Continued on back end papers

*Now available in a lower priced paperback edition in the Wiley Classics Library.

Biostatistics

Biostatistics

A Methodology for the Health Sciences

LLOYD D. FISHER

GERALD VAN BELLE
The University of Washington

A Wiley-Interscience Publication
JOHN WILEY & SONS, INC.
New York · Chichester · Brisbane · Toronto · Singapore

Copyright © 1993 by John Wiley & Sons, Inc.

All rights reserved. Published simultaneously in Canada.

Reproduction or translation of any part of this work
beyond that permitted by Section 107 or 108 of the
1976 United States Copyright Act without the permission
of the copyright owner is unlawful. Requests for
permission or further information should be addressed to
the Permissions Department, John Wiley & Sons, Inc.

Library of Congress Cataloging-in-Publication Data:
Fisher, Lloyd, 1939–
   Biostatistics : a methodology for the health sciences / Lloyd D.
Fisher, Gerald van Belle.
       p.   cm. -- (Wiley series in probability and mathematical
statistics. Applied probability and statistics)
   "A Wiley–Interscience publication."
   Includes bibliographical references and index.
   ISBN 0-471-58465-7
   1. Biometry.   I. Van Belle.   II. Title.   III. Series.
QH323.5.F57   1993
610'.1'5195--dc20   92-24336

Printed in the United States of America

10 9 8 7 6 5

Ad majorem Dei gloriam

Preface

The purpose of this text is for individuals, such as yourself, to learn how to apply statistical methods to the biomedical sciences. The book is written so that individuals with no prior training in statistics and a mathematical knowledge through algebra can follow the text—although the more mathematical training one has, the easier the learning. This text is written for people in a wide variety of biomedical fields, including (alphabetically) biologists, biostatisticians, dentists, epidemiologists, health services researchers, health administrators, nurses, and physicians. The text appears to have a daunting amount of material. Indeed there is a great deal of material but most students will not cover it all. Also, over 30% of the text is devoted to notes, problems, and references so that there is not as much material as seems at first sight. In addition to not covering entire chapters, the following are optional materials: asterisks (*) after a section number denote more advanced material that the instructor may want to skip; the notes at the end of each chapter contain material for extending and enriching the primary material of the chapter, but this may be skipped.

With the exception of a few minor references and comments the following figure gives dependencies between the chapters that allow chapters to be skipped.

While the order of authorship may appear alphabetical, in fact it is random (we tossed a fair coin to determine the sequence) and the book is an equal collaborative effort of the authors. We have many individuals to thank. Our families have been helpful and long-suffering during the writing of this text: for LF, Ginny, Brad, and Laura; for GvB, Johanna, Loeske, William John, Gerard, Christine, Louis, and Bud and Stacy. The many students who were taught with various versions of portions of this material were very helpful. We are also grateful to the many collaborating investigators who taught us much about science as well as the joys of collaborative research. Among those deserving thanks are for LF: Ed Alderman, Christer Allgulander, Fred Applebaum, Michele Battie, Stan Bigos, Tom Bigger, Jeff Borer, Martial Bourassa, Raleigh Bowden, Bob Bruce, Bernie Chaitman, Reg Clift, Rollie Dickson, Kris Doney, Eric Foster, Bob Frye, Bernard Gersh, Karl Hammermeister, Dave Holmes, Mel Judkins, George Kaiser, Ward Kennedy, Tom Killip, Ray Lipicky, Paul Martin, George McDonald, Joel Meyers, Bill Myers, Michael Mock, Gene Passamani, Don Peterson, Bill Rogers, Tom Ryan, Jean Sanders, Lester Sauvage, Rainer Storb, Keith Sullivan, Bob Temple, Don Thomas, Don Weiner, Bob Witherspoon, and a large number of others. For GvB: Ralph Bradley, Richard Cornell, Polly Feigl, Pat Friel, Al Heyman, Myles Hollander, Jim Hughes, Dave Kalman, Jane Koenig, Tom Koepsell, Bud Kukull, Eric Larson, Will Longstreth, Dave Luthy, Duane Meeter, Lorene Nelson,

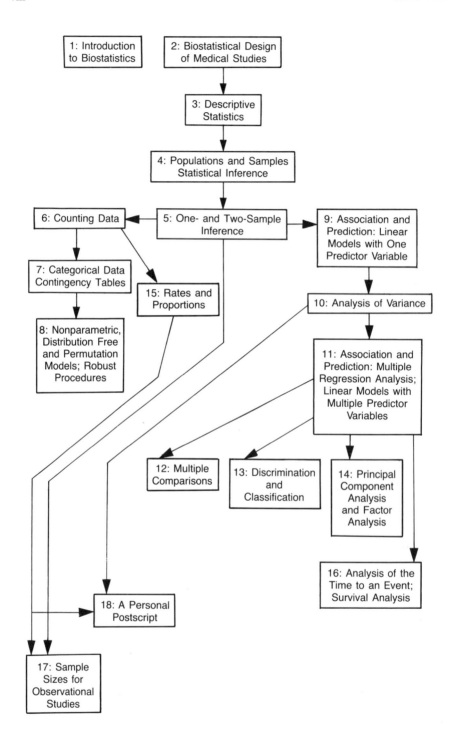

Don Martin, Gil Omenn, Don Peterson, Gordon Pledger, Richard Savage, Kirk Shy, Nancy Temkin and many others. In addition GvB acknowledges the secretarial and moral support of Sue Goleeke. There were many excellent and able typists over the years; special thanks to Myrna Kramer, Pat Coley and Jan Alcorn. We owe a special thanks to Amy Plummer for superb work in tracking down authors and publishers for permission to cite their work. We thank Robert Fisher for help with numerous figures. Rob Christ did an excellent job of using LATEX for the final version of the text. Finally several individuals assisted with running particular examples and creating the tables; we thank Barry Storer, Margie Jones, and Gary Schoch.

Our initial contact with Wiley was the indefatigable Beatrice Shube. Her enthusiasm for our effort carried over to her successor, Kate Roach. The associate managing editor, Rose Ann Campise, was of great help during the final preparation of this manuscript.

With a work this size there are bound to be some errors, inaccuracies and ambiguous statements. We would appreciate receiving your comments. We have set up a special electronic-mail account for your feedback:

fvb@dehpost.sphcm.washington.edu

LLOYD D. FISHER
GERALD VAN BELLE

Contents

Biostatistics

CHAPTER 1

Introduction to Biostatistics

1.1 INTRODUCTION

We welcome the reader who wishes to learn biostatistics. This chapter introduces you to the subject. We define statistics and biostatistics.

Then examples are given where biostatistical techniques are useful. These examples show that biostatistics is an important tool in advancing our biological knowledge; biostatistics helps evaluate many life-and-death issues in medicine.

We urge you to read the examples carefully. Ask yourself, "what can be inferred from the information presented?" How would you design a study or experiment to investigate the problem at hand? What would you do with the data after they are collected? We want you to realize that biostatistics is a tool that can be used to benefit you and society.

The chapter closes with a description of what you may accomplish through use of this text. To paraphrase Pythagoras, there is no royal road to biostatistics. You need to be involved. You need to work hard. You need to think. You need to analyze actual data. The end result will be a tool that has immediate practical uses. As you thoughtfully consider the material presented here you will develop thought patterns that are useful in evaluating information in all areas of your life.

1.2 WHAT IS THE FIELD OF STATISTICS?

Much of the joy and grief in life arises in situations that involve considerable uncertainty. Here are a few such situations.

1. Parents of a child with a genetic defect consider whether or not they should have another child. They will base the decision upon the chance that the next child will have the same defect.

2. A physician must compare the prognosis, or future course, of a patient under several therapies, in order to choose the best therapy. A therapy may be a success, a failure, or somewhere in between; the evaluation of the chance of each occurrence necessarily enters into the decision.

3. In an experiment to investigate whether a food additive is carcinogenic (that is, causes or at least enhances the possibility of having cancer), the Food and

1

Drug Administration has animals treated with and without the additive. Often cancer will develop in both the treated and untreated groups of animals. In both groups there will be animals that do not develop cancer. There is a need for some method of determining whether the group treated with the additive has "too much" cancer.

4. It is well known that "smoking causes cancer." Smoking does not cause cancer in the same manner that striking a billiard ball with another causes the second billiard ball to move. There are many individuals who smoke heavily for long periods of time and do not develop cancer. The formation of cancer subsequent to smoking is not an invariable consequence but occurs only a fraction of the time. Data collected to examine the association between smoking and cancer must be analyzed with recognition of an uncertain and variable outcome.

5. In designing and planning medical care facilities, planners take into account differing needs for medical care. Needs change because there are new modes of therapy, as well as demographic shifts that may increase or decrease the need for facilities. All of the uncertainty associated with the future health of the population and its future geographic and demographic patterns should be taken into account.

Inherent in all of these examples is the idea of uncertainty. Similar situations do not always result in the same outcome. Statistics deals with this variability. This somewhat vague formulation will become clearer in this text. Many definitions of statistics explicitly bring in the idea of variability. Some definitions of statistics are given in the Notes at the end of this chapter.

1.3 WHY BIOSTATISTICS?

Biostatistics is the study of statistics as applied to biological areas. Biological laboratory experiments, medical research (including clinical research) and health services research all use statistical methods. Many other biological disciplines rely upon statistical methodology.

Why should one study biostatistics rather than statistics, since the methods have wide applicability? There are three reasons for focusing on biostatistics.

1. Some statistical methods are more heavily used in biostatistics than in other fields. For example, a general statistical textbook would not discuss the life-table method of analyzing survival data—of importance in many biostatistical applications. The topics in this book are tailored to the applications in mind.

2. Examples are drawn from the biological, medical, and health care areas; this helps you maintain motivation. It also helps you understand how to apply statistical methods.

3. A third reason for a biostatistical text is to teach the material to an audience of health professionals. In this case, the interaction between the students and teacher, but especially among the students themselves, is of great value in learning and applying the subject matter.

1.4 GOALS OF THIS BOOK

Suppose that we wanted to learn something about drugs; we can think of four different levels of knowledge. At the first level, a person may merely know that drugs act chemically when introduced into the body and produce many different effects. A second, higher, level of knowledge is to know that a specific drug is given in certain situations; but we have no idea why the particular drug works. We do not know whether a drug might be useful in a situation we have not seen before. At the next, third level, we have a good idea why things work and also know how to administer drugs. At this level we do not have a complete knowledge of all the biochemical principals involved; but we do have considerable knowledge about the activity and workings of the drug.

Finally, at the fourth and highest level, we have a detailed knowledge of all of the interactions of the drug; we know the current research. This level is appropriate for researchers, those seeking to develop new drugs and understand further the mechanisms of existent drugs. Think of the field of biostatistics in analogy to the drug field above. It is our goal that individuals who complete the material of this book should be on the third level. This book is written to enable you to do more than apply statistical techniques mindlessly.

The greatest danger is in statistical analysis untouched by the human mind. We have the following objectives:

1. You should understand specified statistical concepts and procedures.
2. You should be able to identify procedures appropriate (and inappropriate) to a given situation. You should also have the knowledge to recognize when you do not know of an appropriate technique.
3. You should be able to carry out appropriate specified statistical procedures.

These are high goals for you, the reader of this text. But experience has shown that professionals in a wide variety of biological and medical areas can and do attain this level of expertise. The material presented in this text is often difficult and challenging; time and effort, however, will result in the acquisition of a valuable and indispensable tool which is useful in our daily lives as well as scientific work.

1.5 STATISTICAL PROBLEMS IN BIOMEDICAL RESEARCH

We conclude this chapter with several examples of situations in which biostatistical design and analysis have been or could have been of use. The examples are placed here to introduce you to the subject, to provide motivation for you if you have not thought about such matters before, and to encourage thought about the need for methods of approaching variability and uncertainty in data.

The examples below deal with clinical medicine, an area which has general interest.

1.5.1 Example 1: King Charles II

This first example deals with the treatment of King Charles II during his terminal illness. The following quote is taken from Haggard [1929]:

Some idea of the nature and number of the drug substances used in the medicine of the past may be obtained from the records of the treatment given King Charles II at the time of his death. These records are extant in the writings of a Dr. Scarburgh, one of the twelve or fourteen physicians called in to treat the king. At eight o'clock on Monday morning of February 2, 1685, King Charles was being shaved in his bedroom. With a sudden cry he fell backward and had a violent convulsion. He became unconscious, rallied once or twice, and after a few days died. Seventeenth-century autopsy records are far from complete, but one could hazard a guess that the king suffered with an embolism—that is, a floating blood clot which has plugged up an artery and deprived some portion of his brain of blood—or else his kidneys were diseased. As the first step in treatment the king was bled to the extent of a pint from a vein in his right arm. Next his shoulder was cut into and the incised area "cupped" to suck out an additional eight ounces of blood. After this homicidal onslaught the drugging began. An emetic and purgative were administered, and soon after a second purgative. This was followed by an enema containing antimony, sacred bitters, rock salt, mallow leaves, violets, beet root, camomile flowers, fennel seeds, linseed, cinnamon, cardamon seed, saphron, cochineal, and aloes. The enema was repeated in two hours and a purgative given. The king's head was shaved and a blister raised on his scalp. A sneezing powder of hellebore root was administered, and also a powder of cowslip flowers "to strengthen his brain." The cathartics were repeated at frequent intervals and interspersed with a soothing drink composed of barley water, licorice and sweet almond. Likewise white wine, absinthe and anise were given, as also were extracts of thistle leaves, mint, rue, and angelica. For external treatment a plaster of Burgundy pitch and pigeon dung was applied to the king's feet. The bleeding and purging continued, and to the medicaments were added melon seeds, manna, slippery elm, black cherry water, an extract of flowers of lime, lily-of-the-valley, peony, lavender, and dissolved pearls. Later came gentian root, nutmeg, quinine, and cloves. The king's condition did not improve, indeed it grew worse, and in the emergency forty drops of extract of human skull were administered to allay convulsions. A rallying dose of Raleigh's antidote was forced down the king's throat; this antidote contained an enormous number of herbs and animal extracts. Finally bezoar stone was given. Then says Scarburgh: "Alas! after an ill-fated night his serene majesty's strength seemed exhausted to such a degree that the whole assembly of physicians lost all hope and became despondent: still so as not to appear to fail in doing their duty in any detail, they brought into play the most active cordial." As a sort of grand summary to this pharmaceutical debauch a mixture of Raleigh's antidote, pearl julep, and ammonia was forced down the throat of the dying king.

From this time and distance there are comical aspects about this observational study describing the "treatment" given to King Charles. It should be remembered that his physicians were doing their best according to the state of their knowledge. Our knowledge has advanced considerably, but it would be intellectual pride to assume that all modes of medical treatment in use today are necessarily beneficial. This example illustrates that there is a need for sound scientific development and verification in the biomedical sciences.

1.5.2   Example 2: Relationship Between the Use of Oral Contraceptives and Thromboembolic Disease

In 1967 in Great Britain, there was concern about higher rates of thromboembolic disease (disease from blood clots) among women using oral contraceptives than among women not using oral contraceptives.

To investigate the possibility of a relationship, Vessey and Doll [1969] studied existing cases with thromboembolic disease. Such a study is called a retrospective study because retrospectively, or after the fact, the cases were identified and data accumulated for analysis. The study began by identifying women aged 16–40 years who had been discharged from one of 19 hospitals with a diagnosis of deep vein thrombosis, pulmonary embolism, cerebral thrombosis, or coronary thrombosis.

The idea of the study was to interview the cases to see if more of them were using oral contraceptives than one would "expect." The investigators needed to know how much oral contraceptive usage to expect assuming such usage does not predispose individuals to thromboembolic disease. This is done by identifying a group of women "comparable" to the cases. The amount of oral contraceptive use in this control, or comparison, group is used as a standard of comparison for the cases. In this study, two control women were selected for each case: the control women had suffered an acute surgical or medical condition, or had been admitted for elective surgery. The controls had the same age, date of hospital admission and parity (number of live births) as the case. The controls were selected to have the absence of any predisposing cause of thromboembolic disease.

If there is no relationship between oral contraception and thromboembolic disease, the cases with thromboembolic disease would be no more likely to use oral contraceptives than the controls. In this study, 42 of 84 cases, or 50%, used oral contraceptives. Twenty-three of the 168 controls, or 14%, of the controls used oral contraceptives. After deciding that such a difference is unlikely to occur by chance, the authors concluded that there is a relationship between oral contraceptive use and thromboembolic disease.

This study is an example of a case-control study. The aim of such a study is to examine potential risk factors (that is, factors that may dispose an individual to have the disease) for a disease. The study begins with the identification of cases with the specified disease. A control group is then selected. The control group is a group of subjects comparable to the cases, except for the presence of the disease and the possible presence of the risk factor(s). The case and control groups are then examined to see if a risk factor occurs more often in the cases than would be expected by chance as compared to the controls.

1.5.3 Example 3: Estrogen Therapy and Endometrial Carcinoma

This is a second example of a case-control study that involved matching of cases and controls.

Estrogen, a steroid compound, is primarily produced in the ovaries, but also in the adrenal cortex and the testes. In the menopause, there is a decline or cessation of ovarian function and an estrogen deficiency often occurs, resulting in a general endocrine imbalance and instability of the autonomic nervous system. Diminished estrogen production is thought to be a factor in age-related disorders such as osteoporosis, atherosclerosis and arthritis. All these potential disorders constitute a rationale for the use of estrogen therapy in menopausal and post-menopausal women.

However, on the basis of previous studies, a specific linkage between exogenous estrogen (i.e., estrogen therapy) and endometrial carcinoma (the endometrium is the lining of the uterus) has been suggested, prompting a study by Smith *et al.* [1975] to evaluate the possibility of this linkage in humans.

The authors defined the group of interest to be all "women 48 years of age or older in whom the diagnosis of adenocarcinoma of the endometrium was made in 1960–1972 after curettage or hysterectomy." A total of 317 patients (cases) was obtained from the two hospitals in the study. Estrogen treatment was defined to be usage for 6 months or more before diagnosis of endometrial carcinoma.

Table 1.1. Classification of 317 Case-Control Pairs.

Endometrial Carcinoma (Case)	Other Gynecological Neoplasms (Matched Control)		Total Use
	Estrogen Use	No Estrogen Use	
Estrogen use	39	113	152
No estrogen use	15	150	165
Total	54	263	317

The question arises, "Is there more estrogen usage than one would in some sense expect?" In order to answer this question, it is desirable to compare the cases with a group of women who are comparable, except possibly with regard to estrogen usage, and also with regard to the fact that they did not develop endometrial carcinoma. The approach, as before, is to find a suitable "control" group.

Controls "were selected at the same institutions from patients with other gynecological neoplasms... (it is assumed that estrogen does not *protect* against other gynecological neoplasms—otherwise these controls would be inappropriate). Cases were matched for age at diagnosis (within 4 years) and year of diagnosis (within 2 years) to a corresponding control." The first factor was included because the incidence of endometrial cancer increases with age; the second, because estrogen therapy is relatively new and its use and indications for use are changing.

The results of the study are summarized in Table 1.1. Of the 317 endometrial carcinoma cases, 152 had used estrogen. Of the controls, only 54 had used estrogen. Table 1.1 presents that data organized by pairs. Each pair can be classified according to estrogen usage into four categories, two of which provide no information about the specific linkage of endometrial carcinoma and estrogen usage. The pairs without information are the ones where both or none used estrogen. This eliminates 189 $(39 + 150)$ pairs. (Such pairs where the usage agrees are called concordant pairs.) This leaves 128 $(15 + 113)$ discordant pairs.

If there is no specific linkage between estrogen usage and endometrial carcinoma, approximately half of these pairs should be discordant in one direction and half in the other. In fact, Table 1.1 has a split of 113 to 15. This situation is very unlikely to occur "by chance." One can conclude that endometrial cancer and estrogen use are associated.

The issue of the *New England Journal of Medicine* containing the paper by Smith *et al.*, contains a second paper by Ziel and Finkle [1975] dealing with the same topic with a similar study design using two matched controls for each patient with endometrial carcinoma. Matching variables in this study were birthdate, area of residence by postal zip code, and duration of Health Plan membership. Findings were similar to that of the Smith *et al.* [1975] paper.

1.5.4 Example 4: Use of Laboratory Tests and the Relation to Quality of Care

An important feature of medical care are laboratory tests. These tests affect both the quality and the cost of care. The frequency with which such tests are ordered varies with the physician. It is not clear how the frequency of such tests influences the quality of medical care. Laboratory tests are sometimes ordered as part of "defensive" medical practice. Some of the variation is due to training. Studies investigating the relationship between use of tests and quality of care need to be carefully designed to measure the quantities of interest reliably, without bias. Given the expense of laboratory tests and limited time and resources, there clearly is a need for evaluation of the relationship between the use of laboratory tests and the quality of care.

The study to be discussed here consisted of 21 physicians serving medical internship as reported by Schroeder *et al.* [1974]. The interns were independently ranked on overall clinical capability (i.e., quality of care) by five faculty internists who had interacted with them during their medical training.

Only patients admitted with uncomplicated acute myocardial infarction or uncomplicated chest pain were considered for the study. "Medical records of all patients hospitalized on the coronary care unit between July 1, 1971 and June 20, 1972, were analyzed and all patients meeting the eligibility criteria were included in the study...." The frequency of laboratory utilization ordered during the first 3 days of hospitalization was translated into cost. Since daily EKGs and enzyme determinations (SGOT, LDH and CPK) were ordered on all patients, the costs of these tests were excluded. Mean costs of laboratory use were calculated for each intern's subset of patients, and the interns were ranked in order of increasing costs on a per-patient basis.

The ranking by the five faculty internists and ranking by cost are given in Table 1.2. There is considerable variation in the evaluations by the five internists; for example,

Table 1.2. Independent Assessment of Clinical Competence of 21 Medical Interns by Five Faculty Internists and Ranking of Cost of Laboratory Procedures Ordered, The George Washington University Hospital, 1971–1972. Data from Schroeder *et al.* [1974]. By permission of Medical Care.

Intern No.	Clinical Competence[a]							Rank of Costs of Procedures Ordered[b]
	I	II	III	IV	V	Total	Rank	
A	1	2	1	2	1	7	1	10
B	2	6	2	1	2	13	2	5
C	5	4	11	5	3	28	3	7
D	4	5	3	12	7	31	4	8
E	3	9	8	9	8	37	5	16
F	13	11	7	3	5	39	7	9
G	7	12	5	4	11	39	7	13
H	11	3	9	10	6	39	7	18
I	9	15	6	8	4	42	9	12
J	16	8	4	7	14	49	10	1
K	17	1	17	11	9	55	11	20
L	6	7	21	16	10	60	12	19
M	8	20	14	6	17	65	13	21
N	18	10	13	13	13	67	14	14
O	12	14	12	18	15	71	15	17
P	19	13	10	17	16	75	16	11
Q	20	16	16	15	12	77	17	4
R	14	18	19	14	19	84	18	15
S	10	19	18	20	20	87	19	3
T	15	17	20	21	21	94	20.5	2
U	21	21	15	19	18	94	20.5	5

[a] 1 = most competent.
[b] 1 = least expensive.

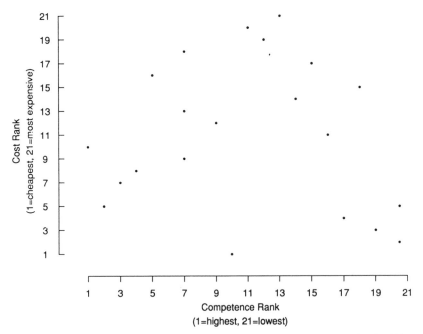

Figure 1.1. Rank order of clinical competence *vs.* rank order of cost of laboratory tests orders for 21 interns. The George Washington University Hospital, 1971–1972. Data from Schroeder *et al.* [1974].

intern K is ranked seventeenth in clinical competence by internists I and III, but first by internist II.

This table still does not clearly answer the question whether there is a relationship between clinical competence and the frequency of use of laboratory tests and their cost. Figure 1.1 shows the relationship between cost and one measure of clinical competence; on the basis of this graph and some statistical calculations the authors conclude that, "at least in the setting measured, no overall correlation existed between cost of medical care and competence of medical care."

This study contains good examples of the kinds of (basically statistical) problems facing a researcher in the health administration area. First of all, what is the population of interest? In other words, what population do the 21 interns represent? Second, there are difficult measurement problems: is level of clinical competence, as evaluated by an internist, equivalent to the level of quality of care? How reliable are the internists? The variation in their assessments has already been noted. Is cost of laboratory use synonymous with cost of medical care, as the authors seem to imply in their conclusion?

1.5.5   Example 5: Salk Poliomyelitis Vaccine Field Trial

This example is taken from the paper by Meier [1989]. In this paper entitled "The biggest public health experiment ever: the 1954 field trial of the Salk poliomyelitis vaccine," the author discusses the evaluation of the Salk vaccine. A more detailed account is in the book by Tanur *et al.* [1989], which also contains many other interesting examples of the use of statistics.

By 1953 the Salk vaccine had been developed. It appeared to be effective against poliomyelitis (polio) and a study was designed to test the effectiveness of the vaccine. The following facts were available:

1. The rate of occurrence of polio during the trial could be expected to be about 50 per 100,000 population.
2. The incidence of polio was related to hygienic status, and hence socioeconomic status; a higher incidence was associated with higher socioeconomic status.
3. The incidence of polio varied greatly from place to place and time to time.
4. Half of polio cases were classified as "paralytic."
5. The effectiveness of the vaccine was estimated to be 50%.
6. Except for "paralytic" polio, the disease is difficult to diagnose; common symptoms of mild polio are fever and weakness.

What kind of study can be designed to test the effectiveness of the vaccine? What are the criteria to judge its effectiveness? How large a sample is needed? What are the ethical constraints? These questions and others had to be answered by the statistician in collaboration with other investigators.

The simplest experimental approach would be to distribute the vaccine as widely as possible, and then to observe the subsequent rate of reported polio. A suitable population might be elementary school pupils. This approach is the vital statistics approach. It is simple and requires a minimum of statistical planning. But there are serious drawbacks to this approach. First, the annual variation of the number of polio cases was considerable: the total number of reported cases between 1947 and 1952 varied from approximately 10,000 cases per year to almost 60,000 cases per year. A small number of cases in the test year could occur "by chance."

Second, as mentioned above, except for the classic "respirator-case," polio is a difficult disease to diagnose. Mild forms of polio can be manifested in sweating and fever. Once the large scale study was introduced, there would be a focusing on the disease and a large number of "false positives" could be expected. A false positive is an individual mistakenly diagnosed as having the disease (positive for the disease).

A second approach uses a control group. A "comparable" group of children would not be given the vaccine. Specifically, it was proposed to make the vaccine available to all children in the second grade (the treatment group) and to make first and third grades the control group. The incidence of polio could then be compared in the two groups. This design can be termed the observed control approach. Some of the problem associated with the vital statistics approach, such as the variation in time, are no longer present but there are others: knowledge of vaccination status can influence diagnosis, families volunteering for the vaccination tend to be better educated and from a higher socioeconomic stratum.

To overcome most of the objections to the vital statistic design and the observed control approach, a randomized placebo control experiment was set up. A randomized experiment is an experiment where treatment assignment is made "by chance." For this example one may think of individuals receiving the vaccine or placebo decided by the flip of a coin. The placebo (from the Latin: "I please") is a preparation designed to look exactly like the polio vaccine but lacking the active ingredient. The random method of selection assured that the treatment and control groups were roughly comparable. Any difference would occur entirely by chance. Preparations were placed in coded vials and neither the subject nor anyone involved in the vaccination and evaluation process knew whether a vial contained the active or inert ingredient. A study with this feature is termed a double blind study. The study is single blind if the subjects receiving the treatment or control do not know which they receive. Thus the participating subjects are blinded with regard to whether they are treatment or control subjects. The study is double blind if the individuals evaluating the outcome variables also do not know whether the subject received treatment or not. Usually half of the preparations are active and half are inert—although this proportion may be varied. (You may want to list the advantages, and disadvantages, of such a study.) One of the objections to this design is that it is considered unethical to administer an inert substance; another is that withholding the best possible treatment is unethical. Of course if the vaccine had been known to be the best treatment, the trial would not have been performed.

Both the observed control design and the randomized placebo control design were used in the 1954 field trial. The main results are shown in Table 1.3. The total number of subjects involved in the two types of studies was close to 2,000,000 with approximately 750,000 subjects in the latter study and 1,000,000 subjects in the former study. All reported cases of polio were reviewed by a blinded panel of experts and classified as "definite polio," "doubtful polio" or "nonpolio."

The definite polio cases were subdivided into paralytic, nonparalytic, and fatal. In Table 1.3, paralytic and fatal cases are grouped. The table illustrates the effectiveness

Table 1.3. Summary of Study Cases by Diagnostic Class and Vaccination Status (Rates per 100,000) in the 1954 Field Trial of the Salk Poliomyelitic Vaccine. From "The Biggest Public Health Experiment Ever: The 1954 Field Trial of the Salk Poliomyelitis Vaccine" by Paul Meier. In *Statistics: A Guide to the Unknown*, 3rd Edition by J. M. Tanur, editor. Copyright © 1989 by Wadsworth & Brooks/Cole Advanced Books & Software, Pacific Grove, CA 93950.

Study Group	Study Population	All Reported Cases No.	Rate	Paralytic[a] No.	Rate	Non-paralytic No.	Rate	Nonpolio No.	Rate
All areas: total	1,829,916	1028	56	700	38	178	10	150	8
Randomized Placebo Control Areas									
Total	749,236	432	58	274	37	88	12	70	9
Vaccinated	200,745	82	41	33	16	24	12	25	12
Placebo	201,229	166	82	119	59	27	13	20	10
Not inoculated[b]	338,778	182	54	121	36	36	11	25	7
Incomplete vaccinations	8,484	2	24	1	12	1	12	–	–
Observed Control Areas									
Total	1,080,680	596	55	426	39	90	8	80	7
Vaccinated	221,998	76	34	38	17	18	8	20	9
Controls[c]	725,173	450	62	341	47	61	8	48	6
Grade 2—not inoculated	123,605	66	53	43	35	11	9	12	10
Incomplete vaccinations	9,904	4	40	4	40	–	–	–	–

[a]"Paralytic" includes fatal polio.
[b]Includes 8577 who received one or two injections of placebo.
[c]First and third grade total population.

of the vaccine with the rate per 100,000 for the vaccinated groups approximately one-half that of the control group.

1.5.6   Example 6: Internal Mammary Artery Ligation

One of the greatest health problems in the world, and especially in the industrialized nations, is coronary artery disease. The coronary arteries are the arteries around the outside of the heart. These arteries bring blood to the heart muscle (myocardium). Coronary artery disease brings a narrowing of the coronary arteries. Such narrowing often results in chest, neck and arm pain (angina pectoris) precipitated by exertion. When arteries completely block off or occlude, a portion of the heart muscle is deprived of its blood supply with the life giving oxygen and nutrients. A myocardial infarction, or heart attack, is the death of a portion of the heart muscle.

As the coronary arteries narrow the body often compensates by building "collateral circulation." Collateral circulation consists of branches from existing coronary arteries that develop to bring blood to an area of restricted blood flow. The internal mammary arteries are arteries that bring blood to the chest. The tributaries of the internal mammary arteries develop collateral circulation to the coronary arteries. It was thus reasoned that by tying off, or ligating the internal mammary arteries, a larger blood supply would be forced to the heart. An operation, internal mammary artery ligation was developed to implement this procedure.

Early results of the operation were most promising. Battezzati et al. [1959] reported on 304 cases who underwent internal mammary artery ligation: 94.8% of the patients reported improvement; 4.9% reported no appreciable change.

It would seem that the surgery gave great improvement. Still, the possibility remained that the improvement resulted from a placebo effect. A placebo effect is a change, or perceived change, resulting from the psychological benefits of having undergone treatment. It is well known that inert tablets will cure a substantial portion of headaches, stomach aches and afford pain relief. The placebo effect of surgery might be even more substantial.

Two studies of internal mammary artery ligation were performed using a sham operation as a control. Both studies were double blind. Neither the patients nor the physicians evaluating the effect of surgery knew whether the ligation had taken place. In each study, incisions were made in the patient's chest and the internal mammary arteries exposed. In the sham operation nothing further was done. For the other patients, the arteries were ligated. Both studies selected the patients having the ligation or sham operation by random assignment.

Cobb et al. [1959] reported upon the subjective patient estimates of "significant" improvement. Patients were asked to estimate the percent improvement after the surgery. Another indication of the amount of pain experienced is the number of nitroglycerin tablets taken for anginal pain. Table 1.4 reports these data.

Dimond et al. [1960] reported a study of eighteen patients of whom five received the sham operation and thirteen received surgery. Table 1.5 presents the patients' opinion of the percentage benefit of surgery.

Both papers conclude that it is unlikely that the internal mammary artery ligation has benefit, beyond the placebo effect, in the treatment of coronary artery disease. Note that 12 of the 14, or 86%, of those receiving the sham operation reported improvement in the two studies. These studies point to the need for appropriate comparison groups when making scientific inferences.

Table 1.4. Subjective Improvement as Measured by Patient Reporting and Number of Nitroglycerin Tablets Taken in the Cobb *et al.* [1959] Study.

	Ligated	Nonligated
Number of patients	8	9
Average % improvement reported	32	43
Subjects reporting 40% or more improvement	5	5
Subjects reporting no improvement	3	2
Nitroglycerin tablets taken:		
Average before operation (No./week)	43	30
Average after operation (No./week)	25	17
Average % decrease in No./week	34	43

Table 1.5. Patient's Opinion of Surgical Benefit in the Dimond *et al.* [1959] Study.

Patient's Opinion of the Benefit of Surgery	Patient Number[a]
Cured (90–100%)	4, 10, 11, 12*, 14*
Definite benefit (50–90%)	2, 3*, 6, 8, 9*, 13*, 15, 17, 18
Improved but disappointed (25–50%)	7
Improved for 2 weeks, now same or worse	1, 5, 16

[a] The numbers (1–18) refer to the individual patients as they occurred in the series, grouped according to their own evaluation of their benefit, expressed as percentage. Those numbers followed by an asterisk indicate a patient on whom the sham operation was performed.

The use of clinical trials has greatly enhanced medical progress. Some examples are given throughout this book; but this is not the primary emphasis of the text. Good references for learning much about clinical trials are Meinert [1986], Friedman *et al.* [1981], and Fleiss [1986].

NOTES

Some Definitions of Statistics:

1. "The science of statistics is essentially a branch of Applied Mathematics, and may be regarded as mathematics applied to observational data ... Statistics may be regarded (i) as the study of populations, (ii) as the study of variation, (iii) as the study of methods of the reduction of data." Fisher [1950]
2. "Statistics is the branch of the scientific method which deals with the data obtained by counting or measuring the properties of populations of natural phenomena." Kendall and Stuart [1963]
3. "The science and art of dealing with variation in such a way as to obtain reliable results." Mainland [1963]
4. "Statistics is concerned with the inferential process, in particular with the planning and analysis of experiments or surveys, with the nature of observational errors and sources of variability that obscure underlying patterns, and with the efficient summarizing of sets of data." Kruskal [1968]
5. "Statistics = Uncertainty and Behavior." Savage [1968]

6. "... the principal object of statistics [is] to make inference on the probability of events from their observed frequencies." von Mises [1957]

7. "The technology of the scientific method." Mood [1950]

8. "The statement, still frequently made, that statistics is a branch of mathematics is no more true than would be a similar claim in respect of engineering ... good statistical practice is equally demanding of appreciation of factors outside the formal mathematical structure, essential though that structure is." Finney [1975]

There is clearly no complete consensus in the definitions of statistics. But certain elements reappear in all of the definitions: variation, uncertainty, inference, science. In the previous sections we have illustrated how the concepts occur in some typical biomedical studies. The need for biostatistics has thus been shown.

PROBLEMS

The problems of Chapter 1 are intended to provoke thought and interest. Some of the papers contain material, both scientific and statistical, that you are not expected to understand at this stage.

N.B.: These problems require outside reading and are only appropriate if the material is available.

1. Gastric Freezing in the Treatment of Duodenal Ulcers: This exercise deals with duodenal ulcers. The following is an oversimplified description of the biology involved. A peptic ulcer is a sharply circumscribed loss of tissue involving the innermost layers of the digestive tract exposed to acid—peptic gastric juice (pepsin is a digestive enzyme). Most commonly, these areas include the lining of the stomach itself on the lesser curvature or in the antrum, and the first 3–4 cm of the duodenum (first part of the small intestine between the pylorus and the opening of the bile and pancreatic ducts). This first 3–4 cm is called the duodenal bulb (see Figure 1.2). Ulcers can be single or multiple with each one usually smaller than 2 cm in diameter. The prevalence has been estimated at 10–12% based on X-ray surveys and autopsy series. Ulcers occur in all age groups, but symptoms usually occur in individuals between the ages of 20–40 years. The diagnosis of peptic ulcer is made with certainty only by X-ray barium studies or at surgery. Symptoms can range from a mild pain, tenderness, or discomfort in the epigastric region to vomiting of bright red or "coffee-grounds" blood. The ulcer can extend through the wall of the stomach (perforate) causing sudden or rapid onset of severe pain and collapse. Topical treatment for peptic ulcer has focused on reducing the gastric acid or its pH by means of antacids, buffers, bland diets, or antivagal drugs (vagal nerves stimulate the production of digestive proteins). Surgical treatment involves removal of stomach tissue or cutting of the vagus nerve. Untreated (and all too frequently, treated) peptic ulcers have a tendency to heal and then relapse, and frequently become chronic.

   a. Read Wangensteen et al. [1962] and answer the following questions:

      i. Does the animal evidence seem reasonable and convincing to you?

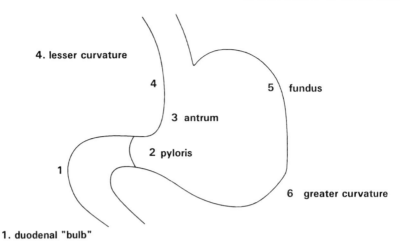

4. lesser curvature

4   5 fundus

3 antrum

2 pyloris

1

6   greater curvature

1. duodenal "bulb"

Figure 1.2. Common locations of duodenal ulcers.

      ii. Combining the animal evidence with the patient data does it seem reasonable to conclude that there is a decrease in the free HCl?

   b. Read the article on the procedure as presented in *Time Magazine* [1962].

      i. Is the *Time* article a reasonable presentation of the results?

      ii. What aspects of the article make you think it is written for general appeal?

   c. Read the article by Hitchcock *et al.* [1966].

      i. How do you reconcile the Wangensteen article with Hitchcock's?

      ii. What are the weaknesses of the Hitchcock study?

   d. Read Ruffin *et al.* [1969].

2. Read the article by Silverman [1977] entitled The Lesson of Retrolental Fibroplasia. Give a brief (less than one page) synopsis of the content of the article.

3. These papers are from the internal mammary artery ligation example.

   a. Read Battezzati *et al.* [1959]. How convincing would the evidence of this paper be if you did not know the results?

   b. Read Ratcliff [1957]. What would the public conclude from this article? What should be the role of the news media in evaluating new techniques?

   c. Read Cobb *et al.* [1959]. Do you think it is ethically justifiable to perform a sham operation? List the benefits and drawbacks for individuals being randomized in a sham versus real surgical trial. Is it ethically necessary to inform the participants of the details of the study? Is it ethical for physicians to prescribe placebos in routine medical practice? (The placebo has been called the most powerful drug in modern medicine.)

4. Read Karlowski *et al.* [1975]. The randomized trial took place among the employees of the National Institutes of Health.

a. Would you have expected this group of people to adhere to the study plan (or study protocol)?

b. What went wrong with the trial? If the blinding is broken by individuals receiving treatment, what may happen to the placebo effect?

c. Do you think an "appropriate statistical analysis" might be able to compensate for a lack of blinding? Why or why not?

5. This question refers to the paper of Vessey and Doll [1969]. The controls in the paper were selected from individuals with acute medical or surgical conditions, or who had been admitted to the hospital for elective surgery. Such "hospital controls" are frequently used in case-control studies. Discuss the good and bad points of choosing a hospital control group. The following questions may be worth considering. If controls are selected from the same hospitals as the cases, do you expect them to be from the same socioeconomic group as the cases? Will they be as healthy as a timely "comparable" population?

REFERENCES

Battezzati, M., Tagliaferro, A., and Cattaneo, A. D. [1959]. Clinical evaluation of bilateral internal mammary artery ligation as treatment of coronary heart disease. *American Journal of Cardiology*, **4:** 180–183.

Cobb, L. A., Thomas, G. I., Dillard, D. H., Merendino, K. A., and Bruce, R. A. [1959]. An evaluation of internal-mammary-artery ligation by a double blind technique. *New England Journal of Medicine*, **260:** 1115–1118.

Dimond, E. G., Kittle, C. F., and Crockett, J. E. [1960]. Comparison of internal mammary artery ligation and sham operation for angina pectoris. *American Journal of Cardiology*, **5:** 483–486.

Finney, D. J. [1975]. Numbers and data. *Biometrics*, **31:** 375–386.

Fisher, R. A. [1950]. *Statistical Methods for Research Workers*, 11th edition. Hafner, New York.

Fleiss, J. L. [1986]. *The Design and Analysis of Clinical Experiments*. Wiley, New York.

Friedman, L. M., Furberg, C. D., and DeMets, D. L. [1981]. *Fundamentals of Clinical Trials*, John Wright, Boston, MA.

Haggard, H. W. [1929]. *Devils, Drugs, and Doctors*. Blue Ribbon Books, New York.

Hitchcock, C. R., Ruiz, E., Sutherland, R. D., and Bitter, J. E. [1966]. Eighteen-month follow-up of gastric freezing in 173 patients with duodenal ulcer. *Journal of the American Medical Association*, **195:** 115–119.

Karlowski, T. R., Chalmers, T. C., Frenkel, L. D., Kapikian, A. Z., Lewis, T. L., and Lynch, J. M. [1975]. Ascorbic acid for the common cold; a prophylactic and therapeutic trial. *Journal of the American Medical Association*, **231:** 1038–1042.

Kendall, M. G., and Stuart, A. [1963]. *The Advanced Theory of Statistics*, Volume 1, 2nd edition. Charles Griffin, London.

Kruskal, W. [1968]. In *International Encyclopedia of the Social Sciences*. D. L. Sills (ed). Macmillan, New York.

Mainland, D. [1963]. *Elementary Medical Statistics,*. 2nd edition. Saunders, Philadelphia.

Meier, P. [1989]. The biggest public health experiment ever: the 1954 field trial of the Salk poliomyelitis vaccine. In *Statistics: A Guide to the Unknown*. Third Edition J. M. Tanur (ed.). Wadsworth & Brooks/Cole Advanced Books & Software, Pacific Grove, CA.

Meinert, C. L. [1986]. *Clinical Trials: Design, Conduct and Analysis*, Oxford University Press, New York.

Mood, A. M. [1950]. *Introduction to the Theory of Statistics.* MacGraw-Hill, New York.

Ratcliff, J. D. [1957]. New surgery for ailing hearts. *Reader's Digest,* **71:** 70–73.

Ruffin, J. M., Grizzle, J. E., Hightower, N. C., McHarcy, G., Shull, H., and Kirsner, J. B. [1969]. A cooperative double-blind evaluation of gastric "freezing" in the treatment of duodenal ulcer. *New England Journal of Medicine,* **281:** 16–19.

Savage, I. R. [1968]. *Statistics: Uncertainty and Behavior.* Houghton-Mifflin, Boston.

Schroeder, S. A., Schliftman, A., and Piemme, T. E. [1974]. Variation among physicians in use of laboratory tests: relation to quality of care. *Medical Care,* **12:** 709–713.

Silverman, W. A. [1977]. The lesson of retrolental fibroplasia. *Scientific American,* **236:** 100–107.

Smith, D. C., Prentice, R. L., Thompson, D. J., and Herrmann, W. L. [1975]. Exogenous estrogen and endometrial carcinoma. *New England Journal of Medicine,* **293:** 1164–1167.

Tanur, J. M., Mosteller, F., Kruskal, W. H., Link, R. F., Pieters, R. S., and Rising, G. R. [1989]. *Statistics: A Guide to the Unknown.* Third Edition. Wadsworth & Brooks/Cole Advanced Books & Software, Pacific Grove, CA.

Time Magazine [1962]. Frozen ulcers. *Time,* May 18: 45–47.

Vessey, M. P. and Doll, R. [1969]. Investigation of the relation of between use of oral contraceptives and thromboembolic disease; a further report. *British Medical Journal,* **2:** 651–657.

von Mises, R. [1957]. *Probability, Statistics and Truth,* 2nd edition. Macmillan, New York.

Wangensteen, O. H., Peter, E. T., Nicoloff, M., Walder, A. I., Sosin, H., and Bernstein, E. F. [1962]. Achieving "physiological gastrectomy" by gastric freezing. *Journal of the American Medical Association,* **180:** 439–444.

Ziel, H. K. and Finkle, W. D. [1975]. Increased risk of endometrial carcinoma among users of conjugated estrogens. *New England Journal of Medicine,* **293:** 1167–1170.

CHAPTER 2

Biostatistical Design of Medical Studies

2.1 INTRODUCTION

In this chapter we introduce some of the principles of biostatistical design. Many of the ideas will be expanded in later chapters. This chapter also serves as a reminder that statistics is not an end in itself but a tool to be used in investigating the world around us. The study of statistics should serve to develop critical, analytical thought and "common sense" as well as to introduce specific tools and methods of processing data.

2.2 PROBLEMS TO BE INVESTIGATED

Biomedical studies arise in many ways. A particular study may result from a sequence of experiments, each one leading naturally to the next. The study may be triggered by observation of an interesting case, or observation of a mold, e.g., penicillin in a petri dish. The study may be instigated by a governmental agency in response to a question of national importance. The basic ideas of the study may be defined by an advisory panel. Many of the critical studies and experiments in biomedical science have come from one individual with an idea for a radical interpretation of past data.

The formulation of the problem to be studied lies outside the realm of statistics *per se*. Statistical considerations may suggest that an experiment is too expensive to conduct, or may suggest a different approach than at first planned. The need to statistically evaluate the data from a study forces an investigator to sharpen the focus of the study. It makes one translate intuitive ideas into an analytical model capable of generating data that may be evaluated statistically.

To answer a given scientific question, many different studies may be considered. Possible studies may range from small laboratory experiments, to large and expensive experiments involving humans, to observational studies. It is worth spending a considerable amount of time thinking about alternatives. In most cases your first idea for a study will not be your best—unless it is your only idea.

In laboratory research, many different experiments may shed light upon a given hypothesis or question. Sometimes less-than-optimal execution of a well-conceived experiment sheds more light than arduous and excellent experimentation unimaginatively designed. One mark of a good scientist is that he or she attacks important problems in a clever manner.

2.3 SOME DIFFERENT TYPES OF STUDIES

A problem may be investigated in a variety of different ways. To decide upon your method of approach, it is necessary to understand the types of studies that might be done. To facilitate the discussion of design, we introduce definitions of commonly used types of studies.

Definition 2.1. An *observational study* collects data from an existing situation. The data collection does not intentionally interfere with the running of the system.

There are subtleties associated with observational studies. The act of observation may introduce change into the system. For example, if physicians know that their behavior is being monitored and charted for study purposes, they may tend to adhere more strictly to procedures than would otherwise be the case. Pathologists performing autopsies guided by a study form may invariably look for a certain finding, not routinely sought. The act of sending out questionnaires about health care may sensitize individuals to the need for health care; this might result in more demand for health care. Constantly asking about a person's health can introduce hypochondria.

A side effect introduced by the act of observation is the *Hawthorne Effect*, after a famous experiment carried out at the Hawthorne works of the Western Electric Company. Employees were engaged in the production of electrical relays. The study was designed to investigate the effect of better working conditions including increased pay, shorter hours, better lighting and ventilation, and pauses for rest and refreshment. All were introduced with "resulting" increased output. As a control, working conditions were returned to original conditions. Production continued to rise! The investigators concluded that increased morale due to the attention and resulting *esprit de corps* among workers resulted in better production. Humans and animals are not machines or passive experimental units (Roethlisberger [1941]).

Definition 2.2. An *experiment* is a study in which the investigator deliberately sets one or more factors to a specific level.

Experiments lead to stronger scientific inferences than observational studies. The "cleanest" experiments exist in the physical sciences; nevertheless, in the biological sciences, particularly with the use of randomization (a topic to be discussed below), strong scientific inferences can be obtained. Experiments are superior to observational studies in part because in an observational study one may not be observing one or more variables that are of crucial importance to interpreting the observations. Observational studies are always open to misinterpretation due to the lack of knowledge in a given field. In an experiment, by seeing the resulting change when a factor is varied, the causal inference is much stronger.

Definition 2.3. A *laboratory experiment* is an experiment which takes place in an environment (called the laboratory) where experimental manipulation is facilitated.

Although this definition is loose, the connotation of the term "laboratory experiment" is that the experiment is run in a situation where most of the variables of interest can be very closely controlled (e.g., temperature, air quality). In laboratory experiments involving animals, the aim (not always reached) is that animals are treated in the same manner in all respects except with regard to the factors varied by the investigator.

Definition 2.4. A *comparative experiment* is an experiment which compares two or more techniques, treatments or levels of some variable.

There are many examples of comparative experiments in biomedical areas. For example, it is common in nutrition to compare laboratory animals on different diets. There are many experiments comparing different drugs. Experiments may compare the effect of a given treatment with no treatment. (From a strictly logical point of view, "no treatment" is in itself a type of treatment.) There are also comparative observational studies. In a comparative *study* one might, for example, observe women using and women not using birth control pills and examine the incidence of complications such as thrombophlebitis. The women themselves would decide whether or not to use birth control pills. The user and nonuser groups would probably differ in a great many other ways. In a comparative *experiment*, one might have women selected by chance to receive birth control pills with the control group using some other method.

Definition 2.5. An *experimental unit* or *study unit* is the smallest unit upon which the experiment or study is performed.

In a clinical study, the experimental units are usually humans. (In some other cases, it may be an eye; for example, one eye may receive treatment, the other being a control.) In animal experiments, the experimental unit is usually an animal. With a study on teaching, the experimental unit may be a class—as the teaching method will usually be given to an entire class. Study units are the object of consideration when one discusses sample size.

Definition 2.6. An experiment is a *cross-over experiment* if the same experimental unit receives more than one treatment, or is investigated under more than one condition of the experiment. The different treatments are given during nonoverlapping time periods.

An example of a cross-over experiment is one in which laboratory animals are treated sequentially with more than one drug and blood levels of certain metabolites are measured for each of the drugs. A major benefit of a cross-over experiment is that each experimental unit serves as its own control (the term "control" to be explained in more detail below) eliminating subject-to-subject variability in response to the treatment or experimental conditions being considered. Major disadvantages of a cross-over experiment are that (a) there may be a carry-over effect of the first treatment continuing into the next treatment period, (b) the experimental unit may change over time, (c) in animal or human experiments, the treatment introduces permanent physiological changes, (d) the experiment may take longer so that investigator and subject enthusiasm wanes, and (e) the chance of dropping out increases.

Definition 2.7. A *clinical study* is one which takes place in the setting of clinical medicine.

A study which takes place in an organizational unit dispensing health care—such as a hospital, psychiatric clinic, well-child clinic, or group practice clinic—is a clinical study.

We now turn to the concepts of prospective studies and retrospective studies, usually involving human populations.

Definition 2.8. A *cohort* of people is a group of people whose membership is clearly defined.

Examples of cohorts are: all individuals enrolling in the Graduate School at the University of Washington for Fall quarter of 1990; all females between the ages of 30 and 35 (as of a certain date) whose residence is within the New York City limits; all smokers in the United States as of January 1, 1953, where an individual is defined to be a smoker if he or she smoked one or more cigarettes during the preceding calendar year. Often cohorts are followed over some time interval.

Definition 2.9. An *endpoint* is a clearly-defined outcome or event associated with an experimental or study unit.

An endpoint may be the presence of a particular disease or 5-year survival after, say, a radical mastectomy. An important characteristic of an endpoint is that it can be clearly defined and observed.

Definition 2.10. A *prospective study* is one in which a cohort of people is followed for the occurrence or nonoccurrence of specified endpoints or events or measurements.

In the analysis of data from a prospective study, the occurrence of the endpoints is often related to characteristics of the cohort measured at the beginning of the study.

Definition 2.11. *Baseline characteristics* or *baseline variables* are values collected at the time of entry into the study.

The Salk polio vaccine trial mentioned earlier is an example of a prospective study, in fact, a prospective experiment. On occasion, you may be able to conduct a prospective study from existing data; that is, some unit of government or other agency may have collected data for other purposes which allows you to analyze the data as a prospective study. In other words, there is a well-defined cohort for which records have already been collected (for some other purpose) and which can be used for your study. Such studies are sometimes called historical prospective studies.

One drawback associated with prospective studies is that the endpoint of interest may occur infrequently. In this case, extremely large numbers of people need to be followed in order that the study have enough endpoints for statistical analysis (*cf.* again the Salk polio vaccine trial). Other designs, as discussed below, help get around this problem.

Definition 2.12. A *retrospective study* is one in which individuals having a particular outcome or endpoint are identified and studied.

These individuals are usually compared to other individuals without the endpoint. The groups are compared to see whether the individuals with the given endpoint have a higher fraction with one or more of the factors that are conjectured to increase the risk of endpoints.

Subjects with particular characteristics of interest are often collected into registries. Such a registry usually covers a well-defined population. In Sweden, for example, there is a twin registry. In the United States there are cancer registries, often defined for a specified metropolitan area. Registries can be used for retrospective as well as prospective studies. A cancer registry can be used retrospectively to compare presence or absence of possible causal factors of cancer after generating appropriate controls—either randomly from the same population or by some matching mechanism. Alternatively, a cancer registry can be used prospectively by comparing survival times of cancer patients having various therapies.

One way of avoiding the large sample sizes needed to prospectively collect enough cases is to use the case control study as seen in the last chapter.

Definition 2.13. A *case control study* selects all cases, usually of a disease, meeting fixed criteria. A comparison group for the cases, called *controls* is also selected. The cases and controls are compared with respect to various characteristics.

The study of oral contraceptives and thromboembolic disease mentioned in Chapter 1 was an example of the case control method. Sometimes controls are selected to match the individual case; in other situations, an entire group of control individuals is selected for comparison with the entire group of cases.

Definition 2.14. In a *matched case control study* controls are selected to match characteristics of individual cases. The case and the control(s) are associated with each other. There may be more than one control for each case.

Suppose that we want to study characteristics of cases of a disease. One way to do this would be to identify new cases appearing during some time interval. A second possibility would be to identify all known cases at some fixed time. The first approach is a longitudinal approach; the second approach is called cross-sectional.

Definition 2.15. A *longitudinal study* collects information on study units over a specified time interval. A *cross-sectional study* collects data on study units at some fixed time.

Figure 2.1 illustrates the difference. The longitudinal study might collect information on the six new cases appearing over the specified time interval. The cross-sectional study would identify the nine available cases at the fixed time point. The cross-sectional study will have proportionately more cases with a long duration. (Why?)

For completeness, we repeat the definitions given informally in Chapter 1.

Definition 2.16. A *placebo* treatment is designed to appear exactly like a comparison treatment, but to be devoid of the active part of the treatment.

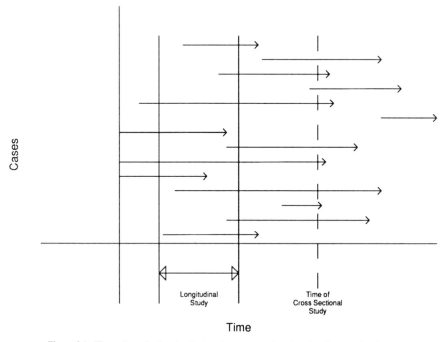

Figure 2.1. Illustration of a longitudinal and a cross-sectional study of cases of a disease.

Definition 2.17. The *placebo effect* results from the knowledge that one is treated, rather than changes due to physical, physiological, and chemical activity of a treatment.

Definition 2.18. A study is *single blind* if treated subjects are unaware of which treatment (including any control) they are receiving. A study is *double blind* if it is single blind and the individuals evaluating the outcome variables are also unaware of which treatment the subjects are receiving.

2.4 STEPS NECESSARY TO PERFORM A STUDY

In this section we outline briefly the steps involved in conducting a study. The steps are interrelated and are oversimplified here in order to isolate different elements of scientific research and to discuss the statistical issues involved.

Step 1. A question or problem area of interest is considered. This does not involve biostatistics *per se*.

Step 2. A study is to be designed to answer the question. The design of the study must consider at least the following elements:

    a. Identify the data to be collected. This includes the variables to be measured as well as the number of experimental units, that is, the size of the study or experiment.

      b. An appropriate analytical model needs to be developed for describing and processing the data.

      c. What are the inferences one hopes to make from the particular study? What conclusions might one draw from the study? To what population(s) is the conclusion applicable?

Step 3. The study is carried out and the data are collected.

Step 4. The data are analyzed and conclusions and inferences are drawn.

Step 5. The results are used. This may involve changing operating procedures, publishing results or planning a subsequent study.

2.5 ETHICS

Many studies and experiments in the biomedical field involve both animal and/or human participants. Moral and legal issues are involved in both areas. Ethics must be of primary concern. In particular, we will mention five points relevant to experimentation with humans.

1. It is our opinion that all investigators involved in a study are responsible for the conduct of an ethical study to the extent that they may be expected to know what is involved in the study. For example, we think that it is unethical to be involved in the analysis of data which has been collected in an unethical manner.

2. Investigators are close to a study and often excited about its potential benefits and advances. It is difficult for them to consider all ethical issues objectively. For this reason, in proposed studies involving humans (or animals), there should be review by individuals not concerned or connected with the study or the investigators. The reviewers should not directly profit in any way if the study is carried out. Implementation of the study should be contingent upon such a review.

3. Individuals participating in an experiment should understand and sign an informed consent form. The principle of informed consent says that a participant should know about the conduct of the study and about any possible harm and/or benefits that may result from participation in the study. For those unable to give an informed consent, appropriate representation may give the consent.

4. Subjects should be free to withdraw at any time, or to refuse initial participation, without being penalized or jeopardized with respect to current and future care and activities.

5. Both the Nuremberg Code and the Helsinki Accord recommend that, when possible, animal studies should be done prior to human experimentation.

References relevant to ethical issues are the US Department of Health, Education and Welfare's statement on *Protection of Human Subjects*, Papworth's book, *Human Guinea Pigs* [1967] and Spicker *et al.* [1988]; Papworth is extremely critical of the conduct of modern biological experimentation. There are also guidelines for studies involving animals. See, for example, *Guide for the Care and Use of Laboratory Animals* [1985].

2.6 DATA COLLECTION: DESIGN OF FORMS

2.6.1 What Data are to be Collected?

In studies involving only one or two investigators, there is often almost complete agreement as to what data are to be collected. In this case, it is very important that good laboratory records be maintained. It is especially important that variations in the experimental procedure (for example, loss of power during a time period, unexpected change in temperature in a room containing laboratory animals) be recorded. If there are peculiar patterns in the data, the detailed notes may point to possible causes. The necessity for keeping detailed notes is even more crucial in large studies or experiments involving many investigators; it is difficult for any one person to have complete knowledge of the study.

In a large collaborative study involving a human population, it is not always easy to decide what data to collect. For example, often there is interest in getting prognostic information. How many potentially prognostic variables should you record? Suppose you are measuring pain relief or quality of life; how many questions do you need to reasonably characterize these abstract ideas? In looking for complications of drugs, should you instruct investigators to enter all complications? This may be an unreliable procedure if you are dependent upon a large, diverse group of observers. In studies with many investigators, each investigator will want to collect data relating to her or his special interests. You can rapidly arrive at large, complex forms. If too much data is collected, there are various "prices" to be paid. One obvious price is the expense of collecting and handling large and complex data. Another is reluctance (especially by volunteer subjects) to fill out long, complicated forms leading to possible biases in subject recruitment. If the study lasts a long time, the investigators may become fatigued by the onerous task of data collection. Fatigue and lack of enthusiasm can affect the quality of the data through a lack of care and effort in its collection.

On the other hand, there are many examples where too little data were collected. One of the most difficult tasks in designing forms is to remember to include all necessary items. The more complex the situation, the more difficult the task. It is easy to look at existing questions and respond to them. If a question is missing, how is one alerted to the fact? One of the authors was involved in the design of a follow-up form where mortality could not be recorded. There was an explanation for this: the patients were to fill out the forms. Nevertheless, it was necessary to include forms that would allow those responsible for follow-up to record mortality, the primary endpoint of the study.

In order to assure that all necessary data are on the form, the following steps are suggested:

1. There should be thorough review of all forms *with a written response* by all participating investigators.
2. Decide upon the statistical analyses beforehand. Check that *specific* analyses involving *specific* variables can be run. Often the analysis is changed during processing of the data or in the course of "interactive" data analysis. This preliminary step is still necessary to ensure that data are available to answer the primary questions.

3. Look at other studies and papers in the area being studied. It may be useful to mimic analyses in the most outstanding of these papers. If they contain variables not recorded in the new study, find out why. The usual reason for excluding variables is that they are not needed to answer the problems addressed.

4. If there is a pilot phase to the study, as suggested below, analyze the data of the pilot phase to see if you can answer the questions of interest when more data become available.

2.6.2 Clarity of Questions

The task of designing clear and unambiguous questions is much greater than is usually realized. The following points are of help in designing such questions.

1. Who is filling out the forms? Forms to be filled out by many people should, as much as possible, be self-explanatory. There should not be another source to which individuals are required to go for explanation—often they would not take the trouble. This need not be done if trained technicians or interviewers are being used in certain phases of the study.

2. The degree of accuracy and the units required should be specified where possible. For example, data on heights should not be recorded in both inches and centimeters in the same place. It may be useful to allow both entries and have a computer adjust to a common unit. In this case have two possible entries, one designated centimeters and the other designated inches.

3. A response should be required on all sections of a form. Then if a portion of the form has no response, this would indicate that the answer was missing. (If an answer is required only under certain circumstances, you cannot determine whether a question was missed or a correct "no answer" response was given; a blank would be a valid answer. For example, in pathology, traditionally the pathologist reports only "positive" findings. If a finding is absent in the data, was the particular finding not considered, and missed, or was the positive outcome not there?)

4. There are many alternatives when collecting data about humans: forms filled out by an individual, an in-person interview by a trained interviewer, a telephone interview, forms filled out by medical personnel after a general discussion with the subject, or forms filled out by direct observation. It is an eye-opening experience to collect the "same" data several different ways. This leads to a healthy respect for the amount of variability in the data. It also may lead to clarification of the data being collected. In collecting subjective opinions of individuals, there is usually interaction between the subject and the method of data collection. This may greatly influence, albeit unconsciously, the subject's response.

The following points should also be noted. A high level of formal education of subjects and/or interviewer is not necessarily associated with greater accuracy or reproducibility of collected data. The personality of the subject and/or interviewer can be more important than the level of education. The effort and attention given to a particular part of a complex data set should be proportional to its importance. Prompt editing of data for mistakes produces higher quality data than when there is considerable delay between collecting, editing, and correction of forms.

2.6.3 Pretesting of Forms and Pilot Studies

If it is extremely difficult, indeed almost impossible, to design a satisfactory form, how is one to proceed? It is necessary to have a pretest of the forms, except in the simplest of experiments and studies. By a pretest, we mean the forms will be filled out by one or more individuals prior to beginning the actual study and data collection.

In this case, several points should be considered. Individuals filling out the forms should be representative of the people who will be filling out the forms during the study. You can be misled by having health professionals fill out forms that are designed for the "average" patient. You should ask the individuals filling out the pretest forms if they have any questions or are not sure about parts of the forms. However, it is important not to interfere while the forms are being used but to let them be used in the same context as the study; then ask the questions. Preliminary data should be analyzed; you should look for differences in responses from different clinics or individuals. Such analyses may indicate that a variable is being interpreted differently by different groups. The pretest forms should be edited by those responsible for the design. Comments written on the forms or answers that are not legitimate can be important in improving the forms. During this phase of the study, one should pursue vigorously the causes of missing data.

A more complete approach is to have a pilot study. A pilot study consists of going through the actual mechanics of the proposed study. Thus, a pilot study not only works out the "bugs" from forms used in the data collection but also operational problems within the study. Where possible, data collected in a pilot study should be compared with examples of the "same" data collected in other studies.

Suppose there is recording of data that are not quantitative but categorical, e.g., the amount of impairment of an animal, whether an animal was losing its hair, whether a patient has improved morale. There is a danger that the investigator(s) may use a convention that would not be readily understood by others. In order to evaluate the extent to which the data collected is understood, it is good procedure to ask others to examine some of the same study units and to record their opinion without first discussing what is meant by the categories being recorded. If there is great variability, this should give appropriate caution in the interpretation of the data. This problem may be most severe when only one individual is involved in the data collection.

2.6.4 Layout and Appearance

The physical appearance of forms is important if many individuals are to fill them out. People attach more importance to a printed page than to a mimeographed page, even though the layout is the same. If one is depending upon voluntary reporting of data, it may be worthwhile to spend the extra money to have forms printed in multiple colors with an attractive logo and appearance.

2.7 DATA EDITING AND VERIFICATION

If a study involves many people filling out forms, it will be necessary to have an individual and/or computer review the content of the forms before beginning analysis. In most studies, there are inexplicably large numbers of mistakes and missing data. If missing and miscoded data can be attacked vigorously *from the beginning* of a study, the quality of data can be vastly improved. Among checks that go into data editing are the following:

1. *Validity Checks.* Check that only allowable values or codes are given for the answers to the questions. For example, a negative weight is not allowed. A simple extension of this idea is to require that most of the data fall within a given range; range checks are set so that a small fraction of the valid data will be outside the range and will be "flagged"; for example, the height of a professional basketball team center (who happens to be a subject in the study) may fall outside the allowed range even though the height is correct. By checking out-of-range values many incorrectly recorded values can be detected.

2. *Consistency Checks.* There should be internal consistency of the data. Following are some examples:

   a. If more than one form is involved, the dates on these forms should be consistent with each other; e.g., a date of surgery should precede the date of discharge for that surgery.

   b. Consistency checks can be built into the study by collecting crucial data in two different ways, e.g., ask for both date of birth and age.

   c. If the data are sequentially collected, it is useful to examine unexpected changes between forms; e.g., changes in height, or drastic changes such as changes of weight by 70%. Occasionally such changes are correct, but they should be investigated.

   d. In some cases there are certain combinations of replies that are mutually inconsistent; checks for these should be incorporated into the editing and verification procedures.

3. *Missing Forms.* In some case control studies, a particular control may refuse to participate in the study. Some preliminary data on this control may already have been collected. Some mechanism should be set up so that it is clear that no further information will be obtained for that control. (It will be useful to keep the preliminary information so that possible selection bias can be detected). If forms are entered sequentially, it will be useful to decide when missing forms will be labeled "overdue" or "missing."

2.8 DATA HANDLING

All except the smallest experiments involve data which are eventually processed or analyzed by computer. Forms should be designed with this fact in mind. It should be easy to enter the form by keyboard. Some forms are called "self-coding": columns are given next to each variable for data entry. Except in cases where the forms are to be entered by a variety of people at different sites, the added cluttering of the form by the self-coding system is not worth the potential ease in data entry. Experienced individuals entering the same type of form over and over soon know which columns to use. Alternatively, it is possible to overlay plastic sheets which give the columns for data entry.

For very large studies, the logistics of collecting the data, putting the data on a computer system, and linking records may hinder the study more than any other factor. While it is not appropriate to discuss these issues in detail here, the reader should be aware of this problem. In any large study, individuals with expertise in data handling and computer management of data should be consulted during the design phase. Inappropriately constructed data files result in unnecessary expense and delay during the analytic phase. In projects extending over a long period of time and requiring periodic reports, it is important that the timing and management of the data collection and management be specified. Experience has shown that there will be inevitable delays even with the best plans. It is useful to allow some slack time between required submission of forms and reports, between final submission and data analysis.

Computer files or tapes will occasionally be accidentally erased. It is necessary to have back-up computer tapes and documentation in the event of such a disaster. If information on individual subject participants is required, there are confidentiality laws to be considered as well as the investigator's ethical responsibility to protect subject interests.

During the design of any study, everyone will underestimate the amount of work involved in accomplishing the task. Experience shows that caution is necessary in estimating time schedules. During a long study, constant vigilance is required to maintain quality of data collection and flow. In laboratory experimentation, technicians may tend to become bored and slack off unless monitored. Clinical study personnel will tire of collecting the data and may try to accomplish this too rapidly unless monitored.

Data collection and handling usually involves almost all participants of the study and should not be underestimated. It is a common experience for research studies to be planned without allowing sufficient time or money for data processing and analysis. It is difficult to give a rule of thumb, but in a wide variety of studies, 15% of the expense has been in data handling, processing, and analysis.

2.9 AMOUNT OF DATA COLLECTED: SAMPLE SIZE

It is part of scientific folklore that one of the tasks of a statistician is to determine an appropriate sample size for a study. Statistical considerations do have a large bearing upon the selection of a sample size. However, there is other scientific input that must be considered in order to arrive at the number of experimental units needed. If the purpose of an experiment is to estimate some quantity, then there is a need to know how precise an estimate is desired and how confident the investigator wishes to be that the estimate is within a specified degree of precision. If the purpose of an experiment is to compare several treatments, it is necessary to know what difference is considered important and how certain the investigator wishes to be of detecting such a difference. Statistical calculation of sample size requires that all these considerations be quantified. (This topic is discussed in subsequent chapters.) In a descriptive observational study, the size of the study is determined by specifying the needed accuracy of estimates of population characteristics.

2.10 INFERENCES FROM A STUDY

2.10.1 Bias

The statistical term "bias" refers to a situation in which the statistical method used does not estimate the quantity thought to be estimated, or does not test the hypothesis thought to be tested. This definition will be made more precise later. In this section, the term will be used on a intuitive level.

Consider some examples of biased statistical procedures.

1. A proposal is made to measure the average amount of health care in the United States by means of a personal health questionnaire, to be passed out at an American Medical Association convention. In this case, the AMA respondents constitute a biased sample of the overall population.

2. A famous historical example involves a telephone poll made during the Roosevelt–Landon presidential contest. At that time—and to some extent today—a large section of the population could not afford a telephone. Consequently, the poll was conducted among the more well-to-do citizens who constituted a biased sample with respect to presidential preference.

3. In a laboratory experiment, animals receiving one treatment are kept on one side of the room and animals on a second treatment on another side. If there is a large differential in lighting and heat between the two sides of the room, one could find "treatment effects" that were in fact ascribable to differences in light and/or heat. Some work by Riley [1975] suggests that level of stress (e.g., bottom cage *vs.* top cage) affects the resistance of animals to carcinogens.

In the examples of Chapter 1, some methods of minimizing bias were considered. Single blind and double blind experiments reduce bias.

2.10.2 Similarity in a Comparative Study

If physicists in Berkeley perform an experiment in electron physics, it is expected that the same experiment could be successfully performed (given the appropriate equipment) in Moscow or London. One expects the same results because the current physical model is that all electrons are precisely the same, that is, they are identical and the experiments are truly similar experiments. In a comparative experiment, we would like to try out experiments on similar units.

We now discuss similarity where it is assumed for the sake of discussion that the experimental units are humans. The ideas and results, however, can be extended to animals and other types of experimental units.

The experimental situations being compared will be called treatments. To get a fair comparison, it is necessary that the treatments be given to similar units. For example, if cancer patients whose disease had not progressed much receive a new treatment and their survival is compared to the standard treatment administered to all types of patients, the comparison would not be justified; the treatments were not given to similar groups.

Of all human beings, *identical twins* are the most alike by having identical genetic background. Often they are raised together, so that they share the same environment. Even in an observational twin study, a strong scientific inference can be made if enough appropriate pairs of identical twins can be found. For example, suppose the two "treatments" are smoking and non-smoking. If one had identical twins raised together where one of the pair smoked and the other did not, then the incidence of lung cancer, the general health, and the survival experience could provide quite strong scientific inferences as to the health effect of smoking. (In Sweden there is a twin registry to aid public health and medical studies.) It is difficult to conduct twin studies because sufficient numbers of identical twins need to be located, such that one member of the pair has one treatment and the other twin another treatment. It is expensive to identify and find them. Since they have the same environment it is most likely, in a smoking study, that either both would smoke or both would not smoke. Such studies are logistically not possible in most circumstances.

A second approach is that of matching or pairing individuals. The rationale behind *matched* or *matched pair* studies is to find two individuals who are identical with regard to all "pertinent" variables under consideration, except the treatment. This may be thought of as an attempt to find a surrogate identical twin. In many studies, individuals are matched with regard to age, sex, race, and some indicator of socio-economic status. In a prospective study, the two matched individuals receive differing treatments. In a retrospective study, the individual with the endpoint is identified first (the person usually has some disease); as we have seen, such studies are called case control studies. One weakness of such studies is that there may not be a sufficient number of subjects to make "good" matches. Matching on too many variables makes it virtually impossible to find a sufficient number of control subjects. No matter how well the matching is done, there is the possibility that the groups receiving the two treatments (or the case and control groups) are not sufficiently similar because of some unrecognized variables.

A third approach is not to match on specific variables but to try to select the subjects on an intuitive basis. For example, such procedures often select the next person entering the clinic, or have the patient select a friend of the same sex. The rationale here is that a friend will tend to belong to the same socioeconomic environment and have the same ethnic characteristics.

Still another approach even farther removed from the "identical twins" approach is to select a group receiving a given treatment and then to select in its entirety a second group as a control. The hope is that by careful consideration of the problem and good intuition, the control group will, in some sense, mirror the first treatment group with regard to "all pertinent characteristics," except the treatment and endpoint. In a retrospective study, the first group usually consists of cases, and a control group selected from the remaining population.

The final approach is to select the two groups in some manner realizing that they will not be similar, and to measure pertinent variables, such as the variables that one had considered matching upon, as well as the appropriate endpoint variables. The idea is to make statistical adjustments to find out what would have happened had the two groups been comparable. Such adjustments are done in a variety of ways. The techniques are discussed in following chapters.

None of the above methods of obtaining "valid" comparisons are totally satisfactory. In the 1920s, Sir Ronald A. Fisher and others made one of the great advances in scientific methodology—they assigned treatments to patients by chance; that is, they assigned treatments *randomly*. The technique is called *randomization*. The statistical or chance rule of assignment will satisfy certain properties that are best expressed by the concepts of probability theory. These concepts are given in Chapter 4. For assignment to two therapies, a coin toss could be used. A head would mean assignment to therapy one; a tail would result in assignment to therapy two. Each patient would have an equal chance of getting each therapy. Assignments to past patients would not have any effect on the therapy assigned to the next patient. By the laws of probability, on the average, treatment groups will be similar. *The groups will even be similar with respect to variables not measured, or not even thought about!* The mathematics of probability allow one to estimate whether differences in the outcome might be due to the chance assignment to the two groups or whether the differences should be ascribed to true differences between the treatments. These points are discussed in more detail later.

2.10.3 Inference to a Larger Population

Usually it is desired to apply the results of a study to a population beyond the experimental units. In an experiment with guinea pigs, the assumption is that if other guinea pigs had been used, the "same" results would have been found. In reporting good results with a new surgical procedure, it is implicit that this new procedure is probably good for a wide variety of patients in a wide variety of clinical settings. In order to extend results to a larger population, experimental units should be "representative" of the larger population. The best way to assure this is to select the experimental units "at random," or by chance, from the larger population. The mechanics and interpretation of such random sampling are discussed in Chapter 4. Random sampling assures, on the average, a representative sample. In other instances, if one is willing to make assumptions, the extension may be valid. There is an implicit assumption in much clinical research that a treatment is good for almost everyone or almost no one. Many techniques are used initially on the subjects available at a given clinic. It is assumed that a result is true for all clinics if it works in the one setting.

Sometimes the results of a technique are compared with "historical" controls; that is, a new treatment is compared with the results of previous patients using an older technique. The use of historical controls can be hazardous; patient populations change with time, often in ways that have much more importance than is generally realized.

Another approach with weaker inference is the use of an animal model. The term "animal model" indicates that the particular animal is susceptible to, or suffers from a disease similar to that experienced by humans. If a treatment works on the animal, it may be useful for humans. There would then be an investigation in the human population to see whether the assumption is valid.

The results of an observational study carried out in one country may be extended to other countries. This is not always appropriate. Much of the "bread and butter" of epidemiology consists of noting that the same risk factor seems to produce different results in different populations, or in noting that the particular endpoint of a disease occurs with differing rates in different countries. There has been considerable advance in medical science by noting different responses among different populations. This is a broadening of the topic of this section: extending inferences in one population to another population.

2.10.4   Precision and Validity of Measurements

Statistical theory leads to the examination of variation in a method of measurement. The variation may be estimated by making repeated measurements on the same experimental unit. If instrumentation is involved, multiple measurements may be taken using more than one of the instruments to note the variation between instruments. If different observers, interviewers or technicians take measurements, a quantification of the variability between different observers may be made.

It is necessary to have information on the precision of a method of measurement in calculating sample size for a study. This information is also used in considering whether or not variables deserve repeated measurements to gain increased precision about the true response of the experimental unit.

Statistics helps in thinking about alternative methods of measuring a quantity. When introducing a new apparatus or new technique to measure a quantity of interest, validation against the old method is useful. In considering subjective ratings by different individuals (even when the subjective rating is given as a numerical scale) it often turns out that a quantity is not measured in the same fashion if the measurement method is changed. A new laboratory apparatus may measure consistently higher than an old one. In two methods of evaluating pain relief, one way of phrasing a question may tend to give a higher percentage of improvement. Methodologic statistical studies are helpful in placing interpretations and inferences in their proper context.

2.10.5   Quantification and Reduction of Uncertainty

Because of variability there is uncertainty associated with the interpretation of study results. Statistical theory allows quantification of the uncertainty. If a quantity is being estimated, the amount of uncertainty in the estimate must be assessed. In considering a hypothesis one may give numerical assessment of the chance of occurrence of the observed results when the hypothesis is true.

Appreciation of statistical methodology often leads to the design of a study with increased precision and consequently a smaller sample size. An example of an efficient technique is the statistical idea of blocking. Blocks are subsets of relatively homogeneous experimental units. The strategy is to apply all the treatments randomly to the units within a particular block. Such a design is called a randomized block design. The advantage of the technique is that comparisons of treatments are intra-block comparisons (that is, comparisons within blocks) and are more precise because of the homogeneity of the experimental units within the blocks, so that it is easier to detect treatment differences. Simple randomization, as discussed before, does ensure similar groups but the variability within the treatment groups will be greater if no blocking of experimental units has been done. For example, if age is important prognostically in the outcome of a comparative trial of two therapies, there are two approaches one may take. If one ignores age and randomizes the two therapies, the therapies will be tested on similar groups, but the variability in outcome due to age will tend to mask the effects of the two treatments. Suppose you place individuals whose ages are close into blocks, and assign each treatment by a chance mechanism within each block. If you then compare the treatments within the blocks, the effect of age on the outcome of the two therapies will be largely eliminated. A more precise comparison of the therapeutic effects can be gained. This increased precision due to statistical design leads to a study requiring a smaller sample size than a completely randomized design.

A good statistical design allows the investigation of several factors at one time with little added cost:

> "No aphorism is more frequently repeated with field trials than we must ask Nature a few questions, or ideally, one question at a time. The writer is convinced that this view is wholly mistaken. Nature, he suggests, will best respond to a logical and carefully thought out questionnaire; indeed if we ask her a single question, she will often refuse to answer until some other topic has been discussed." (Sir R. A. Fisher as quoted by Yates [1964].)

PROBLEMS

1. Consider the following terms defined in Chapters 1 and 2: single blind, double blind, placebo, observational study, experiment, laboratory experiment, comparative experiment, cross-over experiment, clinical study, cohort, prospective study, retrospective study, case-control study, matched case-control study. In the examples of Chapter 1, which terms apply to which parts of these examples?

2. List possible advantages and disadvantages of a double blind study. Give some examples where a double blind study clearly cannot be carried out; suggest how virtues of "blinding" can still be retained.

3. Discuss the ethical aspects of a randomized placebo controlled experiment. Can you think of situations where it would be extremely difficult to carry out such an experiment?

4. Discuss the advantages of randomization in a randomized placebo controlled experiment. Can you think of alternative, possibly better, designs? Consider (at least) the aspects of bias and efficiency.

5. This problem involves design of two questions on "stress" to be used on a data collection form for the population of a group practice health maintenance organization. After a few years of follow-up it is desired to assess the effect of physical and psychological stress.
   a. Design a question which classifies jobs by the amount of physical work involved. Use eight or fewer categories. Assume that the answer to the question is to be based on job title. That is, someone will code the answer given a job title.
   b. Same as (a), but now the classification should pertain to the amount of psychological stress.
   c. Have yourself and (independently) a friend answer your two questions for the following occupational categories: student, college professor, plumber, waitress, homemaker, salesperson, unemployed, retired, unable to work (due to illness), physician, hospital administrator, grocery clerk, prisoner.
   d. What other types of questions would you need to design to capture the total amount of stress in the individual's life?

6. In designing a form, careful distinction must be made between the following categories of nonresponse to a question: (1) not applicable, (2) not noted, (3) don't know, (4) none, and (5) normal. If nothing is filled in, someone has to determine which of the five categories applies—and often this cannot be done after the interview or the records have been destroyed. This is particularly troublesome when medical records are abstracted. Suppose you are checking medical records to record the number of pregnancies (gravidity) of a patient. Unless the gravidity is specifically given you have a problem. If no number is given, any one of the four categories above could apply. Give two other examples of questions with ambiguous interpretation of "blank" responses. Devise a scheme for interview data that is unambiguous and does not require further editing.

REFERENCES

Meier, P., Free, Jr., S. M., and Jackson, G. L. [1968]. Reconsideration of methodology in studies of pain relief. *Biometrics*, **14:** 330–342.

Papworth, M. H. [1967]. *Human Guinea Pigs*. Beacon Press, Boston, MA.

Riley, V. [1975]. Mouse mammary tumors; alteration of incidence as apparent function of stress. *Science*, **189:** 465–467.

Roethlisberger, F. S. [1941]. *Management and Morals*. Harvard University Press, Cambridge, MA.

Spicker *et al.* (eds) [1988]. *The Use of Human Beings in Research, with Special Reference to Clinical Trials*. Kluwer Academic, Boston, MA.

US Department of Agriculture [1989]. Animal Welfare; proposed rules, part III. *Federal Register*, March 15, 1989.

US Department of Health, Education and Welfare [1975]. Protection of human subjects, part III. *Federal Register*, August 8, 1975, **40:** 11854.

US Department of Health, Education and Welfare [1985]. *Guide for the Care and Use of Laboratory Animals*. DHEW Publication No. (NIH) 86–23. US Government Printing Office, Washington, DC.

Yates, F. [1964]. Sir Ronald Fisher and the design of experiments. *Biometrics*, **20:** 307–321. Used wth permission from The Biometric Society.

CHAPTER 3

Descriptive Statistics

3.1 INTRODUCTION

The beginning of an introductory statistics textbook usually contains a few paragraphs placing the subject matter in its encyclopedic order, discussing the limitations or wide ramifications of the topic and tends to the more philosophical rather than the substantive-scientific. Briefly, we consider science to be a study of the world emphasizing qualities of permanence, order, and structure. Such a study involves a drastic reduction of the real world and often numerical aspects only are considered. If there is no obvious numerical aspect or ordering, an attempt is made to impose it. For example, quality of medical care is not an immediately numerically scaled phenomenon, but some scale is often induced or imposed. Statistics is concerned with the estimation, summarization and obtaining of reliable numerical characteristics of the world. It will be seen that this is in line with some of the definitions given in the Notes in Chapter 1.

It may be objected that a characteristic such as the sex of a newborn baby is not numerical, but it can be coded (arbitrarily) in a numerical way, for example: $0 =$ male, $1 =$ female. Many such characteristics can be numerically *labeled*, and as long as the code, or the dictionary, is known, it is possible to go back and forth.

Consider a set of measurements of head circumferences of term infants born in a particular hospital. We have a quantity of interest—head circumference—which varies from baby to baby, and a collection of actual values of head circumferences.

Definition 3.1. A *variable* is a quantity that may vary from object to object.

Definition 3.2. A *sample* (or data set) is a collection of values of one or more variables. A member of the sample is called an *element*.

We distinguish between a variable and the value of a variable in the same way that the label, "title of a book in the library" is distinguished from the title, *Gray's Anatomy*. A variable will usually be represented by a capital letter, say, Y and a value of the variable by a lower case letter, say, $Y = y$.

In this chapter, we discuss briefly the types of variables typically dealt with in statistics. We then go on to discuss ways of *describing* samples of values of variables, both numerically and graphically. A key concept is that of a *frequency distribution*. Such presentations can be considered part of *descriptive* statistics. Finally, we discuss one of the earliest challenges to statistics, how to *reduce* samples to a few summarizing numbers. This will be considered under the heading descriptive *statistics*.

3.2 TYPES OF VARIABLES

3.2.1 Qualitative (Categorical) Variables

Some examples of qualitative (or categorical) variables and their values are:

1. Color of an individual's hair (black, grey, red, ..., brown).
2. Sex of child (male, female).
3. Province of residence of a Canadian citizen (Newfoundland, Nova Scotia, ..., British Columbia).
4. Cause of death of newborn (congenital malformation, asphyxia, ...).

Definition 3.3. A *qualitative variable* has values that are intrinsically nonnumerical (categorical).

As suggested in the introduction, the values of a qualitative variable can always be put into numerical form. The simplest numerical form is consecutive labeling of the values of the variable. The values of a qualitative variable are also referred to as *outcomes* or *states*.

Note that examples 3 and 4 are ambiguous. In example 3, what shall we do with Canadian citizens living outside of Canada? We could arbitrarily add another "province" with the label "Outside Canada." Example 4 is ambiguous because there may be more than one cause of death. Both of these examples show that it is not always easy to anticipate all the values of a variable. Either the list of values must be changed or the variable must be redefined.

The arithmetic operation associated with the values of qualitative variables is usually that of counting. Counting is perhaps the most elementary—but not necessarily simple—operation that organizes or abstracts characteristics. A count is an answer to the question: "How many?" (Counting assumes that whatever is counted shares some characteristics with the other "objects." Hence it disregards what is unique and reduces the objects under consideration to a common category or class.) Counting leads to statements such as, "the number of births in Ontario in 1979 was 121,655."

Qualitative variables can often be ordered or ranked. Ranking or ordering places a set of objects in a sequence according to a specified scale. In Chapter 2, clinicians ranked interns according to the quality of medical care delivered. The "objects" were the interns and the scale was "quality of medical care delivered." The interns could also be ranked according to their height from shortest to tallest—the "objects" are again the interns and the scale is "height." The provinces of Canada could be ordered by their population sizes from lowest to highest. Another possible ordering is by the latitudes of, say, the capitals of each of the provinces. Even hair color could be ordered by the wavelength of the dominant color. Two points should be noted in connection with ordering or qualitative variables. First, as indicated by the example of the provinces, there is more than one ordering that can be imposed on the outcomes of a variable, i.e., there is no natural ordering; the kind of ordering imposed will depend on the nature of the variable and the purpose for which it is studied—if we wanted to study the impact of crowding or pollution in Canadian provinces, we might want to rank them by population size. If we wanted to study rates of melanoma as related to amount of ultraviolet radiation, we might want to rank them by the latitude of the provinces as summarized, say by the latitudes of the capitals or most populous areas. Second, the ordering need not be complete, that is, we may not be able to rank each outcome above or below another. For example, two of the Canadian provinces may have virtually identical populations so that it is not possible to order them. Such orderings are called *partial*.

3.2.2 Quantitative Variables

Some examples of quantitative variables (with scale of measurement; values) are the following:

1. Height of father (half inch units; 0.0, 0.5, 1.0, 1.5, ..., 99.0, 99.5, 100.0).
2. Number of particles emitted by a radioactive source (counts per minute; 0, 1, 2, 3, ...).
3. Total body calcium of a patient with osteoporosis (nearest gram; 0, 1, 2, ..., 9999, 10,000).
4. Survival time of a patient diagnosed with lung cancer (nearest day; 0, 1, 2, ..., 19,999, 20,000).
5. Apgar score of infant 60 seconds after birth (counts; 0, 1, 2, ..., 8, 9, 10).
6. Number of children in a family (counts; 0, 1, 2, 3, ...).

Definition 3.4. A *quantitative variable* has values that are intrinsically numerical.

As illustrated by the above examples, we must specify two aspects of a variable: the scale of measurement and the values the variable can take on. Some quantitative variables have numerical values that are integers, or discrete. Such variables are referred to as *discrete* variables. The variable "number of particles emitted by a radioactive source" is such an example; there are "gaps" between the successive values of this variable. It is not possible to observe 3.5 particles. (It is sometimes a source of amusement when discrete numbers are manipulated to produce values that cannot occur—for example, "the average American family" has 2.125 children.) Other quantitative variables have values that are potentially associated with real numbers—such variables are called *continuous* variables. For example, the survival time of a patient diagnosed with lung cancer may be expressed to the nearest day, but this phrase implies that there has been rounding. We could refine the measurement to, say, hours, or even more precisely, to minutes or seconds. The exactness of the values of such a variable is determined by the precision of the measuring instrument as well as the usefulness of extending the value. Usually, a reasonable unit is assumed and it is considered *pedantic* to have a unit that is too refined, or *rough* to have a unit that does not permit distinction between the objects on which the variable is measured. Examples 1, 3, and 4 above deal with continuous variables; those in the other examples are discrete. Note that with quantitative variables there is a natural ordering, e.g., from lowest to highest value (see also Note 3.6 for another taxonomy of data).

In each illustration of qualitative and quantitative variables, we listed all of the possible values of a variable. (Sometimes the values could not be listed; usually indicated by inserting three dots "..." into the sequence.) This leads to:

Definition 3.5. The *sample space* or *population* is the set of all possible values of a variable.

The definition or listing of the sample space is not a trivial task. In the examples of qualitative variables, we already discussed some ambiguities associated with the definitions of a variable and the sample space associated with the variable. Your definition must be reasonably precise without being "picky." Consider again the variable "Province of residence of a Canadian citizen" and the sample space (Newfoundland, Nova Scotia, . . . , British Columbia). Some questions that can be raised include:

1. What about citizens living in the Northwest Territories? (reasonable question)
2. Are landed immigrants who are not citizens as yet to be excluded? (reasonable question)
3. What time point is intended? Today? January 1, 1990? (reasonable question)
4. If January 1, 1990 is used, what about citizens who died on that day? Are they to be included? (This is becoming somewhat "picky.")

3.3 *DESCRIPTIVE* STATISTICS

3.3.1 Tabulations and Frequency Distributions

One of the simplest ways to summarize data is by tabulation. John Graunt, in 1652, published his observations on Bills of Mortality, excerpts of which can be found in Newman [1956]. Table 3.1 is a condensation of Graunt's list of 63 diseases and casualties.

Several things should be noted about the table. To make up the table, three ingredients are needed: (1) a *collection* of objects (in this case, humans); (2) a *variable* of interest (the cause of death); (3) *frequency* of occurrence of each of the categories.

Table 3.1. Diseases and Casualties for the Year 1632, City of London. A Selection from Graunt's Tables.

Disease	Casualties
Abortive and stillborn	445
Affrighted	1
Aged	628
Ague	43
⋮	⋮
Crisomes and infants	2268
⋮	⋮
Tissick	34
Vomiting	1
Worms	27
In all	9535

Table 3.2. Rearrangement of Graunt's Data (Table 3.1), by Ten Most Common Causes of Death.

Disease	Casualties
Crisomes and infants	2268
Consumption	1797
Fever	1108
Aged	628
Flocks and smallpox	531
Teeth	470
Abortive and stillborn	445
Bloody flux, scowring, and flux	348
Dropsy and swelling	267
Convulsion	241
Childbed	171
Total	8274

These will be defined more precisely later. Secondly, we note that the disease categories are arranged alphabetically (ordering number 1). This may not be too helpful if we want to look at the most common causes of death. Let us rearrange Graunt's table by listing disease categories by greatest frequencies (ordering number 2).

Table 3.2 contains a listing of the ten most common disease categories in Graunt's table. This table summarizes $8274/9535 = 87\%$ of the data in Table 3.1.

From Table 3.2 we see at once that "Crisomes" is the most frequent cause of death. (A "Crisome" is an infant dying within 1 month of birth. Gaunt lists the number of "Christenings" [births] as 9584, so a crude estimate of neonatal mortality is $2268/9584 \doteq 24\%$. The symbol "$\doteq$" means "approximately equal to.") Finally, we note that data for 1633 almost certainly would not have been identical to that of 1632. However, the number in the category "Crisomes" probably would have remained the largest. An example of a statistical question is whether this predominance of "Crisomes and Infants" has a quality of permanence from one year to the next.

A second example of a tabulation involves keypunching errors made by a data entry operator. To be entered were 156 lines of data, each line containing data on the number of crib deaths for a particular month in King County, Washington, for the years 1965–1977. Other data on a line consisted of meteorological data as well as the total number of births for that month for King County. Each line required the punching of 47 characters, excluding the spaces. The numbers of errors per line starting with January, 1965, and ending with December, 1977, are listed in Table 3.3.

One of the problems with this table is its bulk. It is difficult to grasp its significance. You would not transmit this table over the phone to explain to someone the number of errors made. One way to summarize this table is to specify how many times a particular combination of errors occurred. One possibility is the following:

Number of Errors per Line	Number of Lines
0	124
1	27
2	5
3 or more	0

Table 3.3. Number of Keypunching Errors Per Line for 156 Consecutive Lines of Data Entered (each digit represents number of errors in a line).

0	0	1	0	2	0	0	0	1	0	0	0
0	0	0	0	1	0	0	1	2	0	0	1
1	0	0	2	0	0	0	0	0	0	0	0
0	1	0	0	0	0	0	0	0	0	0	0
1	0	0	0	0	0	0	0	0	0	0	0
0	0	0	0	0	0	0	0	1	0	0	0
0	1	1	1	1	0	0	0	0	0	0	1
0	1	0	0	1	0	0	0	0	2	0	0
1	0	0	0	2	0	0	0	0	0	0	0
1	0	0	0	1	0	1	0	0	0	0	0
1	1	1	0	0	0	0	0	0	0	0	0
0	1	0	1	1	0	0	0	0	0	0	0
0	0	0	0	0	0	1	0	0	0	0	0

This list is based on three ingredients as before: a *collection* of lines of data, a *variable* (the number of errors per line) and the *frequency* with which values of the variable occur. Have we lost something in going to this summary? Yes—we have lost the order in which the observations occurred. That could be important if we wanted to find out whether errors came "in bunches" or whether there was a learning process so that fewer errors occurred as practice was gained. The original data are already a condensation. The "number of errors per line" does not give information about the location of the errors in the line nor the type of error. (For educational purposes the latter might be very important.)

A difference between the variables of Tables 3.2 and 3.3 is that the variable in the second example was *numerically valued*, i.e., took on numerical values, in contrast with the *categorically valued* variable of the first example. Statisticians typically mean the former when *variable* is used by itself and we will specify *categorical variable* when appropriate. (As discussed before, a categorical variable can always be made numerical by (as in Table 3.1) alphabetically arranging the values and numbering the observed categories 1, 2, 3,.... This is not biologically meaningful because the ordering is a function of the language used.)

The data of the above two examples were discrete. A different type of variable is represented by the age at death of crib death or SIDS (Sudden Infant Death Syndrome) cases. Table 3.4 displays ages at death in days of 78 cases of SIDS in King County, Washington, during the years 1976–1977. The variable, age at death, is continuous. However, there is rounding to the nearest whole day. Thus, "68 days"

Table 3.4. Age at Death (in days) of 78 Cases of Crib Death (SIDS) Occurring in King County, Washington, 1976–1977.

225	174	274	164	130	96	102	80	81	148	130	48
68	64	234	24	187	117	42	38	28	53	120	66
176	120	77	79	108	117	96	80	87	85	61	65
68	139	307	185	150	88	108	60	108	95	25	80
143	57	53	90	76	99	29	110	113	67	22	118
47	34	206	104	90	157	80	171	23	92	115	87
42	77	65	45	32	44						

Table 3.5. Frequency Distribution of Age at Death of 78 SIDS Cases Occurring in King County Washington, 1976–1977.

Age Interval (Days)	Number of Deaths
1–30	6
31–60	13
61–90	23
91–120	18
121–150	7
151–180	5
181–210	3
211–240	1
241–270	0
271–300	1
301–330	1
Total	78

could represent 68.438... or 67.8873..., where the three dots indicate an unending decimal sequence.

Again, the table staggers us by its bulk. Unlike the previous example, it will not be too helpful to list the number of times a particular value occurs: there are just too many different ages. One way to reduce the bulk is to define intervals of days and count the number of observations that fall in each interval. Table 3.5 displays the data grouped into 30-day intervals (months). Now the data makes more sense. We note, for example, that many deaths occur between the ages of 61 and 90 days (2–3 months) and that very few deaths occur after 180 days (6 months). Somewhat surprisingly, there are relatively few deaths in the first month of life. This age distribution pattern is unique to SIDS.

We again note the three characteristics on which Table 3.5 is based: (1) a *collection* of 78 objects—SIDS cases; (2) a *variable* of interest—age at death; and (3) the *frequency* of occurrence of values falling in specified intervals. We are now ready to define these three characteristics more explicitly.

Definition 3.6. An *empirical frequency distribution (EFD)* of a variable is a listing of the values or ranges of values of the variable together with the frequencies with which these values or ranges of values occur.

The adjective "empirical" emphasizes that an *observed* set of values of a variable is being discussed; if this is obvious we may use just "frequency distribution" (as in the heading of Table 3.5).

The choice of interval width and interval endpoint is somewhat arbitrary. They are usually chosen for convenience. In Table 3.5, a "natural" width is 30 days (1 month) and convenient endpoints are 1 day, 31 days, 61 days, and so on. A good rule is to try to produce between seven and ten intervals. To do this, divide the range of the values (*largest–smallest*) by 7, and then adjust to make a simple interval. For example suppose the variable is "weight of adult male" (expressed to the nearest kilogram) and the values vary from 54 kg to 115 kg. The range is 115–54 = 61 kg, suggesting intervals of width $61/7 \doteq 8.7$ kg. This is clearly not a very good width; the closest "natural" width is 10 kg (producing a slightly coarser grid). A reasonable starting point is 50 kg so that the intervals have endpoints 50 kg, 60 kg, 70 kg, and so on.

To compare several EFDs it is useful to make them comparable with respect to the total number of subjects. To make them comparable we need:

Definition 3.7.   The *size* of a sample is the number of elements in the sample.

Definition 3.8.   An *empirical relative frequency distribution (ERFD)* is an empirical frequency distribution where the frequencies have been divided by the sample size.

Equivalently, the relative frequency of the value of a variable is the proportion of times that value of the variable occurs. (The context often makes it clear that an *empirical* frequency distribution is involved. Similarly, many authors omit the adjective *relative* so that "frequency distribution" is shorthand for "empirical relative frequency distribution.")

To illustrate ERFDs, consider the data from Winkelstein *et al.* [1975], in Table 3.6, consisting of systolic blood pressures of three groups of Japanese men: native Japanese, first-generation immigrants to the United States (Issei) and second-generation Japanese in the United States (Nisei). The sample sizes are 2232, 263, and 1561, respectively.

It is difficult to compare these distributions because the sample sizes differ. The *relative* frequencies (proportions) are obtained by dividing each frequency by the corresponding sample size. The ERFD is presented in Table 3.7. For example, the (empirical) relative frequency of native Japanese with systolic blood pressure less than 106 mmHg is $218/2232 = 0.098$.

It is still difficult to make comparisons. One of the purposes of the study was to determine how much variables such as blood pressure were affected by environmental conditions. To see if there is a *shift* in the blood pressures, we could consider the proportion of men with blood pressures less than some specified value and compare the groups that way. Consider, for example, the proportion of men with systolic blood pressures less than or equal to 134 mmHg. For the native Japanese this is (Table 3.7): $(0.098 + 0.122 + 0.151 + 0.162) = 0.533$, or 53.3%. For the Issei and Nisei

Table 3.6. Empirical Frequency Distribution of Systolic Blood Pressure of Native Japanese, First Generation and Second Generation Immigrants to the United States. Males aged 45–69 years. Data from Winkelstein *et al.* [1975].

Blood Pressure (mmHg)	Native Japanese	Issei	California Nisei
< 106	218	4	23
106–114	272	23	132
116–124	337	49	290
126–134	362	33	347
136–144	302	41	346
146–154	261	38	202
156–164	166	23	109
> 166	314	52	112
Totals	2232	263	1561

Table 3.7. Empirical Relative Frequency Distribution of Systolic Blood Pressure of Native Japanese, First Generation and Second Generation Immigrants to the United States. Males aged 45–69 years.

Blood Pressure (mmHg)	Native Japanese	Issei	California Nisei
< 106	0.098	0.015	0.015
106–114	0.122	0.087	0.085
116–124	0.151	0.186	0.186
126–134	0.162	0.125	0.222
136–144	0.135	0.156	0.222
146–154	0.117	0.144	0.129
156–164	0.074	0.087	0.070
> 166	0.141	0.198	0.072
Totals	1.000	0.998	1.001
Sample size	(2232)	(263)	(1561)

these figures are 0.413 and 0.508, respectively. These last two figures are somewhat lower than the first suggesting that there has been a shift to higher systolic blood pressure among the immigrants. Whether this shift represents sampling variability or a genuine shift in these groups can be determined by methods developed in the next three chapters.

The concept discussed above is formalized in the empirical cumulative distribution (ECD).

Definition 3.9. The *empirical cumulative distribution (ECD)* of a variable is a listing of values of the variable together with the *proportion* of observations less than or equal to that value (cumulative proportion).

Before we construct the ECD for a sample, we need to clear up one problem associated with rounding of values of continuous variables. Consider the age of death of the SIDS cases of Table 3.4. The first age listed is 225 days. Any value between 224.5+ and 225.5− is rounded off to 225 (224.5+ indicates a value greater than 224.5 by some arbitrarily small amount, and similarly, 225.5− indicates a value less than 225.5). Thus, the upper endpoint of the interval 1–30 days in Table 3.5 is 30.4$\overline{9}$, or 30.5.

The ECD associated with the data of Table 3.5 is presented in Table 3.8 which contains: (1) the age intervals, (2) endpoints of the intervals, (3) EFD, (4) ERFD, and (5) ECD.

Two comments are in order: (1) there is a slight rounding error in the last column because the relative frequencies are rounded to three decimal places—if we had calculated from the frequencies rather than the relative frequencies this problem would not have occurred; and (2) given the cumulative proportions the original proportions can be recovered. For example, consider the following endpoints and their cumulative frequencies:

$$150.5 \quad 0.860$$
$$180.5 \quad 0.924$$

Table 3.8. Frequency Distribution of Age at Death of 78 SIDS Cases Occurring in King County Washington, 1976–1977.

Age Interval (Days)	Endpoint of Interval (Days)	Number of Deaths	Relative Frequency (Proportion)	Cumulative Proportion
1–30	30.5	6	0.077	0.077
31–60	60.5	13	0.167	0.244
61–90	90.5	23	0.295	0.539
91–120	120.5	18	0.231	0.770
121–150	150.5	7	0.090	0.860
151–180	180.5	5	0.064	0.924
181–210	210.5	3	0.038	0.962
211–240	240.5	1	0.013	0.975
241–270	270.5	0	0.000	0.975
271–300	300.5	1	0.013	0.988
301–330	330.5	1	0.013	1.001
Total		78	1.001	

Subtracting, $0.924 - 0.860 = 0.064$ produces the proportion in the interval 151–180. Mathematically, the ERFD and the ECD are equivalent.

3.3.2 Graphs

Graphical displays frequently provide very effective descriptions of samples. In this section, we discuss some very common ways of doing this and close with some examples that are innovative. Graphs can also be used to enhance certain features of data as well as distort them. A good discussion can be found in Huff [1982].

One of the most common ways of pictorially describing a sample is to plot on one axis values of the variable and on another axis the frequency of occurrence of a value or some measure related to it. Consider the data of Table 3.8 which consists of the age at death of 78 SIDS cases. Figure 3.1 represents the empirical frequency distribution by means of a *histogram*. The area of a bar is proportional to the frequency of occurrence of values in the interval. Since the intervals are all equal, the height of a bar is proportional to the frequency.

The values of a variable are usually plotted on the abscissa (X-axis), the frequencies on the ordinate (Y-axis). The ordinate on the left hand side of Figure 3.1 contains the relative frequencies of occurrence of the values.

Figure 3.2 displays the empirical cumulative distribution (ECD). This is a *step-function* with jumps at the endpoints of the interval. The height of the jump is equal to the relative frequency of the observations in the interval. The ECD is nondecreasing and is bounded above by 1. Both Figures 3.1 and 3.2 emphasize the discreteness of data. A *frequency polygon* and *cumulative frequency polygon* are often used with continuous variables to emphasize the continuity of the data. A frequency polygon is obtained by joining the heights of the bars of the histogram at their midpoints. The frequency polygon for the data of Table 3.8 is displayed in Figure 3.3. A question arises: where is the midpoint of the interval? To calculate the midpoint for the interval 31–60 days, we note that the limits of this interval are 30.5–60.5. The mid-

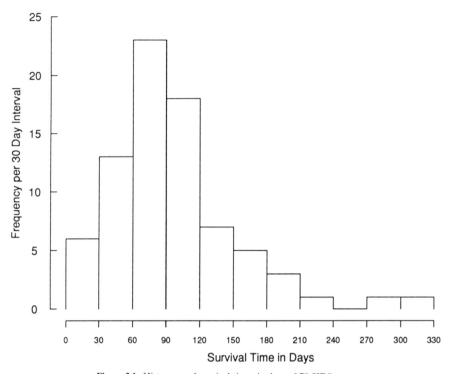

Figure 3.1. Histogram of survival times in days of 78 SIDS cases.

point is half-way between these endpoints, hence, *midpoint* = $(30.5 + 60.5)/2 = 45.5$ days.

All midpoints are spaced in intervals of 30 days so that the midpoints are 15.5, 45.5, 75.5, and so on. To close the polygon, the midpoints of two additional intervals are needed: one to the left of the first interval (1–30) and one to the right of the last observed interval (301–330), both of these with zero observed frequencies.

The cumulative frequency polygon is constructed by joining the cumulative relative frequencies observed at the endpoints of their respective intervals. Figure 3.4 displays the cumulative relative frequency of the SIDS data of Table 3.8. The curve has the value 0.0 below 0.5 and the value 1.0 to the right of 330.5. Both the histograms and the cumulative frequency graphs implicitly assume that the observations in our interval are evenly distributed over that interval.

One advantage of the cumulative frequency polygon is that the proportion (or percentage) of observations less than a specified value can be read off easily from the graph. For example, from Figure 3.4, it can be seen that 50% of the observations have a value less than 88 days (this is the median of the sample). See Section 3.4.2 for further discussion.

EFDs can often be graphed in an innovative way to illustrate a point. Consider the data in Figure 3.5 which contains the frequency of births per day as related to phases of the moon. Data was collected by Schwab [1975] on the number of births for 2 years, grouped by each day of the 29-day lunar cycle, and is presented here

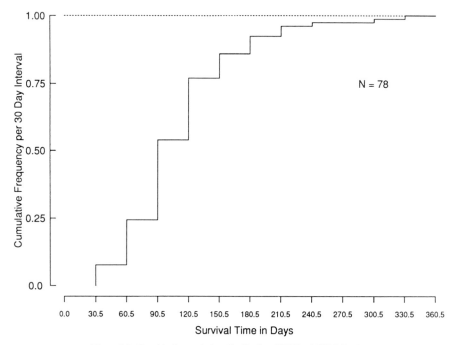

Figure 3.2. Empirical cumulative distribution (ECD) of SIDS deaths.

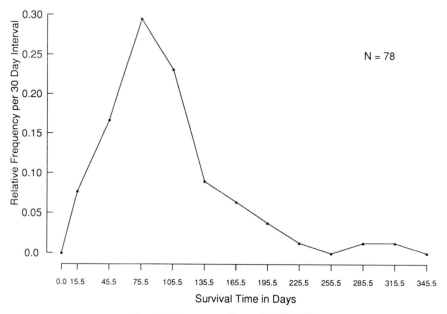

Figure 3.3. Frequency polygon of SIDS deaths.

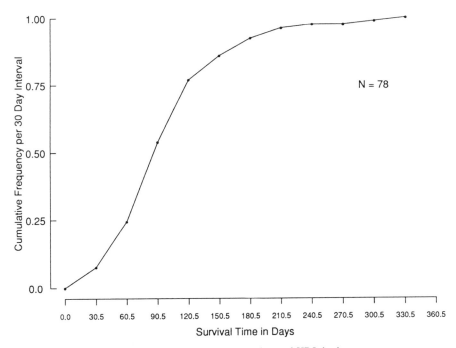

Figure 3.4. Cumulative frequency polygon of SIDS deaths.

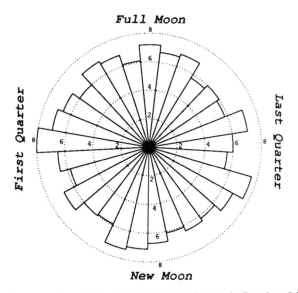

Figure 3.5. Average number of births per day over 29-day lunar cycle. Data from Schwab [1975].

as a circular distribution where the lengths of the sectors are proportional to the frequencies. (There is clearly no evidence supporting the hypothesis that the cycle of the moon influences the birth rate.)

Sometimes more than one variable is associated with each of the objects under study. Data arising from such situations are called *multivariate data*. A moment's reflection will convince you that most biomedical data is multivariate in nature. For example, the variable "blood pressure of a patient" is usually expressed by two numbers, systolic and diastolic blood pressure. We often specify age and sex of the patient to characterize the blood pressure more accurately. In the multivariate situation, in addition to describing the frequency with which each value of each variable occurs, we may also want to study the relationships among the variables. For example, Table 1.2 and Figure 1.1 attempt to assess the relationship between the variables "Clinical Competence" and "Cost of Laboratory Procedures Ordered" of interns. Graphs of multivariate data will be found throughout the text. Here we present a few examples of visually displaying values of several variables at the same time. Figure 3.6 is a presentation from Fish *et al.* [1976] and depicts the following three variables simultaneously:

1. Mother's education (years).
2. Socioeconomic index.
3. IQ of offspring at 7 years.

Note that each cell in the table contains the number of children whose mothers had

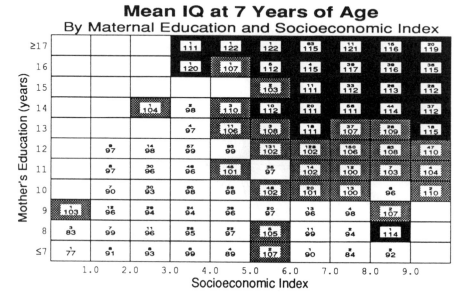

Figure 3.6. Mean IQ at 7 years of age by maternal education and socioeconomic index. From Fish *et al.* [1976]. Copyright 1976 Amercan Medical Association.

the specified combination of education and socioeconomic index, the average IQ for these children and, in addition, shading of intervals of IQs. A frequency distribution for the mother's years of education could be obtained by summing the bracketed numbers in a cell in each row. For example, the number of mothers with years of education greater than or equal to 17 years is $1 + 1 + 1 + 63 + 11 + 15 + 20 = 112$. Such a frequency distribution is called a *marginal* frequency distribution because the frequencies are the marginal totals for the table.

Figure 3.7 compares the site of occurrences of familial melanoma with nonfamilial, or sporadic, cases. These data, from Wallace and Exton [1972], do not suggest an obvious difference between the two types. In part this is due to the relatively rarer occurrence of familial cases. There are ways of quantifying these data. You might think of some way of doing this.

For beautiful books on the visual display of data see Tufte [1983] and [1990]. A very readable compendium of graphical methods is contained in Moses [1987].

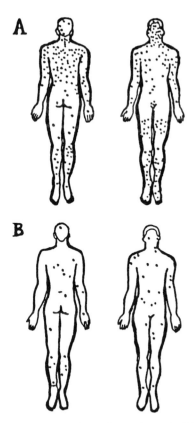

Figure 3.7. Sites of occurrences of familial melanoma in nonfamilial, or sporadic, cases.

3.3.3 Stem and Leaf Diagrams

An elegant way of describing data consists of stem-and-leaf diagrams. (A phrase coined by J. W. Tukey—see his book for some additional innovative methods of describing data; Tukey [1977].) The following data deal with the aflatoxin levels of raw peanut kernels as described by Quesenberry *et al.* [1976]. Approximately 560 g of ground meal was divided among 16 centrifuge bottles and analyzed. One sample was lost so that only 15 readings are available (measurement units are not given). The values were:

$$30, 26, 26, 36, 48, 50, 16, 31, 22, 27, 23, 35, 52, 28, 37.$$

How can this sample best be described? One way is by means of a stem-and-leaf diagram. In this case the number of 10s form the "stem" and the unit digits represent the "leaves."

Stem (Tens)	Leaf (Units)
1	6
2	6 6 2 7 3 8
3	0 6 1 5 7
4	8
5	0 2

For example, the row 3 | 06157 is a description of the observations 30, 36, 31, 35 and 37. The most frequently occurring category is the 20s. The smallest value is 16, the largest value, 52.

A nice feature of this stem-and-leaf diagram is that all of the values can be recovered (but not in the sequence in which the observations were made). Another

Table 3.9. Stem-and-Leaf Diagram of Age at Death of 78 SIDS Cases. Data from Table 3.4.

Stem (20s)	Leaf (units)
2	4 8 8 5 9 2 4 3 2
4	8 2 3 7 3 7 2 5 4
6	8 4 6 7 9 1 5 8 0 6 7 7 5
8	6 0 1 6 0 7 5 8 5 0 0 9 0 0 2 7
10	2 7 8 7 8 8 0 3 8 4 5
12	0 0 0 0 9
14	8 0 3 7
16	4 4 6 1
18	7 5
20	6
22	5 4
24	
26	4
28	
30	7

useful feature which will be used is that a quick ordering of the observations can be obtained by use of the stem-and-leaf diagram.

The reason this example worked well is that the number of 10s categories was small: five. But consider the data of Table 3.4—the age at death in days of 78 SIDS cases. The smallest value is 22 days, the largest value is 307 days. There will be 28 stems if we use the above method. One alternative is to convert the days to months, producing 11 stems. Another alternative uses the stems in intervals of 20, say, 0, 20, 40, ... A value of 23 is represented by a 2 | 3 and a value of 33 as 2 | 3 and so on. Table 3.9 is constructed in this way. The diagram illustrates very clearly the pattern of ages at death.

Another suggestion uses intervals of 20 and represents each observation by two digits. This is somewhat wasteful but works well with small sets of data.

3.4 DESCRIPTIVE *STATISTICS*

3.4.1 Introduction

In the previous section the emphasis was on tabular and visual display of data. It is clear that these techniques can be used to great advantage summarizing and highlighting data. However, even a table or a graph takes up quite a bit of space, cannot be summarized in the mind too easily, and, particularly for a graph, represents the data with some imprecision. For these and other reasons, numerical characteristics of data are calculated routinely.

Definition 3.10.   A *statistic* is a numerical characteristic of a sample.

One of the functions of statistics as a field of study is to describe samples by as few numerical characteristics as possible. Most numerical characteristics can be classified broadly into statistics derived from percentiles of a frequency distribution and those statistics derived from moments of a frequency distribution—both terms will be explained below. Roughly speaking, the former approach tends to be associated with a statistical methodology usually termed nonparametric, the latter with parametric methods. The two classes will be used, contrasted, and evaluated throughout this text.

3.4.2 Statistics Derived from Percentiles

A percentile has an intuitively simple meaning—for example, the 25th percentile is that value of a variable such that 25% of the observations are less than that value and 75% of the observations are greater. You can supply a similar definition for, say, the 75th percentile. However, when we apply these definitions to a particular sample we may run into three problems: (1) a small sample size, (2) tied values, or (3) nonuniqueness of a percentile. Consider the following sample of four observations:

$$22, 22, 24, 27$$

How can we define the 25th percentile for this sample? There is no value of the variable with this property. But for the 75th percentile there is an infinite number of values—for example, 24.5, 25, and 26.9378 all satisfy the definition of the 75th percentile. For large samples, these problems disappear and we will define percentiles

for small samples in a way that is consistent with the intuitive definition. To find
a particular percentile in practice, we would rank the observations from smallest
to largest and count until the specified proportion had been reached. For example,
to find the 50th percentile of the above four numbers, we want to be somewhere
between the second and third largest observation (between the values for ranks two
and three). Usually this value is taken to be halfway between the two values. This
could be thought of as the value with rank 2.5—call this a "half rank." Note that

$$2.5 = (50/100)(1 + \text{sample size}).$$

You can verify that the following definition is consistent with your intuitive understanding of percentiles:

Definition 3.11. The *Pth percentile* of a sample of n observations is that value of
the variable with rank $(P/100)(1 + n)$. If this rank is not an integer it is rounded to
the nearest half rank.

Consider the 15 aflatoxin levels listed at the beginning of Section 3.3.3. The stem-
and-leaf diagram rearranged to put the values in increasing order is:

Stem (Tens)	Leaf (Units)
1	6
2	2 3 6 6 7 8
3	0 1 5 6 7
4	8
5	0 2

The 50th percentile is that value with rank $(50/100)(1 + 15) = 8$. The 8th largest
(or smallest) observation is 30. The 25th percentile is the observation with rank
$(25/100)(1 + 15) = 4$, and this is 26. Similarly, the 75th percentile is 37. The 10th
percentile (or decile) is that value with rank $(10/100)(1 + 15) = 1.6$, so we take
the value halfway between the smallest and second smallest observation, which is
$(1/2)(16+22) = 19$. The 90th percentile is the value with rank $(90/100)(1+15) = 14.4$;
this is rounded to the nearest half rank of 14.5. The value with this half rank is
$(1/2)(49 + 52) = 50.5$.
Certain percentile or functions of percentiles have specific names:

Percentile	Name
50	Median
25	Lower quartile
75	Upper quartile

All these statistics tell something about the location of the data. If we want to
describe how spread out the values of a sample are, we can use the range of values
(largest minus smallest), but a problem is that this statistic is very much dependent
on the sample size. A better statistic is given by:

Definition 3.12. The *interquartile range* is the difference between the 75th and 25th percentiles.

For the aflatoxin example, the interquartile range is $37 - 26 = 11$. Recall the *range* of a set of numbers is the largest value minus the smallest value. The data can be summarized as follows:

$$
\left. \begin{array}{ll}
\text{Median} & 30 \\
\text{Minimum} & 16 \\
\text{Maximum} & 52
\end{array} \right\} \text{Measures of location}
$$

$$
\left. \begin{array}{ll}
\text{Interquartile range} & 11 \\
\text{range} & 36
\end{array} \right\} \text{Measures of spread}
$$

The first three measures describe the location of the data, the last two give a description of their spread. If we were to add 100 to each of the observations the median, minimum and maximum would be shifted by 100 but the interquartile range and range would be unaffected.

These data can be graphically summarized by means of a *box plot*—a plot of the median, quartiles, and the minimum and maximum values. Note that a rectangle with upper and lower edges at the 25th and 75th percentiles is drawn with a line in the rectangle at the median (50th percentile). Lines are drawn from the rectangle (box) to the minimum and maximum values. The box plot for these data (Figure 3.8) indicates that the 50% of the data between the lower and upper quartiles is distributed over a much narrower range than the remaining 50% of the data.

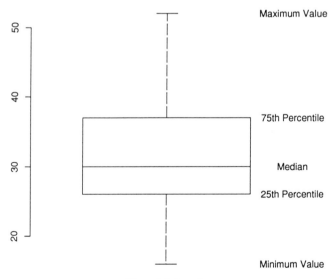

Figure 3.8. Box plot

3.4.3 Statistics Derived from Moments

The statistics discussed in the previous section have primarily dealt with describing the location and the variation of a sample of values of a variable. In this section, we introduce another class of statistics that have a similar purpose. In this class are the ordinary average, or arithmetic mean, and standard deviation. The reason these statistics are said to be derived from *moments* is that they are based on powers or moments of the observations.

Definition 3.13. The *arithmetic mean* of a sample of values of a variable is the average of all the observations.

Consider the aflatoxin data mentioned at the beginning of Section 3.3.3. The arithmetic mean of the data is

$$\frac{(30 + 26 + 26 + \cdots + 28 + 37)}{15} = \frac{487}{15} = 32.4\overline{6} \doteq 32.5.$$

A reasonable rule is to express the mean with one more significant digit than the observations, hence we round 32.4$\overline{6}$—a nonterminating decimal—to 32.5. (See also Note 3.2 on significant digits and rounding.)

Notation: The specification of some of the statistics to be calculated can be simplified by use of notation. We will use a capital letter for the name of a variable and the corresponding lowercase letter for a value. For example, $Y =$ *aflatoxin level* (the name of the variable); $y = 30$ (the value of aflatoxin level for a particular specimen). We will use the Greek symbol $\sum$ to mean "sum all the observations." Thus for the aflatoxin example, $\sum y$ is shorthand for the statement "sum all the aflatoxin levels." Finally, we will use the symbol $\overline{y}$ to denote the arithmetic mean of the sample. The arithmetic mean of a sample of n values of a variable can now be written as

$$\overline{y} = \frac{\sum y}{n}.$$

For example, $\sum y = 487$, $n = 15$ and $\overline{y} = 487/15 \doteq 32.5$. Consider now the variable of Table 3.3: the number of keypunching errors per line. Suppose we want the average number of errors per line. By definition this is $(0+0+1+0+2+\cdots+0+0+0+0)/156 = 37/156 \doteq 0.2$ errors per line. But this is a tedious way to calculate the average. A simpler way utilizes the frequency distribution or relative frequency distribution.

The total number of errors is $(124 \times 0) + (27 \times 1) + (5 \times 2) + (0 \times 3) = 37$, that is, there are 124 lines without errors, 27 lines each of which contains one error for a total of 27 errors for these kinds of lines and 5 lines with two errors for a total of 10 errors for these kinds of lines, and finally, no lines with 3 errors (or more). So the arithmetic mean is

$$\overline{y} = \frac{\sum fy}{\sum f} = \frac{\sum fy}{n},$$

since the total of the frequencies, f, add up to n, the sample size. Here, the sum $\sum fy$ is over observed values of y, each value appearing once.

Table 3.10. Calculation of Arithmetic Average from Empirical Frequency and Empirical Relative Frequency Distribution. Data from Table 3.3.

No. of Errors per Line, y	No. of Lines, f	Proportion of Lines, p	p times y
0	124	0.79487	0.00000
1	27	0.17308	0.17308
2	5	0.03205	0.06410
3	0	0.00000	0.00000
Total	156	1.00000	0.23718

The arithmetic mean can also be calculated from the empirical relative frequencies. We use the following algebraic property:

$$\bar{y} = \frac{\sum fy}{n} = \sum \frac{fy}{n} = \sum \frac{f}{n}y = \sum py.$$

The f/n are precisely the empirical relative frequencies or proportions, p. The calculations using proportions are given in Table 3.10. The value obtained for the sample mean is the same as before. The formula $\bar{y} = \sum py$ will be used extensively in the next chapter when we come to probability distributions. If the values y represent the midpoints of intervals in an empirical frequency distribution, the mean of the grouped data can be calculated in the same way.

Analogous to the interquartile range there is a measure of spread based on sample moments.

Definition 3.14. The *standard deviation* of a sample of n values of a variable Y is

$$s = \sqrt{\frac{\sum(y-\bar{y})^2}{n-1}}.$$

Roughly, the standard deviation is the square root of the average of the square of the deviations from the sample mean. The reason for dividing by $n-1$ is explained in Note 3.3. Before giving an example, we note the following properties of the standard deviation:

1. The standard deviation has the same units of measurement as the variable. If the observations are expressed in centimeters, the standard deviation is expressed in centimeters.
2. If a constant value is added to each of the observations, the value of the standard deviation is unchanged.
3. If the observations are multiplied by a positive constant value, the standard deviation is multiplied by the same constant value.
4. The following two formulæ are sometimes computationally more convenient in calculating the standard deviation:

$$s = \sqrt{\frac{\sum y^2 - n\bar{y}^2}{n-1}} = \sqrt{\frac{\sum y^2 - (\sum y)^2/n}{n-1}}.$$

Figure 3.9. Variation is important: statistician drowning in a river of average depth of 10.634 inches.

5. The square of the standard deviation is called the variance.
6. In many situations the standard deviation can be approximated by

$$s \doteq \frac{\text{Interquartile range}}{1.35}$$

7. In many cases it is true that approximately 68% of the observations fall within one standard deviation of the mean; approximately 95% within two standard deviations.

To illustrate the calculation of the standard deviation, we use the aflatoxin data of Section 3.3.3 as summarized in Table 3.11.

This table lists the data for calculating the standard deviation in each of three ways defined above. First consider the information in column (5) of Table 3.11. The quantity at the bottom of this column is $\sum(y - \bar{y})^2$, so that

$$s \doteq \sqrt{\frac{1581.7332}{15 - 1}} \doteq \sqrt{112.0909} \doteq 10.63.$$

It is clear that considerable computational labor is involved in calculating this quantity. The arithmetic mean, $\bar{y}$, in this example has been expanded to four decimal places to minimize round-off error. The deviations, $y - \bar{y}$, are also expressed as four-place decimals. These numbers must then be squared and added. A simpler method uses

Table 3.11. Three Approaches for Calculating the Standard Deviation of the Aflatoxin Data from Section 3.3.3.

(1) Obs. No.	(2) Value y	(3) $\bar{y}$	(4) $y - \bar{y}$	(5) $(y - \bar{y})^2$	(6) y^2	(7)[a] $\|y - \bar{y}\|$
1	30	32.4667	2.4667	6.0846	900	2.4667
2	26	32.4667	6.4667	41.8182	676	6.4667
3	26	32.4667	6.4667	41.8182	676	6.4667
4	36	32.4667	−3.5333	12.4842	1296	3.5333
5	48	32.4667	−15.5333	241.2834	2304	15.5333
6	50	32.4667	−17.5333	307.4166	2500	17.5333
7	16	32.4667	16.4667	271.1522	256	16.4667
8	31	32.4667	1.4667	2.1512	961	1.4667
9	22	32.4667	10.4667	109.5518	484	10.4667
10	27	32.4667	5.4667	29.8848	729	5.4667
11	23	32.4667	9.4667	89.6184	529	9.4667
12	35	32.4667	−2.5333	6.4176	1225	2.5333
13	52	32.4667	−19.5333	381.5498	2704	19.5333
14	28	32.4667	4.4667	19.9514	784	4.4667
15	37	32.4667	−4.5333	20.5508	1369	4.5333
Total	487	487.0005	0.0005	1581.7332	17393	126.4001

[a]See Section 3.4.4 for a discussion of this column.

the last of the alternative formulæ for the standard deviation, based on the algebraic equivalence of

$$\sum (y - \bar{y})^2 = \sum y^2 - \frac{(\sum y)^2}{n}$$
$$= 17393 - \frac{(487)^2}{15}$$
$$= 17393 - 15811.2\bar{6}$$
$$\doteq 1581.73$$

and

$$s \doteq \sqrt{\frac{1581.73}{15 - 1}} \doteq \sqrt{112.9810} \doteq 10.63,$$

the same answer as before.

The quantity $\sum y^2$ is obtained from column (6) in Table 3.11 and $\sum y$ from column (2). The second formula differs from the third only in the way the value $15811.2\bar{6}$ is calculated. It can be shown algebraically that

$$\frac{(\sum y)^2}{n} = n\bar{y}^2.$$

For example $\bar{y} \doteq 32.4667$ so that $n\bar{y}^2 \doteq 15(32.4667)^2 \doteq 15811.2991$. Note that errors have crept in at the second decimal place, because the mean is expressed to six significant digits whereas $n\bar{y}^2$ is expressed to nine digits.

3.4.4 Other Measures of Location and Spread

There are many other measures of location and spread. In the former category we mention the mode and the geometric mean.

Definition 3.15. The *mode* of a sample of values of a variable Y is that value that occurs most frequently.

The mode is usually calculated for large sets of discrete data. Consider the data in Table 3.12 which contains the distribution of the number of boys per family of eight children.

The most frequently occurring value of the variable Y, the number of boys per family of eight children, is 4.

There are more families with that number of boys than any other specified number of boys. For data arranged in histograms, the mode is usually associated with the midpoint of the interval having the highest frequency. For example, the mode of the systolic blood pressure of the native Japanese men listed in Table 3.6 is 130 mmHg, the modal value for Issei is 120 mmHg.

Definition 3.16. The *geometric mean* of a sample of nonnegative values of a variable Y is the nth root of the product of the n values, where n is the sample size.

Equivalently, it is the antilogarithm of the arithmetic mean of the logarithms of the values. (See Note 3.1 for a brief discussion of logarithms.)

Consider the following four observations of systolic blood pressure in mmHg:

$$118, 120, 122, 160.$$

The arithmetic mean is 130 mmHg which is larger than the first three values, the reason is that the 160 mmHg blood pressure "pulls" the mean to the right. The geometric mean is $(118 \times 120 \times 122 \times 160)^{1/4} \doteq 128.9$ mmHg. The geometric mean is less affected by the extreme value of 160 mmHg. The median is 121 mmHg. If the value of

Table 3.12. Number of Boys in Families of Eight Children (Geissler data). Data reprinted in Fisher [1958].

No. of Boys Per Family of Eight Children	Empirical Frequency (No. of Families)	Empirical Relative Frequency (Proportion) of Families
0	215	0.0040
1	1485	0.0277
2	5331	0.0993
3	10649	0.1984
4	14959	0.2787
5	11929	0.2222
6	6678	0.1244
7	2092	0.0390
8	342	0.0064
Total	53680	1.0000

160 mmHg is changed to a more extreme value, the mean will be affected the most, the geometric mean somewhat less, and the median not at all.

One other measure of spread is the average deviation. In column (7) of Table 3.11 are listed the deviations from the mean, $y - \bar{y}$, disregarding the sign of negative deviations. Such values are called absolute deviations and denoted by $|y - \bar{y}|$.

Definition 3.17. The *average deviation* of a sample of values of a variable is the arithmetic average of the absolute values of the deviations about the sample mean.

Using symbols the average deviation can be written as:

$$\text{Average deviation} = \frac{\sum |y - \bar{y}|}{n}.$$

For the data of Table 3.11 the average deviation is $126.4001/15 \doteq 8.43$.

3.4.5 Which Statistics?

Table 3.13 lists the statistics that have been defined so far, categorized by their usage.

The question arises: which statistic should be used for a particular situation? There is no simple answer because the choice depends on the data and the needs of the investigator. Statistics derived from percentiles and those derived from moments can be compared with respect to:

1. **Robustness.** The robustness of a statistic is related to its resistance to being affected by extreme values. In Section 3.4.4, it was shown that the mean—as compared to the median and geometric mean—is most affected by extreme values. The median is said to be more robust.
2. **Summarizing Capability.** The arithmetic mean is more appropriate if the data can be described by a particular mathematical model: the normal or Gaussian frequency distribution, which is the basis for a large part of the theory of statistics. This is described in the next chapter.
3. **Computational Ease.** Percentiles are easily calculated for small sets of data, say $n \leq 50$. For larger sets, the ranking of the observations can become a laborious, time-consuming task unless done by computer.
4. **Similarity.** In many samples, the mean and median are not too different. If the empirical frequency distribution of the data is almost symmetrical, the mean and the median tend to be close to each other.

Table 3.13. Statistics Defined in this Chapter.

Location	Spread
Median	Interquartile range
Percentile	Range
Arithmetic mean	Standard deviation
Geometric mean	Average deviation
Mode	

As a summary, it is suggested that the median and mean be calculated as measures of location and the interquartile range and standard deviation as measures of spread. The other statistics have limited or specialized use. There is additional discussion of robustness in Chapter 8.

NOTES

Note 3.1: Logarithms

A logarithm is an exponent on a base. The base is usually 10 or e (2.71828183...). Logarithms with base 10 are called common logarithms; logarithms with base e are called natural logarithms. To illustrate these concepts, consider:

$$100 = 10^2 = (2.71828183\ldots)^{4.605170\ldots} = e^{4.605170\ldots}$$

That is, the logarithm to the base 10 of 100 is 2, usually written

$$\log_{10}(100) = 2$$

and the logarithm of 100 to the base e is

$$\log_e(100) = 4.605170\ldots$$

The three dots indicate that the number is an unending decimal expansion. Unless otherwise stated in this text, logarithms will always be natural logarithms. Other bases are sometimes useful—in particular, the base 2. In determining hemagglutination levels, a series of dilutions of serum are set, each dilution being half of the previous one. The dilution series may be $1:1, 1:2, 1:4, 1:8, 1:16, 1:32$, etc.

The logarithm of the dilution factor using the base 2 is then simply

$$\log_2(1) = 0$$
$$\log_2(2) = 1$$
$$\log_2(4) = 2$$
$$\log_2(8) = 3$$
$$\log_2(16) = 4 \text{ etc.}$$

The following properties of logarithms are the only ones needed in this text. For simplicity, we use the base e, but the operations are valid for any base.

1. Multiplication of numbers is equivalent to adding logarithms ($e^a \times e^b = e^{a+b}$).
2. The logarithm of the reciprocal of a number is the negative of the logarithm of the number ($1/e^a = e^{-a}$).
3. Rule (2) is a special case of this rule: division of numbers is equivalent to subtracting logarithms ($e^a/e^b = e^{a-b}$).

Most pocket calculators permit rapid calculations of logarithms and antilogarithms. Tables are also available. You should verify that you can still use logarithms by working a few problems both ways.

Note 3.2: "Significant Digits": Rounding and Approximation

In working with numbers that are used to estimate some quantity, we are soon faced with the question of the number of "significant digits" to carry, or to report. A typical rule is to report the mean of a set of observations to one more place and the standard deviation to two more places than the original observation. But this is merely a guideline—which may be wrong. Following DeLury [1958], we can think of two ways in which approximation to the value of a quantity can arise:

1. Through arithmetical operations only, or
2. Through measurement.

If we express the mean of the three numbers 140, 150 and 152 as 147.3, we have approximated the exact mean, $147\frac{1}{3}$, so that there is *rounding error*. This error arises purely as the result of the arithmetical operation of division. The rounding error can be calculated exactly: $147.\overline{3} - 147.3 = 0.0\overline{3}$.

But this is not the complete story. If the above three observations are the weights of three teenage boys—measured to the nearest pound—the true average weight can vary all the way from $146.8\overline{3}$ to $147.8\overline{3}$ pounds; that is, the recorded weights (140, 150, 152) could vary from the three lower values (139.5, 149.5, 151.5) to the three high values (140.5, 150.5, 152.5), producing the two averages above. This kind . of rounding can be called *measurement rounding*. Knowledge of the measurement operation is required to assess the extent of the measurement rounding error: if the above three numbers represent systolic blood pressure readings in mmHg expressed to the nearest *even* number you can verify that the actual arithmetic mean of these three observations can vary from 146.33 to 148.33 so that even the third "significant" digit could be in error.

Unfortunately, we are not quite done yet with assessing the extent of an approximation. If the weights of the three boys are a sample from populations of boys and the population mean is to be estimated, we will also have to deal with *sampling variability* (a second aspect of the measurement process). And the effect of sampling variability is likely to be much larger than the effect of rounding error and measurement roundings. Assessing the extent of sampling variability is discussed in the next chapter.

For the present time we give you the following guidelines:

1. Minimize or eliminate rounding errors in intermediate arithmetical calculations. So, for example, instead of calculating

$$\sum (y - \bar{y})^2$$

in the process of calculating the standard deviation, use the equivalent relationship

$$\sum y^2 - \frac{(\sum y)^2}{n}.$$

You should also note that we are more likely to use approximations with the arithmetical operations of division and the taking of square roots, less likely with addition, multiplication, and subtraction. So if you can sequence the cal-

culations with division and square root being last, rounding errors due to arithmetical calculations will have been minimized.

2. The rule stated above is reasonable. In the next chapter you will learn of a
   better way of assessing the extent of approximation in measuring a quantity of
   interest.

Note 3.3: Degrees of Freedom

The concept of "degrees of freedom" appears again and again in this text. To make
the concept clear, we need the idea of a linear constraint on a set of numbers; this
is illustrated by several examples. Consider the numbers of girls, X, and the number
of boys, Y, in a family. (Note that X and Y are variables.) The numbers X and Y
are free to vary and we say that there are two degrees of freedom associated with
these variables. However, suppose the total number of children in a family, as in
the example, is specified to be precisely 8. Then, given the number of girls is 3, the
number of boys is fixed—namely, $8 - 3 = 5$. Given the constraint on the total number
of children, the two variables X and Y are no longer both free to vary, but fixing
one determines the other. That is, now there is only one degree of freedom. The
constraint can be expressed as

$$X + Y = 8 \quad \text{so that} \quad Y = 8 - X.$$

Constraints of the above type are called linear constraints.

A second example is based on Mendel's work in plant propagation. Mendel [1911],
reported results of many genetic experiments. One data set related two variables: form
and color. Table 3.14 summarizes these characteristics for 556 garden peas.

Let A, B, C, D be the numbers of peas as follows:

	Form	
Color	**Round**	**Wrinkled**
Yellow	A	B
Green	C	D

For example, A is the number of peas that are round and yellow. Without any
restrictions, the numbers A, B, C and D can be any nonnegative integers: there are
four degrees of freedom. Suppose now the total number of peas is fixed at 556 (as in
Table 3.14). That is, $A + B + C + D = 556$. Now only three of the numbers are free to
vary. Suppose, in addition, that the number of yellows peas is fixed at 416. Now only
two numbers can vary; for example, fixing A determines B, and fixing C determines
D. Finally, if the numbers of round peas is also fixed, only one number in the table
can be chosen. If, instead of the last constraint on the number of round peas, the
number of green peas had been fixed, two degrees would have remained since the
constraints "number of yellow peas fixed" and "number of green peas fixed" are not
independent, given that the total number of peas is fixed.

These results can be summarized in the following rule: given a set of N quantities
and $M \, (\leq N)$ linear, independent constraints, the number of degrees of freedom
associated with the N quantities is $N - M$.

Calculations of averages will almost always involve the number of degrees of freedom associated with the statistic rather than its number of components. For example,

Table 3.14. Frequency Distribution of Form and Color of 556 Garden Peas. Data from Mendel [1911].

	Variable 1: Form		
Variable 2: Color	Round	Wrinkled	Total
Yellow	315	101	416
Green	108	32	140
Total	423	133	556

the quantity $\sum(y - \bar{y})^2$ used in calculating the standard deviation of a sample of, say, n values of a variable Y has $n - 1$ degrees of freedom associated with it because $\sum(y - \bar{y}) = 0$. That is, the sum of the deviations about the mean is zero (cf. column 4, Table 3.11).

Note 3.4: Mean, Median, and Mode

There is an approximate relationship between mean, median, and mode. If the empirical frequency distribution has one peak ("unimodal") and is only moderately asymmetric or skewed, the following approximate relationship holds:

$$\text{Mean} - \text{Mode} \doteq 3(\text{Mean} - \text{Median}).$$

Note 3.5: Moments

Given a sample of observations $y_1, y_2, \ldots, y_n$ of a variable Y the rth sample moment about zero, m_r^* is defined to be

$$m_r^* = \frac{\sum y^r}{n}, \quad \text{for } r = 1, 2, 3, \ldots.$$

For example, $m_1^* = \sum y^1/n = \sum y/n = \bar{y}$ is just the arithmetic mean. The rth sample moment about the mean, m_r, is defined to be

$$m_r = \frac{\sum(y - \bar{y})^r}{n}, \quad \text{for } r = 1, 2, 3, \ldots.$$

The value of m_1 is zero (see Problem 15). It is clear that m_2 and s^2 (the sample variance) are closely connected. For a large number of observations m_2 will be approximately equal to s^2. One of the earliest statistical procedures (about 1900) was the method of moments by Karl Pearson. The method specified that all estimates derived from a sample should be based on sample moments. Some properties of moments are:

1. $m_1 = 0$.
2. Odd-numbered moments about the mean of symmetric frequency distributions are equal to zero.
3. A unimodal frequency distribution is skewed to the right if the mean is greater than the mode; it is skewed to the left if the mean is less than the mode. For distributions skewed to the right $m_3 > 0$; for distributions skewed to the left $m_3 < 0$.

The latter property is used to characterize the "skewness of a distribution," defined by

$$a_3 = \frac{\sum(y - \bar{y})^3}{\left[\sum(y - \bar{y})^2\right]^{3/2}} = \frac{m_3}{(m_2)^{3/2}}.$$

The division by $(m_2)^{3/2}$ is to standardize the statistic which now is unitless. Thus, a set of observations expressed in degrees Fahrenheit will have the same value of a_3 when expressed in degrees Celsius. Values of $a_3 > 0$ indicate positive skewness, skewness to the right, while value of $a_3 < 0$ indicate negative skewness. Some typical curves and corresponding values for the skewness statistics are illustrated in Figure 3.10. Note that all but the last two frequency distributions are symmetric, the last figure with skewness $a_3 = -2.71$ is a mirror image of the penultimate figure with skewness $a_3 = 2.71$.

The fourth moment about the mean is involved in the characterization of the flatness or peakedness of a distribution, labeled kurtosis (degree of archedness); a measure of kurtosis is defined by

$$a_4 = \frac{\sum(y - \bar{y})^4}{\left[\sum(y - \bar{y})^2\right]^2} = \frac{m_4}{(m_2)^2}.$$

Again, as in the case of a_3, the statistic is unitless. The following terms are used to characterize values of a_4.

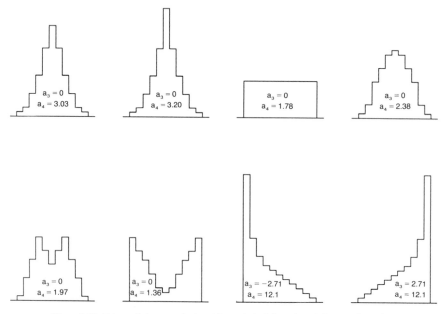

Figure 3.10. Values of skewness (a_3) and kurtosis (a_4) for selected data configurations.

$a_4 = 3$	Mesokurtic: this is the value for a bell-shaped distribution (Gaussian or normal distribution).
$a_4 < 3$	Leptokurtic: thin or peaked shape (or small "tails").
$a_4 > 3$	Platykurtic: flat shaped (or large "tails").

Values of this statistic associated with particular frequency distribution configurations are illustrated in Figure 3.10. The first figure is similar to a bell-shaped curve and has a value $a_4 = 3.03$, very close to 3. Other frequency distributions have values as indicated. It is only meaningful to speak of kurtosis for symmetric distributions.

Note 3.6: Taxonomy of Data

Social scientists have thought hard about the kinds of data. The following table summarizes a fairly standard taxonomy of data.

The four scales: nominal, ordinal, interval and ratio are characterized in the following table:

Scale	Characteristic Question	Statistic	Statistic to be used
Nominal	Do A and B differ?	List of diseases; marital status	mode
Ordinal	Is A bigger (better) than B?	Quality of teaching: (unacceptable/acceptable)	median
Interval	How much do A and B differ?	Temperatures, dates of birth	mean
Ratio	How many times is A bigger than B?	Distances, ages, heights	mean

This table is to be used as a guide only. You can be too rigid in applying this scheme. Frequently ordinal data are coded in increasing numerical order and averages are taken. Or, interval and ratio measurements are ranked (i.e., reduced to ordinal status) and averages taken at that point. Even with ordinal data, we sometimes calculate averages. For example: coding male $= 0$, female $= 1$ in a class of 100 students, then the average is the proportion of females in the class. For further discussion see Luce and Narens [1987].

PROBLEMS

1. Characterize the following variables and classify them as qualitative or quantitative. If qualitative can the variable be ordered? If quantitative is the variable discrete or continuous? In each case define the values of the variable:

(1) Race
(2) Date of birth
(3) Systolic blood pressure
(4) Intelligence quotient

(5) Apgar score
(6) White blood count
(7) Weight
(8) Quality of medical care

2. For each of the variables listed in Problem 1, define a suitable sample space. For two of the sample spaces so defined, explain how you would draw a sample. What statistics could be used to summarize such a sample?

3. Many variables of medical interest are derived from (functions of) several other variables. For example, as a measure of obesity there is the "ponderal index," which is given by weight/height^2. Another example is the dose of an anticonvulsant to be administered usually calculated on the basis of milligram of medicine per kilogram of body weight. What are some assumptions when these kinds of variables are used? Give two additional examples.

4. Every row of 12 observations in Table 3.3 (number of keypunching errors per month of data) can be summed to form the number of keypunching errors per year of data. Calculate the 13 values for this variable. Make a stem and leaf diagram. Calculate the (sample) mean and standard deviation. How does this mean and standard deviation compare with the mean and standard deviation for the number of keypunching errors per line of data?

5. The precise specification of the value of a variable is not always easy. Consider the data dealing with keypunching errors in Table 3.3. How is an error defined? A fairly frequent occurrence was the transposition of two digits—for example, a value of "63" might have been entered as "36." Does this represent one or two errors? Sometimes a zero was omitted, changing, for example, 0.0317 to 0.317. Does this represent four errors or one? Consider the list of qualitative variables at the beginning of Section 3.2, and name some problems you might encounter in defining the values of some of the variables.

6. For the data in Figure 3.6 calculate the marginal frequency distribution for the mother's education (in years). Construct a histogram for this distribution. What is your definition of the endpoints of the intervals?

7. Give three examples of frequency distributions from areas of your own research interest. Be sure to specify (a) what constitutes the sample; (b) the variable of interest; and (c) the frequencies of values or ranges of values of the variables.

8. A constant is added to each observation in a set of data (relocation). Describe the effect on the median, lower quartile, range, interquartile range, minimum, mean, variance and standard deviation. What is the effect on the above statistics if each observation is multiplied by a constant (rescaling)? Relocation and rescaling are called linear transformations. Such transformations are frequently used; for example, converting from °C to °F, defined by: °F $= 1.8 \times$ °C $+ 32$. What is the rescaling constant? Give two more examples of rescaling and relocation. An example of nonlinear transformation is going from the radius of a circle to its area: $A = \pi r^2$. Give two more examples of nonlinear transformations.

9. The following data are from a paper by Vlachakis and Mendlowitz [1976], and deal with the treatment of essential hypertension ("essential" is a technical term

meaning that the cause is unknown; a synonym is "idiopathic"). Seventeen patients received treatments C, A, and B where

$C =$ Control period,

$A =$ Propranolol + phenoxybenzamine,

$B =$ Propranolol + phenoxybenzamine, + hydrochlorothiazide.

Each patient received C first, then either A or B and finally B or A. The data consist of the systolic blood pressure in the recumbent position. (Note that in this example blood pressures are not always even-numbered.)

	C	A	B		C	A	B
1	185	148	132	10	180	132	136
2	160	128	120	11	176	140	135
3	190	144	118	12	200	165	144
4	192	158	115	13	188	140	115
5	218	152	148	14	200	140	126
6	200	135	134	15	178	135	140
7	210	150	128	16	180	130	130
8	225	165	140	17	150	122	132
9	190	155	138				

a. Construct stem-leaf diagrams for each of the three treatments. Can you think of some innovative way of displaying the three diagrams together to highlight the data?

b. Graph the ECDFs for each of the treatments C, A, and B on the same graph.

c. Construct box plots for each of the treatments C, A, and B. State your conclusions with respect to the systolic blood pressures associated with the three treatments.

d. Consider the difference between treatments A and B for each patient. Construct a box plot for the difference. Compare this result with that of Part (b).

e. Calculate the mean and standard deviation for each of the three treatments C, A, and B.

f. Consider, again, the difference between treatment A and B for each patient. Calculate the mean and standard deviation for the difference. Relate the mean to the means obtained in Part (d). How many standard deviations is the mean away from zero?

10. The following is a frequency distribution of fasting serum insulin ($\mu U/ml$) of males and females in a rural population of Jamaican adults; data from Florey *et al.* [1977]. (Serum insulin levels are expressed as whole numbers, so that "7−" represents the values 7 and 8.)

(μU/ml)	Males	Females	(μU/ml)	Males	Females
7–	1	3	29–	8	14
9–	9	3	31–	8	11
11–	20	9	33–	4	10
13–	32	21	35–	4	8
15–	32	23	37–	3	7
17–	22	39	39–	1	2
19–	23	39	41–	1	3
21–	19	23	43–	1	1
23–	20	27	$\geq$ 45	6	11
25–	13	23			
27–	8	19	Total	235	296

The last frequencies are associated with levels greater than 45. Assume that these represent the levels 45 and 46.

a. Plot both frequency distributions as histograms.

b. Plot the relative frequency distributions.

c. Calculate the ECDF.

d. Construct box plots for males and females. State your conclusions.

e. Assume that all the observations are concentrated at the midpoints of the intervals. Calculate the mean and standard deviation for males and females.

f. The distribution is obviously skewed. Transform the levels for males to logarithms and calculate the mean and standard deviation. The transformation can be carried in at least two ways: (a) consider the observations to be centered at the midpoints, transform the midpoints to logarithms and group into 6–8 intervals, (b) set up 6–8 intervals on the logarithmic scale, transform to the original scale and estimate by interpolation the number of observations in the interval. The antilogarithm of the logarithmic mean is what kind of mean? Compare it with the median and arithmetic mean.

11. There has been a long-held belief that births occur more frequently in the "small hours of the morning" than at any other time of the (24-h) day. Sutton [1945] collected the time of birth at the King George V Memorial Hospital, Sydney, for 2654 consecutive births. (Note: The total number of observations listed is 2650, not 2654 as stated by Sutton.) The frequency of births by hour is as follows:

Time	Births	Time	Births	Time	Births
6–7pm	92	2am	151	10am	101
7pm	102	3am	110	11am	107
8pm	100	4am	144	12pm	97
9pm	101	5–6am	136	1pm	93
10pm	127	6–7am	117	2pm	100
11pm	118	7am	80	3pm	93
12am	97	8am	125	4pm	131
1am	136	9am	87	5–6pm	105

a. Sutton states that the data "confirmed the belief ... that more births occur in the small hours of the morning than at any other time in the 24 hours." Develop a graphical display that illustrates this point.

b. Is there evidence of Sutton's statement, "An interesting point emerging was the relatively small number of births during the meal hours of the staff; this suggested either hastening or holding back of the second stage during meal hours"?

c. The data points in fact represent frequencies of values of a variable that has been divided into intervals. What is the variable?

12. At the International Health Exhibition in Britain, in 1884, Francis Galton, a scientist with strong statistical interests, obtained data on the strength of pull. Here is his data for "519 males ages 23–26."

Pull Strength	Cases Observed	Pull Strength	Cases Observed
Under 50 pounds	10	Under 90 pounds	113
Under 60 pounds	42	Under 100 pounds	22
Under 70 pounds	140	Above 100 pounds	24
Under 80 pounds	168		
		Total	519

Assume the smallest and largest categories are spread uniformly over a 10 pound interval.

a. The description of the data is exactly as in Galton [1889]. What are the intervals assuming strength of pull is measured to the nearest pound?

b. Calculate the median, 25th and 75th percentiles.

c. Graph the ECDF.

d. Calculate the mean and standard deviation assuming the observations are centered at the midpoints of the intervals.

e. Calculate the proportion of observations within one standard deviation of the mean.

13. The aflatoxin data cited at the beginning of Section 3.3.3 were taken from a larger set in the paper by Quesenberry *et al.* [1976]. The authors state:

"Aflatoxin is a toxic material that can be produced in peanuts by the fungus *Aspergillus flavus*. As a precautionary measure all commercial lots of peanuts in the United States (approximately 20,000 each crop year) are tested for aflatoxin.... Because aflatoxin is often highly concentrated in a small percentage of the kernels, variation among aflatoxin determinations is large.... Estimation of the distribution (of levels) is important.... About 6200 g of raw peanut kernels contaminated with aflatoxin were comminuted (ground up). The ground meal was then divided into 11 subsamples (lots) weighing approximately 560 g each. Each subsample was blended with 2800 ml methanol-water-hexane solution for two minutes, and the homogenate divided equally among 16 centrifuge bottles. One observation was lost from each of three subsamples leaving eight subsamples with 16 determinations and three subsamples with 15 determinations."

The original data were given to two decimal places—they are shown here rounded off to the nearest whole number. The data are as follows (asterisks indicate lost observations):

Lot

1	2	3	4	5	6	7	8	9	10	11
121	95	20	22	30	11	29	34	17	8	53
72	56	20	33	26	19	33	28	18	6	113
118	72	25	23	26	13	37	35	11	7	70
91	59	22	68	36	13	25	33	12	5	100
105	115	25	28	48	12	25	32	25	7	87
151	42	21	27	50	17	36	29	20	7	83
125	99	19	29	16	13	49	32	17	12	83
84	54	24	29	31	18	38	33	9	8	65
138	90	24	52	22	18	29	31	15	9	74
83	92	20	29	27	17	29	32	21	14	112
117	67	12	22	23	16	32	29	17	13	98
91	92	24	29	35	14	40	26	19	11	85
101	100	15	37	52	11	36	37	23	5	82
75	77	15	41	28	15	31	28	17	7	95
137	92	23	24	37	16	32	31	15	4	60
146	66	22	36	*	12	*	32	17	12	*

a. Make stem-and-leaf diagrams of the data of lots 1, 2, and 10. Make box plots for these three lots and discuss differences among these lots with respect to location and spread.

b. The data are analyzed by means of the MINITAB computer program. The data are entered by columns and the command DESCRIBE is used to give standard descriptive statistics for each lot. The output from the program (slightly modified) is given below:

```
MTB > desc c1-c11
           N    N*      MEAN   MEDIAN   STDEV    MIN   MAX        Q1        Q3
   C1     16     0    109.69   111.00   25.62     72   151     85.75    134.00
   C2     16     0     79.25    83.50   20.51     42   115     60.75     94.25
   C3     16     0    20.687   21.500   3.860     12    25     19.25     24.00
   C4     16     0     33.06    29.00   12.17     22    68     24.75     36.75
   C5     15     1     32.47    30.00   10.63     16    52     26.00     37.00
   C6     16     0    14.688   14.500   2.651     11    19     12.25     17.00
   C7     15     1     33.40    32.00    6.23     25    49     29.00     37.00
   C8     16     0    31.375   32.000   2.849     26    37     29.00     33.00
   C9     16     0     17.06    17.00    4.19      9    25     15.00     19.75
  C10     16     0     8.438    7.500   3.076      4    14      6.25     11.75
  C11     15     1     84.00    83.00   17.74     53   113     70.00     98.00
```

c. Verify that the statistics for Lot number 1 are correct in the printout. ($N*$ = number of missing observations, $Q1$ and $Q3$ are the 25th and 75th percentiles, respectively.)

d. There is an interesting pattern between the means and their standard deviation. Make a plot of the means versus standard deviation. Describe the pattern.

e. One way of describing the pattern between the means and the standard deviations is to calculate the ratio of the standard deviation to the mean. This ratio is called the coefficient of variation. It is usually multiplied by 100 and expressed as the percent coefficient of variation. Calculate the coefficients of variation in percentages for each of the 11 lots and make a plot of their value with the associated means. Do you see any pattern now? Verify that the average of the coefficients of variation is about 24%. A reasonable number to keep in mind for many biological measurements is that the variability as measured by the standard deviation is about 30% of the mean.

14. A paper by Robertson *et al.* [1976] discusses the level of plasma prostaglandin E (iPGE in pg/ml) in patients with cancer with and without hypercalcemia. The data are given below:

PATIENTS WITH HYPERCALCEMIA

Patient Number	Mean Plasma iPGE (pg/ml)	Mean Serum Calcium (ml/dl)
1	500	13.3
2	500	11.2
3	301	13.4
4	272	11.5
5	226	11.4
6	183	11.6
7	183	11.7
8	177	12.1
9	136	12.5
10	118	12.2
11	60	18.0

PATIENTS WITHOUT HYPERCALCEMIA

Patient Number	Mean Plasma iPGE (pg/ml)	Mean Serum Calcium (ml/dl)
12	254	10.1
13	172	9.4
14	168	9.3
15	150	8.6
16	148	10.5
17	144	10.3
18	130	10.5
19	121	10.2
20	100	9.7
21	88	9.2

Note that the variables are "Mean Plasma iPGE" and "Mean Serum Ca" levels—presumably more than one assay was carried out for each patient's level. The number of such tests for each patient is not indicated, nor the criterion for the number.

a. Calculate the mean and standard deviation of plasma iPGE level for the patients with hypercalcemia; do the same for patients without hypercalcemia.

b. Make box plots for plasma iPGE levels for each group. Can you draw any conclusions from these plots? Do they suggest that the two groups differ in plasma iPGE levels?

c. The article states that normal limits for serum calcium levels are 8.5–10.5 mg/dl. It is clear that patients were classified as "hypercalcemic" if their serum calcium levels exceeded 10.5 mg/dl. Without classifying patients it may be postulated that high plasma iPGE levels tend to be associated with high serum calcium levels. Make a plot of the plasma iPGE and serum calcium levels to determine if there is a suggestion of a pattern relating these two variables.

15. Prove or verify the following for the observations $y_1, y_2, \ldots, y_n$.

a. $\sum 2y = 2 \sum y.$

b. $\sum (y - \bar{y}) = 0.$

c. By means of an example, show that $\sum y^2 \neq \left(\sum y \right)^2.$

d. If a is a constant, $\sum ay = a \sum y.$

e. If a is a constant, $\sum (a + y) = na + \sum y.$

f. $\sum \dfrac{y}{n} = \dfrac{1}{n} \sum y.$

g. $\sum (a + y)^2 = na^2 + 2a \sum y + \sum y^2$

h. $\sum (y - \bar{y})^2 = \sum y^2 - \dfrac{\left(\sum y \right)^2}{n}.$

i. $\sum (y - \bar{y})^2 = \sum y^2 - n\bar{y}^2$

16. A variable Y is grouped into intervals of width h and represented by the midpoint of the interval. What is the maximum error possible in calculating the mean of all the observations?

17. Prove that the two definitions of the geometric mean are equivalent.

18. Calculate the average number of boys per family of eight children for the data given in Table 3.12.

19. The formula $\bar{Y} = \sum py$ is also valid for observations not arranged in a frequency distribution as follows: if we let $1/N = p$ we get back to the formula $\bar{Y} = \sum py$. Show that this is so for the following four observations: 3, 9, 1, 7.

20. Calculate the average systolic blood pressure of native Japanese men using the fre-

quency data of Table 3.6. Verify that the same value is obtained using the relative frequency data of Table 3.7.

21. Using the taxonomy of data described in Note 3.6, classify each of the variables in Problem 1 according to the scheme described in the note.

REFERENCES

DeLury, D. B. [1958]. Computations with approximate numbers. *The Mathematics Teacher*, **51:** 521–530. Reprinted in Ku, H. H. (ed.) [1969]. *Precision Measurement and Calibration*. NBS Special Publication 300. US Government Printing Office, Washington DC.

Fisch, R. O., Bilek, M. K., Horrobin, J. M., and Chang, P. [1976]. Children with superior intelligence at 7 years of age. *American Journal of Diseases of Children*, **130:** 481–487.

Fisher, R. A. [1958]. *Statistical Methods for Research Workers*, 13th edition. Oliver and Boyd, London.

Florey, C. du V., Milner, R. D. G., and Miall, W. E. [1977]. Serum insulin and blood sugar levels in a rural population of Jamaican adults. *Journal of Chronic Diseases*, **30:** 49–60. Used with permission from Pergamon Press, Inc.

Galton, F. [1889]. *Natural Inheritance*. Macmillan, London.

Graunt, J. [1622]. Natural and political observations mentioned in a following index and made upon the Bills of Mortality. In J. R. Newman [1956]. *The World of Mathematics*. **3:** 1421–1435. Simon and Schuster, New York.

Huff, D. [1982]. *How to Lie with Statistics*. W. W. Norton, New York.

Ku, H. H. (ed.) [1969]. *Precision, Measurement and Calibration*. NBS Special Publication 300, Volume I. Superintendent of Documents, US Government Printing Office, Washington DC.

Luce, R. D. and Narens, L. [1987]. Measurement scales on the continuum. *Science*, **236:** 1527–1532.

Mendel, G. [1911]. *Versuche Uber Pflanzenhybriden*. Wilhelm Engelmann, Leipzig. p. 18.

Meyer, M. A., Broome, F. R. and Schweitzer, R. H., Jr. [1975]. Color statistical mapping by the US Bureau of the Census. *American Cartographer*, **2:** 100–117. Reproduced in Fienberg S. E. [1979]. Graphical methods in statistics. *American Statistician*, **33:** 165–178.

Moses, L.E. [1987]. Graphical methods in statistical analysis. *Annual Reviews of Public Health*, **8:** 309–353.

Newman, J. R. (ed.) [1956]. *The World of Mathematics*. **3:** 1421–1435. Simon and Schuster, New York.

Quesenberry, P. D., Whitaker, T. B. and Dickens, J. W. [1976]. On testing normality using several samples: an analysis of peanut aflatoxin data. *Biometrics*, **32:** 753–759. With permission of the Biometric Society.

Robertson, R. P., Baylink, D. J., Metz, S. A. and Cummings, K. B. [1976]. Plasma prostaglandin E in patients with cancer with and without hypercalcemia. *Journal of Clinical Endocrinology and Metabolism*, **43:** 1330–1335.

Schwab, B. [1975]. Delivery of babies and full moon, a letter to the editor. *Canadian Medical Association Journal*, **113:** 489 and 493.

Sutton, D. H. [1945]. Gestation period. *The Medical Journal of Australia*, Volume I, **32:** 611–613. Used with permission.

Tufte, E. R. [1983]. *The Visual Display of Quantitative Information*. Graphics Press, Cheshire, CT.

Tufte, E. R. [1990]. *Envisioning Information*. Graphics Press, Cheshire, CT.

Tukey, J. W. [1972]. Some graphic and semi-graphic displays. In *Statistical Papers in Honor of George W. Snedecor.* T. A. Bancroft, Editor. Iowa State University Press, Ames, IA.

Tukey, J. W. [1977]. *Exploratory Data Analysis.* Addison-Wesley. Reading, MA.

Vlachakis, N. D. and Mendlowitz, M. [1976]. Alpha- and beta-adrenergic receptor blocking agents combined with a diuretic in the treatment of essential hypertension. *Journal of Clinical Pharmacology,* **16:** 352–360.

Wallace, D. C. and Exton, L. A. [1972]. Genetic predisposition to development of malignant melonoma. In *Melonoma and Skin Cancer.* McCarthy, Editor. V. C. N. Blight, Government Printer, Sydney, Australia, p. 65–81.

Wilk, M. B. and Gnanadesikan, R. [1968]. Probability plotting methods for the analysis of data. *Biometrika,* **55:** 1–17.

Winkelstein, W., Jr., Kagan, A., Kato, H. and Sacks, S. T. [1975]. Epidemiological studies of coronary heart disease and stroke in Japanese men living in Japan, Hawaii and California: blood pressure distributions. *American Journal of Epidemiology,* **102:** 502–513.

CHAPTER 4

Statistical Inference: Populations and Samples

4.1 INTRODUCTION

Statistical inference has been defined as "the attempt to reach a conclusion concerning all members of a class from observations of only some of them" [Runes, 1959]. In statistics, "all members of a class" form the *population* or *sample space*, and the observed subset a *sample*; we discussed this in Sections 3.1 and 3.2. We now discuss the *process* of obtaining a valid sample from a population, i.e., when is it valid to make a statement about a population on the basis of a sample? One of the assumptions in any scientific investigation is that valid inferences can be made—that the results of a study can apply to a larger population. For example, we can assume that a new therapy developed at the Memorial Sloan-Kettering Cancer Center in New York is applicable to cancer patients in Great Britain. You can easily supply additional examples.

In the next section, we note which characteristics of a population are of interest, and illustrate this with two examples. Section 4.3 introduces probability theory as a way by which we can define valid sampling procedures. In Section 4.4 we apply the theory to a well-known statistical model for a population—the normal frequency distribution, which has practical as well as theoretical interest. One reason for the importance of the normal distribution is given in Section 4.5, which discusses the concept of sampling distribution. The next three sections discuss inferences about population means and variances on the basis of a single sample. Following this section are notes, problems and references.

4.2 POPULATION AND SAMPLE

4.2.1 Definition and Examples

You should review the concepts of *variable*, *sample space* or *population*, and *statistic*, as discussed in Chapter 3.

Definition 4.1. A *parameter* is a numerical characteristic of a population.

Analogous to numerical characteristics of a sample (statistics) we will be interested in numerical characteristics of populations (parameters). The population characteristics are usually unknown because the whole population cannot be enumerated or studied. The problem of statistical inference can then be stated as follows: on the basis of a sample from a population, what can be said about the population from which the sample came. In this section we illustrate the four concepts of population and its corresponding parameters, and sample and its corresponding statistics.

Example 4.1.   We illustrate those four concepts with an example from Chapter 3, systolic blood pressure for Japanese men, aged 45–69, living in Japan. The "population" can be considered to be the collection of blood pressures of all Japanese men. The blood pressures are assumed to have been taken under standardized conditions. Clearly, Winkelstein *et al.* [1975] could not possibly measure all Japanese men, but a subset of 2232 eligible men were chosen. This is the sample. A numerical quantity of interest could be the average systolic blood pressure. This average for the population is a parameter; the average for the sample is the statistic. Since the total population cannot be measured, the parameter value is unknown. The statistic, the average for the sample, can be calculated. You are probably assuming now that the sample average is a good estimate of the population average. You may be correct. Later in this chapter we specify under what conditions this is true, but note for now that all the elements of inference are present.   □

Example 4.2.   Consider this experimental situation. We want to assess the effectiveness of a new special diet for children with phenylketonuria (PKU). One effect of this condition is that untreated children become mentally retarded. The diet is used with a set of PKU children and their IQs are measured when they reach four years of age. What is the population? It is hypothetical in this case: all PKU children who could potentially be treated with the new diet. The variable of interest is the IQ associated with each child. The sample is the set of children actually treated. A parameter could be the median IQ of the hypothetical population; a statistic might be the median IQ of the children in the sample. The question to be answered is whether the median IQ of this treated hypothetical population is the same or comparable to that of non-PKU children.   □

A sampling situation has the following components: a population of measurement is specified, a sample is taken from the population and measurements are made. A statistic is calculated which—in some way—makes a statement about the corresponding population parameter. Some practical questions that come up are:

* **1.** Is the population defined unambiguously?
 2. Is the variable clearly observable?
 3. Is the sample "valid"?
 4. Is the sample "big enough"?

The first two questions have been discussed in previous chapters. In this chapter we will begin to answer the last two.

Conventionally, parameters are indicated by Greek letters, and the estimate of the parameter by the corresponding Roman letter. For example, μ is the population mean, and m is the sample mean. Similarly, the population standard deviation will be indicated by σ, and the corresponding sample estimate by s.

4.2.2 Estimation and Hypothesis Testing

Two approaches are commonly used in making statements about population parameters: *estimation* and *hypothesis testing*. Estimation, as the name suggests, attempts to estimate values of parameters. As discussed before, the sample mean is thought to estimate, in some way, the mean of the population from which the sample was drawn. In Example 4.1, the mean of the observed blood pressures is considered an estimate of the corresponding population value. Hypothesis testing makes inferences about (population) parameters by supposing they have certain values, and then testing whether the observed data is consistent with the hypothesis. Example 4.2 illustrates this framework: is the mean IQ of the population of PKU children treated with the special diet the same as that of the population of non-PKU children? We could hypothesize that it is and determine, in some way, whether the data are inconsistent with this hypothesis.

You could argue that in the second example we are also dealing with estimation. If one could estimate the mean IQ of the treated population then the hypothesis can be dealt with. This is quite true. In Section 4.7, we will see that in many instances hypothesis testing and estimation are but two sides of the same coin.

One additional comment about estimation: a distinction is usually made between *point estimate* and *interval estimate*. A sample mean is a point estimate. An interval estimate is a range of values that is reasonably certain to contain the value of the parameter of interest.

4.3 VALID INFERENCE THROUGH PROBABILITY THEORY

4.3.1 The Precise Specification of Our Ignorance

Everyone "knows" that the probability of heads coming up in the toss of a coin is one-half; that the probability of a three in the toss of a die is one-sixth. More subtly, the probability that a randomly selected patient has systolic blood pressure less than the population median is one-half—although some may claim, after the measurement is made, it is either "zero" or "one"—that is, the systolic blood pressure of the patient is either below the median or greater than or equal to the median.

What do we mean by the phrase "the probability of"? Consider one more situation. We toss a thumbtack on a hard, smooth surface such as a table, if the outcome is ⊥, we call it "up"; if the outcome is ⼈, we call it "down." What is the probability of "up"? It is clear that in this example we do not know, *a priori*, the probability of "up"—it depends on the physical characteristics of the thumbtack. How would you *estimate* the probability of "up"? Intuitively, you would toss the thumbtack a large number of times and observe the proportion of times the thumbtack landed "up." And that is the way we define probability. Mathematically, we define the probability of "up" as the relative frequency of the occurrence of "up" as the number of tosses become indefinitely large. This is an illustration of the *relative frequency* concept of probability. Some of its ingredients are: (1) a trial or experiment has a set of specified outcomes; (2) the outcome of one trial does not influence the outcome of another trial; (3) the trials are identical; (4) the probability of a specified outcome is the limit of its relative frequency of occurrence as the number of trials becomes indefinitely large.

Probabilities provide a link between a population and samples. A probability can be thought of as a numerical statement about what we know and do not know: a precise specification of our ignorance (Fisher [1956]). In the thumbtack-tossing experiment, we know that the relative frequency of occurrences of "up" will approach some number: the probability of "up." What we do not know is what the outcome will be on the next toss. A probability, then, is a characteristic of a population of outcomes. When we say that the probability of a head in a coin toss is one-half, we are making a statement about a population of tosses. For alternate interpretations of probability see Note 4.1.

On the basis of the relative frequency interpretation of probability we deduce that probabilities are numbers between zero and one (including zero and one).

The outcome of a trial such as a coin toss will be denoted by a capital letter, for example,

$$H = \text{"coin toss results in head"},$$

$$T = \text{"coin toss results in tail."}$$

Frequently the letter can be chosen as a mnemonic for the outcome.

The probability of an outcome, O, in a trial will be denoted by $P[O]$. Thus, in the coin-tossing experiment, we have $P[H]$ and $P[T]$ for the probabilities of "head" and "tail," respectively.

4.3.2   Working with Probabilities

Outcomes of trials can be categorized by two criteria: *statistical independence* and *mutual exclusiveness*.

Definition 4.2. Two outcomes are *statistically independent* if the probability of their joint occurrence is the product of the probabilities of occurrence of each outcome.

Using notation, let C be one outcome and D be another outcome; $P[C]$ is the probability of occurrence of C, and $P[D]$ is the probability of occurrence of D. Then C and D are statistically independent if

$$P[CD] = P[C]P[D],$$

where $[CD]$ means both C and D occur.

Statistically independent events are the model for events that "have nothing to do with each other." In other words, the occurrence of one event does not change the probability of the other occurring. Later this will be explained in more detail.

Models of independent outcomes are the outcomes of successive tosses of a coin, die, or the spinning of a roulette wheel. For example, suppose the outcomes of two tosses of a coin are statistically independent. Then the probability of two heads, $P[HH]$, by statistical independence is:

$$P[HH] = P[H]P[H] = \frac{1}{2} \times \frac{1}{2} = \frac{1}{4}.$$

Similarly,

$$P[HT] = \frac{1}{2} \times \frac{1}{2} = \frac{1}{4}$$

$$P[TH] = \frac{1}{2} \times \frac{1}{2} = \frac{1}{4}$$

and

$$P[TT] = \frac{1}{2} \times \frac{1}{2} = \frac{1}{4}.$$

Note that the outcome *HT* means *head on toss 1 and tail on toss 2.*

You may wonder why we refer to coin-tossing and dice throws so much. One reason has been given already: these activities form patterns of probabilistic situations. Secondly, they can be models for many experimental situations. Suppose we consider the Winkelstein *et al.* study dealing with blood pressures of Japanese men. What is the probability that each of two men has a blood pressure less than the median of the population? We can use the coin-toss model: by definition, half of the population has blood pressure less than the median. The populations can then be thought of as a very large collection of trials each of which has two outcomes: less than the median, greater than or equal to the median. If the selection of two men can be modeled by the coin tossing experiment, then the probability that both men have blood pressures less than the median is $1/2 \times 1/2 = 1/4$. This we now formalize:

Definition 4.3. Outcomes of a series of repetitions of a trial are a *random sample* of outcomes if the probability of their joint occurrence is the product of the probabilities of each occurring separately. If every possible sample of *k* outcomes has the same probability of occurrence, the sample is called a *simple random sample.*

Suppose we are dealing with the outcomes of trials. We label the outcomes O_k, where the subscript is used to denote the order in the sequence; O_1 is the specified outcome for the first trial, O_2 is the outcome for the second trial, and so on. Then the outcomes form a random sample if

$$P[O_1 O_2 O_3 \cdots O_k] = P[O_1]P[O_2]P[O_3] \cdots P[O_k].$$

The phrase "a random sample" is therefore not so much a statement about the sample as a statement about the method that produced the sample. The randomness of the sample allows us to make valid statements about the population from which it came. It also allows us to quantify what we know and do not know.

How can we draw a random sample? For the coin tosses and dice throws this is fairly obvious. But how do we draw a random sample of Japanese men? Theoretically, we could have their names on slips of paper in a very large barrel. The contents are stirred and slips of paper drawn out—the random sample. Clearly, this is not done in practice. In fact, many times a sample is claimed to be random by default: "there is no reason to believe it is not random." Thus, college students taking part in a experiment are implicitly assumed to be a "random sample of people." Sometimes this is reasonable; as mentioned in the introduction, cancer patients treated in New York are considered very similar—with respect to cancer—to cancer patients in California. There is a gradation in the seriousness of nonrandomness of samples: "red blood cells from healthy adult volunteers" are apt to be similar in many respects the world over (and dissimilar in others); "diets of teenagers," on the other hand, will vary from region to region.

Obtaining a truly random sample is a difficult task which is rarely carried out successfully. A standard criticism of any study is that the sample of data is not a random sample so that the inference is not valid. Some of the problems in sampling have already been discussed in Chapter 2, here we list a few additional problems:

1. The population or sample space is not defined.
2. Part of the population of interest is not available for study.
3. The population is not identifiable or changes with time.
4. The sampling procedure is faulty, due to limitations in time, money, and effort.
5. Random allocation of members of a group to two or more treatments does not imply that the group itself is necessarily a random sample.

Most of these problems are present in any study, sometimes in an unexpected way. For example, in an experiment involving rats, the animals were "haphazardly" drawn from a cage for assignment to one treatment and the remaining rats were given another treatment. "Differences" between the treatments were due to the fact that the more agile and larger animals evaded "haphazard" selection and wound up in the second treatment. For some practical ways of drawing random samples, see Note 4.8.

Now we consider probabilities of mutually exclusive events:

Definition 4.4. Two outcomes are *mutually exclusive* if at most one of them can occur at a time; that is, the outcomes do not overlap.

Using notation, let C be one outcome and D another; then it can be shown (using the relative frequency definition) that $P[C \text{ or } D] = P[C] + P[D]$ if the outcomes are mutually exclusive. Here, the connective "or" is used in its inclusive sense, "either or, or both."

Some examples of mutually exclusive outcomes are H and T on a coin toss; the race of a person for purposes of a study can be defined as "black," "white," or "other," and each subject can belong to only one category; the method of delivery can be either "vaginal" or by means of a "cesarean section."

Example 4.3. We now illustrate outcomes that are not mutually exclusive. Suppose that the Japanese men in the Winkelstein data are categorized by weight: "Reasonable Weight" or "Overweight," and their blood pressures by "Normal" or "High." Suppose we have the following table:

	Blood Pressure		
	Normal (*N*)	High (*H*)	
Reasonable weight (*R*)	0.6	0.1	0.7
Overweight (*O*)	0.2	0.1	0.3
	0.8	0.2	1.0

The entries in the table are the probabilities of outcomes for an individual randomly selected from the population, so that, for example, 20% of Japanese men are considered overweight and have normal blood pressure. Consider the outcomes "Overweight" and "High Blood Pressure." What is the probability of the outcome [O or H] (overweight, high blood pressure, or both)? This corresponds to the portion of the table boldfaced as follows:

	N	H	
R	0.6	**0.1**	0.7
O	**0.2**	**0.1**	0.3
	0.8	0.2	1.0

$$P[O \text{ or } H] = 0.2 + 0.1 + 0.1 = 0.4.$$

But $P[O] + P[H] = 0.2 + 0.3 = 0.5$. Hence O and H are not mutually exclusive. In terms of calculation, we see that we have added in the outcome $P[OH]$ twice:

	N	H	
R		0.1	
O	0.2	0.1	0.3
		0.2	

The correct value is obtained if we subtract $P[OH]$ as follows:

$$P[O \text{ or } H] = P[O] + P[H] - P[OH]$$
$$= 0.3 + 0.2 - 0.1$$
$$= 0.4.$$

This example is an illustration of the addition rule of probabilities.

Definition 4.5 (Addition Rule). For any two outcomes, the probability of occurrence of either outcome or both is the sum of the probabilities of each occurring minus the probability of their joint occurrence.

Using notation, for any two outcomes C and D,

$$P[C \text{ or } D] = P[C] + [D] - P[CD].$$

Two outcomes, C and D, are mutually exclusive if they cannot occur together. In this case, $P[CD] = 0$ and $P[C \text{ or } D] = P[C] + P[D]$ as stated previously.

We conclude this section by briefly discussing dependent outcomes. The outcomes O and H in the above example were not mutually exclusive. Were they independent? By Definition 4.2, O and H are statistically independent if $P[OH] = P[O]P[H]$.

From the table, we get $P[OH] = 0.1$, $P[O] = 0.3$, and $P[H] = 0.2$ so that

$$0.1 \neq (0.3)(0.2).$$

Of subjects with reasonable weight, only 1 in 7 has high blood pressure, but among overweight individuals, 1 in 3 has high blood pressure. Thus, the probability of high blood pressure in overweight subjects is greater than the probability of high blood pressure in subjects with normal weight. The reverse statement can also be made: two out of eight persons with normal blood pressure are overweight, 1 out of 2 persons with high blood pressure is overweight.

The statement "of subjects with reasonable weight, only 1 in 7 has high blood pressure" can be stated as a probability: "the probability that a person with reasonable weight has high blood pressure is 1/7." Formally, this is written as.

$$P[H \mid R] = \frac{1}{7}.$$

or P[High blood pressure *given* a reasonable weight] $= 1/7$. The probability $P[H \mid R]$ is called a *conditional* probability. You can verify that $P[H \mid R] = P[HR]/P[R]$.

Definition 4.6.   For any two outcomes C and D, the *conditional probability* of the occurrence of C *given* the occurrence of D, $P[C \mid D]$, is given by

$$P[C \mid D] = \frac{P[CD]}{P[D]}.$$

For completeness we now state the multiplication rule of probability (which will be discussed in more detail in Chapter 6).

Definition 4.7 (Multiplication Rule).   For any two outcomes C and D, the probability of the joint occurrence of C and D, $P[CD]$, is given by

$$P[CD] = P[C]P[D \mid C],$$

or equivalently

$$P[CD] = P[D]P[C \mid D].$$

Continuing Example 4.3, what is the probability that a randomly selected person is overweight and has high blood pressure? In our notation we want $P[OH]$. By the multiplication rule, this probability is

$$P[OH] = P[O]P[H \mid O].$$

Using Definition 4.6,

$$P[H \mid O] = \frac{P[OH]}{P[O]} = \frac{0.1}{0.3} = \frac{1}{3},$$

so that

$$P[OH] = 0.3 \left(\frac{1}{3} \right) = 0.1.$$

Alternatively we could have calculated $P[OH]$ by

$$P[OH] = P[H]P[O \mid H],$$

which becomes

$$P[OH] = (0.2) \left(\frac{0.1}{0.2} \right) = 0.1.$$

We can also state the criterion for statistical independence in terms of conditional probabilities. From Definition 4.2, two outcomes C and D are statistically independent if $P[CD] = P[C]P[D]$, i.e., the probability of the joint occurrence of C and D is the product of the probability of C and the probability of D. The multiplication rule states that for *any* two outcomes C and D,

$$P[CD] = P[C]P[D \mid C].$$

Under independence,

$$P[CD] = P[C]P[D].$$

Combining the two, we see that C and D are independent if (and only if) $P[D \mid C] = P[D]$. In other words, the probability of occurrence of D is not altered by the occurrence of C. This has intuitive appeal.   □

When do we use the addition rule; when the multiplication rule? Use the addition rule to calculate the probability that either one or both events occur. Use the multiplication rule to calculate the probability of the joint occurrence of two events.

4.3.3   Random Variables and Distributions

Basic to the field of statistics is the concept of a random variable:

Definition 4.8.   A *random variable* is a variable associated with a random sample.

The only difference between a *variable* defined in Chapter 3 and a *random variable* is the process that generates the value of the variable. If this process is random we speak of a random variable. All the examples of variables in Chapter 3 can be interpreted in terms of random variables if the samples are random samples. The empirical relative frequency of occurrence of a value of the variable becomes an estimate of the probability of occurrence of that value. For example, the relative frequencies of the values of the variable "number of boys in families with eight children" in Table 3.12 become estimates of the probabilities of occurrence of these values.

The distinction between discrete and continuous variables carries over to random variables. Also, as with variables, we denote the label of a random variable by capital letters (say X, Y, V, ...) and a value of the random variable by the corresponding lower case letter (x, y, v, ...).

We are interested in describing the probabilities with which values of a random variable occur. For discrete random variables, this is straightforward. For example, let Y be the outcome of the toss of a die. Then Y can take on the values 1, 2, 3, 4, 5, 6 and we write

$$P[Y = 1] = \frac{1}{6}, \quad P[Y = 2] = \frac{1}{6}, \quad \ldots, \quad P[Y = 6] = \frac{1}{6}.$$

This leads to the following definition:

Definition 4.9. A *probability function* is a function that for each possible value of a discrete random variable takes on the probability of that value occurring. The function is usually presented as a listing of the values with the probabilities of occurence of the values.

Consider again the data of Table 3.12, the number of boys in families with eight children. The observed empirical relative frequencies can be considered estimates of probabilities if the 53,680 families are a random sample. The probability distribution is then estimated as follows:

Number of Boys	Probability
0	0.0040
1	0.0277
2	0.0993
3	0.1984
4	0.2787
5	0.2222
6	0.1244
7	0.0390
8	0.0064
Total	1.0000

The estimated probability of observing precisely two boys in a family of eight children, is 0.0993 or, approximately, 1 in 10. Since the sample is very large we will treat—in this discussion—the estimated probabilities as if they were the actual probabilities. If Y represents the number of boys in a family with eight children, then we write

$$P[Y = 2] = 0.0993.$$

What is the probability of two boys or fewer? This can be expressed as

$$P[Y \leq 2] = P[Y = 2 \text{ or } Y = 1 \text{ or } Y = 0].$$

Since these are mutually exclusive outcomes

$$P[Y \leq 2] = P[Y = 2] + P[Y = 1] + P[Y = 0]$$
$$= 0.0993 + 0.0277 + 0.0040$$
$$= 0.1310.$$

Approximately 13% of families with eight children will have two or fewer boys. A probability function can be graphically represented by a plot of the values of the variable against the probability of the value. The probability function for the Geissler data is presented in Figure 4.1.

How can we describe probabilities associated with continuous random variables? Somewhat paradoxically, the probability of a specified value for a continuous random variable is zero! For example, the probability of finding anyone with height 63.141592654 inches—and not 63.141592653 inches—is virtually zero. If we were to continue the decimal expansion, the probability becomes smaller yet. But we do find people with height, say, 63 inches. When we write 63 inches, however, we do not mean 63.000... inches (and we are almost certain not to find anybody with that height), but we have in mind *an interval* of values of height, anyone with height between 62.500... and 63.500... inches. We could then divide the values of the continuous random variable into intervals, treat the midpoints of the intervals as the values of a discrete variable and list the probabilities associated with these values. Table 3.7 illustrates this approach with the division of the systolic blood pressure of Japanese men into discrete intervals.

We will start with the histogram and the *relative* frequencies associated with the

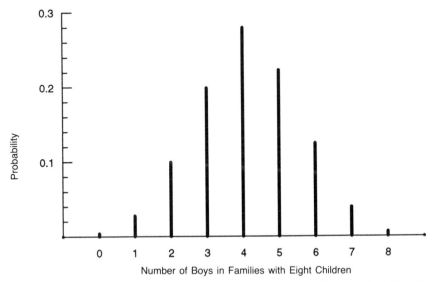

Figure 4.1. Probability function of the random variable "number of boys in families with eight children." Data from Table 3.12.

intervals of values in the histogram. The area under the "curve" is equal to 1 if the width of each interval is one; or if we *normalize* (that is, multiply by a constant so the area is equal to one). Suppose now that the interval widths are made smaller and smaller, and simultaneously the number of cases increased. Normalize so the area under the curve remains equal to one, then the curve is assumed to take on a smooth shape. Such shapes are called *probability density functions* or, more briefly, *densities*:

Definition 4.10. *A probability density function* is a curve which specifies by means of the area under the curve over an interval the probability that a continuous random variable falls within the interval. The total area under the curve is one.

Some simple densities are illustrated in Figure 4.2.

Figures 4.2(a) and (b) represent uniform densities on the intervals $(-1, 1)$ and $(0, 1)$, respectively. Figure 4.2(c) illustrates a triangular density, and Figure 4.2(d) an exponential density. The latter curve is defined over the whole positive axis. (It requires calculus to show that the area under this curve is one.) The probability that a continuous random variable takes on a value in a specified interval is equal to the area over the interval. For example, the probability that the random variable in Figure 4.2(a) falls in the interval 0.2–0.6 is equal to the area over the interval. This is, $(0.6 - 0.2)(0.5) = 0.20$, so that we expect 20% of values of this random variable to fall in this interval. One of the most important probability density function is the normal distribution; it will be discussed in detail in Section 4.4.

How can we talk about a random sample of observations of a continuous variable? The simplest way is to consider the drawing of an observation as a trial and the probability of observing an arbitrary (but specified) value or smaller of the random variable. Definition 4.3 can then be applied.

Before turning to the normal distribution we introduce the concept of averages

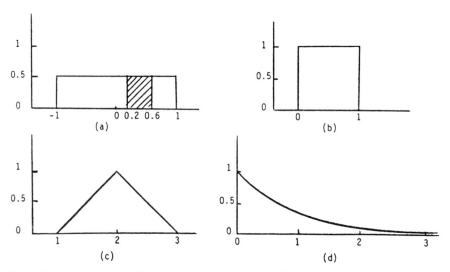

Figure 4.2. Examples of probability density functions. In each case, the area under the curve is equal to one.

of random variables. In Section 3.4.3, we discussed the average of a discrete variable based on the empirical relative frequency distribution. The average of a discrete variable Y with values $y_1, y_2, \ldots, y_k$ occurring with relative frequencies $p_1, p_2, \ldots, p_k$, respectively, was shown to be

$$\bar{y} = \sum py.$$

(We leave off the subscripts since it is clear that we are summing over all the values.) Now, if Y is a *random* variable and $p_1, p_2, \ldots, p_k$ are the *probabilities* of occurrence of the values $y_1, y_2, \ldots, y_k$, we give the quantity $\sum py$ a special name:

Definition 4.11. The *expected value of a discrete random variable* Y, denoted by $E(Y)$, is

$$E(Y) = \sum py,$$

where $p_1, \ldots, p_k$ are the probabilities of occurrence of the k possible values $y_1, \ldots, y_k$ of Y. The quantity $E(Y)$ is usually denoted by μ.

To calculate the expected value for the data of Table 3.12, the number of boys in families with eight children we proceed as follows. Let $p_1, p_2, \ldots, p_k$ represent the probabilities $P[Y = 0]$, $P[Y = 1]$, $\ldots$, $P[Y = 8]$. Then the expected value is

$$\begin{aligned}
E(Y) &= p_0 \times 0 + p_1 \times 1 + \cdots + p_8 \times 8 \\
&= (0.0040)(0) + (0.0277)(1) + (0.0993)(2) + \cdots + (0.0064)(8) \\
&= 4.1179 \\
&\doteq 4.12 \text{ boys.}
\end{aligned}$$

This leads to the statement, "a family with eight children will have an average of 4.12 boys."

Corresponding to the sample variance, s^2, is the variance associated with a discrete random variable:

Definition 4.12. The *variance of a discrete random variable* Y is

$$E(Y - \mu)^2 = \sum p(y - \mu)^2.$$

where $p_1, \ldots, p_k$ are the probabilities of occurrence of the k possible values $y_1, \ldots, y_k$ of Y.

The quantity $E(Y - \mu)^2$ is usually denoted by σ^2, where σ is the Greek letter *sigma*. For the example above, we calculate

$$\begin{aligned}
\sigma^2 &= (0.0040)(0 - 4.1179)^2 + (0.0277)(1 - 4.1179)^2 + \cdots + (0.0064)(1 - 4.1179)^2 \\
&= 2.0666.
\end{aligned}$$

Several comments about $E(Y - \mu)^2$ can be made:

1. Computationally, it is equivalent to calculating the sample variance using a divisior of n rather than $n-1$, and probabilities rather than relative frequencies.
2. The square root of σ^2 (σ) is called the (population) *standard deviation* of the random variable.
3. It can be shown that $\sum p(y - \mu)^2 = \sum py^2 - \mu^2$. The quantity $\sum py^2$ is called the *second moment about the origin*, and can be defined as the average value of the squares of Y, or the expected value of Y^2. This then can be written as $E(Y^2)$, so that $E(Y - \mu)^2 = E(Y^2) - E^2(Y) = E(Y^2) - \mu^2$. See Note 4.9 for further development of the algebra of expectations.

What about the mean and variance of a continuous random variable? As before, we could divide the range of the continuous random variable into a number of intervals, calculate the associated probabilities of the variable, assume that the values are concentrated at the midpoints of the intervals and proceed with Definitions 4.8 and 4.9. This is precisely what is done with one additional step: the intervals are made narrower and narrower. The mean is then the limit of a sequence of means calculated in this way, and similarly the variance. In these few sentences, we have crudely summarized the mathematical process known as integration. We will only state the results of such processes but will not actually derive or demonstrate them. For the densities presented in Figure 4.2, the following results can be stated:

Figure	Name	μ	σ^2
4.2(a)	Uniform on $(-1, 1)$	0	1/3
4.2(b)	Uniform on $(0, 1)$	1/2	1/12
4.2(c)	Triangular on $(1, 3)$	2	1/6
4.2(d)	Exponential	1	1

The first three densities in Figure 4.2 are examples of *symmetric* densities. A symmetric density always has equality of mean and median. The exponential density is not symmetric, it is "skewed to the right." Such a density has a mean that is larger than the median; for Figure 4.2(d), the median is about 0.69.

It is useful at times to state the functional form for the density. If Y is the random variable then for a value $Y = y$, the height of the density is given by $f(y)$. The densities in Figure 4.2 have the following functional forms:

Figure	Name of Density	Function	Range of Y
4.2(a)	Uniform on $(-1, 1)$	$f(y) = 0.5$	$(-1, 1)$
		$f(y) = 0$	elsewhere
4.2(b)	Uniform on $(0, 1)$	$f(y) = 1$	$(0, 1)$
		$f(y) = 0$	elsewhere
4.2(c)	Triangular on $(1, 3)$	$f(y) = y - 1$	$(1, 2)$
		$f(y) = 3 - y$	$(2, 3)$
		$f(y) = 0$	elsewhere
4.2(d)	Exponential	$f(y) = e^{-y}$	$(0, \infty)$
		$f(y) = 0$	elsewhere

The letter e in $f(y) = e^{-y}$ is the base of the natural logarithms. The symbol ∞ stands for positive infinity.

4.4 NORMAL DISTRIBUTIONS

4.4.1 Introduction and Motivation

Statistically, a population is the set of all possible values of a variable; random selection of objects of the population makes the variable a random variable and the population is completely described ("modeled") if the probability function or the probability density function is specified. A statistical challenge is to find models of populations that use a few parameters (say, two or three) and yet have wide applicability to real data. The normal or Gaussian distribution is one such statistical model.

The term "Gaussian" refers to Carl Friedrich Gauss, who developed and applied this model. The term "normal" appears to have been coined by Francis Galton. It is important to remember that there is nothing normal or abnormal about the normal distribution! A given data set may or may not be adequately modeled by the normal distribution. However, the normal distribution often proves to be a satisfactory model for data sets. The first and most important reason is that it "works" as will be indicated below. Secondly, there is a mathematical reason suggesting that a Gaussian distribution may adequately represent many data sets—the famous Central Limit Theorem discussed in Section 4.5. Finally, there is a matter of practicality. The statistical theory and methods associated with the normal distribution work in a nice fashion and have many desirable mathematical properties. But, no matter how convenient the theory, the assumptions that a data set is modeled adequately by a normal curve should be verified when looking at a particular data set. One such method is presented in Section 4.4.4.

4.4.2 Examples of Data that Might be Modeled by Normal Distribution

The first example is taken from a paper by Golubjatnikov *et al.* [1972], entitled, "Serum cholesterol levels of Mexican and Wisconsin school children." Figure 4.3 shows serum

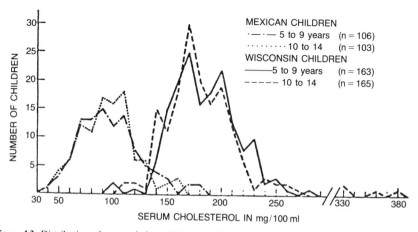

Figure 4.3. Distribution of serum cholesterol levels in Mexican and Wisconsin school children. Data from Golubjatnikov *et al.* [1972].

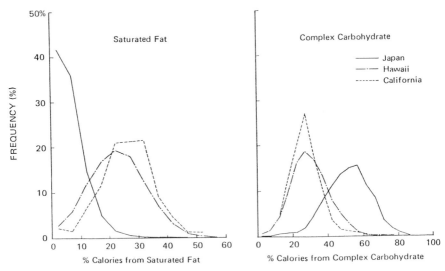

Figure 4.4. Frequency distribution of dietary saturated fat and dietary complex carbohydrate intake. Data from Kato *et al.* [1973].

cholesterol levels of Mexican and Wisconsin children in two different age groups. In each case there is considerable fluctuation in the graphs, probably due to the small numbers of individuals considered. However, it might be possible to model such data with a normal curve. Note, that there seem to be possibly too many values in the right tail to model the data by a normal curve; such curves are symmetric about their center point.

Figure 4.4 is taken from a paper by Kato *et al.* [1973] dealing with epidemiologic studies of coronary heart disease and stroke in Japanese men living in Japan, Hawaii and California. The curves present the frequency distribution of the percentage of calories from saturated fat and from complex carbohydrate in the three groups of men. Such percentages necessarily lie on the interval from 0 to 100. For the Hawaiian and Californian men with regard to the saturated fat, the bell-shaped curve might be a reasonable model. Note, however, that for the Japanese men, with a very low percentage of the diet from saturated fat, a bell-shaped curve would obviously be inappropriate.

A third example from Kesteloot and van Houte [1973] examines blood pressure measurements on 42,000 members of the Belgian army and territorial police. Figure 4.5 gives two different age groups. Again, particularly in the graphs of the diastolic pressures, it appears that a bell-shaped curve might not be a bad model.

Another example of data which do not appear to be modeled very well by a symmetric bell-shaped curve is from the paper by Hagerup *et al.* [1972], dealing with serum cholesterol, serum triglyceride, and ABO blood groups in a population of 50-year-old Danish men and women. Figure 4.6 shows the distribution of serum triglycerides. There is a notable asymmetry to the distribution, there being too many values to the right of the peak of the distribution as opposed to the left.

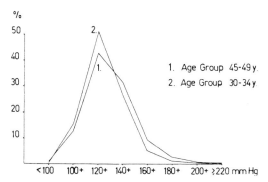

FIGURE 1. Distribution of SBP according to age. (Distribution is slightly skewed towards the higher values.)

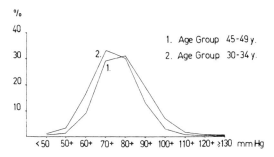

FIGURE 2. Distribution of DBP according to age. (Distribution is slightly skewed towards the higher values.)

Figure 4.5. Distributions of systolic and diastolic blood pressures according to age. Data from Kesteloot and van Houte [1973].

Another example of data which are not normally distributed are the 2-h plasma glucose levels (mg/100 ml) in Pima Indians. The data in Figure 4.7, from Rushforth *et al.* [1971] are the plasma glucose levels for male Pima Indians for each decade of age. The data become clearly bimodal (two modes) with increasing decade of age. Note also that the overall curve is shifting to the right with increasing decade: the first mode shifts from approximately 100 mg/100 ml in the 5–14 year decade to about 170 mg/100 ml in the 65–74 year decade.

4.4.3 Calculating Areas under the Normal Curve

A normal distribution is completely specified by its mean, μ, and standard deviation, σ. Figure 4.8 illustrates some normal distributions with specific means and standard deviations. Note that two normal distributions with the same standard deviation but different means have the same shape and are merely shifted; similarly two normal

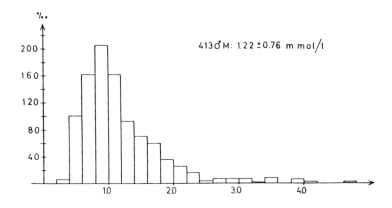

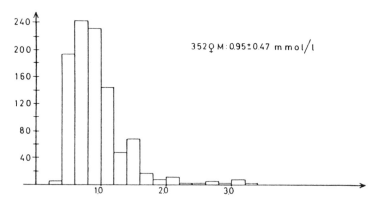

Figure 4.6. Serum triglycerides: 50-year survey in Glostrup. Fasting blood samples were drawn for determination of serum triglyceride by the method of Laurell. Data from Hagerup *et al.* [1972].

distributions with the same means but different standard deviations are centered in the same place but have different shapes. Consequently, μ is called a location parameter and σ a shape parameter.

The standard deviation is the distance from the mean to the point of inflection of the curve. This is the point where a tangent to the curve switches from being over the curve to under the curve.

As with any density, the probability that a normally distributed random variable takes on a value in a specified interval is equal to the area over the interval. So we need to be able to calculate these areas in order to know the desired probabilities. Unfortunately, there is no simple algebraic formula that gives these areas, so that tables must be used (see Note 4.14). Fortunately, we only need one table. For any normal distribution, we can calculate areas under its curve using a table for a normal distribution with mean $\mu = 0$ and standard deviation $\sigma = 1$ by expressing the variable in the number of standard deviations from the mean. Using algebraic notation we get the following:

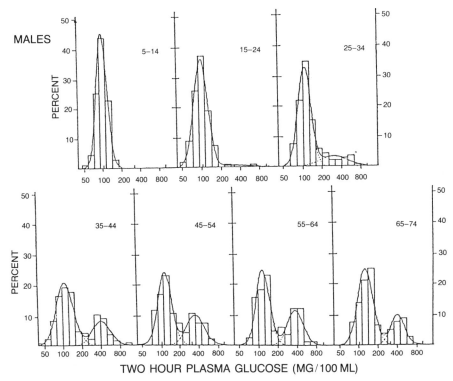

Figure 4.7. Distribution of 2-h plasma glucose levels (mg/100 ml) in male Pima Indians by decade. Data from Rushforth *et al.* [1971].

Definition 4.13. For a random variable Y with mean μ and standard deviation σ, the associated *standard score*, Z, is

$$Z = \frac{Y - \mu}{\sigma}.$$

Given values for μ and σ, we can go from the "Y scale" to the "Z scale" and vice versa. Algebraically, we can solve for Y and get $Y = \mu + \sigma Z$. This is also the procedure that is used to get from degrees Centigrade (°C) to degrees Fahrenheit (°F). The relationship is:

$$°C = \frac{°F - 32}{1.8}.$$

Similarly,

$$°F = 32 + 1.8 \times °C.$$

Definition 4.14. A *standard normal distribution* is a normal distribution with mean $\mu = 0$ and standard deviation $\sigma = 1$.

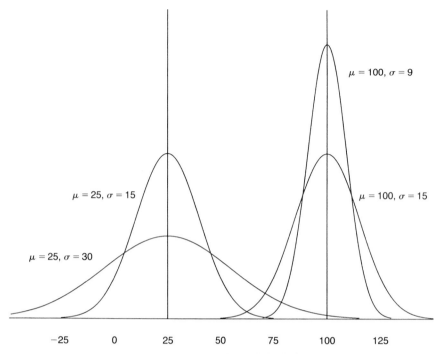

Figure 4.8. Examples of normal distributions.

Table A.1 in the Appendix gives standard normal probabilities. The table lists the area to the left of the stated value of the standard normal deviate under the columns headed "cum. dist." For example, the area to the left of $Z = 0.10$ is 0.5398 as illustrated.

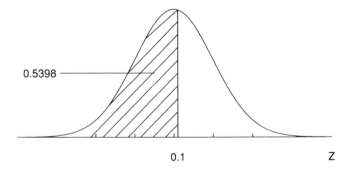

In words, 53.98% of normally distributed observations have values less than 0.10 standard deviations above the mean. We use the notation, $P[Z \leq 0.10] = 0.5398$, or in general, $P[Z \leq z]$. To indicate a value of Z associated with a specified area, p, to its left we will use a subscript on the value, Z_p. For example, $P[Z \leq z_{0.1}] = 0.10$; that is, we want that value of Z such that 0.1 of the area is to its left (call it $z_{0.1}$), or equivalently, such a proportion 0.1 of Z values are less than or equal to $z_{0.1}$. By symmetry, we note that $z_{1-p} = -z_p$.

Since the total area under the curve is 1, we can get areas in the right hand tail by subtraction. Formally,

$$P[Z > z] = 1 - P[Z \leq z].$$

In terms of the above example, $P[Z > 0.10] = 1 - 0.5398 = 0.4602$. By symmetry, areas to the left of $Z = 0$ can also be obtained. For example, $P[Z \leq -0.10] = P[Z > 0.10] = 0.4602$. These values are indicated in the following pictures.

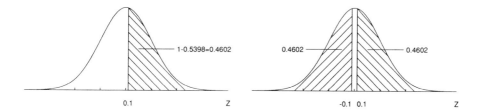

We now illustrate the use of the standard normal table with two word problems. When calculating areas under the normal curve you will find it helpful to draw a rough normal curve and shade in the required area.

Example 4.4. Suppose IQ is normally distributed with mean $\mu = 100$ and standard deviation $\sigma = 15$. A person with IQ > 115 has a "high IQ." What proportion of the population has high IQs? We sketch the required area.

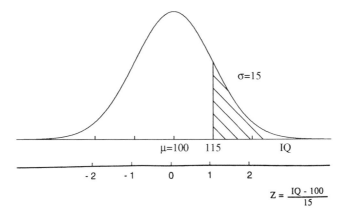

It is clear that IQ $= 115$ is one standard deviation above the mean, so the statement $P[IQ > 115]$ is equivalent to $P[Z > 1]$. This can be obtained from Table A.1 using the relationship $P[Z > 1] = 1 - P[Z \leq 1] = 1 - 0.8413 = 0.1587$. Thus 15.87% of the population has a high IQ. By the same token, if an IQ below 85 is labeled "low IQ," then 15.87% of the population has a low IQ.   □

Example 4.5.   Consider the serum cholesterol levels of Wisconsin children as pictured in Figure 4.3. Suppose the population mean is 175 mg/100 ml and the population standard deviation is 30 mg/100 ml. Suppose a "normal cholesterol value" is taken to be a value within two standard deviations of the mean. What are the "normal limits" and what proportion of Wisconsin children will be within normal limits? Again, on the basis of a sketch:

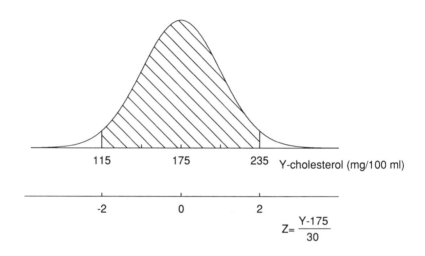

We want the area within ±2 standard deviations of the mean. This can be expressed as $P[-2 \leq Z \leq +2]$. By symmetry and the property that the area under the normal curve is 1.0, we can express this as

$$P[-2 \leq Z \leq 2] = 1 - 2P[Z > 2].$$

(You should draw a picture to convince yourself.) From Table A.1, $P[Z \leq 2] = 0.9772$, so that $P[Z > 2] = 1 - 0.9772 = 0.0228$. (Note that this value is computed for you in the column labeled "one sided.") The desired probability is

$$P[-2 \leq Z \leq 2] = 1 - 2(0.0228)$$
$$= 0.9544.$$

In words, 95.44% of the population of Wisconsin school children have cholesterol values within "normal limits."

Suppose we change the question: instead of defining normal limits and calculating the proportion within these limits, we define the limits such that, say, 95% of the population has cholesterol values within the stated limits. Before, we went from cholesterol level to Z-value to area, now we want to go from area to Z-value to cholesterol values. In this case, Table A.2 will be useful. Again, we begin with a picture:

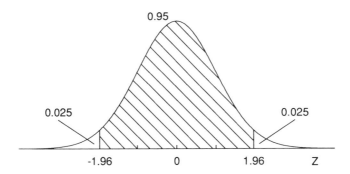

From Table A.2, we get $P[Z > 1.96] = 0.025$ so that $P[-1.96 \le Z \le 1.96] = 0.95$; in words, 95% of normally distributed observations are within ± 1.96 standard deviations of the mean. Or, translated to cholestrol values by the formula $Y = 175 + 30Z$. For $Z = 1.96$, $Y = 175 + (30)(1.96) = 233.8 \doteq 234$, and for $Z = -1.96$, $Y = 175 + (30)(-1.96) = 116.2 \doteq 116$. On the basis of the model, 95% of cholesterol values of Wisconsin children are between 116 and 234 mg/100 ml. If the mean and standard deviation of cholesterol values of Wisconsin children are 175 mg/100 ml and 30 mg/100 ml, respectively, then the 95% limits $(116, 234)$ are called 95% *tolerance limits*.   □

Example 4.6. **(Normal Laboratory Values).** Many times it is useful to know the range of "normal values" of a substance (variable) in a "normal population." A laboratory test can then be carried out to determine whether a subject's values are high, low, or within normal limits. An article by Zervas *et al.* [1970] provides a list of "normal values" for more than 150 substances ranging from ammonia to Vitamin B12. These values have been reprinted in *The Merck Manual of Diagnosis and Therapy 15th Edition, Berkow (Ed.)*, [1987]. The term "normal values" does *not* imply that variables are normally distributed (i.e., follow a Gaussian or bell-shaped curve). A paper by Elveback *et al.* [1970] already indicated that, of seven common substances (calcium, phosphate, total protein, albumin, urea, magnesium, and alkaline phosphatase), only albumin values can be adequately summarized by a normal distribution. All the other substances had distributions of values that were skewed. The authors (correctly) conclude that "the distributions of values in healthy persons *cannot* be assumed to be normal." Admittedly, this leaves an unsatisfactory situation: what then do we mean by "normal limits"? What proportion of "normal values" will fall outside the "normal limits" as the result of random variation? None of these—and other—critical questions can now be answered because a statistical model is not available. But that appears to be the best we can do at this point; as the authors point out, "good limits are hard to get, and bad limits hard to change."   □

4.4.4 Probability Paper

How can we know whether the normal distribution model fits a particular set of data? There are many tests for normality, some graphical, some numerical. In this

Table 4.1. Frequency Distribution of Stature of 928 Adult Children, from Galton [1889].

Endpoint (inches)	Frequency	Cumulative Frequency	Cumulative Percentage
60.7	0	0	0.0
62.7	12	12	1.3
64.7	91	103	11.1
66.7	165	268	28.9
68.7	258	526	56.7
70.7	266	792	85.3
72.7	105	897	96.7
74.7[a]	31	928	100.0

[a]Assumed endpoint.

subsection, we want to discuss a simple graphical test which uses *normal probability paper*, (often abbreviated *probability paper*). This paper is structured in such a way that the cumulative relative frequencies of data from a normal distribution fall on a straight line. If the data are in the form of an empirical frequency distribution, the cumulative percentages (or proportions) are plotted against the endpoints of the intervals of values of the variable.

A famous book by Galton [1889] contains data on the stature of parents and their adult children. Table 4.1 gives the frequency distributions of heights of 928 adult children. The cumulative percentages plotted against the endpoints of the intervals in Figure 4.9 produce the usual sigmoid-shaped curve.

These data are now plotted on normal probability paper in Figure 4.10. The vertical

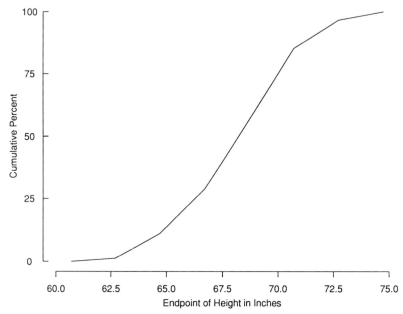

Figure 4.9. Empirical cumulative frequency polygon of heights of 928 adult children. Data from Galton [1889].

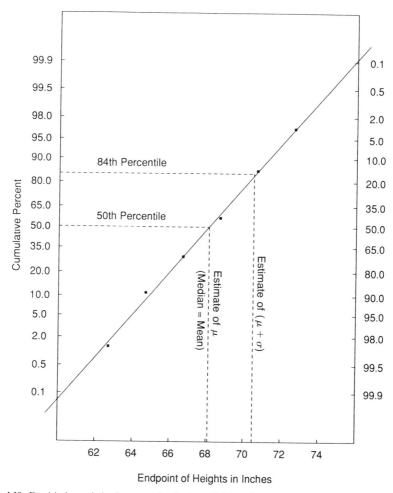

Figure 4.10. Empirical cumulative frequency distribution of heights of 928 adult children plotted on probability paper. Data from Galton [1889].

scale has been stretched near 0% and 100% in such a way that data from a normal distribution should fall on a straight line. Clearly, the data are consistent with a normal distribution model.

Note that the right vertical scale decreases from 99.9% to 0.01%, while the left vertical increases from 0.01% to 99.99%. The cumulative percentages are plotted using the left hand scale.

Another feature of such a graph is that both mean and standard deviation can be read from the graph using the following facts:

1. The mean is equal to the median for the normal distribution; and

2. From Table A.1, we see that 84.13% of the observations fall below $\mu + \sigma$.

Hence, reading these percentiles off the graph, we can get estimates of μ and $\mu + \sigma$. From the graph, we estimate μ to be 68.1 inches and $\mu + \sigma$ to be 72.5 inches, so that an estimate of σ is $72.5 - 68.1 = 4.4$ inches.

4.5 SAMPLING DISTRIBUTIONS

4.5.1 Statistics Are Random Variables

Consider a large multi-center collaborative study of the effectiveness of a new cancer therapy. A great deal of care is taken to standardize the treatment from center to center, but it is obvious that the average survival time on the new therapy (or increased survival time if compared to a standard treatment) will vary from center to center. This is an illustration of basic statistical fact: sample statistics vary from sample to sample. The key idea is that a statistic associated with a random sample is a random variable. What we want to do in this subsection is to relate the variability of a statistic based on a random sample to the variability of the random variable on which the sample is based.

Definition 4.15. The probability (density) function of a statistic is called the *sampling distribution of the statistic*.

What are some of the characteristics of the sampling distribution? In this subsection, we state some results about the sample mean. In Section 4.8 some properties of the sampling distribution of the sample variance are discussed.

4.5.2 Properties of Sampling Distribution

Result 4.1. If a random variable Y has population mean μ and population variance σ^2, then the sampling distribution of sample means (of samples of size n) has population mean μ, and population variance σ^2/n. Note that this result does not assume normality of the "parent" population.

Definition 4.16. The standard deviation of the sampling distribution is called the *standard error*.

Example 4.7. Suppose that IQ is a random variable with mean $\sigma = 100$ and standard deviation $\mu = 15$. Now consider the average IQ of classes of 25 students. What are the population mean and variance of these class averages? By Result 4.1, the class averages have population mean $\mu = 100$ and population variance, $\sigma^2/n = 15^2/25 = 9$. Or, the standard error is $\sqrt{\sigma^2/n} = \sqrt{15^2/25} = \sqrt{9} = 3$.
To summarize:

	Population		
	Mean	**Variance**	$\sqrt{\textbf{Variance}}$
Single observation, Y	100	$15^2 = 225$	$15 = \sigma$
Mean of 25 observations, $\overline{Y}$	100	$15^2/25 = 9$	$3 = \sigma/\sqrt{n}$

The standard error of the sampling distribution of the sample mean $\overline{Y}$ is indicated by $\sigma_{\overline{y}}$ to distinguish it from the standard deviation, σ, associated with the random variable Y. It is instructive to contemplate the formula for the standard error, $\sigma/(\sqrt{n})$. This formula makes clear that a reduction in variability by, say, a factor of two, requires a fourfold increase in sample size. Consider Example 4.7. How large must a class be to reduce the standard error from 3 to 1.5? We want $\sigma/(\sqrt{n}) = 1.5$. Given that $\sigma = 15$ and solving for n, we get $n = 100$. This is a fourfold increase in class size, from 25 to 100. In general, if we want to reduce the standard error by a factor of k we must increase the sample size by a factor of k^2. This suggests that if a study consists of, say, 100 observations, and with a great deal of additional effort (out of proportion to the effort of getting the 100 observations), another ten observations can be obtained, that the additional ten may not be worth the effort.

The standard error based on 100 observations is $\sigma/(\sqrt{100})$. The ratio of these standard errors is

$$\frac{\sigma/\sqrt{100}}{\sigma/\sqrt{110}} = \frac{\sqrt{100}}{\sqrt{110}} = 0.95.$$

Hence a 10% increase in sample size produces only a 5% increase in precision. Of course, precision is not the only criterion we are interested in; if the 110 observations are randomly selected persons to be interviewed, it may be that the last 10 are very hard to locate or difficult to persuade to take part in the study and not including them may introduce a serious *bias*. But with respect to *precision* there is not much difference between means based on 100 observations and means based on 110 observations (see Note 4.10).

4.5.3 The Central Limit Theorem

Although Result 4.1 gives some characteristics of the sampling distribution, it does not permit us to calculate probabilities because we do not know the form of the sampling distribution. To be able to do this we need the following:

Result 4.2. If Y is normally distributed with mean μ and variance σ^2, then $\overline{Y}$, based on a random sample of n observations, is *normally* distributed with mean μ and variance σ^2/n.

Result 4.2 basically states that if Y is normally distributed then $\overline{Y}$, the mean of a random sample, is normally distributed. Result 4.1 then specifies the mean and variance of the sampling distribution. Result 4.2 implies that as the sample size increases that the (normal) distribution of the sample mean becomes more and more "pinched." Figure 4.11 shows three sampling distributions for means of random samples of size one, two, and four.

What is the probability that the average IQ of a class of 25 students exceeds 106? By Result 4.2, $\overline{Y}$, the average of 25 IQs, is normally distributed with mean $\mu = 100$, and standard error $\sigma/(\sqrt{n}) = 15/(\sqrt{25}) = 3$. Hence the probability that $\overline{Y} > 106$ can be calculated as

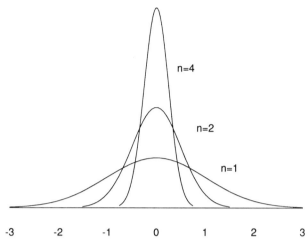

Figure 4.11. Three sampling distributions for means of random samples of size 1, 2, and 4 from a $N(0, 1)$ population.

$$P[\overline{Y} \geq 106] = P\left[Z \geq \frac{106 - 100}{3}\right]$$
$$= P[Z \geq 2]$$
$$= 1 - 0.9772$$
$$= 0.0228.$$

So, approximately 2% of average IQs of classes of 25 students will exceed 106. This can be compared with the probability that a single person's IQ exceeds 106:

$$P[Y > 106] = P\left[Z > \frac{6}{15}\right] = P[Z > 0.4] = 0.3446.$$

The final result we want to state is known as the Central Limit Theorem.

Result 4.3. If a random variable Y has population mean μ and population variance σ^2, then the sample mean $\overline{Y}$, based on n observations, is approximately normally distributed with mean μ and variance σ^2/n, for sufficiently large n.

This is a remarkable result and the most important reason for the central role of the normal distribution in statistics. What this basically states is that means of random samples from *any* distribution (with mean and variance) will tend to be normally distributed as the sample size becomes sufficiently large. How large is "large"? Consider the distributions of Figure 4.2. Samples of six or more from the first three distributions will have means that are virtually normally distributed. The fourth distribution will take somewhat larger samples before approximate normality is obtained; n must be around 25 or 30. Figure 4.12 shows the sampling distributions of means of samples of various sizes drawn from Figure 4.2(d).

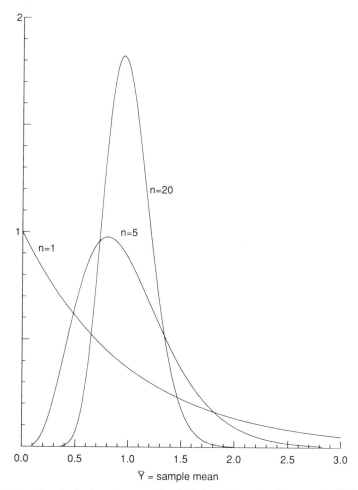

Figure 4.12. Sampling distributions of means of 5 and 20 observations when the parent distribution is exponential.

The Central Limit Theorem provides some reassurance when we are not certain whether observations are normally distributed. The means of reasonably sized samples will have a distribution that is approximately normal. So inference procedures based on the sample means can often use the normal distribution. But you must be careful not to impute normality to the original observations.

4.6 INFERENCE ABOUT THE MEAN OF A POPULATION

4.6.1 Point and Interval Estimates

In this section we want to discuss inference about the mean of a population when the population variance is known. The assumption may seem artificial but sometimes this situation will occur. For example, it may be that a new treatment alters the level of a response variable but not its variability, so that the variability can be assumed to be known from previous experiments. (In Section 4.8, we discuss a method for comparing the variability of an experiment with previous established variability; in Chapter 5, the problem of inference when both population mean and variance are unknown is considered.)

To put the problem more formally, we have a random variable Y with unknown population mean μ. A random sample of size n is taken and inferences about μ are to be made on the basis of the sample. We assume that the population variance is known; denote it by σ^2. Normality will also be assumed; even when the population is not normal, we may be able to appeal to the Central Limit Theorem.

A "natural" estimate of the population mean μ is the sample mean $\overline{Y}$. It is a "natural" estimate of μ because we know that $\overline{Y}$ is normally distributed with the same mean, μ, and variance σ^2/n. Even if Y is not normal, $\overline{Y}$ is approximately normal on the basis of the Central Limit Theorem. The statistic $\overline{Y}$ is called a *point estimate* since we estimate the parameter μ by a single value or point.

Now the question arises: How precise is the estimate? How can we distinguish between two samples of, say, 25 and 100 observations? Both may give the same—or approximately the same—sample mean, but we know that the mean based on the 100 observations is more accurate, that is, has a smaller standard error. One possible way of summarizing this information is to give the sample mean and its standard error. This would be useful for *comparing* two samples. But this does not seem to be a useful approach in considering one sample and its information about the parameter. To use the information in the sample, we set up an *interval* estimate as follows: consider the quantity $\mu \pm (1.96)\sigma/\sqrt{n}$. It describes the spread of sample means; in particular, 95% of means of samples of size n will fall in the interval $[\mu - 1.96\sigma/\sqrt{n}, \mu + 1.96\sigma/\sqrt{n}]$. The interval has the property that as n increases, the width decreases (refer to Section 4.5 for further discussion). Suppose we now replace μ by its point estimate, $\overline{Y}$. How can we interpret the resulting interval? Since the sample mean, $\overline{Y}$, varies from sample to sample it cannot mean that 95% of the sample means will fall in the interval for a specific sample mean. The interpretation is that the probability is 0.95 that the interval *straddles* the population mean. Such an interval is referred to as a 95% confidence interval for the population mean, μ. We now formalize this definition.

Definition 4.17. A $100(1 - \alpha)\%$ *confidence interval* for the mean μ of a normal population (with variance known) based on a random sample of size n is

$$\overline{Y} \pm z_{1-\alpha/2}\frac{\sigma}{\sqrt{n}},$$

where $z_{1-\alpha/2}$ is the value of the standard normal deviate such that $100(1 - \alpha)\%$ of the area falls within $\pm z_{1-\alpha/2}$.

Strictly speaking, we should write

$$\left(\overline{Y} + z_{\alpha/2}\frac{\sigma}{\sqrt{n}}, \overline{Y} + z_{1-\alpha/2}\frac{\sigma}{\sqrt{n}} \right),$$

but by symmetry, $z_{\alpha/2} = -z_{1-\alpha/2}$, so that it is quicker to use the above expression.

Example 4.8. In Section 3.3.1, we discussed the age at death of 78 cases of crib death (SIDS) occurring in King County, Washington, in 1976–1977. Birth certificates were obtained for these cases and birthweights were tabulated. Let $Y =$ birthweight in grams. Then, for these 78 cases, $\overline{Y} = 2993.6 \doteq 2994$ g. From a listing of all the birthweights, it is known that the standard deviation of birthweight is about 800 g, i.e., $\sigma = 800$ g. A 95% confidence interval for the mean birthweight of SIDS cases is calculated to be

$$2994 \pm (1.96)\left(\frac{800}{\sqrt{78}}\right) \quad \text{or} \quad 2994 \pm (1.96)(90.6) \quad \text{or} \quad 2994 \pm 178,$$

producing a lower limit of 2816 g, and an upper limit of 3172 g.

Thus, on the basis of these data, we are 95% confident that we have straddled the population mean, μ, of birthweight of SIDS infants by the interval (2816, 3172).

Suppose we had wanted to be more confident; say, a level of 99%. The value of Z now becomes 2.58 (from Table A.2) and the corresponding limits are $2994 \pm (2.58)\left(800/\sqrt{78}\right)$, or (2760, 3228). The width of the 99% confidence interval is greater than that of the 95% confidence interval (468 g *vs.* 356 g); the price we paid for being more sure that we have straddled the population mean.

Several comments should be made about confidence intervals:

1. Since the population mean μ is fixed, it is not correct to say that the probability is $1 - \alpha$ that μ is in the confidence interval *once it is computed*; that probability is zero or one. Either the mean is in the interval and the probability is equal to one, or the mean is not in the interval and the probability is zero.

2. We can increase our confidence that the interval straddles the population mean by decreasing α, hence increasing $Z_{1-\alpha/2}$. We can take values from Table A.2 to construct the following confidence levels:

Confidence Level	Z-Value
90%	1.64
95%	1.96
99%	2.58
99.9%	3.29

The effect of increasing the confidence level will be to increase the width of the confidence interval.

3. To decrease the width of the confidence interval we can either decrease the confidence level or increase the sample size. The width of the interval is $2z_{1-\alpha/2}\,\sigma/\sqrt{n}$. For a fixed confidence level the width is essentially a function of $\sigma/\sqrt{n}$, the standard error of the mean. To decrease the width by a factor of, say, two, the sample size must be increased by a factor of four, analogous to the discussion in Section 4.5.2.

4. Confidence levels are usually taken to be 95% or 99%. These levels are a matter of convention; there are no theoretical reasons for choosing these values. A rough rule to keep in mind is that a 95% confidence interval is defined by the sample mean ± 2 standard errors. (Note: *not* standard deviations.)

4.6.2 Hypothesis Testing

In estimation, we start with a sample statistic and make a statement about the population parameter: a confidence interval makes a probabilistic statement about straddling the population parameter. In hypothesis testing, we start by assuming a value for a parameter and a probability statement is made about the value of the corresponding statistic. In this section, as in Section 4.6.1, we assume that the population variance is known and that we want to make inferences about the mean of a normal population on the basis of a sample mean. The basic strategy in hypothesis testing is to measure how far an observed statistic is from a hypothesized value of the parameter. If the distance is "great," we would argue that the hypothesized parameter value is inconsistent with the data and we would be inclined to reject the hypothesis (we could be wrong, of course; rare events do happen).

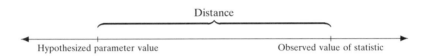

To interpret the distance, we must take into account the basic variability (σ^2) of the observations, and the size of the sample (n) upon which the statistic is based. As a rough rule of thumb that will be explained below, if the observed value of the statistic is more than two standard errors from the hypothesized parameter value, we question the truth of the hypothesis.

To continue Example 4.8, the mean birthweight of the 78 SIDS cases was 2994 g. The standard deviation σ_0 was assumed to be 800 g, and the standard error $\sigma/\sqrt{n} = 800/\sqrt{78} = 90.6$ g. One question that comes up in the study of SIDS is whether SIDS cases tend to have a different birth weight than the general population. For the general population, the average birthweight is about 3300 g. Is the *sample* mean value of 2994 g consistent with this value?

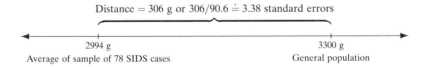

The diagram shows that the distance between the two values is 306 g. The standard error is 90.6, so that the observed value is $306/90.6 = 3.38$ standard errors from the hypothesized population mean. By the rule we stated, the distance is so great that we would conclude that the mean of the *sample* of SIDS births is inconsistent with the mean value in the general population. Hence, we would conclude that the SIDS births come from a population with mean birthweight somewhat less than that of the general population. (This raises more questions, of course: are the gestational ages comparable? what about the racial composition? and so on.) The best estimate we have of the mean birthweight of the population of SIDS cases is the sample mean: in this case, 2994 g; about 300 g lower than that for the normal population.

Before introducing some standard hypothesis testing terminology, two additional points should be made:

1. We have expressed "distance" in terms of number of standard errors from the hypothesized parameter value. Equivalently, we can associate a tail probability with the observed value of the statistic. For the sampling situation described above we know that the sample mean $\overline{Y}$ is normally distributed with standard error $\sigma/\sqrt{n}$.

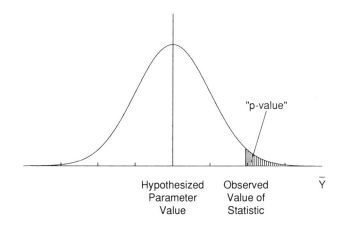

Hypothesized   Observed $\overline{Y}$
Parameter   Value of
Value   Statistic

As the picture indicates, the farther away the observed value of the statistic from the hypothesized parameter value, the smaller the area (probability) in the tail. This tail probability is usually called the *p-value*. For example (using Table A.2), the area to the right of 1.96 standard errors is 0.025; the area to the right of 2.58 standard errors is 0.005. Conversely, if we specify the area, the number of standard errors will be determined.

2. Suppose we planned before doing the statistical test that we would not question the hypothesized parameter value if the observed value of the statistic fell within, say, two standard errors of the parameter value. We could divide the sample space for the statistic (i.e., the real line) into three regions as follows:

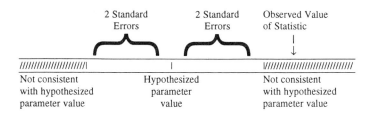

These regions could have been set up before the value of the statistic was observed. All that needs to be determined then is in which region the observed value of the statistic falls to determine if it is consistent with the hypothesized value.

We now formalize some of these concepts:

Definition 4.18.   A *null hypothesis* specifies a hypothesized real value, or values, for a parameter (see Note 4.15 for further discussion).

Definition 4.19.   The *rejection region* consists of the set of values of a statistic for which the null hypothesis is rejected. The values of the boundaries of the region are called the *critical values*.

Definition 4.20.   A *Type I error* occurs when the null hypothesis is rejected when, in fact, it is true.

Definition 4.21.   An *alternative hypothesis* specifies a real value or range of values for a parameter which will be considered when the null hypothesis is rejected.

Definition 4.22.   A *Type II error* occurs when the null hypothesis is not rejected when it is false.

Definition 4.23.   The *power of a test* is the probability of rejecting the null hypothesis when it is false.

Definition 4.24.   The *p-value* in a hypothesis testing situation is that value of p, $0 \leq p \leq 1$ such that for $\alpha > p$ the test rejects the null hypothesis at significance level α, and for $\alpha < p$ the test does not reject the null hypothesis. Intuitively, the *p*-value is the probability under the null hypothesis of observing a value as unlikely or more unlikely, than the value of the test statistic. The *p*-value is a measure of the distance from the observed statistic to the value of the parameter specified by the null hypothesis.

Testing some hypotheses can be tricky. From *American Scientist*, March/April 1976.

"It may very well bring about immortality, but it will take forever to test it."

© 1976 by Sidney Harris — *American Scientist* Magazine

Notation

1. The null hypothesis is denoted by H_0. The alternative hypothesis by H_A.
2. The probability of a Type I error is denoted by α; the probability of a Type II error by β. The power is then

$$\text{Power} = 1 - \text{Probability of Type II error}$$
$$= 1 - \beta$$

Continuing Example 4.8, we can think of our assessment of the birthweight of SIDS babies as a kind of decision problem illustrated in the following layout:

Decision SIDS Birthweights	State of Nature SIDS Birthweights	
	Same as Normal	**Not the same**
Same as normal	Correct $(1 - \alpha)$	Type II error (β)
Not the same	Type I error (α)	Correct $(1 - \beta)$

The table illustrates the two types of errors that can be made depending on our decision and the "state of nature." The null hypothesis for this example can be written as

$$H_0 : \mu = 3300 \text{ g}$$

and the alternative hypothesis written as

$$H_A : \mu \neq 3300 \text{ g}.$$

Suppose we want to reject the null hypothesis when the sample mean $\overline{Y}$ is more than two standard errors from the H_0 value of 3300 grams. The standard error is 90.6 g. The rejection region is then determined by $3300 \pm (2)(90.6)$ or 3300 ± 181.

We can then set up the hypothesis testing framework as indicated by the following diagram:

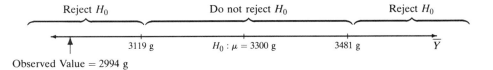

The rejection region consists of values to the left of 3119 g (i.e., $\mu - 2\sigma/\sqrt{n}$) and to the right of 3481 g (i.e., $\mu + 2\sigma/\sqrt{n}$). The observed value of the statistic, $\overline{Y} = 2994$ g, falls in the rejection region, and we therefore reject the null hypothesis that SIDS cases have the same mean birthweight as normal children. On the basis of the observed sample value, we conclude that SIDS babies tend to weigh less than normal babies.

The probability of a Type I error is the probability that the mean of a sample of 78 observations from a population with mean 3300 g is less than 3119 g, or greater than 3481 g:

$$P[3119 \leq \overline{Y} \leq 3481] = P\left[\frac{3119 - 3300}{90.6} \leq Z \leq \frac{3481 - 3300}{90.6}\right]$$
$$= P[-2 \leq Z \leq +2],$$

where Z is a standard normal deviate.

From Table A.1,

$$P[Z \leq 2] = 0.9772,$$

so that

$$1 - P[-2 \leq Z \leq 2] = (2)(0.0228) = 0.0456,$$

the probability of a Type I error. The probability is 0.0455 from the two-sided p-value of Table A.1. The difference relates to rounding.

The probability of a Type II error can be computed when a value for the parameter under the alternative hypothesis is specified. Suppose for these data that the alternative hypothesis is

$$H_A : \mu = 3000 \text{ g},$$

this value being suggested from previous studies. To calculate the probability of a Type II error—and the power—we assume that $\overline{Y}$, the mean of the 78 observations, comes from a normal distribution with mean 3000 g, and standard error as before, 90.6 g.

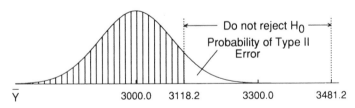

Y = mean birthweight (grams) of 78 SIDS cases

As the picture indicates, the probability of a Type II error is the area over the interval $(3119, 3481)$. This can be calculated:

$$P[\text{Type II error}] = P[3119 \leq \overline{Y} \leq 3481]$$
$$= P\left[\frac{3119 - 3000}{90.6} \leq Z \leq \frac{3481 - 3000}{90.6}\right]$$
$$\doteq P[1.31 \leq Z \leq 5.31]$$
$$\doteq 1 - 0.905$$
$$\doteq 0.095.$$

So, $\beta = 0.095$ and the power is $1 - \beta = 0.905$. Again, these calculations can be made before any data are collected, and they say that if the SIDS population mean birthweight were 3000 g and the normal population birthweight 3300 g, then the probability is 0.905 that a mean from a sample of 78 observations will be declared signficantly different from 3300 g.

Let us summarize the analysis of this example:

$$\text{Hypothesis testing setup (no data taken)} \begin{cases} H_0 : \mu = 3300 \text{ g,} \\ H_A : \mu = 3000 \text{ g,} \\ \sigma = 800 \text{ g (known),} \\ n = 78, \\ \text{Rejection region: } \pm 2 \text{ standard} \\ \quad \text{errors from 3000 g,} \\ \alpha = 0.0456, \\ \beta = 0.095, \\ 1 - \beta = 0.905, \end{cases}$$

$$\text{Observe: } \overline{Y} = 2994,$$
$$\text{Conclusion: Reject } H_0.$$

The value of α is usually specified beforehand: The most common value is 0.05, somewhat less common values are 0.01 or 0.001. Corresponding to the confidence level in interval estimation, we have the *significance level* in hypothesis testing. The significance level is often expressed as a percent and defined to be $100\alpha\%$. Thus, for $\alpha = 0.05$, the hypothesis test is carried out at the 5% significance level, or 0.05 significance level.

The use of a single symbol β for the probability of a Type II error is standard, but a bit misleading. We expect β to stand for one number in the same way that α stands for one number. In fact, β is a function whose argument is the assumed true value of the parameter being tested. For example, in the context of $H_A : \mu = 3000$ g, β is a function of μ, and could be written $\beta(\mu)$. It follows that the power is also a function of the true parameter: power $= 1 - \beta(\mu)$. Thus one must specify a value of μ to compute the power.

We finish this introduction to hypothesis testing with a discussion of the one-tail test and two-tail test. These are related to the choice of the rejection region. Even if α is specified, there is an infinity of rejection regions such that the area over the region is equal to α. Usually only two types of regions are considered. A picture will make this clear:

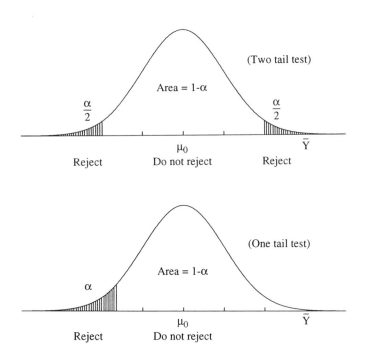

A two-tail test is associated with a rejection region that extends both to the left and to the right of the hypothesized parameter value. A one-tail test is associated with a region to one side of the parameter value. The alternative hypothesis determines the type of test to be carried out. Consider again the birthweight of SIDS cases. Suppose we know that if the mean birthweight of these cases is not the same as that of normal infants (3300 g), then it must be less; it is not possible for it to be more. In that case, if the null hypothesis is false, we would expect the sample mean to be below 3300 g, and we would reject the null hypothesis for values of $\bar{Y}$ below 3300 g. We could then write the null hypothesis and alternative hypothesis as follows:

$$H_0 : \mu = 3300 \text{ g,}$$
$$H_A : \mu < 3300 \text{ g.}$$

We would want to carry out a one-tail test in this case by setting up a rejection region to the left of the parameter value. Suppose we want to test at the 0.05 level, and we only want to reject for values of $\bar{Y}$ below 3300 g. From Table A.2, we see that we must locate the start of the rejection region 1.64 standard errors to the left of $\mu = 3300$ g:

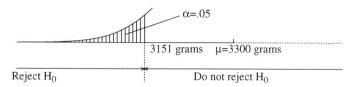

$\alpha = .05$

3151 grams $\mu = 3300$ grams

Reject H_0 Do not reject H_0

The value is $3300 - (1.64)(800/\sqrt{78})$ or $3300 - (1.64)(90.6) = 3151$ g.

Suppose we want a two-tail test at the 0.05 level. The Z-value (Table A.2) is now 1.96, which distributes 0.025 in the left tail and 0.025 in the right tail. The corresponding values for the critical region are $3300 \pm (1.96)(90.6)$ or $(3122, 3478)$, producing a region very similar to the one calculated earlier.

The question is, when should you do a one-tail test and when a two-tail test? As was stated, the alternative hypothesis determines this. An alternative hypothesis of the form $H_A : \mu \neq \mu_0$ is called two-sided, and will require a two-tail test. Similarly, the alternative $H_A : \mu < \mu_0$ is called one-sided, and will lead to a one-tail test. So should the alternative hypothesis be one-sided or two-sided? The experimental situation will determine this. For example, if nothing is known about the effect of a proposed therapy, the alternative hypothesis should be made two-sided. However, if it is suspected that a new therapy will do nothing, or increase a response level and if there is no reason to distinguish between no effect and a decrease in the response level, then the test should be one-tailed. The general rule is, the more specific you can make the experiment, the greater the power of the test (see Fleiss [1981], Section 2.4). (See Exercise 33 to convince yourself that the power of a one-tailed test is greater, *if* the alternative hypothesis correctly specifies the situation.)

4.7 CONFIDENCE INTERVALS *VS.* TESTS OF HYPOTHESES

You may have noticed that there is a very close connection between the confidence intervals and the tests of hypotheses we have constructed. In both approaches we have used the standard normal distribution and the quantity α.

In confidence intervals, we:

1. Specify the confidence level $(1 - \alpha)$

2. Read $z_{1-\alpha/2}$ from a standard normal table

3. Calculate $\overline{Y} \pm z_{1-\alpha/2}\sigma/\sqrt{n}$.

In hypothesis testing, we:

1. Specify the null hypothesis $(H_0 : \mu = \mu_0)$

2. Specify α, the probability of a Type I error

3. Read $z_{1-\alpha/2}$ from a standard normal table

4. Calculate $\mu_0 \pm z_{1-\alpha/2}\sigma/\sqrt{n}$

5. Observe $\overline{Y}$; reject or accept H_0.

The two approaches can be represented pictorially as follows:

Confidence
Interval

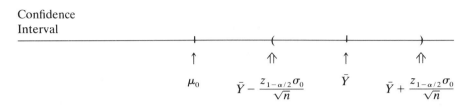

Hypothesis
Test

It is easy to verify that if the confidence interval does not straddle μ (as is the case in the above illustration), then $\overline{Y}$ will fall in the rejection region and vice versa. Will this always be the case? The answer is "yes." When we are dealing with inference about the value of a parameter, the two approaches will give the same answer. To show the equivalence algebraically, we start with the key inequality

$$P\left[-z_{1-\alpha/2} \leq \frac{\overline{Y} - \mu}{\sigma/\sqrt{n}} \leq z_{1-\alpha/2}\right] = 1 - \alpha.$$

If we solve the inequality for $\overline{Y}$, we get

$$P[\mu - z_{1-\alpha/2}\sigma/\sqrt{n} \leq \overline{Y} \leq \mu + z_{1-\alpha/2}/\sqrt{n}] = 1 - \alpha.$$

Given a value $\mu = \mu_0$, the statement produces a region $(\mu_0 \pm z_{1-\alpha/2}\sigma/\sqrt{n})$ within which $100(1 - \alpha)\%$ of sample means fall. If we solve the inequality for μ, we get

$$P[\overline{Y} - z_{1-\alpha/2}\sigma/\sqrt{n} \leq \mu \leq \overline{Y} + z_{1-\alpha/2}\sigma/\sqrt{n}] = 1 - \alpha.$$

This is a confidence interval for the population mean μ.

In the next chapter, we examine this approach in more detail and present a general methodology.

If confidence intervals and hypothesis testing are but two sides of the same coin, why bother with both? The answer is (to continue the analogy) that the two sides of the coin are not the same; there is different information. The confidence interval approach emphasizes the precision of the estimate by means of the width of the interval and provides a point estimate for the parameter, regardless of any hypothesis. The hypothesis testing approach deals with the consistency of observed (new) data with the hypothesized parameter value. It gives a probability of observing the value of the statistic or a more extreme value. In addition, it will provide a method for estimating sample sizes. Finally, by means of power calculations, we can decide beforehand whether a proposed study is feasible, i.e., what is the probability that the study will demonstrate a difference if a (specified) difference exists?

You should become familiar with both approaches to statistical inference. Do not use one to the exclusion of another. In some research fields, hypothesis testing has been elevated to the only "proper" way of doing inference; all scientific questions have to be put into a hypothesis testing framework. This is absurd and stultifying, particularly in pilot studies or investigations into uncharted fields. On the other hand, not to consider *possible* outcomes of an experiment and the chance of picking up differences is also unbalanced. Many times it will be useful to specify very carefully what is known about the parameter(s) of interest *and* to specify, in perhaps a crude way, alternative values or ranges of values for these parameters. If it is a matter of emphasis, you should stress hypothesis testing before carrying out a study and estimation after the study has been done.

4.8 INFERENCE ABOUT THE VARIANCE OF A POPULATION

4.8.1 Distribution of the Sample Variance

In the previous sections we assumed that the population variance of a normal distribution was known. In this section we want to make inferences about the population variance on the basis of a sample variance.

In making inferences about the population mean, we needed to know the sampling distribution of the sample mean. Similarly, we need to know the sampling distribution of the sample variance in order to make inferences about the population variance; analogous to the statement that for a normal random variable, Y, with sample mean $\overline{Y}$, the quantity

$$\frac{\overline{Y} - \mu}{\sigma/\sqrt{n}}$$

has a normal distribution with mean 0 and variance 1. We now state a result about the quantity $(n-1)s^2/\sigma^2$. The basic information is contained in the following statement:

Result 4.4. If a random variable Y is normally distributed with mean μ and variance σ^2 then for a random sample of size n the quantity $(n-1)s^2/\sigma^2$ has a chi-square distribution with $(n-1)$ degrees of freedom.

Each distribution is indexed by $(n-1)$ the degrees of freedom. Recall that the sample variance is calculated by dividing $\sum(y-\overline{y})^2$ by $(n-1)$, the degrees of freedom.

The chi-square distribution is skewed; the amount of skewness decreases as the degrees of freedom increases. Since $(n-1)s^2/\sigma^2$ can never be negative, the sample space for the chi-square distribution is the nonnegative part of the real line. Several chi-square distributions are shown in Figure 4.13. The mean of a chi-square distribution is equal to the degrees of freedom and the variance is twice the degrees of the

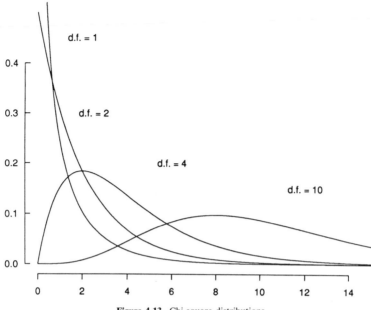

Figure 4.13. Chi-square distributions.

freedom. Formally,

$$E\left[\frac{(n-1)s^2}{\sigma^2}\right] = n-1, \tag{4.1}$$

$$\text{var}\left[\frac{(n-1)s^2}{\sigma^2}\right] = 2(n-1). \tag{4.2}$$

It may seem somewhat strange to talk about the variance of the sample variance, but under repeated sampling the sample variance will vary from sample to sample, and the chi-square distribution describes this variation if the observations are from a normal distribution.

Unlike the normal distribution, a tabulation of the chi-square distribution requires a separate listing for each degree of freedom. In Table A.3, a tabulation is presented of percentiles of the chi-square distribution. For example, 95% of chi-square random variables with ten degrees of freedom have values less than or equal to 18.31. Note that the median (50th percentile) is very close to the degrees of freedom when the number of the degrees of freedom is ten or more.

The symbol for a chi-square random variable is χ^2, the Greek letter chi, to the power of two. So we usually write $\chi^2 = (n-1)s^2/\sigma^2$. The degrees of freedom are usually indicated by the Greek letter ν *(nu)*. Hence, χ^2_ν is a symbol for a chi-square random variable with ν degrees of freedom. It is not possible to maintain the notation of using a capital letter for a variable and the corresponding lower case for the value of the variable.

4.8.2 Inference About a Population Variance

We begin with hypothesis testing. We have a sample of size n from a normal distribution, the sample variance s^2 has been calculated, and we want to know whether the observed value of s^2 is consistent with a hypothesized population value σ_0^2, perhaps known from previous research. Consider the quantity

$$\chi^2 = \frac{(n-1)s^2}{\sigma^2}.$$

If s^2 is very close to σ^2, the ratio s^2/σ^2 is close to 1; if s^2 differs very much from σ^2, the ratio is either very large or very close to 0: this implies that $\chi^2 = (n-1)s^2/\sigma^2$ is either very large or very small, and we would want to reject the null hypothesis. This procedure is analogous to a hypothesis test about a population mean; we measured the distance of the observed sample mean from the hypothesized value in units of standard errors; in this case we measure the "distance" in units of the hypothesized variance.

Example 4.9. The SIDS cases discussed in Section 3.3.1 were assumed to come from a normal population with variance $\sigma^2 = (800)^2$. To check this assumption the variance, s^2, is calculated for the first 11 cases occurring in 1969. The birthweights (in grams) were

3374, 3515, 3572, 2977, 4111, 1899, 3544, 3912, 3515, 3232, 3289.

The sample variance is calculated to be

$$s^2 = (574.3126 \text{ g})^2.$$

The observed value of the chi-square quantity is

$$\chi^2 = (11-1)(574.3126)^2/(800)^2,$$
$$= 5.15 \text{ with 10 degrees of freedom.}$$

Figure 4.14 pictures the chi-square distribution with 10 degrees of freedom. The 2.5th and 97.5th percentiles are 3.25 and 20.48 (see Table A.3). Hence, 95% of chi-square values will fall between 3.25 and 20.48.

If we follow the usual procedure of setting our significance level at $\alpha = 0.05$, we will not reject the null hypothesis that $\sigma^2 = (800 \text{ g})^2$, since the observed value $\chi^2 = 5.15$ is less extreme than 3.25. Hence, there is not sufficient evidence for using a value of σ^2 not equal to 800 g.

An alternative to formally setting up the rejection regions, we could have noted, using Table A.3, that the observed value of $\chi^2 = 5.15$ is between the 5th and 50th percentiles, and therefore the corresponding two-sided p-value is greater than 0.10.

A $100(1-\alpha)\%$ confidence interval is constructed using the the approach of Section 4.7. The key inequality is

$$P[\chi_{\alpha/2}^2 \leq \chi^2 \leq \chi_{1-\alpha/2}^2] = 1 - \alpha.$$

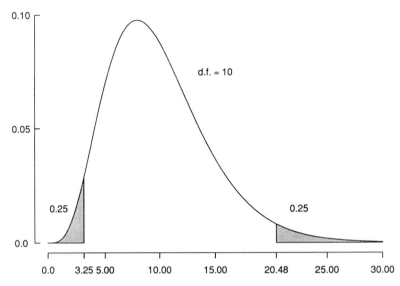

Figure 4.14. Chi-square distribution with 10 degrees of freedom.

The degrees of freedom are not indicated but assumed to be $(n - 1)$. The values $\chi^2_{\alpha/2}$ and $\chi^2_{1-\alpha/2}$ are chi-square values such that $(1 - \alpha)$ of the area is between them. (In Figure 4.14, these values are 3.25 and 20.48 for $1 - \alpha = 0.95$.)

The quantity χ^2 is now replaced by its equivalent $(n - 1)s^2/\sigma^2$, so that

$$P\left[\chi^2_{\alpha/2} \leq \frac{(n - 1)s^2}{\sigma^2} \leq \chi^2_{1-\alpha/2}\right] = 1 - \alpha.$$

If we solve for σ^2, we obtain a $100(1 - \alpha)\%$ confidence interval for the population variance. A little algebra shows that this is

$$P[(n - 1)s^2/\chi^2_{1-\alpha/2} \leq \sigma^2 \leq (n - 1)s^2/\chi^2_{\alpha/2}] = 1 - \alpha.$$

Given an observed value of s^2, the required confidence interval can now be calculated.

To continue our example, the variance for the eleven SIDS cases above is $s^2 = (574.3126 \text{ g})^2$. For $1 - \alpha = 0.95$, the values of χ^2 are (see Figure 4.14)

$$\chi^2_{0.025} = 3.25, \quad \chi^2_{0.095} = 20.48.$$

We can write the key inequality then as

$$P[3.25 \leq \chi^2 \leq 20.48] = 0.95.$$

The 95% confidence interval for σ^2 can then be calculated:

$$\frac{(10)(574.3126)^2}{20.48} \leq \sigma^2 \leq \frac{(10)(574.3126)^2}{3.25}$$

and simplifying,

$$161{,}052 \le \sigma^2 \le 1{,}014{,}877.$$

The corresponding values for the population standard deviation are

$$\text{Lower 95\% limit for } \sigma = \sqrt{161{,}052} = 401 \text{ g,}$$

$$\text{Upper 95\% limit for } \sigma = \sqrt{1{,}014{,}877} = 1007 \text{ g.}$$

These are rather wide limits. Note that they include the null hypothesis value of $\sigma = 800$ g. Thus, the confidence interval approach leads to the same conclusion as the hypothesis testing approach.   □

NOTES

Note 4.1: Definition of Probability

The relative frequency definition of probability was advanced by von Mises, Fisher and others (see Hacking [1965]). A radically different view is held by the so-called personal or subjective school, exemplified in the work of De Finetti, L. J. Savage, and I. R. Savage. According to this school, probability reflects subjective ignorance, and knowledge which can be quantified in terms of betting behavior. I. R. Savage [1968] states:

"My probability for the event A under circumstances H is the amount of money I am indifferent to betting on A in an elementary gambling situation."

What does Savage mean? Consider the thumbtack experiment discussed in Section 4.3.1. Let the event A be that the thumbtack in a single toss falls ⌐. The other possible outcome is ∠. Call this event B. You are to bet a dollars on A and b dollars on B, such that you are indifferent to betting either on A or on B (you must bet). You clearly would not want to put all your money on A, then you would prefer outcome A; there is a split then in the total amount, $a + b$, to be bet so that you are indifferent to either outcome A or B. Then *your* probability of A, $P[A]$, is

$$P[A] = \frac{b}{a+b}.$$

If the total amount to be bet is one unit, then you would split it $1 - P$, P, where $0 \le P \le 1$, so that

$$P[A] = \frac{P}{1 - P + P} = P.$$

Note that Savage is very careful to require the estimate of the probability to be made under specified circumstances. (If the thumbtack could land, say, ⊤ on a soft surface, you would clearly want to modify your probability.)

This definition of probability is also called *personal probability*. An advantage of this view is that it can discuss more situations than the relative frequency definition, for example: the probability (rather, my probability) of life on Mars, or my probability that a cure for cancer will be found. You should not identify personal probability with the irrational or whimsical. Personal probabilities do utilize empirical evidence such as the behavior of a tossed coin.

But personal probabilities are still personal; it is not clear how helpful or useful they are, nor will it always be possible to assess a situation in terms of betting behavior; in the business world this may be possible, but in scientific research it seems rather artificial. Perhaps this is the reason that texts in business statistics are more likely to view probabilities as personal.

There are other views of probability. For a survey see the book by Hacking [1965] and references therein.

Note 4.2: Probability Inequalities

For the normal distribution, approximately 68% of observations are within one standard deviation of the mean, and 95% of observations are within two standard deviations of the mean. If the distribution is not normal, a weaker statement can be made: the proportion of observations within K standard deviations of the mean is greater than or equal to $(1 - 1/K^2)$; notationally, for a variable Y,

$$P\left[-K \leq \frac{Y - E(Y)}{\sigma} \leq K\right] \leq 1 - 1/K^2,$$

where K is the number of standard deviations from the mean. This is a version of Chebyshev's inequality. For example, this inequality states that at least 75% of the observations fall within two standard deviations of the mean (compared to 95% for the normal distribution). This is not nearly as stringent as the first result stated, but it is more general.

If the variable Y can take on only positive values and the mean of Y is μ, the following inequality holds:

$$P[Y \leq y] \leq 1 - \mu/y.$$

This inequality is known as the Markov inequality.

Note 4.3: Inference vs. Decision

The hypothesis tests discussed in Sections 4.6 and 4.7 can be thought of as decisions that are made with respect to a value of a parameter (or "state of nature"). There is a controversy in statistics whether the process of inference is equivalent to a decision process. It seems that a "decision" is sometimes not possible in a field of science. For example, it is not possible, at this point, to decide whether better control of insulin levels will reduce the risk of neuropathy in diabetes mellitus. In this case and others, the kinds of inferences we can make are more tenuous, and cannot really be called decisions. For an interesting discussion, see Moore [1985], pp. 328–332. This is an excellent book covering a variety of statistical topics ranging from ethical issues in experimentation to formal statistical reasoning.

Note 4.4: "Representative" Samples

A random sample from a population was defined in terms of repeated independent trials or drawings of observations. We want to make a distinction between a random and a representative sample. A random sample has been defined in terms of repeated independent sampling from a population. However (see Section 4.3.2), cancer patients treated in New York are clearly not a random sample of all cancer patients in the world, or even the United States. They will differ from cancer patients in, for instance, Great Britain in many ways. Yet, we do frequently make the assumption that if a cancer treatment worked in New York, patients in Great Britain can also benefit. The experiment in New York has wider applicability. We consider that with respect to the outcome of interest in the New York cancer study (e.g., increased survival time), the New York patients, although not a random sample, constitute a representative sample. That is, the survival times are a random sample from the population of survival times.

It is easier to disprove randomness than representativeness. A measure of scientific judgment is involved in determining the latter. For an interesting discussion of the use of the word "representative," see the papers by Kruskal and Mosteller [1979a–c].

Note 4.5: Multivariate Populations

Usually we study more than one variable. The Winkelstein, *et al.* study (see Example 4.1) measured diastolic and systolic blood pressures, height, weight and cholesterol levels. In the study suggested in Example 4.2, in addition to IQ, we would measure physiological and psychological variables to obtain a more complete picture of the effect of the diet. For completeness we therefore define a *multivariate population* as the set of all possible values of a specified set of variables (measured on the objects of interest). A second category of topics then comes up: relationships among the variables. Words such as *association* and *correlation* come up in this context. A discussion of these topics will be started in Chapter 9.

Note 4.6: Sampling Without Replacement

We want to select two patients "at random" from a group of four patients. The same patient cannot be chosen twice. How can this be done? One procedure is to write each name on a slip of paper, put the four slips of paper in a hat, stir the slips of paper and—without looking—draw out two slips. The patients whose names are on the two slips are then selected. This is known as sampling without replacement. (For the above procedure to be "fair," we require that the slips of paper are indistinguishable and well-mixed.) The events "outcome on first draw" and "outcome on second draw" are clearly not independent. If Patient A is selected in the first draw she is no longer available for the second draw. Let the patients be labelled A, B, C, and D. Let the symbol AB mean "Patient A is selected in the first draw and Patient B in the second draw." Write down all the possible outcomes; there are twelve of them as follows:

$$\begin{array}{llll} AB & BA & CA & DA \\ AC & BC & CB & DB \\ AD & BD & CD & DC. \end{array}$$

We define the selection of two patients to be random if each of the above twelve outcomes is equally likely; that is, the probability that a particular pair is chosen is $1/12$.

This definition has intuitive appeal: we could have prepared twelve slips of paper each with one of the twelve pairs recorded and drawn one slip of paper out. If the slip of paper is drawn randomly, the probability is $1/12$ that a particular slip will be selected.

One further comment. Suppose we only want to know which two patients have been selected, i.e., we are not interested in the order. For example, what is the probability that patients C and D are selected? This can happen in two ways: CD or DC. These events are mutually exclusive so that the required probability is $P[CD \text{ or } DC] = P[CD] + P[DC] = 1/12 + 1/12 = 1/6$.

Note 4.7: Pitfalls in Sampling

It is very important to define carefully the population of interest. Two illustrations of rather subtle pitfalls are *Berkson's fallacy* and *length biased sampling*. Berkson's fallacy is discussed in Murphy [1979] as follows: In many studies, hospital records are reviewed or sampled in order to determine relationships between diseases and/or exposures. Suppose a review of hospital records is made with respect to two diseases, A and B, which are so severe that they always lead to hospitalization. Let their frequencies in the population at large be p_1 and p_2. Then, assuming independence, the probability of the joint occurrence of the two diseases is $p_1 p_2$. Suppose now that a healthy proportion p_3 of subjects (H) never go to the hospital, that is, $P[H] = p_3$. Now write $\overline{H}$ as that part of the population that will enter a hospital at some time; then $P[\overline{H}] = 1 - p_3$. By the rule of conditional probability $P[A \mid \overline{H}] = P[A\overline{H}]/P[\overline{H}] = p_1/(1 - p_3)$. Similarly, $P[B \mid \overline{H}] = p_2/(1 - p_3)$ and $P[AB \mid \overline{H}] = p_1 p_2/(1 - p_3)$, and this is not equal to $P[A \mid \overline{H}]P[B \mid \overline{H}] = [p_1/(1 - p_3)][p_2/(1 - p_3)]$, which must be true in order for the two diseases to be unrelated in the hospital population. Now, you can show that $P[AB \mid \overline{H}] < P[AB]$, and, quoting Murphy,

> "the hospital observer will find that they occur together less commonly than would be expected if they were independent. This is known as Berkson's fallacy. It has been a source of embarrassment to many an elegant theory. Thus, cirrhosis of the liver and common cancer are both reasons for admission to the hospital. *A priori*, we would expect them to be less commonly associated in the hospital than in the population at large. In fact, they have been found to be negatively correlated."

(Murphy's book contains an elegant, readable exposition of probability in medicine; it will be worth your while to read it.)

A second pitfall deals with the area of "length biased sampling." This means that for a particular sampling scheme, some objects in the population may be more likely to be selected than others. A paper by Shepard and Neutra [1977] illustrates this phenomenon in sampling medical visits. Our discussion is based on that paper. The problem arises when we want to make a statement about a population of patients which can only be identified by a sample of patient visits. Therefore, frequent visitors will be more likely to be selected. Consider the following data from Shepard and Neutra [1977] ("Expected Composition of Visit-Based Sample in a Hypothetical Population"):

Variable	Type of Patient		
	Hypertensive	**Other**	**Total**
Number of patients	200	800	1000
Visits per patient per year	12	1	13
Visits contributed	2400	800	3200
Expected number of patients in a 3% sample of visits	72	24	96
Expected % of sample	75	25	100

This table illustrates that, although the hypertensive patients make up 20% of the total patient population, a sample based on visits would consist of 75% hypertensive patients and 25% other.

There are other areas, particularly screening procedures in chronic diseases, that are at risk for this kind of problem. See the article for some suggested solutions as well as references to other papers.

Note 4.8: How to Draw a Random Sample

In Note 4.6, we discussed drawing a random sample without replacement. How can we draw samples with replacement? Simply, of course, the slips could be put back in the hat. However, in some situations we cannot collect the total population to be sampled from, due to its size, for example.

One way to sample populations is to use a table of random numbers. These numbers are really "pseudo-random": they have been generated by a formula. Use of such a table can be illustrated by the following problem: a random sample of 100 patient charts is to be drawn from a hospital record room containing 45,850 charts. Assume that the charts are numbered in some fashion from 1 to 45,850. (It is not necessary that they be numbered consecutively or that the numbers start with 1 and end with 45,850. All that is required is that there is some unique way of numbering each chart.) We enter the random number table randomly by selecting a page and a column on the page at random. Suppose that the first 5-digit numbers are

06812, 16134, 15195, 84169, and 41316.

The first three charts chosen would be chart 06812, 16134 and 15195, in that order. Now what do we do with the 84169? We can skip it and simply go to 41316, realizing that if we follow this procedure, we will have to throw out approximately half of the numbers selected.

A second example: A group of 40 animals is to be assigned at random to one of four treatments A, B, C, and D, with an equal number in each of the treatments. Again, enter the random number table randomly. The first ten-digit numbers between 1 and 40 will be the numbers of the animals assigned to treatment A, the second set of 10-digit numbers to treatment B, the third set to treatment C and the remaining animals are assigned to treatment D. If a random number reappears in a subsequent treatment it can simply be omitted. (Why is this reasonable?)

Pseudo-random numbers are usually generated by a computer. Even microcomputers generally have "random number generators" available.

Note 4.9: Algebra of Expectations

Section 4.3.3 discusses random variables, distributions and expectations of random variables. We defined $E(Y) = \sum py$ for a discrete random variable. A similar definition, involving integrals rather than sums, can be made for continuous random variables. We will now state some rules for working with expectations.

1. If a is a constant, $E(aY) = aE(Y)$.
2. If a and b are constants, $E(aY + b) = aE(Y) + b$.
3. If X and Y are two random variables, $E(X + Y) = E(X) + E(Y)$.
4. If a and b are constants, $E(aX + bY) = E(aX) + E(bY) = aE(X) + bE(Y)$.

You can demonstrate the first three rules by using some simple numbers and calculating their average. For example, let $y_1 = 2$, $y_2 = 4$, and $y_3 = 12$. The average is

$$E(Y) = \frac{1}{3} \times 2 + \frac{1}{3} \times 4 + \frac{1}{3} \times 12 = 6.$$

Two additional comments:

1. The second formula makes sense. Suppose we measure temperature in °C. The average is calculated for a series of readings. The average can be transformed to °F by the formula

$$\text{Average in °F} = \frac{9}{5} \times \text{average in °C} + 32.$$

An alternative approach consists of transforming each original reading to °F and then taking the average. It is intuitive that the two approaches should provide the same answer.

2. It is not true that $E(Y^2) = [E(Y)]^2$. Again, a small example will verify this. Use the same three values ($y_1 = 2$, $y_2 = 4$, and $y_3 = 12$). By definition,

$$E(Y^2) = \frac{2^2 + 4^2 + 12^2}{3} = \frac{4 + 16 + 144}{3} = \frac{164}{3} = 54.\overline{6},$$

but

$$[E(Y)]^2 = 6^2 = 36.$$

Can you think of a special case where the equation $E(Y^2) = [E(Y)]^2$ is true?

Note 4.10: Bias, Precision and Accuracy

Using the algebra of expectations we define a statistic T to be a biased estimate of a parameter τ if $E(T) \neq \tau$. Two typical kinds of bias are $E(T) = \tau + a$, where a is some constant; this is called *location bias*; a second kind $E(T) = b\tau$, where b is a positive constant; this is called *scale bias*. A simple example involves the sample variance, s^2. A more "natural" estimate of σ^2 might be

$$s_*^2 = \frac{\sum (y - \bar{y})^2}{n}.$$

This statistic differs from the usual sample variance in division by n rather than $n-1$. It can be shown (you can try it) that

$$E(s_*^2) = \frac{n-1}{n} \sigma^2;$$

hence, s_*^2 is a biased estimate of σ^2. The statistic s_*^2 can be made unbiased by multiplying s_*^2 by $n/(n-1)$, (see Rule 1 in Note 4.9); that is,

$$E \left[\frac{n}{n-1} s_*^2 \right] = \left(\frac{n}{n-1} \right) \left(\frac{n-1}{n} \right) \sigma^2 = \sigma^2.$$

But $n/(n-1)s_*^2 = s^2$, so s^2 rather than s_*^2 is an unbiased estimate of σ^2.

We can now discuss precision and accuracy. *Precision* refers to the degree of closeness to each other of a set of values of a variable; *accuracy* refers to the degree of closeness of these values to the quantity (parameter) being measured. Thus, precision is an internal characteristic of a set of data while accuracy relates the set to an external standard. For example, a thermometer that consistently reads a temperature five degrees too high may be very precise but will not be very accurate. A second example of the distribution of hits on a target illustrates these two concepts:

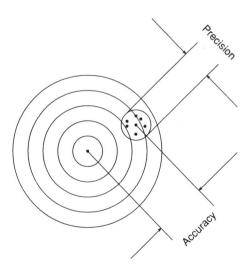

The picture illustrates that accuracy involves the concept of bias. Together with Note 4.9, we can now make these concepts more precise. For simplicity we will refer only to location bias.

Suppose a statistic T estimates a quantity τ in a biased way; $E[T] = \tau + a$. The variance, in this case, is defined to be $E[T - E(T)]^2$. What is the quantity $E[T - \tau]^2$? This can be written as

$$E[T - \tau]^2 = E[T - (\tau + a) + a]^2 = E[T - E[T] + a]^2,$$

$$\underset{\text{(mean square error)}}{E[T - \tau]^2} \quad = \quad \underset{\text{(variance)}}{E[T - E[T]]^2} \quad + \quad \underset{\text{(bias)}}{a^2}.$$

The quantity $E[T - \tau]^2$ is called the *mean square error*. If the statistic is unbiased (i.e., $a = 0$), then the mean square error is equal to the variance (σ^2).

Note 4.11: Use of the Word *Parameter*

We have defined *parameter* as a numerical characteristic of a population of values of a variable. One of the basic tasks of statistics is to estimate values of the unknown parameter on the basis of a sample of values of a variable. There are two other usages of this word. Many clinical scientists use *parameter* for *variable*, as in "we measured the following three parameters, blood pressure, amount of plaque and degree of patient satisfaction." You should be aware of this pernicious usage and valiantly strive to eradicate it from scientific writing. However, we are not sanguine about its ultimate success. A second incorrect usage confuses *parameter* and *perimeter*, as in "the parameters of the study did not allow us to include patients under 12 years of age." A better choice would have been to use the word "limitations."

Note 4.12: "Significant Digits" (continued)

This note continues the discussion of "significant digits" of Note 3.2. We discussed approximations to a quantity due to arithmetical operations, measurement rounding, and finally, sampling variability. Consider the data on SIDS cases of Example 4.11. The mean birthweight of the 78 cases was 2994 g. The probability was 95% that the interval 2994 ± 178 straddles the unknown quantity of interest: the mean birthweight of the population of SIDS cases. This interval turned out to be 2816–3172 g–although the last digits in the two numbers are not very useful. In this case we have carried enough places so that the rule mentioned in Note 3.2 is not applicable. The biggest source of approximation turns out to be due to sampling. The approximations introduced by the arithmetical operation is minimal; you can verify that if we had carried more places in the intermediate calculations, the final confidence interval would have been 2816–3171 g.

Note 4.13: A Matter of Notation

What do we mean by 18 ± 2.6? In many journals you will find this notation. What does it mean? Is it mean plus or minus the standard deviation, or mean plus or minus the standard error? You may have to read a paper carefully to find out. Both meanings are used, and thus need to be clearly specified.

Note 4.14: Formula for the Normal Distribution

The formula for the normal probability density function for a normal random variable Y with mean μ and variance σ^2 is

$$f(y) = \frac{1}{\sqrt{2\pi}\,\sigma} \exp\left[-\frac{1}{2}\left(\frac{y-\mu}{\sigma}\right)^2\right].$$

Here, $\pi = 3.14159\ldots$, and e is the base of the natural logarithm, $e = 2.71828\ldots$.

A standard normal distribution has $\mu = 0$ and $\sigma = 1$. The formula for the standard normal random variable, Z, is

$$f(z) = \frac{1}{\sqrt{2\pi}} \exp\left[-\frac{1}{2}z^2\right].$$

The heights of the curve can be calculated using a hand calculator. By symmetry, only one half of the range of values has to be computed, i.e., $f(z) = f(-z)$. For completeness we give enough points to enable you to graph $f(z)$.

z	$f(z)$	z	$f(z)$	z	$f(z)$	z	$f(z)$	z	$f(z)$
0.0	0.3989	0.5	0.3521	1.0	0.2420	1.5	0.1295	2.0	0.0540
0.1	0.3970	0.6	0.3332	1.1	0.2179	1.6	0.1109	2.1	0.0440
0.2	0.3910	0.7	0.3123	1.2	0.1942	1.7	0.0940	2.2	0.0355
0.3	0.3814	0.8	0.2897	1.3	0.1714	1.8	0.0790	2.3	0.0283
0.4	0.3683	0.9	0.2661	1.4	0.1497	1.9	0.0656	2.4	0.0224

Given any normal variable y with mean μ and variance σ^2, you can calculate $f(y)$ by using the relationships

$$Z = \frac{Y-\mu}{\sigma},$$

and plotting the corresponding heights

$$f(y) = \frac{1}{\sigma}f(z),$$

where Z is defined by the above relationship. For example, suppose we want to graph the curve for IQ where we assume that IQ is normal with mean $\mu = 100$ and standard deviation $\sigma = 15$. What is the height of the curve for an IQ of 109? In this case, $Z = (109 - 100)/15 = 0.60$ and $f(IQ) = (1/15)f(z) = (1/15)(0.3332) = 0.0222$. The height for an IQ of 91 is the same.

Note 4.15: Null hypothesis and Alternative Hypothesis

How do you decide which of the two hypothesis is the null and which is the alternative? Sometimes the advice is to make the null hypothesis the hypothesis of "indifference". This is not helpful; indifference is a poor scientific attitude. We have three suggestions: (1) In many situations there is a prevailing view of the science that is accepted; it will continue to be accepted unless "definitive" evidence to the contrary is produced. In this instance the prevailing view would be made operational in the null hypothesis. The null hypothesis is often the "straw man" that we wish to reject. (Philosophers of science tell us that we never conclusively prove things; we can only disprove theories.) (2) An excellent guide is Occam's razor which states: do not multiply hypotheses beyond necessity. Thus in comparing a new treatment with a standard treatment, the simpler hypothesis is that the treatments have the same effect. To postulate that the treatments are different requires an additional operation. (3) Frequently the null hypothesis is one that allows you to calculate the p-value. Thus if two treatments are assumed the same, we can calculate a p-value for the observed result. If we hypothesize that they are not the same, then without further specification we cannot compute a p-value.

PROBLEMS

1. Give examples of populations with number of elements finite, virtually infinite, potentially infinite and infinite. Define a sample from each of these populations.

2. Give an example from a study in a research area of interest to you that clearly assumes that results are applicable to, as yet, untested subjects.

3. Illustrate the concepts of *population*, *sample*, *parameter*, and *statistic* by two examples from a research area of your choice.

4. In light of the material discussed in this chapter, now review the definitions of statistics presented at the end of Chapter 1, especially the definition by Fisher.

5. In Section 4.3.1, probabilities are defined as long-run relative frequencies. How would you interpret the probabilities in the following situations?
   a. The probability of a genetic defect in a child born to a mother over 40 years of age.
   b. The probability of you, the reader, dying of leukemia.
   c. The probability of life on Mars.
   d. The probability of rain tomorrow. What does the meteorologist mean?

6. Take a thumbtack and throw it onto a hard surface such as a table top. It can come to rest in two ways; and label them as follows:

$$\underline{\perp} = \text{"up"} = U,$$
$$\mathcal{K} = \text{"down"} = D.$$

   a. Guess the probability of U. Record your answer.
   b. Now toss the thumbtack 100 times and calculate the proportion of times the outcome is U. How does this agree with your guess? The observed proportion is an estimate of the probability of U. (Note the implied distinction between "guess" and "estimate.")

c. In a class situation split the class in half. Let each member of the first half of the class toss a thumbtack 10 times and record the outcomes as a histogram:

i. The number of times U occurs in 10 tosses,

ii. The proportion of times U occurs in 10 tosses. Each member of the second half of the class will toss a thumbtack 50 times. Record the outcomes in the same way. Compare the histograms. What conclusions do you draw?

7. The estimation of probabilities and the proper combination of probabilities present great difficulties, even to experts. The best we can do in this text is warn you and point you to some references. A good starting point is the paper by Tversky and Kahneman [1974] reprinted in Kahneman *et al.* [1982]. They categorize the various errors that people make in assessing and working with probabilities. Two examples from this book will test your intuition:

a. In tossing a coin six times, is the sequence HTHHTT more likely than the sequence HHHHHH? Give your "first impression" answer, then calculate the probability of occurrence of each of the two sequences using the rules stated in the chapter.

b. The following is taken directly from the book:

> A certain town is served by two hospitals. In the larger hospital, about 45 babies are born each day, and in the smaller hospital about 15 babies are born each day. As you know, about 50% of all babies are boys. However, the exact percentage varies from day to day. Sometimes it may be higher than 50%, sometimes lower. For a period of one year, each hospital recorded the days on which more than 60% of the babies born were boys. Which hospital do you think recorded more such days? The larger hospital, the smaller hospital, [or were they] about the same (that is within 5% of each other)?

Which of the rules and results stated in this chapter have guided your answer?

8. This problem deals with the "gambler's fallacy," which states, roughly, if an event has not happened for a long time it is "bound to come up." Thus, for example, the probability of a head on the fifth toss of a coin is assumed to be greater if the preceding four tosses all resulted in tails than if the preceding four tosses were all heads. This is incorrect.

a. What statistical property associated with coin tosses is violated by the fallacy?

b. Give some examples of the occurrence of the fallacy from your own area of research.

c. Why do you suppose the fallacy is so ingrained in people?

9. Human blood can be classified by the ABO blood grouping system. The four groups are A, B, AB, or O, depending on whether antigens labelled A and B are present on red blood cells. Hence, the AB blood group is one where both A and B antigens are present; the O group has none of the antigens present. For three US populations the following distributions exist:

	Blood Group				
	A	B	AB	O	Total
Caucasian	0.44	0.08	0.03	0.45	1.00
American Black	0.27	0.20	0.04	0.49	1.00
Chinese	0.22	0.25	0.06	0.47	1.00

For simplicity, consider only the population of American Blacks in the following question. The table shows that for a person randomly selected from this population: $P[A] = 0.27$, $P[B] = 0.20$, $P[AB] = 0.04$ and $P[O] = 0.49$.

a. Calculate the probability that a person is *not* of blood Group A.

b. Calculate the probability that a person is either A *or* O. Are these mutually exclusive events?

c. What is the probability that a person carries A antigens?

d. What is the probability that in a marriage both husband and wife are of blood group O? What rule of probability did you use? (What assumption did you need to make?)

10. This problem continues with the discussion of ABO blood groups of the previous problem. We now consider the black and Caucasian population of the United States. Approximately 20% of the US population is black. This produces the two-way classification of race and blood type as follows:

	Blood Group				
	A	B	AB	O	Total
Caucasian	0.352	0.064	0.024	0.360	0.80
American Black	0.054	0.040	0.008	0.098	0.20
	0.406	0.104	0.032	0.458	1.00

This table specifies, for example, that the probability is 0.352 that a person selected at random is both Caucasian and blood group A.

a. Are the events "Blood Group A" and "Caucasian Race" statistically independent?

b. Are the events "Blood Group A" and "Caucasian Race" mutually exclusive?

c. Assuming statistical independence, what is the "expected probability" of the event "Blood Group A and Caucasian Race"?

d. What is the conditional probability of "Blood Group A" given that the race is Caucasian?

11. The distribution of the Rh factor in a Caucasian population is as follows:

Rh Positive (Rh^+, Rh^+)	Rh Positive (Rh^+, Rh^-)	Rh Negative
0.35	0.48	0.17

Rh$^-$ subjects have two Rh$^-$ genes, while Rh$^+$ subjects have two Rh$^+$ genes or one Rh$^+$ gene and one Rh$^-$ gene. A potential problem occurs when a Rh$^+$ male mates with an Rh$^-$ female.

a. Assuming random mating with respect to the Rh factor, what is the probability of an Rh$^-$ female mating with an Rh$^+$ male?

b. Since each person contributes one gene to an offspring what is the probability of Rh incompatibility given such a mating? (Incompatibility occurs when the fetus is Rh$^+$ and the mother is Rh$^-$.)

c. What is the probability of incompatibility in a population of such matings?

12. The following data for 20–25-year-old white males list four primary causes of death together with a catch-all fifth category, and the probability of death within 5 years:

Cause	Probability
Suicide	0.00126
Homicide	0.00063
Auto accident	0.00581
Leukemia	0.00023
All other causes	0.00788

a. What is the probability of a white male aged 20–25 years dying from *any* cause of death? Which rule did you use to determine this?

b. Out of 10,000 white males in the above age group, how many deaths would you expect in the next 5 years? How many for each cause?

c. Suppose that an insurance company sells insurance to 10,000 white male drivers in the above age bracket. Suppose also that each driver is insured for $100,000 for accidental death. What annual rate would the insurance company have to charge to break even? (Assume a fatal accident rate of 0.00581.) List some reasons why your estimate will be too low or too high.

d. Given that a white male aged 20–25 years has died, what is the most likely cause of death; assume nothing else is known. Can you explain your statement?

13. If $Y \sim N(0,1)$, find
   a. $P[Y \leq 2]$
   b. $P[Y \leq -1]$
   c. $P[Y > 1.645]$
   d. $P[0.4 < Y \leq 1]$
   e. $P[Y \leq -1.96 \text{ or } Y \geq 1.96] = P[|Y| \geq 1.96]$

14. If $Y \sim N(2,4)$, find
   a. $P[Y \leq 2]$
   b. $P[Y \leq 0]$
   c. $P[1 \leq Y < 3]$
   d. $P[0.66 < Y \leq 2.54]$

15. From the paper by Winkelstein *et al.* [1975], glucose data for the 45–49 age group of California Nisei as presented by percentile are:

Percentile	90	80	70	60	50	40	30	20	10
Glucose (mg/100 ml)	218	193	176	161	148	138	128	116	104

    a. Plot these data on normal probability paper connecting the data points by straight lines. Do the data seem normal?

    b. Estimate the mean and standard deviation from the plot.

    c. Calculate the median and the interquartile range.

16. In a sample of size 1000 from a normal distribution, the sample mean $\overline{Y}$ was 15, and the sample variance s^2 was 100.

    a. How many values do you expect to find between 5 and 45?

    b. How many values less than five or greater than 45 do you expect to find?

17. Plot the data of Table 3.8 on probability paper. Do you think that age at death for these SIDS cases is normally distributed? Can you think of some *a priori* reason why this variable, age at death, is not likely to be normally distributed?

18. Plot the aflatoxin data of Section 3.3.3 on normal probability paper by graphing the cumulative proportions against the individual ordered values. Ignoring the last two points on the graph draw a straight line through the remaining points and estimate the median. On the basis of the graph would you consider the last three points in the data set "outliers"? Do you expect the arithmetic mean to be larger or smaller than the median? Why?

19. Plot the data of Table 3.12 (number of boys per family of eight children) on normal probability paper. Consider the "endpoints" of the intervals to be 0.5, 1.5, ..., 8.5. What is your conclusion about the normality of this variable? Estimate the mean and the standard deviation from the graph and compare it with the calulated values of 4.12 and 1.44, respectively.

20. The random variable Y has a normal distribution with mean 1.0 and variance 9.0. Samples of size nine are taken and the sample means, $\overline{Y}$, are calculated.

    a. What is the sampling distribution of $\overline{Y}$?

    b. Calculate $P[1 < \overline{Y} \leq 2.85]$

    c. Let $W = 4\overline{Y}$. What is the sampling distribution of W?

21. The sample mean and standard deviation of a set of temperature observations are 6.1°F and 3.0°F, respectively.

    a. What will be the sample mean and standard deviation of the observations expressed in °C?

    b. Suppose the original observations are distributed with population mean μ°F and standard deviation σ°F. Suppose also that the sample mean of 6.1°F is based on 25 observations. What is the approximate sampling distribution of the mean? What are its parameters?

22. The frequency distributions in Figure 3.10 were based on the following eight sets of frequencies.

				Graph No.				
Y	**1**	**2**	**3**	**4**	**5**	**6**	**7**	**8**
-1	1	1	8	1	1	14	28	10
-2	2	2	8	3	5	11	14	24
-3	5	5	8	8	9	9	10	14
-4	10	9	8	11	14	6	8	10
-5	16	15	8	14	11	3	7	9
-6	20	24	8	15	8	2	6	7
-7	16	15	8	14	11	3	5	6
-8	10	9	8	11	14	6	4	4
-9	5	5	8	8	9	9	3	2
-10	2	2	8	3	5	11	2	1
-11	1	1	8	1	1	14	1	1
Total	88	88	88	88	88	88	88	88
a_4	3.03	3.20	1.78	2.38	1.97	1.36	12.1	5.78

(The numbers are used to label the graph for purposes of this exercise.) Obtain the probability plots associated with graphs 1 and 6.

23. Suppose that the height of male freshmen is normally distributed with mean 69 inches and standard deviation of 3 inches. Suppose also (contrary to fact) that such subjects apply and are accepted at a college without regard to their physical stature.

   a. What is the probability that a randomly selected (male) freshman is 6'6" (78 inches) or more?

   b. How many of such men do you expect to see in a college freshman class of 1000 men?

   c. What is the probability that this class has at least one man 78 inches or more?

24. A normal distribution has mean $\mu = 100$ and standard deviation $\sigma = 15$ (for example, IQ). Give limits within which 95% of the following would lie:

   a. Individual observations

   b. Means of 4 observations

   c. Means of 16 observations

   d. Means of 100 observations

   e. Plot the width of the interval as a function of the sample size. Join the points with an appropriate freehand line.

   f. Using the graph constructed for (e), estimate the width of the 95% interval for means of 36 observations.

25. If the standard error is the measure of the precision of a sample mean, how many observations must be taken to double the precision of a mean of ten observations?

26. The duration of gestation in healthy humans is approximately 280 days with a standard deviation of 10 days.

   a. What proportion of (healthy) pregnant women will be more than 1 week "overdue"? 2 weeks?

   b. The gestation periods for a set of four women suffering from a particular condition are 240, 250, 265, and 280 days. Is this evidence that a shorter gestation period is associated with the condition?

   c. Is the sample variance consistent with the population variance of $10^2 = 100$? (We assume normality.)

   d. In view of (c), do you want to reconsider the answer to question (b)? Why or why not?

27. The mean height of adult men is approximately 69 inches; the mean height of adult women approximately 65 inches. The variance of height for both is 4^2 inches. Assume that husband/wife pairs occur without relation to height, and that heights are approximately normally distributed.

   a. What is the sampling distribution of the mean height of a couple? What are its parameters? (The variance of two statistically independent variables is the sum of the variances.)

   b. What proportion of couples is expected to have a mean height which exceeds 70 inches?

   c. In a collection of 200 couples, how many average heights would be expected to exceed 70 inches?

   *d. In what proportion of couples do you expect the wife to be taller than the husband?

28. A pharmaceutical firm claims that a new analgesic drug relieves mild pain under standard conditions for 3 h, with standard deviation 1 h. Sixteen patients are tested under the same conditions and have an average pain relief of 2.5 h. The hypothesis that the population mean of this sample is actually 3 h is to be tested against the hypothesis that the population mean is in fact less than 3 h; $\alpha = 0.05$.

   a. What is an appropriate test?

   b. Set up the appropriate critical region.

   c. State your conclusion.

   d. Suppose the sample size is doubled. State precisely how the region where the null hypothesis is not rejected is changed.

*29. For Y, from a normal distribution with mean μ and variance σ^2, the variance of $\overline{Y}$, based on n observations, is σ^2/n. It can be shown that the sample median $\widetilde{Y}$ in the above situation has a variance of approximately $1.57\sigma^2/n$. Assume that the standard error of $\widetilde{Y}$ equal to the standard error of $\overline{Y}$ is desired, based on

$n = 10, 20, 50$, and 100 observations. Calculate the corresponding sample sizes needed for the median.

*30. To determine the strength of a digitalis preparation, a continuous intra-jugular perfusion of a tincture is made and the dose required to kill an animal is observed. The lethal dose varies from animal to animal such that its logarithm is normally distributed. One cubic centimeter of the tincture kills 10% of all animals, 2 cm^3 kills 75%. Determine the mean and standard deviation of the distribution of the logarithm of the lethal dose.

31. There were 48 SIDS cases in King County, Washington, during the years 1974 and 1975. The birthweights (in grams) of these 48 cases were:

2466	3941	2807	3118	2098	3175	3515
3317	3742	3062	3033	2353	2013	3515
3260	2892	1616	4423	3572	2750	2807
2807	3005	3374	2722	2495	3459	3374
1984	2495	3062	3005	2608	2353	4394
3232	2013	2551	2977	3118	2637	1503
2438	2722	2863	2013	3232	2863	

a. Calculate the sample mean and standard deviation for this set.

b. Construct a 95% confidence interval for the population mean birthweight assuming the population standard deviation is 800 g. Does this confidence interval include the mean birthweight of 3300 g for normal children?

c. Calculate the p-value of the observed sample mean, assuming the population mean is 3300 g and the population standard deviation is 800 g. Do the results of this part and Part (b) agree?

d. Is the sample standard deviation consistent with a population standard deviation of 800? Carry out a hypothesis test comparing the sample variance with population variance $(800)^2$. The critical values for a chi-square variable with 47 degrees of freedom are as follows:

$$\chi^2_{0.025} = 29.96, \quad \chi^2_{0.975} = 67.82.$$

e. Set up a 95% confidence interval for the population standard deviation. Do this by first constructing a 95% confidence interval for the population variance and then taking square roots.

32. In a sample of 100 patients who had been hospitalized recently, the average cost of hospitalization was $5000, the median cost was $4000, and the modal cost was $2500.

a. What was the total cost of hospitalization for all 100 patients? Which statistic did you use? Why?

b. List one practical use for *each* of the three statistics.

c. Considering the ordering of the values of the statistics what can you say about the distribution of the raw data? Will it be skewed or symmetric? If skewed, which way will the skewness be?

33. For Example 4.8 as discussed in Section 4.6.2:

   a. Calculate the probability of a Type II error and the power if α is fixed at 0.05.

   b. Calculate the power associated with a one tailed-test.

   c. What is the price paid for the increased power in Part (b)?

34. The theory of hypothesis testing can be used to determine statistical characteristics of laboratory tests, keeping in mind the provision mentioned in connection with Example 4.6. Suppose that albumin has a normal (Gaussian) distribution in a healthy population with mean $\mu = 3.75 \, \text{mg}/100 \, \text{ml}$ and $\sigma = 0.50 \, \text{mg}/100 \, \text{ml}$. The normal range of values will be defined as $\mu \pm 1.96\sigma$, so that values outside these limits will be classified as "abnormal." Patients with advanced chronic liver disease have reduced albumin levels; suppose the mean for patients from this population is $2.5 \, \text{mg}/100 \, \text{ml}$ and the standard deviation is the same as the normal population.

   a. What are the critical values for the rejection region? (Here we work with an individual patient, $n = 1$.)

   b. What proportion of patients with advanced chronic liver disease (ACLD) will have "normal" albumin test levels?

   c. What is the probability that a patient with ACLD will be correctly classified on a test of albumin level?

   d. Give an interpretation of Type I error, Type II error and power for this example.

   e. Suppose we only consider low albumin levels to be "abnormal." We want the same Type II error as above. What is the critical value now?

   f. In part (e) above, what is the associated power?

35. This problem illustrates the power of probability theory.

   a. Two SIDS infants are selected at random from the population of SIDS infants. We note their birthweights. What is the probability that both birthweights are (1) below the population median, (2) above the population median (3) straddle the population median? The last interval is a nonparametric confidence interval.

   b. Do the same as in Part (a) for four SIDS infants. Do you see the pattern?

   c. How many infants are needed to have interval (3) in Part (a) have a probability > 0.95?

REFERENCES

Bednarek, F. J. and Roloff, D. W. [1976]. Treatment of apnea of prematurity with aminophylline. *Pediatrics*, **58:** 335–339.

Berkow, R. (Ed.). [1987]. *The Merck Manual of Diagnosis and Therapy*, 15th Edition. Merck, Rahway, NJ.

Case Records of the Massachusetts General Hospital [1970]. Normal laboratory values. *New England Journal of Medicine*, **283:** 1276–1285.

Chen, J. R., Francisco, R. B., and Miller, T. E. [1977]. Legionnaires' disease: nickel levels. *Science*, **196:** 906–908.

Dobson, J. C., Kushida, E., Williamson, M., and Friedman, E. G. [1976]. Intellectual performance of 36 phenylketonuria patients and their non-affected siblings. *Pediatrics*, **58:** 53–58.

Elveback, L. R., Guillier, L. and Keating, Jr., F. R. [1970]. Health, normality and the ghost of Gauss. *Journal of the American Medical Association*, **211:** 69–75.

Fisher, R. A. [1956]. *Statistical Methods and Scientific Inference*. Oliver and Boyd, London.

Fleiss, J. L. [1981]. *Statistical Methods for Rates and Proportions*, 2nd edition. Wiley, New York.

Galton, F. [1889]. *Natural Inheritance*. Macmillan, London.

Golubjatnikov, R., Paskey, T., and Inhorn, S. L. [1972]. Serum cholesterol levels of Mexican and Wisconsin school children. *American Journal of Epidemiology*, **96:** 36–39.

Hacking, I. [1965]. *Logic of Statistical Inference*. Cambridge University Press, London.

Hagerup, L., Hansen, P. F., and Skov, F. [1972]. Serum cholesterol, serum-triglyceride and ABO blood groups in a population of 50-year-old Danish men and women. *American Journal of Epidemiology*, **95:** 99–103.

Kato, H., Tillotson, J., Nichaman, M. Z., Rhoads, G. G., and Hamilton, H. B. [1973]. Epidemiologic studies of coronary heart disease and stroke in Japanese men living in Japan, Hawaii and California: serum lipids and diet. *American Journal of Epidemiology*, **97:** 372–385.

Kahneman, D., Slovic, P., and Tversky, A. (eds.) [1982]. *Judgement Under Uncertainty: Heuristics and Biases*. Cambridge University Press, Cambridge.

Kesteloot, H. and van Houte, O. [1973]. An epidemiologic study of blood pressure in a large male population. *American Journal of Epidemiology*, **99:** 14–29.

Kruskal, W. and Mosteller, F. [1979a]. Representative sampling I: non-scientific literature. *International Statistical Review*, **47:** 13–24.

Kruskal, W. and Mosteller, F. [1979b]. Representative sampling II: scientific literature excluding statistics. *International Statistical Review*, **47:** 111–127.

Kruskal, W. and Mosteller, F. [1979c]. Representative sampling III: the current statistical literature. *International Statistical Review*, **47:** 245–265.

Moore, D. S. [1985]. *Statistics: Concepts and Controversies,*. 2nd edition. W. H. Freeman, New York.

Murphy, E. A. [1979]. *Biostatistics in Medicine*. The Johns Hopkins University Press, Baltimore, MD.

Runes, D. D. [1959]. *Dictionary of Philosophy*. Littlefield, Adams, Ames, IA.

Rushforth, N. B., Bennet, P. H., Steinberg, A. G., Burch, T. A., and Miller M. [1971]. Diabetes in the Pima Indians: evidence of bimodality in glucose tolerance distribution. *Diabetes*, **20:** 756–765. Copyright © 1971 by the American Diabetic Association.

Savage, I. R. [1968]. *Statistics: Uncertainty and Behavior*. Houghton Mifflin, Boston.

Shapiro, S., Goldberg, J. D., and Hutchinson, G. B. [1974]. Lead time in breast cancer detection and implications for periodicity of screening. *American Journal of Epidemiology*, **100:** 357–366.

Shepard, D. S. and Neutra, R. [1977]. Pitfalls in sampling medical visits. *American Journal of Public Health*, **67:** 743–750. Copyright © by the American Public Health Association.

Tversky, A. and Kahneman, D. [1974]. Judgment under uncertainty: heuristics and biases. *Science*, **185:** 1124–1131. Copyright © by the AAAS.

Winkelstein, Jr., W., Kazan, A., Kato, H., and Sachs, S. T. [1975]. Epidemiologic studies of coronary heart disease and stroke in Japanese men living in Japan, Hawaii and California: blood pressure distributions. *American Journal of Epidemiology*, **102:** 502–513.

Zervas, M., Hamacher, H., Holmes, O., and Rieder, S. V. [1970]. Normal laboratory values. *New England Journal of Medicine*, **283:** 1276–1285.

CHAPTER 5

One- and Two-Sample Inference

5.1 INTRODUCTION

In the previous chapter, we laid the groundwork for statistical inference. The following steps were involved:

1. Define the population of interest;
2. Specify the parameter(s) of interest;
3. Take a random sample from the population;
4. Make statistical inferences about the parameter(s):
   a. estimation;
   b. hypothesis testing.

A good deal of "behind-the-scenes" work was necessary, such as specifying what is meant by a *random* sample, but you will recognize that the above four steps summarize the process. In this chapter, we will (1) formalize the inferential process by defining "pivotal quantities" and their uses (Section 5.2); (2) consider normal distributions for which *both* the mean and the variance are unknown; this will involve the use of the famous "Student *t*-distribution" (Sections 5.3 and 5.4); (3) extend the inferential process to the comparison of two normal populations, including comparison of the variances (Sections 5.5–5.7); and (4) finally begin to answer the question frequently asked of statisticians: "how many observations should I take?" (Section 5.9). The chapter concludes with notes, problems, and references.

5.2 PIVOTAL VARIABLES

5.2.1 Definitions

In Chapter 4, confidence intervals and tests of hypotheses were introduced in a somewhat *ad hoc* fashion as inference procedures about population parameters. To be able to make inferences, we needed the sampling distributions of the statistics which estimated the parameters. To make inferences about the mean of a normal distribution (with variance known), we needed to know that the sample mean of a random sample was normally distributed; to make inferences about the variance of a normal distribution, we used the chi-square distribution. A pattern also emerged in the development of estimation and hypothesis testing procedures. We will now discuss the unifying scheme. This will greatly simplify the understanding of the statistical procedures so that attention can be focused on the assumptions and appropriateness of such procedures, rather than the understanding of the mechanics.

In Chapter 4, we basically used two quantities in making inferences:

$$Z = \frac{\overline{Y} - \mu}{\sigma/\sqrt{n}} \quad \text{and} \quad \chi^2 = \frac{(n-1)s^2}{\sigma^2}.$$

What are some of their common features?

1. Each of these expressions involve *at least* a statistic *and* a parameter for the statistic estimated; for example, s^2 and σ^2 in the second formula.

2. The distribution of the quantity was tabulated: standard normal table, chi-square table.

3. The distribution of the quantity was not dependent on a value of the parameter. Such a distribution is called a fixed distribution.

4. Both confidence intervals and tests of hypotheses were derived from a probability inequality involving either Z or χ^2.

Formally, we define:

Definition 5.1. A *pivotal variable* is a function of statistic(s) and parameter(s) having the same fixed distribution (usually tabulated) for all values of the parameter(s).

The quantities Z and χ^2 are pivotal variables. One of the objectives of theoretical statistics is to develop appropriate pivotal variables for experimental situations that cannot be adequately modeled by existing variables.

In Table 5.1 are listed eight pivotal variables and their use in statistical inference. In this chapter, we introduce pivotal variables two, five, six, and eight. Pivotal variables three and four are introduced in Chapter 6. For each variable, the fixed or tabulated distribution is given as well as the formula for a $100(1 - \alpha)\%$ confidence interval. The corresponding test of hypothesis is obtained by replacing the statistic(s) by the hypothesized parameter value(s). The table also lists the assumptions underlying the test. Most of the time, the minimal assumption is that of normality of the underlying observations or appeal is made to the Central Limit Theorem.

Pivotal variables are used primarily in inferences based on the normal distribution. They provide a methodology for estimation and hypothesis testing. The aim of estimation and hypothesis testing is to make probabilistic statements about parameters. For example, confidence intervals and p-values make statements about parameters that have probabilistic aspects. In Chapters 6–8, we discuss inferences that do not depend as explicitly on pivotal variables; however, even in these procedures, the methodology associated with pivotal variables is used.

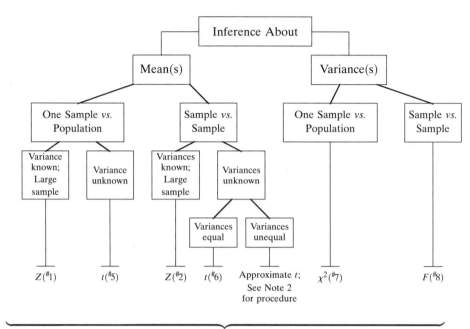

$$\underbrace{\hspace{8cm}}$$

Pivotal Variable

5.3 WORKING WITH PIVOTAL VARIABLES

We have already introduced the manipulation of pivotal variables in Section 4.7. Table 5.1 summarizes the end result of the manipulations. In this section, we again outline the process for the case of one sample from a normal population with the variance known. We have a random sample of size n from a normal population with mean μ and variance σ^2 (known). We start with the basic probabilistic inequality

$$P[z_{\alpha/2} \le Z \le z_{1-\alpha/2}] = 1 - \alpha.$$

We substitute $Z = (\overline{Y} - \mu)/(\sigma_0/\sqrt{n})$, writing σ_0 to indicate that the population variance is assumed to be known:

$$P\left[z_{\alpha/2} \le \frac{\overline{Y} - \mu}{\sigma_0/\sqrt{n}} \le z_{1-\alpha/2}\right] = 1 - \alpha.$$

Table 5.1. Pivotal Variables and Their Use in Statistical Inference.

	Pivotal Variable	Model	Other[a]
			Assumptions
1.	$\dfrac{\overline{Y} - \mu}{\sigma/\sqrt{n}} = Z$	$N(0,1)$	(i) and (iii); or (ii)
2.	$\dfrac{(\overline{Y}_1 - \overline{Y}_2) - (\mu_1 - \mu_2)}{\sqrt{\sigma_1^2/n_1 + \sigma_2^2/n_2}} = Z$	$N(0,1)$	(i) and (iii); or (ii)
3.	$\dfrac{p - \pi}{\sqrt{p(1-p)/n}} = Z$	$N(0,1)$	(ii)
4.	$\dfrac{(p_1 - p_2) - (\pi_1 - \pi_2)}{\sqrt{p_1 q_1/n_1 + p_2 q_2/n_2}} = Z$	$N(0,1)$	(ii)
5.	$\dfrac{\overline{Y} - \mu}{s/\sqrt{n}} = t$	t_{n-1}	(i)
6.	$\dfrac{(\overline{Y}_1 - \overline{Y}_2) - (\mu_1 - \mu_2)}{s_p\sqrt{1/n_1 + 1/n_2}} = t$	$t_{n_1+n_2-2}$	(i) and (iv)
7.	$\dfrac{(n-1)s^2}{\sigma^2} = \chi^2$	χ^2_{n-1}	(i)
8.	$\dfrac{s_1^2/\sigma_1^2}{s_2^2/\sigma_2^2} = F$	$F_{n_1-1,\,n_2-1}$	(i)

[a]**Assumptions (other):** **(i)** Observations (for paired data, the differences) are independent, normally distributed; **(ii)** large sample result; **(iii)** variance(s) known; **(iv)** population variances equal.

Solving for μ produces a $100(1-\alpha)\%$ confidence interval for μ; solving for $\overline{Y}$ and substituting a hypothesized value, μ_0, for μ, produces the nonrejection region for a $100(\alpha)\%$ test of the hypothesis:

$100(1-\alpha)\%$ confidence interval for μ:

$$\left[\overline{Y} + z_{\alpha/2}\sigma_0/\sqrt{n}, \quad \overline{Y} + z_{1-\alpha/2}\sigma_0/\sqrt{n}\right];$$

$100(\alpha)\%$ hypothesis test of $\mu = \mu_0$; reject if $\overline{Y}$ is not in:

$$\left[\mu_0 + z_{\alpha/2}\sigma_0/\sqrt{n}, \quad \mu_0 + z_{1-\alpha/2}\sigma_0/\sqrt{n}\right].$$

Notice again the similarity between the two intervals.

These intervals can be written in an abbreviated form using the fact that $z_{\alpha/2} = -z_{1-\alpha/2}$,

Table 5.1. Pivotal Variables and Their Use in Statistical Inference (continued).

$100(1-\alpha)\%$ Confidence Interval[a]	Inference/Comments
1. $\overline{Y} \pm z_* \, \sigma/\sqrt{n}$	μ or $\mu = \mu_1 - \mu_2$ based on paired data $z_* = z_{1-\alpha/2}$
2. $(\overline{Y}_1 - \overline{Y}_2) \pm z_* \sqrt{\dfrac{\sigma_1^2}{n_1} + \dfrac{\sigma_2^2}{n_2}}$	$\mu_1 - \mu_2$ based on independent data $z_* = z_{1-\alpha/2}$
3. $p \pm z_* \sqrt{p(1-p)/n}$	π $z_* = z_{1-\alpha/2}$
4. $(p_1 - p_2) \pm z_* \sqrt{\dfrac{p_1 q_1}{n_1} + \dfrac{p_2 q_2}{n_2}}$	$\pi_1 - \pi_2$ based on independent data $z_* = z_{1-\alpha/2}$ $q_1 = 1 - p_1;\quad q_2 = 1 - p_2$
5. $\overline{Y} \pm t_* s/\sqrt{n}$	μ or $\mu = \mu_1 - \mu_2$ based on paired data $t_* = t_{n-1,1-\alpha/2}$
6. $(\overline{Y}_1 - \overline{Y}_2) \pm t_* s_p \sqrt{\dfrac{1}{n_1} + \dfrac{1}{n_2}}$	$\mu_1 - \mu_2$ based on independent data $t_* = t_{n_1+n_2-2,\,1-\alpha/2}$ $s_p^2 = \dfrac{(n_1-1)s_1^2 + (n_2-1)s_2^2}{(n_1 + n_2 - 2)}$
7. $\dfrac{(n-1)s^2}{\chi_*^2},\quad \dfrac{(n-1)s^2}{\chi_{**}^2}$	σ^2 $\chi_*^2 = \chi_{n-1,1-\alpha/2}^2$ $\chi_{**}^2 = \chi_{n-1,\alpha/2}^2$
8. $\dfrac{s_1^2/s_2^2}{F_*},\quad \dfrac{s_1^2/s_2^2}{F_{**}}$	σ_1^2/σ_2^2 $F_* = F_{n_1-1,n_2-1,1-\alpha/2}$ $F_{**} = F_{n_1-1,n_2-1,\alpha/2}$

[a]To determine appropriate critical region in a test of hypothesis, replace statistic(s) by hypothesized values of parameter(s).

$$\overline{Y} \pm z_{1-\alpha/2}\sigma_0/\sqrt{n} \quad \text{and} \quad \mu_0 \pm z_{1-\alpha/2}\sigma_0/\sqrt{n}$$

for the confidence intervals and tests of hypothesis, respectively.

To calculate the p-value associated with a test statistic, again use is made of the pivotal variable. The null hypothesis value of the parameter is used to calculate the probability of the observed value of the statistic or an observation more extreme. As an illustration, suppose that a population variance is claimed to be 100 ($\sigma_0^2 = 100$) vs. a larger value ($\sigma^2 > 100$). From a random sample of size 11, we are given $s^2 = 220$. What is the p-value for this value (or more extreme)? We use the pivotal quantity $(n-1)s^2/\sigma_0^2$, which under the null hypothesis is chi-square with ten degrees of freedom.

The one-sided *p*-value is the probability of a value of $s^2 \geq 220$. Using the pivotal variable, we get

$$P\left[\chi^2 \geq \frac{(11-1)(220)}{100}\right] = P[\chi^2 \geq 22.0],$$

where χ^2 has $11 - 1 = 10$ degrees of freedom. Using the chi-square *p*-value table, Table A.4 in the Appendix, we see that the calculated chi-square one-sided *p*-value is 0.0151.

Additional examples in the use of pivotal variables will occur throughout this and later chapters. See Note 5.1 for some additional comments on the pivotal variable approach.

5.4   THE *t*-DISTRIBUTION

For a random sample from a normal distribution with mean μ and variance σ^2 (known), the quantity $Z = (\overline{Y} - \mu)/(\sigma/\sqrt{n})$ is a pivotal quantity which has a normal $(0, 1)$ distribution. What if the variance is unknown? Suppose we replace the variance σ^2 by its estimate s^2 and consider the quantity $(\overline{Y} - \mu)/(s/\sqrt{n})$. What is its sampling distribution?

This problem was solved by the statistician W. S. Gossett, in 1908, who published the result under the pseudonym "Student" using the notation

$$t = \frac{\overline{Y} - \mu}{s/\sqrt{n}}.$$

The distribution of this variable is now called Student's *t*-distribution. Gossett showed that the distribution of *t* was similar to that of the normal distribution, but somewhat more "heavy-tailed" (see below), and that for each sample size there is a different distribution. The distributions are indexed by $(n - 1)$; the degrees of freedom identical to that of the chi-square distribution. The *t*-distribution is symmetrical, and as the degrees of freedom become infinite, the standard normal distribution is reached.

A picture of the *t*-distribution for various degrees of freedom, as well as the limiting case of the normal distribution, is given in Figure 5.1. Note that, like the standard normal distribution, the *t*-distribution is bell-shaped and symmetrical about zero. The *t*-distribution is "heavy-tailed": the area to the right of a specified positive value is greater than for the normal distribution; in other words, the *t*-distribution is less "pinched." This is reasonable; unlike a standard normal deviate where only the mean $(\overline{Y})$ can vary (μ and σ are fixed), the *t* statistic can vary with *both* $\overline{Y}$ and *s*, so that *t* will vary even if $\overline{Y}$ is fixed.

Percentiles of the *t*-distribution are denoted by the symbol $t_{\nu, \alpha}$, where ν indicates the degrees of freedom, and α the 100αth percentile. This is indicated by the following graph:

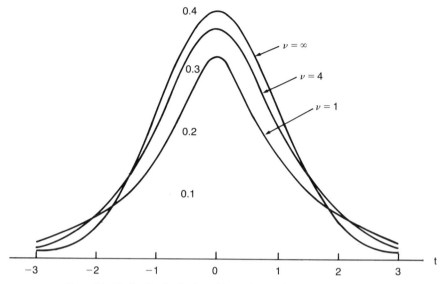

Figure 5.1. "Student" t-distribution with one, four, and ∞ degrees of freedom.

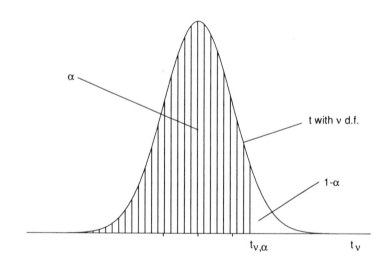

In Table 5.1, rather than writing all the subscripts on the t variate, an asterisk is used and explained in the comment part of the table.

Table A.5 lists the percentiles of the t-distribution for each degree of freedom to 30, by fives to 100, and values for 200, 500, and ∞ degrees of freedom. This table lists the t values such that the percent to the left is as specified by the column heading. For example, for an area of 0.975 (97.5%), the t value for six degrees of freedom is 2.45. The last row in this column corresponds to a t with an infinite number of degrees of freedom, and the value of 1.96 is identical to the corresponding value of Z, that is,

$P[Z \leq 1.96] = 0.975$. You should verify that the last row in this table corresponds precisely to the normal distribution values, i.e., $t_\infty = Z$. What are the mean and the variance of the t-distribution? The mean will be zero, and the variance is $\nu/(\nu - 2)$. In the symbols used in the previous chapter, $E(t) = 0$ and $\text{var}(t) = \nu/(\nu - 2)$.

Table A.6 lists areas greater in absolute value than specified t values. Thus, the probability of a t value greater than one in absolute value for one degree of freedom is 0.500; the corresponding areas for seven, 30, and ∞ degrees of freedom are 0.351, 0.325, and 0.317, respectively. Thus, at 30 degrees of freedom, the t-distribution is, for most practical purposes, indistinguishable from a normal distribution. The term "heavy-tailed" can now be made precise: for a specified value (for example, with an abscissa value of 1), $P[t_1 \geq 1] > P[t_7 \geq 1] > P[t_{10} \geq 1] > P[Z \geq 1]$.

5.5 ONE-SAMPLE INFERENCE: LOCATION

5.5.1 Estimation and Testing

We begin this section with an example.

Example 5.1. In Example 4.9, we considered the birthweight in grams of the first eleven SIDS cases occurring in King County in 1969. In this example, we consider the birthweights of the first 15 cases born in 1977. The birthweights for the latter group are:

$$2013 \quad 3827 \quad 3090 \quad 3260 \quad 4309 \quad 3374 \quad 3544 \quad 2835$$
$$3487 \quad 3289 \quad 3714 \quad 2240 \quad 2041 \quad 3629 \quad 3345$$

The mean and standard deviation of this sample are 3199.8 g and 663.00 g, respectively. Without assuming that the population standard deviation is known, can we obtain an interval estimate for the population mean or test the null hypothesis that the population birthweight average of SIDS cases is 3300 g (the same as the general population)?

We can now use the t-distribution. Assuming that birthweights are normally distributed, the quantity

$$\frac{\overline{Y} - \mu}{s/\sqrt{15}}$$

has a t-distribution with $15 - 1 = 14$ degrees of freedom.

Using the estimation procedure, the point estimate of the population mean birthweight of SIDS cases is $3199.8 \doteq 3200$ g. A 95% confidence interval can be constructed on the basis of the t-distribution. For a t-distribution with $15 - 1 = 14$ degrees of freedom, the critical values are ± 2.14, that is, $P[-2.14 \leq t_{14} \leq 2.14] = 0.95$. Using Table 5.1, a 95% confidence interval is constructed using pivotal variable no. 5,

$$3200 \pm (2.14)(663.0)/\sqrt{15} = 3200 \pm 366,$$
$$\text{Lower limit: 2834 g,} \quad \text{Upper limit: 3566 g.}$$

Several comments are in order:

1. This interval includes 3300 g, the average birthweight in the non-SIDS population. If the analysis had followed a hypothesis testing procedure, we could not have rejected the null hypothesis on the basis of a two-tailed test.

2. The standard error, $633.0/\sqrt{15}$, is multiplied by 2.14 rather than the critical value 1.96 using a normal distribution. Thus, the confidence interval is wider by approximately 9%. This is the price paid for our ignorance about the value of the population standard deviation.   □

5.5.2   *t*-Tests for Paired Data

A second illustration of the one sample *t*-test involves the application to "paired data." What is "paired data"? Typically, this involves repeated or multiple measurements on the same subjects. For example, we may have a measurement of the level of pain before and after administration of an analgesic drug. A somewhat different experiment might consider the level of pain in response to each of *two* drugs. One of these could be a placebo. The first experiment has the weakness that there may be a spontaneous reduction in level of pain, e.g., postoperative pain level, and thus, the difference in the responses (after/before) may be made up of two effects: an effect of the drug as well as the spontaneous reduction. These are some experimental design considerations that are discussed further in Chapter 10. The point we want to make with these two illustrations is that the basic data consists of pairs, and what we want to look at is the differences within the pairs. If, in the second illustration, the treatments are to be compared, a common null hypothesis is that the effects are the same and therefore the differences in the treatments should be centered around zero. A natural approach then tests whether the mean of the *sample differences* could have come from a population of differences with mean zero. If we assume that the means of the sample differences are normally distributed we can then apply the *t*-test (under the null hypothesis), and estimate the variance of the population of differences σ^2, by the variance of the *sample differences*, s^2.

PEANUTS reprinted by permission of UFS, Inc.

Example 5.2.   The procedure is illustrated with the following data from Bednarek and Roloff [1976] dealing with the treatment of apnea (a transient cessation of respiration) using a drug, aminophylline, in premature infants. The variable of interest was "average number of apneic episodes per hour," and was measured before and after treatment with the drug. An episode was defined as the absence of spontaneous breathing for more than 20 s or less if associated with bradycardia or cyanosis.

Patients who had "six or more apneic episodes on each of two consecutive 8-h shifts were admitted to the study." For purposes of the study, consider only the difference between the average number of episodes 24 h before the treatment and

Table 5.2. Response of 13 Patients to Aminophylline Treatment at 16 h Compared with 24 h Before Treatment (Apneic Episodes per Hour). Data from Bednarek and Roloff [1976].

Patient	24 h Before	16 h After	Before–After (Difference)
1	1.71	0.13	1.58
2	1.25	0.88	0.37
3	2.13	1.38	0.75
4	1.29	0.13	1.16
5	1.58	0.25	1.33
6	4.00	2.63	1.37
7	1.42	1.38	0.04
8	1.08	0.50	0.58
9	1.83	1.25	0.58
10	0.67	0.75	−0.08
11	1.13	0.00	1.13
12	2.71	2.38	0.33
13	1.96	1.13	0.83
Total	22.76	12.79	9.97
Mean	1.751	0.984	0.767
Variance	0.7316	0.6941	0.2747
Standard Deviation	0.855	0.833	0.524

16 h after the treatment. This difference is given in the fourth column of Table 5.2. The average difference for the 13 patients is 0.767 episodes/h. That is, there is a change from 1.751 episodes/h before treatment to 0.984 episodes/h at 16 h after the treatment.

The standard deviation of the differences is $s = 0.524$. The pivotal quantity to be used is no. 5 from Table 5.1. The argument is as follows: the basic statement about the pivotal variable t with $13 - 1 = 12$ degrees of freedom is $P[-2.18 \leq t_{12} \leq 2.18] = 0.95$ using Table A.5. The form taken for this example is

$$P\left[-2.18 \leq \frac{\overline{Y} - \mu}{0.524/\sqrt{13}} \leq 2.18\right] = 0.95.$$

To set up the region to test some hypothesis, we solve for $\overline{Y}$ as before. The region then is

$$P[\mu - 0.317 \leq \overline{Y} \leq \mu + 0.317] = 0.95.$$

What is a "reasonable" value to hypothesize for μ? The usual procedure in this type of situation is to assume that the treatment has "no effect." That is, the average difference in the number of apneic episodes from before to after treatment represents random variation. If there is no difference in the population average number of episodes before and after treatment, we can write this as

$$H_0 : \mu = 0.$$

We can now set up the hypothesis testing region as illustrated in Figures 5.2 and 5.3.

Figure 5.2 indicates that the sample space can be partitioned without knowing the

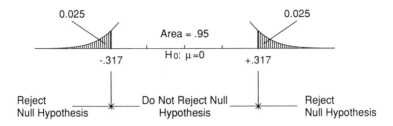

Figure 5.2. Partitioning of sample space of $\overline{Y}$ into two regions: (a) region where the null hypothesis is not rejected, and (b) region where it is rejected. Data from Table 5.2.

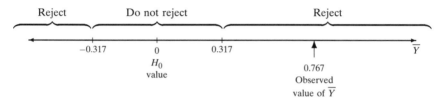

Figure 5.3. Observed value of $\overline{Y}$ and location on the sample space. Data from Table 5.2.

observed value of $\overline{Y}$. Figure 5.3 indicates the observed value of $\overline{Y} = 0.767$ episodes per hour; it clearly falls into the rejection region. Note that the scale has been changed from Figure 5.2 to accommodate the observed value.

Hence, the null hypothesis is rejected and it is concluded that the observed average number of apneic episodes at 16 h after treatment differs significantly from the observed average number of apneic episodes 24 h before treatment.

The above kind of test is often used when two treatments are applied to the same experimental unit, or when the experimental unit is observed over time and a treatment is administered so that it is meaningful to speak of "pre-treatment" and "post-treatment" situations. As mentioned before, there is the possibility that changes, if observed, are in fact due to changes over time and not related to the treatment.

To construct a confidence interval, we solve the inequality for μ so that we get

$$P[\overline{Y} - 0.317 \leq \mu \leq \overline{Y} + 0.317] = 0.95.$$

Again, this interval can be set up to this point without knowing the value of $\overline{Y}$. The value of $\overline{Y}$ is observed to be 0.767 episodes/h, so that the 95% confidence interval becomes

$$[0.767 - 0.317 \leq \mu \leq 0.767 + 0.317] \quad \text{or} \quad [0.450 \leq \mu \leq 1.084].$$

This interval is displayed in Figure 5.4. Two things should be noted:

1. The width of the confidence interval is the same as the width of the region where the null hypothesis is not rejected (cf. Figure 5.3).
2. The 95% confidence interval does not include zero, the null hypothesis value of μ. □

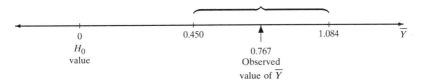

Figure 5.4. A 95% confidence interval for the difference in number of apneic episodes per hour. Data from Table 5.2.

5.6 TWO-SAMPLE STATISTICAL INFERENCE: LOCATION

5.6.1 Independent Random Variables

A great deal of research activity involves the comparison of two or more groups. For example, two cancer therapies may be investigated: one group of patients receives one treatment and a second group the other. The experimental situation can be thought of in two ways: (1) there is one population of subjects, and the treatments induce two sub-populations; or (2) we have two populations which are identical except in their responses to their respective treatments. If the assignment of treatment is random, the two situations are equivalent.

Before exploring this situation, we need to state a definition and a statistical result:

Definition 5.2. Two random variables Y_1 and Y_2 are *statistically independent* if for all fixed values of numbers (say y_1 and y_2),

$$P[Y_1 \le y_1, Y_2 \le y_2] = P[Y_1 \le y_1]P[Y_2 \le y_2].$$

The notation $[Y_1 \le y_1, Y_2 \le y_2]$ means that Y_1 takes on a value less than or equal to y_1, and Y_2 takes on a value less than or equal to y_2. If we define an event A to have occurred when Y_1 takes on a value less than or equal to y_1, and an event B when Y_2 takes on a value less than or equal to y_2, then Definition 5.2 is equivalent to the statistical independence of events $P[AB] = P[A]P[B]$ as defined in Chapter 4. So the difference between statistical independence of random variables and statistical independence of events is that the former in effect describes a relationship between many events (since the definition has to be true for *any* set of values of y_1 and y_2). A basic result can now be stated:

Result 5.1. If Y_1 and Y_2 are statistically independent random variables, then for any two constants a_1 and a_2, the random variable $W = a_1 Y_1 + a_2 Y_2$ has mean and variance

$$E(W) = a_1 E(Y_1) + a_2 E(Y_2),$$

$$\text{var}(W) = a_1^2 \text{var}(Y_1) + a_2^2 \text{var}(Y_2).$$

The only new aspect of this result is that of the variance. In Note 4.9, the expectation of W was already derived. Before giving an example, we also state:

Result 5.2. If Y_1 and Y_2 are statistically independent random variables which are normally distributed, then $W = a_1 Y_1 + a_2 Y_2$ is normally distributed with mean and variance given by Result 5.1.

Example 5.3. Let Y_1 be normally distributed with mean $\mu_1 = 100$ and variance $\sigma_1^2 = 225$; let Y_2 be normally distributed with mean $\mu_2 = 50$ and variance $\sigma_2^2 = 175$. If Y_1 and Y_2 are statistically independent, then $W = Y_1 + Y_2$ is normally distributed with mean $100 + 50 = 150$ and variance $225 + 175 = 400$. This and additional examples are given in the following summary:

$$Y_1 \sim N(100, 225), \quad Y_2 \sim N(50, 175);$$

W	Mean of W	Variance of W
$Y_1 + Y_2$	150	400
$Y_1 - Y_2$	50	400
$2Y_1 + Y_2$	250	1075
$2Y_1 - 2Y_2$	100	1600

Note that the variance of $Y_1 - Y_2$ is the same as the variance of $Y_1 + Y_2$; this is because the coefficient of Y_1, -1, is squared in the variance formula, and $(-1)^2 = (+1)^2 = 1$. In words, the variance of a sum of independent random variables is the same as the variance of a difference of independent random variables.   □

Example 5.4. The following example is more interesting and indicates the usefulness of the two results stated. Heights of females and males are normally distributed with means 162 cm and 178 cm and variances $(6.4 \text{ cm})^2$ and $(7.5 \text{ cm})^2$, respectively. Let $Y_1 = $ height of female; let $Y_2 = $ height of male. Then we can write

$$Y_1 \sim N\left(162, (6.4)^2\right) \quad \text{and} \quad Y_2 \sim N\left(178, (7.5)^2\right).$$

Now consider husband/wife pairs. Suppose (probably somewhat contrary to societal *mores*) that husband/wife pairs are formed independent of stature. That is, we interpret this statement to mean that Y_1 and Y_2 are statistically independent. The question is, on the basis of this model, what is the probability that the wife is taller than the husband? We formulate the problem as follows: construct the new variable $W = Y_1 - Y_2$. From Result 5.2, it follows that

$$W \sim N\left(-16, (6.4)^2 + (7.5)^2\right).$$

Now the question can be translated into a question about W, namely, if the wife is taller than the husband, then $Y_1 > Y_2$, or $Y_1 - Y_2 > 0$, or $W > 0$. Thus, the question is reformulated as $P[W > 0]$. Hence,

$$P[W > 0] = P\left[Z > \frac{0 - (-16)}{\sqrt{(6.4)^2 + (7.5)^2}}\right]$$
$$\doteq P[Z > 16/9.86]$$
$$\doteq P[Z > 1.62]$$
$$\doteq 0.053.$$

so that under the model in 5.3% of husband/wife pairs, the wife will be taller than the husband. The sketch below indicates the area of interest.   □

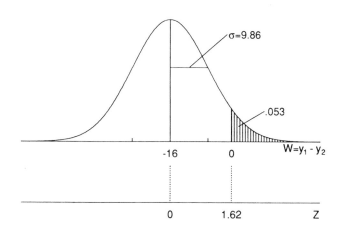

5.6.2   Estimation and Testing

The most important application of Result 5.1 involves the distribution of the difference of two sample means. If $\overline{Y}_1$ and $\overline{Y}_2$ are the means from two random samples of size n_1 and n_2, respectively, and Y_1 and Y_2 are normally distributed with means μ_1 and μ_2, and variances σ_1^2 and σ_2^2, then by Result 5.2,

$$\overline{Y}_1 - \overline{Y}_2 \sim N\left(\mu_1 - \mu_2, \frac{\sigma_1^2}{n_1} + \frac{\sigma_2^2}{n_2}\right),$$

so that

$$\frac{(Y_1 - Y_2) - (\mu_1 - \mu_2)}{\sqrt{\sigma_1^2/n_1 + \sigma_2^2/n_2}} = Z$$

has a standard normal distribution. This, again, is a pivotal variable, number 2 in Table 5.1. We are now in a position to construct confidence intervals for the quantity $(\mu_1 - \mu_2)$ or to do hypothesis testing. In many situations, it will be reasonable to assume (null hypothesis) that $\mu_1 = \mu_2$ so that $\mu_1 - \mu_2 = 0$; although the values of the two parameters are unknown, it is reasonable for testing purposes to assume that they are equal, and hence, the difference will be zero. For example, in a study involving two treatments, we could assume that the treatments were equally effective (or ineffective) and differences between the treatments should be centered at zero.

How do we determine whether or not random variables are statistically independent? This will be discussed in Chapter 9. For the present time, we will either assume that the variables we are dealing with are either statistically independent, or if not (as in the case of the paired t-test discussed in Section 5.5.2), use aspects of the data that can be considered statistically independent.

Example 5.5. Zelazo *et al.* [1972] studied the age at which children walked as related to "walking exercises" given newborn infants. They state "if a newborn infant is held under his arms and his bare feet are permitted to touch a flat surface, he will perform well-coordinated walking movements similar to those of an adult." This reflex disappears by about 8 weeks. They placed 24 white male infants into one of four "treatment" groups. For purposes of this example, we consider only two of the four groups: "Active Exercise Group" and "Eight-Week Control Group." The active group received daily stimulation of the walking reflex for 8 weeks. The control group was tested at the end of the 8-week treatment period, but there was no intervention. The age at which the child subsequently began to walk was then reported by the mother. The data and basic calculations are shown in Table 5.3.

For purposes of this example, we will assume that the sample standard deviations are, in fact, population standard deviations so that Result 5.2 can be applied. In Example 5.6, we will reconsider this example using the two-sample *t*-test. For this example, we have

$$n_1 = 6, \qquad\qquad\qquad n_2 = 5,$$
$$\overline{Y}_1 = 10.125 \text{ months}, \qquad\qquad \overline{Y}_2 = 12.350 \text{ months},$$
$$\sigma_1 = 1.4470 \text{ months (assumed)}, \qquad \sigma_2 = 0.9618 \text{ months (assumed)}.$$

For purposes of this example, the quantity

$$Z = \frac{(\overline{Y}_1 - \overline{Y}_2) - (\mu_1 - \mu_2)}{\sqrt{(1.4470)^2/6 + (0.9618)^2/5}} = \frac{(\overline{Y}_1 - \overline{Y}_2) - (\mu_1 - \mu_2)}{0.7307}$$

has a standard normal distribution and is based on pivotal variable number 2 of Table 5.1. Let us first set up a 95% confidence interval on the difference $(\mu_1 - \mu_2)$ in the population means. The 95% confidence interval is

$$(\overline{Y}_1 - \overline{Y}_2) \pm 1.96(0.7307),$$

Table 5.3. Distribution of Ages (in months) in Infants for Walking Alone. Data from Zelazo
***et al.* [1972].**

	Age in Months for Walking Alone	
	Active Exercise Group	Eight-Week Control Group
	9.00	13.25
	9.50	11.50
	9.75	12.00
	10.00	13.50
	13.00	11.50
	9.50	_[a]
n	6	5
Mean	10.125	12.350
Standard deviation	1.4470	0.9618

[a]One observation is missing from the paper by Zelazo *et al.*

with

$$\text{Upper limit} = (10.125 - 12.350) + 1.4322 = -0.79 \text{ months,}$$
$$\text{Lower limit} = (10.125 - 12.350) - 1.4322 = -3.66 \text{ months.}$$

Drawing a time line, we get

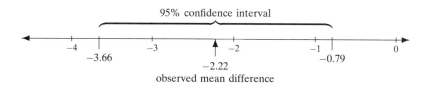

The 95% confidence interval does not straddle zero, so we would conclude that there is a real difference in age in months when the baby first walked in the exercise group as compared to the control group. The best estimate of the difference is $10.125 - 12.350 = -2.22$ months, that is, the age at first walking is about 2 months earlier than the control group.

Note the flow of the argument: the babies were a homogenous group before treatment. Allocation to the various groups was on a random basis (assumed but not stated explicitly in the article); the only subsequent differences between the groups were the treatments, hence significant differences between the groups must be attributable to the treatments. (Can you think of some reservations you may want checked before accepting the conclusion?)

Formulating the problem as a hypothesis testing problem is done as follows: a reasonable null hypothesis is that $\mu_1 - \mu_2 = 0$, in this case, the hypothesis of no effect. Comparable to the 95% confidence interval, a test at the 5% level will be carried out. Conveniently, $\mu_1 - \mu_2 = 0$, so that the nonrejection region is simply $0 \pm 1.96(0.7307)$ or 0 ± 1.4322. Plotting this on a line, we get:

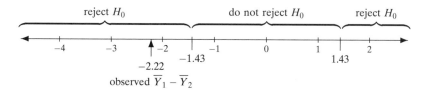

We would reject the null hypothesis, $H_0 : \mu_1 - \mu_2 = 0$, and accept the alternative hypothesis $H_A : \mu_1 \neq \mu_2$; in fact, on the basis of the data, we conclude that $\mu_1 < \mu_2$.

To calculate the (one-sided) p-value associated with the observed difference, we again use the pivotal variable

$$P\left[[\overline{Y}_1 - \overline{Y}_2] \leq -2.225\right] \doteq P\left[Z \leq \frac{-2.225 - 0}{0.7307}\right]$$
$$\doteq P\left[Z \leq -3.05\right]$$
$$\doteq 0.0011.$$

The p-value is 0.0011, much less than 0.05, and again, we would reject the null hypothesis. To make the p-value comparable to the two-sided confidence and hypothesis testing procedure, we must multiply it by two, to give a p-value

$$p\text{-value} = 2(0.0011) = 0.0022. \qquad \square$$

We conclude this section by considering the two sample location problem when the population variances are not known. For this we need:

Result 5.3. If $\overline{Y}_1$ and $\overline{Y}_2$ are based on two independent random samples of size n_1 and n_2 from two normal distributions with means μ_1 and μ_2 and the same variances $\sigma_1^2 = \sigma_2^2 = \sigma^2$, then

$$\frac{(\overline{Y}_1 - \overline{Y}_2) - (\mu_1 - \mu_2)}{s_p \sqrt{1/n_1 + 1/n_2}}$$

has a t-distribution with $n_1 + n_2 - 2$ degrees of freedom. Here s_p^2 is "the pooled estimate of common variance σ^2" as defined below.

This result is summarized by pivotal variable 6 in Table 5.1. Result 5.3 assumes that the population variances are the same, $\sigma_1^2 = \sigma_2^2 = \sigma^2$. There are then two estimates of σ^2: s_1^2 from the first sample, and s_2^2 from the second sample. How can these estimates be combined to provide the best possible estimate of σ^2? If the sample sizes, n_1 and n_2, differ, then the variance based on the larger sample should be given more weight; the pooled estimate of σ^2 provides this. It is defined by

$$s_p^2 = \frac{(n_1 - 1)s_1^2 + (n_2 - 1)s_2^2}{n_1 + n_2 - 2}.$$

If $n_1 = n_2$, then $s_p^2 = \frac{1}{2}(s_1^2 + s_2^2)$, just the arithmetic average of the variances. For $n_1 \neq n_2$, the variance with the larger sample size receives more weight. See Note 5.2 for a further discussion. How can we test whether two sample variances, s_1^2 and s_2^2, estimate the same population variance σ^2? In the next section, we outline a procedure for testing this assumption.

Example 5.6. Continuing Example 5.5, consider again the data in Table 5.3 on the age at which children first walk. We will now take the more realistic approach by treating the standard deviations as sample standard deviations, as they should be.
The pooled estimate of the (assumed) common variance is

$$s_p^2 \doteq \frac{(6-1)(1.4470)^2 + (5-1)(0.9618)^2}{6+5-2} \doteq \frac{14.1693}{9} \doteq 1.5744$$

$$s_p \doteq 1.2547 \text{ months.}$$

A 95% confidence interval for the difference $\mu_1 - \mu_2$ is first constructed. From Table A.5, the critical t-value for 9 degrees of freedom, is $t_{9,\,0.975} = 2.26$. The 95% confidence interval is calculated to be

$$(10.125 - 12.350) \pm (2.26)(1.2547)\sqrt{1/6 + 1/5} \doteq -2.225 \pm 1.717;$$

$$\text{Lower limit} = -3.94 \text{ months} \quad \text{and} \quad \text{Upper limit} = -0.51 \text{ months.}$$

Notice that these limits are wider than the limits $(-3.66, -0.79)$ calculated on the assumption that the variances are known. The wider limits are the price for the additional uncertainty.

The same effect is observed in testing the null hypothesis that $\mu_1 - \mu_2 = 0$. The rejection region, using a 5% significance level, is outside

$$0 \pm (2.26)(2.2547)\sqrt{1/6 + 1/5} \doteq 0 \pm 1.72.$$

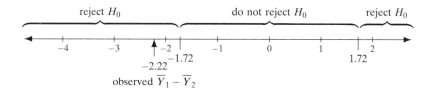

The observed value -2.22 months also falls in the rejection region. Compared to the regions constructed when the variances were assumed known, the region where the null hypothesis is *not* rejected in this case is wider. □

5.7 TWO-SAMPLE INFERENCE: SCALE

5.7.1 The F-distribution

The final inference procedure to be discussed in this chapter deals with the equality of variances of two normal populations. The two-sample t-test assumed that the population variances were equal. We will now present a procedure that will allow us to test this hypothesis. The emphasis will be on testing rather than estimation.

Result 5.4. Given two random samples of size n_1 and n_2, with sample variances s_1^2 and s_2^2, from two normal populations with variances σ_1^2 and σ_2^2, then the variable

$$F = \frac{s_1^2/\sigma_1^2}{s_2^2/\sigma_2^2}$$

has an F-distribution with $(n_1 - 1)$ and $(n_2 - 1)$ degrees of freedom.

The F-distribution (named in honor of Sir R. A. Fisher) is a tabulated distribution; percentiles of the F-distribution are presented in Tables A.7a–A.7f. The distribution is indexed by the degrees of freedom associated with s_1^2 (the numerator degrees of freedom) and the degrees of freedom associated with s_2^2 (the denominator degrees of freedom). A picture of the F-distribution is presented in Figure 5.5. The distribution is skewed; the extent of skewness depends upon the degrees of freedom. As *both* increase, the distribution becomes more symmetric.

We write $F_{\nu_1, \nu_2, \alpha}$ to indicate the 100αth percentile value of an F statistic with ν_1 and ν_2 degrees of freedom. The mean of an F-distribution is $\nu_2/(\nu_2 - 2)$, for $\nu_2 > 2$; the variance is given in Note 5.3. In this note, you will also find a brief discussion of the relationship between the four tabulated distributions we have now discussed:

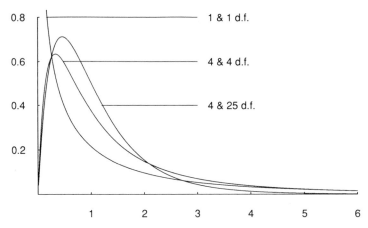

Figure 5.5. F-Distribution for three sets of degrees of freedom.

normal, chi-square, Student t, and F. Since the F-distribution needs to be tabulated for sets of degrees of freedom, the tables are much more extensive. To save space, only the upper percentage points are usually tabulated, and the relationship

$$F_{\nu_1, \nu_2, \alpha} = \frac{1}{F_{\nu_2, \nu_1, 1-\alpha}}$$

is used to calculate the corresponding lower percentage points. For example, suppose we want $F_{5, 4, 0.05}$. This can be calculated using the tabulated value $F_{4, 5, 0.95} = 5.19$:

$$F_{5, 4, 0.05} = \frac{1}{5.19} \doteq 0.19.$$

It is clear that

$$F = \frac{s_1^2 / \sigma_1^2}{s_2^2 / \sigma_2^2}$$

is a pivotal variable, listed as number 8 in Table 5.1. Inferences can be made on the *ratio* σ_1^2 / σ_2^2. (To make inferences about σ_1^2 (or σ_2^2) by itself, we would use the chi-square distribution and the procedure outlined in Chapter 4.) Conveniently, if we want to test whether the variances are equal, that is, $\sigma_1^2 = \sigma_2^2$, then the ratio σ_1^2 / σ_2^2 is equal to 1, and "drops out" of the pivotal variable, which can then be written

$$F = \frac{s_1^2 / \sigma_1^2}{s_2^2 / \sigma_2^2} = s_1^2 / s_1^2.$$

We would reject the null hypothesis of equality of variances if the observed ratio s_1^2 / s_1^2 is "very large" or "very small"; how large or small to be determined by the F-distribution.

5.7.2 Testing and Estimation

Continuing Example 5.5, the sample variances in Table 5.3 were $s_1^2 = (1.4470)^2 = 2.0938$ and $s_2^2 = (0.9618)^2 = 0.9251$. Associated with s_1^2 are $6 - 1 = 5$ degrees of freedom, and with s_2^2, $5 - 1 = 4$ degrees of freedom. Under the null hypothesis of equality of population variances, the ratio s_1^2/s_2^2 has an F-distribution with $(5, 4)$ degrees of freedom. For a two-tail test at the 10% level we need $F_{5,4,0.05}$ and $F_{5,4,0.95}$. From Table A.7, the value for $F_{5,4,0.95}$ is 6.26. Using the relationship $F_{\nu_1,\nu_2,\alpha} = 1/F_{\nu_2,\nu_1,1-\alpha}$, we obtain $F_{5,4,0.05} = 1/F_{4,5,0.95} = 0.19$. The observed value of F is $F_{5,4} = s_1^2/s_2^2 = 2.0938/0.9251 \doteq 2.26$.

From the following diagram, it is clear that the null hypothesis of equality of variances is not rejected:

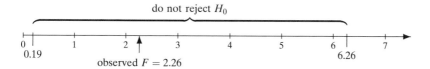

Notice that the rejection region is not symmetric about 1, due to the zero bound on the left hand side. It is instructive to consider F ratios for which the null hypothesis would have been rejected. On the right hand side, $F_{5,4,0.95} = 6.26$; this implies that s_1^2 must be 6.26 times as large as s_2^2 before the 10% significance level is reached. On the left hand side, $F_{5,4,0.05} = 0.19$, so that s_1^2 must be 0.19 times as small as s_2^2 before the 10% significance level is reached. These are reasonably wide limits (even at the 10% level) and therefore, some difference in the sample variances can be tolerated before tests such as the two sample t-tests are demonstrably inappropriate.

What should you do with a test such as the t-test when the variances are not equal? See Note 5.2. Also look at Problem 5.5 for an example of such a situation.

A few comments about terminology: sample variances that are (effectively) the same are called *homogenous*, and those that are not are called *heterogeneous*. A test for equality of population variances then, is a test for homogeneity or heterogeneity. In the more technical statistical literature, you will find the equivalent terms *homoscedasticity* and *heteroscedasticity tests*.

A confidence interval on the ratios of the population variances σ_1^2/σ_2^2 can be constructed using the pivotal variable approach once more. To set up a $100(1 - \alpha)\%$ confidence interval, we need the $100(\alpha/2)$ percentile and $100(1 - \alpha/2)$ percentile of the F-distribution.

Continuing with Example 5.5, suppose we want to construct a 90% confidence interval on σ_1^2/σ_2^2 on the basis of the observed sample. Values for the 5th and 95th percentiles have already been obtained: $F_{5,4,0.05} = 0.19$ and $F_{5,4,0.95} = 6.26$. A 90% confidence interval on σ_1^2/σ_2^2 is then determined by

$$\left(\frac{s_1^2/s_2^2}{F_{5,4,0.95}}, \frac{s_1^2/s_2^2}{F_{5,4,0.05}} \right).$$

For the observed data, this is

$$\left(\frac{2.0938/0.9251}{6.26}, \frac{2.0938/0.9251}{0.19} \right) = (0.36, 11.9).$$

Thus, on the basis of the observed data, we can be 90% confident that the interval $(0.36, 11.9)$ straddles or covers the ratio σ_1^2/σ_2^2 of the population variances. This interval includes 1.0. So, also on the basis of the estimation procedure, we conclude that $\sigma_1^2/\sigma_2^2 = 1$ is not unreasonable.

A 90% confidence interval on the ratio of the standard deviations, σ_1/σ_2, can be obtained by taking square roots of the points $(0.36, 11.9)$, producing $(0.60, 3.45)$ for the interval.

5.8 SAMPLE SIZE CALCULATIONS

One of the questions most frequently asked of a statistician is "How big must my n be?" Stripped of its pseudo-jargon, a valid question is being asked: "how many observations are needed in this study?" Unfortunately, the question cannot be answered before additional information is supplied. We first put the requirements in words in the context of a study comparing two treatments, then we introduce the appropriate statistical terminology. To determine sample size, you need to specify or know:

1. How variable the data are;
2. The chance you are willing to tolerate of incorrectly concluding there is an effect when the treatments are equivalent;
3. The magnitude of the effect to be detected; and
4. The certainty with which you wish to detect the effect.

Each of these considerations are clearly relevant. The more variation in the data, the more observations are needed to "pin down" a treatment effect; when there is no difference, there is a chance that a difference will be observed, which, due to sampling variability, is declared significant. The more certain you want to be of detecting an effect, the more observations you will need, everything else remaining equal. Finally, if the difference in the treatments is very large, a rather economical experiment can be run; conversely, a very small difference in the treatments will require very large sample sizes to detect.

We now phrase the problem in statistical terms: the model we want to consider involves two normal populations with equal variances, σ^2, differing at most in their means, μ_1 and μ_2. To determine the sample size, we must specify

1. σ^2;
2. The probability, α, of a Type I error;
3. The magnitude of the difference $\mu_1 - \mu_2$ to be detected; and
4. The power, $1 - \beta$, or equivalently the probability of a Type II error, β.

Sample sizes are calculated as a function of

$$\Delta = \frac{|\mu_1 - \mu_2|}{\sigma},$$

which is defined to be the standardized distance between the two populations. For a two-sided test, the formula for the required sample size *per group* is

$$n = \frac{2(z_{1-\alpha/2} + z_{1-\beta})^2}{\Delta^2}.$$

It is instructive to contemplate this formula. The standardized difference enters as a square. Thus, to detect a treatment different *half* as small as perhaps initially considered will require *four* times as many observations per group. Decreasing the probabilities of the Type I and Type II errors has the same effect on the sample size; it increases it. However, the increment is not as drastic as it is with Δ. For example, to reduce the probability of a Type I error from 0.05 to 0.025 changes the Z-value from $Z_{0.975} = 1.96$ to $Z_{0.9875} = 2.24$; even though $Z_{1-\alpha/2}$ is squared, the effect will not even be close to doubling the sample size. Finally, the formula assumes that the difference $\mu_1 - \mu_2$ can either be positive or negative. If the direction of the difference can be specified beforehand, a one-tail value for Z can be used. This will result in a reduction of the required sample sizes for each of the groups since $z_{1-\alpha}$ would be used.

Example 5.7.   At a significance level of $1 - \alpha = 0.95$ (one-tail) and power $1 - \beta = 0.80$, a difference $\Delta = 0.3$ is to be detected. The appropriate Z values are

$$Z_{0.95} = 1.645 \quad \text{(a more accurate value than given in Table A.2)}$$
$$Z_{0.80} = 0.84.$$

The required sample size per group is

$$n = \frac{2(1.645 + 0.84)^2}{(0.3)^2} = 137.2.$$

The value is rounded up to 138, so that at least 138 observations *per group* are needed to detect the specified difference at the specified significant level and power.   □

Suppose the variance σ^2 is not known, how can we estimate the sample size needed to detect a standardized difference Δ? One possibility is to have an estimate of the variance σ^2 based on a previous study or sample. Unfortunately, no explicit formulae can be given when the variance is estimated; many statistical texts suggest adding between two and four observations per group to get a reasonable approximation to the sample size (see below).

Finally, suppose *one group*—as in a paired experiment—is to be used to determine whether a populations mean μ differs from a hypothesized mean μ_0. Using the same standardized difference $\Delta = |\mu - \mu_0|/\sigma$, it can be shown that the appropriate number in the group is

$$n = \frac{(z_{1-\alpha/2} + z_{1-\beta})^2}{\Delta^2},$$

or one-half the number needed in one group in the two-sample case. This is why tables for sample sizes in the one-sample case tell you, in order to apply the table to the two-sample case, to (1) double the number in the table, and (2) use that number *for each group*.

Example 5.8. Consider data involving PKU children. Assume that IQ in the general population has mean $\mu = 100$ and standard deviation is 15. Suppose that a sample of eight PKU children whose diet has been terminated has an average IQ of 94, which is not significantly different from 100. How large would the sample have to be in order to detect a difference of six IQ points (i.e., the population mean is 94)? The question cannot be answered yet. (Before reading on: what else must be specified?) Additionally, we need to specify the Type I and Type II errors. Suppose $\alpha = 0.05$ and $\beta = 0.10$. We make the test one-tail because the alternative hypothesis is that the IQ for PKU children is less than that of children in the general population. A value of $\beta = 0.10$ implies that the power is $1 - \beta = 0.90$. We first calculate the standardized distance

$$\Delta = \frac{|94 - 100|}{15} = \frac{6}{15} = 0.40.$$

Then, $z_{1-0.05} = z_{0.95} = 1.645$ and $z_{1-0.10} = z_{0.90} = 1.28$. Hence,

$$n = \frac{(1.645 + 1.28)^2}{(0.40)^2} = 53.5.$$

Rounding up, we estimate that it will take a sample of 54 observations to detect a difference of $100 - 94 = 6$ IQ points (or greater) with probabilities of Type I and Type II errors as specified.

If the variance is not known, and estimated by s^2, say $s^2 = 15^2$, then statistical tables (not included in this text) indicate the sample size is 55; not much higher than the 54 we calculated. A summary outline for calculating sample sizes is given in Figure 5.6.   □

There is something artificial and circular about all of these calculations. If the difference Δ is known, there is no need to perform an experiment to estimate the difference. Calculations of this type are used primarily to make the researcher aware of the kinds of differences that can be detected. Often, a calculation of this type will convince a researcher *not* to carry out a piece of research; or at least, to think very carefully about the possible ways of increasing precision, perhaps even contemplating a radically different attack on the problem. In addition, the size of a sample may be limited by considerations such as cost, recruitment rate, or time constraints beyond control of the investigations.

In the next chapter, we consider questions of sample size for discrete variables.

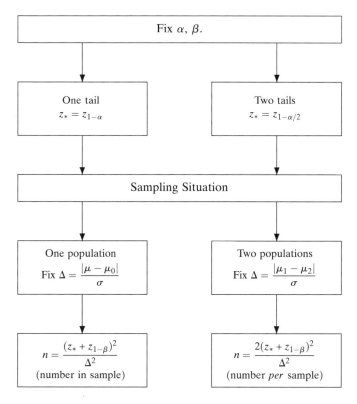

Comments:

1. In the case of two populations, if $\sigma_1^2 \neq \sigma_2^2$, define $\sigma^2 = (\sigma_1^2 + \sigma_2^2)/2$ and proceed as before.

2. If σ is to be estimated from the data, add to the calculated values the following values for an approximate sample size:

		One population	Two populations
One tail	$\alpha = 0.05$	Add 2	Add 1
	$\alpha = 0.01$	Add 4	Add 2
Two tails	$\alpha = 0.05$	Add 2	Add 1
	$\alpha = 0.01$	Add 3	Add 2

Figure 5.6. Sample size calculations for measurement data.

NOTES

Note 5.1: Inference by Means of Pivotal Variables: Some Comments

1. The problem of finding pivotal variables is a task for statisticians. Part of the problem is that such variables are not always unique. For example, when working with the normal distribution, why not use the sample median rather than the sample

mean? After all, the median is admittedly more robust. However, the variance of the sample median is larger than that of the sample mean, so that a less precise probabilistic statement would result.

2. The principal advantage of using the pivotal variable approach is that it gives you a unified picture of a great number of procedures you will need.

3. There is a philosophical problem about the interpretation of a confidence interval. For example, consider the probability inequality

$$P[-1.96 \leq Z \leq 1.96] = 0.95,$$

which leads to a 95% confidence interval for the mean of a normal population on the basis of a random sample of size n,

$$P[\overline{Y} - 1.96\sigma/\sqrt{n} \leq \mu \leq \overline{Y} + 1.96\sigma/\sqrt{n}] = 0.95.$$

It is argued that once $\overline{Y}$ is observed, *this* interval either covers the mean or not, that is, P is either zero or one. One answer is that probabilities are not associated with particular events—whether they have occured or may occur at some future time—but with a population of events. For this reason, we say *after the fact* that we are 95% confident the mean is in the interval; *not* that the probability is 0.95 the mean is in the interval.

4. Given two possible values for a parameter, which one will be designated as the null hypothesis value and which one as the alternative hypothesis value in a hypothesis testing situation? If nothing else is given, then the designation will be arbitrary. Usually, there are at least four considerations in designating the "null value" of a parameter:

   a. Often, the "null value" of the parameter permits the calculation of a p-value. For example, if there are two hypotheses, $\mu = \mu_0$ and $\mu \neq \mu_0$, only under $\mu = \mu_0$ can we calculate the probability of an occurrence of the observed value or a more extreme value.

   b. Past experience or previous work may suggest a specified value. The new experimentation or treatment then has a purpose, rejection of the previously established value, or assessment of the magnitude of the change.

   c. Occam's Razor can be appealed to. It states: "Do not multiply hypotheses beyond necessity," meaning in this context that we usually start from the value of a parameter we would assume if no new data were available or to be produced.

   d. Often, the null hypothesis is a "straw man" we hope to reject, for example, a new drug has the same effect as a placebo.

5. Sometimes it is argued that the smaller the p-value, the stronger the treatment effect. You now will recognize that this cannot be asserted baldly. Consider the two-sample t-test. A p-value associated with this test will depend on the quantities $(\mu_1 - \mu_2)$, s_p, n_1, and n_2. Thus, differences in p-values between two experiments may simply reflect differences in sample size or differences in background variability (as measured by s_p).

Note 5.2: Additional Notes on the t-Test

1. Heterogeneous variances in the two-sample t-test. Suppose the assumption of homogeneity of variances in the two-sample t-test is not tenable. What can be done? At least three avenues are open:

   a. Use an approximation to the t procedure;
   b. Transform the data;
   c. Use another test, such as a nonparametric test.

With respect to the last point, alternative approaches are discussed in Chapter 8. With respect to the first point, one possibility is to rank the observations from smallest to largest (disregarding group membership) and then carry out the t-test on the ranks. This is a surprisingly good *test*, but does not allow us to estimate the magnitude of the difference between the two groups. See Conover and Iman [1981] for an interesting discussion and Thompson [1991] for some cautions. Another approach adjusts the degrees of freedom of the two-sample t-test. The procedure is as follows: let $Y_1 \sim N(\mu_1, \sigma_1^2)$ and $Y_2 \sim N(\mu_2, \sigma_2^2)$, and samples of size n_1 and n_2 are taken, respectively. The variable

$$\frac{(\overline{Y}_1 - \overline{Y}_2) - (\mu_1 - \mu_2)}{\sqrt{\sigma_1^2/n_1 + \sigma_2^2/n_2}}$$

has a standard normal distribution. However, the analogous quantity with the population variances σ_1^2 and σ_2^2 replaced by the sample variances s_1^2 and s_2^2 does not have a t-distribution. The problem of finding the distribution of this quantity is known as the Behrens–Fisher problem. The solution to the problem is still controversial. One possible approach is to treat the statistic

$$\frac{(\overline{Y}_1 - \overline{Y}_2) - (\mu_1 - \mu_2)}{\sqrt{s_1^2/n_1 + s_2^2/n_2}}$$

as a standard normal deviate for $n_1 + n_2 > 30$. As argued before, this is apt to be a reasonable approximation. A second approach adjusts the degrees of freedom of this statistic in relation to the extent of dissimilarity of the two sample variances. The t-table is entered not with $n_1 + n_2 - 2$ degrees of freedom, but with

$$\text{Degrees of freedom} = \frac{\left(s_1^2/n_1 + s_2^2/n_2\right)^2}{\dfrac{(s_1^2/n_1)^2}{n_1 + 1} + \dfrac{(s_2^2/n_2)^2}{n_2 + 1}} - 2.$$

This value need not be an integer, and the critical t-value may have to be obtained by linear interpolation, or conservatively rounding down the degrees of freedom.

2. The two-sample t-test, and design of experiments. Given that a group has to be divided into two subgroups, the arrangement that minimizes the standard error of the difference is that of equal sample sizes in each group when there is a common σ^2. To illustrate, suppose ten objects are to be partitioned into two groups; consider the multiplier $\sqrt{1/n_1 + 1/n_2}$ which determines the relative size of the standard errors.

n_1	n_2	$\sqrt{1/n_1 + 1/n_2}$
5	5	0.63
6	4	0.65
7	3	0.69
8	2	0.79

This listing indicates that small deviations from a $5:5$ ratio do not affect the multiplier very much. It is sometimes claimed that sample sizes must be equal for a valid t-test: except for giving the smallest standard error of the difference there is no such constraint.

3. The "wrong" t-test. What is the effect of carrying out a two-sample t-test on paired data, that is, data that should have been analyzed by a paired t-test? Usually, the level of significance is reduced. On the one hand, the degrees of freedom are *increased* from $(n-1)$, assuming n pairs of observations, to $2(n-1)$, but at the same time, additional variation is introduced, so that the standard error of the difference is now larger. In any event the assumption of statistical independence between "groups" is usually inappropriate.

4. Robustness of the t-test. The t-test tends to be sensitive to "outliers." An "outlier" is an unusually small or large value of a variable. Chapter 8 discusses other methods of analysis. As a matter of routine, you should always graph the data in some way. A simple stem-and-leaf diagram will reveal much about the structure of the data.

Note 5.3: Relationships and Characteristics of the Fixed Distributions in this Chapter

We have already suggested some relationships between the fixed distributions. The connection is more remarkable yet, and illustrates the fundamental role of the normal distribution. The basic connection is between the standard normal and the chi-square distribution. Suppose we randomly draw ten independent values from a standard normal distribution, square each value and sum them. This sum is a random variable. What is its sampling distribution? It turns out to be chi-square with ten degrees of freedom. Using notation, let $Z_1, Z_2, \ldots, Z_{10}$ be the values of Z obtained in drawings one to ten. Then, $Z_1^2 + \cdots + Z_{10}^2$ has a chi-square distribution with ten degrees of freedom: $\chi_{10}^2 = Z_1^2 + \cdots + Z_{10}^2$. This generalizes the special case $\chi_1^2 = Z^2$.

The second connection is between the F-distribution and the chi-square distribution. Suppose we have two independent chi-square random variables with ν_1 and ν_2 degrees of freedom. The ratio

$$\frac{\chi_{\nu_1}^2 / \nu_1}{\chi_{\nu_2}^2 / \nu_2} = F_{\nu_1, \nu_2}$$

has an F-distribution with ν_1 and ν_2 degrees of freedom. Finally, the square of a t-variable with ν degrees of freedom is $F_{1,\nu}$. Summarizing:

$$\chi_\nu^2 = \sum_{i=1}^{\nu} Z_i^2, \quad t_\nu^2 = F_{1,\nu} = \frac{\chi_1^2/1}{\chi_\nu^2/\nu}.$$

A special case connects all four pivotal variables:

$$Z^2 = t_\infty^2 = \chi_1^2 = F_{1,\infty}.$$

Thus, given the F-table, all the other tables can be generated from it.

For completeness, we now summarize the mean and variance of the four fixed distributions:

Distribution	Symbol	Mean	Variance	
Normal	Z	0	1	
Student t	t_ν	0	$\dfrac{\nu}{\nu - 2}$	$(\nu > 2)$
Chi-square	χ_ν^2	ν	2ν	
Fisher's F	F_{ν_1,ν_2}	$\dfrac{\nu_2}{\nu_2 - 2}$	$\dfrac{2\nu_2^2(\nu_1 + \nu_2 - 2)}{\nu_1(\nu_2 - 2)^2(\nu_2 - 4)}$	$(\nu_2 > 4)$

Note 5.4: One-sided Tests and One-sided Confidence Intervals

Corresponding to one-sided (one-tailed) tests are one-sided confidence intervals. A one-sided confidence interval is derived from a pivotal quantity in the same was as a two-sided confidence interval. For example, in the case of a one-sample t-test, a pivotal equation is

$$P\left[-\infty \le \frac{\bar{x} - \mu}{s/\sqrt{n}} \le t_{n-1,1-\alpha} \right] = 1 - \alpha.$$

Solving for μ produces a $100(1 - \alpha)\%$ *upper* one-sided confidence interval for μ: $(\bar{x} - t_{n-1,1-\alpha} s/\sqrt{n}, \infty)$. Similar intervals can be constructed for all the pivotal variables.

PROBLEMS

1. Rickman et al. [1974] made a study of changes in serum cholesterol and trigly-ceride levels of subjects following the "Stillman diet." The diet consists primarily of protein and animal fats, restricting carbohydrate intake. The subjects followed the diet with length of time varying from 3 to 17 days. The mean cholesterol level increased significantly from 215 mg/100 ml at baseline to 248 mg/100 ml at the end of the diet. In this problem, we deal with the triglyceride level.

Subject	Days on Diet	Weight (in kg) Initial	Final	Triglyceride (in mg/100 ml) Baseline	Final
1	10	54.6	49.6	159	194
2	11	56.4	52.8	93	122
3	17	58.6	55.9	130	158
4	4	55.9	54.6	174	154
5	9	60.0	56.7	148	93
6	6	57.3	55.5	148	90
7	3	62.7	59.6	85	101
8	6	63.6	59.6	180	99
9	4	71.4	69.1	92	183
10	4	72.7	70.5	89	82
11	4	49.6	47.1	204	100
12	7	78.2	75.0	182	104
13	8	55.9	53.2	110	72
14	7	71.8	68.6	88	108
15	7	71.8	66.8	134	110
16	14	70.5	66.8	84	81

a. Make a stem-and-leaf diagram of the *changes* in triglyceride levels.

b. Calculate the average change in triglyceride level. Calculate the standard error of the difference.

c. Test the significance of the average change.

d. Construct a 90% confidence interval on the difference.

e. The authors indicate that subjects (5, 6), (7, 8), (9, 10) and (15, 16) were "repeaters"; that is, the same subjects who followed the diet for two sequences. Do you think it is reasonable to include their data on the "second time around" with that of the other subjects? Supposing not, how would you now analyze the data? Carry out the analysis. Does it change your conclusions?

2. The following data are from Dobson *et al.* [1976]. Thirty-six patients with a confirmed diagnosis of phenylketonuria (PKU) were identified and placed on dietary therapy before reaching 121 days of age. The children were tested for IQ (Stanford–Binet) between the ages of four and six; subsequently, their normal siblings of closest age were also tested with the Stanford–Binet. The following are the first 15 pairs listed in the paper:

Pair	1	2	3	4	5	6	7	8
IQ of PKU case	89	98	116	67	128	81	96	116
IQ of sibling	77	110	94	91	122	94	121	114

	9	10	11	12	13	14	15
IQ of PKU case	110	90	76	71	100	108	74
IQ of sibling	88	91	99	93	104	102	82

a. State a suitable null and an alternative hypotheses with regard to these data.

  b. Test the null hypothesis.

  c. State your conclusions.

  d. What are your assumptions?

  e. Discuss the concept of power with respect to this set of data using the fact that PKU invariably led to mental retardation until the cause was found and treatment consisting of restricted diet was instituted.

  f. The mean difference (PKU case – sibling) in IQ for the full 36 pairs was -5.25; the standard deviation of the difference was 13.18. Test the hypothesis of no difference in IQ for this whole set of data.

3. Data by Mazze et al. [1971] deals with the pre-operative and post-operative creatinine clearance (ml/min) of six patients anesthetized by halothane:

	Patient					
	1	**2**	**3**	**4**	**5**	**6**
Preoperative	110	101	61	73	143	118
Postoperative	149	105	162	93	143	100

  a. Why is the paired t-test preferable to the two-sample t-test in this case?

  b. Carry out the paired t-test and test the significance of the difference.

  c. What is the model for your analysis?

  d. Set up a 99% confidence interval on the difference.

  e. Graph the data by plotting the pairs of values for each patient.

4. Some of the physiological effects of alcohol are well known. A paper by Squires *et al.* [1978] assessed the acute effects of alcohol on auditory brainstem potentials in humans. Six volunteers (including the three authors) participated in the study. The latency (delay) in response to an auditory stimulus was measured before and after an intoxicating dose of alcohol. Seven different peak responses were identified. In this exercise, we only discuss latency peak no. 3. The measurements of the latency of peak (in milliseconds after the stimulus onset) in the six subjects were as follows:

	Latency of Peak					
	1	**2**	**3**	**4**	**5**	**6**
Before alcohol	3.85	3.81	3.60	3.68	3.78	3.83
After alcohol	3.82	3.95	3.80	3.87	3.88	3.94

  a. Test the significance of the difference at the 0.05 level.

  b. Calculate the p-value associated with the observed result.

  c. Is your p-value based on a one-tail test or a two-tail test? Why?

  d. As in the previous problem, graph these data and state your conclusion.

e. Carry out an (incorrect) two-sample test and state your conclusions.

f. Using the observed sample variances s_1^2 and s_2^2 associated with the set of readings before and after, calculate the variance of the difference *assuming* independence (call this Variance 1). How does this value compare with the variance of the difference calculated in the first part of the problem? (Call this Variance 2.) Why do you suppose Variance 1 is so much bigger than Variance 2? The *average* of the differences is the same as the difference in the averages. Show this. Hence, the two-sample t-test differed from the paired t-test only in the divisor. Which of the two tests in more powerful in this case, that is, declares a difference significant when in fact there is one?

5. The following data from Schecter *et al.* [1973] deals with sodium chloride preference as related to hypertension. Two groups, 12 normal and 10 hypertensive subjects, were isolated for a week and compared with respect to Na^+ intake. The following are the average daily Na^+ intakes:

Normal	10.2	2.2	0.0	2.6	0.0	43.1
Hypertensive	92.8	54.8	51.6	61.7	250.8	84.5

Normal	45.8	63.6	1.8	0.0	3.7	0.0
Hypertensive	34.7	62.2	11.0	39.1		

a. Compare the average daily Na^+ intake of the hypertensive subjects with that of the normal volunteers by means of an appropriate t-test.

b. State your assumptions.

c. Assuming that the population variances are not homogenous, carry out an appropriate t-test (see Note 5.2).

6. Kapitulnick *et al.* [1976] compared the metabolism of a drug, zoxazolamine, in placentas from 13 women who smoked during pregnancy and 11 who did not. The purpose of the study was to investigate the presence of the drug as a possible proxy for the rate at which benzo[a]pyrene (a byproduct of cigarette smoke) is metabolized. The following data were obtained in the measurement of zoxazolamine hydroxylase production (nmol $3H_2O$ formed/g/h):

Nonsmoker	0.18	0.36	0.24	0.50	0.42	0.36	0.50
Smoker	0.66	0.60	0.96	1.37	1.51	3.56	3.36

Nonsmoker	0.60	0.56	0.36	0.68		
Smoker	4.86	7.50	9.00	10.08	14.76	16.50

a. Calculate the sample mean and standard deviation for each of the two groups.

b. Test the assumption that the two sample variances came from a population with the same variance.

c. Carry out the t-test using the approximation to the t procedure discussed in Note 5.2. What are your conclusions?

d. Suppose we agree that the variability (as measured by the standard deviations) is proportional to the level of the response. Statistical theory then suggests that the logarithms of the responses should have roughly the same variability. Take logarithms of the data and test, once more, the homogeneity of the variances.

7. Sometime you may be asked to do a two-sample t-test knowing only the mean, standard deviation, and sample sizes. A paper by Holtzman et al. [1975] illustrates the problem. This paper dealt with terminating the phenylalanine-restricted diet in 4-year-old children with phenylketonuria (PKU). The purpose of the diet is to reduce the phenylalanine level. A high level is associated with mental retardation. After obtaining informed consent, eligible children of 4 years of age were randomly divided into two groups. Children in one group had their restricted diet terminated while children in the other group were continued on the restricted diet. At 6 years of age, the phenylalanine levels were tested in all children, and the following data reported:

	Diet Terminated	Diet Continued
Number of children	5	4
Mean phenylalanine level (mg/dl)	26.9	16.7
Standard deviation	4.1	7.3

   a. State a reasonable null hypothesis and alternative hypothesis.
   b. Calculate the pooled estimate of the variance s_p^2.
   c. Test the null hypothesis of Part (a) of this problem. Is your test one-tail, or two? Why?
   d. Test the hypothesis that the sample variances came from two populations with the same variance.
   e. Construct a 95% confidence interval on the difference in the population phenylalanine levels.
   f. Interpret the interval constructed in Part (e).
   g. "This set of data has little power," someone says. What does it mean? Interpret the implications of a Type II error in this example.
   h. What is the interpretation of a Type I error in this example? Which, in your opinion, is more serious in this example: a Type I error, or a Type II error?
   i. On the basis of this data, what would you recommend to a parent with a 4-year-old PKU child?
   j. Can you think of some additional information that would make the analysis more precise?

8. Several population studies have demonstrated an inverse correlation of Sudden Infant Death Syndrome (SIDS) rate with birthweight. The occurrence of SIDS in one of a pair of twins provides an opportunity to test the hypothesis that birthweight is a major determinant of SIDS. The following set of data collected by Dr. D. R. Peterson of the Department of Epidemiology, University of Washington, consists of the birthweights (in grams) of each of 22 dizygous twins, and each of 19 monozygous twins:

Dizygous Twins		Monozygous Twins	
SID	**Non-SID**	**SID**	**Non-SID**
1474	2098	1701	1956
3657	3119	2580	2438
3005	3515	2750	2807
2041	2126	1956	1843
2325	2211	1871	2041
2296	2750	2296	2183
3430	3402	2268	2495
3515	3232	2070	1673
1956	1701	1786	1843
2098	2410	3175	3572
3204	2892	2495	2778
2381	2608	1956	1588
2892	2693	2296	2183
2920	3232	3232	2778
3005	3005	1446	2268
2268	2325	1559	1304
3260	3686	2835	2892
3260	2778	2495	2353
2155	2552	1559	2466
2835	2693		
2466	1899		
3232	3714		

    a. With respect to the dizygous twins, test the above hypothesis. State the null hypothesis.

    b. Make a similar test on the monozygous twins.

    c. Discuss your conclusions.

9. A pharmaceutical firm claims that a new analgesic drug relieves mild pain under standard conditions for 3 h with a standard deviation of 1 h. Sixteen patients are tested under the same conditions and have an average pain relief of 2.5 h. The hypothesis that the population mean of this sample is also 3 h is to be tested against the hypothesis that the population mean is in fact less than 3 h; $\alpha = 0.5$.

    a. What is an appropriate test?

    b. Set up the appropriate critical region.

    c. State your conclusion.

    d. Suppose the sample size is doubled. State precisely how the non-rejection region for the null hypothesis is changed.

10. Consider Problem 3.9, dealing with the treatment of essential hypertension. Compare treatments A and B by means of an appropriate t-test. Set up a 99% confidence interval on the reduction of blood pressure under treatment B as compared to treatment A.

11. During July and August, 1976, a large number of Legionnaires attending a convention died of mysterious and unknown cause. Epidemiologists talked of "an outbreak of Legionnaires' disease." One possible cause was thought to be toxins; in particular, nickel. Chen *et al.* [1977] examined the nickel levels in the lungs of nine of the cases, and selected nine controls. All specimens were coded by the Center for Disease Control in Atlanta before being examined by the investigators. The data are as follows ($\mu g/100\,g$ dry weight):

Legionnaire cases	65	24	52	86	120	82	399	87	139
Control cases	12	10	31	6	5	5	29	9	12

Note that there was no attempt to match cases and controls.

a. State a suitable null hypothesis and test it.

b. We now know that Legionnaires' disease is caused by a bacterium, genus *Legionella*, of which there are several species. How would you explain the "significant" results obtained in part (a)? (Chen *et al.* [1977] consider various explanations also.)

12. Review Note 5.3. Generate a few values for the normal, *t*, and chi-square tables from the *F*-table.

13. It is claimed that a new drug treatment can substantially reduce blood pressure. For purposes of this exercise, assume that only diastolic blood pressure is considered. A certain population of hypertensive patients has a mean blood pressure of 96 mmHg. The standard deviation of diastolic blood pressure (variability from subject to subject) is 12 mmHg. To be biologically meaningful, the new drug treatment should lower the blood pressure to at least 90 mmHg. A random sample of patients from the hypertensive population will be treated with the new drug.

a. Assuming $\alpha = 0.05$ and $\beta = 0.05$, calculate the sample size required to demonstrate the specified effect.

b. Considering the labile nature of blood pressure, it might be argued that any "treatment effect" will merely be a "put-on-study effect." So, the experiment is redesigned to consider two random samples from the hypertensive population, one of which will receive the new treatment, and the other, a placebo. Assuming the same specifications as above, what is the required sample size per group?

c. As mentioned above, blood pressure readings are notoriously variable. Suppose that a subject's diastolic blood pressure varies randomly from measurement period to measurement period with a standard deviation of 4 mmHg. Assuming that measurement variability is independent of subject-to-subject variability, what is the overall variance or the total variability in the population? Recalculate the sample sizes for the situation described in (a) and (b).

d. Suppose that the *change* in blood pressure from baseline is used. Suppose the standard deviation of the change is 6 mmHg. How will this change the sample sizes of (a) and (b)?

14. In a paper in the *New England Journal of Medicine*, Rodeheffer *et al.* [1983] assessed the effect of a medication, nifedipine, on the number of painful attacks in patients with Raynaud's phenomenon. This phenomenon causes severe digital pain and functional disability, particularly in patients with underlying connective tissue disease. The drug causes "vascular smooth-muscle relaxation and relief of arterial vasospasm." In this study, 15 patients were selected and randomly assigned to one of two treatment sequences: placebo–nifedipine, or nifedipine–placebo. The data in Table 5.5 were obtained.

a. Why were patients *randomly* assigned to one of the two sequences? What are the advantages?

b. The data of interest are in the columns marked "Placebo" and "Nifedipine" (columns 7 and 9). State a suitable null hypothesis and alternative hypothesis for these data. Justify your choices. Test the significance of the difference in total number of attacks in two weeks on placebo with that of treatment. Use a *t*-test on the differences in the response. Calculate the *p*-value.

c. Construct a 95% confidence interval for the difference. State your conclusions.

d. Make a scatter plot of the placebo response (*x*-axis) *vs.* the nifedipine response (*y*-axis). If there was no significant difference between the treatments, about what line should the observations be scattered?

e. Suppose a statistician considers only the placebo readings and calculates a 95% confidence interval on the population mean. Similarly, the statistician calculates a 95% confidence interval on the nifedipine mean. A graph is made to see if the intervals overlap. Do this for these data. Compare your results with that of part (c). Is there a contradiction? Explain.

f. One way to get rid of outliers is to carry out the following procedure: take the differences of the data in columns 7 (placebo) and 9 (nifedipine), and rank them disregarding the signs of the differences. Put the sign of the difference on the rank. Now, carry out a paired *t*-test on the signed ranks. What would be an appropriate null hypothesis? What would be an appropriate alternative hypothesis? Name one advantages and one disadvantage of this procedure. (It is one form of a nonparametric test discussed in detail in Chapter 8.)

15. Rush *et al.* [1973] reported the design of a randomized controlled trial of nutritional supplementation in pregnancy. The trial was to be conducted in a poor American black population. The variable of interest was the birthweight of infants born to study participants; study design called for the random allocation of participants to one of three treatment groups. The authors then state: "The required size of the treatment groups was calculated from the following statistics: the standard deviation of birthweight ... is of the order of 500 g. An increment of 120 g in birthweight was arbitrarily taken to constitute a biologically meaningful gain. Given an expected difference between subjects and controls of 120 g, the required sample size for each group, in order to have a 5% risk of falsely rejecting, and a 20% risk of falsely accepting the null hypothesis, is about 320."

a. What are the values for α and β?

b. What is the estimate of Δ, the standardized difference?

c. The wording in the paper suggests that sample size calculations are based on a two-sample test. Is the test one-tail or two-tail?

Table 5.4. Effect of Nifedipine on Patients with Raynaud's Phenomenon. Data from Rodeheffer *et al.* [1983].

Case	Age(yr)/ Sex	Diagnosis[a]	History of Digital Ulcer	ANA[d]	Duration of Raynaud's Phenomenon (Years)	Placebo		Nifedipine	
						Total Number of Attacks in 2 weeks	Patient Assessment of Therapy[b]	Total Number of Attacks in 2 weeks	Patient Assessment of Therapy[b]
1	49/F	R[c]	No	20	4	15	0	0	3+
2	20/F	R	No	Neg	3	3	1+	5	0
3	23/F	R	No	Neg	8	14	2+	6	2+
4	33/F	R	No	640	5	6	0	0	3+
5	31/F	R[c]	No	2560	2	12	0	2	3+
6	52/F	PSS	No	320	3	6	1+	1	0
7	45/M	PSS[c]	Yes	320	4	3	1+	2	2+
8	49/F	PSS	Yes	320	4	22	0	30	1+
9	29/M	PSS	Yes	1280	7	15	0	14	1+
10	33/F	PSS[c]	No	2560	9	11	1+	5	1+
11	36/F	PSS	Yes	2560	13	7	2+	2	3+
12	33/F	PSS[c]	Yes	2560	11	12	0	4	2+
13	39/F	PSS	No	320	6	45	0	45	0
14	39/M	PSS	Yes	80	6	14	1+	15	2+
15	32/F	SLE[c]	Yes	1280	5	35	1+	31	2+

[a]R denotes Raynaud's phenomenon without systemic disease; PSS Raynaud's phenomenon with progressive systemic sclerosis; and SLE Raynaud's phenomenon with systemic lupus erythematosus (in addition, this patient had cryoglobulinemia).

[b]The Wilcoxon signed rank test, two-tailed, was performed on the patient assessment of placebo versus nifedipine therapy: P = 0.02. Global assessment scale: 1− = worse; 0 = no change; 1+ = minimal improvement; 2+ = moderate improvement; and 3+ = marked improvement.

[c]Previous unsuccessful treatment with prazosin.

[d]Reciprocal of antinuclear antibody titers.

d. Using a one-tail test, verify that the sample size per group is $n = 215$. The number 320 reflects adjustments for losses and, perhaps, "multiple comparisons" since there are three groups (see Chapter 12).

16. This problem deals with the data of Problem 5.14. In column 4 of the tabulated data, patients are divided into those with a history of digital ulcers and those without. We want to compare these two groups. There are seven patients with a history and eight without.

    a. Consider the total number of attacks (in column 9) on the active drug. Carry out a two-sample t-test. Comparing the group with a digital ulcer history with the group without this history. State your assumptions and conclusions.

    b. Rank all the observations in column 9, then separate the ranks into the two groups defined in Part (a). Now carry out a two-sample t-test on the ranks. Compare your conclusions with those of Part (b). Name an advantage to this approach. Name a disadvantage to this approach.

    c. We now do the following: take the difference between the "Placebo" and "Nifedipine" columns, and repeat the procedures of Parts (a) and (b). Supposing the conclusions of Part (a) are not the same as those in this part, how would you interpret such discrepancies?

    d. The test carried out in Part (c) is often called a *test for interaction*. Why do you suppose this is so?

REFERENCES

Bednarek, F. J. and Roloff, D. W. [1976]. Treatment of apnea of prematurity with aminophylline. *Pediatrics*, **58:** 335–339. Used with permission.

Chen, J. R., Francisco, R. B., and Miller, T. E. [1977]. Legionnaires' disease: nickel levels. *Science*, **196:** 906–908. Copyright © 1977 by the AAAS

Conover, W. J. and Iman, R. L. [1981]. Rank transformations as a bridge between parametric and nonparametric statistics. *American Statistician*, **35:** 124–129.

Dobson, J.C., Kushida, E., Williamson, M., and Friedman, E.G. [1976]. Intellectual performance of 36 phenylketonuria patients and their non-affected siblings. *Pediatrics*, **58:** 53–58. Used with permission.

Holtzman, N. A., Welcher, D. M., and Mellits, E. D. [1975]. Termination of restricted diet in children with phenylketonuria: a randomized controlled study. *New England Journal of Medicine*, **293:** 1121–1124.

Kapitulnik, J., Levin, W., Poppers, J., Tomaszewski, J. E., Jerina, D. M., and Conney, A. H. [1976]. Comparison of the hydroxylation of zoxazolamine and benzo[a]pyrene in human placenta: effect of cigarette smoking. *Clinical Pharmaceuticals and Therapeutics*, **20:** 557–564.

Mazze, R. I., Shue, G. L., and Jackson, S. H. [1971]. Renal dysfunction associated with methoxyflurane anesthesia. *Journal of the American Medical Association*, **216:** 278–288. Copyright © 1971 by The American Medical Association.

Rickman, R., Mitchell, N., Dingman, J., and Dalen, J. E. [1974]. Changes in serum cholesterol during the Stillman Diet. *Journal of the American Medical Association*, **228:** 54–58. Copyright © 1974 by The American Medical Association.

Rodeheffer, R. J., Romner, J. A., Wigley, F., and Smith, C. R. [1983]. Controlled double-blind trial of Nifedipine in the treatment of Raynaud's phenomenon. *New England Journal of Medicine*, **308:** 880–883.

Rush, D., Stein, Z., and Susser, M. [1973]. The rationale for, and design of, a randomized controlled trial of nutritional supplementation in pregnancy. *Nutritional Reports International*, **7:** 547–553. Used with permission of the publisher, Butter–Heinemann.

Schechter, P. J., Horwitz, D., and Henkin, R. I. [1973]. Sodium chloride preference in essential hypertension. *Journal of the American Medical Association*, **225:** 1311–1315. Copyright © 1973 by The American Medical Association.

Squires, K. C., Chen, N. S., and Starr, A. [1978]. Acute effects of alcohol on auditory brainstem potentials in humans. *Science*, **201:** 174–176.

Thompson, G. L. [1991]. A unified approach to rank tests for multivariate and repeated measures designs. *Journal of the American Statistical Association*, **86:** 410–419.

Zelazo, P. R., Zelazo, N. A., and Kolb, S. [1972]. "Walking" in the newborn. *Science*, **176:** 314–315.

CHAPTER 6

Counting Data

6.1 INTRODUCTION

From the previous chapters, recall the basic ideas of statistics. *Descriptive statistics* present data, usually in summary form. Appropriate *models* concisely describe data. The model *parameters* are *estimated* from the data. *Standard errors* and *confidence intervals* quantify the precision of estimates. Scientific hypotheses may be tested. A *formal hypothesis test* involves four things: (1) planning an experiment, or recognizing an opportunity, to collect appropriate data; (2) selecting a *significance level* and *critical region*; (3) collecting the data; and (4) rejecting the *null hypothesis* being tested if the value of the test statistic falls into the critical region. A less formal approach is to compute the *p-value*, a measure of how plausibly the data agrees with the null hypothesis under study. The remainder of this text shows you how to apply these concepts in different situations.

Our position will be illustrated by an analogy with medicine. Suppose that someone outlines for you the primary means of therapy. You are told that drugs have many uses involving the biochemistry of our bodies. You learn of surgical interventions. Further, you are introduced to psychological treatment, physical therapy, radiation, and prosthetic devices as well as dietary measures. Are you ready to practice medicine? No. Before beginning to practice medicine you would need specific knowledge in three areas:

1. You must learn a large number of specific treatments. It is not enough to know that drug therapy is useful; you must know specific drugs.

2. You need the ability to diagnose diseases. It is not enough to know that different diseases exist; faced with a patient you need to diagnose that patient's individual malady.

3. Once the diagnosis is made, you apply the treatment appropriate for the diagnosis.

If any of the three elements is weak or lacking, you would be giving less than optimal treatment.

You have analogous requirements to become qualified as a practicing applied statistician.

1. You need a large number of methods of analyzing data. As we proceed you will learn many different methods of summarizing data. You will learn different methods of statistical testing. As you study this text, you will have more data analytic methods at your disposal.

2. As the physician diagnoses disease, you must know which statistical model may be appropriate for your data. Physicians order more tests when in doubt about a diagnosis; you will learn statistical tests to see if a model fits data. Physicians learn to organize signs and symptoms in searching for a diagnosis; you will learn to examine characteristics of data that suggest appropriate models.

3. When you have selected (diagnosed) the model for a data set, you proceed to describe and analyze the data using the appropriate methods.

The brief description of medicine above is over simplified because: (1) often a diagnosis is uncertain; (2) more than one treatment may be tried to see which treatment works best; (3) occasionally a patient appears whose symptoms defy diagnosis; (4) for certain diagnoses there is no treatment; and finally, (5) one concludes that medicine is an art based upon science.

There are statistical analogies: (1) there may be more than one possible model for a data set; (2) different methods of analyzing a data set may be tried; (3) for some data sets no analysis you know will appear appropriate; (4) some data sets have no useful analysis (this may be true, for example, if there are few observations or a large amount of missing data); and finally, (5) statistical analysis is an art based upon science. The science underlying statistical analysis is probability theory and associated models. Data analysis is an interactive procedure using steps based on statistical theory. You will see examples of this as we progress.

No analogy is perfect. Often the statistician may choose the diagnosis before the disease! By designing and carrying out careful data collection you may assure being faced with a disease that you can treat.

No one wishes to be treated by individuals who have only "book learning." To become proficient at data analysis you *must* gain hands-on experience. Work the problems. If possible, gain access to computer software for statistical analysis. Critically read literature in your field for statistical content.

Most importantly, learn your limitations. Do not force data into an inappropriate method of analysis. Know when to consult a specialist.

6.2 COUNTING DATA

Throughout recorded history people have been able to count. The word *statistics* comes from the Latin word for "state"; early statistics were counts used for the purposes of the state. Censuses were conducted for military and taxation purposes. Modern statistics is often dated from the 1662 comments on the Bills of Mortality in London. The Bills of Mortality counted the number of deaths due to each cause. John Graunt [1662] noticed patterns of regularity in the Bills of Mortality (see Section 3.3.1). Such vital statistics are important today for assessing the public health. This chapter returns to the origin of statistics by dealing with data that arise by counting the number of occurrences of some event.

Count data leads to many different models. The following sections present examples of count data. The different types of count data will each be presented in three steps. First, you learn to recognize count data that fit a particular model. (This is the diagnosis phase.) Second, you examine the model to be used. (You learn about the illness.) Third, you learn the methods of analyzing data using the model. (At this stage you learn how to treat the disease.)

6.3 BINOMIAL RANDOM VARIABLES

6.3.1 Recognizing Binomial Random Variables

Four conditions characterize binomial data:

1. A response or trait takes on one and only one of two possibilities. Such a response is called a binary response. Examples are:
   a. In a survey of the health system, individuals are asked whether or not they have hospitalization insurance.
   b. Blood samples are tested for the presence or absence of an antigen.
   c. Rats fed a potential carcinogen are examined for tumors.
   d. Individuals are classified as having or not having cleft lip.
   e. Injection of a compound does or does not cause cardiac arrhythmia in dogs.
   f. Newborn children are classified as having or not having Down's Syndrome.

2. The response is observed a known number of times. Each observation of the response is sometimes called a *Bernoulli trial*. In (a) above, the number of trials is the number of individuals questioned. In (b), each blood sample is a trial. Each newborn child constitutes a trial in (f).

3. The chance, or probability, that a particular outcome occurs is the same for each trial. In a survey such as (a), people are sampled at random from the population. Since each person has the same chance of being interviewed, the probability that the person has hospitalization insurance is the same in each case. In a laboratory receiving blood samples, the samples could be considered to have the same probability of having an antigen *if* the samples arise from members of a population who submit tests when "randomly" seeking medical care. The samples would not have the same probability if batches of samples arrive from different environments; for example, from school children, a military base, and a retirement home.

4. The outcome of one trial must not be influenced by the outcome of the other trials. Using the terminology of Chapter 5, the trials outcomes are independent random variables. In (b), the trials would not be independent if there was contamination between separate blood samples. The newborn children of (f) might be considered independent trials for the occurrence of Down's Syndrome if each child has different parents. If multiple births are in the data set, the assumption of independence would not be appropriate.

We illustrate and reinforce these ideas by examples that may be modeled by the binomial distribution.

Example 6.1. Weber *et al.* [1976] studied the irritating effects of cigarette smoke. Sixty subjects sat, in groups of five to six, in a climatic chamber 30 m². Tobacco smoke was produced by a smoking machine. After 10 cigarettes had been smoked, 47 of the 60 subjects reported that they wished to leave the room.

Let us consider the appropriateness of the binomial model for these data. Condition 1 is satisfied. Each subject was to report whether or not he or she desired to leave the room. The answer gives one of two possibilities: yes or no. Sixty trials are observed to take place (i.e., condition 2 is satisfied).

The third condition requires that each subject have the same probability of "wishing to leave the room." The paper does not explain how the subjects were selected. Perhaps the authors advertised for volunteers. In this case, the subjects might be considered "representative" of a larger population who would volunteer. The probability would be the *unknown* probability a person selected at random from this larger population would wish to leave the room.

As we will see below, the binomial model is often used to make inferences about the unknown probability of an outcome in the "true population." Many would say that an experiment such as this shows that cigarette smoke irritates people. The extension from the ill-defined population of this experiment to humankind in general does *not* rest on this experiment. It must be based upon other formal or informal evidence that humans do have much in common; in particular, one would need to assume that if one portion of humankind is irritated by cigarette smoke, so will other segments. Do you think such inferences are reasonable?

The fourth condition needed is that the trials are independent variables. The authors report in detail that the room was cleared of all smoke between uses of the climatic chamber. There should not be a "carry-over" effect here. Recall that subjects were tested in groups of five or six. How do you think one person's response would be changed if another person were coughing? Rubbing their eyes? Complaining? It seems possible that condition four is not fulfilled; that is, it seems possible that the responses were not independent.

In summary, a binomial model might be used for these data, but with some reservation. The overwhelming majority of data collected on human populations is collected under less than ideal conditions; a subjective evaluation of the worth of an experiment often enters in. ☐

Example 6.2. Karlowski *et al.* [1975] reported on a controlled clinical trial of ascorbic acid (vitamin C) for the common cold. Placebo and vitamin C were randomly assigned to the subjects; the experiment was to be a double-blind trial. It turned out that some subjects were testing their capsules and claimed to know the medication. Of 64 individuals who tested the capsule and guessed at the treatment, 55 were correct. Could such a split arise by chance if testing did not help one to guess correctly?

One thinks of using a binomial model for these data since there is a binary response (correct or incorrect guess) observed on a known number of individuals. Assuming that individuals tested only their own capsules, the guesses should be statistically independent. Finally, if the guesses are "at random," each subject should have the same probability—one-half—of making a correct guess since half the participants receive vitamin C and half a placebo. This binomial model would lead to a test of the hypothesis that the probability of a correct guess was one-half. ☐

Example 6.3. Bucher *et al.* [1976] studied the occurrence of hemolytic disease in newborns resulting from ABO incompatability among the parents. Parents are said to be incompatible if the father has antigens that the mother lacks. This provides the opportunity for production of maternal antibodies from fetal-maternal stimulation. Low-weight immune antibodies that cross the placental barrier apparently cause the disease (Cavalli-Sforza and Bodmer [1971]). The authors reviewed 7464 consecutive infants born at North Carolina Hospital. Of 3584 "Black births," 43 had ABO hemolytic disease. What can be said about the true probability that a Black birth has ABO hemolytic disease?

It seems reasonable to consider the number of ABO hemolytic disease cases to be binomial. The presence of disease among the 3584 trials should be independent (assuming no parents had more than one birth during the period of case recruitment— October, 1965 to March, 1973—and little or no effect from kinship of parents). The births may conceptually be thought of as a sample of the population of "potential" Black births during the given time period at the hospital.   □

6.3.2 The Binomial Model

In speaking about a Bernoulli trial, without reference to a particular example, it is customary to label one outcome as a "success" and the other outcome as a "failure." The mathematical model for the binomial distribution depends upon two parameters: n, the number of trials, and π, the probability of a success in one trial. A binomial random variable, say Y, is the count of the number of successes in the n trials. Of course, Y can only take on the values $0, 1, 2, \ldots, n$. If π, the probability of a success, is large (close to one), then Y, the number of successes, will tend to be large. Conversely, if the probability of success is small (near zero), Y will tend to be small.

In order to do statistical analysis with binomial variables we need the probability distribution of Y. Let k be an integer between 0 and n inclusive. We need to know $P[Y = k]$. In other words, we want the probability of k successes in n independent trials when π is the probability of success. The symbol $b(k; n, \pi)$ will be used to denote this probability. The answer involves the *binomial coefficient*. The binomial coefficient $\binom{n}{k}$ is the number of different ways that k objects may be selected from n objects. (Problem 6.24 helps you to derive the value of $\binom{n}{k}$.) For each positive integer n, n factorial (written $n!$) is defined to be $1 \times 2 \times \cdots \times n$. So $6! = 1 \times 2 \times 3 \times 4 \times 5 \times 6 = 720$. $0!$, zero factorial, is defined to be 1. With this notation the binomial coefficient may be written

$$\binom{n}{k} = \frac{n!}{(n-k)!\,k!} = \frac{n(n-1)\cdots(k+1)}{(n-k)(n-k-1)\cdots 1}. \tag{6.1}$$

Example 6.4. This is illustrated with the following two cases:

1. Of ten residents, three are to be chosen to cover a hospital service on a holiday. In how many ways may the residents be chosen? The answer is:

$$\binom{10}{3} = \frac{10!}{7!\,3!} = \frac{1 \times 2 \times 3 \times 4 \times 5 \times 6 \times 7 \times 8 \times 9 \times 10}{(1 \times 2 \times 3 \times 4 \times 5 \times 6 \times 7)(1 \times 2 \times 3)} = 120.$$

2. Of eight consecutive patients, four are to be assigned to drug A and four to drug B. In how many ways may the assignments be made? Think of the eight

positions as eight objects; we need to choose four for the drug A individuals. The answer is

$$\binom{8}{4} = \frac{8!}{4!\,4!} = \frac{1 \times 2 \times 3 \times 4 \times 5 \times 6 \times 7 \times 8}{(1 \times 2 \times 3 \times 4)(1 \times 2 \times 3 \times 4)} = 70. \qquad \square$$

The binomial probability, $b(k; n, \pi)$ may be written

$$b(k; n, \pi) = \binom{n}{k} \pi^k (1 - \pi)^{n-k}. \tag{6.2}$$

Example 6.5. Ten individuals are treated surgically. For each individual there is a 70% chance of successful surgery (that is, $\pi = 0.7$). What is the probability of only five or fewer successful surgeries?

$$
\begin{aligned}
P[\text{five or fewer successful cases}] &= \\
P[\text{five successful cases}] &+ P[\text{four successful cases}] \\
+ P[\text{three successful cases}] &+ P[\text{two successful cases}] \\
+ P[\text{one successful case}] &+ P[\text{no successful case}] \\
= b(5; 10, 0.7) + b(4; 10, 0.7) &+ b(3; 10, 0.7) + b(2; 10, 0.7) \\
+ b(1; 10, 0.7) &+ b(0; 10, 0.7) \\
= 0.1029 + 0.0368 + 0.0090 &+ 0.0014 + 0.0001 + 0.0000 \\
= 0.1502.
\end{aligned}
$$

[note: actual value is 0.1503, round off error]. $\qquad \square$

The binomial probabilities may be calculated directly, found by a computer program, or looked up in tables. Table A.8 in the Appendix contains binomial probabilities. Note that this table also gives cumulative probabilities. For the example the value is 0.1503 (using $\pi = 0.3$ as described below), rather than 0.1502, due to round off error. If there are k successes in n trials, then there are $n - k$ failures. The probability of a failure is 1 minus the probability of a success. Thus, in n trials the probability of k events with probability π occurring is the same as the probability of $n - k$ events with probability $1 - \pi$ occurring. That is,

$$b(k; n, \pi) = b(n - k; n, 1 - \pi). \tag{6.3}$$

Because of Equation (6.3), tables of binomial probabilities usually only have probabilities for $\pi \leq 0.5$. For values greater than 0.5, Equation (6.3) may be used to give the probability.

The mean and variance of a binomial random variable with parameters π and n are given by

$$E(Y) = n\pi,$$

$$\text{var}(Y) = n\pi(1 - \pi). \tag{6.4}$$

From Equation (6.4), it follows that Y/n has the expected value π.

$$E\left(\frac{Y}{n}\right) = \pi. \tag{6.5}$$

In other words, the proportion of successes in n binomial trials is an unbiased estimate of the probability of success.

6.3.3 Hypothesis Testing for Binomial Variables

The hypothesis testing framework established in Chapter 4 may be used for the binomial distribution. There is one minor complication. The binomial random variable only can take on a finite number of values. Because of this, it may not be possible to find hypothesis tests such that the significance level is precisely some fixed value. If this is the case, we construct regions so that the significance level is close to the desired significance level.

In most situations involving the binomial distribution, the number of trials (n) is known. We consider statistical tests about the true value of π. Let $p = Y/n$. If π is hypothesized to be π_0, an observed value of p close to π_0 reinforces the hypothesis; a value of p differing greatly from π_0 makes the hypothesis seem unlikely.

Procedure 6.1.   To construct a significance test of $H_0 : \pi = \pi_0$ against $H_A : \pi \neq \pi_o$, at significance level α:

1. Find the smallest c such that $P\left[|p - \pi_0| \geq c\right] \leq \alpha$ when H_0 is true.
2. Compute the *actual* significance level of the test; the actual significance level is $P\left[|p - \pi_0| \geq c\right]$.
3. Observe p, call it $\hat{p}$; reject H_0 if $|\hat{p} - \pi_0| \geq c$.

The quantity c is used to determine the critical value (see Definition 4.19), that is, determine the bounds of the rejection region which will be $\pi_0 \pm c$. Equivalently, working in the Y scale, the region is defined by $n\pi_0 \pm nc$.

Example 6.6.   For $n = 10$, we want to construct a test of the null hypothesis $H_0 : \pi = 0.4$ versus the alternative hypothesis $H_A : \pi \neq 0.4$. Thus, we want a two-sided test. The significance level is to be as close to $\alpha = 0.05$ as possible. We work in the $Y = np$ scale. Under H_0, Y has mean $n\pi = (10)(0.4) = 4$. We want to find a value C such that $P[|Y - 4| \geq C]$ is as close to $\alpha = 0.05$ (and less than α) as possible. The quantity C is the distance Y is from the null hypothesis value 4. Using Table A.8, we construct the following table:

C	$4-C$	$C+4$	$P[|Y-4| \geq C] =$	α
6	—	10	0.0001	$= P[Y = 10]$
5	—	9	0.0017	$= P[Y \geq 9]$
4	0	8	0.0183	$= P[Y = 0] + P[Y \geq 8]$
3	1	7	0.1012	$= P[Y \leq 1] + P[Y \geq 7]$
2	2	6	0.3335	$= P[Y \leq 2] + P[Y \geq 6]$
1	3	5	0.7492	$= P[Y \leq 3] + P[Y \geq 5]$

The closest α-value to 0.05 is $\alpha = 0.0183$; the next value is 0.1012. Hence we choose $C = 4$; we reject the null hypothesis $H_0 : n\pi = 4$ if $Y = 0$ or $Y \geq 8$. Equivalently if $p = 0$ or $p \geq 0.8$. Or, in the original formulation, if $|p - 0.4| \geq 0.4$ since $C = 10c$. □

Procedure 6.2. To find the *p-value* for testing the hypothesis $H_0 : \pi = \pi_0$ versus $H_A : \pi \neq \pi_0$:

1. Observe p: $\widehat{p}$ is now fixed, where $\widehat{p} = y/n$.
2. Let $\widetilde{p}$ be a binomial random variable with parameters n and π_0. The *p*-value is $P\left[|\widetilde{p} - \pi_0| \geq |\widehat{p} - \pi_0|\right]$.

Example 6.7. Find the *p*-value for testing $\pi = 0.5$ if $n = 10$ and we observe $p = 0.2$. $|\widetilde{p} - 0.5| \geq |0.2 - 0.5| = 0.3$ only if $\widetilde{p} = 0.0, 0.1, 0.2, 0.8, 0.9$ or 1.0. Thus, the *p*-value is: $0.0010 + 0.0098 + 0.0439 + 0.0439 + 0.0098 + 0.0010 = 0.1094$ using Table A.8. □

The appropriate one-sided hypothesis test and calculation of a one-sided *p*-value is given in Problem 6.25.

6.3.4 Confidence Intervals

Confidence intervals for a binomial proportion can be found by computer or by looking up the confidence limits in a table. Such tables are not included in this text, but are available in any standard handbook of statistical tables, for example, Odeh *et al.* [1977], Owen [1962], and Beyer [1968].

6.3.5 Large Sample Hypothesis Testing

The Central Limit Theorem holds for binomial random variables. If Y is binomial with parameters n and π, then for "large n,"

$$\frac{Y - E(Y)}{\sqrt{\mathrm{var}(Y)}} = \frac{Y - n\pi}{\sqrt{n\pi(1 - \pi)}}$$

has approximately the same probability distribution as an $N(0, 1)$ random variable. Equivalently, since $Y = np$, the quantity $(p - \pi)/\sqrt{\pi(1 - \pi)/n}$ approaches a normal distribution. We will work interchangeably in the p scale or the Y scale. For large n, hypothesis tests and confidence intervals may be formed by using critical values of the standard normal distribution.

The closer π is to $1/2$, the better the normal approximation will be. If $n \leq 50$, it is preferable to use tables for the binomial distribution and hypothesis tests as outlined above. A reasonable rule of thumb is that n is "large" if $n\pi(1 - \pi) \geq 10$.

In using the Central Limit Theorem, we are approximating the distribution of a discrete random variable by the continuous normal distribution. The approximation can be improved by using a *continuity correction*. The normal random variable with continuity correction is given by

$$
Z_c = \begin{cases}
\dfrac{Y - n\pi - 1/2}{\sqrt{n\pi(1 - \pi)}} & , \quad \text{if } Y - n\pi > 1/2, \\[3ex]
\dfrac{Y - n\pi}{\sqrt{n\pi(1 - \pi)}} & , \quad \text{if } |Y - n\pi| \leq 1/2, \\[3ex]
\dfrac{Y - n\pi + 1/2}{\sqrt{n\pi(1 - \pi)}} & , \quad \text{if } Y - n\pi < -1/2.
\end{cases}
$$

For $n\pi(1 - \pi) \geq 100$, or quite large, the factor of $1/2$ is usually ignored.

Procedure 6.3. Let Y be binomial n, π, with a large n. A hypothesis test of $H_0 : \pi = \pi_0$ versus $H_A : \pi \neq \pi_0$ at significance level α is given by computing Z_c with $\pi = \pi_0$. The null hypothesis is rejected if $|Z_c| \geq z_{1-\alpha/2}$.

Example 6.8. In Example 6.2, of the 64 individuals who tested their capsules, 55 guessed the treatment correctly. Could so many individuals have guessed the correct treatment "by chance"? In the example, we saw that chance guessing would correspond to $\pi_0 = 1/2$. At a 5% significance level, is it plausible that $\pi_0 = 1/2$?

As $n\pi_0(1 - \pi_0) = 64 \times 1/2 \times 1/2 = 16$, a large sample approximation is reasonable. $y - n\pi_0 = 55 - 64 \times 1/2 = 23$ so that

$$
Z_c = \frac{Y - n\pi_0 - 1/2}{\sqrt{n\pi_0(1 - \pi_0)}} = \frac{22.5}{\sqrt{64 \times 1/2 \times 1/2}} = 5.625.
$$

As $|Z_c| = 5.625 > 1.96 = z_{0.975}$, the null hypothesis that the correct guessing occurs purely by chance must be rejected.   □

Procedure 6.4. The large-sample two-sided p-value for testing $H_0 : \pi = \pi_0$ versus $H_A : \pi \neq \pi_0$ is given by $2(1 - \Phi(|Z_c|))$. $\Phi(x)$ is the probability an $N(0, 1)$ random variable is less than x. $|Z_c|$ is the absolute value of Z_c.

6.3.6 Large Sample Confidence Intervals

Procedure 6.5. For large n, say $n\widehat{p}(1 - \widehat{p}) \geq 10$, an approximate $100(1 - \alpha)\%$ confidence interval for π is given by

$$
\left(\widehat{p} - z_{1-\alpha/2}\sqrt{\frac{\widehat{p}(1 - \widehat{p})}{n}}, \quad \widehat{p} + z_{1-\alpha/2}\sqrt{\frac{\widehat{p}(1 - \widehat{p})}{n}} \right), \tag{6.6}
$$

where $\widehat{p} = y/n$ is the observed proportion of successes.

Example 6.9. Find a 95% confidence interval for the true fraction of black children having ABO hemolytic disease in the population represented by the data of

Example 6.3. Using the above formula, the confidence interval is

$$\frac{43}{3584} \pm 1.96 \sqrt{\frac{(43/3584)(1 - 43/3584)}{3584}} \quad \text{or} \quad (0.0084, 0.0156). \qquad \square$$

6.4 COMPARING TWO PROPORTIONS

Often one is not interested in only one proportion, but wants to compare two proportions. A Health Services researcher may want to see whether one of two races has a higher percentage of prenatal care. A clinician may wish to discover which of two drugs has a higher proportion of cures. An epidemiologist may be interested in discovering whether women on oral contraceptives have a higher incidence of thrombophlebitis than those not on oral contraceptives. This section considers the statistical methods appropriate for comparing two proportions.

6.4.1 Fisher's Exact Test

Data to estimate two different proportions will arise from observations upon two populations. Call the two sets of observations Sample 1 and Sample 2. Often the data are presented in 2×2 (verbally, "two by two") tables as follows:

	Success	Failure
Sample 1	n_{11}	n_{12}
Sample 2	n_{21}	n_{22}

The first sample has n_{11} successes in $n_{11} + n_{12}$ trials; the second sample has n_{21} successes in $n_{21} + n_{22}$ trials. Often the null hypothesis of interest is that the probability of success in the two populations is the same. Fisher's Exact Test is a test of this hypothesis for small samples.

The test uses the row and column totals. Let $n_{1\bullet}$ denote summation over the second index; that is, $n_{1\bullet} = n_{11} + n_{12}$. Similarly define $n_{2\bullet}$, $n_{\bullet1}$, and $n_{\bullet2}$. Let $n_{\bullet\bullet}$ denote summation over both indices; that is, $n_{\bullet\bullet} = n_{11} + n_{12} + n_{21} + n_{22}$. Writing the table with row and column totals gives:

	Success	Failure	
Sample 1	n_{11}	n_{12}	$n_{1\bullet}$
Sample 2	n_{21}	n_{22}	$n_{2\bullet}$
	$n_{\bullet1}$	$n_{\bullet2}$	$n_{\bullet\bullet}$

Suppose that the probabilities of success in the two populations are the same. Further suppose that we are given the row and column totals, but *not* n_{11}, n_{12}, n_{21}, and n_{22}. What is the probability distribution of n_{11}?

Consider the $n_{\bullet\bullet}$ trials as $n_{\bullet\bullet}$ objects; for example, $n_{1\bullet}$ purple balls and $n_{2\bullet}$ gold balls. Since each trial has the same probability of success any subset of $n_{\bullet1}$ trials (balls) has the same probability of being chosen as any other. Thus, the probability that n_{11} has the value k is the same as the probability that there are k purple balls among $n_{\bullet1}$ balls chosen without replacement from an urn with $n_1.$ purple balls and $n_2.$ gold balls. The probability distribution of n_{11} is called the *hypergeometric* distribution.

The mathematical form of the hypergeometric probability distribution is derived in Problem 6.26. Note 6.1 explains in detail how to use tables for Fisher's Exact Test. Table A.9 has critical values.

Example 6.10. Kennedy *et al.* [1981] consider patients who have undergone coronary artery bypass graft surgery (CABG). CABG takes a saphenous vein from the leg and connects the vein to the aorta, where blood is pumped from the heart, and to a coronary artery, an artery that supplies the heart muscle with blood. The vein is placed beyond a narrowing, or stenosis, in the coronary artery. If the artery would close at the narrowing, the heart muscle would still receive blood. There is, however, some risk to this open heart surgery. Among patients with moderate narrowing (50–74%) of the left main coronary artery emergency cases have a high surgical mortality rate. The question is whether emergency cases have a surgical mortality different from that of nonemergency cases. The in-hospital mortality figures for emergency surgery and other surgery were:

	Discharge Status	
Surgical Priority	**Dead**	**Alive**
Emergency	1	19
Other	7	369

From the hypergeometric distribution, the probability of an observation this extreme is $0.3419 = P[n_{11} \geq 1] = P[n_{11} = 1] + \cdots + P[n_{11} = 8]$. (Values for $n_{..}$ this large are not tabulated and need to be computed directly.) These data do not show any difference beyond that expected by chance. □

Example 6.11. Sudden infant death syndrome (SIDS), or crib death, results in the unexplained death of approximately two of every 1000 infants during their first year of life. In order to study the genetic component of such deaths, Peterson *et al.* [1980] examined sets of twins with at least one SIDS child. If there is a large genetic component, the probability of both twins dying will be larger for identical twin sets than for fraternal twin sets. If there is no genetic component, but only an environmental component, the probabilities should be the same. The following table gives the data.

	SIDS Children	
Type of Twin	**One**	**Both**
Monozygous (identical)	23	1
Dizygous (fraternal)	35	2

The Fisher's exact test one-sided p-value for testing that the probability is higher for monozygous twins is $p = 0.784$. Thus, there is no evidence for a genetic component in these data. □

6.4.2 Large Sample Tests and Confidence Intervals

As mentioned above, in many situations one wishes to compare proportions as estimated by samples from two populations to see if the true population parameters might be equal or if one is larger than the other. Examples of such situations are: a drug and placebo trial comparing the percentage of patients experiencing pain relief; the percentage of rats developing tumors under diets involving different doses of a food additive; and an epidemiologic study comparing the percentage of infants suffering from malnutrition in two countries.

Suppose that the first binomial variable (the sample from the first population) is of size n_1 with probability π_1, estimated by the sample proportion p_1. The second sample estimates π_2 by p_2 from a sample of size n_2.

It is natural to compare the proportions by the difference $p_1 - p_2$. The mean and variance are given by

$$E(p_1 - p_2) = \pi_1 - \pi_2,$$

$$\text{var}(p_1 - p_2) = \frac{\pi_1(1 - \pi_1)}{n_1} + \frac{\pi_2(1 - \pi_2)}{n_2}.$$

A version of the Central Limit Theorem shows that for large n_1 and n_2 (say both $n_1\pi_1(1 - \pi_1)$ and $n_2\pi_2(1 - \pi_2)$ greater than 10),

$$\frac{p_1 - p_2 - (\pi_1 - \pi_2)}{\sqrt{\dfrac{p_1(1 - p_1)}{n_1} + \dfrac{p_2(1 - p_2)}{n_2}}} = z$$

is an approximately normal pivotal variable. From this, hypothesis tests and confidence intervals develop in the usual manner as illustrated below.

Example 6.12. The paper by Bucher et al. [1976], Example 6.3, examines racial differences in the incidence of ABO hemolytic disease by examining records for infants born at North Carolina Memorial Hospital. In this paper a variety of possible ways of defining hemolytic disease are considered. Using their Class I definition, the samples of black and white infants have the following proportions with hemolytic disease:

$$\text{Black infants,} \quad n_1 = 3584, \quad p_1 = 43/3584;$$

$$\text{White infants,} \quad n_2 = 3831, \quad p_2 = 17/3831.$$

It is desired to perform a two-sided test of the hypothesis $\pi_1 = \pi_2$ at the $\alpha = 0.05$ significance level. The test statistic is

$$Z = \frac{(43/3584) - (17/3831)}{\sqrt{\dfrac{(43/3584)(1 - 43/3584)}{3584} + \dfrac{(17/3831)(1 - 17/3831)}{3831}}} \doteq 3.58.$$

The two-sided p-value is $P\left[|Z| \geq 3.58\right] = 0.0003$ from Table A.1. As $0.0003 < 0.05$, the null hypothesis of equal rates, $\pi_1 = \pi_2$, is rejected at the significance level 0.05.

The pivotal variable may also be used to construct a confidence interval for $\pi_1 - \pi_2$. Algebraic manipulation shows that the end points of a symmetric (about $p_1 - p_2$) confidence interval are given by

$$p_1 - p_2 \pm z_{1-\alpha/2} \sqrt{\frac{p_1(1-p_1)}{n_1} + \frac{p_2(1-p_2)}{n_2}}.$$

For a 95% confidence interval $z_{1-\alpha/2} = 1.96$ and the interval for this example is

$$0.00756 \pm 0.00414 \quad \text{or} \quad (0.00342, 0.01170). \qquad \square$$

A second statistic for testing for equality in two proportions is the χ^2 (chi-square) statistic. This statistic will be considered in more general situations in the next chapter. Suppose that the data are:

	Sample 1	Sample 2	
Success	$n_1 p_1 = n_{11}$	$n_2 p_2 = n_{12}$	$n_{1\bullet}$
Failure	$n_1(1-p_1) = n_{21}$	$n_2(1-p_2) = n_{22}$	$n_{2\bullet}$
	$n_1 = n_{\bullet 1}$	$n_2 = n_{\bullet 2}$	$n_{\bullet\bullet}$

A statistic for testing $H_0 : \pi_1 = \pi_2$ is the χ^2 statistic with one degree of freedom. It is calculated by

$$X^2 = \frac{n_{\bullet\bullet}(n_{11}n_{22} - n_{12}n_{21})^2}{n_{1\bullet}n_{2\bullet}n_{\bullet 1}n_{\bullet 2}}.$$

For technical reasons (Note 6.2) the chi-square distribution with continuity correction, designated by X_c^2, is used by some individuals. The formula for X_c^2 is

$$X_c^2 = \frac{n_{\bullet\bullet}\left(|n_{11}n_{22} - n_{12}n_{21}| - \frac{1}{2}n_{\bullet\bullet}\right)^2}{n_{1\bullet}n_{2\bullet}n_{\bullet 1}n_{\bullet 2}}.$$

For the Bucher *et al.* [1976] data, the values are

	Race		
ABO Hemolytic Disease	**Black**	**White**	**Total**
Yes	43	17	60
No	3541	3814	7355
Total	3584	3831	7415

$$X^2 = \frac{7415\,(43 \cdot 3814 - 17 \cdot 3541)^2}{60 \cdot 7355 \cdot 3584 \cdot 3831} = 13.19$$

$$X_c^2 = \frac{7415\left(|43 \cdot 3814 - 17 \cdot 3541| - 7415/2\right)^2}{60 \cdot 7355 \cdot 3584 \cdot 3831} = 12.26.$$

These statistics, for large n, have a chi-square (χ^2) distribution with one degree of freedom under the null hypothesis of equal proportions. If the null hypothesis is not true X^2 or X_c^2 will tend to be large. The null hypothesis is rejected for large values of X^2 or X_c^2. Table A.3 has χ^2 critical values. The Bucher data have $p < 0.001$ since the 0.001 critical value is 10.83, and require rejection of the null hypothesis of equal proportions.

From Note 5.3 we know that $\chi_1^2 = Z^2$. For this example the value of $Z^2 = 3.58^2 = 12.82$ is close to the value $X^2 = 13.19$. The two values would have been identical (except for rounding) if we had used in the calculation of Z an estimate of the standard error of $\sqrt{pq(1/n_1 + 1/n_2)}$ where $p = 60/7415$ is the pooled estimate of π under the null hypothesis $\pi_1 = \pi_2 = \pi$.

6.4.3   Finding Sample Sizes Needed for Testing the Difference Between Proportions

Consider a study planned to test the equality of the proportions π_1 and π_2. Only studies in which both populations are sampled the same number of times, $n = n_1 = n_2$, will be considered here. There are five quantities that characterize the performance and design of the test:

1. π_1, the proportion in the first population.

2. π_2, the proportion in the second population under the alternative hypothesis.

3. n, the number of observations to be obtained from *each* of the two populations.

4. The significance level α at which the statistical test will be made. α is the probability of rejecting the null hypothesis when it is true. The null hypothesis is that $\pi_1 = \pi_2$.

5. The probability, β, of accepting the null hypothesis when it is not true, but the alternative is true. Here we will have $\pi_1 \neq \pi_2$ under the alternative hypothesis (π_1 and π_2 as specified in 1 and 2 above).

These quantities are interrelated. It is not possible to change one of them without changing at least one of the others. The actual determination of sample size is usually an iterative process; the usual state of affairs is that the desire for precision and the practicality of obtaining an appropriate sample size are in conflict. In practice, one usually considers various possible combinations and arrives at a "reasonable" sample size or decides that it is not possible to perform an adequate experiment within the constraints involved.

The "classical" approach is to specify π_1, π_2 (for the alternative hypothesis), α and β. These parameters determine the sample size n.

Table A.10 gives some sample sizes for such binomial studies using one-sided hypothesis tests (see Problem 6.27). An approximation for n is

$$n = 2 \left(\left(\frac{z_{1-\alpha} + z_{1-\beta}}{\pi_1 - \pi_2} \right) \Big/ \sqrt{\frac{1}{2}(\pi_1(1 - \pi_1) + \pi_2(1 - \pi_2))} \right)^2,$$

where $\alpha = 1 - \Phi(z_{1-\alpha})$; that is, $z_{1-\alpha}$ is the value such that a $N(0, 1)$ variable Z has $P[Z > z_{1-\alpha}] = \alpha$. In words, $z_{1-\alpha}$ is the one-sided normal α critical value. Similarly, $z_{1-\beta}$ is the one-sided normal β critical value.

Figure 6.1 contains a flow diagram for calculating sample sizes for discrete (binomial) as well as continuous variables. It illustrates that the format is the same for both: first, values of α and β are selected. A one-sided or two-sided test determines $z_{1-\alpha}$ or $z_{1-\alpha/2}$ and the quantity NUM, respectively. For discrete data, the quantities π_1 and π_2 are specified, and $\Delta = |\pi_1 - \pi_2| / \sqrt{\frac{1}{2}(\pi_1(1-\pi_1) + \pi_2(1-\pi_2))}$ is calculated. This corresponds to the standardized differences $\Delta = |\mu_1 - \mu_2|/\sigma$ associated with normal or continuous data. The quantity $n = 2(\text{NUM}/\Delta)^2$ then produces the sample size needed for a two-sample procedure. For a one sample procedure, the sample size is $n/2$. Hence a two-sample procedure requires a total of *four* times the number of observations of the one-sample procedure. Various refinements are available in Figure 6.1. A listing of the most common z-values is provided. If a one-sample test is wanted, the values of μ_2 and π_2 can be considered the null hypothesis values. Finally, the equation for the sample size in the discrete case is an approximation, and a more

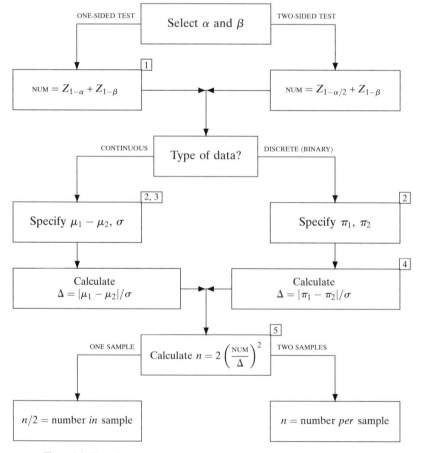

Figure 6.1. Flow chart for sample size calculations (continuous and discrete variables).

precise estimate, n^*, can be obtained from

$$n^* = n + 2/\Delta.$$

This formula is reasonably accurate. For further discussion and references, see Note 17.2.

Other approaches are possible. For example, one might specify the largest feasible sample size n, α, π_1, and π_2 and then determine the power β. The following graph, Figure 6.2, Feigl [1978], for $\alpha = 0.05$ and $\beta = 0.10$, relates π_1, $\Delta = \pi_2 - \pi_1$ and n for two-sided tests.

Finally we note that in certain situations where sample size is necessarily limited, for example, a rare cancer site with several competing therapies, trials with $\alpha = 0.10$ and $\beta = 0.50$ have been run.

In practice, it is difficult to arrive at a sample size. To one unacquainted with statistical ideas, there is a natural tendency to expect too much from an experiment. In answer to the question, "what difference would you like to detect?" the novice will often reply, "any difference," leading to an infinite sample size!

6.4.4 Relative Risk and the Odds Ratio

In this section, we consider studies looking for an association between two binary variables; that is, variables which take on only two outcomes. For definiteness we will speak of the two variables as disease and exposure, since the following techniques are often used in epidemiologic and public health studies. In effect we are comparing two proportions (the proportions with disease) in two populations: those with and without exposure. This section considers methods of summarizing the association.

Figure 6.1. (*Continued*)

1 Values of Z_c for various values of c are:

c	0.500	0.800	0.900	0.950	0.975	0.990	0.995
Z_c	0.000	0.842	1.282	1.645	1.960	2.326	2.576

2 If one sample, μ_2 and π_2 are null hypothesis values.

3 If $\sigma_1^2 \neq \sigma_2^2$, calculate $\sigma^2 = \frac{1}{2}(\sigma_1^2 + \sigma_2^2)$.

4 $\sigma = \sqrt{\frac{1}{2}\left(\pi_1(1 - \pi_1) + \pi_2(1 - \pi_2)\right)}$.

5 Sample size for discrete case is an approximation. For an improved estimate, use $n^* = n + 2/\Delta$.

Note: Two sample case, unequal sample sizes. Let n_1 and kn_1 be the required sample sizes. Calculate n as before. Then calculate $n_1 = n(k + 1)/2k$ and $n_2 = kn_1$. (Total sample size will be larger.) If also, $\sigma_1^2 \neq \sigma_2^2$ calculate n using σ_1; then calculate $n_1 = (n/2)\left(1 + \sigma_2^2/(k\sigma_1^2)\right)$ and $n_2 = kn_1$.

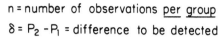

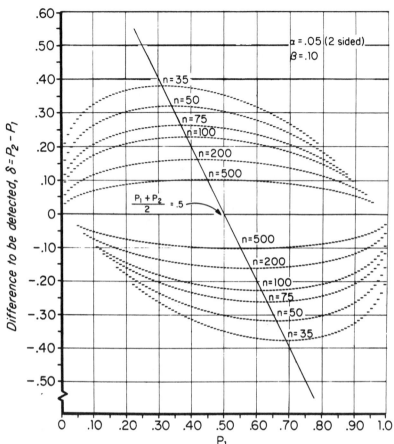

Figure 6.2. Sample sizes required for testing two proportions, π_1 and π_2 with 90% probability of obtaining a significant result at the 5% (two-sided) level.

Suppose that one had a complete enumeration of the population at hand and the true proportions in the population. The probabilities may be presented in a 2×2 table:

Exposure	Disease	
	+ (Yes)	**− (No)**
+ (Yes)	π_{11}	π_{12}
− (No)	π_{21}	π_{22}

where $\pi_{11} + \pi_{12} + \pi_{21} + \pi_{22} = 1$.

There are subtleties glossed over here. For example, by disease (for a human population) does one mean that the person develops the disease at some time before death, has the disease at a particular point in time, or develops it by some age? This ignores the problems of accurate diagnosis, a notoriously difficult problem. Similarly, exposure needs to be carfully defined as to time of exposure, length of exposure, and so on.

What might be a reasonable measure of the effect of exposure? A plausible comparison is $P[\text{disease+} \mid \text{exposure+}]$ with $P[\text{disease+} \mid \text{exposure-}]$. In words, it makes sense to compare the probability of having disease among those exposed with the probability of having the disease among those not exposed.

Definition 6.1. A standard measure of the strength of the exposure effect is the *relative risk*. The relative risk is defined to be

$$\rho = \frac{P[\text{disease+} \mid \text{exposure+}]}{P[\text{disease+} \mid \text{exposure-}]} = \frac{\pi_{11}/(\pi_{11} + \pi_{12})}{\pi_{21}/(\pi_{21} + \pi_{22})} = \frac{\pi_{11}(\pi_{21} + \pi_{22})}{\pi_{21}(\pi_{11} + \pi_{12})}.$$

Thus, a relative risk of 5 means that an exposed person is 5 times as likely to have the disease. The following tables of proportions or probabilities each have a relative risk of two.

Exposure	Disease +	Disease −
+	0.50	0.00
−	0.25	0.25

Exposure	Disease +	Disease −
+	0.25	0.25
−	0.125	0.375

Exposure	Disease +	Disease −
+	0.10	0.40
−	0.05	0.45

Exposure	Disease +	Disease −
+	0.00010	0.49990
−	0.00005	0.49995

We see that many patterns may give rise to the same relative risk. This is not surprising as one number is being used to summarize four numbers. In particular, information on the amount of disease and/or exposure is missing.

Definition 6.2. Given one has the exposure, the *odds* (or betting odds) of getting the disease are

$$\frac{P[\text{disease+} \mid \text{exposure+}]}{P[\text{disease-} \mid \text{exposure+}]}.$$

Similarly, one may define the odds of getting the disease given no exposure. Another measure of the amount of association between the disease and exposure is the *odds ratio*. The odds ratio is defined to be

$$\omega = \frac{P[\text{disease}+ \mid \text{exposure}+]/P[\text{disease}- \mid \text{exposure}+]}{P[\text{disease}+ \mid \text{exposure}-]/P[\text{disease}- \mid \text{exposure}-]}$$

$$= \frac{\left(\pi_{11}/(\pi_{11} + \pi_{12})\right)/\left(\pi_{12}/(\pi_{11} + \pi_{12})\right)}{\left(\pi_{21}/(\pi_{21} + \pi_{22})\right)/\left(\pi_{22}/(\pi_{21} + \pi_{22})\right)}$$

$$= \frac{\pi_{11}\,\pi_{22}}{\pi_{12}\,\pi_{21}}.$$

The odds ratio is also called the cross-product ratio on occasion; this name is suggested by the following scheme:

$$\pi_{11} \qquad \pi_{12}$$

$$\pi_{21} \qquad \pi_{22}$$

Consider now how much the relative risk and odds ratio may differ by looking at the ratio of the two terms, ρ and ω,

$$\frac{\rho}{\omega} = \left(\frac{\pi_{21} + \pi_{22}}{\pi_{22}}\right)\left(\frac{\pi_{12}}{\pi_{11} + \pi_{12}}\right).$$

Suppose that the disease affects a small segment of the population. Then π_{11} is small compared to π_{12}, so that $\pi_{12}/(\pi_{11} + \pi_{12})$ is approximately equal to one. Also, π_{21} will be small compared to π_{22} so that $(\pi_{21} + \pi_{22})/\pi_{22}$ is approximately one. Thus, in this case, $\rho/\omega \doteq 1$. Restating this: if the disease affects a small fraction of the population (in both exposed and unexposed groups), the odds ratio and the relative risk are approximately equal. For this reason the odds ratio is often called the *approximate relative risk*. If the disease affects less than 5% in each group, the two quantities can be considered approximately equal.

The data for looking at the relative risk or the odds ratio usually arises in one of three ways. The three patterns are given below with an illustration of each.

The observed numbers in each of the four cells will be denoted as follows:

	Disease	
Exposure	**+**	**−**
+	n_{11}	n_{12}
−	n_{21}	n_{22}

As before, a dot will replace a subscript when the entries for that subscript are summed. For example,

$$n_{1\bullet} = n_{11} + n_{12},$$

$$n_{\bullet 2} = n_{12} + n_{22},$$

$$n_{\bullet\bullet} = n_{11} + n_{12} + n_{21} + n_{22}.$$

Pattern 6.1: Cross-Sectional Studies—Prospective Studies of a Sample of the Population. There is a sample of size $n_{\bullet\bullet}$ from the population; both traits (exposure and disease) are measured on each subject. This is called cross-sectional data when the status of the two traits is measured at some fixed cross-section in time. In this case the expected number in each cell is the expectation:

$$n_{\bullet\bullet}\pi_{11} \quad n_{\bullet\bullet}\pi_{12}$$
$$n_{\bullet\bullet}\pi_{21} \quad n_{\bullet\bullet}\pi_{22}$$

Example 6.13. The following data are from Meyer *et al.* [1976]. This study collected information on all births in ten Ontario (Canada) teaching hospitals during 1960–1961. A total of 51,490 births was involved, including fetal deaths and neonatal deaths (perinatal mortality). The paper considers the association of perinatal events and maternal smoking during pregnancy. Data relating perinatal mortality and smoking are:

	Perinatal Mortality		
Maternal Smoking	Yes	No	Total
Yes	619	20443	21062
No	634	26682	27316
Total	1253	47125	48378

The estimation of the relative risk and odds ratio is discussed below.

Pattern 6.2: Prospective Study, Groups Based Upon Exposure. In a prospective study of exposure, fixed numbers—say $n_{1\bullet}$ and $n_{2\bullet}$—of individuals with and without the exposure are followed. The endpoints are then noted. In this case the expected number of observations in the cells are

$$
\begin{array}{cc|c}
n_{1\bullet}\dfrac{\pi_{11}}{\pi_{11} + \pi_{12}} & n_{1\bullet}\dfrac{\pi_{12}}{\pi_{11} + \pi_{12}} & n_{1\bullet} \\[3mm]
n_{2\bullet}\dfrac{\pi_{21}}{\pi_{21} + \pi_{22}} & n_{2\bullet}\dfrac{\pi_{22}}{\pi_{21} + \pi_{22}} & n_{2\bullet}
\end{array}
$$

Note that as the sample sizes of the exposure and nonexposure groups are determined by the experimenter, the data will not allow estimates of the proportion exposed, but only the conditional probability of disease given exposure or nonexposure.

Example 6.14. As an example, consider a paper by Shapiro *et al.* [1974] entitled, "Lead time in breast cancer detection and implications for periodicity of screening." In the paper they state that "by the end of this [5-year] period, there were 40 deaths in the [screened] study group of about 31,000 women as compared with 63 such deaths in a comparable group of women." Placing this in a 2 × 2 table and considering the screening to be the exposure, the data are:

| | Breast Cancer Death | | |
On Study (Screened)	Yes	No	Total
Yes	40	30960	31000
No	63	30937	31000

Pattern 6.3: Retrospective Studies. The third way of commonly collecting the data is the retrospective study. Usually, cases and an appropriate control group are identified. (Matched or paired data are *not* being discussed here.) In this case, the size of the disease and control groups, $n_{\cdot 1}$ and $n_{\cdot 2}$ are specified. From such data, one cannot estimate the probability of disease, but rather the probability of being exposed given that an individual has the disease and the probability of exposure given that an individual does not thave the disease. The expected number of observations in each cell is:

$$n_{\cdot 1}\frac{\pi_{11}}{\pi_{11}+\pi_{21}} \qquad n_{\cdot 2}\frac{\pi_{12}}{\pi_{12}+\pi_{22}}$$

$$n_{\cdot 1}\frac{\pi_{21}}{\pi_{11}+\pi_{21}} \qquad n_{\cdot 2}\frac{\pi_{22}}{\pi_{12}+\pi_{22}}$$

$$n_{\cdot 1} \qquad\qquad n_{\cdot 2}$$

Example 6.15. Kelsey and Hardy [1975] studied the driving of motor vehicles as a risk factor for acute herniated lumbar intervertebral disc. Their cases were people between the ages of 20–64; the studies were conducted in the New Haven metropolitan area at three hospitals or in the office of two private radiologists. The cases had low back X-rays and were interviewed and given a few simple diagnostic tests. A control group was composed of those with low back X-rays who were not classified as surgical, probable or possible cases for herniated disc and who had not had their symptoms for more than one year. The in-patients, cases, and controls, of the Yale–New Haven hospital were asked if their job involved driving a motor vehicle. The data were:

| | Herniated Disc? | |
Motor Vehicle Job?	Yes (Cases)	No (Controls)
Yes	8	1
No	47	26
Total	55	27

Consider a two-way layout of disease and exposure to an agent thought to influence the disease.

| | Disease | |
Exposure	+	−
+	n_{11}	n_{12}
−	n_{21}	n_{22}

Table 6.1. Characterization of Cross-Sectional, Prospective and Retrospective Studies and Relationship to Possible Estimation of Relative Risk and Odds Ratio.

Type of Study	Totals for		Can one estimate the	
	Column	Row	Relative Risk?	Odds Ratio?
Cross-sectional or prospective sample	Random	Random	Yes	Yes
Prospective on exposure	Fixed	Random	Yes	Yes
Retrospective	Random	Fixed	No	Yes

The three types of studies discussed above can be thought of as involving conditions on the marginal totals indicated in Table 6.1.

For example, a prospective study can be thought of as a situation where the totals for "Exposure+" and "Exposure−" are fixed by the experimenter and the column totals will vary randomly depending on the association between the disease and the exposure.

For each of these three types of table, how might one estimate the relative risk and/or the odds ratio? From our tables of expected numbers of observations, it is seen that for tables of the types 1 and 2,

$$\frac{E(n_{11})/\big(E(n_{11}) + E(n_{12})\big)}{E(n_{21})/\big(E(n_{21}) + E(n_{22})\big)} = \frac{E(n_{11})/E(n_{1\bullet})}{E(n_{21})/E(n_{2\bullet})} = \frac{\pi_{11}/(\pi_{11} + \pi_{12})}{\pi_{21}/(\pi_{21} + \pi_{22})} = \rho.$$

Thus one estimates the relative risk ρ by replacing the expected value of n_{11} by the observed value of n_{11}, etc., giving

$$\hat{\rho} = \frac{n_{11}/n_{1\bullet}}{n_{21}/n_{2\bullet}}.$$

For retrospective studies of type 3 it is not possible to estimate ρ unless the disease is rare, in which case the estimate of the odds ratio gives a reasonable estimate of the relative risk.

For all three types of tables, one sees that

$$\frac{E(n_{11})E(n_{22})}{E(n_{12})E(n_{21})} = \frac{\pi_{11}\pi_{22}}{\pi_{12}\pi_{12}} = \omega.$$

Therefore we estimate the odds ratio by

$$\hat{\omega} = \frac{n_{11}n_{22}}{n_{12}n_{21}}.$$

It is clear from the definition of relative risk that if exposure has no association with the disease then $\rho = 1$. That is, both "exposed" and "nonexposed" have the same probability of disease. We verify this mathematically, and also that under the null hypothesis of no association, the odds ratio ω is also one. Under H_0:

$$\pi_{ij} = \pi_{i\bullet}\pi_{\bullet j} \quad \text{for} \quad i = 1, 2 \quad \text{and} \quad j = 1, 2.$$

Thus,

$$\rho = \frac{\pi_{11}/\pi_{1\bullet}}{\pi_{21}/\pi_{2\bullet}} = \frac{\pi_{1\bullet}\pi_{\bullet1}/\pi_{1\bullet}}{\pi_{2\bullet}\pi_{\bullet1}/\pi_{2\bullet}} = 1 \quad \text{and} \quad \omega = \frac{\pi_{11}\pi_{22}}{\pi_{12}\pi_{21}} = \frac{\pi_{1\bullet}\pi_{\bullet1}\pi_{2\bullet}\pi_{\bullet2}}{\pi_{1\bullet}\pi_{\bullet2}\pi_{2\bullet}\pi_{\bullet1}} = 1.$$

If ρ or ω are greater than one, the exposed group has an increased risk of the disease. If ρ or ω are less than one, the group not exposed has an increased risk of the disease. Note that an increased or decreased risk may, or may not, be due to a causal mechanism.

For the three examples above, let us calculate the estimated relative risk and odds ratio where appropriate.

For the smoking and perinatal mortality data,

$$\hat{\rho} = \frac{619/21062}{634/27316} \doteq 1.27, \quad \hat{\omega} = \frac{619 \cdot 26682}{634 \cdot 20443} \doteq 1.27.$$

From this data we estimate that smoking during pregnancy is associated with an increased risk of perinatal mortality that is 1.27 times as large. (Note: We have not concluded that smoking causes the mortality, only that there is an association.)

The data relating screening for early detection of breast cancer and 5-year breast cancer mortality gives estimates;

$$\hat{\rho} = \frac{40/31000}{63/31000} \doteq 0.63, \quad \hat{\omega} = \frac{40 \cdot 30937}{63 \cdot 30960} \doteq 0.63.$$

Thus, in this study, screening was associated with a risk of dying of breast cancer within 5 years only 0.63 times as great as the risk among those not screened.

In the unmatched case-control study, only ω can be estimated.

$$\hat{\omega} = \frac{8 \cdot 26}{1 \cdot 47} \doteq 4.43.$$

It is estimated that driving jobs increase the risk of a herniated lumbar intervertebral disc by a factor of 4.43.

Might there really be no association in the above tables and the estimated $\hat{\rho}$'s and $\hat{\omega}$'s differ from 1 merely by chance? You may test the hypothesis of no association by using Fisher's exact test (small samples) or the chi-squared test (for large samples).

For the three examples, using the table of χ^2 critical values with one degree of freedom, we test the statistical significance of the association by using the chi-square statistic with continuity correction.

Smoking–Perinatal mortality:

$$X_c^2 = \frac{48378\left(|619 \times 26682 - 634 \times 20443| - \frac{1}{2}(48378)\right)^2}{21062 \times 27316 \times 1253 \times 47125} = 17.76.$$

From Table A.3, $p < 0.001$, and there is significant association. (Equivalently, for one degree of freedom, $Z = \sqrt{X_c^2} = 4.21$ and Table A.1 shows $p < 0.0001$.)

Breast cancer and screening:

$$X_c^2 = \frac{62000\left(|40 \times 30937 - 63 \times 30960| - \frac{1}{2}(62000)\right)^2}{31000 \times 31000 \times 103 \times 61897} = 4.71.$$

From the table, $0.01 < p < 0.05$ and the association is statistically significant at the 0.05 level.

Motor-vehicle job and herniated disc: $X_c^2 = 1.21$. From the χ^2 table, $p > 0.25$, and there is *not* a statistical association using only the Yale–New Haven data. In the next section, we return to this data set.

If there is association, what can one say about the accuracy of the estimates?

For the first two examples, where there is a statistically significant association, we turn to the construction of confidence intervals for ω. The procedure is to construct a confidence interval for $\ln \omega$, the natural log of ω, and to "exponentiate" the endpoints to find the confidence interval for ω. Our logarithms are natural logarithms, that is, to the base e. Recall e is a number; $e = 2.71828...$.

The estimate of $\ln \omega$ is $\ln \widehat{\omega}$. The standard error of $\ln \widehat{\omega}$ is estimated by

$$\sqrt{1/n_{11} + 1/n_{12} + 1/n_{21} + 1/n_{22}}.$$

The estimate is approximately normally distributed; thus normal critical values are used in constructing the confidence intervals. A $100(1 - \alpha)\%$ confidence interval for $\ln \omega$ is given by

$$\ln \widehat{\omega} \pm z_{1-\alpha/2} \sqrt{1/n_{11} + 1/n_{12} + 1/n_{21} + 1/n_{22}}$$

where an $N(0, 1)$ variable has probability $\alpha/2$ of exceeding $z_{1-\alpha/2}$.

Upon finding the endpoints of this confidence interval, we exponentiate the values of the endpoints to find the confidence interval for ω.

We find a 99% confidence interval for ω with the smoking and perinatal mortality data. First we construct the confidence interval for $\ln \omega$:

$$\ln(1.27) \pm 2.576 \sqrt{1/619 + 1/20443 + 1/26682 + 1/634}$$

or 0.2390 ± 0.1475 or $(0.0915, 0.3865)$. The confidence interval for ω is

$$\left(e^{0.0915}, e^{0.3865}\right) = (1.10, 1.47).$$

To find a 95% confidence interval for the breast cancer-screening data,

$$\ln(0.63) \pm 1.96 \sqrt{1/40 + 1/30960 + 1/30937 + 1/63}$$

or -0.4620 ± 0.3966 or $(-0.8586, -0.0654)$. The 95% confidence interval for the odds ratio, ω, is $(0.424, 0.937)$.

The reason for using logarithms in constructing the confidence intervals is that $\ln \widehat{\omega}$ is more normally distributed than ω. The standard error of ω may be directly estimated by

$$\widehat{\omega} \sqrt{1/n_{11} + 1/n_{12} + 1/n_{21} + 1/n_{22}}$$

(see Note 6.3 for the rationale). However, confidence intervals should be constructed as illustrated above.

6.4.5 Combination of 2 × 2 Tables

In this section we consider methods of combining 2×2 tables. The tables arise in one of two ways. In the first situation, we are interested in investigating an association between disease and exposure. There is, however, a third variable taking a finite number of values. We wish to "adjust" for the effect of the third variable. The values of the "confounding" third variable sometimes arise by taking a continuous variable and grouping by intervals; thus the values are sometimes called *strata*. A second situation in which we will deal with several 2×2 tables is when the study of association and disease is made in more than one group. In some reasonable way, one would like to consider the combination of the 2×2 tables from each group.

Why combine 2 × 2 tables?

To see why one needs to worry about such things, suppose that there are two strata. In our first example there is no association between exposure and disease in each stratum, but if we ignore strata and "pool" our data (that is, add it all together), an association appears.

STRATUM 1

Exposure	Disease	
	+	**−**
+	5	50
−	10	100

$$\widehat{\omega}_1 = \frac{5 \cdot 100}{10 \cdot 50} = 1.$$

STRATUM 2

Exposure	Disease	
	+	**−**
+	40	60
−	40	60

$$\widehat{\omega}_2 = \frac{40 \cdot 60}{40 \cdot 60} = 1.$$

In both tables the odds ratio is one, and there is no association. Combining tables, the combined table and its odds ratio are:

Exposure	Disease	
	+	**−**
+	45	110
−	50	160

$$\widehat{\omega}_{\text{combined}} = \frac{45 \times 160}{50 \times 110} \doteq 1.31.$$

When combining tables with no association, or odds ratios of one, the combination may show association. For example, one would expect to find a positive relationship between breast cancer and being a homemaker. Possibly tables given separately for each sex would not show such an association. If the inference to be derived were that homemaking might be causally related to breast cancer, it is clear that one would need to adjust for sex.

On the other hand, there can be an association within each stratum that disappears in the pooled data set. The following numbers illustrate this:

STRATUM 1

	Disease	
Exposure	+	−
+	60	100
−	10	50

$$\widehat{\omega}_1 = \frac{60 \times 50}{10 \times 100} = 3$$

STRATUM 2

	Disease	
Exposure	+	−
+	50	10
−	100	60

$$\widehat{\omega}_2 = \frac{50 \times 60}{100 \times 10} = 3$$

COMBINED DATA

	Disease	
Exposure	+	−
+	110	110
−	110	110

$$\widehat{\omega}_{\text{combined}} = 1$$

Thus, ignoring a confounding variable may "hide" an association that exists within each stratum, but not observed in the combined data.

Formally, our two situations are the same if we identify the stratum with differing groups. Also, note that there may be more than one confounding variable, that each strata of the "third" variable could correspond to a different combination of several other variables.

Questions of Interest in Multiple 2 × 2 Tables

In examining more than one 2 × 2 table, one or more of three questions is usually asked. This is illustrated by using the data of the study involving cases of acute herniated lumbar disc and controls (not matched) in Example 6.15 which compares the proportions with jobs driving motor vehicles. There are seven different hospital services involved, although only one of them was presented in Example 6.15. Numbering the sources from 1 to 7 and giving the data as 2 × 2 tables, the tables and the seven odds ratios are:

[1]

	Herniated Disc	
Motor Vehicle Job	+	−
+	8	1
−	47	26

$$\widehat{\omega} = 4.43$$

[2]	+	−	
+	5	0	$\widehat{\omega} = \infty$
−	17	21	

[5]	+	−	
+	1	3	$\widehat{\omega} = 0.67$
−	5	10	

[3]	+	−	
+	4	4	$\widehat{\omega} = 5.92$
−	13	77	

[6]	+	−	
+	1	2	$\widehat{\omega} = 1.83$
−	3	11	

[4]	+	−	
+	2	10	$\widehat{\omega} = 1.08$
−	12	65	

[7]	+	−	
+	2	2	$\widehat{\omega} = 3.08$
−	12	37	

The seven odds ratios are 4.43, ∞, 5.92, 1.08, 0.67, 1.83, and 3.08. The ratios vary so much that one might wonder whether each hospital service has the same degree of association (Question 1). If they do not have the same degree of association, one might question whether the controls are appropriate, the patient populations are different, etc.

One also would like an estimate of the overall or "average" association (Question 2). From the previous examples it is seen that it might not be wise to sum all the tables and compute the association based on the pooled tables.

Finally, another question, related to the first two, is whether there is any evidence of any association, either overall or in some of the groups (Question 3).

Two Approaches to Estimating an Overall Odds Ratio

If the seven different tables come from populations with the same odds ratio, how do we estimate the common or overall odds ratio? We will consider two approaches.

The first technique is to work with the natural logarithm, log to the base e, of the estimated odds ratio, $\widehat{\omega}$.

Let $a_i = \ln \widehat{\omega}_i$, where $\widehat{\omega}_i$ is the estimated odds ratio in the ith of k 2×2 tables. The standard error of a_i is estimated by

$$s_i = \sqrt{1/n_{11} + 1/n_{12} + 1/n_{21} + 1/n_{22}},$$

where n_{11}, n_{12}, n_{21}, and n_{22} are the values from the ith 2×2 table. How do we investigate the problems mentioned above? To do this, one needs to understand a little of how the χ^2 distribution arises. The square of a standard normal variable has a chi-square distribution with one degree of freedom (d.f.). If independent chi-square variables are added, the result is a chi-square variable whose degrees of freedom is the sum of the degrees of freedom of the variables that were added (see Note 5.3 also).

We now apply this to the problem at hand. Under the null hypothesis of no association in any of the tables, each a_i/s_i is approximately a standard normal value. If there is no association, $\omega = 1$ and $\ln \omega = 0$. Thus, log $\widehat{\omega}_i$ has a mean of approximately zero. Its square, $(a_i/s_i)^2$, is approximately a χ^2 variable with one degree of freedom. The sum of all k of these independent, approximately chi-square variables, is approximately a chi-square variable with k degrees of freedom. The sum is

$$X^2 = \sum_{i=1}^{k} \left(\frac{a_i}{s_i} \right)^2 ;$$

and under the null hypothesis it has approximately a χ^2 distribution with k degrees of freedom.

It is possible to partition this sum into two parts. One part tests whether the association might be the same in all k tables (i.e., it tests for homogeneity). The second part will test to see whether on the basis of all the tables there is any association.

Suppose that one wants to "average" the association from all of the 2×2 tables, it seems reasonable to give more weight to the better estimates of association. That is, one wants the estimates with higher variances to get less weight. An appropriate weighted average is

$$\bar{a} = \sum_{i=1}^{k} (a_i/s_i^2) \Big/ \sum_{i=1}^{k} (1/s_i^2).$$

The χ^2 statistic then is partitioned, or broken down, into two parts:

$$X^2 = \sum_{i=1}^{k} \left(\frac{a_i}{s_i} \right)^2 = \sum_{i=1}^{k} \left(\frac{1}{s_i^2} \right) (a_i - \bar{a})^2 + \sum_{i=1}^{k} \left(\frac{1}{s_i^2} \right) \bar{a}^2.$$

On the right hand side, the first sum is approximately a χ^2 random variable with $k - 1$ degrees of freedom if all k groups have the same degree of association. It tests for the homogeneity of the association in the different groups. That is, if χ^2 for homogeneity is too large, we reject the null hypothesis that the degree of association (whatever it is) is the same in each group. The second term tests whether there is association on the average. This has approximately a χ^2 distribution with one degree of freedom if there is no association in each group. Thus, define

$$\chi_H^2 = \sum_{i=1}^{k} \left(\frac{1}{s_i^2} \right) (a_i - \bar{a})^2 = \sum_{i=1}^{k} \frac{a_i^2}{s_i^2} - \bar{a}^2 \sum_{i=1}^{k} \left(\frac{1}{s_i^2} \right),$$

and

$$\chi_A^2 = \bar{a}^2 \sum_{i=1}^{k} \left(\frac{1}{s_i^2} \right).$$

Of course, if we decide that there are different degrees of association in different groups, this means that at least one of the groups must have some association.

Consider now the data given above. A few additional points are introduced. We use the log of the odds ratio, but the second group has $\hat{\omega} = \infty$. What shall we do about this?

With small numbers, this may happen due to a zero in a cell. The bias of the method is reduced by adding 0.5 to each cell in each table:

[1]	+	−
+	8.5	1.5
−	47.5	26.5

[2]	+	−
+	5.5	0.5
−	17.5	21.5

[5]	+	−
+	1.5	3.5
−	5.5	10.5

[3]	+	−
+	4.5	4.5
−	13.5	77.5

[6]	+	−
+	1.5	2.5
−	3.5	11.5

[4]	+	−
+	2.5	10.5
−	12.5	65.5

[7]	+	−
+	2.5	2.5
−	12.5	37.5

Now,

$$\widehat{\omega}_i = \frac{(n_{11}+0.5)(n_{22}+0.5)}{(n_{12}+0.5)(n_{21}+0.5)}, \quad s_i = \sqrt{\frac{1}{n_{11}+0.5}+\frac{1}{n_{22}+0.5}+\frac{1}{n_{12}+0.5}+\frac{1}{n_{21}+0.5}}.$$

Performing the above calculations:

Table i	$\widehat{\omega}_i$	$a_i = \log \widehat{\omega}_i$	s_i^2	$1/s_i^2$	a_i^2/s_i^2	a_i/s_i^2
1	3.16	1.15	0.843	1.186	1.571	1.365
2	13.51	2.60	2.285	0.438	2.966	1.139
3	5.74	1.75	0.531	1.882	5.747	3.289
4	1.25	0.22	0.591	1.693	0.083	0.375
5	0.82	−0.20	1.229	0.813	0.033	−0.163
6	1.97	0.68	1.439	0.695	0.320	0.472
7	3.00	1.10	0.907	1.103	1.331	1.212
Total				7.810	12.051	7.689

Then

$$\bar{a} = \sum_{i=1}^{k}\left(\frac{a_i}{s_i^2}\right) \Big/ \sum_{i=1}^{k}\left(\frac{1}{s_i^2}\right) = \frac{7.689}{7.810} \doteq 0.985,$$

$$X_A^2 = (0.985)^2(7.810) \doteq 7.57,$$

$$X_H^2 = \sum \frac{a_i^2}{s_i^2} - X_A^2 = 12.05 - 7.57 = 4.48.$$

X_H^2 with $7 - 1 = 6$ degrees of freedom, has an $\alpha = 0.05$ critical value of 12.59 from Table A.3. We do *not* conclude that the association differs between groups.

Moving to the X_A^2, we find that $7.57 > 6.63$, the χ^2 critical value with one degree of freedom at the 0.010 level. We conclude that there *is* some overall association.

The odds ratio is estimated by $\hat{\omega} = e^{\bar{a}} = e^{0.985} = 2.68$. The standard error of $\bar{a}$ is estimated by

$$\frac{1}{\sqrt{\sum_{i=1}^{k}(1/s_i^2)}}.$$

To find a confidence interval for ω, first find one for $\ln \omega$ and "exponentiate" back. To find a 95% confidence interval, the calculation is

$$\bar{a} \pm \frac{z_{0.975}}{\sqrt{\sum(1/s_i^2)}} = 0.985 \pm \frac{1.96}{\sqrt{7.810}} \quad \text{or} \quad 0.985 \pm 0.701 \quad \text{or} \quad (0.284, 1.696).$$

Taking exponentials, the confidence interval for the overall odds ratio is $(1.33, 5.45)$.

The second method of estimation is due to Mantel and Haenszel [1959]. Their estimate of the odds ratio is

$$\hat{\omega} = \sum_{i=1}^{k}\left(n_{11}(i)n_{22}(i)/n_{\bullet\bullet}(i)\right) \bigg/ \sum_{i=1}^{k}\left(n_{12}(i)n_{21}(i)/n_{\bullet\bullet}(i)\right),$$

where $n_{11}(i)$, $n_{22}(i)$, $n_{12}(i)$, $n_{21}(i)$, and $n_{\bullet\bullet}(i)$ are n_{11}, n_{22}, n_{12}, n_{21}, and $n_{\bullet\bullet}$ for the ith table.

In this problem,

$$\hat{\omega} = \frac{\frac{8\times26}{82} + \frac{5\times21}{43} + \frac{4\times77}{98} + \frac{2\times65}{89} + \frac{1\times10}{19} + \frac{1\times11}{17} + \frac{2\times37}{53}}{\frac{47\times1}{82} + \frac{17\times0}{43} + \frac{13\times4}{98} + \frac{12\times10}{89} + \frac{5\times3}{19} + \frac{3\times2}{17} + \frac{12\times2}{53}}$$

$$\doteq 12.1516/4.0473 \doteq 3.00.$$

A test of association is given by the following statistic, X_A^2, which is approximately a chi-square random variable with one degree of freedom.

$$X_A^2 = \frac{\left(\left|\sum_{i=1}^{k}n_{11}(i) - \sum_{i=1}^{k}n_{1\bullet}(i)n_{\bullet1}(i)/n_{\bullet\bullet}(i)\right| - \frac{1}{2}\right)^2}{\sum_{i=1}^{k}\left(\frac{n_{1\bullet}(i)n_{2\bullet}(i)n_{\bullet1}(i)n_{\bullet2}(i)}{n_{\bullet\bullet}(i)^2(n_{\bullet\bullet}(i) - 1)}\right)}.$$

The herniated disc data yield $X_A^2 = 7.92$ so that as above there is a significant

($p < 0.01$) association between an acute herniated lumbar intervertebral disc and whether or not a job requires driving a motor vehicle.

See Schlesselman [1982] and Breslow and Day [1980] for methods of setting confidence intervals for ω using the Mantel–Haenszel estimate.

In most circumstances, combining 2×2 tables will be used to adjust for other variables that define the strata, i.e., that define the different tables. The homogeneity of the odds ratio is usually of less interest, unless the odds ratio differs widely among tables. Before testing for homogeneity of the odds ratio one should be certain that this is what is desired (see Note 6.4).

6.4.6 Screening and Diagnosis: Sensitivity, Specificity, Bayes' Theorem

In clinical medicine, and also in epidemiology, tests are often used to screen for the presence or absence of a disease. In the simplest case the test will simply be classified as having a positive (disease likely) or negative (disease unlikely) finding. Further, suppose that there is a "gold standard" that tells us whether or not a subject actually has the disease. The definitive classification might be based upon data from follow-up, invasive radiographic or surgical procedures, or autopsy results. In many cases the gold standard itself will only be relatively correct, but nevertheless the best classification available. This section will discuss summarization of the prediction of disease (as measured by our gold standard) by the test being considered. Ideally, those with the disease should all be classified as having disease, and those without disease should be classified as nondiseased. For this reason, two indices of the performance of a test consider how often such correct classification occurs.

Definition 6.3. The *sensitivity* of a test is the percentage of individuals with disease who are classified as having disease. A test is sensitive to the disease if it is positive for most individuals having the disease. The *specificity* of a test is the percentage of individuals without the disease who are classified as not having the disease. A test is specific if it is positive for a small percentage of those without the disease.

Further terminology associated with screening and diagnostic tests are true positive, true negative, false positive and false negative tests.

Definition 6.4. A test is a *true positive test* if it is positive and the subject has the disease. A test is a *true negative test* if the test is negative and the subject does not have the disease. A *false positive test* is a positive test of a person without the disease. A *false negative test* is a negative test of a person with the disease.

Definition 6.5. The *predictive value of a positive test* is the percentage of subjects with a positive test who have the disease; the *predictive value of a negative test* is the percentage of subjects with a negative test who do not have the disease.

Suppose that data are collected on a test and presented in a 2×2 table as follows:

Screening Test Result	Disease Category	
	Diseased (+)	Nondiseased (−)
Positive (+) test	a (true +s)	b (false +s)
Negative (−) test	c (false −s)	d (true −s)

The sensitivity is estimated by $100a/(a + c)$; the specificity by $100d/(b + d)$. If the subjects are representative of a population, the predictive value of positive and negative tests are estimated by $100a/(a + b)$ and $100d/(c + d)$, respectively. These predictive values are only useful when the proportions with and without the disease in the study group are approximately the same as in the population where the test will be used to predict or classify (see below).

Example 6.16. Remein and Wilkerson [1961] considered a number of screening tests for diabetes. They had a group of consultants establish criteria, their "gold standard," for diabetes. On each of a number of days they recruited patients being seen in the outpatient department of the Boston City Hospital for reasons other than suspected diabetes. The table below presents results on the Folin–Wu Blood Test used 1 h after a test meal and using a blood sugar level of 150 mg/100 ml of blood sugar as a positive test.

Test	Diabetic	Nondiabetic	Total
+	56	49	105
−	14	461	475
Total	70	510	580

From this table note there are 56 true positive tests as compared to 14 false negative tests. The sensitivity is $100(56)/(56 + 14) = 80.0\%$. The 49 false positive tests and 461 true negative tests give a specificity of $100(461)/(49 + 461) = 90.4\%$. The predictive value of a positive test is $100(56)/(56+49) = 53.3\%$. The predictive value of a negative test is $100(461)/(14 + 461) = 97.1\%$. □

If a test has a fixed value for its sensitivity and specificity, the predictive values will change depending upon the prevalence of the disease in the population being tested. The values are related by *Bayes' Theorem*. This theorem tells us how to update the probability of an event A; for example, the event of a subject having disease. If the subject is selected at random from some population, the probability of A is the fraction of individuals having the disease. Suppose additional information becomes available; for example, the results of a diagnostic test might become available. In the light of this new information we would like to update or change our assessment of the probability that A occurs (that the subject has disease). The probability of A before receiving additional information is called the *a priori* or *prior* probability. The updated probability of A after receiving new information is called the *a posteriori* or *posterior* probability. Bayes' Theorem tells how to find the posterior probability.

Bayes' Theorem uses the concept of a conditional probability. In Example 6.17 we review this concept.

Example 6.17. Comstock and Partridge [1972] conducted an informal census of Washington County, Maryland in 1963. There were 127 arteriosclerotic heart disease deaths in the follow-up period. Of the deaths, 38 occurred among individuals whose usual frequency of church attendence was once or more per week. There were 24,245 such individuals as compared to 30,603 individuals whose usual attendance was less than once weekly. What is the probability of an arteriosclerotic heart disease death (event A) in 3 years given church attendance usually once or more per week (event B)?

From the data

$$P[A] = 127/(24245 + 30603) = 0.0023,$$
$$P[B] = 24245/(24245 + 30603) = 0.4420,$$
$$P[A \& B] = 38/(24245 + 30603) = 0.0007,$$
$$P[A \mid B] = P[A \& B]/P[B] = 0.0007/0.4420 = 0.0016.$$

If you knew someone attended church once or more per week the prior estimate of 0.0023 of the probability of an arteriosclerotic heart disease death in 3 years would be changed to a posterior estimate of 0.0016.   □

Using the conditional probability concept Bayes' Theorem may be stated.

Fact 6.1: Bayes' Theorem.   Let $B_1, \ldots, B_k$ be events such that one and only one of them must occur. Then for each i,

$$P[B_i \mid A] = \frac{P[A \mid B_i]P[B_i]}{P[A \mid B_1]P[B_1] + \cdots + P[A \mid B_k]P[B_k]}.$$

Example 6.18. We use the data of Example 6.16 and Bayes' Theorem to show that the predictive power of the test is related to the prevalence of the disease in the population. Suppose that the prevalence of the disease were not 70/580 (as in the data given), but rather 6%. Also suppose the sensitivity and specificity of the test were 80.0% and 90.4% as in the example. What is the predictive value of a positive test?

We want $P[\text{disease+} \mid \text{test+}]$. Let B_1 be the event that the patient has disease and B_2 be the event of no disease. Let A be the occurrence of a positive test. A sensitivity of 80.0% is the same as $P[A \mid B_1] = 0.800$. A specificity of 90.4% is equivalent to $P[\text{not } A \mid B_2] = 0.904$. It is easy to see that

$$P[\text{not } A \mid B] + P[A \mid B] = 1$$

for any A and B. Thus, $P[A \mid B_2] = 1 - 0.904 = 0.096$. By assumption, $P[\text{disease+}] = P[B_1] = 0.06$, and $P[\text{disease}-] = P[B_2] = 0.94$. By Bayes' Theorem,

$$P[\text{disease+} \mid \text{test+}] =$$

$$\frac{P[\text{test+} \mid \text{disease+}]P[\text{disease+}]}{P[\text{test+} \mid \text{disease+}]P[\text{disease+}] + P[\text{test+} \mid \text{disease}-]P[\text{disease}-]}$$

Using our definitions of A, B_1 and B_2, this is

$$P[B_1 \mid A] = \frac{P[A \mid B_1]P[B_1]}{P[A \mid B_1]P[B_1] + P[A \mid B_2]P[B_2]}$$
$$= (0.800 \times 0.06)/(0.800 \times 0.06 + 0.096 \times 0.94)$$
$$= 0.347.$$

If the disease prevalence is 6%, the predictive value of a positive test is 34.7% rather than 53.3% when the disease prevalence is $70/580$ (12.1%). $\square$

Problems 6.15 and 6.28 illustrate the importance of disease prevalence in assessing the results of a test. See Note 6.8 for relationships between sensitivity, specificity, prevalence and predictive values of a positive test.

6.5 MATCHED OR PAIRED OBSERVATIONS

6.5.1 Motivation

The comparisons among proportions in the preceding sections dealt with samples from different populations, or from different subsets of a specified population. In many situations, the estimates of the proportions are based on the same objects or come from closely related, matched or paired observations. You have seen matched or paired data used with a one-sample t-test.

A standard epidemiological tool is the retrospective paired case-control study. An example was given in Chapter 1. Let us recall the rationale for such studies. Suppose that one wants to see whether or not there is an association between a risk factor (say, use of oral contraceptives), and a disease (say, thromboembolism). Because the incidence of the disease is low, an extremely large prospective study would be needed to collect an adequate number of cases. One strategy is to *start* with the cases. The question then becomes one of finding appropriate controls for the cases. In a matched pair study, one control is identified for each case. The control, not having the disease, should be identical to the case in all relevant ways except, possibly, for the risk factor (see Note 6.6).

Example 6.19.   This example is a retrospective matched pair case-control study by Sartwell *et al.* [1969], to study thromboembolism and oral contraceptive use. The cases were 175 women of reproductive age (15–44), discharged alive from 43 hospitals in five cities after initial attacks of idiopathic (i.e., of unknown cause) thrombophlebitis (blood clots in the veins with inflammation in the vessel walls), pulmonary embolism (a clot carried through the blood and obstructing lung blood flow), or cerebral thrombosis or embolism. The controls were matched with their cases for hospital, residence, time of hospitalization, race, age, marital status, parity, and pay status. More specifically, the controls were female patients from the same hospital during the same 6-month interval. The controls were within 5 years of age and matched on parity (0, 1, 2, 3, or more prior pregnancies). The hospital pay status (ward, semi-private, or private) was the same. The data for oral contraceptive use are

| | Control use? | |
Case use?	Yes	No
Yes	10	57
No	13	95

The question of interest: are cases more likely than controls to use oral contraceptives?   □

6.5.2 Matched-Pair Data: McNemar's Test and Estimation of the Odds Ratio

The 2×2 table of Example 6.19 does not satisfy the assumptions of previous sections. The proportions using oral contraceptives among cases and controls cannot be considered samples from two populations since the cases and controls are paired; that is, they come together. Once a case is selected the control for the case is constrained to be one of a small subset of individuals who match the case in various ways.

Suppose there is no association between oral contraceptive use and thromboembolism after taking into account relevant factors. Suppose a case and control are such that only one of the pair uses oral contraceptives. Which one is more likely to use oral contraceptives? They may both be likely or unlikely to use oral contraceptives depending upon a variety of factors. Since the pair have the same values of such factors neither member of the pair is more likely to have the risk factor! That is, in the case of disagreement, or discordant pairs, the probability the case has the risk factor is $1/2$. More generally, suppose the data are:

| | Control has risk factor? | |
Case has risk factor?	Yes	No
Yes	a	b
No	c	d

If there is no association between disease (that is, case or control) and presence or absence of the risk factor the number b is binomial with $\pi = 1/2$ and $n = b + c$. To test for association we test $\pi = 1/2$ as shown previously. For large n, say $n \geq 30$,

$$X^2 = \frac{(b-c)^2}{b+c}$$

has a chi-square distribution with one degree of freedom if $\pi = 1/2$. For Example 6.19,

$$X^2 = \frac{(57-13)^2}{57+13} = 27.66.$$

From the chi-square table, $p < 0.001$ so that there is a statistically significant association between thromboembolism and oral contraceptive usage. This statistical test is called McNemar's Test.

Procedure 6.6. For retrospective matched pair data, the odds ratio is estimated by

$$\widehat{\omega}_{\text{paired}} = \frac{b}{c}.$$

The standard error of the estimate is estimated by

$$(1 + \widehat{\omega}_{\text{paired}}) \sqrt{\frac{\widehat{\omega}_{\text{paired}}}{b + c}}.$$

In Example 6.19, we estimate the odds ratio by

$$\widehat{\omega} = \frac{57}{13} \doteq 4.38.$$

The standard error is estimated by

$$(1 + 4.38) \sqrt{\frac{4.38}{70}} \doteq 1.35.$$

An approximate 95% confidence interval is given by

$$4.38 \pm (1.96)(1.35) \quad \text{or} \quad (1.74, 7.02).$$

More precise intervals may be based upon the use of confidence intervals for a binomial proportion and the fact that $\widehat{\omega}_{\text{paired}} / (\widehat{\omega}_{\text{paired}} + 1) = b/(b + c)$ is a binomial proportion (see Fleiss[1981]). See Note 6.6 for a further discussion of the chi-square analysis of paired data.

6.6 POISSON RANDOM VARIABLES

The Poisson distribution occurs primarily in two closely related situations. The first is a situation in which one counts discrete events in space or time, or some other continuous situation. For example, one might note the time of arrival (considered as a particular point in time) at an emergency medical service over some fixed time period. One may count the number of discrete occurrences of arrivals over this continuum of time. Conceptually, we may get any nonnegative integer, no matter how large, as our answer. A second example occurs when counting numbers of red blood cells that occur in a specified rectangular area marked off in the field of view. In a diluted blood sample where the distance between cells is such that they do not tend to "bump into each other," we may idealize the cells as being represented by points in the plane. Thus, within the particular area of interest, we are counting the observed number of points. A third example where one would expect to model the number of counts by a Poisson distribution would be a situation in which one is counting the number of particle emissions from a radioactive source. If the time period of observation is such that the radioactivity of the source does not significantly decrease (that is, the time period is small compared to the half life of a particle), then the counts (which may be considered as coming at discrete time points) would again be appropriately modeled by a Poisson distribution.

The second major use of the Poisson distribution is as an approximation to the binomial distribution. If n is large and π is small in a binomial situation, then the number of successes is very closely modeled by the Poisson distribution. The closeness of the approximation is specified by a mathematical theorem. As a rough rule of thumb, for most purposes the Poisson approximation will be adequate if π is less than or equal to 0.1, and n is greater than or equal to 20.

In order for the Poisson distribution to be an appropriate model for counting discrete points occurring in some sort of a continuum, the following two assumptions must hold:

1. The number of events occurring in one part of the continuum should be statistically independent of the number of events occurring in another part of the continuum. For example, in the emergency room, if we measure the number of arrivals during the first half hour, this event could reasonably be considered statistically independent of the number of arrivals during the second half hour. If there has been some cataclysmic event such as an earthquake, the assumption will not be valid. Similarly, in counting red blood cells in a diluted blood solution, the number of red cells in one square might reasonably be modeled as statistically independent of the number of red cells in another square.

2. The second assumption is that the expected number of counts in a given part of the continuum should approach zero as its size approaches zero. Thus, in observing blood cells one does not expect to find any in a very small area of a diluted specimen.

6.6.1 Examples of Poisson Data

Example 6.3 (Bucher *et al.* [1976]) examines racial differences in the incidence of ABO hemolytic disease by examining records for infants born at the North Carolina Memorial Hospital. The samples of black and white infants gave the following estimated proportions with hemolytic disease:

$$\text{Black infants,} \quad n_1 = 3584, \quad p_1 = 43/3584;$$

$$\text{White infants,} \quad n_2 = 3831, \quad p_2 = 17/3831.$$

The observed number of cases might reasonably be modeled by the Poisson distribution. (Note: the n is large and π is small in a binomial situation.) In this paper, studying the incidence of ABO hemolytic disease in black and white infants, the observed fractions for black and white infants of having the disease were 43/3584 and 17/3831. The 43 and 17 cases may be considered values of Poisson random variables.

A second example that would be appropriately modeled by the Poisson distribution is the number of deaths resulting from a large scale vaccination program. In this case, n will be very large and π will be quite small. One might use the Poisson distribution in investigating the simultaneous occurrence of a disease and its association within a vaccination program. How likely is it that the particular "chance occurrence" might actually occur by chance?

Example 6.20. As a further example, a paper by Fisher *et al.* [1922] considers the accuracy of the plating method of estimating the density of bacterial populations. The process we are speaking about consists in making a suspension of a known mass of soil in a known volume of salt solution, and then diluting the suspension to a known degree. The bacterial numbers in the diluted suspension are estimated by plating a known volume in a nutrient gel medium and counting the number of colonies that develop from the plate. The estimate was made by a calculation which takes into account the mass of the soil taken and the degree of dilution. If we consider the colonies to be points occurring in the volume of gel, a Poisson model for the number of counts would be appropriate. The following is a table of counts from seven different plates with portions of soil taken from a sample of Barnfield soil assayed in four parallel dilutions:

Plate	I	II	III	IV
1	72	74	78	69
2	69	72	74	67
3	63	70	70	66
4	59	69	58	64
5	59	66	58	62
6	53	58	56	58
7	51	52	56	54
Mean	60.86	65.86	64.29	62.86

☐

Example 6.21. A famous example of the Poisson distribution is data by von Bortkiewicz [1898] showing the chance of a cavalryman being killed by a horse-kick in the course of a year. The data are from recordings of ten corps over a period of 20 years supplying 200 readings. A question of interest here might be whether a Poisson model is appropriate. Was the corps with four deaths an "unlucky" accident or might there have been negligence of some kind?

HORSEKICK FATALITY DATA

Number of Deaths per Corps per Year	Frequency
0	109
1	65
2	22
3	3
4	1
5	0
6	0

☐

6.6.2 The Poisson Model

The Poisson probability distribution is characterized by one parameter, λ. For each nonnegative integer k, if Y is a variable with the Poisson distribution with parameter λ,

$$P[Y = k] = \frac{e^{-\lambda}\lambda^k}{k!}.$$

The parameter λ is both the mean and variance of the Poisson distribution,

$$E(Y) = \text{var}(Y) = \lambda.$$

Bar graphs of the Poisson probabilities are given in Figure 6.3 for selected values of λ. As the mean (equal to the variance) increases, the distribution moves to the right and becomes more spread out and more symmetrical.

In using the Poisson distribution to approximate the binomial distribution, the parameter λ is chosen to equal $n\pi$, the expected value of the binomial distribution. Poisson and binomial probabilities are given in the table below for comparison. This table gives an idea of the accuracy of the approximation (table entry is $P[Y = k]$, $\lambda = 2 = n\pi$) for the first seven values of three distributions.

	Binomial Probabilities			
k	$n = 10$ $\pi = 0.20$	$n = 20$ $\pi = 0.10$	$n = 40$ $\pi = 0.05$	**Poisson Probabilities**
0	0.1074	0.1216	0.1285	0.1353
1	0.2684	0.2702	0.2706	0.2707
2	0.3020	0.2852	0.2777	0.2707
3	0.2013	0.1901	0.1851	0.1804
4	0.0881	0.0898	0.0901	0.0902
5	0.0264	0.0319	0.0342	0.0361
6	0.0055	0.0089	0.0105	0.0120

A fact that is often useful is that a sum of independent Poisson variables is itself a Poisson variable. The parameter for the sum is the sum of the individual parameter values.

The parameter λ of the Poisson distribution is estimated by the sample mean when a sample is available. For example, the horsekick data leads to an estimate of λ—say l—given by

$$l = \frac{0 \times 109 + 1 \times 65 + 2 \times 22 + 3 \times 3 + 4 \times 1}{109 + 65 + 22 + 3 + 1} = 0.61.$$

Now, we consider the construction of confidence intervals for a Poisson parameter. Consider the case of one observation, Y, and a small result, say, $Y \leq 100$. Table A.11 gives endpoints for 90%, 95%, and 99% confidence intervals. From this we find a

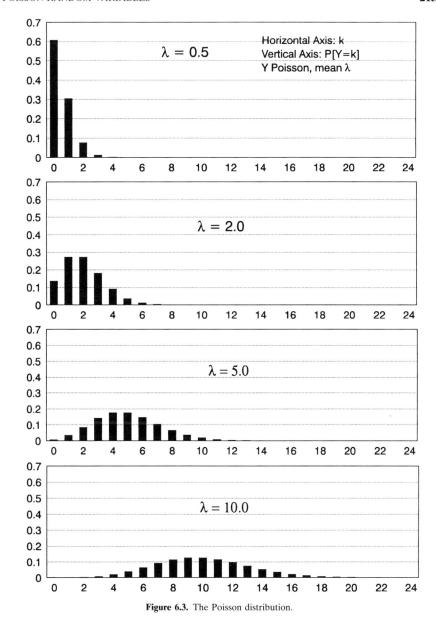

Figure 6.3. The Poisson distribution.

95% confidence interval for the proportion of black infants having ABO hemolytic disease, in the Bucher *et al.* [1976] study. The approximate Poisson variable is the binomial variable, which in this case is equal to 43; thus, a 95% confidence interval for $\lambda = n\pi$ is $(31.12, 57.92)$. The equation $\lambda = n\pi$ equates the mean values for the Poisson and binomial models. Now $n\pi$ is in $(31.12, 57.92)$ if and only if π is in the interval

$$\left(\frac{31.12}{n}, \frac{57.92}{n}\right).$$

In this case, $n = 3584$, so the confidence interval is

$$\left(\frac{31.12}{3584}, \frac{57.92}{3584}\right) \quad \text{or} \quad (0.0087, 0.0162).$$

These results are comparable with the 95% binomial limits obtained in Example 6.9: (0.0084, 0.0156).

6.6.3 Large Sample Statistical Inference for the Poisson Distribution

The Normal Approximation to the Poisson Distribution

The Poisson distribution has the property that the mean and variance are equal. For the mean large, say ≥ 100, the normal approximation can be used. That is, let $Y \sim \text{Poisson}(\lambda)$ and $\lambda \geq 100$. Then, approximately, $Y \sim N(\lambda, \lambda)$. An approximate $100(1 - \alpha)\%$ confidence interval for λ can be formed from

$$Y \pm z_{1-\alpha/2}\sqrt{Y},$$

where $z_{1-\alpha/2}$ is a standard normal deviate at two-sided significance level α. This formula is based on the fact that Y estimates the mean as well as the variance. Consider, again, the data of Bucher *et al.* [1976] (Example 6.3) dealing with the incidence of ABO hemolytic disease. The observed value of Y, the number of black infants with ABO hemolytic disease, was 43. A 95% confidence interval for the mean, λ, is $(31.12, 57.92)$ from Table A.11. Even though $Y \leq 100$, let us use the normal approximation. The estimate of the variance, σ^2, of the normal distribution is $Y = 43$, so that the standard deviation is 6.56. An approximate 95% confidence interval is $43 \pm (1.96)(6.56)$, producing $(30.1, 55.9)$, which is close to the values $(31.12, 57.92)$ tabled.

Suppose that instead of one Poisson value, there is a random sample of size n, $Y_1, Y_2, \ldots, Y_n$ from a Poisson distribution with mean λ. How should one construct a confidence interval for λ based on this data? The sum $Y = Y_1 + Y_2 + \cdots + Y_n$ is Poisson with mean $n\lambda$. Construct a confidence interval for $n\lambda$ as above, say (L, U). Then, an appropriate confidence interval for λ is $(L/n, U/n)$. Consider Example 6.20, which deals with estimating the bacterial density of soil suspensions. The results for Sample I were 72, 69, 63, 59, 59, 53, and 51. We want to set up a 95% confidence interval for the mean density using the seven observations. For this example, $n = 7$.

$$Y = Y_1 + Y_2 + \cdots + Y_7 = 27 + 69 + \cdots + 51 = 426.$$

A 95% confidence interval for 7λ is $426 \pm 1.96\sqrt{426}$.

$$L = 385.55, \quad L/7 = 55.1,$$
$$U = 466.45, \quad U/7 = 66.6,$$
$$\overline{Y} = 60.9.$$

The 95% confidence interval is $(55.1, 66.6)$.

The Square Root Transformation

It is often considered a disadvantage to have a distribution with a variance not "stable" but dependent on the mean in some way, as, for example, the Poisson distribution. The question is, whether there is a transformation, $g(Y)$, of the variable such that the variance is no longer dependent on the mean. The answer is "yes." For the Poisson distribution, it is the square root transformation. It can be shown for "reasonably large" λ, say $\lambda \geq 30$, that if $Y \sim \text{Poisson}(\lambda)$, then $\text{var}(\sqrt{Y}) \doteq 0.25$.

A side benefit is that the distribution of $\sqrt{Y}$ is more "nearly normal," that is, for specified λ, the difference between the sampling distribution of $\sqrt{Y}$ and the normal distribution is smaller for most values as compared to the difference between the distribution of Y and the normal distribution.

For the above situation, it is approximately true that

$$\sqrt{Y} \sim N(\sqrt{\lambda}, 0.25).$$

Consider Example 6.20 again. A confidence interval for $\sqrt{\lambda}$ will be constructed and then converted to an interval for λ. Let $X = \sqrt{Y}$.

Y	72	69	63	59	59	53	51
$X = \sqrt{Y}$	8.49	8.31	7.94	7.68	7.68	7.28	7.14

The sample mean and variance of X are $\overline{X} = 7.7886$ and $s_x^2 = 0.2483$. The sample variance is very close to the variance predicted by the theory $\sigma_x^2 = 0.2500$. A 95% confidence interval on $\sqrt{\lambda}$ can be set up from

$$\overline{X} \pm 1.96 \frac{s_x}{\sqrt{7}} \quad \text{or} \quad 7.7886 \pm (1.96)\sqrt{\frac{0.2483}{7}},$$

producing lower and upper limits in the X scale.

$$L_x = 7.4195, \quad U_x = 8.1577,$$

$$L_x^2 = 55.0, \quad U_x^2 = 66.5,$$

which are remarkably close to the values given previously.

The Poisson Homogeneity Test

In Chapter 4, the question of a test of normality was discussed and a graphical procedure was suggested. Fisher et al. [1922], in the paper of Example 6.20, derived an approximate test for determining whether or not a sample of observations could have come from a Poisson distribution with the same mean. The test does not determine "Poissonness," but rather, equality of means. If the experimental situations are identical (that is, we have a random sample), the test is a test for "Poissonness."

The test is the "Poisson Homogeneity Test" and is based on the property that— for the Poisson distribution—the mean equals the variance. The test is the following: suppose $Y_1, Y_2, \ldots, Y_n$ are a random sample from a Poisson distribution with mean λ. Then, for a large λ—say, $\lambda \geq 50$—the quantity

$$X^2 = \frac{(n-1)s^2}{\overline{Y}}$$

has approximately a chi-square distribution with $n - 1$ degrees of freedom, where s^2 is the sample variance.

Consider again, the data in Example 6.20. The mean and standard deviation of the seven observations are

$$n = 7, \quad \overline{Y} = 60.86, \quad s_y = 7.7552,$$

$$X^2 = \frac{(7-1)(7.7552)^2}{60.86} = 5.93.$$

Under the null hypothesis that all the observations are from a Poisson distribution with the same mean, the statistic $X^2 = 5.93$ can be referred to a chi-square distribution with six degrees of freedom. What will the rejection region be? This is determined by the alternative hypothesis. In this case it is reasonable to suppose that the sample variance will be greater than expected if the null hypothesis is not true. Hence, we want to reject the null hypothesis when χ^2 is "large"; "large" in this case means $P[X^2 \geq \chi^2_{1-\alpha}] = \alpha$.

Suppose $\alpha = 0.05$; the critical value for $\chi^2_{1-\alpha}$ with 6 d.f. is 12.59. The observed value $X^2 = 5.93$ is much less than that and the the null hypothesis is not rejected.

6.7 GOODNESS-OF-FIT TESTS

The use of appropriate mathematical models has made advances possible in biomedical science; the key word is "appropriate." An inappropriate model can lead to false or inappropriate ideas. In some situations the appropriateness of a model is clear. A random sample of a population will lead to a binomial variable for the response to a yes or no question. In other situations the issue may be in doubt. In such cases one would like to examine the data to see if the model used seems to fit the data. Tests of this type are called *goodness of fit tests*. This section examines some tests where the tests are based upon count data. The count data may arise from continuous data. One may count the number of observations in different intervals of the real line; examples are given in Sections 6.7.2 and 6.7.4.

6.7.1 Multinomial Random Variables

Binomial random variables count the number of successes in n independent trials where one and only one of two possibilities must occur. *Multinomial random variables* generalize this to allow more than two possible outcomes. In a multinomial situation, outcomes are observed which take one and only one of two or more, say k, possibilities. There are n independent trials each with the same probability of a particular outcome. Multinomial random variables count the number of occurrences of a particular outcome. Let n_i be the number of occurrences of outcome i. Thus, n_i is an integer taking a value among $0, 1, 2, \ldots, n$. There are k different n_i which add up to n since one and only one outcome occurs on each trial:

$$n_1 + n_2 + \cdots + n_k = n.$$

Let us focus on a particular outcome, say the ith. What are the mean and variance of n_i? We may classify each outcome into one of two posssibilities, the ith outcome or anything else. There are then n independent trials with two outcomes. We see that n_i is a binomial random variable when considered alone. Let π_i, where $i = 1, \ldots, k$, be the probability that the ith outcome occurs. Then

$$E(n_i) = n\pi_i, \quad \text{var}(n_i) = n\pi_i(1 - \pi_i), \tag{6.7}$$

for $i = 1, 2, \ldots, k$.

Often, multinomial outcomes are visualized as placing the outcome of each of the n trials into a separate *cell* or box. The probability π_i is then the probability an outcome lands in the ith cell.

The remainder of this section deals with multinomial observations. Tests are presented to see if a specified multinomial model holds.

6.7.2 Known Cell Probabilities

In this subsection, the cell probabilities $\pi_1, \ldots, \pi_k$ will be specified. We will use the specified values as a null hypothesis to be compared with the data $n_1, \ldots, n_k$. Since $E(n_i) = n\pi_i$, it is reasonable to examine the differences $n_i - n\pi_i$. The statistical test is given by the following fact.

Fact 6.2. Let n_i, where $i = 1, \ldots, k$, be multinomial. Under $H_0 : \pi_i = \pi_i^0$,

$$X^2 = \sum_{i=1}^{k} \left((n_i - n\pi_i^0)^2 / n\pi_i^0 \right)$$

has approximately a chi-square distribution with $k - 1$ degrees of freedom. If some π_i are not equal to π_i^0, X^2 will tend to be too large.

The distribution of X^2 is well approximated by the chi-square distribution if all of the expected values, $n\pi_i^0$, are at least five, except possibly for one or two of the values. When the null hypothesis is not true the null hypothesis is rejected for X^2 too large. At significance level α, reject H_0 if $X^2 \geq \chi^2_{1-\alpha, k-1}$ where $\chi^2_{1-\alpha, k-1}$ is the $1 - \alpha$ percentage point for a χ^2 random variable with $k - 1$ degrees of freedom.

Since there are k cells, one might expect the labeling of the degrees of freedom to be k instead of $k - 1$. However, since the n_i add up to n we only need to know $k - 1$ of them to know all k values. There are really only $k - 1$ quantities that may vary at a time; the last quantity is specified by the other $k - 1$ values.

The form of X^2 may be kept in mind by noting that we are comparing the observed, n_i, and expected values, $n\pi_i^0$. Thus,

$$X^2 = \sum (\text{observed} - \text{expected})^2 / \text{expected}.$$

Example 6.22. Are births uniformly spread throughout the year? The following data gives the number of births in King County, Washington from 1968 through 1979 by month. The estimated probability of a birth in a given month is found by taking the number of days in that month and dividing by the total number of births (leap years are included in the table).

Month	Births	Days	π_i^0	$n\pi_i^0$	$\dfrac{(n_i - n\pi_i^0)^2}{n\pi_i^0}$
January	13,016	310	0.08486	13,633	27.92
February	12,398	283	0.07747	12,446	0.19
March	14,341	310	0.08486	13,633	36.77
April	13,744	300	0.08212	13,193	23.01
May	13,894	310	0.08486	13,633	5.00
June	13,433	300	0.08212	13,193	4.37
July	13,787	310	0.08486	13,633	1.74
August	13,537	310	0.08486	13,633	0.68
September	13,459	300	0.08212	13,193	5.36
October	13,144	310	0.08486	13,633	17.54
November	12,497	300	0.08212	13,193	36.72
December	13,404	310	0.08486	13,633	3.85
Totals	160,654 (n)	3653	0.99997		163.15 (X^2)

Testing the null hypothesis, using the χ^2 Table A.3, we see that $163.15 > 31.26 = \chi^2_{0.001,11}$, so that $p < 0.001$. We reject the null hypothesis that births occur uniformly throughout the year. With this large sample size ($n = 160654$) it is not surprising that the null hypothesis can be rejected. We can examine the magnitude of the effect by comparing the ratio of observed to expected numbers of births.

Month	Observed/Expected Births
January	0.955
February	0.996
March	1.052
April	1.042
May	1.019
June	1.018
July	1.011
August	0.993
September	1.020
October	0.964
November	0.947
December	0.983

There is an excess of births in the spring (March and April) and a deficit in the late fall and winter (October through January). Note that the difference from expected is small. The maximum "excess" of births occurred in March and was only 5.2% above the number expected. A plot of the ratio vs. month would show a distinct sinusoidal pattern. □

Example 6.23. Mendel [1866] is justly famous for his theory and experiments on the principles of heredity. Sir R. A. Fisher [1936] reviewed Mendel's work and found a surprisingly good fit to the data. Consider two parents heterozygous for a dominant-recessive trait. That is, each parent has one dominant gene and one recessive gene. Mendel hypothesized that all four combinations of genes would be equally likely in the offspring. Let "A" denote the dominant gene and "a" denote the recessive gene. The two parents are Aa. The offspring should be:

Genotype	Probability
AA	1/4
Aa	1/2
aa	1/4

The Aa combination has probability 1/2 since one cannot distinguish between the two cases where the dominant gene comes from one parent and the recessive gene from the other parent. In one of Mendel's experiments he examined whether a seed was wrinkled, denoted by a, or smooth, denoted by A. By looking at offspring of these seeds Mendel classified the seeds as aa, Aa or AA. The results were:

	AA	Aa	aa	Total
Number	159	321	159	639

as presented in Table II of Fisher [1936]. Do these data support the hypothesized 1:2:1 ratio? The chi-square statistic is

$$X^2 = \frac{(159 - 159.75)^2}{159.75} + \frac{(321 - 319.5)^2}{319.5} + \frac{(159 - 159.75)^2}{159.75} = 0.014.$$

For the χ^2 distribution with two degrees of freedom, $p > 0.95$ from Table A.3 (in fact $p = 0.993$), so that the result has more agreement than would be expected by chance. We will return to these data in Example 6.24. ☐

6.7.3 Addition of Independent Chi-Square Variables: Mean and Variance of the Chi-Square Distribution

Chi-square random variables occur so often in statistical analysis that it will be useful to know more facts about chi-square variables. In this subsection facts are presented and then applied to an example (see Note 5.3 also).

Fact 6.3. Chi-square variables have the following properties:

1. Let X^2 be a chi-square random variable with m degrees of freedom. Then

$$E(X^2) = m \quad \text{and} \quad \text{var}(X^2) = 2m.$$

2. Let $X_1^2, \ldots, X_n^2$ be independent chi-square variables with $m_1, \ldots, m_n$ degrees of freedom. Then $X^2 = X_1^2 + \cdots + X_n^2$ is a chi-square random variable with $m = m_1 + m_2 + \cdots + m_n$ degrees of freedom.

3. Let X^2 be a chi-square random variable with m degrees of freedom. If m is large, say $m \geq 30$,

$$\frac{X^2 - m}{\sqrt{2m}}$$

is approximately a $N(0, 1)$ random variable.

Example 6.24. We considered Mendel's data, reported by Fisher [1936], in Example 6.23. As Fisher examined the data, he became convinced that the data fit the hypothesis too well (Box [1978], pp. 195, 300). Fisher comments: "although no explanation can be expected to be satisfactory, it remains a possibility among others that Mendel was deceived by some assistant who knew too well what was expected."

One reason Fisher arrived at his conclusion was by combining χ^2 values from different experiments by Mendel. The following table presents the data.

Experiments	X^2	Degrees of Freedom
3:1 Ratios	2.14	7
2:1 Ratios	5.17	8
Bifactorial experiments	2.81	8
Gametic ratios	3.67	15
Trifactorial experiments	15.32	26
Total	29.11	64

If all the null hypotheses are true, by the above facts, $X^2 = 29.11$ should look like a χ^2 with 64 degrees of freedom. An approximate normal variable,

$$Z = \frac{29.11 - 64}{\sqrt{128}} = -3.08$$

has less than 1 chance in 1000 of being this small ($p = 0.99995$). One can only conclude that something peculiar occurred in the collection and reporting of Mendel's data.   □

6.7.4 Chi-Square Tests for Unknown Cell Probabilities

Above we considered tests of the goodness-of-fit of multinomial data when the probability of being in an individual cell was precisely specified; for example, by a genetic model of how traits are inherited. In other situations, the cell probabilities are not known, but may be estimated. First we motivate the techniques by presenting a possible use; next, we present the techniques, and finally, we illustrate the use of the techniques by example.

Consider a sample of n numbers that may come from a normal distribution. How might we check the assumption of normality? One approach is to divide the real number line into a finite number of intervals. The observed number of points in each interval may then be counted. The numbers in the different intervals or cells are multinomial random variables. If the sample were normal with known mean μ and known standard deviation σ, the probability, π_i, that a point falls between the endpoints of the ith interval—say Y_1 and Y_2—is known to be

$$\pi_i = \Phi\left(\frac{Y_2 - \mu}{\sigma}\right) - \Phi\left(\frac{Y_1 - \mu}{\sigma}\right),$$

where Φ is the distribution function of a standard normal random variable. In most cases, μ and σ are not known, so μ and σ, and thus π_i, must be estimated. Now π_i depends upon two variables, μ and σ: $\pi_i = \pi_i(\mu, \sigma)$ where the notation $\pi_i(\mu, \sigma)$ means π_i is a function of μ and σ. It is natural if we estimate μ and σ by, say, $\widehat{\mu}$ and $\widehat{\sigma}$, to estimate π_i by $p_i(\widehat{\mu}, \widehat{\omega})$. That is,

$$p_i(\widehat{\mu}, \widehat{\sigma}) = \Phi\left(\frac{Y_2 - \widehat{\mu}}{\widehat{\sigma}}\right) - \Phi\left(\frac{Y_1 - \widehat{\mu}}{\widehat{\sigma}}\right).$$

From this, a statistic (X^2) can be formed as above. If there are k cells,

$$X^2 = \sum \frac{(\text{observed} - \text{expected})^2}{\text{expected}} = \sum_{i=1}^{k} \frac{\left(n_i - np_i(\widehat{\mu}, \widehat{\sigma})\right)^2}{np_i(\widehat{\mu}, \widehat{\sigma})}.$$

Does X^2 now have a chi-square distribution? The following facts describe the situation.

Fact 6.4. Suppose n observations are grouped or placed into k categories or cells such that the probability of being in cell i is $\pi_i = \pi_i(\Theta_1, \ldots, \Theta_s)$, where π_i depends upon s parameters Θ_i, and where $s < k - 1$. Suppose that none of the s parameters are determined by the remaining $s - 1$ parameters. Then

1. If $\widehat{\Theta}_1, \ldots, \widehat{\Theta}_s$, the parameter estimates, are chosen to minimize X^2, the distribution of X^2 is approximately a chi-square random variable with $k - s - 1$ degrees of freedom for large n. Estimates chosen to minimize the value of X^2 are called *minimum chi-square estimates*.

2. If estimates of $\Theta_1, \ldots, \Theta_s$ other than the minimum chi-square estimates are used, then for large n the distribution function of X^2 lies between the distribution functions of chi-square variables with $k - s - 1$ degrees of freedom and $k - 1$ degrees of freedom. More specifically, let $X^2_{1-\alpha, m}$ denote the α significance level critical value for a chi-square distribution with m degrees of freedom. The significance level α critical value of X^2 is less than or equal to $\chi^2_{1-\alpha, k-1}$. A conservative test of the multinomial model is to reject the null hypothesis that the model is correct if $X^2 \geq \chi^2_{1-\alpha, k-1}$.

These complex statements are best understood by applying them to an example.

Example 6.25. Table 3.4 of Section 3.3.1 gives the age in days at death of 78 SIDS cases. Test for normality at the 5% significance level using a χ^2-test.

Before performing the test, one needs to divide the real number line into intervals or cells. The usual approach is to:

1. Estimate the parameters involved. In this case the unknown parameters are μ and σ. We estimate by $\overline{Y}$ and s.
2. Decide on k, the number of intervals. Let there be n observations. A good approach is to choose k as follows:

   a. For $20 \le n \le 100$, $k \doteq n/5$.

   b. For $n > 300$, $k \doteq 3.5n^{2/5}$ (here, $n^{2/5}$ is n raised to the 2/5 power).
3. Find the endpoints of the k intervals so that each interval has probability $1/k$. The k intervals are:

$$
\begin{array}{ll}
(-\infty, a_1] & \text{interval one} \\
(a_1, a_2] & \text{interval two} \\
\vdots & \vdots \\
(a_{k-2}, a_{k-1}] & \text{interval } (k-1) \\
(a_{k-1}, \infty) & \text{interval } k
\end{array}
$$

Let Z_i be the value such that a standard normal random variable takes a value less than Z_i with probability i/k. Then

$$
a_i = \overline{X} + sZ_i.
$$

(In testing for another distribution than the normal distribution, other methods of finding cells of approximately equal probability need to be used.)

4. Compute the statistic

$$
X^2 = \sum_{i=1}^{k} \frac{(n_i - n/k)^2}{n/k}
$$

where n_i is the number of data points in cell i.

To apply steps 1–4 to the data at hand, one computes $n = 78$, $\overline{X} = 97.85$, and $s = 55.66$. As $78/5 = 15.6$, we will use $k = 15$ intervals. From tables of the normal distribution, we find Z_i, $i = 1, 2, \ldots, 14$, so that a standard normal random variable has probability $i/15$ of being less than Z_i. The values of Z_i and a_i are:

i	Z_i	a_i	i	Z_i	a_i	i	Z_i	a_i
1	-1.50	12.8	6	-0.25	84.9	11	0.62	135.0
2	-1.11	35.3	7	-0.08	94.7	12	0.84	147.7
3	-0.84	50.9	8	0.08	103.9	13	1.11	163.3
4	-0.62	63.5	9	0.25	113.7	14	1.50	185.8
5	-0.43	74.5	10	0.43	124.1			

The observed number of observations in the 15 cells, from left to right, are 0, 8, 7, 5, 7, 9, 7, 5, 6, 6, 2, 2, 3, 5, and 6. In each cell, the expected number of observations is $np_i = n/k$ or $78/15 = 5.2$. Then,

$$
X^2 = \frac{(0 - 5.2)^2}{5.2} + \frac{(8 - 5.2)^2}{5.2} + \cdots + \frac{(6 - 5.2)^2}{5.2} = 16.62.
$$

We know that the 0.05 critical values are between the chi-square critical values with 12 and 14 degrees of freedom. The two values are 21.03 and 23.68. Thus, we do not reject the hypothesis of normality. (If the X^2 value had been greater 23.68 we would have rejected the null hypothesis of normality. If X^2 were in between 21.03 and 23.68 the answer would be in doubt. In that case, it would advisable to compute the minimum chi-square estimates so that a known distribution results.)

Note that the largest observation, 307, is $(307 - 97.85)/55.6 = 3.76$ sample standard deviations from the sample mean. In using a chi-square goodness-of-fit test all large observations are placed into a single cell. The magnitude of the value is lost. If one is worried about large outlying values there are better tests of the fit to normality.

□

NOTES

Note 6.1: Use of Tables for Fisher's Exact Test

Tables for Fisher's exact test are the hardest of all statistical tables to use.

1. The data are presented with the following notation:

$$
\begin{array}{cc|c}
a & A - a & A \\
b & B - b & B \\
\hline
k & N - k & N
\end{array}
$$

2. The tricky part begins! In order to save printed space, the rows and/or columns are interchanged until two conditions hold: (i) $A \geq B$, and (ii) $a/A \geq b/B$.

Examples

a.

$$
\begin{array}{cc|c}
10 & 8 & 18 \\
3 & 4 & 7 \\
\hline
13 & 12 & 25
\end{array}
$$

In this case, both conditions ($18 \geq 7$ and $10/18 \geq 3/7$) are filled; hence, no rearrangement is necessary.

b.

$$
\begin{array}{cc|c}
3 & 4 & 7 \\
10 & 8 & 18 \\
\hline
13 & 12 & 25
\end{array}
$$

Here, $7 \ngeq 18$, so we interchange rows to get

$$
\begin{array}{cc|c}
10 & 9 & 18 \\
3 & 4 & 7 \\
\hline
13 & 12 & 25
\end{array}
$$

Stop as in the first example.

c. Even if the first condition is fulfilled, you might have something like

$$
\begin{array}{cc|c}
8 & 10 & 18 \\
4 & 3 & 7 \\
\hline
12 & 13 & 25
\end{array}
$$

Condition (i) is fulfilled ($18 \geq 7$), but condition (ii) ($8/18 \not\geq 4/7$) is not, so we interchange columns to get

$$
\begin{array}{cc|c}
10 & 8 & 18 \\
3 & 4 & 7 \\
\hline
13 & 12 & 25
\end{array}
$$

Stop as in the first example.

d. Finally, let us take

$$
\begin{array}{cc|c}
4 & 3 & 7 \\
8 & 10 & 18 \\
\hline
12 & 13 & 25
\end{array}
$$

First, $7 \not\geq 18$, so we need to interchange the rows to get

$$
\begin{array}{cc|c}
8 & 10 & 18 \\
4 & 3 & 7 \\
\hline
12 & 13 & 25
\end{array}
$$

Then, since $8/18 \not\geq 4/7$, we must interchange the columns to get

$$
\begin{array}{cc|c}
10 & 8 & 18 \\
3 & 4 & 7 \\
\hline
13 & 12 & 25
\end{array}
$$

3. **Two-sided Tests.** Table A.9 is constructed to give one-sided hypothesis tests. To do a two-sided test at significance level α, locate the column corresponding to $\alpha/2$ and the line corresponding to the given A, B and a. *The null hypothesis is rejected if b is less than or equal to the integer in bold type.* A dash in the space for b means that the numbers are not appropriate to ever reject at the given significance level. The number following the critical value is the one-sided p-value when the number in bold type occurs. If that number does occur, the two-sided p-value is two times the number in the table. (A better method if a computer is available to compute the probability of each possible value of b is to sum the probabilities over all possibilities where the probability is not larger then the probability for the observed value of b.) The following examples (shown after rows and columns are appropriately interchanged) use Table A.9.

Examples

a. Perform a two-sided test at significance level 0.05. Use the $0.05/2 = 0.025$ columns.

$$
\begin{array}{cc|c}
7 & 2 & 9 \\
3 & 6 & 9 \\
\hline
10 & 8 & 18
\end{array}
$$

Here $A = 9$, $B = 9$, $a = 7$, and $b = 3$. The critical value for b is 1. As $3 > 1$, we do *not* reject the null hypothesis at the 0.05 significance level.

b. Test at the 0.02 significance level (two-sided test) for equal proportions in the following 2×2 table.

$$
\begin{array}{cc|c}
7 & 3 & 10 \\
1 & 8 & 9 \\
\hline
8 & 11 & 19
\end{array}
$$

Here $A = 10$, $B = 9$, $a = 7$, and $b = 1$. Use the $0.02/2 = 0.01$ column. The critical value is 0. As $1 > 0$, we do *not* reject the null hypothesis. Note that the 0.025 column has a critical value of 1 as observed, thus we reject at the 0.05 significance level. Since our observed value is the tabulated critical value, the p-value is $2 \cdot 0.015 = 0.030$.

c. Do a two-sided Fisher's exact test at the 0.10 significance level. We use the $0.10/2 = 0.05$ probability column.

$$
\begin{array}{cc|c}
9 & 0 & 9 \\
2 & 4 & 6 \\
\hline
11 & 4 & 15
\end{array}
$$

Here $A = 9$, $B = 6$, $a = 9$, and $b = 2$. The critical value is 3. As $2 \le 3$, we reject at the 0.10 significance level. We can, however, go further than this. Going across the row corresponding to $A = 9$, $B = 6$, and $a = 9$, we see that 2 appears as a critical value with the probability 0.011. The p-value is $2 \cdot 0.011 = 0.022$.

4. One-sided tests. The hardest part in using Fisher's Exact Test for one-sided tests is to know whether the appropriate one-sided rejection region is to reject when b is large or small. The confusion results because of the necessity of rearranging the rows and columns.

One procedure that works is to select one number (or cell) that you think will be too large if the null hypothesis does not hold and to label the number with a plus sign, for example, 7+. Fisher's exact test is a conditional test with the row and column sums being fixed. If one number is expected to be too large (if the alternative hypothesis holds), the other number in the row is expected to be too small (since the row sum is fixed). Label the second number with a minus sign, for example, 5−. A similar argument holds for the columns. The +, − pattern has one of two possibilities:

+	−
−	+

−	+
+	−

Each time the rows and columns interchange, the pattern changes. For example, suppose the original pattern, including the entry that will be too large under the alternative, is

6	4	10
1	12+	13
7	16	23

$\Longrightarrow$

6+	4−	10
1−	12+	13
7	16	23

$\Longrightarrow$

1−	12+	13
6+	4−	10
7	16	23

$\Longrightarrow$

12+	1−	13
4−	6+	10
16	7	23

Here $A = 13$, $B = 10$, $a = 12+$, and $b = 4−$. To perform the one-sided test at a given significance level α, use the column with probability equal to α. Find the critical value. *If the sign of the b entry is +, do not reject the null hypothesis.* If the sign on the b entry is minus, reject the null hypothesis if b is less than or equal to the critical value. If b equals the critical value, the following probability is the p-value. If b is less than the critical value, the p-value is less than the probability in small type after the critical value.

As an example, at the $\alpha = 0.05$ level test, perform a one-sided test of the 2×2 tables below:

	Rash Cured		
	Yes	**No**	**Total**
Treatment	12	3	15
Placebo	5	10	15
Total	17	13	30

One expects the treatment/yes cell to be too large under the alternative. Thus, we have

12+	3−	15
5−	10+	15
17	13	30

Here, $A = 15$, $B = 15$, $a = 12$, and $b = 5−$. The critical value (not given here) is 6. As the b entry has a − sign, we compare it to the critical value of 6. As $5 < 6$, reject the null hypothesis of the same cure rates at the 0.05 significance level.

Suppose, however, the table had been:

| | **Rash Cured** | | |
	Yes	**No**	**Total**
Treatment	5+	10−	15
Placebo	12−	3+	15
Total	17	13	30

The table becomes

12−	3+	15
5+	10−	15
17	13	30

In this case, $A = 15$, $B = 15$, $a = 12$, and $b = 5+$. As b has a + sign, we do *not* reject the null hypothesis, even though $5 < 6$, the critical value.

Tables with values extending beyond those of standard tables are given in Finney *et al.* [1963] and Bennett and Horst [1966]. In addition, many statistical computer software systems allow calculation of p-values using Fisher's Exact Test for 2×2 tables.

Note 6.2: Continuity Correction for 2×2 Table Chi-Square Values

There has been controversy about the appropriateness of the continuity correction for 2×2 tables (Conover [1974]). The continuity correction makes the *actual* significance levels under the null hypothesis closer to the hypergeometric (Fisher's Exact Test) actual significance levels. When compared to the chi-square distribution, the *actual* significance levels are too low (Conover [1974]; Starmer *et al.* [1974]; and Grizzle [1967]). The *uncorrected* "chi-square" value referred to chi-square critical values gives actual and nominal significance levels that are close. For this reason, the authors recommend that the continuity correction *not* be used. Use of the continuity correction would be correct, but over-conservative. For arguments on the opposite side, see Mantel and Greenhouse [1968]. A good summary can be found in Little [1989].

Note 6.3: Standard Error of $\widehat{\omega}$ as Related to the Standard Error of $\log \widehat{\omega}$

Let X be a positive variate with mean μ_x and standard deviation σ_x. Let $Y = \log_e X$. Let the mean and standard deviation of Y be μ_y and σ_y, respectively. It can be shown that under certain conditions

$$\frac{\sigma_x}{\mu_x} \doteq \sigma_y.$$

The quantity σ_x/μ_x is known as the coefficient of variation. Another way of writing this is

$$\sigma_x \doteq \mu_x \sigma_y.$$

If the parameters are replaced by the appropriate statistics, the expression becomes

$$s_x \doteq \bar{x}s_y,$$

and the standard deviation of $\hat{\omega}$ then follows from this relationship.

Note 6.4: Some Limitations of the Odds Ratio

The odds ratio uses one number to summarize four numbers and some information about the relationship is necessarily lost. The following example shows one of the limitations.

Fleiss [1981] discusses the limitations of the odds ratio as a measure for public health. He presents the following data taken from Smoking and Health [1964].

MORTALITY RATES PER 100,000 PERSON-YEARS FROM LUNG CANCER AND CORONARY ARTERY DISEASE FOR SMOKERS AND NONSMOKERS OF CIGARETTES

	Smokers	Nonsmokers	Odds Ratio	Difference
Cancer of the Lung	48.33	4.49	10.8	43.84
Coronary Artery Disease	294.67	169.54	1.7	125.13

The point is that although the risk ω is increased much more for cancer, the added number dying of coronary artery disease is higher, and in some sense smoking has more effect in this case.

Note 6.5: The Mantel–Haenszel Test for Association

The chi-square test of association given in conjunction with the Mantel–Haenszel Test of Section 6.4.5 arises from the approach of the section by choosing a_i and s_i appropriately (Fleiss [1981]). The corresponding chi-square test for homogeneity does *not* make sense and should not be used. Mantel et al. [1977] give the problems associated with using the approach to look at homogeneity.

Note 6.6: Matched Pair Studies

One of the difficult aspects in the design and execution of matched pair studies is to decide upon the matching variables, and then find matches to the degree desired. In practice, many decisions are made for logistic and monetary reasons; these factors will not be discussed here. The primary purpose of matching is to have a *valid* comparison. Variables are matched to increase the validity of the comparison. Inappropriate matching can hurt the statistical power of the comparison. Breslow and Day [1980] and Miettinen [1970] give some fundamental background. Fisher and Patil [1974] further elucidate the matter (see also Problem 6.30).

Note 6.7: More on the Chi-Square Goodness-of-Fit Test

The presentation of the goodness-of-fit test as presented in this chapter did not mention some of the subtleties associated with the subject. A few arcane points, with appropriate references, are given in this note.

1. In Fact 6.4, the estimate used should be maximum likelihood estimates or equivalent estimates (Chernoff and Lehmann [1954]).

2. The initial chi-square limit theorems were proved for fixed cell boundaries. Limiting theorems where the boundaries were random (depending upon the data) were proved later (Kendall and Stuart [1967], Sections 30.20 and 30.21).

3. The number of cells to be used (as a function of the sample size) has its own literature. More detail is given in Kendall and Stuart [1967], Sections 30.28 to 30.30. The recommendations for k in this text are based upon this material.

Note 6.8: The Predictive Value of a Positive Test

The predictive value of a positive test, PV^+, is related to the prevalence (PREV), sensitivity (SENS) and specificity (SPEC) of a test by the following equation:

$$PV^+ = \frac{1}{1 + \left[\dfrac{1 - \text{SPEC}}{\text{SENS}}\right]\left[\dfrac{1 - \text{PREV}}{\text{PREV}}\right]}.$$

Here the PREV, SENS and SPEC are on a scale of 0 to 1 of proportions instead of percents. If we define $\text{logit}(p) = \log[p/(1 - p)]$, then the predictive value of a positive test is very simply related to the prevalence as follows:

$$\text{logit}[PV^+] = \log\left[\frac{\text{SENS}}{1 - \text{SPEC}}\right] + \text{logit}(\text{PREV}).$$

This is a very informative formula. For rare diseases (i.e., low prevalence), the term $\text{logit}(\text{PREV})$ will dominate the predictive value of a positive test. So no matter what the sensitivity or the specificity of a test, the predictive value will be low. A similar relationship can be established for the predictive value of a negative test.

Frequently a plot is made of SENS vs. $(1 - \text{SPEC})$ as some cut-off criterion is varied. Such a plot is called an ROC curve (**R**eceiver **O**perating **C**haracteristic). Then the quantity $(\text{SENS})/(1 - \text{SPEC})$ is associated with a point in the curve, and defines the tangent of the angle the point makes with the origin and x-axis. Since an ROC curve does not depend upon the prevalence, it does not provide enough information to discriminate between tests.

Note 6.9: Rule of Threes

Table A.11 lists as upper 90% confidence interval for a Poisson random variable the value 3. This has led to the "rule of threes" which states that if in n trials zero events of interest are observed, a 95% confidence bound on the underlying rate is $3/n$. For a fuller discussion, see Hanley [1983]. See also Problem 6.29.

PROBLEMS

1. In a randomized trial of surgical and medical treatment a clinic finds eight of nine patients randomized to medicine. They complain that the randomization must not be working, that is, π cannot be one-half.

a. Is their argument reasonable from their point of view?

*__*b.__ With 15 clinics in the trial, what is the probability that *all* 15 clinics have fewer than eight individuals randomized to each treatment, of the first nine individuals randomized? Assume independent binomial distributions with $\pi = 1/2$ at each site.

2. In a dietary study, 14 of 20 subjects lost weight. If weight is assumed to fluctuate by chance, with probability $1/2$ of losing weight, what is the exact two-sided p-value for testing the null hypothesis $\pi = 1/2$?

3. Edwards and Fraccaro [1960] present Swedish data about the sex of a child and the parity. These data are:

Sex	**Order of Birth**							Total
	1	**2**	**3**	**4**	**5**	**6**	**7**	
Males	2846	2554	2162	1667	1341	987	666	12223
Females	2631	2361	1996	1676	1230	914	668	11476
Total	5477	4915	4158	3343	2571	1901	1334	23699

a. Find the p-value for testing the hypothesis that a birth is equally likely to be of either sex using the combined data and binomial assumptions.

b. Construct a 90% confidence interval for the probability that a birth is a female child.

c. Repeat (a) and (b) using only the data for birth order 6.

4. Ounsted [1953] presents data about cases with convulsive disorders. Among the cases there were 82 females and 118 males. At the 5% significance level, test the hypothesis that a case is equally likely to be of either sex. The siblings of the cases were 121 females and 156 males. Test at the 10% significance level the hypothesis that the siblings represent 53% or more male births.

5. Smith, Delgado and Rutledge [1976] report data on ovarian carcinoma (cancer of the ovaries). Individuals had different numbers of courses of chemotherapy. The 5-year survival data for those with 1–4 and 10 or more courses of chemotherapy are:

Courses	**Five-Year Status**	
	Dead	**Alive**
1–4	21	2
≥ 10	2	8

Using Fisher's exact test, is there a statistically significant association ($p \leq 0.05$) in this table? (In this problem and the next you will need to compute the hypergeometric probabilities using the results of Problem 6.26.)

6. Borer *et al.* [1980], study 45 individuals after an acute myocardial infarction (heart attack). They measure the ejection fraction (EF); the EF is the percent of the blood pumped from the left ventricle (the pumping chamber of the heart) during a heart beat. A low EF indicates damaged or dead heart muscle (myocardium). During follow-up four patients died. Dividing EF into low ($< 35\%$) and high ($\geq 35\%$) EF groups gave the following table.

	Vital Status	
EF	Dead	Alive
$< 35\%$	4	9
$\geq 35\%$	0	32

Is there reason to suspect, at a 0.05 significance level, that death is more likely in the low EF group? Use a one-sided p-value for your answer, since biological plausibility (and prior literature) indicates low EF is a risk factor for mortality.

7. Using the data of Problem 6.4, test the hypothesis that the proportions of male births among those with convulsive disorders and among their siblings are the same.

8. Lawson and Jick [1976] compare drug prescription in America and Scotland.

   a. In patients with congestive heart failure, two or more drugs were prescribed in 257 of 437 American patients. In Scotland, 39 of 179 patients had two or more drugs prescribed. Test the null hypothesis of equal proportions giving the resulting p-value. Construct a 95% confidence interval for the difference in the proportions.

   b. Patients with dehydration received two or more drugs in 55 of 74 Scottish cases as compared to 255 of 536 in America. Answer the questions of (a).

9. A randomized study among patients with angina (heart chest pain) is to be conducted with 5-year follow-up. Individuals are to be randomized to medical and surgical treatment. Suppose that the estimated 5-year medical mortality is 10%, and it is hoped that the surgical mortality will be only half as much (5%) or less. If a test of binomial proportions at the 5% significance level is to be performed, and we want to be 90% certain of detecting a difference of 5% or more, what sample sizes are needed for the two (equal sized) groups?

10. A cancer with poor prognosis, a 3-year mortality of 85%, is studied. A new mode of chemotherapy is to be evaluated. Suppose when testing at the 0.10 significance level, one wishes to be 95% certain of detecting a difference if survival has been increased to 50% or more. The randomized clinical trial will have equal numbers of individuals in each group. How many patients should be randomized?

11. Comstock and Partridge [1972], Example 6.17, show data giving an association between church attendance and health. From the data of Example 6.17, which was collected from a prospective study:

a. Compute the relative risk of an arteriosclerotic death in the 3-year follow-up period if one usually attends church less than one a week as compared to once a week or more.

b. Compute the odds ratio and a 95% confidence interval.

c. Find the percent error of the odds ratio as an approximation to the relative risk, that is, compute $100(OR - RR)/RR$.

d. The data in this population on deaths from cirrhosis of the liver are:

	Cirrhosis fatality?	
Usual church attendance	**Yes**	**No**
≥ 1 per week	5	24240
< 1 per week	25	30578

Repeat (a), (b) and (c) for these data.

12. Peterson *et al.* [1979] studied the patterns of infant deaths (especially SIDS) in King County, Washington during the years 1969–1977. They compared the SIDS deaths with a 1% sample of all births during the specified time period. Tables relating the occurence of SIDS with maternal age less than or equal to 19 years of age, and to birth order greater than one follow for those with single births.

	Child	
Birth order	**SIDS**	**Control**
> 1	201	689
$= 1$	92	626

	Child	
Maternal age	**SIDS**	**Control**
≤ 19	76	164
> 19	217	1151

	Child	
	SIDS	**Control**
Birth order > 1 *and* maternal age ≤ 19	26	17
Birth order $= 1$ *or* maternal age > 19	267	1298

	Child	
	SIDS	**Control**
Birth order > 1 *and* maternal age ≤ 19	26	17
Birth order $= 1$ *and* maternal age > 19	42	479

a. Compute odds ratios and 95% confidence intervals for the four tables.

b. Which table of the last two do you think best reflects the risk of both risk factors at once? Why? (There is not a definitely correct answer.)

*c. The control data represent a 1% sample of the population data. Knowing this, how would you directly estimate the relative risk?

13. Rosenberg *et al.* [1980] studied the relationship between coffee drinking and myocardial infarction in young women, aged 30–49 years. This retrospective study included 487 cases hospitalized for the occurrence of a myocardial infarction [MI]. Nine hundred eighty controls hospitalized for an acute condition (trauma, acute cholecystitis, acute respiratory diseases and appendicitis) were selected. Data for consumption of five or more cups of coffee containing caffeine were:

Cups per Day	MI	Control
≥ 5	152	183
< 5	335	797

Compute the odds ratio of a MI for heavy (≥ 5 cups per day) coffee drinkers versus nonheavy coffee drinkers. Find the 90% confidence interval for the odds ratio.

14. The data of Problem 6.13 were considered to be possibly confounded with smoking. The 2×2 tables by smoking status, in cigarettes per day, are given below.

NEVER SMOKED

Cups per Day	MI	Control
≥ 5	7	31
< 5	55	269

FORMER SMOKER

Cups per Day	MI	Control
≥ 5	7	18
< 5	20	112

25–34 CIGARETTES PER DAY

Cups per Day	MI	Control
≥ 5	34	24
< 5	50	55

1–14 CIGARETTES PER DAY

Cups per Day	MI	Control
≥ 5	7	24
< 5	33	114

35–44 CIGARETTES PER DAY

Cups per Day	MI	Control
≥ 5	27	24
< 5	55	58

15–24 CIGARETTES PER DAY

Cups per Day	MI	Control
≥ 5	40	45
< 5	88	172

45+ CIGARETTES PER DAY

Cups per Day	MI	Control
≥ 5	30	17
< 5	34	17

a. Compute the Mantel–Haenszel estimate of the odds ratio and the chi-square statistic for association. Would you reject the null hypothesis of no association between smoking and myocardial infarction at the 5% significance level?

b. Using the log odds ratio as the measure of association in each subtable, compute the chi-square statistic for association. Find the estimated overall odds ratio and a 95% confidence interval for this quantity.

15. The paper of Remein and Wilkerson [1961] Example 6.16, considers screening tests for diabetes. The Somogyi–Nelson (venous) blood test (data at 1 h after a test meal and using 130 mg 100 ml as the blood sugar cutoff level) gives the following table:

Test	Diabetic	Nondiabetic	Total
+	59	48	107
−	11	462	473
Total	70	510	580

    a. Compute the sensitivity, specificity, predictive value of a positive test and predictive value of a negative test.

    b. Using the sensitivity and specificity of the test as given in (a) plot curves of the predictive values of the test versus the percent of the population with diabetes (0–100%). The first curve will give the probability of diabetes given a positive test. The second curve will give the probability of diabetes given a negative test.

16. Remein and Wilkerson [1961], in the paper referenced in Problem 6.15, present tables showing the trade-off between sensitivity and specificity that arises by changing the cutoff value for a positive test. For blood samples collected 1 h after a test meal, three different blood tests gave the following data:

Blood Sugar (mg/100 ml)	Somogyi–Nelson SENS	Somogyi–Nelson SPEC	Folin–Wu SENS	Folin–Wu SPEC	Anthrone SENS	Anthrone SPEC
70	—	—	100.0	8.2	100.0	2.7
80	—	1.6	97.1	22.4	100.0	9.4
90	100.0	8.8	97.1	39.0	100.0	22.4
100	98.6	21.4	95.7	57.3	98.6	37.3
110	98.6	38.4	92.9	70.6	94.3	54.3
120	97.1	55.9	88.6	83.3	88.6	67.1
130	92.9	70.2	78.6	90.6	81.4	80.6
140	85.7	81.4	68.6	95.1	74.3	88.2
150	80.0	90.4	57.1	97.8	64.3	92.7
160	74.3	94.3	52.9	99.4	58.6	96.3
170	61.4	97.8	47.1	99.6	51.4	98.6
180	52.9	99.0	40.0	99.8	45.7	99.2
190	44.3	99.8	34.3	100.0	40.0	99.8
200	40.0	99.8	28.6	100.0	35.7	99.8

    a. Plot three curves, one for each testing method, on the same graph. Let the vertical axis be the sensitivity and the horizontal axis (1 − specificity) of the test. The curves are generated by the changing cutoff values.

    b. Which test looks most promising, if any? Why? (see also Note 6.8).

17. Example 6.19 discusses the data of Sartwell *et al.* [1969], examining the relationship between thromboembolism and oral contraceptive use. The data are presented below for several subsets of the population. For each subset:

a. Perform McNemar's test for a case-control difference (5% significance level).

b. Estimate the relative risk.

c. Find an appropriate 90% confidence interval for the relative risk.
Nonwhite

	Case	
Control	**Yes**	**No**
Yes	3	3
No	11	9

Married

	Case	
Control	**Yes**	**No**
Yes	8	10
No	41	46

Age 15–29

	Case	
Control	**Yes**	**No**
Yes	5	33
No	7	57

18. Janerich *et al.* [1980] compared oral contraceptive usage among mothers of malformed infants and matched controls who gave birth to healthy children. The controls were matched for maternal age and race of the mother. For each of the following, estimate the odds ratio and form a % confidence interval for the odds ratio.

a. Women who conceived while using the pill or immediately following pill use.

	Case	
Control	**Yes**	**No**
Yes	1	33
No	49	632

b. Women who experienced at least one complete pill-free menstrual period prior to conception.

	Case	
Control	**Yes**	**No**
Yes	38	105
No	105	467

c. Cases restricted to major structural anatomical malformations. Use of oral contraceptives after the last menstrual period or in the menstrual cycle prior to conception.

	Case	
Control	**Yes**	**No**
Yes	0	21
No	45	470

d. As in (c) but restricted to mothers of age 30 or more.

	Case	
Control	**Yes**	**No**
Yes	0	1
No	6	103

19. Robinette *et al.* [1980] studied the effects upon health of occupational exposure to microwave radiation (radar). The study looked at groups of enlisted naval personnel who were enrolled during the Korean war period. Using Table A.11 of Poisson confidence limits, find 95% confidence intervals for the percent of men dying of causes as given in the data below. Deaths were recorded that occurred during 1950–1974.

a. Eight of 1412 Aviation Electronics Technicians died of malignant neoplasms.

b. Six of the 1412 Aviation Electronics Technicians died of suicide, homicide, or other trauma.

c. Nineteen of 10,116 Radarmen died by suicide.

d. Sixteen of 3298 Fire Control Technicians died of malignant neoplasms.

e. Three of 9253 Radiomen died of infective and parasitic disease.

f. None of 1412 Aviation Electronics Technicians died of infective and parasitic disease.

20. The following data are from the paper mentioned in Problem 6.19. Find 95% confidence intervals for the population percent dying based on the data given below.

a. Use the normal approximation to the Poisson distribution (which is approximating a binomial distribution).

i. 199 of 13,078 Electronics Technicians died of disease.

ii. 100 of 13,078 Electronics Technicians died of circulatory disease.

iii. 308 of 10,116 Radarmen died (of any cause).

iv. 441 of 13,078 Electronics Technicians died (of any cause).

v. 103 of 10,116 Radarmen died of an accidental death.

b. Use the large sample binomial confidence intervals (of Section 6.3.6) on Parts (i)–(v). Do you think the intervals are similar to those calculated in Part (a)?

21. Infant deaths in King County, Washington were grouped by season of the year. The number of deaths by season, for selected causes of death, were:

	Season			
	Winter	**Spring**	**Summer**	**Autumn**
Asphyxia	50	48	46	34
Immaturity	30	40	36	35
Congenital malformations	95	93	88	83
Infection	40	19	40	43
Sudden infant death syndrome	78	71	87	86

a. At the 5% significance level test, the hypothesis that SIDS deaths are uniformly ($p = 1/4$) spread among season.

b. At the 10% significance level test the hypothesis that the deaths due to infection are uniformly spread among season.

c. What can you say about the p-value for testing that asphyxia deaths are uniformly spread among seasons? Immaturity deaths?

22. Fisher [1958] after Carver [1927]: The following 3839 seedlings are progeny of self-fertilized heterozygotes. (Each seedling can be classified as either Starchy or Sugary and as either Green or White):

Number of Seedlings	Green	White	Total
Starchy	1997	906	2903
Surgary	904	32	936
Total	2901	938	3839

*a. On the assumption that the Green and Starchy genes are dominant and that the factors are independent, show that by Mendel's law the ratio of expected frequencies (starchy green, starchy white, sugary green, sugary white) should be 9:3:3:1.

b. Calculate the expected frequencies under the hypothesis that Mendel's law holds and assuming 3839 seedlings.

c. The data are multinomial with parameters π_1, π_2, π_3, and π_4, say. What does Mendel's law imply about the relationships among the parameters?

d. Test the goodness-of-fit.

23. Goodness-of-fit for the Binomial Distribution. Fisher [1958] presents data of Geissler [1889] of the number of male births in German families with eight offspring. One model one might consider for these data is the binomial distribution. This problem requires a goodness-of-fit test.

a. Estimate π, the probability that a birth is male. This is done by using the estimate $p =$ (total no. of male births)/(total number of births). The data are in Table 3.12.

b. Using the p of (a), find the binomial probabilities for number of boys $= 0, 1, 2, 3, 4, 5, 6, 7,$ and 8. Estimate the expected number of observations in each cell, if the binomial distribution is correct.

c. Compute the X^2 value.

d. The X^2 distribution lies between chi-square distributions with what two degrees of freedom? (Section 6.7.4).

e. Test the goodness-of-fit by finding the two critical values of (d). What can you say about the p-value for the goodness-of-fit test?

*24. The Binomial Coefficient and Binomial Probabilities.

a. Let $R(n)$ be the number of ways to arrange n distinct objects in a row. Show that $R(n) = n! = 1 \cdot 2 \cdot 3 \cdots n$. By definition, $R(0) = 1$. Hint: Clearly $R(1) = 1$. Use *mathematical induction*. That is, show that if $R(n-1) = (n-1)!$, then $R(n) = n!$. This would show that, for all positive integers n, $R(n) = n!$. Why? (To show $R(n) = n!$, suppose $R(n-1) = (n-1)!$. Argue that you may choose any of the n objects for the first position. For each such choice, the remaining $n-1$ objects may be arranged in $R(n-1) = (n-1)!$ different ways.)

b. Show that the number of ways to select k objects from n objects, denoted by $\binom{n}{k}$ (the binomial coefficient), is $n!/\left((n-k)!\,k!\right)$. Hint: we will choose the k objects by arranging the n objects in a row; the first k objects will be the ones we select. There are $R(n)$ ways to do this. When we do this, we get the *same* k objects many times. There are $R(k)$ ways to arrange the *same* k objects in the first k positions. For each such arrangement, the other $n-k$ objects may be arranged in $R(n-k)$ ways. The number of ways to arrange these same objects is $R(k)R(n-k)$. Since each k objects are counted $R(k)R(n-k)$ times in the $R(n)$ arrangements the number of different ways to select k objects is

$$\frac{R(n)}{R(k)R(n-k)} = \frac{n!}{k!\,(n-k)!}$$

from (a). Then check that $\binom{n}{n} = \binom{n}{0} = 1$.

c. Consider the binomial situation: n independent trials each with probability π of success. Show that the probability of k successes

$$b(k; n, \pi) = \binom{n}{k} \pi^k (1 - \pi)^{n-k}.$$

Hint: Think of the n trials as ordered. There are $\binom{n}{k}$ ways to choose the k trials that give a success. Using the independence of the trials argue the probability of those k trials being a success is $\pi^k (1 - \pi)^{n-k}$.

d. Compute from the definition of $b(k; n, \pi)$.

    i. $b(3; 5, 0.5)$,

    ii. $b(3; 3, 0.3)$,

    iii. $b(2; 4, 0.2)$,

    iv. $b(1; 3, 0.7)$,

    v. $b(4; 6, 0.1)$.

25. One-Sided Tests and p-values for the Binomial Distribution. Section 6.3.3 presented procedures for two-sided hypothesis tests with the binomial distribution. This problem deals with one-sided tests. We will present the procedures for a test of $H_0 : \pi \geq \pi_0$ vs. $H_A : \pi < \pi_0$. (The same procedures would be used for $H_0 : \pi = \pi_0$ vs. $H_A : \pi < \pi_0$. For $H_0 : \pi \leq \pi_0$ vs. $H_A : \pi > \pi_0$, the procedure would be modified [see below].)

Procedure A: To construct a significance test of $H_0 : \pi \geq \pi_0$ vs. $H_a : \pi < \pi_0$ at significance level α:

a. Let Y be binomial n, π_0 and $p = Y/n$. Find the largest c such that $P[p \leq c] \leq \alpha$.

b. Compute the actual significance level of the test as $P[p \leq c]$.

c. Observe p. Reject H_0 if $p \leq c$.

Procedure B: The p-value for the test if we observe p is $P[\widetilde{p} \leq p]$ where p is the *fixed* observed value and $\widetilde{p}$ equals $\widetilde{Y}/n$ where $\widetilde{Y}$ is binomial n, π_0.

a. In Problem 6.2 above let π be the probability of losing weight.

    i. Find the critical value c for testing $H_0 : \pi \geq 1/2$ vs. $H_A : \pi < 1/2$ at the 10% significance level.

    ii. Find the one-sided p-value for the data of Problem 6.2.

b. Modify procedures A and B for the hypotheses $H_0 : \pi \leq \pi_0$ vs. $H_A : \pi > \pi_0$.

*26. Hypergeometric Probabilities. The terminology and notation of Section 6.4.1 is used. We consider proportions of success from two samples of size $n_{1\bullet}$ and $n_{2\bullet}$; suppose we are told there are $n_{\bullet 1}$ total success. That is, we observe

	Success	**Failures**	
Sample 1	?		$n_{1\bullet}$
Sample 2			$n_{2\bullet}$
	$n_{\bullet 1}$	$n_{\bullet 2}$	$n_{\bullet\bullet}$

If both populations are equally likely to have a success, what can we say about n_{11}, the number of successes in population one, which goes in the cell with the question mark?
Show that

$$P[n_{11} = k] = \binom{n_{1\bullet}}{k}\binom{n_{2\bullet}}{n_{\bullet 1} - k} \bigg/ \binom{n_{\bullet\bullet}}{n_{\bullet 1}},$$

for $k \leq n_{1\bullet}$, $k \leq n_{\bullet 1}$ and $n_{\bullet 1} - k \leq n_{2\bullet}$. Note: $P[n_{11} = k]$, which has the parameters $n_{1\bullet}$, $n_{2\bullet}$ and $n_{\bullet 1}$, is called a *hypergeometric* probability. Hint: As suggested in Section 6.4.1, think of each trial (in sample one or two) as a ball (purple $(n_{1\bullet})$ or gold $(n_{2\bullet})$). Since successes are equally likely in either population, any ball is as likely as any other to be drawn in the $n_{\bullet 1}$ successes. All subsets of size $n_{\bullet 1}$ are equally likely, so the probability of k successes is the number of subsets with k purple balls divided by the total number of subsets of size $n_{\bullet 1}$. Argue the first number is $\binom{n_{1\bullet}}{k}\binom{n_{2\bullet}}{n_{\bullet 1}-k}$ and the second is $\binom{n_{\bullet\bullet}}{n_{\bullet 1}}$.

27. This problem gives more practice in finding sample sizes needed to test for a difference in two binomial populations.

    a. Use Figure 6.2 to find *approximate* two-sided sample sizes *per group* for $\alpha = 0.05$ and $\beta = 0.10$ when

      i. $P_1 = 0.5$, $P_2 = 0.6$;

      ii. $P_1 = 0.20$, $P_2 = 0.10$;

      iii. $P_1 = 0.70$, $P_2 = 0.90$.

    b. For each of the following, find one-sided sample sizes *per group* as needed both from the table and (*) from the formula of Section 6.4.3. To test π_1 vs. π_2, we need the same sample size as we would to test $1 - \pi_1$ vs. $1 - \pi_2$. Why?

      i. $\alpha = 0.05$, $\beta = 0.10$, $P_1 = 0.25$, $P_2 = 0.10$;

      ii. $\alpha = 0.05$, $\beta = 0.05$, $P_1 = 0.60$, $P_2 = 0.50$;

      iii. $\alpha = 0.01$, $\beta = 0.01$, $P_1 = 0.15$, $P_2 = 0.05$;

      iv. $\alpha = 0.01$, $\beta = 0.05$, $P_1 = 0.85$, $P_2 = 0.75$.

28. You are examined by an excellent screening test (sensitivity and specificity of 99%) for a rare disease (0.1% or 1/1000 of the population). Unfortunately, the test is positive. What is the probability you have the disease?

*29. **a.** Derive the "rule of threes" mentioned in Note 6.9.

    b. Can you find a similar constant to set up a 99% confidence interval?

*30. Consider the data of Problem 17; matched pair data. What null hypothesis does the usual chi-square test for a 2 × 2 table test on these data? What would you decide about the matching if this chi-square was not significant? (For example the table for marital status).

REFERENCES

Bennett, B. M. and Horst, C. [1966]. *Supplement to Tables for Testing Significance in a 2 × 2 Contingency Table*. Cambridge University Press, Cambridge.

Beyer, W. H., Ed. [1968]. *CRC Handbook of Tables for Probability and Statistics*. CRC Press, Cleveland, OH.

Borer, J. S., Rosing, D. R., Miller, R. H., Stark, R. M., Kent, K. M., Bacharach, S. L., Green, M. V., Lake, C. R., Cohen, H., Holmes, D., Donahue, D., Baker, W., and Epstein, S. E. [1980]. Natural history of left ventricular function during 1 year after acute myocardial infarction:

comparison with clinical, electrocardiographic and biochemical determinations. *American Journal of Cardiology*, **46:** 1–12.

Box, J. F. [1978]. *R. A. Fisher, The Life of a Scientist*. Wiley, New York.

Breslow, N.E. and Day, N.E. [1980]. *Statistical Methods in Cancer Research*. Volume 1—The analysis of case-control studies. International Agency for Research in Cancer, Lyon, France. IARC Publications No. 32.

Bucher, K. A., Patterson, A. M., Elston, R. C., Jones, C. A., and Kirkman, H. N. Jr. [1976]. Racial difference in incidence of ABO hemolytic disease. *American Journal of Public Health*, **66:** 854–858. Copyright © 1976 by the American Public Health Association.

Cavalli-Sforza, L. L. and Bodmer, W. F. [1971]. *The Genetics of Human Populations*, W. H. Freeman, San Francisco, CA.

Carver, W. A. [1927]. A genetic study of certain chlorophyll deficiencies in maize. *Genetics*, **12:** 415–440.

Chernoff, H. and Lehmann, E. L. [1954]. The use of maximum likelihood estimates in χ^2 tests for goodness of fit. *Annals of Mathematical Statistics*, **25:** 579–586.

Comstock, G. W. and Partridge, K. B. [1972]. Church attendance and health, *Journal of Chronic Diseases*, **25:** 665–672. Used with permission of Pergamon Press, Inc.

Conover, W. J. [1974]. Some reasons for not using the Yates continuity correction on 2×2 contingency tables. (With discussion.) *Journal of the American Statistical Association*, **69:** 374–382.

Edwards, A. W. F. and Fraccaro, M. [1960]. Distribution and sequences of sex in a selected sample of Swedish families. *Annals of Human Genetics, London*, **24:** 245–252. Cambridge University Press.

Feigl, P. [1978]. A graphical aid for determining sample size when comparing two independent proportions, *Biometrics*, **34:** 111–122.

Finney, D. J., Latscha, R., Bennett, B. M., and Hsu, P. [1963]. *Tables for Testing Significance in a 2 × 2 Contingency Table*. Cambridge University Press, Cambridge.

Fisher, L. and Patil, K. [1974]. Matching and unrelatedness. *American Journal of Epidemiology*, **100:** 347–349.

Fisher, R. A. [1936]. Has Mendel's work been rediscovered? *Annals of Science*, **1:** 115–137.

Fisher, R. A. [1958]. *Statistical Methods for Research Workers,* 13th edition. Oliver and Boyd, London.

Fisher, R. A., Thornton, H. G., and MacKenzie, W. A. [1922]. The accuracy of the plating method of estimating the density of bacterial populations. *Annals of Applied Biology*, **9:** 325–359.

Fleiss, J. L. [1981]. *Statistical Methods for Rates and Proportions,* 2nd edition. Wiley, New York.

Geissler, A. [1889]. Beitrage zur Frage des Geschlechts Verhaltnisses der Geborenen. *Zeitschrift des K. Sachsischen Statistischen Bureaus.*

Graunt, J. [1662]. *Natural and Political Observations Mentioned in a Following Index and Made Upon the Bills of Mortality.* Given in part in Newman, J. R., *The World of Mathematics*, Volume 3, Simon and Schuster, New York.

Grizzle, J. E. [1967]. Continuity correction in the χ^2-test for 2×2 tables. *American Statistician*, **21:** 28–32.

Hanley, J.A. and Lippman-Hand, A. [1983]. If nothing goes wrong, is everything alright? *Journal of the American Medical Association*, **249:** 1743–1745.

Janerich, D. T., Piper, J. M., and Glebatis, D. M. [1980]. Oral contraceptives and birth defects. *American Journal of Epidemiology*, **112:** 73–79.

Karlowski, T. R., Chalmers, T. C., Frenkel, L. D., Zapikian, A. Z., Lewis, T. L., and Lynch, J. M. [1975]. Ascorbic acid for the common cold. A prophylactic and therapeutic trial. *Journal of the American Medical Association*, **231:** 1038–1042.

Kelsey, J. L. and Hardy, R. J. [1975]. Driving of motor vehicles as a risk factor for acute herniated lumbar intervertebral disc. *American Journal of Epidemiology*, **102:** 63–73.

Kendall, M. G. and Stuart, A. [1967]. *The Advanced Theory of Statistics, Volume 2. Inference and Relationship.* Hafner, New York.

Kennedy, J. W., Kaiser, G. W., Fisher, L. D., Fritz, J. K., Myers, W., Mudd, J. G., and Ryan, T. J. [1981]. Clinical and angiographic predictors of operative mortality from the collaborative study in coronary artery surgery (CASS). *Circulation,* **63:** 793–802.

Lawson, D. H. and Jick, H. [1976]. Drug prescribing in hospitals: An international comparison. *American Journal of Public Health,* **66:** 644–648.

Little, R. J. A. [1989]. Testing the equality of two independent binomial proportions. *American Statistician,* **43:** 283–288.

Mantel, N., Brown, C., and Byar, D. P. [1977]. Tests for homogeneity of effect in an epidemiologic investigation. *American Journal of Epidemiology,* **106:** 125–129.

Mantel, N. and Greenhouse, S. W. [1968]. What is the continuity correction? *American Statistician,* **22:** 27–30.

Mantel, N. and Haenszel, W. [1959]. Statistical aspects of the analysis of data from retrospective studies of disease. *Journal of the National Cancer Institute,* **22:** 719–748.

Mendel, G. [1866]. Versuche uber Pflanzenhybriden. *Verhandlungen Naturforschender Vereines in Brunn,* **10:** 1.

Meyer, M. B., Jonas, B. S., and Tonascia, J. A. [1976]. Perinatal events associated with maternal smoking during pregnancy. *American Journal of Epidemiology,* **103:** 464–476.

Miettinen, O. S. [1970]. Matching and design efficiency in retrospective studies. *American Journal of Epidemiology,* **91:** 111–118.

Odeh, R. E., Owen, D. B., Birnbaum, Z. W., and Fisher, L. D. [1977]. *Pocketbook of Statistical Tables,* Marcel Dekker, New York.

Ounsted, C. [1953]. The sex ratio in convulsive disorders with a note on single-sex sibships. *Journal of Neurology, Neurosurgery and Psychiatry,* **16:** 267–274.

Owen, D. B. [1962]. *Handbook of Statistical Tables,* Addison-Wesley, Reading, MA.

Peterson, D. R., Chinn, N. M., and Fisher, L. D. [1980]. The sudden infant death syndrome: repetitions in families. *Journal of Pediatrics,* **97:** 265–267.

Peterson, D. R., van Belle, G., and Chinn, N. M. [1979]. Epidemiologic comparisons of the sudden infant death syndrome with other major components of infant mortality. *American Journal of Epidemiology,* **110:** 699–707.

Remein, Q. R. and Wilkerson, H. L. C. [1961]. The efficiency of screening tests for diabetes. *Journal of Chronic Diseases,* **13:** 6–21. Used with permission of Pergamon Press, Inc.

Robinette, C. D., Silverman, C., and Jablon, S. [1980]. Effects upon health of occupational exposure to microwave radiation (radar). *American Journal of Epidemiology,* **112:** 39–53.

Rosenberg, L., Slone, D., Shapiro, S., Kaufman, D. W., Stolley, P. D., and Miettinen, O. S. [1980]. Coffee drinking and myocardial infarction in young women. *American Journal of Epidemiology,* **111:** 675–681.

Sartwell, P. E., Masi, A. T., Arthes, F. G., Greene, G. R., and Smith, H. E. [1969]. Thromboembolism and oral contraceptives: an epidemiologic case-control study. *American Journal of Epidemiology,* **90:** 365–380.

Schlesselman, J. J. [1982]. *Case-Control Studies. Design, Conduct, Analysis.* Monographs in Epidemiology and Biostatistics. Oxford University Press, New York.

Shapiro, S., Goldberg, J. D., and Hutchinson, G. B. [1974]. Lead time in breast cancer detection and implications for periodicity of screening. *American Journal of Epidemiology,* **100:** 357–366.

Smith, J. P., Delgado, G., and Rutledge, F. [1976]. Second-look operation in ovarian cancer. *Cancer,* **38:** 1438–1442. Used with permission from J. B. Lippincott Company.

Smoking and Health: Report of the Advisory Committee to the Surgeon General of the Public Health Service, [1964]. US Department of Health, Education and Welfare.

Starmer, C. F., Grizzle, J. E., and Sen, P. K. [1974]. Comment. *Journal of the American Statistical Association*, **69:** 376–378.

von Bortkiewicz, L. [1898]. *Das Gesetz der Kleinen Zahlen*. Teubner, Leipzig.

Weber, A., Jermini, C., and Grandjean, E. [1976]. Irritating effects on man of air pollution due to cigarette smoke. *American Journal of Public Health*, **66:** 672–676.

CHAPTER 7

Categorical Data: Contingency Tables

7.1 INTRODUCTION

In the last chapter, *discrete variables* came up by counting the number of times specific outcomes occurred. In looking at the presence or absence of a risk factor and a disease, the *odds ratio* and *relative risk* were introduced. In doing this, we looked at the relationship between two discrete variables; each variable took on one of two possible states (i.e., risk factor present or absent and disease present or absent). This chapter shows how to analyze more general discrete data. Two types of generality are presented.

The first generalization considers two jointly distributed discrete variables. Each variable may take on more than two possible values. Some examples of discrete variables with three or more possible values might be: smoking status (which might take on the values "never smoked," "former smoker," and "current smoker"); employment status (which could be coded as "full-time," "part-time," "unemployed," "unable to work due to medical reason," "retired," "quit," and "other"); and clinical judgment of improvement (classified into categories of "considerable improvement," "slight improvement," "no change," "slight worsening," "considerable worsening," and "death").

The second generalization allows us to consider three or more discrete variables (rather than just two) at the same time. For example, method of treatment, sex and employment status may be analyzed jointly. With three or more variables to investigate, it becomes difficult to obtain a "feeling" for the interrelationships among the variables. If the data fit a relatively simple mathematical model, our understanding of the data may be greatly increased.

In this chapter, our first *multivariate statistical model* is encountered. The model is the *log-linear model* for multivariate discrete data. The remainder of this text depends upon a variety of models for analyzing data; this chapter is an exciting, important, and challenging introduction to such models!

7.2 TWO-WAY CONTINGENCY TABLES

Let two or more discrete variables be measured on each unit in an experiment or observational study. In this chapter, methods of examining the relationship among the variables are studied. Most of the chapter will study the relationship of two discrete variables. In this case, we count the number of occurrences of each pair of possibilities and enter them in a table. Such tables are called contingency tables. Example 7.1 presents two contingency tables.

Example 7.1. Problem 1.1 dealt with freezing for gastric ulcer. Please read that introduction before continuing.

In 1962, Wangensteen et al., published a paper in the *Journal of the American Medical Association* advocating gastric freezing. A balloon was lowered into a subject's stomach, and coolant at a temperature of $-17°C$ to $-20°C$ was introduced through tubing connected to the balloon. Freezing was continued for approximately 1 h. The rationale was that gastric digestion could be interrupted and it was thought that a duodenal ulcer might heal if treatment could be continued over a period of time. The authors advanced three reasons for the interruption of gastric digestion: (1) interruption of vagal secretory responses; (2) "rendering of the central mucosa nonresponsive to food ingestion, ... "; and (3) "impairing the capacity of the parietal cells to secrete acid and the chief cells to secrete pepsin." Table 7.1 was presented as evidence for the effectiveness of gastric freezing. It shows a decrease in acid secretion.

On the basis of this table and other data, the authors state, "These data provide convincing objective evidence of significant decreases in gastric secretory responses attending effective gastric freezing," and conclude, "When profound gastric hypothermia is employed with resultant freezing of the gastric mucosa, the method becomes a useful agent in the control of many of the manifestations of peptic ulcer diathesis. Symptomatic relief is the rule, followed quite regularly by X-ray evidence of healing of duodenal ulcer craters and evidence of effective depression of gastric secretory responses." *Time* magazine [1962] reported that, "All [the patients'] ulcers healed within two to six weeks."

However, careful studies attempting to confirm the above conclusion failed. Two studies in particular failed to confirm the evidence, one by Hitchcock et al. [1966], the other by Ruffin et al. [1969]. The later study used an elaborate sham procedure (control) to simulate gastric freezing, to the extent that the tube entering the patient's mouth was cooled to the same temperature as in the actual procedure, but the coolant entering the stomach was at room temperature so that no freezing took place. The authors defined an "endpoint" to have occurred if one of the following criteria was met: "perforation; ulcer pain requiring hospitalization for relief; obstruction, partial or complete, two or more weeks after hyperthermia; hemorrhage, surgery for ulcer; repeat hypothermia; or X-ray therapy to the stomach."

Several institutions cooperated in the study, and to insure objectivity and equal

Table 7.1. Gastric Response of Ten Patients with Duodenal Ulcer Whose Stomachs Were Frozen at $-17°C$ to $-20°C$ for an Hour. Data from Wangensteen et al. [1962].

| | | Average Percent Decrease in HCl after Gastric Freezing | | |
| | Patients | | | |
Patients	with Decrease of Free HCl	Overnight Secretion	Peptone Stimulation	Insulin
10	10^a	87	51	71

[a] All patients, except one, had at least a 50% decrease in free HCl in overnight secretion.

Table 7.2. Causes of Endpoints. Data from Ruffin *et al.* [1969].

Group	Patients	With Hemorrhage	With Operation	With Hospitalization	Not reaching endpoint
F (Freeze)	69	9	17	9	34
S (Sham)	68	9	14	7	38

numbers, random allocations to treatment and sham were balanced within groups of eight. At the termination of the study patients were classified as in Table 7.2. The authors conclude: "... [the] results of this study demonstrate conclusively that the 'freezing' procedure was not better than the sham in the treatment of duodenal ulcer, confirming the work of others.... It is reasonable to assume that the relief of pain and subjective improvement reported by early investigators was probably due to the psychological effect of the procedure."   □

Contingency tables set up from two variables are called two-way tables. Let the variable corresponding to rows have r (for **r**ow) possible outcomes which we index by i ($i = 1, 2, \ldots, r$). Let the variables corresponding to the column headings have c (for **c**olumn) possible states indexed by j ($j = 1, 2, \ldots, c$). One speaks of an $r \times c$ contingency table. Let n_{ij} be the number of observations corresponding to the ith state of the row variable and the jth state of the column variable. In the example above, $n_{11} = 9$, $n_{12} = 17$, $n_{13} = 9$, $n_{14} = 34$, $n_{21} = 9$, $n_{22} = 14$, $n_{23} = 7$, and $n_{24} = 38$. In general, the data are presented as:

	j			
i	1	2	$\cdots$	c
1	n_{11}	n_{12}	$\cdots$	n_{1c}
2	n_{21}	n_{22}	$\cdots$	n_{2c}
$\vdots$	$\vdots$	$\vdots$	$\ddots$	$\vdots$
r	n_{r1}	n_{r2}	$\cdots$	n_{rc}

Such tables usually arise in one of two ways:

1. A sample of observations is taken. On each unit we observe the values of two traits. Let π_{ij} be the probability that the row variable takes on level i and the column variable takes on level j. Since one of the combinations must occur,

$$\sum_{i=1}^{r} \sum_{j=1}^{c} \pi_{ij} = 1. \tag{7.1}$$

2. Each row corresponds to a sample from a different population. In this case let π_{ij} be the probability the column variable takes on state j when sampling from the ith population. Thus, for each i,

$$\sum_{j=1}^{c} \pi_{ij} = 1. \tag{7.2}$$

If the samples correspond to the column variable the π_{ij} are the probabilities that the row variable takes on state i when sampling from population j. In this circumstance, for each j,

$$\sum_{i=1}^{r} \pi_{ij} = 1. \tag{7.3}$$

Table 7.2 of Example 7.1 comes from the second model since the treatment is assigned by the experimenter; it is not a trait of the experimental unit. Examples for the first model are given below.

The usual null hypothesis in a Model 1 situation is that of independence of row and column variables. That is (assuming row variable $= i$ and column variable $= j$), $P[i \text{ and } j] = P[i]P[j]$,

$$H_0 : \pi_{ij} = \pi_{i\bullet} \pi_{\bullet j}.$$

In the Model 2 situation, suppose that the row variable identifies the population. The usual null hypothesis is that all r populations have the same probabilities of taking on each value of the column variable. That is, for any two rows, denoted by i and i', say, and all j,

$$H_0 : \pi_{ij} = \pi_{i'j}.$$

If one of these hypotheses holds, we say there is *no association*; otherwise the table is said to have *association* between the categorical variables.

We will use the following notation for the sum over the elements of a row and/or column: $n_{i\bullet}$ is the sum of the elements of the ith row; $n_{\bullet j}$ is the sum of the elements of the jth column.

$$n_{i\bullet} = \sum_{j=1}^{c} n_{ij}, \quad n_{\bullet j} = \sum_{i=1}^{r} n_{ij}, \quad n_{\bullet\bullet} = \sum_{i=1}^{r}\sum_{j=1}^{c} n_{ij}.$$

It is shown in Note 7.1 that either under Model 1 or Model 2, the null hypothesis is reasonably tested by comparing n_{ij} with

$$\frac{n_{i\bullet}n_{\bullet j}}{n_{\bullet\bullet}}.$$

The latter is the value expected in the ijth cell given the observed marginal configuration and assuming either of the null hypotheses under Model 1 or Model 2. This is illustrated:

$n_{11} = 9$	$n_{12} = 17$	$n_{13} = 9$	$n_{14} = 34$	$n_{1\bullet} = 69$
$n_{21} = 9$	$n_{22} = 14$	$n_{23} = 7$	$n_{24} = 38$	$n_{2\bullet} = 68$
$n_{\bullet 1} = 18$	$n_{\bullet 2} = 31$	$n_{\bullet 3} = 16$	$n_{\bullet 4} = 72$	$n_{\bullet\bullet} = 137$

Under the null hypothesis, the table of expected values $n_{i\bullet}n_{\bullet j}/n_{\bullet\bullet}$ is

$$69 \times 18/137 \quad 69 \times 31/137 \quad 69 \times 16/137 \quad 69 \times 72/137$$
$$68 \times 18/137 \quad 68 \times 31/137 \quad 68 \times 16/137 \quad 68 \times 72/137$$

or

$$9.07 \quad 15.61 \quad 8.06 \quad 36.26$$
$$8.93 \quad 15.39 \quad 7.94 \quad 35.74$$

It is a remarkable fact that both null hypotheses above may be tested by the χ^2 statistic,

$$X^2 = \sum_{i=1}^{r} \sum_{j=1}^{c} \frac{\left(n_{ij} - n_{i\bullet}n_{\bullet j}/n_{\bullet\bullet}\right)^2}{n_{i\bullet}n_{\bullet j}/n_{\bullet\bullet}}.$$

Note that n_{ij} is the observed cell entry; $n_{i\bullet}n_{\bullet j}/n_{\bullet\bullet}$ is the expected cell entry, so this statistic may be remembered as

$$X^2 = \sum (\text{observed} - \text{expected})^2/\text{expected}.$$

For example, the above table gives

$$X^2 = (9 - 9.07)^2/9.07 + (17 - 15.61)^2/15.61 + (9 - 8.06)^2/8.06$$
$$+ (34 - 36.26)^2/36.26 + (9 - 8.93)^2/8.93 + (14 - 15.39)^2/15.39$$
$$+ (7 - 7.94)^2/7.94 + (38 - 35.76)^2/35.76 = 0.752.$$

Under the null hypothesis, the X^2 statistic has approximately a χ^2 distribution with $(r - 1)(c - 1)$ degrees of freedom. This approximation is for large samples and is appropriate when all of the *expected* values, $n_{i\bullet}n_{\bullet j}/n_{\bullet\bullet}$, are five or greater. There is some evidence to indicate the approximation is valid if all the expected values, except possibly one, are five or greater.

For our example, the degrees of freedom for the example are $(2 - 1)(4 - 1) = 3$. The rejection region is for X^2 too large. The 0.05 critical value is 7.81. As $0.752 < 7.81$, we do *not* reject the null hypothesis at the 0.05 significance level.

Example 7.2. Robertson [1975] examines seatbelt use in automobiles with starter interlock and buzzer/light systems. The use or nonuse of safety belts by drivers in their vehicles was observed at 138 sites in Baltimore, Maryland; Houston, Texas; Los Angeles, California; the New Jersey suburbs; New York City; Richmond, Virginia and Washington, D.C. during late 1973 and early 1974. The sites were such that observers could see whether or not seatbelts were being used. The sites were freeway entrances and exits, traffic jam areas, and other points where vehicles usually slowed to less than 15 miles/h. The observers dictated onto tape the sex, estimated age, and racial appearance of the driver of the approaching car; as the vehicles slowed alongside, the observer recorded whether or not the lap-belt and/or shoulder-belt was in use, not in use, or could not be seen. The license plate numbers were subsequently sent to the appropriate motor vehicle administration, where they were matched to records from which the manufacturer and year were determined. In the 1973 models, there was a buzzer/light system that came on when the seatbelt was not being used. The buzzer was activated for at least 1 min when the driver's seat was occupied, the ignition switch was on, the transmission gear selector was in a forward position, and the driver's lap-belt was not extended at least 4 inches from its normal resting position. Effective on August 15, 1973, a federal standard required that the automobile could be started only under certain conditions. In this case, when the driver was seated, the belts had to be extended more than 4 inches from their normally stored position and/or latched. Robertson states that as a result of the strong negative public reaction to the interlock system, federal law has banned the interlock system. The data on the buzzer/light-equipped models and the interlock-equipped models is given in Table 7.3. As can be seen from the table, column percentages were presented to aid assimilation of the information of the table.

Percentages in two-way contingency tables are useful in aiding visual comprehension of the contents. There are three types of percent tables.

1. *Column percent* tables give the percentages for each column (the columns add to 100%, except possibly for rounding errors). This is best for comparing the distributions of different columns.

2. *Row percent* tables give the percentages for each row (the rows add to 100%). This is best for comparing the distributions of different rows.

3. The *total percent* table gives percentages, so that all the entries in a table add to 100%. This aids investigation of the proportions in each combination.

The column percentages in Table 7.3 facilitate comparison of the seatbelt usage in the 1973 buzzer/light models and the 1974 interlock models. They illustrate that there are strategies for getting around the interlock system, such as disabling it, connecting the seatbelt and leaving it connected on the seat, as well as possible other strategies, so that even with the interlock system, not everyone uses it. The computed value of the chi-square statistic for this table is 1751.6 with 2 degrees of freedom. The p-value is effectively zero, as seen in Table A.3 in the Appendix.

Given that we have a statistically significant association, the next question that arises is: "to what may we attribute this association?" In order to decide why the association occurs, it is useful to have an idea of which entries in the table differ more than would be expected by chance from their value under the null hypothesis of no association. Under the null hypothesis, for each entry in the table, the following *adjusted residual value* is approximately distributed as a standard normal distribution. The term *residual* is used since it looks at the difference between the observed value

Table 7.3. Robertson [1975] Seatbelt Data.

Belt Use	1973 Models (Buzzer/light)		1974 Models (Interlock)		Total
	%	Number	%	Number	
Lap and Shoulder	7	432	48	1007	1439
Lap Only	21	1262	11	227	1489
None	72	4257	41	867	5124
Total	100	5951	100	2101	8052

and the value expected under the null hypothesis. This difference is then standardized by its standard error,

$$z_{ij} = \frac{(n_{ij} - (n_{i\bullet}n_{\bullet j}/n_{\bullet\bullet}))}{\sqrt{\dfrac{n_{i\bullet}n_{\bullet j}}{n_{\bullet\bullet}}\left(1 - \dfrac{n_{i\bullet}}{n_{\bullet\bullet}}\right)\left(1 - \dfrac{n_{\bullet j}}{n_{\bullet\bullet}}\right)}}. \tag{7.4}$$

For example, for the $(1,1)$ entry in the table, a standardized residual is given by

$$\frac{(432 - 1439 \times 5951/8052)}{\sqrt{\dfrac{1439 \cdot 5951}{8052}\left(1 - \dfrac{1439}{8052}\right)\left(1 - \dfrac{5951}{8052}\right)}} = 41.83.$$

The matrix of the observed residual values with the corresponding normal probability p-values is given in Table 7.4. Note that the values add to zero for the residuals across each row. This occurs because there are only two columns. The observed adjusted residual values are so far from zero that the normal p-values are miniscule.

In general, there is a problem in looking at a contingency table with many cells. Because there are a large number of residual values in the table, it may be that one or more of them differs by chance from zero at the 5% significance level. Even *under the null hypothesis*, because of the many possibilities examined, *this would occur much more than 5% of the time.* One conservative way to deal with this problem is to multiply the p-values by the number of rows minus one and the number of columns minus one. If the corresponding p-value is less than 0.05, one can conclude the entry is different from that expected by the null hypothesis at the 5% significance level *even after looking at all of the different entries.* (This problem of looking at many possibilities is called the *multiple comparison problem*, and is dealt with in considerable detail in Chapter 12.) For this example, even after multiplying by the number of rows minus one and the number of columns minus one, all of the entries differ from those expected under the null hypothesis. Thus, one can conclude, using the sign of the residual to tell us whether the percent is too high or too low, that in the 1973 models there is less lap and shoulder usage than in the 1974 models. Further, if we look at the "none" category, there are fewer individuals without any belt usage in the 1974 interlock models than in the 1973 buzzer/light-equipped models. One

Table 7.4. Adjusted Residual Values (Example 7.2).

i	j	Residual (Z_{ij})	p-value	p-value $\times (r-1) \times (c-1)$
1	1	−41.83	0+	0+
1	2	41.83	0+	0+
2	1	10.56	3×10^{-22}	6×10^{-22}
2	2	−10.56	3×10^{-22}	6×10^{-22}
3	1	24.79	9×10^{-53}	2×10^{-52}
3	2	−24.79	9×10^{-53}	2×10^{-52}

would conclude that the interlock system, although a system disliked by the public, was successful as a public health measure in increasing the amount of seatbelt usage.

□

Suppose we decide there is an association in a contingency table. We can interpret the table by using residuals (as we have done above) to help to find out whether particular entries differ more than expected by chance.

Another approach to interpretation is to characterize numerically the amount of association between the variables, or proportions in different groups, in the contingency table. To date, there has been no one measure of the amount of association in contingency tables that has gained widespread acceptance. There have been many proposals, all of which have some merit. Note 7.2 presents some measures of the amount of association.

7.3 THE CHI-SQUARE TEST FOR TREND IN $2 \times k$ TABLES

There are a variety of techniques for improving the statistical power of χ^2 tests. Recall that power is a function of the alternative hypothesis. One weakness of the chi-square test is that it is an "omnibus" test; it tests for independence versus dependence without specifying the nature of the latter. In some cases, a small subset of alternative hypotheses may be specified to increase the power of the chi-square test by defining a special test. One such situation occurs in $2 \times k$ tables when the alternative hypothesis is that there is an ordering in the variable producing the k categories. For example, exposure categories can be ordered, and the alternative hypothesis may be that the probability of disease *increases* with increasing exposure.

In this case the row variable takes on one of two states (say + or − for definiteness). For each state of the column variable ($j = 1, 2, \ldots, k$), let π_j be the conditional probability of a positive response. The test for trend is designed to have statistical power against the alternatives:

$$H_1 : \pi_1 \leq \pi_2 \leq \cdots \leq \pi_k, \quad \text{with at least one strict inequality,}$$

$$H_2 : \pi_1 \geq \pi_2 \geq \cdots \geq \pi_k, \quad \text{with at least one strict inequality.}$$

That is, the alternatives of interest are that the proportion of + responses increases or decreases with the column variable. For these alternatives to be of interest, the column variable will have a "natural" ordering. To compute the statistic, a "score" needs to be assigned to each state j of the column variable. The scores x_j are assigned so that they increase or decrease. Often the x_j are consecutive integers.

The data are laid out as follows:

i	1	2	$\cdots$	k	Total
1+	n_{11}	n_{12}	$\cdots$	n_{1k}	$n_{1\bullet}$
2−	n_{21}	n_{22}	$\cdots$	n_{2k}	$n_{2\bullet}$
Total	$n_{\bullet 1}$	$n_{\bullet 2}$	$\cdots$	$n_{\bullet k}$	$n_{\bullet\bullet}$
Score	x_1	x_2	$\cdots$	x_k	

(Column header: j)

Before stating the test, we define some notation. Let

$$[n_1 x] = \sum_{j=1}^{k} n_{1j} x_j - \frac{n_{1\bullet} \sum n_{\bullet j} x_j}{n_{\bullet\bullet}}$$

and

$$[x^2] = \sum_{j=1}^{k} n_{\bullet j} x_j^2 - \frac{\left(\sum n_{\bullet j} x_j\right)^2}{n_{\bullet\bullet}}$$

and

$$p = \frac{n_{1\bullet}}{n_{\bullet\bullet}}.$$

Then the chi-square test for trend is defined to be

$$X_{trend}^2 = \frac{[n_1 x]^2}{[x^2] p (1 - p)}$$

and when there is no association this quantity has approximately a chi-square distribution with one degree of freedom. (In the terminology of Chapter 9, this is a chi-square test for the slope of a weighted regression line with dependent variable $p_j = n_{1j}/n_{\bullet j}$, predictor variable x_j, and weights $n_{1j}/p(1 - p)$, where $j = 1, 2, \ldots, k$.)

Example 7.3. For an example of this test we use data of Maki *et al.* [1977], relating risk of catheter-related infection to the duration of catheterization. An infection was considered to be present if there were 15 or more colonies of micro-organisms present in a culture associated with the withdrawn catheter. A part of the data dealing with the number of positive cultures as related to duration of catheterization is given in Table 7.5. A somewhat natural set of values of the scores x_i is the duration of catheterization in days. The designation ≥ 4 is, somewhat arbitrarily, scored 4.

Before carrying out the analysis, note that a graph of proportion of positive cultures *vs.* duration, as in Figure 7.1, clearly suggests a trend. The general chi-square test on the 2×4 table produces a value of $X^2 = 6.99$ with 3 degrees of freedom and a significance level between 0.0752 and 0.0658 from Table A.4.

Table 7.5. Relations of Results of Semi-Quantitative Culture and Catheterization. Data from Maki *et al.* [1977].

Culture	Duration of Catheterization (Days)				
	1	2	3	≥ 4	Total
Positive[a]	1[b]	5	5	14	25
Negative	46	64	39	76	225
Total	47	69	44	90	250

[a]Culture positive if 15 or more colonies on primary plate.
[b]Numbers in body of table are the numbers of catheters.

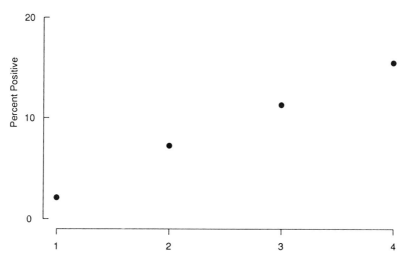

Figure 7.1. Graph of percentage of cultures positive *vs.* duration of catheterization. Data from Table 7.5. The fractions $1/47$, etc., are the number of positive cultures to the total number of cultures for a particular day.

To calculate the chi-square test for trend, we calculate the quantities $[n_1x]$, $[x^2]$ and p as defined above.

$$[n_1x] = 82 - (25 \cdot 677)/250 = 14.3,$$

$$[x^2] = 2159 - (677^2/250) \doteq 325.6840,$$

$$p = 25/250 = 0.1, \quad (1 - p) = 0.9,$$

$$X^2_{\text{trend}} = \frac{[n_1x]^2}{[x^2]p(1-p)} \doteq \frac{14.3^2}{325.6840 \cdot 0.1 \cdot 0.9} \doteq 6.98.$$

This statistic has one degree of freedom associated with it, and from the chi-square Table A.3, it can be seen that $0.005 < p < 0.01$; hence there is a significant linear trend. □

Note two things about the chi-square test for trend. First, the degrees of freedom are one, *regardless* of how large the value k. Secondly, the values of the scores chosen (x_j) are not too crucial, and evenly-spaced scores will give more statistical power against a trend than the usual χ^2 test. The example above indicates one type of contingency table in which ordering is clear; when the categories result from grouping a continuous variable.

7.4 THE MEASUREMENT OF AGREEMENT: *kappa* (κ)

It often happens in measuring or categorizing objects that the variability of the measurement or categorization is investigated. For example, one might have two physicians independently judge a patient's status as "improved," "remained the same," or "worsened." A study of psychiatric patients might have two psychiatrists independently classifying patients into diagnostic categories. When we have two discrete classifications of the same object, we may put the entries into a two-way *square* ($r = c$) contingency table. The chi-square test of this chapter may then be used to test for association. Usually, when two measurements are taken of the same objects, there is not much trouble showing association; rather the concern is to study the degree or amount of agreement in the association. This section deals with a statistic, *kappa* (κ), designed for such situations. We will see that the statistic has a nice interpretation; the value of the statistic can be taken as a measure of the degree of agreement. As we develop this statistic, we shall illustrate it with the following example.

Example 7.4. Fisher *et al.* [1982], studied the reproducibility of coronary arteriography. In the coronary artery surgery study (CASS), coronary arteriography is the key diagnostic procedure. In this procedure, a tube is inserted into the heart, and fluid injected that is opaque to X-rays. By taking X-ray motion pictures, the coronary arteries may be examined for possible narrowing, or stenosis. The three major arterial systems of the heart were judged with respect to narrowing. Narrowing was significant if it was 70% or more of the diameter of the artery. Because the angiographic films are a key diagnostic tool and are important in the decision about the appropriateness of bypass surgery, the quality of the arteriography was monitored and the amount of agreement was ascertained.

Table 7.6 presents the results for randomly selected films with two readings. One reading was that of the patient's clinical site and was used for therapeutic decisions. The angiographic film was then sent to another clinical site designated as a Quality Control site. The Quality Control site read the films blindly; that is, without knowledge of the clinical site's reading. From these readings, the amount of disease was classified as "none" (entirely normal), "zero-vessel disease but some disease," and one-, two-, and three-vessel disease.

We wish to study the amount of agreement. One possible measure of this is the proportion of the pairs of readings which are the same. This quantity is estimated by adding up the numbers on the diagonal of the table; those are the numbers where both the clinical site and the quality control site read the same quantity. In such a situation

Table 7.6. Agreement with Respect to Number of Diseased Vessels.

Quality Control Site Reading	Clinical Site Reading					Total
	Normal	Some	One	Two	Three	
Normal	13	8	1	0	0	22
Some	6	43	19	4	5	77
One	1	9	155	54	24	243
Two	0	2	18	162	68	250
Three	0	0	11	27	240	278
Total	20	62	204	247	337	870

as this, the contingency table will be square. Let r be the number of categories (in the table of this example, $r = 5$). The proportion of cases with agreement is given by

$$P_A = \frac{n_{11} + n_{22} + \cdots + n_{rr}}{n_{\bullet\bullet}} = \sum_{i=1}^{r} \frac{n_{ii}}{n_{\bullet\bullet}}.$$

For this table, the proportion with agreement is given by $P_A = (13 + 43 + 155 + 162 + 240)/870 = 613/870 \doteq 0.7046$.

The proportion of agreement is limited because it is heavily determined by the proportions of people in the different categories. Consider, for example, a situation where each of two judges places 90% of the measurements in one category and 10% in the second category, such as in the following table:

81	9		90
9	1		10
90	10		100

In this table, there is no association whatsoever between the two measurements. In fact, the chi-square value is precisely zero by design; there is no more agreement between the individuals than that expected by chance. Nevertheless, because both individuals have a large proportion of the cases in the first category, in 82% of the cases there is agreement; that is, $P_A = 0.82$. We have a paradox: on the one hand the agreement seems good (there is an agreement 82% of the time); on the other hand, the agreement is no more than can be expected by chance. In order to have a more useful measure of the amount of agreement, the *kappa* statistic was developed to adjust for the amount of agreement that one expects purely by chance.

If one knows the totals of the different rows and columns, the proportion of observations expected to agree by chance is given by the following equation:

$$P_C = \frac{n_{1\bullet}n_{\bullet 1} + \cdots + n_{r\bullet}n_{\bullet r}}{n_{\bullet\bullet}^2} = \sum_{i=1}^{r} \frac{n_{i\bullet}n_{\bullet i}}{n_{\bullet\bullet}^2}.$$

For the example, the proportion of agreement expected by chance is given by

$$P_C = \frac{22 \times 20 + 77 \times 62 + 243 \times 204 + 250 \times 247 + 278 \times 337}{870^2} \doteq 0.2777.$$

The *kappa* statistic uses the fact that the best possible agreement is one, and that by chance, one expects an agreement P_C. A reasonable measure of the amount of agreement is the proportion of difference between one and P_C that can be accounted for by actual observed agreement. That is, *kappa* is the ratio of the agreement actually observed minus the agreement expected by chance, divided by 1 (which corresponds to perfect agreement) minus the agreement expected by chance:

$$\kappa = \frac{P_A - P_C}{1 - P_C}.$$

For our example, the computed value of *kappa* is

$$\kappa = \frac{0.7046 - 0.2777}{1 - 0.2777} \doteq 0.59.$$

The *kappa* statistic runs from $-P_C/(1 - P_C)$ to 1. If the agreement is totally by chance, the expected value is zero. *Kappa* is equal to one if and only if there is complete agreement between the two categorizations.

Since the *kappa* statistic is usually used where it is clear that there will be statistically significant agreement, the real issue is the amount of agreement. κ is a measure of the amount of agreement. In our example, one can state that 59% of the difference between perfect agreement and the agreement expected by chance is accounted for by the agreement between the clinical and quality control reading sites.

Now that we have a parameter to measure the amount of agreement, we need to consider the effect of the sample size. For small samples, the estimation of κ will be quite variable; for larger samples it should be quite good. For relatively large samples, when there is no association, the variance of the estimate is estimated as follows:

$$
\mathrm{var}_o(\kappa) = \frac{P_C + P_C^2 - \displaystyle\sum_{i=1}^{r} \left(\frac{n_{i\bullet}^2 n_{\bullet i} + n_{i\bullet} n_{\bullet i}^2}{n_{\bullet\bullet}^3} \right)}{n_{\bullet\bullet}(1 - P_C)^2}.
$$

The subscript on $\mathrm{var}_o(\kappa)$ indicates that it is the variance under the null hypothesis. The standard error of the estimate is the square root of this quantity. κ divided by the standard error is approximately a standard normal variable when there is no association between the quantities. This may be used as a statistical test for association in lieu of the chi-square test.

A more useful function of the general standard error is construction of a confidence interval for the true κ. A $100(1 - \alpha)\%$ confidence interval for the population value of κ, for large samples, is given by the following equation:

$$
\left(\kappa - z_{1-\alpha/2}\sqrt{\mathrm{var}(\kappa)}, \quad \kappa + z_{1-\alpha/2}\sqrt{\mathrm{var}(\kappa)} \right).
$$

The estimated standard error, allowing for association, is the square root of

$$
\mathrm{var}(\kappa) = \frac{\left(\displaystyle\sum_{i=1}^{n} \frac{n_{ii}}{n_{\bullet\bullet}} \left(1 - \left(\frac{n_{i\bullet} + n_{\bullet i}}{n_{\bullet\bullet}}\right)(1-\kappa)\right)^2 + \sum\sum_{i \neq j} \frac{n_{ij}}{n_{\bullet\bullet}} \left(\left(\frac{n_{\bullet i} + n_{j\bullet}}{n_{\bullet\bullet}}\right)(1-\kappa)\right)^2 - (\kappa - P_C(1-\kappa))^2 \right)}{n_{\bullet\bullet}(1 - P_C)^2}.
$$

For our particular example, the estimated variance of κ is

$$
\mathrm{var}(\kappa) = 0.000449.
$$

The standard error of κ is approximately 0.0212. The 95% confidence interval is

$$
(0.57 - 1.96 \times 0.0212, \ 0.57 + 1.96 \times 0.0212) \doteq (0.55, 0.63). \qquad \square
$$

Two texts which have information on κ are Fleiss [1981] and Reynolds [1977]. κ has been generalized several ways. One may argue that some disagreements are more severe than others, and a weight should be assigned to the particular amount of agreement or disagreement. This is done by Cohen [1968]. The large sample standard error of κ is discussed in Fleiss *et al.* [1969]. κ has also been extended to the case where more than two measurements are made on the same objects (Light [1971] and Ross [1977]). Also, the *kappa* statistic ignores the ordering of the categories (see Maclure and Willet [1987]). Finally, summary articles touching on many of the issues involved with κ and related topics are given by Fleiss [1975] and Landis and Koch [1975].

7.5 *PARTITION OF CHI-SQUARE

In studying contingency tables, the chi-square test gives us a test of association. If we cannot show there is an association, the matter may rest at that point. However, if there is association, then one would like to explain it. We have already seen one method of looking at the association. This method is to take the p-values resulting from the adjusted residuals and to identify particular cells that have too many or too few observations under the hypothesis of independence. In this section, we consider another method of looking at the patterns of association. We first consider a method of breaking the overall chi-square statistic for the contingency table down into two parts. The different parts each have their own chi-square test. Then we consider a general technique for looking at tables resulting from the original contingency table. Thus, both parts of this section show how to take a contingency table and to look at parts of the table, or tables made from combinations of the rows and columns within the original table.

In discussing these new tables generated from an original table, it will be useful to have a term for the tables being formed from the original table.

Definition 7.1. A *partial table* of a contingency table is a new contingency table that results from omitting and/or adding together rows and/or columns of the original table.

This definition is not in general usage; it is given here for ease of exposition. The following will demonstrate what is meant. Consider the following 3×5 contingency table:

$$
\begin{array}{ccccc}
21 & 14 & 116 & 2 & 10 \\
17 & 2 & 15 & 4 & 12 \\
23 & 0 & 101 & 8 & 16
\end{array}
$$

We shall consider the partial table that results by deleting the second row as well as the third and fourth columns and then adding the first two remaining columns. The first step would be to delete the second row and third and fourth columns, giving the following table:

$$
\begin{array}{ccc}
21 & 14 & 10 \\
23 & 0 & 16
\end{array}
$$

We now add the first two columns, giving the following:

$$
\begin{array}{cc}
35 & 10 \\
23 & 16
\end{array}
$$

This is a partial table of the first table.

We now turn to our general technique for partitioning a contingency table into two parts. The particular partition into two parts comes from two partial tables of the original contingency table. The two partial tables in a sense divide up the original table. The division has three properties.

1. The degrees of the freedom of the original table is equal to the sum of the degrees of the freedom of the two partial tables.

2. The chi-square statistic for the original contingency table is equal to the sum of the two chi-square statistics for the two partial tables.

3. Under the null hypothesis of independence in the large table, the two chi-square statistics for the two partial tables are asymptotically independent each with chi-square distributions with their given degrees of freedom.

The method selects a subset of the rows (or a subset of the columns) and constructs one table by deleting all of the other rows (or columns). The rows and columns that were deleted in the first table are included in the second table. This complex verbal explanation becomes easier when illustrated graphically. We will consider the case where we wish to have the first partial table consist of the first J columns out of the total of c columns, and suppose the original table is as given ($J \geq 2$):

$$
\begin{array}{|cccc|} n_{1,1} & \cdots & n_{1,J-1} & n_{1,J} \\ n_{2,1} & \cdots & n_{2,J-1} & n_{2,J} \\ \vdots & \ddots & \vdots & \vdots \\ n_{r,1} & \cdots & n_{r,J-1} & n_{r,J} \end{array}
\begin{array}{|cccc|} n_{1,J+1} & \cdots & n_{1,c} \\ n_{2,J+1} & \cdots & n_{2,c} \\ \vdots & \ddots & \vdots \\ n_{r,J+1} & \cdots & n_{r,c} \end{array}
$$

The first partial table will be

$$
\begin{array}{|cccc|} n_{1,1} & n_{1,2} & \cdots & n_{1,J} \\ n_{2,1} & n_{2,2} & \cdots & n_{2,J} \\ \vdots & \vdots & \ddots & \vdots \\ n_{r,1} & n_{r,2} & \cdots & n_{r,J} \end{array}
$$

To get the second table, we take the last columns, corresponding to the indices $J+1, J+2, \ldots, c$. In addition, we add a first column which is the sum of all the entries in each row of the table above; that is, define new elements m_1 up to m_r as follows:

$$ m_1 = n_{1,1} + n_{1,2} + \cdots + n_{1,J} $$

$$ m_2 = n_{2,1} + n_{2,2} + \cdots + n_{2,J} $$

$$ \vdots $$

$$ m_r = n_{r,1} + n_{r,2} + \cdots + n_{r,J} $$

With this notation, the second table is the following:

$$
\begin{array}{c|cccc}
m_1 & n_{1,J+1} & n_{1,J+2} & \cdots & n_{1,c} \\
m_2 & n_{2,J+1} & n_{2,J+2} & \cdots & n_{2,c} \\
\vdots & \vdots & \vdots & \ddots & \vdots \\
m_r & n_{r,J+1} & n_{r,J+2} & \cdots & n_{r,c}
\end{array}
$$

That is, the second table is composed of two parts. The first part is from the original table. To that, one more column is added, where each element in the column is the sum across the rows of the first table.

The first partial table is $r \times J$ and has $(r-1)(J-1)$ degrees of freedom. The second table has r rows and $c - J + 1$ columns; thus, there are $(r-1)(c-J)$ degrees of freedom. Note that the following holds:

$$
\begin{array}{ccccc}
\dfrac{\text{Degrees of freedom}}{\text{Original table}} & = & \dfrac{\text{Degrees of freedom}}{\text{Table One}} & + & \dfrac{\text{Degrees of freedom}}{\text{Table Two}} \\
(r-1)(c-1) & = & (r-1)(J-1) & + & (r-1)(c-J)
\end{array}
$$

In other words, point 1 above has been shown: the degrees of freedom in these two tables add up to the degrees of freedom of the original table. The second statement above was that the chi-square distribution broke up into two parts which added up to the original chi-square. This is not strictly speaking true if we compute the chi-square statistic X^2 as given above in this chapter. There is a chi-square statistic called the *likelihood ratio chi-square statistic* (which we shall note by LRX^2) that satisfies point 2 above. As the sample size becomes large, the likelihood ratio chi-square statistic and the X^2 chi-square statistic approach the same value; the two statistics are effectively the same. We shall use the likelihood ratio chi-square statistic here, so that the value of the statistic for the entire table is exactly equal to the sum of the two parts. In the example below, and in the problems where both statistics are computed, it will be seen that the values are very close in general. The formula for computing the likelihood ratio chi-square statistic is given by the next equation.

$$
\text{LRX}^2 = 2 \sum_{i=1}^{r} \sum_{j=1}^{c} n_{ij} \ln \left(\frac{n_{ij} n_{\bullet\bullet}}{n_{i\bullet} n_{\bullet j}} \right) = 2 \sum_{ij} (\text{observed}) \ln \left(\frac{\text{observed}}{\text{expected}} \right).
$$

In this equation, "ln" refers to the logarithm to the base e. The natural log function is available on many hand calculators. (However, the partitions of chi-square given below can effectively be done using the X^2 chi-square statistic as well. In that case, the chi-square statistic for the table will not exactly equal the sum of the chi-square statistics for the two partial tables.) The second point above can be expressed algebraically as the following equation:

$$
\text{LRX}^2(\text{original table}) = \text{LRX}^2(\text{table one}) + \text{LRX}^2(\text{table two})
$$

The third important point: under the null hypothesis of independence the two likelihood ratio chi-square statistics are approximately independent statistical tests. That is, these two tests are like statistics from separate studies or experiments. We illustrate this by an example.

Example 7.5. In a paper by Weiner *et al.* [1979], patients who undergo maximal exercise testing and coronary angiography for suspected coronary artery disease are studied. Among the men studied with chest pain, there were three clinical groups. In heart disease, narrowing of the coronary arteries often prevents an adequate blood supply from getting to the heart muscle. This causes chest pain called *angina*. This is especially true when subjects are exerting themselves and have a need for higher output from the heart, and thus more oxygen for the heart muscle. Lack of oxygen to the muscle is called myocardial ischemia, so such chest pain is also called ischemic chest pain. The individuals studied in this paper, based upon their clinical patterns, had their chest pain categorized in one of three categories. Two of the categories referred to angina chest pain: "definite angina" or "probable angina." In addition, there was a third set of individuals whose chest pain syndrome did not seem to be ischemic in origin—this was called "nonischemic chest pain." In addition, these individuals underwent coronary angiography where their coronary arteries were studied for narrowing by X-ray motion pictures. Depending upon whether none, one, two, or three of the major coronary arteries had significant disease (defined as a 70% or greater narrowing), the individuals were categorized as having zero-vessel, one-vessel, or two- or three-vessel disease. The data from this study is presented in the following 3×3 contingency table:

	Vessels Diseased		
Chest Pain	**0**	**1**	**2 or 3**
Definite angina	66	135	419
Probable angina	179	139	276
Nonischemic	197	39	15

There is reason to believe that the two anginal chest pain syndromes may tend to have the same pattern of disease, but that they both will differ from the nonischemic chest pain. One might propose before seeing the data to partition this table into two partial tables. The first table would compare the definite and probable angina groups, that is, the first two rows. The second table would compare those patients with definite or probable angina (that is, the presumed ischemic chest pain patients) versus the nonischemic patients. The resulting partial tables are the following:

	Vessels Diseased					**Vessels Diseased**		
Chest Pain	**0**	**1**	**2 or 3**		**Chest Pain**	**0**	**1**	**2 or 3**
Definite	66	135	419		Angina (D or P)	245	274	695
Probable	179	139	276		Nonischemic	197	39	15

If we then compute the degrees of freedom and the likelihood ratio chi-square statistics and normal chi-square statistics for each table, we obtain the following:

	DF	LRX^2	*p*-Value	X^2
Table one	2	83.28	0+	81.08
Table two	2	353.62	0+	352.00
Original table	4	436.90	0+	418.48

The first $\text{LR}X^2$ value is given by twice the following expression:

$$66\ln(66/125.12) + 135\ln(135/139.93) + 419\ln(419/354.94)$$
$$+ 179\ln(179/119.88) + 139\ln(139/134.07) + 276\ln(276/340.06).$$

The values 125.12, 139.93, 354.94, 119.88, 134.07, and 340.06 are the expected values under the null hypothesis of no association.

From this, we see that both of the partial tables are very highly statistically significant, although as measured by the chi-square value, the combined table has a larger value for the chi-square statistic. In part, this will occur because the second table involves the sum of the rows and has a larger total sample size than the first table.

After arriving at this point, it should be noted that further options are available. Instead of looking at only two tables, we could have looked at further tables by partitioning each of the two partial tables already at hand. By continuing the process of division using this partitioning method any chi-square statistic may be decomposed into individual one degree of freedom chi-square values (see Problem 7.31). Further, on the two partial tables, we may use the weapons already at our disposal. Once we note that they are statistically significant, we may further investigate the association by calculating the adjusted residuals and p-values as in Section 7.2, and interpreting the findings based upon these numbers. The reason that these two tables have statistics that are asymptotically independent under the null hypothesis is developed further in Problem 7.30.

We now turn briefly to the second method of partitioning contingency tables. Sometimes one wishes to divide the contingency table into partial tables that do not fall under the pattern given above. In general, there is no assurance that the individual statistics will be asymptotically independent. The subject of general partitioning is quite complex; further information is given in Lancaster [1969] and in Goodman [1968]. A conservative approach that can always be used is the following: decide before seeing the data how many hypotheses to investigate. Suppose that at significance level α, we want to be sure that any associations that we identify are actually there: that is, we do not want to make the mistake of rejecting the null hypothesis in *any* of the tables all with a probability of α or less. The conservative approach is to compute the chi-square statistics and p-values for each of the partial tables, then multiply these p-values by the number of partial tables examined. Assert that there is association in a table only when the p-value for the table multiplied by the number of different tables is less than α. (This is called the Bonferroni approach and is discussed further in Chapter 12.)

7.6 *LOG-LINEAR MODELS

For the first time we will examine statistical methods that deal with more than two variables at one time. Such methods are important for the following reasons: in one dimension, we have been able to summarize data with the normal distribution and its two parameters, the mean and the variance, or equivalently the mean and the standard deviation. Even when the data did not appear normally distributed, we could get a feeling for our data by histograms and other graphical methods in one dimension. When we observe two numbers at the same time, or are working with two-dimensional data, we can plot the points and examine the data visually. (This will be discussed further in Chapter 9. Even in the case of two variables, we shall see that it is useful to have models summarizing the data.) When we move to three variables, however, it is much harder to get a "feeling" for the data. Possibly, in three dimensions, we could construct visual methods of examining the data, although this would be difficult. With more than three variables, such physical plots cannot be obtained; although mathematicians may think of space and time as being a four-dimensional space, we, living in a three-dimensional world, cannot readily grasp what the points mean. In this case, it becomes very important to simplify our understanding of the data by fitting a model to the data. *If* the model fits, it may summarize the complex situation very succinctly. In addition, the model may point out relationships that may reasonably be understood in a simple way. The fitting of probability models or distributions to many variables at one time is an important topic.

The models are necessarily mathematically complex; thus, the reader needs discipline and perseverance to work through and understand the methods. It is a very worthwhile task. Such methods are especially useful in the analysis of observational biomedical data. We now proceed to our first model for multiple variables, the log-linear model.

Before beginning the details of the actual model, we define some terms that we will be using. The models we will investigate are for *multivariate categorical data*. We already know the meaning of *categorical data*; it is values of a variable or variables that put the individuals into one of a finite number of categories. The term *multivariate* comes from the prefix "multi-," meaning "many," and "variate," referring to variables; the term refers to multiple variables at one time.

Definition 7.2. *Multivariate data* are data for which each observation consists of values for more than one random variable on each experimental unit. *Multivariate statistical analysis* consists of data analysis of multivariate data.

The majority of data collected is, in fact, multivariate data. If one measures systolic and diastolic blood pressure on each subject, there are two variables; thus, multivariate data. If we administer a questionnaire on the specifics of brushing teeth, flossing, and so on, the response of a person to each question is a separate variable, and thus one has multivariate data. Strictly speaking, some of the two-way contingency table data we have looked at is multivariate data since it cross-classifies by two variables. On the other hand, the tables which arose from looking at one quantity in different sub-groups is not multivariate when the observation of the group was not observed on experimental units picked from some population, but rather the group was part of a data collection or experimental procedure.

Additional terminology is included in the term *log-linear models*. We already have an idea of the meaning of a model. Let us consider the two terms *log* and *linear*. The logarithm was discussed in connection with the likelihood ratio chi-square statistics. (In this section, and indeed throughout this book, the logarithm will be to the base e.) Briefly, recall some of the properties of the logarithm. Of most importance to us is that the log of the product of terms is the sum of the individual logs. For example, if we have three numbers, a, b, and c (all positive), then

$$\ln(abc) = \ln a + \ln b + \ln c.$$

Here, "ln" represents the *natural logarithm*, or the log to the base e. Recall that by the definition of the natural log, if one exponentiates the logarithm—that is, takes the number e to the power represented by the logarithm—one gets the original number back:

$$e^{\ln a} = a.$$

Inexpensive hand calculators compute both the logarithm and the exponential of a number. If you are rusty with such manipulations, Problem 7.26 gives practice in the use of logarithms and exponentials.

The second term we have used is the term *linear*. It is associated with a straight line or a linear relationship. For two variables x and y, y is a linear function of x if $y = a + bx$, where a and b are constants. For three variables, x, y and z, z is a linear function of x and y, if $z = a + bx + cy$, where a, b, and c are constant. In general, in a linear relationship, one *adds* a constant multiple for each of the variables involved. The linear models we use will look like the following: Let

$$g_{ij}^{IJ}$$

be the logarithm of the probability that an observation falls into the ijth cell in the two-dimensional contingency table. Let there be I rows and J columns. One possible model would be

$$g_{ij}^{IJ} = u_i^I + u_j^J.$$

(For more detail on why the term "linear" is used for such models, see Note 7.4.)

We first consider the case of two-way tables. Suppose that we want to fit a model for independence. We know that independence in terms of the cell probabilities π_{ij} is equivalent to the following equation:

$$\pi_{ij} = \pi_{i\bullet} \pi_{\bullet j}.$$

If we take logarithms of this equation and use the notation g_{ij} for the natural log of the cell probability, the following results:

$$g_{ij} = \ln \pi_{ij} = \ln \pi_{i\bullet} + \ln \pi_{\bullet j}.$$

When we denote the natural logs of $\pi_{i\bullet}$ and $\pi_{\bullet j}$ by the quantities h_i^I and h_j^J, we have

$$g_{ij} = h_i^I + h_j^J.$$

The quantities h_i^I and h_j^J are not all independent. They come from the marginal probabilities for the I row variables and the J column variables. For example, the h_i^I's satisfy the equation.

$$e^{h_1^I} + e^{h_2^I} + \cdots + e^{h_I^I} = 1.$$

This equation is rather awkward and unwieldy to work with; in particular, given $I - 1$ of the h_i's, the determination of the other coefficient takes a bit of work. It is possible to choose a different normalization of the parameters if we add a constant. Rewrite the equation above as follows:

$$g_{ij} = \left(\sum_{i'=1}^{I} \frac{h_{i'}^I}{I} \right) + \left(\sum_{j'=1}^{J} \frac{h_{j'}^J}{J} \right) + \left(h_i^I - \sum_{i'=1}^{I} \frac{h_{i'}^I}{I} \right) + \left(h_j^J - \sum_{j'=1}^{J} \frac{h_{j'}^J}{J} \right).$$

The two quantities in parentheses farthest to the right both add to zero when we sum over the indices i and j, respectively. In fact, that is why those terms were added and subtracted. Thus, we can rewrite the equation for g_{ij} as follows:

$$g_{ij} = u + u_i^I + u_j^J, \quad i = 1, \ldots, I; \ j = 1, \ldots, J,$$

where

$$\sum_{i=1}^{I} u_i^I = 0, \quad \sum_{j=1}^{J} u_j^J = 0.$$

It is easier to work with this normalization. Note that this is a linear model for the log of the cell probability π_{ij}; that is, this is a log-linear model.

Recall that the estimates for the $\pi_{i\bullet}$ and $\pi_{\bullet j}$ were $n_{i\bullet}/n_{\bullet\bullet}$ and $n_{\bullet j}/n_{\bullet\bullet}$, respectively. If one follows through all of the mathematics involved, estimates for the parameters in the log-linear model result. At this point, we shall slightly abuse our notation by using the same notation for both the population parameter values and the estimated parameter values from the sample at hand. The estimates are

$$u = \frac{1}{I} \sum_{i=1}^{I} \ln \frac{n_{i\bullet}}{n_{\bullet\bullet}} + \frac{1}{J} \sum_{j=1}^{J} \ln \frac{n_{\bullet j}}{n_{\bullet\bullet}},$$

$$u_i^I = \ln \frac{n_{i\bullet}}{n_{\bullet\bullet}} - \frac{1}{I} \sum_{i'=1}^{I} \ln \frac{n_{i'\bullet}}{n_{\bullet\bullet}},$$

$$u_j^J = \ln \frac{n_{\bullet j}}{n_{\bullet\bullet}} - \frac{1}{J} \sum_{j'=1}^{I} \ln \frac{n_{\bullet j'}}{n_{\bullet\bullet}}.$$

From these estimates, we get fitted values for the number of observations in each cell. This is done as follows: by inserting the estimated parameters from the log-linear model and then taking the exponential, we have an estimate of the probability an observation falls into the ijth cell. Multiplying this by $n_{\bullet\bullet}$, we have an estimate of the number of observations we should see in the cell if the model is correct. In this particular case, the fitted value for the ijth cell turns out to be the expected value from the chi-square test presented earlier in this chapter, that is, $n_{i\bullet}n_{\bullet j}/n_{\bullet\bullet}$.

Let us illustrate these complex formulas by finding the estimates for one of the examples above.

Example 7.6. Continuing Example 7.1, we know that for the 2×4 table, we have the following values:

$$n_{\bullet 1} = 18, \quad n_{\bullet 2} = 31, \quad n_{\bullet 3} = 16, \quad n_{\bullet 4} = 72,$$

$$n_{1\bullet} = 69, \quad n_{2\bullet} = 68, \quad n_{\bullet\bullet} = 137;$$

$$\ln(n_{1\cdot}/n_{\cdot\cdot}) \doteq -0.6859, \quad \ln(n_{2\cdot}/n_{\cdot\cdot}) \doteq -0.7005,$$

$$\ln(n_{\cdot1}/n_{\cdot\cdot}) \doteq -2.0296, \quad \ln(n_{\cdot2}/n_{\cdot\cdot}) \doteq -1.4860,$$

$$\ln(n_{\cdot3}/n_{\cdot\cdot}) \doteq -2.1474, \quad \ln(n_{\cdot4}/n_{\cdot\cdot}) \doteq -0.6433.$$

With these numbers, we may compute the parameters for the log-linear model. They are

$$u \doteq \frac{-0.6859 - 0.7005}{2} + \frac{-2.0296 - 1.4860 - 2.1474 - 0.6433}{4}$$

$$\doteq -0.6932 - 1.5766 = -2.2698,$$

$$u_1^J \doteq -2.0296 - (-1.5766) \doteq -0.4530,$$

$$u_1^I \doteq -0.6859 - (-0.6932) \doteq 0.0073, \quad u_2^J \doteq -1.4860 - (-1.5766) \doteq 0.0906,$$

$$u_2^I \doteq -0.7004 - (-0.6932) \doteq -0.0073, \quad u_3^J \doteq -2.1474 - (-1.5766) \doteq -0.5708,$$

$$u_4^J \doteq -0.6433 - (-1.5766) \doteq 0.9333.$$

The larger the value of the coefficient, the larger will be the cell probability. For example, looking at the two values indexed by i, the second state having a minus sign will lead to a slightly smaller contribution to the cell probability than the term with the plus sign. (This is also clear from the marginal probabilities, which are $68/137$ and $69/137$.) The small magnitude of the term means that the difference between the two I state values has very little effect on the cell probability. We see that of all the contributions for the j variable values, $j = 4$ has the biggest effect, 1 and 3 have fairly large effects (tending to make the cell probability small), while 2 is intermediate. $\qquad\square$

The chi-square goodness of fit and the likelihood ratio chi-square statistics which may be applied to this setting are

$$X^2 = \sum(\text{observed} - \text{fitted})^2/\text{fitted},$$

$$\text{LRX}^2 = 2\sum(\text{observed}\ln(\text{observed}/\text{fitted})).$$

Finally, if the model for independence does not hold, we may add more parameters. We can find a log-linear model which will fit any possible pattern of cell probabilities. The equation for the log of the cell probabilities is given by the following:

$$g_{ij} = u + u_i^I + u_j^J + u_{ij}^{IJ}, \quad i = 1, \dots, I; \ j = 1, \dots, J,$$

where

$$\sum_{i=1}^I u_i^I = 0, \quad \sum_{j=1}^J u_j^J = 0, \quad \sum_{i=1}^I u_{ij}^{IJ} = 0, \quad \sum_{j=1}^J u_{ij}^{IJ} = 0.$$

It seems rather paradoxical, or at least needlessly confusing, to take a value indexed by i and j, and to set it equal to the sum of four values, including some indexed by i and j; the right hand side is much more complex than the left hand side. The reason for doing this is that, usually, the full (or saturated) model, which can give any possible pattern of cell probabilities, is not desirable. It is hoped during the modeling effort that the data will allow a simpler model. This would allow a simpler interpretation of the data. In the case at hand, we examine the possibility of the simpler interpretation that the two variables are independent. If they are not, the particular model is not too useful.

Note two properties of the fitted values. First, in order to fit the independence model, where each term depends upon at most one factor or one variable, we only needed to know the marginal values of the frequencies, the $n_{i\bullet}$ and $n_{\bullet j}$. We did not need to know the complete distribution of the frequencies to find our fitted values. Second, when we had fit values to the frequency table, the fitted values summed to the marginal value used in the estimation; that is, if we sum across i or j, the sum of the expected values is equal to the actually observed sum.

At this point, it seems that we have needlessly confused a relatively easy matter: the analysis of two-way contingency tables. If only two-way contingency tables were involved, this would be a telling criticism; however, the strength of log-linear models appears when we have more than two cross-classified categorical variables. We shall now discuss the situation for three cross-classified categorical variables. The analyses may be extended to any number of variables; such extensions will not be done in this text.

Suppose that the three variables are labeled X, Y, and Z, where the index i is used for the X variable, j for the Y variable, and k for the Z variable. (This is to say that X will take values $1, \ldots, I$, Y will take on $1, \ldots, J$, and so forth.) The methods of this section will be illustrated by the following example.

Example 7.7. The study of Weiner *et al.* [1979] has been discussed previously in this chapter. The study involves exercise treadmill tests for men and women. Among men with chest pain thought probably to be angina, a three-way classification of the data is as follows: one variable looks at the resting electrocardiogram and tells whether or not certain parts of the electrocardiogram (called the ST- and T-waves) are normal or abnormal. Thus $J = 2$. A second variable considers whether or not the exercise test was positive or negative ($I = 2$). A positive exercise test shows evidence of an ischemic response (that is, lack of appropriate oxygen to the heart muscles for the effort being exerted). A positive test is thought to be an indicator of coronary artery disease. The third variable was an evaluation of the coronary artery disease as determined by coronary ateriography. The disease is classified as normal or minimal disease called "zero vessel disease," "one vessel disease" and "a multiple vessel disease" category ($K = 3$). The data are presented in Table 7.7.   □

The most general log-linear model for the three factors is given by the following extension of the two-factor work:

$$g_{ijk} = u + u_i^I + u_j^J + u_k^K + u_{ij}^{IJ} + u_{ik}^{IK} + u_{jk}^{JK} + u_{ijk}^{IJK},$$

Table 7.7. Exercise Test Data from Weiner *et al.* [1979]

Exercise Test Response (I)	Resting Electrocardiogram ST and T-waves (J)	Number of Vessels Diseased (K)		
		0 ($k = 1$)	1 ($k = 2$)	2 or 3 ($k = 3$)
+ ($i = 1$)	normal ($j = 1$)	30	64	147
	abnormal ($j = 2$)	17	22	80
− ($i = 2$)	normal ($j = 1$)	118	46	38
	abnormal ($j = 2$)	14	7	11

where

$$\sum_{i=1}^{I} u_i^I = \sum_{j=1}^{J} u_j^J = \sum_{k=1}^{K} u_k^K = 0,$$

$$\sum_{i=1}^{I} u_{ij}^{IJ} = \sum_{j=1}^{J} u_{ij}^{IJ} = \sum_{i=1}^{I} u_{ik}^{IK} = \sum_{k=1}^{K} u_{ik}^{IK} = \sum_{j=1}^{J} u_{jk}^{JK} = \sum_{k=1}^{K} u_{jk}^{JK} = 0,$$

$$\sum_{i=1}^{I} u_{ijk}^{IJK} = \sum_{j=1}^{J} u_{ijk}^{IJK} = \sum_{k=1}^{K} u_{ijk}^{IJK} = 0.$$

In other words, there is a u term for every possible combination of the variables, including no variables at all. For each term involving one or more variables, if we sum over any one variable, the sum is equal to zero. The term involving I, J, and K is called a *three-factor term*, or a *second-order interaction term*; in general, if a coefficient involves M variables, it is called an M-factor term or an $M - 1$ order interaction term.

With this notation we may now formulate a variety of simpler models for our three-way contingency table. For example, the model might be any one of the following simpler models:

$$H_1 : g_{ijk} = u + u_i^I + u_j^J + u_k^K,$$

$$H_2 : g_{ijk} = u + u_i^I + u_j^J + u_k^K + \iota.$$

$$H_3 : g_{ijk} = u + u_i^I + u_j^J + u_k^K + u_{ij}^{IJ} + u_{ik}^{IK} + u_{jk}^{JK}.$$

The notation has become so formidable that it is useful to introduce a shorthand notation for the hypotheses. One or more capitalized indices contained in brackets will indicate a hypothesis where the terms involving that particular set of indices as well as any terms involving subsets of the indices are to be included in the model. Any terms not specified in this form are assumed not to be in the model. For example,

$$[IJ] \longrightarrow u + u_i^I + u_j^J + u_{ij}^{IJ},$$

$$[K] \longrightarrow u + u_k^K,$$

$$[IJK] \longrightarrow u + u_i^I + u_j^J + u_k^K + u_{ij}^{IJ} + u_{ik}^{IK} + u_{jk}^{JK} + u_{ijk}^{IJK}.$$

The formulation of the three hypotheses given above in this notation would be simplified as follows:

$$H_1 : [I][J][K],$$

$$H_2 : [IJ][K],$$

$$H_3 : [IJ][IK][JK].$$

This notation describes a *hierarchical hypothesis*, that is, if we have two factor terms containing, say, variables I and J, then we also have the one factor terms for the same variables. The hypothesis would not be written $[IJ][I][J]$, for example, because the last two parts would be redundant, as already implied by the first. Using this bracket notation for the three factor model, there are eight possible hypotheses of interest. All except the most complex one have a simple interpretation in terms of the probability relationships between the factors X, Y, and Z. This is given in Table 7.8.

Hypotheses 5, 6, and 7 are of particular interest. Take, for example, Hypothesis 5. This hypothesis states that if you take into account the X variable, then there is no association between Y and Z. In particular, if one only looks at the two-way table of Y and Z, an association may be seen, because in fact they are associated. However, if Hypothesis 5 holds, one could then conclude that the association is due to interaction with the variable X and could be "explained away" by taking into account the values of X.

There is a relationship between hypotheses involving the bracket notation and the corresponding tables that one gets from the higher dimensional contingency table. For example, consider the term $[IJ]$. This is related to the contingency table one gets by summing over K (that is, over the Z variable). In general, a contingency table that results from summing over the cells for one or more variables in a higher dimensional contingency table is called a *marginal table*. Very simple examples of marginal tables are the marginal total column and the marginal total row along the bottom of the two-way table.

Using the idea of marginal tables, we can discuss some properties of fits of the various hierarchical hypotheses for log-linear models. Three facts are important:

Table 7.8. Three Factor Hypotheses and Their Interpretation.

Hypothesis	Meaning in Words	The hypothesis restated in terms of the π_{ijk}'s
1. $[I][J][K]$	X, Y, and Z are independent	$\pi_{ijk} = \pi_{i\bullet\bullet}\,\pi_{\bullet j\bullet}\,\pi_{\bullet\bullet k}$
2. $[IJ][K]$	Z is independent of X and Y	$\pi_{ijk} = \pi_{ij\bullet}\,\pi_{\bullet\bullet k}$
3. $[IK][J]$	Y is independent of X and Z	$\pi_{ijk} = \pi_{i\bullet k}\,\pi_{\bullet j\bullet}$
4. $[I][JK]$	X is independent of Y and Z	$\pi_{ijk} = \pi_{i\bullet\bullet}\,\pi_{\bullet jk}$
5. $[IJ][IK]$	For X known, Y and Z are independent; that is, Y and Z are conditionally independent given X	$\pi_{ijk} = \pi_{ij\bullet}\,\pi_{i\bullet k}\,/\,\pi_{i\bullet\bullet}$
6. $[IJ][JK]$	X and Z are conditionally independent given Y	$\pi_{ijk} = \pi_{ij\bullet}\,\pi_{\bullet jk}\,/\,\pi_{\bullet j\bullet}$
7. $[IK][JK]$	X and Y are conditionally independent given Z	$\pi_{ijk} = \pi_{i\bullet k}\,\pi_{\bullet jk}\,/\,\pi_{\bullet\bullet k}$
8. $[IJ][IK][JK]$	No three factor interaction	No simple form

1. The fit is estimated using only the marginal tables associated with the bracket terms that state the hypothesis. For example, consider Hypothesis 1, the independence of the X, Y, and Z variables. To compute the estimated fit, one only needs the one-dimensional frequency counts for the X, Y, and Z variables individually and does not need to know the joint relationship between them.

2. Suppose one looks at the fitted estimates for the frequencies and sums the *fitted* values to give marginal tables. The marginal sum for the fit is equal to the marginal table for the actual data set when the marginal table is involved in the fitting.

3. The chi-square and likelihood ratio chi-square tests discussed above, using the observed and fitted values still hold.

We consider fitting Hypothesis 5 to the data of Example 7.7. The hypothesis stated that if one knows the response to the maximal treadmill test, then the resting electrocardiogram ST- and T-wave abnormalities are independent of the number of vessels diseased. The observed frequencies and the fitted frequencies, as well as the values of the u-parameters for this model are given in Table 7.9.

The relationship between the fitted parameter values and the expected, or fitted, number of observations in a cell is given by the following equations:

$$\widehat{\pi}_{ijk} = e^{u + u_i^I + u_j^J + u_k^K + u_{ij}^{IJ} + u_{ik}^{IK}},$$

The fitted value $= n_{\cdots}\,\widehat{\pi}_{ijk}$,   where $n_{\cdots}$ is the total number of observations.

Table 7.9. Fitted Model for the Hypothesis that the Resting Electrocardiogram ST- and T-Wave (normal or abnormal) is Independent of the Number of Vessels Diseased (0, 1 and 2–3) Conditionally Upon Knowing the Exercise Response (+ or −).

Cell (i,j,k)	Observed	Fitted	u-Parameters
(1,1,1)	30	31.46	$u = -2.885$
(1,1,2)	64	57.57	$u_1^I = -u_2^I = 0.321$
(1,1,3)	147	151.97	$u_1^J = -u_2^J = 0.637$
(1,2,1)	17	15.54	$u_1^K = -0.046,\ u_2^K = -0.200$
(1,2,2)	22	28.43	$u_3^K = 0.246$
(1,2,3)	80	75.04	$u_{1,1}^{IJ} = -0.284,\ u_{1,2}^{IJ} = 0.284$
(2,1,1)	118	113.95	$u_{2,1}^{IJ} = 0.284,\ u_{2,2}^{IJ} = -0.284$
(2,1,2)	46	45.75	$u_{1,1}^{IK} = -0.680,\ u_{1,2}^{IK} = 0.078$
(2,1,3)	38	42.30	$u_{1,3}^{IK} = 0.602$
(2,2,1)	14	18.05	$u_{2,1}^{IK} = 0.680,\ u_{2,2}^{IK} = -0.078$
(2,2,2)	7	7.25	$u_{2,3}^{IK} = -0.602$
(2,2,3)	11	6.70	

For these data, we compute the right hand side of the first equation for the $(1,1,1)$ cell. In this case,

$$\widehat{\pi}_{111} = \exp(-2.885 + 0.321 + 0.637 - 0.046 - 0.284 - 0.680)$$

$$= e^{-2.937} \doteq 0.053,$$

The fitted value $\doteq 594 \times 0.053 \doteq 31.48$

where exp(argument) is equal to the number e raised to a power equal to the argument. The computed value of 31.48 differs slightly from the tabulated value, because the tabulated value came from computer output that carried more accuracy than the accuracy used in this computation.

We may test whether the hypothesis is a reasonable fit by computing the chi-square value under this hypothesis. The likelihood ratio chi-square value is computed as follows:

$$\text{LRX}^2 = 2\big(30\ln(30/31.46) + \cdots + 11\ln(11/6.70)\big) \doteq 6.86.$$

In order to assess the statistical significance we need the degrees of freedom to examine the chi-square value. For the log-linear model the degrees of freedom is given by the following rule:

Rule 7.1. The chi-square statistic for model fit of a log-linear model has degrees of freedom equal to the total number of cells in the table $(I \times J \times K)$ minus the number of independent parameters fitted. By "independent parameters" we mean the following: the number of parameters fitted for the X variable is $I - 1$ since the u_i^I terms sum to zero. For each of the possible terms in the model the number of independent parameters is given in the Table 7.10.

For the particular model at hand, the number of independent parameters fitted is the sum of the last column in Table 7.11.

Table 7.10. Degrees of Freedom for Log-Linear Model Chi-Square.

Term	Number of Parameters
u	1
u_i^I	$I - 1$
u_j^J	$J - 1$
u_k^K	$K - 1$
u_{ij}^{IJ}	$(I - 1)(J - 1)$
u_{ik}^{IK}	$(I - 1)(K - 1)$
u_{jk}^{JK}	$(J - 1)(K - 1)$
u_{ijk}^{IJK}	$(I - 1)(J - 1)(K - 1)$

Table 7.11. Parameters for Example 7.6.

Model Terms	Number of Parameters	
	General	Example 7.6
u	1	1
u_i^I	$I - 1$	1
u_j^J	$J - 1$	1
u_k^K	$K - 1$	2
u_{ij}^{IJ}	$(I - 1)(J - 1)$	1
u_{ik}^{IK}	$(I - 1)(K - 1)$	2

There are twelve cells in the table, so that the number of degrees of freedom is $12 - 8$, or 4 degrees of freedom. The p-value for a chi-square of 6.86 for 4 degrees of freedom is 0.14, so that we cannot reject the hypothesis that this particular model fits the data.

We are now faced with a new consideration. Just because this one model fits the data, there may be other models that fit the data as well, including some simpler model. In general, one would like as simple a model as possible (Occam's razor); however, models with more parameters generally give a better fit. In particular, a simpler model may have a p-value much closer to the significance level one is using. For example, if one model has a p of 0.06 and is simple, and a slightly more complicated model has a p of 0.78, which is to be preferred? If the sample size is small, the p of 0.06 may correspond to estimated cell values that differ considerably from the actual values. For a very large sample, the fit may be excellent. There is no hard-and-fast rule in the trade-off between the simplicity of the model and the goodness of the fit. In order to understand the data, we are happy with the simple model that fits fairly well although, presumably, it is not precisely the probability model that would fit the entirety of the population values. Here we would hope for considerable scientific understanding from the simple model.

For this example, Table 7.12 shows for each of the eight possible models the degrees of freedom, the LRX^2 value (with its corresponding p-value for reference), and the "usual" goodness of fit chi-square value.

Table 7.12. Chi-Square Goodenss-of-Fit Statistics for Example 6 Exercise Data.

Model	DF	LRX^2	p-Value	X^2
$[I][J][K]$	7	184.21	< 0.0001	192.35
$[IJ][K]$	6	154.35	< 0.0001	149.08
$[IK][J]$	5	36.71	< 0.0001	34.09
$[I][JK]$	5	168.05	< 0.0001	160.35
$[IJ][IK]$	4	6.86	0.14	7.13
$[IJ][JK]$	4	138.19	< 0.0001	132.30
$[IK][JK]$	3	20.56	0.0001	21.84
$[IJ][IK][JK]$	2	2.96	0.23	3.03

We see that there are only two possible models if we are to simplify at all rather than using the entire data set as representative. They are the model fit above and the model which contains each of the three two-factor interactions. The model fit above is simpler, while the other model below has a larger *p*-value, possibly indicating a possibly better fit. One way of approaching this is through what are called *nested* hypotheses.

Definition 7.3. One hypothesis is *nested* within another if it is the special case of the other hypothesis. That is, whenever the nested hypothesis holds it necessarily implies that the hypothesis it is nested in also holds.

If nested hypotheses are considered, then one takes the difference between the likelihood ratio chi-square statistic for the more restrictive hypothesis, minus the likelihood ratio chi-square statistic for the more general hypothesis. This difference will itself be a chi-square statistic if the special case holds. The degrees of freedom of the difference is equal to the difference of freedom for the two hypotheses. In this case, the chi-square statistic for the difference is 6.86 - 2.96 = 3.90. The degrees of freedom are $4 - 2 = 2$. This corresponds to a *p*-value of more than 0.10. At the 5% significance level, there is marginal evidence that the more general hypothesis does fit the data better than the restrictive hypothesis. In this case, however, because of the greater simplicity of the restrictive hypothesis, one might choose it to fit the data. Once again, there is no hard and fast answer to the payoff between fit of the data and the simplicity of interpretation of a hypothesis.

This material is an introduction to log-linear models. There are many extensions, some of which are mentioned briefly in the notes at the end of this chapter. An excellent introduction to log-linear models is given in Fienberg [1977]. Other elementary books on log-linear models are those by Everitt [1992] and Reynolds [1977]. A more advanced and thorough treatment is given by Haberman [1978] and Haberman [1979]. A text touching on this subject and many others is Bishop *et al.* [1975].

NOTES

Note 7.1: Testing Independence in Model 1 and Model 2 Tables

This Note refers to Section 7.2.

1. Model 1. The usual null hypothesis is that the results are statistically independent. That is (assuming row variable = i and column variable = j):

$$P[i \text{ and } j] = P[i]P[j].$$

The probability on the left hand side of the equation is π_{ij}. From Section 7.2, the marginal probabilities are found to be

$$\pi_{i\bullet} = \sum_{j=1}^{c} \pi_{ij} \quad \text{and} \quad \pi_{\bullet j} = \sum_{i=1}^{r} \pi_{ij}.$$

The null hypothesis of statistical independence of the variables is

$$H_0 : \pi_{ij} = \pi_{i\bullet} \pi_{\bullet j}.$$

Consider how one might estimate these probabilities under two circumstances:

a. Without assuming the variables are independent.

b. Assuming the variables are independent.

In the first instance (a), we are in a binomial situation. Let a success be the occurence of the ijth pair. Let

$$n_{\bullet\bullet} = \sum_{i=1}^{r} \sum_{j=1}^{c} n_{ij}.$$

The binomial estimate for π_{ij} is the number of successes divided by the number of trials:

$$p_{ij} = n_{ij}/n_{\bullet\bullet}.$$

If we assume independence, the natural approach is to estimate $\pi_{i\bullet}$ and $\pi_{\bullet j}$. But the occurrence of state i for the row variable is also a binomial event. The estimate of $\pi_{i\bullet}$ is the number of occurences of state i for the row variable ($n_{i\bullet}$) divided by the sample size ($n_{\bullet\bullet}$). Thus,

$$p_{i\bullet} = n_{i\bullet}/n_{\bullet\bullet}.$$

Similarly, $\pi_{\bullet j}$ is estimated by

$$p_{\bullet j} = n_{\bullet j}/n_{\bullet\bullet}.$$

Under the hypothesis of statistical independence, the estimate of $\pi_{i\bullet}\pi_{\bullet j} = \pi_{ij}$ is

$$\frac{n_{i\bullet}n_{\bullet j}}{n_{\bullet\bullet}^2}.$$

The chi-square test will involve comparing estimates of the expected number of observations with and without assuming independence. With independence, we expect to observe $n_{\bullet\bullet}\pi_{ij}$ entries in the ijth cell. This is estimated by

$$n_{\bullet\bullet}p_{i\bullet}p_{\bullet j} = \frac{n_{i\bullet}n_{\bullet j}}{n_{\bullet\bullet}}.$$

2. Model 2. Suppose that the row variable identifies the population. The null hypothesis is that all r populations have the same probabilities of taking on each value of the column variable. That is, for any two rows, denoted by i and i', say, and all j,

$$H_0 : \pi_{ij} = \pi_{i'j}.$$

As in the first part above, we want to estimate these probabilities in two cases:

a. Without assuming anything about the probabilities.

b. Under H_0, that is, assuming that each population has the same distribution of the column variable.

Under (a), if no assumptions are made, π_{ij} is the probability of obtaining state j for the column variable in the $n_{i\bullet}$ trials from the ith population. Again the binomial estimate holds

$$p_{ij} = n_{ij}/n_{i\bullet}.$$

If the null hypothesis holds, we may "pool" all our $n_{\bullet\bullet}$ trials to get a more accurate estimate of the probabilities. Then the proportion of times the column variable takes on state j is

$$p_j = n_{\bullet j}/n_{\bullet\bullet}.$$

As in the first part, let us calculate the numbers we expect in the cells under (a) and (b). If (a) holds, the expected number of successes in the $n_{i\bullet}$ trials of the ith population is $n_{i\bullet}\pi_{ij}$. We estimate this by

$$n_{i\bullet}(n_{ij}/n_{i\bullet}) = n_{ij}.$$

Under the null hypothesis, the expected number $n_{i\bullet}\pi_{ij}$ is estimated by

$$n_{i\bullet}p_j = \frac{n_{i\bullet}n_{\bullet j}}{n_{\bullet\bullet}}.$$

In summary, under either Model 1 or Model 2, the null hypothesis is reasonably tested by comparing n_{ij} with $n_{i\bullet}n_{\bullet j}/n_{\bullet\bullet}$.

Note 7.2: Measures of Association in Contingency Tables

Suppose we reject the null nypothesis of no association between the row and column categories in a contingency table. It is useful then to have a measure of the degree of association. Goodman and Kruskal [1979] in a series of papers argue that no one measure of association for contingency tables is best for all purposes. Measures must be chosen to help with the problem at hand. Among the measures they discuss are the following seven.

1. The Measure λ_C. Call the row variable or row categorization R and the column variable or column categorization C. Suppose we wish to use the value of R to predict the value of C. The measure λ_C is an estimate of the proportion of the errors in classification made if we do not know R, that can be eliminated by knowing R before making a prediction. From the data, λ_C is given by

$$\lambda_C = \frac{\left(\sum_{i=1}^{r} \max_j n_{ij}\right) - \max_j n_{\bullet j}}{n_{\bullet\bullet} - \max_j n_{\bullet j}}$$

λ_R is defined analogously.

2. **The Symmetric Measure λ.** λ_C does not treat the row and column classifications symmetrically. A symmetric measure may be found by assuming that the chances are $1/2$ and $1/2$ of needing to predict the row and column variables, respectively. The proportion of the errors in classification that may be reduced by knowing the other (row or column variable) when predicting is estimated by λ.

$$\lambda = \frac{\left(\sum_{i=1}^{r} \max_{j} n_{ij}\right) + \left(\sum_{j=1}^{c} \max_{i} n_{ij}\right) - \max_{i} n_{i\bullet} - \max_{j} n_{\bullet j}}{2n_{\bullet\bullet} - \left(\max_{i} n_{i\bullet} + \max_{j} n_{\bullet j}\right)}.$$

3. **The Measure γ for Ordered Categories.** In many applications of contingency tables the categories have a natural order; for example: last grade in school, age categories, number of weeks hospitalized. Suppose the orderings of the variables correspond to the indices i and j for the rows and columns. The γ measure is the difference in the proportion of the time the two measures have the same ordering minus the proportion of the time they have the opposite ordering, when there are no ties. Suppose the indices for the two observations are i, j and $\underline{i}$, $\underline{j}$. The indices have the same ordering if

$$(1)\ i < \underline{i}\ \text{and}\ j < \underline{j}\quad \text{or}\quad (2)\ i > \underline{i}\ \text{and}\ j > \underline{j}.$$

They have the opposite ordering if

$$(1)\ i < \underline{i}\ \text{and}\ j > \underline{j}\quad \text{or}\quad (2)\ i > \underline{i}\ \text{and}\ j < \underline{j}.$$

There are ties if $i = \underline{i}$ and/or $j = \underline{j}$. The index is

$$\gamma = (2S - 1 + T)/(1 - T),$$

where

$$S = 2 \sum_{i=1}^{r} \sum_{j=1}^{c} \left(n_{ij} \sum_{\underline{i}>i} \sum_{\underline{j}>j} n_{\underline{i}\underline{j}}\right) \bigg/ n_{\bullet\bullet}^2.$$

and

$$T = \left(\sum_{i=1}^{r}\left(\sum_{j=1}^{c} n_{ij}\right)^2 + \sum_{j=1}^{c}\left(\sum_{i=1}^{r} n_{ij}\right)^2 - \sum_{i=1}^{r}\sum_{j=1}^{c} n_{ij}^2\right) \bigg/ n_{\bullet\bullet}^2.$$

4. **Karl Pearson's Contingency Coefficient, C.** Since the chi-square statistic (X^2) is based on the square of the difference between the observed values in the contingency table and the estimated values, if association does not hold it is reasonable to base a measure of association on X^2. However, chi-square increases as the sample

size increases. One would like a measure of association that estimated a property of the total population. For this reason $X^2/n_{\bullet\bullet}$ is used in the next three measures. Karl Pearson proposed the measure C.

$$C = \sqrt{\frac{X^2/n_{\bullet\bullet}}{1 + X^2/n_{\bullet\bullet}}}.$$

5. Cramer's V. Harold Cramer proposed a statistic with values between 0 and 1. The coefficient can actually attain both values.

$$V = \sqrt{\frac{X^2/n_{\bullet\bullet}}{\text{minimum}(r - 1, c - 1)}}.$$

6. Tshuprow's T, and the Φ^2 Coefficient. The two final coefficients based on X^2 are:

$$T = \sqrt{\frac{X^2/n_{\bullet\bullet}}{\sqrt{(r - 1)(c - 1)}}} \quad \text{and} \quad \Phi = \sqrt{X^2/n_{\bullet\bullet}}.$$

We compute these measures of association for two contingency tables. The first table comes from the Robertson [1975] seat belt paper discussed in the text. The data are taken for 1974 cars with the interlock system. They relate age to seat belt usage. The data and the column percents are:

Belt Use	Age (Years)			Column Percents (Age)		
	< 30	30–49	≥ 50	< 30	30–49	≥ 50
Lap and shoulder	206	580	213	45	50	45
Lap only	36	125	65	8	11	14
None	213	459	192	47	39	41
				100	100	100

Although the chi-square value is 14.06 with $p = 0.007$, we can see from the column percentages that the relationship is weak. The coefficients of association are:

$$\lambda_C = 0, \qquad \lambda = 0.003, \qquad C = 0.08, \qquad T = 0.04,$$
$$\lambda_R = 0.006, \qquad \gamma = -0.03, \qquad V = 0.06, \qquad \Phi = 0.08.$$

All these coefficients lie beween −1 or 0, and +1, in general. They are zero if the variables are not associated at all. These values are small, indicating little association.

Consider Example 7.5 (Weiner *et al.* [1979]).

Chest Pain	Frequency (Vessels Diseased)			Row Percents (Vessels Diseased)			Total
	0	**1**	**2 or 3**	**0**	**1**	**2 or 3**	
Definite angina	66	135	419	11	22	68	101
Probable angina	179	139	276	30	23	46	99
Nonischemic	197	39	15	78	16	6	100

The chi-square statistic is 418.48 with a p-value of effectively zero. Note that those with definite angina were very likely (89%) to have disease, and even the probability of having multivessel disease was 68%. Chest pain thought to be nonischemic was associated with "no disease" 78% of the time. Thus, there is a strong relationship. The measures of association are:

$$\lambda_C = 0.24, \quad \lambda = 0.20, \quad C = 0.47, \quad T = 0.38,$$

$$\lambda_R = 0.16, \quad \gamma = -0.64, \quad V = 0.38, \quad \Phi = 0.53.$$

More information on these measures of association and other potentially useful measures is available in Reynolds [1977], and Goodman and Kruskal [1979].

Note 7.3: Testing for Symmetry in a Contingency Table

In a square table, one sometimes wants to test the table for symmetry. For example, when examining two alternative means of classification, one may be interested not only in the amount of agreement (κ), but also in seeing that the pattern of mis-classification is the same. In this case estimate the expected value in the ijth cell by $(n_{ij} + n_{ji})/2$. The usual chi-square value is appropriate with $r(r-1)/2$ degrees of freedom, where r is the number of rows (and columns). See van Belle and Cornell [1971].

Note 7.4: The Use of the Term "Linear" in Log-linear Models

Linear equations are equations of the form $y = c + a_1 X_1 + a_2 X_2 + \cdots + a_n X_n$ for some variables $X_1, \ldots, X_n$ and constants c and $a_1, \ldots, a_n$. The log-linear model equations can be put into this form. For concreteness, consider the model $[IJ][K]$ where $i = 1, 2$, $j = 1, 2$, and $k = 1, 2$. Define new variables as follows:

$$X_1 = \begin{cases} 1, & \text{if } i = 1, \\ 0, & \text{if } i = 2; \end{cases} \qquad X_2 = \begin{cases} 1, & \text{if } i = 2, \\ 0, & \text{if } i = 1; \end{cases} \qquad X_3 = \begin{cases} 1, & \text{if } j = 1, \\ 0, & \text{if } j = 2; \end{cases}$$

$$X_4 = \begin{cases} 1, & \text{if } j = 2, \\ 0, & \text{if } j = 1; \end{cases} \qquad X_5 = \begin{cases} 1, & \text{if } k = 1, \\ 0, & \text{if } k = 2; \end{cases} \qquad X_6 = \begin{cases} 1, & \text{if } k = 2, \\ 0, & \text{if } k = 1; \end{cases}$$

$$X_7 = \begin{cases} 1, & \text{if } i = 1, j = 1, \\ 0, & \text{otherwise;} \end{cases} \qquad X_8 = \begin{cases} 1, & \text{if } i = 1, j = 2, \\ 0, & \text{otherwise;} \end{cases}$$

$$X_9 = \begin{cases} 1, & \text{if } i = 2, j = 1, \\ 0, & \text{otherwise;} \end{cases} \qquad X_{10} = \begin{cases} 1, & \text{if } i = 2, j = 2, \\ 0, & \text{otherwise.} \end{cases}$$

Then the model is

$$\log \pi_{ijk} = u + u_1^I X_1 + u_2^I X_2 + u_1^J X_3 + u_2^J X_4 + u_1^K X_5 + u_2^K X_6$$
$$+ u_{1,1}^{IJ} X_7 + u_{1,2}^{IJ} X_8 + u_{2,1}^{IJ} X_9 + u_{2,2}^{IJ} X_{10}.$$

Thus the log-linear model is a linear equation, of the same form as $y = c + a_1 X_1 + a_2 X_2 + \cdots + a_n X_n$. The multiple regression chapter discusses such equations. Variables created to pick out a certain state (e.g., $i = 2$) by taking the value 1 when the state occurs, and taking the value 0 otherwise, are called *indicator* or *dummy variables*.

Note 7.5: Variables of Constant Probability in Log-linear Models

Consider the three-factor X, Y, and Z log-linear model. Suppose Z terms are entirely "omitted" from the model, for example, $[IJ]$ or

$$\log \pi_{ijk} = u + u_i^I + u_j^J + u_{ij}^{IJ}.$$

The model then fits the situation where Z is uniform on its state; that is,

$$P[Z = k] = 1/k, \quad k = 1, \ldots, K.$$

Note 7.6: Log-linear Models with Zero Cell Entries

Zero values in the contingency tables used for log-linear models are of two types. Some arise as *sampling zeros* (values could have been observed, but were not in the sample). In this case, if zeros occur in marginal tables used in the estimation:

1. Only certain u-parameters may be estimated,
2. The chi-square goodness-of-fit statistic has reduced degrees of freedom.

Some zeros are necessarily *fixed*; for example, some genetic combinations are fatal to offspring and will not be observed in a population. Log-linear models can be used in the analysis (see Bishop *et al.* [1975], Haberman [1979], and Fienberg [1977]).

Note 7.7: The GSK Approach to Higher Dimensional Contingency Tables.

The second major method of analyzing multivariate contingency tables is due to Grizzle *et al.* [1969]. They present an analysis method closely related to multiple regression (Chapter 11). References that consider this method are Reynolds [1977] and Kleinbaum *et al.* [1988].

PROBLEMS

In Problems 7.1–7.9, do the following (1–8) as well as any other requested work. Problems 7.1–7.5 are taken from the seatbelt paper of Robertson [1975].

1. If a table of expected values is given with one or more missing values, compute the missing values.
2. If the chi-square value is not given, compute the value of the chi-square statistic.
3. State the degrees of freedom.
4. State whether the chi-square p-value is less than or greater than 0.01, 0.05, and 0.10 using Appendix Table 3 or 4.
5. When tables are given with missing values for the adjusted residual values, p-values and $(r-1) \cdot (c-1) \cdot p$-values, fill in the missing values. (For the p-values, Appendix Tables 3 and 4 will not always allow exact answers; place a reasonable estimate.)
6. When percent tables are given with missing values, fill in the missing percentages for the row percent table, column percent table and total percent table, as applicable.
7. Using the 0.05 significance level interpret the findings. (Exponential notation is used for some numbers, e.g., $34000 = 3.4 \times 10^4 = 3.4E4$; $0.0021 = 2.1 \times 10^{-3} = 2.1E{-}3$.)
8. Describe verbally what the row and column percents mean. That is, "of those with zero vessels diseased, ...", etc.

1. In 1974 vehicles, the seatbelt usage was considered in association with the ownership of the vehicle. ("L/S" means "both lap and shoulder belt.")

Belt Use	Ownership			
	Individuals	**Rental**	**Lease**	**Other corporate**
L/S	583	145	86	182
Lap Only	139	24	24	31
None	524	59	74	145

Expected			
615.6	112.6	90.9	176.9
134.7	24.7	19.9	38.7
495.7	90.7	?	?

Adjusted Residuals			
−2.99	?	−0.76	0.60
0.63	−0.15	1.02	−1.44
2.65	−4.55	?	0.31

p-Values			
0.0028	5E−6	0.4481	0.5497
0.5291	0.8821	?	?
0.0080	5.3E−6	0.8992	0.7586

$(r-1) \times (c-1) \times p$-Values			
0.017	3E−5	1+	1+
1+	1+	1+	0.8869
0.048	3E−5	1+	1+

Column Percents			
47	?	47	51
11	11	13	9
42	?	?	41

DF =	?
$X^2 = 26.72$	

2. In 1974 cars, belt usage and manufacturer were also examined. One hundred eighty-nine cars from "other" manufacturers are not entered into the table.

	Manufacturer					
Belt Use	**GM**	**Toyota**	**AMC**	**Chrysler**	**Ford**	**VW**
L/S	498	25	36	74	285	33
Lap only	102	5	12	29	43	11
None	334	18	30	67	259	51

Adjusted Residuals					
3.06	0.33	−0.65	−1.70	−0.69	−3.00
0.49	−0.03	?	2.89	−3.06	0.33
−3.43	−0.32	−0.23	−0.08	2.63	?

p-Values					
0.0022	0.7421	0.5180	0.0898	0.4898	0.0027
0.6208	0.9730	?	0.0039	0.0022	0.7415
0.0006	0.7527	?	0.9366	0.0085	0.0043

Column Percents				
5352	46	44	49	?
1110	15	17	7	?
3638	38	39	?	?

DF = ?
$X^2 = 34.30$

3. The relationship between belt usage and racial appearance in the 1974 models is given here. Thirty-four cases whose racial appearance was "other" are excluded from this table.

	Racial Appearance	
Belt Use	**White**	**Black**
L/S	866	116
Lap only	206	20
None	757	102

Expected	
868.9	113.1
?	26.0
?	98.9

Adj Residuals	
−0.40	0.40
1.33	−1.33
?	?

p-Values	
0.69	0.69
?	?
0.67	0.67

DF = ?
$X^2 = ?$

4. The following data are given as the first example in Note 7.2. In the 1974 cars, belt usage and age were cross-tabulated.

Expected		
217.59	556.64	?
49.22	125.93	?
?	481.42	194.39

Adjusted Residuals		
?	2.06	−1.23
−2.26	−0.13	?
2.67	−2.00	−0.25

p-Values		
0.219	?	0.217
0.024	0.895	0.017
0.007	?	0.799

$(r-1) \times (c-1) \times$ p-Values		
0.88	0.16	0.87
?	?	?
0.03	0.18	1+

Column %s		
45	?	45
?	?	14
47	39	41

Row %s		
?	?	?
16	55	29
25	53	22

DF = ?
$X^2 = 14.06$

5. In the 1974 cars, seatbelt usage and sex of the driver were related as follows:

Belt Use	Sex	
	Female	Male
L/S	267	739
Lap only	85	142
None	261	606

Expected	
?	?
66.3	160.7
253.1	613.9

Adj Residuals	
?	?
2.90	−2.90
0.77	−0.77

p-Values	
0.0104	0.0104
0.0038	0.0038
?	?

$(r-1) \times (c-1) \times$ p-Values	
0.02	0.02
0.01	0.01
?	?

Col %s	
44	50
14	?
43	?

Total %s	
13	35
4	?
?	?

$$DF = \ ?$$
$$X^2 = \ ?$$

6. The data are given in the second example of Note 7.2. The association of chest pain classification and amount of coronary artery disease was examined.

Adjusted Residuals		
−13.95	0.33	12.54
?	1.57	−1.27
18.32	−2.47	−14.80

$(r-1) \times (c-1) \times p$-Values		
1.0E−30	1+	4.3E−27
1+	0.47	0.82
1.4E−40	0.05	8.8E−33

Row %s		
11	22	68
30	23	46
?	?	?

Column %s		
?	43	59
?	44	39
?	12	2

$$DF = \ ?$$
$$X^2 = 418.48$$

7. Peterson *et al.* [1979] studied the age at death of children who died from the Sudden Infant Death Syndrome, SIDS. The deaths from a variety of causes, including SIDS, were cross-classified by the age at death, per the following table (taken from death records in King County, Washington, over the years 1969–1977):

	Age at Death				
Cause	0 Days	1–6 Days	2–4 Weeks	5–26 Weeks	27–51 Weeks
Hyaline membrane disease	19	51	7	0	0
Respiratory distress syndrome	68	191	46	0	3
Asphyxia of the newborn	105	60	7	4	2
Immaturity	104	34	3	0	0
Birth injury	115	105	17	2	0
Congenital malformation	79	101	72	75	32
Infection	7	38	36	43	18
SIDS	0	0	24	274	24
All other	60	51	28	58	35

$$DF = \ ?$$
$$X^2 = 1504.18$$

a. The values of $(r - 1) \times (c - 1) \times p$-value for the adjusted residual are given, here, multiplied by -1 if the adjusted residual is negative and multiplied by $+1$ if the adjusted residual is positive.

$-1+$	$1.4E-9$	$-1+$	$-3.8E-5$	-0.89
-0.43	$4.6E-26$	$1+$	$-3.3E-20$	$-3.2E-3$
$3.0E-18$	$1+$	-0.02	$-4.5E-10$	-0.18
$2.3E-26$	$-1+$	$-5.8E-3$	$-1.2E-9$	-0.08
$1.1E-11$	$3.9E-4$	-0.42	$-3.8E-15$	$-1.6E-3$
-0.20	$-1+$	$7.7E-6$	$-1+$	0.12
$-1.1E-8$	$-1+$	$1.3E-5$	0.90	$6.5E-3$
$-31.2E-25$	$-3.4E-28$	-0.19	$1.7E-57$	$1+$
$-1+$	-0.03	$1+$	$1+$	$2.9E-9$

What is the distribution of the SIDS cases under the null hypothesis that all causes have the same distribution?

b. What percent display (row, column, or total) would best emphasize the difference? (See Problem 7.23 for a better way of handling these data.)

8. Morehead [1975] studied the relationship between the retention of intrauterine devices (IUDs) and other factors. The study participants were from New Orleans, Louisiana. Tables relating retention to the subjects' age and to parity (the number of pregnancies) are studied in this problem (one patient had a missing age).

a. Was age related to IUD retention?

Age	Continuers	Terminators
19–24	41	48
25–29	50	40
30+	63	27

Expected		Adj Residuals		p-Values	
50.95	?	-2.61	2.61	0.0091	0.0091
51.52	38.5	-0.40	0.40	?	?
51.52	38.5	?	?	0.0027	0.0027

Column %s		Row %s			
26.6	41.7	46.1	53.9	DF $=$?	
?	34.8	?	?	$X^2 =$?	
?	23.5	70.0	30.0		

b. The relationship of parity and IUD retention gave these data:

Parity	Continuers	Terminators
1–2	59	53
3–4	39	34
5+	57	28

Adj Residuals		Total %s	
−1.32	1.32	?	19.6
−0.81	0.81	14.4	?
?	?	21.1	?

$$\text{DF} = \ ?$$
$$X^2 = 4.74$$

9. McKeown *et al.* [1952] investigate evidence that the environment is involved in infantile pyloric stenosis. The relationship between the age at onset of the symptoms in days, and the rank of the birth (first child, second child, and so on) was given as follows:

Birth Rank	Age at Onset of Symptoms (days)						
	0–6	7–13	14–20	21–27	28–34	35–41	≥ 42
1	42	41	116	140	99	45	58
2	28	35	63	53	49	23	31
≥ 3	26	21	39	48	39	14	23

a. Find the expected value (under independence) for cell ($i = 2, j = 3$). For this cell compute (observed − expected)^2/expected.

b. The chi-square statistic is 13.91. What are the degrees of freedom? What can you say about the *p*-value?

c. The authors present, in the paper, not the frequencies as above, but the column percents. Fill in the missing values in both tables below. The arrangement is the same as the first table.

44	42	53	58	53	55	52
29	36	29	?	26	28	28
?	?	18	20	21	17	21

The adjusted residual *p*-values are

0.076	0.036	0.780	0.042	0.863	0.636	0.041
0.667	0.041	0.551	0.035	0.710	0.874	0.734
0.084	0.734	?	0.856	0.843	0.445	0.954

What can you conclude?

d. The authors note the first 2 weeks appear to have different patterns. They present the data also as:

Birth rank	Age at onset (days)	
	0–13	$\geq$ 14
1	83	458
2	63	219
≥ 3	47	163

For this table, $X^2 = 8.35$. What are the degrees of freedom? What can you say about the p-value?

e. Fill in the missing values in the adjusted residual table, p-value table, and column percent table. Interpret the data.

Adj Residuals	
−2.89	2.89
?	?
1.54	−1.54

p-Values	
0.0039	0.0039
0.065	0.065
?	?

Col %s	
43	55
33	?
24	?

f. Why is it crucial to know whether prior to seeing these data the investigators had hypothesized a difference in the parity distribution between the first 2 weeks and the remainder of the time period?

Problems 7.10–7.16 deal with the chi-square test for trend. The data are from a paper by Kennedy *et al.* [1981] relating operative mortality during coronary bypass operations to various risk factors. For each of the tables let the scores for the chi-square test for trend be consecutive integers. For each of the tables:

1. Compute the chi-square statistic for trend. Using Table A.3, give the strongest possible statement about the p-value.

2. Compute, where not given, the percentage of operative mortality, and plot the percent for the different categories using equally-spaced intervals.

3. The usual chi-square statistic (with $k - 1$ degrees of freedom) is given with its p-value. When possible, from Table A.3 or the chi-square values, tell which statistic is more highly significant (has the smallest p-value). Does your figure in (2) suggest why?

10. The amount of anginal (coronary artery disease) chest pain is categorized by the Canadian Heart Classification from mild (class I) to severe (class IV).

Surgical Mortality	Anginal Pain Classification				Usual
	I	II	III	IV	
Yes	6	19	47	59	$X^2 = 31.19$
No	242	1371	2494	1314	$p = 7.7\text{E}{-}7$
% surgical mortality	2.4	1.4	1.8	?	

11. Congestive heart failure occurs when the heart is not pumping sufficient blood. A heart damaged by a myocardial infarction, heart attack, can incur congestive heart failure. A score from 0 (good) to 4 (bad) for congestive heart failure is related to operative mortality.

Operative Mortality	Congestive Heart Failure Score					Usual
	0	1	2	3	4	
Yes	73	50	13	12	4	$X^2 = 46.45$
No	4480	1394	404	164	36	$p = 1.8E-9$
% operative mortality	1.6	3.4	?	6.8	10.0	

12. A measure of left ventricular performance, or the pumping action of the heart, is the *ejection fraction*, which is the percent of the blood in the left ventricle that is pumped during the beat. A high number indicates a more efficient performance.

Operative Mortality	Ejection Fraction (%)					Usual
	< 19	20–29	30–39	40–49	≥ 50	
Yes	1	4	5	22	74	$X^2 = 8.34$
No	14	88	292	685	3839	$p = 0.080$
% operative mortality	6.7	?	?	3.1	1.9	

13. A score was derived from looking at how the wall of the left ventricle moved while the heart was beating (details in CASS [1981]). A score of 5 was normal; the larger the score, the worse the motion of the left ventricular wall looked. The relationship to operative mortality is given here.

Operative Mortality	Wall Motion Score					Usual
	5–7	8–11	12–15	16–19	≥ 20	
Yes	65	36	32	10	2	$X^2 = 28.32$
No	3664	1605	746	185	20	$p = 1.1E-5$
% operative mortality	1.7	2.2	?	5.1	9.1	

What do you conclude about the relationship? That is, if you were writing a paragraph to describe this finding in a medical journal, what would you say?

14. After the blood has been pumped from the heart, and the pressure is at its lowest

point, a low blood pressure in the left ventricle is desirable. This "left ventricular end diastolic pressure" [LVEDP] is measured in millimeters of mercury (mmHg).

| Operative Mortality | LVEDP | | | | Usual |
	0–12	13–18	19–24	≥ 24	
Yes	56	43	22	26	$X^2 = 34.49$
No	3452	1692	762	416	$p = 1.6\text{E}{-7}$
% operative mortality	?	2.5	2.8	5.9	

15. The number of diseased vessels and operative mortality are given by:

| Operative Mortality | Diseased Vessels | | | Usual |
	1	2	3	
Yes	17	43	91	$X^2 = 7.95$
No	1196	2018	3199	$p = 0.019$
% operative mortality	1.4	2.1	?	

16. The left main coronary artery, if occluded (that is, totally blocked), blocks two of the three major arterial vessels to the heart. Such an event almost always leads to death. Thus, people with much narrowing of the left main coronary artery usually receive surgical therapy. Is this narrowing also associated with higher surgical mortality?

| Operative Mortality | Percentage Narrowing | | | | Usual |
	0–49	50–74	75–89	≥ 90	
Yes	116	8	10	19	$X^2 = 37.75$
No	5497	486	268	222	$p = 3.2\text{E}{-8}$
% operative mortality	2.1	1.6	?	7.9	

17. In Robertson's [1975] seatbelt study, the observers (unknown to them) were checked by sending cars through with a known seatbelt status. The agreement between the observers and the known status were:

| Belt Use Reported | Belt Use in Vehicles Sent | | |
	S/L	Lap only	No Belt
Shoulder and Lap	28	2	0
Lap Only	3	33	6
No Belt	0	15	103

 a. Compute P_A, P_C, and κ.

 b. Construct a 95% confidence interval for κ.

 c. Find the two-sided p-value for testing $\kappa = 0$ (for the entire population) by using $Z = \kappa/\text{SE}_0(\kappa)$.

18. The following table is from Example 7.4 (Fisher *et al.* [1982] copyright Wiley-Liss). The coronary artery tree has considerable biological variability. If the right coronary artery is normal-sized and supplies its usual share of blood to the heart, the circulation of blood is called right dominant. As the right coronary artery becomes less important, the blood supply is characterized as balanced, and then left dominant. The data for the clinical site and quality control site joint readings of angiographic films is given here.

Dominance (QC site)	Dominance (Clinical Site)		
	Left	**Balanced**	**Right**
Left	64	7	4
Balanced	4	35	32
Right	8	21	607

 a. Compute P_A, P_C, and κ (Section 7.4).

 b. Find $\text{var}(\kappa)$ and construct a 90% confidence interval for the population value of κ.

19. Example 7.4 discusses the quality control data for the CASS arteriography (films of the arteries). A separate paper (Wexler *et al.* [1982] copyright Wiley-Liss), examines the study of the left ventricle. Problem 7.12 describes the ejection fraction. Clinical site and quality control site readings of ejection gave the following table.

Ejection Fraction (Clinical Site)	Ejection Fraction (QC site)		
	$\geq 50\%$	**30–49%**	**<30%**
$\geq 50\%$	302	27	5
30–49%	40	55	9
<30%	1	9	18

 a. Compute P_A, P_C, and κ.

 b. Find $\text{SE}(\kappa)$ and construct a 99% confidence interval for the population value of κ.

20. The value of κ depends upon how we construct our categories. Suppose in Example 7.4, we combine normal and other zero vessel disease to create a zero-vessel disease category. Suppose also we combine two- and three-vessel disease into a multivessel-disease category. Then the table becomes:

Vessels Diseased (QC Site)	Vessels Diseased (Clinical Site)		
	0	**1**	**Multi-**
0	70	20	9
1	10	155	78
Multi-	2	29	497

a. Compute P_A, P_C, and κ.

b. Is this *kappa* value greater than or less than the value in Example 7.4? Will this always occur? Why?

c. Construct a 95% confidence interval for the population value of κ.

21. Zeiner-Henriksen [1972a] compared personal interview and postal inquiry methods of assessing infarction. His introduction follows:

"The questionnaire developed at the London School of Hygiene and Tropical Medicine and later recommended by the World Health Organization for use in field studies of cardiovascular disease has been extensively used in various populations. While originally developed for personal interviews, this questionnaire has also been employed for postal inquiries. The postal inquiry method is of course much cheaper than personal interviewing and is without interviewer error.

A Finnish-Norwegian lung cancer study offered an opportunity to evaluate the repeatability at interview of the cardiac pain questionnaire, and to compare the interview symptom results with those of a similar postal inquiry. The last project, confined to a postal inquiry of the chest pain questions in a sub-sample of the 4092 men interviewed, was launched in April 1965, $2\frac{1}{2}$ to 3 years after the original interviews.

The objective was to compare the postal inquiry method with the personal interview method as a means of assessing the prevalence of angina and possible infarction, ... "

The data are (where PI stands for "possible infarction," and AP means "angina pectoris"):

Postal Inquiry	Interview						
	PI + AP	PI only	AP only	PI/AP Negative Nonspecific	Other	Incomplete	Total
PI + AP	23	15	9	6	—	1	54
PI only	14	18	14	24	8	—	78
AP only	3	5	20	12	17	3	60
PI/AP Negative							
Nonspecific	2	8	8	54	24	5	101
Other	2	3	5	62	279	1	352
Incomplete	—	2	—	22	37	—	61
Total	44	51	56	180	365	10	706

a. Compute P_A, P_C, and κ.

b. Construct a 90% confidence interval for the population value of κ ($\sqrt{\text{var}(\kappa)} = 0.0231$).

c. Group the data in three categories by:

    i. combining PI + AP, PI only and AP only;

    ii. combining the two PI/AP negatives categories;

    iii. leaving "incomplete" as a third category.

Recompute P_A, P_C, and κ. (This new grouping has categories, "cardiovascular symptoms," "no symptoms," and "incomplete.")

22. As discussed in the last problem (Zeiner-Henriksen [1972b]) an ischemic chest pain (angina) and infarction (heart attack) interview was evaluated for reproducibility by having reinterviews. The following table shows the results (where I+ means a positive infarction, I− means a negative infarction, and A+ and A− represent a positive or negative indication of angina):

	Interview					
				I−A−		
Postal Inquiry	I+A+	I+A−	I−A+	**Nonspecific**	**Other**	**Total**
I+A+	11	3	1	1	—	16
I+A−	2	14	—	4	—	20
I−A+	5	2	7	1	1	16
I−A−						
Nonspecific	1	4	5	39	9	58
Other	1	8	6	40	72	127
Total	20	31	19	85	82	237

a. Compute P_A, P_C, and κ for these data.

b. Construct a 95% confidence interval for the population value of *kappa*. $\text{se}(\kappa) = 0.043$.

c. What is the value of the Z statistic, for testing no association, that is computed from *kappa* and its estimated standard error $\sqrt{\text{var}_0(\kappa)} = 0.037$?

*23. Consider the SIDS data of Problem 7.7. The main interest of the contingency table is in showing that the time-of-death pattern of SIDS children differs from other causes of death. It is of secondary interest to know whether the other causes of death differ among themselves.

a. Show by using the partitioning method of Section 7.5 that these two comparisons may be analyzed by a partition of chi-square.

b. Complete the following table. (Hint: Recall that the LRX^2 tables add up.)

Source	DF	LRX^2	p
SIDS *vs.* the rest	?	877.75	< 0.001
Comparison among the rest	?	?	< 0.001
Whole Table	?	1606.86	< 0.001

c. The table for SIDS *vs.* the rest and associated table follow. Fill in the blanks and interpret these data.

	0 days	1–6 days	2–4 weeks	5–26 weeks	27–51 weeks
Rest	557	?	216	182	90
SIDS	0	0	?	274	24

Adjusted Residuals				
12.18	13.31	2.75	?	−1.48
−12.18	−13.31	−2.75	?	1.48

$(r - 1) \times (c - 1) \times p$-Values				
0+	0+	?	0+	0.56
0+	0+	?	0+	0.56

Row Percents				
33.2	?	?	10.9	5.4
0.0	0.0	?	?	7.5

*24. Weiner *et al.* [1979] studied men and women with suspected coronary disease. They were studied by a maximal exercise treadmill test. A positive test (≥ 1 mm of ST-wave depression on the exercise electrocardiogram) is thought to be indicative of coronary artery disease. Disease was classified into zero-, one- (or single-), and multivessel disease. Among people with chest pain thought probably anginal (that is, due to coronary artery disease) the following data are found.

	Vessels Diseased		
Category	0	1	Multi-
Males, + test	47	86	227
Males, − test	132	53	49
Females, + test	62	28	44
Females, − test	83	14	9

We want to compare males and females for prevalence of disease. We also want to see whether the exercise test is related to disease separately for men and women.

a. Show that the three comparisons can be evaluated by two applications of the partition of chi-square of Section 7.5.

b. Complete the following table on the partition of chi-square. (The LRX^2 column does not add up correctly due to round-off error.)

Source	DF	LRX^2	p-Value
Male vs. female	?	68.85	0+
Males, + vs. −	?	147.50	0+
Females, + vs. −	?	29.72	0+
Whole table	?	246.06	0+

c. Terms of the form $n_{ij} \ln(n_{ij} n_{..}/n_{i.} n_{.j})$ were used in computing the LRX^{2}s for the whole table. What is the value of this for $i = 1$ and $j = 1$?

d. With all three comparisons so highly statistically significant, one can proceed to the analysis of each of the subtables. For the males, the relationship of + or − test and disease give the data below. Fill in the missing values, interpret these data and answer the questions.

	Vessels Diseased		
Exercise Test	**0**	**1**	**Multi-**
+	47	86	?
−	132	?	49

Expected		
108.5	84.2	167.3
70.5	54.8	?

Adjusted Residuals		
0+	0.73	0+
0+	0.73	0+

Row Percents		
?	?	63.1
56.4	22.6	20.9

Column Percents		
26.3	61.9	?
73.7	38.1	?

Formulate a question for which the row percents would be a good method of presenting the data. Formulate a question where the column percents would be more appropriate.

*25. Consider the contingency table of Problem 7.1. Suppose that we wish to compare seatbelt usage in cars owned by individuals to all other types of ownership. We also then would like to compare rental cars (where there is not time to disable the interlock system) to the other cars not owned by individuals.

a. Show this can be done by two applications of the partitioning method of Section 7.5.

b. Complete the following table partitioning LRX^2 for the whole table.

Source	DF	LRX^2	p-Values
Individual vs. other	?	9.12	?
Rental vs. lease and corporate	?	?	?
Lease vs. corporate	2	?	0.27
Whole table	6	27.58	0.0001

c. The individuals *vs.* other categories' analysis results in the following tables. Fill in the missing values and interpret the data.

	Individuals	Rest
L/S	583	413
L	139	79
None	524	?

Adj Residuals	
2.99	−2.99
?	?
−2.65	2.65

p-Values	
?	?
0.529	0.529
0.008	0.008

Column %s	
?	46.8
?	11.2
36.1	42.1

*26. Exponentials and natural logarithms.

a. Find the natural logarithms, $\ln x$, of the following x: 1.24, 0.63, 0.78, 2.41, 2.7182818, 1.00, 0.10. For what values do you think $\ln x$ is positive? For what values do you think $\ln x$ is negative? (A plot of the values may help.)

b. Find the exponential, e^x, of the following x: −2.73, 5.62, 0.00, −0.11, 17.3, 2.45. When is e^x less than one? When is e^x greater than one?

c. $\ln(a \times b) = \ln a + \ln b$. Verify this for the following pairs of a and b:

a	2.00	0.36	0.11	0.62
b	0.50	1.42	0.89	0.77

d. $e^{a+b} = e^a \cdot e^b$. Verify this for the following pairs of numbers.

a	−2.11	0.36	0.88	−1.31
b	2.11	1.59	−2.67	−0.45

*27. Example 7.7 uses Weiner *et al.* [1979] data for the cases with probable angina. The results for the cases with definite angina are as follows. (I, J, and K refer to variables as in Example 7.7.)

Model	DF	LRX^2	p-Value
$[I][J][K]$	7	114.41	0+
$[IJ][K]$	6	103.17	0+
$[IK][J]$	5	26.32	0+
$[I][JK]$	5	94.89	0+
$[IJ][IK]$	4	15.08	0.0045
$[IJ][JK]$	4	83.65	0+
$[IK][JK]$	3	6.80	0.079
$[IJ][IK][JK]$	2	2.50	0.286

a. Which models are at all plausible?

b. The data for the fit of the $[IJ][IK][JK]$ hypothesis are:

Cell (i,j,k)	Observed	Fitted	u-Parameters
(1,1,1)	17	18.74	$u = -3.37$
(1,1,2)	86	85.01	$u_1^I = -u_2^I = 0.503$
(1,1,3)	244	243.25	$u_1^J = -u_2^J = 0.886$
(1,2,1)	5	3.26	$u_1^K = -0.775,\ u_2^K = -0.128,\ u_3^K = 0.903$
(1,2,2)	14	14.99	$u_{1,1}^{IJ} = -u_{1,2}^{IJ} = -u_{2,1}^{IJ} = u_{2,2}^{IJ} = -0.157$
(1,2,3)	99	99.75	$u_{1,1}^{IK} = -u_{2,1}^{IK} = -0.728$
(2,1,1)	42	40.26	$u_{1,2}^{IK} = -u_{2,2}^{IK} = 0.143$
(2,1,2)	31	31.99	$u_{1,3}^{IK} = -u_{2,3}^{IK} = 0.586$
(2,1,3)	37	37.75	$u_{1,1}^{JK} = -u_{2,1}^{JK} = 0.145$
(2,2,1)	2	3.74	$u_{1,2}^{JK} = -u_{2,2}^{JK} = 0.138$
(2,2,2)	4	3.01	$u_{1,3}^{JK} = -u_{2,3}^{JK} = -0.283$
(2,2,3)	9	8.25	

Using the u-parameters, compute the fitted value for the (1,2,3) cell, showing it is (approximately) equal to 99.75 as given.

c. Using the fact that Hypothesis 7 is nested within Hypothesis 8, compute the chi-square statistic for the additional gain in fit between the models. What is the p-value (as best as you can tell from the tables)?

*28. As in the last problem, the cases of Example 7.7, but with chest pain thought not to be due to heart disease (nonischemic), gave the following goodness-of-fit likelihood ratio chi-square statistics.

Model	DF	LRX^2	p-Value
$[I][J][K]$	7	35.26	0+
$[IJ][K]$	6	28.45	0+
$[IK][J]$	5	11.68	0.039
$[I][JK]$	5	32.46	0+
$[IJ][IK]$	4	4.87	0.30
$[IJ][JK]$	4	25.65	0+
$[IK][JK]$	3	8.89	0.031
$[IJ][IK][JK]$	2	2.47	0.29

a. Which model would you prefer? Why?

b. For model $[IJ][IK]$, here is the information on the fit:

Cell (i,j,k)	Observed	Fitted	u-Parameters
(1,1,1)	33	32.51	$u = -3.378$
(1,1,2)	13	12.01	$u_1^I = -u_2^I = 0.115$
(1,1,3)	7	8.48	$u_1^J = -u_2^J = 0.658$
(1,2,1)	13	13.49	$u_1^K = 1.364,\ u_2^K = -0.097,\ u_3^K = -1.267$
(1,2,2)	4	4.99	$u_{1,1}^{IJ} = -u_{1,2}^{IJ} = -u_{2,1}^{IJ} = u_{2,2}^{IJ} = -0.218$
(1,2,3)	5	3.52	$u_{1,1}^{IK} = -u_{2,1}^{IK} = -0.584$
(2,1,1)	126	128.69	$u_{1,2}^{IK} = -u_{2,2}^{IK} = -0.119$
(2,1,2)	21	18.75	$u_{1,3}^{IK} = -u_{2,3}^{IK} = 0.703$
(2,1,3)	3	2.56	
(2,2,1)	25	22.31	
(2,2,2)	1	3.25	
(2,2,3)	0	0.44	

Using the u-parameter values verify the fitted value for the (2,1,1) cell.

c. Interpret the probabilistic meaning of the model in words for the variables of this problem.

*29. Willkens et al. [1981], study possible diagnostic criteria for Reiter's Syndrome. This rheumatic disease was considered in the context of other rheumatic diseases. Eighty-three Reiter's Syndrome cases were compared with 136 cases with one of the following four diagnoses: ankylosing spondylitis, seronegative definite rheumatoid arthritis, psoriatic arthritis, and gonococcal arthritis. A large number of potential diagnostic criteria were considered. Here we consider two factors. One factor is the presence or absence of urethritis and/or cervicitis (for females). Secondly, the duration of the initial attack was evaluated as greater than or equal to 1 month, or less than 1 month. The data and the goodness-of-fit statistics follow:

		Initial attack $[J]$	
Urethritis and/or Cervicitis $[I]$	**Disease $[K]$**	**< 1 month**	**≥ 1 month**
Yes	Reiter's	2	70
	Other	11	3
No	Reiter's	1	10
	Other	20	132

Model	DF	LRX^2	p-Value
$[I][J][K]$	?	200.65	?
$[IJ][K]$	?	200.41	?
$[IK][J]$	?	40.63	?
$[I][JK]$	?	187.78	?
$[IJ][IK]$	?	40.39	?
$[IJ][JK]$	?	187.55	?
$[IK][JK]$	?	27.76	?
$[IJ][IK][JK]$	?	5.94	?

a. Fill in the question marks in the above table.

b. Which model(s) seem plausible (at the 0.05 significance level)?

c. Since we are looking for criteria to differentiate between Reiter's Syndrome and the other diseases, one strategy that makes sense is to assume independence of the disease category ($[K]$), and then to look for the largest departures from the observed and fitted cells. The model we want is then $[IJ][K]$. The fit is given in the next table.

Cell (i, j, k)	Observed	Fitted
(1,1,1)	70	24.33
(1,1,2)	3	48.67
(1,2,1)	2	4.33
(1,2,2)	11	8.67
(2,1,1)	10	47.33
(2,1,2)	132	94.67
(2,2,1)	1	7.00
(2,2,2)	20	14.00

Which cell of Reiter's Syndrome cases has the largest excess of observed minus fitted?

d. If you use the cell found in (c) as your criteria for Reiter's Syndrome, what is the specificity and sensitivity of this diagnostic criteria for these cases?

*30. We here show why the chi-square partition of Section 7.5 is reasonable. Rather than thinking of the table entries as frequencies, consider the π_{ij}'s, the probabilities of an entry falling into the ijth cell under Model 1. Under statistical independence, the table entries are specified by the marginal probabilities $\pi_{i\bullet}$ and $\pi_{\bullet j}$, since $\pi_{ij} = \pi_{i\bullet}\pi_{\bullet j}$. Let the $\pi_{i\bullet}$ and $\pi_{\bullet j}$ be fixed.

a. Show that for *any* pattern of π_{ij}'s in Table One (with the fixed marginal probabilities), there is a pattern of π_{ij}'s for Table Two that satisfy $\pi_{ij} = \pi_{i\bullet}\pi_{\bullet j}$.

b. Show that for *any* pattern of π_{ij}'s in Table Two (with the fixed marginals), there is a pattern of π_{ij}'s for Table One that satisfy $\pi_{ij} = \pi_{i\bullet}\pi_{\bullet j}$.

From this, we see that the two tables represent different situations with no necessary connection between the two.

*31. Show by repeated application of the partitioning of chi-square presented in Section 7.5 that the $(r-1)(c-1)$ degrees of freedom in an $r \times c$ contingency table may be divided into $(r-1)(c-1)$ individual degrees of freedom which correspond to 2×2 tables. The 2×2 tables have lower right entry n_{ij} for $i \geq 2$ and $j \geq 2$. The other entries are given by adding the terms in the upper left, upper right and lower left of the four quadrants given by the lines:

$$
\begin{array}{ccc|c}
n_{1,1} & \cdots & n_{1,j-1} & n_{1,j} \\
\vdots & \ddots & \vdots & \vdots \\
n_{i-1,1} & \cdots & n_{i-1,j-1} & n_{i-1,j} \\
\hline
n_{i,1} & \cdots & n_{i,j-1} & n_{i,j}
\end{array}
$$

Hint: You might use mathematical induction:

a. Show this holds for the 2×2 table (true by default).

b. If it holds for $r \times c$ tables, show that it holds for $(r+1) \times c$ tables and $r \times (c+1)$ tables.

c. Conclude it holds for all tables.

*32. The text claims that the three-factor log-linear model $[IJ][IK]$ means that the J and K variables are independent conditionally upon the I variable. Prove this by showing the following steps:

a. By definition Y and Z are independent conditionally upon X if

$$P[Y = j \text{ and } Z = k \mid X = i] = P[Y = j \mid X = i]P[Z = k \mid X = i].$$

Using the probabilities π_{ijk}, show that this is equivalent to

$$\frac{\pi_{ijk}}{\pi_{i\bullet\bullet}} = \left(\frac{\pi_{ij\bullet}}{\pi_{i\bullet\bullet}}\right)\left(\frac{\pi_{i\bullet k}}{\pi_{i\bullet\bullet}}\right).$$

b. If the above equation holds true, show that

$$\ln \pi_{ijk} = u + u_i^I + u_j^J + u_k^K + u_{ij}^{IJ} + u_{ik}^{IK},$$

where

$$u_{ij}^{IK} = \ln(\pi_{ij\bullet}) - \frac{1}{I}\sum_{i=1}^{I}\ln(\pi_{ij\bullet}) - \frac{1}{J}\sum_{j=1}^{J}\ln(\pi_{ij\bullet}) + \frac{1}{IJ}\sum_{i=1}^{I}\sum_{j=1}^{J}\ln(\pi_{ij\bullet})$$

$$u_{ik}^{IK} = \ln(\pi_{i\bullet k}) - \frac{1}{I}\sum_{i=1}^{I}\ln(\pi_{i\bullet k}) - \frac{1}{K}\sum_{k=1}^{K}\ln(\pi_{i\bullet k}) + \frac{1}{IK}\sum_{i=1}^{I}\sum_{k=1}^{K}\ln(\pi_{i\bullet k})$$

$$u_i^I = \frac{1}{J}\sum_{j=1}^{J}\ln(\pi_{ij\bullet}) + \frac{1}{K}\sum_{k=1}^{K}\ln(\pi_{i\bullet k}) - \ln(\pi_{i\bullet\bullet}) + \frac{1}{IJ}\sum_{i=1}^{I}\sum_{j=1}^{J}\ln(\pi_{ij\bullet})$$

$$+ \frac{1}{IK}\sum_{i=1}^{I}\sum_{k=1}^{K}\ln(\pi_{i\bullet k}) - \frac{1}{I}\sum_{i=1}^{I}\ln(\pi_{i\bullet\bullet})$$

$$u_j^J = \frac{1}{I}\sum_{i=1}^{I}\ln(\pi_{ij\bullet}) - \frac{1}{IJ}\sum_{j=1}^{J}\sum_{i=1}^{I}\ln(\pi_{ij\bullet})$$

$$u_k^K = \frac{1}{I}\sum_{i=1}^{I}\ln(\pi_{i\bullet k}) - \frac{1}{IK}\sum_{k=1}^{K}\sum_{i=1}^{I}\ln(\pi_{i\bullet k})$$

$$u = -\frac{1}{IJ}\sum_{i=1}^{I}\sum_{j=1}^{J}\ln(\pi_{ij\bullet}) - \frac{1}{IK}\sum_{i=1}^{I}\sum_{k=1}^{K}\ln(\pi_{i\bullet k}) - \frac{1}{I}\sum_{i=1}^{I}\ln(\pi_{i\bullet\bullet})$$

c. If the above equation holds, use $\pi_{ijk} = e^{\ln \pi_{ijk}}$ to show that the first equation then holds.

*33. The notation and models for the three-factor log-linear model extend to larger numbers of factors. For example, for variables W, X, Y, and Z (denoted by the indices i, j, k and l, respectively), the following notation and model correspond:

$$[IJK][L] = u + u_i^I + u_j^J + u_k^K + u_l^L + u_{ij}^{IJ} + u_{ik}^{IK} + u_{jk}^{JK} + u_{ijk}^{IJK}$$

a. For the four-factor model, write the log-linear u-terms corresponding to the following model notations:

   i. $[IJ][KL]$,

   ii. $[IJK][IJL][JKL]$,

   iii. $[IJ][IK][JK][L]$.

b. Give the bracket notation for the models corresponding to the u- parameters:

   i. $u + u_i^I + u_j^J + u_k^K + u_l^L$,

   ii. $u + u_i^I + u_j^J + u_k^K + u_l^L + u_{ij}^{IJ} + u_{kl}^{KL}$,

   iii. $u + u_i^I + u_j^J + u_k^K + u_l^L + u_{ij}^{IJ} + u_{ik}^{IK} + u_{il}^{IL} + u_{jk}^{JK} + u_{ijk}^{IJK}$.

*34. Verify the values of the contingency coefficients, or measures of association, given in the first example of Note 7.2.

*35. Verify the values of the measures of association given in the second example of Note 7.2.

*36. Prove the following properties of some of the measures of association, or contingency coefficients, presented in Note 7.2.

a. $0 \leq \lambda_C \leq 1$. Show by example that 0 and 1 are possible values.

b. $0 \le \lambda \le 1$. Show by example that 0 and 1 are possible values. What happens if the two traits are independent in the sample $n_{ij} = n_{i\bullet}n_{\bullet j}/n_{\bullet\bullet}$?

c. $-1 \le \gamma \le 1$. Can γ be -1 or $+1$? If the traits are independent in the sample, show $\gamma = 0$. Can $\gamma = 0$ otherwise? If yes, give an example.

d. $0 < C < 1$.

e. $0 \le V \le 1$.

f. $0 \le T \le 1$ (use (e) to show this).

g. Show by example that ϕ^2 can be larger than 1.

*37. Compute the contingency coefficients of Note 7.2, omitting γ, for the data of:

a. Problem 7.1.

b. Problem 7.5.

REFERENCES

Bishop, Y. M. M., Fienberg, S. E., and Holland, P. W. [1975]. *Discrete Multivariate Analysis: Theory and Practice*. MIT Press, Cambridge, MA.

CASS [1981] (principal investigators of CASS and their associates; T. Killip (Ed.), L. Fisher and M. Mock, (associate eds). National Heart, Lung and Blood Institute Coronary Artery Surgery Study. *Circulation*, **63**: part II, I-1 to I-81.

Cohen, J. [1968]. Weighted *kappa*: nominal scale agreement with provision for scaled disagreement or partial credit. *Psychological Bulletin*, **70**: 213–220.

Everitt, B. S. [1992]. *The Analysis of Contingency Tables*, 2nd edition. Halstad Press, New York.

Fienberg, S. E. [1977]. *The Analysis of Cross-Classified Categorical Data*. MIT Press, Cambridge, MA.

Fisher, L. D., Judkins, M. P., Lesperance, J., Cameron, A., Swaye, P., Ryan, T. J., Maynard, C., Bourassa, M., Kennedy, J. W., Gosselin, A., Kemp, H., Faxon, D., Wexler, L., and Davis, K. [1982]. Reproducibility of coronary arteriographic reading in the Coronary Artery Surgery Study (CASS). *Catheterization and Cardiovascular Diagnosis*, **8**: 565–575. Copyright © 1982 by Wiley-Liss.

Fleiss, J. L. [1981]. *Statistical Methods for Rates and Proportions*, 2nd edition. Wiley, New York.

Fleiss, J. L. [1975]. Measuring agreement between two judges on the presence or absence of a trait. *Biometrics*, **31**: 651–659.

Fleiss, J. L., Cohen, J. and Everitt, B. S. [1969]. Large sample standard errors of *kappa* and weighted *kappa*. *Psychological Bulletin*, **72**: 323–327.

Goodman, L. A. [1968]. The analysis of cross-classified data: independence, quasi-independence, and interactions in contingency tables with or without missing entries. *Journal of the American Statistical Association*, **63**: 1091–1131.

Goodman, L. A. and Kruskal, W. H. [1979]. *Measures of Association for Cross-Classifications*. Springer-Verlag, New York.

Grizzle, J. E., Starmer, C. F., and Koch, G. G. [1969]. Analysis of categorical data by linear models. *Biometrics*, **25**: 489–504.

Haberman, S. J. [1978]. *Analysis of Qualitative Data, Volume 1: Introductory Topics*. Academic Press, New York.

Haberman, S. J. [1979]. *Analysis of Qualitative Data, Volume 2: New Developments*. Academic Press, New York.

Hitchcock, C. R., Ruiz, E., Sutherland, D., and Bitter, J. E. [1966]. Eighteen-month follow-up of gastric freezing in 173 patients with duodenal ulcer. *Journal of the American Medical Association*, **195:** 115–119.

Kennedy, J. W., Kaiser, G. C., Fisher, L. D., Fritz, J. K., Myers, W., Mudd, J. G., and Ryan, T. J. [1981]. Clinical and angiographic predictors of operative mortality from the collaborative study in coronary artery surgery (CASS). *Circulation*, **63:** 793–802.

Kleinbaum, D. G., Kupper, L. L., and Muller, K. G. [1988]. *Applied Regression Analysis and Other Multivariable Methods*, 2nd edition. Dros-Kent, Boston, MA.

Lancaster, H. O. [1969]. *The Chi-Squared Distribution*. Wiley, New York.

Landis, J. L. and Koch, G. G. [1975]. A review of statistical methods in the analysis of data arising from observer reliability studies, parts I and II. *Statistica Neerlandia*, **29:** 101–123, 151–161.

Light, R. J. [1971]. Measures of response agreement for qualitative data: some generalizations and alternatives. *Psychological Bulletin*, **76:** 365–377.

Maclure, M. and Willett, W. C. [1987]. Misinterpretation and misuses of the kappa statistic. *American Journal of Epidemiology*, **126:** 161–169.

Maki, D. G., Weise, C. E., and Sarafin, H. W. [1977]. A semi-quantitative culture method for identifying intravenous-catheter-related infection. *New England Journal of Medicine*, **296:** 1305–1309.

McKeown, T., MacMahon, B., and Record, R. G. [1952]. Evidence of post-natal environmental influence in the aetiology of infantile pyloric stenosis. *Archives of Diseases in Children*, **58:** 386–390.

Morehead, J. E. [1975]. Intrauterine device retention: a study of selected social-psychological aspects. *American Journal of Public Health*, **65:** 720–730.

Peterson, D. R., van Belle, G., and Chinn, N. M. [1979]. Epidemiologic comparisons of the Sudden Infant Death Syndrome with other major components of infant mortality. *American Journal of Epidemiology*, **110:** 699-707.

Reynolds, H. T. [1977]. *The Analysis of Cross-Classifications*. The Free Press, New York.

Robertson, L. S. [1975]. Safety belt use in automobiles with starter-interlock and buzzer-light reminder systems. *American Journal of Public Health*, **65:** 1319–1325. Copyright © 1975 by the American Health Association.

Ross, D. C. [1977]. Testing patterned hypotheses in multi-way contingency tables using weighted *kappa* and weighted chi-square. *Educational and Psychological Measurement*, **37:** 291–307.

Ruffin, J. M., Grizzle, J. E., Hightower, N. C., McHarcy, G., Shull, H., and Kirsner, J. B. [1969]. A cooperative double-blind evaluation of gastric "freezing" in the treatment of duodenal ulcer. *New England Journal of Medicine*, **281:** 16–19.

Time Magazine. [1962]. Frozen Ulcers. *Time*, May **18:** 45–47.

van Belle, G. and Cornell, R. G. [1971]. Strengthening tests of symmetry in contingency tables. *Biometrics*, **27:** 1074–1078.

Wangensteen, C. H., Peter, E. T., Nicoloff, M., Walder, A. I., Sosin, H., and Bernstein, E. F. [1962]. Achieving "physiologic gastrectomy" by gastric freezing. *Journal of the American Medical Association*, **180:** 439–444. Copyright © 1962 by the American Medical Association.

Weiner, D. A., Ryan, T. J., McCabe, C. H., Kennedy, J. W., Schloss, M., Tristani, F., Chaitman, B. R., and Fisher, L. D. [1979]. Correlations among history of angina, ST-segment response and prevalence of coronary-artery disease in the Coronary Artery Surgery Study (CASS). *New England Journal of Medicine*, **301:** 230–235.

Wexler, L., Lesperance, J., Ryan, T. J., Bourassa, M. G., Fisher, L. D., Maynard, C., Kemp, H. G., Cameron, A., Gosselin, A. J., and Judkins, M. P. [1982]. Interobserver variability in interpreting contrast left ventriculograms (CASS). *Catheterization and Cardiovascular Diagnosis*, **8:** 341–355.

Willkens, R. F., Arnett, F. C., Bitter, T., Calin, A., Fisher, L., Ford, D. K., Good, A. E., and Masi, A. T. [1981]: Reiter's Syndrome: evaluation of preliminary criteria. *Arthritis and Rheumatism*, **24:** 844–849. Used with permission from J. B. Lippincott Company.

Zeiner-Henriksen, T. [1972a]. Comparison of personal interview and inquiry methods for assessing prevalences of angina and possible infarction. *Journal of Chronic Diseases*, **25:** 433–440. Used with permission of Pergamon Press, Inc.

Zeiner-Henriksen, T. [1972b]. The repeatability at interview of symptoms of angina and possible infarction. *Journal of Chronic Diseases*, **25:** 407–414. Used with permission of Pergamon Press, Inc.

CHAPTER 8

Nonparametric, Distribution-Free and Permutation Models: Robust Procedures

In Chapter 4, we worked with the normal distribution, noting the fact that many populations have distributions that are approximately normal. Chapter 5 presented elegant one- and two-sample methods for estimating the mean of a normal distribution, or the difference of the means, and constructing confidence intervals. In the same chapter, we examined the corresponding tests about the mean(s) from normally distributed populations. The techniques that we learned are very useful. Suppose, however, that the population under consideration is not normal. What should we do? If the population is not normal, is it appropriate to use the same t-statistic that applies when the sample comes from a normally distributed population? If not, is there some other approach that can be used to analyze such data?

In this chapter, we consider such questions. Section 8.1 introduces terminology associated with statistical procedures needing few assumptions. Section 8.3 notes that some of the statistical methods we have already looked at require very few assumptions.

The majority of this chapter is devoted to specific statistical methods that require weaker assumptions than that of normality. Statistical methods are presented that apply to a wide range of situations. Methods of constructing statistical tests for specific situations, including computer simulation, are also discussed. We conclude with

1. An indication of newer research in the topics of this chapter, and

2. Suggestions for additional reading if you wish to learn more about the subject matter.

8.1 ROBUSTNESS: NONPARAMETRIC AND DISTRIBUTION-FREE PROCEDURES

This section presents terminology associated with statistical procedures that require few assumptions for their validity.

304

The first idea we consider is *robustness*:

Definition 8.1. A statistical procedure is *robust* if it performs well when the needed assumptions are not violated "too badly", or if the procedure performs well for a large family of probability distributions.

By a *procedure*, we mean an estimate, a statistical test, or a method of constructing a confidence interval.

We elaborate upon this definition to give the reader a better idea of the meaning of the term. The first thing to note is that the definition is *not* a mathematical definition. We have talked about a procedure performing "well," but have not given a precise mathematical definition of what "well" means. The term *robust* is analogous to beauty: things may be considered more or less beautiful. Depending upon the specific criteria for beauty, there may be greater or lesser agreement about the beauty of an object. Similarly, different statisticians may disagree about the robustness of a particular statistical procedure, depending upon the probability distributions of concern, and use of the procedure. Nevertheless, as the concept of beauty is useful, the concept of robustness also proves to be useful conceptually, and in discussing the range of applicability of statistical procedures.

We discuss some of the ways a statistical test may be robust. Suppose we have a test statistic whose distribution is derived for some family of distributions (for example, normal distributions). Suppose also that the test is to be applied at a particular significance level, which we designate the *nominal* significance level. When other distributions are considered, the *actual* probability of rejecting the null hypothesis when it holds may differ from the *nominal* significance level if the distribution is not one of those used to derive the statistical test. For example, in testing for a specific value of the mean with a normally distributed sample, the t-test may be used. Suppose, however, the distribution considered is not normal. Then, if testing at the 5% significance level, the actual significance level (the true probability of rejecting under the *null* hypothesis that the population mean has the hypothesized value) may not be 5%; it may vary. A statistical test would be robust over a larger family of distributions if the true significance level and nominal significance level were close to each other. Also, a statistical test is robust if under specific alternatives, the probability of rejecting the null hypothesis tends to be large even when the alternatives are in a more extensive family of probability distributions.

A statistical test may be robust for large samples, but not for small samples. For example, for most distributions, if one uses the t-test for the mean when the sample size becomes quite large, the central limit theory shows that the nominal significance level is approximately the same as the true significance level when the null hypothesis holds. On the other hand, if one takes a sample from one normal population with a large probability and with a small probability takes the sample from another normal distribution with the same mean but a much larger variance, the properties of the t-test are poor for small samples (Tukey [1960]).

A technique of constructing confidence intervals is robust to the extent that the nominal confidence level is maintained over a larger family of distributions. For example, return to the t-test. If we construct 95% confidence intervals for the mean, the method is robust to the extent that samples from a nonnormal distribution straddle the mean about 95% of the time. Alternatively, a method of constructing confidence intervals is nonrobust if the confidence with which the parameters are in the interval differs greatly from the nominal confidence level.

An estimate of a parameter is robust to the extent that the estimate is close to the true parameter value over a large class of probability distributions.

Turning to a new topic, the normal distribution model is useful for summarizing data, because two parameters (in this case, the mean and variance, or equivalently, the mean and the standard deviation) describe the entire distribution. Such a set or family of distribution functions with each member described (or indexed) by a few parameters is called a *parametric family*. The distributions used for test statistics are also parametric families. For example, the t-distribution, the F-distribution, and the χ^2-distribution depend on one or two integer parameters: the degrees of freedom. Other examples of parametric families are the binomial distribution, with its two parameters n and π, and the Poisson distribution, with its parameter λ.

By contrast, *nonparametric families* of distributions are families which cannot be conveniently characterized, or indexed, by a few parameters. For example, if one looked at all the possible continuous distributions, it is not possible to find a few parameters that characterize all these distributions.

Definition 8.2. A family of probability distributions is *nonparametric* if the distributions of the family cannot be conveniently characterized or indexed by a few parameters. A family of probability distributions that can be characterized by a few parameters is a *parametric* family.

The t-test holds for the family of normal distributions, that is, for a parametric family. It would be nice to have a test statistic whose distribution was valid for a larger family of distributions.

Definition 8.3. Statistical procedures that hold, or are valid for a nonparametric family of distributions, are called *nonparametric statistical procedures*.

(It should be noted that there is not universal agreement about the definition of nonparametrical statistical procedures; see Note 8.1.) The t-test is *not* nonparametric, but rather the t-distribution is based on sampling from a normal distribution. We shall see many examples in this chapter of test statistics that are nonparametric.

The usefulness of the t-distribution results from the fact that samples from a normal distribution give the same t-distribution for all normal distributions under the null hypothesis. More generally, it is very useful to construct a test statistic whose distribution is the same for all members of some family of distributions. That is, assuming the sample comes from some member of the family, and the null hypothesis holds, the statistic has a known distribution; in other words, the distribution does not depend upon, or is *free*, of which member of the underlying family of distributions is sampled. This leads to our next definition.

Definition 8.4. A statistical procedure is *distribution-free* over a specified family of distributions if the statistical properties of the procedure do not depend upon (or are free of) the underlying distribution being sampled.

A test statistic is distribution-free if, under the null hypothesis, it has the same distribution for all members of the family. A method of constructing confidence intervals is distribution-free if the nominal confidence level holds for all members of the underlying family of distributions.

In practice, one selects statistical procedures that hold over a wide class of distributions. Often, the wide class of distributions is nonparametric, and the resulting statistical procedure is distribution-free for the family. The procedure then would be both nonparametric and distribution-free. The terms *nonparametric* and *distribution-free* are used somewhat loosely, and are often considered interchangeable. The term *nonparametric* is used much more often than the term *distribution-free*.

One would expect that a nonparametric procedure would not have as much statistical power as a parametric procedure *if* the observed sample comes from the parametric family. One method of comparing procedures is to look at their relative efficiency. *Relative efficiency* is a complex term when defined precisely (see Note 8.2), but the essence is contained in the following definition:

Definition 8.5. The *relative efficiency* of statistical procedure A to statistical procedure B is the ratio of the sample size needed for B to the sample size needed for A in order that both procedures have the same statistical power.

For example, if the relative efficiency of A to B is 1.5, then B needs 50% more observations than A in order to get the same amount of statistical power.

8.2 REVIEW OF THE χ^2-TEST FOR CONTINGENCY TABLES: FISHER'S EXACT TEST AND McNEMAR'S TEST

Consider the χ^2-test for association in contingency tables. The limiting χ^2-distribution did not depend upon the original probabilities under the null hypothesis of independence. However, recall that the χ^2-distribution is actually a limiting distribution; we required that the expected number of observations in the various cells be five or more so that the distribution of the test statistic was approximately χ^2. Thus, the χ^2-test is an *asymptotic distribution-free test*. Is the test a nonparametric test? The answer to this is both yes and no. The probability distribution we are sampling from can be characterized by the probability of falling into each of the cells in the contingency table. In fact, since the sum of the probabilities add up to one, we only need to know the probability for all of the cells except one. Thus, we could consider the probability distribution for the contingency table as parametric because it is specified by these parameters. If so, we would call the test parametric, *not* nonparametric. On the other hand, the number of parameters is quite large. We might consider the family of distributions nonparametric because of the large number of parameters needed. In this case, the χ^2-test would be a nonparametric statistical test. Since the χ^2-distribution holds asymptotically against all possible probability distributions under the null hypothesis, the distribution is asymptotically nonparametric, distribution-free, and robust.

Similarly, Fisher's Exact Test (if the marginal values in a 2×2 table are known) is a nonparametric and distribution-free test; under the null hypothesis of independence, its distribution is known. Finally, note that McNemar's Test holds regardless of the underlying distributions; this is another distribution-free statistical test. Since the different pairs for McNemar's Test might have different probabilities involved, there would be a large number of parameters involved, and one would consider McNemar's Test a nonparametric test as well as being distribution-free.

8.3   THE SIGN TEST

Suppose we are testing a drug to reduce blood pressure using a crossover design with a placebo. We might analyze the data by taking the blood pressure while not on the drug and subtracting it from the blood pressure while on the drug. These differences resulting from the matched or paired data will have an expected mean of zero if the drug under consideration had no more effect than the placebo effect. If we want to assume normality, a one-sample t-test with a hypothesized mean of zero is appropriate. Suppose, however, we knew from past experience that there were occasional large fluctuations in blood pressure due to biological variability. We might then be hesitant to use the t-test because of the known fact that one or two large observations, or "outliers," destroyed the probability distribution of the test (see Problem 8.20). What should we do?

An alternative nonparametric way of analyzing the data is the following. Suppose there is no treatment effect. All of the difference between the blood pressures measured on-drug and on-placebo will be due to biological variability. Thus, the difference between the two measurements will be due to symmetric random variability; the number is equally likely to be positive or negative. The *sign test* is appropriate for the null hypothesis that observed values have the same probability of being positive or negative: if we look at the number of positive numbers among the differences (and exclude values equal to zero), under the null hypothesis of no drug effect, this number has a binomial distribution, with $\pi = 1/2$. *A test of the null hypothesis could be a test of the binomial parameter $\pi = 1/2$.* This was discussed in Chapter 6, when we considered McNemar's Test. Such tests are called *sign tests*, since we are looking at the sign of the difference.

Definition 8.6.   Tests based upon the sign of an observation (that is, plus or minus), and which test the hypothesis that the observation is equally likely to be a plus or minus, are called *sign test procedures*.

Note that it is possible to use a sign test in situations where numbers are not observed, but there is only a rating. For example, one could have a blinded evaluation of patients as worse on-drug than on-placebo, the same on-drug as on-placebo, and better on-drug than on-placebo. By considering only those who were better or worse on the drug, the null hypothesis of no effect is equivalent to testing that each outcome is equally likely; that is, the binomial probability is $1/2$, the sign test may be used. Ratings of this type are useful in evaluating drugs when numerical quantification is not available.

As tests of $\pi = 1/2$ for binomial random variables were discussed in Chapter 6, we will not elaborate here. Problems 8.1–8.3 use the sign test.

Suppose that the distribution of blood pressures *did* follow a normal distribution: how much would be lost in the way of efficiency by using the sign test? The relative efficiency of the sign test with respect to the t-test when the normal assumptions are satisfied is 0.64; that is, compared to analyzing the data using the t-test, 36% of the samples are effectively thrown away. Alternatively, one needs $1/0.64$, or 1.56 times as many observations for the sign test as one would need using the t-test to have the same statistical power.

The sign test is useful in many situations. It is a "quick-and-dirty" test that one may compute mentally without the use of computational equipment; provided that statistical tables are available, you can get a quick estimate of the statistical significance of an appropriate null hypothesis.

8.4 RANKS

Many of the nonparametric, distribution-free tests are based upon one simple and brilliant idea. The approach is motivated by an example.

Example 8.1. The following data are for individuals who are exercised on a treadmill to their maximum capacity. There were five individuals in a group that underwent heavy distance running training and five control individuals who were sedentary and not trained. The maximum oxygen intake rate adjusted for body weight is measured in ml/kg/min. The quantity is called VO_2 MAX. The values for the untrained individuals were 45, 38, 48, 49, and 51. The values for the trained individuals were 63, 55, 59, 65, and 77. Because of the larger spread among the trained individuals and especially one extremely large VO_2 MAX (as can be seen from Figure 8.1), the values do not look like they are normally distributed. On the other hand, it certainly appears that the training has some benefits, since the five trained individuals all exceed the treadmill times of the five sedentary individuals. Although we do not want to assume that the values are normally distributed, we should somehow use the fact that the larger observations come from one group, and the smaller observations come from the other group. We desire a statistical test whose distribution can be tabulated under the null hypothesis that the probability distributions are the same in the two groups.

The crucial idea is the rank of the observation, which is the position of the observation among the other observations when they are arranged in order.

Definition 8.7. The *rank* of an observation, among a set of observations, is its position when the observations are arranged from the smallest to the largest. The smallest observation has rank 1, the next smallest has rank 2, and so on. If observations are tied, the rank assigned is the average of the ranks appropriate to the equal numbers.

For example, the ranks of the ten observations given above would be found as follows: first, order the observations from the smallest to largest; then, number them from left to right, beginning at 1.

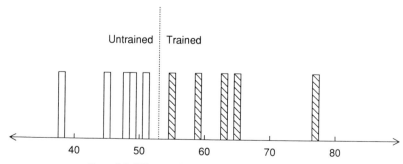

Figure 8.1. VO_2 MAX in trained and untrained individuals.

Observation	38	45	48	49	51	55	59	63	65	77
Rank	1	2	3	4	5	6	7	8	9	10

We now consider several of the benefits of using ranks. In the example above, suppose there was no difference in the VO_2 MAX between the two populations. Then we have ten independent samples (five from each population). Since there would be nothing to distinguish between observations, the five observations from the set of individuals who experienced training would be equally likely to be any five of the given observations. That is, if we consider the ranks from 1 to 10, all subsets of size five would be equally likely to represent the ranks of the five trained individuals. This is true regardless of the underlying distribution of the ten observations.   □

We repeat for emphasis: *If we consider continuous probability distributions (so that there are no ties) under the null hypothesis that two groups of observations come from the same distribution the ranks have the same distribution*! Thus, tests based upon the ranks will be nonparametric tests over the family of continuous probability distributions. Another way of making the same point: any test that results from using the ranks will be distribution-free, because the distribution of the ranks does not depend upon the underlying probability distribution under the null hypothesis.

There is a price to be paid in using rank tests. If we have a small number of observations, say two in each group, even if the two observations in one group are larger than both observations in the other group a rank test will not allow rejection of the null hypothesis that the distributions are the same. On the other hand, if one knows that the data are approximately normally distributed if the two large observations are considerably larger than the smaller observations, the *t*-test would allow one to reject the null hypothesis that the distributions are the same. However, this increased statistical power *critically* depends upon the normality assumptions. With small sample sizes, one cannot check the adequacy of the assumptions. One may reject the null hypothesis incorrectly (when in fact the two distributions are the same) because a large outlying value is observed.

Many nonparametric statistical tests can be devised using the simple idea of ranks. The next three sections of this chapter present specific rank tests of certain hypotheses.

8.5   THE WILCOXON SIGNED RANK TEST

In this section, we consider our first rank test. The test is an alternative to the one-sample *t*-test. Whenever the one-sample *t*-test of Chapter 5 is appropriate, this test may also be used, as its assumptions will be satisfied. However, since the test is a nonparametric test, its assumptions will be satisfied much more generally than under the assumptions needed for the one-sample *t*-test. In this section, we first discuss the needed assumptions and null hypothesis for this test. The test itself is then presented and illustrated by an example. For large sample sizes, the value of the test statistic may be approximated by a standard normal distribution; the appropriate procedure for this is also presented.

8.5.1 Assumptions and Null Hypotheses

The signed rank test is appropriate for statistically independent observations. The null hypothesis to be tested is that each observation comes from a distribution that is symmetric with a mean of zero. That is, for any particular observation, the value is equally likely to be positive or negative.

For the one-sample t-test, we have independent observations from a normal distribution; suppose the null hypothesis to be tested has a mean of zero. When the mean is zero, the distribution is symmetric about zero, and positive or negative values are equally likely. Thus, the signed rank test may be used wherever the one-sample t-test of mean zero is appropriate. For large sample sizes, the signed rank test has an efficiency of 0.955 relative to the t-test; the price paid for using this nonparametric test is equivalent to losing only 4.5% of the observations. In addition, when the normal assumptions for the t-test hold and the mean is not zero, the signed rank test has equivalent statistical power.

An example where the signed rank test is appropriate is a crossover experiment with a drug and a placebo. Suppose that individuals have the sequence "placebo, then drug" or "drug, then placebo" each assigned at random, with a probability of 0.5. The null hypothesis of interest is that the drug has the same effect as the placebo. If one takes the difference between measurements taken on the drug and on the placebo, and if the treatment has no effect, then the distribution of the difference will not depend upon whether the drug was given first or second. The probability is one-half that the placebo was given first and that the observation being looked at is the second observation minus the first observation. The probability is also one-half that the observation being examined came from an individual who took the drug first. In this case, the observation being used in the signed rank test would be the first observation minus the second observation. Since under the null hypothesis, these two differences have the same distribution except for a minus sign, the distribution of observations under the null hypothesis of "no treatment effect" is symmetric about zero.

8.5.2 Alternative Hypotheses Tested with Power

To use the test, we need to know what type of alternative hypotheses may be detected with some statistical power. For example, suppose one is measuring blood pressure, and the drug supposedly lowers the blood pressure compared to a placebo. The difference between the measurements on the drug and the blood pressure will tend to be negative. If we look at the observations, two things will occur. First, there will tend to be more observations which have a negative value (that is, a minus sign) than expected by chance. Second, if we look at the values of the data, the largest absolute values will tend to be negative values. The differences that are positive will usually have smaller absolute values. The signed rank test is designed to use both sorts of information. The signed rank statistic is designed to have power where the alternatives of interest correspond roughly to a shift of the distribution (for example, the median, rather than being zero, is positive or negative).

8.5.3 Computation of the Test Statistic

We compute the signed rank statistic as follows:

1. Rank the absolute values of the observations from smallest to largest. Note that we do *not* rank the observations themselves, but rather the absolute values; that is, we ignore minus signs. Drop observations equal to zero.

2. Add up the values of the ranks assigned to the positive observations. Do the same to the negative observations. The smaller of the two values is the value of the Wilcoxon Signed Rank statistic used in Table A.12 in the Appendix.

The procedure is illustrated in the following example.

Example 8.2. Brown and Hurlock [1975] investigated three methods of preparing the breasts for breastfeeding. The methods were:

1. Toughening the skin of the nipple by nipple friction or rolling.

2. Creams to soften and lubricate the nipple.

3. Prenatal expression of the first milk secreted before or after birth (colostrum).

Each subject had one randomly chosen treated breast, and one untreated breast. Nineteen different subjects were randomized to each of three treatment groups that is, each subject received the three treatments in random order. The purpose of the study was to evaluate methods of preventing postnatal nipple pain and trauma. The effects were evaluated by the mothers filling out a subjective questionnaire rating nipple sensitivity from "comfortable" (1) to "painful" (2) after each feeding. The data are presented in Table 8.1.

Table 8.1. Mean Subjective Difference Between Treated and Untreated Breasts. Data from Brown and Hurlock [1975].

Nipple Rolling	Masse Cream	Expression of Colostrum	Nipple Rolling	Masse Cream	Expression of Colostrum
−0.525	0.026	−0.006	0.007	0.000	0.048
0.172	0.739	0.000	−0.122	0.000	0.300
−0.577	−0.095	−0.257	−0.040	0.060	0.182
0.200	−0.040	−0.070	0.000	−0.180	−0.378
0.040	0.006	0.107	−0.100	0.000	−0.075
−0.143	−0.600	0.362	0.050	0.040	−0.040
0.043	0.007	−0.263	−0.575	0.080	−0.080
0.010	0.008	0.010	0.031	−0.450	−0.100
0.000	0.000	−0.080	−0.060	0.000	−0.020
−0.522	−0.100	−0.010			

We use the signed rank test to examine the statistical significance of the nipple-rolling data. The first step is to rank the absolute values of the observations, omitting zero values. The observations ranked by absolute value and their ranks are:

Observation	0.007	0.010	0.031	0.040	−0.040	0.043
Rank	1	2	3	4.5	4.5	6

Observation	0.050	−0.060	−0.100	−0.122	−0.143	0.172
Rank	7	8	9	10	11	12

Observation	0.200	−0.522	−0.525	−0.575	−0.577	
Rank	13	14	15	16	17	

Note the tied absolute values corresponding to ranks 4 and 5. The average rank 4.5 is used for both observations. Also note that two zero observations were dropped.

The sum of the ranks of the positive numbers is $S = 1+2+3+4.5+6+7+12+13 = 48.5$. This is less than the sum of the negative ranks. For a sample size of 17, Table A.12 shows that the two-sided p-value is ≥ 0.10. If there are no ties, Owen [1962] shows that $P[S \geq 48.5] = 0.1$ and the two-sided p-value is 0.2. No treatment effect has been shown. □

8.5.4 Large Samples

When the number of observations is large, we may compute a statistic that has approximately a standard normal distribution under the null hypothesis. We do this by subtracting the mean under the null hypothesis from the observed signed rank statistic, and dividing by the standard deviation under the null hypothesis. Here we do not take the minimum of the sums of positive and negative ranks; the usual one- and two-sided normal procedures can be used. The mean and variance under the null hypothesis are given in the following two equations:

$$E(S) = n(n + 1)/4, \tag{8.1}$$

$$\text{var}(S) = n(n + 1)(2n + 1)/24. \tag{8.2}$$

From this, one gets the following statistic that is approximately normally distributed for large sample sizes:

$$Z = \frac{S - E(S)}{\sqrt{\text{var}(S)}}. \tag{8.3}$$

Sometimes, data are recorded on such a scale that ties can occur for the absolute values. In this case, tables for the signed rank test are conservative; that is, the probability of rejecting the null hypothesis when it is true is *less* than the nominal significance level. The asymptotic statistic may be adjusted for the presence of ties. The effect of ties is to reduce the variance in the statistic. The rank of a term involved in a tie is replaced by the average of the ranks of those tied observations. Consider, for example, the following data:

$$6, -6, -2, 0, 1, 2, 5, 6, 6, -3, -3, -2, 0$$

Note that there are not only some ties, but zeros. In the case of zeros, the zero observations are omitted from the computation as noted before. These data, ranked by absolute value, with average ranks replacing the given rank when the absolute values are tied, are shown below. The first row (A) represents the data ranked by absolute value, omitting zero values; the second row (B) gives the ranks; and the third row (C) gives the ranks, with ties averaged (in this row, ranks of positive numbers are in bold type):

A	1	−2	2	−2	−3	−3	5	6	−6	6	6
B	1	2	3	4	5	6	7	8	9	10	11
C	**1**	3	**3**	3	5.5	5.5	**7**	**9.5**	9.5	**9.5**	**9.5**

Note that the ties are with respect to the absolute value (without regard to sign). Thus the three ranks corresponding to observations of −2 and +2 are 2, 3, and 4, the average of which is 3. The S-statistic is computed by adding the ranks for the positive values. In this case,

$$S = 1 + 3 + 7 + 9.5 + 9.5 + 9.5 = 39.5.$$

Before computing the asymptotic statistic, the variance of S must be adjusted because of the ties. To make this adjustment, we need to know the number of groups that have ties, and the number of ties in each group. In looking at the above data, we see that there are three sets of ties corresponding to absolute values 2, 3, and 6. The number of ties corresponding to observations of absolute value 2 (the "2 group") is 3; the number of ties in the "3 group" is 2; and the number of ties in the "6 group" is 4. In general, let q be the number of groups of ties, and let t_i, where i goes from 1 to q, be the number of observations involved in the particular group. In this case,

$$t_1 = 3, \quad t_2 = 2, \quad t_3 = 4, \quad q = 3.$$

In general, the variance of S is reduced according to the following equation:

$$\text{var}(s) = \frac{n(n+1)(2n+1) - \frac{1}{2}\sum_{i=1}^{q} t_i(t_i - 1)(t_i + 1)}{24}. \tag{8.4}$$

For the data that we are working with, we started with 13 observations, but the n used for the test statistic is 11, since two zeros were eliminated. In this case, the expected mean and variance are

$$E(S) = 11 \times (12/4) = 33,$$

$$\text{var}(S) = \frac{11 \times 12 \times 23 - \frac{1}{2}(3 \times 2 \times 4 + 2 \times 1 \times 3 + 4 \times 3 \times 5)}{24} \doteq 135.6.$$

Using test statistic S, gives

$$Z = \frac{(S - E(S))}{\sqrt{\text{var}(S)}} = \frac{(39.5 - 33)}{\sqrt{135.6}} \doteq 0.56.$$

With a Z-value of only 0.56, one would not reject the null hypothesis for commonly used values of the significance level. For testing at a 0.05 significance level, if n is 15 or larger with few ties, the normal approximation may reasonably be used. Note 8.4 and Problem 8.21 have more information about the distribution of the signed rank test.

To continue with Example 8.2, we compute the asymptotic Z-statistic for the signed rank test using the data given. In this case,, $n = 17$ after eliminating zero values. We have one set of two tied values, so that $q = 1$ and $t_1 = 2$. The null hypothesis mean is $17 \times 18/4 = 76.5$. This variance is $\left(17 \times 18 \times 35 - (1/2) \times 2 \times 1 \times 3\right)/24 = 446.125$. Therefore, $Z = (48.5 - 76.5)/21.12 \doteq -1.326$. Table A.1 shows that a two-sided p is about 0.186. This agrees with $p = 0.2$ as given above from tables for the distribution of S.

8.6 THE WILCOXON (MANN–WHITNEY) TWO-SAMPLE TEST

Our second example of a rank test is designed for use in the two-sample problem. Given samples from two different populations, the statistic tests the hypothesis that the distributions of the two populations are the same. The test may be used whenever the two-sample t-test is appropriate. Since the test given depends upon the ranks, it is nonparametric and may be used more generally. In this section, we discuss the null hypothesis to be tested, and the efficiency of the test relative to the two-sample t-test. The test statistic is presented and illustrated by two examples. The large-sample approximation to the statistic is given. Finally, the relationship between two equivalent statistics, the Wilcoxon statistic and the Mann–Whitney statistic, is discussed.

8.6.1 Null Hypothesis, Alternatives, and Power

The null hypothesis tested is that each of two independent samples has the same probability distribution. Table A.13 for the Mann–Whitney two-sample statistic assumes that there are no ties. Whenever the two-sample t-test may be used, the Wilcoxon statistic may also be used. The statistic is designed to have statistical power in situations where the alternative of interest has one population with generally larger values than the other. This occurs, for example, when the two distributions are normally distributed, but the means differ. For normal distributions with a shift in the mean, the efficiency of the Wilcoxon Test relative to the two-sample t-test is 0.955; a small price is paid for using the nonparametric test of far greater applicability.

8.6.2 The Test Statistic

The test statistic itself is easy to compute. The combined sample of observations from both populations are ordered from the smallest observation to the largest. The sum of the ranks of the population with the smaller sample size (or in the case of equal sample sizes, an arbitrarily designated first population) gives the value of the Wilcoxon statistic.

To evaluate the statistic, we use some notation. Let m be the number of observations for the smaller sample, and n the number of observations in the larger sample. The Wilcoxon statistic W is the sum of the ranks of the m observations when both sets of observations are ranked together.

The computation is illustrated in the following example:

Example 8.3. This example deals with a small subset of data from the Coronary Artery Surgery Study [CASS, 1981]. Patients were studied for suspected or proven coronary artery disease. The disease was diagnosed by coronary angiography. In coronary angiography, a tube is placed into the aorta (where the blood leaves the heart) and a dye is injected into the arteries of the heart, allowing X-ray motion pictures (cine angiograms) of the arteries. If an artery is narrowed by 70% or more, the artery is considered significantly diseased. The heart has three major arterial systems, and so the disease (or lack thereof) is classified as zero-, one-, two- or three-vessel disease (to be abbreviated 0VD, 1VD, 2VD, and 3VD). Narrowed vessels do not allow as much blood to give oxygen and nutrients to the heart. This leads to chest pain (angina) and total blockage of arteries, killing a portion of the heart (called a *heart attack* or *myocardial infarction*). For those reasons, one does not expect people with disease to be able to exercise vigorously. Some individuals in CASS were evaluated by running on a treadmill to their maximal exercise performance. The treadmill increases in speed and slope according to a set schedule. The total time on the treadmill is a measure of exercise capacity. The data in Table 8.2 present treadmill time in seconds for men with normal arteries (but suspected coronary artery disease) and men with three-vessel disease.

Table 8.2. Treadmill times in seconds. Data from CASS [1981].

Normal	1014	684	810	990	840	978	1002	1110		
3VD	864	636	638	708	786	600	1320	750	594	750

Note that $m = 8$ (normal arteries) and $n = 10$ (three-vessel disease). The first step is to rank the combined sample and assign ranks.

Value	Rank	Group	Value	Rank	Group	Value	Rank	Group
594	1	3VD	750	7.5	3VD	978	13	Normal
600	2	3VD	750	7.5	3VD	990	14	Normal
636	3	3VD	786	9	3VD	1002	15	Normal
638	4	3VD	810	10	Normal	1014	16	Normal
684	5	Normal	840	11	Normal	1111	17	Normal
708	6	3VD	864	12	3VD	1320	18	3VD

The sum of the ranks of the smaller normal group is 101. Table A.13, for the closely related Mann–Whitney statistic of Section 8.6.4, shows that we reject the null hypothesis of equal population distributions at a 5% significance level.

Under the null hypothesis, the expected value of the Wilcoxon statistic is

$$E(W) = \frac{m(m+n+1)}{2}. \tag{8.5}$$

In this case, the expected value is 76. As we conjectured (*before* seeing the data) that the normal individuals would exercise longer (that is, W would be large), a one-sided test which rejects the null hypothesis if W is too large might have been used. Table A.12 shows that at the 5% significance level, we would have rejected the null hypothesis using the one-sided test. (This is also clear, since the more-stringent two-sided test rejected the null hypothesis.) □

8.6.3 Large Sample Approximation

This is a large-sample approximation to the Wilcoxon statistic (W) under the null hypothesis that the two samples come from the same distribution. The mean and variance of W, with or without ties, is given by Equations 8.5 through 8.7. In these equations, m is the size of the smaller group (the number of ranks being added to give W); n is the number of observations in the larger group; q is the number of groups of tied observations (as per the last section); and t_i is the number of ranks that are tied in the ith set of ties. First, without ties,

$$\text{var}(W) = \frac{mn(m+n+1)}{12},\tag{8.6}$$

and with ties,

$$\text{var}(W) = \frac{mn(m+n+1)}{12} - \left(\sum_{i=1}^{q} t_i(t_i - 1)(t_i + 1)\right)\frac{mn}{12(m+n)(m+n-1)}.\tag{8.7}$$

Using these values, an asymptotic statistic with an approximately standard normal distribution is

$$Z = \frac{W - E(W)}{\sqrt{\text{var}(W)}}.\tag{8.8}$$

Example 8.3 (continued). The normal approximation is best used when $n \geq 15$ and $m \geq 15$. Here, however, we compute the asymptotic statistic for the date of Example 8.3.

$$E(W) = \frac{8(10 + 8 + 1)}{2} = 76,$$

$$\text{var}(W) = \frac{8 \cdot 10(8 + 10 + 1)}{12} - 2(2 - 1)(2 + 1)\left(\frac{8 \cdot 10}{12(8 + 10)(8 + 10 + 1)}\right)$$

$$= 126.67 - 0.12 = 126.55,$$

$$Z = \frac{101 - 76}{\sqrt{126.55}} \doteq 2.22.$$

The one-sided p-value is 0.013, and the two-sided p-value is $2(0.013) = 0.026$. In fact, the exact one-sided p-value is 0.013. Note that the correction for ties leaves the variance virtually unchanged. □

Example 8.4. The Wilcoxon Test may be used for data that is ordered and ordinal. Consider the angiographic findings from the CASS [1981] study for men and women in Table 8.3.

Let us test whether the distribution of disease is the same in the men and women studied in the CASS registry.

You probably recognize that this is a contingency table, and the χ^2-test may be applied. If we want to examine the possibility of a trend in the proportions, the χ^2-test

Table 8.3. Extent of Coronary Artery Disease by Sex. Data from CASS [1981].

Extent of Disease	Male	Female	Total
None	2157	2360	4517
Mild	824	572	1396
Moderate	656	291	947
Significant			
1VD	3887	1020	4907
2VD	4504	835	5339
3VD	6115	882	6997
Total	18143	5960	24103

for trend could be used. That test assumes that the proportion of females changes in a linear fashion between categories. Another approach is to use the Wilcoxon Test as described here.

The observations may be ranked by the six categories (none, mild, moderate, 1VD, 2VD, and 3VD). There are many ties: 4517 ties for the lowest rank, 1396 ties for the next rank, and so on. We need to compute the average rank for each of the six categories. If J observations have come before a category with K tied observations, the average rank for the k tied observations is

$$\text{Average rank} = \frac{2J + K + 1}{2}. \tag{8.9}$$

For these data, the average ranks are computed as

K	J	Average
4517	0	2259
1396	4517	5215.5
947	5913	6387
4907	6860	9314
5339	11767	14437
6997	17106	20605

Now our smaller sample of females has 2360 observations with rank 2259, 572 observations with rank 5215.5, and so forth. Thus, the sum of the ranks is

$$W = 2360(2259) + 572(5215.5) + 291(6387) + 1020(9314) + 835(14437) + 882(20605)$$

$$= 49,901,908.$$

The expected value from Equation (8.5) is

$$E(W) = \frac{5960(5960 + 18143 + 1)}{2} = 71,829,920.$$

From Equation (8.7), the variance, taking into account ties, is

$$\text{var}(W) = 5960 \times 18143 \times \frac{5960 + 18143 + 1}{12}$$

$$- (4517 \times 4516 \times 4518 + \cdots + 6997 \times 6996 \times 6998)\frac{5960 \times 18143}{12 \times 20103 \times 20102}$$

$$= 2.06 \times 10^{11}.$$

From this,

$$z = \frac{W - E(W)}{\sqrt{\text{var}(W)}} \doteq -48.29.$$

The p-value is extremely small and the population distributions clearly differ.

8.6.4 The Mann–Whitney Statistic

Mann and Whitney developed a test statistic that is equivalent to the Wilcoxon test statistic. To obtain the value for the Mann–Whitney Test, which we denote by U, one arranges the observations from the smallest to the largest. The statistic U is obtained by counting the number of times an observation from the group with the smallest number of observations precedes an observation from the second group. With no ties, the statistics U and W are related by the following equation:

$$U + W = \frac{m(m + 2n + 1)}{2}. \tag{8.10}$$

Since the two statistics add to a constant, using one of them is equivalent to using the other. We have used the Wilcoxon statistic because it is easier to compute by hand. The values of the two statistics are so closely related that books of statistical tables contain tables for only one of the two statistics, since the transformation from one to the other is almost immediate. Table A.13 is for the Mann–Whitney statistic.

To use the table for Example 8.3, the Mann–Whitney statistic would be

$$U = \frac{8(8 + 2(10) + 1)}{2} - 101 = 116 - 101 = 15.$$

From Table A.13, the two-sided 5% significance levels are given by the tabulated values and mn minus the tabulated value. The tabulated two-sided value is 63, and $8 \times 10 - 63 = 17$. We do reject for a two-sided 5% test. For a one-sided test, the upper critical value is 60; we want the lower critical value of $8 \times 10 - 60 = 20$. Clearly, again we reject at the 5% significance level.

8.7 THE KOLMOGOROV–SMIRNOV TWO-SAMPLE TEST

Definition 3.9 showed one method of describing the distributions of values from a population: the *empirical cumulative distribution*. For each value on the real line, the empirical cumulative distribution gives the proportion of observations less than or equal to that value. One visual way of comparing two population samples would be a graph of the two empirical cumulative distributions. If the two empirical cumulative distributions differ greatly, one would suspect that the populations being sampled were not the same. If the two curves were quite close, then it would be reasonable to assume that the underlying population distributions were essentially the same.

The *Kolmogorov–Smirnov statistic* is based upon this observation. The value of the statistic is the maximum absolute difference between the two empirical cumulative distribution functions. Note 8.5 discusses the fact that the Kolmogorov–Smirnov statistic is a rank test. Consequently, the test is a nonparametric test of the null hypothesis that the two distributions are the same. The Kolmogorov–Smirnov statistic is not efficient when the *t*-test applies, but it is an omnibus test of the difference between two populations; it is a test for any possible departure between the population distributions.

The procedure is illustrated in the following example:

Example 8.4 (continued). The data of Example 8.3 are used to illustrate the statistic. Using the method of Chapter 3, Figure 8.2 was constructed with both distribution functions.

From Figure 8.2, we see that the maximum difference is 0.675 between 786 and 810. Tables of the statistic are usually tabulated not in terms of the maximum absolute difference D, but in terms of $(mn/d)D$ or mnD, where m and n are the two sample sizes, and d is the lowest common denominator of m and n. The benefit of this is that $(mn/d)D$ or mnD is always an integer. In this case, $m = 8$, $n = 10$, and $d = 2$. Thus, $(mn/d)D = (8 \cdot 10/2)0.675 = 27$ and $mnD = 54$. Table 44 of Odeh *et al.* [1977] gives

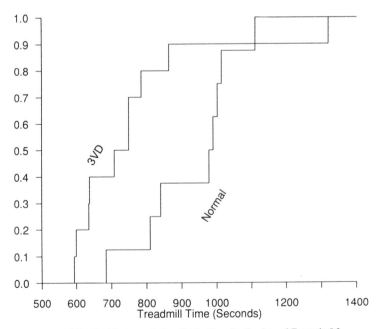

Figure 8.2. Empirical cumulative distributions for the data of Example 8.3.

the 0.05 critical value for mnD as 48. Since $54 > 48$, we reject the null hypothesis at the 5% significance level. Tables of critical values are not given in this text, but are available in standard tables (e.g., Odeh *et al.* [1977], Owen [1962], and CRC Tables [1968]).

The tables are designed for the case with no ties. If there are ties, the test is conservative; that is, the probability of rejecting the null hypothesis when it is true is even less than the nominal significance level.

The large sample distribution of D is known. Let n and m both be large, say, both 40 or more. The large sample test rejects the null hypothesis according to the following table.

Significance Level	Reject the Null Hypothesis if ...
0.001	$KS \geq 1.95$
0.01	$KS \geq 1.63$
0.05	$KS \geq 1.36$
0.10	$KS \geq 1.22$

Let KS be defined as

$$KS = \max_x \sqrt{\frac{nm}{n+m}} \left| F_n(x) - G_m(x) \right| = \sqrt{\frac{nm}{n+m}} \, D, \qquad (8.11)$$

where F_n and G_m are the two empirical cumulative distributions.

8.8 NONPARAMETRIC ESTIMATION AND CONFIDENCE INTERVALS

Many nonparametric tests have associated estimates of parameters. Confidence intervals for these estimates are also often available. In this section, we present one estimate associated with the Wilcoxon (or Mann–Whitney) two-sample test statistic. We also show how to construct a confidence interval for the median of a distribution.

In considering the Mann–Whitney test statistic described in Section 8.6, let us suppose that the sample from the first population was denoted by X's, and the sample from the second population by Y's. Suppose we observe m X's and n Y's. The Mann–Whitney test statistic U is the number of times an X was less than a Y among the nm X and Y pairs. As shown in Equation(8.12), the Mann–Whitney test statistic U, when divided by mn, gives an unbiased estimate of the probability that X is less than Y.

$$E(U/mn) = P[X < Y]. \qquad (8.12)$$

Further, an approximate $100(1-\alpha)\%$ confidence interval for the probability that X is less than Y may be constructed using the asymptotic normality of the Mann–Whitney test statistic. The confidence interval is given by the following equation:

$$\frac{U}{mn} \pm \frac{Z_{1-\alpha/2}}{m} \sqrt{\frac{U}{mn}\left(1 - \frac{U}{mn}\right)}. \qquad (8.13)$$

Example 8.5.   This example illustrates the use of the Mann–Whitney test statistic to estimate the probability that X is less than Y, and to find a 95% confidence interval for $P[X < Y]$.

Examine the data of Table 8.2 (Example 8.3). We shall estimate the probability that the treadmill time of a randomly chosen individual with normal arteries is less than that of a three-vessel disease patient.

Note that 1014 is less than one three-vessel treadmill time; 684 is less than 6 of the three-vessel treadmill times, and so forth. Thus,

$$U = 1 + 6 + 2 + 1 + 2 + 1 + 1 + 1 = 15.$$

We also could have found U by using Equation (8.9) and $W = 101$ from Example 8.3. Our estimate of $P[X < Y]$ is $15/(8 \times 10) = 0.1875$. The confidence interval is

$$0.1875 \pm (1.96/8)(0.1875)(1 - 0.1875) \doteq 0.1875 \pm 0.0373. \qquad \square$$

In Chapter 3, we saw how to estimate the median of a distribution. We now show how to construct a confidence interval for the median that will hold for any distribution. To do this, we use *order statistics*.

Definition 8.8. Suppose that one observes a sample. Arrange the sample from the smallest to the largest number. The smallest number is the *first-order statistic*, the second smallest is the **second-order statistic,** and so on; in general, the *ith-order statistic* is the ith number in line.

The notation used for an order statistic is to put the subscript corresponding to the particular order statistic in parentheses. That is,

$$X_{(1)} \leq X_{(2)} \leq \cdots \leq X_{(n)}.$$

To find a $100(1-\alpha)\%$ confidence interval for the median, we first find from tables of the binomial distribution with $\pi = 0.5$, the largest value of k such that the probability of k or fewer successes is less than or equal to $\alpha/2$. That is, we choose k to be the largest value of k such that

$$P[\text{number of heads in } n \text{ flips of a fair coin} = 0 \text{ or } 1 \text{ or } \dots \text{ or } k] \leq \frac{\alpha}{2}.$$

Given the value of k, the confidence interval for the median is the interval between the $k + 1$- and $n - k$-order statistics. That is, the interval is

$$(X_{(k+1)}, X_{(n-k)}).$$

Example 8.6. The treadmill times of 20 females with normal or minimal coronary artery disease in the CASS study are

570, 618, 30, 780, 630, 738, 900, 750, 750, 540, 660,
780, 720, 750, 936, 900, 762, 840, 816, 690.

We estimate the median time and construct a 90% confidence interval for the median of this population distribution. The order statistics (ordered observations) from one to twenty are,

30, 540, 570, 618, 630, 660, 690, 720, 738, 750, 750,
750, 762, 780, 780, 816, 840, 900, 900, 936.

Since we have an odd number of observations,

$$\text{Median} = \frac{(X_{(10)} + X_{(11)})}{2} = \frac{750 + 750}{2} = 750.$$

Table A.8 shows that if X is binomial $n = 20$ and $\pi = 0.5$, $P[X \le 5] = 0.0207$ and $P[X \le 6] = 0.0577$. Thus, $k = 5$. Now, $X_{(6)} = 690$ and $X_{(15)} = 780$. Hence, the confidence interval is $(690, 780)$. The actual confidence is $100(1 - 2 \times 0.0207)\% \doteq 95.9\%$. Because of the discrete nature of the data, the nominal 90% confidence interval is also a 95.9% confidence interval.

8.9 *PERMUTATION AND RANDOMIZATION TESTS

This section presents a method that may be used to generate a wide variety of statistical procedures. The arguments involved are subtle; you need to pay careful attention to understand the logic. We illustrate the idea by working from an example.

Suppose one had two samples, one of size n and one of size m. Consider the null hypothesis that the distributions of the two populations are the same. Let us suppose that, in fact, this null hypothesis is true; the combined $n + m$ observations are independent and sampled from the same population. Suppose now that you are told that one of the $n + m$ observations is equal to 10. Which of the $n + m$ observations is most likely to have taken the value 10? There is really nothing to distinguish the observations, since they are all taken from the same distribution or population. Thus, any of the $n + m$ observations is equally likely to be the one that was equal to 10. More generally, suppose our samples are taken in a known order; for example, the first n observations come from the first population, and the next m from the second. Let us suppose that the null hypothesis still holds. Suppose you are now given the observed values in the sample, all $n + m$ of them, but not told which value was obtained from which ordered observation. Which arrangement is most likely? Since all the observations come from the same distribution, and the observations are independent, there is nothing that would tend to associate any one sequence or arrangement of the numbers with a higher probability than any other sequence. In other words, every assignment of the observed numbers to the $n + m$ observations is equally likely. This is the idea underlying a class of tests called *permutation tests*. To understand why they are called this, we need the definition of a permutation:

Definition 8.9. Given a set of $(n + m)$ objects arranged or numbered in a sequence, a *permutation* of the objects is a rearrangement of the objects into the same or a different order. The number of permutations is $(n + m)!$

What we said above is that if the null hypothesis holds in the two-sample problem, all permutations of the observed numbers are equally likely. Let us illustrate this with a small example. Suppose that we have two observations from the first group, and also two observations from the second group. Suppose that we know that the four observations take on the values 3, 7, 8, and 10. Listed below are the possible permutations where the first two observations would be considered to come from the first group, and the second two from the second group. (Note that x represents the first group, and y represents the second.)

x		y		$\bar{x} - \bar{y}$		x		y		$\bar{x} - \bar{y}$	
3	7	8	10			7	8	3	10		
3	7	10	8			7	8	10	3		
7	3	8	10	$\Big\}-4$		8	7	3	10	$\Big\}1$	
7	3	10	8			8	7	10	3		
3	8	7	10			7	10	3	8		
3	8	10	7			7	10	8	3		
8	3	7	10	$\Big\}-3$		10	7	3	8	$\Big\}3$	
8	3	10	7			10	7	8	3		
3	10	7	8			8	10	3	7		
3	10	8	7			8	10	7	3		
10	3	7	8	$\Big\}-1$		10	8	3	7	$\Big\}4$	
10	3	8	7			10	8	7	3		

If we only know the four values 3, 7, 8, and 10, but do not know in which order they came, any of the 24 possible arrangements listed above are equally likely. If we wanted to perform a two-sample test, we could generate a statistic and calculate its value for each of the 24 arrangements. We could then order the values of the statistic according to some alternative hypothesis so that the more extreme values were more likely under the alternative hypothesis. By looking at what sequence *actually occurred*, we can get a p-value for this set of data. The p-value is determined by the position of the statistic among the possible values. The p-value is the number of possibilities as extreme or more extreme than that observed divided by the number of possibilities.

Suppose, for example, that with the above data, we decided to use the difference in means between the two groups, $\bar{x} - \bar{y}$, as our test statistic. Suppose also that our alternative hypothesis is that group 1 has a larger mean than group 2. Then, if any of the last four rows of the above table had occurred, the one-sided p-value would be 4/24, or 1/6. Note that this would be the most extreme finding possible. On the other hand, if the data had been 8, 7, 3, and 10, with an $\bar{x} - \bar{y} = 1$, then the p-value would be 12/24, or 1/2.

The tests we have been discussing are called *permutation tests*. They are possible when a permutation of all or some subset of the data is considered equally likely under the null hypothesis; the test is based upon this fact. These tests are sometimes also called *conditional tests*, because the test takes some portion of the data as fixed or known. In the case above, we assume that we know the actual observed values, although we do not know in which order they occurred. We have seen an example of a conditional test before: Fisher's Exact Test in Chapter 6 treated the row and column totals as known; conditionally, upon that information, the test considered what happened to the entries in the table. The permutation test can be used to calculate appropriate p-values for tests such as the t-test when, in fact, normal assumptions do not hold. To do this, proceed as in the next example.

Example 8.7. Given two samples, a sample of size n of X observations, and a sample of size m of Y observations, it can be shown (Problem 8.24) that the two-sample t-test is a monotone function of $\bar{x} - \bar{y}$; that is, as $\bar{x} - \bar{y}$ increases, t also increases. Thus, if we perform a permutation test on $\bar{x} - \bar{y}$, we are in fact basing our test upon extreme values of the t-statistic. The illustration above is equivalent to a t-test on the four values given. Consider now the data

$$x_1 = 1.3, \quad x_2 = 2.3, \quad x_3 = 1.9, \quad y_1 = 2.8, \quad y_2 = 3.9.$$

The 120 permutations $(3 + 2)!$ ordered by the values of $\bar{x} - \bar{y}$ are found in Table 8.4.

The permutations fall into ten groups of twelve permutations with the same value of $\bar{x} - \bar{y}$. The observed value of $\bar{x} - \bar{y}$ is -1.52, the lowest possible value. A one-sided test of $E(Y) < E(X)$ would have $p = 0.1 = 12/120$. The two-sided p-value is 0.2.

The Wilcoxon test may be considered a permutation test, where the values used are the ranks and not the observed values. For the Wilcoxon test, we know what the values of the ranks will be; thus, one set of statistical tables may be generated that may be used for the entire sample. For the general permutation test, since the computation depends upon the numbers actually observed, it cannot be calculated until we have the sample in hand. Further, the computations for large sample sizes are very time-consuming. If n is equal to 20, there are over 2×10^{18} possible permutations. Thus, the computational work for permutation tests becomes large rapidly; this limits their use.

We now turn to *randomization tests*. Randomization tests proceed in a similar manner to permutation tests. In general, one assumes that some aspects of the data are known. If certain aspects of the data are known (for example, we might know the numbers that were observed, but not which group they are in), one can calculate a number of equally likely outcomes for the complete data. For example, in the permutation test, if we know the actual values, then all possible permutations of the values are equally likely. In other words, it is as if a permutation were to be selected at random; the permutation tests are examples of randomization tests.

Here we consider another example. This idea is the same as that used in the discussion of the signed rank test above. Suppose that under the null hypothesis, the numbers observed are independent and symmetric about zero. Suppose also that we are given the absolute values of the numbers observed, but not whether they are positive or negative. Take a particular number "a". Is it more likely to be positive or negative? Because the distribution is symmetric about zero, it is not more likely to be either one. It is equally likely to be $+a$ or $-a$. Extending this to all the observations, every pattern of assigning pluses or minuses to our absolute values is equally likely to occur under the null hypothesis that all observations are symmetric above zero. We can then calculate the value of a test statistic for all the different patterns for pluses and minuses. A test basing the p-value on these values would be called a randomization test.

Example 8.8. One can perform a randomization one-sample t-test, taking advantage of the absolute values observed rather than introducing the ranks. For example, consider the first four paired observations of Example 8.2. The values are -0.0525, 0.172, 0.577, and 0.200. Assign all sixteen patterns of pluses and minuses to the four absolute values (0.0525, 0.172, 0.577, and 0.200) and calculate the values of the paired or one-sample t-test. The sixteen computed values, in increasing order, are -3.47, -1.63, -1.49, $-\mathbf{0.86}$, -0.46, -0.34, -0.08, -0.02, 0.02, 0.08, 0.34, 0.46, 0.86, 1.48, 1.63, and 3.47. The observed t-value (in bold type) is -0.86. It is the fourth of sixteen values. The two-sided p-value is $2(4/16) = 0.5$. $\qquad\square$

Table 8.4. Data for Example 8.7.

x_1	x_2	x_3	y_1	y_2	$\bar{x}-\bar{y}$	x_1	x_2	x_3	y_1	y_2	$\bar{x}-\bar{y}$
1.9	2.3	1.3	3.9	2.8		3.9	2.3	1.3	2.8	1.9	
1.9	2.3	1.3	2.8	3.9		3.9	2.3	1.3	1.9	2.8	
1.9	1.3	2.3	3.9	2.8		3.9	1.3	2.3	2.8	1.9	
1.9	1.3	2.3	2.8	3.9		3.9	1.3	2.3	1.9	2.8	
2.3	1.9	1.3	3.9	2.8		2.3	3.9	1.3	2.8	1.9	
2.3	1.9	1.3	2.8	3.9	−1.52	2.3	3.9	1.3	1.9	2.8	0.15
2.3	1.3	1.9	3.9	2.8		2.3	1.3	3.9	2.8	1.9	
2.3	1.3	1.9	2.8	3.9		2.3	1.3	3.9	1.9	2.8	
1.3	2.3	1.9	3.9	2.8		1.3	3.9	2.3	2.8	1.9	
1.3	2.3	1.9	2.8	3.9		1.3	3.9	2.3	1.9	2.8	
1.3	1.9	2.3	3.9	2.8		1.3	2.3	3.9	2.8	1.9	
1.3	1.9	2.3	2.8	3.9		1.3	2.3	3.9	1.9	2.8	
2.8	1.9	1.3	3.9	2.3		3.9	2.8	1.3	1.9	2.3	
2.8	1.9	1.3	2.3	3.9		3.9	2.8	1.3	2.3	1.9	
2.8	1.3	1.9	3.9	2.3		3.9	1.3	2.8	1.9	2.3	
2.8	1.3	1.9	2.3	3.9		3.9	1.3	2.8	2.3	1.9	
1.9	2.8	1.3	3.9	2.3		2.8	3.9	1.3	1.9	2.3	
1.9	2.8	1.3	2.3	3.9	−1.10	2.8	3.9	1.3	2.3	1.9	0.57
1.9	1.3	2.8	3.9	2.3		2.8	1.3	3.9	1.9	2.3	
1.9	1.3	2.8	2.3	3.9		2.8	1.3	3.9	2.3	1.9	
1.3	2.8	1.9	3.9	2.3		1.3	3.9	2.8	1.9	2.3	
1.3	2.8	1.9	2.3	3.9		1.3	3.9	2.8	2.3	1.9	
1.3	1.9	2.8	3.9	2.3		1.3	2.8	3.9	1.9	2.3	
1.3	1.9	2.8	2.3	3.9		1.3	2.8	3.9	2.3	1.9	
2.8	2.3	1.3	3.9	1.9		3.9	1.9	2.3	2.8	1.3	
2.8	2.3	1.3	1.9	3.9		3.9	1.9	2.3	1.3	2.8	
2.8	1.3	2.3	3.9	1.9		3.9	2.3	1.9	2.8	1.3	
2.8	1.3	2.3	1.9	3.9		3.9	2.3	1.9	1.3	2.8	
2.3	2.8	1.3	3.9	1.9		1.9	3.9	2.3	2.8	1.3	
2.3	2.8	1.3	1.9	3.9	−0.77	1.9	3.9	2.3	1.3	2.8	0.65
2.3	1.3	2.8	3.9	1.9		1.9	2.3	3.9	2.8	1.3	
2.3	1.3	2.8	1.9	3.9		1.9	2.3	3.9	1.3	2.8	
1.3	2.8	2.3	3.9	1.9		2.3	3.9	1.9	2.8	1.3	
1.3	2.8	2.3	1.9	3.9		2.3	3.9	1.9	1.3	2.8	
1.3	2.3	2.8	3.9	1.9		2.3	1.9	3.9	2.8	1.3	
1.3	2.3	2.8	1.9	3.9		2.3	1.9	3.9	1.3	2.8	
2.8	1.9	2.3	3.9	1.3		1.9	3.9	2.8	2.3	1.3	
2.8	1.9	2.3	1.3	3.9		1.9	3.9	2.8	1.3	2.3	
2.8	2.3	1.9	3.9	1.3		1.9	2.8	3.9	2.3	1.3	
2.8	2.3	1.9	1.3	3.9		1.9	2.8	3.9	1.3	2.3	
1.9	2.8	2.3	3.9	1.3		3.9	2.8	1.9	2.3	1.3	
1.9	2.8	2.3	1.3	3.9	−0.27	3.9	2.8	1.9	1.3	2.3	1.07
1.9	2.3	2.8	3.9	1.3		3.9	1.9	2.8	2.3	1.3	
1.9	2.3	2.8	1.3	3.9		3.9	1.9	2.8	1.3	2.3	
2.3	2.8	1.9	3.9	1.3		2.8	3.9	1.9	2.3	1.3	
2.3	2.8	1.9	1.3	3.9		2.8	3.9	1.9	1.3	2.3	
2.3	1.9	2.8	3.9	1.3		2.8	1.9	3.9	2.3	1.3	
2.3	1.9	2.8	1.3	3.9		2.8	1.9	3.9	1.3	2.3	
3.9	1.9	1.3	2.8	2.3		2.3	3.9	2.8	1.9	1.3	
3.9	1.9	1.3	2.3	2.8		2.3	3.9	2.8	1.3	1.9	
3.9	1.3	1.9	2.8	2.3		2.3	2.8	3.9	1.9	1.3	
3.9	1.3	1.9	2.3	2.8		2.3	2.8	3.9	1.3	1.9	
1.9	3.9	1.3	2.8	2.3		3.9	2.8	2.3	1.9	1.3	
1.9	3.9	1.3	2.3	2.8	−0.18	3.9	2.8	2.3	1.3	1.9	1.40
1.9	1.3	3.9	2.8	2.3		3.9	2.3	2.8	1.9	1.3	
1.9	1.3	3.9	2.3	2.8		3.9	2.3	2.8	1.3	1.9	
1.3	3.9	1.9	2.8	2.3		2.8	3.9	2.3	1.9	1.3	
1.3	3.9	1.9	2.3	2.8		2.8	3.9	2.3	1.3	1.9	
1.3	1.9	3.9	2.8	2.3		2.8	2.3	3.9	1.9	1.3	
1.3	1.9	3.9	2.3	2.8		2.8	2.3	3.9	1.3	1.9	

8.10 *MONTE CARLO TECHNIQUES

8.10.1 Evaluation of Statistical Significance

To compute statistical significance, we need to compare the observed values to something else. In the case of symmetry about the origin, we have seen it is possible to compare the observed value to the distribution where the plus and minus signs are independent with probability $1/2$. In cases where we do not know a prior appropriate comparison distribution, as in a drug trial, the distribution without the drug is found by either using the same individuals in a crossover trial or forming a control group by a separate sample of people who are not treated with the drug. There are cases where one can conceptually write down the probability structure that would generate the distribution under the null hypothesis, but in practice could not calculate the distribution. One example of this would be the permutation test. As we mentioned previously, if there are twenty different values in the sample, there are more than 2×10^{18} different permutations. To generate them all would not be feasible even with modern electronic computers. However, one could evaluate the particular value of the test statistic by generating a second sample from the null distribution with all permutations being equally likely. If there were some way to randomly generate permutations and compute the value of the statistic, one could take the observed statistic (thinking of this as a sample of size one) and compare it to the randomly generated value under the null hypothesis, the second sample. One would then order the observed and generated values of the statistic and decide which values are more extreme; this would lead to a rejection region for the null hypothesis. From this, a p-value could be computed. These abstract ideas are illustrated by the following examples.

Example 8.9: The Permutation Test. As mentioned above, for fixed observed values, the two-sample t-test is a monotone function of the value of $\bar{x} - \bar{y}$, the difference in the means of the two samples. Suppose that we have the observed $\bar{x} - \bar{y}$. One might then generate random permutations and compute the values of $\bar{x} - \bar{y}$. Suppose we generate n such values. For a two-sided test, let us order the *absolute* values of the statistic including both our random sample under the null hypothesis and the actual observation, giving us $n + 1$ values. Suppose that the actual observed value of the statistic from the data is the kth-order statistic, where we have ordered the absolute values from the smallest to the largest. Larger values tend to give more evidence against the null hypothesis of equal means. Suppose we would reject for all observations as large as the kth-order statistic or larger. This corresponds to a p-value of $(n + 2 - k)/(n + 1)$. □

One problem that we have not discussed yet is the method for generating the random permutation and $\bar{x} - \bar{y}$ values. This is usually done by computer. The computer generates random permutations by using what are called *random number generators* (see Note 8.9). A study using the generation of random quantities by computer is called a *Monte Carlo study*, for the gambling establishment at Monte Carlo with its random gambling devices and games. Note that by using Monte Carlo permutations, we can avoid the need to generate all possible permutations! This makes permutation tests feasible for large numbers of observations.

Another type of example comes about when one does not know how to theoretically compute the distribution under the null hypothesis.

Example 8.10. This example will not give all the data, but will describe how a Monte Carlo test was used. In the Coronary Artery Surgery Study (CASS, Alderman *et al.* [1982]), a study was made of the reasons people were treated by coronary bypass surgery or medical therapy. Among fifteen different institutions, it was found that many characteristics affected the assignments of patients to surgical therapy. A multivariate statistical analysis of a type described later in this book (linear discriminant analysis) was used to identify factors related to choice of therapy and to estimate the probability someone would have surgery or not. It was clear that the sites differed in the percentage of people assigned to surgery, but it was also clear that the clinical sites had patient populations with different characteristics. Thus, one could not immediately conclude the clinics had different philosophies of assignment to therapy merely by running a χ^2 test. Conceivably, the differences between clinics could be accounted for by the different characteristics of the patient populations. Using the estimated probability that each patient would have surgery or not, the total number of surgical cases were distributed among the clinics using a Monte Carlo technique. The corresponding χ^2 test for the observed and expected values was computed for each of these randomly generated assignments under the null hypothesis of no clinical difference. This was done 1000 times. The actual observed value for the statistic turned out to be larger than any of the 1000 simulations. Thus, the estimated *p*-value for the significance of the conjecture that the clinics had different methods of assigning people to therapy was less than $1/1001$. It was thus concluded that the clinics had different philosophies by which they assigned people to medical or surgical therapy. We now turn to some other possible uses of the Monte Carlo technique.

8.10.2 Empirical Evaluation of the Behavior of Statistics: Modeling and Evaluation

Monte Carlo generation on computer is also useful for studying the behavior of statistics. For example, we know that the χ^2 test for contingency tables, as discussed in Chapter 7, has approximately a χ^2-distribution for large samples. But is the distribution approximately χ^2 for smaller samples? In other words, is the statistic fairly robust with respect to sample size? What happens when there are small numbers of observations in the cells? One way to evaluate small-sample behavior is a Monte Carlo study (also called a simulation study). One can generate multinomial samples with the two traits independent, compute the χ^2-statistic, and observe, for example, how often one would reject at the 5% significance level. The Monte Carlo simulation would allow evaluation of how large the sample needs to be for the asymptotic χ^2 critical value to be useful.

Another use of the Monte Carlo method is to model very complex situations. For example, you might need to design a hospital communications network with many independent inputs. If you knew roughly the distribution of calls from the possible inputs, you could simulate by Monte Carlo techniques the activity of a proposed network if it were built. In this manner, you could see whether or not the network was often overloaded. As another example, you could model the hospital system of an area under the assumption of new hospitals being added and various assumptions about the case load. You could also model what might happen in catastrophic circumstances (*provided* realistic assumptions could be made). In general, the modeling and simulation approach gives one method of evaluating how changes in an environment might affect other factors without going through the expensive and potentially catastrophic exercise of actually building whatever is to be simulated. Of course, such modeling depends *heavily* upon the skill of the individuals constructing the model, the realism of the assumptions they make, and whether or not the probabilistic assumptions used correspond approximately to the real-life situation.

A starting reference for learning about Monte Carlo ideas is a small booklet by Hoffman [1979]. Advanced mathematical references are Hammersley and Handscomb [1964] and Shreider [1966]. A text on the generation of pseudorandom numbers is Newman and Odell [1971]. For a recent text containing many applications and references, see Manly [1991]. A more theoretical text is Edgington [1987].

The idea of splitting a sample to estimate the effect of a model in an unbiased manner will be discussed in Chapter 11 and 13 among others. Systematically omitting part of a sample, estimating values and testing on the omitted part is used; if one does this, say for all subsets of a certain size, a *jackknife* procedure is being used (see Efron [1982]). For statistical procedures that are somewhat intractable from an analytical point of view, construction of confidence intervals and tests may be performed by taking new Monte Carlo samples from the observed values (!) and performing the procedure many times; for large samples in a variety of situations this procedure, called the *bootstrap*, may be used. A good introduction to bootstrap methods is Efron and Tibshirani [1986].

8.11 *ROBUST TECHNIQUES

Robust techniques cover more than the field of nonparametric and distribution-free statistics. In general, distribution-free statistics give robust techniques, but it is possible to make more classical methods robust against certain violations of assumptions.

We illustrate with three approaches to making the sample mean robust. Another approach discussed earlier, which we shall not discuss again here, is to use the sample median as a measure of location. The three approaches are modifications of the traditional mean statistic $\bar{x}$. Of concern in computing the sample mean is the effect that an "outlier" will have. An observation far away from the main data set can have an enormous effect upon the sample mean. One would like to eliminate or lessen the effect of such outlying and possibly spurious observations.

An approach that has been suggested is the α-trimmed mean. With the α-trimmed mean, we take some of the largest and smallest observations and drop them from each end. We then compute the usual sample mean on the remaining data.

Definition 8.10. *The α-trimmed mean* of n observations is computed as follows: Let k be the smallest integer greater than or equal to αn. Let $X_{(i)}$ be the order statistics of the sample. The α-trimmed mean drops approximately a proportion α of the observations from both ends up the distribution. That is,

$$\alpha\text{-trimmed mean} = \frac{1}{n - 2k} \sum_{i=k+1}^{n-k} X_{(i)}.$$

We move on to the two other ways of modifying the mean, and then illustrate all three with a data set.

The second method of modifying the mean is called Winsorization. The α-trimmed mean drops the largest and smallest observations from the samples. In the Winsorized mean, such observations are included, but the large effect is reduced. The approach is to shrink the smallest and largest observations to the next remaining observations, and count them as if they had those values. This will become clearer with the example below.

Definition 8.11. *The α-Winsorized mean* is computed as follows. Let k be the smallest integer greater than or equal to αn. The α-Winsorized mean is

$$\alpha\text{-Winsorized mean} = \frac{1}{n}\left((k+1)\left(X_{(k+1)} + X_{(n-k)}\right) + \sum_{i=k+2}^{n-k-1} X_{(i)}\right).$$

The third method is to weight observations differentially. In general, we would want to weight the observations at the ends or tails less, and those in the middle more. Thus, we will base the weights on the order statistics where the weights for the first few order statistics and the last few order statistics are typically small. In particular, we define the weighted mean to be

$$\text{Weighted mean} = \frac{\displaystyle\sum_{i=1}^{n} W_i X_{(i)}}{\displaystyle\sum_{i=1}^{n} W_i}, \quad \text{where } W_i \geq 0.$$

Problem 8.26 shows that the α-trimmed mean and the α-Winsorized mean are examples of weighted means with appropriately-chosen weights.

Example 8.11. We compute the mean, median, 0.1-trimmed mean, and 0.1-Winsorized mean for the female treadmill data of Example 8.16.

$$\text{Mean} = \bar{x} = \frac{30 + \cdots + 936}{20} = 708,$$

$$\text{Median} = \frac{X_{(10)} + X_{(11)}}{2} = 750.$$

Now, $0.1 \times 20 = 2$, so $k = 2$.

$$\alpha\text{-trimmed mean} = \frac{570 + \cdots + 900}{16} = 734.625,$$

$$\alpha\text{-Winsorized mean} = \frac{1}{20}\left(3(570 + 900) + 618 + \cdots + 840\right) = 734.7. \qquad \square$$

Note that the median and both robust mean estimates are considerably higher than the sample mean $\bar{x}$. This is because of the small outlier of 30.

Robust techniques apply in a much more general context than shown here, and indeed are more useful in other situations. In particular, for regression and multiple regression (subjects of subsequent chapters in this book), a large amount of statistical theory has been developed for making the procedures more robust (Huber [1981]).

8.12   *FURTHER READING AND DIRECTIONS

There are several books dealing with nonparametric statistics. Among these are
Lehmann [1975] and Kraft and van Eeden [1968]. Other books deal exclusively with
nonparametric statistical techniques. Three which are accessible in mathematical level
for the readers of this text are Marascuilo and McSweeney [1977], Bradley [1968], and
Siegel [1956].

A book that gives more of a feeling for the mathematics involved at a level above
this test, but which does not require calculus is Hajek [1969]. Another very comprehensive text which outlines much of the theory of statistical tests, but is on a
somewhat more advanced mathematical level, is Hollander and Wolfe [1973]. Finally,
a comprehensive text on robust methods, written at a very advanced mathematical
level, is Huber [1981].

Other sections of this text will give nonparametric and robust techniques in more
general settings. They may be identified by one of the words "nonparametric," "distribution-free," or "robust" in the title of the section.

NOTES

Note 8.1: Definitions of Nonparametric and Distribution-free

The definitions given in this chapter are close to those of Huber [1981]. Bradley
[1968] states that, "roughly speaking, a nonparametric test is a test which makes no
hypothesis about the value of a parameter in a statistical density function, whereas a
distribution-free test is one which makes no assumptions about the precise form of
the sampled population."

Note 8.2: Relative Efficiency

The statements about relative efficiency in this chapter refer to asymptotic relative efficiency (Bradley [1968], Hollander and Wolfe [1973], and Marascuilo and McSweeney
[1977]). For two possible *estimates*, the asymptotic relative efficiency of A to B is the
limit of the ratio of the variance of B to the variance of A as the sample size increases.
For two possible *tests*, first select a sequence of alternatives such that as n becomes
large, the power (probability of rejecting the null hypothesis) for test A converges
to a fixed number greater than zero and less than one. Let this number be C. For
each member of the sequence, find sample sizes n_A and n_B such that both tests have
(almost) power C. The limit of the ratio n_B to n_A is the asymptotic relative efficiency.

Since the definition is for large sample sizes (asymptotic), for smaller sample sizes,
the efficiency may be more or less than the figures we have given. Both Bradley [1968]
and Hollander and Wolfe [1973] have considerable information on the topic.

Note 8.3: Crossover Designs for Drugs

These are subject to a variety of subtle differences. There may be carryover effects
from the drugs. Changes over time—for example, extreme weather changes—may
make the second part of the crossover design different than the first. Some drugs may

permanently change the subjects in some way. Peterson and Fisher [1980] give many references germane to randomized clinical trials.

Note 8.4: The Signed Rank Test

The values of the ranks are known; for n observations, they are the integers $1 - n$. The only question is the sign of the observation associated with each rank. Under the null hypothesis, the sign is equally likely to be plus or minus. Further, knowing the rank of an observation based on the absolute values does not predict the sign, which is still equally likely to be plus or minus independently of the other observations. Thus, all 2^n patterns of plus and minus signs are equally likely. For $n = 2$, the four patterns are:

Ranks	1	2	1	2	1	2	1	2
Signs	−	−	+	−	−	+	+	+
S	0		1		2		3	

So, $P[S \leq 0] = 1/4$, $P[S \leq 1] = 1/2$, $P[S \leq 2] = 3/4$, and $P[S \leq 3] = 1$.

Note 8.5: The Kolmogorov–Smirnov Statistic is a Rank Statistic

We illustrate one technique used to show that the Kolmogorov–Smirnov statistic is a rank test. Looking at Figure 8.2, we could slide both curves along the X-axis without changing the value of the maximum difference, D. Since the curves are horizontal, we can stretch them along the axis (as long as the order of the jumps does not change) and not change the value of D. Place the first jump at 1, the second at 2, and so forth. We have placed the jumps then at the ranks! The height of the jumps depend on the sample size. Thus, we can compute D from the ranks (and knowing which group have the rank) and the sample sizes. Thus, D is nonparametric and distribution-free.

Note 8.6: One-sample Kolmogorov–Smirnov Tests and One-sided Kolmogorov–Smirnov Tests

It is possible to compare one sample to a hypothesized distribution. Let F be the empirical cumulative distribution function of a sample. Let H be a hypothesized distribution function. The statistic

$$\max_x |F(x) - H(x)|$$

is the one-sample statistic. If H is continuous, critical values are tabulated for this nonparametric test in the tables already referenced in this chapter.

The Kolmogorov–Smirnov two-sample statistic was based upon the largest difference between two empirical cumulative distribution functions. That is,

$$D = \max |F(x) - G(x)|,$$

where F and G are the two empirical cumulative distribution functions. Since the absolute value is involved, we are not differentiating between F being larger and G

being larger. If we had hypothesized as an alternative that the F population took on larger values in general, then F would tend to be less than G, and we could use

$$\max\big(G(x) - F(x)\big).$$

Such one-sided Kolmogorov–Smirnov statistics are used and tabulated. They also are nonparametric rank tests for use with one-sided alternatives.

Note 8.7: Runs Tests

Given observations from two groups, suppose we rank them and use the symbol † when the observation is from group one, and the symbol ‡ when the observation is from group two. If the group one observations were 0.5, 1.2, 2.7, 3.5, 3.6, 4.1, and 4.2, and the group two observations were 2.1, 2.5, 2.6, 3.1, 5.8, and 6.0, the sequence would be

$$† † ‡ ‡ † ‡ † † † † ‡ ‡.$$

We define a *run* to be a set of adjacent symbols of one type that cannot be made longer. In the example there are three †-runs and three ‡-runs, for a total of six runs. The *length of a run* is the number of symbols in the set. The †-runs have lengths 2, 1, and 4; the ‡-runs have lengths 3, 1, and 2. The longest run is the last †-run of length four.

Under the two-sample null hypothesis, if there are n †s and m ‡s, all $\binom{n+m}{m}$ arrangements are equally likely.

Two common test statistics based on runs are (1) the total number of runs, and (2) the length of the longest run. Under most reasonable alternatives, the total number of runs will tend to be too small, and the length of the longest run will tend to be too large.

Tables for the total number of runs test are available in Odeh *et al.* [1977], Owen [1962] and the CRC Handbook [1968]. For large n and m, an asymptotic standard normal statistic may be calculated using the fact that

$$E(\text{number of runs}) = \frac{2nm}{n+m} + 1,$$

$$\text{var(number of runs)} = \frac{2nm(2nm - n - m)}{(n+m)^2(n+m-1)}.$$

Bradley [1968] has considerable material and references about runs tests. For most purposes, runs tests are not very efficient; however, they are easy to compute, and have considerable intuitive appeal.

Note 8.8: More General Rank Tests

The theory of tests based upon ranks is well-developed (Hajek [1969], Hajek and Sidak [1967], and Huber [1981]). Consider the two-sample problem with groups of size n and m, respectively. Let R_i $(i = 1, 2, \ldots, n)$ be the ranks of the first sample. Statistics of the following form, with a a function of R_i, have been extensively studied.

$$S = \frac{1}{n} \sum_{i=1}^{n} a(R_i).$$

The $a(R_i)$ may be chosen to be efficient in particular situations. For example, let $a(R_i)$ be such that a standard normal variable has probability $R_i/(n+m+1)$ of being less than or equal to this value. Then, when the usual two-sample t-test normal assumptions hold the relative efficiency is one. That is, this rank test is as efficient as the t-test for large samples. This test is called the *normal scores test* or *van der Waerden* test.

Note 8.9: The Monte Carlo Technique and Pseudorandom Number Generators

The term Monte Carlo technique was introduced by the mathematician Stanislaw Ulam (Ulam [1976]) while working on the Manhattan atomic bomb project.

Computers do not actually generate random numbers; rather, the numbers are generated in a sequence by a specific computer algorithm. Thus, the numbers are called *pseudorandom numbers*. Although not random, the sequence of numbers need to appear random. Thus, they are tested by statistical tests. For example, a program to generate random integers from zero to nine may have a sequence of generated integers tested by the χ^2 goodness-of-fit test to see that the "probability" of each outcome is $1/10$. A generator of uniform numbers on the interval $(0,1)$ can have its empirical distribution compared to the uniform distribution by the one-sample Kolmogorov–Smirnov test (Note 8.6).

The subject of pseudorandom number generators is very deep both philosophically and mathematically.

Many popular games on personal computers and TV attachments use pseudorandom number generation.

PROBLEMS

1. The following data deals with the treatment of essential hypertension ("essential" is a technical term meaning that the cause is unknown; a synonym is "idiopathic"), and is from a paper by Vlachakis and Mendlowitz [1976]. Seventeen patients received treatments C, A, and B, where

   C = Control period

   A = Propranolol + phenoxybenzamine

   B = Propranolol + phenoxybenzamine + hydrochlorothiazide

   Each patient received C first, then either A or B and finally B or A. The data consist of the systolic blood pressure in the recumbent position.

Patient	C	A	B	Patient	C	A	B
1	185	148	132	10	180	132	136
2	160	128	120	11	176	140	135
3	190	144	118	12	200	165	144
4	192	158	115	13	188	140	115
5	218	152	148	14	200	140	126
6	200	135	134	15	178	135	140
7	210	150	128	16	180	130	130
8	225	165	140	17	150	122	132
9	190	155	138				

a. Take the differences between the systolic blood pressures on treatments A and C. Use the sign test to test for a treatment A effect (two-sided test, give the p-value).

b. Take the differences between treatments B and C. Use the sign test to test for a treatment B effect (one-sided test, give the p-value).

c. Take the differences between treatments B and A. Test for a treatment difference using the sign test (two-sided test, give the p-value).

2. Several population studies have demonstrated an inverse correlation of Sudden Infant Death Syndrome (SIDS) rate with birthweight. The occurrence of SIDS in one of a pair of twins provides an opportunity to test the hypothesis that birthweight is a major determinant of SIDS. The following set of data collected by Dr. D. R. Peterson of the Department of Epidemiology, University of Washington, consists of the birthweights of each of 22 dizygous twins, and each of 19 monozygous twins.

Dizygous Twins		Monozygous Twins	
SIDS	**Non-SIDS**	**SIDS**	**Non-SIDS**
1474	2098	1701	1956
3657	3119	2580	2438
3005	3515	2750	2807
2041	2126	1956	1843
2325	2211	1871	2041
2296	2750	2296	2183
3430	3402	2268	2495
3515	3232	2070	1673
1956	1701	1786	1843
2098	2410	3175	3572
3204	2892	2495	2778
2381	2608	1956	1588
2892	2693	2296	2183
2920	3232	3232	2778
3005	3005	1446	2268
2268	2325	1559	1304

Dizygous Twins (*continued*)		Monozygous Twins (*continued*)	
SIDS	**Non-SIDS**	**SIDS**	**Non-SIDS**
3260	3686	2835	2892
3260	2778	2495	2353
2155	2552	1559	2466
2835	2693		
2466	1899		
3232	3714		

a. For the dizygous twins test the alternative hypothesis that the SIDS child of each pair has the lower birth weight by taking differences and using the sign test. Find the one-sided p-value.

b. As in (a), but do the test for the monozygous twins.

c. As in (a), but do the test for the combined data set.

3. The following data are from Dobson *et al.* [1976]. Thirty-six patients with a confirmed diagnosis of phenylketonuria (PKU) were identified and placed on dietary therapy before reaching 121 days of age. The children were tested for IQ (Stanford–Binet) between the ages of four and six; subsequently, their normal siblings of closest age were also tested with the Stanford–Binet. The following fifteen pairs are the first fifteen listed in the paper.

Pair	1	2	3	4	5	6	7	8
IQ of PKU case	89	98	116	67	128	81	96	116
IQ of sibling	77	110	94	91	122	94	121	114

Pair	9	10	11	12	13	14	15
IQ of PKU case	110	90	76	71	100	108	74
IQ of sibling	88	91	99	93	104	102	82

The null hypothesis is that the PKU children, on the average, have the same IQ as their siblings. Using the sign test, find the two-sided p-value for testing against the alternative hypothesis that the IQ levels differ.

4. Repeat Problem 8.1 using the signed rank test rather than the sign test. Test at the 0.05 significance level.

5. Repeat Problem 8.2, Parts (a) and (b), using the signed rank test rather than the sign test. Test at the 0.05 significance level.

6. Repeat Problem 8.3, using the signed rank test rather than the sign test. Test at the 0.05 significance level.

7. Bednarek and Roloff [1976] deal with the treatment of apnea (a transient cessation of breathing) in premature infants using a drug called aminophylline. The variable of interest was "average number of apneic episodes per hour," and was measured before and after treatment with the drug. An episode was defined as the absence of spontaneous breathing for more than 20 seconds, or less if associated with bradycardia or cyanosis. The following table details the response of thirteen patients to aminophylline treatment at 16 h compared with 24 h before treatment (in apneic episodes per hour):

Patient	24 h Before	16 h After	Before – After (difference)
1	1.71	0.13	1.58
2	1.25	0.88	0.37
3	2.13	1.38	0.75
4	1.29	0.13	1.16
5	1.58	0.25	1.33
6	4.00	2.63	1.37
7	1.42	1.38	0.04
8	1.08	0.50	0.58
9	1.83	1.25	0.58
10	0.67	0.75	−0.08
11	1.13	0.00	1.13
12	2.71	2.38	0.33
13	1.96	1.13	0.83

a. Use the sign test to examine a treatment effect (give the two-sided p-value).

b. Use the signed rank test to examine a treatment effect (two-sided test at 0.05 significance level).

8. The following data from Schechter *et al.* [1973] deals with sodium chloride preference as related to hypertension. Two groups, twelve normal and ten hypertensive subjects, were isolated for a week and compared with respect to Na^+ intake. The following are the average daily Na^+ intakes:

Normal	10.2	2.2	0.0	2.6	0.0	43.1
Hypertensive	92.8	54.8	51.6	61.7	250.8	84.5

Normal	45.8	63.6	1.8	0.0	3.7	0.0
Hypertensive	34.7	62.2	11.0	39.1		

Compare the average daily Na^+ intake of the hypertensive subjects with that of the normal volunteers by means of the Wilcoxon two-sample test at the 5% significance level.

9. During July and August, 1976, a large number of Legionnaires attending a convention died of mysterious and unknown cause. Epidemiologists have talked of "an outbreak of Legionnaires' disease." Chen *et al.* [1977] examined the hypothesis of nickel contamination as a toxin. They examined the nickel levels in the lungs of nine cases and nine controls. The authors point out that contamination at autopsy is a possibility. The data are as follows (μg/100 g dry weight):

Legionnaire cases	65	24	52	86	120	82	399	87	139
Control cases	12	10	31	6	5	5	29	9	12

Note that there was no attempt to match cases and controls. Use the Wilcoxon test at the one-sided 5% level to test the null hypothesis that the numbers are samples from similar populations.

10. A paper by Robertson *et al.* [1976] discusses the level of plasma prostaglandin E (iPGE in pg/ml) in patients with cancer and without hypercalcemia. The data are given below:

PATIENTS WITH HYPERCALCEMIA

Patient Number	Mean Plasma iPGE (pg/ml)	Mean Serum Calcium (ml/dl)
1	500	13.3
2	500	11.2
3	301	13.4
4	272	11.5
5	226	11.4
6	183	11.6
7	183	11.7
8	177	12.1
9	136	12.5
10	118	12.2
11	60	18.0

PATIENTS WITHOUT HYPERCALCEMIA

Patient Number	Mean Plasma iPGE (pg/ml)	Mean Serum Calcium (ml/dl)
12	254	10.1
13	172	9.4
14	168	9.3
15	150	8.6
16	148	10.5
17	144	10.3
18	130	10.5
19	121	10.2
20	100	9.7
21	88	9.2

Note that the variables are "Mean Plasma iPGE" and "Mean Serum Ca" levels; presumably more than one assay was carried out for each patient's level. The number of such tests for each patient is not indicated, nor the criterion for the number. Using the Wilcoxon two-sample test, test for differences between the two groups in:

a. Mean plasma iPGE.

b. Mean serum Ca.

11. Sherwin and Layfield [1976] present data about protein leakage in the lungs of male mice exposed to 0.5 parts per million of nitrogen dioxide (NO_2). Serum fluorescence data were obtained by sacrificing animals at various intervals. Use the two-sided Wilcoxon test, 0.05 significance level, to look for differences between controls and exposed mice.

   a. At 10 days:

Controls	143	169	95	111	132	150	141
Exposed	152	83	91	86	150	108	78

   b. At 14 days:

Controls	76	40	119	72	163	78
Exposed	119	104	125	147	200	173

12. Using the data of Problem 8.8:

   a. Find the value of the Kolmogorov–Smirnov statistic.

   b. Plot the two empirical distribution functions.

   c. Do the curves differ at the 5% significance level? For sample sizes ten and twelve, the 10%, 5%, and 1% critical values for mnD are 60, 66, and 80, respectively.

13. Using the data of Problem 8.9:

   a. Find the value of the Kolmogorov–Smirnov statistic.

   b. Do you reject the null hypothesis at the 5% level? For $m = 9$ and $n = 9$, the 10%, 5%, and 1% critical values of mnD are 54, 54, and 63, respectively.

14. Using the data of Problem 8.10:

   a. Find the value of the Kolmogorov–Smirnov statistic.

   b. What can you say about the p-value? For $m = 10$ and $n = 11$, the 10%, 5%, and 1% critical values of mnD are 57, 60, and 77, respectively.

15. Using the data of Problem 8.11:

   a. Find the value of the Kolmogorov–Smirnov statistic.

   b. Do you reject at 10%, 5%, and 1%, respectively? Do this for Parts (a) and (b) of Problem 8.11. For $m = 7$ and $n = 7$, the 10%, 5%, and 1% critical values of mnD are 35, 42, and 42, respectively. The corresponding critical values for $m = 6$ and $n = 6$ are 30, 30, and 36.

16. Test at the 0.05 significance level for a significant improvement with the cream treatment of Example 8.2.

   a. Use the sign test.

   b. Use the signed rank test.

c. Use the t-test.

17. Use the expression of colostrum data of Example 8.2, and test at the 0.10 significance level the null hypothesis of no treatment effect.

   a. Use the sign test.

   b. Use the signed rank test.

   c. Use the usual t-test.

18. For each of (a), (b), and (c) below, test the null hypothesis of no treatment difference from Example 8.2 using

   a. The Wilcoxon two-sample test.

   b. The Kolmogorov–Smirnov two-sample test. For $m = n = 19$, the 20%, 10%, 5%, 1%, and 0.1% critical values for mnD are 133, 152, 171, 190, and 228, respectively.

   c. The two-sample t-test.

   Compare the two-sided p-values to the extent possible. Using the data of Example 8.2, examine treatments:

   d. Nipple-rolling *vs.* masse cream.

   e. Nipple-rolling *vs.* expression of colostrum.

   f. Masse cream *vs.* expression of colostrum.

19. As discussed in Chapter 3, Winkelstein *et al.* [1975] studied systolic blood pressures of three groups of Japanese men: native Japanese, first-generation immigrants to the United States (Issei) and second-generation Japanese in the United States (Nisei).

Blood Pressure (mmHg)	Native Japanese	Issei	Nisei
< 106	218	4	23
106–114	272	23	132
116–124	337	49	290
126–134	362	33	347
136–144	302	41	346
146–154	261	38	202
156–164	166	23	109
> 166	314	52	112

Use the asymptotic Wilcoxon two-sample statistic to test:

   a. Native Japanese *vs.* California Issei.

   b. Native Japanese *vs.* California Nisei.

   c. California Issei *vs.* California Nisei.

*20. An outlier is an observation far from the rest of the data. This may represent valid data or a mistake in experimentation, data collection, or data entry. At any rate, a few outlying observations may have an extremely large effect. Consider a one-sample t-test of mean zero based upon ten observations with

$$\bar{x} = 10 \quad \text{and} \quad s^2 = 1.$$

Suppose now that one observation of value x is added to the sample.

a. Show that the value of the new sample mean, variance, and t-statistic are

$$\bar{x} = \frac{100 + x}{11},$$

$$s^2 = \frac{10x^2 - 200x + 1099}{11 \times 10},$$

$$t = \frac{100 + x}{\sqrt{x^2 - 20x + 109.9}}.$$

b. Graph t as a function of x.

c. For which values of x would one reject the null hypothesis of mean zero? What does the effect of an outlier (large absolute value) do in this case?

d. Would you reject the null hypothesis without the outlier?

*21. Using the ideas of Note 8.4 about the signed rank test, verify the following values from Owen [1962], by permission of © Addison-Wesley Publishing Company, when $n = 4$.

s	0	1	2	3	4	5
$P[S \leq s]$	0.062	0.125	0.188	0.312	0.438	0.562

s	6	7	8	9	10
$P[S \leq s]$	0.688	0.812	0.875	0.938	1.000

*22. The Wilcoxon two-sample test depends upon the fact that, under the null hypothesis, if two samples are drawn without ties, all $\binom{n+m}{n}$ arrangements of the n ranks from the first sample, and the m ranks from the second sample, are equally likely. That is, if $n = 1$ and $m = 2$, the three arrangements

$$\mathbf{1}\ 2\ 3; \quad W = 1$$
$$1\ \mathbf{2}\ 3; \quad W = 2$$
$$1\ 2\ \mathbf{3}; \quad W = 3$$

are equally likely. Here, the rank from population one is in bold type.

a. If $n = 2$ and $m = 4$ graph the distribution function of the Wilcoxon two-sample statistic when the null hypothesis holds.

b. Find $E(W)$. Does it agree with Equation (8.5)?

c. Find var(W). Does it agree with Equation (8.6)?

*23. (Runs test; Note 8.7.) For each of the following two-sample data sets:

a. Compute the number of runs.

b. Test statistical significance at the 5% and 1% level.

c. Compute the expected value and variance under the null hypothesis.

    d. Compute the asymptotic Z-statistic and give its two-sided p-value.

    Use the data set of:

    d. Example 8.3. The 5% and 1% critical values for sample sizes 8 and 10 are 6 and 4, respectively.

    e. Problem 8.9 (mean plasma iPGE). The 5% and 1% critical values for sample sizes 9 and 9 are 6 and 4, respectively.

    f. Problem 8.11(b). The 5% and 1% critical values for sample sizes 6 and 6 are 3 and 2, respectively.

*24. (Permutation two-sample t-test.) To use the permutation two-sample t-test, the text (in Section 8.9) used the fact that for $n + m$ fixed values, the t-test was a monotone function of $\bar{x} - \bar{y}$. To show this, prove the following equality:

$$t = \cfrac{1}{\sqrt{\cfrac{(n+m)\left(\sum_i x_i^2 + \sum_i y_i^2\right) - \left(\sum_i x_i + \sum_i y_i\right)^2 - nm(\bar{x} - \bar{y})^2}{nm(n+m-2)(\bar{x} - \bar{y})^2}}}.$$

Note that the first two terms in the numerator of the square root are constant for all permutations, so t is a function of $\bar{x} - \bar{y}$.

*25. (One-sample randomization t-test.) For the randomization one-sample t-test, the paired x_i, y_i values give $\bar{x} - \bar{y}$ values. Assume that the $|x_i - y_i|$ are known, but the signs are random, independently $+$ or $-$ with probability $1/2$. The 2^n ($i = 1, 2, \ldots, n$) patterns of pluses and minuses are equally likely.

    a. Show that the one-sample t-statistic is a monotone function of $\bar{x} - \bar{y}$ when the $|x_i - y_i|$ are known. Do this by showing

$$t = \cfrac{\bar{x} - \bar{y}}{\sqrt{\cfrac{-n(\bar{x} - \bar{y})^2 + \sum_i (x_i - y_i)^2}{n(n-1)}}}.$$

    b. For the data

i	X_i	Y_i
1	1	2
2	3	1
3	1	5

compute the eight possible randomization values of t. What is the two-sided randomization p-value for the observed t?

*26. (Section 8.11 [Robust Estimation of the Mean].) Show that the α-trimmed mean and the α-Winsorized mean are weighted means by explicitly showing the weights W_i that are given the two means.

*27. (Section 8.11 [Robust Estimation of the Mean].)

    a. For the combined data for SIDS in Problem 8.2, compute

        i. The 0.05 trimmed mean.

        ii. The 0.05 Winsorized mean.

        iii. The weighted mean with weights $W_i = i(n + 1 - i)$, where n is the number of observations.

    b. The same as in Problem 8.27(a), but do this for the non-SIDS twins.

REFERENCES

Alderman, E., Fisher, L. D., Maynard, C., Mock, M. B., Ringgvist, I., Bourassa, M. G., Kaiser, G. C., and Gillespie, M. J. [1982]. Determinants of coronary surgery in a consecutive patient series from geographically dispersed medical centers; the Coronary Artery Surgery Study. *Circulation*, **66:** 562–568.

Bednarek, E., and Roloff, D. W. [1976]. Treatment of apnea of prematurity with aminophylline. *Pediatrics*, **58:** 335–339.

Beyer W. H. (ed.) [1968]. *CRC Handbook of Tables for Probability and Statistics.* CRC Press, Boca Raton, FL.

Bradley, J. V. [1968]. *Distribution-free Statistical Tests.* Prentice Hall, Englewood Cliffs, NJ.

Brown, M. S., and Hurlock, J. T. [1975]. Preparation of the breast for breastfeeding. *Nursing Research*, **24:** 448–451.

CASS [1981] (principal investigators of CASS and their associates; T. Killip (ed.), L. D. Fisher and M. Mock (associate eds.). National Heart, Lung and Blood Institute Coronary Artery Surgery Study. *Circulation*, **63:** part II, I–1 to I–81. Used with permission from the American Heart Association.

Chen, J. R., Francisco, R. B., and Miller, T. E. [1977]. Legionnaires' disease: nickel levels. *Science*, **196:** 906–908.

Dobson, J. C., Kushida, E., Williamson, M., and Friedman, E. [1976]. Intellectual performance of 36 phenylketonuria patients and their nonaffected siblings. *Pediatrics*, **58:** 53–58.

Edgington, E. S. [1987] *Randomization Tests.* Marcel Dekker, New York.

Efron, B. [1982]. *The Jackknife, Bootstrap and Other Resampling Plans.* Society for Industrial and Applied Mathematics, Philedelphia, PA.

Efron, B. and Tibshirani, R. [1986]. The Bootstrap (with Discussion). *Statistical Science*, **1:** 54–77.

Hajek, J. [1969]. *A Course in Nonparametric Statistics.* Holden-Day, San Francisco, CA.

Hajek, J. and Sidak, Z. [1967]. *Theory of Rank Tests.* Academic Press, New York.

Hammersley, J. M. and Handscomb, D. C. [1964]. *Monte Carlo Methods.* Wiley, New York; Methuen, London.

Hoffman, D. T. [1979]. *Monte Carlo: The Use of Random Digits to Simulate Experiments.* Models and Monographs in Undergraduate Mathematics and its Applications, Unit 269. EDC/UMAP, Newton, MA.

Hollander, M. and Wolfe, D. A. [1973]. *Nonparametric Statistical Methods.* Wiley, New York.

Huber, P. J. [1981]. *Robust Statistics.* Wiley, New York.

Kraft, C. H. and van Eeden, C. [1968]. *A Nonparametric Introduction to Statistics.* Macmillan, New York.

Lehmann, E. L. [1975]. *Nonparametrics: Statistical Methods Based on Ranks.* Holden-Day, San Francisco, CA.

Manly, B. F. [1991] *Randomization and Monte Carlo Methods in Biology.* Chapman and Hall, London.

Marascuilo, L. A. and McSweeney, M. [1977]. *Nonparametric and Distribution-free Methods for the Social Sciences.* Brooks/Cole, Scituate, MA.

Newman, T. G. and Odell, P. L. [1971]. *The Generation of Random Variates.* Number 29 of Griffin's Statistical Monographs and Courses. Hafner, New York.

Odeh, R. E., Owen, D. B., Birnbaum, Z. W., and Fisher, L. D. [1977]. *Pocketbook of Statistical Tables.* Marcel Dekker, New York.

Owen, D. B. [1962]. *Handbook of Statistical Tables*, Addison-Wesley, Reading, MA.

Peterson, A. P. and Fisher, L. D. [1980]. Teaching the principles of clinical trials design. *Biometrics*, **36:** 687–697.

Robertson, R. P., Baylink, D. J., Metz, S. A. and Cummings, K. B. [1976]. Plasma prostaglandin in patients with cancer with and without hypercalcemia. *Journal of Clinical Endocrinology and Metabolism*, **43:** 1330–1335.

Schechter, P. J., Horwitz, D., and Henkin, R. I. [1973]. Sodium chloride preference in essential hypertension. *Journal of the American Medical Association*, **225:** 1311–1315.

Sherwin, R. P. and Layfield, L. J. [1976]. Protein leakage in the lungs of mice exposed to 0.5 ppm nitrogen dioxide; a fluorescence assay for protein. *Archives of Environmental Health*, **31:** 116–118.

Shreider, Yu A. (ed.) [1966]. *The Monte Carlo Method.* Pergamon Press, Oxford, UK.

Siegel, S. [1988]. *Nonparametric Statistics for the Behavioral Sciences.* 2nd edition. McGraw-Hill, New York.

Tukey, J. W. [1960]. A survey of sampling from contaminated distributions. In *Contributions to Probability and Statistics; Essays in Honor of Harold Hotelling.* Olkin, I. *et al.* (eds.), Stanford University Press, Stanford, CA, Chapter 39.

Ulam, S. M. [1976]. *Adventures of a Mathematician.* Charles Scribner, New York.

Vlachakis, N. D. and Mendlowitz, M. [1976]. Alpha- and beta-adrenergic receptor blocking agents combined with a diuretic in the treatment of essential hypertension. *Journal of Clinical Pharmacology*, **16:** 352–360.

Winkelstein, Jr., W., Kazan, A., Kato, H., and Sachs, S. T. [1975]. Epidemiologic studies of coronary heart disease and stroke in Japanese men living in Japan, Hawaii, and California: blood pressure distributions. *American Journal of Epidemiology*, **102:** 502–513.

CHAPTER 9

Association and Prediction: Linear Models with One Predictor Variable

9.1 INTRODUCTION

Motivation for the methods of this chapter is aided by the use of examples. For this reason, we first consider three data sets. These data are used to motivate the methods to follow. The data are also used to illustrate the methods used in Chapter 11. After the three examples are presented, we return to this introduction.

Example 9.1. Table 9.1 and Figure 9.1 contain data on mortality due to malignant melanoma of the skin of white males during the period 1950–1969, for each state in the United States as well as the District of Columbia. No mortality data are available for Alaska and Hawaii for this period. It is well known that the incidence of melanoma can be related to the amount of sunshine and, somewhat equivalently, the latitude of the area. The table contains the latitude as well as the longitude for each of the states. These numbers were simply obtained by estimating the center of the state and reading off the latitude as given in a standard atlas. Finally, the 1965 population and contiguity to an ocean are noted, where "1" indicates contiguity: the state borders one of the oceans.

In the next section, we shall be particularly interested in the relationship between the melanoma mortality and the latitude of the states. These data are presented in Figure 1.

Definition 9.1. When two variables are collected for each data point, a plot is very useful. Such plots of the two values for each of the data points are called *scatter diagrams*, or *scattergrams*.

Note several things about the scattergram of malignant melanoma rates *vs.* latitude. There appears to be a rough relationship. As the latitude increases, the melanoma rate decreases. Nevertheless, there is no one-to-one relationship between the values. There is considerable scatter in the picture. One problem is to decide whether or not the scatter could be due to chance or whether there is some relationship. It might

Table 9.1. Mortality rate (per 10 million [10^7]) of White Males due to Malignant Melanoma of the Skin for the Period 1950–1959 by State and Some Related Variables. Mortality Data from: US Cancer Mortality by County: 1950–1959.

State	Mortality per 10,000,000	Latitude (degrees)	Longitude (degrees)	Population (million, 1965)	Ocean State[a]
Alabama	219	33.0	87.0	3.46	1
Arizona	160	34.5	112.0	1.61	0
Arkansas	170	35.0	92.5	1.96	0
California	182	37.5	119.5	18.60	1
Colorado	149	39.0	105.5	1.97	0
Connecticut	159	41.8	72.8	2.83	1
Delaware	200	39.0	75.5	0.50	1
Washington, DC	177	39.0	77.0	0.76	0
Florida	197	28.0	82.0	5.80	1
Georgia	214	33.0	83.5	4.36	1
Idaho	116	44.5	114.0	0.69	0
Illinois	124	40.0	89.5	10.64	0
Indiana	128	40.2	86.2	4.88	0
Iowa	128	42.2	93.8	2.76	0
Kansas	166	38.5	98.5	2.23	0
Kentucky	147	37.8	85.0	3.18	0
Louisiana	190	31.2	91.8	3.53	1
Maine	117	45.2	69.0	0.99	1
Maryland	162	39.0	76.5	3.52	1
Massachusetts	143	42.2	71.8	5.35	1
Michigan	117	43.5	84.5	8.22	0
Minnesota	116	46.0	94.5	3.55	0
Mississippi	207	32.8	90.0	2.32	1
Missouri	131	38.5	92.0	4.50	0
Montana	109	47.0	110.5	0.71	0
Nebraska	122	41.5	99.5	1.48	0
Nevada	191	39.0	117.0	0.44	0
New Hampshire	129	43.8	71.5	0.67	1
New Jersey	159	40.2	74.5	6.77	1
New Mexico	141	35.0	106.0	1.03	0
New York	152	43.0	75.5	18.07	1
North Carolina	199	35.5	79.5	4.91	1
North Dakota	115	47.5	100.5	0.65	0
Ohio	131	40.2	82.8	10.24	0
Oklahoma	182	35.5	97.2	2.48	0
Oregon	136	44.0	120.5	1.90	1
Pennsylvania	132	40.8	77.8	11.52	0
Rhode Island	137	41.8	71.5	0.92	1
South Carolina	178	33.8	81.0	2.54	1
South Dakota	86	44.8	100.0	0.70	0
Tennessee	186	36.0	86.2	3.84	0
Texas	229	31.5	98.0	10.55	1
Utah	142	39.5	111.5	0.99	0
Vermont	153	44.0	72.5	0.40	1
Virginia	166	37.5	78.5	4.46	1
Washington	117	47.5	121.0	2.99	1
West Virginia	136	38.8	80.8	1.81	0
Wisconsin	110	44.5	90.2	4.14	0
Wyoming	134	43.0	107.5	0.34	0

[a]1 = state borders on ocean.

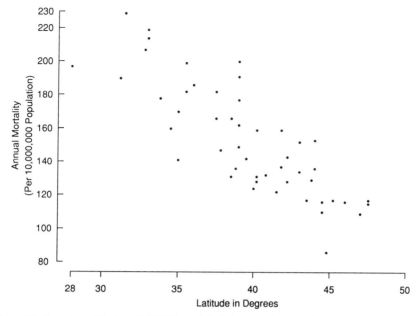

Figure 9.1. Annual mortality (per 10,000,000 population) due to malignant melanoma of the skin for white males by state and latitude of the center of the state for the period 1950–1959.

be of interest to estimate the melanoma rate for various latitudes. In this case, how would we estimate the relationship? In order to convey the relationship to others, it would also be useful to have some simple way of summarizing the relationship. There are two aspects of the relationship which might be summarized. One is how the melanoma rate changes with latitude; it would also be useful to summarize the variability of the scattergram. □

Example 9.2. To assess physical conditioning in normal individuals, it is useful to know how much energy they are capable of expending. Since the process of expending energy requires oxygen, one way to evaluate this is to look at the rate at which they use oxygen at peak physical activity. To examine the peak physical activity, tests have been designed where the individual runs on a treadmill. At specified time intervals, the speed at which the treadmill moves and the grade of the treadmill both increase. The individual is then systematically run to maximum physical capacity. The maximum capacity is determined by the individual; the person stops when unable to go further. Data from Bruce *et al.* [1973] are discussed.

The oxygen consumption was measured in the following way. The patient's nose was blocked off by a clip. Expired air was collected from a silicone rubber mouthpiece fitted with a very low resistance valve. The valve was connected by plastic tubes into a series of evacuated neoprene balloons. The inlet valve for each balloon was opened for sixty seconds to sample the expired air. Measurements were made of the volumes of expired air and the oxygen content was obtained using a paramagnetic analyzer capable of measuring the oxygen. From this, the rate at which oxygen was used in mm/min was calculated. Physical conditioning, however, is relative to the size of the individual involved. Smaller individuals need less oxygen to perform at the same speed. On the other hand, smaller individuals have smaller hearts; so relatively, the same level of effort may be exerted. For this reason, the maximum oxygen content is normalized by body weight; a quantity, VO_2 MAX, is computed by looking at the volume of oxygen used per minute per kilogram of body weight. Of course, the effort expended to go further on the treadmill increases with the duration of time on the treadmill, so there should be some relationship between VO_2 MAX and duration on the treadmill. This relationship is presented below.

Other pertinent variables which are used in the problems and in additional chapters are recorded below in Table 9.2, including the maximum heart rate during exercise, the subject's age, height, and weight. The 44 individuals listed in Table 9.2 were all healthy. They were classified as active if they usually participated at least three times per week in activities vigorous enough to raise a sweat.

The duration of the treadmill exercise and VO_2 MAX data are presented in Figure 9.2.

In this scattergram, we see that as the treadmill time increases, by and large, the VO_2 MAX increases. There is, however, some variability. The increase is not an infallible rule. There are individuals who run longer but have less oxygen consumption than someone else who has exercised for a shorter time period. Because of the expense and difficulty in collecting the expired air volumes, it is useful to evaluate oxygen consumption and conditioning by having individuals run on the treadmill and recording the duration. As we can see from Figure 9.2, this would not be a perfect solution

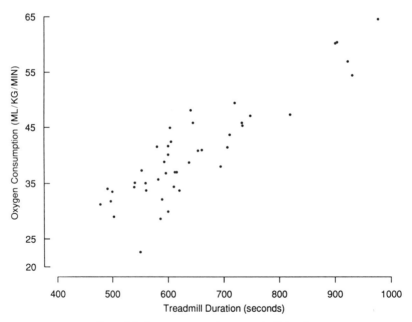

Figure 9.2. Oxygen consumption *vs.* treadmill duration.

Table 9.2. Exercise Data for Healthy Active Males. Data from Bruce _et al._ [1973].

Case	Duration (s)	VO$_2$ MAX	Heart Rate (beats/min)	Age	Height (cm)	Weight (kg)
1	706	41.5	192	46	165	57
2	732	45.9	190	25	193	95
3	930	54.5	190	25	187	82
4	900	60.3	174	31	191	84
5	903	60.5	194	30	171	67
6	976	64.6	168	36	177	78
7	819	47.4	185	29	174	70
8	922	57.0	200	27	185	76
9	600	40.2	164	56	180	78
10	540	35.2	175	47	180	80
11	560	33.8	175	46	180	81
12	637	38.8	162	55	180	79
13	593	38.9	190	50	161	66
14	719	49.5	175	52	174	76
15	615	37.1	164	46	173	84
16	589	32.2	156	60	169	69
17	478	31.3	174	49	178	78
18	620	33.8	166	54	181	101
19	710	43.7	184	57	179	74
20	600	41.7	160	50	170	66
21	660	41.0	186	41	175	75
22	644	45.9	175	58	173	79
23	582	35.8	175	55	160	79
24	503	29.1	175	46	164	65
25	747	47.2	174	47	180	81
26	600	30.0	174	56	183	100
27	491	34.1	168	82	183	82
28	694	38.1	164	48	181	77
29	586	28.7	146	68	166	65
30	612	37.1	156	54	177	80
31	610	34.5	180	56	179	82
32	539	34.4	164	50	182	87
33	559	35.1	166	48	174	72
34	653	40.9	184	56	176	75
35	733	45.4	186	45	179	75
36	596	36.9	174	45	179	79
37	580	41.6	188	43	179	73
38	550	22.7	180	54	180	75
39	497	31.9	168	55	172	71
40	605	42.5	174	41	187	84
41	552	37.4	166	44	185	81
42	640	48.2	174	41	186	83
43	500	33.6	180	50	175	78
44	603	45.0	182	42	176	85

to the problem. Duration would not totally determine the VO_2 MAX level. Nevertheless, it would give us considerable information. When we do this, how should we predict what the VO_2 MAX level would be from the duration? Clearly, such a predictive equation should be developed from the data at hand. When we do this, we want to characterize the accuracy of such predictions and succinctly summarize the relationship between the two variables.   □

Example 9.3.   Dern and Wiorkowski [1969] collected data dealing with the erythrocyte adenisone triphosphate (ATP) levels in youngest and older sons in 17 families. The purpose of the study was to determine the effect of storage of the red blood cells on the ATP level. The level is important because it determines the ability of the blood to carry energy to the cells of the body. The study found considerable variation in the ATP levels, even before storage. Some of the variation could be explained on the basis of variation by family (genetic variation). The data for the oldest and the youngest son are extracted from the more complete data set in the paper. Table 9.3 presents the data for 17 pairs of brothers along with the ages of the brothers.

Figure 9.3 is a scattergram of the values in Table 9.3. Again, there appears to be some relationship between the two values with both brothers tending to have high or low values at the same time. Again, we would like to consider whether or not such variability might occur by chance. If chance is not the explanation, how could we summarize the pattern of variation for the pairs of numbers?   □

Table 9.3. Erythrocyte adenisone triphosphate (ATP) levels in youngest and oldest sons in seventeen families together with age (before storage). Data from Dern and Wiorkowski [1969]. ATP levels expressed as micromoles per gram of hemoglobin.

Family	Youngest		Oldest	
	Age	ATP level	Age	ATP level
1	24	4.18	41	4.81
2	25	5.16	26	4.98
3	19	4.85	27	4.48
4	28	3.43	32	4.19
5	22	4.53	25	4.27
6	7	5.13	23	4.87
7	21	4.10	24	4.74
8	17	4.77	25	4.53
9	25	4.12	26	3.72
10	24	4.65	25	4.62
11	12	6.03	25	5.83
12	16	5.94	24	4.40
13	9	5.99	22	4.87
14	18	5.43	24	5.44
15	14	5.00	26	4.70
16	24	4.82	26	4.14
17	20	5.25	24	5.30

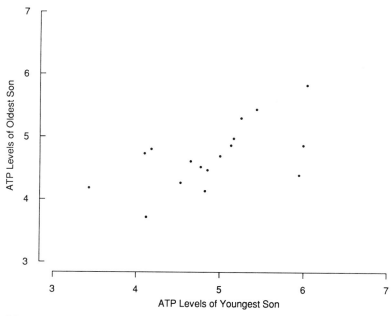

Figure 9.3. ATP Levels (μmol/g of hemoglobin) of youngest and oldest sons in 17 families. Data from Dern and Wiorkowski [1969].

The three scattergrams have certain features in common:

1. Each scattergram refers to a situation where two quantities are associated with each experimental unit. In the first example, the melanoma rate for the state and the latitude of the state are plotted. The state is the individual unit. In the second example, for each person studied on the treadmill, the VO_2 MAX vs. the treadmill time in seconds were plotted. In the third example, the experimental unit was the family and the ATP values of the youngest and oldest sons were plotted.

2. In each of the three diagrams, there appears to be a rough trend or association between the variables. In the melanoma rate date, as the latitude increases, the melanoma rate tends to decrease. In the treadmill data, as the duration on the treadmill increased, the VO_2 MAX also increased. In the ATP data, both brothers tended to have either a high or low value for ATP.

3. Although increasing and decreasing trends were evident, there was not a one-to-one relationship between the two quantities. It was not true that every state with a higher latitude had a lower melanoma rate in comparison with a state at a lower latitude. It was not true that in each case when individual A ran on the treadmill a longer time than individual B that individual A had a higher VO_2 MAX value. There were some pairs of brothers for which one pair did not have the two highest values when compared to the other pair. This is in contrast to certain physical relationships. For example, if one plotted the volume of a cube as a function of the length of a side, there is the one-to-one relationship: the volume increases as the length of the side increases. In the data

we are considering, there is a rough relationship, but there is still considerably variability or scatter.

4. In order to effectively use and summarize such scattergrams, there is a need for a method to quantitate how much of a change the trends represent. For example, if we consider two states where one has a latitude five degrees south of the other, how much difference is expected in the melanoma rates? Suppose that we train an individual to increase the duration of treadmill exercise by 70 s; how much of a change in VO_2 MAX capacity is likely to occur?

5. Suppose that we have some method of quantitating the overall relationship between the two variables in the scattergram. Since the relationship is not precisely one-to-one, there is a need to summarize how much of the variability the relationship explains. Another way of putting this is we need a summary quantity which tells us how closely the two variables are related in the scattergram.

6. If we have methods of quantifying these things, we need to know whether or not any estimated relationships might occur by chance. If not, we still want to be able to quantify the uncertainty in our estimated relationships.

The remainder of this chapter deals with the issues we have just raised. The next section uses a linear equation (a straight line) to summarize the relationship between two variables in a scattergram.

9.2   A SIMPLE LINEAR REGRESSION MODEL

9.2.1   Summarizing the Data by a Linear Relationship

The three scattergrams above have a feature in common: the overall relationship is roughly linear; that is, a straight line that characterizes the relationships between the two variables could be placed through the data. In this and subsequent chapters, we will be looking at linear relationships. A linear relationship is one expressed by a linear equation. For variables $U, V, W, \ldots$, and constants $a, b, c, \ldots$, a linear equation for Y is given by

$$Y = a + bU + cV + dW + \cdots.$$

In the scattergrams for the melanoma data and the exercise data, let X denote the variable on the horizontal axis (abscissa), and Y be the notation for the variable on the vertical axis (ordinate). Let us summarize the data by fitting the straight line equation $Y = a + bX$ to the data. In each case, let us think of the X variable as predicting a value for Y. In the first two examples, that would mean that, given the latitude of the state, we would predict a value for the melanoma rate; given the duration of the exercise test, we would predict the VO_2 MAX for the individual.

There is terminology associated with this procedure. The variable being predicted is called the *dependent variable*, or the *response variable*; the variable we are using to predict is called the *independent variable*, the *predictor variable*, or the *covariate variable*. For a particular value, say, X_i of the predictor variable, our predicted value for Y is given by

$$\widehat{Y}_i = a + bX_i. \tag{9.1}$$

The fit of our predicted values to the observed values (X_i, Y_i) may be summarized by the difference between the observed value Y_i and the predicted value $\widehat{Y}_i$. This difference is called a *residual value*.

$$\text{Residual value} = y_i - \widehat{y}_i = \text{observed value} - \text{predicted value.} \qquad (9.2)$$

It is reasonable to fit the line by trying to make the residual values as small as possible. The *principle of least squares* chooses a and b to minimize the sum of squares of the residual values. This is given in the following definition:

Definition 9.2. Given data (x_i, y_i), $i = 1, 2, \ldots, n$, the *least squares fit* to the data chooses a and b to minimize

$$\sum_{i=1}^{n}(y_i - \widehat{y}_i)^2,$$

where $\widehat{y}_i = a + bx_i$.

The values a and b that minimize the sum of squares are described below. At this point, we introduce some notation similar to that of Section 7.3:

$$[y^2] = \sum_i (y_i - \bar{y})^2,$$

$$[x^2] = \sum_i (x_i - \bar{x})^2,$$

$$[xy] = \sum_i (x_i - \bar{x})(y_i - \bar{y}).$$

We decided to choose values a and b so that the quantity

$$\sum_i (y_i - \widehat{y}_i)^2 = \sum_i (y_i - a - bx_i)^2$$

is minimized. It can be shown that the values for a and b that minimize the quantity are given by

$$b = \frac{\sum (x_i - \bar{x})(y_i - \bar{y})}{\sum (x_i - \bar{x})^2} = \frac{[xy]}{[x^2]}$$

and

$$a = \bar{y} - b\bar{x}.$$

For the melanoma data, we have the following quantities:

$$\bar{x} = 39.533, \quad \bar{y} = 159.878,$$

$$\sum_i (x_i - \bar{x})(y_i - \bar{y}) = [xy] = -6100.171,$$

$$\sum_i (x_i - \bar{x})^2 = [x^2] = 1020.499,$$

$$\sum_i (y_i - \bar{y})^2 = [y^2] = 53637.265,$$

The least squares slope b is

$$b = \frac{-6100.171}{1020.499} = -5.9776,$$

and the least squares intercept a is

$$a = 159.878 - (-5.9776 \times 39.533) = 389.190.$$

Figure 9.4 presents the melanoma data with the line of least squares fit drawn in.

Because of the method of selecting the line, the line of course goes through the data. The least squares line always has the property that it goes through the point in

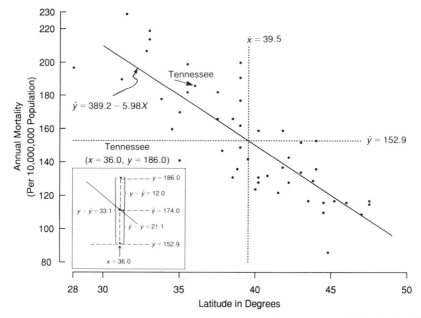

Figure 9.4. Annual mortality (per 10,000,000 population) due to malignant melanoma of the skin for white males by state and latitude of the center of the state for the period 1950–1959 (least squares regression line is given).

Table 9.4. Predicted Mortality Rates by Latitude for the Data of Table 9.1. For the quantities s_2 and s_3, see Section 9.2.3.

Latitude (x)	Predicted mortality (y)	s_1	s_2	s_3
30	209.9	19.12	6.32	20.13
35	180.0	19.12	3.85	19.50
39.5 (mean)	152.9 (mean)	19.12	2.73	19.31
40	150.1	19.12	2.74	19.31
45	120.2	19.12	4.26	19.58
50	90.3	19.12	6.83	20.30

the scattergram corresponding to the sample mean of the two variables. The sample means of the variables are located by the intersection of dotted lines. Further, the point for Tennessee is detailed in the box in the lower left-hand corner. The predicted value from the equation was 174, whereas the actual melanoma rate for this state was 186. Thus, the residual value is the difference, 12. We see that the predicted value, 174, is closer to the observed value than to the overall Y mean, which is 152.9.

For the melanoma data, the line of least squares fit is $Y = 389.19 - 5.9776X$. For each state's observed mortality rate, there is then a predicted mortality rate based on knowledge of the latitude. Some predicted values are listed in Table 9.4. The further north the state, the lower the mortality due to malignant melanoma; but now we have quantified the change.

Note that the predicted mortality at the mean latitude (39.5°) is exactly the mean value of the *observed* mortalities; as noted above, the regression line goes through the point $(\bar{x}, \bar{y})$.

9.2.2 Linear Regression Models

With the line of least squares fit, we shall associate a mathematical model. This *linear regression model* takes the predictor or covariate observation as being fixed. Even if it is sampled at random, the analysis is conditional upon knowing the value of X. In the first example above, the latitude of each state is fixed. In the second example, the healthy people may be considered to be a representative—although not random— sample of a larger population; in this case, the duration may be considered a random quantity. In the linear regression analysis of this chapter, we know X and are interested in predicting the value of Y. The regression model assumes that for a fixed value of X, the expected value of Y is some function. In addition to this expected value, a random error term is added. It is assumed that the error has a mean value of zero. We shall restrict ourselves to situations where the expected value of Y for known X is a linear function. Thus, our linear regression model is the following:

$$\text{Expected value of } Y \text{ knowing } X = E(Y \mid X) = \alpha + \beta X$$

$$Y = \alpha + \beta X + e, \quad \text{where } e \text{ (error) has } E(e) = 0.$$

The parameters α and β are population parameters. Given a sample of observations, the estimates a and b that we found above are estimates of the population parameters. In the mortality rates of the states, the random variability arises both because of the randomness of the rates in a given year and random factors associated with the state, other than latitude. These factors make the observations during a particular time period reasonably modeled as a random quantity. For the exercise test data, we may consider the normal individuals tested as a random sample from a population of active normal males who might have been tested.

Definition 9.3.   The line $E(Y \mid X) = \alpha + \beta X$ is called the *population regression line*. Here, $E(Y \mid X)$ is the expected value of Y when X is known. The coefficients α and β are called *population regression coefficients*. The line $Y = a + bX$ is called the *estimated regression line*, and a and b are called *estimated regression coefficients*. The term *estimated* is often dropped, and *regression line* and *regression coefficients* are used for these estimated quantities.

For each X, $E(Y \mid X)$ is the mean of a population of observations. Figure 9.5(a) shows a linear regression situation. Figure 9.5(b) shows a situation where the regression $E(Y \mid X)$ is not linear.

In order to perform statistical inference, another assumption is added: that the error term is normally distributed with mean zero and variance σ_1^2.

Note that the variance of the error term is *not* the variance of the Y variable. It is the variance of the Y variable *when* the value of the X variable is known.

Given data, the variance σ_1^2 is estimated by the quantity $s_{y \cdot x}^2$, where this quantity

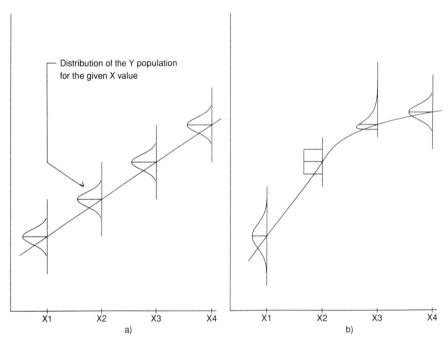

Figure 9.5. Linear regression assumptions and violations. On the left, the expected values of Y for given X values fall on a straight line. The variation about the line is normal with the same variance at each X. On the right, the expected values fall on a curve, not a straight line. The distribution of Y is different for different X values. The distributions are not all normal, and the variances differ.

is defined as

$$s_{y.x}^2 = \sum \frac{(Y_i - \widehat{Y}_i)^2}{n-2}.$$

Recall that the usual sample variance was divided by $n-1$. The $n-2$ occurs because two parameters, α and β, are estimated in fitting the data rather than one parameter, the sample mean, that was estimated before.

9.2.3 Inference

We have the model

$$Y = \alpha + \beta X + e, \quad \text{where } e \sim N(o, \sigma_1^2).$$

On the basis of n pairs of observations we presented estimates a and b of α and β, respectively. To test hypotheses regarding α and β, we need to assume normality of the term e.

Figure 9.5(a) shows a situation where these assumptions are satisfied. Note that:

1. $E(Y \mid X)$ is linear.
2. For each X, the Y-distribution is normal.
3. For each X, the normal Y-distribution has the same variance.

Figure 9.5(b) shows a situation where all these assumptions don't hold.

1. $E(Y \mid X)$ is not a straight line; it curves.
2. At X_2 and X_3, the Y-distribution is not normal.
3. Even at points X_1 and X_4 where the Y-distributions are normal, they have different variances.

It can be shown, under the correct normal model, that

$$b \sim N\left(\beta, \frac{\sigma_1^2}{[x^2]}\right) \quad \text{and} \quad a \sim N\left(\alpha, \sigma_1^2\left[\frac{1}{n} + \frac{\bar{x}^2}{[x^2]}\right]\right).$$

Recall that σ_1^2 is estimated by $s_{y.x}^2 = \sum (Y_i - \widehat{Y}_i)^2/(n-2)$. Note that the divisor is $n-2$: the number of degrees of freedom. The reason, as just mentioned, is that now *two* parameters are estimated: α and β. Given these facts, we can now either construct confidence intervals or tests of hypotheses after constructing appropriate pivotal variables:

$$\frac{b-\beta}{\sigma_1/\sqrt{[x^2]}} \sim N(0,1), \quad \frac{b-\beta}{s_{y.x}/\sqrt{[x^2]}} \sim t_{n-2},$$

and similar terms involving the intercept a are discussed below.

Returning to Example 9.1, the melanoma data by state, the following quantities are known or can be calculated:

$$a = 389.190,$$
$$b = -5.9776,$$
$$s_{y.x}^2 = \sum_i \frac{(Y_i - \widehat{Y}_i)^2}{n-2} = \frac{17173.1}{47} = 365.3844,$$

$$[x^2] = 1020.499, \quad s_{y.x} = 19.1150.$$

On the assumption that there is no relationship between latitude and mortality, that is, $\beta = 0$, the variable b has mean zero. A t-test yields

$$t_{47} \doteq \frac{-5.9776}{19.1150/\sqrt{1020.499}} \doteq \frac{-5.9776}{0.59837} \doteq -9.99.$$

From Table A.5, the critical value for a t-variable with 47 degrees of freedom, at the 0.0001 level (two-tail) is approximately 4.25; hence, the hypothesis is rejected and we conclude that there is a relationship between latitude and mortality; the mortality increases about 6.0 persons per 10,000,000 for every degree further south. This, of course, comes from the value of $b = -5.9776 \doteq -6.0$. Similarily, a 95% confidence interval for β can be constructed using the t-value of 2.01, and the standard error of the slope, $0.59837 = s_{y.x}/\sqrt{[x^2]}$.

A 95% confidence interval is $-5.9776 \pm (2.01 \times 0.59837)$, producing lower and upper limits of -7.18 and -4.77, respectively. Again, the confidence interval does not include zero, and the same conclusion is reached as in the case of the hypothesis test.

The inference has been concerned with the slope β and intercept α up to now. We now want to consider two additional situations:

1. Inference about population means, $\alpha + \beta X$, for a fixed value of X, and
2. Inference about a future observation at a fixed value of X.

To distinguish between the two cases, let $\widehat{\mu}_x$ and $\widehat{y}_x$ be the predicted mean and a new random single observation at the point x, respectively.

First, then, inference about the population mean at a fixed X value: it is natural to estimate $\alpha + \beta X$ by $a + bx$; the predicted value of Y at the value of $X = x$. Rewrite this quantity as

$$\widehat{\mu}_x = \bar{y} + b(x - \bar{x}).$$

It can be shown that $\bar{y}$ and b are statistically independent so that the variance of the quantity is

$$\mathrm{var}[\bar{y} + b(x - \bar{x})] = \mathrm{var}(\bar{y}) + (x - \bar{x})^2 \mathrm{var}(b)$$

$$= \frac{\sigma_1^2}{n} + (x - \bar{x})^2 \frac{\sigma_1^2}{[x^2]}$$

$$= \sigma_1^2 \left[\frac{1}{n} + \frac{(x - \bar{x})^2}{[x^2]} \right] = \sigma_2^2, \quad \text{say.}$$

Tests and confidence intervals for $E(Y \mid X)$ at a fixed value of x may be based upon the t-distribution.

The quantity σ_2^2 reduces to the variance for the intercept, a, at $X = 0$. It is useful to study this quantity carefully; there are important implications for design (see Note 9.3). The variance, σ_2^2, is not constant, but depends on the value of x. The more x differs from $\bar{x}$, the greater the contribution of $(x - \bar{x})^2/[x^2]$ to the variance of $a + bx$. The contribution is zero at $x = \bar{x}$. At $x = \bar{x}$, $y = \bar{y}$ the slope is not used. Regardless of the slope the line goes through mean point $(\overline{X}, \overline{Y})$. Consider again Example 9.1. We need the following information:

$$s_{y \cdot x} = 19.1150,$$

$$n = 49,$$

$$\bar{x} = 39.533,$$

$$[x^2] = 1020.499.$$

Let

$$s_2^2 = s_{y \cdot x}^2 \left[\frac{1}{n} + \frac{(x - \bar{x})^2}{[x^2]} \right],$$

that is, s_2^2 estimates σ_2^2. Values of s_2 as related to latitude are given in Table 9.4. Confidence interval bands for the mean, $\alpha + \beta X$, (at 95% level) are given in Figure 9.6

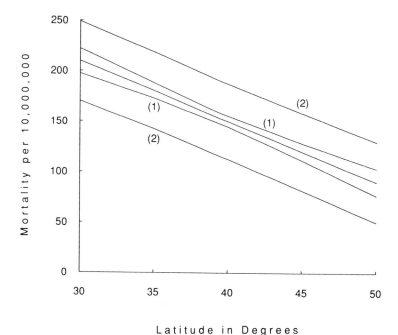

Figure 9.6. Data of Figure 9.1: 95% confidence bands for population means (1) and 95% confidence bands for a future observation (2).

by the narrower bands. The curvature is slight due to the large value of $[x^2]$ and the relatively narrow range of prediction.

We now turn to the second problem: predicting a future observation on the basis of the observed data. The variance is given by

$$s_3^2 = s_{y.x}^2 \left[1 + \frac{1}{n} + \frac{(x - \bar{x})^2}{[x^2]} \right].$$

This is reasonable in view of the following argument: at the point $\alpha + \beta X$ an observation has variance σ_2^2 (estimated by $s_{y.x}^2$). In addition, there is uncertainty in the true value $\alpha + \beta X$. This adds variability to the estimate. A future observation is assumed to be independent of past observations. Hence the variance can be added and the quantity s_3^2 results when σ_1^2 is estimated by $s_{y.x}^2$. Confidence interval bands for future observations (95% level) are represented by outer lines in Figure 9.6. This band means that we are 95% certain the next observation at the fixed point x, will be within the given bands. Note that the curvature is not nearly as marked.

9.2.4 Analysis of Variance

Consider Example 9.1 and the data for Tennessee, as graphed in Figure 9.4. The basic data for this state are (omitting subscripts):

$$y = 186.0 = \text{observed mortality,}$$

$$x = 36.0 = \text{latitude of center of state,}$$

$$\widehat{y} = 174.0 = \text{predicted mortality using latitude of 36.0,}$$

$$\bar{y} = 152.9 = \text{average mortality for United States.}$$

Partition the data as follows:

$$(y - \bar{y}) = (\widehat{y} - \bar{y}) + (y - \widehat{y})$$

$$\begin{array}{ccc} \text{Total} & \text{Attributable} & \text{Residual} \\ \text{variation} = & \text{to regression} + & \text{from regression} \end{array}$$

$$186.0 - 152.9 = (174.0 - 152.9) + (186.0 - 174.0)$$

$$33.1 = 21.1 + 12.0$$

Note that the quantity

$$\widehat{y} - \bar{y} = a + bx - \bar{y}$$

$$= \bar{y} - b\bar{x} + bx - \bar{y}$$

$$= b(x - \bar{x}).$$

The quantity is zero if $b = 0$, that is, if there is no regression relationship between Y and X. In addition, it is zero if prediction is made at the point $x = \bar{x}$.

These quantities can be calculated for each of the states, as indicated in abbreviated form in Table 9.5.

Table 9.5. Deviations from Mean and Regression. Data of Table 9.1.

Case	State	Observed Mortality (y)	Latitude (x)	Predicted Mortality[a]	Variation		
					Total $y - \bar{y}$ =	Regression $\hat{y} - \bar{y}$ +	Residual $y - \hat{y}$
1	Alabama	219.0	33.0	191.9	66.1 =	39.0 +	27.1
2	Arizona	160.0	34.5	183.0	7.1 =	30.1 +	−23.0
⋮	⋮			⋮		⋮	
41	Tennesee	186.0	36.0	174.0	33.1 =	21.1 +	12.0
⋮	⋮			⋮		⋮	
48	Wisconsin	110.0	44.5	123.2	−42.9 =	−29.7 +	−13.2
49	Wyoming	134.0	43.0	132.2	−18.9 =	−20.7 +	1.8
Total					0 =	0 +	0
Mean		152.9	39.5	152.9	0 =	0 +	0
Sum of Squares					53637.3 =	36464.2 +	17173.1

[a]Predicted mortality based on regression line $y = 389.19 − 5.9776x$ where x is latitude of center of state.

The sums of squares of these quantities are given at the bottom of Table 9.5. The remarkable fact is that

$$\sum (y_i - \bar{y})^2 = \sum (\hat{y}_i - \bar{y})^2 + \sum (y_i - \hat{y}_i)^2,$$

$$53637.3 = 36464.2 + 17173.1,$$

that is, the total variation as measured by $\sum (y_i - \bar{y})^2$ has been additively partitioned into a part attributable to regression and the residual from regression. The quantity $\sum (\hat{y}_i - \bar{y})^2 = \sum b^2(x_i - \bar{x})^2 = b^2[x^2]$. (But, since $b = [xy]/[x^2]$, this becomes $\sum (\hat{y}_i - \bar{y})^2 = [xy]^2/[x^2]$). Associated with each sum of squares is a degree of freedom (DF) which can also be partitioned as follows:

$$\text{Total variation} = \text{Attributable to regression} + \text{Residual variation}$$

$$\text{DF} \quad n - 1 = 1 + n - 2$$

$$49 = 1 + 48$$

The total variation has $n - 1$ DF, not n, since we adjusted Y about the mean $\bar{Y}$. These quantities are commonly entered into an analysis of variance table as follows:

Source of Variation	DF	SS	MS	**F-Ratio**
Regression	1	36464.2	36464.2	99.80
Residual	47	17173.1	365.384	
Total	48	53637.3		

The quantity 365.384 is precisely $s_{y.x}^2$. The F-ratio is discussed below. The mean square is the sum of squares divided by the degrees of freedom.

The analysis of variance table of any set of n pairs of observations (x_i, y_i), $i = 1, \ldots, n$, is

Source of Variation	DF	SS	MS	**F-Ratio**
Regression	1	$[xy]^2/[x^2]$	$[xy]^2/[x^2]$	$\dfrac{[xy]^2/[x^2]}{s_{y.x}^2}$
Residual	$n-2$	By subtraction	$s_{y.x}^2$	
Total	$n-1$	$[y^2]$		

Several points should be noted about this table and the regression procedure:

1. Only five quantities need to be calculated from the raw data to completely determine the regression line and sums of squares: $\sum x_i$, $\sum y_i$, $\sum x_i^2$, $\sum y_i^2$, and $\sum x_i y_i$. From these quantities one can calculate:

$$[x^2] = \sum (x_i - \bar{x})^2 = \sum x_i^2 - \left(\sum x_i\right)^2/n,$$

$$[y^2] = \sum (y_i - \bar{y})^2 = \sum y_i^2 - \left(\sum y_i\right)^2/n,$$

$$[xy] = \sum (x_i - \bar{x})(y_i - \bar{y}) = \sum x_i y_i - \left(\sum y_i \sum x_i\right)/n.$$

2. The greater the slope the greater the ss due to regression. That is,

$$\text{ss(Regression)} = b^2 \sum (x_i - \bar{x})^2 = [xy]^2/[x^2].$$

If the slope is "negligible", the ss(Regression) will tend to be "small".

3. The proportion of the total variation attributable to regression is usually denoted by r^2. That is,

$$r^2 = \frac{\text{Variation attributable to regression}}{\text{Total variation}}$$

$$= \frac{[xy]^2/[x^2]}{[y^2]}$$

$$= \frac{[xy]^2}{[x^2][y^2]}.$$

It is clear that $0 \le r^2 \le 1$ (why?). If $b = 0$, then $[xy]^2/[x^2] = 0$ and the variation attributable to regression is zero. If $[xy]^2/[x^2]$ is equal to $[y^2]$, then all of the variation can be attributed to regression; to be more precise, to *linear* regression; that is, all the observations fall on the line $a + bx$. Thus r^2 measures the degree of *linear* relationship between X and Y. r, called the *correlation coefficient*, is studied in Section 9.3. For the data in Table 9.4,

$$r^2 = \frac{36464.2}{53637.3} = 0.67983.$$

That is, approximately 68% of the variation in mortality can be attributed to variation in latitude. Equivalently, the variation in mortality can be reduced 68% knowing the latitude.

4. Now consider the ratio

$$F = \frac{[xy]^2/[x^2]}{s_{y.x}^2}.$$

Under the assumption of the model, i.e. $y \sim N(\alpha + \beta X, \sigma_1^2)$, the ratio F tends to be near 1 if $\beta = 0$ and tends to be larger than 1 if $\beta \neq 0$ (either positively or negatively). F has the F distribution as introduced in Chapter 5. In the example $F_{1,47} = 99.80$, the critical value at the 0.05 level $F_{1,47} = 4.03$ (by interpolation). The critical value at the 0.001 level is $F_{1,47} = 12.4$ (by interpolation). Hence, we reject the hypotheses that $\beta = 0$. We tested the significance of the slope using a t-test given the value

$$t_{47} = -9.9898.$$

The F-value we obtained was

$$F_{1,47} = 99.80.$$

In fact,

$$(-9.9898)^2 = 99.80.$$

That is,

$$t_{47}^2 = F_{1,47}.$$

Recall that,

$$t_\nu^2 = F_{1,\nu}.$$

Thus, the t-test and the F-test for the significance of the slope are equivalent.

9.2.5 Appropriateness of the Model

In Chapter 5, we considered the appropriateness of the model $y \sim N(\mu, \sigma^2)$ for a set of data and discussed briefly some ways of verifying the appropriateness of this model. In this section, we have the model

$$y \sim N(\alpha + \beta X, \sigma_1^2)$$

and want to consider its validity. At least three questions can be asked:

1. Is the relationship between Y and X linear?

2. The variance σ_1^2 is assumed to be constant for all values of X (homogeneity of variable), is this so?

3. Does the normal model hold?

Two very simple graphical procedures, both utilizing the residuals from regression $y_i - \hat{y}_i$, can be used to verify the above assumptions. Also one computation upon the residuals is useful. The two graphical procedures are considered first.

	To Test For	**Graphical Procedures**
1	Linearity of regression and homogeneity of variance	Plot $(y_i - \hat{y}_i)$ against $\hat{y}_i$, $i = 1, \ldots, n$
2	Normality	Normal probability plot of $y_i - \hat{y}_i$, $i = 1, \ldots, n$

Linearity of Regression and Homogeneity of Variance

Given only one predictor variable, X, the graph of Y vs. X will suggest nonlinearity or heterogeneity of variance, see the top row of regression patterns in Figure 9.7. But, if there is more than one predictor variable, as in Chapter 11, the simple two-dimensional graph is not possible. But there is a way of detecting such patterns by considering residual plots $y - \hat{y}$ against a variety of variables. A common practice is to plot $y - \hat{y}$ against $\hat{y}$; this graph is usually referred to as a *residual plot*. The advantage is, of course, that no matter how many predictor variables are used, it is always possible to plot $y - \hat{y}$. The second row of graphs in Figure 9.7 indicate the residual patterns associated with the regression patterns of the top row. Pattern I indicates a reasonable linear trend, Pattern II indicates that the observation ($X = 90$, $Y = 50$) differs from the pattern established by the first eight points. Also, note that this single point "pulls down" the regression line a considerable amount. Pattern III is curvilinear and the residual plot confirms the nonlinearity. Pattern IV indicates that the variation from the line increases as X (and Y) increase, thus providing evidence for the heterogeneity of variance. Again, the residual plot confirms the pattern.

Before turning to the questions of normality of the data, consider the same kind of analysis carried out on the melanoma data. The residuals are plotted in Figure 9.8, part (a). There is no evidence that there is nonlinearity or heterogeneity of variance.

Normality

One way of detecting gross deviations from normality is to graph the residuals from regression on normal probability paper as introduced in Chapter 4. The last row of patterns in Figure 9.6 are the normal probability plots of the deviations from linear regression. A minor alteration in calculating the cumulative percentages is to associate with the ith ordered observation (among the n observations) the cumulative proportion

$$\frac{i - 1/2}{n}$$

rather than i/n. There are two reasons for this: it does not produce any infinite values at $i = n$ and it introduces symmetry into the situation, that is, the largest value

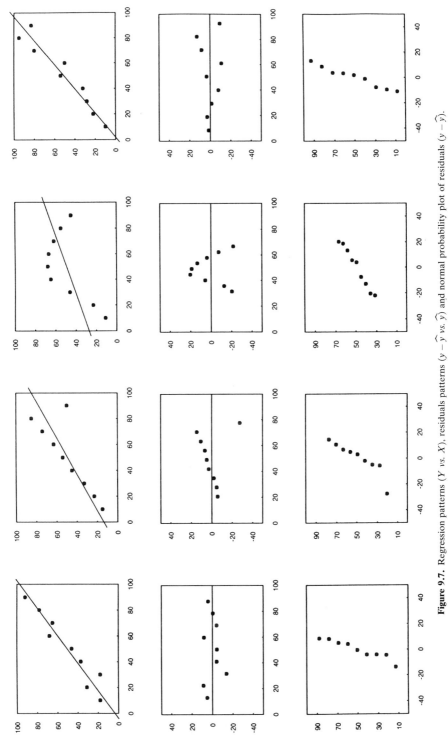

Figure 9.7. Regression patterns (Y vs. X), residuals patterns ($y - \hat{y}$ vs. $\hat{y}$) and normal probability plot of residuals ($y - \hat{y}$).

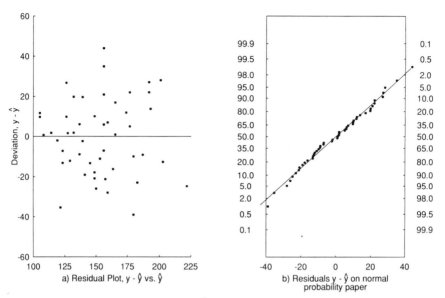

Figure 9.8. Melanoma data (plot a) residuals $(y-\widehat{y})$ from regression lines $Y = 389.19 - 589.8X$ plotted against $\widehat{y}$ and (plot b) normal probability plot of residuals, $y - \widehat{y}$.

is represented by a point as well as the smallest value. The last row in Figure 9.7 indicates that a normal probability plot indicates outliers clearly but is not useful in detecting heterogeneity of variance or curvilinearity.

Of particular concern are points not fit closely by the data. The upper right and lower left points often tail in toward the center in least squares plot. Points on the top far to the right and on the bottom far to the left (as in pattern 2) are of particular concern.

The normal probability plot asssociated with the residuals of the melanoma are plotted in Figure 9.8(b). There is no evidence against the normality assumption.

Normal Deviates

For given X, the variability of Y from $\alpha + \beta X$ is $N(0, \sigma_1^2)$. Thus the values of $(Y_i - (\alpha + \beta X_i))/\sigma_1$ are a sample from a standard normal distribution. We can estimate these quantities by $(y_i - \widehat{y}_i)/s_{y \cdot x}$. That is, we divide the residual values by the estimated standard deviation about the regression line. Values too large in absolute value indicate either a poor fit or outlier observations. Values above three are certainly worth examining.

9.2.6   The Two Sample t-Test as a Regression Problem

In this subsection we will show the usefulness of the linear model approach by illustrating how the two sample t-test can be considered a special kind of linear model. For an example, we again return to the data on mortality rates due to melanoma contained in Table 9.1. This time we consider the rates in relationship to contiguity to an ocean; there are two groups of states: those that border on an ocean and those that

Table 9.6. Comparison by Two Sample t-Test of Mortality Rates Due to Melanoma (Y) by Contiguity to Ocean.

Contiguity to ocean		No =	0	Yes =	1
Number of states		$n_1 =$	27	$n_2 =$	22
Mean mortality		$\bar{y}_1 =$	138.741	$\bar{y}_2 =$	170.227
Variance[a]		$s_1^2 =$	697.97	$s_2^2 =$	1117.70
Pooled variance		$s_p^2 =$	885.51		
Standard error of difference			$s_p\sqrt{\frac{1}{n_1}+\frac{1}{n_2}}= 8.5468$		
Mean difference			$\bar{y}_2 - \bar{y}_1 = 31.487$		
t-Value			$t = 3.684$		
Degrees of freedom			$DF = 47$		
p-Value			$p < 0.001$		

[a]Subscripts on variances denote group membership in this table.

do not. The question is whether the average mortality rate for the first groups differs from that of the second group. The t-test and analysis are contained in Table 9.6.

The mean difference, $\bar{y}_1 - \bar{y}_2 = 31.486$, has a standard error of 8.5468 so that the calculated t-value is $t = 3.684$ with 47 degrees of freedom which exceeds the largest value in the t-table at 40 or 60 degrees of freedom and consequently, $p < 0.001$. The conclusion then is that the mortality rate due to malignant melanoma is appreciably higher in states contiguous to an ocean as compared to "inland" states, the difference being approximately 31 deaths per 10^7 population per year.

Now consider the following (equivalent) regression problem. Let Y be the mortality rate and X the predictor variable; "X = contiguity to ocean" and X takes on only two values, 0, 1. (For simplicity, we again label all the variables and parameters, Y, X, α, β and σ_1^2, but except for Y, they obviously are different from the way they were defined in earlier sections.) The model is

$$Y \sim N(\alpha + \beta X, \sigma_1^2).$$

The data are graphed in Figure 9.9. The calculations for the regression line are as follows:

$$n = 49, \qquad\qquad b = [xy]/[x^2] = 31.487,$$
$$[y^2] = 53637.265, \qquad a = 138.741,$$
$$[xy] = 381.6939, \qquad \text{Regression line}$$
$$[x^2] = 12.12245, \qquad Y = 138.741 + 31.487X,$$
$$\bar{y} = 152.8776, \qquad (n-2)s_{y.x}^2 = [y^2] - [xy]^2/[x^2]$$
$$\bar{x} = 0.44898, \qquad\qquad\quad = 41619.0488,$$
$$s_{y.x}^2 = 885.51.$$

The similarity to the t-test becomes obvious, the intercept, $a = 138.741$, is precisely the mean mortality for the "inland" states. The "slope", $b = 31.487$ is the mean

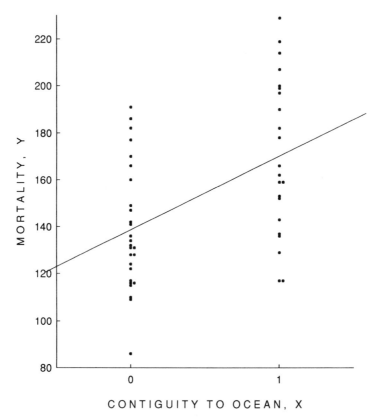

Figure 9.9. Melanoma data: regression of rate (Y) on contiguity to ocean (X). $X = 0$ not contiguous to ocean; $X = 1$ contiguous to ocean.

difference between the two groups of states, and $s^2_{y.x}$, the residual variance is the pooled variance. The t-test for the slope is equivalent to the t-test for the difference in the two means.

$$\text{Variance of slope} = s^2_b = s^2_{y.x}/[x^2]$$

$$= (885.51)/(12.12245)$$

$$= 73.0471$$

$$s_b = 8.5468$$

$$t = 31.487/8.5468$$

$$= 3.684.$$

The t-test for the slope has 47 degrees of freedom, as does the two-sample t-test. Note also that s_b is the standard error of the differences in the two-sample t-test.

Finally, the regression analysis can be put into analysis of variance form as dis-

Table 9.7. Regression Analysis of Mortality and Contiguity to Ocean.

Source of Variation	DF	SS	MS	F-Ratio
Regression	1	12018.22	12018.22	13.57[a]
Residual	47	41619.04	885.51	
Total	48	53637.26		

[a]Significant at the 0.001 level.

played in Table 9.7:

$$ss(\text{Regression}) = [xy]^2/[x^2]$$

$$= (381.6939)^2/(12.12245)$$

$$= 12018.22,$$

$$ss(\text{Residual}) = [y^2] - [xy]^2/[x^2]$$

$$= 53637.26 - 12018.22$$

$$= 41619.04.$$

We note that the proportion of variation in mortality rates attributable to "contiguity to ocean" is

$$r^2 = \frac{[xy]^2/[x^2]}{[y^2]}$$

$$= \frac{12018.22}{53637.06}$$

$$= 0.2241.$$

Approximately 22% of the variation in mortality can be attributed to the predictor variable: "contiguity to ocean."

In Chapter 11 we deal with the relationship between the three variables: mortality, latitude, and contiguity to an ocean.

The predictor variable "contiguity to ocean" which takes on only two values, 0, 1 in this case, is called a *dummy variable* or *indicator variable*. In Chapter 11 more use is made of such variables.

9.3 CORRELATION AND COVARIANCE

9.3.1 Introduction

In the last section the method of least squares was used to find a line for predicting one variable from the other. The reponse variable Y, or dependent variable Y, was random for given X. Even if X and Y were jointly distributed so that X was a random variable, the model only had assumptions about the distribution of Y given the value of X. There are cases, however, where both variables vary jointly, and there is a considerable amount of symmetry. In particular, there does not seem to be a reason to predict one variable from the other. The ATP Example 9.3 is of that type. As another example, we may want to characterize the length and weight relationship of newborn infants. The basic sampling unit is an infant, and two measurements are made, both of which vary. There is a certain symmetry in this situation: there is no "causal direction," length does not cause weight or vice versa. Both variables vary together in some way and are probably related to each other through several other underlying variables which determine (cause) length and weight. In this section, we shall provide a quantitative measure of the strength of the relationship between the two variables, and discuss some of the properties of this measure. The measure (the correlation coefficient) is a measure of the strength of the linear relationship between two variables.

9.3.2   Correlation and Covariance

We would like to develop a measure (preferable one number) which summarizes the strength of any linear relationship between two variables X and Y. Consider Example 9.2, the exercise test data. The X variable is measured in seconds and the Y variable is measured in milliliters per minute per kilogram. When totally different units are used on the two axes, one can change the units for one of the variables, and the picture seems to change. For example, if we went from seconds to minutes where one minute was graphed over the interval of one second in Figure 9.2, the data of Figure 9.2 would go almost straight up in the air. Whatever measure we use should not depend on the choice of units for the two variables. We already have one technique of adjusting for or removing the units involved: to standardize the variables. We have done this for the t-test, and we often had to do it for the construction of test statistics in the previous chapters of this text. Further, since we are just concerned with how closely the family of points is related, if we shift our picture (that is, change the means of the X and Y variables) the strength of the relationship between the two variables should not change. For that reason, we will subtract the mean of each variable, so that the pictures will be centered about zero. In order that we have a solution which does not depend on units, we will standardize each variable by dividing by the standard deviation. Thus, we are now working with two new variables, say U and V which are related to X and Y as follows:

$$U_i = (X_i - \overline{X})/s_x, \quad V_i = (Y_i - \overline{Y})/s_y,$$

where

$$s_x^2 = \sum (X_i - \overline{X})^2/(n-1) \quad \text{and} \quad s_y^2 = \sum (Y_i - \overline{Y})^2/(n-1).$$

Let us consider how the variables U_i and V_i vary together.

In Figure 9.10 we see three possible types of association. Panel A presents a positive relationship, or association between, the variables. As one increases, the other tends to increase. Panel B represents a tighter, negative relationship. As one decreases, the other tends to increase, and vice versa. By the word tighter, we mean

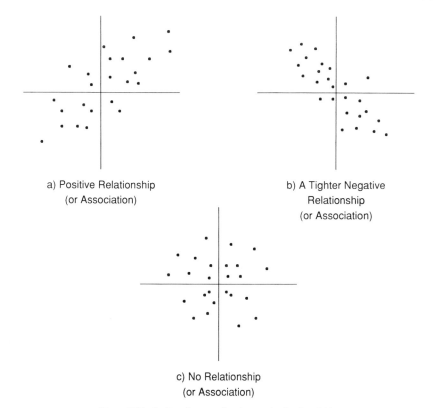

a) Positive Relationship
(or Association)

b) A Tighter Negative
Relationship
(or Association)

c) No Relationship
(or Association)

Figure 9.10. Scatter diagrams for the standardized variables.

that the variability about a fitted regression line would not be as large. Pattern C represents little or no association, with a somewhat circular distribution of points.

One mathematical function that would capture these aspects of the data results from multiplying U_i and V_i. If the variables tend to be positive or negative together, then the product will always be positive. If we add up those multiples, we would get a positive number. On the other hand, if one variable tends to be negative when the other is positive and vice versa, when we multiply the U_i and V_i together, the product will be negative; when we add them, we will get a negative number of substantial absolute value.

On the other hand, if there is no relationship between U and V, when we multiply them half the time the product will be positive and half the time the product will be negative; if we sum them, the positive and negative terms will tend to cancel out, and we will get something close to zero. Thus, adding the products of the standardized variables seems to be a reasonable method of characterizing the association between the variables. This gives us our definition of the correlation coefficient.

Definition 9.4. The *sample Pearson product moment correlation coefficient*, denoted by r, or r_{XY}, is defined to be

$$r = [xy]/\sqrt{[x^2][y^2]} = \frac{\sum (x_i - \bar{x})(y_i - \bar{y})}{\sqrt{\sum (x_i - \bar{x})^2 \sum (y_i - \bar{y})^2}} = \left(\frac{1}{n-1}\right) \sum u_i v_i.$$

This quantity is usually called the *correlation coefficient.*

Note that the denominator looks like the product of the sample standard deviations of X and Y except for a factor of $n - 1$. If we define the sample covariance by the following equation, then we could define the correlation coefficient according to the second alternative definition.

Definition 9.5. The *sample covariance,* s_{xy}, is defined by

$$s_{xy} = \sum_i (x_i - \bar{x})(y_i - \bar{y})/(n - 1).$$

Alternate Definition 9.4. *The sample Pearson Product moment correlation coefficient* is defined by

$$r = s_{xy}/s_x s_y = [xy]/\sqrt{[x^2][y^2]}.$$

The prefix co- is a prefix meaning "with", "together" and "in association", occurring in words derived from Latin. Thus, the co-talks about the two variables varying together or in association. The term covariance has the same meaning as the variance of one variable, how spread out or variable things are. Thus the term co-variance. It is hard to interpret the value of the covariance alone because it is composed of two parts; the variability of the individual variables, and also their linear association. A small covariance can occur because X and/or Y has small variability. It can also occur because the two variables are not associated. Thus, in interpreting the covariance, one usually needs to have some idea of the variability in both of the variables.
A large covariance, however, does imply that at least one of the two variables has a large variance.
The correlation coefficient is a rescaling of the covariance by the standard deviations of X and Y. The motivation for the construction of the covariance and correlation coefficient is the following: s_{xy} is the average of the product of the deviations about the means of X and Y. If X tends to be large when Y is large, both deviations will be positive and the product will be positive. Similarly, if X is small when Y is small, both deviations will be negative but their products will still be positive. Hence, the average of the products for all of the cases will tend to be positive. If there is no relationship between X and Y, a positive deviation in X may be paired with a positive or negative deviation in Y and the product will either be positive or negative, and on the average will tend to center around zero. In the first case X and Y are said to be positively correlated, in the second case there is no correlation between X and Y. A third case results when large values of X tend to be associated with small values of Y and vice versa. In this situation, the product of deviations will tend to be negative and the variables are said to be negatively correlated. The statistic r rescales the average of the product of the deviations about the means by the standard deviations of X and Y.

The statistic r has the following properties:

1. r has value between -1 and 1.

2. $r = 1$ if and only if all the observations are on a straight line with positive slope.

3. $r = -1$ if and only if all observations are on a straight line with negative slope.

4. r takes on the same value if X, or Y, changes units or has a constant added or subtracted.

5. r measures the extent of *linear* association between two variables.

6. r tends to be close to zero if there is no linear association between X and Y.

Some typical scattergrams and associated values of r are given in Figure 9.11. Figures 9.11(a) and 9.11(b) indicate perfect linear relationships between two variables. Figure 9.11(c) indicates no correlation. Figures 9.11(d) and 9.11(e) indicate typical patterns representing less than perfect correlation. Figures 9.11(f) – 9.11(i) portray various pathological situations. Figure 9.11(f) indicates that although there is an explicit relationship between X and Y, the linear relationship is zero; thus $r = 0$ does not imply that there is no relationship between X and Y. In statistical terminology:

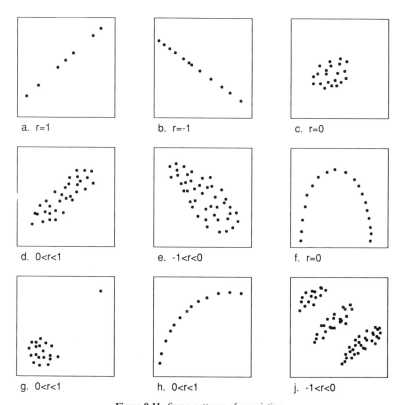

Figure 9.11. Some patterns of association.

$r = 0$ does not imply that the variables are statistically independent. There is one important exception to this statement which is discussed in Section 9.3.4. Figure 9.11(g) indicates that except for the one extreme point there is no correlation. The coefficient of correlation is very sensitive to such outliers and in Section 9.3.8 we discuss correlations that are not as sensitive, that is, more robust. Figure 9.11(h) indicates that an explicit relationship between X and Y is not identified by the correlation coefficient if the relationship is not linear. Finally, Figure 9.11(i) suggest that there are three subgroups of cases; within each subgroup there is a positive correlation but the correlation is negative when the subgroups are combined. The reason is that the subgroups have different means and care must be taken when combining data. For example, natural subgroups defined by sex or race may differ in their means in a direction opposite to the correlation within each subgroup.

Now consider Example 9.3. The scattergram in Figure 9.3 suggests positive association between the ATP level of the youngest son (X) and that of the oldest son (Y). The data for this example produces the following summary statistics (the subscripts on the values of X and Y have been suppressed, for example $\sum x_i = \sum x$).

$$n = 17,$$

$$\sum x = 83.38, \qquad\qquad\qquad \bar{x} = 4.90,$$

$$\sum y = 79.89, \qquad\qquad\qquad \bar{y} = 4.70,$$

$$\sum x^2 = 417.1874, \qquad \sum(x - \bar{x})^2 = 8.233024, \quad s_x = 0.717331,$$

$$\sum y^2 = 379.6631, \qquad \sum(y - \bar{y})^2 = 4.227094, \quad s_y = 0.513997,$$

$$\sum xy = 395.3612, \quad \sum(x - \bar{x})(y - \bar{y}) = 3.524247, \quad s_{xy} = 0.220265.$$

$$r = \frac{0.220265}{(0.717331)(0.513997)} = 0.597.$$

In practice r will simply be calculated from the equivalent formula

$$r = \frac{[xy]}{\sqrt{[x^2][y^2]}} = \frac{3.524247}{\sqrt{(8.233024)(4.227094)}} = \frac{3.524247}{5.899302} = 0.597.$$

The sample correlation coefficient and covariance estimate population parameters. The expected value of the covariance is

$$E(s_{xy}) = E((X - \mu_x)(Y - \mu_y))$$

$$= \sigma_{xy}$$

where

$$\mu_x = E(X) \quad \text{and} \quad \mu_y = E(Y).$$

The population covariance is the average of the product of X about its mean times Y about its mean.

The sample correlation coefficient estimates the population correlation coefficient ρ, defined as follows:

Definition 9.6. Let (X, Y) be two jointly distributed random variables. The (*population*) *correlation coefficient* is

$$\rho = \sigma_{xy}/\sigma_x\sigma_y = \frac{E((X - \mu_x)(Y - \mu_y))}{\sqrt{\text{var}(X)\text{var}(Y)}},$$

where σ_{xy} is the covariance of X and Y, σ_x is the standard deviation of X, and σ_y is the standard deviation of Y. ρ is zero if X and Y are statistically independent variables.

There is now the question about statistical "significance" of a value r. In practical terms, suppose we have sampled 17 families and calculated the correlation coefficient in ATP levels between the youngest son and the oldest son. How much variation could we have expected relative to the value observed for this set? Could the population correlation coefficient $\rho = 0$? In the next two subsections we will deal with this question.

9.3.3 Relationship between Correlation and Regression

In Section 9.2.4, r^2 was presented indicating a close connection between correlation and regression. In this section, the connection will be made explicit in several ways. Formally, one of the variables X, Y could be considered the dependent variable and the other the predictor variable and the techniques of Section 9.2 applied. It is easy to see that in most cases the slope of the regression of Y on X will not be the same as that of X on Y. To keep the distinction clear, the following notation will be used:

$b_{yx} =$ slope of the regression of the "dependent" variable Y
on the "predictor" variable X,

$a_y =$ intercept of the regression of Y on X.

Similarly,

$b_{xy} =$ slope of the regression of the "dependent" variable X
on the "predictor" variable Y,

$a_x =$ intercept of the regression of X on Y.

These quantities are calculated as follows:

	Regress Y on X	Regress X on Y
Slope	$b_{yx} = \dfrac{[xy]}{[x^2]}$	$b_{xy} = \dfrac{[xy]}{[y^2]}$
Intercept	$a_y = \bar{y} - b_{yx}\bar{x}$	$a_x = \bar{x} - b_{xy}\bar{y}$
Residual variance	$s_{y.x}^2 = \dfrac{[y^2] - [xy]^2/[x^2]}{n-2}$	$s_{x.y}^2 = \dfrac{[x^2] - [xy]^2/[y^2]}{n-2}$

From these quantities, the following relationships can be derived:

a. Consider the product:

$$b_{yx}b_{xy} = \frac{[xy]^2}{[x^2][y^2]}$$

$$= r^2.$$

Hence

$$r = \pm\sqrt{b_{yx}b_{xy}}.$$

In words, r is the geometric mean of the slope of the regression of Y on X and the slope of the regression of X on Y.

b.

$$b_{yx} = r\frac{s_y}{s_x}, \quad b_{xy} = r\frac{s_x}{s_y},$$

where s_x and s_y are the sample standard deviations of X and Y, respectively.

c. Using the relationships in (b), the regression line of Y on X

$$\widehat{Y} = a_y + b_{yx}X$$

can be transformed to

$$\widehat{Y} = a_y + \left(\frac{rs_y}{s_x}\right)X$$

$$= \bar{y} + \left(\frac{rs_y}{s_x}\right)(X - \bar{x}).$$

d. Finally, the t-test for the slope, in the regression of Y on X,

$$t_{n-2} = \frac{b_{yx}}{s_{by.x}}$$

$$= \frac{b_{yx}}{s_{y.x}/\sqrt{[x^2]}}$$

is algebraically equivalent to

$$r\bigg/\sqrt{\frac{1 - r^2}{n - 2}}.$$

Hence, testing the significance of the slope is equivalent to testing the significance of the correlation.

Consider again Example 9.3. The data are summarized in Table 9.8. This table indicates that the two regression lines are not the same but that the t-tests for testing the significance of the slopes produce the same observed value, and this value is identical to the test of significance of the correlation coefficient. If the corresponding analyses of variance are carried out, it will be found that the F-ratio in the two analyses are identical and give an equivalent statistical test.

Table 9.8. Regression Analyses of ATP Levels of Oldest and Youngest Sons. Data from Table 9.3.

	Regression Analyses of ATP Levels	
Dependent variable	Y^a	X^a
Predictor variable	X^b	Y^b
Slope	$b_{yx} = 0.42806$	$b_{xy} = 0.83373$
Intercept	$a_y = 2.59989$	$a_x = 0.98668$
Regression line	$\widehat{Y} = 2.600 + 0.428X$	$\widehat{X} = 0.987 + 0.834Y$
Variance about mean	$s_y^2 = 0.26419$	$s_x^2 = 0.51456$
Residual variance	$s_{y.x}^2 = 0.18123$	$s_{x.y}^2 = 0.35298$
Standard error of slope	$s_{b_{y.x}} = 0.14837$	$s_{b_{x.y}} = 0.28897$
Test of significance	$t_{15} = \dfrac{0.42806}{0.14837} = 2.885$	$t_{15} = \dfrac{0.83373}{0.28897} = 2.885$
Correlation	$r_{xy} = r_{yx} = r = 0.597401$	
Test of significance	$t_{15} = \dfrac{0.597401}{\sqrt{\dfrac{1 - (0.597401)^2}{17 - 2}}}$	
	$= \dfrac{0.597401}{0.20706}$	
	$= 2.885$	

[a] ATP level of oldest son.
[b] ATP level of youngest son.

9.3.4 Bivariate Normal Distribution

The statement that a random variable Y has a normal distribution with mean μ and variance σ^2 is a statement about the distribution of the values of Y, and is written in a shorthand way as

$$Y \sim N(\mu, \sigma^2).$$

Such a distribution is called a univariate distribution.

Definition 9.7. A specification of the distribution of two (or more) variables is called a *bivariate* (or *multivariate*) *distribution*.

The definition of such a distribution will require the specification of the numerical characteristics of each of the variables separately as well as the relationships among the variables. The most common bivariate distribution is the normal distribution. The equation for the density of this distribution as well as additional properties are given in Note 9.6.

We write that (X, Y) have a bivariate normal distribution as

$$(X, Y) \sim N(\mu_x, \mu_y, \sigma_x^2, \sigma_y^2, \rho).$$

Here, μ_x, μ_y, σ_x^2 and σ_y^2 are the means and variances of X and Y, respectively. The quantity ρ is the (population) correlation coefficient. If we assume this model, it is this quantity, ρ, that is estimated by the sample correlation, r.

The following considerations may help to give you some feeling for the bivariate normal distribution. A continuous distribution of two variables, X and Y, may be modeled as follows. Pour 1 pound of sand on a floor (the X-Y plane). The probability that a pair (X, Y) falls into an area, say A, on the floor is the weight of the sand on the area A. For a bivariate normal distribution, the sand forms one mountain, or pile, sloping down from its peak at (μ_x, μ_y), the mean of (X, Y). Cross sections of the sand at constant heights are all ellipses. Figure 9.12 shows a bivariate normal distribution. Panel (a) is a view of the sand pile. Panel (b) is a topographical map of the terrain.

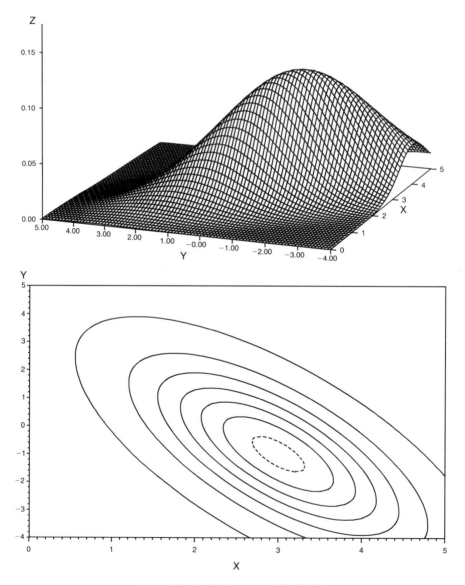

Figure 9.12. The bivariate normal distribution.

The bivariate normal distribution has the property that at every fixed value of X (or Y) the variable Y (or X) has a univariate normal distribution. In particular, write

$$Y_x = \text{the random variable } Y \text{ at a fixed value of } X = x.$$

It can be shown that at this fixed value of $X = x$,

$$Y_x \sim N\left(\alpha_y + \frac{\sigma_y}{\sigma_x}\rho x, \ \sigma_y^2(1 - \rho^2)\right).$$

This is the regression model discussed previously:

$$Y_x \sim N(\alpha + \beta x, \ \sigma_1^2)$$

where

$$\alpha = \mu_y - \beta\mu_x, \quad \beta = \frac{\sigma_y}{\sigma_x}\rho, \quad \sigma_1^2 = \sigma_y^2(1 - \rho^2).$$

Similarly, for

$$X_y = \text{the random variable } X \text{ at a fixed value of } Y = y.$$

it can be shown that

$$X_y \sim N\left(\alpha_x + \frac{\sigma_x}{\sigma_y}\rho y, \ \sigma_x^2(1 - \rho^2)\right).$$

The null hypothesis $\beta_{yx} = 0$ (or, $\beta_{xy} = 0$) is equivalent then to the hypothesis $\rho = 0$, and the t-test for $\beta = 0$ can be applied.

Suppose now that the null hypothesis is

$$\rho = \rho_0,$$

where ρ_0 is some arbitrary, but specified value. The sample correlation coefficient r does not have a normal distribution and the usual normal theory cannot be applied. However, R.A. Fisher showed that the quantity

$$Z_r = \frac{1}{2}\ln\left(\frac{1 + r}{1 - r}\right)$$

has approximately the normal distribution as follows:

$$Z_r \sim N\left(\frac{1}{2}\ln\left(\frac{1 + \rho}{1 - \rho}\right), \frac{1}{n - 3}\right)$$

where n is the number of pairs of values of X and Y from which r is computed. This is graphically illustrated in Figure 9.13.

Z_r may be used to test hypotheses about ρ and to construct a confidence interval for ρ. This is illustrated below. Table A.14 in the Appendix tabulates the Z transformation. (It may also be performed on hand calculators. The inverse, or reverse,

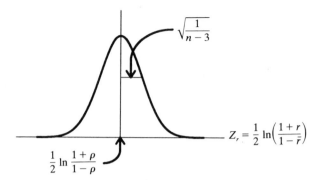

Figure 9.13. Sampling distribution of transformed correlation coefficient, r.

function to r is $(e^{2Z} - 1)/(e^{2Z} + 1)$. Z_r is also the inverse of the hyperbolic tangent, tanh. To "undo" the operation tanh used. If one has a hand calculator, with tanh and its inverse, the Z_r transformations may be performed quickly.)

Consider again Example 9.3 involving the ATP levels of youngest and oldest sons in the 17 families. The correlation coefficient was calculated to be

$$r = 0.5974.$$

This value was significantly different from zero, that is, the null hypothesis $\rho = 0$ was rejected. However, the authors show in the paper that genetic theory predicts the correlation to be $\rho = 0.5$. Does the observed value differ significantly from this value? To test this hypothesis we use the Fisher Z_r transformation. Under the genetic theory, the null hypothesis stated in terms of Z_r is

$$Z_r \sim N\left(\frac{1}{2}\ln\left(\frac{1+0.5}{1-0.5}\right), \frac{1}{17-3}\right)$$

$$\sim N(0.5493, 0.07143).$$

The observed value is (from Table A.14 or a calculator):

$$Z_r = \frac{1}{2}\ln\left(\frac{1+0.5974}{1-0.5974}\right) = 0.6891.$$

The corresponding standard normal deviate is

$$z = \frac{0.6891 - 0.5493}{\sqrt{0.07143}} = \frac{0.1398}{0.2673} = 0.5231.$$

This value does not exceed the critical values at, say, the 0.05 level, and there is no evidence to reject the null hypothesis.

Confidence intervals for ρ may be formed by first using Z_r to find a confidence interval for $1/2\ln((1+\rho)/(1-\rho))$. We then transform back to find the confidence interval for ρ. To illustrate: a $100(1-\alpha)\%$ confidence interval for $1/2\ln((1+\rho)/(1-\rho))$ is given by

$$Z_r \pm z_{1-\alpha/2}\sqrt{\frac{1}{n-3}}.$$

For a 90% confidence interval with these data, the interval is $(0.6891 - 1.645\sqrt{1/14},$ $0.6891 + 1.645\sqrt{1/14}) = (0.249, 1.13)$. From Table A.14 or a calculator, when $Z_r = 0.249$, $r = 0.244$, and when $Z_r = 1.13$, $r = 0.811$. Thus the 90% confidence interval for ρ is $(0.244, 0.811)$. This value straddles 0.5.

9.3.5 Critical Values and Sample Size

We discussed the t-test for testing the hypothesis $\rho = 0$. The formula was

$$t_{n-2} = r \bigg/ \sqrt{\frac{1 - r^2}{n - 2}}.$$

This formula is very simple and can be used for finding critical values and for sample size calculations: given that the number of observation-pairs is specified, the critical value for t with $n - 2$ degrees of freedom is determined and hence, the r critical value can be calculated. For simplicity, write $t_{n-1} = t$; solving the above equation for r^2 yields

$$r^2 = \frac{t^2}{t^2 + n - 2}.$$

For example, suppose $n = 20$, the corresponding t-value with 18 degrees of freedom at the 0.05 level is $t_{18} = 2.101$. Hence,

$$r^2 = \frac{(2.101)^2}{(2.101)^2 + 18} = 0.1969$$

and the corresponding value for r is $r \pm 0.444$; that is, with 20 observations the value of r must exceed 0.444 or be less than -0.444 to be significant at the 0.05 level. Table A.15 lists critical values for r, as a function of sample size.

Another approach is to determine the sample size needed to "make" an observed value of r significant. Algebraic manipulation of the formula gives

$$n = (t^2/r^2) - t^2 + 2.$$

A useful approximation can be derived if it is assumed that we are interested in reasonably small values of r, say $r < 0.5$; in this case $t \doteq 2$ at the 0.05 level and the formula becomes

$$n = (2/r)^2 - 2.$$

For example, suppose $r = 0.3$, the sample size needed to make this value significant is

$$n = (2/0.3)^2 - 2 = 44 - 2 = 42.$$

A somewhat more refined calculation yields $n = 43$, so the approximation works reasonably well.

9.3.6   Using the Correlation Coefficient as a Measure of Agreement for Two Methods of Measuring the Same Quantity

We have seen that for X and Y jointly distributed random variables, the correlation coefficient ρ is a population parameter value: ρ is a measure of how closely X and Y have a linear association, ρ^2 is the proportion of the Y variance that can be explained by linear prediction from X, and vice versa.

Suppose that the regression holds and we may choose X. Figure 9.14 shows data from a regression model with three different patterns of X variables chosen. The same errors were added in each figure. The X values were spread out over larger and larger intervals. Since the spread *about* the regression line remains the same and the range of Y increases as the X range increases the proportion of Y variability explained by X increases: 0.50 to 0.68 to 0.79. For the same random errors and population regression line, r can be anywhere between 0 and 1 depending upon which X values are used! In this case the correlation coefficient depends not only upon the model, but also upon experimental design, where the X's are taken. For this reason some authors say that the r should never be used unless one has a *bi*variate sample: otherwise we do not know what r means; another experimenter with the same regression model could choose different X values and obtain a radically different result.

We discuss these ideas in the context of the exercise data of Example 9.2. Suppose we were strong supporters of maximal treadmill stress testing and wanted to show how closely treadmill duration and VO$_2$ MAX are related. Our strategy for obtaining a large correlation coefficient will be to obtain a large spread of X values, duration. We may know that some of the largest duration and VO$_2$ MAX values were obtained by world

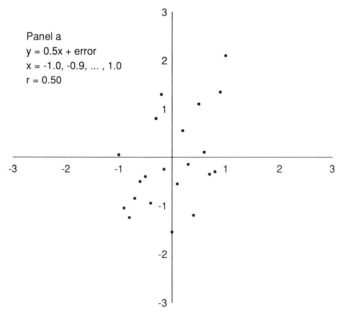

Figure 9.14. The regression model $Y = 0.5X + e$ was used. Twenty-one random $N(0, 1)$ errors were generated by computer (see Section 8.10). The same errors were used in each panel.

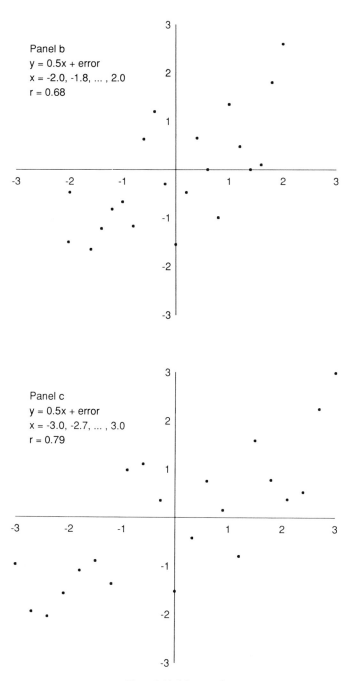

Panel b
y = 0.5x + error
x = -2.0, -1.8, ... , 2.0
r = 0.68

Panel c
y = 0.5x + error
x = -3.0, -2.7, ... , 3.0
r = 0.79

Figure 9.14. (*Continued*)

class cross-country skiers; so we would recruit some. For low values we might search for elderly overweight and deconditioned individuals. Taking a combined group of these two types of subjects should result in a large value of r.

If the same experiment is run using only very old, very overweight and very deconditioned individuals, the small range will produce a small, statistically insignificant r value.

Since the same treadmill test procedure is associated with large and small r values what does r mean? A *better* summary indicator is the estimate, $s_{y.x}$ of the residual standard deviation σ_1. If the linear regression model holds, this would be estimated to be the same in each case.

Is it wrong to calculate or present r when a bivariate sample is not obtained? Our answer is a qualified no; that is, it is all right to present r in regression situations provided that:

1. The limitations are kept in mind and discussed. Possible comments on the situation for other sorts of X values might be appropriate.
2. The standard deviation of the residuals should be estimated and presented.

In Chapter 7, the kappa statistic was presented. This was a measure of the amount of agreement when two categorical measurements of the same objects were available. If the two measurements were continuous, the correlation coefficient r is often used as a measure of the agreement of the two techniques. Such use of r is subject to the comments above.

9.3.7 Errors in Both Variables

An assumption in the linear regression model has been that the predictor variable could be measured without error and that the variation in the dependent variable was of one kind only and could be modelled completely if the value of the predictor variable was fixed. In almost all cases, these assumptions do not hold. For example, in measuring psychological characteristics of individuals, there is (1) variation in the characteristics from individual to individual; (2) error in the measurement of these psychological characteristics. It is almost certainly true that this problem is present in all scientific work. However, it may be that the measurement error is "small" relative to the variation of the individuals and, hence, the former can be neglected.

The problem is difficult and we will not discuss it beyond the effect of errors on the correlation coefficient. For a more complete treatment, consult Acton [1959] or Kendall and Stuart [1967] Volume 2.

Suppose we are interested in the correlation between two random variables X and Y which are measured with errors so that instead of X and Y, we observe

$$W = X + d, \quad V = Y + e,$$

where d and e are errors. The sampling we have in mind is the following: a "case" is selected at random from the population of interest. The characteristics X and Y are measured but with random independent errors d and e. It is assumed that these errors have mean zero and variances σ_1^2 and σ_2^2, respectively. Another "case" is then selected and the measurement process is repeated with error. Of interest is the

Table 9.9. Effect of Errors of Measurement on the Correlation Between Two Random Variables.

$\dfrac{\sigma_1^2}{\sigma_X^2}$	$\dfrac{\sigma_2^2}{\sigma_Y^2}$	ρ_{VW}
0	0	$1\ \rho_{XY}$
0.05	0.05	$0.95\rho_{XY}$
0.10	0.10	$0.91\rho_{XY}$
0.20	0.10	$0.87\rho_{XY}$
0.20	0.20	$0.83\rho_{XY}$
0.30	0.30	$0.77\rho_{XY}$

correlation ρ_{XY} between X and Y, but the correlation ρ_{VW} is estimated. What is the relationship between these two correlations? The correlation ρ_{XY} can be written

$$\rho_{XY} = \frac{\sigma_{XY}}{\sqrt{\sigma_X^2 \sigma_Y^2}}.$$

The reason for writing the correlation this way can be understood when the correlation between V and W is considered:

$$\rho_{VW} = \frac{\sigma_{XY}}{\sqrt{(\sigma_X^2 + \sigma_1^2)(\sigma_Y^2 + \sigma_2^2)}}$$

$$= \frac{\sigma_{XY}}{\sigma_X \sigma_Y \sqrt{\left(1 + \dfrac{\sigma_1^2}{\sigma_X^2}\right)\left(1 + \dfrac{\sigma_2^2}{\sigma_Y^2}\right)}}$$

$$= \frac{\rho_{XY}}{\sqrt{\left(1 + \dfrac{\sigma_1^2}{\sigma_X^2}\right)\left(1 + \dfrac{\sigma_2^2}{\sigma_Y^2}\right)}}.$$

The last two formulae indicate that the correlation between V and W is smaller in absolute value than the correlation between X and Y by an amount determined by the ratio of the measurement errors to the variance in the population. Table 9.9 gives the effect on ρ_{XY} as related to the ratios of σ_1^2/σ_X^2 and σ_2^2/σ_Y^2.

A 10% error of measurment in the variables X and Y produces a 9% reduction in the correlation coefficient. The conclusion is that errors of measurement reduce the correlation between two variables; this phenomenon is called *attenuation*.

9.3.8 Nonparametric Estimates of Correlation

As indicated in previous discussions, the correlation coefficient is quite sensitive to outliers. There are many ways of getting estimates of correlation that are more robust; the paper by Devlin *et al.* [1975] contains a description of some of these methods. In this section we want to discuss two methods of testing correlations derived from the ranks of observations.

The procedure leading to the Spearman rank correlation is as follows; given a set of n observations on the variables X, Y, the values for X are replaced by their ranks, and similarly, the values for Y. Ties are simple assigned the average of the ranks asoociated with the tied observations. The following scheme illistrates the procedure:

Case	X	Rank(X)	Y	Rank(Y)	d = Rank(X) − Rank(Y)
1	x_1	R_{x_1}	y_1	R_{y_1}	$d_1 = R_{x_1} - R_{y_1}$
2	x_2	R_{x_2}	y_2	R_{y_2}	$d_2 = R_{x_2} - R_{y_2}$
3	x_3	R_{x_3}	y_3	R_{y_3}	$d_3 = R_{x_3} - R_{y_3}$
⋮	⋮	⋮	⋮	⋮	⋮
n	x_n	R_{x_n}	y_n	R_{y_n}	$d_n = R_{x_n} - R_{y_n}$

The correlation is then calculated between R_x and R_y. In practice, the *Spearman rank correlation formula* is used

$$r_s = r_{R_x R_y} = 1 - \frac{6 \sum d_i^2}{n^3 - n}.$$

It can be shown that the usual Pearson product-moment correlation formula reduces to this formula when the calculations are made on the ranks, if there are no ties. Note: For one or two ties, the results are virtually the same. It is possible to correct the Spearman formula for ties but a simpler procedure is to calculate r_s by application of usual product-moment formula to the ranks. Table A.16 gives percentile points for testing the hypothesis that X and Y are independent.

Example 9.4.   Consider again the data in Table 9.3 dealing with the ATP levels of the oldest and youngest sons. This data is reproduced in Table 9.10 together with the ranks, the ATP levels being ranked from lowest to highest.

Note that the oldest sons in Families 6 and 13 had the same ATP levels; they would have been assigned ranks 12 and 13 if the values had been recorded more accurately; consequently, they are both assigned a rank of 12.5.

For this example,

$$n = 17,$$

$$\sum d_i^2 = 298.5,$$

$$r_s = 1 - \frac{(6)(298.5)}{17^3 - 17} = 1 - 0.3658 = 0.6342.$$

This value compares reasonably well with the value $r_{xy} = 0.597$ calculated on the actual data. If the usual Pearson product-moment formula is applied to the ranks, the value $r_s = 0.6340$ is obtained. The reader may verify that this is the case. The reason for the slight difference is due to the tie in values for two of the oldest sons.

Table A.16 shows statistical significance at the two-sided 0.05 level since $r_s = 0.6342 > 0.490.$ □

The second nonparametric correlation coefficient is the *Kendall rank correlation coefficient*. Recall our motivation for the correlation coefficient r. If there is positive association, increase in X will tend to correspond to increase in Y. That is, given

Table 9.10. Rank Correlation Analysis of ATP Levels in Youngest and Oldest Sons in Seventeen Families.

Family	Youngest ATP level	Rank(X)	Oldest ATP level	Rank(Y)	d^a
1	4.18	4	4.81	11	-7
2	5.16	12	4.98	14	-2
3	4.85	9	4.48	6	3
4	3.43	1	4.19	3	-2
5	4.53	5	4.27	4	1
6	5.13	11	4.87	12.5	-1.5
7	4.10	2	4.74	10	-8
8	4.77	7	4.53	7	0
9	4.12	3	3.72	1	2
10	4.65	6	4.62	8	-2
11	6.03	17	5.83	17	0
12	5.94	15	4.40	5	10
13	5.99	16	4.87	12.5	3.5
14	5.43	14	5.44	16	-2
15	5.00	10	4.70	9	1
16	4.82	8	4.14	2	6
17	5.25	13	5.30	15	-2

$$\sum d = 0$$

$$\sum d^2 = 298.5$$

^{a}Rank(X) − rank(Y).

two data points (X_1, Y_1) and (X_2, Y_2) if $X_1 - X_2$ is positive then $Y_1 - Y_2$ is positive. In this case, $(X_1 - X_2)(Y_1 - Y_2)$ is usually positive. If there is negative association $(X_1 - X_2)(Y_1 - Y_2)$ will usually be negative. If X and Y are independent the expected value is zero. Kendall's rank correlation coefficient is based on this observation.

Definition 9.8. Consider a bivariate sample of size n, $(X_1, Y_1), \ldots, (X_n, Y_n)$. For each pair, count 1 if $(X_i - X_j)(Y_i - Y_j) > 0$. Count -1 if $(X_i - X_j)(Y_i - Y_j) < 0$. Count zero if $(X_i - X_j)(Y_i - Y_j) = 0$. Let κ be the sum of these $n(n-1)/2$ counts. Kendall's τ is

$$\tau = \frac{\kappa}{n(n-1)/2}.$$

1. The value of τ is between -1 and 1. Under the null hypothesis of independence τ is symmetric about zero. Table A.17 gives critical values when there are no ties. The table is in terms of $K = 2\kappa + n(n-1)/4 = n(n-1)(1+\tau)/4$ for no ties. Note that this κ is not related to the kappa of Chapter 7.

2. Note that $(R_{X_i} - R_{X_j})(R_{Y_i} - R_{Y_j})$ has the same sign as $(X_i - X_j)(Y_i - Y_j)$. That is, both are positive, or both are negative or both are zero. If we calculated τ from the ranks of the (X_i, Y_i), we get the same number. Thus τ is a nonparametric quantity based upon ranks; it does not depend upon normality assumptions.

3. The expected value of τ is

$$P\left((X_i - X_j)(Y_i - Y_j) > 0\right) - P\left((X_i - X_j)(Y_i - Y_j) < 0\right).$$

4. For large n and no or few ties an approximate standard normal test statistic is

$$Z = \frac{\kappa}{\sqrt{n(n-1)(2n+5)/18}}.$$

More information where there are ties is given in Note 9.7.

5. If $(X_i - X_j)(Y_i - Y_j) > 0$ the pairs are said to be concordant. If $(X_i - X_j)(Y_i - Y_j) < 0$ the pairs are discordant. The quantity K is then the number of concordant pairs.

Return to the ATP data of Table 9.10. $(X_1 - X_2)(Y_1 - Y_2) = (4.18 - 5.16)(4.81 - 4.98) > 0$, so we count +1. Comparing each of the $17 \times 16/2 = 136$ pairs gives the +1's, 0's and −1's in Table 9.11. Adding these numbers $\kappa = 67$, $K = 101.5$ and $\tau = 67/(17 \times 16/2) = 0.493$. From Table A.17, this is statistically significant with $p < 0.005$ or 0.01 for a two sided test. The asymptotic Z-value is

$$Z = \frac{67}{\sqrt{17 \times 16 \times 39/18}} = 2.67$$

with $p = 0.0076$ (two-sided). Note that we used Table A.17 with a tie, or zero value for data $i = 13$, $j = 6$. With ties the table is conservative; that is, the probability of rejecting the null hypothesis is even less than the table value.

Table 9.11. Data for Example 9.4 (continued).[a]

i	1	2	3	4	5	6	7	8	9	10	11	12	13	14	15	16
2	1															
3	−1	1														
4	1	1	1													
5	−1	1	1	1												
6	1	1	1	1	1											
7	1	1	−1	1	−1	1										
8	−1	1	−1	1	1	1	−1									
9	1	1	1	−1	1	1	−1	1								
10	−1	1	−1	1	1	1	−1	−1	1							
11	1	1	1	1	1	1	1	1	1	1						
12	−1	−1	−1	1	1	−1	−1	−1	1	−1	1					
13	1	−1	1	1	1	0	1	1	1	1	1	1				
14	1	1	1	1	1	1	1	1	1	1	1	−1	−1			
15	−1	1	1	1	1	1	−1	1	1	1	1	−1	1	1		
16	−1	1	1	−1	−1	1	−1	−1	1	−1	1	1	1	1	1	
17	1	1	1	1	1	1	1	1	1	1	1	−1	−1	1	1	1

[a]Consider $(X_i - X_j)(Y_i - Y_j)$: the entries are 1 if this is positive, 0 if this equals 0 and −1 if this is negative.

9.3.9 Change and Association

Consider two continuous measurements of the same quantity on the same individuals at different times or under different circumstances. The two times might be before and after some treatment. They might be for a person taking a drug and not taking a drug. If we want to see if there there is a difference in the means at the two times or under the two circumstances, we have several statistical tests: the paired t-test, the signed rank test and the sign test. Note that we have observed pairs of numbers on each individual.

We now have new methods when pairs of numbers are observed: linear regression and correlation. Which technique should be used in a given circumstance?

The first set of techniques looks for *changes between the two measurements*. The second set of techniques look for association and sometimes *the ability to predict*. The two concepts are different ideas:

i. Consider two independent length measurements from the same X-rays of a sample of patients. Presumably there is a "true" length. The measurements should fluctuate about the true length. Since the true length will fluctuate from patient to patient, the two readings should be associated, hopefully highly correlated. Since both measurements are of the same quantity, there should be little or no change. This would be a case where one expects association, but no change.

ii. Consider cardiac measurements on patients before and after a heart transplant. The initial measurements refer to a failing heart. After heart transplant the measurements refer to the donor heart. There will be little or no association because the measurements of output, and so on refer to different (somewhat randomly paired) hearts.

There are situations where both change and prediction or association are relevant. After observing a change, one might like to investigate how the new changed values relate to the original values.

9.4 COMMON MISAPPLICATION OF REGRESSION AND CORRELATION METHODS

In this section, we discuss some of the pitfalls of regression and correlation methods.

9.4.1 Regression to the Mean

Consider Figure 9.15, which has data points with approximately zero correlation or association. Suppose that we select for some purpose cases by using measurement A. For example, we might select all those in a certain interval such as that denoted by I or such as that denoted by II. Since there is no association, if we take a second measurement, measurement B, regardless of how we choose our region, I or II for example, the mean of measurement B will be the overall mean of the population. Suppose A and B are measurements on the same quantity (for example, a test measured before and after treatment). If cases are selected for low initial values such as region II, after treatment the average score would improve because it would tend to

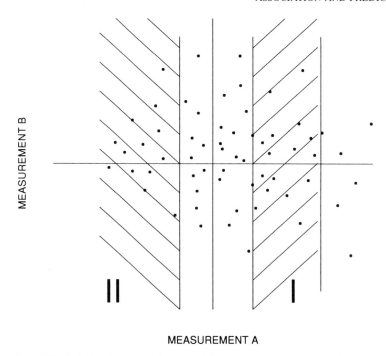

Figure 9.15. Variables with no association; shaded areas represent regression to the mean.

be close to the mean of the population values. It would appear there was improvement; in fact the change would be due to random variability and the case selection. This phenomenon is called *regression to the mean*.

As another example, consider individuals in a quantitative measurement of the amount of rash due to an allergy. Individuals will have considerable variability due to biology variability and environmental variability. Over time, in a random fashion, perhaps related to the season, the severity of the rash will ebb and flow. Such individuals will naturally tend to seek medical help when things are in a particularly bad state. Following the soliciting of help, biological variability will give improvement with or without treatment. Thus, if the treatment is evaluated (using before and after values) there would be a natural drop in the amount of rash simply because medical help was solicited during particularly bad times. This phenomenon again is *regression to the mean*. The phenomenon of regression to the mean is one reason that control groups are used in clinical studies.

9.4.2   Spurious Correlation

Consider a series of population units, for example, states. Suppose that we wish to relate the occurrence of death from two distinct causes, for example, cancer at two different sites on the body. If we take all the states and plot a scatter diagram of the number of deaths from the two causes, there will be a relationship simply because states with many more people such as California or New York will have a large number of deaths compared to a smaller state such as Wyoming or New Hampshire.

It is necessary to somehow adjust for or take into account the population. The most natural thing to do is to take the death rate from certain causes, that is divide the number of deaths by the population of the state. This would appear to be a good solution to the problem. This, however, introduces another problem. If we have two variables, X and Y, which are *not related* and we divide them by a third variable Z, which is random, the two ratios X/Z and Y/Z *will be related*. Suppose Z is the true denominator measured with error. The reason for the relationship is that when Z is on the low side, since we are dividing by Z, we will increase both numbers at the same time; when Z is larger than it should be and we divide X and Y by Z, we decrease both numbers. The introduction of correlation due to computing rates using the same denominator is called *spurious correlation*. For further discussion on this, see Neyman [1952], and Kronmal [1993] who gives a superb, readable review. The best way to adjust for population size is to use the techniques of multiple regression which will be discussed in Chapter 11.

9.4.3 Extrapolation Beyond the Range of the Data

For many data sets, including the three of this chapter, the linear relationship does a reasonable job of summarizing the association between two variables. In other situations, the relationship may be reasonably well modeled as linear over a part of the range of X, but not over the entire range of X. Suppose however, data had only been collected on a small range of X. Then a linear model might fit the accumulated data quite well. If one takes the regression line and uses it as an indication of what would happen for data values *outside the range covered by the actual data*, trouble can result. In order to have confidence in such extrapolation, one needs to know that indeed the linear relationship holds over a broader range than the range associated with the actual data. Sometimes this assumption is valid, but often it is quite wrong. There is no way of knowing in general to what extent extrapolation beyond the data gives problems. Some of the possibilities are indicated graphically in Figure 9.16. Note that virtually any of these patterns of curves, when data are observed over a short range can reasonably be approximated by a linear function. Over a wider range, a linear approximation is not adequate. But if one does not have data over the wide range this cannot be seen.

9.4.4 Inferring Causality from Correlation

Because two variables are associated does not necessarily mean that there is any causal connection between them. For example, if one recorded for each year two numbers: the numbers of hospital beds and the total attendance at major league baseball games, there would be a positive association because both of these variables have increased as the population increased. The direct connection is undoubtedly slight at best. Thus, regression and correlation approaches show observed relationships which may or may not represent a causal relationship. In general, the strongest inference for causality comes from experimental data; in this case, factors are changed by the experimenter to observe change in a response. Regression and correlation from observational data may be very suggestive but do not definitively establish any causal relationships.

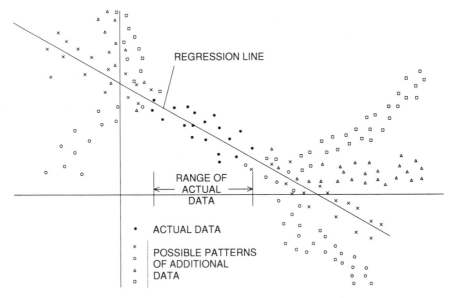

Figure 9.16. The danger of extrapolating beyond observed data.

9.4.5 Interpretation of the Slope of the Regression Line

During the discussion, we have noted that the regression equation implies that if the predictor or independent variable X increases by an amount ΔX, then on the average, Y increases by an amount $\Delta Y = b\Delta X$. This is sometimes interpreted to mean that if we can modify a situation such that the X variable is changed by ΔX, then the Y variable will change correspondingly; this may or may not be the case. For example, if we look at a scatter diagram of adults' height and weight, it does not follow if we induce a change in an individual's weight, either by dieting or by excess calories that the individual's height will change correspondingly. Nevertheless, there is an association between height and weight. Thus, the particular inference depends on the science involved. Earlier in this chapter, it was noted that from the relation between VO_2 MAX and the duration of the exercise test that if an individual is trained to have an increased duration, then the VO_2 MAX will also increase. This particular inference is correct and has been documented by serial studies recording both variables. It follows from other data and scientific understanding. It is *not* a logical consequence of the observed association.

9.4.6 Outlying Observations

As noted above, outlying observations can have a large effect upon the actual regression line (see Figure 9.7, for example). If one examines these scattergrams or residual plots the problem should be recognized. In many situations, however, individuals look at large numbers of correlations and do not have the time, the wherewithal, or possibly the knowledge to examine all of the necessary visual presentations. In such a case an outlier can be missed and data may be interpreted inappropriately.

NOTES

Note 9.1: Origin of the Term Regression

Sir Francis Galton first used the term in 1885. He studied the heights of parents and offspring. He found (on the average) that children of tall parents were closer to the average height (were shorter); children of short parents were taller and closer to the average height. The children's height *regressed* to the average.

Note 9.2: Maximum Likelihood Estimation of Regression and Correlation Parameters

For a data set from a continuous probability density, the probability of observing the data is proportional to the probability density function. It makes sense to *estimate the parameters by choosing parameters to make the probability of the observed data as large as possible*. Such estimates are called *maximum likelihood estimates* (MLE's). Knowing $X_1, \ldots, X_n$ in the regression problem, the likelihood function for the observed $Y_1, \ldots, Y_n$ is (assuming normality)

$$\prod_{i=1}^{n} \frac{1}{\sqrt{2\pi}\sigma} \exp\left[-\frac{1}{2\sigma^2}(Y_i - (\alpha + \beta X_i))^2\right].$$

The maximum likelihood estimates of α, β are the least squares estimates a and b. For the bivariate normal distribution the MLE of ρ is r.

Note 9.3: Notes on the Variance of a, Variance of $a + bx$ and Choice of x for Small Variance (Experimental Design)

1. The variance of a in the regression equation $y = a + bx$, can be derived as follows: $a = \bar{y} + b\bar{x}$; it is true that $\bar{y}$ and b are statistically independent hence,

$$\text{var}(a) = \text{var}(\bar{y} + b\bar{x})$$

$$= \text{var}(\bar{y}) + \bar{x}^2 \text{var}(b)$$

$$= \frac{\sigma_1^2}{n} + \bar{x}^2 \frac{\sigma_1^2}{[x^2]}$$

$$= \sigma_1^2 \left(\frac{1}{n} + \frac{\bar{x}^2}{[x^2]}\right).$$

2. Consider the variance of the estimate of the mean of y at some arbitrary fixed point X;

$$\sigma_1^2 \left(\frac{1}{n} + \frac{(x - \bar{x})^2}{[x^2]}\right).$$

a. Given a choice of x, the quantity is minimized at $x = \bar{x}$.

b. For values of x close to $\bar{x}$ the contribution to the variance is minimal.

c. The contribution increase as the *square* of the distance the predictor variable x is from $\bar{x}$.

d. If there was a choice in the selection of the predictor variables, the quantity $[x^2] = \sum(x_i - \bar{x})^2$ is maximized if the predictor variables are spaced as far apart as possible. If X can have a range of values say, X_{min} to X_{max}, the quantity $[x^2]$ is maximized if half the observations are placed at X_{min} and the other half at X_{max}. The quantity $(x - \bar{x})^2/[x^2]$ will then be as small as possible. Of course, there is a price paid for this design: it is not possible to check the linearity of the relationship between Y and X.

Note 9.4: Regression Lines Through the Origin

Suppose we want to fit the model $Y \sim N(\beta X, \sigma^2)$, that is, the line goes through the origin. In many situations this is an appropriate model, e.g., in relating body weight to height, it is reasonable to assume that the regression line must go through the origin. However, the regression relationship may not be linear over the whole range and often the interval of interest is quite far removed from the origin.

Given n pairs of observation (x_i, y_i) $i = 1, \ldots, n$, and a regression line through the origin is desired, it can be shown that the least squares estimate, b, of β, is

$$b = \frac{\sum x_i y_i}{\sum x^2}.$$

The residual sum of squares is based on the quantity

$$\sum (y_i - \widehat{y}_i)^2 = \sum (y_i - bx_i)^2,$$

and has associated with it, $n - 1$ degrees of freedom, since only one parameter, β, is estimated.

Note 9.5: Robust Regression Models

The least squares regression coefficients result from minimizing

$$\sum_{i=1}^{n} g(Y_i - a - bX_i),$$

where the function $g(z) = z^2$. For large z (large residuals) this term is very large. In the second column of figures in Figure 9.7 we saw that one outlying value could heavily modify an otherwise nice fit.

One way to give less importance to large residuals is to choose the function g to put less weight on outlying values. Many robust regression techniques take this approach. We can choose g so that for most z, $g(z) = z^2$ as in the least squares estimates, but for very large $|z|$, $g(z)$ is less than z^2, even zero for extreme z! See Draper and Smith [1981], pp. 342–344 and Huber [1981], Chapter 7.

Note 9.6: Bivariate Normal Density Function

The formula for the density of the bivariate normal distribution is

$$f_{X,Y}(x,y) = \frac{1}{2\pi\sigma_X\sigma_Y\sqrt{1-p^2}} \exp\left[-\frac{1}{2(1-p^2)}\left(Z_X^2 - 2\rho Z_X Z_Y + Z_Y^2\right)\right]$$

where

$$Z_X = \frac{x - \mu_X}{\sigma_X} \quad \text{and} \quad Z_Y = \frac{y - \mu_Y}{\sigma_Y}.$$

The quantities μ_X, μ_Y, σ_X, σ_Y are, as usual, the means and standard deviations of X and Y, respectively.

Several characteristics of this distribution can be deduced from this formula:

1. If $\rho = 0$, the equation becomes,

$$f_{X,Y}(x,y) = \frac{1}{2\pi\sigma_X\sigma_Y} \exp\left[-\frac{1}{2}\left(Z_X^2 + Z_Y^2\right)\right]$$

and can be written as,

$$= \frac{1}{\sqrt{2\pi}\sigma_X} \exp\left(-\frac{1}{2}Z_X^2\right) \frac{1}{\sqrt{2\pi}\sigma_Y} \exp\left[-\frac{1}{2}Z_Y^2\right]$$

$$= f_X(x)f_Y(y).$$

Thus in the case of the bivariate normal distribution, $\rho = 0$, that is the correlation is zero, implies the random variables X and Y are statistically independent.

2. Suppose $f_{X,Y}(x,y)$ is fixed at some specified value, this implies that the expression in the exponent of the density $f_{X,Y}(x,y)$ has a fixed value, say, K.

$$K = \frac{-1}{2(1-\rho^2)}\left(\left(\frac{x-\mu_X}{\sigma_X}\right)^2 - 2\rho\left(\frac{x-\mu_X}{\sigma_X}\right)\left(\frac{y-\mu_Y}{\sigma_Y}\right) + \left(\frac{y-\mu_Y}{\sigma_Y}\right)^2\right).$$

This is the equation of an ellipse centered at (μ_X, μ_Y).

Note 9.7: Ties in Kendall's Tau

When there are ties in the X_i and/or Y_i values for Kendall's tau the variability is reduced. The asymptotic formula needs to be adjusted accordingly (Hollander and Wolfe [1973]). Let the X_i values have g distinct values with ties with t_j tied observations at the jth tied value. Let the Y_i values have h distinct tied values with u_k tied observations at the kth tied value.

Under the null hypothesis of independence between the X and Y values, the variance of K is

$$\text{var}(K) = (n(n-1)(2n+5)/18)$$

$$- \left(\sum_{j=1}^{g} t_j(t_j - 1)(2t_j + 5)/18 \right)$$

$$- \left(\sum_{j=1}^{h} u_k(u_k - 1)(2u_k + 5)/18 \right)$$

$$+ \left(\frac{\left(\sum_{j=1}^{g} t_j(t_j - 1)(t_j - 2) \right) \left(\sum_{k=1}^{h} u_k(u_k - 1)(u_k - 2) \right)}{(9n(n-1)(n-2))} \right)$$

$$+ \left(\frac{\left(\sum_{j=1}^{g} t_j(t_j - 1) \right) \left(\sum_{k=1}^{h} u_k(u_k - 1) \right)}{(2n(n-1))} \right).$$

The asymptotic normal Z value is

$$Z = K/\sqrt{\text{var}(K)}.$$

Note 9.8: Weighted Regression Analysis

Weighted Linear Regression. In certain cases the assumption of homogeneity of variance of the dependent variable, Y, at all levels of X is not tenable. Suppose that the precision of value $Y = y$ is proportional to a value W, the weight. Usually the precision is the reciprocal of the variance at X_i. The data then can be modeled as follows:

Case	X	Y	W
1	x_1	y_1	w_1
2	x_2	y_2	w_2
⋮	⋮	⋮	⋮
i	x_i	y_i	w_i
⋮	⋮	⋮	⋮
n	x_n	y_n	w_n

Define $\sum w_i(x_i - \bar{x}_i)^2 = [wx^2]$, $\sum w(x_i - \bar{x})(y_i - \bar{y}) = [wxy]$. It can be shown that the *weighted* least squares line has slope and intercept,

$$b = [wxy]/[wx^2] \quad \text{and} \quad a = \bar{y} - b\bar{x},$$

where

$$\bar{y} = \frac{\sum w_i y_i}{\sum w_i} \quad \text{and} \quad \bar{x} = \frac{\sum w_i x_i}{\sum w_i}.$$

It is a weighted least squares solution in that the quantity $\sum w_i(y_i - \hat{y}_i)^2$ is minimized. If all the weights are the same, say equal to 1, the ordinary least squares solutions are obtained.

PROBLEMS

In most of the problems below you are asked to perform some subset of the following tasks:

a. Plot the scatter diagram for the data.

b. Compute for $\overline{X}$, $\overline{Y}$, $[x^2]$, $[y^2]$, and $[xy]$ those quantities not given.

c. Find the regression coefficients a and b.

d. Place the regression line on the scatter diagram.

e. Give $s_{y.x}^2$ and $s_{y.x}$.

f. Compute the missing predicted values, residuals and normal deviates for the given portion of the table.

g. Plot the residual plot.

h. Interpret the residual plot.

i. Plot the residual normal probability plot.

j. Interpet the residual normal probability plot.

k.   **i.** Construct the 90% confidence interval for β.

   ii. Construct the 95% confidence interval for β.

   iii. Construct the 99% confidence interval for β.

   iv. Compute the t-statistic for testing $\beta = 0$. What can you say about its p-value?

l.   **i.** Construct the 90% confidence interval for α.

   ii. Construct the 95% confidence interval for α.

   iii. Construct the 99% confidence interval for α.

m. Construct the ANOVA table and use Table A.7 to give information about the p-value.

n. Construct the 95% confidence interval for $\alpha + \beta X$ at the specified X value(s).

o. Construct the interval such that one is 95% certain a new observation at the specified X value(s) will fall into the interval.

p. Compute the correlation coefficient r.

q.   **i.** Construct the 90% confidence interval for ρ.

   ii. Construct the 95% confidence interval for ρ.

   iii. Construct the 99% confidence interval for ρ.

r. Test the independence of X and Y using Spearman's rank correlation coefficient. Compute the coefficient.

s. Test the independence of X and Y using Kendall's rank correlation coefficient. Compute the value of the coefficient.

t. Compute student's paired t-test for the data, if not given; in any case, interpret.

u. Compute the signed rank statistic, if not given; in any case, interpret.

The first set of problems, problems 1 to 4 come from the exercise data Example 9.2.

1. Suppose we use duration, X, to predict VO_2 MAX, Y. The scatter diagram is Figure 9.2. The $\overline{X} = 647.4$, $\overline{Y} = 40.57$, $[x^2] = 673496.4$, $[y^2] = 3506.2$, $[xy] = $

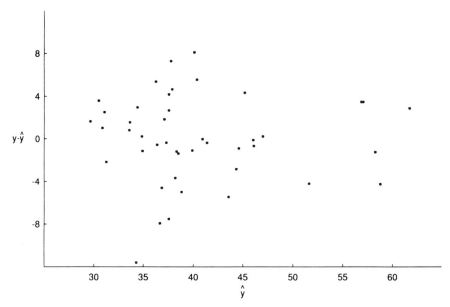

Figure 9.17. Residual plot for the data of Example 9.2, VO$_2$ MAX predicted from duration.

43352.5. Do (c), (e), (f), (h), (k-ii), (k-iv), (l-ii), (m), (n) at $x = 650$, (p) and (q-ii) (the residual plot is Figure 9.17).

X	Y	$\widehat{Y}$	$Y - \widehat{Y}$	Normal Deviate
706	41.5	44.5	−3.0	−0.80
732	45.9	46.13	−0.23	−0.06
930	54.5	?	?	?
900	60.3	?	3.59	0.96
903	60.5	56.90	3.60	0.97
976	64.6	61.50	3.10	0.83
819	47.4	?	−4.21	−1.13
922	57.0	58.10	−1.10	−0.29
600	40.2	37.82	?	0.64
540	35.2	?	1.16	0.31
560	33.8	35.30	−1.50	?
637	38.8	40.15	−1.35	−0.36
593	38.9	?	1.52	0.41
719	49.5	45.31	?	1.23
615	37.1	38.77	−1.67	−0.45
589	32.2	37.13	?	−1.32
478	31.3	30.14	1.16	0.31
620	33.8	39.08	−5.28	?
710	43.7	44.75	−1.05	−0.28
600	41.7	37.82	3.88	1.04
660	41.0	41.60	−0.60	−0.16

What proportion of the Y variance is explained by X? (In practice, duration is used as a reasonable approximation to VO_2 MAX.)

2. One expects exercise performance to reduce with age. In this problem, $X =$ age and $Y =$ duration.
$\overline{X} = 47.2$, $\overline{Y} = 647.4$, $[x^2] = 4303.2$, $[y^2] = 673496.4$, $[xy] = -36538.5$. Do (c), (e), (k-i), (l-i), (p), (q-i).

3. To see if maximum heart rate changes with age, the following numbers are found where $X =$ age, $Y =$ maximum heart rate.
$\overline{X} = 47.2$, $\overline{Y} = 174.8$, $[x^2] = 4303.2$, $[y^2] = 5608.5$, $[xy] = -2915.4$. Do (c), (e), (k-iii), (k-iv), (m), (p), and (p-iii).

4. The relationship between height and weight was examined in these active healthy males.
$X =$ height, $Y =$ weight, $\overline{X} = 177.7$, $\overline{Y} = 77.8$, $[x^2] = 1985.2$, $[y^2] = 3154.5$, $[xy] = 1845.6$. Do (c), (e), (m), (p), (q-i). How do the p-values for the F-test (in part (m)) and for the transformed Z for r compare?
There were two normal deviates of values 3.44 and 2.95. If these two individuals were removed from the calculation $\overline{X} = 177.5$, $\overline{Y} = 76.7$, $[x^2] = 1944.5$, $[y^2] = 2076.12$, and $[xy] = 1642.5$. How much do the regression coefficients a and b, and correlation coefficient r, change?

Problems 5–8 also refer to the Bruce *et al.* [1973] paper, as did Example 9.2 and Problems 1–4. The data for 43 active females are given in Table 9.12.

5. The duration and VO_2 MAX relationship for the active females is studied in this problem. $\overline{X} = 541.9$, $\overline{Y} = 29.1$, $[x^2] = 251260.4$, $[y^2] = 1028.7$, $[xy] = 12636.5$. Do (c), (e), (f), (g), (h), (i), (j), (k-iv), (m), (p), and (q-ii). Table 9.13 contains the residuals.
If the data are rerun with the sixth case omitted, the values of $\overline{X}$, $\overline{Y}$, $[x^2]$, $[y^2]$, and $[xy]$ are changed to 512.9, 29.2, 243843.1, 1001.5, and 13085.6, respectively. Find the new estimates a, b, and r. By what percent are they changed?

6. With $X =$ age and $Y =$ duration; $\overline{X} = 45.1$, $\overline{Y} = 514.9$, $[x^2] = 4399.2$, $[y^2] = 251260.4$. $[xy] = -22911.3$. For each 10-year increase in age, how much does duration tend to change? What proportion of the variability in VO_2 MAX is accounted for by age? Do (m) and (q-ii).

7. With $X =$ age and $Y =$ maximum heart rate, $\overline{X} = 45.1$, $\overline{Y} = 180.6$, $[x^2] = 4399.2$, $[y^2] = 5474.6$. $[xy] = -2017.3$. Do (c), (e), (k-i), (k-iv), (l-i), (m), (n) at $X = 30$ and $X = 50$, (o) at $X = 45$, (p) and (q-ii).

8. $X =$ height and $Y =$ weight, $\overline{X} = 164.7$, $\overline{Y} = 61.3$, $[x^2] = 1667.1$, $[y^2] = 2607.4$. $[xy] = 1006.2$. Do (c), (e), (h), (k-iv), (m), and (p). Check that $t^2 = F$. The residual plot is Figure 9.18.

For Problems 9–12 additional Bruce *et al.* [1973] data is used.
Tables 9.14 and 9.15 have the data for 94 *sedentary males*.

Table 9.12. Exercise Data for Healthy Active Females. Data from Bruce *et al.* [1973]. For Problems 5–8.

Duration	VO$_2$ MAX	HR	Age	Height	Weight
660	38.1	184	23	177	83
628	38.4	183	21	163	52
637	41.7	200	21	174	61
575	33.5	170	42	160	50
590	28.6	188	34	170	68
600	23.9	190	43	171	68
562	29.6	190	30	172	63
495	27.3	180	49	157	53
540	33.2	184	30	178	63
470	26.6	162	57	161	63
408	23.6	188	58	159	54
387	23.1	170	51	162	55
564	36.6	184	32	165	57
603	35.8	175	42	170	53
420	28.0	180	51	158	47
573	33.8	200	46	161	60
602	33.6	190	37	173	56
430	21.0	170	50	161	62
508	31.2	158	65	165	58
565	31.2	186	40	154	69
464	23.7	166	52	166	67
495	24.5	170	40	160	58
461	30.5	188	52	162	64
540	25.9	190	47	161	72
588	32.7	194	43	164	56
498	26.9	190	48	176	82
483	24.6	190	43	165	61
554	28.8	188	45	166	62
521	25.9	184	52	167	62
436	24.4	170	52	168	62
398	26.3	168	56	162	66
366	23.2	175	56	159	56
439	24.6	156	51	161	61
549	28.8	184	44	154	56
360	19.6	180	56	167	79
566	31.4	184	40	165	56
407	26.6	156	53	157	52
602	30.6	194	52	161	65
488	27.5	190	40	178	64
526	30.9	188	55	162	61
524	33.9	164	39	166	59
562	32.3	185	57	168	68
496	26.9	178	46	156	53

Table 9.13. Data for Problem 5.

X	Y	$\widehat{Y}$	Residual	Normal Deviate
660	38.1	36.35	1.75	0.56
628	38.4	34.74	3.66	1.18
637	41.7	35.19	6.51	2.10
575	33.5	32.08	1.42	0.46
590	28.6	32.83	−4.23	−1.37
600	23.9	?	?	?
562	29.6	31.42	−1.82	−0.59
495	27.3	28.05	−0.75	−0.24
540	33.2	?	2.88	0.93
470	26.6	26.80	−0.20	−0.06
408	23.6	23.68	−0.07	−0.02
387	23.1	22.62	0.48	0.15
564	36.6	31.52	5.08	1.64
603	35.8	33.49	2.21	0.75
420	28.0	24.28	3.72	1.20
573	33.8	?	?	0.59
602	33.6	33.43	0.17	0.05
430	21.0	24.78	−3.78	?
508	31.2	28.71	2.49	?
565	31.2	31.57	−0.37	−0.12
464	23.7	26.49	−2.79	−0.90
495	24.5	28.05	−3.55	−1.10
461	30.5	26.34	4.16	1.34
540	25.9	30.32	−4.42	−1.43
588	32.7	?	−0.03	−0.00
498	26.9	?	−1.30	−0.42
483	24.6	27.45	−2.85	−0.92
554	28.8	31.02	−2.22	−0.72
521	25.9	29.36	−3.46	−1.12
436	24.4	25.09	−0.69	−0.22
398	26.3	23.18	3.12	1.01
366	23.2	21.57	1.63	0.53
439	24.6	25.24	−0.64	−0.21
549	28.8	30.77	−1.97	−0.64
360	19.6	21.26	−1.66	−0.54
566	31.4	31.62	−0.22	−0.07
407	26.6	23.63	2.97	0.96
602	30.6	33.43	−2.83	−0.92
488	27.5	27.70	−0.20	−0.06
526	30.9	29.61	1.29	0.42
524	33.9	29.51	4.39	1.42
562	32.3	31.42	0.88	0.28
496	26.9	28.10	−1.20	−0.39

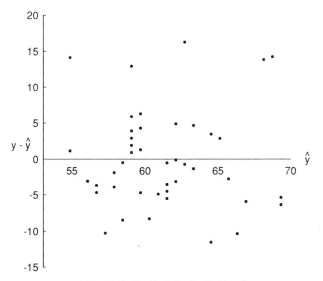

Figure 9.18. Residual plot for Problem 8.

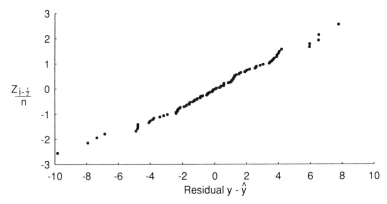

Figure 9.19. Normal probability plot for Problem 9.

9. The duration, X, and VO$_2$ MAX, Y, give $\overline{X} = 577.1$, $\overline{Y} = 35.6$, $[x^2] = 1425990.9$, $[y^2] = 5245.3$, $[xy] = 78280.1$. Do (c), (e), (j), (k-i), (k-iv), (l-i), (m), and (p). The normal probability plot is Figure 9.19.

10. X = age is related to Y = duration. $\overline{X} = 49.8$, $\overline{Y} = 577.1$, $[x^2] = 11395.7$, $[y^2] = 1425990.9$, $[xy] = -87611.9$. Do (c), (e), (m), (p), and (q-ii).

11. The prediction of age by maximal heart rate for sedentary males is considered here. $\overline{X} = 49.8$, $\overline{Y} = 18.6$, $[x^2] = 11395.7$, $[y^2] = 32146.4$, $[xy] = -12064.1$.

Table 9.14. Exercise Data for Sedentary Males. Data from Bruce *et al.* [1973]. For Problems 9–12.

Duration	VO₂ MAX	HR	Age	Height	Weight
360	24.7	168	40	175	96
770	46.8	190	25	168	68
663	41.2	175	41	187	82
679	31.4	190	37	176	82
780	45.7	200	26	179	73
727	47.6	210	28	185	84
647	38.6	208	26	177	77
675	43.2	200	42	162	72
735	48.2	196	30	188	85
827	50.9	184	21	178	69
760	47.2	184	33	182	87
814	41.8	208	31	182	82
778	42.9	184	29	174	73
590	35.1	174	42	188	93
567	37.6	176	40	184	86
648	47.3	200	40	168	80
730	44.4	204	44	183	78
660	46.7	190	44	176	81
663	41.6	184	40	174	78
589	40.2	200	43	193	92
600	35.8	190	41	176	68
480	30.2	174	44	172	84
630	38.4	164	39	181	72
646	41.3	190	39	187	90
630	31.2	190	42	173	69
630	42.6	190	53	181	53
624	39.4	172	57	172	57
572	35.4	164	58	181	58
622	35.9	190	61	168	61
209	16.0	104	74	171	74
536	29.3	175	57	181	57
602	36.7	175	49	175	49
727	43.0	168	53	172	53
260	15.3	112	75	170	75
622	42.3	175	47	185	47
705	43.7	174	51	169	51
669	40.3	174	65	170	65
425	28.5	170	56	167	56
645	38.0	175	50	177	50
576	30.8	184	48	188	48
605	40.2	156	46	187	46
458	29.5	148	61	185	61
551	32.3	188	49	182	49
607	35.5	179	53	179	53
599	35.3	166	55	182	55
453	32.3	160	69	182	69
337	23.8	204	68	176	68

Table 9.15. Data for Problems 9–12 (continued).

Duration	VO₂ MAX	HR	Age	Height	Weight
663	41.4	182	47	171	47
603	39.0	180	48	180	48
610	38.6	190	55	180	55
472	31.5	175	53	192	85
458	25.7	166	58	178	81
446	24.6	160	50	178	77
532	30.0	160	51	175	82
656	42.0	186	52	176	73
583	34.4	175	52	172	77
595	34.9	180	48	179	78
552	35.5	156	45	167	89
675	38.7	162	58	183	85
622	38.4	186	45	175	76
591	32.4	170	62	175	79
582	33.6	156	63	171	69
518	30.0	166	57	174	75
444	28.9	170	48	180	105
473	29.5	175	52	177	77
490	30.4	168	59	173	74
596	34.4	192	46	190	92
529	37.0	175	54	168	82
652	43.4	156	54	180	85
714	46.0	175	46	174	77
646	43.0	184	45	178	80
551	29.3	160	54	172	86
601	36.8	184	48	169	82
579	35.0	170	54	180	80
325	21.9	140	61	175	76
392	25.4	168	60	180	89
659	40.7	178	45	181	81
631	33.8	184	48	173	74
405	28.8	170	63	168	79
560	35.8	180	60	181	82
615	40.3	190	47	178	78
580	33.4	180	66	173	68
530	39.0	174	47	169	64
495	23.2	145	69	171	84
330	20.5	138	60	185	87
600	36.4	200	50	182	81
443	23.5	166	50	175	84
508	29.7	188	61	188	80
596	43.2	168	57	174	66
461	30.4	170	47	171	65
583	34.7	164	46	187	83
620	37.1	174	61	165	71
620	41.4	190	45	171	79
180	19.8	125	71	185	80

Do (c), (m), and (p). Verify (to accuracy given) that $(\overline{X}, \overline{Y})$ lies on the regression line.

12. The height and weight data give $\overline{X} = 177.3$, $\overline{Y} = 79.0$, $[x^2] = 4030.1$, $[y^2] = 7060.0$, $[xy] = 2857.0$. Do (c), (e), (k-iv), (n) at $X = 160$, 170 and 180, and (p).

Mehta *et al.* [1981] studied the effect of the drug dipyridamole on blood platelet function in eight patients with at least 50% narrowing of one or more coronary arteries. Active platelets are sequestered in the coronary arteries giving reduced platelet function in the coronary venous blood, that is, in blood leaving the heart muscle after delivering oxygen and nutrients. More active platelets in the coronary arteries can lead to thrombosis, blood clots and a heart attack. Drugs lessening the chance of thrombosis may be useful in treatment.

Platelet aggregation measures the extent to which platelets aggregate, or cluster together in the presence of some chemical which stimulates clustering or aggregation. The measure used was the percent increase in light transmission after an aggregating agent was added to plasma. (The clustering of the cells make more "holes" in the plasma to let light through.) Two aggregating agents, adenosine diphosphate (ADP) and epinephrine (EPI), were used in this experiment. A second measure taken from the blood count was the count of platelets.

Blood was sampled from two sites, the aorta (blood being pumped from the heart) and the coronary sinus (blood returning from nourishing the heart muscle). Control samples as well as samples after intravenous infusion of 100 milligrams of dipyridiamole were taken. The data are given in tabular form.

	Platelet Aggregation (%)[a]							
	Control				**Dipyridamole**			
	Aorta		**Coronary Sinus**		**Aorta**		**Coronary Sinus**	
Case	**EPI**	**ADP**	**EPI**	**ADP**	**EPI**	**ADP**	**EPI**	**ADP**
1	87	75	89	23	89	75	89	35
2	70	23	42	14	45	16	47	18
3	96	75	96	31	96	83	96	84
4	65	51	70	33	70	55	70	57
5	85	16	79	4	69	13	53	22
6	98	83	98	80	83	70	94	88
7	77	14	97	13	84	35	73	67
8	98	50	99	40	85	50	91	48
Mean	85	48	84	30	78	50	77	52
±SEM	5	10	7	8	6	9	7	9

[a]From Mehta *et al.* [1981].

	Platelet Counts ($\times 1000/mm^3$)[a]			
	Control		**Dipyridamole**	
Case	**Aorta**	**Coronary Sinus**	**Aorta**	**Coronary Sinus**
1	390	355	455	445
2	240	190	248	205
3	135	125	150	145
4	305	268	285	290
5	255	195	230	220
6	283	307	291	312
7	435	350	457	374
8	290	250	301	284
Mean	292	255	302	284
±SEM	32	29	38	34

[a]From Mehta *et al.* [1981].

Problems 13–22 refer to these data.

13. Relate the control platelet counts in the aorta, X, and coronary sinus, Y. Do (a), (b), (c), (d), (e), compute the (X, Y, $\hat{Y}$, residual, normal deviate) table, (g), (h), (i), (j), (k-i), (k-iv), (l), (m), (p), (r), and (s).

14. Look at the association between the platelet counts in the aorta, X, and coronary sinus, Y, when being treated with dipyridamole. Do (a), (b), (c), (d), (m), (r), and (s).

15. Examine the control platelet aggregation percent for EPI, X, and ADP, Y, in the aorta. Do (a), (b), (c), (d), (e), and (m).

16. Examine the association between the EPI, X, and ADP, Y, in the control situation at the coronary sinus. Do (a), (b), (c), (d), (e), (m), (p), (r), and (s).

17. Interpret at the 5% significance level. Look at the platelet aggregation % for epinephrine in the aorta and coronary sinus under the control data. Do (m), (p) and (t), (u). Explain in words how there can be association but no (statistical) difference between the values at the two locations.

18. Under dipyridamole treatment study the platelet aggregation % for EPI in the aorta, X, and coronary sinus, Y. Do (a), (b), (c), (d), (e), (g), (h), (m), (p), (r), (s), (t), and (u).

19. The control aggregation % for ADP is compared in the aorta, X, and coronary sinus, Y, in this problem. Do (a), (b), (c), (d), (e), (f), (g), (h), (i), (j), (m), (p), and (q-ii).

20. Under dipyridamole, the aggregation % for ADP in the aorta, X, and coronary sinus, Y, is studied here. Do (b), (c), (e), (k-ii), (k-iv), (l-ii), (m), (p), (q-ii), (r), and (s).

21. The aortic platelet counts under the control, X, and dipyridamole, Y, are compared in this problem. Do (b), (c), (e), (m), (p), (q-ii), (t), and (u). Do the platelet counts differ under the two treatments (use $\alpha = 0.05$)? Are the platelet counts associated under the two treatments ($\alpha = 0.05$)?

22. The coronary sinus ADP aggregation % was studied during the control period, the X variable, and on dipyridamole, the Y variable. Do (b), (c), (d), (e), (m), and (t). At the 5% significance level is there a change between the treatment and control periods? Can you show association between the two values? How do you reconcile these findings?

Problems 23–29 deal with the data in Tables 9.16 and 9.17. Jensen *et al.* [1980] studied nineteen patients with coronary artery disease. Thirteen had a prior myocardial infarction (heart attack); three had coronary bypass surgery. The patients were evaluated before and after 3 months or more on a structured supervised training program.

Table 9.16. Resting and Maximal Ejection Fraction Measured by Radionuclide Ventriculography, Maximal Heart Rate, Systolic Blood Pressure, Rate Pressure Product and Estimate VO$_2$ MAX Before (Pre) and After (Post) Training.

Case	Resting EF		Maximal EF		Maximal HR	
	Pre	Post	Pre	Post	Pre	Post
1	0.39	0.48	0.46	0.48	110	119
2	0.57	0.49	0.51	0.57	120	125
3	0.77	0.63	0.70	0.82	108	105
4	0.48	0.50	0.51	0.51	85	88
5	0.55	0.46	0.45	0.55	107	103
6	0.60	0.50	0.52	0.54	125	115
7	0.63	0.61	0.75	0.68	170	166
8	0.73	0.61	0.53	0.71	160	142
9	0.70	0.68	0.80	0.79	125	114
10	0.66	0.68	0.54	0.43	131	150
11	0.40	0.31	0.42	0.30	135	174
12	0.48	0.46	0.48	0.30	97	94
13	0.63	0.78	0.60	0.75	135	132
14	0.41	0.37	0.41	0.44	127	162
15	0.75	0.54	0.76	0.57	126	148
16	0.58	0.64	0.62	0.72	102	112
17	0.50	0.58	0.54	0.65	145	140
18	0.71	0.81	0.65	0.60	152	145
19	0.37	0.38	0.32	0.31	155	170
Mean	0.57	0.55	0.56	0.56	127	132
±SD	0.13	0.13	0.13	0.16	23	26

Table 9.17. Resting and Maximal Ejection Fraction Measured by Radionuclide Ventriculography, Maximal Heart Rate, Systolic Blood Pressure, Rate Pressure Product and Estimate VO₂ MAX Before (Pre) and After (Post) Training (continued).

Case	Maximal SBP		Maximal RPP		Est. VO$_2$ MAX (cm^3/kg-min)	
	Pre	Post	Pre	Post	Pre	Post
1	148	156	163	186	24	30
2	180	196	216	245	28	44
3	185	200	200	210	28	28
4	150	148	128	130	34	38
5	150	156	161	161	20	28
6	164	172	205	198	30	36
7	180	210	306	349	64	54
8	182	176	291	250	44	40
9	186	170	233	194	30	28
10	220	230	288	345	30	30
11	188	205	254	357	28	44
12	120	165	116	155	22	20
13	175	160	236	211	20	36
14	190	180	241	292	36	38
15	140	170	176	252	36	44
16	200	230	204	258	28	36
17	215	185	312	259	44	44
18	165	190	251	276	28	34
19	165	200	256	340	44	52
Mean	174	184	223	246	31	37
±SD	25	24	57	69	8	9

The cardiac performance was evaluated using radionuclide studies while the patients were at rest and also exercising with bicycle pedals (while lying supine). Variables measured included (1) ejection fraction (EF) the fraction of the blood in the left ventricle ejected during a heart beat, (2) heart rate (HR) at maximum exercise in beats per minute, (3) systolic blood pressure (SBP) in millimeters of mercury, (4) the rate pressure product (RPP) maximum heart rate times the maximum systolic blood pressure divided by 100, and (5) the estimated maximum oxygen consumption in cubic centimeters of oxygen per kilogram of body weight per minute.

23. The resting ejection fraction is measured before, X, and after Y, training. $\overline{X} = 0.574$, $\overline{Y} = 0.553$, $[x^2] = 0.29886$, $[y^2] = 0.32541$, $[xy] = 0.23385$, Paired $t = -0.984$. Do (c), (e), (k-iv), (m), (p). Is there a change in resting ejection fraction demonstrated with 6 months of exercise training? Are the two ejection fractions associated?

24. The ejection fraction at maximal exercise was measured before, X, and after, Y, training. $\overline{X} = 0.556$, $\overline{Y} = 0.564$, $[x^2] = 0.30284$, $[y^2] = 0.46706$, $[xy] = 0.2809$. Is there association ($\alpha = 0.05$) between the two ejection fractions? If yes, do (c), (k-iii), (l-iii), (p), and (q-ii). Is there a change ($\alpha = 0.05$) between the two ejection fractions? If yes, find a 95% confidence interval for the average difference.

25. The maximum systolic blood pressure was measured before, X, and after, Y, training. $\overline{X} = 173.8$, $\overline{Y} = 184.2$, $[x^2] = 11488.5$, $[y^2] = 10458.5$, $[xy] = 7419.5$, paired $t = 2.263$. Do (a), (b), (c), (d), (e), (m), (p), and (t). Does the exercise training produce a change? How much? Can we predict individually the maximum SBP after training from that before? How much of the variability in maximum SBP after exercise is accounted for by knowing the value before exercise?

26. The before, X, and after, Y, rate pressure product give $\overline{X} = 223.0$, $\overline{Y} = 245.7$, $[x^2] = 58476$, $[y^2] = 85038$, $[xy] = 54465$, paired $t = 2.256$. Do (c), (e), (f), (g), (h), (m). Find the large sample p-value for Kendall's tau for association.

Maximal SBP

Pre X	Post Y	$\widehat{Y}$	$Y - \widehat{Y}$	Normal Deviate
163	186	189.90	−3.80	−0.08
216	245	239.16	?	?
200	210	224.26	−14.26	−0.32
128	130	157.20	−27.20	−0.61
161	161	?	−26.94	?
205	198	228.92	−30.92	−0.69
306	349	322.99	26.01	?
291	250	309.02	−59.02	−1.31
233	194	255.00	−61.00	−1.36
288	345	306.22	38.77	0.86
254	357	?	?	?
116	155	146.02	8.98	0.20
236	211	257.79	−46.79	−1.04
241	292	262.45	29.55	0.66
176	252	201.91	50.09	1.12
204	258	227.99	30.01	0.67
312	259	328.58	−69.58	−1.55
251	276	271.76	4.24	0.09
256	340	276.42	63.58	1.42

27. The maximum oxygen consumption, VO$_2$ MAX, is measured before, X, and after, Y. Here $\overline{X} = 32.53$, $\overline{Y} = 37.05$, $[x^2] = 2030.7$, $[y^2] = 1362.9$, $[xy] = 54465$, paired $t = 2.811$. Do (c), (k-ii), (m), (n), at $x = 30, 35, 40$, (p), (q-ii), and (t).

28. The ejection fractions at rest, X, and at maximum exercise, Y, before training is used in this problem. $\overline{X} = 0.574$, $\overline{Y} = 0.556$, $[x^2] = 0.29886$, $[y^2] = 0.30284$, $[xy] = 0.24379$, paired $t = -0.980$. Analyze these data, including a scatter diagram, and write a short paragraph describing the change and/or association seen.

29. The ejection fractions at rest, X, and after exercises, Y, for the subjects after training:

i. is associated

ii. does not change on the average

iii. explain about 52% of the variability in each other.

Justify statements (i)–(iii). $\overline{X} = 0.553$, $\overline{Y} = 0.564$, $[x^2] = 0.32541$, $[y^2] = 0.4671$, $[xy] = 0.28014$, paired $t = 0.424$.

Problems 30–33 refer to the following study. Boucher et al. [1981] studied patients before and after surgery for isolated aortic regurgitation and isolated mitral regurgitation. The aortic valve is in the heart valve between the left ventricle where blood is pumped from the heart and the aorta, the large artery beginning the arterial system. When the valve is not functioning and closing properly some of the blood pumped from the heart returns (or regurgitates) as the heart relaxes before its next pumping action. To compensate for this, the heart volume increases to pump more blood out (since some of it returns). To correct for this, open heart surgery is performed and an artificial valve is sewn into the heart. Data on 20 patients with aortic regurgitation and corrective surgery are given in Tables 9.18 and 9.19.

NYHA Class measures the amount of impairment in daily activities that the patient

Table 9.18. Data for 20 Patients with Aortic Regurgitation: Table for Introduction to Problems 30–33.

| Case | Age (years) and Sex | NYHA Class | Preoperative | | | | | |
			HR beats/min	SBP (mmHG)	EF	EDVI (ml/m^2)	SVI (ml/m^2)	ESVI (ml/m^2)
1	33M	I	75	150	0.54	225	121	104
2	36M	I	110	150	0.64	82	52	30
3	37M	I	75	140	0.50	267	134	134
4	38M	I	70	150	0.41	225	92	133
5	38M	I	68	215	0.53	186	99	87
6	54M	I	76	160	0.56	116	65	51
7	56F	I	60	140	0.81	79	64	15
8	70M	I	70	160	0.67	85	37	28
9	22M	II	68	140	0.57	132	95	57
10	28F	II	75	180	0.58	141	82	59
11	40M	II	65	110	0.62	190	118	72
12	48F	II	70	120	0.36	232	84	148
13	42F	III	70	120	0.64	142	91	51
14	57M	III	85	150	0.60	179	107	30
15	61M	III	66	140	0.56	214	120	94
16	64M	III	54	150	0.60	145	87	58
17	61M	IV	110	126	0.55	83	46	37
18	62M	IV	75	132	0.56	119	67	52
19	64M	IV	80	120	0.39	226	88	138
20	65M	IV	80	110	0.29	195	57	138
Mean	49		75	143	0.55	162	85	77
±SD	14		14	25	0.12	60	26	43

Table 9.19. Data for 20 Patients with Aortic Regurgitation. Table for Introduction to Problems 30–33 (continued).

Case	Age (years) and Sex	NYHA Class	Postoperative					
			HR beats/min	SBP (mmHG)	EF	EDVI (ml/m^2)	SVI (ml/m^2)	ESVI (ml/m^2)
1	33M	I	80	115	0.38	113	43	43
2	36M	I	100	125	0.58	56	32	24
3	37M	I	100	130	0.27	93	25	68
4	38M	I	85	110	0.17	160	27	133
5	38M	I	94	130	0.47	111	52	59
6	54M	I	74	110	0.50	83	42	42
7	56F	I	85	120	0.56	59	33	26
8	70M	I	85	130	0.59	68	40	28
9	22M	II	120	136	0.33	119	39	80
10	28F	II	92	160	0.32	71	23	48
11	40M	II	85	110	0.47	70	33	37
12	48F	II	84	120	0.24	149	36	113
13	42F	III	84	100	0.63	55	35	20
14	57M	III	86	135	0.33	91	72	61
15	61M	III	100	138	0.34	92	31	61
16	64M	III	60	130	0.30	118	35	83
17	61M	IV	88	130	0.62	63	39	24
18	62M	IV	75	126	0.29	100	29	71
19	64M	IV	78	110	0.26	198	52	147
20	65M	IV	75	90	0.26	176	46	130
Mean	49		87	123	0.40	102	38	65
±SD	14		13	15	0.14	41	11	39

suffers: (I) is least impairment, II is mild impairment, III is moderate impairment and (IV) is severe impairment; **HR**, heart rate in beats/minute; **SBP**, the systolic (pumping or maximum) blood pressure in millimeters of mercury; **EF**, the ejection fraction; that is, the fraction of blood in the left ventricle pumped out during a beat; **EDVI**, the volume (in milliliters) of the left ventricle after the heart relaxes, adjusted for physical size; the adjustment is to divide by an estimate of the patient's body surface area, BSA, in square meters; **SVI**, the volume (in milliliters) of the left ventricle after the blood is pumped out adjusted for BSA; **ESVI**, the volume (in milliliters) of the left ventricle pumped out during one cycle, adjusted for BSA; ESVI = EDVI − SVI.

These values were measured before and after valve replacement surgery. The patients in this study were selected to have left ventricular volume overload; that is, expanded EDVI.

Another group of 20 patients with mitral valve disease and left ventricular volume overload were studied. The mitral valve is the valve allowing oxygenated blood from the lungs into the left ventricle for pumping to the body. Mitral regurgitation allows blood to be pumped "backward" and to be mixed with "new" blood coming from the lungs. The data for these patients are given in Tables 9.20 and 9.21.

Table 9.20. Data for 20 Patients with Mitral Regurgitation.

Case	Age (years) and Sex	NYHA Class	HR beats/min	SBP (mmHG)	EF	EDVI (ml/m^2)	SVI (ml/m^2)	ESVI (ml/m^2)
					Preoperative			
1	23M	II	75	95	0.69	71	49	22
2	31M	II	70	150	0.77	184	142	42
3	40F	II	86	90	0.68	84	57	30
4	47M	II	120	150	0.51	135	67	66
5	54F	II	85	120	0.73	127	93	34
6	57M	II	80	130	0.74	149	110	39
7	61M	II	55	120	0.67	196	131	65
8	37M	III	72	120	0.70	214	150	64
9	52M	III	108	105	0.66	126	83	43
10	52F	III	80	115	0.52	167	70	97
11	52M	III	80	105	0.76	130	99	31
12	56M	III	80	115	0.60	136	82	54
13	58F	III	65	110	0.62	146	91	56
14	59M	III	102	90	0.63	82	52	30
15	66M	III	60	100	0.62	76	47	29
16	67F	III	75	140	0.71	94	67	27
17	71F	III	88	140	0.65	111	72	39
18	55M	IV	80	125	0.66	136	90	46
19	59F	IV	115	130	0.72	96	69	27
20	60M	IV	64	140	0.60	161	97	64
Mean	53		81	121	0.66	131	86	45
±SD	12		17	17	0.09	40	30	19

30. a. The preoperative, X, and postoperative, Y, ejection fraction in the patients with aortic valve replacement gave $\overline{X} = 0.549$, $\overline{Y} = 0.396$, $[x^2] = 0.26158$, $[y^2] = 0.39170$, $[xy] = 0.21981$, paired $t = -6.474$. Do (a), (c), (d), (e), (m), (p), and (t). Is there a change? Are ejection fractions before and after surgery related?

b. The mitral valve cases had $\overline{X} = 0.662$, $\overline{Y} = 0.478$, $[x^2] = 0.09592$, $[y^2] = 0.24812$, $[xy] = 0.04458$, paired $t = -7.105$. Perform the same tasks as in (a).

c. When the emphasis is on the change, rather than possible association and predictive value the type of figure of Figure 9.20 (from Boucher *et al.* [1981]) may be preferred to a scatter diagram. Plot the scatter diagram for the aortic regurgitation data and comment upon the relative merits of the two graphics.

31. a. For the mitral valve cases, we use the end systolic volume index (ESVI) before surgery to try to predict the end diastolic volume index (EDVI) after surgery. $\overline{X} = 45.25$, $\overline{Y} = 77.9$, $[x^2] = 6753.8$, $[y^2] = 16885.5$, $[xy] = 7739.5$. Do (c), (d), (e), (f), (h), (j), (k-iv), (m), and (p).

Table 9.21. Data for 20 Patients with Mitral Regurgitation (continued).

| Case | Age (years) and Sex | NYHA Class | Postoperative | | | | | |
			HR beats/min	SBP (mmHG)	EF	EDVI (ml/m^2)	SVI (ml/m^2)	ESVI (ml/m^2)
1	23M	II	90	100	0.60	67	40	27
2	31M	II	95	110	0.64	64	41	23
3	40F	II	80	110	0.77	59	45	14
4	47M	II	90	120	0.36	96	35	61
5	54F	II	100	110	0.41	59	24	35
6	57M	II	75	115	0.54	71	38	33
7	61M	II	140	120	0.41	165	68	97
8	37M	III	95	120	0.25	84	21	63
9	52M	III	100	125	0.43	67	29	38
10	52F	III	90	90	0.44	124	55	69
11	52M	III	98	116	0.55	68	37	31
12	56M	III	61	108	0.56	112	63	49
13	58F	III	88	120	0.50	76	38	38
14	59M	III	100	100	0.48	40	19	21
15	66M	III	85	124	0.51	31	16	15
16	67F	III	84	120	0.39	81	32	49
17	71F	III	100	100	0.44	76	33	43
18	55M	IV	108	124	0.43	63	27	36
19	59F	IV	100	110	0.49	62	30	32
20	60M	IV	90	110	0.36	93	34	60
Mean	53		93	113	0.48	78	36	42
±SD	12		15	9	0.11	30	14	21

X	Y	$\widehat{Y}$	Residuals	Normal Deviate
22	67	51.26	15.74	0.75
42	64	74.18	−10.18	−0.48
30	59	60.42	−1.42	−0.06
66	96	101.68	−5.68	−0.27
34	59	65.01	−6.01	−0.28
39	71	70.74	0.26	0.01
65	165	?	?	?
64	84	99.39	15.29	−0.73
43	67	75.32	?	−0.39
97	124	137.20	−13.20	?
31	68	61.57	?	?
54	112	87.93	24.07	1.14
56	76	?	?	−0.67
30	40	?	−20.42	−0.97
29	31	?	?	?
27	81	56.99	24.01	1.14
39	76	70.74	5.26	0.25
46	63	78.76	−15.76	−0.75
27	62	56.99	5.01	0.24
64	93	99.39	−6.39	−0.30

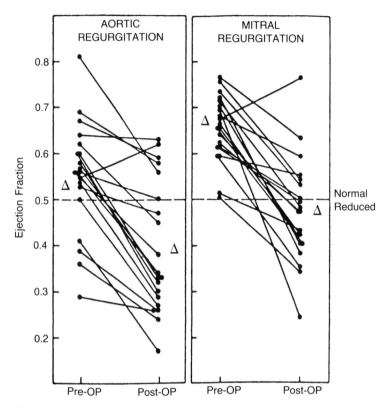

Figure 9.20. Figure for Problem 30(c). Individual values for ejection fraction before (Pre-OP) and early after (Post-OP) surgery are plotted; mean values (Δ) are also shown. Preoperatively, only four patients with aortic regurgitation had an ejection fraction below normal. After operation, 13 patients with aortic regurgitation and 9 with mitral regurgitation had an ejection fraction below normal. The lower limit of normal (0.50) is represented by a broken line.

The residual plot and normal probability plot are found in Figures 9.21 and 9.22.

b. If subject 7 is omitted, $\overline{X} = 44.2$, $\overline{Y} = 73.3$, $[x^2] = 6343.2$, $[y^2] = 8900.1$, $[xy] = 5928.7$. Do (c), (m), and (p). What are the changes in (a), (b), and (r) from Part (a)?

c. For the aortic cases; $\overline{X} = 75.8$, $\overline{Y} = 102.3$, $[x^2] = 35307.2$, $[y^2] = 32513.8$, $[xy] = 27076$. Do (c), (k-iv), (p), and (q-ii).

32. We want to investigate the predictive value of the preoperative ESVI to predict the postoperative ejection fraction, EF. For each part do (a), (c), (d), (k-i), (k-iv), (m), and (p).

a. The aortic cases have $\overline{X} = 75.8$, $\overline{Y} = 0.396$, $[x^2] = 35307.2$, $[y^2] = 0.39170$, $[xy] = 84.338$.

b. The mitral cases have $\overline{X} = 45.3$, $\overline{Y} = 0.478$, $[x^2] = 6753.8$, $[y^2] = 0.24812$, $[xy] = -18.610$.

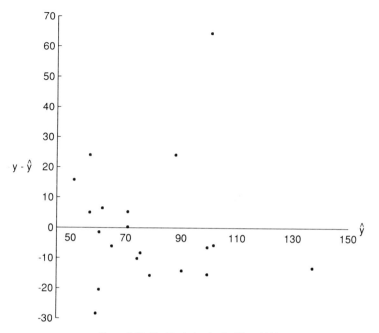

Figure 9.21. Residual plot for Problem 31(a).

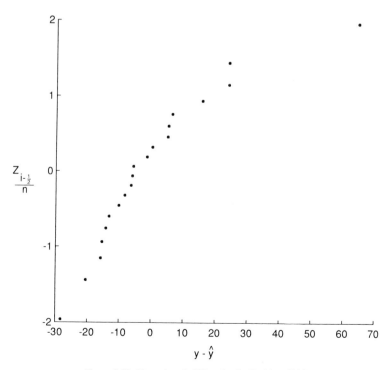

Figure 9.22. Normal probability plot for Problem 31(a).

33. Investigate the relationship between the preoperative heart rate and the postoperative heart rate. If there are outliers eliminate (their) effect. Specifically address these questions:

> **i.** Is there an overall change from preop to postop HR?
>
> **ii.** Are the preop and postop HR's associated? If there is an association, summarize it.

a. For the aortic cases: $\sum X = 1502$, $\sum Y = 17.30$, $\sum X^2 = 116446$, $\sum Y^2 = 152662$, $\sum XY = 130556$.

X	Y	$\widehat{Y}$	Residuals	Normal Deviate
75	80	86.48	−6.48	−0.51
110	100	92.56	7.44	0.59
75	100	86.48	13.52	1.06
70	85	85.61	0.61	−0.04
68	94	85.27	8.73	0.69
76	74	86.66	−12.66	−1.00
60	85	83.88	1.12	0.08
70	85	85.61	0.61	−0.04
68	120	85.27	34.73	2.73
75	92	86.48	5.52	0.43
65	85	84.75	0.25	0.02
70	84	85.61	−1.61	−0.13
70	84	85.61	−1.61	−0.13
85	86	88.22	−2.22	−0.17
66	100	84.92	15.08	1.19
54	60	82.84	−22.84	−1.80
110	88	92.56	−4.56	0.36
75	75	86.48	−11.48	−0.90
80	78	87.35	−9.35	−0.74
80	75	87.35	−12.35	−0.97

b. For the mitral cases: $\sum X = 1640$, $\sum Y = 1869$, $\sum X^2 = 140338$, $\sum Y^2 = 179089$, $\sum XY = 152860$.

X	Y	$\widehat{Y}$	Residuals	Normal Deviate
75	90	93.93	−3.93	−0.25
70	95	94.27	0.73	0.04
86	80	93.18	−13.18	−0.84
120	90	90.87	−0.87	−0.05
85	100	93.25	6.75	0.43
80	75	93.59	−18.59	−1.19
55	140	95.28	44.72	2.86
72	95	94.13	0.87	0.05

(*continued*)

X	Y	$\widehat{Y}$	Residuals	Normal Deviate
108	100	91.68	8.32	0.53
80	90	93.59	−3.59	−0.23
80	98	93.59	4.41	0.28
80	61	93.95	−32.59	−2.08
65	88	94.61	−6.61	0.42
102	100	92.09	7.91	0.51
60	85	94.94	−9.94	−0.64
75	84	93.93	−9.93	−0.63
88	100	93.04	6.96	0.44
80	108	93.59	14.41	0.92
115	100	91.21	8.79	0.56
64	90	94.67	−4.67	−0.30

REFERENCES

Acton, F. S. [1959]. *Analysis of Straight-Line Data*. Dover Publications, New York.

Boucher, C. A., Bingham, J. B., Osbakken, M. D., Okada, R. D., Strauss. H. W., Block, P. C., Levine, F. H., Phillips, H. R., and Phost, G. M. [1981]. Early changes in left ventricular volume overload. *American Journal of Cardiology*, **47**: 991–1004.

Bruce, R. A., Kusumi, F., and Hosmer, D. [1973]. Maximal oxygen intake and nomographic assessment of functional aerobic impairment in cardiovascular disease. *American Heart Journal*, **65**: 546–562.

Dern, R. J. and Wiorkowski, J. J. [1969]. Studies on the preservation of human blood, IV. the hereditary component of pre- and post storage erythrocyte adenosine triphosphate levels. *Journal of Laboratory and Clinical Medicine*, **73**: 1019–1029.

Devlin, S. J., Gnanadesikan, R., and Kettenring, J. R. [1975]. Robust estimation and outlier detection with correlation coefficients. *Biometrics*, **62**: 531–545.

Draper, N. R. and Smith, H. [1981]. *Applied Regression Analysis*, 2nd edition. Wiley, New York.

Hollander, M. and Wolfe, D. A. [1973]. *Nonparametric Statistical Methods*. Wiley, New York.

Huber, P. J. [1981]. *Robust Statistics*. Wiley, New York.

Jensen, D., Atwood, J. E., Frolicher, V., McKirnan, M. D. Battler, A., Ashburn, W., and Ross, Jr., J. [1980]. Improvement in ventricular function during exercise studied with radionuclide ventriculography after cardiac rehabilitation. *American Journal of Cardiology*, **46**: 770–777.

Kendall, M. G. and Stuart, A. [1967]. *The Advanced Theory of Statistics, Volume 2. Inference and Relationships*, 2nd edition. Hafner, New York.

Kronmal, R. A. [1993]. *Spurious correlation and the fallacy of the ratio standard revisited*. (To appear).

Mehta, J., Mehta, P., Pepine, C. J., and Conti, C. R. [1981]. Platelet function studies in coronary artery disease. X. Effects of dipyridamole. *American Journal of Cardiology*, **47**: 1111–1114.

Neyman, J. [1952]. On a most powerful method of discovering statistical regularities. *Lectures and Conferences on Mathermatical Statistics and Probability*. US Department of Agriculture, Washington, DC, pp. 143–154.

U.S. Cancer Mortality by County: 1950–59 [1974]. DHEW Publication No. (NIH) 74-615, Bethesda, MD.

CHAPTER 10

Analysis of Variance

10.1 INTRODUCTION

The phrase "analysis of variance" was coined by Fisher [1950], who defined it as, "The separation of variance ascribable to one group of causes from the variance ascribable to other groups." Another way of stating this is to consider it as a partitioning of total variance into component parts. One illustration of this procedure is contained in the previous chapter on linear regression where the total variability of the dependent variable was partitioned into two components: one associated with regression and the other associated with (residual) variation about the regression line. Analysis of variance models are a special class of linear models.

Definition 10.1. An *analysis of variance model* is a linear regression model in which the predictor variables are classification variables. The categories of a variable are called the *levels* of the variable.

The meaning of this definition will become clearer as you read this chapter.

The topics of analysis of variance and design of experiments are closely related. This has already been evident in earlier chapters. For example, use of a paired t-test implies that the data are paired and thus may indicate a certain type of experiment. Similarly, a partitioning of total variation in a regression situation implies that two variables measured are linearly related. A general principle is involved: the analysis of a set of data should be appropriate for the design. We will indicate the close relationship between design and analysis throughout this chapter.

The chapter begins with the one-way analysis of variance. Total variability is partitioned into a variance between groups and a variance within groups. The groups could consist of different treatments or different classifications. In Section 10.2, we develop the construction of an analysis of variance from group means and standard deviations, and consider the analysis of variance using ranks. Section 10.3 discusses the two-way analysis of variance: a special two-way analysis involving "randomized blocks" and the corresponding rank analysis are discussed, and then two kinds of classification variables (random and fixed) are covered. Special but common designs are presented in Sections 10.4 and 10.5. Finally, Section 10.6 discusses the testing of the assumptions of the analysis of variance including ways of transforming the data to make the assumptions valid. Notes and specialized topics conclude our discussion.

418

A few comments about notation and computations: the formulas for the analysis of variance look formidable, but follow a logical pattern. The following rules are followed or held (we remind you on occasion):

1. Indices for groups follow a mnemonic pattern. For example, the subscript i runs from $1, \ldots, I$; the subscript j from $1, \ldots, J$; k from $1, \ldots, K$, and so forth.
2. Sums of values of the random variables are indicated by replacing the subscript by a dot. For example,

$$Y_{i\bullet} = \sum_{j=1}^{J} Y_{ij}, \quad Y_{\bullet jk} = \sum_{i=1}^{I} Y_{ijk}, \quad Y_{\bullet j\bullet} = \sum_{i=1}^{I}\sum_{k=1}^{K} Y_{ijk}.$$

3. It is expensive to print subscripts and superscripts on $\sum$ signs. A very simple rule is that summations are always over the given subscripts. For example,

$$\sum Y_i = \sum_{i=1}^{I} Y_i, \quad \sum Y_{ijk} = \sum_{i=1}^{I}\sum_{j=1}^{J}\sum_{k=1}^{K} Y_{ijk}.$$

We may write expressions initially with the subscripts and superscripts, but after the patterns have been established, we omit them. See Table 10.5 for an example.

4. The symbol n_{ij} denotes the number of Y_{ijk} observations, and so on. The total sample size is denoted by n rather than $n_{\bullet\bullet\bullet}$; it will be obvious from the context that the total sample size is meant.
5. The means are indicated by $\overline{Y}_{ij\bullet}$, $\overline{Y}_{\bullet j\bullet}$, and so on. The number of observations associated with a mean is always n with the same subscript, e.g., $\overline{Y}_{ij\bullet} = Y_{ij\bullet}/n_{ij}$, or $\overline{Y}_{\bullet j\bullet} = Y_{\bullet j\bullet}/n_{\bullet j}$.
6. The analysis of variance is an analysis of variability associated with a single observation. This implies that sums of squares of subtotals or totals must always be divided by the number of observations making up the total; for example, $\sum Y_{i\bullet}^2/n_i$, if $Y_{i\bullet}$ is the sum of n_i observations. The rule is then that the divisor is always the number of observations represented by the dotted subscripts. Another example: $Y_{\bullet\bullet}^2/n_{\bullet\bullet}$, since $Y_{\bullet\bullet}$ is the sum of $n_{\bullet\bullet}$ observations.
7. Similar to rules 5 and 6, a sum of squares involving means always have as weighting factor the number of observations on which the mean is based. For example,

$$\sum_{i=1}^{I} n_i \left(\overline{Y}_{i\bullet} - \overline{Y}_{\bullet\bullet}\right)^2$$

because the mean $\overline{Y}_{i\bullet}$ is based on n_i observations.

8. The ANOVA models are best expressed in terms of means and deviations from means. The computations are best carried out in terms of totals to avoid unnecessary calculations and prevent rounding error. (This is similar to the definition and calculation of the sample standard deviation.) For example,

$$\sum n_i \left(\overline{Y}_{i\bullet} - \overline{Y}_{\bullet\bullet}\right)^2 = \sum \frac{Y_{i\bullet}^2}{n_i} - \frac{Y_{\bullet\bullet}^2}{n_{\bullet\bullet}}.$$

See Problem 10.25.

10.2 ONE-WAY ANALYSIS OF VARIANCE

10.2.1 Motivating Example

Example 10.1. To motivate the one-way analysis of variance, we return to the data of Zelazo *et al.* [1972] which deals with the age at which children first walked (see Chapter 5). The experiment involved reinforcement of the walking and placing reflexes in newborns. The walking and placing reflexes disappear by about 8 weeks of age. In this experiment, newborn children were randomly placed into one of four treatment groups: active exercise; passive exercise; no exercise; or an 8-week control group. Infants in the active-exercise group received walking and placing stimulation four times a day for 8 weeks, infants in the passive-exercise group received an equal amount of gross motor stimulation, infants in the no-exercise group were tested along with the first two groups at weekly intervals, and the 8-week control group consisted of infants observed only at 8 weeks to control for possible effects of repeated examination. The response variable was age (in months) at which the infant first walked. The data are presented in Table 10.1. For purposes of this example we have added the mean of the fourth group to that group to make the sample sizes equal; this will not change the mean of the fourth group. Equal sample sizes are not required for the one-way analysis of variance.

Assume that the age at which an infant first walks alone is normally distributed with variance σ^2. For the four treatment groups let the means be μ_1, μ_2, μ_3, and μ_4. Since σ^2 is unknown, we could calculate the sample variance for each of the four groups and come up with a pooled estimate, s_p^2, of σ^2. For this example, since the sample sizes per group are assumed to be equal, this is

Table 10.1. Distribution of Ages (in months) at Which Infants First Walked Alone. Data from Zelazo *et al.* [1972].

	Active Group	Passive Group	No-Exercise Group	8-Week Control Group
	9.00	11.00	11.50	13.25
	9.50	10.00	12.00	11.50
	9.75	10.00	9.00	12.00
	10.00	11.75	11.50	13.50
	13.00	10.50	13.25	11.50
	9.50	15.00	13.00	12.35[a]
Mean	10.125	11.375	11.708	12.350
Variance	2.0938	3.5938	2.3104	0.7400
$Y_{i\bullet}$	60.75	68.25	70.25	74.10

[a]This observation is missing from the original data set. For purposes of this illustration, it is estimated by the sample mean. See text for further discussion.

$$s_p^2 = \frac{1}{4}(2.0938 + 3.5938 + 2.3104 + 0.7400) = 2.1845.$$

But we have one more estimate of σ^2. If the four treatments do not differ ($H_0 : \mu_1 = \mu_2 = \mu_3 = \mu_4 = \mu$), then the sample means are normally distributed with variance $\sigma^2/6$. The quantity $\sigma^2/6$ can be estimated by $s_{\bar{Y}}^2$, the variance of the sample means. For this example it is,

$$s_{\bar{Y}}^2 = 0.87439.$$

Hence, $6s_{\bar{Y}}^2 = 5.2463$ is also an estimate of σ^2. Under the null hypothesis, $6s_{\bar{Y}}^2/s_p^2$ will follow an F-distribution. How many degrees of freedom are involved? The quantity $s_{\bar{Y}}^2$ has three degrees of freedom associated with it (since it is a variance based on four observations). The quantity s_p^2 has 20 degrees of freedom (since each of its four component variances has 5 degrees of freedom). So, the quantity $6s_{\bar{Y}}^2/s_p^2$ under the null hypothesis has an F-distribution with 3 and 20 degrees of freedom. What if the null hypothesis is not true, i.e., the μ_1, μ_2, μ_3, and μ_4 are not all equal? It can be shown that $6s_{\bar{Y}}^2$ then estimates $\sigma^2 + positive\ constant$, so that the ratio $6s_{\bar{Y}}^2/s_p^2$ tends to be larger than 1. The usual hypothesis-testing approach is to reject the null hypothesis if the ratio is "too large," with the critical value selected from an F-table. The analysis is summarized in an "analysis of variance" table as in Table 10.2.

The variances $6s_{\bar{Y}}^2/s_p^2$ and s_p^2 are called *mean* squares for reasons to be explained later. It is clear that the first variance measures the variability between groups, and the second measures the variability within groups. The F-ratio of 2.40 is referred to an F-table. The critical value at the 0.05 level is $F_{3,20,0.95} = 3.10$, the observed value 2.40 is smaller, and we do not reject the null hypothesis at the 0.05 level. The data are displayed in Figure 10.1.

From the graph, it can be seen that the active group had the lowest mean value. The nonsignificance of the F-test suggests that the active group mean is not significantly lower than that of the other three groups. $\square$

10.2.2 Using the Normal Distribution Model

Basic Approach
The one-way analysis of variance is a generalization of the t-test. As in the above motivating example, it can be used to examine the age at which groups of infants first walk alone, each group receiving a different treatment; or we may compare patient

Table 10.2. Simplified Analysis of Variance Table of Data of Table 10.1.

Source of Variation	DF	Mean Squares	F-Ratio
Between Groups	3	$6s_{\bar{Y}}^2 = 5.2463$	$\dfrac{6s_{\bar{Y}}^2}{s_p^2} = \dfrac{5.2463}{2.1845} = 2.40$
Within Groups	20	$s_p^2 = 2.1845$	

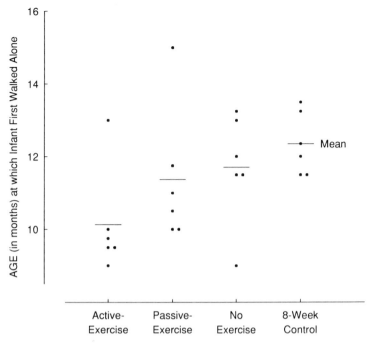

Figure 10.1. Distribution of ages at which infants first walked alone. Data from Table 10.1.

costs (in dollars per day) in a sample of hospitals from a metropolitan area. (There is a subtle distinction between the two examples; see Section 10.3.4 for a further discussion).

Definition 10.2. An analysis of variance of observations, each of which belongs to one of I disjoint groups, is a *one-way analysis of variance on I groups.*

Suppose that samples are taken from I normal populations that differ at most in their means; the observations can be modeled by

$$Y_{ij} = \mu_i + \epsilon_{ij}, \quad i = 1,\ldots,I, \quad j = 1,\ldots,n_i. \tag{10.1}$$

The mean for normal population i is μ_i; we assume that there are n_i observations from this population. Also, by assumption, the ϵ_{ij} are independent $N(0, \sigma^2)$ variables. In words: Y_{ij} denotes the jth sample from a population with mean μ_i and variance σ^2. If $I = 2$, you can see that this is precisely the model for the two-sample t-test.

The only difference between the situation now and that of Section 10.2.1 is that we allow the number of observations to vary from group to group. The within-group estimate of the variance σ^2 now becomes a weighted sum of sample variances. Let s_i^2 be the sample variance from group i, where $i = 1,\ldots,I$. The within-group estimate

σ^2 is

$$\frac{\sum(n_i - 1)s_i^2}{\sum(n_i - 1)} = \frac{\sum(n_i - 1)s_i^2}{n - I},$$

where $n = n_1 + n_2 + \cdots + n_I$ is the total number of observations.

Under the null hypothesis $H_0 : \mu_1 = \mu_2 = \cdots = \mu_I = \mu$, the variability among the group of sample means also estimates σ^2. We will show below that the proper expression is

$$\frac{\sum n_i(\overline{Y}_{i\bullet} - \overline{Y}_{\bullet\bullet})^2}{I - 1},$$

where

$$\overline{Y}_{i\bullet} = \sum_{j=1}^{n_i} Y_{ij}/n_i$$

is the sample mean for group i, and

$$\overline{Y}_{\bullet\bullet} = \sum_{i=1}^{I} \sum_{j=1}^{n_i} \frac{Y_{ij}}{n} = \sum \frac{n_i \overline{Y}_{i\bullet}}{n}$$

is the grand mean. These quantities can again be arranged in an analysis of variance table, as displayed in Table 10.3.

Under the null hypothesis, $H_0 : \mu_1 = \mu_2 = \cdots = \mu_I = \mu$, the quantity A/B in Table 10.3 follows an F-distribution with $(I - 1)$ and $(n - I)$ degrees of freedom.

We now reanalyze our first example in Section 10.2.1, deleting the sixth observation, 12.35, in the 8-week control group. The means and variances for the four groups are now:

	Active	Passive	No-Exercise	Control	Overall
Mean $(\overline{Y}_{i\bullet})$	10.125	11.375	11.708	12.350	11.348
Variance (s_i^2)	2.0938	3.5938	2.3104	0.925	—
n_i	6	6	6	5	23

Table 10.3. One-Way Analysis of Variance Table for I Groups and n_i Observations Per Group; $i = 1, \ldots, I$.

Source of Variation	DF	Mean Square	F-Ratio
Between groups	$I - 1$	$A = \dfrac{\sum n_i(\overline{Y}_{i\bullet} - \overline{Y}_{\bullet\bullet})^2}{I - 1}$	A/B
Within groups	$n - I$	$B = \sum \dfrac{(n_i - 1)s_i^2}{n - I}$	

Table 10.4. Analysis of Variance of Data of Example 10.1 Omitting the Last Observation.

Source of Variation	DF	Mean Square	F-Ratio
Between Groups	3	4.9253	2.14
Within Groups	19	2.2994	

Therefore,

$$\sum n_i (\overline{Y}_{i\bullet} - \overline{Y}_{\bullet\bullet})^2 = 6(10.125 - 11.348)^2 + 6(11.375 - 11.348)^2$$
$$+ 6(11.708 - 11.348)^2 + 5(12.350 - 11.348)^2$$
$$= 14.776.$$

The between-group mean square is $14.776/(4-1) = 4.9253$. The within-group mean square is

$$\frac{1}{23-4}\left(5(2.0938) + 5(3.5938) + 5(2.3104) + 4(0.925)\right) = 2.2994.$$

The analysis of variance table is displayed in Table 10.4.

The critical value $F_{3,19,0.95} = 3.13$ so, again, the four groups do not differ significantly.

Linear Model Approach

In this subsection, we approach the analysis of variance using linear models. The model $Y_{ij} = \mu_i + \epsilon_{ij}$ is usually written as

$$Y_{ij} = \mu + \alpha_i + \epsilon_{ij}, \quad i = 1, \ldots, I, \ j = 1, \ldots, n_i. \tag{10.2}$$

The quantity μ is defined as

$$\mu = \sum_{i=1}^{I} \sum_{j=1}^{n_i} \frac{\mu_i}{n},$$

where $n = \sum n_i$ (the total number of observations). The quantity α_i is defined as $\alpha_i = \mu - \mu_i$. This implies that

$$\sum_{i=1}^{I} \sum_{j=1}^{n_i} \alpha_i = \sum n_i \alpha_i = 0. \tag{10.3}$$

Definition 10.3. The quantity $\alpha_i = \mu - \mu_i$ is the *main effect* of the ith population.

Comments:

1. The symbol α with a subscript will denote an element of the analysis of variance model, not the Type I error. The context will make it clear which meaning is intended.

2. The equation $\sum n_i \alpha_i = 0$ is a constraint. It implies that fixing any $(I - 1)$ of the main effects determines the remaining value.

If we hypothesize that the I populations have the same means,

$$H_0 : \mu_1 = \mu_2 = \cdots = \mu_I = \mu,$$

then an equivalent statement is:

$$H_0 : \alpha_1 = \alpha_2 = \cdots = \alpha_I = 0 \quad \text{or} \quad H_0 : \alpha_i = 0, \quad i = 1, \ldots, I.$$

How are the quantities μ_i, $i = 1, \ldots, I$ and σ^2 to be estimated from the data? (Or, equivalently, μ, α_i, $i = 1, \ldots, I$ and σ^2.) Basically, we follow the same strategy as in Section 10.2.1. The variances within the I groups are pooled to provide an estimate of σ^2 and the variability between groups provides a second estimate under the null hypothesis. The data can be displayed as follows:

	Sample				
	1	**2**	**3**	$\cdots$	**I**
	Y_{11}	Y_{21}	Y_{31}	$\cdots$	Y_{I1}
	Y_{12}	Y_{22}	Y_{32}	$\cdots$	Y_{I2}
	$\vdots$	$\vdots$	$\vdots$	$\vdots$	$\vdots$
	Y_{1n_1}	Y_{2n_2}	Y_{3n_3}	$\cdots$	Y_{In_I}
Observations	n_1	n_2	n_3	$\cdots$	n_I
Means	$\overline{Y}_{1\bullet}$	$\overline{Y}_{2\bullet}$	$\overline{Y}_{3\bullet}$	$\cdots$	$\overline{Y}_{I\bullet}$
Totals	$Y_{1\bullet}$	$Y_{2\bullet}$	$Y_{3\bullet}$	$\cdots$	$Y_{I\bullet}$

For this set of data, a partitioning can be set up that mimics the model defined by Equation (10.2):

$$\left.\begin{array}{ll} \text{Model:} & Y_{ij} = \mu + \alpha_i + \epsilon_{ij}, \\[2mm] \text{Data:} & Y_{ij} = \overline{Y}_{\bullet\bullet} + a_i + e_{ij}, \end{array}\right\} \quad i = 1, \ldots, I, \; j = 1, \ldots, n_i, \quad (10.4)$$

where $a_i = \overline{Y}_{i\bullet} - \overline{Y}_{\bullet\bullet}$ and $e_{ij} = Y_{ij} - \overline{Y}_{i\bullet}$ for $i = 1, \ldots, I$ and $j = 1, \ldots, n_i$. It is easy to verify that the condition $\sum n_i \alpha_i = 0$ is mimicked by $\sum n_i a_i = 0$. Each data point is partitioned into three component estimates:

$$Y_{ij} = \overline{Y}_{\bullet\bullet} + (\overline{Y}_i - \overline{Y}_{\bullet\bullet}) + (Y_{ij} - \overline{Y}_{i\bullet}) = \text{mean} + i\text{th main effect} + \text{error}.$$

The expression on the right side of Y_{ij} is an algebraic identity. It is a remarkable property of this partitioning that the sum of squares of the Y_{ij} is equal to the sum of the three sums of squares of the elements on the right side:

$$\sum_{i=1}^{I}\sum_{j=1}^{n_i} Y_{ij}^2 = \sum_{i=1}^{I}\sum_{j=1}^{n_i} \overline{Y}_{..}^2 + \sum_{i=1}^{I}\sum_{j=1}^{n_i}(\overline{Y}_{i.} - \overline{Y}_{..})^2 + \sum_{i=1}^{I}\sum_{j=1}^{n_i}(Y_{ij} - \overline{Y}_{i.})^2$$

$$= n\overline{Y}_{..}^2 + \sum_{i=1}^{I} n_i(\overline{Y}_{i.} - \overline{Y}_{..})^2 + \sum_{i=1}^{I}\sum_{j=1}^{n_i}(Y_{ij} - \overline{Y}_{i.})^2 \tag{10.5}$$

and the degrees of freedom can also be partitioned: $n = 1 + (I - 1) + (n - I)$. You will recognize the terms on the right side as the ingredients needed for setting up the analysis of variance table as discussed in the previous subsection. It should also be noted that the quantities on the right side are random variables (since they are based on statistics). It can be shown that their expected values are

$$E\left(\sum n_i(\overline{Y}_{i.} - \overline{Y}_{..})^2\right) = \sum n_i\alpha_i^2 + (I - 1)\sigma^2 \tag{10.6}$$

and

$$E\left(\sum_{i=1}^{I}\sum_{j=1}^{n_i}(Y_{ij} - \overline{Y}_{i.})^2\right) = (n - I)\sigma^2. \tag{10.7}$$

If the null hypothesis $H_0 : \alpha_1 = \alpha_2 = \cdots = \alpha_I = 0$ is true (i.e., $\mu_1 = \mu_2 = \cdots = \mu_I = \mu$), then $\sum n_i\alpha_i^2 = 0$ and both of the above terms provide an estimate of σ^2 (after division by $(I - 1)$ and $(n - I)$, respectively). This layout and analysis is summarized in Table 10.5.

The quantities making up the component parts of Equation (10.5) are called *sums of squares* (ss). "Grand Mean" is usually omitted; it is used to test the null hypothesis that $\mu = 0$. This is rarely of very much interest, particularly if the null hypothesis $H_0 : \mu_1 = \mu_2 = \cdots = \mu_I$ is rejected (but see Example 10.7). "Between Groups" is used to test the latter null hypothesis, or the equivalent hypothesis, $H_0 : \alpha_1 = \alpha_2 = \cdots = \alpha_I = 0$.

Before returning to Example 1, we give a few computational notes.

Computational Notes
As in the case of calculating standard deviations, the computations usually are not based on the means but rather the group totals. Only three quantities have to be calculated for the one-way ANOVA. Let

$$Y_{i.} = \sum_{j=1}^{n_i} Y_{ij} = \text{total in the } i\text{th treatment group} \tag{10.8}$$

and

$$Y_{..} = \sum Y_{i.} = \text{grand total.} \tag{10.9}$$

Table 10.5. Layout for the One-Way Analysis of Variance.

Source of Variation	DF	SS[a]	MS	F-ratio	DF of F-ratio	E(MS)	Hypothesis Tested
Grand mean	1	$SS_\mu = n\overline{Y}_{\bullet\bullet}^2$	$MS_\mu = SS_\mu$	$\dfrac{MS_\mu}{MS_\epsilon}$	$(1, n-1)$	$n\mu^2 + \sigma^2$	$\mu = 0$
Between groups (main effects)	$I-1$	$SS_\alpha = \sum n_i(\overline{Y}_{i\bullet} - \overline{Y}_{\bullet\bullet})^2$	$MS_\alpha = \dfrac{SS_\alpha}{I-1}$	$\dfrac{MS_\alpha}{MS_\epsilon}$	$(I-1, n-I)$	$\sum \dfrac{n_i\alpha_i^2}{I-1} + \sigma^2$	$\alpha_1 = \cdots = \alpha_I$ or $\mu_1 = \cdots = \mu_I$
Within groups (residuals)	$n-I$	$SS_\epsilon = \sum\sum(Y_{ij} - \overline{Y}_{i\bullet})^2$	$MS_\epsilon = \dfrac{SS_\epsilon}{n-I}$	—	—	σ^2	σ^2
Total	n	$\sum\sum Y_{ij}^2$					

[a]Summation is over all displayed subscripts.

Model: $Y_{ij} = \mu_i + \epsilon_{ij}$ $\qquad \epsilon_{ij} \sim$ iid $N(0, \sigma^2)$

$\qquad\qquad = \mu + \alpha_i + \epsilon_{ij}$ $\qquad i = 1,\ldots,I,\; j = 1,\ldots,n_i$

Data: $Y_{ij} = \overline{Y}_{\bullet\bullet} + (\overline{Y}_{i\bullet} - \overline{Y}_{\bullet\bullet}) + (Y_{ij} - \overline{Y}_{i\bullet})$

(iid= independent and identically distributed)

Equivalent model:

$\qquad Y_{ij} \sim N(\mu_i, \sigma^2),$ where Y_{ij}s are independent.

The three quantities that have to be calculated are

$$\sum_{i=1}^{I}\sum_{j=1}^{n_i} Y_{ij}^2 = \sum\sum Y_{ij}^2, \quad \sum_{i=1}^{I}\frac{Y_{i\bullet}^2}{n_i} = \sum\frac{Y_{i\bullet}^2}{n_i}, \quad \frac{Y_{\bullet\bullet}^2}{n},$$

where $n = \sum n_i =$ total observations. It is easy to establish the following relationships:

$$\text{SS}_\mu = \frac{Y_{\bullet\bullet}^2}{n}, \tag{10.10}$$

$$\text{SS}_\alpha = \sum\frac{Y_{i\bullet}^2}{n_i} - \frac{Y_{\bullet\bullet}^2}{n}, \tag{10.11}$$

$$\text{SS}_\epsilon = \sum\sum Y_{ij}^2 - \sum\frac{Y_{i\bullet}^2}{n_i}. \tag{10.12}$$

The subscripts are omitted.

We have an algebraic identity in $\sum\sum Y_{ij}^2 = \text{SS}_\mu + \text{SS}_\alpha + \text{SS}_\epsilon$. Defining SS_{TOTAL} as $\text{SS}_{\text{TOTAL}} = \sum\sum Y_{ij}^2 - \text{SS}_\mu$, we get $\text{SS}_{\text{TOTAL}} = \text{SS}_\alpha + \text{SS}_\epsilon$, and degrees of freedom $(n-1) = (I-1) + (n-I)$.

This formulation is a simplified version of Equation (10.5). Note that the original data are needed only for $\sum\sum Y_{ij}^2$; all other sums of squares can be calculated from group or overall totals.

Continuing Example 10.1, omitting again the last observation (12.35):

$$\sum\sum Y_{ij}^2 = 9.00^2 + 9.50^2 + \cdots + 11.50^2 = 3020.2500$$

$$\sum\frac{Y_{i\bullet}^2}{n_i} = \frac{60.75^2}{6} + \frac{68.25^2}{6} + \frac{70.25^2}{6} + \frac{61.75^2}{5} = 2976.5604$$

$$\frac{Y_{\bullet\bullet}^2}{n} = \frac{261.00^2}{23} = 2961.7826.$$

The ANOVA table omitting rows for SS_μ and SS_{TOTAL} becomes:

Source of Variation	DF	SS	MS	F-Ratio
Between Groups	3	14.7778	4.9259	2.14
Within Groups	19	43.6896	2.2995	

The numbers in this table are not subject to rounding error and differ slightly from those in Table 10.4.

Estimates of the components of the expected mean squares of Table 10.5 can now be obtained. The estimate of σ^2 is $\hat{\sigma}^2 = 2.2995$, and the estimate of $\sum n_i \alpha_i^2/(I-1)$ is

$$\frac{\sum n_i \hat{\alpha}_i^2}{I-1} = 4.9259 - 2.2995 = 2.6264.$$

How is this quantity to be interpreted in view of the nonrejection of the null hypothesis? Theoretically, the quantity can never be less than zero (all the terms are positive). The best interpretation looks back to MS_α, which is a random variable which (under the null hypothesis) estimates σ^2. Under the null hypothesis, MS_α and MS_ϵ both estimate σ^2, and $\sum n_i \alpha_i^2/(I-1)$ is zero.

10.2.3 One-Way ANOVA From Group Means and Standard Deviation

In many research papers, the raw data are not presented but rather the means and standard deviations (or variances) for each of the, say, I treatment groups under consideration. It is instructive to construct an analysis of variance from these data and see how the assumption of the equality of the population variances for each of the groups enters in. Advantages of constructing the ANOVA table are

1. Pooling the sample standard deviations (variances) of the groups produces a more precise estimate of the population standard deviation, this becomes very important if the sample sizes are small;
2. A simultaneous comparison of all group means can be made by means of the F-test, rather than a series of two-sample t-tests. The analysis can be modeled on the layout in Table 10.3.

Suppose that for each of I groups the following quantities are available:

Group	Sample Size	Sample Mean	Sample Variance
i	n_i	$\overline{Y}_{i\bullet}$	s_i^2

The quantities $n = \sum n_i$, $Y_{i\bullet} = n_i \overline{Y}_{i\bullet}$ and $Y_{\bullet\bullet} = \sum Y_{i\bullet}$ can be calculated. The "Within Groups" ss is the quantity B in Table 10.3 times $n - I$, and the "Between Groups" ss can be calculated as

$$\mathrm{ss}_\alpha = \sum \frac{Y_{i\bullet}^2}{n_i} - \frac{Y_{\bullet\bullet}^2}{n}.$$

Example 10.2. Barboriak *et al.* [1972] studied risk factors in patients undergoing coronary bypass surgery for coronary artery disease. The authors looked for an association between cholesterol level (a putative risk factor) and the number of diseased blood vessels. The data are:

Diseased vessels (*i*)	Sample size (*n_i*)	Mean Cholesterol Level ($\overline{Y}_{i\bullet}$)	Standard Deviation (*s_i*)
1	29	260	56.0
2	49	289	87.5
3	76	295	72.4

Using Equations (10.8)–(10.12), we get $n = 29 + 49 + 76 = 154$,

$$Y_{1\bullet} = n_1 \overline{Y}_{1\bullet} = 29 \cdot 260 = 7540, \qquad Y_{3\bullet} = n_3 \overline{Y}_{3\bullet} = 76 \cdot 295 = 22420,$$

$$Y_{2\bullet} = n_2 \overline{Y}_{2\bullet} = 49 \cdot 289 = 14161, \qquad Y_{\bullet\bullet} = \sum n_i \overline{Y}_{i\bullet} = \sum Y_{i\bullet} = 44121.$$

$$\mathrm{ss}_\alpha = \frac{7540^2}{29} + \frac{14161^2}{49} + \frac{22420^2}{76} - \frac{44121^2}{154} = 12666829.0 - 12640666.5 = 26162.5,$$

$$\mathrm{ss}_\epsilon = \sum (n_i - 1)s_i^2 = 28 \times 56.0^2 + 48 \times 87.5^2 + 75 \times 72.4^2 = 848440.$$

Table 10.6. Analysis of Variance of Data of Example 10.2

Source	DF	SS	MS	F-Ratio
Main Effects (disease status)	2	26162.50	13081.2	2.33
Residual (error)	151	848440.0	5618.5	—

The analysis of variance table can now be constructed as in Table 10.3. (There is no need to calculate the total ss.)

The critical value for F at the 0.05 level with 2 and 120 degrees of freedom is 3.07; the observed F-value does not exceed this critical value and the conclusion is that the average cholesterol levels do not differ significantly.   □

10.2.4   One-Way Analysis of Variance Using Ranks

In this section the rank procedures discussed in Chapter 8 are extended to the one-way analysis of variance. For three or more groups, Kruskal and Wallis [1952] have given a one-way ANOVA based on ranks. The model is

$$Y_{ij} = \mu_i + \epsilon_{ij}, \quad i = 1, \ldots, I, \ j = 1, \ldots, n_i.$$

The only assumption about the ϵ_{ij} is that they are independently and identically distributed, not necessarily normal. It is assumed that there are no ties among the observations. For a small number of ties in the data, the average of the ranks for the tied observations is usually assigned (see Note 10.1). The test procedure will be conservative in the presence of ties, i.e., the p-value will be smaller when adjustment for ties is made.

The null hypothesis of interest is

$$H_0 : \mu_1 = \mu_2 = \cdots = \mu_I = \mu.$$

The procedure for obtaining the ranks is similar to that for the two-sample Wilcoxon rank sum procedure: the $n_1 + n_2 + \cdots + n_I = n$ observations are ranked without regard to which group they belong, let $R_{ij} = $ rank of observation j in group i.

$$T_{KW} = \frac{12 \sum n_i (\overline{R}_{i\bullet} - \overline{R}_{\bullet\bullet})^2}{n(n+1)}, \tag{10.13}$$

where $\overline{R}_{i\bullet}$ is the average of the ranks of the observations in group i:

$$\overline{R}_{i\bullet} = \sum_{j=1}^{n_i} \frac{R_{ij}}{n_i},$$

and $\overline{R}_{\bullet\bullet}$ is the grand mean of the ranks. The value of the mean $(\overline{R}_{\bullet\bullet})$ must be $(n+1)/2$ (why?) and this provides a partial check on the arithmetic. Large values of T_{KW} imply that the average ranks for the group differ, so that the null hypothesis is rejected for large values of this statistic. If the null hypothesis is true and all the n_i become large, the distribution of the statistic T_{KW} approaches a χ^2-distribution with $I-1$ degrees of freedom. Thus, for large sample sizes, critical values for T_{KW} can be read from a χ^2-table. For small values of n_i, say, in the range of two to five, exact critical values have been tabulated (see for example CRC Table X.9 [1968]). Such tables are available for three or four groups.

An equivalent formula for T_{KW} as defined by Equation (10.13) is

$$T_{KW} = \frac{12 \sum R_{i\bullet}^2 / n_i}{n(n+1)} - 3(n+1), \tag{10.14}$$

where $R_{i\bullet}$ is the total of the ranks for the ith group.

Example 10.3. Chikos *et al.* [1977] studied errors in the reading of chest X-rays. The opinion of 10 radiologists about the status of the left ventricle of the heart ("normal" *vs.* "abnormal") was compared to data obtained by ventriculography (which consists of the insertion of a catheter into the left ventricle, injection of a radiopague fluid and the taking of a series of X-rays). The ventriculography data were used to classify a subject's left ventricle as "normal" or "abnormal." Using this "gold standard," the percentage of errors for each radiologist was computed. The authors were interested in the effect of experience and for this purpose the radiologists were classified into one of three groups: senior staff, junior staff, and residents. The data for these three groups are:

	Senior Staff	**Junior Staff**	**Residents**
i	1	2	3
n_i	2	4	4
Y_{ij}	7.3	13.3	14.7
	7.4	10.6	23.0
(Percent error)		15.0	22.7
		20.7	26.6

To compute the Kruskal–Wallis statistic T_{KW}, the data are ranked disregarding groups:

Observation	7.3	7.4	10.6	13.3	14.7	15.0	20.7	22.7	23.0	26.6
Rank	1	2	3	4	5	6	7	8	9	10
Group	1	1	2	2	3	2	2	3	3	3

The sums and means of the ranks for each group are calculated to be

$$R_{1\bullet} = 1 + 2 = 3, \qquad \overline{R}_{1\bullet} = 1.5,$$
$$R_{2\bullet} = 3 + 4 + 6 + 7 = 20, \qquad \overline{R}_{2\bullet} = 5.0,$$
$$R_{3\bullet} = 5 + 8 + 9 + 10 = 32, \qquad \overline{R}_{3\bullet} = 8.0.$$

(The sum of the ranks is $R_1 + R_2 + R_3 = 55 = (10 \times 11)/2$ providing a partial check of the ranking procedure.)

Using Equation (10.14), the T_{KW} statistic has a value of

$$T_{KW} = \frac{12(3^2/2 + 20^2/4 + 32^2/4)}{10(10+1)} - 3(10+1) = 6.33.$$

This value can be referred to as a χ^2-table with two degrees of freedom. The p-value is $0.025 < p < 0.05$. The exact p-value can be obtained from, for example, Table X.9 of the CRC tables [1968]. (This table does not list the critical values of T_{KW} for $n_1 = 2$, $n_2 = 4$, $n_3 = 4$; however, the order in which the groups are labeled does not matter so that the values $n_1 = 4$, $n_2 = 4$, $n_3 = 2$ may be used.) From this table it is seen that $0.011 < p < 0.046$, indicating that the chi-square approximation is satisfactory even for these small sample sizes. The conclusion from both analyses is that among staff levels there are significant differences in the accuracy of reading left ventricular abnormality from a chest X-ray.   □

10.3   TWO-WAY ANALYSIS OF VARIANCE

10.3.1   Using the Normal Distribution Model

In this section we consider data that arise when a response variable can be classified in two ways. For example, the response variable may be blood pressure and the classification variables type of drug treatment and sex of the subject; another example arises from classifying individuals by type of health insurance and race, the response variable could be number of physician contacts per year.

Definition 10.4.   An analysis of variance of observations each of which can be classified in two ways is called a *two-way analysis of variance.*

The data are usually displayed in "cells" with the row categories the values of one classification variable and the columns representing values of the second classification variable.

A completely general two-way ANOVA model with each cell mean any value could be

$$Y_{ijk} = \mu_{ij} + \epsilon_{ijk},   \tag{10.15}$$

where $i = 1, \ldots, I$, $j = 1, \ldots, J$, and $k = 1, \ldots, n_{ij}$. By assumption, the ϵ_{ijk} are iid $N(0, \sigma^2)$: independently and identically distributed $N(0, \sigma^2)$. This model could be treated as a one-way ANOVA with IJ groups with a test of the hypothesis that all μ_{ij} are the same, implying that the classification variables are not related to the response variable. However, if there is a significant difference among the IJ group means we want to know whether these differences can be attributed to:

1. One of the classification variables;
2. Both of the classification variables acting separately (no interaction); or
3. Both of the classification variables acting separately and jointly (interaction).

In many situations involving classification variables, the mean μ_{ij} may be modeled as the sum of two terms, an effect of variable one plus an effect of variable two:

$$\mu_{ij} = u_i + v_j, \quad i = 1, \ldots, I, \ j = 1, \ldots, J. \tag{10.16}$$

Here μ_{ij} depends, in an additive fashion, on the ith level of the first variable and the jth level of the second variable. One problem is that u_i and v_j are not uniquely defined; for any constant C if $\mu_i^* = u_i + C$ and $v_j^* = v_j - C$ then $\mu_{ij} = u_i^* + v_j^*$. Thus, the values of u_i and v_j can be pinned down to within a constant. The constant is specified by convention and is associated with the experimental setup. Suppose there are n_{ij} observations at the ith level of variable one, and jth level of variable two. The frequencies of observations can be laid out in a contingency table as follows:

Levels of Variable 1	Levels of Variable 2						Total
	1	**2**	$\cdots$	j	$\cdots$	J	**Total**
1	n_{11}	n_{12}	$\cdots$	n_{1j}	$\cdots$	n_{1J}	$n_{1\bullet}$
2	n_{21}	n_{22}	$\cdots$	n_{2j}	$\cdots$	n_{2J}	$n_{2\bullet}$
$\vdots$	$\vdots$	$\vdots$	$\ddots$	$\vdots$	$\ddots$	$\vdots$	$\vdots$
i	n_{i1}	n_{i2}	$\cdots$	n_{ij}	$\cdots$	n_{iJ}	$n_{i\bullet}$
$\vdots$	$\vdots$	$\vdots$	$\ddots$	$\vdots$	$\ddots$	$\vdots$	$\vdots$
I	n_{I1}	n_{I2}	$\cdots$	n_{Ij}	$\cdots$	n_{IJ}	$n_{I\bullet}$
Total	$n_{\bullet 1}$	$n_{\bullet 2}$	$\cdots$	$n_{\bullet j}$	$\cdots$	$n_{\bullet J}$	$n_{\bullet\bullet}$

The experiment has a total of $n_{\bullet\bullet}$ observations. The notation is identical to that used in a two-way contingency table layout. (A major difference is that all the frequencies are usually chosen by the experimenter, we shall return to this point when talking about a balanced ANOVA design). Using the model of Equation (10.16), the value of μ_{ij} is defined as

$$\mu_{ij} = \mu + \alpha_i + \beta_j, \tag{10.17}$$

where $\mu = \sum \sum n_{ij} \mu_{ij} / n_{\bullet\bullet}$, $\sum n_{i\bullet} \alpha_i = 0$, and $\sum n_{\bullet j} \beta_j = 0$. This is similar to the constraints put upon the one-way ANOVA model; see Equations (10.2) and (10.3), and Problem 10.25d.

Example 10.4. An experimental setup involves two explanatory variables each at three levels. There are 24 observations distributed as follows:

Levels of Variable 1	Levels of Variable 2			Total
	1	**2**	**3**	**Total**
1	2	2	2	6
2	3	3	3	9
3	3	3	3	9
Total	8	8	8	24

The effects of the first variable are assumed to be: $\alpha_1 = 3$, $\alpha_2 = 6$, and $\alpha_3 = -8$; the effects of the second variable are: $\beta_1 = 1$, $\beta_2 = -3$, and $\beta_3 = 2$. The overall level is $\mu = 20$. If the model defined by Equation (10.17) holds, the cell means μ_{ij} are completely specified as follows:

Effects of the First Variable	Effects of the Second Variable			Total
	$\beta_1 = 1$	$\beta_2 = -3$	$\beta_3 = 2$	
$\alpha_1 = 3$	$\mu_{11} = 24$	$\mu_{12} = 20$	$\mu_{13} = 25$	$\mu_{1\bullet} = 23$
$\alpha_2 = 6$	$\mu_{21} = 27$	$\mu_{22} = 23$	$\mu_{23} = 28$	$\mu_{2\bullet} = 26$
$\alpha_3 = -8$	$\mu_{31} = 13$	$\mu_{32} = 9$	$\mu_{33} = 14$	$\mu_{3\bullet} = 12$
Total	$\mu_{\bullet 1} = 21$	$\mu_{\bullet 2} = 17$	$\mu_{\bullet 3} = 22$	$\mu = 20$

For example, $\mu_{11} = 20+3+1 = 24$ and $\mu_{33} = 20-8+2 = 14$. Note that $\sum n_{i\bullet}\alpha_i = 6 \cdot 3 + 9 \cdot 6 + 9(-8) = 0$ and, similarly, $\sum n_{\bullet j}\beta_j = 0$. Note also that $\mu_{1\bullet} = \sum n_{1j}\mu_{1j} / \sum n_{1j} = \mu + \alpha_1 = 20+3 = 23$, that is, a marginal mean is just the overall mean plus the effect of the variable associated with that margin. The means are graphed in Figure 10.2. The points have been joined by dashed lines to make the pattern clear; there need not be any continuity between the levels. A similar graph could be made with the level of the second variable plotted on the abscissa and the lines indexed by the levels of the first variable.

Definition 10.5. A two-way ANOVA model satisfying Equation (10.17) is called an *additive model*.

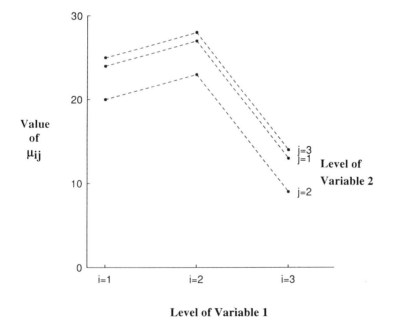

Figure 10.2. Graph of additive ANOVA model (Example 10.4).

Some implications of this model are discussed. You will find it helpful to refer to Example 10.4 and Figure 10.2 in understanding the following:

1. The statement of Equation (10.17) is equivalent to saying that "changing the level of variable 1 while the level of the second variable remains fixed, changes the value of the mean by the same amount regardless of the (fixed) level of the second variable."
2. Statement 1 holds with variable 1 and 2 interchanged.
3. If the values of μ_{ij} ($i = 1, \dots, I$) are plotted for the different levels of the second variable, the curves are parallel (see Figure 10.2).
4. Statement 3 holds with the roles of variable 1 and variable 2 interchanged.
5. The model defined by Equation (10.17) imposes $1 + (I - 1) + (J - 1)$ constraints on the IJ means μ_{ij}, leaving $(I - 1)(J - 1)$ degrees of freedom.

We now want to define a nonadditive model, but before doing so, we must introduce one other concept.

Definition 10.6. A two-way ANOVA has a *balanced (orthogonal) design* if, for every i and j,

$$n_{ij} = \frac{n_{i\bullet}n_{\bullet j}}{n_{\bullet\bullet}}.$$

That is, the cell frequencies are functions of the product of the marginal totals. The reason this characteristic is needed is that only for balanced designs can the total variability be partitioned in an additive fashion. In Section 10.5, we introduce a discussion of unbalanced or nonorthogonal designs, the topic is treated in terms of multiple regression models in the next chapter.

Definition 10.7. A *balanced two-way* ANOVA *model with interaction* (a nonadditive model) is defined by

$$Y_{ijk} = \mu + \alpha_i + \beta_j + \gamma_{ij} + \epsilon_{ijk}, \quad \begin{matrix} i = 1, \dots, I, \\ j = 1, \dots, J, \\ k = 1, \dots, n_{ij}, \end{matrix} \qquad (10.18)$$

subject to the following conditions:

1. $n_{ij} = n_{i\bullet}n_{\bullet j}/n_{\bullet\bullet}$ for every i and j.
2. $\sum n_{i\bullet}\alpha_i = \sum n_{\bullet j}\beta_j = 0$.
3. $\sum n_{i\bullet}\gamma_{ij} = 0$ for all $j = 1, \dots, J$, $\sum n_{\bullet j}\gamma_{ij} = 0$ for all $i = 1, \dots, I$.
4. The ϵ_{ijk} are iid $N(0, \sigma^2)$. This assumption implies homogeneity of variances among the IJ cells.

If the γ_{ij} are zero, the model is equivalent to the one defined by Equation (10.17), there is no interaction and the model is additive.

As in Section (10.2), Equations (10.4) and (10.5), a set of data as defined at the beginning of this section can be partitioned into parts each of which estimates the component part of the model:

$$Y_{ijk} = \overline{Y}_{\bullet\bullet\bullet} + a_i + b_j + g_{ij} + e_{ijk}, \tag{10.19}$$

where

$\overline{Y}_{\bullet\bullet\bullet}$ = the effect of the mean,

$a_i = \overline{Y}_{i\bullet\bullet} - \overline{Y}_{\bullet\bullet\bullet}$ = main effect of ith level of variable 1,

$b_j = \overline{Y}_{\bullet j\bullet} - \overline{Y}_{\bullet\bullet\bullet}$ = main effect of jth level of variable 2,

$g_{ij} = \overline{Y}_{ij\bullet} - \overline{Y}_{i\bullet\bullet} - \overline{Y}_{\bullet j\bullet} + \overline{Y}_{\bullet\bullet\bullet}$ = interaction of ith and jth levels of variables 1 and 2,

$e_{ijk} = \overline{Y}_{ijk} - \overline{Y}_{ij\bullet}$ = residual effect (error).

The quantities $\overline{Y}_{i\bullet\bullet}$ and $\overline{Y}_{\bullet j\bullet}$ are the means of the ith level of variable 1 and the jth level of variable 2. In symbols,

$$\overline{Y}_{i\bullet\bullet} = \sum_{j=1}^{J}\sum_{k=1}^{n_{ij}} \frac{Y_{ijk}}{n_{i\bullet}} \quad \text{and} \quad \overline{Y}_{\bullet j\bullet} = \sum_{i=1}^{I}\sum_{k=1}^{n_{ij}} \frac{Y_{ijk}}{n_{\bullet j}}.$$

The interaction term, g_{ij} can be rewritten as

$$g_{ij} = (\overline{Y}_{ij\bullet} - \overline{Y}_{\bullet\bullet\bullet}) - (\overline{Y}_{i\bullet\bullet} - \overline{Y}_{\bullet\bullet\bullet}) - (\overline{Y}_{\bullet j\bullet} - \overline{Y}_{\bullet\bullet\bullet}),$$

which is the overall deviation of the mean of the ijth cell from the grand mean minus the main effects of variable 1 and variable 2. If the data can be fully explained by main effects, the term g_{ij} will be zero. Hence, g_{ij} measures the extent to which the data deviate from an additive model.

For a balanced design the total sum of squares, $\text{ss}_{\text{TOTAL}} = \sum\sum\sum(Y_{ijk} - \overline{Y}_{\bullet\bullet\bullet})^2$, and degrees of freedom can be partitioned additively into four parts:

$$\begin{aligned} \text{ss}_{\text{TOTAL}} &= \text{ss}_\alpha + \text{ss}_\beta + \text{ss}_\gamma + \text{ss}_\epsilon, \\ n_{\bullet\bullet} - 1 &= (I-1) + (J-1) + (I-1)(J-1) + (n_{\bullet\bullet} - IJ). \end{aligned} \tag{10.20}$$

Let

$$Y_{ij\bullet} = \sum_{k=1}^{n_{ij}} Y_{ijk} = \text{total for cell } ij,$$

$$Y_{i\bullet\bullet} = \sum_{j=1}^{J} Y_{ij\bullet} = \text{total for row } i,$$

$$Y_{\bullet j\bullet} = \sum_{i=1}^{I} Y_{ij\bullet} = \text{total for column } j.$$

Then the equations for the sums of squares together with computationally simpler formulæ are

$$\text{SS}_\alpha = \sum n_{i\bullet}(\overline{Y}_{i\bullet\bullet} - \overline{Y}_{\bullet\bullet\bullet})^2 = \sum \frac{Y_{i\bullet\bullet}^2}{n_{i\bullet}} - \frac{Y_{\bullet\bullet\bullet}^2}{n_{\bullet\bullet}},$$

$$\text{SS}_\beta = \sum n_{\bullet j}(\overline{Y}_{\bullet j\bullet} - \overline{Y}_{\bullet\bullet\bullet})^2 = \sum \frac{Y_{\bullet j\bullet}^2}{n_{\bullet j}} - \frac{Y_{\bullet\bullet\bullet}^2}{n_{\bullet\bullet}},$$

$$\text{SS}_\gamma = \sum\sum n_{ij}(\overline{Y}_{ij\bullet} - \overline{Y}_{i\bullet\bullet} - \overline{Y}_{\bullet j\bullet} + \overline{Y}_{\bullet\bullet\bullet})^2 = \sum\sum \frac{Y_{ij\bullet}^2}{n_{ij}} - \frac{Y_{\bullet\bullet\bullet}^2}{n} - \text{SS}_\alpha - \text{SS}_\beta,$$

$$\text{SS}_\epsilon = \sum\sum\sum (Y_{ijk} - \overline{Y}_{ij\bullet})^2 = \sum\sum\sum Y_{ijk}^2 - \sum\sum \frac{Y_{ij\bullet}^2}{n_{ij}}.$$

(10.21)

The partition of the sum of squares, the mean squares and the expected mean squares are given in Table 10.7.

A series of F-tests can be carried out to test the significance of the components of the model specified by Equation (10.18). The first test carried out is usually the test for interaction: $\text{MS}_\gamma/\text{MS}_\epsilon$. Under the null hypothesis $H_0 : \gamma_{ij} = 0$ for all i and j, this ratio has an F-distribution with $(I-1)(J-1)$ and $n-IJ$ degrees of freedom. The null hypothesis is rejected for large values of this ratio. Interaction is indicated by nonparallelism of the treatment effects. In Figure 10.3, some possible patterns are indicated. The expected results of F-tests are given at the top of each graph. For example, Pattern 1 shows NO–YES–NO, implying that the test for the main effect of variable 1 was not significant, the test for main effect of variable 2 was significant, and the test for interaction was not significant. It now becomes clear that if interaction is present, main effects are going to be difficult to interpret. For example, Pattern 4 in Figure 10.3 indicates significant interaction but no significant main effects. But the significant interaction implies that at level 1 of variable 1 there is a significant difference in the main effect of variable 2. What is happening is that the effect of variable 2 is in the opposite direction at the second level of variable 1. This pattern is extreme. A more common pattern is that of Pattern 6. How is this pattern to be interpreted? First of all, there is interaction; second, above the interaction there are significant main effects.

There are substantial practical problems associated with significant interaction patterns. For example, suppose that the two variables represent two drugs for pain-relief administered simultaneously to a patient. With Pattern 2, the inference would be that the two drugs together are more effective than either one acting singly. In Pattern 4 (and Pattern 3), the drugs are said to act antagonistically. In Pattern 6, the drugs are said to act synergistically; the effect of both drugs combined is greater than the sum of each acting alone. (For some subtle problems associated with these patterns, see the discussion of transformations in Section 10.6.)

If interaction is not present, the main effects can be tested by means of the F-tests $\text{MS}_\alpha/\text{MS}_\epsilon$ and $\text{MS}_\beta/\text{MS}_\epsilon$ with $(I-1, n-IJ)$ and $(J-1, n-IJ)$ degrees of freedom, respectively. If a main effect is significant, the question arises: which levels of the main effect differ significantly? At this point, a visual inspection of the levels may be sufficient to establish the pattern; in Chapter 12 we establish a more formal approach.

Table 10.7. Layout for the Two-Way Analysis of Variance.

Source of Variation	DF	SS[a]	MS
Grand mean	1	$SS_\mu = n\overline{Y}_{...}^2$	$MS_\mu = SS_\mu$
Row main effects	$I - 1$	$SS_\alpha = \sum n_{i\bullet}(\overline{Y}_{i\bullet\bullet} - \overline{Y}_{...})^2$	$MS_\alpha = \dfrac{SS_\alpha}{I-1}$
Column main effects	$J - 1$	$SS_\beta = \sum n_{\bullet j}(\overline{Y}_{\bullet j\bullet} - \overline{Y}_{...})^2$	$MS_\beta = \dfrac{SS_\beta}{J-1}$
Row×column interaction	$(I-1)(J-1)$	$SS_\gamma = \sum n_{ij}(\overline{Y}_{ij\bullet} - \overline{Y}_{i\bullet\bullet} - \overline{Y}_{\bullet j\bullet} + \overline{Y}_{...})^2$	$MS_\gamma = \dfrac{SS_\gamma}{(I-1)(J-1)}$
Residual	$n - IJ$	$SS_\epsilon = \sum(Y_{ijk} - \overline{Y}_{ij\bullet})^2$	$MS_\epsilon = \dfrac{SS_\epsilon}{n-IJ}$
Total	n	$\sum Y_{ijk}^2$	—

Source of Variation	F-Ratio	DF of F-Ratio	$E(MS)$
Grand mean	$\dfrac{MS_\mu}{MS_\epsilon}$	$(1, n-IJ)$	$\sigma^2 + n\mu^2$
Row main effects	$\dfrac{MS_\alpha}{MS_\epsilon}$	$(I-1, n-IJ)$	$\sigma^2 + \dfrac{\sum n_{i\bullet}\alpha_i^2}{I-1}$
Column main effects	$\dfrac{MS_\beta}{MS_\epsilon}$	$(J-1, n-IJ)$	$\sigma^2 + \dfrac{\sum n_{\bullet j}\beta_j^2}{J-1}$
Row×column interaction	$\dfrac{MS_\gamma}{MS_\epsilon}$	$\left((I-1)(J-1), n-IJ\right)$	$\sigma^2 + \dfrac{\sum n_{ij}\gamma_{ij}^2}{(I-1)(J-1)}$
Residual	—	—	σ^2

Source of Variation	Hypothesis Being Tested
Grand mean	$\mu = 0$
Row main effects	$\alpha_i = 0$ for all i
Column main effects	$\beta_j = 0$ for all j
Row × column interaction	$\gamma_{ij} = 0$ for all i and j, or $\mu_{ij} = u_i + v_j$

[a]Summation is over all displayed subscripts.

Model: $Y_{ijk} = \mu_{ij} + \epsilon_{ijk}$ [where $\epsilon_{ijk} \sim$ iid $N(0, \sigma^2)$] $= \mu + \alpha_i + \beta_j + \gamma_{ij} + \epsilon_{ijk}$. Data: $Y_{ijk} = \overline{Y}_{...} + (\overline{Y}_{i\bullet\bullet} - \overline{Y}_{...}) + (\overline{Y}_{\bullet j\bullet} - \overline{Y}_{...}) + (\overline{Y}_{ij\bullet} - \overline{Y}_{i\bullet\bullet} - \overline{Y}_{\bullet j\bullet} + \overline{Y}_{...}) + (Y_{ijk} - \overline{Y}_{ij\bullet})$. Equivalent model: $Y_{ijk} \sim N(\mu_{ij}, \sigma^2)$ where Y_{ijk}s are independent.

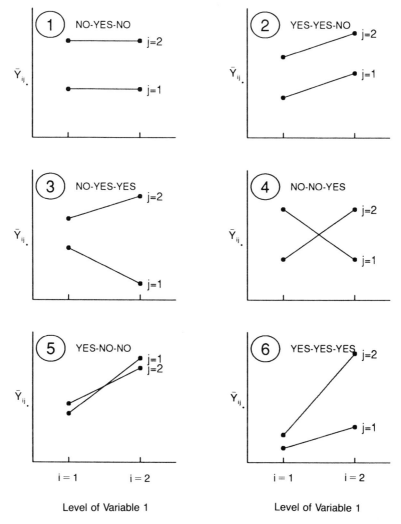

Figure 10.3. Some possible patterns for observed cell means in two-way ANOVA with two levels for each variable. Results of F-tests for main effects variable 1, variable 2 and interaction are indicated by YES or NO. See text for discussion.

As usual, the test MS_μ / MS_ϵ is of little interest and this line is frequently omitted in an analysis of variance table.

Example 10.5. Nitrogen dioxide (NO_2) is an automobile emission pollutant, but less is known about its effects than other pollutants such as particulate matter. Several animal models have been studied to gain an understanding of the effects of NO_2. Sherwin and Layfield [1976] studied protein leakage in the lungs of mice exposed to 0.5 parts per million (ppm) NO_2 for 10, 12, and 14 days. Half of a total group of 44 animals was exposed to the NO_2; the other half served as controls. Control and experimental animals were matched on the basis of weight, but this aspect will be ignored in the analysis since the matching did not appear to influence the results. Thirty-eight animals were available for analysis; the raw data and some basic statistics are listed in Table 10.8.

The response is the percent of serum fluorescence. High serum fluorescence values indicate a greater protein leakage and some kind of insult to the lung tissue. The authors carried out t-tests and state that with regard to serum fluorescence, "no significant differences" were found.

The standard deviations are very similar, suggesting that the homogeneity of variance assumption is probably valid. It is a good idea again to graph the results to get some "feel" for the data and this is done in Figure 10.4. We can see from this

Table 10.8. Serum Fluorescence Readings of Mice Exposed to Nitrogen Dioxide (NO_2) for 10, 12 and 14 Days Compared with Control Animals. Data for Example 10.5.

	Serum Fluorescence		
	10 Days ($j = 1$)	12 Days ($j = 2$)	14 Days ($j = 3$)
Control ($i = 1$)	143	179	76
	169	160	40
	95	87	119
	111	115	72
	132	171	163
	150	146	78
	141	—	—
Exposed ($i = 2$)	152	141	119
	83	132	104
	91	201	125
	86	242	147
	150	209	200
	108	114	178
	75	—	—

n_{ij}

		j	
i	1	2	3
1	7	6	6
2	7	6	6

$Y_{ij\bullet}$

		j	
i	1	2	3
1	941	858	548
2	745	1039	873

$\overline{Y}_{ij\bullet}$

		j	
i	1	2	3
1	134.4	143.0	91.3
2	106.4	173.2	145.5

s_{ij}

		j	
i	1	2	3
1	24.7	35.5	43.2
2	32.1	51.0	37.1

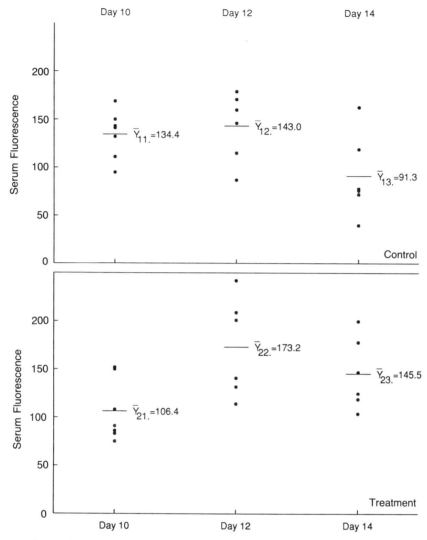

Figure 10.4. Serum fluorescence of mice exposed to nitrogen dioxide. Data from Example 10.5.

figure that there are no outlying observations that would invalidate the normality assumption of the two-way ANOVA model.

To obtain the entries for the two-way ANOVA table, we basically need six quantities:

$$n, \quad Y_{\cdots}, \quad \sum Y_{ijk}^2, \quad \sum \frac{Y_{i\cdot\cdot}^2}{n_{i\cdot}}, \quad \sum \frac{Y_{\cdot j\cdot}^2}{n_{\cdot j}}, \quad \sum \frac{Y_{ij\cdot}^2}{n_{ij}}.$$

With these quantities, and using Equations (10.20) and (10.21), the entire table can be computed. The values are as follows:

$$n = 38, \qquad Y_{\cdots} = 5004, \qquad \sum Y_{ijk}^2 = 730828,$$

$$\sum \frac{Y_{i\cdots}^2}{n_{i\cdot}} = 661476.74, \qquad \sum \frac{Y_{\cdot j\cdot}^2}{n_{\cdot j}} = 671196.74, \qquad \sum \frac{Y_{ij\cdot}^2}{n_{ij}} = 685472.90.$$

Sums of squares can now be calculated:

$$\text{ss}_\alpha = \text{ss}_{\text{TREATMENT}} = 661476.74 - \frac{5004^2}{38} = 2528.95,$$

$$\text{ss}_\beta = \text{ss}_{\text{DAYS}} = 671196.74 - \frac{5004^2}{38} = 12248.95,$$

$$\text{ss}_\gamma = \text{ss}_{\text{TREATMENT} \times \text{DAYS}} = 685472.90 - \frac{5004^2}{38} - 2528.95 - 12248.95 = 11747.21,$$

$$\text{ss}_\epsilon = \text{ss}_{\text{RESIDUAL}} = 730828 - 685472.90 = 45355.10.$$

(It can be shown that $\text{ss}_\epsilon = \sum(n_{ij} - 1)s_{ij}^2$. You can verify this for these data). The ANOVA table is presented in Table 10.9.

The MS for interaction is significant at the 0.05 level ($F_{2,32} = 4.14$, $p < 0.05$). How is this to be interpreted? The means $\overline{Y}_{ij\cdot}$ are graphed in Figure 10.5. There clearly is nonparallelism and the model is not an additive one. But more should be said in order to interpret the results; particularly the role of the control animals. Clearly, control animals were used to provide a measurement of background variation. The differences in mean fluorescence levels among the control animals indicate that the baseline response level changed from Day 10 to Day 14. If we consider the response of the animals exposed to nitrogen dioxide standardized by the control level, a different picture emerges. In Figure 10.5, the differences in means between exposed and unexposed animals is plotted as a dashed line with scale on the right hand side of the graph. This line indicates that there is an increasing effect of exposure with time. The interpretation of the significant interaction effect then is, possibly, that exposure did induce increased protein leakage with greater leakage attributable to longer exposure. This contradicts the authors' analysis of the data using t-tests. If the matching by weight was retained, it would have been possible to consider the differences between exposed and control animals and carry out a one-way ANOVA on the differences. See Problem 10.5.   □

Table 10.9. Analysis of Variance of Serum Fluorescence Levels of Mice Exposed to Nitrogen Dioxide (NO_2). Data From Example 10.5.

Source of Variation	DF	SS	MS	F	p-
Treatment	1	2528.95	2528.95	1.78	> 0.10
Days	2	12248.95	6124.48	4.32	< 0.05
Treatment × days	2	11747.21	5873.60	4.14	< 0.05
Residual	32	45355.10	1417.35	—	—
Total	37	71880.21	—	—	—

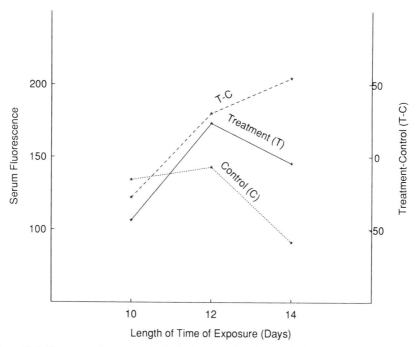

Figure 10.5. Mean serum fluorescense level of mice exposed to nitrogen dioxide, treatment *vs.* control. Data from Example 10.5. The difference (*treatment – control*) is given by dashed lines.

Two-Way ANOVA *From Means and Standard Deviations*

As in the one-way ANOVA, a two-way ANOVA can be reconstructed from means and standard deviations. Let $\overline{Y}_{ij\bullet}$ be the mean, s_{ij} the standard deviation, and n_{ij} the sample size associated with cell ij $(i = 1,\ldots,I, j = 1,\ldots,J)$, assuming a balanced design. Then,

$$Y_{\bullet\bullet\bullet} = \sum_{i=1}^{I}\sum_{j=1}^{J} n_{ij}\overline{Y}_{ij}, \quad Y_{i\bullet\bullet} = \sum_{j=1}^{J} n_{ij}\overline{Y}_{ij\bullet}, \quad Y_{\bullet j\bullet} = \sum_{i=1}^{I} n_{ij}\overline{Y}_{ij\bullet}.$$

Using Equation (10.21), ss_α and ss_β can now be calculated. The term $\sum Y_{ij\bullet}^2/n_{ij}$ in ss_γ is equivalent to

$$\sum \frac{Y_{ij\bullet}^2}{n_{ij}} = \sum n_{ij}\overline{Y}_{ij\bullet}^2.$$

Finally, ss_ϵ can be calculated from

$$\mathrm{ss}_\epsilon = \sum (n_{ij} - 1)s_{ij}^2. \tag{10.22}$$

Problems 10.22 and 10.23 deal with data presented in terms of means and standard deviations. There will be some round-off error in the two-way analysis constructed in this way but it will not affect the conclusion.

It is easy to write a computer subroutine that produces such a table upon input of means, standard deviations and sample sizes.

10.3.2 Randomized Block Design (RBD)

In Chapter 2, we discussed the statistical concept of blocking. A block consists of a subset of homogeneous experimental units. The background variability among blocks is usually much greater than within blocks, and the experimental strategy is to assign randomly all treatments to the units of a block. A simple example of blocking is illustrated by the paired t-test. Suppose two anti-epileptic agents are to be compared. One possible (valid) design is to assign randomly half of a group of patients to one agent and half to the other. By this randomization procedure, the variability among patients is "turned" into error. Appropriate analyses are the two-sample t-test, the one-way analysis of variance, or a two-sample nonparametric test. However, if possible, a better design would be to test both drugs on the same patient; this would eliminate patient to patient variability and comparisons are made within patients. The patients in this case act as "blocks." A paired t-test or analogous nonparametric test is now appropriate. In order for this design to work we would want to assign the drugs randomly within a patient. This would eliminate a possible additive sequence effect; hence, the term "randomized block design." In addition, we would want to have a reasonably large time interval between drugs to eliminate possible carry-over effects; that is, we cannot permit a treatment $\times$ period interaction. Other examples of naturally occurring blocks are animal litters, families, and classrooms. Constructed blocks could be made up of sets of subjects matched on age, race and sex.

Blocking is done for two purposes:

1. To obtain smaller residual variability; and
2. To examine treatments under a wide range of conditions.

A basic design principle is to partition a population of study units in such a way that background variability between blocks is maximized, and consequently background variability within blocks is minimized.

Definition 10.8. In a *randomized block design*, each treatment is given once and only once in each block. Within a block, the treatments are assigned randomly to the experimental units.

Note that a randomized block design, by definition, is a balanced design: this is somewhat restrictive. For example, in animal experiments it would require litters to be of the same size.

The statistical model associated with the randomized block design is

$$Y_{ij} = \mu + \beta_i + \tau_j + \epsilon_{ij}, \quad i = 1, \ldots, I, \ j = 1, \ldots, J; \qquad (10.23)$$

and

a. $\sum \beta_i = \sum \tau_j = 0,$
b. ϵ are iid $N(0, \sigma^2)$.

In this model, β_i is the effect of block i, and τ_j the effect of treatment j. In this model, as indicated, we assume no interaction between blocks and treatments, i.e., if there is a difference between treatments, the magnitude of this effect does not vary from block to block except for random variation. In Section 10.6, we discuss a partial check on the validity of the assumption of no interaction.

The analysis of variance table for this design is a simplified version of Table 10.7: the number of observations is the same in each block and for each treatment. In addition, there is no ss for interaction; another way of looking at this is that the ss for interaction is the error ss. The calculations are laid out in Table 10.10.

Tests of significance proceed in the usual way. The expected mean squares can be derived from Table 10.7, making use of the simpler design.

The computations for the randomized block design are very simple. You can verify that

$$
ss_\mu = \frac{Y_{\bullet\bullet}^2}{n}, \quad ss_\beta = \frac{\sum Y_{i\bullet}^2}{J} - \frac{Y_{\bullet\bullet}^2}{n}, \quad ss_\tau = \frac{\sum Y_{\bullet j}^2}{I} - \frac{Y_{\bullet\bullet}^2}{n},
$$
$$
ss_\epsilon = \sum Y_{ij}^2 - \frac{Y_{\bullet\bullet}^2}{n} - ss_\beta - ss_\tau. \tag{10.24}
$$

Example 10.6. The pancreas, a large gland, secretes digestive enzymes into the intestine. Lack of this fluid results in bowel absorption problems (steatorrhea); this can be diagnosed by excess fat in feces. Commercial pancreatic enzyme supplements are available in three forms: capsule, tablets, and enteric-coated tablets. The enteric-coated tablets have a protective shell to prevent gastrointestinal reaction. Graham [1977] investigated the effectiveness of these three formulations in six patients with steatorrhea; the three randomly assigned treatments were preceded by a control period. For purposes of this example, we will consider the control period as a treatment, even though it was not randomized. The data are displayed in Table 10.11.

To use Equation 10.4, we will need the quantities

$$
Y_{\bullet\bullet} = 618.6, \quad \frac{\sum Y_{i\bullet}^2}{4} = 21532.80, \quad \frac{\sum Y_{\bullet j}^2}{6} = 17953.02, \quad \sum Y_{ij}^2 = 25146.8.
$$

The analysis of variance table, omitting ss_μ, is displayed in Table 10.12.

The treatment effects are highly significant. A visual inspection of Table 10.11 suggests that capsules and tablets are the most effective, enteric-coated tablets less effective. The author points out that the "normal" amount of fecal fat is less than $6\,g/24\,h$, suggesting that, at best, the treatments are palliative. The F-test for patients is also highly significant, indicating that the levels among patients varied considerably: Patient 4 had the lowest average level at $6.1\,g/24\,h$; Patient 5 had the highest level with $47.1\,g/24\,h$. □

10.3.3 Analyses of Randomized Block Designs Using Ranks

A nonparametric analysis of randomized block data using only the ranks was developed by Friedman [1937]. The model is that of Equation (10.23), but the ϵ_{ij} are no longer required to be normally distributed. We assume that there are no ties in the data; for a small number of ties the average ranks may be used. The idea of the test is

Table 10.10. Layout for the Randomized Block Design.

Source of Variation	DF	SS[a]	MS
Grand mean	1	$SS_\mu = n\overline{Y}_{\bullet\bullet}^2$	$MS_\mu = SS_\mu$
Blocks	$I-1$	$SS_\beta = J\sum(\overline{Y}_{i\bullet} - \overline{Y}_{\bullet\bullet})^2$	$MS_\beta = \dfrac{SS_\beta}{I-1}$
Treatments	$J-1$	$SS_\tau = I\sum(\overline{Y}_{\bullet j} - \overline{Y}_{\bullet\bullet})^2$	$MS_\tau = \dfrac{SS_\tau}{J-1}$
Residual	$(I-1)(J-1)$	$SS_\epsilon = \sum(Y_{ij} - \overline{Y}_{i\bullet} - \overline{Y}_{\bullet j} + \overline{Y}_{\bullet\bullet})^2$	$MS_\epsilon = \dfrac{SS_\epsilon}{(I-1)(J-1)}$
Total	IJ	$\sum Y_{ij}^2$	—

Source of Variation	F-Ratio	DF of F-Ratio	$E(MS)$
Grand mean	$\dfrac{MS_\mu}{MS_\epsilon}$	$\left(1, (I-1)(J-1)\right)$	$\sigma^2 + IJ\mu^2$
Blocks	$\dfrac{MS_\beta}{MS_\epsilon}$	$\left(I-1, (I-1)(J-1)\right)$	$\sigma^2 + \dfrac{J\sum\beta_i^2}{I-1}$
Treatments	$\dfrac{MS_\tau}{MS_\epsilon}$	$\left(J-1, (I-1)(J-1)\right)$	$\sigma^2 + \dfrac{I\sum\tau_j^2}{J-1}$
Residual	—	—	σ^2

Source of Variation	Hypothesis Being Tested
Grand mean	$\mu = 0$
Blocks	$\beta_i = 0$ for all i
Treatments	$\tau_j = 0$ for all j

[a]Summation is over all displayed subscripts.
Model: $Y_{ij} = \mu + \beta_i + \tau_j + \epsilon_{ij}$ [where $\epsilon_{ij} \sim$ iid $N(0, \sigma^2)$]. Data: $Y_{ij} = \overline{Y}_{\bullet\bullet} + (\overline{Y}_{i\bullet} - \overline{Y}_{\bullet\bullet}) + (\overline{Y}_{\bullet j} - \overline{Y}_{\bullet\bullet}) + (\overline{Y}_{ij} - \overline{Y}_{i\bullet} - \overline{Y}_{\bullet j} + \overline{Y}_{\bullet\bullet})$. Equivalent model: $Y_{ij} \sim N(\mu_{ij}, \sigma^2)$ where Y_{ij}s are independent.

simple: if there are no treatment effects ($\tau_j = 0$ for all j), the ranks of the observations within a block are randomly distributed. For block i, let

$$R_{ij} = \text{rank of } Y_{ij} \text{ among } Y_{i1}, Y_{i2}, \ldots, Y_{iJ}.$$

The Friedman statistic for testing the null hypothesis $H_0 : \tau_j = 0$ (where $j = 1, \ldots, J$) is

Table 10.11. Effectiveness of Pancreatic Supplements on Fat Absorption in Patients with Steatorrhea. Data from Example 10.6.

Case	None (Control)	Tablet	Capsule	Enteric-Coated Tablet	$Y_{i\bullet}$	$\overline{Y}_{i\bullet}$
	Fecal Fat (g/24 hours)					
1	44.5	7.3	3.4	12.4	67.6	16.9
2	33.0	21.0	23.1	25.4	102.5	25.6
3	19.1	5.0	11.8	22.0	57.9	14.5
4	9.4	4.6	4.6	5.8	24.4	6.1
5	71.3	23.3	25.6	68.2	188.4	47.1
6	51.2	38.0	36.0	52.6	177.8	44.4
$Y_{\bullet j}$	228.5	99.2	104.5	186.4	618.6	—
$\overline{Y}_{\bullet j}$	38.1	16.5	17.4	31.1	$\overline{Y}_{\bullet\bullet} = 25.8$	

Table 10.12. Randomized Block Analysis of Fecal Fat Excretion of Patients with Steatorrhea. Data from Table 10.11; see Example 10.6.

Source of Variation	DF	SS	MS	F-Ratio	p-Value
Patients (Blocks)	5	5588.38	1117.68	10.44	< 0.001
Treatments	3	2008.60	669.53	6.26	< 0.01
Residual	15	1605.40	107.03	—	—
Total	23	9202.38	—	—	—

$$T_{FR} = 12I \sum_{j=1}^{J} \frac{(\overline{R}_{\bullet j} - \overline{R}_{\bullet\bullet})^2}{J(J+1)}. \tag{10.25}$$

Computationally, the following formula is easier:

$$T_{FR} = \frac{12}{IJ(J+1)} \sum_{j=1}^{J} R_{\bullet j}^2 - 3(I)(J+1). \tag{10.26}$$

The null hypothesis is rejected for large values of T_{FR}. For small randomized block designs the critical values of T_{FR} are tabulated; see, for example, Table 39 in Odeh *et al.* [1977] which goes up to $I = J = 6$. As the number of blocks becomes very large, the distribution of T_{FR} approaches that of a χ^2-distribution with $(J-1)$ degrees of freedom. See also Notes 10.1 and 10.2.

Example 10.6 (continued). Replacing the observations for each *individual* by their ranks produces the following table:

Case	Control	Tablet	Capsule	Enteric-Coated Tablet
		Treatment		
1	4	2	1	3
2	4	1	2	3
3	3	1	2	4
4	4	1.5	1.5	3
5	4	1	2	3
6	3	2	1	4
$R_{\cdot j}$	22	8.5	9.5	20

For individual 4, the two tied observations are replaced by the average of the two ranks. (As a check, the total $R_{\cdot\cdot}$ of ranks must be $R_{\cdot\cdot} = IJ(J+1)/2$. (Why?) For this example $I = 6, J = 4, IJ(J+1)/2 = (6 \cdot 4 \cdot 5)/2 = 60$, and $R_{\cdot\cdot} = 22+8.5+9.5+20 = 60$). The Friedman statistic, using Equation (10.26), has the value

$$T_{FR} = \frac{12}{6 \times 4 \times 5}(22^2 + 8.5^2 + 9.5^2 + 20^2) - (3 \times 6 \times 5) = 104.65 - 90 = 14.65.$$

This quantity is referred to Table A.4 in the Appendix with $3\,\mathrm{DF}$ ($14.65/3 = 4.88$); the p-value is $p = 0.0021$, which is less than 0.005. From exact tables such as Odeh et al. [1977], the exact p-value is $p < 0.001$. The conclusion is the same as that of the analysis of variance in Section 10.3.2. Note also that the ranking of treatments in terms of the total ranks is the same as in Table 10.11. For an alternative rank analysis of these data see Problem 10.20.

10.3.4 Types of ANOVA Models

In Section 10.2.2, two examples were mentioned of one-way analyses of variance. The first dealt with the age at which children begin to walk as a function of various training procedures; the second example dealt with patient hospitalization costs, based upon an examination of some hospitals (treatments) randomly selected from all the hospitals in a large metropolitan area (from each selected hospital a specified number of patient records are selected for cost analysis). The experimental design associated with the first example differs from the second: in a repetition of the first study, the same set of treatments could be used; in the second study, a new set of hospitals could presumably be selected; that is, the "treatment levels" are randomly selected from a larger set of treatment levels.

Definition 10.9. If the levels of a classification variable in an ANOVA situation are selected at random from a population, the variable is said to be a *random factor* or *random effect*. Factors with the levels fixed by those conducting the study or which are fixed classifications (e.g., sex) are called *fixed factors* or *fixed effects*.

Definition 10.10. ANOVA situations with all classification variables fixed are called *fixed effects models* (Model I). If all the classification variables are random effects the design is a *random effects model* (Model II). If both random and fixed effects are present, the design is a *mixed effects model*.

Historically, no distinction was made between Model I and II designs, in part due to identical analyses in simple situations and similar analyses in more complicated situations. Eisenhart [1947] was the first to describe systematically the differences between the two models. Some other examples of random effects models are:

1. A manufacturer of spectrophotometers randomly selects five instruments from its production line and obtains a series of replicated readings on each machine.
2. To estimate the maximal exercise performance in a healthy adult population twenty subjects are randomly selected and 10 independent estimates of maximal exercise performance for each individual are obtained.
3. To determine knowledge about the effect of drugs among 6th graders a researcher randomly selects five sixth grade classes from among the 100 sixth grade classes in a large school district. Each child in the selected 6th grades fills out a questionnaire.

How can we determine whether a design is Model I or Model II? The basic criterion deals with the population to which inferences are to be made. Another way of looking at this is to consider the number of times randomness is introduced (ideally). In Example 2 above there are two sources of randomness: subjects and observations within subjects. If more than one "layer of randomness" has to be passed through in order to reach the population of interest then we have a random effects model.

An example of a mixed model is Example 2 above with a further partitioning of subjects into male and female. The factor, sex, is fixed.

Sometimes a set of data can be modeled either by a fixed or random effects model. Consider Example 1 again. Suppose a cancer research center has bought the five instruments and is now running some standardization experiments. For the purpose of the research center the effects of machines are fixed effects.

To distinguish a random effects model from a fixed effects model the components of the model are written as random variables. The two-way random effects ANOVA model with interaction is written as

$$Y_{ijk} = \mu + A_i + B_j + G_{ij} + e_{ijk}, \quad i = 1, \ldots, I, \; j = 1, \ldots, J, \; k = 1, \ldots, n_{ij}. \quad (10.27)$$

The assumptions are:

1. e_{ijk} are iid $N(0, \sigma^2)$, as before,
2. A_i are iid $N(0, \sigma_\alpha^2)$,
3. B_j are iid $N(0, \sigma_\beta^2)$,
4. G_{ij} are iid $N(0, \sigma_\gamma^2)$.

The total variance can now be partitioned into several components (hence another term for these models: components of variance models). Assume that the experiment is balanced with $n_{ij} = m$ for all i and j. The difference between the fixed effect and random effect model is in the expected mean squares. Table 10.13 compares the EMS for both models, taking the EMS for the fixed effect model from Table 10.7.

The test for interaction is the same in both models. However, if interaction is present the test for main effects in the random effects model must use MS_γ in the denominator rather than MS_ϵ in order to obtain a valid test.

Table 10.13. Comparison of Expected Mean Squares in the Two Way ANOVA, **Fixed Effect** *vs.*
Random Effect Models. m **Observations in Each Cell.**

Source of Variation	DF	EMS	
		Fixed Effect	Random Effect
Row Main Effects	$I - 1$	$\sigma^2 + \dfrac{Jm \sum \alpha_i^2}{I - 1}$	$\sigma^2 + m\sigma_\gamma^2 + mJ\sigma_\alpha^2$
Column Main Effects	$J - 1$	$\sigma^2 + \dfrac{Im \sum \beta_j^2}{J - 1}$	$\sigma + m\sigma_\gamma^2 + mI\sigma_\beta^2$
Row × column Interaction	$(I - 1)(J - 1)$	$\sigma^2 + \dfrac{IJm \sum \gamma_{ij}^2}{(I - 1)(J - 1)}$	$\sigma^2 + m\sigma_\gamma^2$
Residual	$n_{\bullet\bullet} - IJ$	σ^2	σ^2

The null hypothesis

$$H_0 : \gamma_{ij} = 0 \quad \text{all } i \text{ and } j$$

in the fixed effect model has as counterpart

$$H_0 : \sigma_\gamma^2 = 0$$

in the random effect model. In both cases the test is carried out using the ratio MS$_\gamma$/MS$_\epsilon$ with $(I - 1)(J - 1)$ and $n - IJ$ degrees of freedom. If interaction is not present, the tests for main effects are the same in both models. However, if H_0 is not rejected the tests for main effects are different in the two models. In the random effects model the expected mean square for main effects now contains a term involving σ_γ^2. Hence the appropriate F-test involves MS$_\gamma$ in the denominator rather than MS$_\epsilon$; the degrees of freedom are changed accordingly.

Several comments can be made:

1. Frequently the degrees of freedom associated with MS$_\gamma$ are fewer than those of MS$_\epsilon$ so that there is a loss of precision if MS$_\gamma$ has to be used to test main effects.

2. From a design point of view, if m, I, J can be chosen it may pay to choose m small and I, J relatively large if a random effects model is appropriate. A minimum of two replicates per treatment combination is needed to obtain an estimate of σ^2. If possible the rest of the observations should be allocated to the levels of the variables. This may not always be possible due to costs or other considerations. If the total cost of the experiment is fixed an algorithm can be developed for choosing the values of m, I, and J.

3. The difference between the fixed and random effects models for the two-way ANOVA designs is not as crucial as it seems. We have indicated caution in proceeding to the tests of main effects if interaction is present in the fixed model (see Figure 10.3 and associated discussion). In the random effects model the same caution holds. It is perhaps too strong to say that main effects should not be tested when interaction is present but you should certainly be able to explain what information you hope to obtain from such tests after a full interpretation of the (significant) interaction.

4. Expected mean squares for an unbalanced random effects model are not derivable or are very complicated. A more useful approach is that of multiple regression discussed in the next chapter. See also Section 10.5.

5. For the randomized block design the MS_ϵ can be considered the mean square for interaction. Hence, in this case the F-tests are appropriate for both models. (Does this contradict the statement made in 3?) Note also that there is little interest in the test of block effects except as a verification that the blocking was effective.

Good discussions about inference in the case of random effects models can be found in Snedecor and Cochran [1980] and Winer [1971].

10.4 REPEATED MEASURES DESIGNS AND OTHER DESIGNS

10.4.1 Repeated Measures Designs

Consider a situation in which blood pressures of two populations are to be compared. One individual is selected at random from each population. The blood pressure of each individual is measured 100 times. How would you react to a data analysis that used the two-sample t-test with two samples of size 100 and showed that the blood pressures differed in the two populations? The idea is ridiculous, but in one form or another appears frequently in the research literature. Where does the fallacy lie? There are two sources of variability: within individuals and among individuals. The variability within individuals is incorrectly assumed to represent the variability among individuals. Another way of saying this is that the 100 readings are not independent samples from the population of interest. They are repeated measurements on the same experimental unit. The repeated measures may be useful in this context in pinning down more accurately the blood pressure of the two individuals, they do not, however, make up for the small sample size. Another feature we want to consider is that the sequence of observations within the individual cannot be randomized, for example, a sequence of measurements of growth. Thus, we typically, do not have a randomized block design.

Definition 10.11. In a *repeated measures design* multiple (two or more) measurements are made sequentially upon the same observational unit.

A repeated measures design usually is an example of a mixed model with the observational unit a random effect, for example persons or animals, and the treatments on the observational units fixed effects. Frequently, data from repeated measure designs are somewhat unbalanced and this makes the analysis more difficult. One approach is to summarize the repeated measures in some meaningful way by single measures and then analyze the single measures in the usual way. This is the way many computer programs analyze such data. We motivate this approach by an example.

Example 10.7. Hillel and Patten [1990] were interested in the effect of accessory nerve injury as result of neck surgery in cancer. The surgery frequently decreases the strength of the arm on the affected side. To assess the potential recovery, the unaffected arm was to be used as a control. But there is a question of the comparability of arms due to dominance, age, gender, and other factors. In order to assess this effect 33 normal volunteers were examined by several measurements. The one discussed here is that of torque, or the ability to abduct (move or pull) the shoulder using a standard machine built for that purpose. The subjects were tested under three consecutive conditions (in order of increasing strenuousness): $90°/s$, $60°/s$, and $30°/s$. The data presented in Table 10.14 are the best of three trials under each condition. For completeness, the age and height of each of the subjects is also presented. The researchers wanted answers to at least five questions, all dealing with differences between dominant and nondominant sides:

Table 10.14. Peak Torque for 33 Subjects by Gender, Dominant Arm and Age Group under Three Conditions. Data for Example 10.7.

Subject		Age	Height (inches)	Weight (pounds)	90°		60°		30°	
					DM[a]	ND[a]	DM	ND	DM	ND
Female	1	20	64	107	17	13	20	17	23	22
	2	23	68	140	25	25	28	29	31	31
	3	23	67	135	27	28	30	31	32	33
	4	23	67	155	23	28	27	29	27	32
	5	25	65	115	15	11	15	13	17	17
	6	26	68	147	27	17	25	21	32	27
	7	31	62	147	25	17	25	21	29	24
	8	31	66	137	19	15	17	17	21	19
	9	33	66	160	28	26	31	27	31	31
	10	36	66	118	23	23	26	27	27	25
	11	56	67	210	23	31	37	44	49	53
	12	59	67	130	15	17	17	19	20	20
	13	60	63	132	17	15	19	21	24	28
	14	60	64	180	15	15	17	19	19	21
	15	67	62	135	13	5	15	8	15	14
	16	73	62	124	11	9	13	13	19	17
Male	1	26	69	140	43	43	44	43	49	41
	2	28	71	175	45	43	48	45	53	52
	3	28	70	125	25	29	29	37	39	41
	4	28	70	175	39	41	49	47	55	44
	5	29	72	150	38	33	40	33	44	37
	6	30	68	145	53	41	51	40	59	44
	7	31	74	240	60	49	71	54	68	53
	8	32	67	168	32	31	37	31	39	30
	9	40	69	174	47	37	43	47	49	53
	10	41	72	190	33	25	29	25	39	27
	11	41	68	184	39	24	43	25	39	33
	12	56	70	200	21	11	23	12	33	24
	13	58	72	168	41	35	45	37	49	39
	14	59	73	170	31	32	31	31	35	38
	15	60	73	225	39	41	47	45	55	49
	16	68	67	140	31	23	33	27	37	33
	17	72	69	125	13	17	17	19	17	25

[a]DM, dominant arm; ND, nondominant arm.

1. Is there a difference between the dominant and nondominant arms.
2. Does the difference vary between men and women.
3. Does the difference depend on age, height or weight.
4. Does the difference depend on treatment condition.
5. Is there interaction between any of the factors or variables mentioned in (1)–(4).

For purposes of this example, we will only address questions (1), (2), (4), and (5), leaving question (3) for the discussion of the analysis of covariance in Chapter 11.

Columns two to four in Table 10.15 contain the differences between the dominant and nondominant arms. Columns five to seven are re-expressions of the three differences as follows. Let d90, d60, and d30 be the differences between the dominant and nondominant arms under each of the three conditions. Then we define:

$$Constant = (d90 + d60 + d30)/\sqrt{3},$$

$$Linear = (d90 - d30)/\sqrt{2},$$

$$Quadratic = (d90 - 2 \cdot d60 + d30)/\sqrt{6}.$$

For example, for the first female subject, rounding off to one decimal place,

$$(4 + 3 + 1)/\sqrt{3} = 4.6,$$

$$(4 - 1)/\sqrt{2} = 9.9,$$

$$(4 - 2 \times (3) + 1)/\sqrt{6} = -0.4.$$

The first component clearly represents an average difference of dominance over the three conditions. The divisor is chosen to make the variance of this term equal to the variance of a single difference. The second term represents a slope within an individual. If the three conditions were considered as values of a predictor variable with values -1 (for 30°), 0 (for 60°), and 1 (for 90°) the slope would be expressed as in the second, or linear, term. The linear term assesses a possible trend in the differences over the three conditions within an individual. The last term, the quadratic term fits a quadratic curve through the data assessing possible curvature or nonlinearity within an individual. This partitioning of the observations within an individual has the property that sums of squares are maintained. For example, for the first female subject,

$$4^2 + 3^2 + 1^2 = 26 = (4.6)^2 + (2.1)^2 + (-0.4)^2,$$

except for rounding. (If you were to calculate these terms to more decimal places you would find that the right side is identical to the left side.) In words, the variability in response within an individual has been partitioned into a constant component, a linear component and a quadratic component. The questions posed can now be answered unambiguously since the three components have been constructed to be "orthogonal" or, uncorrelated. An analysis of variance is carried out on the three terms; unlike the usual analysis of variance a term for the mean is included; results are summarized in

Table 10.15. Differences in Torque Under Three Conditions and Associated Orthogonal Contrasts. Data from Table 10.14, Example 10.7. See Table 10.14 for Notation.

		DM – ND			Orthogonal Contrasts		
		90°	60°	30°	Constant	Linear	Quadratic
Female	1	4	3	1	4.6	2.1	−0.4
	2	0	−1	0	−0.6	0.0	0.8
	3	−1	−1	−1	−1.7	0.0	0.0
	4	−5	−2	−5	−6.9	0.0	−2.4
	5	4	2	0	3.5	2.8	0.0
	6	10	4	5	11.0	3.5	2.9
	7	8	4	5	9.8	2.1	2.0
	8	4	0	2	3.5	1.4	2.4
	9	2	4	0	3.5	1.4	−2.4
	10	0	−1	2	0.6	−1.4	1.6
	11	−8	−7	−4	−11.0	−2.8	0.8
	12	−2	−2	0	−2.3	−1.4	0.8
	13	2	−2	−4	−2.3	4.2	0.8
	14	0	−2	−2	−2.3	1.4	0.8
	15	8	7	1	9.2	4.9	−2.0
	16	2	0	2	2.3	0.0	1.6
Male	1	0	1	8	5.2	−5.7	2.4
	2	2	3	1	3.5	0.7	−1.2
	3	−4	−8	−2	−8.1	−1.4	4.1
	4	−2	2	11	6.4	−9.2	2.0
	5	5	7	7	11.0	−1.4	−0.8
	6	12	11	15	21.9	−2.1	2.0
	7	11	17	15	24.8	−2.8	−3.3
	8	1	6	9	9.2	−5.7	−0.8
	9	10	−4	−4	1.2	9.9	5.7
	10	8	4	12	13.9	−2.8	4.9
	11	15	18	6	22.5	6.4	−6.1
	12	10	11	9	17.3	0.7	−1.2
	13	6	8	10	13.9	−2.8	0.0
	14	−1	0	−3	−2.3	1.4	−1.6
	15	−2	2	6	3.5	−5.7	0.0
	16	8	6	4	10.4	2.8	0.0
	17	−4	−2	−8	−8.1	2.8	−3.3

Table 10.16. We start with discussing the analysis of the quadratic component. The analysis indicates that there are no significant differences between males and females in terms of the quadratic, or nonlinear component. Nor is there an overall effect. Next, conclusions are similar for the linear effect. We conclude that there is no linear trend for abductions at 90°, 60°, and 30°. This leaves the constant term which indicates (1) that there is a significant gender effect of dominance ($F_{1,31} = 6.48$, $p < 0.05$) and an overall dominance effect. The average of the constant term for females is 1.31, for males is 8.6. One question that can be raised is whether the difference between female and male is a true gender difference or can be attributed to differences is size. An analysis of covariance can answer this question (see Problem 11.40). ☐

Table 10.16. Analysis of Variance and Means of the Data in Table 10.15, Example 10.7.

Source of Variation		DF	SS	MS	F-Ratio
			Analysis of Variance		
Constant	Mean	1	900.7	900.7	13.3
	Gender	1	438.5	438.5	6.48
	Error 1	31	2099.2	67.72	
Linear	Mean	1	0.33	0.33	0.02
	Gender	1	33.43	33.43	2.43
	Error 2	31	426.0	13.74	
Quadratic	Mean	1	3.09	3.09	0.50
	Gender	1	0.70	0.70	0.11
	Error 3	31	191.2	6.17	

Means

		Constant	Linear	Quadratic
Female ($n = 16$)	Mean	1.306	1.138	0.456
	Standard Deviation	5.920	2.121	1.609
Male ($n = 17$)	Mean	8.600	−0.876	0.165
	Standard Deviation	9.917	4.734	3.085

Data from a repeated measures design often look like those of a randomized block design. The major difference is the way the data are generated. In the randomized block, the treatments are allocated randomly to a block. In the repeated measures design this is not the case; not being possible as in the case of observations over time, or because of experimental constraints as in the above example. If the data are analyzed as a randomized block, care must be taken that the assumptions of the randomized block design are satisfied. The key assumption is that of *compound symmetry*: the sample correlations among treatments over subjects must all estimate the same population correlation. The randomization ensures this in the randomized block design. For example, for the data of Example 10.6 in Table 10.11 the correlations are as follows:

	Control	**Tablet**	**Capsule**
Tablet	0.658		
Capsule	0.599	0.960	
Coated Tablet	0.852	0.784	0.833

These correlations are reasonably comparable. If the correlations are not assumed equal, a conservative F-test can be carried out by referring the observed value of F for treatments to an F-table with 1 and $(I-1)$ degrees of freedom (rather than $(J-1)$

and $(I-1)(J-1)$ degrees of freedom). Alternatives to the above two approaches include multivariate analyses. There is a huge literature on repeated measures analysis. The psychometric literature contains many papers on this topic. To explore this area consult recent issues of journals such as The American Statistician. One example is a paper by Looney and Stanley [1989].

10.4.2 Factorial Designs

An experimental layout that is very common in agricultural and nutritional studies is the balanced factorial design. It is less common in medical research due to the ever present risk of missing observations and ethical constraints.

Definition 10.12. In a *factorial design* each level of a factor occurs with every level of every other factor. Experimental units are assigned randomly to treatment combinations.

Suppose there are three factors with levels $I = 3$, $J = 2$, and $K = 4$. Then there are $3 \times 2 \times 4 = 24$ treatment combinations. If there are three observations per combination, 72 experimental units are needed. Factorial designs, if feasible, are very economical and permit assessment of joint effects of treatments that are not possible with experiments dealing with one treatment at a time. The two-way analysis of variance can be thought of as dealing with a two-factor experiment. The generalization to three or more factors does not require new concepts or strategies, just increased computational complexity.

10.4.3 Hierarchical or Nested Designs

A hierarchical or nested design is illustrated by the following example. As part of a program to standardize measurement of the blood level of phenytoin, an antiepileptic drug, samples with known amounts of active ingredients are sent to four commercial laboratories for analysis. Each laboratory employs a number of technicians who make one or more determinations of the blood level. A possible layout is the following:

Laboratory	1	2	3	4
Technician	1 2	3 4 5	6 7	8 9
Assay	∧ ∧	∧ ∧ ∧	⋀⋀	∧ ∧

In this example Laboratory 2 employs three technicians who routinely do this assay, all other laboratories use two technicians. In Laboratory 3, each technician runs three assays, in the other laboratories each technician runs two assays. There are three factors: laboratories, technicians, and assays; the arrangement is *not* factorial: there is no reason to match Technician 1 with any technician from another laboratory.

Definition 10.13. In a *hierarchical or nested design* levels of one or more factors are subsampled within one or more other factors. In other words, the levels of one or more factors are not crossed with one or more other factors.

In the above example the factors, "Technicians" and "Assay," are not "crossed" with the first factor but rather nested within that factor. For the factor "Technician" to be "crossed," its levels would have to repeat within each level of "Laboratory." That is why we deliberately labeled the levels of "Technician" consecutively and introduced some imbalance. Determining whether a design is factorial or hierarchical is not always easy. If the first of the two technicians within a laboratory was the senior technician and the second (or second and third) a junior technician then "Technician" could perhaps be thought of as having two levels "Senior" and "Junior" which could then be crossed with "Laboratory." A second reason is that designs are sometimes mixed, having both factorial and hierarchical components. In the above example, if "Technician" occurred at two levels then "Technician" and "'Laboratory" could be crossed or factorial but "Assay" would continue to be nested within "Technician."

10.4.4 Split Plot Designs

A related experimental design is the Split-Plot design. We will illustrate it with an example. We want to test the effect of physiotherapy in conjunction with drug therapy on the mobility of patients with arthritis. Patients are randomly assigned to physiotherapy and each patient is given a standard drug and a placebo in random order. The experimental layout is as follows:

| | | Physiotherapy | | | |
| | | $i = 1$ (Yes) | | $i = 2$ (No) | |
	Patient	1	$2 \cdots J$	1	$2 \cdots J$
$k = 1$	Drug	Y_{111}	$— \cdots —$	Y_{211}	$— \cdots —$
$k = 2$	Placebo	Y_{112}	$— \cdots —$	Y_{212}	$— \cdots —$

The patients form the "whole plots" and the drug administration the "split plot." These designs are characterized by almost separate analyses of specified effects. To illustrate in this example, let

$$D_{ij} = Y_{ij1} - Y_{ij2} \quad \text{and} \quad T_{ij} = Y_{ij1} + Y_{ij2}, \quad i = 1, 2, \, j = 1, \ldots, J.$$

In words, D_{ij} is the difference between Drug and Placebo for patient j receiving physiotherapy level i; T_{ij} is the sum of readings for Drug and Placebo. Now carry out an analysis of variance (or two-sample t-test) on each of these variables. The tables are as follows:

| One-Way ANOVA | DF | Interpretation of Split-Plot Analyses | |
		Differences	Sums
Mean	1	Mean differences	Mean sums
Between groups	1	Differences × physiotherapy	Sums × physiotherapy
Within groups	$2(J - 1)$	Differences within physiotherapy	Sums within physiotherapy
Total	$2J$	"Total"	"Total"

An analysis of variance of the sums is, in effect, an assessment of physiotherapy (averaged or summed over drug and placebo), that is, a comparison of the $\overline{T}_{1\bullet}$ and $\overline{T}_{2\bullet}$.

The analysis of differences is very interesting. The assessment of the significance of "Between Groups" is a comparison of the average differences between Drug and Placebo with physiotherapy and without physiotherapy, that is, $\overline{D}_{1\bullet} - \overline{D}_{2\bullet}$ is a test for interaction. Additionally, the "Mean Differences" term can be used to test the hypothesis that $\overline{D}_{\bullet\bullet}$ comes from a population with mean zero, that is, it is a comparison of Drug and Placebo. This test only makes sense if the null hypothesis of no interaction is not rejected.

These remarks are intended to give you an appreciation for these designs. For more details consult a text on design of experiments such as Winer [1971].

10.5 UNBALANCED OR NONORTHOGONAL DESIGNS

In the previous sections we have discussed balanced designs. The balanced design is necessary to obtain an additive partition of the sum of squares. If the design is not balanced there are basically three strategies available; the first is to try to restore balance. If only one or two observations are "missing" this is a possible strategy, but if more than two or three are missing a second or third alternative will have to be used. The second alternative is to use an unweighted means analysis. The third strategy is to use a multiple regression approach, this will be discussed in detail in Section 11.8.

10.5.1 Causes of Imbalance

Perhaps the most important thing you can do in the case of unbalanced data is to reflect on the reason(s) for the imbalance. If the imbalance is due to some random mechanism unrelated to the factors under study the procedures to be discussed below are appropriate. If the imbalance is due to some specific reason, perhaps related to the treatment, then it will be profitable to think very carefully about the implications. Usually, such imbalance suggests a bias in the treatment effects. For example, if a drug has major side effects which causes patients to drop out of a study then the effect of the drug may be inappropriately estimated if only the remaining patients are used in the analysis; if one does the analysis only on patients for whom "all data are available" biased estimates may result.

10.5.2 Restoring Balance

Missing Data in the Randomized Block Design
Suppose that the ijth observation is missing in a randomized block design consisting of I blocks and J treatments. The usual procedure is to:

1. Estimate the missing data point by least squares using the formula:

$$\widehat{Y}_{ij} = \frac{(IY_{i\bullet} + JY_{\bullet j} - Y_{\bullet\bullet})}{(I-1)(J-1)} \tag{10.28}$$

where the row, column and grand totals are those for the values present.

2. Carry out the usual analysis of variance on this augmented data set.

3. Reduce the degrees of freedom for MS_ϵ by 1.

If more than one observation is missing, say two or three, values are guessed for all but one, the latter is estimated by Equation (10.28), a second missing value deleted and the process repeated until convergence. The degrees of freedom for MS_ϵ are now reduced by the number of observations that are missing.

Example 10.7 (continued). We return to Example 10.6, Table 10.11. Suppose observation $Y_{31} = 19.1$ is missing and we want to estimate it. For this example, $I = 6, J = 4, Y_{3\bullet} = 38.8, Y_{\bullet 1} = 209.4,$ and $Y_{\bullet\bullet} = 599.5$. We estimate Y_{31} by

$$\widehat{Y}_{31} = \frac{(6(38.8) + 4(209.4) - 599.5)}{(6-1)(4-1)} = 31.4$$

This value appears to be drastically different from 19.1. It is. It also indicates that there is no substitute for real data. The analysis of variance is not altered a great deal.

Source of Variable	DF	SS	MS	F
Patients(Blocks)	5	5341.93	1068.39	9.90
Treatments	3	2330.30	776.77	7.20
Residual	14	1510.94	107.92	—
Total	22	9183.17	—	—

The F-ratios have not changed much from those in Table 10.12. So in this case, the conclusions are unchanged. Note that the degrees of freedom for residual are reduced by 1. This means that the critical values of the F-statistics are increased slightly. Therefore this experiment has less power than the one without missing data. □

Missing Data in Two-Way and Factorial Designs
If a cell in a two-way design has a missing observation, it is possible to replace the missing point by the mean for that cell, carry out the analysis as before and subtract one degree of freedom for MS_ϵ. A second approach is to carry out an unweighted means analysis. We will illustrate both procedures by means of an example.

Example 10.8. These data are part of data used in Wallace *et al.* [1977]. The observations are from a patient with prostatic carcinoma. The question of interest is whether the immune system of such a patient differs from that of noncarcinoma subjects. One way of assessing this is to stimulate in vitro the patient's lymphocytes with phytohemagglutinin (PHA). This causes blastic transformation. Of interest is the amount of blastogenic generation as measured by DNA incorporation of a radioactive compound. The observed data are the mean radioactive counts/minute both when stimulated with PHA and when not stimluated by PHA. As a control, the amount of PHA stimulation in a pooled sera of normal blood donors was used. To examine the response of a subject's lymphocytes, the quantity,

$$\frac{\dfrac{\text{Subject's mean count/minute stimulated with PHA}}{\text{Subject's mean count/minute without PHA}}}{\dfrac{\text{Normal sera mean count/minute stimulated with PHA}}{\text{Normal sera mean count/minute without PHA}}} = \frac{(X_{11}/X_{12})}{(X_{21}/X_{22})} \quad (10.29)$$

Table 10.17. DNA Incorporation of Sera of Patient with Prostatic Carcinoma as Compared to Sera From Normal Blood Donors. Data for Example 10.17. ($\log_e$ of Counts in Parentheses).

	Radioactivity (counts/min)	
Subject	With PHA	Without PHA
	129594 (11.772)	301 (5.707)
	143687 (11.875)	333 (5.808)
Patient Sera	115953 (11.661)	295 (5.687)
	103098 (11.543)	285 (5.652)
	98125 (11.494)	
	43125 (10.672)	247 (5.509)
	46324 (10.743)	298 (5.697)
Blood Donor Sera	42117 (10.648)	387 (5.958)
	45482 (10.725)	
	31192 (10.348)	

was used. If the lymphocytes responded in the same way to the subject's sera and the pooled sera, the ratio should be approximately equal to one. The data are displayed in Table 10.17.

There is a great deal of variability in the counts/minute as related to level. In Section 10.6.3, we suggest that logarithms are appropriate for stabilization of the variability. There is a bonus involved in this case. Under the null hypothesis of no difference in patient and blood donor sera the ratio in Equation (10.28) is 1; that is

$$H_0 : \frac{E(X_{11})/E(X_{12})}{E(X_{21})/E(X_{22})} = 1.$$

This is equivalent to

$$H_0 : \log_e \frac{E(X_{11})/E(X_{12})}{E(X_{21})/E(X_{22})} = \log_e 1 = 0$$

or

$$\log_e E(X_{11}) - \log_e E(X_{12}) - \log_e E(X_{21}) + \log_e E(X_{22}) = 0. \tag{10.30}$$

Now define

$$Y_{ijk} = \log_e X_{ijk}, \quad i = 1, 2, \quad j = 1, 2, \ k = 1, \dots, n_{ij}.$$

It can be shown that Equation (10.30) is zero only if the true interaction term is zero. Thus, the hypothesis that the patient's immune system does not differ from that of noncarcinoma subjects is translated into a null hypothesis about interaction involving the logarithms of the radioactive counts.

We finally get to the "missing data" problem. The data are not balanced: $n_{ij} \neq n_{i\bullet}n_{\bullet j}/n_{\bullet\bullet}$ (we could delete one observation from the (1,2) cell but considering the small numbers, we want to retain as much information as possible). One strategy is to add an observation to cell (2,2) equal to the mean for that cell and adjust the degrees of freedom for interaction. The mean $\overline{Y}_{22\bullet}$ is 5.721. The analysis of variance becomes:

Source	DF	SS	MS	*F*-Ratio	*p*-Value
Subject	1	1.4893	1.4893	—	—
PHA	1	131.0722	131.0722	—	—
PHA × subject	1	1.2247	1.2247	50.0	$p < 0.001$
Error	13	0.3184	0.02449	—	—
Total	16	—	—	—	—

Note that the MS for error has 13 degrees of freedom, not 14. The MS for error will be the correct estimate using this procedure but the MS for interaction (and main effects) will not be the same as the one obtained by techniques of the next chapter. However, it should be close.

10.5.3 Unweighted Means Analysis

The second approach is that of unweighted mean analysis. Again, assuming that the unequal cell frequencies are not due to treatment effects, the cell means are used and an average sample size calculated for each cell. The appropriate average sample size is given by the harmonic mean. In the context of our example, the harmonic mean is defined to be

$$\widetilde{n} = \frac{IJ}{1/n_{11} + 1/n_{12} + 1/n_{21} + 1/n_{22}},$$

where $n_{ij} =$ number of observations in cell (i,j). The harmonic mean is used because the standard error of the mean of cell (i,j) is proportional to $1/n_{ij}$. All calculations for row and column effects are now based on cell means and the harmonic mean of the cell sample sizes. Write the cell means and marginal means as follows:

$\overline{Y}_{11\bullet}$	$\overline{Y}_{12\bullet}$	$\widehat{M}_{1\bullet}$
$\overline{Y}_{21\bullet}$	$\overline{Y}_{22\bullet}$	$\widehat{M}_{2\bullet}$
$\widehat{M}_{\bullet 1}$	$\widehat{M}_{\bullet 2}$	$\widehat{M}_{\bullet\bullet}$

The marginal and overall means are just the arithmetic average of the cell means, that is, the unweighted average (hence, the name "unweighted mean" analysis). The row and column sums of squares are calculated as follows:

$$\mathrm{ss}_\alpha = \widetilde{n}J \sum (\overline{M}_{i\bullet} - \overline{M}_{\bullet\bullet})^2,$$

$$\mathrm{ss}_\beta = \widetilde{n}I \sum (\overline{M}_{\bullet j} - \overline{M}_{\bullet\bullet})^2,$$

$$\mathrm{ss}_\gamma = \widetilde{n} \sum (\overline{Y}_{ij\bullet} - \overline{M}_{i\bullet} - \overline{M}_{\bullet j} + \overline{M}_{\bullet\bullet})^2.$$

ss_ϵ is calculated in the usual way: $ss_\epsilon = \sum_i (Y_{ijk} - \overline{Y}_{ij\bullet})^2$. For the example, the calculations are:

Means

11.669000	5.713500	8.691250
10.627200	5.721333	8.174266
11.148100	5.717416	8.432758

The harmonic mean $\widetilde{n}$ is

$$\widetilde{n} = \frac{(2)(2)}{1/5 + 1/4 + 1/5 + 1/3} = 4.067797.$$

$$ss_\mu = (4.067797)(2)\left((8.691250 - 8.432758)^2 + (8.174266 - 8.432758)^2\right) = 1.0872,$$

$$ss_\beta = (4.067797)(2)\left((11.148100 - 8.432758)^2 + (5.717416 - 8.432758)^2\right) = 119.6888,$$

$$ss_\gamma = (4.067797)\left((4)(0.262408)^2\right) = 1.1204,$$

making use of the fact that all the interaction deviations are equal in absolute value:

$$\overline{Y}_{11\bullet} - \overline{M}_{1\bullet} - \overline{M}_{\bullet 1} + \overline{M}_{\bullet\bullet} = 0.262408,$$

$$\overline{Y}_{12\bullet} - \overline{M}_{1\bullet} - \overline{M}_{\bullet 2} + \overline{M}_{\bullet\bullet} = -0.262408, \ldots.$$

The analysis of variance table based on the unweighted means is:

Source	DF	SS	MS	*F*-Ratio	*p*-Value
Subject	1	1.0872	1.0872		
PHA	1	119.688	119.6888	1	
PHA × subject	1	1.1204	1.1204	45.7	< 0.001
Error	13	0.3184	0.02449		
Total	16				

The conclusion remains unchanged. It turns out in this case that the test for interaction is identical to the multiple regression procedure of the next chapter.

10.6 VALIDITY OF ANOVA MODELS

10.6.1 Assumptions in ANOVA Models

All of the models considered in this chapter have assumed at least the following:

1. Homogeneity of variance
2. Normality of the residual error

3. Statistical independence of the residual errors
4. Linearity of the model.

For example, consider again the model associated with the one-way analysis of variance (leaving off the subscripts):

$$Y = \mu + \alpha + \epsilon.$$

We assumed that the error term ϵ: (a) had constant variance for all values of μ and α, (b) was normally distributed, (c) values of ϵ were randomly (independently) selected and (d) the response Y was linearly related to μ, α and ϵ.

In addition, the random effects and repeated measures models made assumptions about the covariances of the random factors and the residual error; other models assumed zero interaction (additivity).

If one or more of the assumptions does not hold one of the following approaches is frequently used:

1. The data are analyzed by a method that makes fewer assumptions, for example, nonparametric analysis.
2. Part of the data is eliminated or not used, for example, extreme values "outliers" are deleted or replaced by less extreme values.
3. The measurement variables are replaced by categorical variables and some kind of analysis of frequencies is carried out; for example, "age at first pregnancy" is replaced by "teenage mother: Yes-No" and the number of observations in various categories is now the outcome variable.
4. A weighted analysis is done; for example, if the variance is not constant at all levels of response, the responses are weighted by the inverse of the variances. The log-linear models of Chapter 7 are an example of a weighting procedure.
5. The data are "transformed" to make the assumptions valid. Typical transformations are: logarithmic, square root, reciprocal, arcsin $\sqrt{\ }$. These transformations are nonlinear. Linear transformations do not alter the analysis of variance tests.
6. Finally, appeal is made to the "robustness" of the ANOVA and the analysis carried out anyway. This is a little bit like riding a bicycle without holding on to the handle bars, it takes experience and courage; if you arrive safely everyone is impressed, if not, they told you so.

The most common approach is to transform the data. There are advantages and disadvantages to transformations. A brief discussion is presented in the next subsection. In the other subsections we present some specific tests of the assumptions of the ANOVA model.

10.6.2   Transformations

Common Transformations
Some statisticians recommend routine transformations of data before any analysis is carried out. We recommend the contrary approach; do not carry out transformations unless necessary and then be very careful, particularly in estimation. We discuss this more fully below but first some common transformations. Table 10.18 lists

Table 10.18. Characteristics of Some Common Transformations of a Random Variable Y.

$W = g(Y)$	Range of Y	Variance of Y	Variance of W	Normalizing	Stabilizing	Linearity	Comments	Uses
$\sqrt{Y}$	$0 \le Y \le \infty$	$\lambda^2 \mu_Y$	$\lambda^2/4$	U	Y	—	For $\mu_Y < 10$ use $W = 1/2(\sqrt{Y} + \sqrt{Y+1})$ (Freeman Tukey transformation)	Poisson
$\log_e Y$	$0 \le Y \le \infty$	$\lambda^2 \mu_Y^2$	λ^2	U	Y	C	Use $\log_e(Y+1)$ if zeroes occur	Wide range of Y, e.g., 1–1000
$\dfrac{1}{Y}$	$0 \le Y \le \infty$	$\lambda^2 \mu_Y^4$	λ^2	U	Y	C	Use $1/(Y+1)$ if zeroes occur	Survival time, response time
Y^b	$0 \le Y \le \infty$	—	1	Y	Y	C	Box–Cox transformation Power transformation	Generalized transformation
$\log_e \dfrac{Y}{1-Y}$	$0 \le Y \le 1$	$\lambda^2 \mu_Y(1 - \mu_Y)$	$\dfrac{\lambda^2}{\mu_Y(1-\mu_Y)}$	U	N	C	Logit transformation	Logistic regression, binomial
$\arcsin\sqrt{Y}$	$0 \le Y \le 1$	$\lambda^2 \mu_Y(1 - \mu_Y)$	$\lambda^2/4$	U	Y	C	"Angle" transformation	Binomial
$1/2\log_e \dfrac{1+Y}{1-Y}$	$-1 \le Y \le 1$	$\lambda^2(1 - \mu_Y^2)^2$	λ^2	Y	Y	C	R. A. Fisher's Z-transformation	Normalize correlation coefficient
$\Phi^{-1}\left(\dfrac{\text{Rank } Y}{n}\right)$	$-\infty \le Y \le \infty$	—	1	Y	Y	—	Normal scores transformation $(\text{Rank}(Y) - 1/2)/n$ is sometimes used	Nonparametric analysis

^{a}C = could; N = no; U = usually; Y = yes.

seven of the most commonly used transformations and one somewhat more specialized one. Each row in the table lists some of the characteristics of the transformation
and its uses. A large number of these transformations are variance stabilizing. For
example, if the variance of Y is $\lambda^2 \mu_Y$ where λ is a constant and μ_Y is the expected
value of Y then $\sqrt{Y}$ tends to have a variance that is constant and equal to $\lambda^2/4$.
Hence, this transformation is frequently associated with a Poisson random variable:
in this case $\lambda = 1$ so that $\sqrt{Y}$ tends to have a variance of $1/4$ regardless of the value
of μ_Y. This result is approximate in that it holds for large values of μ_Y. However, the
transformation works remarkably well even for small μ_Y, say, equal to 10. Freeman
and Tukey [1950] have proposed a modification of the square root transformation
which stabilizes the variance for even smaller values of μ_Y. Variance stabilizing transformations tend to be normalizing as well, and can be derived explicitly as a function
of the variance of the original variable.

The logarithmic transformation is used to stabilize the variance and/or change
a multiplicative model into an linear model. When the standard deviation of Y is
proportional to μ_Y the logarithmic transformation tends to stabilize the variance. The
reciprocal transformation (1/observation) is used when the variance is proportional to
μ_Y^4. These first three transformations deal with a progression in the dependence of the
variance of Y on μ_Y: from μ_Y to μ_Y^4. The transformations consist of raising Y to an
exponent from $Y^{1/2}$ to Y^{-1}. If we define the limit of Y^b to be $\log_e Y$ as b approaches
0, then these transformations represent a gradation in exponents. A further logical
step is to let the data determine the value of b. This transformation, Y^b, is an example
of a power transformation. (Power, here, does not imply "powerful" but simply that
Y is raised to the bth power). See Note 10.4 for some additional comments.

The next two transformations are used with proportions or rates. The first one of
these is the ubiquitous logistic transformation, which is not variance stabilizing but
does frequently induce linearity (cf. Section 7.6: log-linear models). The angle transformation is variance stabilizing but has a finite range; it is not used much anymore
because computational power is now available to use the more complex but richer
logistic transformation.

The Fisher Z-transformation is used to transform responses whose range is between -1 and $+1$. It was developed specifically for the Pearson product-moment
correlation coefficient and discussed in Chapter 9. Finally, we mention one transformation via ranks, the normal scores transformation. This transformation is used
extensively in nonparametric analyses and discussed in Chapter 8.

Benefits of Transformation
There are benefits to using transformations. It is well to state them explicity since we
also have some critical comments. The benefits include the following:

1. Methods using the normal distribution can be used.
2. Tables, procedures, and computer programs are available.
3. A transformation derived for one purpose tends to achieve some other purposes
   as well. But not always.
4. Inferences (especially relating to hypothesis testing) can be made more easily.
5. Confidence intervals in the transformed scale can be "transformed back." (But
   estimates of standard errors cannot).

Risk of Transformations

Transformations are more useful for testing purposes than for estimation. The following drawbacks of transformations should be kept in mind:

1. The order of statistics may not be preserved. Consider the following two sets of data: Sample 1: 1, 10; Sample 2: 5, 5. The arithmetic means are 5.5 and 5.0, respectively. The geometric means (i.e., the antilogarithms of the arithmetic mean of the logarithms of the observations) are 3.2 and 5.0, respectively. Hence, the ordering of the *means* is not preserved by the transformation (the ordering of the *raw* data is preserved).

2. Contrary to some, we think that there may be a "natural scale" of measurement. Some examples of variables with a natural scale of measurement are "life expectancy" measured in years, days or months; cost of medical care in dollars; number of accidents attributable to alcoholism. Administrators or legislators may not be impressed with, or willing to think about the cost of medical care in terms of "square root of millions of dollars expended."

3. Closely related is the problem of bias. An obvious approach to the criticism in (2) is to do the analysis in the transformed units and then transform back to the original scale. Unfortunately, this introduces bias as mentioned in (1). Formally, if Y is the variable of interest and $W = g(Y)$ its transform then it is usually the case that

$$E(W) \neq g(E(Y)).$$

There are ways of assessing this bias and eliminating it but such methods are cumbersome and require an additional layer of computations, something the transformation was often designed to reduce!

4. Finally, many of the virtues of transformations are asymptotic virtues; they are approached as the sample size becomes very large. This should be kept in mind when analyzing relatively small data sets.

10.6.3 Testing of Homogeneity of Variance

It is often the case that the variance or standard deviation is proportional to the mean level of response. There are two common situations where this occurs. First, where the range of response varies over two or more orders of magnitude; second, in situations where the range of response is bounded, on the left, the right or both. Examples of the former are Poisson random variables; examples of the latter, responses such as proportions, rates or random variables that cannot be negative.

The simplest verification of homogeneity of variance is provided by a graph, plotting the variance or standard deviation versus the level of response.

Example 10.5 (continued).   In Table 10.8, the means and standard deviations of serum fluorescence readings of mice exposed to nitrogen dioxide are given. In Figure 10.6 the standard deviations are plotted against the means of the various treatment combinations.

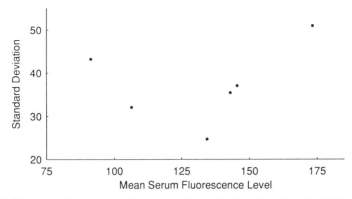

Figure 10.6. Mean serum fluorescence level and standard deviation. Data from Example 10.5, Section 10.3.1.

Table 10.19. Aflatoxin Levels in Peanut Kernels. Means and Standard Deviations for Eleven Samples Using Transformations. Data for Example 10.8.

		\multicolumn{8}{c}{Mean and Standard Deviation of Aflatoxin Level}
		\multicolumn{2}{c}{Y}	\multicolumn{2}{c}{$W = Y^{1/4}$}	\multicolumn{2}{c}{$W = \sqrt{Y}$}	\multicolumn{2}{c}{$W = \log Y$}
Sample	n	Mean	SD	Mean	SD	Mean	SD	Mean	SD
1	16	110	25.6	3.2	0.192	10.4	1.24	4.7	0.240
2	16	79	20.6	3.0	0.204	8.8	1.19	4.3	0.281
3	16	21	3.9	2.1	0.109	4.5	0.45	3.0	0.213
4	16	33	12.2	2.4	0.192	5.7	0.96	3.4	0.311
5	15	32	10.6	2.4	0.194	5.6	0.92	3.4	0.328
6	16	15	2.7	2.0	0.089	3.8	0.35	2.7	0.183
7	15	33	6.2	2.4	0.111	5.8	0.54	3.5	0.183
8	16	31	2.8	2.4	0.054	5.6	0.26	3.4	0.092
9	16	17	4.2	2.0	0.129	4.1	0.51	2.8	0.261
10	16	8	3.1	1.7	0.143	2.9	0.49	2.1	0.339
11	15	84	17.7	3.0	0.164	9.1	0.98	4.4	0.221

This example does not demonstrate any pattern between the standard deviation and the cell means. It would not be expected because the range of the cell means is fairly small.   □

Example 10.8.   A more interesting example is the data of Quesenberry *et al.* [1976] discussed and referenced in Chapter 3, Problem 13. Samples of peanut kernels were analyzed for aflatoxin levels. Each sample was divided into 15 or 16 subsamples. There was considerable variability in mean levels and corresponding standard deviations.

A plot of means versus standard deviations displays an increasing pattern suggesting a logarithmic transformation to stabilize the variance. This transformation as well as two other transformations ($\sqrt{Y}, Y^{1/4}$) are summarized in Table 10.19.

The means versus the standard deviations are plotted in Figure 10.7. The first pattern clearly indicates a linear trend, the plot for the data expressed as logarithms sug-

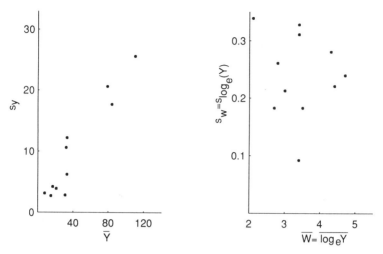

Figure 10.7. Means *vs.* standard deviation, arithmetic and logarithmic scale. Data from Example 10.8.

gests very little pattern. This does not prove that the lognormal model is appropriate. Quesenberry *et al.* [1976] in fact, considered two classes of models: the 11 samples are from normal distributions with means and variances $\mu_i, \sigma_i^2, i = 1, \ldots, 11$; the second class of models assumes that the logarithms of the aflatoxin levels for the 11 samples come from normal distributions with means and variances $\gamma_i, \theta^2, i = 1, \ldots, 11$.

On the basis of their analysis, they conclude that the normal models are more appropriate. The cost is, of course, that 10 more parameters have to be estimated. Graphs of mean versus standard deviation for the $\sqrt{Y}$ and $Y^{1/4}$ scale still suggest a relationship. □

The tests of homogeneity of variance developed here are graphical. There are more formal tests. All of the tests assume normality and are sensitive to departure from normality. In view of the robustness of the analysis of variance to heterogeneity of variance, Box [1953] remarked that:

... "to make the preliminary tests on variances is rather like putting to sea in a rowing boat to find out whether conditions are sufficiently calm for an ocean liner to leave port."

There are four common tests of homogeneity of variance, associated with the names of Hartley, Cochran, Bartlett, and Scheffé. Only the first two are described here, they will be adequate for most purposes. For a description of the other tests see, for example, Winer [1971]. Suppose there are k samples with sample size n_i and sample variance s_i^2, $i = 1, \ldots, k$. For the moment assume that all n_i are equal to n. Hartley's test calculates

$$F_{\text{MAX}} = \frac{s^2_{\text{maximum}}}{s^2_{\text{minimum}}},$$

Cochran's test calculates

$$C = \frac{s^2_{\text{maximum}}}{\sum s_i^2}.$$

Both statistics are referred to appropriate tables, the F_{MAX} statistic to Table A.19, the C statistic to Table A.20. If the sample sizes are not equal, the tables can be entered with the minimum sample size to give a conservative test and with the maximum sample size to give a "liberal" test, i.e., the null hypothesis is rejected more frequently than the nominal significance level.

Example 10.8 (continued). For the transformations considered, the F_{MAX} test and C test statistics are:

Scale	F_{MAX}	C
Y	$\left(\dfrac{25.6}{2.7}\right)^2 = 89.9$	$\dfrac{(25.7)^2}{1758.1} = 0.38$
$\sqrt{Y}$	$\left(\dfrac{1.24}{0.26}\right)^2 = 22.7$	$\dfrac{(1.24)^2}{9.787} = 0.16$
$Y^{1/4}$	$\left(\dfrac{0.204}{0.054}\right)^2 = 14.1$	$\dfrac{(0.204)^2}{0.252} = 0.16$
$\log_e Y$	$\left(\dfrac{0.339}{0.092}\right)^2 = 13.6$	$\dfrac{(0.339)^2}{0.694} = 0.17$
Critical value at 0.05 level	5.8	0.15

The critical values have been obtained by interpolation. The F_{MAX} test indicates that none of the transformations achieve satisfactory homogeneity of variance, validating one of Quesenberry *et al.*'s conclusions. The Cochran test suggests that there is little to choose between the three transformations.

A question remains: How valid is the analysis of variance under heterogeneity of variance? Box [1953] indicates that for three treatments a ratio of 3 in the maximum to minimum *population* variance does not alter the significance level of the test appreciably. (One-way ANOVA model with $n_{i\bullet} = 5, I = 3$). The analysis of variance is therefore reasonably robust with respect to deviation from homogeneity of variance.

□

10.6.4 Testing of Normality in ANOVA

Tests of normality are not as common or well developed as tests of homogeneity of variance. There are at least two reasons: first, they are not as crucial because even if the underlying distribution of the data is not normal appeal can be made to the central limit theorem. Second, it turns out that fairly large sample sizes are needed (say, $n > 50$) to discriminate between distributions. Again, most tests are graphical.

Consider for simplicity, the one-way analysis of variance model

$$Y_{ij} = \mu + \alpha_i + \epsilon_{ij}, \quad i = 1, \ldots, I; \ j = 1, \ldots, n_i.$$

By assumption the ϵ_{ij} are iid $N(0, \sigma^2)$. The ϵ_{ij} are estimated by

$$\epsilon_{ij} = Y_{ij} - \overline{Y}_{i\bullet}.$$

The e_{ij} are normally distributed with population mean 0; $\sum e_{ij}^2/(n - I)$ is an unbiased estimate of σ^2 but the e_{ij} are not statistically independent. They can be made statistically independent but it is not worthwhile for testing the normality. Some kind of normal probability plot is usually made and a decision made based upon a visual inspection. Frequently such a plot is used to identify outliers. Before giving an example we give a simple procedure for use when probability paper is not available.

Definition 10.14.   Given a sample of n observations, $Y_1, Y_2, \ldots, Y_n$, the *order statistics* $Y_{(1)}, Y_{(2)}, \ldots, Y_{(n)}$ are the values ranked from lowest to highest.

Now suppose we generate samples of size n from an $N(0,1)$ distribution and average the values of the order statistics.

Definition 10.15.   *Rankits* are the expected values of the order statistics of a sample of size n from an $N(0,1)$ distribution. That is, let $Z_{(1)}, \ldots, Z_{(n)}$ be the order statstics from an $N(0,1)$ population then the rankits are $E(Z_{(1)}), E(Z_{(2)}), \ldots, E(Z_{(n)})$.

Rankits have been tabulated in Table A.21. A plot of the order statistics of the residuals against the rankits is equivalent to a normal probability plot. A reasonable approximation for the ith rankit is given by the formula

$$E(Z_{(i)}) \doteq 4.91 \left[p^{0.14} - (1 - p)^{0.14} \right], \tag{10.31}$$

where

$$p = \frac{i - 3/8}{n + 1/4}.$$

For a discussion see Joiner and Rosenblatt [1971].

We now return to the first example in this chapter, Example 10.1. A one-way analysis of variance was constructed for these data and we now want to test the normality assumption.

Example 10.1 (continued).   The distribution of ages at which infants first walked discussed in Section 10.2.1 (see Table 10.1) is now analyzed for normality. The residuals $Y_{ij} - \overline{Y}_{i\bullet}$ for the 23 observations are:

−1.125	−0.375	−0.208	0.900
−0.625	−1.375	0.292	−0.850
−0.375	−1.375	−2.708	−0.350
−0.125	0.375	−0.208	1.150
2.875	−0.875	1.542	−0.850
−0.625	3.625	1.292	

Note that the last observation has been omitted again so that we are working with the 23 observations given in the paper. These observations are now ranked from smallest to largest to be plotted on probability paper. To illustrate the use of rankits, we will calculate the expected values of the 23 normal (0,1) order statistics using Equation (10.31). Below are given the 23 order statistics for e_{ij}, $e_{(ij)}$ and the corresponding rankits:

$e_{(ij)}$	−2.708	−1.375	−1.375	−1.125	−0.875	−0.850	−0.850	−0.625
$E(Z_{(ij)})$	−1.93	−1.48	−1.21	−1.01	−0.84	−0.70	−0.57	−0.44

$e_{(ij)}$	−0.625	−0.375	−0.375	−0.350	−0.208	−0.208	−0.125	−0.292
$E(Z_{(ij)})$	−0.33	−0.22	−0.11	0.0	0.11	0.22	0.33	0.44

$e_{(ij)}$	0.375	0.900	1.150	1.292	1.542	2.875	3.625
$E(Z_{(ij)})$	0.57	0.70	0.84	1.01	1.21	1.48	1.93

For example the largest deviation is -2.708 the expected value of $Z_{(1)}$ associated with this deviation is calculated as follows:

$$p = \frac{1 - 3/8}{23 + 1/4} = 0.02688,$$

$$E(Z_{(1)}) = 4.91 \left((0.02688)^{0.14} - (1 - 0.02688)^{0.14}\right)$$

$$= -1.93.$$

The rankits and the ordered residuals are plotted in Figure 10.8. What do we do with this graph? Is there evidence of nonnormality?

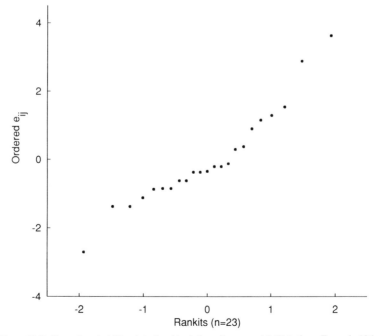

Figure 10.8. Normal probability plot of residuals from linear model. Data from Example 10.1.

There does seem to be some excessive deviation in the tails, the question is, how important is it? One way to judge this would be to generate many plots for normal and nonnormal data and compare the plots to develop a visual "feel" for the data. This has been done by Daniel and Wood [1971] and Daniel [1976]. Comparison of this plot with the plots in Daniel and Wood suggests that these data deviate moderately from normality. For a further discussion, see Section 11.7.1.   □

More formal tests of normality can be carried out using the Kolmogorov-Smirnov test of Chapter 8. A good test is based on the Pearson product-moment correlation of the order statistics and corresponding rankits. If the residuals are normally distributed there should be a very high correlation between the order statistics and the rankits. The (null) hypothesis of normality is rejected when the correlation is *not large enough*. Weisberg and Bingham [1975] show that this is a very effective procedure. The critical values for the correlation have been tabulated; see, for example, Ryan et al. [1980]. For $n \geq 15$, the critical value is of the order of 0.95 or more. This is a simple number to remember. For Example 10.1, discussed in this section, the correlation between the order statistics of the residuals, $e_{(ij)}$ and the rankits $E(Z_{(ij)})$ is $r = 0.9128$ for $n = 23$. This is somewhat lower than the critical value of 0.95 again, suggesting that the residuals are "not quite" normally distributed.

10.6.5 Independence

One of the most difficult assumptions to verify is that of statistical independence of the residuals. There are two problems. First, tests of independence for continuous variables are difficult to implement. Frequently, such tests are, in fact, tests of no correlation among the residuals so that if the errors are normally distributed and uncorrelated then they are independent. Secondly, the observed residuals in the analysis of variance have a built-in dependence due to the constraints on the linear model. For example, in the one-way analysis of variance with I treatments and, say, $n_{i \cdot} = m$ observations per treatment, there are mI residuals but only $(m-1)I$ degrees of freedom; this induces some correlation among the residuals. This is not an important dependence and can be taken care of.

Tests for dependence usually are tests for serial correlation, i.e., correlation among adjacent values. This assumes that the observations can be ordered in space or time. The most common test statistic for serial correlation is the Durbin–Watson statistic. See, for example, Draper and Smith [1981]. Computer packages frequently will print this statistic assuming the observations are entered in the same sequence in which they were obtained. This, of course, is rarely the case and the statistic and its value should not be used. Such "free information" is sometimes hard to ignore; the motto for computer output is *caveat lector* (let the reader beware).

10.6.6 Linearity in ANOVA

Linearity, like independence, is difficult to verify. Example 10.7 of Section 10.5 illustrated a multiplicative model. The model was transformed to a linear (nonadditive) model by considering the logarithm of the original observations. Other types of nonlinear models are discussed in Chapters 11–15. Evidence for a nonlinear model may consist of heterogeneity of variance or interaction. However, this need not always be the case. Scheffé [1959] gives the following example. Suppose there are $I+J+1$ independent Poisson variables defined as follows: $U_1, U_2, \ldots, U_I$ have means $\alpha_1, \alpha_2, \ldots, \alpha_I$; $V_1, V_2, \ldots, V_J$ have means $\beta_1, \beta_2, \ldots, \beta_J$ and W has mean γ. Let $Y_{ij} = W + U_i + V_j$, then $E(Y_{ij}) = \gamma + \alpha_i + \beta_j$, that is, we have an additive, linear model. But $\text{var}(Y_{ij}) = \gamma + \alpha_i + \beta_j$ so that there is heterogeneity of variance (unless all the α_i are equal and all the β_j are equal). The square root transformation destroys the linearity and the additivity. Scheffé [1959] states:

> "It is not obvious whether Y or $\sqrt{Y}$ is more nearly normal ... but in the present context it hardly matters."

A linear model is frequently assumed to be appropriate for a set of data without any theoretical basis. It may be a reasonable first-order approximation to the "state of nature" but should be recognized as such.

Sometimes a nonlinear model can be derived from theoretical principles. The form of the model may then suggest a transformation to linearity. But as the above example illustrates, the transformation need not induce other required properties of ANOVA models, or may even destroy them.

Another strategy for testing linearity is to add some nonlinear terms to the model and then test their significance. Sections 11.6 and 11.7 elaborate on this strategy.

10.6.7 Additivity

The term "additivity" is used somewhat ambiguously in the statistical literature. It is sometimes used to describe the transformation of a multiplicative model to a linear model. The effects of the treatment variables become "additive" rather than multiplicative. We have called such a transformation a linearizing transformation. It is not always possible to find such a transformation (see Section 11.9.5). We have reserved the term "additivity" for the additive model illustrated by the two-way analysis of variance model (see Definition 10.4). A test for additvity then becomes a test for "no interaction." Scheffé [1959] proves that transformations to additivity exists for a very broad class of models.

The problem is that the existence of interaction may be of key concern. Consider Example 10.8 of Section 10.5. The existence of interaction in this example is taken as evidence that the immune system of a patient with prostatic carcinoma differed from that of normal blood donors. This finding has important implications for a theory of carcinogenesis. These data are an example of the importance of expressing observations in an appropriate scale. Of course, what evidence is there that the logarithms of the radioactive count is the appropriate scale? There is some arbitrariness but the original model was stated in terms of percentage changes and this implies constant changes on a logarithmic scale.

So the problem has been pushed back one step: Why state the original problem in terms of percentage changes? The answer must be found in the experimental situation and the nature of the data. Ultimately the researcher will have to provide justification for the initial model used.

This discussion has been rather philosophical. One other situation will be considered: the randomized block design. There is no test for interaction because there is only one observation per cell. Tukey [1949] suggested a procedure that is an example of a general class of procedures. The validity of a model is evaluated by considering an enlarged model and testing the significance of the terms in the enlarged model. To be specific, consider the randomized block design model of Equation (10.23):

$$Y_{ij} = \mu + \beta_i + \tau_j + \varepsilon_{ij}, \quad i = 1, \ldots, I, \ j = 1, \ldots, J.$$

Tukey [1949] embedded this model in the "richer" model

$$Y_{ij} = \mu + \beta_i + \tau_j + \lambda \beta_i \tau_j + \varepsilon_{ij}, \quad i = 1, \ldots, I, \ j = 1, \ldots, J. \tag{10.32}$$

He then proposed to test the null hypothesis,

$$H_0 : \lambda = 0,$$

as a test for nonadditivity. Why this form? It is the simplest nonlinear effect involving both blocks and treatments. The term λ is estimated and tested as follows. Let the model without interaction be estimated by

$$Y_{ij} = \overline{Y}_{\bullet\bullet} + b_i + t_j + e_{ij},$$

where

$$b_i = \overline{Y}_{i\bullet} - \overline{Y}_{\bullet\bullet}, \ t_j = \overline{Y}_{\bullet j} - \overline{Y}_{\bullet\bullet}, \quad \text{and} \quad e_{ij} = Y_{ij} - \overline{Y}_{\bullet\bullet} - b_i - t_j.$$

We have the usual constraints,

$$\sum b_i = \sum t_j = 0$$

and

$$\sum_i e_{ij} = \sum_j e_{ij} = 0 \quad \text{for all } i \text{ and } j.$$

Now define

$$X_{ij} = b_i t_j, \quad i = 1, \ldots, I, \ j = 1, \ldots, J. \tag{10.33}$$

It can be shown that the least-squares estimate, $\widehat{\lambda}$, of λ is

$$\widehat{\lambda} = \frac{\sum X_{ij} Y_{ij}}{\sum X_{ij}^2}. \tag{10.34}$$

Since $\overline{X} = 0$ (why?), the quantity $\widehat{\lambda}$ is precisely the regression of Y_{ij} on X_{ij}. The Sum of Squares for regression is the Sum of Squares for nonadditivity in the analysis of variance table:

$$\text{ss}_\lambda = \text{ss}_{\text{nonadditivity}} = \frac{\left(\sum X_{ij} Y_{ij}\right)^2}{\sum X_{ij}^2}. \tag{10.35}$$

The analysis of variance table for the randomized block design including the test for nonadditivity is displayed in Table 10.20. As expected the ss_λ has 1 degree of freedom

Table 10.20. Analysis of Variance of Randomized Block Design Incorporating Tukey Test for Additivity.

Source of Variation	DF	SS[a]	MS
Grand mean	1	$SS_\mu = n\overline{Y}_{\bullet\bullet}^2$	$MS_\mu = SS_\mu$
Blocks	$I-1$	$SS_\beta = J\sum(\overline{Y}_{i\bullet} - \overline{Y}_{\bullet\bullet})^2$	$MS_\beta = \dfrac{SS_\beta}{I-1}$
Treatments	$J-1$	$SS_\tau = I\sum(\overline{Y}_{\bullet j} - \overline{Y}_{\bullet\bullet})^2$	$MS_\tau = \dfrac{SS_\tau}{J-1}$
Non additivity	1	$SS_\lambda^b = \dfrac{\left(\sum X_{ij}Y_{ij}\right)^2}{\left(\sum X_{ij}^2\right)}$	$MS_\lambda = SS_\lambda$
Residual	$IJ-I-J$	$SS_\epsilon = $ by subtraction	$MS_\epsilon = \dfrac{SS_\epsilon}{IJ-I-J}$
Total	IJ	$\sum Y_{ij}^2$	

Source of Variation	F-Ratio	DF	$E(MS)$	Hypothesis Tested
Grand mean	$\dfrac{MS_\mu}{MS_\epsilon}$	$(1, IJ-I-J)$	$n\mu^2 + \sigma^2$	$\mu = 0$
Blocks	$\dfrac{MS_\beta}{MS_\epsilon}$	$(I-1, IJ-I-J)$	$\dfrac{J\sum\beta_i^2}{I-1} + \sigma^2$	$\beta_i = 0$ all i
Treatments	$\dfrac{MS_\tau}{MS_\epsilon}$	$(J-1, IJ-I-J)$	$\dfrac{I\sum\tau_j^2}{J-1} + \sigma^2$	$\tau_i = 0$ all j
Nonadditivity	$\dfrac{MS_\lambda}{MS_\epsilon}$	$(1, IJ-I-J)$	$\lambda^2 C^c + \sigma^2$	$\lambda = 0$

[a]Summation is over all displayed subscripts.
[b]$X_{ij} = (\overline{Y}_{i\bullet} - \overline{Y}_{\bullet\bullet})(\overline{Y}_{\bullet j} - \overline{Y}_{\bullet\bullet})$.
[c]C is a constant which depends on the cell means; it is zero if the additive model holds.

Model: $Y_{ij} = \mu + \beta_i + \tau_j + \lambda\beta_i\tau_j + \epsilon_{ij}$ $\epsilon_{ij} \sim$ iid $N(0, \sigma^2)$. Data: $Y_{ij} = \overline{Y}_{\bullet\bullet} + (\overline{Y}_{i\bullet} - \overline{Y}_{\bullet\bullet}) + (Y_{\bullet j} - Y_{\bullet\bullet}) + \widehat{\lambda}X_{ij} + \widetilde{e}_{ij}$
(residual obtained by subtraction).

since we are estimating a slope. But who "pays" for the 1 degree of freedom? A little reflection indicates that it must come out of the error term; the number of constraints on the block and treatment effects remain the same.

A graph of Y_{ij} vs. X_{ij} (or equivalently, e_{ij} vs. X_{ij}) will indicate whether there is any pattern.

The idea of testing models within larger models as a way of testing the validity of the model is discussed further in Section 11.7.2.

Example 10.6 (continued). We now apply the Tukey test for additivity to the experiment assessing the effect of pancreatic supplements on fat absorption in patients with steatorrhea, discussed in Section 10.3.2. We need to calculate ss_λ from Equation (10.35) and this involves the regression of Y_{ij} on X_{ij} where X_{ij} is defined by Equation (10.33). To save space we will calculate only a few of the X_{ij}. For example,

$$X_{11} = \left(\overline{Y}_{1\bullet} - \overline{Y}_{\bullet\bullet}\right)\left(\overline{Y}_{\bullet1} - \overline{Y}_{\bullet\bullet}\right)$$
$$= (16.9 - 25.775)(38.083 - 25.775)$$
$$= -109.2$$

and

$$X_{23} = \left(\overline{Y}_{2\bullet} - \overline{Y}_{\bullet\bullet}\right)\left(\overline{Y}_{\bullet3} - \overline{Y}_{\bullet\bullet}\right)$$
$$= (25.625 - 25.775)(17.417 - 25.775)$$
$$= 1.3.$$

(Note that a few more decimal places for the means are used here as compared to Table 10.10). A graph of Y_{ij} vs. X_{ij} is presented in Figure 10.9.

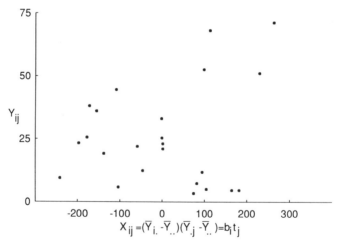

Figure 10.9. Plot for Tukey test for additivity. See text for explanation.

Table 10.21. Randomized Block Analysis with Tukey Test for Additivity of Fecal Fat Excretion of Patients With Steatorrhea. Data from Table 10.10, Example 10.6.

Source of Variation	DF	SS	MS	F	p-Value
Patients	5	5588.38	1117.68	13.1	< 0.001
Treatments	3	2008.60	669.53	7.83	< 0.01
Additivity	1	407.60	407.60	4.76	$0.025 < p < 0.05$
Residual	14	1197.80	85.557		
Total	23	9202.38			

The estimate of the slope is

$$\hat{\lambda} = \frac{\sum X_{ij} Y_{ij}}{\sum X_{ij}^2}$$

$$= \frac{(-109.2)(44.5) + (82.0)(7.3) + \cdots + (98.8)(52.6)}{(-109.2)^2 + (82.0)^2 + \cdots + (98.8)^2}$$

$$= \frac{13807}{467702}$$

$$= 0.029521.$$

ss_λ is

$$ss_\lambda = \frac{(13807)^2}{467702} = 407.60.$$

The analysis of variance is tabulated in Table 10.21.

The test for additivity indicates significance at the 0.05 level ($p = 0.047$), thus there is some evidence that the data cannot be represented by an additive model. Tukey [1949] related the constant "a" in Y^a (power transformation) to the degree of non-additivity by the following formula:

$$\hat{a} = 1 - \hat{\lambda} \overline{Y}_{\bullet \bullet}.$$

The quantity $\hat{a}$ is a statistic and hence a random variable. For a particular set of data the confidence interval on $\hat{a}$ will tend to be fairly wide, hence, a "nice" value of "a" is usually chosen. For the example, $\hat{a} = 1 - (0.029521)(25.775) = 0.239$. A "nice" value for "$a$" is thus 0.25 or even 0.20. □

10.6.8 Strategy for Analysis of Variance

It is useful to have a checklist in carrying out an ANOVA. Not every item on the list needs to be considered, nor necessarily in the order given but you will find it useful to be reminded of these items:

1. Describe how the data were generated; from what population? To what population will inferences be made? State explicitly at what steps in the data generation randomness entered.

2. Specify the ANOVA null hypotheses, alternative hypotheses; whether the model is fixed, random or mixed.

3. Graph the data to get some idea of treatment effects, variability and possible outliers.

4. If necessary, test the homogeneity of variance and the normality.

5. If ANOVA is inappropriate on the data as currently expressed, consider alternatives. If transformations are used, repeat steps 2 and 4.

6. Carry out the ANOVA. Calculate F-ratios. Watch out for F-ratios much less than 1; they usually indicate an inappropriate model.

7. State conclusions and limitations.

8. If null hypotheses are not rejected, consider the power of the study and of the analysis.

9. For more detailed analyses and estimation procedures see Chapter 12.

NOTES

Note 10.1: Ties in Nonparametric Analysis of Variance (One-Way and Randomized Block)

As indicated, both the Kruskal–Wallis and the Friedman tests are conservative in the presence of ties. The adjustment procedure is similar to those used in Chapter 8, Equation 8.4. For the Kruskal–Wallis situation let

$$C_{KW} = \frac{\sum_{\ell=1}^{L} \left(t_\ell^3 - t_\ell\right)}{(n^3 - n)},$$

where L is the number of groups of tied ranks and t_ℓ is the number of ties in group $\ell, \ell = 1, \ldots, L$. Then the statistic T_{KW} (Equation (10.13)) is adjusted to $T_{ADJ} = T_{KW}/(1 - C_{KW})$. Since $0 \leq C_{KW} \leq 1$, $T_{ADJ} \geq T_{KW}$. Hence, if the null hypothesis is rejected with T_{KW} it will certainly be rejected with T_{ADJ} since the degrees of freedom remain unchanged. Usually C_{KW} will be fairly small: suppose there are 10 tied observations in an ANOVA of 20 observations; in this case $C_{KW}(10^3 - 10)/(20^3 - 20) = 0.1241$ so that $T_{ADJ} = T_{KW}/(1 - 0.1241) = 1.14 T_{KW}$. The adjusted value is only 14% larger than the value of T_{KW} even in this extreme situation. (If the 10 ties are made up of 5 groups of 2 ties each the adjustment is less than 0.5%).

A similar adjustment is made for the Friedman statistic, given by Equations (10.25) and (10.26). In this case,

$$C_{FR} = \frac{\sum_{i=1}^{I} \sum_{\ell=1}^{L_i} \left(t_{i\ell}^3 - t_{i\ell}\right)}{I(J^3 - J)},$$

where $t_{i\ell} = $ the number of ties in group ℓ within block i and untied values within a block are counted as a group of size 1.

(Hence $\sum_{\ell=1}^{L_1} t_{i\ell} = J$ for every i.) The adjusted Friedman statistic, T_{ADJ}, is $T_{ADJ} = T_{FR}/(1 - C_{FR})$. Again, unless there are very many ties the adjustment factor, C_{FR} will be relatively small.

Note 10.2: Nonparametric Analyses With Ordered Alternatives

All the tests considered in this chapter have been "omnibus tests," that is, the alternative hypotheses have been general. In the one-way ANOVA the null hypothesis is $H_0 : \mu_1 = \mu_2 = \cdots = \mu_I = \mu$, the alternative hypothesis $H_1 : \mu_i \neq \mu_{i'}$ for at least one i and i'. Since the power of a test is determined by the alternative hypothesis we should be able to "do better" using more specific alternative hypotheses. One such hypothesis involves ordered alternatives. For the one-way ANOVA (see Section 10.2) let $H_1 : \mu_1 \leq \mu_2 \leq \cdots \leq \mu_I$ with at least one strict inequality. A regression-type parametric analysis could be carried out by coding the categories $X = 1, X = 2, \ldots, X = I$.

A nonparametric test of H_0 against an ordered alternative H_1 was developed by Terpstra and Jonckheere (see, for example, Hollander and Wolfe [1973]). The test is based on the Mann–Whitney statistic (see Section 8.6). The Terpstra–Jonckheere statistic is

$$T_{TJ} = \sum_{i=1}^{I-1} \sum_{k=i+1}^{I} M_{ik} = \sum_{i<k} M_{ik},$$

where $M_{ik} =$ number of pairs with the observation in Group i less than that of Group k $(i < k)$ among the $n_i n_k$ pairs.

Under the null hypothesis, $H_0 : \mu_1 = \mu_2 = \cdots = \mu_I = \mu$ the statistic T_{TJ} has a distribution that approaches a normal distribution as n becomes large, with mean and variance given by

$$E[T_{TJ}] = \frac{\left(n^2 - \sum n_i^2\right)}{4}$$

and

$$\mathrm{var}[T_{TJ}] = \frac{\left[n^2(2n+3) - \sum n_i^2(2n_i + 3)\right]}{72},$$

where $n = n_1 + n_2 + \cdots + n_I$. See Problems 10.3 and 10.11 for an application.

In Section 10.3.3, a nonparametric analysis of randomized block design was presented to test the null hypothesis $H_0 : \tau_1 = \tau_2 = \cdots = \tau_J = 0$. Again, we consider an ordered alternative, $H_1 : \tau_1 \leq \tau_2 \leq \cdots \leq \tau_J$ with at least one strict inequality. Using the notation of Section 10.3.3, let $R_{\bullet j} =$ sum of ranks for treatment j. Page [1963] developed a nonparametric test of H_0 against H_1. The statistic $T_{\mathrm{PAGE}} = \sum_{j=1}^{J} jR_{\bullet j}$ under the null hypothesis approaches a normal distribution (as I become large) with mean and variance

$$E[T_{\mathrm{PAGE}}] = \frac{IJ^2(J+1)}{4}$$

and

$$\mathrm{var}[T_{\mathrm{PAGE}}] = \frac{I(J^3 - J)^2}{144(J-1)}.$$

Note 10.3: Alternative Rank Analyses

Conover and Iman [1981] in a series of papers have advocated a very simple rank analysis: replace observations by their ranks and then carry out the usual parametric analysis. These procedures must be viewed with caution when models are nonadditive (Akritas [1990]), and discussion in Chapter 8. Hettmansperger and McKean [1978] provide an illustration of another class of rank-based analytical procedures that can be developed. There are three steps in this type of approach:

1. Define a robust or nonparametric estimate of dispersion;
2. State an appropriate statistical model for the data; and
3. Given a set of data, estimate the values of the parameters of the model to minimize the robust estimate of dispersion.

A drawback of such procedures is that estimates cannot be written explicitly and, more important, the estimation procedure is nonlinear, requiring a computer to carry it out. However, with the increasing availability of microcomputers it will only be a matter of time until software will be developed, making such procedures widely accessible.

It is possible to run routinely a parametric analysis of the raw data and compare it with some alternative rank analysis. If the two analyses do not agree, the data should be examined more carefully to determine the cause of the discrepant results; usually it will be due to "nonnormality" of the data. The researcher then has two choices: if the "nonnormality" is thought to be a characteristic of the biological system from which the data came the rank analysis would be preferred. On the other hand if the "nonnormality" is due to "outliers" (see Chapter 8), there are other options available, all the way from redoing the experiment (more carefully this time), to removing the outliers, to sticking with the analysis of the ranks. Clearly, there are financial, ethical, and professional costs and risks. What should *not* be done in the case of disagreement is to pick the analysis that conforms, in some sense, to the researcher's preconceptions or desires.

Note 10.4: Power Transformation

Let Y^δ be a transformation of Y. The assumption is that Y^δ is normally distributed with mean μ (which will depend on the experimental model) and variance σ^2. The ss_ε will now be a function of δ. It can be shown that the appropriate quantity to be minimized is

$$L(\delta) = \frac{n}{2}\mathrm{ss}_\varepsilon - \sum \ln(\delta y^\delta),$$

and defined to be

$$= \frac{n}{2}\mathrm{ss}_\epsilon - \sum \ln y$$

for $\sigma = 0$ (corresponding to the logarithmic transformation). Typically, this equation is solved by trial and error. With a microcomputer this can be done quickly. Usually there will be a range of values of σ over which the values of $L(\delta)$ will be close to the minimum; it is customary then to pick a value of σ that is simple. For example, if the minimum of $L(\delta)$ occurs at $\sigma = 0.49$, the value chosen will be $\sigma = 0.50$ to correspond to the square root transformation. For an example, see Weisberg [1980]. Empirical evidence suggests that the value of δ derived from the data is frequently close to some "natural" rescaling of the data. (This may just be a case of perfect 20/20 hindsight).

PROBLEMS

For Problems 1–23 below carry out one or more of the following tasks. Additional tasks will be indicated at each problem.

  a. State an appropriate ANOVA model, null hypotheses, alternative hypotheses. State whether the model is fixed, random, or mixed. Define the population to which inferences are to be made.

  b. Test the assumption of homogeneity of variance.

  c. Test the assumption of normality using a probability plot.

  d. Test the assumption of normality correlating residuals and rankits.

  e. Graph the data. Locate the cell means on the graph.

  f. Transform the data. Give a rationale for transformation.

  g. Carry out the analysis of variance. State conclusions and reservations. Compare with the conclusions of the author(s) of the paper. If possible, estimate the power if the results are not significant.

  h. Carry out a nonparametric analysis of variance. Compare with conclusions of parametric analysis.

  i. Partition each observation into its component parts (see, for example, Equations (10.4) and (10.19)) and verify that sum of squares of each component is equal to one of the sums of squares in the ANOVA table.

  j. Construct the ANOVA table from means and standard deviations (or standard errors). Do relevant parts of (g).

1. Olsen *et al.* [1975] studied "morphine and phenytoin binding to plasma proteins in renal and hepatic failure." Twenty-eight subjects with uremia were classified into four groups. The percentage of morphine that was bound is the endpoint.

$$\text{Chronic}(n_1 = 18): \quad 31.5, 35.1, 32.1, 34.2, 26.7, 31.9, 30.8,$$
$$27.3, 27.3, 29.0, 30.0, 36.4, 39.8, 32.0,$$
$$35.9, 29.9, 32.2, 31.8$$
$$\text{Acute}(n_2 = 2): \quad 31.6, 28.5$$
$$\text{Dialysis}(n_3 = 3): \quad 29.3, 32.1, 26.9$$
$$\text{Anephric}(n_4 = 5): \quad 26.5, 22.7, 27.5, 24.9, 23.4$$

  a. Do Tasks (a)–(e) and (g)–(i).

b. In view of the nature of the response variable (percent of morphine bound) explain why strictly speaking the assumption of homogeneity of variance cannot hold.

2. Graham [1977] assayed 16 commercially available pancreatic extracts for six kinds of enzyme activity. See also, Example 10.6, Section 10.3.2. Data for one of these enzymes, proteolytic activity, are presented here. The 16 products were classified by packaging form: capsule, tablet, and enteric-coated tablets. The following data were obtained:

	Proteolytic Activity (U/Unit)				
Tablet ($n = 5$)	6640	4440	240	990	410
Capsule ($n = 4$)	6090	5840	110	195	
Coated ($n = 7$)	1800	1420	980	1088	2200
Tablet	870	690			

a. Do Tasks (a)–(e) and (g)–(i).

b. Is there a transformation that would make the variance more homogeneous? Why is this unlikely to be the case? What is peculiar about the values for the coated tablets?

3. The following data from Rifkind *et al.* [1976] consist of antipyrine clearance of males suffering from β-thalassemia, a chronic type of anemia. In this disease, abnormally thin red blood cells are produced. The treatment of the disease has undesirable side effects including liver damage. Antipyrine is a drug used to assess liver function with a high clearance rate indicating satisfactory liver function. These data deal with the antipyrine clearance rate of 10 male subjects classified according to pubertal stage. The question is whether there is any significant difference in clearance rate among the pubertal stages (I = Infant; V = Adult).

Pubertal Stage	Clearance Rate (Half-Life in Hours)				
I	7.4	5.6	3.7	6.6	6.0
IV	10.9	12.2			
V	11.3	10.0	13.3		

a. Do Tasks (a)–(e) and (g)–(i).

*b. Assuming that the antipyrine clearance rate increases with age, carry out a nonparametric test for trend (see Note 10.2). What is the alternative hypothesis in this case?

4. It is known that organisms react to stress. A more recent discovery is that the immune system's response is altered as a function of stress. In a paper by Keller *et al.* [1981], the immune response of rats as measured by the number of circulating lymphocytes (cells per milliliter $\times 10^{-6}$) was related to the level of stress. The following data are taken from this paper:

Group	Number of Rats	Mean Number of Lymphocytes	Standard Error
Home-cage control	12	6.64	0.80
Apparatus control	12	4.84	0.70
Low-shock	12	3.98	1.13
High-shock	12	2.92	0.42

a. Do Tasks (a), (b), (e), and (j).

b. The authors state: "a significant lymphocytopenia $[F(3, 44) = 3.86, p < 0.02]$ was induced by the stressful conditions." Does your F-ratio agree with theirs?

c. Sharpen the analysis by considering a trend in the response levels as a function of increasing stress level.

5. This problem deals with the data in Table 10.8 of Example 10.5. The authors of the paper state that the animals were matched on the basis of weight but that there was no correlation with weight. Assume that the data are presented in the order in which the animals were matched, that is, $Y_{111} = 143$ is matched with $Y_{211} = 152$; in general, Y_{1jk} is matched with Y_{2jk}.

a. Construct a table of differences $D_{jk} = Y_{2jk} - Y_{1jk}$.

b. Carry out a one-way ANOVA on the differences, include ss_μ in your table.

c. Interpret ss_μ for these data.

d. State your conclusions and compare them with the conclusions of Example 10.5.

e. Relate the MS(Between Groups) in the one-way ANOVA to one of the MS terms in Table 10.8. Can you identify the connection and the reason for it?

*f. We want to correlate the Y_{1jk} observations with the Y_{2jk} observations but the problem is that the response level changes from day to day, which would induce a correlation. So we will use the following "trick." Calculate $Y^*_{ijk} = Y_{ijk} - \overline{Y}_{ij\bullet}$; and correlate Y^*_{1jk} with Y^*_{2jk}. Test this correlation using a t-test with $16 - 1 = 15$ degrees of freedom. Why $16 - 1$? There are $7 - 1 = 6$ independent pairs for Day 10, 5 each for Day 12 and Day 14 for a total of 16. Since the observations sum to zero already we subtract one more degree of freedom for estimating the correlation. If matching was not effective this correlation should be zero.

6. Ross and Bras [1975] investigated the relationship between diet and length of life in 121 randomly bred rats. After 21 days of age, each rat was given a choice of several diets *ad libitum* for the rest of its life. The daily food intake (grams/day) was categorized into one of 6 intervals so that an equal number of rats (except for the last interval) appeared in each interval. The response variable was life-span in days. The following data were obtained:

Mean food intake						
(g/day)	18.3	19.8	20.7	21.6	22.4	24.1
Food intake category	1	2	3	4	5	6
Number of rats	20	20	20	20	20	21
Mean life span(days)	733	653	630	612	600	556
Standard error	117	126	111	115	113	106

a. Carry out Tasks (a), (b), (e), and (j).

b. Can this be thought of as a regression problem? How would the residual MS from Regression be related to the MS error of the analysis of variance?

*c. Can you relate in detail the ANOVA procedure and the regression analysis; particularly an assessment of a nonlinear trend?

7. The following data from Florey *et al.* [1977] are the fasting serum insulin levels (μU/ml) for adult Jamaican females after an overnight fast:

Fasting Serum Insulin Level (μU/ml)

Age	25–34	35–44	45–54	55–64
Number	73	97	74	53
Mean	22.9	26.2	22.0	23.8
SD	10.3	13.0	7.4	10.0

a. Do Tasks (a), (b), (e), and (j).

b. Why did the authors partition the ages of the subjects into intervals? Are there other ways of summarizing and analyzing the data? What advantages or disadvantages are there to your alternatives?

8. The assay of insulin was one of the earliest topics in bioassay. A variety of methods have been developed over the years. In the mid 1960s an assay was developed based on the fact that insulin stimulates glycogen synthesis in mouse diaphragm tissue, in vitro. A paper by Wardlaw and van Belle [1964] describes the statistical aspects of this assay. The data in this problem deal with a qualitative test for insulin activity. A pool of 36 hemidiaphragms was collected for each day's work and the tissues incubated in tubes containing medium with or without insulin. Each tube contained three randomly selected diaphragms. For parts of this problem we will ignore tube effects and assume that each treatment was based on six hemidiaphragms. Four unknown samples were assayed. Since the diaphragms synthesize glycogen in medium, a control preparation of medium only was added as well as a standard insulin preparation. The following data adapted from Wardlaw and van Belle [1964] were obtained:

GLYCOGEN CONTENT (OPTICAL DENSITY IN ANTHRONE TEST × 1000)

Medium Only		Standard Insulin (0.5 mu/ml)		Test Preparation							
				A		B		C		D	
280	290	460	465	470	480	430	300	510	505	310	290
240	275	400	460	440	390	385	505	610	570	350	330
225	350	470	470	425	445	380	485	520	570	250	300

a. Carry out Tasks (a)–(e) and (g)–(i). (To simplify the arithmetic if you are using a calculator, divide the observations by 100).

b. Each column in the data layout represents one tube in which the three hemidiaphragms were incubated so that the design of the experiment is actually hierarchical. To assess the effect of tubes we partition the ss_ε (with 30 DF) into two parts: ss(Between Tubes Within Preparations) = $ss_{BT(WP)}$ with 6 degrees of freedom (why?) and ss(Within Tubes) = ss_{WT} with 24 degrees of freedom (why?). The latter ss can be calculated by considering each tube as a treatment. The former can then be calculated as $ss_{BT(WP)} = ss_\varepsilon - ss_{WT}$. Carry out this analysis and test the null hypothesis that the variability between tubes within preparations is the same as the within tube variability.

9. Schizophrenia is one of the psychiatric illnesses which is thought to have a definite physiological basis. Lake *et al.* [1980] assayed the concentration of norepinephrine in the cerebrospinal fluid of patients (NE in CSF) with one of three types of schizophrenia and controls. They reported the following means and *standard errors*:

NE in CSF (pg/ml)	Control Group	Schizophrenic Group		
		Paranoid	**Undifferentiated**	**Schizoaffective**
N	29	14	10	11
Mean	91	144	101	122
Standard error	6	20	11	21

Carry out Tasks (a), (b), (e), and (j).

10. Corvilain *et al.* [1971] studied the role of the kidney in the catabolism (conversion) of insulin by comparing the metabolic clearance rate in 7 control subjects, 8 patients with chronic renal failure, and 7 anephric (without kidneys) patients. The data for this problem consist of the plasma insulin concentrations (ng/ml) at 45 and 90 min after the start of continuous infusion of labelled insulin. A low plasma concentration is associated with a high metabolic clearance rate.

PLASMA INSULIN CONCENTRATION (ng/ml)

Control			Renal Failure			Anephric		
Patient	**45**	**90**	**Patient**	**45**	**90**	**Patient**	**45**	**90**
1	3.7	3.8	1	3.0	4.2	1	6.7	9.6
2	3.4	4.2	2	3.1	3.9	2	2.6	3.4
3	2.4	3.1	3	4.4	6.1	3	3.4	—[a]
4	3.3	4.4	4	5.1	7.0	4	4.0	5.1
5	2.4	2.9	5	1.9	3.5	5	3.1	4.2
6	4.8	5.4	6	3.4	5.7	6	2.7	3.8
7	3.2	4.1	7	2.9	4.3	7	5.3	6.6
			8	3.8	4.8			

[a]Missing observation

a. Consider the plasma insulin concentration at 45 min. Carry out Tasks (a)–(e) and (g)–(i).

b. Consider the plasma insulin concentration at 90 min. Carry out Tasks (a)–(e) and (g)–(i).

c. Calculate the difference in concentrations between 90 and 45 min for each patient. Carry out Tasks (a)–(e) and (g)–(i). Omit Patient 3 in the anephric group.

d. Graph the means for the three groups at 45 and 90 min on the same graph. What is the overall conclusion that you draw from the three analyses? Were all three analyses necessary? Would two of three have sufficed? Why or why not?

11. We return to the data of Zelazo *et al.* [1972] one more time. Carry out the Terpstra–Jonckeere test for ordered alternatives as discussed in Note 10.2. Justify the use of an ordered alternative hypothesis. Discuss in terms of power the reason that this analysis does indicate a treatment effect; in contrast to previous analyses.

12. One of the problems in the study of SIDS is the lack of a good animal model. Baak and Huber [1974] studied the guinea pig as a possible model observing the effect of lethal histamine shock on the guinea pig thymus. The purpose was to determine if changes in the thymus of the guinea pig correspond to pathological changes observed in SIDS victims. In the experiment 40 animals (20 male, 20 female) were randomly assigned either to "control" or "histamine shock." On the basis of a Wilcoxon two-sample test—which ignored possible sex differences—the authors concluded that the variable Medullar Blood Vessel Surface (mm^2/mm^3) did not differ significantly between "control" and "histamine shock." The data below have been arranged to keep track of sex differences.

MEDULLAR BLOOD VESSEL SURFACE (mm^2/mm^3)

	Control					Histamine Shock				
Female	6.4	6.2	6.9	6.9	5.4	8.4	10.2	6.2	5.4	5.5
	7.5	6.1	7.3	5.9	6.8	7.3	5.2	5.1	5.7	9.8
Male	4.3	7.5	5.2	4.9	5.7	7.5	6.7	5.7	4 9	6.8
	4.3	6.4	6.2	5.0	5.0	6.6	6.9	11.8	6.7	9.0

a. Do Tasks (a)–(e), (g), and (i).

b. Replace the observations by their ranks and repeat the analysis of variance. Compare your conclusions with those of Part (a).

13. In tumor metastasis, tumor cells spread from the original site to other organs. Usually a particular tumor will spread preferentially to specific organs. There are two possibilities as to how this may occur: the tumor cells gradually adapt to the organ to which they have spread or, tumor cells that grow well at this organ are preferentially selected. Nicolson and Custead [1982] studied this problem by comparing the metastatic potential of melanoma tumor cells mechanically lodged in the lungs of mice or injected intravenously and allowed to metastasize to the lung. Each of these cell lines was then harvested and injected subcutaneously. The numbers of pulmonary tumor colonies were recorded for each of three treatments: original line (control); mechanical placement (adaptation); and selection. The following data were obtained in three experiments involving 84 mice:

	Number of Pulmonary Tumor Colonies											
	Control				**Adaption**				**Selection**			
Experiment 1	0	4	20	32	0	3	20		7	92	141	
	0	9	22		0	6	24		64	96	149	
	1	11	31		2	14	29		79	100	151	
Experiment 2	0	8	31	41	0	10	13		0	101	132	
	3	8	32		0	11	14		52	109	136	
	6	22	39		5	12	14		89	110	140	
Experiment 3	0	4	36	49	0	11	21		30	79	111	
	0	18	39		0	13	27		46	89	114	
	2	29	42		3	13	28		51	100	114	

a. Carry out Tasks (a)–(g). You may want to try several transformations, for example, $\sqrt{\ }$, $Y^{1/4}$. An appropriate transformation is logarithmic. In order to avoid problems with zero values use $\log(Y + 1)$.

b. How would you interpret a significant "Experiment by Treatment" interaction?

14. A paper by Gruber [1976] evaluated interactions between two analgesic agents: fenoprofen and propoxyphene. The design of the study was factorial with respect to drug combinations. Propoxyphene (P) was administered in doses of 0, 5, 100, and 150 mg.; fenoprofen (F) in doses of 0, 200, 400, and 600 mg. Each combination of the two factors was studied. In addition, post-episiotomy postpartum patients were categorized into one of four pain classes: "Little," "Some," "Lot," and "Terrible" pain; for each of the 16 medication combinations 8, 10, 10, and 2 patients in the four pain classes were used. The layout of the number of patients could be constructed as follows:

Pain Level	**Medication Combination**					
	$(0P, 0F)$	$(0P, 200F)$	$\cdots$ $(0P, 600F)$	$(50P, 0F)$	$\cdots$	$(150P, 600F)$
"Little"	8	8	$\cdots$ 8	8	$\cdots$	8
"Some"	10	10	$\cdots$ 10	10	$\cdots$	10
"Lot"	10	10	$\cdots$ 10	10	$\cdots$	10
"Terrible"	2	2	$\cdots$ 2	2	$\cdots$	2

a. One response variable was "analgesic score" for a medication combination. The partial ANOVA table for this variable taken from the paper is as follows:

Source	DF	SS	MS	F	p
Pain class	—	3,704	—	—	—
Medications	—	9,076	—	—	—
Interaction	—	3,408	—	—	—
Residual	—	—	—		
Total	479	41,910			

Fill in the lines in the table completing the table.

b. The total analgesic score for the 16 sets of 30 patients classified by the two drug levels is:

TOTAL ANALGESIA SCORE

Propoxyphene Dose (mg)	Fenoprofen Calcium Dose (mg)			
	0	**200**	**400**	**600**
0	409	673	634	756
50	383	605	654	785
100	496	773	760	755
150	496	723	773	755

Carry out a "randomized block analysis" on these total scores dividing the sums of squares by 30 to return the analysis to a single reading status. Link this analysis with the table in part (a). You have, in effect, partitioned the SS for medications in that table into 3 parts. Test the significance of the *three* mean squares.

c. Graph the Mean Analgesia Score (per patient) by plotting the dose on the X-axis for fenoprofen, indicating the levels of the propoxyphene dose in the graph. State your conclusions.

15. While the prescription, "Take two aspirins, drink lots of fluids and go to bed," is usually good advice, it is known that aspirin induces "microbleeding" in the gastrointestinal system as evidenced by minute amounts of blood in the stool. Hence, there is constant research to develop other anti-inflammatory and antipyretic (fever combatting) agents. Arsenault *et al.* [1976] reported on a new agent, R-803, studying its effect in a Latin square design, comparing it to placebo and aspirin (900 mg, q.i.d). For purposes of this exercise the data are extracted in the form of a randomized block design. Each subject received each of three treatments for a week. We will assume the order was random. The variable measured is the amount of blood lost in ml/day as measured over a week.

Subject	Mean Blood Loss (ml/day)								
	1	**2**	**3**	**4**	**5**	**6**	**7**	**8**	**9**
Placebo	0.45	0.54	0.69	0.53	3.03	0.78	0.14	0.82	0.96
R-803	0.82	0.39	0.67	1.19	1.18	1.07	0.49	0.14	0.80
Aspirin	18.00	6.46	6.19	6.52	7.18	9.39	6.93	1.57	4.03

a. Do Tasks (a)–(e) and (g)–(i).

b. Carry out the Tukey test for additivity. What are your conclusions?

16. Occupational exposures to toxic substances are being investigated more and more carefully. Ratney *et al.* [1974] studied the effect of daily exposure of 180 to 200 ppm of methylene chloride on carboxyhemoglobin (COHb) measured during the work day. The following are COHb data (% COHb) for 7 subjects measured 5 times during the day.

% COHb

No. of Hours Since Beginning of Exposure

Subject	0	2	4	6	8
1	4.4	4.9	5.2	5.7	5.7
2	3.3	5.3	6.9	7.0	8.8
3	5.0	6.4	7.2	7.7	9.3
4	5.3	5.3	7.4	7.0	8.3
5	4.1	6.8	9.6	11.5	12.0
6	5.0	6.0	6.8	8.3	8.1
7	4.6	5.2	6.6	7.4	7.1

a. Carry out Tasks (a), (c)–(e), and (g)–(i).

b. Suppose the observation for Subject 3 at time 6 ($Y_{34} = 7.7$) is missing. Estimate its value and redo the ANOVA.

c. Carry out the Tukey test for additivity.

d. Carry out the Page test for trend (see Note 10.2).

e. Why do the data not form a randomized block design?

f. Could this problem be treated by regression methods; where X = hours since exposure, and Y = % COHb? Why or why not?

g. Calculate all 10 pairwise correlations between the treatment combinations. Do they look "reasonably close?"

17. Wang *et al.* [1976] studied the effects on sleep of four hypnotic agents and a placebo. The preparations were: lorazepam 2 and 4 mg, flurazepam 15 and 30 mg. Each of 15 subjects received all five treatments in a random order in 5-night units. The analysis of variance of length of sleep is presented here.

Source	DF	SS	MS	F	p
Treatments	—	—	12.0	—	—
Patients	—	—	14.8	—	—
Residual	—	—	2.2		
Total	74	—			

a. Do Task (a).

b. Fill in the missing values in the ANOVA table.

c. State your conclusions.

d. The article does not present any raw data or means. How satisfactory is this in terms of clinical significance?

18. High blood pressure is a known risk factor for cardiovascular disease and many drugs are now on the market that provide symptomatic as well as therapeutic relief. One of these drugs is propranolol. Hamet *et al.* [1973] in an early study investigated the effect of propranolol in labile hypertension. Among the variables studied was mean blood pressure measured in mmHg (diastolic + 1/3 pulse pressure). A placebo treatment was included in a double blind fashion. The blood pressure was measured in the recumbent and upright positions. The following are the blood pressure data:

	Blood Pressure (mmHg)			
	Recumbent		**Upright**	
Patient	**Placebo**	**Propranolol**	**Placebo**	**Propranolol**
N.F.	96	71	73	87
A.C.	96	85	104	76
P.D.	92	89	83	90
J.L.	97	110	101	85
G.P.	104	85	112	94
A.H.	100	73	101	93
C.L.	93	81	88	85

a. Assuming the treatments are just four treatments, carry out Tasks (a)–(e) and (g)–(i) (i.e., assume a randomized block design).

b. The sum of squares for treatments (3 DF) can be additively partitioned into three parts: $SS_{POSITION}$, SS_{DRUG}, and $SS_{DRUG \times POSITION}$, each with one degree of freedom. To do this construct an "interaction table" of treatment totals.

	Drug		
Position	**Placebo**	**Propranolol**	
Recumbent	678	594	1272
Upright	662	610	1272
	1340	1204	2544

$$SS_{DRUGS} = \frac{1340^2}{14} + \frac{1204^2}{14} - \frac{2544^2}{28} = 660.57,$$

$$SS_{POSITION} = \frac{1272^2}{14} + \frac{1272^2}{14} - \frac{2544^2}{28} = 0 \ (sic),$$

$$SS_{DRUGS \times POSITION} = \frac{678^2}{7} + \frac{594^2}{4} + \frac{662^2}{7} + \frac{610^2}{7} - \frac{2544^2}{28}$$

$$- SS_{DRUGS} - SS_{POSITION} = 36.57.$$

Expand the ANOVA table to include these terms. (The $ss_{POSITION} = 0$ is most unusual; the raw data are as reported in the table.)

c. This analysis could have been carried out as a series of three paired t-tests as follows: for each subject calculate the following 3 quantities "$+ + --$," "$+ - +-$," and "$+ - -+$." For example, for subject N.F. "$+ + --$" $= 96 + 71 - 73 - 87 = 7$, "$+ - +-$" $= 96 - 71 + 73 - 87 = 11$, and "$+ - -+$" $= 96 - 71 - 73 + 87 = 39$. These quantities represent effects of position, drug treatment, and interaction, respectively, and are called contrasts (see Chapter 12 for more details). Each contrast can be tested against 0 by means of a one-sample t-test. Carry out these t-tests. Compare the variances for each contrast; one assumption in the analysis of variance is that these contrast variances all estimate the same variance. How is the sum of the contrast variances related to the ss_e in the ANOVA?

d. Let d_1 = the sum of the observations associated with the pattern $+ + --$, d_2 = the sum of the observations associated with the pattern $+ - +-$, and d_3 = the sum of the observations associated with the pattern $+ - -+$. How is $\left(d_1^2 + d_2^2 + d_3^2\right)$ related to $ss_{TREATMENT}$?

19. Consider the data in Example 10.5. Rank all 38 observations from lowest to highest and carry out the usual analysis of variance on these ranks. Compare your p-values with the p-values of Table 10.9. In view of Note 10.3 does this analysis give you some concern?

20. Consider the data of Table 10.11 dealing with the effectiveness of pancreatic supplements on fat absorption. Rank all of the observations from 1 to 24 (i.e., ignoring both treatment and block categories).

a. Carry out an analysis of variance on the ranks obtained above.

b. Compare your analysis with the analysis using the Friedman statistic. What is a potential drawback in the analysis of Part (a).

c. Return to the "Friedman ranks" in Section 10.3.3. Carry out an analysis of variance on the Friedman ranks. How is the Friedman statistic related to ss_r of the ANOVA of the Friedman ranks?

21. These data are from the same source as in those in Problem 3. We now add data for females to generate a two-way layout:

Antipyrine Clearance (Half-Life in Hours)

	Stage I			IV	V	
Males	7.4	5.6	3.7	10.9	11.3	13.3
	6.6	6.0		12.2	10.0	
Females	9.1	6.3	7.1	11.0	8.3	
	11.3	9.4	7.9		4.3	

a. Do Tasks (a)–(d).

b. Graph the data. Is there any suggestion of interaction? Of main effects?

c. Carry out a weighted means analysis.

d. Partition each observation into its component parts and verify that the sums of squares are *not* additive.

22. Fuertes-de la Haba *et al.* [1976] measured intelligence in offspring of oral and non-oral contraceptive users in Puerto Rico. In the early 1960s subjects were randomly assigned to oral conceptive use or other methods of birth control. Subsequently, mothers with voluntary pregnancies were identified and offspring between ages 5 and 7 were administered a Spanish-Puerto Rican version of the Wechsler Intelligence Scale for Children (WISC). The following data for boys only are taken from the article.

		Age Groups (Years)		
		5	**6**	**7–8**
Oral contraceptive WISC score	*n*	9	18	14
	Mean	81.44	88.50	76.00
	SD	9.42	11.63	9.29
Other birth control WISC score	*n*	11	28	21
	mean	82.91	87.75	83.24
	SD	10.11	10.85	9.60

a. Carry out Tasks (a), (b), and (e).

b. Do an unweighted means analysis. Interpret your findings.

c. The age categories have obviously been "collapsed." What effect could such a collapsing have on the analysis? (Why introduce Age as a variable since IQ is standardized for age?)

d. Suppose we carried out a contingency table analysis on the cell frequencies. What could such an analysis show?

23. The following data have been extracted from the same article as the data in Problem 22 but the data have been "collapsed over age" and are presented by treatment (type of contraceptive) by sex. The response variable is, again, Wechsler IQ score.

		Sex	
		Boys	**Girls**
Oral Contraceptive WISC Score	*n*	41	55
	Mean	82.68	86.87
	SD	11.78	14.66
Other Birth Control WISC Score	*n*	60	54
	Mean	85.28	85.83
	SD	10.55	12.22

a. Carry out Tasks (a), (b), and (e).

b. Do an unweighted means analysis.

    c. Compare your conclusions with those of Problem 22.

*24. This problem considers some implications of the additive model for the two-way ANOVA as defined by Equation (10.18), and illustrated in Example 10.4.

    a. Graph the means of Example 10.4 by using the level of the second variable for the abscissa. Interpret the difference in the patterns.

    b. How many degrees of freedom are left for the means assuming the model defined by Equation (10.18) holds?

    c. We now want to define a nonadditive model retaining the values of the α's, β's and μ, equivalently, retaining the same marginal and overall means. You are free to vary any of the cell means subject to the above constraints. Verify that you can manipulate only four cell means. After changing the cell means calculate for each cell ij the quantity $Y_{ij} = \mu - \alpha_i - \beta_j$. What are some characteristics of these quantities?

    d. Graph the means derived in Part (c) and compare the pattern obtained with that of Figure 10.2.

*25. This problem is designed to give you some experience with algebraic manipulation. It is not designed to teach you algebra but to provide additional insight into the mathematical structure of analysis of variance models. You will want to take this medicine in small doses.

    a. Show that Equation (10.5) follows from the model defined by Equation (10.4).

    b. Prove Equations (10.6) and (10.7).

    c. Prove Equations (10.10)–(10.12) starting with the components of Equation (10.5).

    d. Consider Equation (10.17). Let $\mu_i = \sum n_{ij}\mu_{ij}/n_{i\bullet}$, etc. Relate α_i and β_j to $\mu_{i\bullet}$ and $\mu_{\bullet j}$.

    e. For the two-way ANOVA model as defined by Equation (10.21) show that $\mathrm{SS}_\varepsilon = \mathrm{SS_{ERROR}} = \sum (n_{ij} - 1)s_{ij}^2$ where s_{ij}^2 is the variance of the observations in cell (i, j).

    f. Derive the expected mean squares for MS_α and MS_γ in the fixed and random effects models as given in Table 10.13.

REFERENCES

Akritas, M. G. [1990]. The rank transform in some two-factor designs. *Journal of the American Statistical Association*, **85**: 73–78.

Arsenault, A., Le Bel, E., and Lussier, E. [1976]. Gastrointestinal microbleeding in normal subjects receiving acetysalicylic acid, placebo and R-803, a new antiinflammatory agent, in a design balanced for residual effects. *Journal of Clinical Pharmacology*, **16**: 473–480. Used with permission from J. B. Lippencott Company.

Baak, J. P. A. and Huber, J. [1974]. Effects of lethal histamine shock in the guinea pig thymus. In *SIDS 1974, Proceedings of the Francis E. Camps International Symposium of Sudden and Unexpected Deaths in Infancy*. R. R. Robertson (ed). The Canadian Foundation for the Study of Infant Death, Toronto, Canada.

Barboriak, J. J., Rimm, A., Tristani, F. E., Walker, J. R., and Lepley, Jr., D. [1972]. Risk factors in patients undergoing aorta-coronary bypass surgery. *Journal of Thoracic and Cardiovascular Surgery*, **64**: 92–97.

Beyer, W. H. (ed.) [1968]. *CRC Handbook of Tables for Probability and Statistics*, CRC Press, Cleveland, OH.

Box, G. E. P. [1953]. Non-normality and tests on variances. *Biometrika*, **40:** 318–335. Used with permission of the Biometrika Trustees.

Chikos, P. M. Figley, M. M. and Fisher, L. D. [1977]. Visual assessment of total heart volume and chamber size from standard chest radiographs. *American Journal of Roentgenology*, **128:** 375–380. Copyright © 1977 by the American Roentgenology Society.

Conover, W. J. and Iman, R. L. [1981]. Rank transformations as a bridge between parametric and nonparametric statistics. *American Statistician*, **35:** 124–133.

Cornfield, J. and Tukey, J. W. [1956]. Average values of mean squares in factorials. *Annals of Mathematical Statistics*, **27:** 907–949.

Corvilain, J., Brauman, H., Delcroix, C., Toussaint, C., Vereerstraeten, P., and Franckson, J. R. M. [1971]. Labeled insulin catabolism in chronic renal failure and the anephric state. *Diabetes*, **20:** 467–475.

Daniel, C. [1976]. *Applications of Statistics to Industrial Experiments*. Wiley, New York.

Daniel, C. and Wood, F. [1971] *Fitting Equations to Data*. Wiley, New York.

Draper, N. R. and Smith, H. [1981] *Applied Regression Analysis,* 2nd edition. Wiley, New York.

Eisenhart, C. [1947]. The assumptions underlying the analysis of variance. *Biometrics*, **3:** 1–21.

Fisher, R. A. [1950]. Statistical Methods for Research Workers, 11th edition. Oliver and Boyd, London.

Florey, C. Du V., Milner, R. D. G., and Miall, W. I. [1977]. Serum insulin and blood sugar levels in a rural population of Jamaican adults. *Journal of Chronic Diseases*, **30:** 49–60. Used with permission of Pergamon Press, Inc.

Freeman, M. F. and Tukey, J. W. [1950]. Transformations related to the angular and the square root. *Annals of Mathematical Statistics*, **21:** 607–611.

Friedman, M. [1937]. The use of ranks to avoid the assumption of normality implicit in the analysis of variance. *Journal of the American Statistical Association*, **32:** 675–701.

Fuertes-de la Haba, A., Santiago, G., and Bangdiwala, I. S. [1976]. Measured intelligence in offspring of oral and nonoral contraceptive users. *American Journal Obstetrics and Gynecology*, **7:** 980–982.

Graham, D. Y. [1977]. Enzyme replacement therapy of exocrine pancreatic insufficiency in man. *New England Journal of Medicine*, **296:** 1314–1317.

Gruber, Jr., C. M. [1976]. Evaluating interactions between fenoprofen and propoxyphene: analgesic and adverse reports by postepisiotomy patients. *Journal of Clinical Pharmacology*, **16:** 407–417. Used with permission from J. J. Lippencott Company.

Hamet, P., Kuchel, O., Cuche, J. L., Boucher, R., and Genest, J. [1973]. Effect of propranolol on cyclic AMP excretion and plasma renin activity in labile essential hypertension. *Canadian Medical Association Journal*, **1:** 1099–1103.

Hettmansperger, T. P. and McKean, J. W. [1978]. Statistical inference based on ranks. *Psychometrika*, **43:** 69–79.

Hillel, A. and Patten, C. [1990]. Effects of age and gender on dominance for lateral abduction of the shoulder. Unpublished data; used by permission.

Hollander, M. and Wolfe, D. A. [1973]. *Nonparametrical Statistical Methods*. Wiley, New York.

Joiner, B. L. and Rosenblatt, J. R. [1971]. Some properties of the range in samples from Tukey's symmetric lambda distributions. *Journal of the American Statistical Association*, **66:** 394–399.

Keller, S. E., Weiss, J. W., Schleifer, S. J., Miller, N. E., and Stein, M. [1981]. Suppression of immunity by stress. *Science*, **213:** 1397–1400. Copyright © 1981 by the AAAS.

Koch, G. G., Amara, I. A., Stokes, M. E., and Gillings, D. B. [1980]. Some views on parametric and non-parametric analysis for repeated measurements and selected bibliography. *International Statistical Review*, **48:** 249–265.

Kruskal, W. H. and Wallis, W. A. [1952]. Use of ranks in one-criterion variance analysis. *Journal of the American Statistical Association*, **47:** 583–621.

Lake, C. R., Sternberg, D. E., van Kammen, D. P., Ballenger, J. C., Ziegler, M. G., Post, R. M., Kopin, I. J., and Bunney, W. E. [1980]. Schizophrenia: elevated cerebrospinal fluid norepinephrine. *Science*, **207:** 331–333. Copyright ⓒ 1980 by the AAAS.

Looney, S. W. and Stanley, W. B. [1989]. Exploratory repeated measures analysis for two or more groups. *The American Statistician*, **43:** 220–225.

Nicolson, G. L. and Custead, S. E. [1982]. Tumor metastasis is not due to adaptation of cells to a new organ environment. *Science*, **215:** 176–178. Copyright ⓒ 1982 by the AAAS.

Odeh, R. E., Owen, D. B., Birnbaum, Z. W., and Fisher, L. D. [1977]. *Pocket Book of Statistical Tables*. Marcel Dekker, New York.

Olsen, G. D, Bennett, W. M., and Porter, G. A. [1975]. Morphine and phenytoin binding to plasma proteins in renal and hepatic failure. *Clinical Pharmaceuticals and Therapeutics*, **17:** 677–681.

Page, E. B. [1963]. Ordered hypotheses for multiple treatments: a significance test for linear ranks. *Journal of the American Statistical Association*, **58:** 216–230.

Quesenberry, P. D., Whitaker, T. B., and Dickens, J. W. [1976]. On testing normality using several samples: an analysis of peanut aflatoxin data. *Biometrics*, **32:** 753–759.

Ratney, R. S., Wegman, D. H., and Elkins, H. B. [1974]. In vivo conversion of methylene chloride to carbon monoxide. *Archives of Environmental Health*, **28:** 223–226. Reprinted with permission of the Helen Dwight Reid Educational Foundation. Published by Heldref Publications, 4000 Albemarle St., N. W., Washington D.C. 20016. Copyright ⓒ 1974.

Rifkind, A. B., Canale, V., and New, M. I. [1976]. Antipyrine clearance in homozygous beta-thalassemia. *Clinical Pharmaceuticals and Therapeutics*, **20:** 476–483.

Ross, M. H. and Bras, G. [1975]. Food preference and length of life. *Science*, **190:** 165–167. Copyright ⓒ 1975 by the AAAS.

Ryan, T. A., Jr., Joiner, B. L., and Ryan, B. F. [1980]. *Minitab Reference Manual*, Release 1/10/80. Statistics Department, Pennsylvania State University, University Park, PA.

Scheffé, H. [1959]. *The Analysis of Variance*. Wiley, New York.

Sherwin, R. P. and Layfield, L. J. [1976]. Protein leakage in lungs of mice exposed to 0.5 ppm nitrogen dioxide; a fluorescence assay for protein. *Archives of Environmental Health*, **31:** 116–118.

Snedecor, G. W. and Cochran, W. G. [1980]. *Statistical Methods*, 7th edition. Iowa State University Press, Ames, IA.

Tukey, J. W. [1949]. One degree of freedom for additivity. *Biometrics*, **5:** 232–242.

Wallace, Fisher, and Tremann [1977]. Unpublished manuscript.

Wang, R. I. H., Stockdale, S. L., and Hieb, E. [1976]. Hypnotic efficacy of lorazepam and flurazepam. *Clinical Pharmaceuticals and Therapeutics*, **19:** 191–195.

Wardlaw, A. C. and van Belle, G. [1964]. Statistical aspects of the mouse diaphragm test for insulin. *Diabetes*, **13:** 622–634.

Weisberg, S. [1980]. *Applied Linear Regression*. Wiley, New York.

Weisberg, S. and Bingham, C. [1975]. Approximate analysis of variance test for non-normality suitable for machine calculation. *Technometrics*, **17:** 133–134.

Winer, B. J. [1971]. *Statistical Principles in Experimental Design*, 2nd edition. McGraw-Hill, New York.

Zelazo, P. R., Zelazo, N. A., and Kalb, S. [1972] "Walking" in the newborn. *Science*, **176:** 314–315.

CHAPTER 11

Association and Prediction: Multiple Regression Analysis, Linear Models With Multiple Predictor Variables

11.1 INTRODUCTION

We looked at the linear relationship between two variables, say X and Y, in Chapter 9. We learned to estimate the regression line of Y on X and to test the significance of the relationship. Summarized by the correlation coefficient; the square of the correlation coeffiecients is the percent of the variability explained.

Often we want to predict or explain the behavior of one variable in terms of more than one variable. We want to explain the behavior of variable Y in terms of k variables $X_1, \ldots, X_k$. In this chapter we look at situations where Y may be explained by a linear relationship with the explanatory or predictor variables $X_1, \ldots, X_k$. This chapter is a generalization of Chapter 9 where only one explanatory variable was considered. Some additional possibilities will arise. With more than one potential predictor variable it will often be desirable to find a simple model that explains the relationship. Thus we will consider how to select a subset of predictor variables from a large number of potential predictor variables to find a reasonable predictive equation. Multiple regression analyses, as the methods of this chapter are called, are one of the most widely used tools in statistics. If the appropriate limitations are kept in mind, they can be useful in understanding complex relationships. Because of the difficulty of computing the estimates involved most computations of multiple regression analyses are performed by computer. For this reason, this chapter includes examples of output from multiple regression computer runs.

11.2 THE MULTIPLE REGRESSION MODEL

In this section we present the multiple regression mathematical model. We discuss the methods of estimation and also the assumptions that are needed for statistical inference. The procedures are illustrated with two examples.

496

11.2.1 The Linear Model

Definition 11.1. A *linear equation* for the variable Y in terms of $X_1, \ldots, X_k$, is an equation of the form

$$Y = a + b_1 X_1 + \cdots + b_k X_k. \tag{11.1}$$

The values of $a, b_1, \ldots, b_k$, are fixed constant values. These values are called coefficients.

Suppose we observe Y and want to model its behavior in terms of independent, predictor, or explanatory, or covariate variables, $X_1, \ldots, X_k$. For a particular set of values of the covariates, the Y value will not be known with certainty. As before, we model the expected value of Y for given or known values of the X_i. Throughout this chapter, we consider the behavior of Y for fixed, known, or observed values for the X_i. We have a multiple linear regression model if the expected value of Y for the known $X_1, \ldots, X_k$ is linear. Stated more precisely:

Definition 11.2. Y has a linear regression upon $X_1, \ldots, X_k$, if the expected value of Y for the known X_i values is linear in the X_i values. That is,

$$E(Y \mid X_1, \ldots, X_k) = \alpha + \beta_1 X_1 + \cdots + \beta_k X_k. \tag{11.2}$$

Another way of stating this is the following. Y is equal to a linear function of the X_i, plus an error term whose expectation is zero:

$$Y = \alpha + \beta_1 X_1 + \cdots + \beta_k X_k + \epsilon \tag{11.3}$$

where

$$E(\epsilon) = 0.$$

We use the Greek letters α and β_i for the population parameter values and Latin letters a and b_i for the estimates to be described below. Analogous to definitions in Chapter 9, the number α is called the intercept of the equation and is equal to the expected value of Y when all of the X_i values are 0. The β_i coefficients are the regression coefficients.

11.2.2 The Least Squares Fit

In Chapter 9 we fitted the regression line by choosing the estimates a and b to minimize the sum of squares of the differences between the observed Y values and the predicted or modeled Y values. These differences were called residuals; another way of explaining the estimates is to say that the coefficients were chosen to minimize the sum of squares of the residual values. We use this same approach, for the same reasons, to estimate the regression coefficients in the multiple regression problem. Because we have more than one predictor or covariate variable and multiple observations, the notation becomes slightly more complex. Suppose that there are n observations, we denote the observed values of Y for the ith observation by Y_i and the observed value of the jth variable X_j by X_{ij}. For example, for two predictor variables we can lay out the data in the following array:

Case	Y	X_1	X_2
1	Y_1	X_{11}	X_{12}
2	Y_2	X_{21}	X_{22}
$\vdots$	$\vdots$	$\vdots$	$\vdots$
i	Y_i	X_{i1}	X_{i2}
$\vdots$	$\vdots$	$\vdots$	$\vdots$
n	Y_n	X_{n1}	X_{n2}

The following definition extends the definition of least squares estimation to the multiple regression situation.

Definition 11.3. Given data $(Y_i, X_{i1}, \ldots, X_{ik})$, $i = 1, \ldots, n$ the *least squares fit* of the regression equation chooses $a, b_1, \ldots, b_k$ to minimize

$$\sum_{i=1}^{n} (Y_i - \widehat{Y}_i)^2$$

where $\widehat{Y}_i = a + b_1 X_{i1} + \cdots + b_k X_{ik}$. The b_i are the *(sample) regression* coefficients, a is the *sample intercept*. The difference $Y_i - \widehat{Y}_i$ is the ith *residual*.

The actual fitting is usually done by computer since the solution by hand can be quite tedious. Some details of the solution are presented in Note 11.1.

Example 11.1. We consider a paper by Cullen and van Belle [1975] dealing with the effect of the amount of anesthetic agent administered during an operation. The work also examines the degree of trauma upon the immune system as measured by the decreasing ability of lymphocytes to transform in the presence of mitogen (a substance which enhances cell division). The variables measured (among others) were

$X_1 = $ duration of anesthesia (in hours),

$X_2 = $ trauma factor (see Table 11.1 for classification),

$Y = $ percentage depression of lymphocyte transformation following anesthesia.

It is assumed that the amount of anesthetic agent administered is directly proportional to the duration of anesthesia. The question of the influence of each of the two

Table 11.1. Classification of Surgical Trauma.

0	Diagnostic or therapeutic regional anesthesia; examination under general anesthesia
1	Joint manipulation; minor orthopedic procedures; cystoscopy; dilatation and curettage
2	Extremity, genitourinary, rectal and eye procedures; hernia repair; laparoscopy
3	Laparotomy; craniotomy; laminectomy; peripheral vascular surgery
4	Pelvic extenteration; jejunal interposition; total cycstectomy

Table 11.2. Effect of Duration of Anesthesia (X_1), Degree of Trauma (X_2) On Percentage Depression of Lymphocyte Transformation Following Anesthesia (Y).

Patient	X_1 Duration	X_2 Trauma	Y Percent Depression	Y Predicted Value	$Y - \widehat{Y}$ Residual
1	4.0	3	36.7	33.0	3.7
2	6.0	3	51.3	35.2	16.1
3	1.5	2	40.8	19.9	20.9
4	4.0	2	58.3	22.6	35.7
5	2.5	2	42.2	21.0	21.2
6	3.0	2	34.6	21.5	13.1
7	3.0	2	77.8	21.5	56.3
8	2.5	2	17.2	21.0	-3.8
9	3.0	3	-38.4	31.9	-70.3
10	3.0	3	1.0	31.9	-30.9
11	2.0	3	53.7	20.8	22.9
12	8.0	3	14.3	37.4	-23.1
13	5.0	4	65.0	44.5	20.5
14	2.0	2	5.6	20.4	-14.8
15	2.5	2	4.4	21.0	-16.6
16	2.0	2	1.6	20.4	-18.8
17	1.5	2	6.2	19.9	-13.7
18	1.0	1	12.2	8.9	3.3
19	3.0	3	29.9	31.9	-2.0
20	4.0	3	76.1	33.0	43.1
21	3.0	3	11.5	32.0	-20.5
22	3.0	3	19.8	31.9	-12.1
23	7.0	4	64.9	46.7	18.2
24	6.0	4	47.8	45.6	2.2
25	2.0	2	35.0	20.4	14.6
26	4.0	2	1.7	22.6	-20.9
27	2.0	2	51.5	20.4	31.1
28	1.0	1	20.2	8.9	11.3
29	1.0	1	-9.3	8.9	-18.2
30	2.0	1	13.9	10.0	3.9
31	1.0	1	-19.0	8.9	-27.9
32	3.0	1	-2.3	11.1	-13.4
33	4.0	3	41.6	33.0	8.6
34	8.0	4	18.4	47.8	-29.4
35	2.0	2	9.9	20.4	-10.5
Total	112.5	83	896.1	896.3	-0.2[a]
Mean	3.21	2.37	25.60	25.60	-0.006

[a]0 except for round-off error.

predictor variables is the crucial one which will not be answered in this section. Here we consider the combined effect. The set of 35 patients considered for this example consisted of those receiving general anesthesia. The basic data are reproduced in Table 11.2. The predicted values and deviations are calculated from the least squares regression equation which was $Y = -2.55 + 1.10X_1 + 10.376X_2$. □

11.2.3 Assumptions for Statistical Inference

Recall that in the simple linear regression models of Chapter 9, we needed assumptions about the distribution of the error terms before we proceeded to statistical inference; that is, before we tested hypotheses about the regression coefficient using the F-test from the analysis of variance table. More specifically, we assumed:

Simple Linear Regression Model. Observe (X_i, Y_i), $i = 1, \ldots, n$. The model is

$$Y_i = \alpha + \beta X_i + \epsilon_i \tag{11.4}$$

or

$$Y_i = E(Y_i \mid X_i) + \epsilon_i,$$

where the "error" terms ϵ_i are statistically independent of each other and all have the same normal distribution with mean zero and variance σ^2; that is, $\epsilon_i \sim N(0, \sigma^2)$.

Using this model, it is possible to set up the analysis of variance table associated with the regression line. The table has the following form:

ANOVA TABLE

Source of Variation	Degrees of Freedom	Sum of Squares	Mean Square	F-Ratio
Regression	1	$\text{SS}_{\text{REG}} = \sum_i (\widehat{Y}_i - \overline{Y})^2$	$\text{MS}_{\text{REG}} = \text{SS}_{\text{REG}}$	$\dfrac{\text{MS}_{\text{REG}}}{\text{MS}_{\text{RESID}}}$
Residual	$n - 2$	$\text{SS}_{\text{RESID}} = \sum_i (Y_i - \widehat{Y}_i)^2$	$\text{MS}_{\text{RESID}} = \dfrac{\text{SS}_{\text{RESID}}}{n - 2}$	
Total	$n - 1$	$\sum_i (Y_i - \overline{Y}_i)^2$		

The mean square for residual is an estimate of the variance σ^2 about the regression line. (In this chapter we change notation slightly from that used in Chapter 9. The quantity σ^2 used here is the variance about the regression line. This was σ_1^2 in Chapter 9.)

The F-ratio is an F statistic having numerator and denominator degrees of freedom of 1 and $n - 2$, respectively. We may test the hypothesis that the variable X has linear predictive power for Y, that is $\beta \neq 0$, by using tables of critical values for the F statistic with 1 and $n - 2$ degrees of freedom. Further, using the estimate of the variance about the regression line MS_{RESID}, it was possible to set up confidence intervals for the regression coefficient β.

For multiple regression equations of the currrent chapter, the same assumptions needed in the simple linear regression analyses carry over in a very direct fashion. More specifically, our assumptions for the multiple regression model are the following.

Multiple Regression Model. Observe $(Y_i, X_{i1}, \ldots, X_{ik})$, $i = 1, 2, \ldots, n$ (n observations). The distribution of Y_i for fixed or known, values of $X_{i1}, \ldots, X_{ik}$, is

$$Y_i = E(Y_i \mid X_{i1}, \ldots, X_{ik}) + \epsilon_i, \tag{11.5}$$

where $E(Y_i \mid X_{i1}, \ldots, X_{ik}) = \alpha + \beta_1 X_{i1} + \cdots + \beta_k X_{ik}$ or $Y_i = \alpha + \beta_1 X_{i1} + \cdots + \beta_k X_{ik} + \epsilon_i$. The ϵ_i are statistically independent and all have the same normal distribution with mean 0 and variance σ^2; that is, $e_i \sim N(0, \sigma^2)$.

With these assumptions, we use a computer program to find the least squares estimate of the regression coefficients. From these estimates we have the predicted value for Y_i given the values of $X_{i1}, \ldots, X_{ik}$. That is,

$$\widehat{Y}_i = a + b_1 X_{i1} + \cdots + b_k X_{ik}. \tag{11.6}$$

Using these values, the analysis of variance table for the one-dimensional case generalizes. The analysis of variance table in the multi-dimensional case is now the following:

MULTIPLE REGRESSION ANOVA TABLE

Source of Variation	Degrees of Freedom	Sum of Squares	Mean Square	F-Ratio
Regression	k	$\text{SS}_{\text{REG}} = \sum_i (\widehat{Y}_i - \overline{Y})^2$	$\text{MS}_{\text{REG}} = \dfrac{\text{SS}_{\text{REG}}}{k}$	$\dfrac{\text{MS}_{\text{REG}}}{\text{MS}_{\text{RESID}}}$
Residual	$n - k - 1$	$\text{SS}_{\text{RESID}} = \sum_i (Y_i - \widehat{Y}_i)^2$	$\text{MS}_{\text{RESID}} = \dfrac{\text{SS}_{\text{RESID}}}{n - k - 1}$	
Total	$n - 1$	$\sum_i (Y_i - \overline{Y}_i)^2$		

For the analysis of variance table and multiple regression model, note the following:

1. If $k = 1$ there is one X variable; the equations and analysis of variance table reduce to that of the simple linear regression case.
2. The F statistic tests the hypothesis that the regression line has no predictive power. That is, it tests the hypothesis

$$H_0 : \beta_1 = \beta_2 = \cdots = \beta_k = 0. \tag{11.7}$$

This hypothesis says that all of the beta coefficients are zero, that is, the X variables do not help to predict Y. The alternative hypothesis is that one or more of the regression coefficients $\beta_1, \ldots, \beta_k$ are nonzero. Under the null hypothesis, H_0, the F statistic has an F distribution with k and $n - k - 1$ degrees of freedom. Under the alternative hypotheses, that one or more of the β_i are nonzero, the F statistic tends to be too large. Thus the hypothesis that the regression line has predictive power is tested by using tables of the F distribution and rejection when F is too large.

3. The residual sum of squares is an estimate of the variability about the regression line; that is, it is an estimate of σ^2. Introducing notation similar to that of Chapter 9, we write

$$\sigma^2 = s_{Y \cdot X_1, \ldots, X_k}^2 = \text{MS}_{\text{RESID}} = \frac{\sum_i (Y_i - \widehat{Y}_i)^2}{n - k - 1}. \tag{11.8}$$

4. Using the estimated value of σ^2, it is possible to find estimated standard errors for the b_i, the estimates of the regression coefficients β_i. The estimated standard error is associated with the t distribution with $n - k - 1$ degrees of freedom. The test of $\beta_i = 0$ and an appropriate $100(1 - \alpha)$ percent confidence interval are given by the following equations. To test $H_j : \beta_j = 0$ at significance level α, use two-sided critical values for the t-distribution with $n - k - 1$ degrees of freedom and the test statistic

$$t = \frac{b_j}{\text{SE}(b_j)}, \tag{11.9}$$

where b_j and $\text{SE}(b_j)$ are taken from computer output. Reject H_j if

$$|t| \geq t_{n-k-1,1-\alpha/2}.$$

A $100(1 - \alpha)\%$ confidence interval for β_j is given by

$$b_j \pm \text{SE}(b_j)t_{n-k-1,1-\alpha/2}. \tag{11.10}$$

These two facts follow from the pivotal variable

$$t = \frac{b_j - \beta_j}{\text{SE}(b_j)},$$

which has a t-distribution with $n - k - 1$ degrees of freedom.

5. Interpretations of the estimated coefficients in a multiple regression equation must be done cautiously. Recall (from the simple linear regression chapter) that we used the example of height and weight; we noted that if we managed to get the individuals to eat and/or diet to change their weight, this would not have any substantial effect on the individual's height despite a relationship between height and weight in the population. Similarly, when we look at the estimated multiple regression equation, we can say that for the observed X values the regression coefficients β_i have the following interpretation. If all of the X variables except for one, say X_i, are kept fixed and if X_i changes by one unit, then the expected value of Y changes by β_i. Let us consider this statement again for emphasis. *If all of the X variables except for one X variable, X_i, are held constant, and the observation has X_i changed by an amount one, then the expected value of Y_i changes by the amount β_i.* This is seen by looking at the difference in the expected values.

$$\alpha + \beta_1 X_1 + \cdots + \beta_i(X_i + 1) + \cdots + \beta_k X_k - (\alpha + \cdots + \beta_i X_i + \cdots + \beta_k X_k) = \beta_i.$$

This does not mean that when the regression equation is estimated, by changing X by a certain amount we can therefore change the expected value of Y. Consider a medical example where X_i might be systolic blood pressure, and other X variables are other measures of physiological performance. Any maneuvers taken to change X_i might also result in changing some or all of the other X's in the population. The change in Y of β_i holds for the distribution of X's in the sampled population. By changing the values of X_i we might change the overall relationship between the Y_i's and the X_i's so that the estimated regression equation no longer holds. (Recall again the height and weight example for simple linear regression). For these reasons, interpretations of multiple regression equations must be made tentatively, especially when the data result from observational studies rather than controlled experiments.

6. If two variables, say X_1 and X_2 are closely related, it is difficult to estimate their regression coefficients because they tend to get confused. Take the extreme case where the variables X_1 and X_2 are actually the same value. Then if we look at $\beta_1 X_1 + \beta_2 X_2$ we can factor out the X_1 variable which is equal to X_2. That is, if $X_1 = X_2$, then $\beta_1 X_1 + \beta_2 X_2 = (\beta_1 + \beta_2)X_1$. We see that β_1 and β_2 are not uniquely determined in this case, but any values for β_1 and β_2 whose sum is the same will give the "same" regression equation. More generally, if X_1 and X_2 are very closely associated in a linear fashion (that is, if their correlation is large), it is very difficult to estimate the betas. This difficulty is referred to as *collinearity*. We will return to this fact in more depth below.

7. In Chapter 9 we saw that the assumptions of the simple linear regression model held if the two variables, X and Y, have a bivariate normal distribution. This fact may be extended to the considerations of this chapter. If the variables $Y, X_1, \ldots, X_k$ have a multivariate normal distribution, then conditionally upon knowing the values of $X_1, \ldots, X_k$ the assumptions of the multiple regression model hold. Note 11.2 has more detail on the multivariate normal distribution. We shall not go into this in detail, but merely mention that if the variables have a multivariate normal distribution, then any one of the variables has a normal distribution; any two of the variables have a bivariate normal distribution; and any linear combination of the variables also has a normal distribution.

These generalizations of the findings for simple linear regression are illustrated in the next section which presents several examples of multiple regression.

11.2.4 Examples of Multiple Regression

Example 11.1 (continued). We modeled the percent depression of lymphocyte transformation following anesthesia by using the duration of the anesthesia in hours and trauma factor. The least squares estimates of the regression coefficients, the estimated standard errors and the analysis of variance table are given below.

Constant or Variable i	b_i	$\text{SE}(b_i)$
Duration of anesthesia	1.105	3.620
Trauma factor	10.376	7.460
Constant	−2.55	12.395

ANALYSIS OF VARIANCE TABLE

Source	DF	SS	MS	*F*-Ratio
Regression	2	4192.94	2096.47	3.18
Residual	32	21070.09	658.44	
Total	34	25263.03		

From tables of the F distribution, we see that at the 5% significance level the critical value for 2 and 30 degrees of freedom is 3.32 while for 2 and 40 degrees of freedom it is 3.23. Thus, $F_{2,32,0.95}$ is between 3.23 and 3.32. Since the observed F ratio is 3.18 which is smaller at the 5% significance level, we would not reject a null hypothesis that the regression equation has no contribution to the prediction. (Why is the double negative appropriate here?) This being the case, it would not pay to proceed further to examine the significance of the individual regression coefficients. (You will note that a standard error for the constant term in the regression is also given. This is also a feature of the computer output for most multiple regression packages.)   □

Example 11.2.   This is a continuation of the malignant melonoma of the skin example. In Chapter 9, we examined data on malignant melanoma of the skin from white males. In that chapter we saw that the mortality was related to latitude by a simple linear regression equation, and that contiguity to an ocean was also related to mortality. We now consider the modeling of the mortality result using a multiple regression equation with both the latitude variable and the contiguity to an ocean variable. When this is done the following estimates result.

Constant or Variable	b_i	$\text{SE}(b_i)$
Latitude in degrees	−5.449	0.551
Contiguity to ocean	18.681	5.079
$1 = $ contiguous to ocean		
$0 = $ does not border ocean		
Constant	360.28	22.572

ANALYSIS OF VARIANCE TABLE

Source	**DF**	**SS**	**MS**	**F**
Regression	2	40366.82	20183.41	69.96
Residual	46	13270.45	288.49	
Total	48	53637.27		

The F critical values at the 0.05 level with 2 and 40 and 2 and 60 degrees of freedom are 3.23 and 3.15, respectively. Thus the F statistic for the regression is very highly statistically significant. This being the case, we might then wonder whether or not the significance came from one variable or whether both of the variables contributed to the statistical significance. We first test the significance of the latitude variable at the 5% significance level and also construct a 95% confidence interval. $t = -5.449/0.551 = -9.89$, $|t| > t_{48,0.975} \doteq 2.01$, reject $\beta_1 = 0$ at the 5% significance level. The 95% confidence interval is given by $-5.449 \pm 2.01 \times 0.551$ or $(-6.56, -4.34)$.

Consider a test of the significance of β_2 at the 1% significance level and a 99% confidence interval for β_2. $t = 18.681/5.079 = 3.68$, $|t| > t_{48,0.995} \doteq 2.68$, reject $\beta_2 = 0$ at the 1% significance level. The 99% confidence interval is given by $18.681 \pm 2.68 \times 5.079$ or $(5.07, 32.29)$.

In this example, from the t statistic we conclude that both latitude in degrees and the contiguity to the ocean variables are each contributing to the statistically significant relationship between the melanoma of the skin mortality rates and the multiple regression equation. □

Example 11.3. The data for this problem come from Chapter 9, Problems 9.5–9.8. These data consider maximal exercise treadmill tests for 43 active women. We will consider two possible multiple regression equations from these data. Suppose that we want to predict or explain the variability in VO_2 MAX by using three variables: X_1, the duration of the treadmill test; X_2, the maximum heart rate attained during the test; X_3, the height of the subject in centimeters. Data resulting from the least squares fit are:

Covariate or Constant	b_i	$SE(b_i)$	t ($t_{39,0.975} \doteq 2.02$)
Duration (seconds)	0.0534	0.00762	7.01
Maximum heart rate (beats/min)	−0.0482	0.05046	−0.95
Height (cm)	0.0199	0.08359	0.24
Constant	6.954	13.810	

ANALYSIS OF VARIANCE TABLE

Source	DF	SS	MS	F ($F_{3,39,0.95} \doteq 2.85$)
Regression	3	644.61	214.87	21.82
Residual	39	384.06	9.85	
Total	42	1028.67		

Note that the overall F test is highly significant, being 21.82 compared to a 5% critical value for the F distribution with 3 and 39 degrees of freedom, of approximately 2.85. When we look at the t statistic for the three individual terms, we see that the t for duration, 7.01, is much larger than the corresponding 0.05 critical value of 2.02. The other two variables have values for the t statistic with absolute value much less than 2.02. This raises the possibility that duration is the only variable of the three contributing to the predictive equation. Perhaps we should consider a model where we predict the maximum oxygen consumption only in terms of duration rather than using all three variables. In sections to follow, we will consider the question of selecting a "best" predictive equation using a subset of a given set of potential explanatory or predictor variables.

As a second part of Example 11.3, we use the same data, but consider the dependent variable to be age. We shall try to model this from three explanatory, or independent, or predictor variables. Let X_1 be the duration of the treadmill test in seconds, let X_2 be VO_2 MAX be the maximal oxygen consumption, and let X_3 be the maximum heart rate during the treadmill test. Analysis of these data lead to the following:

Covariate or Constant	b_i	$SE(b_i)$	t ($t_{39,0.975} \doteq 2.02$)
Duration	−0.0524	0.0268	−1.96
VO_2 MAX	−0.633	0.378	−1.67
Maximum heart rate	−0.0884	0.119	−0.74
Constant	106.51	18.63	

ANALYSIS OF VARIANCE TABLE

Source	DF	SS	MS	F ($F_{3,39,0.95} \doteq 2.85$)
Regression	3	2256.97	752.32	13.70
Residual	39	2142.19	54.93	
Total	42	4399.16		

The overall F of 13.7 is very highly statistically significant, indicating that if one has the results of the treadmill test including duration, VO_2 MAX, and the maximum heart rate, one can gain a considerable amount of knowledge about the subject's age. Note however that when we look at the p-values for the individual variables, not one of them is statistically significant! How can it be that the overall regression equation is very highly statistically significant but none of the variables individually can be shown to have contributed at the 5% significance level? This paradox results because the predictive variables are highly correlated among themselves; they are *collinear*, as mentioned above. For example, we already know from Chapter 9 that the duration and the VO_2 MAX are highly correlated variables; there is much overlap in their predictive information. We have trouble showing that the prediction comes from one or the other of the two variables.   □

11.3 LINEAR ASSOCIATION: MULTIPLE, PARTIAL, AND CANONICAL CORRELATION

The simple linear regression equation was very closely associated with the correlation coefficient between the two variables; the square of the correlation coefficient was the proportion of the variability in one variable that could be explained by the other variable using a linear predictive equation. In this section we consider a generalization of the correlation coefficient.

11.3.1 The Multiple Correlation Coefficient

In considering simple linear regression, we saw that r^2 was the proportion of the variability of the Y_i about the mean that could be explained from the regression equation. We generalize this to the case of multiple regression.

Definition 11.4. The *squared multiple correlation coefficient*, denoted by R^2, is the proportion of the variability in the dependent variable Y that may be accounted for by the multiple regression equation. Algebraically,

$$R^2 = \frac{\text{regression sum of squares}}{\text{total sum of squares}}.$$

Since

$$\sum_i (Y_i - \overline{Y}_i)^2 = \sum_i (Y_i - \widehat{Y}_i)^2 + \sum_i (\widehat{Y}_i - \overline{Y}_i)^2$$

$$R^2 = \frac{\text{SS}_{\text{REG}}}{\text{SS}_{\text{TOTAL}}} = \frac{\sum_i (\widehat{Y}_i - \overline{Y})^2}{\sum_i (Y_i - \overline{Y})^2}. \qquad (11.11)$$

Definition 11.5. The positive square root of R^2 is denoted by R. R is the *multiple correlation coefficient*.

The multiple correlation coefficient may also be computed as the correlation between the Y_i and their estimated best linear predictor, $\widehat{Y}_i$. If the data come from a multivariate sample rather than having the X's fixed by experimental design, the quantity R is an estimate of the correlation between Y and the best linear predictor for Y in terms of $X_1, \ldots, X_k$, that is, the correlation between Y and $a+b_1 X_1+\cdots+b_k X_k$. The population correlation will be zero if and only if all of the regression coefficients $\beta_1, \ldots, \beta_k$ are equal to zero. Again, the value of R^2 is an estimate (for a multivariate sample) of the square of the correlation between Y and the best linear predictor for Y in the overall population. Since the population value for R^2 will be zero if and only if the multiple regression coefficients are equal to zero, a test of the statistical significance of R^2 is the F test for the regression equation. R^2 and F are related (as given by the definition of R^2 and the F test in the analysis of variance table). It is easy to show that

$$R^2 = \frac{kF}{kF + n - k - 1}, \quad F = \frac{(n - k - 1)R^2}{k(1 - R^2)}. \qquad (11.12)$$

The multiple correlation coefficient thus has associated with it the same degrees of freedom as the F distribution: k and $n - k - 1$. Statistical significance testing for R^2 is based upon the statistical significance test of the F statistic of regression.

At significance level α reject the null hypothesis of the no linear association between Y and $X_1, \ldots, X_k$ if

$$R^2 \geq \frac{kF_{k,n-k-1,1-\alpha}}{kF_{k,n-k-1,1-\alpha} + n - k - 1},$$

where $F_{k,n-k-1,1-\alpha}$ is the $1 - \alpha$ percentile for the F distribution with k and $n - k - 1$ degrees of freedom.

For any of the examples considered above, it is easy to compute R^2. Consider the last part of the last example, the active female exercise test data where duration, VO_2 MAX and the maximal heart rate were used to "explain" the subject's age. The value for R^2 is given by $2256.97/4399.16 = 0.51$, that is, 51% of the variability in Y(age) is explained by the three explanatory or predictor variables. The multiple regression coefficient, or positive square root, is 0.72.

The multiple regression coefficient has the same limitations as the simple correlation coefficient. In particular, if the explanatory variables take values picked by an experimenter and the variability about the regression line is constant, the value of R^2 may be increased by taking a large spread among the explanatory variables $X_1, \ldots, X_k$. The value for R^2, or R, may be presented when the data do *not* come from a multivariate sample; in this case it is an indicator of the amount of the variability in the dependent variable explained by the covariates. *It is then necessary to remember that the values do not reflect something inherent in the relationship between the dependent and independent variables, but rather reflect a quantity that is subject to change according to the value selection for the independent or explanatory variables.*

Example 11.4. Gardner [1973] considered using environmental factors to explain and predict mortality. He studied the relationship between a number of socioenvironmental factors and mortality in county boroughs of England and Wales. Rates for all sizeable causes of death in the ages 45–74 were considered separately. Four social and environmental factors were used as independent variables in a multiple regression analysis of each death rate. The variables included social factor score, "domestic" air pollution, latitude, and the level of water calcium. He then examined the residuals from this regression model, and considered relating the residual variability to other environmental factors. The only factors showing sizeable and consistent correlation were the long period average rainfall and latitude, with rainfall being the more significant variable for all causes of death. When rainfall was included as a fifth regressor variable, no new factors were seen to be important. Tables 11.3 and 11.4 give the regression coefficients, not for the raw variables but for standardized variables.

These data were developed for 61 English county boroughs, and then used to predict the values for 12 other boroughs. In addition to taking the square of the multiple correlation coefficient for the data used for the prediction, the correlation between observed and predicted values for *the other 12 boroughs* were calculated. Table 11.4 gives the results of these data.

This example has several striking features. Note that Gardner tried to fit a variety of models. This is often done in multiple regression analysis, and we discuss it in more detail in Section 11.5. Also note the dramatic drop (!) in the amount of variability in the death rate that can be explained between the data used to fit the model and the data used to predict values for other boroughs. This may be due to several sources.

Table 11.3. Multiple Regression[a] of Local Death Rates on Five Socioenvironmental Indices in the County Boroughs.

Sex and Age Group	Period	Social Factor Score	"Domestic" Air Pollution	Latitude	Water Calcium	Long Period Average Rainfall
Males	1948–1954	0.16	0.48***	0.10	−0.23	0.27***
45–64	1958–1964	0.19*	0.36***	0.21**	−0.24**	0.30***
Males	1950–1954	0.24*	0.28*	0.02	−0.43***	0.17
65–74	1958–1964	0.39**	0.17	0.13	−0.30**	0.21
Females	1948–1954	0.16	0.20	0.32**	−0.15	0.40***
45–64	1958–1964	0.29*	0.12	0.19	−0.22*	0.39***
Females	1950–1954	0.39***	0.02	0.36***	−0.12	0.40***
65–74	1958–1964	0.40***	−0.05	0.29***	−0.27**	0.29**

[a]A standardized partial regression coefficients given, that is, the variables are reduced to the same mean (zero) and variance (one) to allow values for the five socioenvironmental indices in each cause of death to be compared. The higher of two coefficients is not necessarily the more significant statistically.
*$p < 0.05$; **$p < 0.01$; ***$p < 0.001$.

Table 11.4. Results of Using Estimated Multiple Regression Equations from 61 County Boroughs to Predict Death Rates in 12 Other County Boroughs.

Sex and Age Group	Period	$\widehat{R}^2$	r^{2a}
Males	1948–1954	0.80	0.12
45–64	1958–1964	0.84	0.26
Males	1950–1954	0.73	0.09
65–74	1958–1964	0.76	0.25
Females	1948–1954	0.73	0.46
45–64	1958–1964	0.72	0.48
Females	1950–1954	0.80	0.53
65–74	1958–1964	0.73	0.41

[a] r is the correlation coefficient in the second sample between the predicted value of the dependent variable and its observed value.

First, the value of R^2 is always nonnegative and can only be zero if variability in Y can be perfectly predicted. In general R^2 tends to be too large. There is a value called *adjusted R^2* which we will denote by R_a^2 which takes into account this effect. This estimate of the population R^2, is given by the following equation:

$$R_a^2 = 1 - (1 - R^2) \frac{n-1}{n-k}.$$ (11.13)

For the Gardner data on males from 45 to 64 during the time period 1948–1954, the adjusted R^2 value is given by

$$R_a^2 = 1 - (1 - 0.80) \frac{61-1}{61-5} = 0.786.$$

We see that this does not account for much of the drop. Another possible effect may be related to the fact that Gardner tried a variety of models; in considering multiple models one may get a very good fit just by chance because of the many possibilities tried. The most likely explanation, however, is that a model fitted in one environment and then used in another setting may lose much predictive power because *variables important to one setting may not be as important in another setting.* As another possibility there could be an important variable which is not even known by the person analyzing the data. If this variable varies between the original data set and the new data set, where one desires to predict, extreme drops in predictive power may occur. As a general rule of thumb, *the more complex the model, the less transportable the model is in time and/or space.* This example illustrates that whenever possible, when fitting a multivariate model including multiple linear regression models, if the model is to be used for prediction it is useful to try the model on an independent sample. Great degradation in predictive power is not an unusual occurrence. □

In one example above, we had the peculiar situation that the relationship between the dependent variable age and independent variables duration, VO₂ MAX, and maximal heart rate were such that there was a very highly statistically significant relationship between the regression equation and the dependent variable, but at the 5% significance level we were not able to demonstrate the statistical significance of the regression coefficients of any of the three independent variables. That is, we could not demonstrate that any of the three predictor variables actually added statistically significant information to the prediction. We mentioned that this may occur because of high correlations between variables. This implies that they contain much of the same predictive information. In this case estimation of their individual contribution is very difficult. This idea may be expressed quantitatively by examining the variance of the estimate for a regression coefficient, say β_j. This variance can be shown to be

$$\text{var}(b_j) = \frac{\sigma^2}{[x_j^2](1 - R_j^2)}. \tag{11.14}$$

In this formula σ^2 is the variance about the regression line; $[x_j^2]$ is the sum of the squares of the difference between the observed values for the jth predictor variable and its mean (this bracket notation was used in Chapter 9). R_j^2 is the square of the multiple correlation coefficient between X_j as dependent variable and the other predictor variables as independent variables. Note that if there is only one predictor R_j^2 is zero; in this case the formula reduces to the formula of Chapter 9 for simple linear regression. On the other hand, if X_j is very highly correlated with other predictor variables, we see that the variance of the estimate of b_j increases dramatically. This again illustrates the phenomenon of *collinearity*. A good discussion of the problem may be found in Mason [1975], as well as Hocking [1976].

In certain circumstances, more than one multiple regression coefficient may be considered at one time. It is then necessary to have notation which explicitly gives the variables used.

Definition 11.6. The multiple correlation coefficient of Y with the set of variables $X_1, \ldots, X_k$ is denoted by

$$R_{Y(X_1, \ldots, X_k)}$$

when it is necessary to explicitly show the variables used in the computation of the multiple correlation coefficient.

11.3.2 The Partial Correlation Coefficient

When two variables are linearly related we have used the correlation coefficient as a measure of the amount of association between the two variables. However, we might suspect that a relationship between two variables occurred because they are both related to another variable. For example, there may be a positive correlation between the density of hospital beds in a geographical area and an index of air pollution. We probably would not conjecture that the number of hospital beds increased the air pollution although the opposite could conceivably be true. More likely, both are more immediately related to population density in the area; thus we might like to examine the relationship between the density of hospital beds and air pollution after controlling or adjusting for the population density. We have previously seen examples where we controlled or adjusted for a variable. As one example this was done in the combining of two-by-two tables, using the different strata as an adjustment. A partial correlation coefficient is designed to measure the amount of linear relationship between two variables after adjusting for or controlling for the effect of some set of variables. The method is appropriate when there are linear relationships between the variables and certain model assumptions such as normality hold.

Definition 11.7. The *partial correlation coefficient* of X and Y adjusting for the variables $X_1, \ldots, X_k$ is denoted by: $\rho_{X,Y \cdot X_1, \ldots, X_k}$. The sample partial correlation coefficient of X and Y adjusting for $X_1, \ldots, X_k$ is denoted by $r_{X,Y \cdot X_1, \ldots, X_k}$. The partial correlation coefficient is the correlation of Y minus its best linear predictor in terms of the X_i variables with X minus its best linear predictor in terms of the X_i variables. That is, letting $\widehat{Y}$ be a predicted value of Y from multiple linear regression of Y on $X_1, \ldots, X_k$ and letting $\widehat{X}$ be the predicted value of X from the multiple linear regression of X on $X_1, \ldots, X_k$, the partial correlation coefficient is the correlation of $X - \widehat{X}$ and $Y - \widehat{Y}$.

If all of the variables concerned have a multivariate normal distibution, then the partial correlation coefficient of X and Y adjusting for $X_1, \ldots, X_k$ is the correlation of X and Y conditionally upon knowing the values of $X_1, \ldots, X_k$. The conditional correlation of X and Y in this multivariate normal case is the same for each fixed set of the values for $X_1, \ldots, X_k$ and is equal to the partial correlation coefficient.

The statistical significance of the partial correlation coefficient is equivalent to testing the statistical significance of the regression coefficient for X if a multiple regression is performed with Y as a dependent variable with $X, X_1, \ldots, X_k$ as the independent or explanatory variables. In the next section on nested hypotheses, we will consider such significance testing in more detail.

Partial regression coefficients are usually estimated by computer, but there is a simple formula for the case of three variables. Let us consider the partial correlation coefficient of X and Y adjusting for a variable Z. In terms of the correlation coefficients for the pairs of variables, the partial correlation coefficient in the population and its estimate from the sample are given by the following two equations.

$$\rho_{X,Y \cdot Z} = \frac{\rho_{X,Y} - \rho_{X,Z}\rho_{Y,Z}}{\sqrt{\left(1 - \rho_{X,Z}^2\right)\left(1 - \rho_{Y,Z}^2\right)}},$$

$$r_{X,Y \cdot Z} = \frac{r_{X,Y} - r_{X,Z}r_{Y,Z}}{\sqrt{\left(1 - r_{X,Z}^2\right)\left(1 - r_{Y,Z}^2\right)}}. \tag{11.15}$$

We illustrate the effect of the partial correlation coefficient by the exercise data for active females discussed above. We know that age and duration are correlated. For the data above the correlation coefficient is -0.68913. Let us consider how much of the linear relationship between age and duration is left if we adjust out the effect of the oxygen consumption, VO_2 MAX, for the same data set. The correlation coefficients for the sample are as follows:

$$r_{\text{AGE,DURATION}} = -0.68913,$$

$$r_{\text{AGE,VO}_2 \text{ MAX}} = -0.65099,$$

$$r_{\text{DURATION,VO}_2 \text{ MAX}} = 0.78601.$$

The partial correlation coefficient of age and duration adjusting VO_2 MAX using the equation above is estimated by

$$r_{\text{AGE,DURATION.VO}_2\ \text{MAX}} = \frac{-0.68913 - (-0.65099) \times (-0.78601)}{\sqrt{(1 - (-0.65099)^2)(1 - (0.78601)^2)}} = -0.37812.$$

If we consider the corresponding multiple regression problem with a dependent variable of age and independent variables duration and VO_2 MAX, the t statistic for duration is -2.58. The two-sided 0.05 critical value is 2.02, while the critical value at significance level 0.01 is 2.70. Thus, we see that the p-value for statistical significance of this partial correlation coefficient is between 0.05 and 0.01.

11.3.3   The Partial Multiple Correlation Coefficient

Occasionally one wants to examine the linear relationship, that is the correlation, between one variable, say Y, and a second group of variables, say $X_1, \ldots, X_k$ while adjusting or controlling for a third set of variables $Z_1, \ldots, Z_p$. If it were not for the Z_i variables, we would simply use the multiple correlation coefficient to summarize the relationship between Y and the X variables. The approach taken is the same as for the partial correlation coefficient. First subtract out for each variable its best linear predictor in terms of the Z_i's. From the remaining residual values compute the multiple correlation between the Y residuals and the X residuals. More formally, we have the following definition.

Definition 11.8.   For each variable let a $\widehat{}$ denote the least squares linear predictor for the variable in terms of the quantities $Z_1, \ldots, Z_p$. The best linear predictor for a sample results from the multiple regression of the variable upon the independent variables $Z_1, \ldots, Z_p$. The *partial multiple correlation coefficient* between the variable Y and the variables $X_1, \ldots, X_k$ adjusting for $Z_1, \ldots, Z_p$ is the multiple correlation between the variable $Y - \widehat{Y}$ and the variables $X_1 - \widehat{X}_1, \ldots, X_k - \widehat{X}_k$. The partial multiple correlation coefficient of Y and $X_1, \ldots, X_k$ adjusting for $Z_1, \ldots, Z_p$ is denoted by

$$R_{Y(X_1, \ldots, X_k).Z_1, \ldots, Z_p}.$$

A significance test for the partial multiple correlation coefficient is discussed in Section 11.4 on nested hypotheses. The coefficient is also called the *multiple partial correlation coefficient*.

11.3.4   Canonical Correlation

The multiple correlation coefficient examined the linear relationship between one variable and a set of variables. In canonical correlation we want to examine the relationship between two distinct sets of variables. To understand that what follows is a generalization of the multiple correlation coefficient, it is useful to characterize in another manner the least square predictor for Y in terms of a set of variables. The multiple linear regression can be characterized as the linear combination of the independent variables that is most highly correlated with the dependent variable. This answer is not unique since multiplication by a nonzero constant would give the same positive correlation. However, speaking loosely we may say that the least squares estimate is the combination of independent variables most highly correlated with the dependent variable.

Similarly, we proceed to look at the linear relationship between two sets of variables. Let the sets of variables be $X_1, \ldots, X_k$ and $Y_1, \ldots, Y_p$.

Definition 11.9. The *first canonical correlation* is the largest correlation that can result from a linear combination of the X variables with linear combination of the Y variables. The linear combinations involved are usually unique up to multiplication by the constant. These linear combinations are usually normalized to have a variance of 1. The two resulting linear combinations, one of the X variables and one of the Y variables, are called the *first canonical variables* or the *first pair of canonical variables*.

Since one candidate for the canonical variables would consist of one variable from the first set and any linear combination from the second set it follows that the multiple correlation coefficient between a fixed variable in the first set and the set of variables in the second set must be less than or equal to the canonical correlation coefficient. More generally, the first canonical correlation coefficient gives an upper bound on the multiple correlation coefficient between one variable from one of the subsets and the variables in the second set or any subset of variables from the second set. Thus, if the first canonical correlation is small, there is no use searching in more detail for strong relationships between a fixed variable in one set and the other set of variables.

Even after determining the first canonical correlation and canonical variables, there still may remain some more residual relationship between two sets of variables. One can subtract off from both sets of variables their best linear predictors in terms of the first canonical variables. After doing this, the linear combinations between the two subsets of variables that have the highest correlation are called the *second pair of canonical variables*; their correlation is a *second canonical correlation coefficient*. After this step the linear combinations of the variables in each subset which give the two second canonical variables are uncorrelated with the first two canonical variables. This procedure may be continued until there is no correlation left between the two sets of variables. More detail on canonical correlation is given in most textbooks of multivariate statistical analysis including Morrison [1976] and Timm [1975].

11.4 NESTED HYPOTHESES

In the second part of Example 11.3, we saw a multiple regression equation where we could not show the statistical significance of individual regression coefficients. This raised the possibility of reducing the complexity of the regression equation by eliminating one or more variables from the predictive equation. When we consider such possibilities, we are considering what is called a *nested hypothesis*. In this section we discuss nested hypotheses in the multiple regression setting. First we define nested hypotheses; we then introduce notation for nested hypotheses in multiple regression. In addition to notation for the hypotheses we need notation for the various sums of squares involved. This leads to appropriate F-statistics for testing nested hypotheses. After we understand nested hypotheses, we shall see how to construct F-tests for the partial correlation coefficient and the partial multiple correlation coefficient. Furthermore, the ideas of nested hypotheses will be used below in stepwise regression.

Definition 11.10. One hypothesis, say hypothesis H_1, is *nested* within a second hypothesis, say hypothesis H_2, if whenever the hypothesis H_1 is true, then the hypothesis H_2 is also true. That is to say, the hypothesis H_1 is a special case of the hypothesis H_2.

In our multiple regression situation most nested hypotheses will consist of specifying that some subset of the regression coefficients β_i have the value 0. For example, the larger first hypothesis might be H_2 as follows:

$$H_2 : Y = \alpha + \beta_1 X_1 + \cdots + \beta_k X_k + \epsilon,$$

$$\epsilon \sim N(0, \sigma^2).$$

The smaller (nested) hypothesis H_1 might specify that some subset of the β's, for example, the last $k - j$ betas corresponding to variables $X_{j+1}, \ldots, X_k$, are all 0. We denote this hypothesis by H_1.

$$H_1 : Y = \alpha + \beta_1 X_1 + \cdots + \beta_j X_j + \epsilon,$$

$$\epsilon \sim N(0, \sigma^2).$$

In other words, H_2 holds *and*

$$\beta_{j+1} = \beta_{j+2} = \cdots = \beta_k = 0.$$

A more abbreviated method of stating the hypothesis is the following:

$$H_1 : \beta_{j+1} = \beta_{j+2} = \cdots = \beta_k = 0 \mid \beta_1, \ldots \beta_j.$$

In order to test such nested hypotheses, it will be useful to have a notation for the regression sum of squares for any subset of independent variables in the regression equation. If variables $X_1, \ldots, X_j$ are used as explanatory or independent variables in a multiple regression equation for Y, we denote the regression sum of squares by

$$\text{SS}_{\text{REG}}(X_1, \ldots, X_j).$$

We denote the residual sum of squares (that is the total sum of squares of the dependent variable Y about its mean minus the regression sum of squares) by

$$\text{SS}_{\text{RESID}}(X_1, \ldots, X_j).$$

If we use more variables in a multiple regression equation, the sum of squares explained by the regression can only increase since one potential predictive equation would set all the regression coefficients for the new variables equal to 0. This will almost never occur in practice, if for no other reason than the random variability of the error term allows the fitting of extra regression coefficients to explain a little more of the variability. The increase in the regression sum of squares, however, may be due to chance. The F-test used to test nested hypotheses looks at the increase in the regression sum of squares and examines whether it is plausible that the increase could occur by chance. Thus we need a notation for the increase in the regression sum of squares. This notation follows:

$$\text{SS}_{\text{REG}}(X_{j+1}, \ldots, X_k \mid X_1, \ldots, X_j) = \text{SS}_{\text{REG}}(X_1, \ldots, X_k) - \text{SS}_{\text{REG}}(X_1, \ldots, X_j).$$

This is the sum of squares attributable to $X_{j+1}, \ldots, X_k$ after fitting the variables $X_1, \ldots, X_j$. With this notation we may proceed to the F-test of the hypothesis that adding the last $k - j$ variables does not increase the sum of squares a statistically significant amount beyond the regression sum of squares attributable to $X_1, \ldots, X_k$.

Assume a regression model with k predictor variables, $X_1, \ldots, X_k$. The F-statistic for testing the hypothesis

$$H_1 : \beta_{j+1} = \cdots = \beta_k = 0 \mid \beta_1, \ldots, \beta_j$$

is

$$F = \frac{\text{SS}_{\text{REG}}(X_{j+1}, \ldots, X_k \mid X_1, \ldots, X_j)/(k - j)}{\text{SS}_{\text{RESID}}(X_1, \ldots, X_k)/(n - k - 1)}.$$

Under H_1, F has an F-distribution with $k - j$ and $n - k - 1$ degrees of freedom. Reject H_1 if $F > F_{k-j,n-k-1,1-\alpha}$, the $1 - \alpha$ percentile of the F-distribution.

The partial correlation coefficient is related to the sums of squares as follows. Let X be a predictor variable in addition to $X_1, \ldots, X_k$.

$$r^2_{X,Y.X_1,\ldots,X_k} = \frac{\text{SS}_{\text{REG}}(X \mid X_1, \ldots, X_k)}{\text{SS}_{\text{RESID}}(X_1, \ldots, X_k)}. \qquad (11.16)$$

The sign of $r_{X,Y.X_1,\ldots,X_k}$ is the same as the sign of the X regression coefficient when Y is regressed on $X, Y.X_1, \ldots, X_k$. The F-test for statistical significance of $r_{X,Y.X_1,\ldots,X_k}$ uses

$$F = \frac{\text{SS}_{\text{REG}}(X \mid X_1, \ldots, X_k)}{\text{SS}_{\text{RESID}}(X, X_1, \ldots, X_k)/(n - k - 2)}. \qquad (11.17)$$

Under the null hypothesis that the partial correlation is zero (or equivalently that $\beta_X = 0 \mid \beta_1, \ldots, \beta_k$) F has an F distribution with 1 and $n - k - 2$ degrees of freedom. F is sometimes called the *partial F statistic*. The t-statistic for the statistical significance of β_X is related to F by

$$t^2 = \frac{\beta_X^2}{\text{SE}(\beta_X)^2} = F.$$

Similar results hold for the partial multiple correlation coefficient. The correlation is always positive and its square is related to the sums of squares by

$$R^2_{Y(X_1,\ldots,X_k).Z_1,\ldots,Z_p} = \frac{\text{SS}_{\text{REG}}(X_1, \ldots, X_k \mid Z_1, \ldots, Z_p)}{\text{SS}_{\text{RESID}}(Z_1, \ldots, Z_p)}. \qquad (11.18)$$

The F-test for statistical significance uses the test statistic

$$F = \frac{\text{ss}_{\text{REG}}(X_1,\ldots,X_k \mid Z_1,\ldots,Z_p)/k}{\text{ss}_{\text{RESID}}(X_1,\ldots,X_k,Z_1,\ldots,Z_p)/(n-k-p-1)}. \qquad (11.19)$$

Under the null hypothesis that the population partial multiple correlation coefficient is zero, F has an F distribution with k and $n - k - p - 1$ degrees of freedom. This test is equivalent to testing the nested multiple regression hypothesis:

$$H : \beta_{X_1} = \cdots = \beta_{X_k} = 0 \mid \beta_{Z_1},\ldots,\beta_{Z_p}.$$

Note that in each case above, the contribution to R^2 after adjusting for additional variables is the increase in the regression sum of squares divided by the residual sum of squares after taking the regression on the adjusting variables. The corresponding F-statistic has a numerator degrees of freedom equal to the number of added predictive variables or equivalently the number of additional parameters being estimated. The denominator degrees of freedom are equal to the number of observations minus the total number of parameters estimated. The reason for the -1 in the denominator degrees of freedom in Equation (11.19) is the estimate of the constant in the regression equation.

Example 11.5. We illustrate some of these ideas by returning to the 43 active females who were exercise tested as discussed in Example 11.3. Let us compute the following quantities:

$$r_{\text{VO}_2 \text{ MAX,DURATION.AGE}}$$

$$R^2_{\text{AGE(VO}_2 \text{ MAX,HEART RATE).DURATION}}$$

To examine the relationship between VO$_2$ MAX and duration adjusting for age, let duration be the dependent or response variable. Suppose we then run two multiple regressions; one predicting duration using only age as the predictive variable and a second regression using both age and VO$_2$ MAX as the predictive variable. These runs give the following data:

Y = duration, X_1 = age:

Covariate or Constant	b_i	SE(b_i)	$t\,(t_{41,0.975} \doteq 2.02)$
Age	−5.208	0.855	−6.09
Constant	749.975	39.564	

ANALYSIS OF VARIANCE TABLE

Source	DF	SS	MS	$F\,(F_{1,41,0.95} \doteq 4.08)$
Regression of Duration on age	1	119324.47	119324.47	37.08
Residual	41	131935.95	3217.95	
Total	42	251260.42		

$Y = $ duration, $X_1 = $ age, $X_2 = $ VO$_2$ MAX:

Covariate or Constant	b_i	SE(b_i)	t ($t_{40,0.975} \doteq 2.09$)
Age	−2.327	0.901	−2.583
VO$_2$ MAX	9.151	1.863	4.912
Constant	354.072	86.589	

ANALYSIS OF VARIANCE TABLE

Source	DF	SS	MS	F ($F_{2,40,0.95} \doteq 3.23$)
Regression of duration on age and VO$_2$ MAX	2	168961.48	84480.74	41.06
Residual	40	82298.94	2057.47	
Total	42	251260.42		

Using Equation (11.16), we find the square of the partial correlation coefficient.

$$r^2_{\text{VO}_2 \text{ MAX,DURATION.AGE}} = \frac{168961.48 - 119324.47}{131935.95}$$

$$= \frac{49637.01}{131935.95}$$

$$= 0.376.$$

Since the regression coefficient for VO$_2$ MAX is positive (when regressed with age) having a value of 9.151 the positive square root gives r.

$$r_{\text{VO}_2 \text{ MAX,DURATION.AGE}} = +\sqrt{0.376} = 0.613.$$

To test the statistical significance of the partial correlation coefficient, Equation (11.17) gives

$$F = \frac{168961.48 - 119324.467}{82298.94/(43 - 1 - 1 - 1)} = 24.125.$$

Note that $t^2_{\text{VO}_2 \text{ MAX}} = 24.127 = F$ within round off error. As $F_{1,40,0.999} = 12.61$ this is highly significant ($p < 0.001$). In other words the duration of the treadmill test and the maximum oxygen consumption are significantly related even after adjustment for the subject's age.

Now we turn to the computation and testing of the partial multiple correlation coefficient. To use Equations (11.18) and (11.19), we need to regress age on duration, and also regress age on duration, VO$_2$ MAX, and the maximum heart rate. The analysis of variance tables follow.

Age regressed upon duration:

ANALYSIS OF VARIANCE TABLE

Source	DF	SS	MS	$F\,(F_{1,41,0.95} \doteq 4.08)$
Regression	1	2089.18	2089.18	37.08
Residual	41	2309.98	56.34	
Total	42	4399.16		

Age regressed upon duration, VO_2 MAX and maximum heart rate:

ANALYSIS OF VARIANCE TABLE

Source	DF	SS	MS	$F\,(F_{3,39,0.95} \doteq 2.85)$
Regression	3	2256.97	752.32	13.70
Residual	39	2142.19	54.93	
Total	42	4399.16		

From Equation (11.18)

$$R^2_{\text{AGE(VO}_2\text{ MAX,HEART RATE).DURATION}} = \frac{2256.97 - 2089.18}{2309.98}$$

$$= 0.0726$$

and $R = \sqrt{R^2} = 0.270$.
 The F-test, by Equation (11.19) is,

$$F = \frac{(2256.97 - 2089.18)/2}{2142.19/(43 - 2 - 1 - 1)} = 1.53.$$

As $F_{2,39,0.90} \doteq 2.44$, we have not shown statistical significance even at the 10% significance level. In words: VO_2 MAX and maximum heart rate have no more additional linear relationship with age, after controlling for the duration, than would be expected by chance variability.   □

11.5 SELECTING A "BEST" SUBSET OF EXPLANATORY VARIABLES

11.5.1 The Problem

Sometimes given a large number of potential explanatory variables, one can select a smaller subset that explains the variability in the dependent variable. We have seen examples above where it appears that one or more of the variables in a multiple regression do not contribute, beyond an amount consistent with chance, to the explanation of the variability in the dependent variable. Thus consider a response variable Y with a large number of potential predictor variables X_i. How should we choose a "best" subset of variables to explain the Y variability? This topic is addressed in this section. If we knew the number of predictor variables we wanted, we could use some criterion for the best subset. One natural criterion from the concepts already presented would be to choose the subset which gives the largest value for R^2. Even then the selection of the subset can be a formidable task. For example, suppose there are 30 predictor variables and a subset of ten variables is wanted; there are

$$\binom{30}{10} = 30,045,015$$

possible regression equations which have ten predictor variables. This is not a routinely manageable number even with modern high speed computers. Furthermore, in many instances, we will not know how many possible variables we should place into our prediction equation. If we consider all possible subsets of thirty variables there are over one billion possible combinations for the prediction. Thus once again one cannot examine all subsets. There has been much theoretical work on selecting the best subset according to some criteria; the algorithms allow one to find the best subset without explicitly looking at all of the possible subsets. Still, for large numbers of variables we need another procedure to select the predictive subset.

A further complication arises when we have a very large number of observations; then we may be able to show statistically that all of the potential predictor variables contribute additional information to explain the variability in the dependent variable Y. However, the large majority of the predictor variables may add so little to the explanation that we would prefer a much smaller subset which explains almost as much of the variability and gives a much simpler model. In general, simple models are desirable because they may be used more readily, and often when applied in a different setting, turn out to be more accurate than a model with a large number of variables.

In summary, the task before us in this section is to consider a means of choosing a subset of predictor variables from a pool of potential predictor variables.

11.5.2 Approaches to the Problem Which Consider All Possible Subsets of Explanatory Variables

We discuss two approaches and then apply both approaches to an example.

The first approach is based upon the following idea: if we have the appropriate predictive variables in a multiple regression equation, plus possibly some other variables that have no predictive power, then the residual mean square for the model will estimate σ^2 the variability about the true regression line. On the other hand, if we do not contain enough predictive variables, the residual mean square will contain additional variability due to the poor muliple regression fit and will tend to be too large. We want to use this fact to allow us to get some idea of the number of variables needed in the model. We do this in the following way. Suppose that we consider all possible predictions for some fixed number, say p, of the total possible number of predictor variables. Suppose that the correct predictive equation has a much smaller number of variables than p. Then when we look at all of the different subsets of p predictor variables, most of them will contain the *correct* variables for the predictive equation plus other variables that are not needed. In this case, the mean square residual will be an estimate of σ^2. If we average all of the mean square residuals for the equations with p variables, since most of them will contain the correct predictive variables, we should get an estimate fairly close to σ^2. We examine the mean square residuals by plotting the average mean square residuals for all the regression equations using p variables versus p. As p becomes large, this average value should tend to level off at the true residual variability. By drawing a horizontal line at approximately the value where things average out, we can get some idea of the residual variability. We would then search for a simple model which has approximately this asymptotic estimate of σ^2. That is, we expect a picture such as Figure 11.1.

The second approach is due to C.L. Mallows and is called Mallow's C_p statistic. In this case, let p equal the number of predictive variables in the model, *plus one*. This is a change from the previous paragraph, where p was the number of predictive variables. The switch to this notation is made because in the literature for Mallow's C_p, this is the value used. The statistic is as follows:

$$C_p(\text{model with } p - 1 \text{ explanatory variables}) =$$

$$\frac{\text{ss}_{\text{RESID}}(\text{model})}{\text{ms}_{\text{RESID}}(\text{using all possible predictors})} - (N - 2p),$$

where ms_{RESID}(using all possible predictors) is the residual mean square when the dependent variable Y is regressed upon all possible independent predictors; ss_{RESID}(model) is the residual sum of squares for the possible model being considered (this model uses $p - 1$ explanatory variables); N is the total number of observations; and p is the number of explanatory variables in the model plus one.

To use Mallow's C_p we compute the value of C_p for each possible subset of explanatory variables. The points (C_p, p) are then plotted for each possible model. The following facts about the C_p statistics are true.

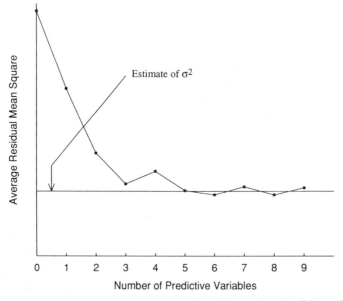

Figure 11.1. Average residual mean square as a function of the number of predicitve variables.

1. If the model fits, the expected value for each C_p is approximately p.

2. If C_p is larger than p, the difference, $C_p - p$, gives approximately the amount of bias in the sum of squares involved in the estimation. The bias occurs because the estimating predictive equation is not the true equation, and thus estimates something other than the correct Y value.

3. The value of C_p itself gives an overall estimate of the sum of the squares of the average difference between correct Y values and the Y values predicted from the model. This difference is composed of two parts, one part due to bias because the estimating equation is not correct (and cannot be correct if the wrong variables are included), and a second part because of variability in the estimate. If the expected value of Y may be modeled by a few variables, there is a cost to adding more variables to the estimation procedure. In this case, statistical noise enters into the estimation of the additional variables, so that by using the more complex estimated predictive equation, future predictions would be off by more.

4. Thus what we would like to look for in our plot is a value C_p which is close to the 45 degree line, $C_p = p$. Such a value would have a low bias. Further, we would like the value of C_p itself to be small, so that the total error sum of squares is not large. The nicest possible case occurs when we can more or less satisfy both demands at the same time.

5. If we have to choose between a C_p value, which is close to p, or one which is smaller but above p, we are choosing between an equation that has a small bias (when $C_p = p$), but in further prediction is likely to have a larger predictive error, and a second equation (the smaller value for C_p) which in the future prediction is more likely to be close to the true value but where we think that the estimated predictive equation is probably biased. Depending upon the use of the model, the trade-off between these two ills may or may not be clear cut.

Example 11.6. In this example, we return to the data of Cullen and van Belle used in Example 11.1. We shall consider the response variable, DPMA, which is the disintegrations per minute of lymphocytes after the surgery. The viability of the lymphocytes was measured in terms of the uptake of nutrients which were radioactively labeled. A large number of disintegrations per minute suggests a high cell division rate, and thus active lymphocytes. The potential predictive variables for explaining the variability in DPMA are trauma factor (as discussed previously), duration (as discussed previously), the disintegrations per minute before the surgery, labeled DPMB, and the lymphocyte count in thousands per cubic millimeter before the surgery, LYMPHB, as well as the lymphocyte count in thousands per cubic millimeter after the surgery, LYMPHA. Let these variables have the following labels:

$$Y = \text{DPMA}, \quad X_1 = \text{Duration}, \quad X_2 = \text{Trauma},$$
$$X_3 = \text{DPMB}, \quad X_4 = \text{LYMPHB}, \quad X_5 = \text{LYMPHA}.$$

Table 11.5 presents the results for the 32 possible regression runs using subsets of the five predictor variables. For each run the value of p, C_p, the residual mean square, the average residual mean square for runs with the same number of variables, the multiple R^2 and the adjusted R^2, R_a^2 are presented. For a given number of variables, the entries are ordered in terms of increasing values of C_p.

Table 11.5. Results from the 32 Regression Runs on the Anesthesia Data of Cullen and van Belle [1975].

Numbers of the Explanatory Variables in the Predictive Equation	p	C_p	Residual Mean Square	Residual Average Mean Square	R^2	R_a^2
None	1	60.75	4047	4047	0	0
3	2	5.98	1645		0.606	0.594
1		49.45	3578		0.142	0.116
2		57.12	3919	3476	0.060	0.032
4		60.48	4069		0.024	−0.005
5		62.70	4168		0.000+	−0.030
2,3	3	2.48	1444		0.664	0.643
1,3		2.82	1459		0.661	0.639
3,5		6.26	1617		0.624	0.600
3,4		6.91	1647		0.617	0.593
1,4		48.37	3549	2922	0.175	0.123
1,2		51.06	3672		0.146	0.093
1,5		51.43	3689		0.142	0.088
2,4		56.32	3914		0.090	0.033
2,5		59.10	4041		0.060	0.001
4,5		62.39	4192		0.024	−0.036
2,3,4	4	3.03	1422		0.680	0.648
1,3,4		3.32	1435		0.677	0.645
1,3,5		3.36	1438		0.676	0.645
2,3,5		3.52	1445		0.674	0.643
1,2,3		3.96	1466	2396	0.670	0.639
3,4,5		7.88	1651		0.628	0.592
1,2,4		50.03	3647		0.178	0.099
1,4,5		50.15	3653		0.177	0.097
1,2,5		52.98	3787		0.146	0.064
2,4,5		57.75	4013		0.096	0.008
1,2,3,4	5	4.44	1440		0.686	0.644
1,3,4,5		4.64	1450		0.684	0.642
2,3,4,5		4.69	1453	1913	0.683	0.641
1,2,3,5		4.83	1460		0.682	0.640
1,2,4,5		51.91	3763		0.180	0.070
1,2,3,4,5	6	6	1468	1468	0.691	0.637

Note several things in Table 11.5. For a fixed number, $p-1$, of predictor variables, if we look at the values for C_p, the residual mean square, R^2 and R_a^2, we see that as C_p increases, the residual mean square increases while R^2 and R_a^2 decrease. This relationship is a mathematical fact. Thus, if we know how many predictor variables, p, we want in our equation, any of the following six criteria for the best subset of predictor variables are equivalent:

1. Pick the predictive equation with a minimum value of C_p.
2. Pick the predictive equation with the minimum value of the residual mean square.
3. Pick the predictive equation with the maximum value of the multiple correlation coefficient, R^2.
4. Pick the predictive equation with the maximum value of the adjusted multiple correlation coefficient, R_a^2.
5. Pick the predictive equation with a maximum sum of squares due to regression.
6. Pick the predictve equation with the minimum sum of squares for the residual variability.

The C_p data are more easily assimilated if we plot them. Figure 11.2 is a C_p plot for these data. The line, $C_p = p$, is drawn for reference. Recall that points near this line have little bias in terms of the fit of the model; for points above this line we have biased estimates of the regression equation. We see that there are a number of models which have little bias. All things being equal, we prefer as small a C_p value as possible, since this is an estimate of the amount of variability between the true values and predicted values, which takes into account two components, the bias in the estimate of the regression line as well as the residual variability due to estimation. For this plot, we are in the fortunate position of the lowest C_p value showing no bias. In addition there are a minimal number of variables involved. This point is circled, and going back to Table 11.5, corresponds to a model with $p = 3$, that is 2 predictor

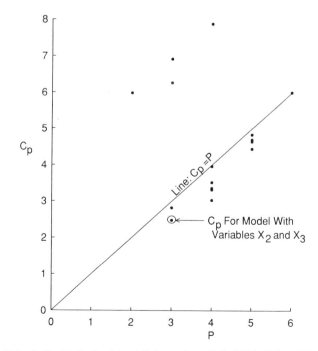

Figure 11.2. Mallow's C_p plot for the data of Cullen and van Belle [1975]. Only points with $C_p < 8$ are plotted.

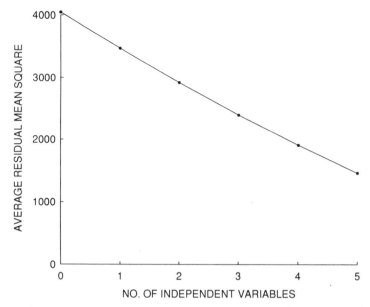

Figure 11.3. Average mean square plot for the Cullen and van Belle data [1975].

variables. They are variables 2 and 3, the trauma variable, as well as DPMB, the lymphocyte count in thousands per cubic millimeters, before the surgery. This is the model we would select using Mallow's C_p approach.

We now turn to the average residual mean square plot to see if that would help us to decide how many variables to use. Figure 11.3 gives this plot. We can see that this plot does not level out, but decreases until we have five variables. Thus this plot does not help us to decide on the number of variables we might consider in the final equation. If we look at Table 11.5 we can see why this happens. Since the final model has two predictive variables, even with three variables many of the subsets, namely four, do not include the most predictive variable, variable number three, and thus have very large mean squares. We have not considered enough variables in the model above and beyond the final model for the curve to level out. With a relatively small number of potential predictor variables, five in this model, the average residual mean square plot is usually not useful.   □

Suppose we have too many predictor variables to consider all combinations. Or, suppose we are worried about the problem of looking at the huge number of possible combinations, because we feel that the multiple comparisons may allow random variability to have too much effect. In this case, how might we proceed? The next section discusses one approach to this problem.

11.5.3 Stepwise Procedures

In this section we consider building a multiple regression model variable by variable. Let us start at the first step. Suppose that we have a dependent variable Y and a set of potential predictor variables, X_i. Suppose we try to explain the variability in Y

by choosing only one of the predictor variables. Which would we want? It is natural to choose the variable that has the largest squared correlation with the dependent variable Y. Because of the relationships among the sums of squares, this is equivalent to the following:

Step 1

1. Choose i to maximize r^2_{Y,X_i}.
2. Choose i to maximize $ss_{REG}(X_i)$.
3. Choose i to minimize $ss_{RESID}(X_i)$.

By renumbering our variables if necessary we can assume that the variable we picked was X_1. Now suppose that we want to add one more variable, say X_i, to X_1, to give us as much predictive power as possible. Which variable shall we add? Again we would like to maximize the correlation between Y and the predicted value of Y, $\widehat{Y}$; equivalently we would like to maximize the multiple correlation coefficient squared. Because of the relationships among the sums of squares, this is equivalent to any of the following at this next step:

Step 2: X_1 is in the model, we now find X_i ($i \neq 1$)

1. Choose i to maximize $R^2_{Y(X_1,X_i)}$.
2. Choose i to maximize $r^2_{Y,X_i.X_1}$.
3. Choose i to maximize $ss_{REG}(X_1, X_i)$.
4. Choose i to maximize $ss_{REG}(X_i \mid X_1)$.
5. Choose i to minimize $ss_{RESID}(X_1, X_i)$.

Our stepwise regression proceeds in this manner. Suppose that j variables have entered. By renumbering our variables if necessary, we can assume without loss of generality that the variables that have entered the predictive equation are $X_1, \ldots, X_j$. If we are to add one more variable to the predictive equation, which variable might we add? As before, we would like to add the variable which makes the correlation between Y and the predictor variables as large as possible. Again, because of the relationships between the sums of squares, this is equivalent to any of the following:

Step $j + 1$: $X_1, \ldots, X_j$ are in the model, we want X_i ($i \neq 1, \ldots, j$)

1. Choose i to maximize $R^2_{Y(X_1,\ldots,X_j,X_i)}$.
2. Choose i to maximize $r^2_{Y,X_i.X_1,\ldots,X_j}$.
3. Choose i to maximize $ss_{REG}(X_1, \ldots, X_j, X_i)$.
4. Choose i to maximize $ss_{REG}(X_i \mid X_1, \ldots, X_j)$.
5. Choose i to minimize $ss_{RESID}(X_1, \ldots, X_j, X_i)$.

If we continue in this manner, eventually we will use all of the potential predictor variables. Recall that our motivation was to select a simple model. Thus we would like a small model; this means that we would like to stop at some step before we have included all of our potential predictor variables. How long shall we go on including predictor variables in this model? There are several mechanisms for stopping. We present the most widely used stopping rule. We would not like to add a new variable if we cannot show statistically that it adds to the predictive power. That is, if in the presence of the other variables already in the model, there is no statistically significant relationship between the response variable and the next variable to be added we will stop adding new predictor variables. Thus, the most common method of stopping is to test the significance of the partial correlation of the next variable and the response variable Y after adjusting for the previously entered variables. We use the partial F test as discussed above. Commonly, the procedure is stopped when the p-value for the F level is greater than some fixed level, often the fixed level is taken to be 0.05. This is equivalent to testing the statistical significance of the partial correlation coefficient. The partial F statistic in the context of regression analysis is also often called the F *to enter*, since the value of F, or equivalently its p-value, is used as a criteria for entering the equation.

Since the F statistic always has numerator degrees of freedom 1 and the denominator degrees of freedom $n - j - 2$, and usually n is much larger than j, the appropriate critical value is effectively the F critical value with 1 and infinity degrees of freedom. For this reason, rather than using a p-value, often the entry criteria is to enter variables as long as the F statistic itself is greater than some fixed amount.

Summarizing, we stop when:

1. The p-value for $r^2_{Y,X_i \cdot X_1,\ldots,X_j}$ is greater than a fixed level.

2. The partial F statistic

$$\frac{\text{SS}_{\text{REG}}(X_i \mid X_1,\ldots,X_j)}{\text{SS}_{\text{RESID}}(X_1,\ldots,X_j,X_i)/(n - j - 2)}$$

is less than some specified value, or its p-value is greater than some fixed level.

All of this is summarized in Table 11.6. We illustrate this by an example.

Example 11.7: Stepwise Multiple Linear Regression. Consider the active female exercise data used above. We shall perform a stepwise regression with VO_2 MAX as the dependent variable and duration, maximum heart rate, age, height, and weight as potential independent variables. Table 11.7 contains a portion of the BMDP computer output for this run.

The 0.05 F critical value with degrees of freedom 1 and 42 is approximately 4.07. Thus at step 0 duration, maximum heart rate, and age are all statistically significantly related to the dependent variable VO_2 MAX.

We see this by examining the F to enter column in the output from step 0. This is the F statistic for the square of the correlation between the individual variable and the dependent variable. In step 0 up on the left, we see the analysis of variance table with only the constant coefficient. Under partial correlation we have the correlation between each individual variable and the dependent variable. At the first step, the computer program scans the possible predictor variables to see which one has the highest absolute value of the correlation with the dependent variable. This is equivalent to choosing the largest F to enter. We see that this variable is duration. In step 1, duration has entered the predictive equation. Up on the left, we see the multiple R which in this case is simply the correlation between the VO_2 MAX and duration

Table 11.6. Stepwise Regression Procedure (Forward) Selection for p Variable Case.

Step	Variable Entered[a]	Intercept and Slopes Calculated[b]	Total ss Attributable to Regression	Contribution of Entered Variable to Regression	F-Ratio to Test Significance of Entered Variable
1	X_1	$a^{(1)}, b_1^{(1)}$	$\text{SS}_{\text{REG}}(X_1)$	$\text{SS}_{\text{REG}}(X_1)$	$\dfrac{\text{ss}(X_1)(n-2)}{\text{SS}_{\text{RESID}}(X_1)} = F_{1,n-2}$
2	X_2	$a^{(2)}, b_1^{(2)}, b_2^{(2)}$	$\text{SS}_{\text{REG}}(X_1, X_2)$	$\text{SS}_{\text{REG}}(X_2 \mid X_1)$	$\dfrac{\text{ss}(X_2 \mid X_1)(n-3)}{\text{SS}_{\text{RESID}}(X_1, X_2)} = F_{1,n-3}$
3	X_3	$a^{(3)}, b_1^{(3)}, b_2^{(3)}, b_3^{(3)}$	$\text{SS}_{\text{REG}}(X_1, X_2, X_3)$	$\text{SS}_{\text{REG}}(X_3 \mid X_1, X_2)$	$\dfrac{\text{ss}(X_3 \mid X_1, X_2)(n-4)}{\text{SS}_{\text{RESID}}(X_1, X_2, X_3)} = F_{1,n-4}$
$\cdots$	$\cdots$	$\cdots$	$\cdots$	$\cdots$	$\cdots$
j	X_j	$a^{(j)}, b_1^{(j)}, b_2^{(j)}, \ldots, b_j^{(j)}$	$\text{SS}_{\text{REG}}(X_1, X_2, \ldots, X_j)$	$\text{SS}_{\text{REG}}(X_j \mid X_1, \ldots, X_{j-1})$	$\dfrac{\text{ss}(X_j \mid X_1, \ldots, X_{j-1})(n-j-1)}{\text{SS}_{\text{RESID}}(X_1, \ldots, X_j)} = F_{1,n-j-1}$
$\cdots$	$\cdots$	$\cdots$	$\cdots$	$\cdots$	$\cdots$
p	X_p	$a^{(p)}, b_1^{(p)}, b_2^{(p)}, \ldots, b_p^{(p)}$	$\text{SS}_{\text{REG}}(X_1, X_2, \ldots, X_p)$	$\text{SS}_{\text{REG}}(X_p \mid X_1, \ldots, X_{p-1})$	$\dfrac{\text{ss}(X_p \mid X_1, \ldots, X_p)(n-p-1)}{\text{SS}_{\text{RESID}}(X_1, \ldots, X_p)} = F_{1,n-p-1}$

[a]To simplify notation, variables are labeled by the step at which they entered the equation.
[b]The superscript notation indicates that the estimate of α changes from step to step, as well as the estimates of $\beta_1, \beta_2, \ldots, \beta_{p-1}$.

Table 11.7. Stepwise Multiple Linear Regression for the Data of Example 11.3.

```
STEP NO.    0
----------------
STD. ERROR OF EST.    4.9489

ANALYSIS OF VARIANCE
                        SUM OF SQUARES    DF    MEAN SQUARE


            RESIDUAL    1028.6670         42    24.49208

                        VARIABLES IN EQUATION FOR VO2MAX

                            STD. ERROR  STD REG                  F
        VARIABLE    COEFFICIENT  OF COEFF    COEFF   TOLERANCE  TO REMOVE  LEVEL
(Y-INTERCEPT        29.05349 )

                VARIABLES NOT IN EQUATION
                    PARTIAL                 F
        VARIABLE    CORR.   TOLERANCE   TO ENTER   LEVEL
    DUR      1  0.78601  1.00000      66.28        1
    HR       3  0.33729  1.00000       5.26        1
    AGE      4 -0.65099  1.00000      30.15        1
    HT       5  0.29942  1.00000       4.04        1
    WT       6 -0.12618  1.00000       0.66        1


STEP NO.    1
----------------
VARIABLE ENTERED    1 DUR

MULTIPLE R              0.7860
MULTIPLE R-SQUARE       0.6178
ADJUSTED R-SQUARE      0.6085

STD. ERROR OF EST.     3.0966

ANALYSIS OF VARIANCE
                        SUM OF SQUARES    DF    MEAN SQUARE   F RATIO
            REGRESSION  635.51730         1     635.5173       66.28
            RESIDUAL    393.15010         41    9.589027

                        VARIABLES IN EQUATION FOR VO2MAX

                            STD. ERROR  STD REG                  F
        VARIABLE    COEFFICIENT  OF COEFF    COEFF   TOLERANCE  TO REMOVE  LEVEL
(Y-INTERCEPT        3.15880 )
    DUR      1      0.05029      0.0062    0.786   1.00000     66.28        1
```

Table 11.7. *(Continued)*

```
                    VARIABLES NOT IN EQUATION
                  PARTIAL                 F
         VARIABLE   CORR.   TOLERANCE  TO ENTER   LEVEL
  HR        3  -0.14731   0.72170      0.89        1
  AGE       4  -0.24403   0.52510      2.53        1
  HT        5   0.01597   0.86364      0.01        1
  WT        6  -0.32457   0.99123      4.71        1


  STEP NO.    2
  ---------------
  VARIABLE ENTERED    6 WT

  MULTIPLE R            0.8112
  MULTIPLE R-SQUARE     0.6581
  ADJUSTED R-SQUARE     0.6410


  STD. ERROR OF EST.    2.9654


  ANALYSIS OF VARIANCE
                      SUM OF SQUARES    DF    MEAN SQUARE    F RATIO
         REGRESSION    676.93490         2    338.4675        38.49
         RESIDUAL      351.73250        40    8.793311


                 VARIABLES IN EQUATION FOR VO2MAX


                                  STD. ERROR  STD REG              F
         VARIABLE   COEFFICIENT   OF COEFF    COEFF   TOLERANCE  TO REMOVE  LEVEL
  (Y-INTERCEPT       10.30026 )
  DUR       1        0.05150       0.0059     0.805    0.99123    75.12       1
  WT        6       -0.12659       0.0583    -0.202    0.99123     4.71       1


                  VARIABLES NOT IN EQUATION


                  PARTIAL                 F
         VARIABLE   CORR.   TOLERANCE  TO ENTER   LEVEL

  HR        3  -0.08377   0.68819      0.28        1
  AGE       4  -0.24750   0.52459      2.54        1
  HT        5   0.20922   0.66111      1.79        1
```

variables, the value for R^2 and the standard error of the estimate; this is the estimated standard deviation about the regression line. This value squared is the mean square for the residual or the estimate for σ^2 if this is the correct model. Below this is the analysis of variance table and below this the value of the regression coefficient, 0.050, for the duration variable. The standard error of the regression coefficient is then given. The standardized regression coefficient is the value of the regression coeffcent if we had replaced duration by its standardized value. The value F to remove in a step-wise regression is the statistical significance of the partial correlation between the variable in the model and the dependent variable when adjusting for other variables in the model. The left hand side lists the variables not already in the equation. Again we have the partial correlations between the potential predictor variables and the dependent variable after adjusting for the variables in the model, in this case one variable, duration. Let us focus on the variable age at step 0 and at step 1. In step 0 there was a very highly statistically significant relationship between VO_2 MAX and age, the F-value being 30.15. After duration enters the predictive equation, in step 1 we see that the statistical significance has disappeared with the F to enter decreasing to 2.53. This occurs because age is very closely related to duration and also highly related to VO_2 MAX. The explanatory power of age may equivalently be explained by the explanatory power of duration. We see that *when a variable does not enter a predicitve model, this does not mean the variable is not related to the dependent variable but possibly that other variables in the model can account for its predictive power.* An equivalent way of viewing this is that the partial correlation has dropped from -0.65 to -0.24. There is another column labeled tolerance. The tolerance is 1 minus the square of the multiple correlation between the particular variable being considered and all of the variables already in the step-wise equation. Recall that if this correlation is large it is very difficult to estimate the regression coefficient (see Equation (11.14)). The tolerance is the term $(1 - R_j^2)$ in Equation (11.14). If the tolerance becomes too small the numerical accuracy of the model is in doubt.

In step 1, scanning the F-to-enter column we see the variable weight which is statistically significantly related to VO_2 MAX at the 5% level. This variable enters at step 2. After this variable has entered there are no statistically significant relationships left between the variables not in the equation and the dependent variable after adjusting for the variables in the model. The stepwise regression would stop at this point unless directed to do otherwise.   □

It is possible to modify the stepwise procedure so that rather than starting with 0 variables and building up, we start with all potential predictive variables in the equation and work down. In this case, at the first step we discard from the model the variable whose regression coefficient has the largest p-value, or equivalently the variable whose correlation with the dependent variable after adjusting for the other variables in the model is as small as possible. At each step, this process continues removing a variable as long as there are variables to remove from the model that are not statistically significantly related to the response variable at some particular level. The procedure of adding in variables that we have discussed in this chapter is called a *step-up stepwise procedure*, while the opposite procedure of removing variables is called a *step-down stepwise procedure*. Further, as the model keeps building, it may be that a variable entered earlier in the stepwise procedure no longer is statistically significantly related to the dependent variable in the presence of the other variables. For this reason most regression programs when performing a step-up regression, at each step have the ability to remove variables that are no longer statistically significant. All of this aims at a simple model (in terms of the number of variables) which explains as much of the variability as possible. The step-up and step-down procedures do not look at as many alternatives as the C_p plot procedure and thus may not be as prone to overfitting the data because of the many models considered. If we perform a step-up or step-down fit for the anesthesia data discussed above, the resulting model is the same as the model picked by the C_p plot.

11.6 POLYNOMIAL REGRESSION

We motivate this section by an example. Consider the data of Bruce *et al.* [1973] for 44 active males with a maximal exercise treadmill test. The oxygen consumption VO_2 MAX was regressed on, or explained by, the age of the participants. Figure 11.4 shows the residual plot.

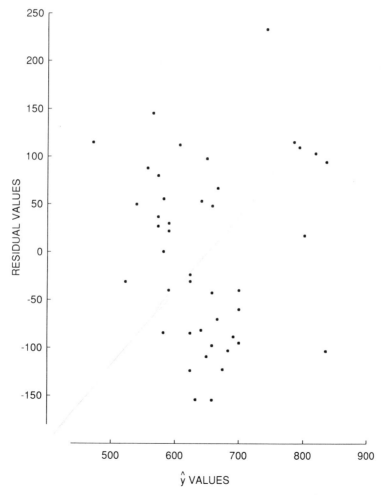

Figure 11.4. Residual plot of the regression of VO_2 MAX on age, active males.

Examination of the residual plot shows the majority of the points on the left are positive with a downward trend. The points on the right have generally higher values with an upward trend. This suggests that possibly the simple linear regression model does not fit the data well. The fact that the residuals come down and then go up suggests that, possibly, the regression curve rather than being linear should be a second order curve, such as

$$Y = a + b_1 X + b_2 X^2 + e.$$

Note that this equation looks like a multiple linear regression equation. We could write this equation as a multiple regression equation,

$$Y = a + b_1 X_1 + b_2 X_2 + e,$$

with $X_1 = X$ and $X_2 = X^2$. This simple observation allows us to fit polynomial equations to data by using multiple linear regression techniques. Observe what we are doing with multiple linear regression: the equation must be linear in the unknown parameters but we may insert *known* functions of an explanatory variable. If we create the new variables $X_1 = X$ and $X_2 = X^2$ and run a multiple regression program we find the following results.

Variable or Constant	b_1	SE(b_i)	t ($t_{41,0.975} \doteq 2.02$)
Age	−35.47	9.19	−3.86
Age^2	0.305	0.02	2.97
Constant	1611.57	199.99	

We note that both terms age and age^2 are statistically significant. Recall that the t-test for the age^2 term is equivalent to the partial correlation of the age squared with VO$_2$ MAX adjusting for the effect of age. This is equivalent to considering the hypothesis of linear regression *nested* within the hypothesis of quadratic regression. Thus we reject the hypothesis of linear regression and could use this quadratic regression formula. A plot of the residuals using the quadratic regression shows no particular trend and is not presented here. One might wonder, now that we have a second order term, whether perhaps a third order term might help the situation. If we run a multiple regression with three variables ($X_3 = X^3$), the following results obtain.

Variable or Constant	b_1	SE(b_i)	t ($t_{40,0.975} \doteq 2.02$)
Age	−18.928	55.905	−0.34
Age^2	−0.0738	1.268	−0.06
Age^3	0.00277	0.00923	0.30
Constant	1384.49	783.15	

Since the age^3 term, which tests the nested hypothesis of the quadratic equation within the cubic equation, is nonsignificant we may accept the quadratic equation as appropriate.

Figure 11.5 presents a scatter diagram of the data as well as the linear and quadratic curves. Note that the quadratic curve is higher at the younger ages and levels off more around 50 to 60. Within the high range of the data the quadratic or second order curve

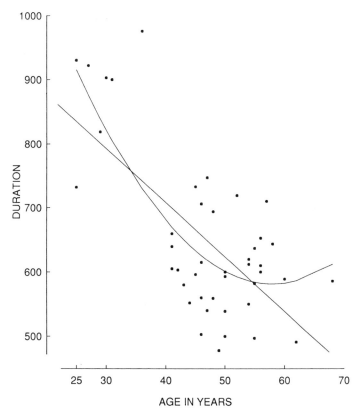

Figure 11.5. Active males with treadmill test (Bruce *et al.* [1973]). linear and quadratic fits.

increases. This may be an artifact of the curve fitting because all physiological knowledge tells us that the capacity for conditioning does not increase with age, although some individuals may improve their exercise performance with extra training. Thus the second order curve would seem to indicate that in a population of healthy active males that the decrease in VO_2 MAX consumption is not as rapid at the higher ages as at the lower ages. This is contrary to the impression one would get from a linear fit. One would not want however to use the quadratic curve to extrapolate beyond or even to the far end of the data in this particular example.

We see that the real restrictions of multiple regression is not that the equation be linear in the observed variables but rather that it be linear in the unknown coefficients. The coefficients may be multiplied by known functions of the observed variables; this makes a variety of models possible. For example, with *two variables* we could also consider as an alternative to a linear fit (as given below) a second order equation or polynomial in two variables. This is also given below.

$$Y = a + b_1X_1 + b_2X_2 + e$$

(linear in X_1 and X_2).

$$Y = a + b_1 X_1 + b_2 X_2 + b_3 X_1^2 + b_4 X_1 X_2 + b_5 X_2^2 + e$$

(a second order polynomial in X_1 and X_2).

Other functions of variables may be used. For example, if we observe a response which we believe is a periodic function of the variable X with a period of length L, we might try an equation of the form

$$Y = a + b_1 \sin(\pi X/L) + b_2 \cos(\pi X/L) + b_3 \sin(2\pi X/L) + b_4 \cos(2\pi X/L) + e.$$

The important point to remember is that not only can polynomials in variables be fit, but any model may be fit where the response is a linear function of known functions of the variables involved.

11.7 GOODNESS-OF-FIT CONSIDERATIONS

As in the one-dimensional case, we need to check the fit of the regression model. We need to see that the form of the model roughly fits the observed data; if we are engaged in statistical inference we need to see that the error distribution looks approximately normal. As in simple linear regression one or two outliers can greatly skew the results; also an inappropriate functional form can give misleading conclusions. In doing multiple regression it is harder then in simple linear regression to check the assumptions, because there are more variables involved. We do not have nice two-dimensional plots that completely display our data. In this section we discuss some of the ways in which multiple regression models may be examined.

11.7.1 Residual Plots and Normal Probability Plots

In the multiple regression situation, a variety of plots may be useful. We discussed in Chapter 9 the residual plots of the predicted value for Y versus the residual. Also useful is a normal probability plot of the residuals. This is useful for detecting outliers and for examining the normality assumption. Plots of the residual as a function of the independent or explanatory variables may point out a need for quadratic terms or for some other functional form. It is useful to have such plots even for potential predictor variables not entered into the predictive equation; they might be omitted because they are related to the response variable in a nonlinear fashion. This might be revealed by such residual plots.

Example 11.8. We return to the healthy normal active females of Example 11.7. Recall that the VO$_2$ MAX in a stepwise regression was predicted by duration and weight. Other variables considered were the maximum heart rate, age, and height. We now examine some of the plots from the BMDP output as well as normal probability plots of the residuals not taken from such output. Figure 11.6 on the left gives the residual plot from the BDMP program. The residuals look fairly good except for the point circled on the right hand margin which lies farther from the value of zero than the rest of the points. The right hand panel gives the square of the residuals. These values will have approximately a chi-square distribution with one degree of freedom

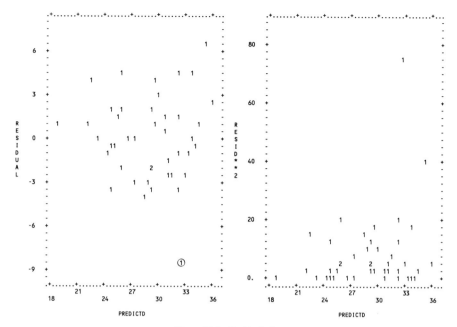

Figure 11.6. Residual plot.

if normality holds. If the model is correct, there will not be a change in the variance with increasing predicted values. There is no systematic change here. However, once again the one value has a large deviation. □

Figure 11.7 gives the normal probability plot for the residuals. In this output, the predicted values are on the horizontal axis rather than on the vertical axis as plotted previously. Again the residuals look quite nice except for the point on the far left; this point corresponds to the circled value in the last figure. This raises the possibility of rerunning the analysis omitting the one "outlier" to see what effect it had upon the analysis. We shall discuss this below after reviewing more graphical data.

Figures 11.8–11.12 deal with the residual values as a function of the five potential predictor variables. In each figure, the left hand panel presents the observed and predicted values for the data points and the right hand panel for the observed values of those data present the residual values. In Figure 11.7, for the duration note that the predicted values are almost linear. This is because most of the predictive power come from the duration variable so that the predicted value is not far removed from a linear function of duration. The residual plot looks nice with the possible exception of the outlier. In Figure 11.8 with respect to weight we have the same sort of behavior as we do in the last three figures for age, maximal heart rate, and height. In no case does there appear to be systematic unexplained variability than might be explained by adding a quadratic term or other terms to the equation.

If we rerun these data removing the potential outlier, the results change as given below.

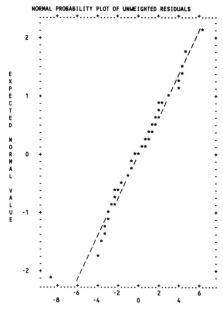

Figure 11.7. Normal residual plot.

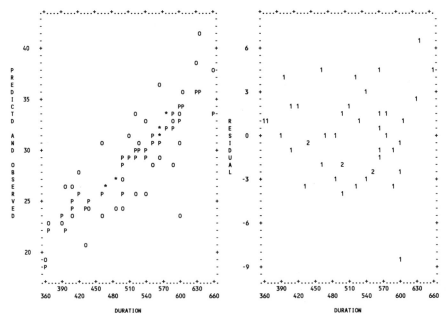

Figure 11.8. Duration *vs.* residual plot.

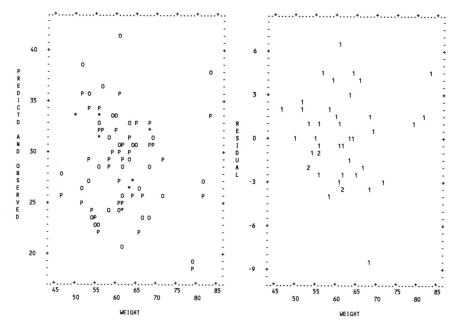

Figure 11.9. Weight *vs.* residual plot.

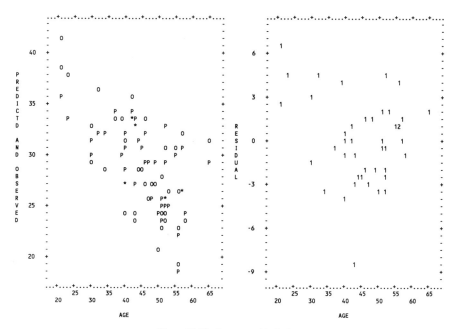

Figure 11.10. Age *vs.* residual plot.

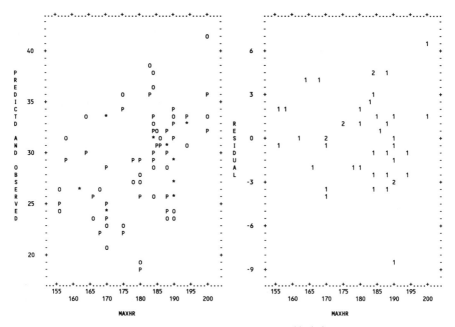

Figure 11.11. Maximum heart rate *vs.* residual plot.

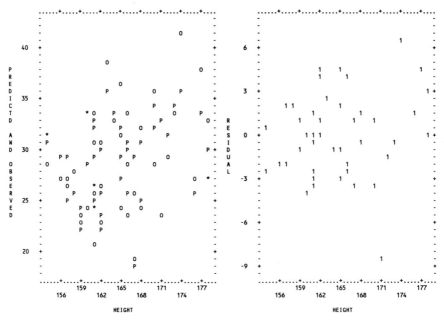

Figure 11.12. Height *vs.* residual plot.

Variable or Constant	All Data		Removing the "Outlier" Point	
	b_i	t	b_i	t
Duration	0.0515	8.67	0.0544	10.17
Weight	−0.127	−2.17	−0.105	−2.02
Constant	10.300		7.704	

We see a moderate change in the coefficient for weight; the change increases the importance of the duration variable. The t statistic for the weight variable is now right on the precise edge of statistical significance of the 0.05 level. Thus although the original model did not mislead us, part of the contribution from weight came from the data point that was removed. This brings up the issue of how such data might be presented in a scientific paper or talk. One possibility would be to present both results and discuss the issue. The removal of outlying values may allow one to get a closer fit to the data, and in this case the residual variability decreased from an estimated σ^2 of 2.97 to 2.64. Still if the outlier is not considered to be due to bad data, but rather due to an exceptional individual, in applying such relationships other exceptional individuals may be expected to appear. In such cases, interpretation necessarily becomes complex. This shows, again, that although there is a nice precision to significance levels, in practice the interpretation of the statistical analysis is an art as well as a science.

11.7.2 Nesting in More Global Hypothesis

Since it is difficult to visually inspect multi-dimensional data, one possibility for testing the model fit is to imbed the model in a more global hypothesis; that is, nest the model used within a more general model. One example of this would be adding quadratic terms and cross-product terms as discussed in Section 11.6. The number of such possible terms goes up greatly as the number of variables increases; this luxury is only available when there is a considerable amount of data.

11.7.3 Splitting the Samples; Jack-Knife Procedures

An estimated equation will fit data better than the true population equation because the estimate is designed to fit the data at hand. One way to get an estimate of the precision in a multiple regression model is to split the sample size into halves at random. One can estimate the parameters from one-half of the data and then predict the values for the remaining unused half of the data. The evaluation of the fit can be performed using the other half of the data. This gives an unbiased estimate of the appropriateness of the fit and the precision. There is however the problem that one-half of the data is "wasted" by not being used for the estimation of the parameters. This may be overcome by estimating the precision in this split-sampling manner but then presenting final estimates based upon the entire dataset.

Another approach which allows more precision in the estimate is to delete subsets of the data and to estimate the model on the remaining data; one then tests the fit on the removed smaller subsets. If this is done systematically, for example by removing one data point at a time, estimating the model using the remaining data and then examining the fit to the omitted data point, the procedure is called a *jack-knife procedure* (see Efron [1982]). Resampling from the observed data, the *bootstrap* method may also be used (Efron and Tibshirani [1986]).

We will not go further into such issues here.

11.8 ANALYSIS OF COVARIANCE

11.8.1 The Need for the Analysis of Covariance

In Chapter 10, we considered the analysis of variance. Associated with categorical classification variables we had a continuous response. Let us consider the simplest case, where we have a one-way analysis of variance consisting of two groups. Suppose that there is a continuous variable X in the background; a covariate. For example, the distribution of the variable X may differ between the groups, or the response may be very closely related to the value for the variable X. Suppose further that the variable X may be considered a more fundamental cause of the response pattern then the grouping variable. We illustrate some of the potential complications by two figures.

In Figure 11.13 on the left hand side, suppose that we have data as shown. The solid circles show the response values for group 1 and the crosses the response value for group 2. There is clearly a difference in response between the two groups. Suppose that we think that it is not the grouping variable that is responsible but the covariate X. On the right hand side, we see a possible pattern which could lead to the response pattern given. In this case, we see that both the observations from group 1 and group 2 have the same response pattern *when the value of X is taken into account*; that is they both fall around one fixed regression line. In this case, the observed difference between the groups may alternatively be explained because they differ in the covariate value X. Thus in certain situations in the analysis of variance one would like to adjust for potential differing values of a covariate. Another way of stating the same thing is: *in certain analysis of variance situations there is a need to remove potential bias due to the fact that categories differ in their values of a covariate X.*

Figure 11.14 shows a pattern of observations on the left for groups 1 and 2. There is no difference between the response in the groups, given the variability of the observations. Consider the same points, however, where we consider the relationship to a covariate X as plotted on the right. The right hand figure shows that the two groups have parallel regression lines which differ by an amount delta. Thus for a fixed value of the covariate X, on the average, the observations from the two groups differ. In this plot, there is clearly a statistically significant difference between the two groups because their regression lines will clearly have different intercepts. Although the two groups have approximately the same distribution of the covariate values, if we consider the covariate we are able to improve the precision of the comparison between the two groups. On the left, most of the variability is not due to intrinsic variability within the groups, but rather is due to the variability in the covariate X. On the right, when the covariate X is taken into account, we can see that there is a difference. Thus a second reason for considering covariates in the analysis of variance is: *consideration of a covariate may improve the precision of the comparison of the categories in the analysis of variance.*

In this section we consider methods which allow one or more covariates to be taken into account when performing an analysis of variance. Because we take into

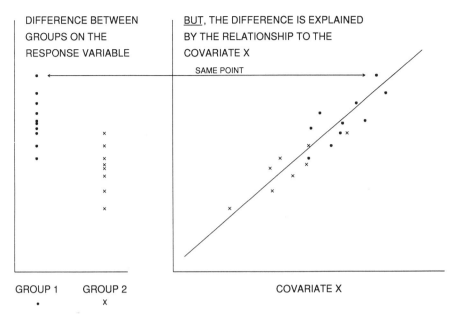

Figure 11.13. One-way analysis of variance with two categories. Group difference because of bias due to different distribution on the covariate X.

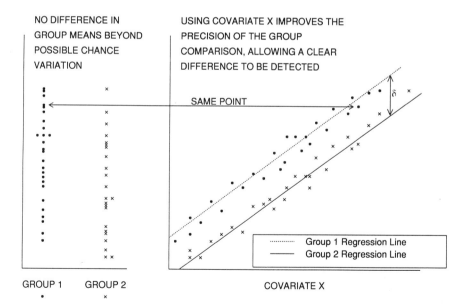

Figure 11.14. Two groups with close distribution on the covariate X by using the relationship of the response to X, separately in each group, a group difference obscured by the variation in X is revealed.

account those variables that vary with the variables of interest, the models and the technique are called the *analysis of covariance*.

11.8.2 The Analysis of Covariance Model

In this section, we consider the one-way analysis of covariance. This is a sufficient introduction to the subject so that more general analysis of variance models with covariates can then be approached.

In the one-way analysis of covariance, we observe a continuous response for each of a fixed number of categories. Suppose that the analysis of variance model is

$$Y_{ij} = \mu + \alpha_i + \epsilon_{ij},$$

where $i = 1, \ldots, I$ indexes the I categories. α_i, the category effect, satisfies $\sum_i \alpha_i = 0$, $j = 1, \ldots, n_i$ indexes the observations in the ith category. The ϵ_{ij} are independent $N(0, \sigma^2)$ random variables.

Suppose now that we wish to take into account the effect of the continuous covariate X. As in Figures 11.13 and 11.14, we suppose that the response is linearly related to X where the slope of the regression line is the same for each of the categories (see Figure 11.15). That is, our analysis of covariance model is

$$Y_{ij} = \mu + \alpha_i + \beta X_{ij} + \epsilon_{ij}, \tag{11.20}$$

with the assumptions as before.

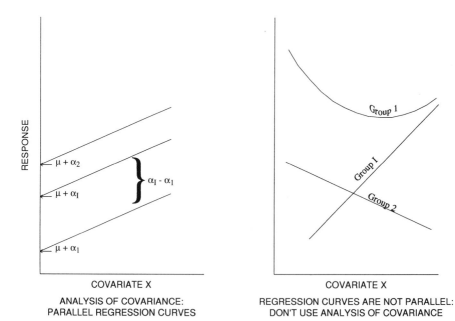

Figure 11.15. Parallel regression curves are assumed in the analysis of covariance.

Although we do not pursue the matter, the analogous analysis of covariance model for the two-way analysis of variance without interaction may be given by

$$Y_{ijk} = \mu + \alpha_i + \beta_j + \beta X_{ijk} + \epsilon_{ijk}.$$

Analysis of covariance models easily generalize to include more than one covariate. For example, if there are p covariates to adjust for, the appropriate equation is

$$Y_{ij} = \mu + \alpha_i + \beta_1 X_{ij}(1) + \beta_2 X_{ij}(2) + \cdots + \beta_p X_{ij}(p) + \epsilon_{ij}.$$

Where $X_{ij}(k)$ is the value for the kth covariate, when the observation comes from the ith category and the jth observation in that category. Further, if the response is not linear, one may model a different form of the response. For example, the following equation models a quadratic response to the covariate X_{ij}:

$$Y_{ij} = \mu + \alpha_i + \beta_1 X_{ij} + \beta_2 X_{ij}^2 + \epsilon_{ij}.$$

In each case in the analysis of covariance, *the assumption is that the response to the covariates is the same within each of the strata or cells for the analysis of covariance.*

It is possible to perform both the analysis of variance and the analysis of covariance by using the methods of multiple linear regression analysis, as given above in this chapter. The trick to thinking of an analysis of variance problem as a multiple regression problem is to use what are called dummy or indicator variables. These variables allow us to consider the unknown parameters in the analysis of variance to be parameters in the multiple regression model.

Definition 11.11. A *dummy variable*, or *indicator variable*, for a category or condition is a variable taking the value one if the observation comes from the category or satisfies the condition. Otherwise the variable takes on the value zero.

We illustrate this definition with two examples. A dummy variable for the male sex is

$$X = \begin{cases} 1, & \text{if the subject is male,} \\ 0, & \text{otherwise.} \end{cases}$$

A series of dummy variables for blood types (A, B, AB, O) are

$$X_1 = \begin{cases} 1, & \text{if the blood type is A,} \\ 0, & \text{otherwise.} \end{cases}$$

$$X_2 = \begin{cases} 1, & \text{if the blood type is B,} \\ 0, & \text{otherwise.} \end{cases}$$

$$X_3 = \begin{cases} 1, & \text{if the blood type is AB,} \\ 0, & \text{otherwise.} \end{cases}$$

$$X_4 = \begin{cases} 1, & \text{if the blood type is O,} \\ 0, & \text{otherwise.} \end{cases}$$

By using dummy variables, analysis of variance models may be turned into multiple regression models. We illustrate this by an example.

Consider a one-way analysis of variance with three groups. Suppose we have two observations in each of the first two groups and three observations in the third group. Our model is

$$Y_{ij} = \mu + \alpha_i + \epsilon_{ij}, \tag{11.21}$$

where i denotes the group and j the observation within the group. Our data are Y_{11}, Y_{12}, Y_{21}, Y_{22}, Y_{31}, Y_{32}, Y_{33}. Let X_1, X_2, and X_3 be indicator variables for the three categories.

$$X_1 = \begin{cases} 1, & \text{if the observation is in group 1,} \\ 0, & \text{otherwise.} \end{cases}$$

$$X_2 = \begin{cases} 1, & \text{if the observation is in group 2,} \\ 0, & \text{otherwise.} \end{cases}$$

$$X_3 = \begin{cases} 1, & \text{if the observation is in group 3,} \\ 0, & \text{otherwise.} \end{cases}$$

Then Equation (11.21) becomes (omitting subscript on Y and e)

$$Y = \mu + \alpha_1 X_1 + \alpha_2 X_2 + \alpha_3 X_3 + \epsilon. \tag{11.22}$$

Note that X_1, X_2, and X_3 are related. If $X_1 = 0$ and $X_2 = 0$, then X_3 must be 1. Hence there are only two independent dummy variables. In general, for k groups there are $(k-1)$ independent dummy variables. This is another illustration of the fact that the k treatment effects in the one-way analysis of variance have $(k-1)$ degrees of freedom. Our data, renumbering the Y_{ij} to be Y_k, $k = 1, \ldots, 7$ are:

Y_k	Y_{ij}	X_1	X_2	X_3
Y_1	Y_{11}	1	0	0
Y_2	Y_{12}	1	0	0
Y_3	Y_{21}	0	1	0
Y_4	Y_{22}	0	1	0
Y_5	Y_{31}	0	0	1
Y_6	Y_{32}	0	0	1
Y_7	Y_{33}	0	0	1

For technical reasons, we do not estimate Equation (11.22). Since

$$\sum_i X_i = 1, \quad R^2_{X_1(X_2, X_3)} = 1.$$

Recall that we cannot estimate regression coefficients well if the multiple correlation is near one. Instead, an equivalent model

$$Y = \delta + \gamma_1 X_1 + \gamma_2 X_2 + \epsilon$$

is used. Here $\delta = \mu + \alpha_3$, $\gamma_1 = \alpha_1 - \alpha_3$, and $\gamma_2 = \alpha_2 - \alpha_3$. That is, all effects are compared relative to group 3. We may now use a multiple regression program to perform the one-way analysis of variance.

To move to an analysis of covariance, we use $Y = \delta + \gamma_1 X_1 + \gamma_2 X_2 + \beta X + \epsilon$ where X is the covariate. If there is no group effect, then we have the same expected value (for fixed X) regardless of the group; that is $\gamma_1 = \gamma_2 = 0$.

More generally, for I groups the model is

$$Y = \delta + \gamma_1 X_1 + \cdots + \gamma_{I-1} X_{I-1} + \beta X + \epsilon.$$

The null hypothesis is $H_0 : \gamma_1 = \gamma_2 = \cdots = \gamma_{I-1} = 0$. This is tested using nested hypotheses. Let $\text{SS}_{\text{REG}}(X)$ be the regression sum of squares for the model $Y = \delta + \beta X + e$. Let

$$\text{SS}_{\text{REG}}(\gamma \mid X) = \text{SS}_{\text{REG}}(X_1, \ldots, X_{I-1}, X) - \text{SS}_{\text{REG}}(X)$$

and

$$\text{SS}_{\text{RESID}}(\gamma, X) = \text{SS}_{\text{TOTAL}} - \text{SS}_{\text{REG}}(X_1, \ldots, X_{I-1}, X).$$

The analysis of covariance table is:

Source	DF	SS	MS	F
Regression on X	1	$\text{SS}_{\text{REG}}(X)$	$\text{MS}_{\text{REG}}(X)$	$\dfrac{\text{MS}_{\text{REG}}(X)}{\text{MS}_{\text{RESID}}}$
Groups adjusted for X	$I - 1$	$\text{SS}_{\text{REG}}(\gamma \mid X)$	$\text{MS}_{\text{REG}}(\gamma \mid X)$	$\dfrac{\text{MS}_{\text{REG}}(\gamma \mid X)}{\text{MS}_{\text{RESID}}}$
Residual	$n - I - 1$	$\text{SS}_{\text{RESID}}(\gamma, X)$	MS_{RESID}	
Total	$n - 1$	SS_{TOTAL}		

The F test for the equality of group means has $I - 1$ and $n - I - 1$ degrees of freedom. If there is a statistically significant group effect, then there is an interest in the separation of the parallel regression lines. The regression lines are:

Group	Line
1	$\widehat{\delta} + \widehat{\gamma}_1 + \widehat{\beta} X$
2	$\widehat{\delta} + \widehat{\gamma}_2 + \widehat{\beta} X$
$\vdots$	$\vdots$
$I - 1$	$\widehat{\delta} + \widehat{\gamma}_{I-1} + \widehat{\beta} X$
I	$\widehat{\delta} + \widehat{\beta} X$

where the hat denotes the usual least square multiple regression estimate. Customarily, these values are calculated for X equal to the average X value over all the observations. These values are called *adjusted means* for the group. This is in contrast to the observed mean for the observations in each group. Note, again, that group I is the reference group. It may sometimes be useful to rearrange the groups to have a specific group be the reference group. For example, suppose there are three treatment groups and one reference group. Then the effects γ_1, γ_2, γ_3 are, naturally, the treatment effects relative to the reference group.

We illustrate these ideas with two examples. In each example there are two groups ($I = 2$) and one covariate for adjustment.

Example 11.9. The data of Cullen and van Belle [1975] is considered again. In this case a larger set of data is used. One group received general anesthesia ($n_1 = 35$) and another group regional anesthesia ($n_2 = 42$). The dependent variable, Y, is the percent depression of lymphocyte transformation following surgery. The covariate, X, is the degree of trauma of the surgical procedure.

Figure 11.16 shows the data with the estimated analysis of covariance regression lines. The top line is the regression line for the general anesthesia group (which had a higher average trauma, 2.4 versus 1.4). The analysis of covariance table is:

Source	DF	SS	MS	F
Regression on trauma	1	4621.52	4621.52	7.65
General versus regional anesthesia adjusted for trauma	1	1249.78	1249.78	2.06
Residual	74	44788.09	605.24	
Total	76	56201.52		

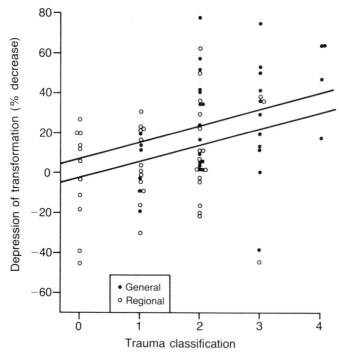

Figure 11.16. Relationship of postoperative depression of lymphocyte transformation to the level of trauma. Each point represents the response of one patient.

Note that trauma is significantly related to the percent depression of lymphocyte transformation, $F = 7.65 > F_{1,74,0.95}$. In testing the adjusted group difference,

$$F = 2.06 < 3.97 = F_{1,74,0.95},$$

so that there is not a statistically significant difference between regional and general anesthesia after adjusting for trauma.

The two regression lines are

$$Y_1 = 25.6000 + 8.4784(X - 2.3714),$$

$$Y_2 = 6.7381 + 8.4784(X - 1.2619).$$

At the average value of $\overline{X} = 1.7552$, the predicted or adjusted means are

$$\widehat{Y}_1 = 25.6000 + (-5.1311) = 20.47,$$

$$\widehat{Y}_2 = 6.7381 + (4.2757) = 11.01.$$

The original difference is $\overline{Y}_{1\bullet} - \overline{Y}_{2\bullet} = 25.6000 - 6.7381 = 18.86$. The adjusted (nonsignificant) difference is $\widehat{Y}_1 - \widehat{Y}_2 = 20.47 - 11.01 = 9.46$, a considerable drop. In fact the unadjusted one-way analysis of variance, or equivalently unpaired t-test, is significant $p < 0.01$. The observed difference may be due to bias in the differing amount of surgical trauma in the two groups.   □

Example 11.10.   Do men and women use the same level of oxygen when their maximal exercise limit is the same? The Bruce *et al.* [1973] maximal exercise data are used. The limit of exercise is expressed by the duration on the treadmill. Thus we wish to know if there is a VO$_2$ MAX difference between the sexes when adjusting for the duration of exercise. The analysis of covariance table is

Source	DF	SS	MS	F
Duration	1	6049.51	6049.51	504.97
Sex adjusting for duration	1	229.83	229.83	19.18
Residual	84	1006.05	11.98	
Total	86	7285.39		

The sex difference is highly statistically significant after adjusting for the treadmill duration. The estimated regression lines are

$$\text{Females:} \quad \text{VO}_2 \text{ MAX} = -1.59 + 0.0595 \times \text{duration},$$
$$\text{Males:} \quad \text{VO}_2 \text{ MAX} = 2.27 + 0.0595 \times \text{duration}.$$

The overall duration mean is 581.89. The means are:

	VO$_2$ MAX Means	
	Observed	Adjusted
Female	29.05	33.03
Male	40.80	36.89

The fact that at maximum exercise normal males use more oxygen per unit of body weight is not entirely accounted for by their average longer duration on the treadmill (647 s *vs.* 515 s). Even when adjusting for duration more oxygen/kilogram/per minute is used.

Model assumptions may be tested by residual plots and normal probability plots as above. One assumption was that the regression lines were parallel. This may be tested by using the model (in the one-way ANOVA)

$$Y = \delta + \gamma_1 X_1 + \cdots + \gamma_{I-1} X_{I-1} + \beta_1 XX_1 + \cdots + \beta_I XX_I + \epsilon.$$

If an observation is in group i $(i = 1, \ldots, I - 1)$, this reduces to

$$Y = \delta + \gamma_i + \beta_i X + \epsilon.$$

Nested within this model is the special $\beta_1 = \beta_2 = \cdots = \beta_I$.

Source	DF	SS
Model with $\gamma_1, \ldots, \gamma_{I-1}, \beta$	I	$\text{SS}_{\text{REG}}(\gamma_1, \ldots, \gamma_{I-1})$
Model with $\gamma_1, \ldots, \gamma_{I-1}, \beta,$		
$\beta_1, \ldots, \beta_I$; extra ss	$I - 1$	$\text{SS}_{\text{REG}}(\beta_1, \ldots, \beta_I \mid \gamma_1, \ldots, \gamma_{I-1}, \beta)$
Residual	$n - 2I$	$\text{SS}_{\text{RESID}}(\gamma_1, \ldots, \gamma_{I-1}, \beta_1, \ldots, \beta_I)$
Total	$n - 1$	SS_{TOTAL}

Source	MS	F
Model with $\gamma_1, \ldots, \gamma_{I-1}, \beta$	$\text{MS}_{\text{REG}}(\gamma\text{'s})$	
Model with $\gamma_1, \ldots, \gamma_{I-1}, \beta,$		
$\beta_1, \ldots, \beta_I$; extra ss	$\text{MS}_{\text{REG}}(\beta_i\text{'s} \mid \gamma_i\text{'s}, \beta)$	$\dfrac{\text{MS}_{\text{REG}}(\beta_i\text{'s} \mid \gamma_i\text{'s}, \beta)}{\text{MS}_{\text{RESID}}(\gamma_i\text{'s}, \beta_i\text{'s})}$
Residual	$\text{MS}_{\text{RESID}}(\gamma_i\text{'s}, \beta_i\text{'s})$	

For the exercise test example, we have:

Source	DF	SS	MS	F
Model with group, equal slopes and duration	2	6279.34	3139.67	
Model with unequal slopes (minus ss for nested equal slope model).	1	29.40	29.40	2.50
Residual	83	976.65	11.77	
Total	86	7285.39		

As $F = 2.50 < F_{1,83,0.95}$, the hypothesis of equal slopes (parallelism) is reasonable and the analysis of covariance was appropriate. This use of a nested hypothesis is an example of the method of Section 11.7.2 for testing goodness-of-fit of a model.

11.9 ADDITIONAL REFERENCES AND DIRECTIONS FOR FURTHER STUDY

11.9.1 There Are now Many References on Multiple Regression Methods

Draper and Smith [1981] present an extensive coverage of the topics of this chapter, plus much more material and a large number of examples with solutions. The text is on a more advanced mathematical level making use of matrix algebra. Kleinbaum and Kupper [1988] present material on a level close to that of this chapter; taking more pages for the topics of this chapter, they have a more leisurely presentation. The text is an excellent supplementary reference to the material of this chapter. Another useful text is Daniel and Wood [1971].

11.9.2 Time Series Data

It would appear that the multiple regression methods of this chapter would apply when one of the explanatory variables is time. This may be true in certain limited cases, but it is not usually true. Analyzing data with time as an independent variable is called *time series analysis*. Often, in time, the errors are dependent at different time points. Box and Jenkins [1970] are one source for time series methods.

11.9.3 Causal Models: Structural Models and Path Analysis

In many studies, especially observational studies of human populations, one might conjecture that certain variables contribute in a causal fashion to the value of another variable. For example, age and sex might be hypothesized to contribute to hospital bed usage, but not vice versa. In a statistical analysis, bed use would be modeled as a linear function of age and sex plus other unexplained variability. If only these three variables were considered, we would have a multiple regression situation. Bed use with other variables might be considered an explanatory variable for number of nursing days used. *Structural models* consist of a series of multiple regression equations; the equations are selected to model conjectured causal pathways. The models do not prove causality, but can examine whether the data are consistent with certain causal pathways.

Three books addressing structural models (from most elementary to more complex) are Li [1975], Duncan [1975], and Goldberger and Duncan [1973]. Issues of causality are addressed in Blalock [1964] and Campbell and Stanley [1963].

11.9.4 Multivariate Multiple Regression Models

In this chapter, we have analyzed the response of one dependent variable as explained by a linear relationship with multiple independent or predictor variables. In many circumstances there are multiple (more than one) dependent variables whose behavior we want to explain in terms of the independent variables. When the models are linear the topic is called *multivariate multiple regression*. The mathematical complexity increases, but in essence each dependent variable is modeled by a separate linear equation. Morrison [1976] and Timm [1975] present such models.

11.9.5 Nonlinear Regression Models

In certain fields it is not possible to express the response of the dependent variable as a linear function of the independent variables. For example, in pharamacokinetics and compartmental analysis, equations such as

$$Y = \beta_1 e^{\beta_2 x} + \beta_3 e^{\beta_4 x} + e$$

and

$$Y = \frac{\beta_1}{x - \beta_2} + e$$

may arise where the β_i's are unknown coefficients and the e is an error (unexplained variability) term. See van Belle *et al.* [1989] for an example of the latter equation. Further examples of *nonlinear* regression equations are given in Chapters 13 (logistic regression) and 16 (the Cox regression model for survival).

There are computer programs for estimating the unknown parameters.

1. The estimation proceeds by trying to get better and better approximations to the "best" (maximum likelihood) estimates. Sometimes the programs do not come up with an estimate; that is, they do not converge.
2. The estimation is much more expensive (in computer time) than the linear models program.
3. The interpretation of the models may be more difficult.
4. It is more difficult to check fit of many of the models visually.

One available package of nonlinear regression programs is given in BMDP [1981].

NOTES

Note 11.1: Least Squares Fit of the Multiple Regression Model

We use the sum of squares notation of Chapter 9. The regression coefficients b_i are solutions to the k equations:

$$[x_1^2]b_1 + [x_1 x_2]b_2 + \cdots + [x_1 x_k]b_k = [x_1 y]$$
$$[x_1 x_2]b_1 + [x_2^2]b_2 + \cdots + [x_2 x_k]b_k = [x_2 y]$$

$$\vdots$$

$$[x_1 x_k]b_1 + [x_2 x_k]b_2 + \cdots + [x_k^2]b_k = [x_k y].$$

For the reader familliar with matrix notation, we give a Y vector and covariate matrix.

$$\underline{Y} = \begin{pmatrix} Y_1 \\ \vdots \\ Y_n \end{pmatrix}, \quad \underline{X} = \begin{pmatrix} X_{11} & \cdots & X_{1k} \\ X_{21} & \cdots & X_{2k} \\ \vdots & \ddots & \vdots \\ X_{n1} & \cdots & X_{nk} \end{pmatrix}$$

The b_i are given by:

$$\begin{pmatrix} b_1 \\ \vdots \\ b_k \end{pmatrix} = (\underline{X}'\underline{X})^{-1}\underline{X}'\underline{Y}$$

where the prime denotes the matrix transpose and the -1 denotes the matrix inverse. Once the b_i's are known, a is given by

$$a = \overline{Y} - (b_i\overline{X}_i + \cdots + b_k\overline{X}_k).$$

Note 11.2: The Multivariate Normal Distribution

The density function for the *multivariate normal distribution* is given for those who know matrix algebra. Consider jointly distributed variables

$$\underline{Z} = \begin{pmatrix} Z_1 \\ \vdots \\ Z_p \end{pmatrix}$$

written as a vector. Let the *mean vector* and *covariance matrix* be given by

$$\underline{\mu} = \begin{pmatrix} E(Z_1) \\ \vdots \\ E(Z_p) \end{pmatrix}, \quad \underline{\Sigma} = \begin{pmatrix} \mathrm{var}(Z_1) & \mathrm{cov}(Z_1, Z_2) & \cdots & \mathrm{cov}(Z_1, Z_p) \\ \vdots & & & \vdots \\ \mathrm{cov}(Z_p, Z_1) & \cdots & \cdots & \mathrm{var}(Z_p) \end{pmatrix}.$$

The density is

$$f(z_1, \ldots, z_p) = (2\pi)^{-p/2}|\underline{\Sigma}|^{-\frac{1}{2}}e^{-\frac{1}{2}(\underline{Z}-\underline{\mu})'\underline{\Sigma}^{-1}(\underline{Z}-\underline{\mu})},$$

where $|\underline{\Sigma}|$ is the determinant of Σ and -1 denotes the matrix inverse. See Graybill [1976] for much more information about the multivariate normal distribution.

Note 11.3: Pure Error

We have seen that it is difficult to test goodness-of-fit without knowing at least one large model that fits the data. This allows estimation of the residual variability. There is a situation where one can get an accurate estimate of the residual variability without any knowledge of an appropriate model. Suppose for some fixed value of the X_i's, there are *repeated* measurements of Y. These Y variables will be multiple independent observations with the same mean and variance. By subtracting the sample mean for the point in question, we can estimate the variance. More generally, if more than one X_i combination has multiple observations we can pool the sum of squares (as in the one-way ANOVA) to estimate the residual variability.

We now show how to partition the sum of squares. Suppose there are K combinations of the covariates X_i for which we observe two or more Y values. Let Y_{ik} denote the ith observation ($i = 1, 2, \ldots, n_k$) at the kth covariate values. Let $\overline{Y}_k$ be the mean of the Y_{ik}:

$$\overline{Y}_k = \sum_{i=1}^{n_k} Y_{ik}/n_k.$$

We define the pure error sum of squares and model of squares as follows:

$$\text{SS}_{\text{PURE ERROR}} = \sum_{k=1}^{K} \sum_{i=1}^{n_k} (Y_{ik} - \overline{Y}_k)^2,$$

$$\text{SS}_{\text{MODEL FIT}} = \text{SS}_{\text{RESID}} - \text{SS}_{\text{PURE ERROR}},$$

Also,

$$\text{MS}_{\text{PURE ERROR}} = \frac{\text{SS}_{\text{PURE ERROR}}}{\text{DF}_{\text{PURE ERROR}}},$$

$$\text{MS}_{\text{MODEL FIT}} = \frac{\text{SS}_{\text{MODEL}}}{\text{DF}_{\text{MODEL}}},$$

where

$$\text{DF}_{\text{PURE ERROR}} = \sum_{k=1}^{K} n_k - K,$$

$$\text{DF}_{\text{MODEL}} = n + K - \sum_{k=1}^{K} n_k - p - 1,$$

n = total number of observations, and p = number of covariates in the multiple regression model.

The analysis of variance table becomes:

Source	DF	SS	MS	F
Regression	p	SS_{REG}	MS_{REG}	$\dfrac{\text{MS}_{\text{REG}}}{\text{MS}_{\text{RESID}}}$
Residual	$n - p - 1$	SS_{RESID}	MS_{RESID}	
*Model	*DF_{MODEL}	SS_{MODEL}	MS_{MODEL}	$\dfrac{\text{MS}_{\text{MODEL}}}{\text{MS}_{\text{PURE ERROR}}}$
*Pure error	*$\text{DF}_{\text{PURE ERROR}}$	$\text{SS}_{\text{PURE ERROR}}$	$\text{MS}_{\text{PURE ERROR}}$	
Total	$n - 1$	SS_{TOTAL}		

The asterisk terms further partition the residual sum of squares. The F statistic $MS_{MODEL}/MS_{PURE\ ERROR}$ with DF_{MODEL} and $DF_{PURE\ ERROR}$ degrees of freedom tests the model fit. If the model is not rejected as unsuitable, then the usual F-statistic tests whether or not the model has predictive power (that is, whether all the $\beta_i = 0$).

PROBLEMS

Problems 1–7 deal with the fitting of one multiple regression equation. Perform each of the following tasks as indicated. Note that various parts are from different sections of the chapter. For example, (e) and (f) are discussed in Section 11.7.

   a. Find the t-value for testing the statistical significance of each of the regression coefficients. Do we reject $\beta_i = 0$ at the 5% significance level? At the 1% significance level?

   b. **i.** Construct a 95% confidence interval for each β_i.

   ii. Construct a 99% confidence interval for each β_i.

   c. Fill in the missing values in the analysis of variance table. Is the regression significant at the 5% significance level? At the 1% significance level.

   d. Fill in the missing values in the partial table of observed, predicted and residual values.

   e. Plot the residual plot of Y versus $Y - \widehat{Y}$. Interpret your plot.

   f. Plot the normal probability plot of the residual values. Do the residuals seem reasonably normal?

1. The 94 sedentary males with treadmill tests of Chapter 9, Problems 9–12 are considered here. The dependent and independent variables were:

$$Y = VO_2 \text{ MAX}, \quad X_1 = \text{duration}, \quad X_2 = \text{maximum heart rate},$$

$$X_3 = \text{height}, \quad X_4 = \text{weight},$$

Constant or Covariate	b_i	$SE(b_i)$
X_1	0.0510	0.00416
X_2	0.0191	0.0258
X_3	−0.0320	0.0444
X_4	0.0089	0.0520
Constant	2.89	11.17

Source	DF	SS	MS	F
Regression	?	4314.69	?	?
Residual	?	?	?	
Total	?	5245.31		

Do Tasks (a), (b-i), and (c). What is R^2?

2. The data of Mehta *et al.* [1981] used in Problems 13–22 of Chapter 9 are used here. The aorta platelet aggregation percent under dipyridamole, using epinehrine, was regressed upon the control values in the aorta and coronary sinus. The results were:

Constant or Covariate	b_i	$\text{SE}(b_i)$
Aorta control	−0.0306	0.301
Coronary sinus control	0.768	0.195
Constant	15.90	

Source	DF	SS	MS	F
Regression	?	?	?	?
Residual	?	231.21	?	
Total	?	1787.88		

Y	$\widehat{Y}$	Residual
89	81.58	7.42
45	?	?
96	86.68	?
70	?	2.34
69	?	?
83	88.15	−5.15
84	88.03	−4.03
85	88.92	−3.92

Do Tasks (a), (b-ii), (c), (d), (e), (f) (with small numbers of points the interpretation in (e) and (f) is problematic).

3. The problem uses the 20 aortic valve surgery cases of Chapter 9; see introduction to Problems 30–33. The response variable is the end diastolic volume adjusted for body size, EDVI. The two predictive variables are the EDVI before surgery and the systolic volume index, SVI, before surgery.

$$Y = \text{EDVI postoperatively,}$$

$$X_1 = \text{EDVI preoperatively,}$$

$$X_2 = \text{SVI preoperatively.}$$

Constant or Covariate	b_i	$\text{SE}(b_i)$
X_1	0.889	0.155
X_2	−1.266	0.337
Constant	65.087	

Source	DF	SS	MS	F
Regression	?	21631.66	?	?
Residual	?	?	?	
Total	?	32513.75		

Y	$\hat{Y}$	Residual	Y	$\hat{Y}$	Residual
111	112.8	0.92	70	84.75	−14.75
56	?	?	149	165.13	−16.13
93	?	−39.99	55	?	?
160	148.78	11.22	91	88.89	2.11
111	?	5,76	118	103.56	−11.56
83	86.00	?	63	?	?
59	?	4.64	100	86.14	13.86
68	93.87	?	198	154.74	43.26
119	62.27	56.73	176	166.39	9.61
71	86.72	?			

Do Tasks (a), (b-i), (c), (d), (f). Find R^2.   □

Problems 4–7 refer to data of Hossack *et al.* [1980] and [1981]. Ten normal men and eleven normal women were studied during a maximal exercise treadmill test. While being exercised they had a catheter (tube) inserted into the pulmonary (lung) artery and a short tube into the left radial or brachial artery. This allowed sampling and observation of arterial pressures and the oxygen content of the blood. From this several parameters as described below were measured or calculated. The data for the eleven women are given in Table 11.8. Descriptions of the variables follow.

Activity: A subject who routinely exercises three or more times per week until perspiring was active (Act), otherwise the subject was sedentary (Sed).

Wt: Weight in kilograms.

Ht: Height in centimeters.

VO_2 MAX: the oxygen used (in milliliters per kilogram of body weigth) in 1 min at maximum exercise.

FAI: functional aerobic impairment. For a patient's age and activity level (active or sedentary) the expected treadmill duration (ED) is estimated from a regression equation. The excess of observed duration (OD) to expected duration (ED) as a percent of ED is the FAI. FAI = $100 \times$ (OD − ED)/ED.

$\dot{Q}$ MAX: the output of the heart in liters of blood per minute at maximum.

HR MAX: the heart rate in beats/minute at maximum exercise.

SV MAX: the volume of blood pumped out of the heart in milliliters during each stroke (at maximum cardial output).

CaO_2: the oxygen content of the arterial system in milliters of oxygen per liter of blood.

$C\bar{v}O_2$: the oxygen content of the venous (vein) system in milliliters of oxygen per liter of blood.

$a\bar{v}O_2D$ MAX: this is the difference in the oxygen content (in milliliters of oxygen per liter of blood) between the arterial system and the venous system (at maximum exercise). Thus, $a\bar{v}O_2D$ MAX = $CaO_2 - C\bar{v}O_2$.

$\bar{P}_{SA}$MAX: average pressure in the arterial system at the end of exercise in milliliters of mercury (mmHg).

$\bar{P}_{PA}$MAX: average pressure in the pulmonary artery at the end of exercise in mmHg.

The data for the 10 normal men are displayed in Table 11.9.

Table 11.8. Physical and Hemodynamic Variables in 11 Normal Women.

Case	Activity	Age (years)	Wt	Ht	$\dot{V}O_2$ MAX	FAI	$\dot{Q}$ MAX	HR MAX	SV MAX	CaO_2	$C\bar{v}O_2$	$a\bar{v}O_2D$ MAX	$\bar{P}_{SA}$MAX	$\bar{P}_{PA}$MAX
1	Sed	45	63.2	163	28.81	−12	12.43	194	64	193	46	147	109	27
2	Sed	52	56.6	166	24.04	−3	12.19	158	87	181	73	108	137	16
3	Sed	43	65.0	155	26.66	−1	11.52	194	59	212	61	151	?	30
4	Sed	51	58.2	161	24.34	−3	10.78	188	63	173	41	132	154	15
5	Sed	61	74.1	167	21.42	−6	11.71	178	66	198	62	136	140	29
6	Sed	52	69.0	161	26.72	−15	12.89	188	72	193	50	143	125	30
7	Sed	60	50.9	166	23.74	−15	10.94	164	68	160	42	118	95	26
8	Sed	56	66.0	158	28.72	−31	13.93	184	81	168	52	136	148	21
9	Sed	56	66.0	165	20.77	6	10.25	166	62	171	53	118	102	27
10	Sed	51	64.3	168	24.77	−4	11.98	176	68	187	54	133	152	38
11	Act	28	55.5	160	47.72	−37	14.36	200	76	202	31	171	132	25
Mean		50.5	62.6	163	27.07	−11	12.09	181	70	187	51	136	129	26
SD		9.3	6.7	4.1	7.34	13	1.27	14	9	15	10	18	21	7

Table 11.9. Physical and Hemodynamic Variables in 10 Normal Men.

Case	Age (yr)	Wt	Ht	VO_2 MAX	FAI	$\dot{Q}$ MAX	HR MAX	SV MAX	$\bar{P}_{SA}$ MAX	$\bar{P}_{PA}$ MAX
1	64	73.6	170	30.3	−4	13.4	156	85	114	24
2	61	90.9	191	27.1	12	17.8	156	115	104	30
3	38	76.8	180	44.4	5	19.4	190	102	100	24
4	62	92.7	185	24.6	18	15.8	173	91	78	33
5	59	92.0	183	41.2	−18	21.1	167	127	133	36
6	47	83.2	185	48.9	−20	22.4	173	132	160	22
7	24	69.8	178	62.1	−2	24.9	188	133	127	25
8	26	78.6	191	50.9	5	20.1	169	119	115	15
9	54	95.9	183	33.2	9	19.2	154	125	108	31
10	20	83.0	176	32.5	34	15.0	196	77	120	18
Mean	46	83.7	182	39.2	4	18.9	169	114	117	26
SD	17	8.9	7	12.0	16	3.5	21	25	22	7

4. For the 10 men, let

$$Y = VO_2 \text{ MAX}, \quad X_1 = \text{weight}, \quad X_2 = \text{HR MAX}, \quad X_3 = \text{SV MAX}.$$

(In practice one would not use three regression variables with only 10 data points. This is done here so that the small data set may be presented in its entirety.)

Constant or Covariate	b_i	SE(b_i)
Weight	−0.699	0.128
HR MAX	0.289	0.078
SV MAX	0.448	0.0511
Constant	−1.454	

Source	DF	SS	MS	F
Regression	?	?	?	?
Residual	?	55.97	?	
Total	?	1305.08		

Y	$\widehat{Y}$	Residual
30.3	30.38	−0.08
27.1	?	−4.64
44.4	45.60	−1.20
24.6	24.65	?
41.2	39.53	1.67
48.9	?	−0.75
62.1	63.80	−1.70
50.9	45.88	?
33.2	32.15	1.05
32.5	?	?

Do Tasks (a), (c), (d), (e), (f).

5. After examining the normal probability plot of residuals, the regression of Problem 4 was rerun omitting cases 2 and 8. In this case we find:

Constant or Covariate	b_i	SE(b_i)
Weight	−0.615	0.039
HR MAX	0.274	0.024
SV MAX	0.436	0.015
Constant	−4.486	

Source	DF	SS	MS	F
Regression	?	1017.98	?	?
Residual	?	?	?	
Total	?	1021.18		

Y	$\widehat{Y}$	Residual
30.3	?	?
44.4	?	−0.45
24.6	25.62	?
41.2	?	1.09
48.9	49.35	?
62.1	?	?
33.2	33.28	−0.08
32.5	31.77	0.73

Do Tasks (a), (b-i), (c), (d), (f). Comment: The very small residual (high R^2) indicates the data are very likely highly "over fit". Compute R^2.

6. The selection of the regression variables of Problems 4 and 5 was based on Mallows' C_p plot. With so few cases, the multiple comparison problem looms large. As an independent verification, we try the result on the data of the 11 normal women.
   We find:

Constant or Covariate	b_i	SE(b_i)
Weight	−0.417	0.201
HR MAX	0.441	0.098
SV MAX	0.363	0.160
Constant	−51.96	

Source	DF	SS	MS	F
Regression	?	419.96	?	?
Residual	?	117.13	?	
Total	?	?		

Y	$\widehat{Y}$	Residual
28.81	?	−1.75
24.04	?	−1.72
26.66	27.99	?
24.34	29.63	?
21.42	?	?
26.72	?	?
23.72	23.89	−0.15
28.72	31.14	−2.42
20.77	16.30	4.46
24.77	23.60	1.17
47.72	40.77	6.95

Do Tasks (a), (b-i), (c), (d), (e), and (f). Do (e) or (f) look suspicious? Why?

7. Another run with the data of Problem 6 omitting the last point.

Constant or Covariate	b_i	$SE(b_i)$
Weight	−0.149	0.074
HR MAX	0.233	0.042
SV MAX	0.193	0.056
Constant	−20.52	

Source	DF	SS	MS	F
Regression	?	?	?	?
Residual	?	?	?	
Total	?	?		

Note the large change in the b_i's omitting the "outlier".

Y	$\widehat{Y}$	Residual
28.81	27.54	1.27
24.04	24.59	−0.55
26.66	?	?
24.34	26.70	−2.36
21.42	?	?
26.72	?	−0.11
23.72	?	0.57
28.72	28.08	?
20.77	20.23	?
24.77	23.96	0.81

Do Tasks (a), (c), and (d). Find R^2. Do you think the female findings roughly support the results for the males?

8. Consider the regression of Y on $X_1, X_2, \ldots, X_6$. Which of the following five hypotheses are *nested* within other hypotheses?

$$H_1 : \beta_1 = \beta_2 = \beta_3 = \beta_4 = \beta_5 = \beta_6 = 0,$$
$$H_2 : \beta_1 = \beta_5 = 0,$$
$$H_3 : \beta_1 = \beta_5,$$
$$H_4 : \beta_2 = \beta_5 = \beta_6 = 0,$$
$$H_5 : \beta_5 = 0.$$

9. Consider a hypothesis H_1 nested within H_2. Let R_1^2 be the multiple correlation coefficient for H_1 and R_2^2 the multiple correlation coefficient for H_2. Suppose there are n observations and H_2 regresses on Y and $X_1, \ldots, X_k$ while H_1 regresses Y only on the first j X_i's $(j < k)$. Show that the F statistic for testing $\beta_{j+1} = \cdots = \beta_k = 0$ may be written as

$$F = \frac{\left(R_2^2 - R_1^2\right)/(k - j)}{\left(1 - R_2^2\right)/(n - k - 1)}. \qquad \square$$

Florey and Acheson [1969] studied blood pressure as it relates to physique, blood glucose, and serum cholesterol separately for males and females, blacks and whites. Table 11.10 presents sample correlation coefficients for black males on the following variables:

Height: in inches.

Weight: in pounds.

Right triceps skinfold: in thickness in centimeters of skin folds on the back of the right arm; this was measured with standard calipers.

Infrascapular skinfold: skinfold thickness on the back below the tip of the right scapula.

Arm girth: circumference of the loose biceps.

Glucose: taken 1 h after a challenge of 50 g of glucose in 250 cm^3 of water.

Total serum cholesterol concentration.

Age: in years.

Systolic blood pressure: (mmHg).

Diastolic blood pressure: (mmHg).

Table 11.10. Simple Correlation Coefficients Between Nine Variables for Black Men: United States, 1960–1962.

Variable	1	2	3	4	5	6	7	8	9
1. Height	—								
2. Weight	0.34	—							
3. Right triceps skinfold	−0.04	0.61	—						
4. Infrascapular skinfold	−0.05	0.72	0.72	—					
5. Arm Girth	0.10	0.89	0.60	0.70	—				
6. Glucose	−0.20	−0.05	0.09	0.10	−0.03	—			
7. Cholesterol	−0.08	0.15	0.17	0.20	0.17	0.12	—		
8. Age	−0.23	−0.09	−0.05	0.02	−0.10	0.37	0.34	—	
9. Systolic blood pressure	−0.18	0.11	0.07	0.12	0.12	0.29	0.20	0.47	—
10. Diastolic blood pressure	−0.09	0.17	0.08	0.16	0.18	0.20	0.17	0.33	0.79

Notes: Number of Observations for samples: $N = 358$ and $N = 349$. Figures which are underlined were derived from persons in the sample for whom glucose and cholesterol measurements were available.

An additional variable considered was the *ponderal index* defined to be the height divided by the cube root of the weight.

Note the samples sizes varied because of a few uncollected blood specimens. For the problem below use $N = 349$.

10. Using the Florey and Acheson [1969] data above, the correlation squared of systolic blood pressure, variable 9 with the age and physical variables (variables 1, 2, 3, 4, 5 and 8) is 0.266. If we add variables 6 and 7, the blood glucose and chloresterol variables, R^2 increases to 0.281. Using the result of Problem 9, is this a statistically significant difference?

11. Suppose that the following description of a series of multiple regression runs was presented. Find any incorrect or inconsistent statements (if they occur).

    Forty-five individuals were given a battery of psychological tests. The dependent variable of self-image was analyzed by multiple regression analysis with five predictor variables: 1, tension index; 2, perception of success in life; 3, IQ; 4, aggression index; and 5, a hypochondriacal index. The multiple correlation with variables 1, 4, and 5 was -0.329, $p < 0.001$. When variables 2 and 3 were added to the predictive equation $R^2 = 0.18$, $p > 0.05$. The relationship of self-image to the variables was complex; the correlation with variables 2 and 3 was low (0.03 and -0.09, respectively), but the multiple correlation of self-image with variables 2 and 3 was higher than expected, $R^2 = 0.22$, $p < 0.01$.

12. Using the definition of R^2 (Definition 11.4) and the multiple regression F test in Section 11.2.3, show that

$$R^2 = \frac{kF}{kF + n - k - 1}$$

and

$$F = \frac{(n - k - 1)R^2}{k(1 - R^2)}. \qquad \square$$

Haynes *et al.* [1978] consider the relationship of psychological factors and coronary heart disease. As part of a long ongoing study of coronary heart disease, the Framingham study, from 1965 to 1967, questionnaires were given to 1822 individuals.

Of particular interest was "type A" behavior. Roughly speaking type A individuals feel considerable time pressure, are very driving and aggressive, and feel a need for perfection. Such behavior has been linked with coronary artery disease. The questions and scales (see end of questions) used in this study are:

Psychosocial Scale and Items Used in the Framingham Study

1. Framingham type A behavior pattern:
   Traits and qualities which describe you:[1]
   > Being hard-driving and competitive
   > Usually pressed for time
   > Being bossy and dominating

       Having a strong need to excel in most things

       Eating too quickly

   Feeling at the end of an average day of work:

       Often felt very pressed for time

       Work stayed with you so you were thinking about it after hours

       Work often stretched you to the very limits of your energy and capacity

       Often felt uncertain, uncomfortable, or dissatisfied with how you were doing

   Do you get upset when you have to wait for anything?

2. Emotional lability:

   Traits and qualities which describe you:[1]

       Having feelings easily hurt

       Getting angry very easily

       Getting easily excited

       Getting easily sad or depressed

       Worrying about things more than necessary

   Do you cry easily?

   Are you easily embarassed?

   Are your feelings easily hurt?

   Are you generally a high strung person?

   Are you usually self-conscious?

   Are you easily upset?

   Do you feel sometimes that you are about to go to pieces?

   Are you generally calm and not easily upset?

3. Ambitiousness:

   Traits and qualities which describe you:[1]

       Being very socially ambitious

       Being financially ambitious

       Having a strong need to excel in most things

4. Noneasygoing:

   Traits and qualities which describe you:[1]

       Having a sense of humor

       Being easygoing

       Having ability to enjoy life

5. Nonsupport from boss:

   Boss (the person directly above you):[2]

       Is a person you can completely trust

       Is cooperative

       Is a person you can rely upon to carry his load

       Is a person who appreciates you

       Is a person who interferes with you or makes it difficult for you to get your work done

       Is a person who generally lets you know how you stand

       Is a person who takes a personal interest in you

6. Marital dissatisfaction:

   Everything considered, how happy would you say that your marriage has been?[3]

   Everything considered, how happy would you say that your spouse has found your marriage to be?[3]

About marriage, are you more satisfied, as satisfied, or less satisfied than most of your close friends are with their marriages?[4]

7. Marital disagreement:

How often do you and your spouse disagree about:[5]
  Handling family finances or money matters
  How to spend leisure time
  Religious matters
  Amount of time that should be spent together
  Gambling
  Sexual relations
  Dealings with in-laws
  On bringing up children
  Where to live
  Way of making a living
  Household chores
  Drinking

8. Work overload:

Regular line of work fairly often involves:[2]
  Working overtime?
  Meeting deadlines or rigid time schedules?

9. Aging worries:

Worry about:[6]
  Growing old
  Retirement
  Sickness
  Death
  Loneliness

10. Personal worries:

Worry about:[6]
  Sexual problems
  Change of life
  Money matters
  Family problems
  Not being a success

11. Tensions:

Often troubled by feelings of tenseness, tightness, restlessness, or inability to relax?[5]
Often bothered by nervousness or shaking?
Often have trouble sleeping or falling asleep?
Feel under a great deal of tension?
Have trouble relaxing?
Often have periods of restlessness so that you cannot sit for long?
Often felt difficulties were piling up too much for you to handle?

12. Reader's daily stress:

At the end of the day I am completely exhausted mentally and physically[1]
There is a great amount of nervous strain connected with my daily activities
My daily activities are extremely trying and stressful
In general I am usually tense and nervous

13. Anxiety symptoms:
    Often become tired easily or feel continuously fatigued?[2]
    Often have giddiness or dizziness or a feeling of unsteadiness?
    Often have palpitations, or a pounding or racing heart?
    Often bothered by breathlessness, sighing respiration or difficulty in getting a deep breath?
    Often have poor concentration or vagueness in thinking?

14. Anger symptoms:
    When really angry or annoyed:[7]
        Get tense or worried
        Get a headache
        Feel weak
        Feel depressed
        Get nervous or shaky

15. Anger-in:
    When really angry or annoyed:[7]
        Try to act as though nothing much happened
        Keep it to yourself
        Apologize even though you are right

16. Anger-out:
    When really angry or annoyed:[7]
        Take it out on others
        Blame someone else

17. Anger-discuss:
    When really angry or annoyed:[7]
        Get it off your chest
        Talk to a friend or relative

Response Sets:

1. Very well, fairly well, somewhat, not at all
2. Yes, no
3. Very happy, happy, average, unhappy, very unhappy
4. More satisfied, as satisfied, less satisfied
5. Often, once in a while, never
6. A great deal, somewhat, a little, not at all
7. Very likely, somewhat likely, not too likely

The correlations between the indices were reported in Table 11.11.

13. Using the Haynes *et al.* [1978] data of Table 11.11. The multiple correlation squared of the Framingham Type A variable with all 16 of the other variables is 0.424. Note the high correlations for variables 2, 3, 14, 15, and 17.

$$R^2_{1(2,3,14,15,17)} = 0.352.$$

   a. Is there a statistically significant ($p < 0.05$) gain in R^2 by adding the remainder of the variables?

Table 11.11. Correlations Among 17 Framingham Psychosocial Scales with Continuous Distributions.

Phychosocial Scales	1	2	3	4	5	6	7	8	9	10	11	12	13	14	15	16	17
1. Framingham Type A		.43	.31	.09	.12	.23	.29	.06	.27	.32	-.04	.19	.11	.47	.42	.24	.34
2. Emotional lability			.12	.26	.08	.05	.21	.12	.37	.31	.10	.23	.11	.43	.61	.42	.60
3. Ambitiousness				-.23	.01	.01	-.05	-.04	.04	.06	.08	.03	.09	.12	.06	-.01	.07
4. Non-easygoing					.05	.03	.15	.22	.18	.17	-.12	.16	.00	.19	.22	.17	.18
5. Non-support from boss						.11	.11	-.01	.09	.10	-.06	-.01	-.02	.12	.10	.06	.06
6. Work overload							.11	-.07	.04	.06	-.03	-.07	.04	.15	.11	.02	.06
7. Marital disagreement								.44	.33	.47	-.08	.15	-.01	.21	.22	.18	.19
8. Marital dissatisfaction									.12	.25	.00	.02	-.02	.11	.12	.13	.13
9. Aging worries										.53	.01	.16	.04	.27	.33	.29	.31
10. Personal worries											-.05	.19	.03	.31	.33	.21	.31
11. Anger-in												-.18	-.07	.06	.11	.12	.18
12. Anger-out													.11	.11	.13	.09	.19
13. Anger-discuss														.08	.10	.06	.12
14. Daily stress															.51	.34	.41
15. Tension																.49	.61
16. Anxiety symptoms																	.45
17. Anger symptoms																	

b. Find the partial correlation of variables 1 and 2 after adjusting for variable 15. That is, what is the correlation of the Framingham Type A index and emotional lability if adjustment is made for the amount of tension? □

Stoudt *et al.* [1970] report upon the relationship between certain body size measurements and anthropometric indices. As one would expect, there is considerable correlation among such measurements. The details of the measurements are reported in the reference above. The correlation for women are given in the following table:

Body Measurement	2	3	4	5	6	7
1. Sitting height, erect	0.907	0.440	0.364	0.585	0.209	0.347
2. Sitting height, normal		0.420	0.352	0.533	0.199	0.327
3. Knee height			0.747	0.023	0.196	0.689
4. Popliteal height				−0.095	−0.141	0.429
5. Elbow rest height					0.293	0.051
6. Thigh clearance height						0.465
7. Buttock-knee length						
8. Buttock-popliteal length						
9. Elbow to elbow breadth						
10. Seat breadth						
11. Biacromial diameter						
12. Chest girth						
13. Waist girth						
14. Right arm girth						
15. Right arm skinfold						
16. Intrascapular skinfold						
17. Height						
18. Weight						
19. Age						

	8	9	10	11	12	13
1. Sitting height, erect	0.231	−0.032	0.204	0.350	0.059	−0.076
2. Sitting height, normal	0.230	−0.029	0.197	0.317	0.045	−0.091
3. Knee height	0.585	0.106	0.254	0.406	0.180	−0.121
4. Popliteal height	0.387	−0.200	−0.101	0.255	−0.126	0.166
5. Elbow rest height	−0.045	0.143	0.275	0.094	0.179	0.111
6. Thigh clearance height	0.352	0.597	0.609	0.370	0.594	0.523
7. Buttock-knee length	0.786	0.413	0.552	0.426	0.441	0.410
8. Buttock-popliteal length		0.326	0.390	0.341	0.371	0.333
9. Elbow to elbow breadth			0.696	0.331	0.878	0.870
10. Seat breadth				0.327	0.680	0.666
11. Biacromial diameter					0.433	0.301
12. Chest girth						0.862

Body Measurement	14	15	16	17	18	19
1. Sitting height, erect	0.052	0.057	−0.063	0.772	0.197	−0.339
2. Sitting height, normal	0.034	0.064	−0.063	0.729	0.165	−0.300
3. Knee height	0.128	0.100	0.041	0.782	0.322	−0.128
4. Popliteal height	−0.219	−0.193	−0.248	0.723	−0.035	−0.196
5. Elbow rest height	0.222	0.191	0.150	0.258	0.253	−0.177
6. Thigh clearance height	0.641	0.539	0.541	0.137	0.693	−0.026
7. Buttock-knee length	0.450	0.343	0.296	0.609	0.620	−0.036
8. Buttock-popliteal length	0.355	0.269	0.243	0.514	0.490	−0.005
9. Elbow to elbow breadth	0.835	0.619	0.751	−0.070	0.844	0.393
10. Seat breadth	0.746	0.614	0.596	0.137	0.805	0.187
11. Biacromial diameter	0.331	0.209	0.243	0.407	0.443	−0.116
12. Chest girth	0.843	0.615	0.762	0.016	0.882	0.317
13. Waist girth	0.803	0.589	0.747	−0.090	0.844	0.432
14. Right arm girth		0.740	0.774	−0.026	0.888	0.272
15. Right arm skinfold			0.755	−0.022	−0.641	0.203
16. Intrascapular skinfold				−0.136	0.729	0.278
17. Height					0.189	−0.289
18. Weight						0.204
19. Age						

14. This problem deals with partial correlations.

   a. For the Stoudt *et al.* [1970] data, the multiple correlation of seat breadth with height and weight is 0.64826. Find

$$r_{\text{seat breadth, height.weight}},$$

$$r_{\text{seat breadth, weight.height}}.$$

   b. The Florey and Acheson [1969] data show that the partial multiple correlation between systolic blood pressure and the two predictor variables glucose and cholesterol adjusting for the weight and measurement variables is

$$R^2_{9(6,7).1,2,3,4,5,8} = 0.207, \quad R = 0.144.$$

   What are the numerator and denominator degrees of freedom for testing statistical significance? What is (approximately) the 0.05 (0.01) critical value? Find F in terms of R^2. Do we reject the null hypothesis of no correlation at the 5% (1%) level?

15. Suppose you want to regress Y on $X_1, X_2, \ldots, X_8$. There are 73 observations. Suppose you are given the following sums of squares:

$$\text{SS}_{\text{TOTAL}}, \text{SS}_{\text{REG}}(X_1), \text{SS}_{\text{REG}}(X_4), \text{SS}_{\text{REG}}(X_1, X_5),$$

$$\text{SS}_{\text{REG}}(X_3, X_6), \text{SS}_{\text{REG}}(X_7, X_8), \text{SS}_{\text{REG}}(X_1, X_5, X_6),$$

$$\text{SS}_{\text{REG}}(X_1, X_3, X_6), \text{SS}_{\text{REG}}(X_4, X_7, X_8), \text{SS}_{\text{REG}}(X_3, X_5, X_6, X_8),$$

$$\text{SS}_{\text{REG}}(X_3, X_4, X_7, X_8).\text{SS}_{\text{REG}}(X_3, X_5, X_6, X_7, X_8).$$

For each of the following: (1) state the quantity cannot be estimated, or (2) show (a) how to compute the quantity in terms of the sums of squares, and (b) give the F statistic in terms of the sums of squares and give the degress of freedom.

a. r^2_{Y,X_3},

b. $R^2_{Y(X_1,X_5,X_6)}$,

c. $R^2_{Y(X_1,X_5,X_6).X_3}$,

d. $R^2_{Y(X_3,X_4,X_7,X_8)}$,

e. $r^2_{Y,X_6.X_1,X_5}$,

f. $R^2_{Y(X_5,X_6).X_3,X_4}$,

g. $R^2_{Y(X_3,X_4).X_7,X_8}$,

h. $R^2_{Y(X_3,X_5,X_6).X_7,X_8}$.

16. Suppose in the Framingham study (Haynes *et al.* [1978] above) we want to examine the relationship between Type A behavior and anger (as given by the four anger variables). We would like to be sure the relationship does not occur because of joint relationships with the other variables; that is, we want to adjust for all the variables other than Type A (variable 1) and the anger variables 11, 12, 13 and 17.

   a. What quantity would you use to look at this?
   b. If the value (squared) is 0.019, what is the value of the F statistic to test for significance? The degrees of freedom?

17. Suppose using the Framingham data we decide to examine emotional lability. We want to see how it is related to four areas characterized by variables as follows:

Work	Variables 5 and 6.
Worry and Anxiety	Variables 9, 10, and 16.
Anger	Variables 11, 12, 13, and 17.
Stress and tension	Variables 14 and 15.

   a. To get a rough idea of how much relationship one might expect, we calculate

   $$R^2_{2(5,6,9,10,16,11,12,13,17,14,15)} = 0.49.$$

   b. To see which group or groups of variables may be contributing the most to this relationship, we find

   $$
   \begin{aligned}
   R^2_{2(5,6)} &= 0.01 && \text{Work} \\
   R^2_{2(9,10,16)} &= 0.26 && \text{Worry/anxiety} \\
   R^2_{2(11,12,13,17)} &= 0.38 && \text{Anger} \\
   R^2_{2(14,15)} &= 0.39 && \text{Stress/tension}
   \end{aligned}
   $$

c. As the two most promising set of variables were the anger and the stress/tension, we compute

$$R^2_{2(11,12,13,17,14,15)} = 0.48.$$

i. Might we find a better relationship (larger R^2) by working with indices such as the average score on variables 11, 12, 13, and 17 for the anger index? Why or why not?

ii. After using the anger and stress/tension variables is there statistical significance left in the relationship of lability and work and work/anxiety? What quantity would estimate this relationship? [Chapter 14 shows some other ways to analyze these data.]

18. The Jensen *et al.* [1980] data of 19 subjects were used in Problems 23–29 of Chapter 9. Here we consider the data before training. The exercise VO₂ MAX is to be regressed upon three variables.

$$Y = \text{VO}_2 \text{ MAX},$$

$$X_1 = \text{maximal ejection fraction},$$

$$X_2 = \text{maximal heart rate},$$

$$X_3 = \text{maximal systolic blood pressure}.$$

The residual mean square with all three variables in the model is 73.40. The residual sums of squares are

$$\text{ss}_{\text{RESID}}(X_1, X_2) = 1101.58,$$

$$\text{ss}_{\text{RESID}}(X_1, X_3) = 1839.80,$$

$$\text{ss}_{\text{RESID}}(X_2, X_3) = 1124.78,$$

$$\text{ss}_{\text{RESID}}(X_1) = 1966.32,$$

$$\text{ss}_{\text{RESID}}(X_2) = 1125.98,$$

$$\text{ss}_{\text{RESID}}(X_3) = 1885.98.$$

a. For each model compute C_p.
b. Plot C_p versus p and select the best model.
c. Compute and plot the average mean square residual versus p.

19. The 20 aortic valve cases of Problem 3 give the following data about the values of C_p and the residual mean square.

Numbers of the Explanatory Variables	p	C_p	Residual Mean Square	Numbers of the Explanatory Variables	p	C_p	Residual Mean Square
None	1	14.28	886.99	2,4,5	4	2.29	468.36
				1,4,5		2.41	472.20
4	2	3.87	578.92	3,4,5		2.69	481.50
5		11.60	804.16	1,3,4		6.91	619.81
3		13.63	863.16	1,2,4		6.91	619.90
2		14.14	877.97	2,3,4		7.80	648.81
1		16.00	932.21	2,3,5		14.14	856.68
				1,3,5		14.40	866.45
4,5	3	0.72	454.10	1,2,5		14.45	866.75
1,4		4.94	584.23	1,2,3		15.21	891.72
2,4		5.82	611.35				
3,4		5.87	612.75	1,2,4,5	5	4.05	491.14
1,5		12.76	825.45	2,3,4,5		4.16	494.92
3,5		12.96	831.53	1,3,4,5		4.41	503.66
2,5		13.17	838.17	1,2,3,4		8.90	660.65
2,3		13.23	839.87	1,2,3,5		15.83	903.14
1,3		15.60	912.88				
1,2		15.96	924.03	1,2,3,4,5	6	6	524.37

a. Plot Mallow's C_p plot and select the "best" model.

b. Plot the average residual mean square versus p. Is it useful in this context? Why or why not?

20. The blood pressure, physique, glucose, and serum cholesterol work of Florey and Acheson [1969] was mentioned above. The authors first tried using a variety of regression analyses. It was known that the relationship between age and blood pressure is often curvilinear so an age^2 term was used as a potential predictor variable. After exploratory analyses, stepwise regression of blood pressure (systolic or diastolic) upon five variables (age, age^2, ponderal index, glucose, and cholesterol) was run. The four regressions (black and white, female and male) for systolic blood pressure are given in Tables 11.12–11.15. The "standard error of the estimate" is the estimate of σ^2 at each stage.

a. For the black men, give the values of the partial F statistics and the degrees of freedom as each variable entered the equation.

b. Are the F values in Part (a) significant at the 5% significance level?

c. For fixed ponderal index of 32 and glucose level of 125 mg%, plot the regression curve for systolic blood pressure for white women with ages 20–70 years.

d. Can you determine the partial correlation of systolic blood pressure and glucose adjusting for age in black women from these data? If so, give the value.

*e. Consider all the multiple regression R^2 values of systolic blood pressure with subsets of the five variables used. For white males and these data give all

Table 11.12. Selected Regression Statistics for Systolic Blood Pressure and Selected Independent Variables of *White* Men: United States, 1960–62.

		Dependent Variable = Systolic Blood Pressure				
Step	Variables Entered	Multiple R	R^2	Increase in R^2	Regression Coefficient	Standard Error of Estimate
1	Age squared	0.439	0.193	0.193	0.0104	17.9551
2	Ponderal index	0.488	0.238	0.045	−6.1775	17.4471
3	Glucose	0.499	0.249	0.011	0.0500	17.3221
4	Cholesterol	0.503	0.253	0.004	0.0351	17.2859
5	Age	0.507	0.257	0.004	−0.5136	17.2386

Note: Constant term = 194.997. $N = 2599$.

Table 11.13. Selected Regression Statistics for Systolic Blood Pressure and Selected Independent Variables of *Black* Men: United States, 1960–62.

		Dependent Variable = Systolic Blood Pressure				
Step	Variables Entered	Multiple R	R^2	Increase in R^2	Regression Coefficient	Standard Error of Estimate
1	Age squared	0.474	0.225	0.225	0.6685	21.9399
2	Ponderal index	0.509	0.259	0.034	−6.4515	21.4769
3	Glucose	0.523	0.273	0.014	0.0734	21.3048

Note: Constant term = 180.252. $N = 349$.

Table 11.14. Selected Regression Statistics for Systolic Blood Pressure and Selected Independent Variables of *White* Women: United States, 1960–62.

		Dependent Variable = Systolic Blood Pressure				
Step	Variables Entered	Multiple R	R^2	Increase in R^2	Regression Coefficient	Standard Error of Estimate
1	Age squared	0.623	0.388	0.388	0.00821	18.9317
2	Ponderal index	0.667	0.445	0.057	−7.3925	18.0352
3	Glucose	0.676	0.457	0.012	0.0650	17.8445

Note: Constant term = 193.260. $N = 2931$.

Table 11.15. Selected Regression Statistics for Systolic Blood Pressure and Selected Independent Variables of *Black* Women: United States, 1960–62.

		Dependent Variable = Systolic Blood Pressure				
Step	Variables Entered	Multiple R	R^2	Increase in R^2	Regression Coefficient	Standard Error of Estimate
1	Age squared	0.590	0.348	0.348	0.9318	24.9930
2	Ponderal index	0.634	0.401	0.053	0.1388	23.9851
3	Glucose	0.656	0.430	0.029	−6.0723	23.4223

Note: Constant term = 153.149. $N = 443$.

possible inequalities that are *not* of the obvious form:

$$R^2_{Y(X_{i_1}, \ldots, X_{i_m})} \leq R^2_{Y(X_{j_1}, \ldots, X_{j_n})},$$

where $X_{i_1}, \ldots, X_{i_m}$ is a subset of $X_{j_1}, \ldots, X_{j_n}$.

21. From a correlation matrix it is possible to compute the order in which variables enter a stepwise multiple regression. The partial correlations, F statistics and regression coefficients for the standardized variables (except for the constant) may be computed. The first 18 women's body dimension variables (as given in Stoudt *et al.* [1970] and mentioned above) were used. The dependent variable was weight, which we are trying to predict in terms of the 17 measured dimension variables. Because of the large sample size, it is "easy" to find statistical significance. In such cases the procedure is sometimes terminated while statistically significant predictor variables yet remain. In this case, the addition of predictor variables was stopped when R^2 would increase by less than 0.01 for the next variable. The variable numbers, the partial correlation with the dependent variable (conditioning upon variables in the predictive equation) for the variables not in the model and the corresponding F-value were:

STEP 0

var	PCORR	F	var	PCORR	F
1	0.1970	144.506	10	0.8050	6589.336
2	?	100.165	11	0.4430	873.872
3	0.3230	?	12	0.8820	12537.104
4	−0.0350	4.390	13	0.8440	8862.599
5	0.2530	244.755	14	0.8880	13346.507
6	0.6930	3306.990	15	0.6410	2496.173
7	0.6200	?	16	0.7290	4059.312
8	0.4900	1130.830	17	0.1890	132.581
9	?	8862.599			

The F statistics have 1 and 3579 DF.

STEP 1

var	PCORR	F	var	PCORR	F
1	0.3284	432.622	9	0.4052	?
2	0.2933	?	10	0.4655	989.824
3	0.4568	943.565	11	0.3435	478.797
4	0.3554	517.351	12	0.5394	1467.962
5	0.1246	56.419	13	0.4778	1058.297
6	?	501.893	15	−0.0521	9.746
7	0.5367	1447.655	16	?	74.882
8	0.4065	708.359	17	0.4614	967.603

The F statistics have 1 and 3578 DF.

STEP 5

var	PCORR	F	var	PCORR	F
1	?	323.056	8	0.0051	0.093
2	0.2285	196.834	9	0.0083	0.252
3	0.1623	96.676	11	0.1253	?
4	0.1157	48.503	15	−0.1298	61.260
5	?	183.520	16	−0.0149	?
6	0.2382	214.989	17	0.3131	388.536

The F statistics have 1 and ? DF.

LAST STEP: STEP ?

var	PCORR	F	var	PCORR	F
1	?	5.600	8	−0.0178	1.143
2	−0.0289	2.994	9	0.0217	1.685
3	−0.0085	0.263	11	0.0043	0.067
4	−0.0172	1.062	15	−0.1607	94.635
5	0.0559	?	16	−0.0034	0.042

The F statistics have 1 and 3572 DF.

a. Fill in the question marks at steps 0 and 1.

b. Fill in the question marks in step 5.

c. Fill in the question marks in the last step.

d. Which variables entered the predictive equation?

*__e.__ What can you say about the proportion of the variability in weight explained by the measurements?

f. What can you say about the p-value of the next variable that would have entered the stepwise equation? [Note that this small p has less than 0.01 gain in R^2 if entered into the predictive equation.]

22. Data from Hossack *et al.* [1980] and [1981] for men and women (Problems 4–7) were combined. The maximal cardiac output, QDOT, was regressed upon the maximal oxygen uptake, VO_2 MAX. From other work, the possibility of a curvilinear relationship was entertained. Polynomials of the 0, 1, 2, and 3 degree (or highest power of X) were considered. Portions of the BMDP output are presented below, with appropriate questions (see Figures 11.17–11.19)

a. Goodness-of-Fit Test: For the polynomial of each degreee, a test is made for additional information in the orthogonal polynomials of higher degree. The numerator sum of squares for each of these tests is the sum of squares attributed to all orthogonal polynomials of higher degree and the denominator sum of squares is the residual sum of squares from the fit to the highest degree polynomial (fit to all orthogonal polynomials). A significant F statistic thus indicates that a higher degree polynomial should be considered.

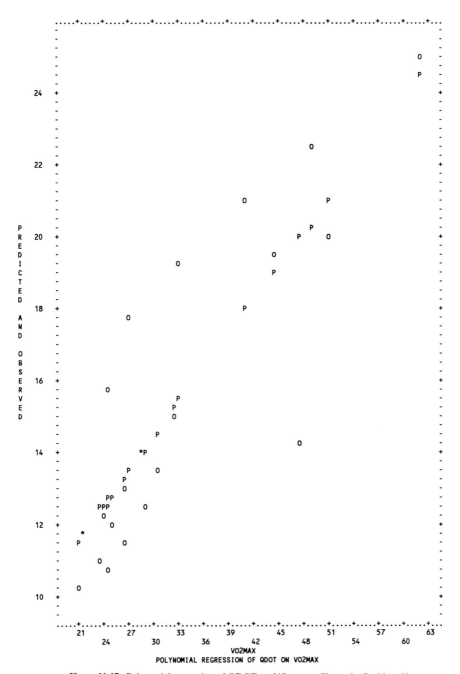

Figure 11.17. Polynomial regression of QDOT on VO$_2$ MAX. Figure for Problem 22.

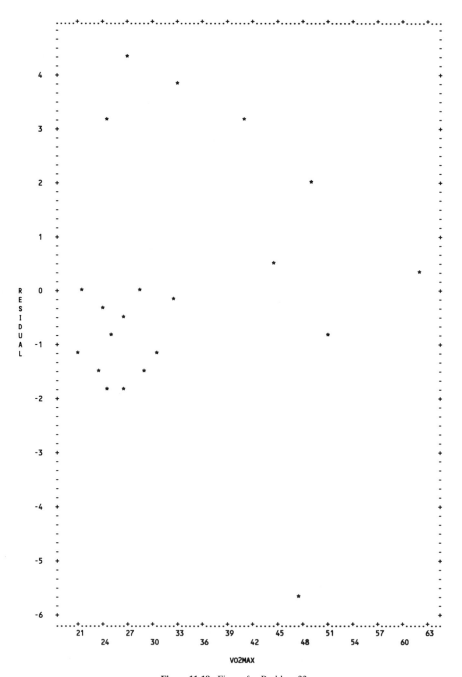

Figure 11.18. Figure for Problem 22.

NORMAL PROBABILITY PLOT OF RESIDUALS FOR DEGREE 1

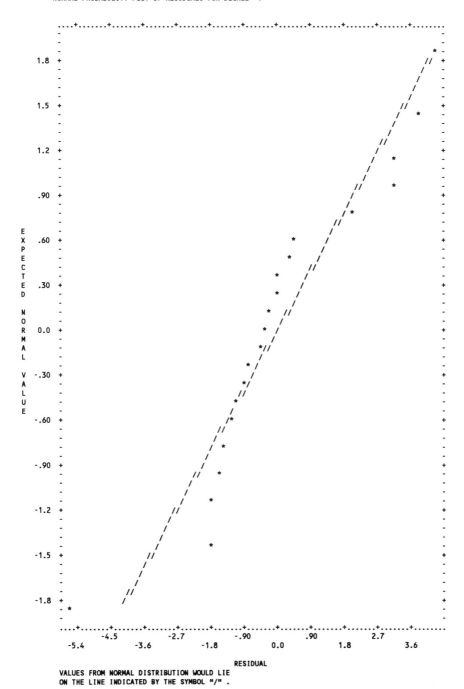

Figure 11.19. Figure for Problem 22.

Degree	SS	DF	MS	F	Tail Probability
0	278.50622	4	69.62656	12.04	0.00
1	12.23208	3	4.07736	0.70	0.56
2	10.58430	2	5.29215	0.91	0.42
3	5.22112	1	5.22112	0.90	0.36
Residual	92.55383	16	5.78461		

What degree polynomial appears most appropriate? Why do the degrees of freedom in this table add up to more than the total number of observations (21)?

b. For a linear equation the coefficients, observed and predicted values, residual plot and normal residual are:

POLYNOMIAL IN X

Degree	Regression Coefficient	Standard Error	T Value
0	4.88737	1.58881	3.08
1	0.31670	0.04558	6.95

(Note that the computer output uses "T" for the t-statistic value.) What would you conclude from the normal probability plot? Is the most outlying point a male or female? Which subject number in its table?

c. For those with access to a polynomial regression program. Rerun the problem removing the outlying point.

Table 11.16. Weight by Height Distribution for Men 25–34 Years of Age: Health Examination Survey, 1960–1962.

Height in Inches	Total	Number of examinees at weight in pounds									
		Under 130	130– 139	140– 149	150– 159	160– 169	170– 179	180– 189	190– 199	200– 209	210+
Total	675	39	50	78	93	92	87	74	56	48	58
<63	11	3	2	2	4	—	—	—	—	—	—
63	11	2	2	1	4	1	1	—	—	—	—
64	34	10	4	5	5	4	3	1	—	1	1
65	28	6	3	—	7	2	6	1	—	2	1
66	67	6	7	8	11	14	9	2	5	2	3
67	70	4	6	17	9	11	5	5	5	5	3
68	120	5	14	18	25	11	13	13	12	5	4
69	80	1	5	9	10	11	14	11	8	8	3
70	103	2	4	9	9	17	16	14	9	8	15
71	48	—	1	5	4	7	7	7	4	5	8
72	57	—	2	2	4	8	8	8	9	5	11
≥73	46	—	—	2	1	6	5	12	4	7	9

Note: Height without shoes; weight partially clothed; clothing weight estimated as averaging 2 pounds.

Table 11.17. Number of Men Aged 25–34 years by weight for height: United States, 1971–1974.

Height in Inches	Total	Number of examinees at weight in pounds									
		Under 130	130– 139	140– 149	150– 159	160– 169	170– 179	180– 189	190– 199	200– 209	210+
Total	804	33	54	86	129	102	103	84	72	42	99
<63	6	1	3	1	—	—	1	—	—	—	—
63	17	4	3	5	3	—	—	—	1	—	1
64	23	3	5	8	2	1	1	1	1	1	—
65	41	5	6	7	11	3	3	1	2	2	1
66	70	5	10	11	11	10	9	5	6	2	1
67	86	3	10	6	19	15	11	9	5	4	4
68	92	5	4	15	12	15	14	13	7	2	5
69	120	3	5	10	26	17	22	8	10	4	15
70	112	2	5	12	15	14	11	18	13	10	12
71	73	2	1	8	14	10	8	7	13	1	9
72	69	—	2	1	10	9	8	9	5	6	19
≥73	95	—	—	2	2	8	15	13	9	10	32

Note: Height without shoes; weight partially clothed; clothing weight estimated as averaging 2 pounds.

23. As in Problem 22, this problem deals with a potential polynomial regression equation. Weight and height were collected from a sample of the US population in surveys done in 1960–1962 (Roberts [1966]) and in 1971–1974 (Abraham et al. [1979]). The data for males 25–34 years of age are given in Tables 11.16 and 11.17. In this problem we use only the 1960–1962 data. Both data sets are used in Problem 38. The weight categories were coded as values 124.5, 134.5, ..., 204.5, 214.5 and the height categories as 62, 63, ..., 72, 73. The contingency table was replaced by 675 "observations". As before, we present some of the results from a BMDP computer output. The height was regressed upon weight.

   a. Goodness-of-Fit Test: For the polynomial of each degree, a test is made for additional information in the orthogonal polynomials of higher degree. The numerator sum of squares attributed to all orthogonal polynomials of higher degree and the denominator sum of squares is the residual sum of squares from the fit to the highest degree polynomial (fit to all polynomials). A significant F statistic thus indicates that a higher degree polynomial should be considered.

Degree	SS	DF	MS	F	Tail Probability
0	900.86747	3	300.28916	54.23	0.00
1	41.69944	2	20.84972	3.77	0.02
2	2.33486	1	2.33486	0.42	0.52
Residual	3715.83771	671	5.53776		

   Which degree polynomial appears most satisfactory?

   b. Coefficients with corresponding t statistics are given for the first, second, and third degree polynomials.

POLYNOMIAL IN X

Degree	Regression Coefficient	Standard Error	T Value
0	61.04225	0.60868	100.29
1	0.04408	0.00355	12.40

POLYNOMIAL IN X

Degree	Regression Coefficient	Standard Error	T Value
0	50.89825	3.85106	13.22
1	0.16548	0.04565	3.62
2	−0.00036	0.00013	−2.67

POLYNOMIAL IN X

Degree	Regression Coefficient	Standard Error	T Value
0	34.30283	25.84667	1.33
1	0.46766	0.46760	1.00
2	−0.00216	0.00278	−0.78
3	0.00000	0.00001	0.65

Does this confirm the results of (a)? How can the second order term be significant for the second degree polynomial, but neither the second or third power has a statistically significant coefficient when a third order polynomial is used?

c. The normal probability plot of residuals for the second degree polynomials is shown in Figure 11.20. What does the tail behavior indicate (as compared to normal tails)? Think about how we obtained those data and how they were generated. Can you explain this phenomenon? This may account for the findings, the original data would be needed to evaluate the extent of this problem.

Most multiple regression analyses (other than examining fit and model assumptions) use sums of squares rather than the original data. Problems 24–29 illustrate this point. The problems are based upon the 20 aortic valve surgery cases of Chapter 9, the introduction to Problems 30–33; Problem 3 above uses these data. We consider the regression sums of squares for all possible subsets of five predictor variables. Here

$$Y = \text{EDVI postoperative}, \quad X_1 = \text{age in years},$$

$$X_2 = \text{heart rate}, \quad X_3 = \text{systolic blood pressure},$$

$$X_4 = \text{EDVI preoperative}, \quad X_5 = \text{SVI preoperative}.$$

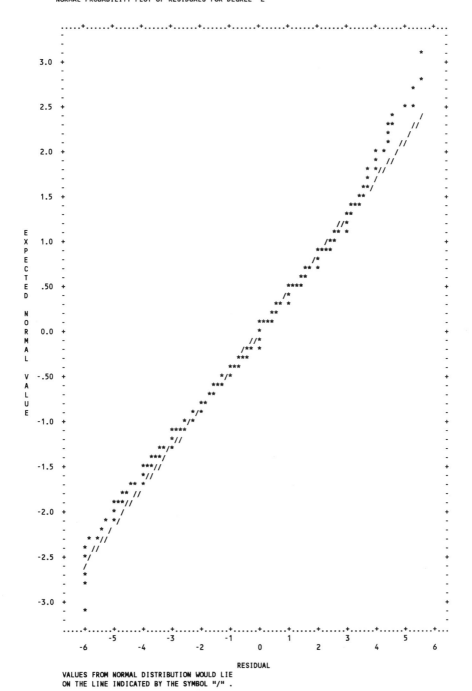

Figure 11.20. Normal probability plot of residuals of degree 2. Figure for Problem 23.

Indices of Variables in the Multiple Regression Equation	Regression Sum of Squares SS_{REG}	Indices of Variables in the Multiple Regression Equation	Regression Sum of Squares SS_{REG}
(SS_{TOTAL})	($SS_{TOTAL} = 32513.75$)	1,2,3	2775.87
1	671.04	1,2,4	14982.64
2	926.11	1,2,5	2585.85
3	1366.28	1,3,4	15013.76
4	12619.27	1,3,5	3064.41
5	658.21	1,4,5	21648.27
1,2	1607.06	2,3,4	13097.36
1,3	1620.17	2,3,5	2743.26
1,4	14973.55	2,4,5	22482.01
1,5	2397.10	3,4,5	21638.86
2,3	2547.67	1,2,3,4	15017.57
2,4	12619.61	1,2,3,5	3412.22
2,5	1145.53	1,2,4,5	22488.97
3,4	13090.47	1,3,4,5	21651.03
3,5	2066.16	2,3,4,5	22511.26
4,5	21631.66	1,2,3,4,5	22530.91

24. From the regression sums of squares compute and plot C_p values for the smallest C_p value for each p (that is, for the largest SS_{REG}). Plot these values. Which model appears best?

25. From the regression sums of squares perform a step-up stepwise regression. Use the 0.05 significance level to stop adding variables. Which variables are in the final model?

*26. From the regression sums of squares perform a *stepdown* stepwise regression. Use the 0.10 significance level to stop removing variables. What is your final model?

27. Compute the following multiple correlation coefficients.

$$R_{Y(X_4,X_5)}, \quad R_{Y(X_1,X_2,X_3,X_4,X_5)}, \quad R_{Y(X_1,X_2,X_3)}.$$

Which are statistically significant at the 0.05 significance level?

28. Compute the following squared partial correlation coefficients and test their statistical significance at the 1% level.

$$r^2_{Y,X_4 \cdot X_1,X_2,X_3,X_5}, \quad r^2_{Y,X_5 \cdot X_1,X_2,X_3,X_4}$$

29. Compute the following partial multiple correlation coefficients and test their statistical significance at the 5% significance level.

$$R_{Y(X_4,X_5) \cdot X_1,X_2,X_3}, \quad R_{Y(X_1,X_2,X_3,X_4) \cdot X_5}.$$

Data on the 94 sedentary males of Chapter 9, Problems 9–12 are used here. The dependent variable was age. The idea is to find an equation that predicted age; this equation might give an approximation to an "exercise age". Subjects might be encouraged, or convinced, to exercise if they heard a statement such as. "Mr. Jones, although you are 28 your exercise performance is that of a 43 year old sedentary man." The potential predictor variables with the regression sum of squares is given below for all combinations.

$$Y = \text{age in years}, \quad X_1 = \text{duration in seconds},$$

$$X_2 = \text{VO}_2 \text{ MAX}, \quad X_3 = \text{heart rate in beats/minute},$$

$$X_4 = \text{height in centimeters}, \quad X_5 = \text{weight in kilograms},$$

$$\text{SS}_{\text{TOTAL}} = 11395.74.$$

Indices of Variables in the Multiple Regression Equation	Regression Sum of Squares SS_{REG}	Indices of Variables in the Multiple Regression Equation	Regression Sum of Squares SS_{REG}
		1,2,3	5988.09
1	5382.81	1,2,4	5658.66
2	4900.82	1,2,5	5777.12
3	4527.51	1,3,4	6097.58
4	295.26	1,3,5	6151.91
5	54.80	1,4,5	5723.50
1,2	5454.48	2,3,4	5851.44
1,3	5953.18	2,3,5	5923.41
1,4	5597.08	2,4,5	5243.27
1,5	5685.88	3,4,5	4630.28
2,3	5731.40	1,2,3,4	6128.27
2,4	5089.15	1,2,3,5	6201.39
2,5	5221.73	1,2,4,5	5805.06
3,4	4628.83	1,3,4,5	6179.52
3,5	4568.73	2,3,4,5	5940.03
4,5	299.81	1,2,3,4,5	6223.12

Problems 30–35 are based on these data.

30. Compute and plot for each p, the smallest C_p value. Which predictive model would you choose?

31. At the 10% significance level perform stepwise regression (do not compute the regression coefficients) selecting variables. Which variables are in the final model? How does this compare to the answer to Problem 30?

*32. At the 0.01 significance level, select variables using a *stepdown* regression equation (no coefficients computed).

33. What are the values of the following correlation and multiple coefficients? Are they significantly nonzero at the 5% significance level?

$$R_{Y(X_1,X_2)}, \quad R_{Y(X_3,X_4,X_5)},$$

$$R_{YX_1}, \quad R_{YX_2}, \quad R_{Y(X_4,X_5)}.$$

34. Compute the following squares of partial correlation coefficients. Are they statistically significant at the 0.10 level?

$$r^2_{Y,X_1 \cdot X_2}, \quad r^2_{Y,X_2 \cdot X_1}, \quad r^2_{Y,X_3 \cdot X_1,X_2}.$$

Describe these quantities in words.

35. Compute the following partial multiple correlation coefficients. Are they significant at the 5% level?

$$R_{Y(X_1,X_2,X_3) \cdot X_4,X_5}, \quad R_{Y(X_1,X_3) \cdot X_2},$$

$$R_{Y(X_2,X_3) \cdot X_1}, \quad R_{Y(X_1,X_2) \cdot X_3}.$$

*36. A canonical correlation analysis was performed upon the measurement data of Stoudt *et al.* [1970] for adult women. One set of variables, Set 1, consisted of variables which would appear to be more indicative of linear dimensions. The other set of variables, Set 2, would appear to be more indicative of bulk and weight. The sets were:

SET 1

Name	Mnemonic
Sitting height, erect	STHTER
Sitting height, normal	STHTNORM
Knee height	KNEEHT
Popliteal height	POPHT
Elbow rest height	ELBWHT
Height	HT

SET 2

Name	Mnemonic
Seat breadth	SEATBRTH
Chest girth	CHESTGRH
Waist girth	WSTGRTH
Right arm girth	RTARMGRH
Right arm skinfold	RTARMSKN
Infrascapular skinfold	INFRASCP
Weight	WT

Results of BDMP canonical correlation analysis follow.

SQUARED MULTIPLE CORRELATIONS OF EACH VARIABLE IN SECOND SET WITH ALL OTHER VARIABLES IN SECOND SET.

Variable		
Number	**Name**	R^2
10	SEATBRTH	0.67381
12	CHESTGRH	0.84267
13	WSTGRTH	0.78835
14	RTARMGRH	0.85049
15	RTARMSKN	0.66139
16	INFRASCP	0.73271
18	WT	0.89021

SQUARED MULTIPLE CORRELATIONS OF EACH VARIABLE IN FIRST SET WITH ALL OTHER VARIABLES IN FIRST SET.

Variable		
Number	**Name**	R^2
1	STHTER	0.88098
2	STHTNORM	0.82552
3	KNEEHT	0.70585
4	POPHT	0.66306
5	ELBWHT	0.46022
17	HT	0.87420

Eigenvalue	**Canonical Correlation**
0.41822	0.64670
0.30149	0.54908
0.06957	0.26377
0.01782	0.13350
0.00246	0.04964
0.00067	0.02589

a. "Eigenvalue" is a mathematical term; the value is the square of the canonical correlation coefficient. Verify this for the first three eigenvalues.

b. List any inequalities (e.g., $R^2 \leq 0.13$) for the multiple correlation coefficients squared (as listed below) that follow from this analysis.

$$R^2_{WT(CHESTGRH,HT)},$$

$$R^2_{HT(SEATBRTH,CHESTGRTH,WT)},$$

$$R^2_{WT(HT,KNEEHT,POPHT)}.$$

c. There is often interest in evaluating the extent to which the different variables enter into the canonical correlations. This is examined in two ways.

    i. The coefficients (of the canonical variables) are given for the standard-ized variables (mean 0, variance 1). This gives an idea, independent of units, of their impact. However, since a variable may be important, but highly correlated with other variables in a set, its contribution may be masked (as in multiple regression). Still, large values indicate influential variables.

    ii. The correlation between the canonical variable (for a set) and each vari-able in its set are given. A high correlation shows that a variable (and its highly correlated "neighbors" in the set) explain the canonical variable. Still, with the correlations among variables, interpretation is an art. For Set 1 and the first canonical variables, what is the most influential vari-able by the standardized regression coefficients? By the correlation of the variable and the canonical variable? Same question for the second most influential variable.

d. Same question as (c) for the second set of variables and the first canonical variables.

e. Using the correlation squared (eigenvalue) to measure the percent of the variablility explained, do the following. Divide each eigenvalue by the sum of the eigenvalues; then multiply by 100. How many pairs of the canonical variables are needed to explain 70% of the joint variability *between* the sets? 90%?

*37. Florey and Acheson [1969] relate blood pressure to several other variables. We perform a canonical correlation analysis relating the blood pressure values to the other variables. The two sets of variables for the canonical correlation analysis are:

SET 1

Name	Mnemonic
Systolic Blood Pressure	SYSBP
Diastolic Blood Pressure	DIABP

SET 2

Name	Mnemonic
Height	HT
Weight	WT
Right triceps, skinfold	RTTRICPS
Infrascapular, skinfold	INFRASCP
Right arm girth	ARMGIRTH
Glucose	GLUCOSE
Cholesterol	CHOL
Age	AGE

The analysis gives the following results for the black males, using a BMDP pro-gram.

**SQUARED MULTIPLE CORRELATIONS OF EACH VARIABLE IN SECOND
SET WITH ALL OTHER VARIABLES IN SECOND SET.**

Variable		
Number	**Name**	R^2
1	HT	0.47506
2	WT	0.89234
3	RTTRICPS	0.55264
4	INFRASCP	0.69101
5	ARMGIRTH	0.84424
6	GLUCOSE	0.17011
7	CHOL	0.16661
8	AGE	0.28618

**SQUARED MULTIPLE CORRELATIONS OF EACH VARIABLE IN FIRST
SET WITH ALL OTHER VARIABLES IN FIRST SET.**

Variable		
Number	**Name**	R^2
9	SYSBP	0.62410
10	DIABP	0.62410

Eigenvalue	**Canonical Correlation**
0.2878	0.53177
0.03147	0.17739

**COEFFICIENTS OF CANONICAL VARIABLES FOR
FIRST SET OF VARIABLES**

		$CNVRF1_1$	$CNVRF2_2$
SYSBP	9	0.109140D+01	−0.121207D+01
DIABP	10	−0.119078D+00	0.162668D+01

**STANDARDIZED COEFFICIENTS FOR CANONICAL VARIABLES
FOR FIRST SET OF VARIABLES (THESE ARE THE COEFFICIENTS
FOR THE STANDARDIZED VARIABLES—MEAN ZERO,
STANDARD DEVIATION ONE)**

		CNVRF1	CNVRF2
SYSBP	9	1.091	−1.212
DIABP	10	−0.119	1.627

COEFFICIENTS FOR CANONICAL VARIABLES FOR SECOND SET OF VARIABLES.

		CNVRS1	CNVRS2
HT	1	−0.331148D+00	0.368504D+00
WT	2	0.580343D+00	−0.190257D+00
RTTRICPS	3	−0.747289D−01	−0.587831D+00
INFRASCP	4	−0.213743D+00	0.606057D+00
ARMGIRTH	5	−0.300343D−02	0.859810D+00
GLUCOSE	6	0.255525D+00	−0.433685D−01
CHOL	7	0.243341D−01	0.121624D+00
AGE	8	0.764213D+00	−0.985981D−01

STANDARDIZED COEFFICIENTS FOR CANONICAL VARIABLES FOR SECOND SET OF VARIABLES (THESE ARE THE COEFFICIENTS FOR THE STANDARDIZED VARIABLES—MEAN ZERO, STANDARD DEVIATION ONE)

		CNVRS1	CNVRS2
HT	1	−0.331	0.369
WT	2	0.580	−0.190
RTTRICPS	3	−0.075	−0.588
INFRASCP	4	−0.214	0.606
ARMGIRTH	5	−0.003	0.860
GLUCOSE	6	0.256	−0.043
CHOL	7	0.024	0.122
AGE	8	0.764	−0.098

CANONICAL VARIABLE LOADINGS (CORRELATIONS OF CANONICAL VARIABLES WITH ORIGINAL VARIABLES)

		CNVRF1	CNVRF2
SYSBP	9	0.997	0.073
DIABP	10	0.743	0.669

		CNVRS1	CNVRS2
HT	1	−0.349	0.405
WT	2	0.188	0.807
RTTRICPS	3	0.126	0.255
INFRASCP	4	0.210	0.647
ARMGIRTH	5	0.206	0.831
GLUCOSE	6	0.550	−0.147
CHOL	7	0.372	0.192
AGE	8	0.891	−0.185

a. Verify that the eigenvalues are the squares of the canonical correlations.

b. If this analysis (computer output) allows any inference about the following multiple correlations squared state the appropriate inequalities.

$$R^2_{\text{SYSBP(HT,WT,AGE)}}, \quad R^2_{\text{AGE(HT,WT)}},$$

$$R^2_{\text{DIABP(SYSBP,HT,WT)}}, \quad R^2_{\text{AGE(DIABP,SYSBP)}}.$$

c. What is the upper bound on the percent of the variability in blood pressure (defined as the average of DIAPB and SYSPB) which may be explained by the other variables?

d. Which variable(s) in Set 1 is most influential in the first canonical variable? (Why do you say this?) Answer the same question for the variables of Set 2.

e. The same question as in (d) except restricted to the second canonical pair of variables.

Problems 38, 39, and 40 are analysis of covariance problems. They use BMDP computer output which is addressed in more detail in the first problem. This problem should be done before Problems 39 and 40.

38. This problem uses the height and weight data of 25–34-year-old men as measured in 1960–1962 and 1971–1974 samples of the US populations. These data are described and presented in Problem 23.

a. The groups are defined by a year variable taking on the value 1 for the 1960 survey and the value 2 for the 1971 survey. Means for the data are:

		Estimates of Means		
		YR1960	**YR1971**	**Total**
HT	1	68.5081	68.9353	68.7403
WT	2	169.3890	171.4030	170.4838

Which survey had the heaviest men? the tallest men? There are at least two possible explanations for weight gain: (1) the weight is increasing due to more overweight and/or building of body muscle; (2) the taller population naturally weighs more.

b. To distinguish between two hypotheses an analysis of covariance adjusting for height is performed. The analysis produced the following output:

DEPENDENT VARIABLE IS WT

Covariate	**Regression Coefficient**	**Standard Error**	**T-Value**
HT	4.22646	0.22742	18.58450

Group	**N**	**Group Mean**	**Adjusted Group Mean**	**Standard Error**
YR1960	675	169.38904	170.37045	0.89258
YR1971	804	171.40295	170.57901	0.91761

ANALYSIS OF VARIANCE

Source	DF	SS	MS	F	Tail Area Probabilty
Equality of adjusted cell means	1	15.7500	15.7500	0.0294	0.8639
Zero slope	1	185086.0000	185086.0000	345.3833	0.0000
Error	1475	790967.3750	535.8857		
Equality of slopes	1	0.1250	0.1250	0.0002	0.9878
Error	1475	790967.2500	536.2490		

SLOPE WITHIN EACH GROUP

		YR1960	YR1971
HT	1	4.2223	4.2298

**T-TEST MATRIX FOR ADJUSTED GROUP MEANS
ON 1476 DEGREES OF FREEDOM**

		YR1960	YR1971
YR1960	1	0.0000	
YR1971	2	0.1720	0.0000

PROBABILITES FOR THE T-VALUES ABOVE

	$YR1960_1$	$YR1971_2$
$YR1960_1$	1.0000	
$YR1971_2$	0.8634	1.0000

   i. Note the "equality of slopes" line of output. This gives the F test for the equality of the slopes with the corresponding p-value. Is the hypothesis of the equality of the slopes feasible? If estimated separately what are the two slopes?

  ii. The test for equal (rather than just parallel) regression lines in the groups corresponds to the line labeled "equality of adjusted cell means". Is there a statistically significant difference between the groups? What are the adjusted cell means? By how many pounds do the adjusted cell means differ? Does hypothesis (1) or (2) seem more plausible with these data?

 iii. A t-test for comparing each pair of groups is presented. The p-value 0.8643 is the same (to roundoff) as the F statistic. This occurs because only two groups are compared.

39. The cases of Bruce *et al.* [1973] are used. We are interested in comparing VO$_2$ MAX, after adjusting for duration and age, in three groups: the active males, sedentary males, and active females. The analysis gives:

**Number of Cases
per Group**

ACTMALE	44
SEDMALE	94
ACTFEM	43
TOTAL	181

Estimates of Means

		ACTMALE	SEDMALE	ACTFEM	TOTAL
VO$_2$ MAX	1	40.8046	35.6330	29.0535	35.3271
Duration	2	647.3864	577.1067	514.8837	579.4091
Age	3	47.2046	49.7872	45.1395	48.0553

DEPENDENT VARIABLE IS VO$_2$ MAX

Covariate	Regression Coefficient	Standard Error	T-Value
Duration	0.05242	0.00292	17.94199
Age	−0.06872	0.03160	−2.17507

Group	N	Group Mean	Adjusted Group Mean	Standard Error
ACTMALE	44	40.80456	37.18298	0.52933
SEDMALE	94	35.63297	35.87268	0.34391
ACTFEM	43	29.05349	32.23531	0.56614

ANALYSIS OF VARIANCE

Source	DF	SS	MS	F	Tail Area Probabilty
Equality of adjusted cell means	2	422.8359	211.4180	19.4336	0.0000
Zero slope	2	7612.9980	3806.4990	349.6947	0.0000
Error	176	1914.7012	10.8790		
Equality of slopes	4	72.7058	18.1765	1.6973	0.1528
Error	172	1841.9954	10.7093		

SLOPES WITHIN EACH GROUP

		ACTMALE	SEDMALE	ACTFEM
Duration	2	0.0552	0.0522	0.0411
Age	3	−0.1439	−0.0434	−0.1007

**T-TEST MATRIX FOR ADJUSTED GROUP MEANS
ON 176 DEGREES OF FREEDOM**

		ACTMALE	SEDMALE	ACTFEM
ACTMALE	1	0.0000		
SEDMALE	2	−2.1005	0.0000	
ACTFEM	3	−5.9627	−5.3662	0.0000

PROBABILITES FOR THE T-VALUES ABOVE

		ACTMALE	SEDMALE	ACTFEM
ACTMALE	1	1.0000		
SEDMALE	2	0.0371	1.0000	
ACTFEM	3	0.0000	0.0000	1.0000

a. Are the slopes of the adjusting variables (covariates) statistically significant?

b. Is the hypothesis of parallel regression equations (equal β's in the groups) tenable?

c. Does the adjustment bring the group means closer together?

d. After adjustment is there a statistically significant difference between the groups?

e. If the answer to (d) is yes, which groups differ at the 10%, 5%, 1% significance level?

40. This problem deals with the data of Example 10.7 presented in Tables 10.14, 10.15, and 10.16.

a. Using the quadratic term of Table 10.15 correlate this term with height, weight, and age for the group of females and also for the group of males. Are the correlations comparable?

b. Do Task (a) by setting up an appropriate regression analysis with dummy variables.

c. Test whether gender makes a significant contribution to the regression model of Part (b).

d. Repeat the analyses for the linear and constant terms of Table 10.15.

e. Do your conclusions differ from those of Example 10.7?

41. This problem examines the heart rate response in normal males and females as reported in Hossack et al. [1980] and [1981]. As heart rate is related to age, and the males were older, this was used as an adjustment covariate. The data are:

**Number of Cases
per Group**

Male	11
Female	10
Total	21

ESTIMATES OF MEANS

		Male	Female	Total
HR	1	180.9091	172.2000	176.7619
Age	2	50.4546	45.5000	48.0952

DEPENDENT VARIABLE IS HR

Covariate	Regression Coefficient	Standard Error	T-Value
Age	-0.75515	0.17335	-4.35610

Group	N	Group Mean	Adjusted Group Mean	Standard Error
Male	11	180.90909	182.69070	3.12758
Female	10	172.19998	170.24017	3.28303

ANALYSIS OF VARIANCE

Source	DF	SS	MS	F	Tail Area Probabilty
Equality of adjusted cell means	1	783.3650	783.3650	7.4071	0.0140
Zero slope	1	2006.8464	2006.8464	18.9756	0.0004
Error	18	1903.6638	105.7591		
Equality of slopes	1	81.5415	81.5415	0.7608	0.3952
Error	17	1822.1223	107.1837		

SLOPES WITHIN EACH GROUP

		Male	Female
Age	2	-1.0231	-0.6687

a. Is it reasonable to assume equal age response in the two groups?

b. Are the adjusted cell means closer or further apart than the unadjusted cell means? Why?

c. After adjustment what is the p-value for a difference between the two groups? Do men or women have a higher heart rate on maximal exercise (after age adjustment) in these data?

REFERENCES

Abraham, S., Johnson, C. L., and Najjar, M. F. [1979]. Weight by Height and Age for Adults 18–74 Years: United States, 1971–1974. Data from the National Health Survey. Series 11, No. 208. DHEW Publications No. (PHS) 79-1656. Washington, DC.

Blalock, H. M., Jr. [1964]. Causal Inferences in Nonexperimental Research. University of North Carolina, Chapel Hill, NC.

BMDP-81, BMDP 1985. Statistical Software Manual. Chief editor, Dixon, W. J. University of California Press, Berkeley, CA, BMDP 1988, volumes 1 and 2.

Boucher, C. A., Bingham, J. B., Osbakken, M. D., Okada, R. D., Strauss, H. W., Block, P. C., Levine, R. B., Phillips, H. R., and Pohost, G. B. [1981]. Early changes in left ventricular size and function after correction of left ventricular volume overload. *American Journal of Cardiology*, **47:** 991–1004.

Box, G. E. P. and Jenkins, G. M. [1970]. Time Series Analysis, Forecasting and Control. Holden-Day, San Francisco, CA.

Bruce, R. A., Kusumi, F., and Hosmer, D. [1973]. Maximal oxygen intake and nomographic assessment of functional aerobic impairment in cardiovascular disease. *American Heart Journal*, **85:** 546–562.

Campbell, D. T. and Stanley, J. C. [1963]. Experimental and quasi-experimental designs for research on teaching. In N. L. Gage (ed.). *Handbook of Research on Teaching*. Rand-McNally, Chicago, pp. 172–246.

Cullen, B. F. and van Belle, G. [1975]. Lymphocyte transformation and changes in leukocyte count: effects of anesthesia and operation. *Anesthesiology*, **43:** 577-583. Used with permission of J. B. Lippincott Company.

Daniel, C. and Wood, F. S. [1971]. *Fitting Equations to Data*. Wiley–Interscience, New York.

Draper, N. R. and Smith, H. [1981]. *Applied Regression Analysis*, 2nd edition. Wiley, New York.

Duncan, O. D. [1975]. *Introduction to Structural Equation Models*. Academic Press, New York.

Efron, B. [1982]. *The Jackknife, Bootstrap and Other Resampling Plans*, Society for Industrial and Applied Mathematics, Philedelphia, PA.

Efron, B. and Tibshirani, R. [1986]. The bootstrap (with discussion), *Statistical Science*, **1:** 54–77.

Florey, C. du V. and Acheson, R. M. [1969]. Blood pressure as it relates to physique, blood glucose and cholesterol. vital and health statistics. data from the national health survey. Public Health Service Publication No. 1000, Series 11, No. 34. Washington, DC.

Gardner, M. J. [1973]. Using the environment to explain and predict mortality. *Journal of the Royal Statistical Society, Series A*, **136:** 421–440.

Goldberger, A. S. and Duncan, O. D. [1973]. *Structural Equation Models in the Social Sciences*. Seminar Press, New York.

Graybill, F. A. [1976]. *Theory and Application of the Linear Model*. Duxbury Press, North Scituate, MA.

Haynes, S. G., Levine, S., Scotch, N., Feinleib, M., and Kannel, W. B. [1978]. The relationship of psychosocial factors to coronary heart disease in the Framingham study. *American Journal of Epidemiology*, **107:** 362–283.

Hocking, R. R. [1976]. The analysis and selection of variables in linear regression. *Biometrics*, **32:** 1–50.

Hossack, K. F., Bruce, R. A., Green, B., Kusumi, F., DeRouen, T. A., and Trimble, S. [1980]. Maximal cardiac output during upright exercise: approximate normal standards and variations with coronary heart disease. *American Journal of Cardiology*, **46:** 204–212.

Hossack, K. F., Kusumi, F., and Bruce, R. A. [1981]. Approximate normal standards of maximal cardiac output during upright exercise in women. *American Journal of Cardiology*, **47:** 1080–1086.

Jensen, D., Atwood, J. E., Frolicher, V., McKirnan, M. D., Battler, A., Ashburn, W., and Ross, Jr., J. [1980]. Improvement in ventricular function during exercise studied with radionuclide ventriculography after cardiac rehabilitation, *American Journal of Cardiology*, **46:** 770–777.

Kleinbaum, D. G. and Kupper, L. L. [1988]. *Applied Regression Analysis and Other Multivariate Methods*, 2nd edition. Duxbury Press, North Scituate, MA.

Li, C. C. [1975]. *Path Analysis: A Primer*. The Boxwood Press, Pacific Grove, CA.

Mason, R. L. [1975]. Regression analysis and problems of multicollinearity. *Communications in Statistics*, **4:** 277–292.

Mehta, J., Mehta, P., Pepine, C. J. and Conti, C. R. [1981]. Platelet function studies in coronary artery disease. X. Effects of dipyridamole. *American Journal of Cardiology*, **47:** 1111–1114.

Morrison, D. F. [1976]. *Multivariate Statistical Methods,* 2nd edition. McGraw-Hill, New York.

Roberts, J. [1966]. Weight by Height and Age of Adults, United States—1960–1962. Vital and Health Statistics. Data from the National Health Survey. Series 11, No. 14, Public Health Service Publication No. 1000, Series 11, No. 14. Washington, DC.

Stoudt, H. W., Damon, A., and McFarland, R. A. [1970]. Skinfolds, Body Girths, Biacromial Diameter, and Selected Anthropometric Indices of Adults: United States, 1960–62. Vital and Health Statistics. Data from the National Health Survey. Public Health Service Publication No. 1000, Series 11, No. 35. Washington, DC.

Timm, N. H. [1975]. *Multivariate Analysis with Applications in Education and Psychology.* Brooks/Cole, Monterey, CA.

van Belle, G., Leurgans, S., Friel, P., Guo, S. and Yerby, M. [1989]. Determination of enzyme binding constants using generalized linear models, with particular reference to Michaelis–Menten models. *Journal of Pharmeceutical Science*, **78:** 413–416.

CHAPTER 12

Multiple Comparisons

12.1 INTRODUCTION

Most of us are aware of the large number of coincidences that appear in our lives. "Imagine meeting you here!" "The ticket number is the same as our street address." One explanation of such phenomenon is statistical. There are so many different things going on in our lives that a few events of small probability (the coincidences) are likely to happen at the same time. See Diaconis and Mosteller [1989] for methods for studying coincidences.

In a more formal setting, the same phenomenon can occur. If many tests or comparisons are carried out at the 0.05 significance level (with the null hypothesis holding in all cases), the probability of deciding that the null hypothesis may be rejected in one or more of the tests is considerably larger. If *many* 95% confidence intervals are set up, there is not 95% confidence that *all* parameters are "in" their confidence intervals. If many treatments are compared, each comparison at a given significance level, the overall probability of a mistake is much larger. If significance tests are done continually while data accumulate, stopping when statistical significance is reached, the significance level is much larger than the nominal "fixed sample size" significance level. The category of problems being discussed is called the *multiple comparison* problem: many (or multiple) statistical procedures are being applied to the same data.

This chapter gives a quantitative feeling for the problem. Statistical methods to handle the situation are also described. We first describe the "multiple testing," or multiple comparison problem in Section 12.2. In Section 12.3, we present three very common methods for obtaining simultaneous confidence intervals for the regression coefficients of a linear model. In Section 12.4, we discuss how to choose between them. In Section 12.5, we briefly introduce you to the ideas behind sequential analysis. The chapter concludes with notes and problems.

12.2 THE MULTIPLE COMPARISON PROBLEM

Suppose that n statistically independent tests are being considered in an experiment. Each test is evaluated at significance level α. Suppose that the null hypothesis holds in each case. What is the probability, α^*, of incorrectly rejecting the null hypothesis in one or more of the tests? For $n = 1$, the probability is α by definition. Table 12.1 gives the probabilities for several values of α and n.

Table 12.1. Probability, α^*, of Rejecting One or More Null Hypotheses When n Tests are Carried Out at Significance Level α and Each Null Hypothesis is True.

Number of Tests n	$\alpha = 0.01$	$\alpha = 0.05$	$\alpha = 0.10$
1	0.01	0.05	0.10
2	0.02	0.10	0.19
3	0.03	0.14	0.27
4	0.04	0.19	0.34
5	0.05	0.23	0.41
6	0.06	0.26	0.47
7	0.07	0.30	0.52
8	0.08	0.34	0.57
9	0.09	0.37	0.61
10	0.10	0.40	0.65
20	0.18	0.64	0.88
50	0.39	0.92	0.99
100	0.63	0.99	1.00
1000	1.00	1.00	1.00

Note that if each test is carried out at a 0.05 level then for 20 tests, the probability is 0.64 of incorrectly rejecting at least one of the null hypotheses.

The table may also be related to confidence intervals. Suppose that each of n $100(1 - \alpha)\%$ confidence intervals comes from an independent data set. The table gives the probability that one or more of the estimated parameters is not straddled by its confidence interval. For example, among five 90% confidence intervals, the probability is 0.41 that at least one of the confidence intervals does not straddle the parameter being estimated.

Now that we see the magnitude of the problem, what shall we do about it? One solution is to use a smaller α level for each test or confidence interval so that the probability of one or more mistakes over all n tests is the desired (nominal) significance level. Table 12.2 shows the α level needed for each test in order that the combined significance level, α^*, be as given at column heading.

The values of α and α^* are related to each other by the equation

$$\alpha^* = 1 - (1 - \alpha)^n \quad \text{or} \quad \alpha = 1 - (1 - \alpha^*)^{1/n}, \tag{12.1}$$

where $(1 - \alpha)^{1/n}$ is the nth root of $1 - \alpha$.

If p-values are being used without a formal significance level, the p-value from an individual test is adjusted by the opposite of Equation 12.1. That is, p^*, the overall p-value, taking into account the fact that there are n tests, is given by

$$p^* = 1 - (1 - p)^n. \tag{12.2}$$

For example, if there are two tests and the p-value of each test is 0.05, the overall p-value is $p^* = 1 - (1 - 0.05)^2 = 0.0975$. For small values of α (or p) and n by the binominal expansion $\alpha^* \doteq \frac{1}{n}\alpha$ (and $p^* = np$), a relationship that also will be derived in the context of the Bonferroni inequality.

Table 12.2. Significance Level, α, Needed for Each Test or Confidence Interval so that the Overall Significance Level (Probability of One or More Mistakes) is α^* When Each Null Hypothesis is True.

n	$\alpha^* = 0.01$	$\alpha^* = 0.05$	$\alpha^* = 0.10$
1	0.010	0.05	0.10
2	0.005	0.0253	0.0513
3	0.00334	0.0170	0.0345
4	0.00251	0.0127	0.0260
5	0.00201	0.0102	0.0209
6	0.00167	0.00851	0.0174
7	0.00143	0.00730	0.0150
8	0.00126	0.00639	0.0131
9	0.00112	0.00568	0.0116
10	0.00100	0.00512	0.0105
20	0.00050	0.00256	0.00525
50	0.00020	0.00103	0.00210
100	0.00010	0.00051	0.00105
1000	0.00001	0.00005	0.00011

Before giving an example, we introduce some terminology and make a few comments. We consider an "experiment" in which n tests or comparisons are made.

Definition 12.1. The significance level at which each test or comparison is carried out in an experiment is called the *per-comparison* error rate.

Definition 12.2. The probability of incorrectly rejecting at least one null hypothesis in an experiment involving one or more tests or comparisons, where the null hypothesis is true in each case, is called the *per-experiment error rate*.

The terminology is less transparent than it seems. In particular, what defines an "experiment"? You could think of your life as an experiment involving many comparisons, if you wanted to restrict your "per-experiment" error level to say, $\alpha^* = 0.05$, you would need to carry out each of the comparisons at ridiculously low values of α. Frequently groups of tests or comparisons form a natural unit and a suitable adjustment can be made. See also Problem 13 for an amusing yet pertinent illustration of the "problem."

Two of the key assumptions in the derivation of Equations 12.1 and 12.2 are statistical independence and the null hypothesis being true for each comparison. In the next two sections, we discuss their relevance and ways of dealing with them.

Example 12.1. To illustrate the methods, consider responses to maximal exercise testing within eight groups by Bruce *et al.* [1974]. The subjects were all males. An indication of exercise performance is functional aerobic impairment (FAI). This index is age and sex adjusted to compare the duration of the maximal treadmill test to that expected for a healthy individual of the subject's age and sex. A larger score indicates more exercise impairment. Working at a 5% significance level, it is desired to compare the average levels in the eight groups. The data are:

Group		N	Mean	Standard Deviation
1	Healthy individuals	1275	0.6	11
2	Hypertensive subjects (HT)	193	8.5	19
3	Post-myocardial infarction (PMI)	97	24.5	21
4	Angina pectoris, chest pain (AP)	306	30.3	24
5	PMI+AP	228	36.9	26
6	HT+AP	138	36.6	23
7	HT+PMI	20	27.6	18
8	PMI+AP+HT	75	44.9	22

Because it was expected that the healthy group would have a smaller variance, a one-way ANOVA was not performed (in the next section you will see how to handle such problems). Instead, we construct eight *simultaneous* 95% confidence intervals. Hence, $\alpha = 1 - (1 - 0.05)^{1/8} \doteq 0.0064$ is to be the α-level for each interval. The intervals are given by

$$\bar{y} \pm \frac{\text{SD}}{\sqrt{n}}\, t_{n-1,1-(0.0064/2)}.$$

The t-values are estimated by interpolation from the table of t-critical values and the normal table ($n > 120$). The eight confidence intervals work out to be:

Group	Critical t-Value	Limits	
		Lower	Upper
1	2.73	−0.2	1.4
2	2.73	4.8	12.2
3	2.79	18.5	30.5
4	2.73	26.6	34.0
5	2.73	32.2	41.6
6	2.77	31.2	42.0
7	3.06	15.3	39.9
8	2.81	37.7	52.1

Displaying these intervals graphically and indicating which group each interval belongs to, gives Figure 12.1.

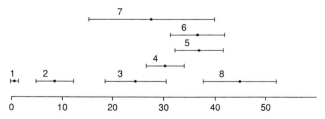

Figure 12.1. Functional aerobic impairment level.

Since all eight groups have a simultaneous 95% confidence interval, it is sufficient (but not necessary) to decide that any two means whose confidence intervals do not overlap are significantly different. Let $\mu_1, \mu_2, \ldots, \mu_8$, be the population means associated with groups $1, 2, \ldots, 8$ respectively. The following conclusions are in order:

1. μ_1 has the smallest mean ($\mu_1 < \mu_i$, $i = 2, \ldots, 8$),
2. μ_2 is the second smallest mean ($\mu_1 < \mu_2 < \mu_i$, $i = 3, \ldots, 8$),
3. $\mu_3 < \mu_5$, $\mu_3 < \mu_6$, $\mu_3 < \mu_8$,
4. $\mu_4 < \mu_8$.

There are seeming paradoxes. We know that $\mu_3 < \mu_5$ but we cannot decide whether μ_7 is larger or smaller than those two means.

Restating the conclusions in words, the healthy group had the best exercise performance, followed by the hypertensive subjects, who were better than the rest. The post-myocardial infarction group performed better than the PMI+AP, PMI+AP+HT, and HT+AR groups. The angina pectoris group had better performance than angina pectoris plus an MI and hypertension. The other orderings were not clear from this data set.

12.3 SIMULTANEOUS CONFIDENCE INTERVALS AND TESTS FOR LINEAR MODELS

12.3.1 Linear Combinations and Contrasts

In the linear models, the estimates of the parameters are usually not independent. Even when the estimates of the parameters are independent, the same error mean square, MS$_e$, is used for each test or confidence interval. Thus, the method of Section 12.2 does not apply. In this section, several techniques dealing with the linear model are considered.

Before introducing the Scheffé method, we need additional concepts of linear combinations and contrasts.

Definition 12.3. A *linear combination of the parameters* $\beta_1, \beta_2, \ldots, \beta_p$ is a sum $\theta = c_1\beta_1 + c_2\beta_2 + \cdots + c_p\beta_p$ where $c_1, c_2, \ldots, c_p$ are known constants.

Associated with any parameter set $\beta_1, \beta_2, \ldots, \beta_p$ is a number which is equal to the number of linearly estimated independent parameters. In ANOVA tables, this is the number of degrees of freedom associated with a particular sum of squares.

A linear combination is a parameter. An estimate of such a parameter is a statistic, a random variable. Let $b_1, b_2, \ldots, b_p$ be unbiased estimates of $\beta_1, \beta_2, \ldots, \beta_p$, then $\widehat{\theta} = c_1b_1 + c_2b_2 + \cdots + c_pb_p$ is an unbiased estimate of θ. If $b_1, b_2, \ldots, b_p$ are jointly normally distributed, then $\widehat{\theta}$ will be normally distributed with mean θ and variance $\sigma_{\widehat{\theta}}^2$. The standard error of $\widehat{\theta}$ is usually quite complex and depends upon possible relationships among the β's as well as correlations among the estimates of the β's. It will be of the form

$$\text{CONSTANT}\sqrt{\text{MS}_e}$$

where MS_e is the residual mean square either from the regression analysis or the analysis of variance. A simple set of linear combinations can be obtained by having only one of the c_i take on the value 1 and all others the value 0.

A particular class of linear combinations that will be very useful is given by:

Definition 12.4. A linear combination $\theta = c_1\beta_1 + c_2\beta_2 + \cdots + c_p\beta_p$ is a *contrast* if $c_1 + c_2 + \cdots + c_p = 0$. The contrast is *simple* if exactly two constants are nonzero and equal to 1 and -1.

The following are examples of linear combinations that are contrasts: $\beta_1 - \beta_2$ (a simple contrast); $\beta_1 - \frac{1}{2}(\beta_2 + \beta_3) = \beta_1 - \frac{1}{2}\beta_2 - \frac{1}{2}\beta_3$; and $(\beta_1 + \beta_8) - (\beta_2 + \beta_4) = \beta_1 + \beta_8 - \beta_2 - \beta_4$. The following are linear combinations that are not contrasts: β_1, $\beta_1 + \beta_6$, $\beta_1 + \frac{1}{2}\beta_2 + \frac{1}{2}\beta_3$. The linear combinations and contrasts have been defined and illustrated using regression notation. They are also applicable to analysis of variance models (which are special regression models) so that the examples can be rewritten as $\mu_1 - \mu_2$; $\mu_1 - \frac{1}{2}(\mu_2 + \mu_3)$, etc. The interpretation is now a bit more transparent: $\mu_1 - \mu_2$ is a comparison of "treatment 1" and "treatment 2"; $\mu_1 - \frac{1}{2}(\mu_2 + \mu_3)$ is a comparison of "treatment 1" with the average of "treatment 2" and "treatment 3."

Since hypothesis testing and estimation are equivalent we will state most results in terms of simultaneous confidence intervals.

12.3.2 The Scheffé Method (S-Method)

A very general method for protecting against a large per-experiment error rate is provided by the Scheffé method. It allows unlimited "fishing", at a price.

Result 12.1. Given a set of parameters $\beta_1, \beta_2, \ldots, \beta_p$, the probability is $(1-\alpha)$ that simultaneously *all* linear combinations of $\beta_1, \beta_2, \ldots, \beta_p$ say $\theta = c_1\beta_1 + c_2\beta_2 + \cdots + c_p\beta_p$ are in the confidence intervals

$$\widehat{\theta} \pm \sqrt{dF_{d,m,1-\alpha}}\, \widehat{\sigma_{\widehat{\theta}}},$$

where the estimate of θ is $\widehat{\theta} = c_1b_1 + c_2b_2 + \cdots + c_pb_p$ with estimated standard error $\widehat{\sigma_{\widehat{\theta}}}$, F is the usual F statistic with (d, m) degrees of freedom, "d" is the number of linearly independent parameters and "m" is the number of degrees of freedom associated with MS_e.

Note that these confidence intervals are of the usual form "Statistic $\pm$ Constant $\times$ Standard Error of Statistic," the only difference is the constant which now depends on the number of parameters involved as well as the degrees of freedom for the error sum of squares. When $d = 1$, for any α,

$$\sqrt{dF_{d,m,1-\alpha}} = \sqrt{F_{1,m,1-\alpha}} = t_{m,1-\alpha},$$

that is, the constant reduces to the usual t-statistic with m degrees of freedom. After discussing some examples, we assess the price paid for the unlimited number of comparisons that can be made.

The easiest way to understand the S-method is to work through some examples.

Example 12.1: Multiple Regression.   In Table 12.3 we present part of the computer output from Cullen and van Belle [1975] discussed in Chapters 9 and 11. We construct simultaneous 95% confidence intervals for the slopes β_i. In this case, the first linear combination is

$$\theta_1 = 1 \times \beta_1 + 0 \times \beta_2 + 0 \times \beta_3 + 0 \times \beta_4 + 0 \times \beta_5;$$

the second linear combination is

$$\theta_2 = 0 \times \beta_1 + 1 \times \beta_2 + 0 \times \beta_3 + 0 \times \beta_4 + 0 \times \beta_5;$$

and so on.

The standard errors of these linear combinations are simply the standard errors of the slopes. There are five slopes $\beta_1, \beta_2, \ldots, \beta_5$ which are linearly independent, but their estimates $b_1, b_2, \ldots, b_5$ are correlated. The MS_e upon which the standard errors of the slopes are based has 29 degrees of freedom. The F-statistic has value $F_{5,29,0.95} = 2.55$.

The 95% *simultaneous* confidence intervals will be of the form

$$b_i \pm \sqrt{(5)(2.55)} s_{b_i}$$

or

$$b_i \pm 3.57 s_{b_i} \quad i = 1, 2, \ldots, 5.$$

For the regression coefficient of DPMB the interval is

$$0.575 \pm (3.57)(0.0834)$$

resulting in 95% confidence limits of $(0.277, 0.873)$.

Computing these values, the confidence intervals are (next page):

Table 12.3. Analysis of Variance, Regression Coefficients, and Confidence Intervals.

		Analysis of Variance			
Source	d.f.	Sum of Squares	Mean Square	F	Significance
Regression	5.0	95827	18965	12.9	0.000
Residual	29.0	42772	1474		

				95% Limits	
Variable	b	Standard Error b	t	Lower	Upper
DPMB	0.575	0.0834	6.89	0.404	0.746
Trauma	−9.21	11.6	−0.792	−33.0	14.6
Lymph B	−8.56	10.2	−0.843	−29.3	12.2
Time	−4.66	5.68	−0.821	−16.3	6.96
Lymph A	−4.55	6.72	−0.677	−18.3	9.19
Constant	−96.3	36.4	2.65	22.0	171

	Limits	
Variable	**Lower**	**Upper**
DPMB	0.277	0.873
Trauma	−50.8	32.3
Lymph B	−44.8	27.7
Time	−24.9	15.6
Lymph A	−28.5	19.4

These limits are much wider than those based on a per-comparison t-statistic. This is due solely to the replacement of $t_{29,0.975} = 2.05$ by $\sqrt{5F_{5,29,0.95}} = 3.57$. Hence, the confidence interval width is increased by a factor of $3.57/2.05 = 1.74$ or 74%.

Example 12.2: One-Way ANOVA. Using the notation of Section 10.2.2 if we wish simultaneous confidence intervals for all I means, then $d = I$, $m = n_\bullet - I$ and the standard error of the estimate of μ_i is

$$\sqrt{\frac{MS_e}{n_i}}, \quad i = 1, \dots, I.$$

Thus, the confidence intervals are of the form

$$\overline{Y}_{i\bullet} \pm \sqrt{IF_{I,n_\bullet - I,1-\alpha}}\sqrt{\frac{MS_e}{n_i}}, \quad i = 1, \dots, I.$$

Suppose we want simultaneous 99% confidence intervals for the morphine binding data of Problem 10.1. The confidence interval for the Chronic group is

$$31.9 \pm \sqrt{(4)\underbrace{(4.22)}_{F_{4,24,0.99}}}\sqrt{\frac{9.825}{18}} = 31.9 \pm 3.0$$

or

$$31.9 \pm 3.0.$$

The four simultaneous 99% confidence intervals are:

	Limits	
Group	**Lower**	**Upper**
$\mu_1 =$ Chronic	28.9	34.9
$\mu_2 =$ Acute	21.0	39.2
$\mu_3 =$ Dialysis	22.0	36.8
$\mu_4 =$ Anephric	19.2	30.8

As all four intervals overlap, we cannot conclude immediately from this approach that the means differ (at the 0.01 level). To compare two means we can also consider confidence intervals for $\mu_i - \mu_i'$. As the Scheffé method allows us to look at all linear combinations, we may also consider the confidence interval for $\mu_i - \mu_i'$.

The formula for the simultaneous confidence intervals is

$$\overline{Y}_{i\bullet} - \overline{Y}_{i'\bullet} \pm \sqrt{IF_{I,n\bullet-1,1-\alpha}}\sqrt{MS_e\left(\frac{1}{n_i} + \frac{1}{n_{i'}}\right)}, \quad i,i' = 1,\ldots,I, \; i \neq i'.$$

In this case, the confidence intervals are:

	Limits	
Contrast	Lower	Upper
$\mu_1 - \mu_2$	-7.8	11.4
$\mu_1 - \mu_3$	-5.5	10.5
$\mu_1 - \mu_4$	0.4	13.4
$\mu_2 - \mu_3$	-11.1	12.5
$\mu_2 - \mu_4$	-5.7	15.9
$\mu_3 - \mu_4$	-5.0	13.8

As the interval for $\mu_1 - \mu_4$ does not contain zero, we conclude that $\mu_1 - \mu_4 > 0$ or $\mu_1 > \mu_4$. This example is typical in that comparison of the linear combination of interest is best done through a confidence interval for that combination.

The comparisons are in the form of contrasts but were not considered so explicitly. Suppose we restrict ourselves to contrasts. This is equivalent to deciding which mean values differ, so that we are no longer considering confidence intervals for a particular mean. This approach gives smaller confidence intervals.

Contrast comparisons among the means μ_i, $i = 1,\ldots,I$ are equivalent to comparisons of α_i, $i = 1, \ldots, I$ in the one-way ANOVA model $Y_{ij} = \mu + \alpha_i + \epsilon_{ij}$, $i = 1, \ldots, I$, $j = 1, \ldots, n_i$; for example, $\mu_1 - \mu_2 = \alpha_1 - \alpha_2$. And there are only $(I-1)$ linearly independent values of α_i since we have the constraint $\sum_i \alpha_i = 0$. This is, therefore, the first example in which the parameters are not linearly independent. (In fact, the main effects are contrasts). Here, we set up confidence intervals for the simple contrasts $\mu_i - \mu_i'$. Here $d = 3$ and the simultaneous confidence intervals are given by

$$\overline{Y}_{i\bullet} - \overline{Y}_{i'\bullet} \pm \sqrt{(I-1)F_{I-1,n\bullet-1,1-\alpha}}\sqrt{MS_e\left(\frac{1}{n_i} + \frac{1}{n_{i'}}\right)}, \quad i,i' = 1,\ldots,I, \; i \neq i'.$$

In the case at hand, the intervals are:

	Limits	
Contrast	Lower	Upper
$\mu_1 - \mu_2$	-7.0	10.6
$\mu_1 - \mu_3$	-4.9	9.9
$\mu_1 - \mu_4$	0.9	12.9
$\mu_2 - \mu_3$	-10.1	11.5
$\mu_2 - \mu_4$	-4.8	15.0
$\mu_3 - \mu_4$	-1.9	10.7

As the $\mu_1 - \mu_4$ interval does not contain zero, we conclude that $\mu_1 > \mu_4$. Note that these intervals are shorter then in the first illustration. If you are interested in comparing each pair of means, this method will occasionally detect differences not found if we require confidence intervals for the mean as well.

Example 12.3: Two-Way Analysis of Variance

1. Main Effects. In two-way ANOVA situations there are many possible sets or linear combinations that may be studied; here we consider a few.

To study all cell means, consider the IJ cells to be part of a one-way ANOVA and use the approach of Example 12.1 or 12.2.

Consider now Example 10.5 in Section 10.3.1. Suppose we want to compare the differences between the means for the different days at a 10% significance level. In this case, we are working with the β_j main effects. The intervals for $\overline{\mu}_{\bullet j} - \overline{\mu}_{\bullet j'} = \beta_j - \beta_{j'}$ are given by

$$\overline{Y}_{\bullet j \bullet} - \overline{Y}_{\bullet j' \bullet} \pm \sqrt{(J-1)F_{J-1,n_{\bullet\bullet}-IJ,1-\alpha}} \sqrt{\mathrm{MS}_e \left(\frac{1}{n_{\bullet j}} + \frac{1}{n_{\bullet j'}} \right)}.$$

The means are 120.4, 158.1, and 118.4, respectively. The following contrasts are of interest:

Contrast	Estimate	90% Limits	
		Lower	**Upper**
$\beta_1 - \beta_2$	-37.7	-70.7	-4.7
$\beta_2 - \beta_3$	39.7	5.5	73.9
$\beta_1 - \beta_3$	2.0	-31.0	35.0

At the 10% significance level, we conclude that $\mu_{\bullet 1} - \mu_{\bullet 2} < 0$ or $\mu_{\bullet 1} < \mu_{\bullet 2}$, and that $\mu_{\bullet 3} < \mu_{\bullet 2}$. Thus, the means (combining cases and controls) of days 10 and 14 are less than the means of day 12.

2. Main Effects Assuming No Interaction. We will illustrate the procedure using Problem 10.12 as an example. This example discussed the effect of histamine shock on medullar blood vessel surface of the guinea pig thymus.

The sex of the animal was used as a covariate. The analysis of variance table is

Source	d.f.	Mean Square	F-Ratio	P-Ratio
Treatment	1	11.56	5.20	< 0.05
Sex	1	1.26	0.57	> 0.05
Treatment by sex	1	5.40	2.43	> 0.05
Error	36	2.225		
Total	39			

There is little evidence of interaction. Suppose we want to fit the model

$$Y_{ijk} = \mu + \alpha_i + \beta_j + \epsilon_{ijk}, \quad \begin{aligned} i &= 1, \ldots, I, \\ j &= 1, \ldots, J, \\ k &= 1, \ldots, n_{ij}. \end{aligned}$$

That is, we ignore the interaction term. It can be shown that the appropriate estimates in the balanced model for the cell means $\mu + \alpha_i + \beta_j$ are

$$\overline{Y}_{\bullet\bullet\bullet} + a_i + b_j, \quad \begin{aligned} i &= 1, \ldots, I, \\ j &= 1, \ldots, J, \end{aligned}$$

or

$$\overline{Y}_{\bullet\bullet\bullet} + (\overline{Y}_{i\bullet\bullet} - \overline{Y}_{\bullet\bullet\bullet}) + (\overline{Y}_{\bullet j\bullet} - \overline{Y}_{\bullet\bullet\bullet}) = \overline{Y}_{i\bullet\bullet} + \overline{Y}_{\bullet j\bullet} - \overline{Y}_{\bullet\bullet\bullet}.$$

The estimates are $\overline{Y}_{\bullet\bullet\bullet} = 6.53$, $\overline{Y}_{1\bullet\bullet} = 6.71$, $\overline{Y}_{2\bullet\bullet} = 6.35$, $\overline{Y}_{\bullet 1\bullet} = 5.99$, $\overline{Y}_{\bullet 2\bullet} = 7.07$. The estimated cell means fitted to the model $E(Y_{ijk}) = \mu + \alpha_i + \beta_j$ by $\overline{Y}_{\bullet\bullet\bullet} + a_i + b_j$ are

	Treatment	
Sex	**Control**	**Shock**
Male	6.17	7.25
Female	5.81	6.89

For multiple comparisons the appropriate formula for simultaneous confidence intervals for each cell mean assuming the interaction term is zero is given by the formula

$$\overline{Y}_{i\bullet\bullet} + \overline{Y}_{\bullet j\bullet} - \overline{Y}_{\bullet\bullet\bullet} \pm \sqrt{(I+J-1)F_{I+J-1, n_{\bullet\bullet}-IJ+1, 1-\alpha}} \sqrt{\mathrm{MS}_e \left(\frac{1}{n_{i\bullet}} + \frac{1}{n_{\bullet j}} - \frac{1}{n_{\bullet\bullet}} \right)}.$$

The degrees of freedom for the F-statistic are $(I+J-1)$ and $(n_{\bullet\bullet} - IJ + 1)$ because there are $I+J-1$ linearly independent cell means and the residual MS_e has $(n_{\bullet\bullet} - IJ + 1)$ degrees of freedom. This MS_e can be obtained by pooling the $\mathrm{SS_{INTERACTION}}$ and $\mathrm{SS_{RESIDUAL}}$ in the ANOVA table. For our example,

$$\mathrm{MS}_e = \frac{1 \times 5.40 + 36 \times 2.225}{37} = 2.311.$$

We will construct the 95% confidence intervals for the four cell means. The confidence interval for the first cell is given by

$$6.17 \pm \sqrt{\underbrace{(2+2-1)F_{3,37,0.95}}_{2.86}} \sqrt{2.311 \left(\frac{1}{20} + \frac{1}{20} - \frac{1}{40} \right)},$$

yielding

$$6.17 \pm 1.22$$

for limits

$$(4.95, 7.39).$$

The four simultaneous 95% confidence limits are:

| | Treatment | |
Sex	Control	Shock
Male	(4.95, 7.39)	(6.03, 8.47)
Female	(4.59, 7.03)	(5.67, 8.11)

Requiring this degree of confidence gives intervals that overlap. However, using the Scheffé method all linear combinations can be examined. With the same 95% confidence, let us examine the sex and treatment differences. The intervals for sex are defined by

$$\overline{Y}_{1\bullet\bullet} - \overline{Y}_{2\bullet\bullet} \pm \sqrt{3F_{3,37,0.95}} \sqrt{MS_e \left(\frac{1}{n_{1\bullet}} + \frac{1}{n_{2\bullet}} \right)},$$

or 0.36 ± 1.41 for limits $(-1.05, 1.77)$. Thus, in these data there is no reason to reject the null hypothesis of no difference in sex. The simultaneous 95% confidence interval for treatment is -1.08 ± 1.41 or $(-2.49, 0.33)$. This confidence interval also straddles zero and at the 95% simultaneous confidence level we conclude that there is no difference in the treatment. This result nicely illustrates a dilemma. The two-way analysis of variance did indicate a significant treatment effect. Is this a contradiction? Not really, we are "protecting" ourselves against an increased Type I error. Since the results are "borderline" even with the analysis of variance, it may be best to conclude that the results are suggestive but not clearly significant. A more substantial point may be made by asking why we should test the effect of sex anyway? It is merely a covariate or blocking factor. This argument raises the question of the appropriate set of comparisons. What do you think?

3. Randomized Block Designs. Usually we are interested in the treatment means only and not the block means. The confidence interval for the contrast $\tau_j - \tau_{j'}$ has the form

$$\overline{Y}_{\bullet j} - \overline{Y}_{\bullet j'} \pm \sqrt{(J-1)F_{J-1,IJ-I-J+1,1-\alpha}} \sqrt{MS_e \frac{2}{I}}.$$

The treatment effect τ_j has confidence interval

$$\overline{Y}_{\bullet j} - \overline{Y}_{\bullet\bullet} \pm \sqrt{(J-1)F_{J-1,IJ-I-J+1,1-\alpha}} \sqrt{MS_e \left(1 - \frac{1}{J} \right) \frac{1}{I}}.$$

Problems 12.16 and 12.18 illustrate the use of these formulae.

12.3.3 The Tukey Method (T-Method)

Another method that holds in nicely balanced ANOVA situations is a method based on an extension of the Student t-test. Recall that in the two-sample t-test, we use

$$t = \sqrt{\frac{n_1 n_2}{n_1 + n_2}} \frac{\overline{Y}_{1\bullet} - \overline{Y}_{2\bullet}}{s},$$

where $\overline{Y}_{1\bullet}$ is the mean of the first sample, $\overline{Y}_{2\bullet}$, is the mean of the second sample, and $s = \sqrt{\mathrm{MS}_e}$ is the pooled standard deviation. The process of dividing by s is called "studentizing" the range.

For more than two means, we are interested in the sampling distribution of the (largest-smallest) mean.

Definition 12.5. Let $Y_1, Y_2, \ldots, Y_k$ be independent and identically distributed (iid) $N(\mu, \sigma^2)$. Let s^2 be an estimate of σ^2 with m degrees of freedom which is independent of the Y_i's. Then the quantity

$$Q_{k,m} = \frac{\mathrm{MAXIMUM}\,(Y_1, Y_2, \ldots, Y_k) - \mathrm{MINIMUM}\,(Y_1, Y_2, \ldots, Y_k)}{s}$$

is called the *studentized range*.

Tukey derived the distribution of $Q_{k,m}$ and showed that it does not depend on μ or σ; a description is given in Miller [1981]. The distribution is tabulated in Table A.18 in the Appendix. Let $q_{k,m,1-\alpha}$ denote the upper critical value, that is

$$P\,[Q_{k,m} \geq q_{k,m,1-\alpha}] = 1 - \alpha.$$

You can verify from the table that for $k = 2$, two groups,

$$q_{2,m,1-\alpha} = \sqrt{2}\, t_{2,m,1-\alpha/2}.$$

We now state the main result for using the T-method of multiple comparisons which will then be specialized and illustrated with some examples.

The result is stated in the analysis of variance context since it is the most common application.

Result 12.2. Given a set of p population means $\mu_1, \mu_2, \ldots, \mu_p$ estimated by p independent sample means $\overline{Y}_1, \overline{Y}_2, \ldots, \overline{Y}_p$ each based on n observations and residual error s^2 based on m degrees of freedom, the probability is $(1 - \alpha)$ that simultaneously all *contrasts* of $\mu_1, \mu_2, \ldots, \mu_p$, say $\theta = c_1\mu_1 + c_2\mu_2 + \cdots + c_p\mu_p$ are in the confidence intervals

$$\widehat{\theta} \pm q_{p,m,1-\alpha}\widehat{\sigma_{\widehat{\theta}}},$$

where

$$\widehat{\theta} = c_1\overline{Y}_1 + c_2\overline{Y}_2 + \cdots + c_p\overline{Y}_p \quad \text{and} \quad \widehat{\sigma_{\widehat{\theta}}} = \frac{s}{\sqrt{n}} \sum_{i=1}^{p} \frac{|c_i|}{2}.$$

The Tukey method is used primarily with pairwise comparisons. In this case, $\hat{\sigma}_{\hat{\theta}}$ reduces to $s/\sqrt{n}$, the standard error of a mean. A requirement is that there be equal numbers of observations in each mean, this implies a balanced design. However, reasonably good approximations can be obtained for some unbalanced situations as will be illustrated next.

One-Way Analysis of Variance

Suppose there are I groups with n observations per group, and means $\mu_1, \mu_2, \ldots, \mu_I$. We are interested in all pairwise comparisons of these means. The estimate of $\mu_i - \mu_{i'}$ is $\overline{Y}_{i\bullet} - \overline{Y}_{i'\bullet}$, the variance of each sample mean estimated by $\text{MS}_e(1/n)$ with $m = I(n-1)$ degrees of freedom. The $100(1 - \alpha)\%$ simultaneous confidence intervals are given by

$$\overline{Y}_{i\bullet} - \overline{Y}_{i'\bullet} \pm q_{I,I(n-1),1-\alpha}\frac{1}{\sqrt{n}}\sqrt{\text{MS}_e}, \quad i, i' = 1, \ldots, I, i \neq i'.$$

This result cannot be applied to the example of Section 12.3.2 since the sample sizes are not equal. However, Dunnett [1980a] has shown that the $100(1-\alpha)\%$ simultaneous confidence intervals can be reasonably approximated by replacing

$$\sqrt{\frac{\text{MS}_e}{n}} \quad \text{by} \quad \sqrt{\text{MS}_e\left(\frac{1}{2}\right)\left(\frac{1}{n_i} + \frac{1}{n_{i'}}\right)},$$

where n_i and $n_{i'}$ are the sample sizes in Groups i and i', respectively, and the degrees of freedom associated with MS_e are the usual ones from the analysis of variance.

We now apply this approximation to the morphine binding data in Section 12.3.2. For this example $1 - \alpha = 0.99$, $I = 4$ and the $\text{MS}_e = 9.825$ has 24 d.f. resulting in $q_{4,24,0.99} = 4.907$. Simultaneous 99% confidence intervals are:

Contrast	n_i	n'_i	$\overline{Y}_{i\bullet} - \overline{Y}_{i'\bullet}$	Estimated Standard Error	99% Limits Lower	Upper
$\mu_1 - \mu_2$	18	2	1.7833	1.6520	−6.32	9.98
$\mu_1 - \mu_3$	18	3	2.4500	1.3822	−4.33	9.23
$\mu_1 - \mu_4$	18	5	6.8833	1.1205	1.39	12.4
$\mu_2 - \mu_3$	2	3	0.6167	2.0233	−9.31	10.5
$\mu_2 - \mu_4$	2	5	5.0500	1.8544	−4.05	14.1
$\mu_3 - \mu_4$	3	5	4.4333	1.6186	−3.51	12.4

We conclude, at a somewhat stringent 99% confidence level, that simultaneously, only one of the pairwise contrasts is significantly different: Group 1 (normal) differing significantly from Group 4 (Anephric).

Two-Way ANOVA with Equal Numbers of Observations Per Cell

Suppose in the two-way ANOVA of Section 10.3.1, there are n observations for each cell. The T-method may then be used to find intervals for either set of main effects (but not both simultaneously). For example, to find intervals for the α_i's, the intervals are:

Contrast	Interval
α_i	$\overline{Y}_{i\bullet\bullet} - \overline{Y}_{\bullet\bullet\bullet} \pm \dfrac{1}{\sqrt{Jn}} q_{I,IJ(n-1),1-\alpha} \sqrt{\mathrm{MS}_e \left(1 - \dfrac{1}{I}\right)}$
$\alpha_i - \alpha_{i'}$	$\overline{Y}_{i\bullet\bullet} - \overline{Y}_{i'\bullet\bullet} \pm \dfrac{1}{\sqrt{Jn}} q_{I,IJ(n-1),1-\alpha} \sqrt{\mathrm{MS}_e}$

We again consider the last example of Section 12.3.2, and want to set up 95% confidence intervals for α_1, α_2, and $\alpha_1 - \alpha_2$. In this example $I = 2$, $J = 2$, and $n = 10$. Using $q_{2,36,0.95} = 2.87$ (by interpolation) the intervals are given below:

Contrast	Estimate	Standard Error	95% Limits Lower	95% Limits Upper
α_1	-0.54	0.2358	-1.22	0.68
α_2	0.54	0.2358	-0.68	1.22
$\alpha_1 - \alpha_2$	-1.08	0.3335	-2.04	-0.12

We have used the MS_e with 36 degrees of freedom, that is, we have fitted a model with interaction. The interpretation of the results is that treatment effects do differ significantly at the 0.05 level; even though there is not enough evidence to reject the null hypothesis that the treatment effects differ from zero.

Randomized Block Designs

Using the notation of Section 12.3.2, suppose that we want to compare contrasts among the treatment means (the $\mu + \tau_j$). The τ_j themselves are contrasts among the means. In this case $m = (I - 1)(J - 1)$. The intervals are:

Contrast	Interval
τ_j	$\overline{Y}_{\bullet j} - \overline{Y}_{\bullet\bullet} \pm \dfrac{1}{\sqrt{I}} q_{J,(I-1)(J-1),1-\alpha} \sqrt{\mathrm{MS}_e \left(1 - \dfrac{1}{J}\right)}$
$\tau_j - \tau_{j'}$	$\overline{Y}_{\bullet j} - \overline{Y}_{\bullet j'} \pm \dfrac{1}{\sqrt{2I}} q_{J,(I-1)(J-1),1-\alpha} \sqrt{\mathrm{MS}_e}$

Consider Example 10.6. We want to compare the effectiveness of pancreatic supplements on fat absorption. The treatment means are

$$\overline{Y}_{\bullet 1} = 38.1, \quad \overline{Y}_{\bullet 2} = 16.5, \quad \overline{Y}_{\bullet 3} = 17.4, \quad \overline{Y}_{\bullet 4} = 31.1.$$

The estimate of σ^2 is $\mathrm{MS}_e = 107.03$ with 15 degrees of freedom. To construct simultaneous 95% T-confidence intervals, we need $q_{4,15,0.95}$; From Table A.18 this value is 4.076. The simultaneous 95% confidence interval for $\tau_1 - \tau_2$ is

$$(38.1 - 16.5) \pm \frac{1}{\sqrt{6}}(4.076)\sqrt{107.03}$$

or

$$21.6 \pm 17.2$$

yielding (4.4, 38.8).

Proceeding similarly, we obtain simultaneous 95% confidence intervals for the six pairwise comparisons:

Contrast	Estimate	95% Limits	
		Upper	Lower
$\mu_1 - \mu_2$	21.6	4.4	38.8
$\mu_1 - \mu_3$	20.7	3.5	37.9
$\mu_1 - \mu_4$	7.0	-10.2	24.2
$\mu_2 - \mu_3$	-0.9	-18.1	16.3
$\mu_2 - \mu_4$	-14.6	-31.8	2.6
$\mu_3 - \mu_4$	-13.7	-30.9	3.5

From this analysis we conclude that Treatment 1 differs from Treatments 2 and 3 but has not been shown to differ from Treatment 4. All other contrasts are not significant.

12.3.4   The Bonferroni Method (B-Method)

In this section, a method is presented which may be used in all situations. The method is conservative and is based on Bonferroni's inequality. It states that the probability of occurrence of one or more of a set of events occurring is less that or equal to the sum of the probabilities. That is, the Bonferroni inequality states

$$P(A_1 U \ldots U A_n) \leq \sum_{i=1}^{n} P(A_i).$$

We know that for disjoint events, the probability of one or more of $A_1, \ldots, A_n$ is equal to the sum of probabilities. If the events are not disjoint, then part of the probability is counted twice or more and there is strict inequality.

Suppose now that n simultaneous tests are to be performed. It is desired to have an overall significance level α. That is, if the null hypothesis is true in all n situations, the probability of incorrectly rejecting one or more of the null hypothesis is less than or equal to α. *Perform each test at significance level α/n, then the overall significance level is less that or equal to α.* Let A_i be the event of incorrectly rejecting in the ith test. Bonferroni's inequality shows that the probability of rejecting one or more of the null hypotheses is less than or equal to $(\alpha/n + \cdots + \alpha/n)$ (n terms) which is equal to α.

We now state the result that makes use of this inequality:

Result 12.3.   Given a set of parameters $\beta_1, \beta_2, \ldots, \beta_p$ and N linear combinations of these parameters, then the probability is greater than or equal to $(1 - \alpha)$ that simultaneously these linear combinations are in the intervals

$$\widehat{\theta} \pm t_{m, 1-\alpha/2N} \widehat{\sigma_{\widehat{\theta}}}.$$

Table 12.4. Variables at Rest and Exercise Before and After Oral Procainamide.[a]

	Procainamide Plasma Level, 1 Hr	Rest						Exercise							
		HR		SP		DP		HR Maximum		SP Maximum		DP Maximum		Arrhythmia Frequency	
		Control	1 Hr	Control	1 Hr	Control	1 Hr	Control	1 Hr	Control	1 Hr	Control	1 Hr	Control	1 Hr
Number of patients	23	23		23		23		23		23		23		23	
Mean ± SD	5.99 ±1.33	73 ±11	87 ±13	129 ±17	118 ±11.8	81 ±11	81 ±9.2	171 ±13.5	170 ±14	187 ±20.6	168 ±20	85 ±12	76 ±10	105 ±108	38 ±69
t		5.053		4.183		0.3796		0.9599		5.225		5.005		3.422	
P[b]		<0.0015		<0.0060		NS		NS		<0.0015		<0.015		<0.0360	

	Severity Index		VO_2 MAX		FAI(%)		Computer ST_B							
							Rest		Maximum		Slope		Zero Recovery	
	Control	1Hr	Control	1Hr	Control	1Hr	Control	1Hr	Control	1Hr	Control	1Hr	Control	1Hr
Number of patients	23		22		23		22		22		22		23	
Mean ±SD	12.9 ±3.0	4.9 ±4.67	33.2 ±5.8	33.0 ±6.0	12.9 ±12.5	13.5 ±11.5	0.036 ±0.044	0.044 ±0.051	-0.190 ±0.126	-0.122 ±0.095	-2.31 ±1.401	-2.05 ±1.29	-0.065 ±0.0003	-0.0302 ±0.077
t	5.870		0.3852		0.5253		0.8861		3.915		1.132		4.320	
P[b]	<0.0015		NS		NS		NS		<0.0120		NS		<0.0045	

[a]Dose, 15 mg per kilogram body weight; HR, heart rate; SP, systolic pressure (mmHg); DP, diastolic pressure (mmHg); VO_2 MAX, maximal oxygen consumption (ml/min); FAI, functional aerobic impairment; ST_B, 100-beat averaged S-T depression, from monitored CB, lead, taken 50–69 ms after nadir of S wave; slope, $\delta Hr/\delta ST_B$; t, paired t-test; NS, not significant; Hr, hour.

[b]Probability mulitiplied by 15 to correct for multiple comparisons (Bonferroni's inequality correction).

The quantity $\widehat{\theta}$ is $c_1 b_1 + c_2 b_2 + \cdots + c_p b_p$, $t_{m,1-\alpha/2N}$ is the $100(1 - \alpha/2N)th$ percentile of a t-statistic with m degrees of freedom, and $\widehat{\sigma}_{\widehat{\theta}}$ is the estimated standard error of the estimate of the linear combination based on m degrees of freedom.

The value of N will vary with the application. In the one-way ANOVA with all the pairwise comparisons among the I treatment means $N = \binom{I}{2}$. Simultaneous confidence intervals, in this case, are of the form

$$\overline{Y}_{i\bullet} - \overline{Y}_{i'\bullet} \pm t_{m,1-\alpha/2\binom{I}{2}} \sqrt{\text{MS}_e \left(\frac{1}{n_i} + \frac{1}{n'_i} \right)}, \quad i, i' = 1, \ldots, I, \ i \neq i'.$$

The value of α need not be partitioned into equal multiples. The simplest is $\alpha = \alpha/N + \alpha/N + \cdots + \alpha/N$ but any partitions of $\alpha = \alpha_1 + \alpha_2 + \cdots + \alpha_N$ is permissible, yielding a per experiment error rate of at most α.

Similarly, when presenting p-values, when N simultaneous tests are being done, multiplication of the p-value for each test by N gives p-values allowing simultaneous consideration of all N tests.

An example of the use of Bonferroni's inequality, is in a paper by Gey et al. [1974]. This paper considers heart beats that have an irregular rhythm (or arrythmia). The study examined the administration of the drug procainamide and evaluated variables associated with the maximal exercise test with and without the drug. Fifteen variables were examined using paired t-tests. All the tests came from data on the *same* 23 patients so the test statistics were not independent. To correct for the multiple comparison values, the p-values were multiplied by 15. Table 12.4 presents 14 of the 15 comparisons. The table shows that even taking into account the multiple comparisons, many of the variables differed when the subject was on the procainamide medication. In particular, the frequency of arrythmic beats was decreased by administration of the drug.

12.4 COMPARISON OF THE THREE PROCEDURES

Of the three methods presented, which should be used? In many situations there is not sufficient balance in the data (for example, equal numbers in each group in a one-way analysis of variance) to use the T-method and the Scheffé method procedure or the Bonferroni inequality should be used. For paired comparisons, the T-method is preferable. For more complex contrasts, the S-method is preferable. A comparison between the Bonferroni or B-method and the S-method is more complicated, depending heavily on the type of application. The Bonferroni method is easier to carry out, and in many situations the critical value will be less that that for the S-method.

In Table 12.5 we compare the critical values for the three methods for the case of the one-way ANOVA with k treatments and 20 degrees of freedom for error MS. With two treatments ($k = 2$ and, therefore, $\nu = 1$) the three methods give identical multipliers (the q statistic has to be divided by $\sqrt{2}$ to have the same scale as the other two statistics).

Hence, if pairwise comparisons are carried out, the Tukey procedure will produce the shortest simultaneous confidence intervals. For the type of situation illustrated in the table the B-method is always preferable to the S-method. It assumes, of course, that the total, N, of comparisons to be made is known. If this is not the case, as in "fishing expeditions," the Scheffé method provides more adequate protection.

Table 12.5. Comparison of the Critical Values for One Way ANOVA with k Treatments. Assume $\binom{k}{2}$ comparisons for the Tukey and Bonferroni Procedures. Based on 20 degrees of freedom for error mean square.

Number of Treatments (k)	Degrees of Freedom $\nu = k - 1$	$\sqrt{\nu F_{\nu,20,0.95}}$	$\frac{1}{\sqrt{2}} q_{\nu,20,0.95}$	$t_{20,1-\alpha/2\binom{k}{2}}$
2	1	2.09	2.09	2.09
3	2	2.64	2.53	2.61
4	3	3.05	2.80	2.93
5	4	3.39	2.99	3.15
11	10	4.85	3.61	3.89
21	20	6.52	4.07	4.46

Dunn [1959] computed tables of the statistic $t_{m,1-\alpha/2N}$ for N up to values of $N = 100$. For an informative discussion of the issues in multiple comparisons, see comments by O'Brien [1983] in *Biometrics*. The same issue contains two additonal papers on the same topic, in particular, a paper by Duncan and Brant [1983]. One of Duncan's approaches has been to modulate the per-comparison critical value on the basis of the overall F-statistic. If the overall F-statistic is large, the critical t-value for a particular comparison can be reduced. This approach has not won universal approval in the statistical community.

12.5 OPTIONAL STOPPING OF EXPERIMENTS

Consider a situation in which experimental data are gathered sequentially. A clinical trial with patients enrolled sequentially is one example. Another situation might be the laboratory experiment in which only one animal can be studied at a time. In the clinical trial context often data must be monitored continuously for ethical reasons. For example in a trial using survival as an endpoint if it should become clear which treatment is best, the trial must be stopped not only to provide all participants with the best treatment, if possible, but also to publicize the results. In the laboratory situation it would seem reasonable to test the null hypothesis after each animal, to see if statistical significance has been achieved. Thus, "needless" experimentation might be avoided.

The problem with the above approach is that multiple tests are being made in the same experiment. If each test is at a fixed significance level α, after multiple tests the probability of incorrectly rejecting the null hypothesis will be larger than α. The order of magnitude of the problem is illustrated by two tables abstracted from tables in Armitage *et al.* [1969].

In the first situation, a binomial sample is accumulated, the distribution being according to the null hypothesis $\pi = 0.5$. After each trial, a two-sided fixed sample size test of the null hypothesis is used. The experiment stops if the null hypothesis is rejected. Table 12.6 presents the probability of incorrectly rejecting $\pi = 0.5$ in the first n trials (called the probability of "being absorbed"). If you were to take a binomial sample of size 100 and test at a 5% significance level after each Bernoulli

Table 12.6. The Probability of Being Absorbed at or Before the nth Observation in Binomial Sampling with Repeated Tests at a Nominal Two-Sided Significance Level α.

n	$\alpha = 0.05$	$\alpha = 0.03$	$\alpha = 0.01$
10	0.05469	0.02930	0.00781
15	0.08191	0.03955	0.01538
20	0.10662	0.05248	0.01840
30	0.13355	0.07631	0.02746
40	0.15351	0.09255	0.03406
50	0.17117	0.10508	0.03931
60	0.18583	0.11616	0.04319
80	0.20889	0.13083	0.05090
100	0.22731	0.14436	0.05586
120	0.24187	0.15515	0.06085
140	0.25503	0.16388	0.06462
150	0.26108	0.16833	0.06619

trial, the probability is 0.227 that the null hypothesis would be incorrectly rejected on or before the 100th trial. Somehow the fact of optional stopping must be taken into account if the significance level of 0.05 is to be maintained.

As a second example, suppose that iid, $N(0,1)$ variables are accumulated and the null hypothesis $H_o : \mu = 0$ is tested. Suppose after each new observation we use a two-sided fixed sample size test. That is, after n observations reject H_o if

$$|\sqrt{n}\,\overline{X}| \geq Z_{1-\alpha/2}$$

where $Z_{1-\alpha/2}$ is the two-sided $N(0,1)$ α critical value. Table 12.7 shows for various n the probability of having rejected H_o on or before the nth observation.

For example, for 100 observations with sequential testing after each observation at a 10% significance level, the probability of incorrectly rejecting the null hypothesis is 0.59 rather that 0.10.

One approach to dealing with the optional stopping dilemma is to use sequential analysis. The mathematics and details of such designs are beyond the scope of the text. A good reference is Whitehead [1983], a classic reference is Armitage [1975]. The basic idea in a sequential trial is that the results are evaluated as they become available. This is particularly applicable to medical studies where patients frequently enter a study sequentially over a fairly long period of time. If the sample size calculations are based on a fixed sample size design (as all the calculations have been in this text) then the researcher basically waits until the end of the study to analyze the data. If the expected effect had appeared already early in the study it would have been clearly more efficient to terminate the study. If the endpoints are mortality or serious morbidity it would not be ethical to continue the study. In the simplest sequential medical trial, a patient's outcome measure is classified as a "success" or a "failure"; or, if two treatments are involved patients might be paired in some manner and a pairwise comparison is made. (Pairs of patients that have the same outcome are not used in this part of the analysis.) We thus "induce" Bernoulli trials on the data. Let S_n and F_n be the number of "successes" and "failures" after the nth trial, $S_n - F_n$ the excess preference.

Table 12.7. The Probability of Rejecting H_o At or Before the nth Observation in Sampling from a Normal Distribution with Known Variance, with Repeated Tests at a Nominal Two-Sided Significance Level α.

n	$\alpha = 0.10$ $Z_{1-\alpha/2} = 1.645$	$\alpha = 0.05$ $Z_{1-\alpha/2} = 1.960$	$\alpha = 0.02$ $Z_{1-\alpha/2} = 2.326$	$\alpha = 0.01$ $Z_{1-\alpha/2} = 2.576$
1	0.10000	0.05000	0.02000	0.01000
2	0.16015	0.08312	0.03453	0.01766
3	0.20207	0.10726	0.04561	0.02366
4	0.23399	0.12617	0.05454	0.02858
5	0.25963	0.14169	0.06201	0.03274
10	0.34169	0.19336	0.08776	0.04738
15	0.38973	0.22509	0.10419	0.05692
20	0.42319	0.24791	0.11628	0.06403
30	0.46896	0.28016	0.13379	0.07444
40	0.50020	0.30293	0.14643	0.08205
50	0.52364	0.32045	0.15633	0.08805
60	0.54223	0.33464	0.16446	0.09300
80	0.57051	0.35674	0.17733	0.10090
100	0.59152	0.37362	0.18732	0.10708
140	0.62292	0.39857	0.20238	0.11647
200	0.65165	0.42429	0.21828	0.12649
500	0.720	0.487	0.259	0.152
750	0.746	0.513	0.276	0.164
1000	0.763	0.530	0.288	0.172

In sequential analysis the quantities k_n in P[for at least one n, $|S_n - F_n| > k_n] = 0.05$ are calculated for each n and plotted on a graph. The value of k_n will depend upon the expected difference between the probabilty of success and failure. At each trial, the excess number of preferences is plotted. A path is thus set up. If the path crosses one of the boundaries defined by k_n the experiment is terminated and the null hypothesis is rejected. The choice of alternative hypothesis depends upon the outcome of the trial.

We illustrate this approach with some data from Brown et al. [1960] which examined the value of a large dose of antitoxin in clinical tetanus. There was no question about the efficacy of antitoxin in *preventing* tetanus but the question of the value of the antitoxin in patients who had developed the disease was not answered. The two treatments were "Antitoxin" and "No Antitoxin." As patients entered the trial they were paired and one of the pair received antitoxin, the other did not. If for any pair, the patient receiving the antitoxin survived while the patient who did not receive the antitoxin died, a preference was established for "antitoxin." If the reverse happened a preference was established for "no antitoxin." If both patients survived, or both died the data was not used in the calculation of $S_n - F_n =$ the excess number of preferences for "antitoxin" over "no antitoxin" after the nth untied pair.

The results were graphed as in Figure 12.2. The boundaries are chosen so that the probability of a type I error remained constant at $\alpha = 0.05$, and the power is $1 - \beta = 0.95$, assuming a probability of preference for antitoxin of 0.75 (compared to the null hypothesis value of 0.50). If the boundary was not crossed by the 62nd untied pair, the study would have been stopped and the null hypothesis not rejected. The upper boundary was crossed at the 18th preference. At that point 15 preferences

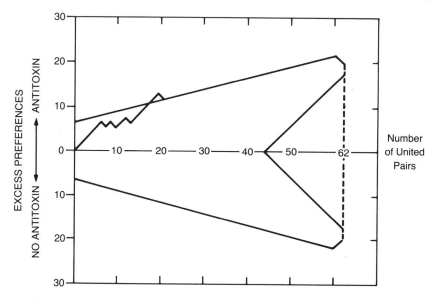

Figure 12.2. Sequential analysis of a trial of antitoxin in the treatment of clinical tetanus. Data from Brown *et al.* [1960].

favored antitoxin and 3 preferences favored no antitoxin. (Approximately 32 pairs had been enrolled by this time so about 14 pairs were tied.) It is interesting to note that the p-value associated with the observed result: ($15\ S$, $3\ F$) is $2 \times 0.00377 = 0.00754$, much less than 0.05. The p-value, however, is not 0.00754 but 0.05, to accommodate the mutiple testing of the null hypothesis.

The maximum sample size could have been 62. Assuming a "fixed sample size" with $\alpha = 0.05$ (two-sided), $\beta = 0.05$, and $\pi_1 = 0.75$, $\pi_0 = 0.50$, a total of 48 preferences would have been needed. This illustrates a characteristic of sequential trials; the number of samples needed *could* be bigger than that based on fixed sample trials. In a world of no treatment effects, this would tend to be the case. The moral would seem to be: If you can't afford to wait, don't – but there may be a price to pay.

There are compromises between "pure" sequential trials and pure "fixed sample" studies that allow periodic inspection of the data. See Whitehead [1983], Pocock [1982] and O'Brien and Fleming [1979].

Another good example of a sequential trial can be found in Lewis *et al.* [1983].

12.6 POST HOC ANALYSIS

12.6.1 The Setting

A particular form of the multiple comparison problem is the *post hoc* analysis. Such an analysis is not explicitly planned at the start of the study but suggested by the data. Other terms associated with such analyses are "data driven" and "sub-group analysis". Aside from the assignment of appropriate p-values there is the more important question of the scientific status of such an analysis. Is the study to be considered exploratory, confirmatory or both? That is, can the *post hoc* analysis only suggest possible connections and associations which have to be confirmed in future studies or can it be considered as confirming them as well? Unfortunately no rigid lines can be drawn here. Every experimenter does, and should do, *post hoc* analyses to ensure that all aspects of the observations are utilized. There is no room for rigid adherence to artificial schema of hypothesis which are laid out row upon boring row. But what is the status of these analyses? Cox [1977] remarks:

> "Some philosophies of science distinguish between exploratory experiments and confirmatory experiments and regard an effect as well established only when it has been demonstrated in a confirmatory experiment. There are undoubtedly good reasons, not specifically concerned with statistical technique, for proceeding this way; but there are many fields of study, especially outside the physical sciences, where mounting confirmatory investigations may take a long time and therefore where it is desirable to aim at drawing reasonably firm conclusions from the same data as used in exploratory analysis."

What statistical approaches and principles can be used? In the following discussion we follow closely suggestions of Cox and Snell [1981], Pocock [1982], and Pocock [1984].

12.6.2   Statistical Approaches and Principles

Analyses Must be Planned
At the start of the study specific analyses must be planned and agreed to. These may be broadly outlined but must be detailed enough to, at least theoretically, answer the questions being asked. Every practicing statistician has met the researcher who has a filing cabinet full of crucial data "just waiting to be analyzed" (by the statistician, who may also feel free to suggest appropriate questions that can be answered by the data).

Planned Analyses Must be Carried Out and Reported
This appears obvious but is not always followed. At worst it becomes a question of scientific integrity and honesty. At best it is potentially misleading to omit reporting such analyses. If the planned analysis is amplified by other analyses which begin to take on more importance, a justification must be provided together with suggested adjustments to the significance level of the tests. The researcher may be compared to the novelist whose minor character develops a life of his own as the novel is written. The development must be rational and believable.

Adjustment for Selection
A *post hoc* analysis is part of a multiple comparisons procedure and appropriate adjustments can be made if the family of comparisons is known.

  Use of the Bonferroni inequality can also have a drastic effect on the significance level. Schweder and Spjøtvoll [1982] provide an ingenious method for assesing the number of sigificant tests in a set of tests. See Note 12.4.

Split-sample Approach
In the split sample approach, the data are randomly divided into two parts, the first part is used to generate the exploratory analyses which are then "confirmed" by the

second part. Cox [1977] says that there are "strong objections on general grounds to procedures where different people analyzing the same data by the same method get different answers". An additional aspect of such analyses is that it does not provide a solution to the problem of subgroup analysis.

Interaction Analysis

The number of comparisons is frequently not defined and most of the above approaches will not work very well. Interaction analysis of subgroups provides valid protection in such *post hoc* analyses. Suppose a treatment effect has been shown for a particular subgroup. To assess the validity of this effect analyze all subgroups jointly and test for an interaction of subgroup and treatment. This procedure embeds the subgroup in a meaningful larger family. If the global test for interaction is significant, it is warranted to focus on the subgroup suggested by the data. Pocock [1984] illustrates this approach with data from the Multiple Risks Factor Intervention Trial Research Group [1982]. This randomized trial of "12,866 men at high risk of coronary heart disease compared to special intervention (SI) aimed at affecting major risk factor (e.g., hypertension, smoking, diet) and usual care (UC). The overall rates of coronary mortality after an average seven year follow-up (1.79% on SI and 1.93% on UC) are not significantly different." The paper presented four subgroups. The extreme right hand column in the table below lists the odds ratio comparing mortality in the Special Intervention and Usual Care groups. The first three subgroups appear homogeneous suggesting a beneficial effect of special intervention. The fourth subgroup (with hypertension and ECG abnormality) appears different. The average odds ratio for the first three subgroups differs significantly from the odds ratio for the fourth group ($p < 0.05$). However, this is a *post hoc* analysis and a test for the homogeneity of the odds ratios over all four subgroups shows no significant differences and, furthermore, the average of the odds ratio does not differ significantly from one. Thus on the basis of the global interaction test there are no significant differences in mortality among the eight groups. (A chi-square analysis of the 2×8 contingency table formed by the two treatment groups and the eight subgroups shows a value of $\chi^2 = 8.65$ with 7 d.f.). Pocock concludes: "taking into account the fact that this was not the only subgroup analysis performed, one should feel confident that there are inadequate grounds for supposing that the special intervention did harm to those with hypertension and ECG abnormalities".

Hypertension	ECG Abnormality	No. of Coronary Death/No. of Men		Odds Ratio
		Special Intervention (%)	Usual Care (%)	
No	No	24/1817 (1.3)	30/1882 (1.6)	0.83
No	Yes	11/592 (1.9)	15/583 (2.6)	0.72
Yes	No	44/2785 (1.6)	58/2808 (2.1)	0.76
Yes	Yes	36/1233 (2.9)	21/1185 (1.8)	1.67

If the overall test of interaction had been significant, or if the comparison had been suggested before the study was started, then the "significant" *p*-value would have had clinical implications.

12.6.3 Summary

Post hoc comparisons usually should be considered exploratory rather than confirmatory; but this rule should not be followed slavishly. It is clear that some adjustment to the significance level must be made to maintain validity of the statistical procedure. In each instance the *p*-value will be adjusted upwards. The question is how much. We have described several techniques for making these adjustments. When reading research reports which include *post hoc* analyses it is prudent to keep in mind that, in all likelihood, many such analyses were tried by the authors but not reported. Thus, scientific caution must be the rule. To be confirmatory, results from such analyses must not only make excellent biological sense but must also satisfy the principle of Occam's razor. That is, there must not be a simpler explanation that is also consistent with the data.

NOTES

Note 12.1: Orthogonal Contrasts

Orthogonal contrasts form a special group of contrasts. Consider two contrasts:

$$\theta_1 = c_{11}\beta_1 + \cdots + c_{1p}\beta_p$$

and

$$\theta_2 = c_{21}\beta_1 + \cdots + c_{2p}\beta_p.$$

The two contrasts are said to be *orthogonal* if

$$\sum_{j=1}^{p} c_{1j}c_{2j} = 0.$$

Clearly, if θ_1, θ_2 are orthogonal then $\widehat{\theta}_1$, $\widehat{\theta}_2$ will be orthogonal since orthogonality is a property of the coefficients. Two orthogonal contrasts are *orthonormal* if, in addition,

$$\sum c_{1j}^2 = \sum c_{2j}^2 = 1.$$

The advantage to considering orthogonal (and orthonormal) contrasts is that they are uncorrelated and, hence, if the observations are normally distributed the contrasts are statistically independent. Hence, the Bonferroni inequality becomes an equality. But there are other advantages. To see those we extend the orthogonality to more than two contrasts. A set of contrasts is orthogonal (orthonormal) if all pairs of contrasts are orthogonal (orthonormal).

Now consider the one-way analysis of variance with I treatments. There are $(I-1)$ degrees of freedom associated with the treatment effect. It can be shown that there are precisely $(I-1)$ orthogonal contrasts to compare the treatment means. The set is not unique, let θ_1, θ_2, $\ldots$, θ_{I-1} form a set of such contrasts. Assume that they are orthonormal, and let $\widehat{\theta}_1$, $\widehat{\theta}_2$, $\ldots$, $\widehat{\theta}_{I-1}$ be the estimate of the orthonormal contrasts. Then it can be shown that

$$\text{SS}_{\text{TREATMENTS}} = \widehat{\theta}_1^2 + \widehat{\theta}_2^2 + \cdots + \widehat{\theta}_{I-1}^2.$$

We have thus partitioned the SS_{TREATMENTS} into $(I-1)$ components (each with one degree of freedom it turns out) and uncorrelated as well. This is a very nice summary of the data.

To illustrate this approach, assume an experiment with four treatments. Let the means be μ_1, μ_2, μ_3, μ_4.

A possible set of contrasts is given by the following pattern

Contrast	μ_1	μ_2	μ_3	μ_4
θ_1	$1/\sqrt{2}$	$-1/\sqrt{2}$	0	0
θ_2	$1/\sqrt{6}$	$1/\sqrt{6}$	$-2/\sqrt{6}$	0
θ_3	$1/\sqrt{12}$	$1/\sqrt{12}$	$1/\sqrt{12}$	$-3/\sqrt{12}$

You can verify that:

1. These contrasts are orthonormal;
2. There are no additional *orthogonal contrasts*;
3. $\theta_1^2 + \theta_2^2 + \theta_3^2 = \sum (\mu_i - \mu)^2$.

The pattern can clearly be extended to any number of means (it is known as the Gram–Schmidt orthogonalization process).

The nonuniqueness of this decomposition becomes obvious from starting the first contrast, say, with

$$\theta_1^* = \frac{1}{\sqrt{2}}\mu_1 - \frac{1}{\sqrt{2}}\mu_4.$$

Sometimes a meaningful set of orthogonal contrasts can be used to summarize an experiment. This approach, using the statistical independence to determine the significance level will minimize the cost of multiple testing. Of course, if these contrasts were carefully specified beforehand you might argue that each one should be tested at level α!

Note 12.2: The Tukey Test

The assumptions underlying the Tukey test include that the variances of the means are equal; this translates into equal sample sizes in the analysis of variance situation. Although the procedure is commonly associated with pairwise comparisons among independent means it can be applied to arbitrary linear combinations and even allows for a common correlation among the means. For further discussion see Miller [1981], pp. 37–48.

Note 12.3: Simultaneous Tests in Contingency Tables

In $r \times c$ contingency tables, there is frequently interest in comparing subsets of the tables. Goodman [1964a,b] derived the large sample form for $100(1 - \alpha)\%$ simultaneous contrasts for all 2×2 comparisons. This is equivalent to examining all $\binom{r}{2}\binom{c}{2}$

possible odds ratios. The intervals are constucted in terms of the logarithms of the ratio. Let

$$\hat{\omega} = \log n_{ij} + \log n_{i'j'} - \log n_{i'j} - \log n_{ij'}$$

be the log odds associated with the indicated frequencies. In Chapter 7 we showed that the approximate variance of this statistic is

$$\hat{\sigma}^2_{\omega} \doteq \frac{1}{n_{ij}} + \frac{1}{n_{i'j'}} + \frac{1}{n_{i'j}} + \frac{1}{n_{ij'}}.$$

Simultaneous $100(1 - \alpha)\%$ confidence intervals are of the form,

$$\hat{\omega} \pm \sqrt{\chi^2_{(r-1)(c-1),(1-\alpha)}}\ \hat{\sigma}_{\omega}.$$

This again is of the same form as the Scheffé approach, but now based upon the chi-square distribution rather that the F-distribution.

The price, again, is fairly steep. At the 0.05 level and a 6×6 contingency table, the critical value of the chi-square statistic is

$$\sqrt{\chi^2_{25,0.95}} = \sqrt{37.65} = 6.14.$$

Of course, there are $\binom{6}{2}\binom{6}{2} = 225$ such tables.

It may be more efficient to use the Bonferroni inequality. In the example above, the corresponding Z-value using the Bonferroni inequality is

$$Z_{1-0.025/225} = Z_{0.999889} \doteq 3.69.$$

So if only 2×2 tables are to be examined the Bonferroni approach will be more economical.

However, the Goodman approach works and is valid for *all* linear contrasts. See Goodman [1964a,b] for additional details.

Note 12.4: *P*-Values for Simultaneous Tests

A paper by Schweder and Spjøtvoll [1982] proposes an ingenious approach to estimating the number of false null hypotheses when a large number of tests are carried out. Suppose N tests are performed, and let N_0 be the unknown number of true null hypotheses. Let N_p be the observed number of p-values greater than p. Since a p-value should be small for a false null hypothesis,

$$E[N_p] \doteq N_0(1 - p),$$

when p is not too small. A plot of N_p against $1 - p$ should suggest a straight line for a value of $1 - p$ small with slope N_0. As more extreme p-values are reached the number N_p should begin to deviate from the line. This may be particularly applicable to "dredging" $r \times c$ contingency tables.

PROBLEMS

For the problems in this chapter the following tasks are defined. Additional tasks are indicated at each problem. Unless otherwise indicated, assume $\alpha^* = 0.05$.

   a. Calculate simultaneous confidence intervals as discussed in Section 12.2. Graph the intervals and state your conclusions.

   b. Apply the Scheffé method. State your conclusions.

   c. Apply the Tukey method. State your conclusions.

   d. Apply the Bonferroni method. State your conclusions.

   e. Compare the indicated methods. Which result is the most reasonable.

1. This problem deals with Problem 10.1. Use a 99% confidence level.

   a. Carry out Task (a).

   b. Compare your results with those obtained in Section 12.3.2.

   c. A more powerful test can be obtained by considering the groups to be ranked in order of increasingly severe disorder. A test for trend can be carried out by coding the groups 1, 2, 3, and 4 and regressing the percentage morphine bound on the regressor variable and testing for significance of the slope. Carry out this test and describe its pros and cons.

   d. Carry out Task (c) using the approximation recommended in Section 12.3.3.

   e. Carry out Task (e).

2. This problem deals with Problem 10.2.

   a. Do Tasks (a)–(e) for pairwise comparisons of all treatment effects.

3. This problem deals with Problem 10.3.

   a. Do Tasks (a)–(d) for all pairwise comparisons.

   b. Do Task (c) defined in Problem 12.1.

   c. Do Task (e).

4. This problem deals with Problem 10.4.

   a. Do Tasks (a)–(e) setting up simultaneous confidence intervals on both main effects and all pairwise comparisons.

   b. A further comparison of interest is control vs. shock. Using the Scheffé approach, test this effect.

   c. Summarize the results from this experiment in a short paragraph.

5. Sometimes we are interested in comparing several treatments against a standard treatment. Dunnett [1954] has considered this problem. If there are I groups, and Group 1 is the Standard Group, $(I-1)$ comparisons can be made at level $(1-\alpha/2(I-1))$ to maintain a per-experiment error rate of α. Apply this approach to the data of Bruce et al. [1974] in Section 12.2 by comparing Groups $2, \ldots, 8$ with Group 1, the healthy individuals. How do your conclusions compare with those of Section 12.2?

6. This problem deals with Problem 10.6.
   a. Carry out Tasks (a)–(e).
   b. Suppose we treat these data as a regression problem (as suggested in Chapter 10). Does it still make sense to test the significance of the difference of adjacent means? Why or why not? What if the trend was nonlinear?

7. This problem deals with Problem 10.7.
   a. Carry out Tasks (a)–(e).

8. This problem deals with Problem 10.8.
   a. Carry out Tasks (b), (c), and (d).
   b. Of particular interest are the comparisons of each of the test preparations A–D with the standard insulin. The "Medium" treatment is not relevant for this analysis. How does this alter Task (d)?
   c. Why would it not be very wise to ignore totally the "Medium" treatment? What aspect of the data for this treatment can be usefully incorporated into the analysis in Part (b).

9. This problem deals with Problem 10.9.
   a. Compare each of the means of the schizophrenic group with the control group using S, T, and B methods.
   b. Which method is preferred?

10. This problem deals with Problem 10.10.
   a. Carry out Tasks (b)–(e) on the plasma concentration of 45 minutes, comparing the two treatments with controls.
   b. Carry out Tasks (b)–(d) on the difference in the plasma concentration at 90 min and 45 min (subtract the 45 min reading from the 90 min reading). Again, compare the two treatments with controls.
   c. Synthesize the conclusions of Parts (a) and (b).
   d. Can you think of a "nice" graphical way of presenting Part (c)?
   e. Consider the combined parts of (a) and (b) above. From a multiple comparison point of view what criticism could you level at this combination? How would you resolve it?

11. Data for this problem are from a paper by Winick et al. [1975]. The paper examines the development of adopted Korean children differing greatly in early nutritional status. The study was a retrospective study of children admitted to the Holt Adoption Service and ultimately placed in homes in the United States. The children were divided into three groups on the basis of how their height, at the time of admission to Holt, related to a reference standard of normal Korean children of the same age:

   Group 1: Designated "*malnourished*"—below the third percentile for both height and weight;

   Group 2: "*Moderately nourished*"—from the third to the twenty-fourth percentile for both height and weight;

Table 12.8. Current Height (Percentiles, Korean Reference Standard) Comparison of Three Nutrition Groups. F Probability is the Probability that the Calculated F from the One-Way ANOVA Ratio Would Occur by Chance. Data for Problem 11.

Group	N	Mean Percentile	SD	F Probability	Contrast Group	t-Test t	t-Test P
1	41	71.32	24.98	0.068	1 vs.2	−1.25	0.264
2	50	76.86	21.25		1 vs.3	−2.22	0.029[a]
3	47	82.81	23.26		2 vs.3	−1.31	0.194
Total	138	77.24	23.41				

[a]Statistically significant.

> Group 3: *"Well-nourished or control"*—at or above the twenty-fifth percentile for both height and weight.

Table 12.8 has data from this paper.

a. Carry out Tasks (a), (b), (c), (d), and (e) for all pairwise comparisons and state your conclusions.

b. Read the paper compare your results with that of the authors.

c. A philosophical point may be raised about the procedure of the paper. Since the overall F-test is not significant at the 0.05 level (see Table 12.8), it would seem inappropriate to "fish" further into the data. Discuss the pros and cons of this argument.

d. Can you suggest alternative more powerful analyses? (What is meant by "more powerful"?)

12. Derive Equation (12.1). Indicate clearly how the independence assumption and the null hypotheses are crucial to this result.

13. A somewhat amusing—but also serious—example of the multiple comparison problem is the following. Suppose that a journal tends to accept only papers that show "significant" results. Now imagine multiple groups of independent researchers (say 20 universities in the United States and Canada) all working on roughly the same topic, and hence, testing the same null hypothesis. If the null hypothesis is true we would expect only one of the researchers to come up with a "significant" result. Knowing the editorial policy of the journal, the 19 researchers with nonsignificant results do not bother to write up their research but the remaining researcher does. The paper is well written, challenging and provocative. The editor accepts the paper and it is published.

a. What is the per-experiment error rate? Assume 20 independent researchers.

b. Define an appropriate editorial policy in view of an unknown number of comparisons.

14. This problem deals with the data of Problem 10.13. The primary interest in these data involves comparisons of three treatments, that is, the experiments represent blocks. Carry out Tasks (a)–(e) focussing on comparison of the means for Tasks (b)–(d).

15. This problem deals with the data of Problem 10.14.

   a. Carry out the Tukey test for pairwise comparisons on the total analgesia score presented in Part (b) of that question. Translate your answers to obtain confidence intervals applicable to single readings.

   *b. The Sum of Squares for Analgesia can be partitioned into three orthogonal contrasts as follows:

	μ_1	μ_2	μ_3	μ_4	**Divisor**
θ_1	-1	-1	-1	3	$\sqrt{12}$
θ_2	1	-1	-1	1	$\sqrt{4}$
θ_3	-1	3	-3	1	$\sqrt{20}$

   Verify that these contrasts are orthogonal. If the coefficients are divided by the divisors at the right, verify that the contrasts are orthonormal.

   *c. Interpret the contrasts θ_1, θ_2, θ_3 defined in Part (b).

   *d. Let $\widehat{\theta}_1$, $\widehat{\theta}_2$, $\widehat{\theta}_3$ be the estimates of the orthonormal contrasts. Verify that

   $$\text{SS}_{\text{treatments}} = \widehat{\theta}_1^2 + \widehat{\theta}_2^2 + \widehat{\theta}_3^2.$$

   Test the significance of each of these contrasts and state your conclusion.

16. This problem deals with Problem 10.15.

   a. Carry out Tasks (b)–(e) on all pairwise comparisons of treatment means.

   b. How would the results in Part (a) be altered if the Tukey test for additivity is used? Is it worth re-analyzing the data?

17. This problem deals with Problem 10.16.

   a. Carry out Tasks (b)–(e) on the treatment effects and on all pairwise comparisons of treatment means.

   *b. Partition the Sums of Squares of Treatments into two pieces, a part attributable to linear regression and the remainder. Test the significance of the regression, adjusting for the multiple comparison problem.

*18. Apply the method of Note 12.4 to the data of Bruce et al. [1974] in Section 12.2, by comparing all pairwise treatment means. How many null true hypotheses do you estimate there are among the $\binom{8}{2} = 28$ comparisons.

*19. This problem deals with the data of Problem 10.18.

   a. We are going to "mold" these data into a regression problem as follows; define six dummy variables I_1 to I_6.

   $$I_i = 1, \quad \text{data from subject } i, i = 1,\ldots,6,$$

   $$0, \quad \text{otherwise.}$$

In addition, define three further dummy variables:

$$I_7 = \begin{cases} 1, & \text{recumbent position,} \\ 0, & \text{otherwise;} \end{cases}$$

$$I_8 = \begin{cases} 1, & \text{placebo,} \\ 0, & \text{otherwise;} \end{cases}$$

$$I_9 = I_7 \times I_8.$$

b. Carry out the regression analyses of Part (a) forcing in the dummy variables I_1 to I_6 first. Group those into one ss with 6 degrees of freedom. Test the significance of the regression coefficients of I_7, I_8, I_9 using the Scheffé procedure.

c. Compare the results of Part (c) of Problem 10.18 with the analysis of Part (b). How can the two analyses be reconciled?

20. This problem deals with the data of Example 10.5 and Problem 10.19.
   a. Carry out Tasks (c) and (d) on pairwise comparisons.
   b. In the context of the Friedman test, suggest a multiple comparison approach.

*21. Apply the method of Note 12.4 to Problem 7.1. There are $\binom{4}{2}\binom{3}{2} = 18$ 2×2 contingency tables. Calculate the chi-squared statistic and associated p-value for each 2×2 table. Estimate the number of true null hypotheses.

22. This problem deals with Problem 11.4.
   a. Set up simultaneous 95% confidence intervals on the three regression coefficients using the Scheffé method.
   b. Use the Bonferroni method to construct comparable 95% confidence intervals.
   c. Which method is preferred?
   d. In regression models, the usual tests involve null hypotheses of the form $H_o : \beta_i = 0$, $i = 1, \ldots, p$. In general, how do you expect the Scheffé method to behave as compared with the Bonferroni method?
   e. Suppose we have another kind of null hypothesis, for example, $H_o : \beta_1 = \beta_2 = \beta_3 = 0$. Does this create a multiple comparison problem? How would you test this null hypothesis?
   f. Suppose we wanted to test, simultaneously, two null hypotheses, $H_o : \beta_1 = \beta_2 = 0$ and $H_o : \beta_3 = 0$. Carry out this test using the Scheffé procedure. State your conclusion. Also use nested hypotheses; how do the two tests compare?

*23. **a.** Verify that the contrasts defined in Problem 10.18, Parts (c), (d), (e) are orthogonal.
   b. Define another set of orthogonal contrasts that is also meaningful. Verify that the ss$_{\text{TREATMENTS}}$ can be partitioned into three sums of squares associated with this set. How do you interpret these contrasts?

REFERENCES

Armitage, P., McPherson, C. K., and Rowe, B. C. [1969]. Repeated significance tests on accumulating data. *Journal of the Royal Statistical Society Series A*, **132:** 235–244.

Armitage, P. [1975]. *Sequential Medical Trials*, 2nd edition. Blackwell, Oxford.

Brown, A., Mohammed, S. D., Montgomery, R. D., Armitage, P., and Lawrence, D. R. [1960]. Value of a large dose of antitoxin in clinical tetanus. *Lancet*, **2:** 227–230.

Bruce, R. A., Gey Jr., G. O., Fisher, L. D., and Peterson, D. R. [1974]. Seattle heart watch: initial clinical, circulatory and electrocardiographic responses to maximal exercise. *American Journal of Cardiology*, **33:** 459-469.

Cox, D. R. [1977]. The role of significance tests. *Scandinavian Journal of Statistics*, **4:** 49-62.

Cox, D. R. and Snell, E. J. [1981]. *Applied Statistics*. Chapman and Hall, London.

Cullen, B. F. and van Belle, G. [1975]. Lymphocyte transformation and changes in leukocyte count: effects of anesthesia and operation. *Anesthesiology*, **43:** 577–583.

Diaconis, P. and Mosteller, F. [1989]. Methods for studying coincidences. *Journal of the American Statistical Association*, **84:** 853–861.

Duncan, D. B. and Brant, L. J. [1983]. Adaptive *t*-tests for multiple comparisons. *Biometrics*, **39:** 790–794.

Dunn, O. J. [1959]. Confidence intervals for the means of dependent, normally distributed variables. *Journal of the American Statistical Association*, **54:** 613–621.

Dunnett, C. W. [1954]. A multiple comparison procedure for comparing several treatments with a control. *Journal of the American Statistical Association*, **50:** 1096–1121.

Dunnett, C. W. [1980a]. Pairwise multiple comparison in the homogeneous variance, unequal sample size case. *Journal of the American Statistical Association*, **75:** 789–795.

Dunnett, C. W. [1980b]. Pairwise multiple comparison in the unequal variance case. *Journal of the American Statistical Association*, **75:** 796–800.

Gey, G. D., Levy, R. H., Fisher, L. D., Pettet, G., and Bruce, R. A. [1974]. Plasma concentration of procainamide and prevalence of exertional arrythmias. *Annals of Internal Medicine*, **80:** 718–722.

Goodman, L. A. [1964a]. Simultaneous confidence intervals for contrasts among multinomial populations. *Annals of Mathematical Statistics*, **35:** 716–725.

Goodman, L. A. [1964b]. Simultaneous confidence limits for cross-product ratios in contingency tables. *Journal of the Royal Statistical Society Series B*, **26:** 86–102.

Lewis, H. D., Davis, J. W., Archibald, D. G., Steinke, W. E., Smitherman, T. C., Doherty, J. E., Schnaper, H. W., LeWinter, M. M., Linares, E., Pouget, J. M., Sabharwai, S. C., Chesler, E., and DeMets, H. [1983]. Protective effects of aspirin against acute myocardial infarction and death in men with unstable angina. *New England Journal of Medicine*, **309:** 396–403.

Miller, R. G. [1981]. *Simultaneous Statistical Inference*, 2nd edition. Springer-Verlag, New York.

Multiple Risks Factor Intervention Trial Research Group [1982]. Multiple risk factor intervention trial: risk factor changes and mortality results. *Journal of the American Medical Association*, **248:** 1465–1477.

O'Brien, P. C. [1983]. The appropriateness of analysis of variance and multiple comparison procedures. *Biometrics*, **39:** 787–794.

O'Brien, P. C. and Fleming T. R. [1979]. A multiple testing procedure for clinical trials. *Biometrics*, **35:** 549–556.

Pocock, S. J. [1982]. Interim analyses for randomised clinical trials: the group sequential approach. *Biometrics*, **36:** 153–162.

Pocock, S. J. [1983]. *Clinical Trials*. Wiley, New York.

Pocock, S. J. [1984]. Current issues in design and interpretation of clinical trials. *Proceedings of the Twelfth International Biometric Conference*, Tokyo, Japan, pp. 31–39.

Schweder, T. and Spjøtvoll, E. [1982]. Plots of P-values to evaluate many test simultaneously. *Biometrika*, **69:** 493–502.

Whitehead, J. [1983]. *The Design and Analysis of Sequential Clinical Trials*. Ellis Horwood, West Sussex, UK.

Winick, M., Meyer, K. K., and Harris, R. C. [1975]. Malnutrition and environmental enrichment by early adoption. *Science*, **190:** 1173–1175. Copyright © 1975 by the AAAS.

CHAPTER 13

Discrimination and Classification

13.1 INTRODUCTION

Discrimination techniques are methods that use measured characteristics or attributes of objects to (a) put the observed units into differing existing classes or subgroups or (b) formulate differing classes or subgroups. Diagnostic procedures try to place individuals into differing groups or disease states. Screening tests are examples of classification techniques. Prognosis may also be considered a discrimination or classification problem; for example, you may try to discriminate between cancer patients who will and who will not survive 5 years.

The subject of diagnostic screening was discussed in Chapter 6, in the context of Bayes' theorem or Bayes procedures. The diagnostic decision was made on the basis of the value of a single variable, for example, the screening procedure for diabetes was based on a measured response in the glucose tolerance test. In this chapter more than one classification variable will be considered. If more than one predictor or classification variable is used, the discrimination procedure is a multivariate procedure. It differs from the multiple regression procedure in that the latter relates a dependent or outcome variable to a set of predictor variables. In discrimination there is not a dependent variable in the usual sense; all the variables are predictor variables. You could argue that the membership in a class is a (discrete) variable. But it may not be possible to order the classes so there is no usual regression relationship.

In Note 13.4, we consider the special case of two classes and its relationship to linear regression.

In this chapter we start with logistic regression as a way of discriminating between two groups; it can be considered the "natural way" of modeling a binary dependent variable. A motivating example begins the chapter, the logistic model is presented in Section 13.2.1, a step-wise procedure is developed in Section 13.2.2. In Section 13.3 the discriminant analysis approach is discussed by applying Bayes' theorem to multivariate normal distributions. Section 13.4 treats special considerations in discrimination and classification, including a comparison of logistic regression and discriminant analysis. In Section 13.5 mathematical models used in prediction are discussed. Section 13.6 briefly treats decision theory in a medical context. Some more specialized topics are discussed in the notes, the chapter closes with problems and references.

13.2 LOGISTIC REGRESSION

13.2.1 A Motivating Example

Example 13.1. Pine *et al.* [1983] followed patients with intra-abdominal sepsis (blood poisoning) severe enough to warrant surgery to determine the incidence of organ failure or death (from sepsis). Those outcomes were correlated with age and pre-existing conditions such as alcoholism and malnutrition. Table 13.1 lists the patients with the values of the associated variables. There are 21 deaths in the set of 106 patients. Survival status is indicated by the variable Y. Five potential predictor variables: shock, malnutrition, alcoholism, age, and bowel infarction were labeled X_1, X_2, X_3, X_4, X_5, respectively. The four variables X_1, X_2, X_3, and X_5 were binary variables, coded "1" if the symptom was present and "0" if absent. The variable X_4 = age in years, was retained as a continuous variable. Consider now just variables Y and X_1: a 2×2 table could be formed as follows:

		Y		
		(Death)	**(Survive)**	
X_1		1	0	
(Shock)	1	7	3	10
(No Shock)	0	14	82	96
		21	85	106

The probability of death given that shock was present is $7/10 = 0.7000$. The probability of death given that no shock was present is $14/96 = 0.1458$ (the odds ratio associated with shock is $(7)(82)/(14)(3) = 13.7$). We want to model the probability of death with shock present or absent. One possible model is

$$\pi = \alpha + \beta_1 X_1,$$

where π is the probability of death (to be estimated), α, β_1 are constants and X_1 is the binary variable associated with shock. There are two problems associated with this approach: there is no guarantee that the predicted estimate of π, say $\hat{\pi}$, will always be in the interval $[0,1]$; it will in this case but need not if X_1 is a continuous variable. Secondly, we cannot use the least squares approach because the variance of $\hat{\pi}$ is no longer constant. The first problem can be solved easily by re-expressing π, but the nonconstancy of variance cannot be eliminated and computers are usually needed to estimate the parameters. The most common re-expression of π leads to the logistic model

$$\log_e \frac{\pi}{1 - \pi} = \alpha + \beta_1 X_1,$$

commonly written as

$$\text{logit}(\pi) = \alpha + \beta_1 X_1. \tag{13.1}$$

Four comments are in order.

Table 13.1. Survival Status of 106 Patients Following Surgery and Associated Pre-Operative Variables. Data from Pine *et al.* [1983]. See Text for Labels.

ID	Y	X_1	X_2	X_3	X_4	X_5	ID	Y	X_1	X_2	X_3	X_4	X_5
1	0	0	0	0	56	0	301	1	0	1	0	50	1
2	0	0	0	0	80	0	302	0	0	0	0	20	0
3	0	0	0	0	61	0	303	0	0	0	0	74	1
4	0	0	0	0	26	0	304	0	0	0	0	54	0
5	0	0	0	0	53	0	305	1	0	1	0	68	0
6	1	0	1	0	87	0	306	0	0	0	0	25	0
7	0	0	0	0	21	0	307	0	0	0	0	27	0
8	1	0	0	1	69	0	308	0	0	0	0	77	0
9	0	0	0	0	57	0	309	0	0	1	0	54	0
10	0	0	1	0	76	0	401	0	0	0	0	43	0
11	1	0	0	1	66	1	402	0	0	1	0	27	0
12	0	0	0	0	48	0	501	1	0	1	1	66	1
13	0	0	0	0	18	0	502	0	0	1	1	47	0
14	0	0	0	0	46	0	503	0	0	0	1	37	0
15	0	0	1	0	22	0	504	0	0	1	0	36	1
16	0	0	1	0	33	0	505	1	1	1	0	76	0
17	0	0	0	0	38	0	506	0	0	0	0	33	0
19	0	0	0	0	27	0	507	0	0	0	0	40	0
20	1	1	1	0	60	1	508	0	0	1	0	90	0
22	0	0	0	0	31	0	510	0	0	0	1	45	0
102	0	0	0	0	59	1	511	0	0	0	0	75	0
103	0	0	0	0	29	0	512	1	0	0	1	70	1
104	0	1	0	0	60	0	513	0	0	0	0	36	0
105	1	1	0	0	63	1	514	0	0	0	1	57	0
106	0	0	0	0	80	0	515	0	0	1	0	22	0
107	0	0	0	0	23	0	516	0	0	0	0	33	0
108	0	0	0	0	71	0	518	0	0	1	0	75	0
110	0	0	0	0	87	0	519	0	0	0	0	22	0
111	1	1	1	0	70	0	520	0	0	1	0	80	0
112	0	0	0	0	22	0	521	1	0	1	0	85	0
113	0	0	0	0	17	0	523	0	0	1	0	90	0
114	1	0	0	1	49	0	524	1	0	0	1	71	0
115	0	1	0	0	50	0	525	0	0	0	1	51	0
116	0	0	0	0	51	0	526	1	0	1	1	67	0
117	0	0	1	1	37	0	527	0	0	1	0	77	0
118	0	0	0	0	76	0	529	0	0	0	0	20	0
119	0	0	0	1	60	0	531	0	0	0	0	52	1
120	1	1	0	0	78	1	532	1	1	0	1	60	0
122	0	0	1	1	60	0	534	0	0	0	0	29	0
123	1	1	1	0	57	0	535	0	0	0	0	30	1
202	0	0	0	0	28	1	536	0	0	0	0	20	0
203	0	0	0	0	94	0	537	0	0	0	0	36	0
204	0	0	0	0	43	0	538	0	0	1	1	54	0
205	0	0	0	0	70	0	539	0	0	0	0	65	0
206	0	0	0	0	70	0	540	1	0	0	0	47	0
207	0	0	0	0	26	0	541	0	0	0	0	22	0
208	0	0	0	0	19	0	542	1	0	0	1	69	0
209	0	0	0	0	80	0	543	1	0	1	1	68	0
210	0	0	1	0	66	0	544	0	0	1	1	49	0
211	0	0	1	0	55	0	545	0	0	0	0	25	0
214	0	0	0	0	36	0	546	0	1	1	0	44	0
215	0	0	0	0	28	0	549	0	0	0	1	56	0
217	0	0	0	0	59	1	550	0	0	1	1	42	0

1. The logit of π has range $(-\infty, \infty)$. The following values can easily be calculated:

$$\text{logit}(1) = +\infty,$$

$$\text{logit}(0) = -\infty,$$

$$\text{logit}(0.5) = 0.$$

2. If we solve for π, the expression that results is

$$\pi = \frac{e^{\alpha + \beta_1 X_1}}{1 + e^{\alpha + \beta_1 X_1}} = \frac{1}{1 + e^{-(\alpha + \beta_1 X_1)}}. \tag{13.2}$$

3. For a given data set, we estimate α, β_1 (and hence π) by a, b_1, and $\hat{\pi}$ from $\text{logit}(\hat{\pi}) = a + b_1 X_1$.

4. What criterion is used to estimate α, β_1, and π? The criterion that is used is known as the conditional maximum likelihood criterion (see Definition 13.2 and Note 13.1). This criterion reduces to the least squares criterion with a normal distribution model.

If we fit this model to the data, we get

$$\text{logit}(\hat{\pi}) = -1.768 + 2.615 X_1.$$

If $X_1 = 0$ (i.e., there is no shock),

$$\text{logit}(\hat{\pi}) = -1.768$$

or

$$\hat{\pi} = \frac{1}{1 + e^{-(-1.768)}} = 0.146.$$

If $X_1 = 1$ (i.e., there is shock),

$$\text{logit}(\hat{\pi}) = -1.768 + 2.615 = 0.847,$$

$$\hat{\pi} = \frac{1}{1 + e^{-0.847}} = 0.700.$$

This is precisely the probability of death given no preoperative shock. The coefficient of X_1, 2.615, also has a special interpretation: it is the logarithm of the odds ratio and the quantity $e^{b_1} = e^{2.615} = 13.7$ is the odds ratio associated with shock (as compared to no shock). This can be shown algebraically to be the case (see Problem 13.1).

As mentioned before, the computations usually require a computer program. This program will ususally print the logistic regression coefficients together with their standard errors. As with the chi-square tests, a large sample normal approximation can be used to test the significance of the regression coefficient. One standard statistical package for fitting the logistic regression model is the generalized linear interactive modeling) GLIM system (Baker and Nelder [1978]). For the above problem the computer program prints:

	Estimate	S.E.	Parameter
1.	−1.768	0.2891	% GM
2.	2.615	0.7463	Shock

The notation "% GM" refers to "Grand Mean" and corresponds to the estimate "a" of α. The parameter coefficient "Shock" corresponds to the estimate "b_1" of β_1. To test the significance of the estimate, b_1, of β_1, a Z-test can be carried out:

$$z = \frac{2.615 - 0}{0.7463} = 3.50, \quad p = 0.0005.$$

This is a test of $H_0 : \beta_1 = 0$, which we reject with a p-value of 0.0005, thus we conclude that there is a significant increased risk of post-operative death associated with shock. A 95% confidence interval on β_1 can be calculated:

$$2.615 \pm (1.96)(0.7463)$$

$$L.L. = 1.152 \quad U.L. = 4.078$$

These limits can be exponentiated to produce 95% confidence limits on the odds ratio (see Problem 1):

$$L = e^{1.152} = 3.16, \quad U = e^{4.078} = 59.0.$$

These limits correspond to the limits obtained in calculating 95% confidence limits of the odds ratio, ω, using the logarithmic approach. The variance in that approach is equal to the reciprocal of the frequencies

$$\text{var}(\ln \hat{\omega}) = \frac{1}{14} + \frac{1}{82} + \frac{1}{7} + \frac{1}{3} = 0.55981,$$

$$\text{standard error}(\ln \hat{\omega}) = \sqrt{0.55982} = 0.7482,$$

which is very close to the standard error, 0.7463, generated by the GLIM program. For one binary risk factor this will usually be the case.

The reasons for considering the logistic approach can now be stated more explicitly; first of all, we are not restricted to one variable and secondly, we can incorporate discrete as well as continuous variables. In the next section we generalize to more than one predictor variable.   □

13.2.2 Logistic Regression for k Predictor Variables

In this section we present the logistic regression model for a binary outcome variable, Y. Typically, Y will be a variable indicating group membership.

Definition 13.1. *A linear logistic regression model* for the binary outcome variable, Y, is an equation of the form,

$$\text{logit}\,(P[Y = 1]) = \alpha + \beta_1 X_1 + \beta_2 X_2 + \cdots + \beta_k X_k. \tag{13.3}$$

The constants $\alpha, \beta_1, \ldots, \beta_k$ are the regression coefficients; the variable Y takes on values 0, 1, say, and define

$$P[Y = 1] = \pi,$$
$$P[Y = 0] = 1 - \pi.$$

The variables $X_1, X_2, \ldots, X_k$ are the predictor variables which can either be discrete or continuous. A typical data set is given by:

Case	Y	X_1	$\cdots$	X_j	$\cdots$	X_k
1	Y_1	X_{11}	$\cdots$	X_{1j}	$\cdots$	X_{1k}
2	Y_2	X_{21}	$\cdots$	X_{2j}	$\cdots$	X_{2k}
$\vdots$	$\vdots$	$\vdots$		$\vdots$		$\vdots$
i	Y_i	X_{i1}	$\cdots$	X_{ij}	$\cdots$	X_{ik}
$\vdots$	$\vdots$	$\vdots$		$\vdots$		$\vdots$
n	Y_n	X_{n1}	$\cdots$	X_{nj}	$\cdots$	$X_{nk}.$

Table 13.1 contains an example of such a data set: for these data $n = 106$ and $k = 5$; four of the predictor variables are discrete (X_1, X_2, X_3, X_5) and one variable is continuous (X_4). The variable Y is a binary variable (taking on values 0, 1) and thereby indicates group membership. The observations can be partitioned then into two groups; one group, the set of X values for which $Y = 0$ and the second group of X values for which $Y = 1$.

Definition 13.2. Given the data structure above and the model specified by Equation (13.11) the coefficients $\alpha_1, \beta_1, \ldots, \beta_k$ are estimated by $a_1, b_1, \ldots, b_k$. These values are obtained by choosing the values for α, β_1, β_2, $\ldots$, β_k that make the probability of the observed $Y_1, Y_2, \ldots, Y_n$ as large as possible for the given X_{ij} values (this is an illustration of the (conditional) maximum likelihood principle).

As in the multiple linear regression case the actual fitting is done by computer since hand computation is no longer feasible. See also Note 13.1 for a simple example of maximization of likelihood.

The estimates $a_1, b_1, \ldots, b_k$ can be tested for statistical significance using standard errors based on large sample inference. The null hypothesis for a test of significance is identical to that for multiple regression: for example, to test the significance of the estimate b_1 of β_1 the null hypothesis is

$$H_0 : \beta_1 = 0 \mid \alpha, \beta_2, \ldots, \beta_k,$$

that is, the significance of β_1 is tested in the presence of all the other estimates. A large sample test statistic for any $\beta_i = 0$ is

$$Z = \frac{b_i - 0}{s_{b_i}},$$

where s_{b_i} is the estimated standard error of b_i.
A more general test of the null hypothesis:

$$H_0 : \beta_i = \beta_i^* \mid \alpha, \beta_1, \ldots, \beta_{i-1}, \beta_{i+1}, \ldots, \beta_k$$

is generated by

$$Z = \frac{b_i - \beta_i^*}{s_{b_i}}. \tag{13.4}$$

Example 13.1 (continued). We now continue the analysis of the data of Pine *et al.*
listed in Table 13.1. Using the GLIM package the following output and calculations
can be generated for all the variables listed.

Variable	Regression Coefficient	Standard Error	Z-Value	p-Value
Intercept	−9.754	2.534	—	—
X_1 (shock)	3.674	1.162	3.16	0.0016
X_2 (malnutrition)	1.217	0.7274	1.67	0.095
X_3 (alcoholism)	3.355	0.9797	3.43	0.0006
X_4 (age)	0.09215	0.03025	3.04	0.0023
X_5 (infarction)	2.798	1.161	2.41	0.016

We would interpret these results as showing that malnutrition, in the presence of
the remaining variables is not an important predictor of survival status. All the other
variables are significant predictors of survival status. All but variable X_4 are discrete,
binary variables. If malnutritrion is dropped from the analysis the following estimates
and standard errors are obtained:

Variable	Regression Coefficient	Standard Error
Intercept	−8.895	2.314
X_1 (shock)	3.701	1.103
X_3 (alcoholism)	3.186	0.9163
X_4 (age)	0.08983	0.02918
X_5 (infarction)	2.386	1.071

If $\widehat{\pi}$ is the predicted probability of death the equation is

$$\text{logit}(\widehat{\pi}) = -8.895 + 3.701X_1 + 3.186X_3 + 0.08983X_4 + 2.386X_5.$$

For each of the values of X_1, X_3, X_5 (a total of eight possible combinations) a
regression curve can be drawn for $\text{logit}(\widehat{\pi})$ versus age. In Figure 13.1 the lines are

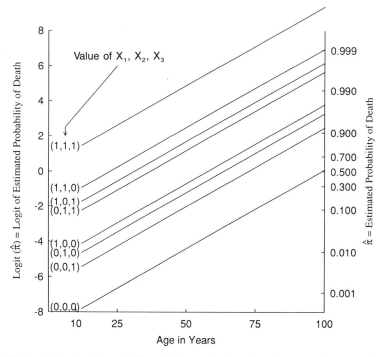

Figure 13.1. Logit of estimated probability of death as a function of age in years and category of status of (X_1, X_3, X_5).

drawn for each one of the eight combinations. For example, corresponding to $X_1 = 1$ (shock present), $X_3 = 0$ (no alcoholism), and $X_5 = 0$ (no infarction) the line

$$\text{logit}(\hat{\pi}) = -8.895 + 3.701 + 0.08983X_4$$

$$= -5.194 + 0.08983X_4$$

is drawn.

This line is indicated by "(100)" as a shorthand way of writing $(X_1 = 1, X_3 = 0, X_5 = 0)$. The eight lines seem to group themselves into four groups: the top line representing all three symptoms present, the next three lines consist of groups with two symptoms present, the next three lines for groups with one symptom present and finally, the group at lowest risk with no symptoms present. The right hand scale of Figure 13.1 consists of probabilities corresponding to the logit scale on the left hand side. In Figure 13.2 the probability of death is plotted in the original scale; only four of the eight groups have been graphed. The group at highest risk is the one with all three binary risk factors present. The curves have been dotted below 15 years and above 90 years since there were no patients in these categories. One of the advantages of the model is that we can draw a curve for the situation with all three risk factors present even though there are no patients in that category; but the estimate depends upon the model. The curve is drawn on the assumption that the risks are additive in the logistic scale (that is what we *mean* by a linear model). This assumption can be

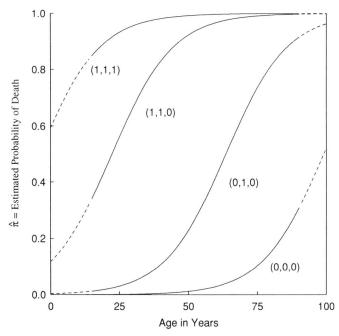

Figure 13.2. Estimated probability of death as a function of age in years and selected values of (X_1, X_3, X_5).

tested by including interaction terms involving these three covariates in the model and testing their significance. When this was done none of the interaction terms were significant, suggesting that the additive model is a reasonable one.

13.2.3   Selecting and Testing an Appropriate Model

Procedures described in Section 11.5 can also be applied to logistic regression procedures in selecting a "best" subset of predictor variables. These include (1) testing the additional reduction in variability due to inclusion of additional variables; (2) a special case of the above: determining the "best" fit of a set of hierarchical models; and (3) the "split-half" technique. In logistic regression, two additional procedures are available. First of all, since the objective is prediction, a good way to test a model is to see how well it predicts. One possible way of doing this is to apply the model to the data. There is a circularity here, of course. Applying the model to the data will tend to make the predictions look much better than they really are. One possible solution to the problem is to use the "split-half" technique: estimate the logistic regression coefficients from the first half of the data and apply the model to the second half. There is still some circularity if the variables are selected (and tested) on all of the data. However, if selection and estimation is on one half and testing on the other, this procedure will produce unbiased estimates of the classification probabilities.

A second test of a logistic regression uses the chi-square goodness-of-fit test discussed in Section 6.7. The idea is the following. Suppose the number of possible combinations of covariate values is small relative to the total number of data points. Then each combination of covariate values can be considered a cell which contains a certain number of (observed) data points. The model then generates an expected number of data points for that cell as well as all other cells. A chi-square goodness-of-fit test of the form $\sum 2O \ln O/E$ can then be carried out with degrees of freedom equal to the number of covariate combinations that actually occur minus the number of parameters estimated (including one degree of freedom for the mean). See Section 7.5 for a discussion of this statistic. We now describe some of these procedures in more detail. A new measure of variability must be defined to allow us to quantify the concepts.

Definition 13.3. Given a set of n data points for a dependent random variable Y and k predictor variables $X_1, X_2, \ldots, X_k$, the deviance $= D(X_1, X_2, \ldots, X_k)$ is a measure of the total variability left after fitting the predictor variables $X_1, X_2, \ldots, X_k$. For the case of Y normally distributed and $E(Y) = \alpha + \beta_1 X_1 + \beta_2 X_2 + \cdots + \beta_k X_k$, the deviance is equal to the residual sum of squares, $\text{ss}_{\text{resid}}(X_1, X_2, \ldots, X_k)$. The degrees of freedom associated with the deviance is $n - k - 1$. The deviance associated with fitting a constant is D.

The deviance is equal to $-2 \log_e(LF)$, where LF is the likelihood function. In Figure 13.10, Note 13.1, the deviance is plotted on the right hand side of the graph.

Result 13.1. Under the assumptions of a logistic model for Y (a binary variable) the difference in deviances of nested models associated with p_1 and $(p_1 + p_2)$ predictor variables is approximately chi-square with p_2 degrees of freedom.

As an example, let $D(X_1, X_2)$ and $D(X_1, X_2, X_3, X_4)$ be two deviances associated with fitting two and four (two additional) variables, respectively. Then the null hypothesis

$$H_0 : \beta_3 = \beta_4 = 0 \mid \alpha, \beta_1, \beta_2,$$

can be tested by $D(X_1, X_2) - D(X_1, X_2, X_3, X_4) = \chi_2^2$. This is approximately a chi-square statistic with 2 degrees of freedom.

Result 13.2. Under the assumptions of a logistic model for Y (a binary variable), let

$$D(X_{p_1+1}, \ldots, X_{p_1+p_2} \mid X_1, \ldots, X_{p_1}) =$$
$$D(X_1, X_2, \ldots, X_{p_1}) - D(X_1, X_2, \ldots, X_{p_1}, \ldots, X_{p_1+p_2}).$$

Then

$$\frac{D(X_{p_1+1}, \ldots, X_{p_1+p_2} \mid X_1, \ldots, X_{p_1})/p_2}{D(X_1, X_2, \ldots, X_{p_1+p_2})/(n - p_1 - p_2 - 1)} = F_{p_2, n-p_1, -p_2-1}$$

has approximately an F distribution with p_2 and $n - p_1 - p_2 - 1$ degrees of freedom under the null hypothesis

$$H_0 : \beta_{p_1+1} = \cdots = \beta_{p_1+p_2} = 0 \mid \alpha, \beta_1, \ldots, \beta_{p_1}.$$

Both Results 13.1 and 13.2 are "approximate" results. They are large sample results in the same sense that the chi-square test is a large sample test. The two tests described are asymptotically equivalent since $F_{p_2, \infty} = \chi_{p_2}^2$. In finite samples they may give different information. Such a difference usually indicates that the logistic assumption does not hold.

A stepwise procedure can now be defined. For a forward stepwise procedure we would proceed as follows:

Step 0. Fit a simple binomial model, that is, without any covariates and calculate the deviance, D.

Step 1.

1. Select that predictor variable that produces the greatest reduction in deviance. Let this be X_1. That is $D - D(X_1) > D - D(X_i), i = 2, \ldots, k$.
2. Test whether $\chi_1^2 = D - D(X_1)$ exceeds a pre-established critical level. If so, proceed to Step 2, otherwise stop.

Step 2.

1. Select a second predictor variable to minimize $D(X_1, X_j)$. Let $j = 2$.
2. Test whether X_2 produces a significant reduction in deviance by calculating

$$\chi_1^2 = D(X_1) - D(X_1, X_2)$$

or by testing

$$F_{1,n-3} = \frac{[D(X_1) - D(X_1, X_2)]/1}{D(X_1, X_2)/(n-3)}.$$

If there is a significant reduction, continue; otherwise stop.

At Step j the variables $X_1, X_2, \ldots, X_j$ are in the model.

Step $j + 1$.

1. Select that predictor variable from $X_{j+1}, \ldots, X_k$ that minimizes $D(X_1, X_2, \ldots, X_j, X_{j'})$, $j + 1 \leq j' \leq k$. Let X_{j+1} be that variable.
2. Test the significance of the addition of X_{j+1} by

$$\chi_1^2 = D(X_1, X_2, \ldots, X_j) - D(X_1, X_2, \ldots, X_j, X_{j+1}) \tag{13.5}$$

or by

$$F_{1,n-j-2} = \frac{[D(X_1, X_2, \ldots, X_j) - D(X_1, X_2, \ldots, X_j, X_{j+1})]/1}{D(X_1, X_2, \ldots, X_j, X_{j+1})/(n-j-2)}. \tag{13.6}$$

The process is continued until either all variables have been entered or the addition of a predictor variable does not significantly decrease the deviance. This is summarised in Table 13.2.

Example 13.1 (continued). To illustrate the stepwise procedure, we continue with the example of Section 13.2.1, the prediction of death associated with intra-abdominal sepsis. One change has been made: the ages of the patients were divided by 10 and truncated to produce a simpler measure of age: AGEC, corresponding to the common way we talk about people as in their twenties, etc. This was done to illustrate the chi-square goodness-of-fit test.

Table 13.2. Stepwise Logistic Regression Procedure (Forward) for the k Variable Case.

Step	Variable Entered	Intercept and Slopes Calculated[a]	Total $-2 \ln LF$	Contribution	Approximate χ^2 With 1 d.f.
0	—	$a^{(0)}$	D		
1	X_1	$a^{(1)}, b_1^{(1)}$	$D(X_1)$	$D(X_1)$	$D - D(X_1)$
2	X_2	$a^{(2)}, b_1^{(2)}, b_2^{(2)}$	$D(X_1, X_2)$	$D(X_2 \mid X_1)$	$D(X_1) - D(X_1, X_2)$
⋮	⋮	⋮	⋯	⋯	⋯
j	X_j	$a^{(j)}, b_1^{(j)}, b_2^{(j)}, \ldots, b_j^{(j)}$	$D(X_1, X_2, \ldots, X_j)$	$D(X_j \mid X_1, \ldots, X_{j-1})$	$D(X_1, \ldots, X_{j-1}) - D(X_1, \ldots, X_j)$
⋮	⋮	⋮	⋯	⋯	⋯
k	X_k	$a^{(k)}, b_1^{(k)}, b_2^{(k)}, \ldots, b_k^{(k)}$	$D(X_1, X_2, \ldots, X_k)$	$D(X_k \mid X_1, \ldots, X_{k-1})$	$D(X_1, \ldots, X_{k-1}) - D(X_1, \ldots, X_k)$

[a]The superscripts indicate the estimates at that step.

The data were re-analyzed using the BMDP Statistical Software (Dixon [1988]) package subroutine: Stepwise Logistic Regression coded as BMDPLR.

Table 13.3 contains the printout of the BMDPLR at Step 0, 1 and Step 4, the last step. We discuss each of these steps.

Step 0. At this step the log likelihood is −52.764. The deviance, D, is −2(−52.764) = 105.528. The goodness-of-fit chi-square is 83.119 with d.f. = 42 and p-value = 0.000. What does this mean? The program has automatically determined that there are 43 nonempty covariate patterns or cells (see Table 13.4). If we assume all cells to have

Table 13.3. BMDPLR Output of Stepwise Logistic Regression (Selected and Edited). Data From Example 13.1

```
STEP NUMBER    0
---------------


                          LOG LIKELIHOOD =    -52.764
GOODNESS OF FIT CHI-SQ    (2*O*LN(O/E))  =     83.119  D.F.=  42  P-VALUE= 0.000
GOODNESS OF FIT CHI-SQ    ( C.C.BROWN )  =      0.000  D.F.=   0  P-VALUE= 1.000


                                         STANDARD
       TERM            COEFFICIENT         ERROR    COEFF/S.E.   EXP(COEFFICIENT)


CONSTANT                  1.3981          0.2437      5.737          4.048


STATISTICS TO ENTER OR REMOVE TERMS
-----------------------------------
                        APPROX.                APPROX.
       TERM             F TO   D.F. D.F.        F TO   D.F. D.F.
                        ENTER                   REMOVE                 P-VALUE


SHOCK                   20.58   1   104                                0.0000
MALN                     6.37   1   104                                0.0131
ETOH                    12.65   1   104                                0.0006
INFARC                   9.95   1   104                                0.0021
AGEC                    13.55   1   104                                0.0004
CONSTANT                                        32.61   1    104       0.0000


STEP NUMBER    1               SHOCK                  IS ENTERED
---------------


                          LOG LIKELIHOOD =    -45.988
IMPROVEMENT CHI-SQUARE    ( 2*(LN(MLR) )  =     13.552  D.F.=   1  P-VALUE= 0.000
GOODNESS OF FIT CHI-SQ    (2*O*LN(O/E))   =     69.567  D.F.=  41  P-VALUE= 0.004
GOODNESS OF FIT CHI-SQ    ( C.C.BROWN )   =      0.000  D.F.=   0  P-VALUE= 1.000

                                         STANDARD
       TERM            COEFFICIENT         ERROR    COEFF/S.E.   EXP(COEFFICIENT)


SHOCK                     1.3075          0.3741      3.495          0.2705
CONSTANT                 -0.46018         0.3741     -1.230          1.584
```

Table 13.3. (*Continued*)

```
STATISTICS TO ENTER OR REMOVE TERMS
-----------------------------------
                APPROX.             APPROX.
     TERM       F TO  D.F. D.F.     F TO  D.F. D.F.
                ENTER               REMOVE              P-VALUE

SHOCK                               11.98  1   103   0.0008
MALN            4.37   1   103                        0.0391
ETOH           20.26   1   103                        0.0000
INFARC          6.80   1   103                        0.0105
AGEC           10.36   1   103                        0.0017
CONSTANT                             1.48  1   103   0.2258


STEP NUMBER   4                INFARC            IS ENTERED
---------------

                 LOG LIKELIHOOD =   -29.992
IMPROVEMENT CHI-SQUARE   ( 2*(LN(MLR) ) =    5.698  D.F.=  1  P-VALUE= 0.017
GOODNESS OF FIT CHI-SQ   (2*O*LN(O/E))  =   37.574  D.F.= 38  P-VALUE= 0.489
GOODNESS OF FIT CHI-SQ (HOSMER-LEMESHOW)=    8.265  D.F.=  6  P-VALUE= 0.219
GOODNESS OF FIT CHI-SQ   ( C.C.BROWN )  =    2.756  D.F.=  2  P-VALUE= 0.252

                                   STANDARD
     TERM         COEFFICIENT       ERROR   COEFF/S.E.  EXP(COEFFICIENT)

SHOCK               1.6624         0.5014      3.316       0.1897
ETOH                1.4750         0.4205      3.508       0.2288
INFARC              1.1069         0.4940      2.240       0.3306
AGEC                0.68841        0.2437      2.825       0.5024
CONSTANT           -2.8238         1.370      -2.061      16.84


STATISTICS TO ENTER OR REMOVE TERMS
-----------------------------------
                APPROX.             APPROX.
     TERM       F TO  D.F. D.F.     F TO  D.F. D.F.
                ENTER               REMOVE              P-VALUE

SHOCK                                8.43  1   100   0.0045
MALN            2.37   1   100                        0.1268
ETOH                                 9.43  1   100   0.0027
INFARC                               3.85  1   100   0.0525
AGEC                                 6.12  1   100   0.0151
CONSTANT                             3.26  1   100   0.0741


SUMMARY OF STEPWISE RESULTS

STEP      TERM              LOG     IMPROVEMENT   GOODNESS OF FIT
NO.   ENTERED REMOVED   DF LIKELIHOOD CHI-SQUARE P-VAL CHI-SQUARE P-VAL
---   ------------------ --- ---------- ---------- ----- ---------- -----
 0                          -52.764                         83.119 0.000
 1 SHOCK                 1  -45.988   13.552 0.000  69.567 0.004
 2 ETOH                  1  -38.485   15.006 0.000  54.561 0.062
 3 AGEC                  1  -32.841   11.289 0.001  43.272 0.294
 4 INFARC                1  -29.992    5.698 0.017  37.574 0.489
```

Table 13.4. Analysis of All Possible Covariate Combinations. Computer Output has been Edited. Summary Description of Cells. Cells are Formed by all Combinations of Values of all Variables.

SHOCK	MALN	ETOH	AGE	INFARC	Number Survived	Number Died	Prob. of Surviving Observed	Prob. of Surviving Predicted	SE Predicted	SE Residual Observed	SE Residual Predicted	Log Odds Predicted	Deviance
0	0	0	0	1	0	1	0.0000	0.3316	0.1442		-0.7400	-0.7008	-0.8977
0	1	1	1	1	0	1	0.0000	0.0974	0.0982		-0.3482	-2.2261	-0.4528
0	1	1	1	1	0	1	0.0000	0.0514	0.0602		-0.2420	-2.9146	-0.3250
1	0	1	0	1	0	1	0.0000	0.8041	0.1403		-2.1660	1.4123	-1.8057
1	1	1	0	1	0	1	0.0000	0.0974	0.0982		-0.3482	-2.2261	-0.4528
1	0	1	0	1	0	1	0.0000	0.0691	0.0780		-0.2863	-2.6010	-0.3784
1	0	1	0	1	0	1	0.0000	0.0359	0.0450		-0.1990	-3.2894	-0.2706
1	1	1	0	1	0	1	0.0000	0.0343	0.0386		-0.1929	-3.3372	-0.2643
1	0	0	0	1	0	1	0.0000	0.5747	0.2074		-1.2807	0.3012	-1.3077
1	1	0	0	1	0	2	0.0000	0.2543	0.1648		-0.9779	-1.0756	-1.0835
1	0	1	0	1	0	1	0.0000	0.0691	0.0780		-0.2863	-2.6010	-0.3784
0	0	1	0	1	1	2	0.3333	0.4969	0.1337		-0.6394	-0.0124	-0.5723
0	1	1	0	1	1	2	0.3333	0.8265	0.0861		-2.4540	1.5609	-1.8889
0	1	1	0	1	1	2	0.3333	0.4969	0.1337		-0.6394	-0.0124	-0.5723
0	0	1	0	1	1	1	0.5000	0.7965	0.0983		-1.1097	1.3645	-0.9309
0	1	0	0	1	1	1	0.5000	0.9497	0.0291		-2.9620	2.9377	-1.8192
0	0	0	0	1	5	1	0.8333	0.9868	0.0121		-3.4103	4.3145	-1.8389
0	0	0	0	0	3	0	1.0000	0.9983	0.0027		0.0718	6.3797	0.1008
0	0	0	0	0	17	0	1.0000	0.9966	0.0045		0.2530	5.6913	0.3385
0	0	0	0	0	7	0	1.0000	0.9933	0.0075		0.2236	5.0029	0.3062
0	0	0	0	0	5	0	1.0000	0.9741	0.0188		0.3783	3.6261	0.5126
0	0	0	0	0	2	0	1.0000	0.9497	0.0291		0.3315	2.9377	0.4545
0	0	0	0	0	6	0	1.0000	0.9046	0.0478		0.8673	2.2493	1.0970
0	0	0	0	0	4	0	1.0000	0.8265	0.0861		1.0291	1.5609	1.2348

0.8357	0.8725	0.6856	0.1519	0.7053	1.0000	0	1	9	0	0	0	0
0.2466	3.4776	0.1801	0.0372	0.9700	1.0000	0	1	2	1	0	0	0
0.3454	2.7891	0.2568	0.0608	0.9421	1.0000	0	1	3	1	0	0	0
1.1437	1.4123	1.0812	0.1403	0.8041	1.0000	0	3	5	1	0	0	0
1.1624	0.0355	1.1019	0.2265	0.5089	1.0000	0	1	7	1	0	0	0
0.4915	2.0529	0.3694	0.0772	0.8862	1.0000	0	1	3	0	1	0	0
1.5707	0.6760	1.3636	0.1156	0.6629	1.0000	0	3	5	1	1	0	0
0.1422	5.6913	0.1016	0.0045	0.9966	1.0000	0	3	2	0	0	0	0
0.1157	5.0029	0.0823	0.0075	0.9933	1.0000	0	1	3	0	0	0	0
0.3242	3.6261	0.2340	0.0188	0.9741	1.0000	0	2	5	0	0	0	0
0.7757	2.2493	0.5863	0.0478	0.9046	1.0000	0	3	7	0	0	0	0
1.1819	0.8725	1.0365	0.1519	0.7053	1.0000	0	2	9	0	1	0	0
0.3454	2.7891	0.2568	0.0608	0.9421	1.0000	0	1	3	1	1	0	0
0.4915	2.0529	0.3694	0.0772	0.8862	1.0000	0	1	3	0	0	0	0
1.1685	1.3645	0.9662	0.0983	0.7965	1.0000	0	3	4	1	1	0	0
0.9069	0.6760	0.7355	0.1156	0.6629	1.0000	0	1	5	0	1	0	0
1.0525	0.3012	0.9476	0.2074	0.5747	1.0000	0	1	5	0	0	1	0
1.3456	-0.3872	1.3268	0.1984	0.4044	1.0000	0	1	6	0	0	1	0
0.7951	0.9896	0.6708	0.1854	0.7290	1.0000	0	1	4	0	0	1	0

equal expectation, that is, the expected frequency $E = 105/43 = 2.442$, then the calculated value of $\chi^2_{42} = \sum 2O \ln O/E = 83.119$. Since the p-value is much less than 0.05, we reject the hypothesis that the frequencies in the cells are uniformly distributed. This is equivalent to rejecting the null hypothesis

$$H_0 : \beta_1 = \beta_2 = \beta_3 = \beta_4 = \beta_5 = 0.$$

We therefore proceed to select a variable to enter the model. The variable with the largest approximate F-to-enter is SHOCK. The F-value is 20.58 with 1 and 104 degrees of freedom and $p < 0.001$. (The F-to-enter in the BMDPLR subroutine is *not* the F statistic of Result 13.2 above, although it should be close. The F-to-enter is based on the large sample estimates of variances and covariances of the logistic regression coefficients). We therefore proceed to Step 1, entering the variable SHOCK.

Step 1. The log likelihood is now -45.988, or the deviance,

$$D(\text{SHOCK}) = -2(-45.988) = 91.976.$$

The improvement chi-square is

$$\chi^2_1 = D - D(\text{SHOCK}) = 105.528 - 91.976 = 13.552$$

with a p-value < 0.0005. This is a test of the hypothesis

$$H_0 : \beta_1 = 0 \mid \alpha.$$

The goodness-of-fit statistic now has $43-2 = 41$ degrees of freedom and the goodness-of-fit chi-square has a p-value of 0.004. The goodness-of-fit chi-square tests the hypothesis

$$H_0 : \beta_2 = \beta_3 = \beta_4 = \beta_5 = 0 \mid \beta_1, \alpha.$$

Again, we reject this null hypothesis: the model does not fit well.

Before selecting the next variable you should note that the regression coefficient, b_1, is 1.3075. The value obtained in Section 13.2.1 is 2.615. What has happened? It illustrates one of the "quirks" of a computer program: the BMDPLR program assumes that a variable is categorical unless told otherwise *and* it automatically rescores the variable as $1, -1$! And we need the variable to be categorical to get a goodness-of-fit. The model fitted at this point is

$$\text{logit}(\widehat{\pi}) = -0.4602 + 1.3075X_1^*,$$

where

$$X_1^* = 2X_1 - 1.$$

To convert back to "our" variable X_1 we substitute for X_1^*

$$\text{logit}(\widehat{\pi}) = -0.4602 + 1.3075(2X_1 - 1)$$
$$= -1.7677 + 2.615X_1.$$

This is the model of Section 13.2.1.

The program then proceeds to Steps 2, 3, and 4. We only reproduce Step 4.

Step 4. After this step the deviance is

$$D(\text{SHOCK,ETOH,AGEC,INFARC}) = -2(-29.992) = 59.984.$$

The deviance at Step 3 was

$$D(\text{SHOCK,ETOH,AGEC}) = -2(-32.841) = 65.682.$$

The improvement chi-square is

$$\chi_1^2 = 65.682 - 59.984 = 5.698, \quad \text{d.f.} = 1, p = 0.017.$$

The F-to-enter for the remaining variable MALN is not significant ($F_{1,100} = 2.37, p = 0.1268$) and the step-wise procedure stops at this point. The goodness-of-fit chi-square is now $\chi_{38}^2 = 37.574, p = 0.489$ indicating no reason, on the basis of this test, to reject the adequacy model using the four predictor variables SHOCK, ETOH, AGEC, and INFARC.

In Table 13.4, the 43 distinct covariate patterns are listed. The first five columns list the observed covariate patterns. Columns 6–9 list the observed numbers of survivors and deaths, the observed proportion surviving and the predicted probability of survival, respectively. The column labelled, "Predicted Log Odds," is calculated from logit(predicted probability of survival); for example -2.226, the first value listed in that column is $\log(0.0974/(1 - 0.0974))$.

Note that the covariance pattern has retained $X_2 = $ MALN, even though it was not significant. If the analysis is run again this time omitting MALN, there remain 27 distinct covariance patterns; the chi-square goodness-of-fit statistic after fitting the other variables has value 21.789 with $27 - 1 - 4 = 22$ degrees of freedom for a p-value equal to 0.473; so the conclusion remains unchanged.

The number of distinct covariate patterns is reasonably small in this example (due to the truncation of AGE). Using the original covariate, AGE, results in 74 distinct patterns. This is too many to be useful for a chi-square goodness-of-fit test since most cells will have frequencies less than 5.

One possible way out is to consider only the number of distinct *categorical* covariate patterns, and calculate observed frequencies for the cells. The expected frequency for a cell is then the summation over the distinct noncategorical covariate patterns of the expected frequencies.

13.3 DISCRIMINANT ANALYSIS

13.3.1 The Framework

In this section we consider the classification procedure using the normal distribution and applying Bayes' Theorem. This approach is an older technique dating back to Fisher [1936] and goes under the name of discriminant analysis. In Section 13.4 we will make some comparisons between the logistic regression approach and discriminant analysis and discuss some issues common to both procedures; for example, the probability of misclassifications.

In this section we assume that there are two groups G_1 and G_2, which occur with proportions π_1 and π_2 such that $\pi_1 + \pi_2 = 1$. One way to think of this is to consider a population made up of a mixture of G_1 and G_2. For each member of this population we know

1. Its group membership, G_1 or G_2,
2. The values of a set of k covariates or predictor variables, $X_1, \ldots, X_k$.

Now multivariate normality needs to be defined.

Definition 13.4. A set of random variables $X_1, \ldots, X_k$, is *multivariate normal* if every linear combination of $X_1, \ldots, X_k$ has a normal distribution.

This definition clearly excludes sets of categorical or discrete random variables from being considered multivariate normal. If we want to classify into one of two groups, G_1 or G_2, the following result can be stated.

Result 13.3. Let the variances and covariances of $X_1, \ldots, X_k$ be the same within the two groups G_1 and G_2, that is, covariance $(X_i, X_j \mid G_1) =$ covariance $(X_i, X_j \mid G_2)$ for all i, j and suppose that the variables are multivariate normal, then by Bayes' Theorem the probability, P_1, that an object comes from group G_1 is

$$P_1 = P[G_1 \mid X_1, \ldots, X_k] = \frac{1}{1 + e^{-(\alpha + \beta_1 X_1 + \cdots + \beta_k X_k)}}, \tag{13.7}$$

where the parameters $\alpha_1, \beta_1, \ldots, \beta_k$ are related to the means, variances, and covariances of the random variables $X_1, \ldots, X_k$; α also depends on π_1 and π_2.

Definition 13.5. The function $\alpha + \beta_1 X_1 + \cdots + \beta_k X_k$ is called the *linear discriminant function*.

The coefficients $\alpha_1, \beta_1, \ldots, \beta_k$ are usually estimated from samples from the two groups. The estimated function $Z = a + b_1 X_1 + \cdots + b_k X_k$ is also called the linear discriminant function. The ambiguity is not dangerous since the context usually makes it clear whether the estimated or population discriminant function is being discussed. The computation of $a_1, b_1, \ldots, b_k$ is usually carried out on a computer and require the same kind of program as that for the multiple linear regression problem.

In classifying into the two groups, G_1, G_2, an object would be classified into G_1 if on the basis of the observed values of $X_1, \ldots, X_k$, $P_1 \geq 1/2$. It is easy to verify that this occurs when the linear discriminant function is greater than or equal to zero. Otherwise, the object is classified into G_2 ($P_1 < 1/2$ implies $P_2 = P(G_2 \mid X_1, \ldots, X_k) > 1/2$).

The coefficients $\beta_1, \ldots, \beta_k$ are not changed when π_1, π_2 change; only α changes.

This means that a change in π_1, π_2 changes the cut-off point but does not change the function.

Before considering an example, we consider another approach that leads to the linear discriminant function. From the chapter on the two sample t-test, recall that the power of a test for discriminating between the two groups depends upon the standardized difference in the means, $\Delta = (\mu_1 - \mu_2)/\sigma$. The same concept can be applied to the discrimination problem as follows. For the predictor variables $X_1, \ldots, X_k$, let

$$\ell = c_0 + c_1 X_1 + \cdots + c_k X_k$$

be a particular linear combination. This linear combination is a random variable because the X_i's are random variables and it is normally distributed by the assumption of multivariate normality. Let $E(\ell \mid G_1)$ be the mean of this normal distribution given that the values of X_i, $i = 1, \ldots, k$ come from G_1, and let $var(\ell \mid G_1)$ be the variance. We will assume, as in Result 13.3, that $var(\ell \mid G_1) = var(\ell \mid G_2) = var(\ell \mid G)$. We now can define:

Definition 13.6. The maximum value of the quantity $\Delta(\ell)$, over all possible values of $c_0, c_1, \ldots, c_k$, defined by

$$\Delta^2(\ell) = \frac{[E(\ell \mid G_1) - E(\ell \mid G_2)]^2}{var(\ell \mid G)} \tag{13.8}$$

is called the *Mahalanobis distance*.

And the following result can be stated:

Result 13.4. Under the conditions of Result 13.3, the linear combination of $X_1, \ldots, X_k$ that produces the Mahalanobis distance is the linear discriminant function (up to the constant).

To provide an intuitive justification for the use of the Mahalanobis distance, consider that the quantity $\Delta^2(\ell)$ is

$$\Delta^2(\ell) = \frac{\text{Variability between } G_1 \text{ and } G_2}{\text{Variability within } G_1 (\text{and } G_2)}. \tag{13.9}$$

The combination ℓ is chosen in order to maximize the ratio of the variability between G_1 and G_2 to the variability within G_1 (and hence G_2 since the variability within G_2 is assumed the same as within G_1).

The function ℓ is called the classification function.

For the case of $k = 1$ (one predictor variable), the Mahalanobis distance is algebraically equivalent to the standardized difference.

Note that logit $(P[G_1 \mid X_1, \ldots, X_k]) = a + b_1 X_1 + \cdots + b_k X_k$, this is the same form as the logistic model. However *this* logit is a linear combination of normal random variables and hence is normally distributed. The quantity logit $(P[G_1 \mid X_1, \ldots, X_k])$ is usually called the *discriminant score* associated with the random variables $X_1, \ldots, X_k$.

13.3.2   An Example

We apply the discriminant analysis to the data of Pine *et al.* [1983] discussed already in Sections 13.1 and 13.2. You will probably raise the objection that the data does not fit the assumptions of the discriminant model (multivariate normality, equal variances, and covariances within each group). This is a valid criticism. But by using this data again we can contrast logistic regression and discriminant analysis. In Section 13.4 we compare these two approaches, and also the example. One of the points that is made in that section is that discriminant analysis is fairly robust against certain kinds of departures from normality. Some reasons and some references are given. We now turn to the example. The data are analyzed using the BMDP program 7M, "Stepwise Discriminant Analysis." As in the previous section, the variable AGE has been recoded to

$$\text{AGEC} = \text{Integer}\,(\text{AGE}/10),$$

which converts AGE to multiples of 10 years.

The output from the BMDP7M program resembles the output from the earlier stepwise regression program; in fact, discrimination between two groups can be thought of as a regression problem (see Note 13.4). We now briefly discuss the first two and the last step in the program. The analysis is summarized in Table 13.5. Here Group 1 = survive; Group 2 = death.

Step 0. Before any of the variables are entered an F-statistic is calculated for each of the variables, the value of this F for a variable is a function of the Mahalanobis distance between the groups, the first variable selected is SHOCK, it has the largest F-value, $F_{1,104} = 20.58$. It is interesting to note that the t-test comparing Group 1 (Survivors) and Group 2 (Deaths) on the variable SHOCK is

$$t_{104} = \frac{0.03529 - 0.33333}{0.26963\sqrt{\frac{1}{85} + \frac{1}{21}}} = -4.53599.$$

The quantities 0.03529, 0.33333 are the means for the variable SHOCK in Groups 1 and 2, respectively. The quantity 0.26963 is the pooled standard deviation which has 104 degrees of freedom associated with it. The square of t_{104}, $t_{104}^2 = (4.53599)^2 = 20.58 = F_{1,104}$.

Step 1. At this step the variable SHOCK has been entered. The F-to-remove is 20.58, the same value as the F-to-enter on the previous step. The "classification functions" are given as

	Group	
Variable	**Survive**	**Death**
SHOCK	0.48548	4.58506
CONSTANT	−0.22935	−2.38309

The classification functions are used to calculate the probabilities P_1 and P_2 as follows:

$$P_1(G_1 \mid \text{SHOCK}) = \frac{e^{-0.22935+0.48548\text{SHOCK}}}{e^{-0.22935+0.48548\text{SHOCK}} + e^{-2.38309+4.58506\text{SHOCK}}}.$$

Table 13.5. Excerpts of BMDP7M Stepwise Discriminant Analysis Output. Data From Example 13.1 (Section 13.2).

```
MEANS
        GROUP =   Survive      Death       ALL GPS.
VARIABLE
   3 SHOCK        0.03529      0.33333      0.09434
   4 MALN         0.24706      0.52381      0.30189
   5 ETOH         0.14118      0.47619      0.20755
   7 INFARC       0.08235      0.33333      0.13208
   8 AGEC         4.34118      6.14286      4.69811


COUNTS             85.          21.         106.


STANDARD DEVIATIONS
        GROUP =   Survive      Death       ALL GPS.
VARIABLE
   3 SHOCK        0.18562      0.48305      0.26963
   4 MALN         0.43386      0.51177      0.44989
   5 ETOH         0.35027      0.51177      0.38660
   7 INFARC       0.27653      0.48305      0.32655
   8 AGEC         2.17427      1.06234      2.00882


STEP NUMBER   0

   VARIABLE     F TO    FORCE TOLERNCE *   VARIABLE     F TO   FORCE TOLERNCE
               REMOVE   LEVEL          *                ENTER  LEVEL
        DF =  1  105                   *        DF =  1  104
                                       *   3 SHOCK      20.58    1   1.00000
                                       *   4 MALN        6.37    1   1.00000
                                       *   5 ETOH       12.65    1   1.00000
                                       *   7 INFARC      9.95    1   1.00000
                                       *   8 AGEC       13.55    1   1.00000
STEP NUMBER   1
VARIABLE ENTERED   3 SHOCK

   VARIABLE     F TO    FORCE TOLERNCE *   VARIABLE     F TO   FORCE TOLERNCE
               REMOVE   LEVEL          *                ENTER  LEVEL
        DF =  1  104                   *        DF =  1  103
   3 SHOCK      20.58    1   1.00000   *   4 MALN        4.43    1   0.99780
                                       *   5 ETOH       19.61    1   0.93533
                                       *   7 INFARC      7.19    1   0.99790
                                       *   8 AGEC        9.82    1   0.99721


U-STATISTIC(WILKS' LAMBDA) 0.8348366  DEGREES OF FREEDOM   1   1   104
APPROXIMATE F-STATISTIC       20.575  DEGREES OF FREEDOM   1.00   104.00

  F - MATRIX        DEGREES OF FREEDOM =   1   104

           Survive
Death      20.58
```

Table 13.5. (*Continued*)

```
CLASSIFICATION FUNCTIONS
        GROUP =   Survive      Death
VARIABLE
  3 SHOCK          0.48548      4.58506


CONSTANT          -0.22935     -2.38309


STEP NUMBER    4


VARIABLE ENTERED    8 AGEC

    VARIABLE      F TO    FORCE TOLERNCE *   VARIABLE      F TO    FORCE TOLERNCE
                 REMOVE   LEVEL           *                ENTER   LEVEL
          DF =  1  101                    *          DF = 1  100
    3 SHOCK       20.30     1   0.93195 *   4 MALN         1.89     1   0.95798
    5 ETOH        18.06     1   0.92704 *
    7 INFARC       7.47     1   0.98939 *
    8 AGEC         7.17     1   0.99677 *


U-STATISTIC(WILKS' LAMBDA) 0.6071519   DEGREES OF FREEDOM   4   1   104
APPROXIMATE F-STATISTIC        16.338   DEGREES OF FREEDOM   4.00   101.00


  F - MATRIX          DEGREES OF FREEDOM =    4  101


            Survive
Death       16.34


CLASSIFICATION FUNCTIONS
        GROUP =   Survive      Death
VARIABLE
  3 SHOCK          0.39886      5.41098
  5 ETOH          1.06069      4.39754
  7 INFARC        0.92113      3.49707
  8 AGEC          1.07260      1.48212


CONSTANT          -2.66880     -8.70284


CLASSIFICATION MATRIX
GROUP      PERCENT   NUMBER OF CASES CLASSIFIED INTO GROUP -
           CORRECT
                     Survive  Death
Survive    83.5       71       14
Death      76.2        5       16


TOTAL      82.1       76       30
```

Table 13.5. (*Continued*)

```
JACKKNIFED CLASSIFICATION^a

GROUP      PERCENT    NUMBER OF CASES CLASSIFIED INTO GROUP -
           CORRECT
                      Survive  Death
 Survive    77.6        66      19
 Death      76.2         5      16


 TOTAL      77.4        71      35
```

```
SUMMARY TABLE

           VARIABLE    F VALUE TO   NO. OF                              DEGREES
 STEP      ENTERED     ENTER        VARIAB.              APPROXIMATE    OF
 NO.       REMOVED     REMOVE       INCLUDED U-STATISTIC F-STATISTIC    FREEDOM
 ---  --------------   ----------   -------- ----------- -----------   -----------
   1   3 SHOCK           20.575        1       0.8348      20.575      1.0   104.0
   2   5 ETOH            19.606        2       0.7013      21.931      2.0   103.0
   3   7 INFARC           8.009        3       0.6503      18.286      3.0   102.0
   4   8 AGEC             7.173        4       0.6072      16.338      4.0   101.0
```

[a]See Section 13.4.3 for discussion.

If SHOCK $= 0$,

$$P_1(G_1 \mid \text{SHOCK} = 0) = \frac{e^{-0.22935}}{e^{-0.22935} + e^{-2.38309}} = 0.89602,$$

that is, the estimated probability of survival given no shock is 0.89602. A simpler way of calculating these probabilities is to note that $P_1(G_1 \mid \text{SHOCK})$ is of the form

$$\frac{e^{\ell_1}}{e^{\ell_1} + e^{\ell_2}} = \frac{1}{1 + e^{-(\ell_1 - \ell_2)}}$$

and $\ell_1 - \ell_2$ is the discriminant function, Z. (The reason that the program prints the classification functions is that this method also works with more than two groups. See Note 13.2 for more details). We recommend that you calculate $\ell_1 - \ell_2$ in order to obtain the discriminant function coefficients, as follows

| | Group | | |
Variable	Survive	Death	Survive – Death
SHOCK	0.48548	4.58506	−4.09958
CONSTANT	−0.22935	−2.38309	2.15374

The discriminant function then has the form

$$Z = 2.15374 - 4.09958\text{SHOCK}.$$

The following values can be verified:

SHOCK	Z	$P_1(G_1 \mid \text{SHOCK})$	$P_2(G_2 \mid \text{SHOCK})$
0	2.15374	0.896	0.104
1	-1.94584	0.125	0.875

A particular value of Z is called the discriminant score, the probabilities can be calculated using $1/(1 + e^{-Z})$ and $1 - 1/(1 + e^{-Z})$ for P_1 and P_2, respectively.

Of the remaining variables ETOH has the highest F-to-enter level; $F_{1,103} = 19.61$. This variable is entered at Step 2. At Step 3 INFARC is entered and at Step 4, AGEC.

Step 4. All the variables that contribute to discrimination have now been entered. The variable MALN has an F-to-enter of $F_{1,100} = 1.892$, which is not significant.

The difference between the two classifications functions at Step 4 gives the discrimination function:

$$Z = 6.03404 - 5.01211\text{SHOCK} - 3.33685\text{ETOH} - 2.57594\text{INFARC} - 0.40952\text{AGEC}.$$

For example, the constant term in the discriminant function, 6.03404, is the difference between the two constants in the classification functions: $-2.66880 - (-8.70284) = 6.03404$. We can now estimate the probability of death for each combination of values of the four predictor variables. For example, let

$$\text{SHOCK} = 1, \quad \text{ETOH} = 0, \quad \text{AGEC} = 5, \quad \text{INFARC} = 1.$$

This would describe a patient arriving in shock, with no history of alcoholism, in his/her fifties and a history of bowel infarction. The discriminant score is

$$Z = 6.03404 - 5.01211 - 2.57594 - 0.40952(5) = -3.60161.$$

This produces a probability of survival (Group G_1)

$$P_1(G_1 \mid Z) = 1/(1 + e^{-3.60161}) = 0.02656,$$

conversely, the probability of death is

$$P_2(G_2 \mid Z) = 1 - P_1(G_1 \mid Z) = 0.9734.$$

On the basis of the discriminant model this patient has a very high probability of dying from sepsis.

The discriminant scores for each of the two groups can be plotted to give some idea of the separation of the two groups: in Figure 13.3 the distribution of the Z-scores is presented. The discriminant scores, Z, range from -4.5 to 5.5. A vertical line indicates the point at which $Z = 0$, a value to the left of $Z = 0$ implies a probability of survival less than 0.5, a value to the right of $Z = 0$ has an associated probability of survival greater than 0.5. Using this discriminant procedure 3 survivors and 10 deaths would be misclassified. These are further discussed in Section 13.4.2.

13.3.3   Selecting and Testing an Appropriate Discriminant Model

The comments made in Section 13.2.3 are also applicable to the discriminant model and you should review them in this context. The discriminant model assumed, in

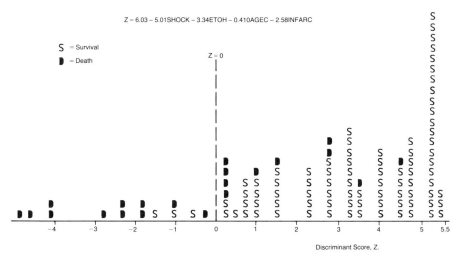

Figure 13.3. Distribution of discriminant scores by survival status. Data from Example 13.1 (Section 13.2).

addition, multivariate normality and homogeneity of variances and covariances within the groups, that is covariance $(X_i, X_j \mid G)$. Although formal tests are available, some estimate of the validity of the homogeneity of the variances can be obtained by considering F-tests for the ratios of variances using degrees of freedom associated with the sample sizes in each of the groups.

To test the normality assumption, the discriminant scores within a group could be plotted on probability paper, or a histogram made. It is clear from Figure 13.3 that the discriminant scores for the example used in this chapter are not normally distributed; not surprising in view of the discrete nature of all the variables used in the model.

A goodness-of-fit test similar to the one described in Section 13.2.3 could be set up by partitioning the continuous variables into a small number of categories, say, three or four, and then applying the chi-square test. See also the next section.

13.3.4 Quadratic Discriminant Analysis

One of the assumptions in the discriminant analysis model is that the variances and covariances are the same in each of the two groups. If this assumption is not made, it can be shown that the discriminant function involves terms of the form X_i^2 and $X_i X_j$ as well as linear terms in the X_i's. The technique is then called *quadratic discriminant analysis*. An example of the application of such procedures is given in Hall *et al.* [1971]. The authors were concerned with the computer diagnosis of rheumatic heart disease. In their study, chest X-rays were digitized and a computer program was written to estimate the cardiac silhouette. From the estimated silhouette, parameters were estimated and used in a discriminant analysis to classify individuals into groups. It was found that a quadratic discriminant analysis gave a better procedure in this sense: adding terms of the form $X_i X_j$ significantly improved prediction by the model. This implies the use of a quadratic discrimination model. In Figure 13.4 the results of the computer generated discriminant procedure are compared with the assessment of the radiologists. The top percentage in each box represents the proportion of diagnoses

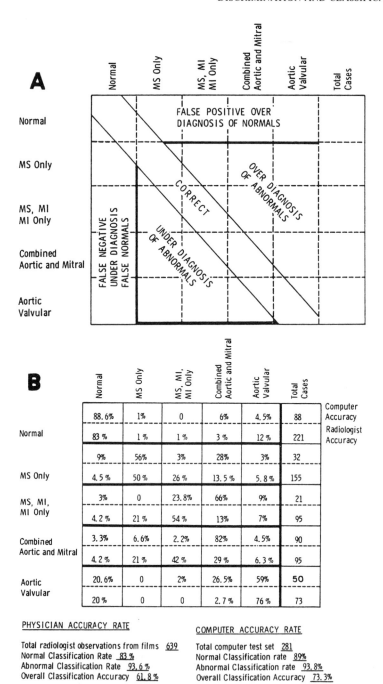

Figure 13.4. Comparison of computer based and radiologist diagnosis of rheumatic heart disease. Computer diagnosis based on a quadratic discriminant function. Data from Hall *et al.* [1971].

placed there by the computer, the bottom percentage represents the assessment by the radiologists. MS refers to stenosis of the mitral valve, MI to mitral valve insufficiency. The percentage correctly classified in each category can be found along the diagonal of the table. All other percentages relate to incorrect diagnosis.

One conclusion from this data is that the computer procedure in this study was more accurate than the radiologists in placing abnormal silhouettes in the correct class. Of course, the two approaches must be compared more carefully, for example, to ascertain the types of diagnostic errors made and the obvious differences between the two procedures. In addition, since the diagnoses were made by 10 radiologists, were some radiologists more accurate than others? See the paper for further discussion.

13.4 CONSIDERATIONS IN DISCRIMINATION AND CLASSIFICATION

In this section, we discuss some issues that are relevant to both logistic regression and discriminant analysis. In addition, we compare these two approaches. Issues common to both procedures include (i) selection of the prior probability; (ii) other criteria for choosing coefficients; (iii) bias due to estimation and testing of the functions on the same set of data; and (iv) comparison of the two methodologies.

13.4.1 Selection and Effect of Prior Probabilities

In Section 6.4.6, we saw that the prevalence of a disease had a marked effect on the predictive value of a positive test. In the context of this chapter we have talked about prior probabilities: the mix of the two groups of interest in the population. Frequently we do not know these probabilities *a priori*. We can then, perhaps, estimate them from the sample or use some arbitrary value. Let $\widehat{\pi}_1 = n_1/(n_1 + n_2)$ be the proportion of observations from Group G_1 in the sample. As before, π_i is the proportion of G_i in the total population with $\pi_1 + \pi_2 = 1$. Let $P[G_1 \mid \widehat{\pi}_1]$ be the posterior probability based on the sample proportions $\widehat{\pi}_1$ and $1 - \widehat{\pi}_1$, and let $P[G_1 \mid \pi_1]$ be the desired posterior probability based on the prior probabilities π_1 and $1 - \pi_1$. Then in the case of logistic regression, the two probabilities are related by

$$\text{logit}\, P[G_1 \mid \pi_1] = (\text{logit}\, \pi_1 - \text{logit}\, \widehat{\pi}_1) + \text{logit}\, P[G_1 \mid \widehat{\pi}_1]. \tag{13.10}$$

Thus the adjustment to the logit is only to the constant term (the intercept). If the "sample mix" is the same as the "population mix" then there is no adjustment. If π_1 is less than $\widehat{\pi}_1$ (common in case-control studies with G_1 the cases and π_1 the prevalence of the disease) the posterior probability is reduced by the amount (logit $\widehat{\pi}_1$ − logit π_1). Equation (13.10) can be expressed explicitly in terms of probabilities:

$$P[G_1 \mid \pi_1] = \cfrac{1}{1 + \left(\cfrac{1 - P[G_1 \mid \widehat{\pi}_1]}{P[G_1 \mid \widehat{\pi}_1]}\right)\left(\cfrac{\widehat{\pi}_1}{1 - \widehat{\pi}_1}\right)\left(\cfrac{1 - \pi_1}{\pi_1}\right)} \tag{13.11}$$

See also Problem 13.11.

The adjustment is identical in discriminant analysis if the model is sample based. That is, it is possible to estimate the discriminate score assuming $\pi_1 = 0.5$ but unequal sample sizes. Most computer programs require you to specify the prior probabilities either as the sample probabilities or other specified values. If no proportions are specified, the priors π_1, $1 - \pi_1$ are assumed equal (in Section 13.3.2 we specified $\widehat{\pi}_1$ and $\widehat{\pi}_2$). If the discriminant score has been calculated assuming equal priors, the adjustment to the posterior probability is simply

$$\text{logit } P[G_1 \mid \pi_1] = -\text{logit } \pi_1 + \text{logit } P[G_1 \mid \pi_1 = 0.5]$$

or

$$P[G_1 \mid \pi_1] = \frac{1}{1 + \left(\dfrac{1 - P[G_1 \mid \pi_1 = 0.5]}{P[G_1 \mid \pi_1 = 0.5]} \right) \left(\dfrac{1 - \pi_1}{\pi_1} \right)} \tag{13.12}$$

If the discriminant function was calculated using the sample proportions $\widehat{\pi}_1$, $1 - \widehat{\pi}_1$ as the priors then the adjustment to the posterior probabilities is as in Equations (13.9) and (13.10).

In terms of the discriminant function, consider only one predictor variable X. Let the prior probabilities be estimated from the data, to produce the function

$$Z = a + bX.$$

This function has the prior probabilities "built-in." To adjust the discriminant score with prior probabilities π_1, $1 - \pi_1$ we get

$$Z_{\text{adj}} = \text{logit } \pi_1 - \text{logit } \widehat{\pi}_1 + Z \tag{13.13}$$

and this is an adjustment to the intercept only. For the example discussed in Section 13.3.2 at Step 1, the estimated probability of survival for a subject in shock was based on the discriminant score

$$Z = 2.15374 - 4.09958(1)$$
$$= -1.94584.$$

Hence, the probability of survival given that SHOCK was present is estimated to be

$$P[\text{Survival} \mid \text{SHOCK}, \widehat{\pi}_1] = \frac{1}{1 + e^{-Z}} = 0.125,$$

as before.

Suppose now that the prior probability is lower and equal to $\pi_1 = 0.60$. From the sample we have $\widehat{\pi}_1 = P[\text{Survival}] = 85/106 = 0.80189$, so that the adjusted discriminant score becomes

$$Z_{\text{adj}} = -1.94584 + \text{logit}(0.60) - \text{logit}(0.80189)$$
$$= -1.94584 - 0.99266$$
$$= -2.93852.$$

The posterior probability is calculated as

$$P[\text{Survival} \mid \text{SHOCK}, \pi_1] = 0.050.$$

Thus the probability of survival of a patient in shock given a prior probability of 0.60 is 0.050 which is considerably lower than the probability of survival given a prior probability of 0.80189. This is intuitively correct since we "know" *a priori* that the survival rate is 60% rather than 80%. Reduction of the prior probability of survival leads, in this case, to a shift to the right of the cut-off line in Figure 13.3.

The choice of values for prior probabilities is crucial. If the event of interest is rare it may be difficult to generate enough information to make the prediction of its occurrence reasonably high. Particularly, in epidemiological screening procedures, if the prevalence (prior probability) of the disease is rare the predictive value of a positive test (posterior probability) may not be high. One way of stating this is that the prior information dominates the information from the experiment. See also Section 13.4.2 for some additional comments. Most computer programs permit a choice of priors in addition to the estimates provided by the sample. The "default" option is often that of equal probabilities.

13.4.2 Criteria for Estimating Coefficients

A common criterion for choice of coefficients in the model is that of maximum likelihood. Note 13.1 discusses conditional maximum likelihood in the context of logistic regression and Section 13.2.1 indicates that this criterion becomes the least squares criterion in the normal case; in Section 13.3.1 we stated that another criterion, in the normal case, is that of choosing that linear combination of the predictor variables that maximizes the Mahalanobis distance.

In the context of discriminant analysis, theory has been developed that allows different costs of misclassification. A standard reference work for this development is Anderson [1988]. He shows that it is only the ratio of the costs that influences the classification rule. Consider the case of only two groups: G_1, G_2; let

$$C(1 \mid 2) = \text{cost of misclassifying an observation from}$$

$$G_2 \text{ as coming from } G_1;$$

similarly,

$$C(2 \mid 1) = \text{cost of misclassifying an observation from}$$

$$G_1 \text{ as coming from } G_2.$$

Let Z be the discriminant function so that for $Z > 0$ the observation is classified into G_1 and for $Z < 0$ the observation is classified into G_2. The discriminant function is modified so that for

$$Z > \ln\left(\frac{C(1 \mid 2)}{C(2 \mid 1)}\right), \tag{13.14}$$

the observation is classified into G_1 and otherwise G_2. If the costs of misclassification are equal, the usual discriminant rule is obtained. If $C(2 \mid 1) > C(1 \mid 2)$, that is, it is more costly to misclassify an observation from G_1, the rule is more likely to classify an observation into G_1. If the costs can be quantified, the above modification will produce the optimal decision rule, that is, we minimize the expected cost of misclassification.

There is one other issue that can be discussed in the context of misclassification. Suppose, again, that there are two groups, G_1, G_2 with prior probabilities π_1, π_2, respectively and $\pi_1 + \pi_2 = 1$. If $\pi_1 > \pi_2$, then we would predict G_1 just on the basis of knowledge of the prior probabilities. Hence, not knowing the values of any of the predictor variables, our success rate will be $100\pi_1\%$. In the context of Example 13.1, there were 85/106 survivors, or $100\widehat{\pi}_1\% \doteq 80\%$. On the basis of this fact alone we can be correct about 80% of the time. Using the discriminant program we classified correctly $92/106 \doteq 87\%$ of cases (see Figure 13.3). On this basis we have not done very much better! However, the following argument can be made: we want to identify *both* survivors and deaths. A discriminant function does classify subjects into the "death" group. This may be considered "bucking the odds." "Forcing" the discriminant function to classify into both groups reduces the chance of correct classification as follows: let p_1 be the proportion classified as G_1 and p_2 the proportion classified as G_2. Then if the classification assigns randomly:

$$P(\text{correct classification}) = P(G_1 \mid \text{classified as } G_1)P(\text{classified } G_1)$$

$$+ P(G_2 \mid \text{classified as } G_2)P(\text{classified } G_2) \qquad (13.15)$$

$$= \pi_1 p_1 + \pi_2 p_2.$$

Using $p_1 = \pi_1$ and $p_2 = \pi_2$, that is, using the same mix in the sample as in the population,

$$P(\text{correct classification}) = \pi_1^2 + \pi_2^2$$

$$= \pi_1^2 + (1 - \pi_1)^2. \qquad (13.16)$$

For our example,

$$\pi_1 = 0.80$$

and

$$P(\text{correct classification}) = 0.80^2 + 0.20^2 = 0.68 = 68\%.$$

Hence, the 87% correct classification may be compared with 68% "chance classification," a somewhat greater improvement. For further discussion see Morrison [1974].

13.4.3 Bias in Estimation of Misclassification Error

As discussed above, there are two types of misclassification errors. For purposes of this section we will consider the two types of errors equally costly. The simplest way to estimate the misclassification errors is to apply the procedure to the actual sample. This will involve a bias because the procedure is applied to the same set of data from which the estimates are obtained. The observed proportion misclassified will usually be lower than the true misclassification rate.

In the case of discriminant analysis another estimate of the true proportion misclassified can be obtained. The estimate makes use of the normality assumptions. Let

$$P(2 \mid 1) = \text{Probability of assigning an individual from Group 1 to Group 2}$$

and

$$P(1 \mid 2) = \text{Probability of assigning an individual from Group 2 to Group 1.}$$

Then an estimate of $P(2 \mid 1)$ based on the data is

$$\widehat{P}(2 \mid 1) = \Phi \left(\frac{\ln(\widehat{\pi}_2/\widehat{\pi}_1) - \Delta^2/2}{\sqrt{\Delta^2}} \right) \tag{13.17}$$

and

$$\widehat{P}(1 \mid 2) = \Phi \left(\frac{-\ln(\widehat{\pi}_2/\widehat{\pi}_1) - \Delta^2/2}{\sqrt{\Delta^2}} \right), \tag{13.18}$$

where Δ^2 is the square of the Mahalanobis distance, and Φ is the symbol for the cumulative normal distribution. For a more detailed analysis see Lachenbruch [1977].

Another approach which works with both logistic regression and discriminant models is to split the sample in half, estimate the model from the first half and apply the model to the second half of the data to estimate the probabilities of misclassification. If the original sample is reasonably large this will be a useful approach. Finally, the jackknife procedures discussed in Chapter 8 can be applied. This is done routinely in some computer programs, see for example, Table 13.5.

13.4.4   Comparison of Logistic Regression and Discriminant Analyses

The two procedures can be compared in several ways. In this section, we discuss some of these and give references for further reading and information.

Assumptions
The linear discriminant model is more stringent in its assumptions: it requires multivariate normality of the predictor variables and equal variances and covariances in each of the groups. These conditions clearly cannot be met by discrete or categorical variables. The logistic model allows both types of variables and assumes only that conditional on the values of the predictor variables (and assuming the model is correct) group membership is binomially distributed. Thus it is a more general model.

Computations
The logistic model requires considerably more computational effort, even on a high-speed computer.

Fitting Criterion
The usual criterion for fitting a logistic model is (conditional) maximum likelihood. This is a generalization of the criterion of maximum likelihood which is equivalent to the principle of least squares in the normal case. Thus, the criteria used in fitting the two models are not as different as might appear at first glance.

Efficiency
Since the logistic regression model can be thought of as a generalization of the discriminant model why not fit this model routinely if there is an appropriate computer program available? Efron [1975] has shown that under the normality assumptions the discriminant model is more efficient than the logistic model in the sense that in large samples the coefficients of the discriminant function have smaller variance. Hence fewer data points are needed to attain a precision comparable with a logistic model.

And this becomes even more important for samples of moderate size where we try to extract as much information as possible from the data. But note that any gain depends upon the multivariate normality of the sample. We suggest using logistic regression unless one is confident of the normality assumptions.

Robustness

Several authors have claimed good robustness properties for discriminant analysis. Truett *et al.* [1967] applied discriminant analysis to the data of the Framingham study. This was a longitudinal study of the incidence of coronary heart disease in Framingham, Massachusetts. In their prediction model the authors used continuous variables such as age (years), serum cholesterol (mg/100 ml) as well as discrete or categorical variables such as: cigarettes per day (0 = never smoked, 1 = less than one pack a day, 2 = one pack a day, 3 = more than a pack a day); ECG (0 = normal, 1 = certain kinds of abnormality). It was found that the linear discriminant model fitted the data reasonably well in estimating the risk or probability of being in the coronary

Halperin *et al.* [1971] re-analyzed these data using a multivariate logistic model. They came to five conclusions which have stood the test of time. If the logistic model holds but the normality assumptions for the predictor variables are violated they concluded that:

1. β_i which are zero will tend to be estimated as zero for large samples by the method of maximum likelihood, but not necessarily by the discrimination function method;

2. If any β_i are nonzero they will tend to be estimated as nonzero by either method, but the discriminant function approach will give asymptotically biased estimates for those β_i and for α; the $|\hat{\beta_i}|$ are too small.

3. Empirically, the assessment of significance for a variable, as measured by the ratio of the estimated coefficient to its estimated standard error, is apt to be about the same whichever method is used;

4. Empirically, the maximum likelihood method usually gives slightly better fits to the model, as evaluated from observed and expected numbers of cases per decile of risk;

5. There is a theoretical basis for the possibility that the discriminant function will give a very poor fit, even if the model holds.

See also Problems 13.5, 13.6, and 13.7.

For example, Fisher and Kennedy [1983] have confirmed that the discriminant model can give estimates of posterior probabilities which are quite unrealistic in some situations. Given the present state of knowledge and the availability of computer programs, it is more appropriate to use the logistic model if there are many categorical variables. See also Knoke [1982] and Gordon [1974] for a critical review of the use of logistic regression.

Example 13.1 (continued). In view of the above comments we want to take a final look at the example of Section 13.2 to which we have applied both models. Figure 13.5 contains a plot of the predicted probabilities of survival using the two models. The top graph contains a plot of the logits of the probabilities, the bottom graph the predicted probabilities. The bottom graph indicates reasonable agreement

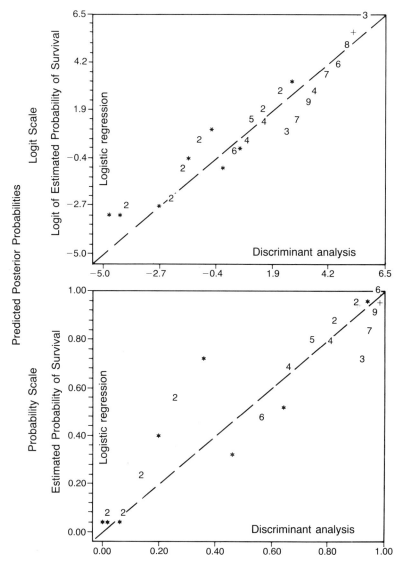

Figure 13.5. Logistic regression *vs.* discriminant analysis. Plots of the posterior probabilities. Data from Example 13.1 (Section 13.2).

in the "tails" but substantial disagreement in the center. And this is the region of importance because we would change our classification depending on the value being below or above 0.5. A review of the predicted probabilities indicates that there are nine cases with this feature. This results in different misclassification proportions for the two procedures. These can be summarized in the following table:

OUTCOME OF CLASSIFICATION PROCEDURE

	Logistic Regression Predicted Status			Discriminant Analysis Predicted Status		
Observed	Death	Survival		Death	Survival	
Death	14	7	21	10	11	21
Survival	3	82	85	3	82	85
Total	17	89	106	13	93	106

The logistic procedure does somewhat better in correctly predicting death, its accuracy is estimated to be $14/21 = 67\%$ compared with $10/21 = 48\%$ for discriminant analysis. The overall proportion correctly classified is $96/106 = 91\%$ for logistic regression and $92/106 = 87\%$ for discriminant analysis. In summary, the logistic model appears to do slightly better than the discriminant model and is certainly more appropriate.   □

Patterns
In Figure 13.6 we illustrate some patterns that may be observed with two predictor variables. The distributions on the axes represent the pattern that would be observed if only one variable was used in discrimination. The line labeled Z represents the distribution of the logits of the estimated probability of misclassification.

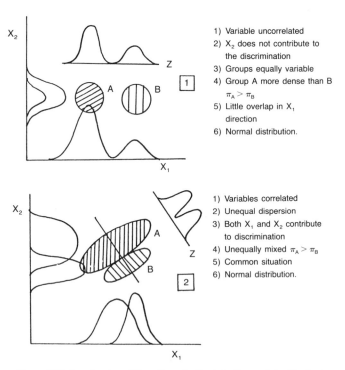

1) Variable uncorrelated
2) X_2 does not contribute to the discrimination
3) Groups equally variable
4) Group A more dense than B
   $\pi_A > \pi_B$
5) Little overlap in X_1 direction
6) Normal distribution.

1) Variables correlated
2) Unequal dispersion
3) Both X_1 and X_2 contribute to discrimination
4) Unequally mixed $\pi_A > \pi_B$
5) Common situation
6) Normal distribution.

Figure 13.6. Sample of possible patterns in discrimination and classification.

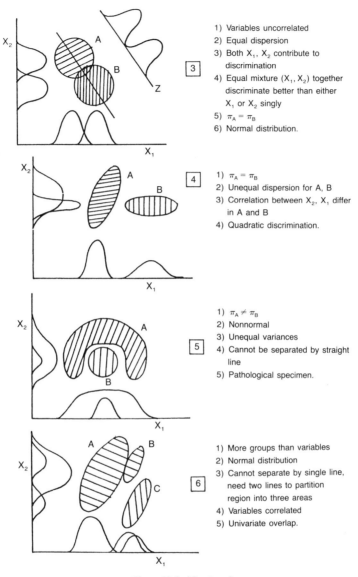

1) Variables uncorrelated
2) Equal dispersion
3) Both X_1, X_2 contribute to discrimination
4) Equal mixture (X_1, X_2) together discriminate better than either X_1 or X_2 singly
5) $\pi_A = \pi_B$
6) Normal distribution.

1) $\pi_A = \pi_B$
2) Unequal dispersion for A, B
3) Correlation between X_2, X_1 differ in A and B
4) Quadratic discrimination.

1) $\pi_A \neq \pi_B$
2) Nonnormal
3) Unequal variances
4) Cannot be separated by straight line
5) Pathological specimen.

1) More groups than variables
2) Normal distribution
3) Cannot separate by single line, need two lines to partition region into three areas
4) Variables correlated
5) Univariate overlap.

Figure 13.6. (*Continued*)

13.5 *MATHEMATICAL MODELS IN COMPUTER-AIDED DIAGNOSIS

13.5.1 Introduction

In Section 13.3.4 (see Figure 13.4) an example was presented of computer diagnosis of rheumatic heart disease. It is an illustration of the use of a mathematical model for prediction. A set of variables is measured and related to a set of outcomes or can be used to discriminate between groups. Routine use of such procedures must be backed by extensive thought, time, and effort. Fisher *et al.* [1976] give an overview of the issues involved in the use of mathematical models, particularly in computer-aided diagnosis. The workstation explosion makes their considerations ever more appropriate. This section contains a part of their paper.

13.5.2   Issues in the Development of Computer-Aided Diagnosis

In spite of the rather high level of mathematical sophistication required for research into computer diagnosis, remarkably little attention has been given to the experimental design of the studies to validate the proposed methods. Without adequate evaluation, of course, the most sophisticated of mathematical models may be of little value.

The first step in any observational study is definition of the "target" population. This is a particularly important aspect of the application of mathematical methods to diagnosis and prognosis, because, unlike many other kinds of medical research projects, the primary purpose is *not* to elucidate the underlying biological phenomenon that is causing the particular desease, but rather to construct a mathematical function that will help decide upon the particular diagnosis or prognosis. While we might be able to assume in the first case that the biological phenomenon is the same from population to population, it is usually not possible to make the same assumption concerning the distribution of the values observed for the variables used by the mathematical model. It is extremely likely that these vary among populations and may even differ due to observer peculiarities. The majority of studies of this type are conducted in university and other medical research facilities, which not only serve populations of wide diversity but also vary considerably in the strengths of their staff. Another point that is frequently overlooked is the likelihood of changes over time in the composition of the populations as well as in medical treatment that may affect the variables observed in the study population. Thus, a mathematical formulation that is successful in classifying individuals when data are originally collected may do a relatively poor job a few years later or at a new location.

Another important aspect of the experimental design is the existence of well-defined groups into which the classification procedure is to classify people. Errors in placing individuals in the groups used for the construction of the classification rule will have a detrimental effect on the performance of this procedure. Further, it is extremely important that the criteria for defining the groups be based on independent objective information not to be used in the mathematical model. Examples of this kind of information might be the results of exploratory surgery, isolation of a particular pathogen, or even the success of a disease-specific treatment. It should be clear that if the variables used to define the disease are also used to construct the classification function then positive results are guaranteed. Yet such results would have little usefulness.

In designing studies for mathematical aids to diagnosis and prognosis, decisions made concerning the selection of variables to be recorded and their definition are quite crucial. Particular attention should be given to the subjective variables, so that, as much as possible, these will be reproducible. If this is not done the performance of the model may depend as much on the individual recording the variables as on the mathematics of the model. Another point that is often overlooked is that the individual recording the information should have no knowledge about the condition of the patient, other than those variables to be used for the classification procedure, that is, the study should be "single-blind." If the person recording the data has other information which points to the correct diagnosis, this may affect his recording of those variables with some judgmental component, thus weighting the mathematical function towards the true diagnosis. This will, of course, falsely inflate the success of the mathematical procedure.

Finally, studies in the construction of mathematical diagnostic aids should always be considered as model-building studies; that is, once the mathematical function is constructed from a particular set of data, it provides only an initial model, and it must be further tested in a new set of independent data before it can be validated. Testing the data on the sample from which the function was computed will invariably give an inflated estimate of the performance of the method. This "internal check" can, however, give the researcher an estimate of the best that he can expect to do in using this function. In order to estimate actual performance, a new sample must be collected and the function used to classify its members ("external check"). Lack of success on this external check may be due to one of several factors. First, the mathematical model may be far from optimal for these particular data; if this is the case, testing other models on the original set and upon the external sample may improve the results. A second difficulty may be that the frequency of the disease in the population may have changed between the time the samples for the internal and external checks were collected, whereas the relationships between the variables and disease have remained constant. Although techniques exist for correcting this problem, under what circumstances these should be applied and how well they work is not known. A more serious difficulty resulting in poor performance of the mathematical model may be caused by changes in the characteristics of the disease or in the manner in which the variables are recorded. As indicated earlier, changes in the personnel involved in the study, or for that matter the physical devices measuring the variables, can strongly influence the performance of these models.

The exceptionally successful use of mathematical models in diagnosis that have been reported (correct classification rates of well over 90%), as well as the probably large number of unpublished poor performances, may well be due to the lack of consideration of the points raised in this section. In addition, of course, the percent classified correctly may be a very poor measure of the effectiveness of a rule, since the false positives and false negatives may have drastically different consequences.

13.5.3 Evaluation of Computer Diagnosis Programs

Before a mathematical model can legitimately be used in a clinical setting, it must be tested with the same rigor and care that optimally is used before a new drug is released. It is obvious that a poorly validated and incorrect computer-aided diagnosis or prognosis program could result in severe medical, legal, and moral problems. Given below is a brief general outline of the steps one would like in the evaluation of a computer diagnosis program. Clearly medical, practical, and mathematical considerations may well modify the details of the study design, but the principles listed below will in general be valid.

Step 1. Define the target population and the personnel who will collect the data (physicians, physical therapists, nurses, physician's assistants, clerks, etc.).

Step 2. Pick the variables that are candidates for use in the model. Carefully define the manner in which these variables are to be recorded and the meaning of the terms used in their definitions.

Step 3. Pick a random sample from the target population. (In many instances it may not be possible to draw a random sample, and we will have to be satisfied with those cases we can conveniently obtain. This, of course, will make inferences concerning the potential usefulness of our mathematical model more tentative and its testing more difficult).

Step 4. Data on these individuals should be collected by a random sample of the type of individual who will collect the data in the everyday medical setting in which we envision its use. (Again, this also may not be possible to accomplish, putting further burden on our testing of the model at a later time).

Step 5. From these data we can now compute the estimates of the parameters of our mathematical model or models. It is useful at this point to estimate our maximum potential performance by using the model to classify this internal sample. We may at this time also try to eliminate those variables that have little influence on our ability to classify the cases correctly. In addition, we evaluate competing mathematical models, discarding those whose "best possible" performance is inadequate.

Step 6. Collect and record data on a new sample and on this sample apply the mathematical models constructed in Step 5. Evaluate the performance of these models.

Step 7. At this point we must decide whether or not the results we have obtained are useful in a general medical context. What information is necessary to make this decision obviously depends on the particular medical setting. One example might be that we would compare the success of our mathematical model with that of a randomly chosen set of physicians given the same information. In any event, only if our model has a demonstrable clinical use will it be accepted by the medical profession.

Step 8. Now we must find a practical, cost effective method for implementing our mathematical model into medical practice. This may be difficult in some settings since some models require access to a computer; usually a personal computer is adequate.

Step 9. We must now design a trial of our mathematical model in real clinical practice under typical operating conditions. If this trial demonstrates the efficacy of our methods, we can at this point release it for general use.

Step 10. We cannot, however, rest, even at this stage, for it is still necessary for us to monitor the continuing performance of our model, particularly keeping in mind that changes in medical practice, personnel, and possibly in the disease itself may invalidate or require us to modify our procedures.

We do not know of any trial resembling the one described in the steps given. This would require resources far in excess of what has been available to most researchers in this area.

13.6 *DECISION THEORY

13.6.1 The Field

Much of life consists of making decisions under conditions that are uncertain. For example, one might decide to major in a field of great interest but with high unemployment. Somehow the judgment of the chances of working in the field after finishing a degree must be weighed into the decision. In medicine, the practitioner must often decide between possible alternative treatments (including no treatment). The prognosis under each alternative treatment involves much uncertainty. Many factors must be taken into account and somehow balanced to arrive at a course of action.

Decision theory is a body of theory and techniques designed to allow the selection of "good" decisions. Typically in a decision theory problem certain elements are present. The elements are:

1. Knowledge of the *possible outcomes* of a situation.
2. Knowledge of the *cost* associated with each possible outcome. (Cost need not be measured in dollars.)
3. Knowledge of *possible decisions, decision rules* or *strategies* that might be employed.
4. Knowledge of personal estimates of the *probability of each outcome* under the different strategies involved.

Using the above elements, a decision theoretic approach chooses a strategy or method of decision-making that is good in some sense. One method is to choose the strategy that minimizes the expected costs. The example below uses this approach. Usually the models applied to a particular patient are advisory since they necessarily are more simplified than the real-life situation.

Decision theory has been heavily used in business, engineering, and the defense industry. The methods are used by systems analysts, operations researchers and some administrators. With the great concern for medical care costs, one would expect more use of such techniques in the future. One benefit of decision theoretic analyses is that they force clarification and definition of a situation. The task of assigning costs to outcomes and procedures makes clearer the cost trade-offs involved. Usually the models applied to a particular patient are advisory since they necessarily are more simplified than the real life situation.

13.6.2 An Example

We now consider the work of Tompkins *et al.* [1977] which examines the cost effectiveness of the management of pharyngitis (sore throat) for acute rheumatic fever (ARF) prevention. The authors present the case as follows:

> "Respiratory infections are the single most prevalent illness in the United States. These illnesses result in more productive time loss and more use of medical services than any other group of acute illnesses. Within the respiratory infections, there are relatively few diseases that can be treated successfully, and even fewer that have any significant long-term morbidity. Streptococcal pharyngitis is the prime example of such an illness. It has been shown to be the predisposing factor for acute rheumatic fever and acute glomerulonephritis; it is one cause of severe local infections (for example, peritonsillar abscess); but prompt treatment

can prevent most of its complications. For these reasons, medical care providers spend much time identifying and treating those cases of pharyngitis caused by Group A streptococci."

The authors consider three possible strategies for treating pharyngitis. It is assumed that those with known penicillin allergies would not be treated with penicillin.

A. The currently recommended strategy: to culture the throat of all patients and to treat with penicillin those whose cultures are positive for Group A streptococci;

B. Treat all patients with penicillin without taking a throat culture; or

C. Neither culture nor treat any patient.

The risk associated with treatment is a possible allergic reaction ranging in severity from mild to fatal. In addition, the treatment may not be effective. The risk of a fatal allergic reaction is very small (the authors estimate the probability is 3×10^{-6}) but the cost is very high. The risk associated with no treatment is that of incurring acute rheumatic fever (ARF).

The authors first traced through the possible paths resulting from each strategy. After arriving at the endpoints, they assigned a cost to each possible endpoint. The three strategies, their possible endpoints and assigned costs are given in Figures 13.7 and 13.8. Tompkins *et al.* discuss in detail the assignment of costs. The specifics will

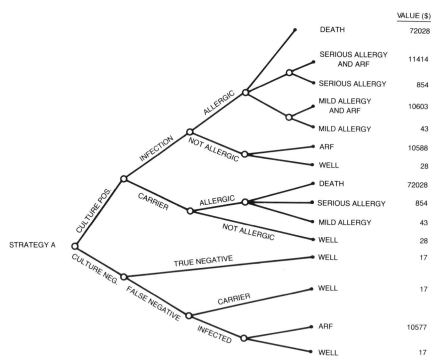

Figure 13.7. Decision tree for strategy A (culture and treat) using benzathine penicillin. ARF, acute rheumatic fever. Tompkins *et al.* [1977].

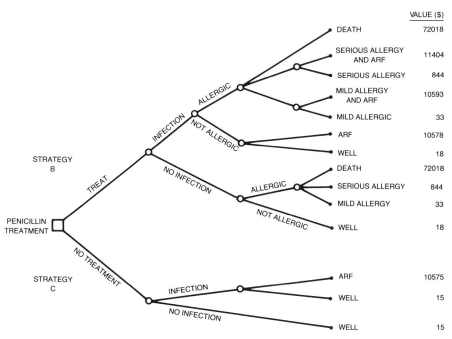

Figure 13.8. Decision tree for strategies B (treat all patients) and C (no culture or treatment). ARF, acute rheumatic fever. Tompkins *et al.* [1977].

not be discussed here other than to note that medical care costs and lost time from work, both due to illness and time off for the physician's appointment(s) are taken into account.

It is worth considering one cost in particular. The authors assign a dollar-value of $72,000 to death. (The remainder of the total for death are medical care costs). The precise dollar amount is not the point of concern here. There is something in the human spirit that rebels against assigning a dollar-value to a death, especially your own and those you love. Yet there are limited resources in society. Dollars not spent in medical care might save lives and improve the quality of life in other areas. For example, providing food for famine areas of the world, public health measures for developing countries, improving air and water quality, and renovation of inner cities are only a few of the alternative uses for resources. Such issues will be faced by default, if in no other way.

In order to proceed, the authors needed the probability of each outcome for each of the strategies. For this purpose they reviewed the literature and found estimates of the needed probabilities.

Consider now the cost associated with the outcome to a random variable. Suppose there are I outcomes with probabilities $P_1, \ldots, P_I$ of occurring and associated costs $c_1, \ldots, c_I$. The expected cost is then

$$E(\text{Cost}) = c_1 P_1 + \cdots + c_I P_I = \sum_{i=1}^{I} c_i P_i. \tag{13.19}$$

For each strategy, A, B, and C, the authors computed the expected costs, E_A(Cost), E_B(Cost) and E_C(Cost), respectively. The strategy (A, B, or C) which has the smallest expected cost is then advocated as the method of minimal cost to society.

At this stage a few complicating factors need to be considered. The fraction of throat cultures positive varies according to whether there is an epidemic or endemic situation. The proportion also varies with the population, geographic location, and time of the year. The analysis was run separately for differing positive throat culture rates (proportions). There are also two methods of penicillin therapy, oral or by injection. The costs vary according to the type of penicillin therapy. Figure 13.9 presents the estimated cost per patient for the three strategies. From this figure the authors summarized their study in the following way:

> "The cost-effectiveness of preventing primary acute rheumatic fever attacks by oral or benzathine penicillin treatment was analyzed for both epidemic and endemic streptococcal pharyngitis situations. Decision analysis was used: the probabilities and the outcome values were calculated from published data. Three penicillin strategies were compared: (A) treating only patients with Group A streptococci-positive throat cultures; (B) treating all patients; (C) treating none of the patients. In the epidemic situation it is medically most effective and least costly to treat all patients with penicillin (Strategy B). In the endemic situation, Strategy B is also most cost-effective when oral penicillin is used in patient populations where the positive throat culture yield is at least 20%. Strategy A is optimal when the yield is between 5% and 20%; below a 5% yield, Strategy C is appropriate. For any individual patient, it is possible that choice of the most cost-effective treatment strategy could be based on the patient's clinical findings."

Note that this analysis considered the total cost from the point of view of society.

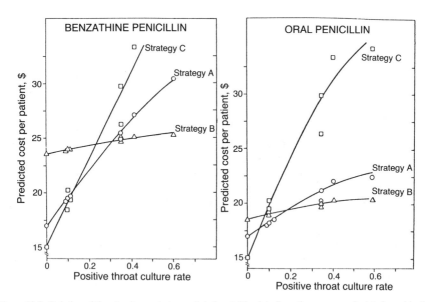

Figure 13.9. Relation of throat culture rate to predicted cost. The data from the paper are depicted graphically, showing the relation between an endemic patient population's throat culture rate and the predicted cost per patient for each clinical strategy. Tompkins *et al.* [1977].

Perhaps the relative cost of the outcomes should be assigned by the patient. This would undoubtedly result in higher values for the adverse outcomes.

Classical texts in this area are by Chernoff and Moses [1959] and Raiffa and Schlaifer [1961].

NOTES

Note 13.1: Conditional Maximum Likelihood and Logistic Regression

The regression coefficients in the logistic regression model are estimated using the conditional maximum likelihood criterion. A full discussion of this topic is beyond the scope of this text but in this note we outline the procedure for the situation involving one covariate. Suppose first that we have a Bernoulli random variable, Y, with probability function,

$$P[Y = 1] = \pi,$$

$$P[Y = 0] = 1 - \pi.$$

We observe n values of Y, $y_1, y_2, \ldots, y_n$ (a sequence of zeros and ones). The probability of observing this sequence is proportional to

$$\prod_{j=1}^{n} \pi^{y_j}(1 - \pi)^{1-y_j} = \pi^{\sum y_j}(1 - \pi)^{n - \sum y_j}. \tag{13.20}$$

This quantity is now considered as a function of π and defined to be the likelihood. To emphasize the dependence on π, we write

$$L(\pi \mid \sum y_j, n) = \pi^{\sum y_j}(1 - \pi)^{n - \sum y_j}. \tag{13.21}$$

Given the value of $\sum y_j$, what is the "best" choice for a value for π? The maximum likelihood principle states that the value of π that maximizes $L(\pi \mid \sum y_j, n)$ should be chosen. It can be shown by elementary calculus that the value of π that maximizes $L(\pi \mid \sum y_j, n)$ is equal to $\sum y_j/n$. You will recognize this as the proportion of the n values of Y that have the value 1. This can also be shown graphically; Figure 13.10 is a graph of $L(\pi \mid \sum y_j, n)$ as a function of π for the situation $\sum y = 6$ and $n = 10$. Note that the graph has one maximum and that it is not quite symmetrical.

In the logistic regression model the probability π is assumed to be a function of an underlying covariate, X, that is, we model

$$\text{logit}(\pi) = \alpha + \beta X,$$

where α, β are constants. Conversely,

$$\pi = \frac{e^{\alpha + \beta X}}{1 + e^{\alpha + \beta X}} = \frac{1}{1 + e^{-(\alpha + \beta X)}}. \tag{13.22}$$

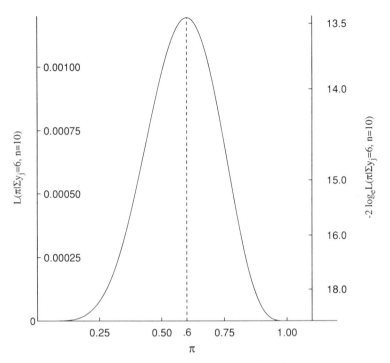

Figure 13.10. Likelihood function, $L(\Pi \mid 6, 10)$.

For fixed values of X the probability π is determined (since α, β are parameters, to be estimated from the data). A set of data now consists of *pairs* of observations: (y_j, x_j), $j = 1, \ldots, n$, where y_j is again a zero-one variable and x_j is an observed value of X for set j. For each outcome, indexed by set j, there is now a probability π_j determined by the value of x_j. The likelihood function is written

$$L(\pi_1, \ldots, \pi_n \mid y_1, \ldots, y_n, x_1, \ldots, x_n, n) = \prod_{j=1}^{n} \pi_j^{y_j} (1 - \pi_j)^{1-y_j}; \qquad (13.23)$$

but π_j can be expressed as

$$\pi_j = \frac{e^{\alpha + \beta X_j}}{1 + e^{\alpha + \beta X_j}}, \qquad (13.24)$$

where x_j is the value of the covariate for subject j.

The likelihood function can then be written and expressed as a function of α, β as follows:

$$L(\alpha, \beta \mid y_1, \ldots, y_n; x_1, \ldots, x_n; n) = \prod_{j=1}^{n} \left(\frac{e^{\alpha+\beta x_j}}{1 + e^{\alpha+\beta x_j}} \right)^{y_j} \left(\frac{1}{1 + e^{\alpha+\beta x_j}} \right)^{1-y_j}$$

$$= \prod_{j=1}^{n} \frac{\left(e^{\alpha+\beta x_j} \right)^{y_j}}{1 + e^{\alpha+\beta x_j}}$$

$$= \frac{e^{\sum_{j=1}^{n} y_j(\alpha+\beta x_j)}}{\prod_{j=1}^{n} (1 + e^{\alpha+\beta x_j})}. \qquad (13.25)$$

The conditional maximum likelihood criterion then requires values for α, β to be chosen so that the above likelihood function is maximized; it is a *conditional* maximization in that it is dependent on the observed values of the covariate $X : x_1, x_2, \ldots, x_n$. Computer programs for logistic regression choose values for α, β by this criterion. For more than one covariate the likelihood function can be similarly deduced.

The above procedure explains the estimation process. To test the significance of the estimates, use is made of "large sample theory." All the tests for the logistic regression model are large sample tests.

Note 13.2: Logistic Discrimination with More than Two Groups

Anderson [1972] and Jones [1975], among others, have considered the case of logistic discrimination with more than two groups. Following Anderson [1972], let for two groups,

$$P(G_1 \mid X) = \frac{\exp\{\alpha_0 + \alpha_1 X_1 + \cdots + \alpha_p X_p\}}{1 + \exp\{\alpha_0 + \alpha_1 X_1 + \cdots + \alpha_p X_p\}}.$$

Then

$$P(G_2 \mid X) = \frac{1}{1 + \exp\{\alpha_0 + \alpha_1 X_1 + \cdots + \alpha_p X_p\}}.$$

This must be so because $P(G_1 \mid X) + P(G_2 \mid X) = 1$, that is, the observation X belongs to either the G_1 or G_2. For k groups, define

$$P(G_s \mid X) = \frac{\exp\{\alpha_{s0} + \alpha_{s1} X_1 + \cdots + \alpha_{sp} X_p\}}{1 + \sum_{j=1}^{k-1} \exp\{\alpha_{j0} + \alpha_{j1} X_1 + \cdots + \alpha_{jp} X_p\}},$$

for groups $s = 1, \ldots, k - 1$, and for group G_k let

$$P(G_k \mid X) = \frac{1}{1 + \sum_{j=1}^{k-1} \exp\{\alpha_{j0} + \alpha_{j1} X_1 + \cdots + \alpha_{jp} X_p\}}. \qquad (13.26)$$

The estimation of coefficients is carried out by computer as before. Unfortunately, most statistical packages do not provide this analysis. The computing time is quite extensive. Similar equations hold for discriminant analysis for more than two groups.

Note 13.3: Logistic Regression and Discriminant Analysis as Multivariate Procedures

Logistic regression and discriminant analysis are examples of multivariate analysis of data. Multiple regression still deals with a (univariate) dependent variable which is related in a quantitative way to a set of predictor variables. The procedures discussed in this chapter are genuinely multivariate. For example, suppose we want to compare two drug treatments but now there is not one response variable but several and the question is whether the *set* of response variables differs between the two drugs, or which of the variables in the set differ. This problem can be handled by the metholodogy of this chapter. Assume a multivariate normal model for a set of data so that discriminant analysis can be applied. Let Δ = Mahalanobis distance (see Definition 13.8) and D_q is the estimate of Δ based upon q variables. Then the null hypothesis $H_0 : \Delta = 0$ can be tested using the statistic

$$F = \frac{(n - q - 1)n_1 n_2}{q(n_1 + n_2)} \frac{D_q^2}{(n - 2)}, \tag{13.27}$$

which has an F distribution with q and $n - q - 1$ degrees of freedom (n_1, n_2 are the sample sizes associated with each group and $n = n_1 + n_2$). It can be shown that $D_q^2 = T^2((1/n_1) + (1/n_2))$ where T^2 is *Hotelling's* multivariate analogue of the two sample (univariate) t-test. Hence, a test of $\Delta = 0$ is equivalent to a multivariate test of no difference between two groups. Most step-wise discriminant programs will print D_q^2 or the associated F statistic at each step. For the example discussed in Section 13.3.2, the F statistic at the fourth and final step is

$$F_{4,101} = 16.34, \quad p \ll 0.001.$$

The estimated Mahalanobis distance is therefore

$$D_4^2 = \frac{q(n_1 + n_2)(n - 2)}{(n - q - 1)(n_1 n_2)} F$$

$$= \frac{4(85 + 21)(104)}{(106 - 4 - 1)(85)(21)}$$

$$= 0.24459$$

or

$$D_4 = 0.495.$$

This suggests considerable overlap, the means of the two populations of linear combinations are separated by approximately 0.5 of a standard deviation.

Note 13.4: Discriminant Analysis of Two Groups and Linear Regression

Given two groups of size n_1, n_2, it was shown already by Fisher [1936] that the discriminant analysis is equivalent to a multiple regression on the dummy variable Y defined as follows:

$$Y = \frac{n_2}{n_1 + n_2} \quad \text{members of Group 1}$$

$$= \frac{-n_1}{n_1 + n_2} \quad \text{members of Group 2.} \tag{13.28}$$

We can now treat this as a regression analysis problem. The multiple regression equation obtained will define the regions in the sample space identical to these defined by the discriminant analysis model.

Note 13.5: Cluster Analysis

This chapter has discussed methods of placing or classifying objects into existing groups. There has been considerable work on the inverse "problem" of finding or constructing groups in an existing set of objects. Much of the work originated in *numerical taxonomy* (the construction of classifications, especially of plants and animals). The work is often called *cluster analysis*. Cluster analysis in taxonomy is presented, for example, in Sokal and Sneath [1963].

Note 13.6: Pattern Recognition

There are a wide variety of techniques used in classification. For example, one approach is to define some measure of distance between observations. A "training set" is stored and new observations are classified into the same class as their nearest neighbor in the training set. In the electrical engineering literature, classification techniques are called *pattern recognition* techniques. Two books that cover a range of approaches (at a high mathematical level) are Patrick [1972] and Meisel [1972].

Note 13.7: Additional Comments

Table 13.6 presents parallels between linear regression (with normal distribution assumptions) and logistic regression. See Hosmer and Lemeshow [1989] for an excellent discussion of logistic regression.

PROBLEMS

1. For the logistic regression model $logit(\pi) = \alpha + \beta X$, where X is a dichotomous 0,1 variable show that e^β is the odds ratio associated with the exposure to X.

2. For the data of Table 13.1 (Section 13.2) the logistic regression model using only the variable X_1, malnutrition, is

$$logit(\hat{\pi}) = -0.646 + 1.210X_1.$$

The 2×2 table associated with these data is:

		Y		
		(Death)	(Survive)	
X_1		1	0	
(Malnutrition)	1	11	21	32
(No malnutrition)	0	10	64	74
		21	85	106

Table 13.6. Comparison of Logistic Regression and "Normal" Regression (One Predictor Variable).

	Logistic Regression	Normal Regression
Dependent variable	Y discrete (binary)	Y Continuous
Covariates	X categorical or continuous	X categorical or continuous
Distribution of Y (given X)	Binomial (n, π)	Normal (μ, σ^2)
Model	$E(Y) = \pi$	$E(Y) = \mu$
Link to X	$\text{logit}(\pi_j) = \alpha + \beta X_j$	$\mu_j = \alpha + \beta X_j$
Data	$y_1, y_2, \ldots, y_n$ $x_1, x_2, \ldots, x_n$	$y_1, y_2, \ldots, y_n$ $x_1, x_2, \ldots, x_n$
Likelihood function (LF)	$\prod_{j=1}^{n} \pi_j^{y_j}(1 - \pi_j)^{1-y_j} =$ $\prod_{j=1}^{n} \left(\dfrac{e^{\alpha + \beta x_j}}{1 + e^{\alpha + \beta x_j}} \right)^{y_j} \left(\dfrac{1}{1 + e^{\alpha + \beta x_j}} \right)^{1-y_j}$	$\prod_{j=1}^{n} \left(\dfrac{1}{\sqrt{2\sigma\pi}} \right)^{n} \exp\left(-1/2 \sum \left(\dfrac{y_j - \mu_j}{\sigma} \right) \right) =$ $\prod_{j=1}^{n} \left(\dfrac{1}{\sqrt{2\sigma\pi}} \right)^{n} \exp\left(-1/2 \sum \left(\dfrac{y_j - \alpha - \beta x_j}{\sigma} \right)^2 \right)$

Fitting criterion (for choosing estimates of α, β)	Maximize LF	Maximize LF
$-2\log$ LF (is proportional to)	$-2\sum y_j(\alpha+\beta X_j)+2\sum\ln(1+e^{\alpha+\beta X_j})$	$\frac{1}{\sigma^2}\sum(y_j-\alpha-\beta X_j)^2$
Equivalent fitting criterion	Minimize $-2\log$ LF (**NOT** least squares)	Minimize $-2\log$ LF (least squares)
Notation	$D(X)=\underset{\text{over }\alpha,\,\beta}{\text{minimum}}(-2\log\text{LF})=$ deviance	$D(X)=\underset{\text{over }\alpha,\,\beta}{\text{minimum}}(-2\log\text{LF})=$ deviance
Testing: $H_0:\beta=0$ in model	$D-D(X)$ is approximately chi-square	$D-D(X)$ is chi-square
Alternative test $H_0:\beta=0$ in model	$\dfrac{D-D(X)}{D(X)/(n-2)}$ is approximately $F_{1,n-2}$	$\dfrac{D-D(X)}{D(X)/(n-2)}=F_{1,n-2}$

a. Verify that the coefficient of X_1 is equal to the logarithm of the odds ratio for malnutrition.

b. Calculate the probability of death given malnutrition using the above model and compare it with the observed probability.

c. The standard error of the regression coefficient is 0.5035, test the significance of the observed value, 1.210. Set up 95% confidence limits on the population value and translate these limits into ones for the population odds ratio.

d. Calculate the standard error of the logarithm of the odds ratio from the 2×2 table and compare it with the value in Part (c).

3. The full model for the data of Table 13.1 is given in Section 13.2.

a. Calculate the logit line for $X_2 = 0$, $X_3 = 1$, and $X_5 = 1$. Plot logit($\hat{\pi}$) versus age in years.

b. Plot $\hat{\pi}$ versus age in years for Part (a).

c. What is the probability of death for a 60 year-old patient with no evidence of shock, but with symptoms of alcoholism and prior bowel infarction?

4. The bottom of Table 13.3 contains a summary of the step-wise logistic regression results of the data dealing with intra-abdominal sepsis. Calculate the deviances going from Step 2 to 3 and from Step 3 to 4. What are the degrees of freedom associated with these deviances?

5. One of the problems in the treatment of acute appendicitis is that perforation of the appendix cannot be predicted accurately. Since the consequences of perforation are serious, surgeons tend to be conservative by removing the appendix. Koepsel *et al.* [1981] attempted to relate the occurrence (or absence) of perforation to a variety of risk factors to enable better assessment of the risk of perforation. A consecutive series of 281 surgery patients was selected initially; of these, 192 were appropriate for analysis, 41 of whom had demonstrable perforated appendices according to the pathology report. The data are listed in Table 13.7. Of the 12 covariates studied six are listed here, with the group indicator Y.

Y = perforation status (1 = yes; 0 = no),

X_1 = sex (1 = male; 0 = female),

X_2 = age in years,

X_3 = duration of symptoms in hours prior to physician contact,

X_4 = time from physician contact to operation (in hours),

X_5 = white blood count in thousands,

X_6 = gangrene (1 = yes; 0 = no).

For X_4 the code 999 is for unknown; for X_5 the code 99 is an unknown code.

a. Compare the means of the continuous variables (X_2, X_3, X_4, X_5) in the two outcome groups $(Y = 0, 1)$ by some appropriate test. Make an appropriate comparison of the association of X_5 and Y. State your conclusion at this point.

Table 13.7. Data for Problem 13.5.

	Y	X_1	X_2	X_3	X_4	X_5	X_6		Y	X_1	X_2	X_3	X_4	X_5	X_6
1	0	0	41	19	1	16	0	49	0	1	15	6	6	19	0
2	1	1	42	48	0	24	1	50	0	0	17	10	4	9	0
3	0	0	11	24	5	14	0	51	0	0	10	72	6	17	0
4	0	1	17	12	2	9	0	52	0	1	9	8	999	15	0
5	1	1	45	36	3	99	1	53	1	1	3	4	2	18	1
6	0	0	15	24	5	14	0	54	0	0	7	16	1	24	0
7	0	1	17	11	24	8	0	55	0	1	60	14	2	11	0
8	0	1	52	30	1	13	0	56	0	1	11	48	3	8	0
9	0	1	15	26	6	13	0	57	0	1	8	48	24	14	0
10	1	1	18	48	2	20	1	58	0	1	9	12	1	12	0
11	0	0	23	48	5	14	0	59	0	1	19	36	1	99	0
12	1	1	9	336	11	13	1	60	1	0	44	24	1	11	1
13	0	0	18	24	3	13	0	61	0	0	46	9	4	12	0
14	0	0	30	8	15	11	0	62	0	1	11	36	2	13	0
15	0	0	16	19	9	10	0	63	0	1	18	8	2	19	0
16	0	1	9	8	2	15	0	64	0	0	21	24	5	12	0
17	0	1	15	48	4	12	0	65	0	0	31	24	8	16	0
18	1	1	25	120	4	8	1	66	0	0	14	7	4	12	0
19	0	0	17	7	17	14	0	67	0	1	17	6	6	19	0
20	0	1	17	12	2	14	0	68	0	0	15	24	1	9	0
21	1	0	63	72	7	11	1	69	0	0	18	24	4	9	0
22	0	0	19	8	1	15	0	70	0	0	38	48	2	99	0
23	0	1	9	48	24	9	0	71	0	1	13	18	4	18	0
24	1	0	9	48	12	14	1	72	1	0	23	168	4	18	0
25	0	0	17	5	1	14	0	73	0	0	15	3	2	14	0
26	0	0	12	48	3	15	0	74	1	0	34	48	3	16	1
27	0	1	6	48	1	26	0	75	0	1	21	24	47	8	1
28	0	0	8	48	3	99	0	76	0	1	50	8	4	12	0
29	1	1	17	30	6	12	1	77	0	0	10	23	6	16	1
30	0	0	11	8	7	15	0	78	0	0	14	48	12	15	0
31	0	1	16	48	2	11	0	79	0	1	26	48	12	13	0
32	0	1	15	10	12	12	0	80	1	0	16	22	1	14	1
33	0	1	13	24	11	15	1	81	1	0	9	24	12	16	1
34	1	1	26	48	4	11	1	82	0	1	26	5	1	16	0
35	0	1	14	7	4	16	0	83	0	1	29	24	1	30	0
36	0	0	44	20	2	13	0	84	0	1	35	408	72	6	0
37	1	1	13	168	999	10	1	85	0	0	18	168	16	12	0
38	0	0	13	14	22	13	0	86	0	1	12	18	4	12	0
39	0	1	24	10	2	19	0	87	0	1	14	7	3	21	0
40	1	0	12	72	2	16	1	88	1	1	45	24	3	18	1
41	0	1	18	15	1	16	0	89	0	1	16	5	21	12	0
42	0	0	19	15	0	9	0	90	0	0	19	240	163	6	0
43	0	0	11	336	20	8	0	91	1	1	9	48	7	23	1
44	0	1	13	14	1	99	0	92	1	1	50	30	5	15	1
45	0	1	25	10	10	11	0	93	0	0	18	2	10	15	0
46	0	1	16	72	5	7	0	94	0	0	27	2	24	17	1
47	0	1	25	72	45	7	0	95	0	1	48	27	5	16	0
48	0	1	42	12	33	19	1	96	0	1	7	18	5	14	0

Table 13.7. (*continued*)

	Y	X_1	X_2	X_3	X_4	X_5	X_6		Y	X_1	X_2	X_3	X_4	X_5	X_6
97	0	1	16	13	1	11	0	145	0	1	41	24	4	14	0
98	0	1	29	5	24	19	1	146	0	0	28	6	1	15	0
99	0	1	18	48	3	11	0	147	1	0	13	48	9	15	1
100	0	1	18	9	2	14	0	148	0	1	10	15	1	99	0
101	1	1	14	14	1	15	1	149	0	1	16	18	4	14	0
102	0	1	32	240	24	7	0	150	0	1	17	18	10	17	0
103	0	1	23	18	2	17	1	151	0	1	38	9	7	11	0
104	0	1	26	16	2	13	0	152	0	1	12	18	2	13	0
105	0	0	30	24	4	20	0	153	0	0	12	72	3	15	0
106	0	1	44	39	15	11	0	154	0	0	27	16	0	14	1
107	1	1	17	24	4	16	1	155	0	1	31	7	8	14	0
108	0	1	30	36	3	15	1	156	0	0	45	20	4	27	0
109	0	1	18	24	2	11	1	157	1	1	52	48	3	15	1
110	0	1	34	96	1	10	0	158	1	1	26	48	13	16	1
111	0	1	15	12	2	10	0	159	0	0	38	15	1	16	0
112	0	1	10	24	4	99	0	160	0	0	19	24	5	99	0
113	0	1	12	14	13	5	0	161	0	1	14	20	2	15	0
114	0	1	10	12	17	17	0	162	0	0	27	22	8	18	0
115	0	1	28	24	2	15	0	163	0	1	20	21	1	99	0
116	0	1	10	96	8	8	0	164	1	1	11	24	8	10	1
117	0	0	22	12	2	12	0	165	0	1	17	72	20	10	0
118	0	0	30	15	5	12	0	166	0	0	27	24	3	9	0
119	0	1	16	36	3	12	0	167	1	0	52	16	4	13	1
120	0	0	16	30	4	15	0	168	1	1	38	48	2	13	1
121	0	1	9	12	12	15	0	169	0	1	16	19	3	12	0
122	1	1	16	144	4	15	1	170	0	1	19	9	4	17	0
123	0	1	17	36	13	6	0	171	0	0	24	24	2	11	0
124	1	1	12	120	2	11	1	172	0	1	12	17	20	6	1
125	0	1	28	17	26	10	0	173	1	1	51	72	2	16	1
126	1	0	13	48	3	21	1	174	1	1	50	72	6	11	1
127	0	0	23	72	3	13	0	175	0	0	28	12	3	13	0
128	1	0	62	72	2	12	1	176	0	0	19	48	8	14	1
129	0	1	17	24	4	14	0	177	0	1	9	24	999	99	0
130	0	0	12	24	12	15	0	178	0	0	40	48	7	14	0
131	0	1	10	12	10	11	0	179	0	0	17	504	7	99	0
132	0	1	47	48	8	9	0	180	0	1	51	24	1	9	1
133	0	1	43	11	8	13	0	181	0	1	31	24	2	10	0
134	1	1	18	36	2	15	1	182	0	0	25	8	9	8	0
135	0	0	6	24	1	9	0	183	0	0	14	24	8	10	0
136	0	0	24	2	22	10	0	184	0	1	7	24	4	15	0
137	0	0	22	11	24	7	0	185	0	1	27	7	2	14	0
138	1	1	39	36	3	15	1	186	0	1	35	72	3	19	1
139	1	1	43	48	2	11	1	187	0	0	11	12	9	11	0
140	0	1	12	7	1	14	0	188	0	1	20	8	6	12	0
141	0	1	14	48	6	16	0	189	0	1	50	48	27	19	0
142	0	1	21	24	1	17	0	190	0	1	16	6	7	7	0
143	1	1	34	48	12	9	1	191	0	1	45	24	4	20	0
144	1	0	60	24	3	14	1	192	1	1	47	336	4	9	1

b. Carry out a stepwise discriminant analysis. Which variables are useful predictors? How much improvement in prediction is there in using the discriminant procedure? How appropriate is the procedure?

c. Carry out a stepwise logistic regression and compare your results with those of Part (b).

d. The authors introduced two additional variables in their analysis: $X_7 = \log(X_2)$ and $X_8 = \log(X_3)$. Test whether these variables improve the prediction scheme. Interpret your findings.

e. Plot the probability of perforation as a function of the duration of symptoms; using the logistic model generate separate curve for subjects aged 10, 20, 30, 40 and 50 years. Interpret your findings.

6. A classic in the use of discriminant analysis is the paper by Truett, *et al.* [1967] which attempted to predict the risk of coronary heart disease using the Framingham data. (The Framingham study was a longitudinal study of the incidence of coronary heart disease in Framingham, Massachusetts; see page 562ff for additional information.) The two groups under consideration were those who did and did not develop coronary heart disease (CHD) in a 12-year follow-up period. There were 2669 women and 2187 men, aged 30–62, involved in the study and free from CHD at their first examination. The variables considered were:

- Age (years)
- Serum cholesterol (mg/100 ml)
- Systolic blood pressure (mm Hg)
- Relative weight (100 × actual weight ÷ median for sex-height group)
- Hemoglobin (g/100 ml)
- Cigarettes per day, coded as

> 0 = never smoked
> 1 = less than a pack a day
> 2 = one pack a day
> 3 = more than a pack a day.

- ECG, coded as

> 0 = for normal,
> 1 = for definite or possible left ventricular hypertrophy, definite nonspecific abnormality and intraventricular block.

Note that the variables cigarettes and ECG cannot be normally distributed as they are discrete variables. Nevertheless, the linear discriminant function model was tried. It was found that the predictions (in terms of the risk or estimated probability of being in the coronary heart disease groups) fitted the data well. The coefficients of the linear discriminant functions for men and women including the standard errors are as follows:

Risk Factors	Women	Men	Standard Errors of Estimated Coefficients	
Constant $(\widehat{\alpha})$	−12.5933	−10.8986		
Age (years)	0.0765	0.0708	0.0133	0.0083
Cholesterol (mg%)	0.0061	0.0105	0.0021	0.0016
Systolic blood pressure (mmHg)	0.0221	0.0166	0.0043	0.0036
Relative weight	0.0053	0.0138	0.0054	0.0051
Hemoglobin (g%)	0.0355	−0.0837	0.0844	0.0542
Cigarettes smoked (see code)	0.0766	0.3610	0.1158	0.0587
ECG abnormality (0.1)	1.4338	1.0459	0.4342	0.2706

a. Determine for both women and men in terms of the p-value the most significant risk factor for CHD in terms of the p-value.

b. Calculate the probability of CHD for a male with the following characteristics: age = 35 years; cholesterol = 220 mg%; systolic blood pressure = 110 mmHg; relative weight = 110; hemoglobin = 130 g%; cigarette code = 3; and ECG code = 0.

c. Calculate the probability of CHD for a female with the above characteristics.

d. How much is the probability in Part (b) changed for a male with all the above characteristics except that he does not smoke (i.e., cigarette code = 0).

e. Calculate and plot the probability of CHD for the woman in Part (c) as a function of age.

7. In a paper which appeared 4 years later, Halperin *et al.* [1971] re-examined the Framingham data analysis (see Problem 13.6) by Truett *et al.* [1967] using a logistic model (for further comparison, see Section 13.4.4). Halperin *et al.* analyzed several subsets of the data; for this problem we abstract the data for men aged 29–39 years, and three variables, cholesterol, systolic blood pressure and cigarette smoking (0 = never smoked; 1 = smoker); cholesterol and systolic blood pressure are measured as in the previous problem. The following coefficients for the logistic and discriminant models (with standard errors in parentheses) were obtained:

	Intercept	Cholesterol (mg/100 ml)	Systolic Blood Pressure	Cigarettes
Logistic	−11.6246	0.0179(0.0036)	0.0277(0.0085)	1.7346(0.6236)
Discriminant	−13.5300	0.0236(0.0039)	0.0302(0.0100)	1.1191(0.3549)

a. Calculate the probability of CHD for a male with relevant characteristics defined in the previous problem Part (b) for both the logistic and discriminant models.

b. Interpret the regression coefficients of the logistic model.

c. The authors in comparing the two methods state, "empirically, the assessment of significance of a variable, as measured by the ratio of the estimated coefficient to its estimated standard error is apt to be about the same whichever method is used." Verify that this is so for this problem. (However, see also the discussion in Section 13.4.4).

8. Both patients and physicians make decisions on the basis of expected length of life. Doubilet and McNeil [1982] provide an example which deals with the treatment of gastric carcinoma without known metastasis. The basic decision is whether to operate (gastrectomy) or not in the face of a fairly high operative mortality. But the mortality depends upon age and sex. The success of surgery depends upon the degree of metastasis. For purposes of this exercise metastasis is divided into two categories: "distant" and "not distant." The decision tree can be laid out as follows:

Operation	Metastasis	Survive Surgery	Outcome	Expectancy
			Cured	16.0
		Yes		
	Not Distant		Not Cured	1.8
		No		0.0
Operate				
		Yes		0.6
	Distant			
		No		0.0
	Not Distant			1.8
Do Not Operate				
	Distant			0.6

For a 60-year-old male, the probability of survival of an operation is 91.2%, the probability of distant metastasis is 43%, the probability of a cure for "not distant" metastasis is 78%. It is assumed that no cure is possible if the disease has metastasized to distant parts of the body, or if no operation is carried out.

a. In each branch of the decision tree, calculate the probabilities and fill them in.

b. The possible life expectancies for a 60-year-old male are given in the column at the right in the figure. Calculate the life expectancy for a 60-year-old male in the case of no operation and in the case of an operation. Hence, calculate the expected gain (loss) in years associated with an operation.

c. All these calculations are based upon averages: the average 60-year-old male; the "average" operation, the average surgeon, etc. Can you think of a way of assessing the robustness of the model? For example, how robust is the expected gain in life (in years) as a result of the operation for a particular 60-year-old male?

d. What variable(s) other than life expectancy should enter into a decision whether to operate or not? How would you, or could you, quantify these variable(s)?

9. In a paper in the *American Statistician*, Hauck [1983] derived confidence bands for the logistic response curve. He illustrated the method with data from the Ontario

Exercise Heart Collaborative Study. The logistic model dealt with the risk of myocardial infarction (MI) during a study period of 4 years. A logistic model based on the two most important variables, smoking (X_1) and serum triglyceride level (X_2) was calculated to be

$$\text{logit}(P) = -2.2791 + 0.7682X_1 + 0.001952(X_2 - 100),$$

where P = probability of an MI during the four year observation period. The variable X_1 had values $X_1 = 0$ (nonsmoker) and $X_1 = 1$ (smoker). As in ordinary regression, the confidence band for the entire line is narrowest at the means of X_1 and $(X_2 - 100)$ and spreads out the further you go from the means. See the paper for more details.

a. The range of values of triglyceride levels is assumed to be from 0 to 550. Graph the probability of MI for smokers and nonsmokers separately.

b. The standard errors of regression coefficients for smoking and serun triglyceride are 0.3137 and 0.001608, respectively. Test their significance.

10. One of the earliest applications of the logistic model to medical screening by Anderson *et al.* [1972] involved the diagnosis of keratoconjunctivitis sicca (KCS) also known as "dry eyes." It is known that rheumatoid arthritic patients are at greater risk but the definitive diagnosis requires an ophthalmologist, hence it would be advantageous to be able to predict the presence of KCS on the basis of symptoms such as a burning sensation in the eye, etc. In this study 40 rheumatoid patients with KCS and 37 patients without KCS were assessed with respect to the presence (scored as 1) or absence (scored as 0) of each of the following symptoms: (1) foreign body sensation; (2) burning; (3) tiredness; (4) dry feeling; (5) redness; (6) difficulty in seeing; (7) itchiness; (8) aches; (9) soreness or pain; and (10) photosensitivity and excess of secretion. The data are reproduced in Table 13.8.

a. Fit a stepwise logistic model to the data. Test the significance of the coefficients.

b. On the basis of the proportions of positive symptoms displayed at the bottom of the table select that variable that should enter the regression model first.

c. Estimate the probability of misclassification.

d. It is known that approximately 12% of patients suffering from rheumatoid arthritis have KCS. On the basis of this information, calculate the appropriate logistic scoring function.

e. Define X = The number of symptoms reported (out of 10). Do a logistic regression using this variable. Test the significance of the regression coefficient. Now do a t-test on the X variable comparing the two groups. Discuss and compare your results.

11. Consider Equation 13.10. A plot of logit $P[G_1 \mid \pi_1]$ vs. logit $P[G_1 \mid \hat{\pi}_1]$ produces straight lines at 45° to the origin with intercepts at logit $\hat{\pi}_1 -$ logit π_1.

Table 13.8. Data for Problem 10

Patient	KCS Patients										Patient	Patients Without KCS									
	1	2	3	4	5	6	7	8	9	10		1	2	3	4	5	6	7	8	9	10
1	1	1	1		1		1			1	1	1						1			
2	1	1	1	1	1			1			2										
3	1	1		1	1	1			1		3		1	1				1			
4	1	1		1	1			1	1		4										
5	1	1	1	1			1			1	5								1		
6	1	1			1					1	6										
7	1	1	1			1		1			7			1				1			
8	1	1		1		1			1		8										
9	1	1	1	1	1		1	1			9										
10	1										10									1	
11	1	1	1	1		1		1			11		1					1			
12	1	1				1	1	1			12										
13	1	1	1	1	1	1	1			1	13							1			
14	1	1	1	1	1		1	1		1	14										
15			1	1			1	1			15							1			
16	1	1		1				1			16										
17		1	1		1		1		1		17							1			
18	1		1	1	1			1			18										
19	1		1	1	1		1				19				1						
20	1	1	1	1			1			1	20										
21				1							21										
22	1	1	1	1	1	1	1				22					1					
23	1	1	1		1					1	23										
24	1	1		1			1	1	1		24				1						
25	1	1	1	1				1			25	1								1	
26			1			1					26										
27	1	1		1	1			1	1	1	27										
28	1	1	1	1						1	28										
29	1		1		1			1			29						1				
30	1	1		1		1				1	30										
31		1	1	1						1	31			1							
32	1	1	1	1	1	1	1			1	32										
33		1	1	1		1		1			33							1			
34	1	1	1	1		1	1			1	34										
35	1		1		1			1			35										1
36	1	1	1	1				1			36										
37	1	1	1	1	1						37							1			1
38	1	1																			
39																					
40		1	1	1			1			1											
Proportion positive	$\frac{32}{40}$	$\frac{30}{40}$	$\frac{26}{40}$	$\frac{28}{40}$	$\frac{19}{40}$	$\frac{10}{40}$	$\frac{16}{40}$	$\frac{15}{40}$	$\frac{9}{40}$	$\frac{15}{40}$	Proportion positive	$\frac{2}{37}$	$\frac{2}{37}$	$\frac{2}{37}$	$\frac{1}{37}$	$\frac{2}{37}$	$\frac{1}{37}$	$\frac{10}{37}$	$\frac{1}{37}$	$\frac{2}{37}$	$\frac{2}{37}$

   a. Make a graph of this relationship for the values of

$$\frac{\widehat{\pi}_1(1 - \pi_1)}{(1 - \widehat{\pi}_1)(\pi_1)}$$

over a reasonable range.

   b. Rescale the graph of Part (a) into probabilities rather than logits. You now have a convenient graph for adjusted predicted probabilities on the basis of various values of priors and sample mixes.

   *c. Make a nomogram to illustrate the above relationships.

   *d. Derive the formulae relating posterior probabilities given different prior probabilities. In the context of Section 13.4.1, link $P[G_1 \mid \pi_1]$ and $P[G_1 \mid \widehat{\pi}_1]$.

12. This problem deals with the data in Section 13.2.1. Calculate the posterior probabilities of survival for a patient in the fourth decade arriving at the hospital in shock and history of myocardial infarction and without other risk factors:

   a. Using the logistic model.

   b. Using the discriminant model.

   c. Graph the two survival curves as a function of age. Use the values 5, 15, 25, ... for the ages in the discriminant model.

   d. Assume the prior probabilities are $\pi_1 = P[\text{Survival}] = 0.60$ and $\pi_2 = 1 - 0.60 = 0.40$. Recalculate the probabilities in Parts (a) and (b).

   e. Define a new variable for the data of Table 13.1 as follows: $X_6 = X_1 + X_2 + X_3 + X_5$. Interpret this variable.

   f. Do a logistic regression and discriminant analysis using variables X_4 and X_6 (defined above). Interpret your results.

   g. Is any information "lost" using the approach of Parts (e) and (f)? If so, what is lost? When is this likely to be important?

13. This problem deals with the data of Problem 15, Chapter 5, comparing the effect of the drug nifedipine on vasospasm attacks in patients suffering from Raynaud's phenomenon. We want to make a multivariate comparison of the seven patients with a history of digital ulcers ("Yes" in column 4) with the eight patients without a history of digital ulcers ("No" in column 4). Variables to be used are age, sex, duration of phenomenon, total number of attacks on placebo and total number of attacks on nifedipine.

   a. Carry out a stepwise logistic regression on these data.

   b. Which variable entered first?

   c. State your conclusion.

   d. Make a scatterplot of the logistic scores and indicate the dividing point.

*14. This problem deals with the data of Problem 10.10, comparing metabolic clearance rates in three groups of subjects.

   a. Use a discriminant analysis on the *three* groups.

   b. Interpret your results.

   c. Graph the data using different symbols to denote the three groups.

d. Suppose you "create" a third variable: concentration at 90 min minus concentration at 45 min. Will this improve the discrimination? Why, or why not?

*15. Using the notation and terminology of Section 13.3 assume two groups G_1 and G_2 (e.g., "death", "survive"; "disease", "no disease"), and a binary covariate, X with values 0, 1 (e.g., "don't smoke", "smoke"; "symptom absent", "symptom present"). The data can be arranged in a 2×2 table:

	Group	
X	G_1	G_2
1		
0		
	π_1	π_2

Here $\pi_1 = $ prior probability of group G_1 membership,

$P(X = i \mid G_1) = $ likelihood of $X = i$ given G_1 membership; $i = 0, 1$.

$P(G_1 \mid X = i) = $ posterior probability of G_1 membership given that $X = i$; $i = 0, 1$.

a. Show that

$$\frac{P(G_1 \mid X = i)}{P(G_2 \mid X = i)} = \frac{\pi_1}{\pi_2} \frac{P(X = i \mid G_1)}{P(X = i \mid G_2)}.$$

Hint: Use Bayes' theorem.

b. The expression in Part (a) can be written as

$$\frac{P(G_1 \mid X = i)}{1 - P(G_1 \mid X = i)} = \frac{\pi_1}{1 - \pi_1} \frac{P(X = i \mid G_1)}{P(X = i \mid G_2)}.$$

In words

$$\begin{array}{c} \text{Posterior odds} \\ \text{of Group 1} \\ \text{membership} \end{array} = \begin{array}{c} \text{Prior odds} \\ \text{of Group 1} \\ \text{membership} \end{array} \times \begin{array}{c} \text{Ratio of likelihoods} \\ \text{of observed values of } X. \end{array}$$

Relate the ratio of likelihoods to the sensitivity and specificity of the procedure.

c. Take logarithms of both sides of the equation in Part (b). Relate your result to Note 6.8 of Chapter 6.

d. The result in Part (b) can be shown to hold for X, continuous or multivariate. What are the assumptions (go back to the simple set-up of Part (a)).

REFERENCES

Anderson, J. A. [1972]. Separate sample logistic regression. *Biometrika*, **59:** 19–35.

Anderson, J. A., Whaley, K., Williamson, J., and Buchanan, W. W. [1972]. A statistical aid to the diagnosis of keratoconjunctivitis sicca. *Quarterly Journal of Medicine, New Series*, **41:** 175–189. Used with permission from Oxford University Press.

Anderson, T. W. [1988]. *An Introduction to Multivariate Statistical Methods*, 2nd edition. Wiley, New York.

Baker, R. J. and Nelder, J. A. [1978]. *Generalized Linear Interactive Modelling (GLIM)*, Release 3. Numerical Algorithms Group, Oxford.

Chernoff, H. and Moses, L. E. [1959]. *Elementary Decision Theory*. Wiley, New York.

Dixon, W. J. (Ed.) [1988]. *BMDP Statistical Software*, Los Angeles, CA.

Doubilet, P. and McNeil, B. J. [1982]. Treatment choice in gastric carcinoma. *Medical Decision Making*, **2:** 261–274.

Efron, B. [1975]. The efficiency of logistic regression compared to normal discriminant analysis. *Journal of the American Statistical Association*, **70:** 892–898.

Fisher, L.D., Kronmal, R., and Diehr, P. [1976]. Mathematical aids to medical decision-making. In *Operations Research in Health Care*, L. J. Shuman, R. D. Speas, Jr., and J. P. Young (eds.). The Johns Hopkins University Press, Baltimore, MD.

Fisher, L. D. and Kennedy, J. W. [1983]. Operative mortality in coronary bypass grafting. *Journal of Thoracic and Cardiovascular Surgery*, **85:** 146–147.

Fisher, R. A. [1936]. The use of multiple measurements in taxonomic problems. *Annals of Eugenics*, **7:** 179–188.

Gordon, T. [1974]. Hazards in the use of the logistic function with special reference to data from prospective cardiovascular studies. *Journal of Chronic Diseases*, **27:** 97–102.

Hall, D. L., Lodwick, G. S., Kruger, R. P., Dwyer, S. J., and Townes, J. R. [1971]. Direct computer diagnosis of rheumatic heart disease. *Radiology*, **101:** 497–509.

Halperin, M., Blackwelder, W. C., and Verter, J. I. [1971]. Estimation of the multivariate logistic risk function: a comparison of the discriminant function and maximum likelihood approaches. *Journal of Chronic Diseases*, **24:** 125–158.

Hosmer, D. W. and Lemeshow, S. [1989]. *Applied Logistic Regression*, Wiley, New York.

Hauck, W. W [1983]. A note on confidence bands for the logistic response curve. *American Statistician*, **37:** 158–160.

Jones, R. H. [1975]. Probability estimation using a multinomial logistic function. *Journal of Statistical Computation and Simulation*, **3:** 315–329.

Knoke, J. D. [1982]. Discriminant analysis with discrete and continuous variables. *Biometrics*, **38:** 191–200. See also correction in *Biometrics*, **38:** 1143.

Koepsel, T. D., Inui, T. S., and Farewell, V. T. [1981]. Factors affecting perforation in acute appendicitis. *Surgery, Gynecology and Obstetrics*, **153:** 508–510. Used with permission.

Lachenbruch, P. A. [1977]. *Discriminant Analysis*, Hafner Press, New York.

Mantel, N. [1966]. Models for complex contingency tables and polychotomous dosage response curves. *Biometrics*, **22:** 83–95.

Meisel, W. S. [1972]. *Computer-Oriented Approaches to Pattern Recognition*. Academic Press, New York.

Morrison, D. G. [1974]. In *Handbook on Marketing Research*, R. Ferber (ed.). McGraw-Hill, New York.

Patrick, E. A. [1972]. *Fundamentals of Pattern Recognition*. Prentice Hall, Englewood Cliffs, NJ.

Pine, R. W., Wertz, M. J., Lennard, E. S., Dellinger, E. P., Carrico, C.J., and Minshew, H. [1983]. Determinants of organ malfunction or death in patients with intra-abdominal sepsis. *Archives of Surgery*, **118:** 242–249. Copyright © 1983 by the American Medical Association.

Raiffa, H. and Schlaifer, R. [1961]. *Applied Statistical Decision Theory*. Harvard Business School, Harvard University, Boston, MA.

Sokal, R. R. and Sneath, P. H. A. [1963]. *Principles of Numerical Taxonomy*. W. H. Freeman, San Francisco, CA.

Tomkins, R. K., Burnes, D. C., and Cable, W. E. [1977]. An analysis of the cost effectiveness of pharyngitis management and acute rheumatic fever prevention. *Annals of Internal Medicine*, **86:** 481–492.

Truett, J., Cornfield, J., and Kannel, W. [1967]. A multivariate analysis of the risk of coronary heart disease in Framingham. *Journal of Chronic Diseases*, **20:** 511-524.

Warner, H. R., Toronto, A. F., and Veasey, L. G. [1964]. Experience with Bayes' theorem for computer diagnosis of congenital heart disease. *Annals of the New York Academy of Sciences*, **115:** 558–567.

Wilson, J. M. G. and Junger, G. [1968]. Principles and practice of screening for disease. *Public Health Papers, No. 34*, World Health Organization, Geneva.

CHAPTER 14

Principal Component Analysis and Factor Analysis

In Chapter 11, Multiple Regression Analysis, and in Chapter 10, The Analysis of Variance, we considered the dependence of a specified response variable upon other variables. The identified response variable played a special role among the variables being considered. This is appropriate in many situations because of the scientific question and/or experimental design. What do you do, however, if you have a variety of variables and desire to examine the relationships between them without identifying one specific response variable?

This chapter presents two methods of examining the relationships among a set of variables without identifying a specific response variable. For these methods no one variable has any more distinguished role or importance than any other variable. The first technique we will examine, principal component analysis, explains as much variability as possible in terms of a few linear combinations of the variables. The second technique, factor analysis, explains the relationships between variables by a few unobserved factors. Both methods depend upon the covariances, or the correlations, between the variables.

14.1 VARIABILITY IN A GIVEN DIRECTION

Consider the 20 observations on two variables X and Y listed in Table 14.1.

These data are such that the original observations had their means subtracted so that the means of the points are zero. Figure 14.1 plots these points, that is, plots the data points about their common mean.

Rather than thinking of the data points as X and Y values, think of the data points as a point in a plane. Consider Figure 14.2(a). When an origin is identified, each point in the plane is identified with a pair of numbers x and y. The x value is found by dropping a line perpendicular to the horizontal axis; while the y value is found by a line perpendicular to the vertical axis. These axes are shown in Figure 14.2(b). There is no necessary reason, however, to use the horizontal and vertical directions to locate our points, although this is traditional. Lines at any angle θ from the horizontal and vertical as shown in Figure 14.2(c) might be used. In terms of these two lines, the data point has values found by dropping perpendicular lines to these two directions;

Table 14.1. Twenty Biometric Observations.

Observation	X	Y	Observation	X	Y
1	−0.52	0.60	11	0.08	0.23
2	0.04	−0.51	12	−0.06	−0.59
3	1.29	−1.19	13	1.25	−1.25
4	−1.12	1.90	14	0.53	−0.45
5	−1.02	0.31	15	0.14	0.47
6	0.10	−1.15	16	0.48	−0.11
7	−0.32	−0.13	17	−0.61	1.04
8	0.08	−0.17	18	−0.47	0.34
9	0.49	0.18	19	0.41	0.29
10	−0.54	0.20	20	−0.22	0.00

Figure 14.2(d) shows the two values. We will call the new values, x' or y' and the old values x and y. It can be shown that x' and y' are linear combinations of x and y. This idea of lines in different directions with perpendiculars to describe the position of points will be used in principal component analysis.

For our data set, the variability in x and y may be summarized by the standard deviation of the x and y values, respectively, as well as the covariance, or equivalently the correlation between them. Consider now the data of Figure 14.1 and Table 14.1. Suppose that we draw a line in a direction of 30° to the horizontal. The 20 observations give 20 x' values in the X' direction when the perpendicular lines are dropped. Figure 14.3 shows the values in the x' direction. Consider now the points along the line in the x' direction corresponding to the feet of the perpendicular lines. We may summarize the variability among these points by our usual measure of variability, the

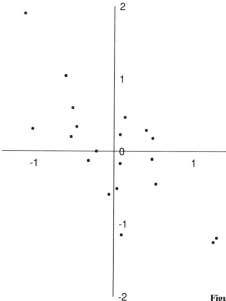

Figure 14.1. Plot of the twenty data points of Table 14.1.

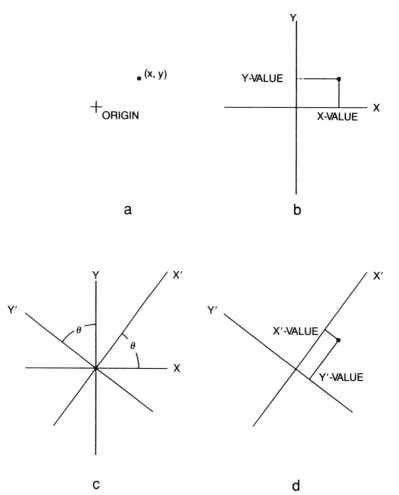

Figure 14.2. Points in the plane, coordinates, and rotation of axes.

standard deviation. This would be computed in our usual manner from the 20 values x'. The variability of the data may be summarized by plotting the standard deviation, say $s(\theta)$, in each direction θ at a distance s from the origin. When we look at the standard deviation in all directions, this results in an egg-shaped curve with dents in the side; or a symmetric curve in the shape of a violin or cello body. For the data at hand, this curve is shown in Figure 14.4; the curve is identified as the standard deviation curve. Note that the standard deviation is not the same in all directions. For our data set, the data are spread out more along a northwest-southeast direction than in the southwest-northeast direction. The standard deviation curve has a minimum distance at about $38°$. The standard deviation steadily increases to a maximum; the maximum is positioned along the line in Figure 14.4 running from the upper left to

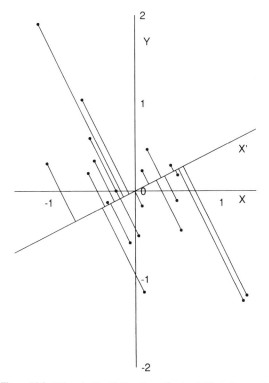

Figure 14.3. Values in the X-direction. X' axis at $30°$ to the x-axis.

the lower right. These two directions are labeled direction 1 and direction 2. If we want to pick one direction which contains as much variability as possible, we would choose direction 1 because the standard deviation is largest in that direction. If all the data points lie on a line then the variability will be a maximum in the direction of the line that contains all the data.

There is some terminology used in finding the value of a data point in a particular direction. The process of dropping a perpendicular line to direction is called *projecting* the point onto the direction. The value in the particular direction (x' in Figure 14.2(d) or Figure 14.3) is called the *projection of the point*. If we know the values x and y, or if we know the values x' and y', we know where the point is in the plane. Two such variables x and y, or equivalently x' and y', which allow us to find the values of the data are called a *basis for the variables*.

These concepts may be generalized when there are more than two variables involved. If we observe three variables x, y, and z, the points may be thought of as points in three dimensions. Suppose that we subtract the means from all the data so the data are centered about the origin of a three-dimensional plot. As you sit reading this material, picture the points suspended about the room. Pick an origin. You may draw a line through the origin in any direction. For any point that you have picked in the room you may "drop" a perpendicular to the line. Given a line, the point on the line where the perpendicular meets the line is the projection of the point onto the

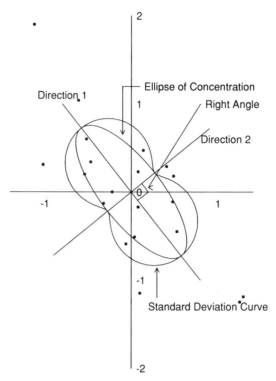

Figure 14.4. The standard deviation in each direction and the ellipse of concentration.

line. We may then calculate the standard deviation for this direction. If the standard deviations are plotted in all directions, a dented egg-shaped surface results. There will be one direction with the greatest variability. When more than three variables are observed, although we cannot picture the situation mentally, mathematically the ideas may be extended; the concept of a direction may be extended in a natural manner. In fact, mathematical statistics is one part of mathematics which heavily uses the geometry of n-dimensional space when there are n variables observed. Fortunately, to understand the statistical methods we do not need to understand the mathematics!

Let us turn our attention again to Figure 14.4. Rather than plotting the standard deviation curve, it is traditional to summarize the variability in the data by an ellipse. The two perpendicular axes of the ellipse lie along the directions of the greatest variability and the least variability. The ellipse, called the *ellipsoid of concentration*, meets the standard deviation curve along its axes at the points of greatest and the least variation. In other directions the standard deviation curve will be larger, that is, further removed from the origin. In three dimensions, rather than plotting an ellipse we plot an egg-shaped surface, the ellipsoid. (One reason the ellipsoid is used: if you have a bivariate normal distribution in the plane; take a very large sample; divide the plane up into small squares, as on graph paper; and place columns whose height is proportional to the number of points, the columns of constant height would lie on an ellipsoid.)

Out of the technical discussion above we want to remember the following ideas:

1. If we observe a set of variables we may think of each data point as a point in a space. In this space, when the points are centered about their mean, there is variability in each direction.
2. The variability is a maximum in one direction. In two dimensions (or more) the minimum lies in a perpendicular direction.
3. The variability is symmetric about each of these particular identified directions.

It is possible to identify the different directions with linear combinations of the variables or coordinates.

Each direction for $X_1, \ldots, X_p$ is associated with a sum

$$Y = a_1 X_1 + a_2 X_2 + \cdots + a_p X_p, \tag{14.1}$$

where

$$a_1^2 + a_2^2 + \cdots + a_p^2 = 1.$$

The constants $a_1, a_2, \ldots, a_p$ are uniquely associated with the direction except that we may multiply at the same time each a by -1. The sum that is given in Equation (14.1) is the value of the projection of the points x_1 to x_p corresponding to the given direction.

14.2 THE PRINCIPAL COMPONENTS

The motivation behind principal component analysis is to find a direction, or a few directions, that explains as much of the variability as possible. Since each direction is associated with a linear sum of the variables, we may say that we want to find a few new variables, which are linear sums of the old variables, which explain as much of the variability as possible. Thus, the first principal component is the linear sum corresponding to the direction of greatest variability:

Definition 14.1. The *first principal component* is the sum

$$Y = a_1 X_1 + \cdots + a_p X_p, \quad a_1^2 + \cdots + a_p^2 = 1, \tag{14.2}$$

corresponding to the direction of greatest variability when variables $X_1, \ldots, X_p$ are under consideration.

Usually, the first principal component will leave much of the variability unexplained. (In the next section, we discuss a method of quantifying the amount of variability explained.) For this reason, we wish to search for a second principal component that explains much of the remaining variability. You might think we would take the next linear combination of variables that explains as much of the variability as possible. But if you examine the picture in Figure 14.4, you see that the closer the direction gets to the first principal component, which would be direction 1 in Figure 14.4, the more variability one would have. Thus, essentially we would be driven to the same variable. Therefore, the search for the second principal component is restricted to variables which are uncorrelated with the first principal component. Geometrically it can be shown that this is equivalent to considering directions which are perpendicular to the direction of the first principal component. In two dimensions such as Figure 14.4, direction 2 would be the direction of the second principal component. However, in three dimensions when we have the line corresponding to the direction of the first principal component, the set of all directions perpendicular to it correspond to a plane and there are a variety of possible directions to search for the second principal component. This leads to the following definition:

Definition 14.2. Suppose that we have the first $k - 1$ principal components for variables $X_1, \ldots, X_p$. The *k*th *principal component* corresponds to the variable or direction which is uncorrelated with the first $k - 1$ principal components and has the largest possible variance.

As a summary of these difficult ideas you should remember the following:

1. Each principal component is chosen to explain as much of the remaining variability as possible after the previous principal components have been chosen.
2. Each principal component is uncorrelated with the other principal components. In the case of a multivariate normal distribution, the principal components are statistically independent.
3. Although it is not clear from the above, the following is true: for each k, the first k principal components explain as much of the variability in a sample as may be explained by any k directions, or equivalently k variables.

14.3 THE AMOUNT OF VARIABILITY EXPLAINED BY THE PRINCIPAL COMPONENTS

Suppose that we want to perform a principal component analysis upon variables $X_1, \ldots, X_p$. If we were dealing with only one variable, say variable X_j, we summarize its variability by the variance. Suppose there are a total of n observations so that for each of the p variables we have n values. Let X_{ij} be the ith observation on the jth variable. Let $\overline{X}_j$ be the mean of the n observations on the jth variable. Then we estimate the variability, that is, the variance, of the variable X_j by

$$\widehat{\text{var}}(X_j) = \sum_{i=1}^{n} \frac{(X_{ij} - \overline{X}_j)^2}{n - 1}. \tag{14.3}$$

A reasonable summary of the variability in the p variables is the sum of the individual variances. This leads us to the next definition.

Definition 14.3. The *total variance, denoted by* V, for variables $X_1, \ldots, X_p$ is the sum of the individual variances. That is,

$$\text{Total variance} = V = \sum_{j=1}^{p} \text{var}(X_j). \tag{14.4}$$

The sample total variance which we will also denote by V since that is the only type of total variance used in this section, is:

$$\text{Sample total variance} = V = \sum_{j=1}^{p} \sum_{i=1}^{n} \frac{(X_{ij} - \overline{X}_j)^2}{n-1}.$$

We now characterize the amount of variability explained by the principal components. Recall that the principal components are themselves variables; they are linear combinations of the X_j variables. Each principal component has a variance itself. It is natural therefore, to compare the variance of the principal components with the variance of the X_j's. This leads us to the following definitions.

Definition 14.4. Let $Y_1, Y_2, \ldots$ be the first, second, and subsequent principal components for the variables $X_1, \ldots, X_p$. In a sample the variance of each Y_k is estimated by

$$\text{var}(Y_k) = \sum_{i=1}^{n} \frac{(Y_{ik} - \overline{Y}_k)^2}{(n-1)} = V_k, \tag{14.5}$$

where Y_{ik} is the value of the kth principal component for the ith observation. That is, we first estimate the coefficients for the kth principal component. The value for the ith observation uses those coefficients and the observed values of the X_j's to compute the value of Y_{ik}. The variance for the kth principal component in the sample is then given by the sample variance for Y_{ik}, $i = 1, 2, \ldots, n$. We denote this variance as seen above by V_k. Using this notation we have the following two definitions.

1. *The percent of variability explained by the kth principal component is*

$$\frac{100 V_k}{V}.$$

2. *The percent of the variability explained by the first m principal components is*

$$100 \sum_{k=1}^{m} \frac{V_k}{V}. \tag{14.6}$$

The following facts about the principal components can be stated:

1. There are exactly p principal components where p is the number of X variables considered. This is because with p uncorrelated variables, there is a one-to-one correspondence between the values of the principal components and the values of the original data; that is, we can go back and forth so that all of the variability is accounted for; the percent of variability explained by the p principal components is 100%.
2. Because we chose the principal components to successively explain more and more of the variance, we have

$$V_1 \geq V_2 \geq \cdots \geq V_p \geq 0.$$

3. The first m principal components explain as much of the total variability as it is possible to explain by m linear functions of the X_j variables.

We now proceed to a geometric interpretation of the principal components. Consider the case where $p = 2$. That is, we observe two variables X_1 and X_2. Plot, as previously in this chapter, the ith data point in the coordinate system which is centered about the means for the X_1 and X_2 variables. Draw a line in the direction of the first principal component and project the data point onto the line. This is done in Figure 14.5.

The square of the distance of the data point from the new origin, which is the sample mean, is given by the following equation, using the Pythagorean theorem:

$$d_i^2 = (X_{i1} - \overline{X}_1)^2 + (X_{i2} - \overline{X}_2)^2 = \sum_{j=1}^{2}(X_{ij} - \overline{X}_j)^2.$$

The square of the distance f_i of the projection turns out to be the difference between the value of the first principal component for the ith observation and the mean of the first principal component squared. That is,

$$f_i^2 = (Y_{i1} - \overline{Y}_1)^2.$$

It is geometrically clear that the distance d_i is larger than f_i. The ith data point will be better represented by its position along the line if it lies closer to the line. That is, if f_i is close to d_i. One way we might judge the adequacy of the variability explained by the first principal component would be to take the ratio of the sum of the lengths

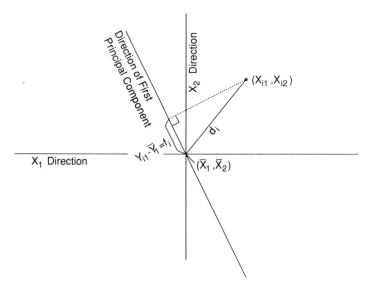

Figure 14.5. Projection of a data point onto the first principal component direction.

of the f_i's squared to the sum of the lengths of the d_i's squared. If we do this, we have the following equation:

$$\frac{\sum_{i=1}^{n} f_i^2}{\sum_{i=1}^{n} d_i^2} = \frac{\sum_{i=1}^{n} (Y_{i1} - \overline{Y}_1)^2}{\sum_{i=1}^{n} \sum_{j=1}^{2} (X_{ij} - \overline{X}_j)^2} = \frac{V_1}{V}. \tag{14.7}$$

That is, we have the proportion of the variability explained. If we multiplied the equation throughout by 100, we would have the percent of the variability explained by the first principal component. This gives us an alternative way of characterizing the first principal component. The direction of the first principal component is the line for which the following holds: when the data are projected onto this line, the sum of the squares of the projections is as large as possible; equivalently the sum of squares is as close as possible to the sum of squares of the lengths of the lines to the original data points from the origin (which is also the mean). From this we see that the percent of variability explained by the first principal component will be 100 if and only if the lengths d_i and f_i are all the same; that is, the first principal component will explain all the variability if and only if all of the data points lie on a single line. The closer all the data points come to lie on a single line, the larger the percent of variability explained by the first principal component.

We now proceed to examine the geometric interpretation in three dimensions. In this case we consider a data point plotted not in terms of the original axes X_1, X_2 and X_3 but rather in terms of the coordinate system given by the principal components Y_1, Y_2 and Y_3. Figure 14.6 presents such a plot for a particular data point. The figure

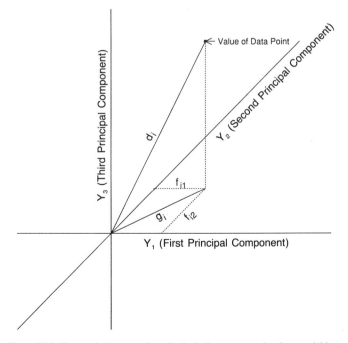

Figure 14.6. Geometric interpretation of principal components for three variables.

is a two-dimensional representation of a three-dimensional situation; two of the axes are vertical and horizontal on the paper. The third axis recedes into the plane formed by the page in this book. Consider the ith data point which lies at a distance d_i from the origin which is at the mean of the data points. This point also turns out to be the mean of the principal component values. Suppose we now drop a line down into the plane which contains the axes corresponding to the first two principal components. This is indicated by the vertical dotted line in the figure. This point in the plane we could now project onto the value for the first principal component and the second principal component. These values with lengths f_{i1} and f_{i2}, are the same as we would get by directly dropping perpendiculars from the point to those two axes. Again, we might assess the adequacy of the characterization of the data point by the first two principal components by comparing the length of its projection in the plane, g_i, with the length of the line from the origin to the original data point, d_i. If we compare the squares of these two lengths, each summed over all of the data points and use the Pythagorean theorem again, the following results hold:

$$\frac{\sum_{i=1}^{n} g_i^2}{\sum_{i=1}^{n} d_i^2} = \frac{\sum_{i=1}^{n} f_{i1}^2 + \sum_{i=1}^{n} f_{i2}^2}{\sum_{i=1}^{n} d_i^2}$$

$$= \frac{\sum_{i=1}^{n} \frac{(Y_{i1} - \overline{Y}_1)^2}{n-1} + \sum_{i=1}^{n} \frac{(Y_{i1} - \overline{Y}_2)^2}{n-1}}{\sum_{i=1}^{n} \frac{d_i^2}{n-1}}$$

$$= \frac{V_1 + V_2}{V}.$$

Using this equation, we see that the percent of the variability explained by the first two principal components is the ratio of the squared lengths of the projections onto the plane of the first two principal components divided by the squared lengths of the original data points about their mean. This also gives us a geometric interpretation of the total variance. It is the sum for all the data points of the squares of the distance between the point corresponding to the mean of the sample and the original data points. In other words, the first two principal components may be characterized as giving a plane for which the projected points onto the plane contain as high a proportion as possible of the squared lengths associated with the original data points. From this we see that the percent of variability explained by the first two principal components will be 100 if and only if all of the data points lie in some plane through the origin, which is the mean of the data.

The coefficients associated with the principal components are usually calculated by computer; in general there is no easy formula to obtain them. Thus the examples in this chapter will begin with the coefficients for the principal components and their variance. (There is an explicit solution when there are only two variables and this is given in Problem 14.9.)

Example 14.1. We turn to the data of Table 14.1. Equations for the principal components are:

$$Y_1 = -0.6245X + 0.7809Y,$$

$$Y_2 = 0.7809X + 0.6245Y.$$

For the first data point $(X, Y) = (-0.52, 0.60)$ the values are

$$Y_1 = -0.6245 \times (-0.52) + 0.7809 \times 0.60 = 0.79,$$

$$Y_2 = 0.7809 \times (-0.52) + 0.6245 \times 0.60 = -0.03.$$

If we compute all of the numbers, we find the values for each of the 20 data points on the principal components are as given in the following table.

Data		Principal Component Values	
X	Y	Y_1	Y_2
−0.52	0.60	0.79	−0.03
0.04	−0.51	−0.42	−0.28
1.29	−1.19	−1.74	0.26
−1.12	1.90	2.19	0.31
−1.02	0.31	0.88	−0.60
0.10	−1.15	−0.96	−0.64
−0.32	−0.13	0.10	−0.33
0.08	−0.17	−0.18	−0.04
0.49	0.18	−0.16	0.50
0.54	0.20	0.49	−0.29
0.08	0.23	0.13	0.21
−0.06	−0.59	−0.42	−0.42
1.25	−1.25	−1.76	0.19
0.53	−0.45	−0.68	0.13
0.14	0.47	0.28	0.40
0.48	−0.11	−0.39	0.31
−0.61	1.04	1.20	0.17
−0.47	0.34	0.56	−0.16
0.41	0.29	−0.02	0.50
−0.22	−0.00	0.13	−0.18

From these data we may compute the sample variance of Y_1 and Y_2 as well as the variance of X and Y. We find the following values:

$$V_1 = 0.861, \quad V_2 = 0.123, \quad \text{var}(X) = 0.411, \quad \text{var}(Y) = 0.573.$$

From these data we may compute the percent of variability explained by the two principal components, individually and together.

1. Percent of variability explained by the first principal component $= 100 \times 0.861/(0.411 + 0.573) = 87.5\%$.
2. Percent of variability explained by the second principal component $= 100 \times 0.123/(0.411 + 0.573) = 12.5\%$.
3. Percent of variability explained by the first two principal components $= 100 \times (0.861 + 0.123)/(0.411 + 0.573) = 100\%$.

We see that the first principal component of the data in Figure 14.4 contains a high proportion of the variability. This may also be seen visually by examining the plot orienting your eyes so that the horizontal line is the direction of the first principal component. Certainly, there is much more variability in that direction than in direction 2, the direction of the second principal component.   □

14.4   USE OF THE COVARIANCE, OR CORRELATION, VALUES, AND PRINCIPAL COMPONENT ANALYSIS

The coefficients of the principal components and their variances can be computed by knowing the covariances between the X_j's. One might think that as a general search for relationships among X_j's the principal component will be appropriate as an exploratory tool. Sometimes this is true. However, consider what happens when we have different scales of measurement. Suppose for example, among our units, one unit is the height in inches while another unit is the systolic blood pressure of individuals in mmHg. In principal component analysis we are adding the variability in the two variables. Suppose now that we change our measurement of height from inches to feet. Then the standard deviation of the height variable will be divided by 12 and the variance will be divided by 144. In the total variance the contribution of height will have dropped greatly. Equivalently, the blood pressure contribution (and any other variables) will become much more important. Recomputing the principal components will produce a different answer. In other words, the measurement units are important in finding the principal component because the variance of any individual variable is directly compared to the variance of another variable without regard to whether or not the units are appropriate for the comparison. We reiterate: *The importance of a variable in principal component analysis changes with a change of scale of one or more of the variables*. For this reason principal component analysis is most appropriate, and probably has its best applications, when all the variables are measured in the same units; for example, the X_j variables may may be measurements of length; in inches with the variables being measurements of different parts of the body, and the covariances between variables such as arm length, leg length and body length.

In some situations with differing units, one still wants to try principal component analyses. In this case, standardized variables are often used, that is, we divide each variable by its standard deviation. Each rescaled variable then has a variance of one and the covariance matrix of the new standardized variables is the correlation matrix of the original variables. The interpretation of the principal components is now less clear. If many of the variables are highly correlated, the first principal component will tend to pick up this fact; for example, with two variables a high correlation means the variables lie along a line. The ellipse of concentration has one axis along the line; that direction gives us the direction of the first principal component. When standardized variables are used, since each variable has a variance of one the sum of the variances is p. In looking at the percent of variability explained there is no need to separately compute the total variance; it is p, the number of variables. We emphasize that when the correlations are used there should be some reason for doing this beside the fact that the variables do not have measurements in comparable units.

14.5 STATISTICAL RESULTS FOR PRINCIPAL COMPONENT ANALYSIS

Suppose that we have a sample of size n from a multivariate normal distribution with unknown covariances. Let $V_i(\text{pop})$ be the true (unknown) population value for the variance of the ith principal component when computed from the (unknown) true variances; let V_i be the variance of the principal components computed from the sample covariances. Then the following are true:

1.

$$\frac{V_i - V_i(\text{pop})}{V_i(\text{pop})\sqrt{2/(n-1)}}, \quad i = 1, \ldots, p, \tag{14.8}$$

for large n is approximately a standard normal, $N(0,1)$, random variable. These variables are approximately statistically independent.

2. $100(1 - \alpha)\%$ confidence intervals for $V_i(\text{pop})$ for large n, are given by

$$\left(\frac{V_i}{1 + z_{1-\alpha/2}\sqrt{2/(n-1)}}, \frac{V_i}{1 - z_{1-\alpha/2}\sqrt{2/(n-1)}} \right) \tag{14.9}$$

where $z_{1-\alpha/2}$ is the $1 - \alpha/2$ percentile value of the $N(0,1)$ distribution.

Further statistical results on principal component analysis are given in Morrison [1976] and in Timm [1975].

Principal component analysis is a least squares technique, as were analysis of variance and multiple linear regression. Outliers in the data can have a large effect on the results (as in other cases where least squares techniques are used).

14.6 PRESENTING THE RESULTS OF A PRINCIPAL COMPONENT ANALYSIS

We have seen that principal component analysis is designed to explain the variability in data. Thus, any presentation should include:

1. The variance of the principal components,
2. The percent of the total variance explained by each individual principal component, and
3. The percent of the total variance explained cumulatively by the first m terms (for each m).

It is also useful to know how closely each variable X_j is related to the values of the principal components Y_i; this is usually done by presenting the correlations between each variable and each of the principal components. Let

$$Y_i = a_{i1}X_1 + \cdots + a_{ip}X_p.$$

The correlation between one of the original variables X_j and the ith principal component Y_i is given by the following equation:

$$r_{ji} = \text{correlation of } X_j \text{ and } Y_i = \frac{a_{ij}\sqrt{V_i}}{s_j}. \tag{14.10}$$

Table 14.2. Summary of a Principal Component Analysis Using the Covariances.

Variables	Correlation of the Principal Components and the X_j's				Standard Deviations of the X_j
	1	2	$\cdots$	p	
X_1	$\dfrac{a_{11}\sqrt{V_1}}{s_1}$	$\cdots$	$\cdots$	$\dfrac{a_{p1}\sqrt{V_p}}{s_1}$	s_1
$\vdots$	$\vdots$	$\vdots$	$\vdots$	$\vdots$	$\vdots$
X_p	$\dfrac{a_{1p}\sqrt{V_1}}{s_p}$	$\cdots$	$\cdots$	$\dfrac{a_{pp}\sqrt{V_p}}{s_p}$	s_p
Variance of principal component	V_1	V_2	$\cdots$	V_p	
% of total variance	$\dfrac{100V_1}{V}$	$\cdots$	$\cdots$	$\dfrac{100V_p}{V}$	
Cumulative % of total variance	$\dfrac{100V_1}{V}$	$\dfrac{100(V_1 + V_2)}{V}$	$\cdots$	1	

In this equation V_i is the variance of the ith principal component while s_j is the standard deviation of X_j. These results are summarized in Table 14.2.

By examining the variables that are highly correlated with a principal component we can see which variables contribute most to the principal component. Alternatively glancing across the rows for each variable X_j we may see which principal component has the highest correlation with the variable. An X_i which has the highest correlations with the first few principal components is contributing more to the overall variability than variables with small correlations with the first few principal components. In Section 14.8, several examples of principal component analysis are given, including an example of the use of such a summary table (Table 14.4).

14.7 USES AND INTERPRETATION OF PRINCIPAL COMPONENT ANALYSIS

Principal component analysis is a technique for explaining variability. Following are some of the uses of principal components.

1. Principal component analysis is a search for linear relationships for explaining variability in a multivariate sample. The first few principal components are important because they may summarize a large proportion of the variability. However, the understanding of which variables contribute to the variability is important only if most of the variance comes about because of important relationships among the variables. After all, we can increase the variance of a variable, say X_1, by increasing the error of measurement. If we have a phenomenally large error of measurement, the variance of X_1 will be much larger than the variances of the rest of the variables. In this case, the first principal component will be approximately equal to X_1 and the amount of variability explained will be close to 1. However, such knowledge is not particularly useful,

since the variability in X_1 does not make X_1 the most important variable, but in this case, reflects a very poorly measured quantity. Thus, to decide that the first few principal components are important summary variables, you must feel that the relationships among them come from linear relationships which may shed some light on the data being studied.

2. We may take the first two principal components and plot the values for the first two principal components of the data points. We know that among all possible plots in only two dimensions this one gives the best fit in a precise mathematical sense.

3. In some situations we have many measurements of somewhat related variables. For example, we might have a large number of size measurements on different portions of the human body. It may be that we want to perform a statistical inference but the large number of variables for the relatively small number of cases involved makes such statistical analysis inappropriate. We may "summarize" the data by using the values on the first few principal components. *If the variability is important* (!), we have then reduced the number of variables without getting involved in multiple comparison problems. We may proceed to statistical analysis. For example, suppose we are trying to perform a discriminant analysis and want to use size as one of the discriminating variables. However, for each of a relatively small number of cases we may have many anthropometric measurements. We might take the first principal component as a variable to summarize all of the size relationships. One of the examples of principal component analysis below gives a principal component analysis of physical size data.

14.8 PRINCIPAL COMPONENT ANALYSIS EXAMPLES

Example 14.2. Stoudt *et al.* [1970] consider measurements taken on a sample of adult females from the United States. The correlations among these measurements (as well as weight and age) are in Chapter 11 before Problem 11.14.

The variance explained for each principal component is presented in Table 14.3.

These data are very highly structured. Only three (of nineteen) principal components explain over 70% of the variance. Table 14.4 summarizes the first three principal components.

The first component, in the direction of greatest variation, is associated heavily with the weight variables. The highest correlation is with weight, 0.957. Other variables associated with size—such as chest and waist measurements, arm girth and skinfolds—also are highly correlated with the first principal component.

The second component is most closely associated with physical length measurements. Height is the most highly correlated variable. Other variables with correlations above 0.7 are the sitting heights (normal and erect), knee height and popliteal height.

Since we are working with a correlation matrix, the total variance is 19, the number of variables. The average "variance," in fact the exact variance, per variable is one. Only these first three principal components have variance greater than one. The other 16 directions correspond to a variance less than one. □

Example 14.3. Reeck and Fisher [1973] performed a statistical analysis of the amino acid composition of protein. The mole percent of the 18 amino acids in a

Table 14.3. Data for Example 14.2.

Principal Component	Variance Explained	Percent of Total Variance	Cumulative Percent of Total Variance
1	7.82	41.1	41.1
2	4.46	23.5	64.6
3	1.91	10.1	74.7
4	0.88	4.6	79.4
5	0.76	4.0	83.3
6	0.56	2.9	86.3
7	0.45	2.4	88.6
8	0.38	2.0	90.7
9	0.35	1.9	92.5
10	0.31	1.6	94.1
11	0.19	1.0	95.1
12	0.18	0.9	96.1
13	0.16	0.8	96.9
14	0.14	0.7	97.7
15	0.13	0.7	98.3
16	0.10	0.5	98.9
17	0.10	0.5	99.4
18	0.06	0.3	99.7
19	0.05	0.3	100.0

sample of 207 proteins was examined. The covariances and correlations are given in Table 14.5. The diagonal entries and numbers above them give the variances and covariances; the lower numbers are the correlations.

The mnemonics are:

Asp	Aspartic acid
Thr	Threonine
Ser	Serine
Glu	Glutamic acid
Pro	Proline
Gly	Glycine
Ala	Alanine
Cys/2	Half-cystine
Val	Valine
Met	Methionine
Ile	Isoleucine
Leu	Leucine
Tyr	Tyrosine
Phe	Phenylalanine
Trp	Tryptophan
Lys	Lysine
His	Histidine
Arg	Arginine

Table 14.4. Example 14.2; First Three Principal Components.

	Correlation of the Principal Components and the Variables		
Variables	1	2	3
SITHTER	0.252	0.772	0.485
SITHTHL	0.235	0.748	0.470
KNEEHT	0.385	0.722	−0.392
POPHT	0.005	0.759	−0.444
ELBOWHT	0.276	0.243	0.783
THIGHHT	0.737	−0.007	0.204
BUTTKN	0.677	0.476	−0.348
BUTTPOP	0.559	0.411	−0.444
ELBOWBR	0.864	−0.325	−0.033
SEATBR	0.832	−0.050	0.096
BIACROM	0.504	0.350	−0.053
CHEST	0.890	−0.228	−0.018
WAIST	0.839	−0.343	−0.106
ARMGTH	0.893	−0.267	0.068
ARMSKIN	0.733	−0.231	0.124
INFRASCA	0.778	−0.371	0.056
HT	0.251	0.923	−0.051
WT	0.957	−0.057	0.001
AGE	0.222	−0.488	−0.289
Variance of principal components	7.82	4.46	1.91
% of total variance	41.1	23.5	10.1
Cumulative % of total variance	41.1	64.6	74.7

The principal component analysis applied to the data gave the following table.

k	1	2	3	4	5	6
C	0.13	0.26	0.37	0.46	0.55	0.61

k	7	8	9	10	11	12
C	0.66	0.70	0.75	0.79	0.83	0.86

k	13	14	15	16	17	18
C	0.90	0.93	0.95	0.98	1.00	1.00

k is the dimension of the subspace used to represent the data. C is the proportion of the total variance accounted for in the best k-dimensional representation.

In contrast to Example 14.2, eight principal components are needed to account for 70% of the variance. In this example there are no simple linear relationships (or directions) that account for most of the variability. In this case the principal component correlations are not presented as the results are not very useful. □

Table 14.5. Example 14.3, Reeck and Fisher [1973] Covariance/Correlation Matrix. Diagonal and Upper Entries are Variances and Covariances. Below the Diagonal are the Correlations.

	Asp	Thr	Ser	Glu	Pro	Gly	Ala	Cys/2	Val	Met	Ile	Leu	Tyr	Phe	Trp	Lys	His	Arg
Asp	6.5649	0.2449	0.7879	-1.5329	-1.9141	-1.8328	-1.7003	-0.4974	-0.1374	0.0810	0.6332	-1.0855	0.6413	0.1879	0.3873	0.7336	0.0041	-1.5633
Thr	0.0517	3.4209	1.3998	-1.3341	-0.3531	-0.7752	-0.6428	0.4468	0.3603	-0.3502	0.1620	-1.2836	0.1804	-0.0978	0.1114	-0.3348	-0.2594	-0.8938
Ser	0.1219	0.2999	6.3687	-1.6465	0.1876	-0.8922	-1.3593	-0.3123	0.6659	-0.6488	-0.3738	-1.1125	0.4403	0.0432	0.2552	-1.6972	-0.3025	-1.4289
Glu	-0.1789	-0.2157	-0.1951	11.1880	-0.5866	-2.1665	-0.7732	-0.1443	-1.5346	0.0002	-0.3804	1.6210	-1.1824	-0.6684	-0.6778	0.0192	-0.3154	0.1169
Pro	-0.3566	-0.0911	-0.0355	-0.0837	4.3891	1.4958	-0.4259	1.0159	-0.7017	-0.4171	-0.8453	-0.9980	-0.0868	-0.1187	0.1163	-0.7021	-0.1612	0.4801
Gly	-0.2324	-0.1362	-0.1149	-0.2105	0.2320	9.4723	1.2857	0.1737	-0.3883	-0.4226	-0.2812	-2.3936	-0.8971	-0.7784	-0.2637	-1.0861	-0.2526	-0.0037
Ala	-0.2417	-0.1266	-0.1962	-0.0842	-0.0741	0.1522	7.5371	-2.1250	0.8498	0.1810	-0.4183	1.2480	-1.3374	-0.4320	-0.5219	-1.1641	-0.2730	0.0701
Cys/2	-0.0717	0.0892	-0.0457	-0.0159	0.1790	0.0208	-0.2857	7.3393	-1.3667	-0.4788	-1.3959	-2.3443	0.5408	-0.6282	0.1136	0.2727	-0.7482	0.1447
Val	-0.0275	0.1001	0.1356	-0.2357	-0.1721	-0.0648	0.1590	-0.2592	3.7885	-0.0632	0.5700	0.2767	-0.1348	-0.2303	-0.2792	-0.7921	-0.0632	-0.8223
Met	0.0294	-0.1759	-0.2388	0.0001	-0.1849	-0.1275	0.0612	-0.1642	-0.0302	1.1589	0.2493	0.2438	-0.1397	0.2060	-0.0159	0.1715	0.1457	0.0945
Ile	0.1426	0.0505	-0.0855	-0.0656	-0.2328	-0.0527	-0.0879	-0.2974	0.1690	0.1337	3.0023	-0.1857	-0.2785	-0.0870	-0.1296	0.2361	-0.0829	-0.3956
Leu	-0.1701	-0.2786	-0.1770	0.1946	-0.1912	-0.3122	0.1825	-0.3474	0.0571	0.0928	-0.0430	6.2047	-1.0362	0.2515	-0.2332	-0.6337	0.3951	1.0593
Tyr	0.1605	0.0625	0.1119	-0.2267	-0.0266	-0.1869	-0.3123	0.1280	-0.0444	-0.0832	-0.1031	-0.2667	1.0362	0.1823	0.9201	-0.5061	0.0855	0.1436
Phe	0.0525	-0.0379	0.0123	-0.1431	-0.0406	-0.1811	-0.1126	-0.1660	-0.0847	0.1370	-0.0360	0.0723	0.1823	1.9512	0.2223	0.1659	0.3434	0.1796
Trp	0.1576	0.0628	0.1054	-0.2113	0.0579	-0.0893	-0.1982	0.0437	-0.1495	-0.0154	-0.0780	-0.0976	0.2262	0.1659	0.9201	-0.5061	0.0855	0.1436
Lys	0.1061	-0.0670	-0.2491	0.0021	-0.1241	-0.1307	-0.1571	0.0373	-0.1507	0.0590	0.0505	-0.0942	0.1823	0.1659	-0.5061	7.2884	-0.1830	-1.0898
His	0.0014	-0.1194	-0.1020	-0.0803	-0.0655	-0.0699	-0.0847	-0.2351	-0.0276	0.1152	-0.0408	0.1350	0.0733	0.0646	0.0855	-0.2030	3.9550	0.0976
Arg	-0.3068	-0.2430	-0.2847	0.0176	0.1152	-0.0006	0.0128	0.0269	-0.2124	0.0441	-0.1148	0.2138	-0.0882	0.0646	0.0753	-0.2030	0.0976	3.9550

14.9 FACTOR ANALYSIS

As in principal component analysis, factor analysis looks at the relationships among variables as expressed by their correlations or covariances. While principal component analysis is designed to model and explain as much of the variability as possible, factor analysis seeks to explain the relationships among the variables. The assumption of the model is that the relationships may be explained by a few unobserved variables which will be called factors. It is hoped that fewer factors than the original number of variables will be needed to explain the relationships among the variables. Thus conceptually one may simplify the understanding of the correlations between the variables.

It is difficult to present the technique without having the model and many of the related issues discussed first. However, it is also difficult to understand the related issues without examples. Thus, it is suggested that you read through the material about the mathematical model, go through the examples, and then with this understanding, reread the material about the mathematical model.

We now turn to the model. We observe jointly distributed random variable $X_1, \ldots, X_p$. The assumption is that each X is a linear sum of the factors plus some remaining residual variability.

That is, the model is the following:

$$
\begin{aligned}
X_1 &= E(X_1) + \lambda_{11}F_1 + \lambda_{12}F_2 + \cdots + \lambda_{1k}F_k + e_1 \\
&\vdots \\
X_p &= E(X_p) + \lambda_{p1}F_1 + \lambda_{p2}F_2 + \cdots + \lambda_{pk}F_k + e_p
\end{aligned} \tag{14.11}
$$

In this model, each X_i is equal to its expected value, plus a linear sum of k factors, plus a term for residual variability. This looks like a series of multiple regression equations; each of the variables X_i is regressed upon the variables $F_1, \ldots, F_k$. There are, however, major differences between this model and the multiple regression model of Chapter 11. Observations and assumptions about this model are the following:

1. The factors F_j are *not* observed; only the $X_1, \ldots, X_p$ are observed, although the X_i variables are expressed in terms of these smaller number of factors F_j.
2. The e_i (which are also unobserved) represent variability in the X_i not explained by the factors. We do *not* assume that these residual variability terms have the same distribution.
3. Usually the number of factors k is unknown, and must be determined from the data. We shall first consider the model and the analysis where the number of factors is known; later we consider how one might search for the appropriate number of factors.

Assumptions made in the model, in addition to the linear equations given above, are the following:

1. The factors F_j are standardized; that is, they have mean zero and variance one.
2. The factors F_j are uncorrelated with each other, and they are uncorrelated with the e_i terms. See Section 14.11 for a relaxation of this requirement.
3. The e_i's have mean zero and are uncorrelated with each other as well as with the F_j's. They may have different variances.

It is a fact that if p factors F are allowed, there is no need for the residual variability terms e_i. One can reproduce any pattern of covariances or correlations using p factors when p variables X_i are observed. This, however, is not very useful because we have summmarized the p variables which were observed with p unknown variables. Thus in general we will be interested in k factors where k is less than p.

Let ψ_i be the variance of e_i. With the assumptions of the model above, the variance of each X_i can be expressed in terms of the coefficients λ_{ij} of the factors and the residual variance ψ_i.

The equation giving the relationship for k factors is

$$\text{var}(X_i) = \lambda_{i1}^2 + \cdots + \lambda_{ik}^2 + \psi_i. \tag{14.12}$$

In words, the variance of each X_i is the sum of the squares of the coefficients of the factors, plus the variance of e_i. The variance of X_i has two parts. The sum of the coefficients λ_{ij} squared depends upon the factors; the factors contribute in common to all of the X_i's. The e_i's correlate only with their own variable X_i, and not with other variables in the model. In particular they are uncorrelated with all of the X_i's except for the one corresponding to their index. Thus we have broken down the variance into a part related to the factors which each variable has in common, and the unique part related to the residual variability term. This leads to the following definition.

Definition 14.5. $c_i = \sum_{j=1}^{k} \lambda_{ij}^2$ is called the *common part of the variance* of X_i, c_i is also called the *communality* of X_i, ψ_i is called the *unique* or *specific part of the variance* of X_i, and ψ_i is also called the *uniqueness* or *specificity*.

Although factor analysis is designed to explain the relationships between the variables and not the variance of the individual variables, if the communalities are large compared to the specificities of the variables, then the model has also succeeded in explaining not only the relationships among the variables but the variability in terms of the common factors.

Not only may the variance be expressed in terms of the coefficients of the factors, but the covariance between any two variables may also be expressed by the following equation:

$$\text{cov}(X_i, X_j) = \lambda_{i1}\lambda_{j1} + \cdots + \lambda_{ik}\lambda_{jk}, \quad \text{for } i \neq j. \tag{14.13}$$

These equations explain the relationships among the variables. If both X_i and X_j have variances equal to 1, then this expression gives the correlation between the two variables. There is a standard name for the coefficients of the common factors.

Definition 14.6. The coefficients λ_{ij} are called the *factor loadings* or *loadings*. λ_{ij} the loading of variable X_i and factor F_j.

In general, $\text{cov}(X_i, F_j) = \lambda_{ij}$. That is, λ_{ij} is the covariance between X_i and F_j. If X_i has variance 1, for example if it is standardized, then since F_j has variance 1, the factor loading is the correlation coefficient between the variable and the factor.

We illustrate the method by two examples.

Example 14.4. We continue with the measurement data of US females of Example 14.2. A factor analysis with three underlying factors was performed on these data. Since we are trying to explain the correlations between the variables, it is useful to examine the fit of the model by comparing the observed and modeled correlations. We do this by examining the residual correlation.

Definition 14.7. The *residual correlation* is the observed correlation minus the fitted correlation from the factor analysis model.

Table 14.6 gives the residual correlations below the diagonal; on the diagonal are

Table 14.6. Residual Correlations for Example 14.4.

		STHTER 1	STHTNORM 2	KNEEHT 3	POPHT 4	ELBWHT 5
STHTER	1	0.034				
STHTNORM	2	0.002	0.151			
KNEEHT	3	−0.001	0.001	0.191		
POPHT	4	0.001	0.002	0.048	0.276	
ELBWHT	5	−0.001	−0.011	0.011	−0.004	0.474
THIGHHT	6	−0.009	0.004	0.003	−0.076	0.035
BUTTKNHT	7	−0.002	0.000	−0.016	−0.056	−0.021
BUTTPOP	8	−0.002	0.011	−0.042	−0.064	−0.035
ELBWELBW	9	0.000	0.013	−0.004	0.014	−0.010
SEATBRTH	10	−0.002	0.013	0.016	−0.041	0.020
BIACROM	11	0.004	−0.005	−0.000	0.014	−0.089
CHESTGRH	12	0.003	0.004	0.003	0.030	−0.015
WSTGRTH	13	0.005	−0.004	0.002	0.032	0.006
RTARMGRH	14	−0.001	−0.004	0.004	−0.009	0.003
RTARMSKN	15	−0.005	0.016	0.025	−0.012	−0.004
INFRASCP	16	−0.002	0.006	0.020	0.016	0.004
HT	17	0.000	−0.001	−0.000	0.003	0.008
WT	18	−0.000	−0.009	−0.004	−0.005	0.008
AGE	19	0.002	0.024	0.003	0.024	−0.042

		THIGH-HT 6	BUTT-KNHT 7	BUTT-POP 8	ELBW-ELBW 9	SEAT-BRTH 10
THIGHHT	6	0.499				
BUTTKNHT	7	0.062	0.251			
BUTTPOP	8	0.040	**0.136**	0.425		
ELBWELBW	9	−0.012	−0.017	−0.016	0.158	
SEATBRTH	10	0.035	0.070	0.010	−0.016	0.338
BIACROM	11	0.049	−0.035	−0.039	0.012	−0.042
CHESTGRH	12	−0.038	−0.044	−0.017	0.036	−0.056
WSTGRTH	13	−0.067	−0.023	−0.021	0.037	−0.029
RTARMGRH	14	0.005	0.005	0.007	−0.014	0.008
RTARMSKN	15	0.048	0.019	0.021	−0.030	0.047
INFRASCP	16	0.004	−0.025	−0.007	−0.003	−0.030
HT	17	−0.003	−0.001	0.001	0.004	−0.014
WT	18	0.017	0.009	−0.004	−0.011	0.019
AGE	19	**−0.172**	−0.056	−0.034	0.078	0.002

Table 14.6. (*Continued*)

		BIA-CROM 11	CHEST-GRH 12	WST-GRTH 13	RTARM-GRH 14	RTARM-SKN 15
BIACROM	11	0.679				
CHESTGRH	12	0.072	0.148			
WSTGRTH	13	−0.008	0.032	0.172		
RTARMGRH	14	−0.014	−0.014	−0.031	0.134	
RTARMSKN	15	−0.053	−0.041	−0.046	0.075	0.487
INFRASCP	16	−0.010	0.013	0.003	0.013	**0.171**
HT	17	0.002	−0.000	−0.002	−0.001	0.003
WT	18	−0.003	0.000	0.004	0.009	−0.030
AGE	19	**−0.106**	0.033	0.105	−0.017	−0.012

		INFRASCP 16	HT 17	WT 18	AGE 19
INFRASCP	16	0.317			
HT	17	0.002	0.056		
WT	18	−0.018	0.001	0.057	
AGE	19	−0.017	0.016	−0.034	0.770

the estimated uniquenesses; the part of the (standardized) variance not explained by the three factors.

A rule of thumb is that the correlation has been reasonably explained when the residual is less than 0.1 in absolute value. This is convenient because it is easy to scan the residual matrix for a zero after a decimal point. Of course, depending upon the purpose more stringent requirements may be considered.

In this example there are four large absolute values of residuals (−0.172, 0.171, 0.136 and −0.106). This suggests that more factors are needed. (Problem 14.10 considers analysis of these data with more factors.) The factor loadings are presented in Table 14.7.

The loadings factors are first presented in Table 14.7, Part A. An easier to read format is given next in Part B. Here the variables are arranged so that high loadings for a factor come first. Low absolute values are dropped and replaced by zeros. A factor which has large absolute values of loadings or small absolute values of loadings is easily associated with the variables for which it loads high.

In this example the first factor has high loadings on weight and bulk measurements (variables 14, 18, 12, 9, 13, 16, 10, 15, and 6) and might be called a *weight* factor. The second factor has high loadings on length or height measurements (variables 3, 17, 4, 7, and 8) and might be considered a *height* factor. The third factor seems to be a *sitting height* factor.

When the covariance matrix has rows arranged in the order of Table 14.7, Part B clusters of correlated variables often appear. This may be appreciated visually by replacing correlations by shaded squares with the amount of shading related to the absolute value of the correlation. Figure 14.7 gives a *shaded correlation matrix* for the correlation data from pages 567, 568.

We know that the variance of each X_i is partially expressed by its communality.

Table 14.7. Factor Loadings for a Three Factor Model: Example 14.4.

Variable	Number	Factor Loadings (Pattern)		
		Factor 1	Factor 2	Factor 3
		A. Rotated		
STHTER	1	0.011	0.346	0.920
STHTNORM	2	−0.003	0.333	0.859
KNEEHT	3	0.082	0.884	0.146
POPHT	4	−0.271	0.801	0.096
ELBWHT	5	0.222	−0.120	0.680
THIGHHT	6	0.672	0.125	0.181
BUTTKNHT	7	0.436	0.742	0.095
BUTTPOP	8	0.339	0.679	−0.006
ELBWELBW	9	0.914	0.050	−0.065
SEATBRTH	10	0.781	0.171	0.151
BIACROM	11	0.344	0.390	0.225
CHESTGRH	12	0.916	0.114	0.007
WSTGRTH	13	0.898	0.072	−0.126
RTARMGRH	14	0.929	0.050	0.028
RTARMSKN	15	0.714	0.010	0.055
INFRASCP	16	0.823	−0.043	−0.060
HT	17	−0.087	0.804	0.538
WT	18	0.929	0.265	0.103
AGE	19	0.328	−0.124	−0.328
		B. Sorted[a]		
RTARMGRH	14	0.929	0.000	0.000
WT	18	0.929	0.265	0.000
CHESTGRH	12	0.916	0.000	0.000
ELBWELBW	9	0.914	0.000	0.000
WSTGRTH	13	0.898	0.000	0.000
INFRASCP	16	0.823	0.000	0.000
SEATBRTH	10	0.781	0.000	0.000
RTARMSKN	15	0.714	0.000	0.000
THIGHHT	6	0.672	0.000	0.000
KNEEHT	3	0.000	0.884	0.000
HT	17	0.000	0.804	0.538
POPHT	4	−0.271	0.801	0.000
BUTTKNHT	7	0.436	0.742	0.000
BUTTPOP	8	0.339	0.679	0.000
STHTER	1	0.000	0.346	0.920
STHTNORM	2	0.000	0.333	0.859
ELBWHT	5	0.000	0.000	0.680
BIACROM	11	0.344	0.390	0.000
AGE	19	0.328	0.000	−0.328
	VP	7.123	3.633	2.627

[a]The factor loading matrix has been rearranged so that the columns appear in decreasing order of variance explained by factors. The rows have been rearranged so that for each successive factor, loadings greater than 0.5000 appear first. Loadings less than 0.2500 have been replaced by zero.

Absolute Values of Correlations in Sorted and Shaded Form

14	RTARMGRH	X
		@
18	WT	XX
		@@
12	CHESTGRH	XXX
		@@@
9	ELBWELBW	XXXX
		@@@@
13	WSTGRTH	XXXXX
		@@@@@
16	INFRASCP	XXXXXX
		00000@
10	SEATBRTH	XXXXXXX
		O@NONN@
15	RTARMSKN	XXXXXXXX
		ONNNNON@
6	THIGHHT	XXXXXXXXX
		NONN N @
3	KNEEHT	.-. . - .X
		@
17	HT	XX
		O@
4	POPHT	. ...- ..XXX
		OO@
7	BUTTKNHT	+X+++-X+XXX+X
		N ON @
8	BUTTPOP	+X+---+-+XX+XX
		N O@
1	STHTER	. . .+X++-X
		O @
2	STHTNORM	. . .+X+--XX
		O @@
5	ELBWHT	.-.. .-.- - XXX
		N @
11	BIACROM	-++----.+++-+++- X
		@
19	AGE	-.-++-.. .-. --..X
		@

The absolute values of the lower diagonal matrix entries have been printed above in shaded form according to the following scheme:

(blank)	Less than or equal to 0.113	X	0.453 up to and including 0.567
.	0.113 up to and including 0.227	X N	0.567 up to and including 0.680
−	0.227 up to and including 0.340	X O	0.680 up to and including 0.794
+	0.340 up to and including 0.453	X @	Greater than 0.794

Figure 14.7. Shaded correlation matrix for Example 14.4. Data from pages 567, 568.

The sum of the squares of loadings for a factor is the portion of the sum of the X_i variances (the total variance) that is explained by the factor. The factors in Table 14.7 Part B were listed by decreasing explained variance.   □

Example 14.5.   As a second example consider coronary artery disease patients with left main coronary artery disease. This patient group was discussed in Chaitman, *et al.* [1981]. In this factor analysis twelve variables were considered and four factors were used with 357 cases. The factor analysis was based on the correlation matrix. The variables and their mnemonics (names) are:

SEX 0 = male, 1 = female.

PREVMI 0 = history of prior myocardial infarction, 1 = no such history.

FEPCHEP Time in weeks since the first episode of anginal chest pain; this analysis was restricted to individuals with anginal chest pain.

CHCLASS Severity of impairment due to angina (chest pain); ranging from I (mildly impaired) to IV (any activity is limited; almost totally bedridden).

LMCA The percent diameter narrowing of the left main coronary artery; this analysis was restricted to 50% or more narrowing.

AGE In years.

SCORE The amount of impairment of the pumping chamber (left ventricle) of the heart; score ranges from 5 (normal) to 30 (not attained).

PS70 The number of proximal (near the beginning of the blood supply) segments of the coronary arteries with 70% or more diameter narrowing.

LEFT This variable (and RIGHT) tells if the right artery of the heart carries as much blood as normal. Left (dominant) implies the right coronary artery carries little blood; 8.8% of these cases fell into this category. Code: LEFT = 1 (left dominant), LEFT = 0 otherwise.

RIGHT There are three types of dominance of the coronary arteries—LEFT above, unbalanced (implicitly coded when LEFT = 0 and RIGHT = 0), and RIGHT. Right dominance is the usual case and occurs when the right coronary artery carries a usual amount of blood. 85.8% of these cases are right dominant—RIGHT = 1; otherwise RIGHT = 0.

NOVESLS This is the number of diseased vessels with $\geq$ 70% stenosis or narrowing of the three major arterial branches, above and beyond the left main disease.

LVEDP The left ventricular end diastolic pressure. This is the pressure in the heart when it is relaxed between beats. A damaged or failing heart has a higher pressure.

Factor analysis is primarily designed for continuous variables. In this example we have many discrete variables, and even dummy or indicator variables. The analysis is considered more descriptive or explanatory in this case.

Examining the residual values in Table 14.8, we see a fairly satisfactory fit; the maximum absolute value of a residual is 0.062, most are much smaller. Examination of the uniqueness diagonal column on top shows that the number of vessels diseased, NOVESLS, is essentially explained by the factors (uniqueness = 0.000) and SCORE is largely explained (uniqueness = 0.021). Some other variables retain almost all of

Table 14.8. Correlations (as the Bottom Entry in Each Cell) and the Residual Correlations (as the Top Entry) in Each Cell. The Diagonal Entry on Top is the Estimated Uniqueness for Each Variable. Four Factors were Used. Example 14.5.

	SEX	PREMI	FEPCHEP	CHCLASS	LMCA	AGE
SEX	0.933					
	1.000					
PREVMI	0.053	0.802				
	0.040	1.000				
FEPCHEP	−0.013	−0.043	0.714			
	−0.002	−0.161	1.000			
CHCLASS	0.056	−0.000	−0.001	0.796		
	0.073	−0.117	0.217	1.000		
LMCA	0.010	0.049	0.005	−0.037	0.989	
	0.012	0.036	0.041	0.004	1.000	
AGE	−0.026	0.019	0.012	−0.001	0.024	0.727
	−0.013	−0.107	0.286	0.227	0.065	1.000
SCORE	0.000	−0.001	−0.000	0.000	0.000	0.000
	0.030	−0.427	0.143	0.185	0.019	0.175
PS70	−0.028	−0.057	−0.027	0.062	−0.016	0.013
	−0.054	−0.188	0.129	0.087	−0.034	0.044
LEFT	0.015	−0.011	−0.015	0.025	0.011	−0.005
	−0.027	−0.022	0.014	0.099	0.063	0.064
RIGHT	0.009	−0.007	−0.009	0.015	0.006	−0.003
	0.054	0.017	−0.033	−0.062	−0.049	−0.077
NOVESLS	0.000	0.000	0.000	−0.000	0.000	0.000
	−0.033	−0.183	0.206	0.014	−0.034	0.130
LVEDP	0.014	0.023	0.001	0.024	0.019	−0.015
	0.020	−0.072	0.119	0.135	0.041	0.109

	SCORE	PS70	LEFT	RIGHT	NOVESLS	LVEDP
SCORE	0.021					
	1.000					
PS70	0.001	0.514				
	0.198	1.000				
LEFT	−0.000	−0.004	0.281			
	0.007	0.004	1.000			
RIGHT	−0.000	−0.004	0.002	0.175		
	−0.041	−0.013	−0.767	1.000		
NOVESLS	0.000	0.000	0.000	0.000	0.000	
	0.284	0.693	−0.071	0.073	1.000	
LVEDP	0.000	−0.025	−0.007	−0.004	0.000	0.930
	0.175	0.029	0.068	−0.086	0.063	1.000

their variability: SEX (uniqueness = 0.933) and LMCA (uniqueness = 0.989). Since we have explained most of the relationships among the variables without using the variability of these factors SEX and LMCA must be weakly related to the other factors. This is readily verified by looking at the correlation matrix; the maximum absolute correlation involving either of the variables is $r = 0.073$, $r^2 = 0.005$. They explain 1/2 of one percent or less of the variability in the other variables.

Table 14.9. Factor Loadings for Example 14.5.

	Factor[a]			
	1	2	3	4
RIGHT	0.898*	0.039	−0.048	−0.120
LEFT	−0.837*	−0.035	0.009	0.129
NOVESLS	0.058	0.980*	0.172	0.078
PS70	−0.023	0.681*	0.118	0.085
SCORE	0.022	0.105	0.971*	0.158
AGE	−0.013	0.077	0.089	0.509x
FEPCHEP	0.037	0.159	0.047	0.507x
CHCLASS	−0.019	−0.041	0.125	0.432x
PREVMI	−0.012	−0.103	−0.401x	−0.163
SEX	0.057	−0.045	0.029	0.030
LMCA	−0.047	−0.040	0.010	0.085
LVEDP	−0.056	0.025	0.143	0.215
$\sum \lambda_{ij}^2$, variance explained	1.523	1.486	1.197	0.853

[a]For emphasis * denotes absolute values ≥ 0.6; x denotes absolute values < 0.6 but ≥ 0.4.

Let us now look at the factor loading (or correlation) values in Table 14.9. The first factor has heavy loadings on the two *dominance* variables. This factor could be labeled a dominance factor. The second factor looks like a *coronary artery disease (CAD) factor*. The third is a heart attack, *ventricular function* factor. The fourth might be labeled a *history* variable.

The first factor exists largely by definition; if LEFT = 1 then RIGHT = 0 and vice versa. The second factor also is expected; if proximal segments are diseased then the arteries are diseased. The third factor makes biological sense. A damaged ventricle often occurs because of a heart attack. The factor with moderate loadings on AGE, FEPCHEP and CHCLASS is not as clear. □

14.10   ESTIMATION

Many methods have been suggested for estimation of the factor loadings and the specificities. That is, the coefficients λ_{ij} and the variance of the residual term e_i. Consider Equation (14.11). Suppose that we change the scale of X_i. Effectively, this is the same as looking at a new variable cX_i; the new value is the old value multiplied by a constant. Multiplying through the equations of Equation (14.11) by the constant, and remembering that we have restricted the factors to have variance one, we see that the factor loading should be multiplied by the same factor as X_i. Only one method of estimation has this property. This method is the maximum likelihood method; it is our method of choice. The method seems to give the best fit, where fit is examined as described below. There are drawbacks to the method. The computer computation is difficult and can be expensive. Sometimes the computations do not converge to an answer. For a review of other methods, we recommend the book by

Gorsuch [1974]. This book, which is cited extensively below, contains a nice review of many of the issues of factor analysis. Two shorter volumes are Kim and Mueller [1978a,b].

14.11 INDETERMINACY OF THE FACTOR SPACE

There appears to be something magical about factor analysis; we are estimating coefficients of variables that are not even observed. It is difficult to imagine that one can estimate this at all. In point of fact, it is not possible to uniquely estimate the F_i, but one can estimate the F_i up to a certain indeterminacy. It is necessary to describe this indeterminacy in mathematical terms.

Mathematically, the factors are unique except for possible linear combinations. Geometrically, suppose that we think of the factors, for example a model with $k = 2$, as corresponding to values in a plane. Let this plane exist in three-dimensional space. For example, the subspace corresponding to the two factors, that is the plane, might be the plane of the paper of this text. Within this three-dimensional space, factor analysis would determine which plane contains the two factors. However, any two perpendicular directions in the factor plane would correspond to factors that equally well fit the data in terms of explaining the covariances or correlations between the variables. Thus, we have the factors identified up to a certain extent, but we are allowed to rotate them within a "subspace".

This indeterminacy allows one to "fiddle" with different combinations of factors, that is, rotations, so that the factors are considered "easy to interpret". As will be discussed at some length below, one of the strengths and weaknesses of factor analysis is the possibility of finding factors that represent some abstract concept. This task is easiest when the factors are associated with some subset of the variables. That is, one would like factors that have high loadings (in terms of absolute value) on some subset of variables and very low (near zero in absolute value) loadings on the rest of the variables. In this case, the factor is closely associated with the subset of the variables that have large absolute loadings. If these variables have something in common conceptually, for example they are all measures of blood pressure, or in a psychological study they all seem to be related to aggressive behavior, one might then identify the specific factor as a blood pressure factor or an aggression factor.

Another complication in the literature of factor analysis is related to the choice of a specific basis in the factor subspace. Suppose for the moment that we are dealing with the correlations among the X_i's. In this case, as we saw before, the loadings on the factors are correlations of the factor with the variable. Thus each loading will be in absolute value less than or equal to 1. It will be easy to interpret our factors if the absolute value is near zero or near 1. Consider Figure 14.8, parts (a) and (b). These figures give us plots of the loadings on factors 1 and 2, with a separate point for each of the variables X_i. In Figure 14.8(a) there is a very nice pattern. The variables corresponding to points on the factor 1 axis of ± 1 or on the factor 2 axis of ± 1 are variables associated with each of the factors. The variables plotted near 0 on both factors have little relationship to the 2 factors; in particular, factor 1 would be associated with the variables having points near ± 1 along its axis, including variables 1 and 10 as labeled. This would be considered a very nice loading pattern, and easy to interpret, having the simple structure as described above. In

Figure 14.8(b) we see that if we look at the original factors 1 and 2, it is hard to interpret the data points, but should we rotate by θ as indicated in the figure, we would have factors easy of interpretation, that is each factor associated with a subset of the X_i variables. By looking at such plots and then drawing lines and deciding upon the angle θ visually, we have what is called *visual rotation*. When the factor subspace contains a variety of factors, that is, $k > 2$, the situation is not as simple. If we rotate factors 1 and 2 to find a simple interpretation, we will have altered the relationship between factors 1 and 2 and the other factors, and thus, in improving the relationship between 1 and 2 to have a simple form, we may weaken the relationship between 1 and 5, for example. Visual rotation of factors is an art which can take days or even weeks. The advantage of such rotation is that the mind can weigh the different tradeoffs. One drawback of visual rotation is that it may be rotated to give factors that conform to some pet hypothesis. Again, the naming and interpretation of factors will be discussed below. Thus, visual rotation can take an enormous amount of time and is subject to the biases of the data analyst (as well as to his or her creativity).

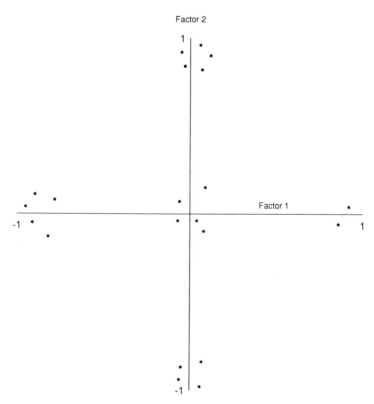

a. Very good loading pattern.
All loadings with absolute value near zero or one.

Figure 14.8. Two factor loading patterns.

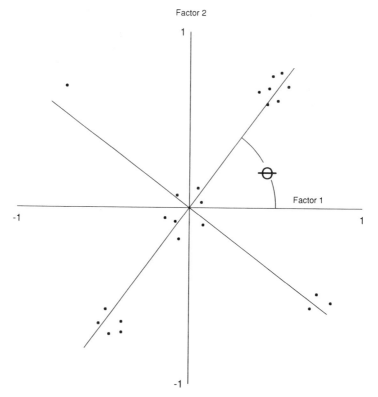

b. This pattern suggests rotating
the factors by the angle ⊖ to have a simple structure.

Figure 14.8. (*Continued*)

Because of the time constraints for analysis, the complexity of the rotation and the potential biases, there has been considerable effort devoted to developing analytic methods of rotating the factors to get the "best rotation". By analytic we mean that there is an algorithm describing whether or not a particular rotation for all of the factors is desirable. The computer software, then, finds the best orientation.

Note 14.1 describes two popular criteria, the varimax method and the quartimax method. A factor analysis is said to have a general factor if there is a factor which is associated with all or almost all of the variables. The varimax method can be useful but does not allow general factors and should not be used when such factors may occur. Otherwise it is considered one of the most satisfactory methods. (In fact, factor analysis was developed in conjunction with the study of intelligence. In particular, one of the issues was: does intelligence consist of one general factor or a variety of uncorrelated factors corresponding to different types of intelligence? Another alternative model for intelligence is a general factor plus other factors associated with some subset of measures of performance thought to be associated with intelligence.)

The second popular method is the quartimax method. This method, in contrast to the varimax method, tends to have one factor with large loadings on all the variables, and not many large loadings among the rest of the factors. In the examples of this chapter we have used the varimax method. We do not have the space to get into all the issues involved in the selection of a rotation method.

Returning to visual rotation, suppose that we have a pattern as given in Figure 14.9. We see that there are no perpendicular axes for which the loadings are 1 or -1, but if we took two axes corresponding to the dotted lines, the interpretation might be simplified. Factors corresponding to the two dotted lines are no longer uncorrelated with each other, and one may wonder to what extent they are "separate" factors. Such factors are called oblique factors, the word oblique coming from the geometric picture and the fact that in geometry oblique lines are lines that do not intersect at a right angle. There are a number of analytic methods for getting oblique rotations, with snappy names such as oblimax, biquartimin, binormamin, and maxplane. References to these are found in Gorsuch [1974]. If oblique axes or bases are used, the formulae for the variance and covariances of the X_i's as given above no longer hold. Again, see Gorsuch for more in-depth consideration of such issues.

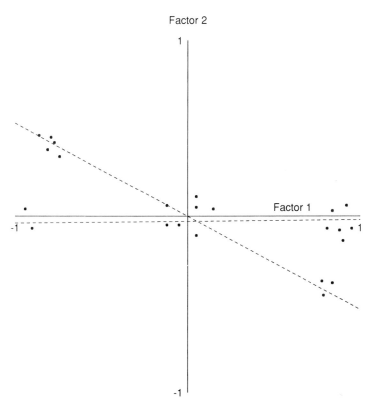

Figure 14.9. Orthogonal and oblique axes for factor loadings.

To try a factor analysis it is not necessary to be expert with every method of estimation and rotation. An exploratory data analysis may be performed to see the extent to which things simplify. We suggest the use of the maximum likelihood estimation method for estimating the coefficients λ_{ij} where the rotation is performed using the varimax method unless one large general factor is suspected to occur.

Example 14.6. We return to Examples 14.4 and 14.5 and examine plots of the correlations of the variables with the factors. Figure 14.10 shows the plots for Example 14.4.

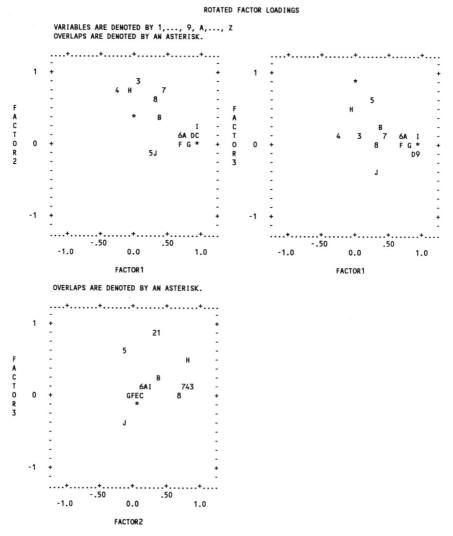

Figure 14.10. Factor loadings for Example 14.4.

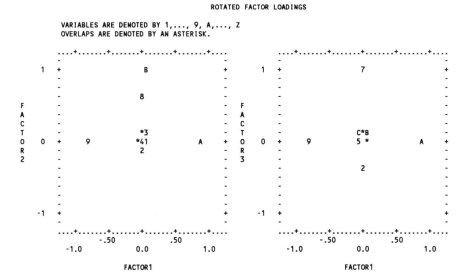

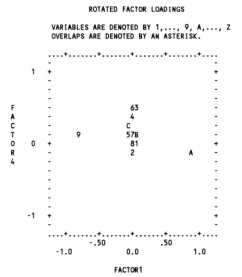

Figure 14.11. Factor loadings for Example 14.5.

The plot for factors 2 and 3 looks reasonable (absolute values near 0 or 1). The other two plots have in-between points making interpretation of the factors difficult. This along with the large residuals mentioned above suggest trying an analysis with a few more factors.

The plots for Example 14.5 are given in Figure 14.11.

These plots suggest factors fairly easy of interpretation with few, if any, points with moderate loadings on several factors. The interpretation of the factors, discussed above in Example 14.5, was fairly straightforward.

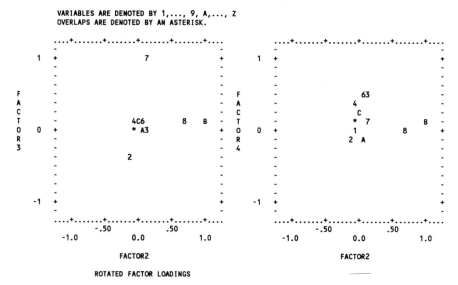

FACTOR2   FACTOR2

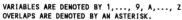

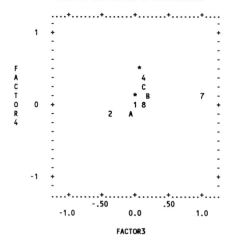

FACTOR3

Figure 14.11. (*Continued*)

14.12   DETERMINING THE NUMBER OF FACTORS

In this section, we consider what to do when the number of factors is unknown. Estimation methods of factor analysis begin with knowledge of k, the number of factors. But this number is usually not known or hypothesized. There is no universal agreement on how to select k; below we examine a number of ways of doing this. The first step is always carried out.

1. Examine the values of the residual correlations. In this section we shall suppose that we are trying to model the correlations between variables rather than their covariances. Recall that with maximum likelihood estimation, fitting one is the same as fitting the other. In looking at the residual correlations, as done in Examples 14.4 and 14.5, we may feel that we have done a good job if all of the correlations have been fit to within some specified difference. If the residual correlations reveal large discrepancies, the model does not fit.

2. Statistical tests of significance. There are statistical tests *if* we can assume that multivariate normality holds and we use the maximum likelihood estimation method. In this case, there is an asymptotic chi-square test for any hypothesized fixed number of factors. Computation of the test statistic is complex and given in Note 14.2. However, it is available in many statistical computer programs. One approach is to look at successively more factors until the statistic is not statistically significant; that is, there are enough factors so that one would not reject at a fixed significance level the hypothesis that the number of factors is as given. This is analogous to a stepwise regression procedure. If we do this we are performing a stepwise procedure and the true and nominal significance levels differ (as usual in a stepwise analysis).

3. Looking at the roots of the correlation matrix.

    a. If the correlations are arranged in a square pattern, or matrix, as usually done, this pattern is called a correlation matrix. Suppose we perform a principal component analysis and examine the variances of the principal components $V_1 \geq V_2 \geq \cdots \geq V_p$. These values are called the eigenvalues or roots of the correlation matrix. If we have the correlation matrix for the entire population, Guttman [1954] showed that the number of factors, k, must be greater than or equal to the number of roots greater than or equal to 1. That is, the number of factors in the factor analytic model must be greater than or equal to the number of principal components whose variance is greater than or equal to 1. Of course in practice we do not have the population correlation matrix but an estimate. The number of such roots greater than or equal to 1 in a sample may turn out to be smaller or larger. However, because of Guttman's result, a reasonable starting value for k is the number of roots greater than or equal to 1 for the sample correlation matrix. For a thorough factor analysis, values of k above and below this number should be tried, and the residual patterns observed. The number of factors in examples four and five was chosen by this method.

    b. *Scree* is the name for the rubble at the bottom of a cliff. The scree test plots the variances of the principle components, that is the roots, or eigenvalues as they are called, of the correlation matrix. If the plot looks somewhat like Figure 14.12, one looks to separate the climb of the cliff from the scree at the bottom of the cliff. We are directed to pick the cliff, roots 1, 2, 3, and possibly 4 rather than the rubble. A clear plastic ruler is laid across the bottom points, and the number of values above the line is the number of important factors. This advice is reasonable when a sharp demarcation can be seen, but often the pattern has no clear break point.

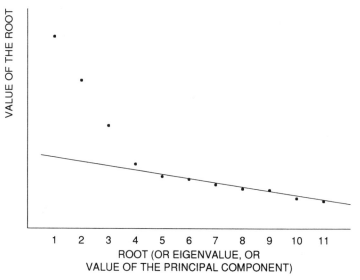

Figure 14.12. Plot for the scree test.

c. Since we are interested in the correlation structure, we might plot as a function of k (the number of factors), the maximum absolute value of all the residuals of the estimated correlations. Another useful plot is the square root of the sum of the squares of all of the residual correlations divided by the number of such residual correlations, which is $p(p-1)/2$. If there is a break in the plots of the curves we would then pick k so that the maximum and the average squared residual correlations are small. For example, in Figure 14.13 we might choose three or four factors. Gorsuch suggests, "In

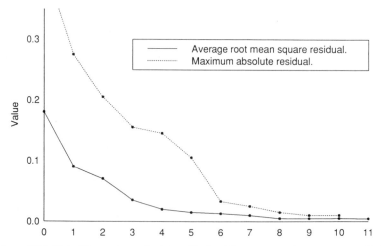

Figure 14.13. Plot of the maximum absolute residual and the average root mean square residual.

the final report, interpretation could be limited to those factors which are well stabilized over the range which the number of factors may reasonably take".

14.13 INTERPRETATION OF FACTORS

Much of the debate about factor analysis stems from the naming and interpretation of factors. Often after a factor analysis is performed, the factors are identified with concepts or objects. *Is a factor an underlying concept, or merely a convenient way of summarizing interrelationships among variables*? A useful word in this context is reify. Reify means to convert into or to regard something as a concrete thing. Should factors be reified?

As Gorsuch states: "A prime use of factor analysis has been in the development of both the theoretical constructs for an area and the operational representatives for the theoretical constructs." In other words, a prime use of factor analysis requires reifying the factors. Also, "The first task of any research program is to establish empirical referents for the abstract concepts embodied in a particular theory."

In psychology, how would one deal with an abstract concept such as aggression? On a questionnaire a variety of possible "aggression" questions might be used. If most or all of them have high loadings on the same factor, and other questions thought to be unrelated to aggression had low loadings, one might identify that factor with aggression. Further, the highest loadings might identify operationally the questions to be used to examine this abstract concept.

Since our knowledge is of the original observations, without a unique set of variables loading a factor, interpretation is difficult. Note well, however, that there is no law saying that one must interpret and name any or all factors.

Gorsuch makes the following points:

1. "The factor can only be interpreted by an individual with extensive background in the substantive area."

2. "The summary of the interpretation is presented as the factor's name. The name may be only descriptive or it may suggest a causal explanation for the occurrence of the factor. Since the name of the factor is all most readers of the research report will remember, it should be carefully chosen." *Perhaps it should not be chosen at all in many cases*.

3. "The widely followed practice of regarding interpretation of a factor as confirmed solely because the post-hoc analysis 'makes sense' is to be deplored. Factor interpretations can only be considered hypotheses for another study."

Interpretation of factors may be strengthened by using cases from other populations. Also, collecting other variables thought to be associated with the factor and including them in the analysis is useful. They should load on the same factor. Taking "marker" variables from other studies is useful in seeing whether an abstract concept has been more or less embodied the same way in two different analyses.

In our opinion, factor analysis is most useful in fields that need to make abstract concepts operationally concrete. It is least useful where physical theories specify the relationship between the variables.

NOTES

Note 14.1: Varimax and Quartimax Methods of Choosing Factors in a Factor Analysis

Many analytic methods of choosing the factors have been developed so that the loading matrix is easy to interpret, that is, has simple structure. These many different methods make the factor analysis literature so complex. We mention two of the methods.

1. **Varimax method.** The varimax method uses the idea of maximizing the sum of the variances of the squares of loadings of the factors. Note the variances are high when the λ_{ij}^2 are near 1 and 0, some of each in each column. In order that variables with large communalities are not overly emphasized, weighted values are used. Suppose we have the loadings λ_{ij} for one selection of factors. Let θ_{ij} be the loadings for a different set of factors, (the linear combinations of the old factors). Define the weighted quantities

$$
\gamma_{ij} = \theta_{ij} \Big/ \sqrt{\sum_{j=1}^{m} \lambda_{ij}^2}.
$$

The method chooses the θ_{ij} to maximize the following:

$$
\sum_{j=1}^{k} \left(\frac{1}{p} \sum_{i=1}^{p} \gamma_{ij}^4 - \frac{1}{p^2} \left(\sum_{i=1}^{p} \gamma_{ij}^2 \right)^2 \right).
$$

Some problems have a factor where all variables load high (e.g., general IQ). Varimax should not be used if a general factor may occur as the low variance discourages general factors. Otherwise it is one of the most satisfactory methods.

2. **Quartimax method.** This works with the variance of the square of all p_k loadings. We maximize over all possible loadings θ_{ij}

$$
\max_{\theta_{ij}} \left(\sum_{i=1}^{p} \sum_{j=1}^{k} \theta_{ij}^4 - \frac{1}{pm} \left(\sum_{i=1}^{p} \sum_{j=1}^{k} \theta_{ij}^2 \right) \right).
$$

Quartimax is used less often since it tends to include one factor with all major loadings and no other major loadings in the rest of the matrix.

Note 14.2: A Statistical Test for the Number of Factors in a Factor Analysis When the $X_1, \ldots, X_p$ are Multivariate Normal and Maximum Likelihood Estimation is Used

This note presupposes familiarity with matrix algebra. Let A be a matrix, A' denotes the transpose of A; if A is square, let $|A|$ be the determinant of A and $\text{Tr}(A)$ be the trace of A. Consider a factor analysis with k factors and estimated *loading matrix*

$$\Lambda = \begin{pmatrix} \lambda_{11} & \cdots & \lambda_{1k} \\ \vdots & \ddots & \vdots \\ \lambda_{n1} & \cdots & \lambda_{nk} \end{pmatrix}.$$

The test statistic is

$$X^2 = \left(n - 1 - \frac{2p+5}{6} - \frac{2k}{3} \right) \log_e \left(\frac{|\Lambda\Lambda' + \psi|}{|S|} \right) \mathrm{Tr} \left(S \left(\Lambda\Lambda' + \psi \right)^{-1} \right) p,$$

where S is the sample covariance matrix and ψ is a diagonal matrix where $\psi_{ii} = s_i - (\Lambda\Lambda')_{ii}$, s_i the sample variance of X_i.

If the true number of factors is less than or equal to k, then X^2 has a chi-square distribution with $\left((p-k)^2 - (p+k) \right) / 2$ degrees of freedom. The null hypothesis of only k factors is rejected if X^2 is too large.

One could try successively more factors until this is not significant. The true and nominal significance levels differ as usual in a stepwise procedure. (For the test to be appropriate the degrees of freedom must be > 0.)

PROBLEMS

The first four problems present principal component analyses using correlation matrices. Portions of computer output (BMDP program 4M) are given. The coefficients for principal components that have a variance of one or more are presented. Because of the connection of principal component analysis and factor analysis mentioned in the text (when the correlations are used), the principal components are also called factors in the output. With a correlation matrix the coefficient values presented are for the standardized variables. You are asked to perform a subset of the following tasks.

  a. Fill in the missing values in the "variance explained" and "cumulative proportion of the total variance" table.

  b. For the principal component(s) specified, give the percent of the total variance accounted for by the principal component(s).

  c. How many principal components are needed to explain 70% of the total variance? 90%? Would a plot with two axes contain most (say $\geq 70\%$) of the variability in the data?

  d. For the case(s) with the value(s) as given, compute the case(s) values on the first two principal components.

1. This problem uses the pyschosocial Framingham data described, with the correlation matrix, in Chapter 11 just before Problem 11.13. The menmonics go in the same order as the presented correlations. The results are presented in Table 14.10 and Table 14.11 pages 735, 736. Perform (a), (b) for principal components 2 and 4, and (c).

2. Measurement data on US females by Stoudt *et al.* [1970] were discussed in this chaper. The same correlation data for adult males were also given (Table 14.12, page 736). The principal component analysis gave the results of Table 14.13, page 738. Perform (a), (b) for principal components 2, 3 and 4, and (c).

3. The Bruce *et al.* [1973] exercise data for 94 sedentary males is used in this problem. These data were used in Problems 9.9–9.12. The exercise variables used are DURAT (duration of the exercise test in seconds), VO_2 MAX (the maximum oxygen consumption (normalized for body weight)), HR (maximum heart rate (beats/min)), AGE (in years), HT (height in centimeters), and WT (weight in kilograms). The correlation values are given in Table 14.14, page 738. The principal component analysis is given in Table 14.15, page 738. Perform (a), (b) for principal components 4, 5, and 6 and (c) (Table 14.16, page 739). Perform (d) for a case with

$$\begin{aligned}
\text{DURAT} &= 600, & VO_2 \text{ MAX} &= 38, \\
\text{HR} &= 185, & \text{AGE} &= 29, \\
\text{HT} &= 165, & \text{WT} &= 71.
\end{aligned}$$

NB: Find the value of the *standardized* variables.

4. The variables are the same as in Problem 14.3. In this analysis 43 active females (whose individual data are given before Problem 9.5) are studied. The correlations are in Table 14.17, page 739. The Principal component analysis in Tables 14.18 and 14.19, page 740. Perform (a), (b) for principal components 1 and 2 and (c). Do (d) for the two cases after Table 14.7 (use std. vars.).

Problems 5–12 (except for 9) consider maximum likelihood factor analysis with varimax rotation (from computer program BMDP4M). Except for Problem 10 the number of factors is selected by Guttman's root criteria (the number of eigenvalues greater than one). Answer the following questions as requested.

   a. Examine the residual correlation matrix. What is the maximum residual correlation? Is it < 0.1? < 0.5?

   b. For the pair(s) of variables, with mnemonics given, find the fitted residual correlation.

   c. Consider the plots of the rotated factors. Dicuss the extent to which the interpretation will be simple.

   d. Discuss the potential for naming and interpreting these factors. Would you be willing to name any? If so, what names?

   e. Give the uniqueness and communality for the variables whose numbers are given.

   f. Is there any reason that you would like to see an analysis with fewer or more factors? If so, why?

   g. If you were willing to associate a factor with variables (or a variable) identify the variables on the shaded form of the correlations. Do the variables cluster (form a dark group), which has little correlation with the other variables?

5. A factor analysis is performed upon the Framingham data of Problem 1. The results are in Tables 14.20–14.23, see pages 741–743, Figures 14.14–14.16, see pages 752–754. Communalities obtained from 5 factors after 17 iterations. The com-

munality of a variable is its squared multiple correlation with the factors; they are in Table 14.21. Perform (a), (b) (TYPEA, EMOTLBLE) and (ANGEROUT, ANGERIN), (c), (d), (e) for variables 1, 5, and 8, (f) and (g). In this study, the TYPEA variable was of special interest. Is it particularly associated with one of the factors?

6. Anderson [1972] studies a causal model of a health services system. (This article performs a path analysis, but that is not discussed here.) Data were obtained for each of 32 counties in New Mexico. The eleven variables used were:

Mnemonic	Meaning
PCENTAG	The percent of the labor force in agriculture 1968.
PCTURBAN	Percent urban 1960.
PCTSPAM	Percent Spanish-American 1960.
PCTNONWH	Percent non-white 1960.
NETMIGR	Net migration 1960–1969.
MEDAGE	Median age 1960.
MEDEDUC	Median education 1960.
PCTUNEMP	Percent unemployed 1968.
INCOME	Per Capita income 1967.
HOSPBEDS	Hospital beds/100000 population.
MORTALTY	Mortality/100000 population due to accidents, suicide, and cirrhosis of the liver 1968. A factor analysis upon causes of death identified this factor—all three causes are known to be related to sociocultural variables.

The correlations are in Table 14.24, page 743. The results are in Tables 14.25–14.27, see pages 744–745, Figures 14.17–14.19, see pages 755–757. Perform (a), (b) (MEDAGE, PCTURBAN) and (MORTALTY, NETMIGR), (c), (d), (e) for 9, 10 and 11, and (f).

7. Starkweather [1970] performed a study entitled Hospital Size, Complexity and Formalization. He states: "Data on 704 United States short-term general hospitals are sorted into a set of dependent variables indicative of organizational formalism and a number of independent variables separately measuring hospital size (number of beds) and various types of complexity commonly associated with size." Here we used his data for a factor analysis of the following variables.

SIZE Number of beds.

CONTROL A hospital was scored:

    1. Proprietary control

    2. Nonprofit community control

    3. Church operated

    4. Public district hospital

    5. City or county control

    6. State control

SCOPE (of patient services) "A count was made of the number of services reported for each sample hospital. Services were weighted 1, 2, or 3 according to their relative impact on hospital operations, as measured by estimated proportion of total operating expenses."

TEACHVOL "The number of students in each of several types of hospital training programs was weighted and the products summed. The number of paramedical students was weighted by 1.5, the number of RN students by 3, and the number of interns and residents by 5.5. These weights represent the average number of years of training typically involved, which in turn constitute a rough measure of the relative impact of students on hospital operations."

TECHTYPE Types of teaching programs. The following scores were summed.

    1. For practical nurse training program

    3. For RN

    4. For medical students

    5. For interns

    7. For residents

NONINPRG Noninpatient programs, Sum the following scores:

    1. For emergency service

    2. For outpatient care

    3. For home care

The results are in Tables 14.28–14.31 see pages 745–746, and Figures 14.20–14.21, see page 758. The factor analytic results follow. Perform (a), (c), (d), (e) for 1, 2, 3, 4, 5, 6, (f), and (g).

8. This factor analysis examines the data used in Problem 14.3, the maximal exercise test data for the sedentary males. The results are given in Tables 14.32–14.34, see page 747; Figures 14.22, 14.23, see page 759. Perform (a), (b) (HR, AGE), (c), (d), (e) for variables 1 and 5, (f), and (g).

*9. Consider two variables, X and Y, with covariances (or correlations) given in the following notation. Prove (a) and (b) below.

	Variable	
Variable	1	2
1	a	c
2	c	b

a. We suppose $c \neq 0$. The variance explained by the first principal component is

$$V_1 = \frac{(a+b) + \sqrt{(a-b)^2 + 4c^2}}{2}.$$

The first principal component is

$$\sqrt{\frac{c^2}{c^2 + (V_1 - a)^2}} X + \frac{c}{|c|} \sqrt{\frac{(V_1 - a)^2}{c^2 + (V_1 - a)^2}} Y.$$

b. If $c = 0$. The first principal component is X if $a \geq b$, and is Y if $a < b$.

c. The introduction to Problems 9.30–9.33 presented data on 20 patients who had their mitral valve replaced. The systolic blood pressure before and after surgery had the following variances and covariance.

	SBP	
	Before	**After**
Before	349.74	21.63
After	21.63	91.94

Find the variance explained by the first and second principal components.

10. The exercise data of the 43 active females of Problem 14.4 are used here. The findings are given in Tables 14.35–14.37 see page 748 and Figures 14.24–14.25, see page 760. Perform (a), (c), (d), (f), and (g). Problem 14.8 examined a similar exercise data for sedentary males. Which factor analysis do you feel was more satisfactory in explaining the relationship among variables? Why? Which analysis had the more interpretable factors? Explain your reasoning.

11. The data on the correlation among male body measurements (of Problem 14.2) are factor analyzed here. The computer output gave the results given in Tables 14.38–14.40 see pages 749–751 and Figure 14.26, see page 761. Perform (a), (b) (POPHT, KNEEHT), (STHTER, BUTTKNHT), (RTARMSKN, INFRASCP), (e) for variables 1 and 11, (f), and (g). Examine the diagonal of the residual values and the communalities. What values are on the diagonal of the residual correlations? (The diagonals are the 1–1, 2–2, 3–3, etc. entries.)

TABLES AND FIGURES FOR PROBLEMS

Table 14.10. Problem 1, Variance Explained by Principal Components.[a]

Factor	Variance Explained	Cumulative Proportion of Total Variance
1	4.279180	0.251716
2	1.633777	0.347821
3	1.360951	?
4	1.227657	0.500092
5	1.166469	0.568708
6	?	0.625013
7	0.877450	0.676627
8	0.869622	0.727782
9	0.724192	0.770381
10	0.700926	0.811612
11	0.608359	?
12	0.568691	0.880850
13	0.490974	0.909731
14	?	0.935451
15	0.386540	0.958189
16	0.363578	0.979576
17	?	?

[a]The variance explained by each factor is the eigenvalue for that factor. Total variance is defined as the sum of the diagonal elements of the correlation (covariance) matrix.

Table 14.11. Problem 1, Principal Components.

		Factor 1	Factor 2	Factor 3	Factor 4	Factor 5
			Unrotated Factor Loadings (Pattern) for Principal Components			
TYPEA	1	0.633	−0.203	0.436	−0.049	0.003
EMOTLBLE	2	0.758	−0.198	−0.146	0.153	−0.005
AMBITIOS	3	0.132	−0.469	0.468	−0.155	−0.460
NONEASY	4	0.353	0.407	−0.268	0.308	0.342
NOBOSSPT	5	0.173	0.047	0.260	−0.206	0.471
WKOVRLD	6	0.162	−0.111	0.385	−0.246	0.575
MTDISSAG	7	0.499	0.542	0.174	−0.305	−0.133
MGDISSAT	8	0.297	0.534	−0.172	−0.276	−0.265
AGEWORRY	9	0.596	0.202	0.060	−0.085	−0.145
PERSONWY	10	0.618	0.346	0.192	−0.174	−0.206
ANGERIN	11	0.061	−0.430	−0.470	−0.443	−0.186
ANGEROUT	12	0.306	0.178	0.199	0.607	−0.215
ANGRDISC	13	0.147	−0.181	0.231	0.443	−0.108
STRESS	14	0.665	−0.189	0.062	−0.053	0.149
TENSION	15	0.771	−0.226	−0.186	0.039	0.118
ANXSYMPT	16	0.594	−0.141	−0.352	0.022	0.067
ANGSYMPT	17	0.723	−0.242	−0.256	0.086	−0.015
	VP[a]	4.279	1.634	1.361	1.228	1.166

[a]The VP for each factor is the sum of the squares of the elements of the column of the factor loading matrix corresponding to that factor. The VP is the variance explained by the factor.

Table 14.12. Problem 2, Correlations.

		STHTER 1	STHT-NORM 2	KNEEHT 3	POPHT 4	ELBWHT 5
STHTER	1	1.000				
STHTNORM	2	0.873	1.000			
KNEEHT	3	0.446	0.443	1.000		
POPHT	4	0.410	0.382	0.798	1.000	
ELBWHT	5	0.544	0.454	−0.029	−0.062	1.000
THIGHHT	6	0.238	0.284	0.228	−0.029	0.217
BUTTKNHT	7	0.418	0.429	0.743	0.619	0.005
BUTTPOP	8	0.227	0.274	0.626	0.524	−0.145
ELBWELBW	9	0.139	0.212	0.139	−0.114	0.231
SEATBRTH	10	0.365	0.422	0.311	0.050	0.286
BIACROM	11	0.365	0.335	0.352	0.275	0.127
CHESTGRH	12	0.238	0.298	0.229	0.000	0.258
WSTGRTH	13	0.106	0.184	0.138	−0.097	0.191
RTARMGRH	14	0.221	0.265	0.194	−0.059	0.269
RTARMSKN	15	0.133	0.191	0.081	−0.097	0.216
INFRASCP	16	0.096	0.152	0.038	−0.166	0.247
HT	17	0.770	0.717	0.802	0.767	0.212
WT	18	0.403	0.433	0.404	0.153	0.324
AGE	19	−0.272	−0.183	−0.215	−0.215	−0.192

Table 14.12. (*Continued*)

		THIGH-HT 6	BUTT-KNHT 7	BUTT-POP 8	ELBW-ELBW 9	SEAT-BRTH 10
THIGHHT	6	1.000				
BUTTKNHT	7	0.348	1.000			
BUTTPOP	8	0.237	0.736	1.000		
ELBWELBW	9	0.603	0.299	0.193	1.000	
SEATBRTH	10	0.579	0.449	0.265	0.707	1.000
BIACROM	11	0.303	0.365	0.252	0.311	0.343
CHESTGRH	12	0.605	0.386	0.252	0.833	0.732
WSTGRTH	13	0.537	0.323	0.216	0.820	0.717
RTARMGRH	14	0.663	0.342	0.224	0.755	0.675
RTARMSKN	15	0.480	0.240	0.128	0.524	0.546
INFRASCP	16	0.503	0.212	0.106	0.674	0.610
HT	17	0.210	0.751	0.600	0.069	0.309
WT	18	0.684	0.551	0.379	0.804	0.813
AGE	19	−0.190	−0.151	−0.108	0.156	0.043

		BIA-CROM 11	CHEST-GRH 12	WST-GRTH 13	RTARM-GRH 14	RTARM-SKN 15
BIACROM	11	1.000				
CHESTGRH	12	0.418	1.000			
WSTGRTH	13	0.249	0.837	1.000		
RTARMGRH	14	0.379	0.784	0.712	1.000	
RTARMSKN	15	0.183	0.558	0.552	0.570	1.000
INFRASCP	16	0.242	0.710	0.727	0.667	0.697
HT	17	0.381	0.189	0.054	0.139	0.060
WT	18	0.474	0.885	0.821	0.849	0.562
AGE	19	−0.261	0.062	0.299	−0.115	−0.039

		INFRASCP 16	HT 17	WT 18	AGE 19
INFRASCP	16	1.000			
HT	17	−0.003	1.000		
WT	18	0.709	0.394	1.000	
AGE	19	0.045	−0.270	−0.058	1.000

Table 14.13. Problem 2, Variance Explained by the Principal Components.[a]

Factor	Variance Explained	Cumulative Proportion of Total Variance
1	7.839282	0.412594
2	4.020110	0.624179
3	1.820741	0.720007
4	1.115168	0.778700
5	0.764398	0.818932
6	?	0.850389
7	0.475083	?
8	0.424948	0.897759
9	0.336247	0.915456
10	?	0.931210
11	0.252205	0.944484
12	?	0.955404
13	0.202398	0.966057
14	0.169678	0.974987
15	0.140613	0.982388
16	0.119548	?
17	0.117741	0.994872
18	0.055062	0.997770
19	0.042365	1.000000

[a]The variance explained by each factor is the eigenvalue for that factor. Total variance is defined as the sum of the diagonal elements of the correlation (covariance) matrix.

Table 14.14. Problem 3, Correlation Matrix.

		DURAT	VO$_2$ MAX	HR	AGE	HT	WT
DURAT	1	1.000					
VO$_2$ MAX	2	0.905	1.000				
HR	3	0.678	0.647	1.000			
AGE	4	−0.687	−0.656	−0.630	1.000		
HT	5	0.035	0.050	0.107	−0.161	1.000	
WT	6	−0.134	−0.147	0.015	−0.069	0.536	1.000

Table 14.15. Problem 3, Variance Explained by the Principal Components.[a]

Factor	Variance Explained	Cumulative Proportion of Total Variance
1	3.124946	0.520824
2	1.570654	?
3	0.483383	0.863164
4	?	0.926062
5	?	0.984563
6	0.092621	1.000000

[a]The variance explained by each factor is the eigenvalue for that factor. Total variance is defined as the sum of the diagonal elements of the correlation (covariance) matrix.

Table 14.16. Problem 3, Principal Components.

		Unrotated Factor Loadings (Pattern) for Principal Components	
		Factor 1	Factor 2
DURAT	1	0.933	−0.117
VO$_2$ MAX	2	0.917	−0.120
HR	3	0.832	0.057
AGE	4	−0.839	−0.134
HT	5	0.128	0.860
WT	6	−0.057	0.884
	VP[a]	3.125	1.571

[a]The VP for each factor is the sum of the squares of the elements of the column of the factor loading matrix corresponding to that factor. The VP is the variance explained by the factor.

		Univariate Summary Statistics	
	Variable	Mean	Standard Deviation
1	DURAT	577.10638	123.83744
2	VO$_2$ MAX	35.63298	7.51007
3	HR	175.39362	18.59195
4	AGE	49.78723	11.06955
5	HT	177.39851	6.58285
6	WT	79.00000	8.71286

Table 14.17. Problem 4, Correlation Matrix.

		DURAT	VO$_2$ MAX	HR	AGE	HT	WT
DURAT	1	1.000					
VO$_2$ MAX	2	0.786	1.000				
HR	3	0.528	0.337	1.000			
AGE	4	−0.689	−0.651	−0.411	1.000		
HT	5	0.369	0.299	0.310	−0.455	1.000	
WT	6	0.094	−0.126	0.232	−0.042	0.483	1.000

Two Cases for Table 14.17

	Subject 1	Subject 2
DURAT	660	628
VO$_2$ MAX	38.1	38.4
HR	184	183
AGE	23	21
HT	177	163
WT	83	52

Table 14.18. Problem 4, Variance Explained by the Principal Components.[a]

Factor	Variance Explained	Cumulative Proportion of Total Variance
1	3.027518	?
2	1.371342	0.733143
3	?	?
4	0.416878	0.918943
5	?	0.972750
6	?	1.000000

[a]The variance explained by each factor is the eigenvalue for that factor. Total variance is defined as the sum of the diagonal elements of the correlation (covariance) matrix.

Table 14.19. Problem 4, Principal Components.

		Unrotated Factor Loadings (Pattern) for Principal Components	
		Factor 1	Factor 2
DURAT	1	0.893	−0.201
VO$_2$ MAX	2	0.803	−0.425
HR	3	0.658	0.162
AGE	4	−0.840	0.164
HT	5	0.626	0.550
WT	6	0.233	0.891
	VP[a]	3.028	1.371

[a]The VP for each factor is the sum of the squares of the elements of the column of the factor loading matrix corresponding to that factor. The VP is the variance explained by the factor.

		Univariate Summary Statistics	
	Variable	Mean	Standard Deviation
1	DURAT	514.88372	77.34592
2	VO$_2$ MAX	29.05349	4.94895
3	HR	180.55814	11.41699
4	AGE	45.13953	10.23435
5	HT	164.69767	6.30017
6	WT	61.32558	7.87921

Table 14.20. Problem 5, Residual Correlations.

		TYPE-A 1	EMOT-LBLE 2	AMBI-TIOS 3	NON-EASY 4	NO-BOSSPT 5	WK-OVRLD 6
TYPEA	1	0.219					
EMOTLBLE	2	0.001	0.410				
AMBITIOS	3	0.001	0.041	0.683			
NONEASY	4	0.003	0.028	−0.012	0.635		
NOBOSSPT	5	−0.010	−0.008	0.001	−0.013	0.964	
WKOVRLD	6	0.005	−0.041	−0.053	−0.008	0.064	0.917
MTDISSAG	7	0.007	−0.010	−0.062	−0.053	0.033	0.057
MGDISSAT	8	0.000	0.000	0.000	0.000	0.000	0.000
AGEWORRY	9	0.002	0.030	0.015	0.017	0.001	−0.017
PERSONWY	10	−0.002	−0.010	0.007	0.007	−0.007	−0.003
ANGERIN	11	0.007	−0.006	−0.028	0.005	−0.018	0.028
ANGEROUT	12	0.001	0.056	0.053	0.014	−0.070	−0.135
ANGRDISC	13	−0.011	0.008	0.044	−0.019	−0.039	0.006
STRESS	14	0.002	−0.032	−0.003	0.018	0.030	0.034
TENSION	15	−0.004	−0.006	−0.016	−0.017	0.013	0.024
ANXSYMPT	16	0.004	−0.026	−0.028	−0.019	0.009	−0.015
ANGSYMPT	17	−0.000	0.018	−0.008	−0.012	−0.006	0.009

		MTDI-SSAG 7	MTDI-SSAT 8	AGE-WORRY 9	PERS-ONWY 10	ANG-ERIN 11	ANG-EROUT 12
MTDISSAG	7	0.574					
MGDISSAT	8	0.000	0.000				
AGEWORRY	9	0.001	−0.000	0.572			
PERSONWY	10	−0.002	0.000	0.001	0.293		
ANGERIN	11	0.010	−0.000	0.015	−0.003	0.794	
ANGEROUT	12	0.006	−0.000	−0.006	−0.001	−0.113	0.891
ANGRDISC	13	−0.029	−0.000	0.000	0.001	−0.086	0.080
STRESS	14	−0.017	−0.000	−0.015	0.013	0.022	−0.050
TENSION	15	0.004	−0.000	−0.020	0.007	−0.014	−0.045
ANXSYMPT	16	0.026	−0.000	0.037	−0.019	0.011	−0.026
ANGSYMPT	17	0.004	−0.000	−0.023	0.006	0.012	0.049

		ANGR-DISC 13	STRESS 14	TENSION 15	ANX-SYMPT 16	ANG-SYMPT 17
ANGRDISC	13	0.975				
STRESS	14	−0.011	0.599			
TENSION	15	−0.005	0.035	0.355		
ANXSYMPT	16	−0.007	0.015	0.020	0.645	
ANGSYMPT	17	0.027	−0.021	−0.004	−0.008	0.398

Table 14.21. Problem 5, Communalities.

1	TYPEA	0.7811
2	EMOTLBLE	0.5896
3	AMBITIOS	0.3168
4	NONEASY	0.3654
5	NOBOSSPT	0.0358
6	WKOVRLD	0.0828
7	MTDISSAG	0.4263
8	MGDISSAT	1.0000
9	AGEWORRY	0.4277
10	PERSONWY	0.7072
11	ANGERIN	0.2063
12	ANGEROUT	0.1087
13	ANGRDISC	0.0254
14	STRESS	0.4010
15	TENSION	0.6445
16	ANXSYMPT	0.3555
17	ANGSYMPT	0.6019

Table 14.22. Problem 5, Factors.

		Factor 1	Factor 2	Factor 3	Factor 4	Factor 5
TYPEA	1	0.331	0.185	0.133	0.753	0.229
EMOTLBLE	2	0.707	0.194	0.054	0.215	−0.046
AMBITIOS	3	0.075	0.012	0.026	0.212	0.515
NONEASY	4	0.215	0.105	0.163	0.123	−0.516
NOBOSSPT	5	0.051	0.101	−0.014	0.142	−0.050
WKOVRLD	6	0.037	0.033	−0.039	0.281	0.003
MTDISSAG	7	0.090	0.474	0.391	0.178	−0.095
MGDISSAT	8	0.093	0.146	0.971	−0.143	−0.085
AGEWORRY	9	0.288	0.576	0.020	0.099	−0.048
PERSONWY	10	0.184	0.799	0.138	0.127	−0.003
ANGERIN	11	0.263	−0.076	−0.025	−0.238	0.272
ANGEROUT	12	0.128	0.179	−0.003	0.196	−0.148
ANGRDISC	13	0.117	−0.004	−0.013	0.102	0.031
STRESS	14	0.493	0.189	0.089	0.337	0.018
TENSION	15	0.753	0.193	0.046	0.190	−0.047
ANXSYMPT	16	0.571	0.138	0.058	0.040	−0.072
ANGSYMPT	17	0.748	0.191	0.044	0.066	0.009
	VP[a]	2.594	1.477	1.181	1.112	0.712

[a]The VP for each factor is the sum of the squares of the elements of the column of the factor pattern matrix corresponding to that factor. When the rotation is orthogonal, the VP is the variance explained by the factor.

Table 14.23. Problem 5, Sorted Rotated Factor Loadings (Pattern).[a]

		Factor 1	Factor 2	Factor 3	Factor 4	Factor 5
TENSION	15	0.753	0.000	0.000	0.000	0.000
ANGSYMPT	17	0.748	0.000	0.000	0.000	0.000
EMOTLBLE	2	0.707	0.000	0.000	0.000	0.000
ANXSYMPT	16	0.571	0.000	0.000	0.000	0.000
PERSONWY	10	0.000	0.799	0.000	0.000	0.000
AGEWORRY	9	0.288	0.576	0.000	0.000	0.000
MTDISSAT	8	0.000	0.000	0.971	0.000	0.000
TYPEA	1	0.331	0.000	0.000	0.753	0.000
NONEASY	4	0.000	0.000	0.000	0.000	−0.516
ABITIOS	3	0.000	0.000	0.000	0.000	0.515
ANGERIN	11	0.263	0.000	0.000	0.000	0.272
ANGEROUT	12	0.000	0.000	0.000	0.000	0.000
ANGRDISC	13	0.000	0.000	0.000	0.000	0.000
STRESS	14	0.493	0.000	0.000	0.337	0.000
MTDISSAG	7	0.000	0.473	0.391	0.000	0.000
WKOVRLD	6	0.000	0.000	0.000	0.281	0.000
NOBOSSPT	5	0.000	0.000	0.000	0.000	0.000
	VP	2.594	1.477	1.181	1.112	0.712

[a]The factor loading matrix has been rearranged so that the columns appear in decreasing order of variance explained by factors. The rows have been rearranged so that for each successive factor, loadings greater then 0.5000 appear first. Loadings less than 0.2500 have been replace by zero.

Table 14.24. Problem 6, Correlation Matrix.

		PCE-NTAG 1	PCTU-RBAN 2	PCT-SPAM 3	PCTN-ONWH 4	NET-MIGR 5	MED-AGE 6
PCENTAG	1	1.000					
PCTURBAN	2	−0.520	1.000				
PCTSPAM	3	0.130	−0.490	1.000			
PCTNONWH	4	−0.360	−0.040	−0.250	1.000		
NETMIGR	5	−0.420	0.420	−0.110	0.050	1.000	
MEDAGE	6	0.300	0.140	−0.320	−0.460	0.240	1.000
MEDEDUC	7	−0.280	0.680	−0.680	−0.180	0.190	0.180
PCTUNEMP	8	−0.010	−0.450	0.750	0.180	−0.220	−0.550
INCOME	9	−0.160	0.590	−0.520	−0.370	0.200	0.400
HOSPBEDS	10	−0.400	0.570	−0.240	0.150	0.190	0.210
MORTALTY	11	−0.200	0.390	0.220	0.390	−0.250	−0.170

		MED-EDUC	PCT-UNEMP	IN-COME	HOSP-BESD	MORTALTY
MEDEDUC	7	1.000				
PCTUNEMP	8	−0.710	1.000			
INCOME	9	0.820	−0.650	1.000		
HOSPBEDS	10	0.200	−0.190	0.240	1.000	
MORTALTY	11	−0.480	0.430	−0.480	0.070	1.000

Table 14.25. Problem 6, Communalities.[a]

1	PCENTAG	0.8112
2	PCTURBAN	0.7020
3	PCTSPAM	0.9999
4	PCTNONWH	0.7988
5	NETMIGR	0.3663
6	MEDAGE	1.0000
7	MEDEDUC	0.9361
8	PCTUNEMP	0.7659
9	INCOME	0.7938
10	HOSPBEDS	0.4002
11	MORTALTY	0.4149

[a]Communalities obtained from 4 factors after 18 iterations. The communality of a variable is its squared multiple correlation with the factors.

Table 14.26. Problem 6, Residual Correlations.

		PCE-NTAG 1	PCTU-RBAN 2	PCT-SPAM 3	PCTN-ONWH 4	NET-MIGR 5	MED-AGE 6
PCENTAG	1	0.189					
PCTURBAN	2	0.017	0.298				
PCTSPAM	3	−0.000	−0.000	0.000			
PCTNONWH	4	0.006	−0.036	−0.000	0.201		
NETMIGR	5	−0.032	0.045	0.000	0.071	0.634	
MEDAGE	6	−0.000	−0.000	0.000	−0.000	0.000	0.000
MEDEDUC	7	−0.006	−0.013	0.000	0.001	−0.006	−0.000
PCTUNEMP	8	0.004	0.029	−0.000	−0.011	−0.056	−0.000
INCOME	9	−0.005	−0.019	0.000	0.007	−0.043	0.000
HOSPBEDS	10	0.037	0.167	0.000	0.024	−0.176	0.000
MORTALTY	11	−0.092	−0.141	0.000	0.003	−0.258	0.000

		MED-EDUC	PCT-UNEMP	IN-COME	HOSP-BESD	MORTALTY
MEDEDUC	7	0.064				
PCTUNEMP	8	−0.008	0.234			
INCOME	9	0.009	0.022	0.206		
HOSPBEDS	10	−0.015	0.012	0.019	0.600	
MORTALTY	11	0.022	0.050	0.011	0.016	0.585

Table 14.27. Problem 6, Factor Loadings (Pattern).

		Factor 1	Factor 2	Factor 3	Factor 4
			Rotated		
PCENTAG	1	−0.025	−0.788	0.059	0.431
PCTURBAN	2	0.507	0.615	−0.253	−0.059
PCTSPAM	3	−0.344	−0.145	0.921	−0.109
PCTNONWH	4	−0.523	0.172	−0.498	−0.497
NETMIGR	5	0.063	0.580	0.015	0.161
MEDAGE	6	0.152	0.132	−0.156	0.967
MEDEDUC	7	0.845	0.267	−0.386	−0.045
PCTUNEMP	8	−0.563	−0.130	0.539	−0.376
INCOME	9	0.793	0.277	−0.199	0.219
HOSPBEDS	10	−0.001	0.604	−0.153	0.110
MORTALTY	11	−0.629	0.108	0.010	−0.090
	VP[a]	2.731	1.945	1.691	1.622
			Sorted[b]		
MEDEDUC	7	0.845	0.267	−0.386	0.000
INCOME	9	0.793	0.277	0.000	0.000
MORTALTY	11	−0.629	0.000	0.000	0.000
PCTUNEMP	8	−0.563	0.000	0.539	−0.376
PCTNONWH	4	−0.523	0.000	−0.498	−0.497
PCENTAG	1	0.000	−0.788	0.000	0.431
PCTURBAN	2	0.507	0.615	−0.253	0.000
HOSPBEDS	10	0.000	0.604	0.000	0.000
NETMIGR	5	0.000	0.580	0.000	0.000
PCTSPAM	3	−0.344	0.000	0.921	0.000
MEDAGE	6	0.000	0.000	0.000	0.967
	VP[a]	2.731	1.945	1.691	1.622

[a]The VP for each factor is the sum of the squares of the elements of the column of the factor pattern matrix corresponding to that factor. When the rotation is orthogonal, the VP is the variance explained by the factor.
[b]This factor loading matrix has been rearranged so that the columns appear in decreasing order of variance explained by factors. The rows have been rearranged so that for each successive factor, loadings greater then 0.5000 appear first. Loadings less than 0.2500 have been replace by zero.

Table 14.28. Problem 7, Correlation Matrix.

		SIZE 1	CON-TROL 2	SCOPE 3	TEACH-VOL 4	TECH-TYPE 5	NON-INPRG 6
SIZE	1	1.000					
CONTROL	2	−0.028	1.000				
SCOPE	3	0.743	−0.098	1.000			
TEACHVOL	4	0.717	−0.040	0.643	1.000		
TECHTYPE	5	0.784	−0.034	0.547	0.667	1.000	
NONINPRG	6	0.523	−0.051	0.495	0.580	0.440	1.000

Table 14.29. Problem 7, Communalities.[a]

1	SIZE	0.8269	
2	CONTROL	0.0055	
3	SCOPE	0.7271	
4	TEACHVOL	0.6443	
5	TECHTYPE	1.0000	
6	NONINPRG	0.3788	

[a]Communalities obtained from 2 factors after 8 iterations. The communality of a variable is its squared multiple correlation with the factors.

Table 14.30. Problem 7, Residual Correlations.

		SIZE 1	CON-TROL 2	SCOPE 3	TEACH-VOL 4	TECH-TYPE 5	NON-INPRG 6
SIZE	1	0.173					
CONTROL	2	0.029	0.995				
SCOPE	3	0.013	−0.036	0.273			
TEACHVOL	4	−0.012	0.012	−0.014	0.356		
TECHTYPE	5	−0.000	0.000	−0.000	−0.000	0.000	
NONINPRG	6	−0.020	−0.008	−0.027	0.094	−0.000	0.621

Table 14.31. Problem 7, Factors.

		Factor Loadings (Pattern)	
		Factor 1	Factor 2
Rotated			
SIZE	1	0.636	0.650
CONTROL	2	−0.016	−0.072
SCOPE	3	0.357	0.774
TEACHVOL	4	0.527	0.605
TECHTYPE	5	0.965	0.261
NONINPRG	6	0.312	0.530
	VP[a]	1.840	1.743
Sorted[b]			
TECHTYPE	5	0.965	0.261
SCOPE	3	0.357	0.774
SIZE	1	0.636	0.650
TEACHVOL	4	0.527	0.605
NONINPRG	6	0.312	0.530
CONTROL	2	0.000	0.000
	VP[a]	1.840	1.743

[a]The VP for each factor is the sum of the squares of the elements of the column of the factor pattern matrix corresponding to that factor. When the rotation is orthogonal, the VP is the variance explained by the factor. [b]This factor loading matrix has been rearranged so that the columns appear in decreasing order of variance explained by factors. The rows have been rearranged so that for each successive factor, loadings greater then 0.5000 appear first. Loadings less than 0.2500 have been replace by zero.

Table 14.32. Problem 8, Residual Correlations.

		DURAT	VO$_2$ MAX	HR	AGE	HT	WT
DURAT	1	0.067					
VO$_2$ MAX	2	0.002	0.126				
HR	3	−0.005	−0.011	0.478			
AGE	4	0.004	0.011	−0.092	0.441		
HT	5	−0.006	0.018	−0.021	0.010	0.574	
WT	6	0.004	−0.004	−0.008	0.007	0.005	0.301

Table 14.33. Problem 8, Communalities.[a]

1	DURAT	0.9331
2	VO$_2$ MAX	0.8740
3	HR	0.5217
4	AGE	0.5591
5	HT	0.4264
6	WT	0.6990

[a]Communalities obtained from 2 factors after 6 iterations. The communality of a variable is its squared multiple correlation with the factors.

Table 14.34. Problem 8, Factors.

			Factor Loadings (Pattern)	
			Factor 1	Factor 2
		Rotated		
DURAT	1		0.962	0.646
VO$_2$ MAX	2		0.930	−0.092
HR	3		0.717	0.089
AGE	4		−0.732	−0.154
HT	5		0.098	0.833
WT	6		−0.071	0.833
	VP[a]		2.856	1.158
		Sorted[b]		
DURAT	1		0.962	0.000
VO$_2$ MAX	2		0.930	0.000
AGE	4		−0.732	0.000
HR	3		0.717	0.000
WT	6		0.000	0.883
HT	5		0.000	0.646
	VP[a]		2.856	1.158

[a]The VP for each factor is the sum of the squares of the elements of the column of the factor pattern matrix corresponding to that factor. When the rotation is orthogonal, the VP is the variance explained by the factor.

[b]This factor loading matrix has been rearranged so that the columns appear in decreasing order of variance explained by factors. The rows have been rearranged so that for each successive factor, loadings greater then 0.5000 appear first. Loadings less than 0.2500 have been replace by zero.

Table 14.35. Problem 10, Residual Correlations.

		DURAT	VO$_2$ MAX	HR	AGE	HT	WT
DURAT	1	0.151					
VO$_2$ MAX	2	0.008	0.241				
HR	3	0.039	−0.072	0.687			
AGE	4	0.015	0.001	−0.013	0.416		
HT	5	−0.045	0.013	−0.007	−0.127	0.605	
WT	6	0.000	0.000	0.000	−0.000	0.000	0.000

Table 14.36. Problem 10, Communalities.[a]

1	DURAT	0.8492
2	VO$_2$ MAX	0.7586
3	HR	0.3127
4	AGE	0.5844
5	HT	0.3952
6	WT	1.0000

[a]Communalities obtained from 2 factors after 10 iterations. The communality of a variable is its squared multiple correlation with the factors.

Table 14.37. Problem 10, Factors.

		Factor Loadings (Pattern)	
		Factor 1	Factor 2
Rotated			
DURAT	1	0.907	0.165
VO$_2$ MAX	2	0.869	−0.059
HR	3	0.489	0.271
AGE	4	−0.758	−0.102
HT	5	0.364	0.513
WT	6	−0.078	0.997
	VP[a]	2.529	1.371
Sorted[b]			
DURAT	1	0.907	0.000
VO$_2$ MAX	2	0.869	0.000
AGE	4	−0.758	0.000
WT	6	0.000	0.997
HT	5	0.364	0.513
HR	3	0.489	0.271
	VP	2.529	1.371

[a]The VP for each factor is the sum of the squares of the elements of the column of the factor pattern matrix corresponding to that factor. When the rotation is orthogonal, the VP is the variance explained by the factor.

[b]This factor loading matrix has been rearranged so that the columns appear in decreasing order of variance explained by factors. The rows have been rearranged so that for each successive factor, loadings greater then 0.5000 appear first. Loadings less than 0.2500 have been replace by zero.

Table 14.38. Problem 11, Residual Correlations.

		STHTER 1	STHT-NORM 2	KNEEHT 3	POPHT 4	ELBWHT 5
STHTER	1	0.028				
STHTNORM	2	0.001	0.205			
KNEEHT	3	0.000	−0.001	0.201		
POPHT	4	0.000	−0.006	0.063	0.254	
ELBWHT	5	−0.001	−0.026	−0.012	0.011	0.519
THIGHHT	6	−0.003	0.026	0.009	−0.064	−0.029
BUTTKNHT	7	0.001	−0.004	−0.024	−0.034	−0.014
BUTTPOP	8	−0.001	0.019	−0.038	−0.060	−0.043
ELBWELBW	9	−0.001	0.008	0.007	−0.009	0.004
SEATBRTH	10	−0.002	0.023	0.015	−0.033	−0.013
BIACROM	11	0.006	−0.009	0.009	0.035	−0.077
CHESTGRH	12	−0.001	0.004	−0.004	0.015	−0.007
WSTGRTH	13	0.001	−0.004	−0.002	0.008	0.006
RTARMGRH	14	0.002	0.011	0.012	−0.006	−0.021
RTARMSKN	15	−0.002	0.025	−0.002	−0.012	0.009
INFRASCP	16	−0.002	0.003	−0.009	−0.002	0.020
HT	17	−0.000	0.001	−0.003	−0.003	0.007
WT	18	0.000	−0.007	0.001	0.004	0.007
AGE	19	−0.001	0.006	0.010	−0.014	−0.023

		THIGH-HT 6	BUTT-KNHT 7	BUTT-POP 8	ELBW-ELBW 9	SEAT-BRTH 10
THIGHHT	6	0.462				
BUTTKNHT	7	0.012	0.222			
BUTTPOP	8	0.016	0.076	0.409		
ELBWELBW	9	0.032	−0.002	0.006	0.215	
SEATBRTH	10	0.023	0.020	−0.017	0.007	0.305
BIACROM	11	−0.052	−0.019	−0.027	0.012	−0.023
CHESTGRH	12	−0.020	−0.013	−0.011	0.025	−0.020
WSTGRTH	13	−0.002	0.006	0.009	−0.006	−0.009
RTARMGRH	14	0.009	0.000	0.013	0.011	−0.017
RTARMSKN	15	0.038	0.039	0.015	−0.019	0.053
INFRASCP	16	−0.025	0.008	−0.000	−0.022	0.001
HT	17	0.005	0.005	0.005	0.000	−0.001
WT	18	−0.004	−0.005	−0.007	−0.006	0.004
AGE	19	−0.012	−0.010	−0.014	0.011	0.007

Table 14.38. (*Continued*)

		BIA-CROM 11	CHEST-GRH 12	WST-GRTH 13	RTARM-GRH 14	RTARM-SKN 15
BIACROM	11	0.684				
CHESTGRH	12	0.051	0.150			
WSTGRTH	13	−0.011	0.000	0.095		
RTARMGRH	14	−0.016	−0.011	−0.010	0.186	
RTARMSKN	15	−0.065	−0.011	0.009	0.007	0.601
INFRASCP	16	−0.024	−0.005	0.014	−0.022	0.199
HT	17	−0.008	0.000	−0.003	−0.005	0.004
WT	18	0.006	0.002	0.002	0.006	−0.023
AGE	19	−0.015	−0.006	−0.002	0.014	−0.024

		INFRASCP 16	HT 17	WT 18	AGE 19
INFRASCP	16	0.365			
HT	17	0.003	0.034		
WT	18	−0.003	0.001	0.033	
AGE	19	−0.022	0.002	0.002	0.311

Table 14.39. Problem 11, Communalities.[a]

1	STHTER	0.9721
2	STHTNORM	0.7952
3	KNEEHT	0.7991
4	POPHT	0.7458
5	ELBWHT	0.4808
6	THIGHHT	0.5379
7	BUTTKNHT	0.7776
8	BUTTPOP	0.5907
9	ELBWELBW	0.7847
10	SEATBRTH	0.6949
11	BIACROM	0.3157
12	CHESTGRH	0.8498
13	WSTGRTH	0.9054
14	RTARMGRH	0.8144
15	RTARMSKN	0.3991
16	INFRASCP	0.6352
17	HT	0.9658
18	WT	0.9671
19	AGE	0.6891

[a]Communalities obtained from 4 factors after 6 iterations. The communality of a variable is its squared multiple correlation with the factors.

Table 14.40. Factors for Problem 11.

		Factor Loadings (Pattern)			
		Factor 1	Factor 2	Factor 3	Factor 4
			Unrotated [a]		
STHTER	1	0.100	0.356	0.908	−0.104
STHTNORM	2	0.168	0.367	0.795	−0.027
KNEEHT	3	0.113	0.875	0.128	−0.062
POPHT	4	−0.156	0.836	0.133	−0.071
ELBWHT	5	0.245	−0.151	0.617	−0.131
THIGHHT	6	0.675	0.131	0.114	−0.230
BUTTKNHT	7	0.308	0.819	0.100	−0.042
BUTTPOP	8	0.188	0.742	−0.063	−0.020
ELBWELBW	9	0.873	0.039	0.058	0.131
SEATBRTH	10	0.765	0.209	0.247	0.070
BIACROM	11	0.351	0.298	0.213	−0.242
CHESTGRH	12	0.902	0.137	0.118	0.062
WSTGRTH	13	0.892	0.064	0.030	0.323
RTARMGRH	14	0.873	0.067	0.097	−0.198
RTARMSKN	15	0.625	−0.001	0.075	−0.053
INFRASCP	16	0.794	−0.053	0.045	0.022
HT	17	0.022	0.836	0.507	−0.098
WT	18	0.907	0.308	0.218	−0.049
AGE	19	0.062	−0.135	−0.160	0.801
	VP [a]	6.409	3.964	2.370	0.978
			Sorted Rotated [b]		
WT	18	0.907	0.308	0.000	0.000
CHESTGRH	12	0.902	0.000	0.000	0.000
WSTGRTH	13	0.892	0.000	0.000	0.323
ELBWELBW	9	0.873	0.000	0.000	0.000
RTARMGRH	14	0.873	0.000	0.000	0.000
INFRASCP	16	0.794	0.000	0.000	0.000
SEATBRTH	10	0.765	0.000	0.000	0.000
THIGHHT	6	0.675	0.000	0.000	0.000
RTARMSKN	15	0.625	0.000	0.000	0.000
KNEEHT	3	0.000	0.875	0.000	0.000
POPHT	4	0.000	0.836	0.000	0.000
HT	17	0.000	0.836	0.507	0.000
BUTTKNHT	7	0.308	0.819	0.000	0.000
BUTTPOP	8	0.000	0.742	0.000	0.000
STHTER	1	0.000	0.356	0.908	0.000
STHTNORM	2	0.000	0.367	0.795	0.000
ELBWHT	5	0.000	0.000	0.617	0.000
AGE	19	0.000	0.000	0.000	0.801
BIACROM	11	0.351	0.298	0.000	0.000
	VP	6.409	3.964	2.370	0.978

[a] The VP for each factor is the sum of the squares of the elements of the column of the factor pattern matrix corresponding to that factor. When the rotation is orthogonal, the VP is the variance explained by the factor.
[b] This factor loading matrix has been rearranged so that the columns appear in decreasing order of variance explained by factors. The rows have been rearranged so that for each successive factor, loadings greater then 0.5000 appear first. Loadings less than 0.2500 have been replace by zero.

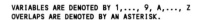

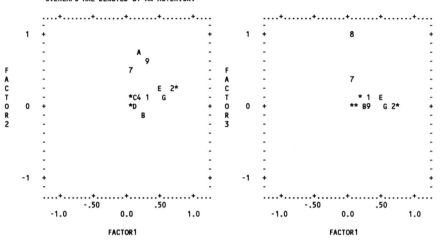

ROTATED FACTOR LOADINGS

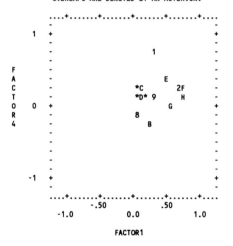

Figure 14.14. Problem 5, plots of factor loadings.

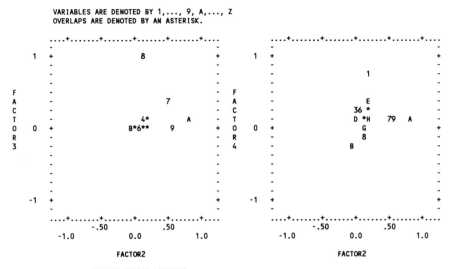

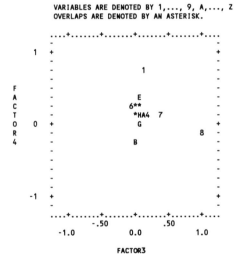

Figure 14.15. Problem 5, plots of factor loadings (continued).

Absolute Values of Correlations
in Sorted and Shaded Form

```
15    TENSION       X
                    @
17    ANGSYMPT      XX
                    @@
2     EMOTLBLE      XXX
                    @@@
16    ANXSYMPT      XXXX
                    ONN@
10    PERSONWY      XXX-X
                      @
9     AGEWORRY      XXX+XX
                      O@
8     MTDISSAT      ....+.X
                      @
1     TYPEA         XXX+X+ X
                    N N    @
4     NONEASY       --+-----.X
                      @
3     AMBITIOS      .    X+X
                      @
11    ANGERIN       .-..   ..X
                      @
12    ANGEROUT      .-+.-- -- -X
                      @
13    ANGRDISC      ...    . ..X
                      @
14    STRESS        XXXXX+.X-. ..X
                    ONN    O    @
7     MTDISSAG      ----XXX+. .. -X
                      O N      @
6     WKOVRLD       .    +    ..X
                      @
5     NOBOSSPT      . ...  .   ...X
                      @
```

The absolute value of the matrix entries have been printed above in shaded form according to the following scheme:

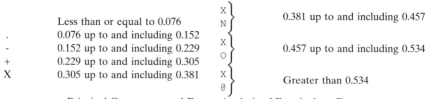

	Less than or equal to 0.076	X N	0.381 up to and including 0.457
.	0.076 up to and including 0.152		
-	0.152 up to and including 0.229	X O	0.457 up to and including 0.534
+	0.229 up to and including 0.305		
X	0.305 up to and including 0.381	X @	Greater than 0.534

Principal Component and Factor Analysis of Framingham Data.

Figure 14.16. Shaded correlation matrix for Problem 5.

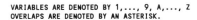

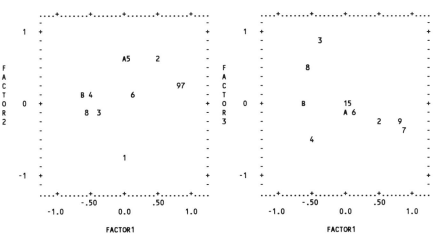

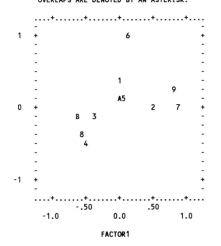

Figure 14.17. Problem 6, plots of factor loadings.

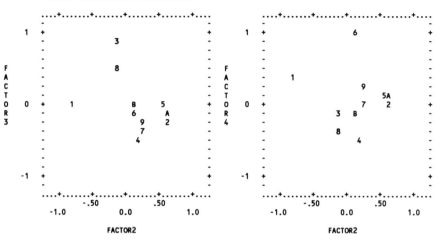

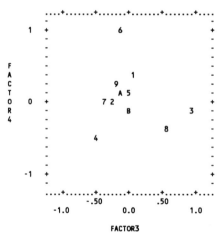

Figure 14.18. Problem 6, plots of factor loadings (continued).

Absolute Values of Correlations
in Sorted and Shaded Form

7	MEDEDUC	X
		@
9	INCOME	XX
		@@
11	MORTALTY	XXX
		@
8	PCTUNEMP	XXXX
		OO @
4	PCTNONWH	.++.X
		@
1	PCENTAG	-.. +X
		@
2	PCTURBAN	XX+X XX
		ON N@
10	HOSPBEDS	.- ..+XX
		N@
5	NETMIGR	.. -- XX.X
		@
3	PCTSPAM	XX-X-.X-.X
		ON @ @
6	MEDAGE	.+.XX-.--+X
		N @

The absolute value of the matrix entries have been printed above in shaded form according to the following scheme:

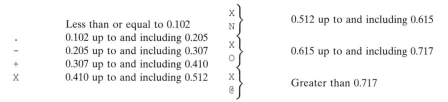

		X	
.	Less than or equal to 0.102	N	0.512 up to and including 0.615
.	0.102 up to and including 0.205		
–	0.205 up to and including 0.307	X	
+	0.307 up to and including 0.410	O	0.615 up to and including 0.717
X	0.410 up to and including 0.512	X	
		@	Greater than 0.717

Figure 14.19. Shaded correlation matrix for Problem 6.

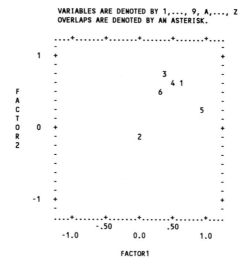

Figure 14.20. Problem 7, plot of factor loadings.

Absolute Values of Correlations
in Sorted and Shaded Form

5	TECHTYPE	X
		@
3	SCOPE	XX
		N@
1	SIZE	XXX
		@@@
4	TEACHVOL	XXXX
		OO@@
6	NONINPRG	XXXXX
		NNN@
2	CONTROL	X
		@

The absolute value of the matrix entries have been printed above in shaded form according to the following scheme:

	Less than or equal to 0.098	X, N	0.490 up to and including 0.588	
.	0.098 up to and including 0.196			
−	0.196 up to and including 0.294	X, O	0.588 up to and including 0.686	
+	0.294 up to and including 0.392			
X	0.392 up to and including 0.490	X, @	Greater than 0.686	

Figure 14.21. Shaded correlation matrix for Problem 7.

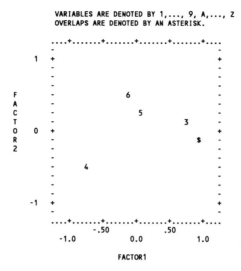

Figure 14.22. Problem 8, plot of factor loadings.

	Absolute Values of Correlations in Sorted and Shaded Form		
1	DURAT	X
		@
2	VO$_2$ MAX	XX
		@@
4	AGE	XXX
		ON@
3	HR	XXXX
		NNN@
6	WT	. . X
		@
5	HT	. XX
		@

The absolute value of the matrix entries have been printed above in shaded form according to the following scheme:

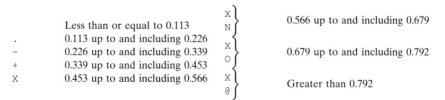

Figure 14.23. Shaded correlation matrix for Problem 8.

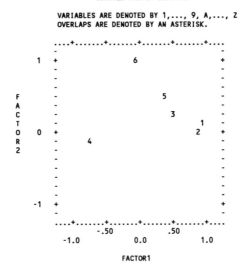

ROTATED FACTOR LOADINGS

VARIABLES ARE DENOTED BY 1,..., 9, A,..., Z
OVERLAPS ARE DENOTED BY AN ASTERISK.

Figure 14.24. Problem 10, plot of factor loadings.

	Absolute Values of Correlations in Sorted and Shaded Form	
1	DURAT	X
		@
2	VO₂ MAX	XX
		@ @
4	AGE	XXX
		@O@
6	WT	. X
		@
5	HT	++XXX
		@
3	HR	X+X−+X
		N @

The absolute value of the matrix entries have been printed above in shaded form according to the following scheme:

		X	0.491 up to and including 0.590
.	Less than or equal to 0.098	N	
	0.098 up to and including 0.197	X	
−	0.197 up to and including 0.295		0.590 up to and including 0.688
+	0.295 up to and including 0.393	O	
X	0.393 up to and including 0.491	X	Greater than 0.688
		@	

Figure 14.25. Shaded correlation matrix for Problem 10.

Absolute Values of Correlations
in Sorted and Shaded Form

```
18    WT           X
                   @
12    CHESTGRH     XX
                   @@
13    WSTGRTH      XXX
                   @@@
9     ELBWELBW     XXXX
                   @@@@
14    RTARMGRH     XXXXX
                   @@OO@
16    INFRASCP     XXXXXX
                   OOOOO@
10    SEATBRTH     XXXXXXX
                   @OOOON@
6     THIGHHT      XXXXXXXX
                   ON NN N@
15    RTARMSKN     XXXXXXXXX
                   NN  NO  @
3     KNEEHT       +-... -- X
                             @
4     POPHT        .   .  . XX
                            @@
17    HT           +.   . -. XXX
                             @O@
7     BUTTKNHT     X+--+.X+-XXXX
                            ONO@
8     BUTTPOP      +-..- --.XXXXX
                           N NO@
1     STHTER       +-  .. +-.X+X+-X
                             O  @
2     STHTNORM     +-..-.+-.X+X+-XX
                            O  @@
5     ELBWHT       --.----..  . .XXX
                             @
19    AGE           -..   . ..-. -..X
                             @
11    BIACROM      X+--+-+-.+-++-++.-X
                             @
```

The absolute value of the matrix entries have been printed above in shaded form according to the following scheme:

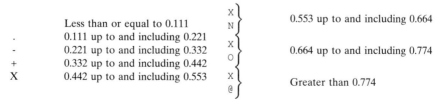

Figure 14.26. Shaded correlation matrix for Problem 11.

REFERENCES

Anderson, J. G. [1972]. Causal model of a health services system. *Health Services Research.* Spring: 23–42. Used with permission from the Hospital and Educational Trust.

Bruce, R. A., Kusumi, F., and Hosmer, D. [1973]. Maximal oxygen intake and nomographic assessment of functional aerobic impairment in cardiovascular disease. *American Heart Journal*, **85:** 546–562.

Chaitman, B. R., Fisher, L., Bourassa, M., Davis, K., Rogers, W., Maynard, C., Tyros, D., Berger, R., Judkins, M., Ringqvist, I., Mack, M. B., Killip, T., and participating CASS Medical Centers [1981]. Effects of coronary bypass surgery on survival in subsets of patients with left main coronary artery disease. Report of the Collaborative Study on Coronary Artery Surgery. *American Journal of Cardiology*, **64:** 360–367.

Gorsuch, R. L. [1974]. *Factor Analysis.* W. B. Saunders, Philadelphia.

Guttman, L. [1954]. Some necessary conditions for common factor analysis. *Psychometrika*, **19(2):** 149–161.

Kim, J-O. and Mueller, C. W. [1978a]. Introduction to Factor Analysis. What It Is and How to Do It. A Sage University Paper, No. 13. Sage Publications, Beverly Hills, CA.

Kim, J-O. and Mueller, C. W. [1978b]. Factor Analysis. Statistical Methods and Practical Issues. A Sage University Paper, No. 14. Sage Publications, Beverly Hills, CA.

Morrison, D. R. [1976]. *Multivariate Statistical Methods,.* 2nd edition. McGraw-Hill, New York.

Reeck, G. R. and Fisher, L. [1973]. A statistical analysis of the amino acid composition of proteins. *International Journal Peptide Protein Research*, **5:** 109–117.

Starkweather, D. B. [1970]. Hospital size, complexity, and formalization. *Health Services Research.* Winter: 330–341. Used with permission from the Hospital and Educational Trust.

Stoudt, H. W., Damon, A., and McFarland, R. A. [1970]. Skinfolds, Body Girths, Biacromial Diameter, and Selected Anthropometric Indices of Adults: United States, 1960–62 Vital and Health Statistics. Data from the National Survey. Public Health Service Publication No. 1000, Series II, No. 35, Washington, DC.

Timm, N. H. [1975]. *Multivariate Analysis with Applications in Education and Psychology.* Brooks/Cole, Monterey, CA.

CHAPTER 15

Rates and Proportions

15.1 INTRODUCTION

In this chapter and the next we want to study in more detail some of the topics dealing with counting data introduced in Chapter 6. In this chapter we want to take an epidemiological approach studying populations by means of describing incidence and prevalence of disease. In a sense this is where statistics began: a numerical description of the characteristics of a state, frequently involving mortality, fecundity, and morbidity. We will call the occurrence of one of those outcomes an "event". In the next chapter, we deal with the more recent developments which have focused on a more detailed modeling of survival (hence also death, morbidity, and fecundity) and, secondly dealt with such data obtained in experiments rather than observational studies. An implication of the latter point is that sample sizes have been much smaller than traditionally used in the epidemiological context. For example, the evaluation of the success of heart transplants has, by necessity, been based on a relatively small set of data.

We begin the chapter with definitions of incidence and prevalence rates and discuss some problems with these "crude" rates. Two methods of standardization, direct and indirect, are then discussed and compared. In Section 15.4, a third standardization procedure is presented to adjust for varying exposure times among individuals. In Section 15.5, a brief tie-in is made to the multiple logistic procedures of Chapter 13. We close the chapter with notes, problems, and references.

15.2 RATES, INCIDENCE, AND PREVALENCE

The term *rate* refers to the amount of change occurring in a quantity with respect to time. In practice, rate refers to the amount of change in a variable over a specified time interval divided by the length of the time interval.

The data used in this chapter to illustrate the concepts come from the Third National Cancer Survey: Incidence Data [1975], henceforth referred to as Third Cancer Survey [1975]. For this reason we discuss the concepts in terms of incidence rates. The *incidence* of a disease in a fixed time interval is the number of new cases diagnosed during the time interval. The *prevalence* of a disease is the number of individuals with the disease at a fixed time point. For a chronic disease, incidence and prevalence may present markedly different ideas of the importance of a disease.

Consider the Third Cancer Survey, National Cancer Institute Monograph 41 [1975]. This survey examined the incidence of cancer (by site) in nine areas during the time period 1969–1971. The areas were the Detroit SMSA (Standard Metropolitan Statistical Area); Pittsburgh SMSA, Atlanta SMSA, Birmingham SMSA, Dallas-Fort Worth SMSA, State of Iowa, Minneapolis-St. Paul SMSA, State of Colorado and the San Francisco-Oakland SMSA. The information used in this chapter refers to the combined data and the Atlanta SMSA and San Francisco-Oakland SMSA. The data are abstracted from tables in the Survey. Suppose we wanted the rate for all sites (of cancer) combined. The rate per year in 1969–1971 time interval would be simply the number of cases divided by three, as the data were collected over a 3-year time interval. The rates are as follows:

$$\text{Combined area:} \quad \frac{181{,}027}{3} = 60{,}342.3,$$

$$\text{Atlanta:} \quad \frac{9341}{3} = 3{,}113.7,$$

$$\text{San Francisco-Oakland:} \quad \frac{30{,}931}{3} = 10{,}310.3.$$

Can we conclude that cancer incidence is worse in the San Francisco-Oakland area than the Atlanta area? The answer is yes and no. Yes, in that there are more cases to take care of in the San Francisco-Oakland area. If we are concerned about the chance of an individual getting cancer, the numbers would not be meaningful. As the San Francisco-Oakland area may have a larger population, the number of cases per number of the population might be less. To make comparisons taking into account the population size, we use

$$\text{Incidence per time interval} = \frac{\text{Number of new cases}}{\text{Total population} \times \text{Time interval}}. \quad (15.1)$$

The result of Equation (15.1) would be quite small so that the number of cases per 100,000 population is used to give a more convenient number. The rate per 100,000 population per year is then

$$\text{Incidence per 100,000 per time interval} = \frac{\text{Number of new cases}}{\text{Total population} \times \text{Time interval}} \times 100{,}000.$$

For these data sets the values are:

$$\text{Combined area:} \quad \frac{181{,}027 \times 100{,}000}{21{,}003{,}451 \times 3} = 287.3 \text{ new cases per 100,000 per year}$$

$$\text{Atlanta:} \quad \frac{9341 \times 100{,}000}{1{,}390{,}164 \times 3} = 224.0 \text{ new cases per 100,000 per year}$$

$$\text{San Francisco-Oakland:} \quad \frac{30{,}931 \times 100{,}000}{3{,}109{,}519 \times 3} = 331.6 \text{ new cases per 100,000 per year.}$$

Even after adjusting for population size, the San Francisco-Oakland area has a higher overall rate.

Note several facts about the estimated rates. The estimates are binomial proportions times a constant (here $100,000/3$). Thus, the rate has a standard error easily estimated. Let N be the total population and n the number of new cases; the rate is $n/N \times C$ ($C = 100,000/3$ in this example) and the standard error is estimated by

$$\sqrt{C^2 \frac{1}{N} \frac{n}{N} \left(1 - \frac{n}{N}\right)}$$

or

$$\text{Standard error of rate per time interval} = C\sqrt{\frac{1}{N} \frac{n}{N} \left(1 - \frac{n}{N}\right)}.$$

For example, the combined area estimate has a standard error of

$$\frac{100,000}{3} \sqrt{\frac{1}{21,003,451} \frac{181,027}{21,003,451} \left(1 - \frac{181,027}{21,003,451}\right)} = 0.67.$$

As the rates are assumed to be binomial proportions, the methods of Chapter 6 may be used to get adjusted estimates or standardized estimates of proportions.

Rates computed by the above methods,

$$\frac{\text{Number of new cases in the interval}}{\text{Population size} \times \text{Time interval}},$$

are called *crude* or *total rates*. This term is used in distinction to *standardized* or *adjusted* rates as discussed below.

Similarly, a *prevalence rate* can be defined as

$$\text{Prevalence} = \frac{\text{Number of cases at a point in time}}{\text{Population size}}.$$

Sometimes a distinction is made between "point prevalence" and "prevalence" to facilitate discussion of chronic disease such as epilepsy and a disease of shorter duration, for example, a common cold or even accidents. It is debatable whether the word prevalence should be used for accidents or illnesses of "short" duration.

15.3 DIRECT AND INDIRECT STANDARDIZATION

15.3.1 Problems With the Use of Crude Rates

Crude rates are useful for certain purposes. For example, the crude rates indicate the load of new cases per capita in a given area of the country. Suppose that we wished to use the cancer rates as epidemiologic indicators. The inference would be that it was likely that environmental or genetic differences were responsible for a difference, if any. There may be simpler explanations, however. Breast cancer rates would likely differ in areas that had differing proportions of the sexes. A retirement community with an older population will tend to have a higher rate. In order to make fair comparisons, we often want to adjust for the differences between populations in one or more factors (covariates). One approach is to find an index that is adjusted in some fashion. The next two sections on direct and indirect standardization discuss two methods of adjustment.

15.3.2 Direct Standardization

In direct standardization, we are interested in adjusting by one or more variables which are divided (or naturally fall) into discrete categories. For example, in Table 15.1 we adjust for sex, and age divided into a total of 18 categories. The idea is to find an answer to the following question: "Suppose that the distribution with regard to the adjusting factors was not as observed, but rather had been the same as this other (reference) population, what would the rate have been?" In other words, we apply the observed risks in our study population to a reference population.

In symbols, the adjusting variable is broken down into I cells. In each cell we know the number of events (the numerator) n_i and the total number of individuals (the denominator) N_i:

Level of adjusting factor, i	1	2	$\cdots$	i	$\cdots$	I
Observed proportion in study population	$\dfrac{n_1}{N_1}$	$\dfrac{n_2}{N_2}$	$\cdots$	$\dfrac{n_i}{N_i}$	$\cdots$	$\dfrac{n_I}{N_I}$

Both numerator and denominator are presented in the table. The crude rate is estimated by

$$C \, \frac{\sum_{i=1}^{I} n_i}{\sum_{i=1}^{I} N_i}.$$

Consider now a *standard or reference* population which instead of having N_i individuals in the ith cell, has M_i individuals.

Reference Population

Level of adjusting factor	1	2	$\cdots$	i	$\cdots$	I
Number in reference population	M_1	M_2	$\cdots$	M_i	$\cdots$	M_I

The question now is, "If the study population has M_i instead of N_i individuals in the ith cell, what would the crude rate have been?" We cannot determine what the crude rate was, but we can estimate what it might have been. In the ith cell the proportion of observed deaths was

$$\frac{n_i}{N_i}.$$

If the same proportion of deaths occurred with M_i individuals we would expect

$$n_i^* = \frac{n_i}{N_i} M_i \text{ deaths.}$$

Thus, if the adjusting variables had been distributed with M_i individuals in the ith cell, we estimate that the data would have been:

Level of adjusting factor	1	2	$\cdots$	i	$\cdots$	I
Expected proportion of cases in reference population	$\dfrac{\left(\dfrac{n_1 M_1}{N_1}\right)}{M_1}$	$\dfrac{\left(\dfrac{n_2 M_2}{N_2}\right)}{M_2}$	$\cdots$	$\dfrac{n_i^*}{M_i}$	$\cdots$	$\dfrac{\left(\dfrac{n_I M_I}{N_I}\right)}{M_I}.$

Finally, the crude rate for this standard population is

$$\text{Adjusted rate} \\ \text{(adjusted to standard population)} = r = \frac{C \sum_{i=1}^{I} \left(\frac{n_i M_i}{N_i} \right)}{\sum_{i=1}^{I} M_i} = \frac{C \sum_{i=1}^{I} n_i^*}{\sum_{i=1}^{I} M_i},$$

r is the *adjusted rate*.

As an example, consider the rate for cancer for all sites for Blacks in the San Francisco-Oakland SMSA, adjusted for sex and age to the total combined sample of the Third Cancer Survey as given by the 1970 census. There are two sex categories, and 18 age categories for a total of 36 cells. The cells are laid out in two columns, rather than in one row of 36 cells. The data are given in Table 15.1 (from the Third Cancer Survey [1975]).

The crude rate for the San Francisco-Oakland Black population is

$$\frac{100,000}{3} \frac{974 + 1188}{169,123 + 160,984} = 218.3.$$

Table 15.2 gives the values of

$$\frac{n_i M_i}{N_i}.$$

Table 15.1. Rate for Cancer of All Sites for Blacks in the San Francisco-Oakland SMSA and Reference Population.

	Study Population n_i / N_i		Reference Population M_i	
Age	Females	Males	Females	Males
< 5	8/16,046	6/16,493	872,451	908,739
5–9	6/18,852	7/19,265	1,012,554	1,053,350
10–14	6/19,034	3/19,070	1,061,579	1,098,507
15–19	7/16,507	6/16,506	971,894	964,845
20–24	16/15,885	9/14,015	919,434	796,774
25–29	27/12,886	19/12,091	755,140	731,598
30–34	28/10,705	18/10,445	620,499	603,548
35–39	46/9,580	25/8,764	595,108	570,117
40–44	83/9,862	47/8,858	650,232	618,891
45–49	109/10,341	108/9,297	661,500	623,879
50–54	125/8,691	131/8,052	595,876	558,124
55–59	120/6,850	189/6,428	520,069	481,137
60–64	102/5,017	158/4,690	442,191	391,746
65–69	119/3,806	159/3,345	367,046	292,621
70–74	75/2,264	154/1,847	300,747	216,929
75–79	44/1,403	72/931	224,513	149,867
80–84	28/765	51/471	139,552	84,360
> 85	25/629	26/416	96,419	51,615
Subtotal	974/169,123	1,188/160,984	10,806,804	10,196,647
Total	2,162/330,107		21,003,451	

Table 15.2. Estimated Number of Casess Per Cell ($n_i M_i / N_i$) if the San Francisco-Oakland Area had the Reference Population Age and Sex Distribution.

Age	Females	Males
< 5	434.97	330.59
5–9	322.26	382.74
10–14	334.64	172.81
15–19	412.14	350.73
20–24	926.09	511.66
25–29	1,582.24	1,149.65
30–34	1,622.98	1,040.10
35–39	2,857.51	1,629.30
40–44	5,472.45	3,283.80
45–49	6,972.58	7,247.38
50–54	8,570.30	9,080.26
55–59	9,110.70	14,146.69
60–64	8,990.13	13,197.41
65–69	11,476.21	13,909.34
70–74	9,962.91	18,087.20
75–79	7,041.03	11,590.14
80–84	5,107.79	9,134.52
> 85	3,832.23	3,225.94
Subtotal	85,029.16	108,470.26
Total		193,499.42

The sex and age-adjusted rate is thus,

$$\frac{100,000}{3} \frac{193,499.42}{21,003,451} = 307.09$$

Note the dramatic change in the estimated rate. This occurs because the San Francisco-Oakland SMSA Black population differs in its age distribution as compared to the overall sample.

The variance is estimated by considering the denominators in the cell as fixed and using the binomial variance of the n_i's. Since the cells constitute independent samples,

$$
\begin{aligned}
\text{var}(r) &= \text{var}\left(C \sum_{i=1}^{I} \frac{n_i M_i}{N_i} \Big/ \sum_{i=1}^{I} M_i \right) \\
&= \frac{C^2}{M_\bullet^2} \sum_{i=1}^{I} \left(\frac{M_i}{N_i} \right)^2 \text{var}(n_i) \\
&\doteq \frac{C^2}{M_\bullet^2} \sum_{i=1}^{I} \left(\frac{M_i}{N_i} \right)^2 N_i \frac{n_i}{N_i} \left(1 - \frac{n_i}{N_i} \right)
\end{aligned}
$$

$$= \frac{C^2}{M_\bullet^2} \sum_{i=1}^{I} \frac{M_i}{N_i} \left(\frac{n_i M_i}{N_i} \right) \left(1 - \frac{n_i}{N_i} \right),$$

where $M_\bullet = \sum_{i=1}^{I} M_i$.

If n_i/N_i is small, then $1 - n_i/N_i \doteq 1$ and

$$\text{var}(r) \doteq \frac{C^2}{M_\bullet^2} \sum_{i=1}^{I} \frac{M_i}{N_i} \left(\frac{n_i M_i}{N_i} \right). \tag{15.2}$$

We use this to compute a 95% confidence interval for the adjusted rate computed above. Using Equation (15.2), the standard error is

$$\text{S.E.}(r) = \frac{C}{M_\bullet} \sqrt{\sum_{i=1}^{I} \frac{M_i}{N_i} \left(\frac{n_i M_i}{N_i} \right)}$$

$$= \frac{100,000}{3} \frac{1}{21,003,451} \left(\frac{872,451}{16,046} 434.97 + \cdots \right)^{1/2}.$$

$$= 7.02.$$

The quantity r is approximately normally distributed so that the interval is

$$307.09 \pm 1.96 \times 7.02 \quad \text{or} \quad (293.3, \; 320.8).$$

If adjusted rates are estimated for two different populations, say r_1 and r_2, with standard errors $\text{SE}(r_1)$ and $\text{SE}(r_2)$, respectively, equality of the adjusted rates may be tested by using

$$z = \frac{r_1 - r_2}{\sqrt{\text{SE}(r_1)^2 + \text{SE}(r_2)^2}}.$$

The $N(0,1)$ critical values are used as z is approximately $N(0,1)$ under the null hypothesis of equal rates.

15.3.3 Indirect Standardization

In indirect standardization, the procedure of direct standardization is used in the opposite direction. That is, we ask the question, "What would the mortality rate have been for the study population if it had the same rates as the population reference?" That is, we apply the observed risks in the reference population to the study population.

Let m_i be the number of deaths in the reference population in the ith cell. The data are:

Level of adjusting factor	1	2	$\cdots$	i	$\cdots$	I
Observed proportion in reference population	$\dfrac{m_1}{M_1}$	$\dfrac{m_2}{M_2}$	$\cdots$	$\dfrac{m_i}{M_i}$	$\cdots$	$\dfrac{m_I}{M_I}$

where both numerator and denominators are presented in the table. Also,

Level of adjusting factor	1	2	$\cdots$	i	$\cdots$	I
Denominators in study population	N_1	N_2	$\cdots$	N_i	$\cdots$	N_I.

The estimate of the rate the study population would have experienced is (analogous to the argument in the previous section):

$$r_{\text{REF}} = \frac{C \sum_{i=1}^{I} N_i \left(\frac{m_i}{M_i}\right)}{\sum_{i=1}^{I} N_i}.$$

The crude rate for the study population is

$$r_{\text{STUDY}} = \frac{C \sum_{i=1}^{I} n_i}{\sum_{i=1}^{I} N_i},$$

where n_i is the observed number of cases in the study population at level i. Usually, there is not much interest in comparing the values r_{REF} and r_{STUDY} as such, because the distribution of the study population with regard to the adjusting factors is not a distribution of much interest. For this reason attention is usually focused on the *standardized mortality ratio*, SMR, when death rates are considered or the *standardized incidence ratio*, SIR, defined to be,

$$\text{Standardized ratio} = s = \frac{r_{\text{STUDY}}}{r_{\text{REF}}} = \frac{\sum_{i=1}^{I} n_i}{\sum_{i=1}^{I} \left(\frac{N_i m_i}{M_i}\right)}. \tag{15.3}$$

The main advantage of the indirect standardization is that the SMR only involves the total number of events so you do not need to know in which cells the deaths occur for the study population. An alternative way of thinking of the SMR is that it is the observed number of deaths in the study population divided by the expected number if the cell specific rates of the reference population held.

As an example, let us compute the SIR of cancer in Black males in the Third Cancer Survey, using White males of the same study as the reference population and adjusting for age. The data are presented in Table 15.3. The standardized incidence ratio is

$$s = \frac{8793}{7,474.16} = 1.17645 = 1.18.$$

One reasonable question is to ask whether this ratio is significantly different from one. An approximate variance can be derived as follows:

$$s = \frac{O}{E}, \quad \text{where } O = \sum_{i=1}^{I} n_i = n_{\bullet} \quad \text{and} \quad E = \sum_{i=1}^{I} N_i \left(\frac{m_i}{M_i}\right).$$

The variance of s is estimated by

$$\text{var}(s) = \frac{\text{var}(O) + s^2 \, \text{var}(E)}{E^2}. \tag{15.4}$$

Table 15.3. Cancer of All Sites, Number of Cases, Black and White Males by Age and Number Eligible by Age. All Areas Combined.

Age	Black Males		White Males		$\dfrac{N_i m_i}{M_i}$	$\left(\dfrac{N_i}{M_i}\right)^2 m_i$
	n_1	N_1	m_1	M_1		
< 5	45	120,122	450	773,459	69.89	10.85
5–9	34	130,379	329	907,543	47.26	6.79
10–14	39	134,313	300	949,669	42.43	6.00
15–19	45	112,969	434	837,614	58.53	7.89
20–24	49	86,689	657	694,670	81.99	10.23
25–29	63	71,348	688	647,304	75.83	8.36
30–34	84	57,844	724	533,856	78.45	8.50
35–39	129	54,752	1,097	505,434	118.83	12.87
40–44	318	57,070	2,027	552,780	209.27	21.61
45–49	582	56,153	3,947	559,241	396.31	39.79
50–54	818	48,753	6,040	503,163	585.23	56.71
55–59	1,170	42,580	8,711	432,982	856.65	84.24
60–64	1,291	33,892	10,966	352,315	1,054.91	101.48
65–69	1,367	27,239	11,913	261,067	1,242.97	129.69
70–74	1,266	17,891	11,735	196,291	1,069.59	97.49
75–79	788	9,827	10,546	138,532	748.10	53.07
80–84	461	4,995	6,643	78,044	425.17	27.21
> 85	244	3,850	3,799	46,766	312.75	25.75
Total	8,793	1,070,700	81,006	8,970,730	7,474.16	708.53

The basic "trick" is to (1) assume that the number of cases in a particular cell follows a Poisson distribution and (2) to note that the sum of independent Poisson random variables is Poisson. Using these two facts

$$\text{var}(O) \doteq \sum_{i=1}^{I} n_i = n \qquad (15.5)$$

and

$$\text{var}(E) \doteq \text{var}\left(\sum_{i=1}^{I} \left(\frac{N_i}{M_i}\right) m_i\right)$$

$$= \sum_{i=1}^{I} \left(\frac{N_i}{M_i}\right)^2 m_i. \qquad (15.6)$$

The variance of s is estimated by using Equations (15.4), (15.5), (15.6).

$$\text{var}(s) = \frac{n_{\bullet} + s^2 \sum \left(\dfrac{N_i}{M_i}\right)^2 m_i}{E^2}.$$

A test of the hypothesis that the population value of s is 1 is obtained from

$$z = \frac{s - 1}{\sqrt{\operatorname{var}(s)}}$$

and $N(0, 1)$ critical values.

For the example,

$$\sum_{i=1}^{I} n_i = n_{\bullet} = 8793,$$

$$E = \sum_{i=1}^{I} \left(\frac{N_i}{M_i} \right) m_i = 7474.16,$$

$$\operatorname{var}(E) \doteq \sum_{i=1}^{I} \left(\frac{N_i}{M_i} \right)^2 m_i = 708.53,$$

$$\operatorname{var}(s) \doteq \frac{8793 + (1.17645)^2 \times 708.53}{(7474.16)^2} = 0.000174957.$$

From this, standard error of $s \doteq 0.013$, and the ratio is significantly different from one using

$$z = \frac{s - 1}{\operatorname{SE}(s)} = \frac{0.17645}{0.013227} = 13.2,$$

and $N(0, 1)$ critical values.

If the reference population is much larger than the study population $\operatorname{var}(E)$ will be much less than $\operatorname{var}(O)$ and you may approximate $\operatorname{var}(s)$ by

$$\frac{\operatorname{var}(O)}{E^2}.$$

15.3.4 Drawbacks to Using Standardized Rates

Any time a complex situation is summarized in one or a few numbers, considerable information is lost. There is always a danger that the lost information is crucial for understanding the situation under study. For example, two populations may have almost the same standardized rates but may differ greatly within the different cells; one population has much larger values in one subset of the cells and the reverse situation in another subset of cells. Even when the standardized rates differ, it is not clear if the difference is somewhat uniform across cells or results mostly from one or a few cells with much larger differences.

The moral of the story is that whenever possible, the rates in the cells used in standardization should be examined individually in addition to working with the standardized rates.

15.4 HAZARD RATES: WHEN SUBJECTS DIFFER IN EXPOSURE TIME

In the rates computed above, each person was exposed (eligible for cancer incidence) over the same length of time (3 years, 1969–1971). (This is not quite true as there is some population mobility, births, and deaths. The assumption that each individual was exposed 3 years is valid to a high degree of approximation.) There are other circumstances where people are observed for varying lengths of time. This happens, for example, when patients are recruited sequentially as they appear at a medical care facility. One approach would be to restrict the analysis to those who had been observed for at least some fixed amount of time (for example, for 1 year). If large numbers of individuals are not observed, this approach is wasteful by throwing away valuable and needed information. This section presents an approach that allows the rates to use all of the available information if certain assumptions are satisfied.

Suppose that we observe subjects over time and look for an event that occurs only once. For definiteness, we will speak about observing individuals where the event is death. Assume that, over the time interval observed, if a subject has survived to some time, t_0, the probability of death in a short interval from t_0 to t_1, is almost $\lambda(t_1 - t_0)$. The quantity λ is called the *hazard rate, force of mortality,* or *instantaneous death rate.* The units of λ are deaths/time unit.

How would we estimate λ from data in a real life situation? Suppose we have n individuals and begin observing the ith individual at time B_i. If the individual dies, let the time of death be D_i. Let the time of last contact be C_i for those individuals still alive. Thus, the time we are observing each individual at risk of death is

$$O_i = \begin{cases} C_i - B_i, & \text{if the subject is alive} \\ D_i - B_i, & \text{if the subject is dead.} \end{cases}$$

An unbiased estimate of λ is

$$\text{Estimated hazard rate} = \widehat{\lambda} = \frac{\text{Number observed deaths}}{\sum_{i=1}^{n} O_i} = \frac{L}{\sum_{i=1}^{n} O_i}. \tag{15.7}$$

As in the earlier sections of this chapter, $\widehat{\lambda}$ is often normalized to have different units. For example, suppose $\widehat{\lambda}$ is in deaths/day of observation. That is, suppose O_i is measured in days. To convert to deaths/100 observation years, we use

$$\widehat{\lambda} \times 365 \, \frac{\text{days}}{\text{year}} \times 100.$$

As an example, consider the paper by Clark *et al.* [1971]. This paper discusses the prognosis of patients who have undergone cardiac (heart) transplantation. They present data on 20 transplanted patients. These data are presented in Table 15.4. To estimate the deaths/year of exposure, we have

$$\frac{12 \text{ deaths}}{3,599 \text{ exposure days}} \frac{365 \text{ days}}{\text{year}} = 1.22 \, \frac{\text{deaths}}{\text{exposure year}}.$$

To compute the variance and standard error of the observed hazard rate we again assume that L, in Equation (15.7) has a Poisson distribution. So, conditional on the

Table 15.4. Stanford Heart Transplant Data.

i	Date of Transplantation	Date of Death	Time at Risk in Days (*if alive)[a]
1	1/6/68	1/21/68	15
2	5/2/68	5/5/68	3
3	8/22/68	10/7/68	46
4	8/31/68	–	608*
5	9/9/68	1/14/68	127
6	10/5/68	12/5/68	61
7	10/26/68	–	552*
8	11/20/68	12/14/68	24
9	11/22/68	8/30/69	281
10	2/8/69	–	447*
11	2/15/69	2/25/69	10
12	3/29/69	5/7/69	39
13	4/13/69	–	383*
14	5/22/69	–	344*
15	7/16/69	11/29/69	136
16	8/16/69	8/17/69	1
17	9/3/69	–	240*
18	9/14/69	11/13/69	60
19	1/3/70	–	118*
20	1/16/70	–	104*

[a]Total exposure days $= 3599$, $L = 12$.

total observation period, the variability of the estimated hazard rate is proportional to the variance of L which is estimated by L itself. Let

$$\widehat{\lambda} = \frac{CL}{\sum_{i=1}^{n} O_i},$$

where C is a constant which standardizes the hazard rate appropriately.

Then the standard error of $\widehat{\lambda}$, $\mathrm{SE}(\widehat{\lambda})$, is approximately

$$\mathrm{SE}(\widehat{\lambda}) \doteq \frac{c}{\sum_{i=1}^{n} O_i} \sqrt{L}.$$

A confidence interval for λ can be constructed by using confidence limits (L_1, L_2) for $E(L)$ (as given in Table A.11 in the Appendix, Confidence Intervals for the Mean of a Poisson Distribution).

$$\text{Confidence interval for } \lambda = \left(\frac{cL_1}{\sum_{i=1}^{n} O_i}, \frac{cL_2}{\sum_{i=1}^{n} O_i} \right).$$

For the example, a 95% confidence interval for the number of deaths is (6.2–21.0). A 95% confidence interval for the hazard rate is then

$$\left(\frac{6.2}{3599} \times 365, \frac{21.0}{3599} \times 365 \right) = (0.63, \ 2.13).$$

Note that this assumes a constant hazard rate from day of transplant; this assumption is suspect. In Chapter 16 some other approaches to analyzing such data are given.

As a second more complicated illustration, consider the work of Bruce *et al.* [1976]. This study analyzed the experience of the Cardiopulmonary Research Institute (CAPRI) in Seattle, Washington. The program provided medically supervised exercise programs for diseased individuals. Over 50% of the participants dropped out of the program. As the individuals who continued participation and those who dropped out had similar characteristics, it was decided to compare the mortality rates for men to see if the training prevented mortality. It was recognized that individuals might drop out because of factors relating to disease and the inference would be weak in the event of an observed difference.

The interest of this example is in the appropriate method of calculating the rates. All individuals, *including the dropouts*, enter into the computation of the mortality for active participants! The reason for this is that had they died during training, they would have been counted as an active participant deaths. Thus training must be credited with the exposure time or observed time when the dropouts were in training. For those who did not die and dropped out, the date of last contact *as an active participant* was the date at which the individuals left the training program. (Topics related to this are dealt with in the next chapter.)

In summary, to compute the mortality rates for active participants all subjects have an observation time. The times are:

1. O_i = (time of death − time of enrollment) for those who died as active participants.
2. O_i = (time of last contact − time of enrollment) for those in the program at last contact.
3. O_i = (Time of dropping the program − time of enrollment) for those who dropped whether or not a subsequent death was observed.

The rate, $\widehat{\lambda}_A$ for active participants is then computed as

$$\widehat{\lambda}_A = \frac{\text{Number of deaths observed during training}}{\sum_{\text{all individuals}} O_i} = \frac{L_A}{\sum O_i}.$$

To estimate the rate for dropouts only those who drop out have time at risk of dying as a dropout. For those who have died, the observed time is

$$O'_i = (\text{time of death} - \text{time the subject dropped out}).$$

For those alive at the last contact,

$$O'_i = (\text{time of last contact} - \text{time the subject dropped out}).$$

The hazard rate for the dropouts, $\widehat{\lambda}_D$, is

$$\widehat{\lambda}_D = \frac{\text{Number of deaths observed during dropout period}}{\sum_{\text{dropouts}} O'_i} = \frac{L_D}{\sum O'_i}.$$

The paper reports rates of 2.7 deaths per 100 person years for the active participants based on 16 deaths. The mortality rate for dropouts was 4.7 based on 34 deaths.

Are rates statistically different at a 5% significance level? For a Poisson variable, L, the variance equals the expected number of observations and is thus estimated by the value of the variable itself. The rates $\widehat{\lambda}$ are of the form

$$\widehat{\lambda} = cL \quad (L \text{ the number of events}).$$

Thus, $\operatorname{var}(\widehat{\lambda}) = c^2 \operatorname{var}(L) \doteq c^2 L = \widehat{\lambda}^2/L$.

To compare the two rates,

$$\operatorname{var}(\widehat{\lambda}_A - \widehat{\lambda}_D) = \operatorname{var}(\widehat{\lambda}_A) + \operatorname{var}(\widehat{\lambda}_D) = \frac{\widehat{\lambda}_A^2}{L_A} + \frac{\widehat{\lambda}_D^2}{L_D}.$$

The approximation is good for large L.

An approximate normal test for the equality of the rates is

$$z = \frac{\widehat{\lambda}_A - \widehat{\lambda}_D}{\sqrt{\dfrac{\widehat{\lambda}_A^2}{L_A} + \dfrac{\widehat{\lambda}_D^2}{L_D}}}.$$

For the example, $L_A = 16$, $\widehat{\lambda}_A = 2.7$, and $L_D = 34$, $\widehat{\lambda}_D = 4.7$ so that

$$z = \frac{2.7 - 4.7}{\sqrt{\dfrac{(2.7)^2}{16} + \dfrac{(4.7)^2}{34}}}$$

$$= -1.90.$$

Thus, the difference between the two groups was not statistically significant at the 5% level.

15.5   THE MULTIPLE LOGISTIC MODEL FOR ESTIMATED RISK AND ADJUSTED RATES

In Chapter 13, the linear discriminant model or multiple logistic model was used to estimate the probability of an event as a function of covariates, $X_1, \ldots, X_n$. Suppose that we want a direct adjusted rate where $X_1(i), \ldots, X_n(i)$ was the covariate value at the midpoints of the ith cell. For the study population, let p_i be the adjusted probability of an event at $X_1(i), \ldots, X_n(i)$. An adjusted estimate of the probability of an event is

$$\widehat{p} = \frac{\sum_{i=1}^{I} M_i p_i}{\sum_{i=1}^{I} M_i},$$

where M_i is the number of reference population individuals in the ith cell. This equation can be written as

$$\widehat{p} = \sum_{i=1}^{I} \left(\frac{M_i}{M_\bullet} p_i \right),$$

where $M_\bullet = \sum_{i=1}^{I} M_i$.

If the study population is small, it is better to estimate the p_i using the approach of Chapter 13 rather than the direct standardization approach of Section 15.3. This will usually be the case when there are several covariates with many possible values.

NOTES

Note 15.1: More Than One Event Per Subject

In some studies, each individual may experience more than one event: for example, seizures in epileptic patients. In this case, each individual could contribute more than once to the numerator in the calculation of a rate. In addition, exposure time or observed time would continue beyond an event as the individual is still at risk for another event. You need to check in this case that there are not individuals with "too many" events; that is, events "cluster" in a small subset of the population. A preliminary test for clustering may then be called for. This is a complicated topic. See Kalbfleisch and Prentice [1980] for references. One possible way of circumventing the problem is to record the time to the second or kth event. This builds a certain robustness into the data, but of course, makes it not possible to investigate the clustering which may be of primary interest.

Note 15.2: Standardization with Varying Observation Time

It is possible to compute standardized rates when the study population has the rate in each cell determined by the method of Section 15.4; that is, individuals are observed for varying lengths of time. In this note we only discuss the method for direct standardization.

Suppose that in each of the i cells the rates in the study population is computed as CL_i/O_i where C is a constant, L_i is the number of events and O_i is the sum of the observed times for individuals in that cell. The adjusted rate is

$$\frac{\sum_{i=1}^{I} \left(\dfrac{M_i}{L_i}\right) O_i}{\sum_{i=1}^{I} M_i} = \frac{C \sum_{i=1}^{I} M_i \widehat{\lambda}_i}{M_\bullet}, \quad \text{where } \widehat{\lambda}_i = \frac{L_i}{O_i}.$$

The standard error is estimated to be

$$\frac{C}{M_\bullet} \sqrt{\sum_{i=1}^{I} \left(\frac{M_i}{O_i}\right) L_i}.$$

Note 15.3: Sources of Demographic and Natural Data

There are many government sources of data in all of the Western countries. Governments of European countries, Canada and the United States regularly publish vital statistics data as well as results of population surveys such as the Third National Cancer Survey [1975] referred to in the text. In the United States, the National Center for Health Statistics publishes more than 20 series of monographs dealing with a variety of topics. For example, Series 20 provides natural data on mortality, Series 21 on natality, marriage, and divorce. These reports are obtainable from the US government.

Note 15.4: Binomial Assumptions

There is some question whether the binomial assumptions (see Chapter 6) always hold. There may be "extra-binomial" variation. In this case standard errors will tend to be underestimated and sample size estimates will be too low, particularly in the case of dependent Bernoulli trials. Such data is not easy to analyze, sometimes a logarithmic transformation is used to stabilize the variance.

PROBLEMS

1. This problem will give practice by asking you to carry out analyses similar to the ones in each of the sections. The numbers from the Third National Cancer Survey [1975] for lung cancer cases for white males in the Pittsburgh and Detroit SMSA's are given in Table 15.5.

   a. Carry out the analyses of Section 15.2 for these SMSA's.

   b. Calculate the direct and indirect standardized rates for lung cancer for White males adjusted for age. Let the Detroit SMSA be the study population and the Pittsburgh SMSA be the reference population.

   c. Compare the rates obtained in Part (b) with those obtained in Part (a).

Table 15.5. Lung Cancer Cases by Age for White Males in the Detroit and Pittsburgh SMSAs.

| Age | Detroit | | Pittsburgh | |
---	Cases	Population Size	Cases	Population Size
< 5	0	149,814	0	82,242
5–9	0	175,924	0	99,975
10–14	2	189,589	1	113,146
15–19	0	156,910	0	100,139
20–24	5	113,003	0	68,062
25–29	1	113,919	0	61,254
30–34	10	92,212	7	53,289
35–39	24	90,395	21	55,604
40–44	101	108,709	56	70,832
45–49	198	110,436	148	74,781
50–54	343	98,756	249	72,247
55–59	461	82,758	368	64,114
60–64	532	63,642	470	50,592
65–69	572	47,713	414	36,087
70–74	473	35,248	330	26,840
75–79	365	25,094	259	19,492
80–84	133	12,577	105	10,987
> 85	51	6,425	52	6,353
Total	3271	1,673,124	2480	1,066,036

2. **a.** Calculate crude rates and standardized cancer rates for the White males of Table 15.5 using Black males of Table 15.3 as the reference population.

   b. Calculate the standard error of the indirect standardized mortality rate and test whether it is different from 1.

   c. Compare the standardized mortality rates for Blacks and Whites.

3. The following data presents the mortality experience for farmers in England and Wales 1949–1953 as compared with national mortality statistics.

Age	National Mortality (1949–1953) rate per 100,000/year	Population of Farmers (1951 census)	Deaths in 1949–1953
20–	129.8	8,481	87
25–	152.5	39,729	289
35–	280.4	65,700	733
45–	816.2	73,376	1,998
55–64	2,312.4	58,226	4,571

   a. Calculate the crude mortality rates.

   b. Calculate the standardized mortality rates.

   c. Test the significance of the standardized mortality rates.

   d. Construct a 95% confidence interval for the standardized mortality rates.

   e. What are the units for the ratios calculated in (a) and (b)?

4. Problems for discussion and thought.

   a. Direct and indirect standardization permit comparison of rates in two populations. Describe in what way this can also be accomplished by multi-way contingency tables.

   b. For calculating standard errors of rates we assumed that events were binomially (or Poisson) distributed. State the assumption of the binomial distribution in terms of say the event death from cancer for a specified population. Which of the assumptions is likely to be valid? Which is not likely to be invalid?

   c. Continuing from (b), we calculate standard errors of rates which are population based, hence the rates are not samples. Why calculate standard errors anyway and do significance testing?

5. This problem deals with a study reported in Bunker et al. [1969]. Halothane, an anesthetic agent, was introduced in 1956. Its early safety record was good but reports of massive hepatic damage and death began to appear. In 1963, a Subcommittee on the National Halothane Study was appointed. Two prominent statisticians, Frederick Mosteller and Lincoln Moses, were members of the Committee. The Committee designed a large cooperative retrospective study, ultimately involving 34 institutions that completed the study. "The primary objective of the study was to compare halothane with other general anesthetics as to incidence of fatal massive hepatic necrosis within six weeks of anesthesia." A 4-year period, 1959–1962, was chosen for the study. One categorization of the patients was

by physical status at the time of the operation. Physical status varies from good (category 1) to moribund (category 7). Another categorization was by mortality level of the surgical procedure having values of low, middle, high. The data below deals with "middle level mortality surgery" and two of the five anesthetic agents studied; the total number of administrations and the number of patients dying within 6 weeks of the operation.

Physical Status	Number of Operations			Number of Deaths		
	Total	Halothane	Cyclo-propane	Total	Halothane	Cyclo-propane
Unknown	69,239	23,684	10,147	1,378	419	297
1	185,919	65,936	27,444	445	125	91
2	104,286	36,842	14,097	1,856	560	361
3	29,491	8,918	3,814	2,135	617	403
4	3,419	1,170	681	590	182	127
5	21,797	6,579	7,423	314	74	101
6	11,112	2,632	3,814	1,392	287	476
7	2,137	439	749	673	111	253
Total	427,400	146,200	68,169	8,783	2,375	2,109

a. Calculate the crude death rates per 100,000 per year for total, halothane, and cyclopropane. Are the crude rates for halothane and cyclopropane significantly different?

b. By direct standardization (relative to the total), calculate standardized death rates for halothane and cyclopropane. Are the standardized rates significantly different?

c. Calculate the standardized mortality rates for halothane and cyclopropane and test the significance of the difference.

d. The calculations of the standard errors of the standardized rates depend on certain assumptions. Which assumptions are likely not to be valid in this example?

6. In 1980, 45 SIDS (Sudden Infant Death Syndrome) deaths were observed in King County. There were 15,000 births.

a. Calculate the SIDS rate per 100,000 births.

b. Construct a 95% confidence interval on the SIDS rate per 100,000 using the Poisson approximation to the binomial.

c. Using the normal approximation to the Poisson, set up the 95% limits.

d. Use the square root transformation for a Poisson random variable to generate a third set of 95% confidence intervals. Are the intervals comparable?

e. The SIDS rate in 1970 in King County is stated to be 250 per 100,000. Someone wants to compare this 1970 rate with the 1980 rate and carries out a test of two proportions $p_1 = 300$ per 100,000 and $p_2 = 250$ per 100,000, using the binomial distributions with $N_1 = N_2 = 100,000$. The large sample normal approximation is used. What part of the Z statistic: $(p_1 - p_2)/$standard error $(p_1 - p_2)$ will be right? What part will be wrong? Why?

7. Annegers *et al.* [1976] investigated ischemic heart disease (IHD) in patients with epilepsy. The hypothesis of interest was whether patients with epilepsy, particularly those on long-term anticonvulsant medication, were at less than expected risk of ischemic heart disease. The study dealt with 516 cases of epilepsy, exposure time was measured from time of diagnosis of epilepsy to time of death or time last seen alive.

   a. For males aged 60–69, the number of years at risk was 161 person-years. In this time interval, four IHD deaths were observed. Calculate the hazard rate for this age group in units of 100,000 persons/year.

   b. Construct a 95% confidence interval.

   c. The expected hazard rate in the general population is 1464 per 100,000 persons/year. How many deaths would you have expected in the 60–69 age group on the basis of the 161 person-years experience?

   d. Do the number of observed and expected deaths differ significantly?

   e. The raw data for the incidence of ischemic heart disease are as follows:

Sex	Age	Epileptics: Person-Years At Risk	New and Nonfatal IHD Cases	Incidence in General Population Per 100,000/year
Male	30–39	354	2	76
	40–49	303	2	430
	50–59	209	3	1,291
	60–69	143	4	2,166
	70+	136	4	1,857
Female	30–39	534	0	9
	40–49	363	1	77
	50–59	218	3	319
	60–69	192	4	930
	70+	210	2	1,087

   Calculate the expected number of deaths for males and the expected number of deaths for females by summing the expected numbers in the age categories (for each sex separately). Treat the total observed as a Poisson random variable and set up a 95% confidence intervals. Do these include the expected number of deaths? State your conclusion.

   *f. Derive a formula for an indirect standardization of these data (see Note 15.2) and apply it to these data.

8. A random sample of 100 subjects from a population is divided into two age groups and for each age group the number of cases of a certain disease is determined. A reference population of 2,000 persons has an age distribution as given below:

	Sample		Reference Population
	Total Number	Number of Cases	Total Number
Age 1	80	8	1,000
Age 2	20	8	1,000

a. What is the crude case rate per 1000 population for the sample?

b. What is the standard error of the crude case rate?

c. What is the age adjusted case rate per 1000 population using direct standard-ization and the reference population above?

d. How would you test the hypothesis that the case rate at Age 1 is not signif-icantly different from the case rate at Age 2?

9. The following data are derived from a paper by Friis *et al.* [1981]. The mortality among male Hispanics and non-Hispanics was as follows:

Age	Hispanic Males		Non-Hispanic Males	
0–4	11,089	0	51,250	0
5–14	18,634	0	120,301	0
15–24	10,409	0	144,363	2
25–34	16,269	2	136,808	9
35–44	11,050	0	106,492	46
45–54	6,368	7	91,513	214
55–64	3,228	8	70,950	357
65–74	1,302	12	34,834	478
75+	1,104	27	16,223	814
Total	79,453	56	772,734	1,920

a. Calculate the crude death rate among Hispanic males.

b. Calculate the crude death rate among non-Hispanic males.

c. Compare (a) and (b) using an appropriate test.

d. Calculate the SMR using non-Hispanic males as the reference population.

e. Test the significance of the SMR as compared with a ratio of 1. Interpret your results.

10. The following data are abstracted from National Center for Health Statistics Se-ries 21, Number 26 [1976] and deals with the mortality experience in poverty and nonpoverty areas of New York and Seattle.

		New York City		Seattle	
Area	Race	Population	Death Rate per 1,000	Population	Death Rate per 1,000
Poverty	White	974,462	9.9	29,016	22.9
	All others	1,057,125	8.5	14,972	12.5
Non-poverty	White	5,074,379	11.6	434,854	11.7
	All other	788,897	6.4	51,989	6.5

a. Using New York City as the "Standard Population" calculate the standardized mortality rates for Seattle taking into account race and poverty area.

b. Estimate the variance of this quantity and calculate 99% confidence limits.

c. Calculate the standardized death rate/100,000 population.

d. Interpret your results.

e. Why would you caution a reviewer of your analysis about the interpretation?

11. In a paper by Foy *et al.* [1983] the risk of getting *Mycoplasma* pneumonia in a 2-year interval was determined on the basis of an extended survey of school children. Of interest was whether children previously exposed to *Mycoplasma pneumoniae* had a smaller risk of recurrence. In the 5–9 year age group the following data were obtained.

	Previously Exposed	Not Previously Exposed
Person-years at risk	680	134
Number with *Mycoplasma* pneumonia	7	8

a. Calculate 95% confidence intervals for the infection rate per 100 person-years for each of the two groups.

b. Test the significance of the difference between the infection rates.

*c. A statistician is asked to calculate the study size needed for a new prospective study between the two groups. He assumes $\alpha = 0.05$, $\beta = 0.20$, and a two tail, two-sample test. He derives the formula

$$\lambda_2 = \left(\sqrt{\lambda_1} - \frac{2.8}{\sqrt{n}} \right),$$

where λ_i is the 2-year infection rate for group i, and n is the number of persons per group. He used the fact that the square root transformation of a Poisson random variable stabilizes the variance (see Section 10.6). Derive the formula and calculate the infection rate in group 2, λ_2 for $\lambda_1 = 10$ or 6 and sample sizes of 20, 40, 60, 80 and 100.

12. In a classic paper dealing with mortality among women first employed before 1930 in the US radium dial-painting industry, Polednak *et al.* [1978] investigated 21 malignant neoplasms among a cohort of 634 women employed between 1915 and 1929. The five highest mortality rates (observed divided by expected deaths) are:

Ranked Cause of Death	Observed Number	Expected Number	Ratio
Bone cancer	22	0.27	81.79
Larynx	1	0.09	11.13
Other sites	18	2.51	7.16
Brain and CNS	3	0.97	3.09
Buccal cavity, pharynx	1	0.47	2.15

a. Test which ratios are significantly different from 1.

b. Assuming that the causes of death were selected without a particular reason adjust the observed p-values using an appropriate multiple comparison procedure.

c. The painters had contact with the radium through the licking of the radium coated paintbrush to make a fine point with which to paint the dial. On the basis of this information would you have "preselected" certain malignant neoplasms? If so, how would you "adjust" the observed p-value?

13. Consider the data in Janerich *et al.* [1974] listing the frequency of infants with Simian creases by sex and maternal smoking status.

Sex of Infant	Maternal Smoking	Birthweight Interval (pounds)			
		< 6	6–6.99	7–7.99	≥ 8
F	No	2/45	5/156	9/242	11/216
	Yes	4/48	8/107	6/110	3/44
M	No	5/40	5/109	23/265	18/278
	Yes	10/55	6/84	10/106	6/74

a. These data can be analyzed by the multidimensional contingency table approach of Chapter 7. However, we can also treat it as a problem in standardization. Describe how indirect standardization can be carried out using the total sample as the reference population, to compare "risk" of Simian creases in smokers and nonsmokers adjusted for birthweight and sex of the infants.

b. Carry out the indirect standardization procedure and compare the standardized rates for smokers and nonsmokers. State your conclusions.

*c. Carry out the logistic model analysis of Chapter 7.

*14. Show that the variance of the standardized mortality ratio, Equation (15.3), is approximately equal to Equation (15.4).

REFERENCES

Annegers, J. F., Elveback, L. R., Labarthe, D. R., and Hauser, W. A. [1976]. Ischemic heart disease in patients with epilepsy. *Epilepsia*, **17:** 11–14.

Bruce, E., Frederick, R., Bruce, R., and Fisher, L. D. [1976]. Comparison of active participants and dropouts in CAPRI cardiopulmonary rehabilitation programs. *American Journal of Cardiology*, **37:** 53–60.

Bunker, J. P., Forest, W. H., Jr., Mosteller, F., and Vandam, L. D. [1969]. *The National Halothane Study. A study of the possible association between halothane anesthesia and postoperative hepatic recrosis.* National Institute of Health. National Institute of Several Medical Sciences. Bethesda, MD.

Clark, D. A., Stinson, E. B., Griepp, R. B., Schroeder, J. S., Shumway, N. E., and Harrison, D. C. [1971]. Cardiac transplantation. VI. Prognosis of patients selected for cardiac transplantation. *Annals of Internal Medicine*, **75:** 15–21. Used with permission.

Foy, H. M., Kenny, G. E., Cooney, M. K., Allan, I. D., and van Belle, G. [1983]. Naturally acquired immunity to mycoplasma pneumonia infections. *Journal of Infectious Diseases.* **147:** 967–973. Used with permission from University of Chigaco Press.

Friis, R., Nanjundappa, G., Prendergast, J. J. Jr., and Welsh, M. [1981]. Coronary heart disease mortality and risk among hispanics and non-hispanics in Orange County, CA. *Public Health Reports*, **96:** 418–422.

Janerich, D. T., Skalko, R. G., and Porter, I. H. (Editors) [1974]. *Congenital Defects: New Directions in Research*. Academic Press, New York.

Kalbfleisch, J. D. and Prentice, R. L. [1980]. *The Statistical Analysis of Failure Time Data*. Wiley, New York.

National Cancer Institute Monograph 41 [1975], Third National Cancer Survey: Incidence Data. DHEW Publication No. (NIH) 75–787.

National Center for Health Statistics Series 21, Number 26 [1976]. Selected Vital and Health Statistics in Poverty and Nonpoverty Areas of 19 Large Cities, United States, 1969–1971. US Government Printing Office, Washington, DC.

Polednak, A. P., Stehney, A. F., and Rowland, R. E. [1978]. Mortality among women first employed before 1930 in the U.S. radium dial-painting industry. *American Journal of Epidemiology*, **107:** 179–195.

CHAPTER 16

Analysis of the Time to an Event: Survival Analysis

16.1 INTRODUCTION

Many biomedical analyses study the time to an event. A cancer study of combination therapy using surgery, radiation, and chemotherapy, may examine the time from the onset of therapy until death. A study of coronary artery bypass surgery may analyze the time from surgery until death. In each of these two cases, the event being used is death. Other events are also analyzed. In some cancer studies, the time from successful therapy (i.e., a patient goes into remission) until remission ends is studied. In cardiovascular studies, one may analyze the time to a heart attack or death whichever event occurs first. A health services project may consider the time from enrollment in a health plan until the first use of the facilities. An analysis of children and their need for dental care may use the time from birth until the first cavity is filled. An assessment of an ointment for contact skin allergies may consider the time from treatment until the rash has cleared up.

In each of the above situations, the data consisted of the time from a fixed or designated initial point until an event occurs. This chapter shows how to analyze such "event data". When the event of interest is death, the subject is called *survival analysis*. Sometimes this name is used as a generic name, even when the endpoint or event being studied is not death but something else. In industrial settings the study of the lifetime of a component (until failure) is called *reliability theory*. For concreteness we will often speak of the event as death and the time as survival time. However, it should always be kept in mind that there are other uses.

In this chapter we will consider the presentation of time to event data, estimation of the time to an event and its statistical variability. We also consider potential predictor or explanatory variables. A third topic is to compare the time to event in several different groups. For example, a study of two alternative modes of cancer therapy may examine which group has the best survival experience.

16.2 THE SURVIVORSHIP FUNCTION OR SURVIVAL CURVE

In previous chapters we examined means of characterizing the distribution of a variable using, for example, the cumulative distribution function and histograms. One

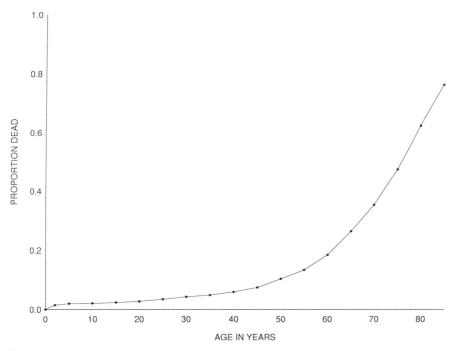

Figure 16.1. Cumulative probability of death, United States 1974. Source: Vital Statistics of the United States 1974, Volume II, Section 5, Life Tables [1976].

might take survival data and present the cumulative distribution function. Figure 16.1 shows an estimate for the United States population in 1974, of the probability of dying before a fixed age. This is an estimate of the cumulative distribution of survival in the United States in 1974. Note that there is an increase in deaths during the first year; after this the rate levels off but then climbs progressively in the later years. This cumulative probability of death is then an estimate of the probability that an individual dies at or before the given time. That is,

$$F(t) = P[\text{Person dies at a time} \leq t].$$

If we had observed the entire survival experience of the 1974 population we would estimate this quantity as we estimated the cumulative distribution function previously. We would estimate it according to the following formula.

$$F(t) = \frac{\text{Number of individuals who die at or before time } t}{\text{Total number observed}}. \quad (16.1)$$

Note, however, that we cannot estimate the survival experience of the 1974 population this way because we have not observed all of its members until death. This is a most fortunate circumstance since the population includes both of the authors of this text as well as many of its readers. In the next section, we will discuss some methods of estimating survival, when one does not observe the true survival of all the individuals.

It is depressing to speak of death; it is more pleasant to speak of life. In analyzing survival data, the custom has grown not of using the cumulative probability of death but using an equivalent function which is called the *survivorship function* or the *survival curve*. This function is merely the percent of individuals who live to a fixed time or beyond.

Definition 16.1.   The *survival curve*, or *survivorship function*, is the proportion or percent of individuals living to a fixed time t or beyond. The curve is then a function of t:

$$S(t) = \text{Percent of individuals surviving to time } t \text{ or beyond if}$$
$$\text{expressed as a percent,}$$

$$S(t) = \text{Proportion of individuals surviving to time } t \text{ or beyond}$$
$$\text{if expressed as a proportion.}$$

(16.2)

If we have a sample from a population, there is a distinction between the population survival curve and the sample or estimated population survival curve. In practice there is no distinct notation unless it is necessary to emphasize the difference. The context will usually show which of the two is meant.

The cumulative distribution function of the survival and the survival curve are closely related. If the two curves are continuous they are related by the following equation:

$$S(t) = 100\,(1 - F(t)) \quad \text{or} \quad S(t) = 1 - F(t).$$

(When we look at the sample curves, the curves are equal at all points except for the points where the curves jump. At these points there is a slight technical problem, because we have used $\leq$ in one instance and $\geq$ in the other instance. But for all practical purposes, the two curves are related by the above equation.)

Figure 16.2 shows the survival curve for the United States population as given in Figure 16.1. As you can see the survival curve results by "flipping over" the cumulative probability of death and using percentages. As mentioned above the estimate of the curve in Figure 16.2 is complicated by the fact that many individuals in the 1974 US population are happily alive. Thus, their true survival is not yet observed. The survival in the overall population is not yet observed. The survival in the overall population is estimated by the method discussed in the next section.

Sometimes the *proportion* surviving to time t or beyond is used. We will use them interchangeably. The two are simply related; to find the percent merely multiply the proportion by 100.

If we observe the survival of all individuals it is easy to estimate the survival curve. In analogy with the estimate of the cumulative distribution function the estimate of the survival curve at a fixed t is merely the percent of individuals whose survival was equal to the value t or greater. That is,

$$S(t) = 100 \left(\frac{\text{Number of individuals who survive to or beyond } t}{\text{Total number observed}} \right). \quad (16.3)$$

In many instances, we are not able to observe all individuals until they reach the event of interest. This makes the estimation problem more challenging. The next section discusses the estimates.

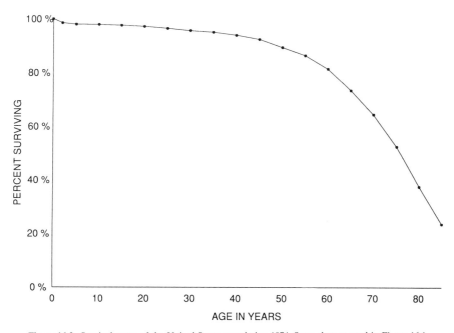

Figure 16.2. Survival curve of the United States population 1974. Same data as used in Figure 16.1.

16.3   ESTIMATION OF THE SURVIVAL CURVE: THE ACTUARIAL OR LIFE TABLE METHOD

Consider a clinical study of a procedure with a high initial mortality rate. For example, very delicate high risk surgery during its development period. Suppose we design a study to follow a group of such individuals for 2 years. Because most of the mortality is expected during the first year, it is decided to concentrate the effort on the first year. Two-thousand individuals are to be entered in the study; half of them will be followed for 2 years while one-half will only be followed for the critical first year. The individuals are randomized into two groups, group 1 to be followed for 1 year and group 2 to be followed for both years. Suppose that the data are as follows:

	Group 1		Group 2	
	Number Observed	**Number Who Died**	**Number Observed**	**Number Who Died**
Year 1	1000	240	1000	200
Year 2	—	—	800	16

We wish to estimate 1-year and 2-year survival. We consider three methods of estimation. The first two methods will not be appropriate but are used to motivate the correct life table method to follow.

One way of estimating survival might be to estimate separately the 1- and 2-year survival. Since it is wasteful to "throw away" data and the reason 2000 people were observed for 1 year was because that year was considered crucial, it is natural to estimate the percent surviving for 1 year by the total population. This percentage is as follows:

$$\text{Percent of 1-year survival} = 100 \times \frac{2000 - 240 - 200}{2000} = 78.0\%.$$

To estimate 2-year survival, we did not observe what happened to the individuals in Group 1 during the second year. Thus we might estimate the survival using only the individuals in Group 2. This estimate is

$$\text{Percent of 2-year survival} = 100 \times \frac{1000 - 200 - 16}{1000} = 78.4\%$$

We now have a problem in that the estimated percent surviving 1 year is less than the percent surviving 2 years! Clearly, as time increases, the percent surviving will decrease. This first method is not a reasonable way to estimate our survival curve.

One way to get around this problem is to use only the individuals from group 2 who are observed for 2 years. Then we have a straightforward estimate of survival at each time period. The percent surviving 1 year or more is 80% while the percent surviving 2 or more years is, as before, 78.4%. This gives a consistent pattern of survival but seems quite wasteful; we deliberately designed the study to allow us to observe more individuals in the first year when the mortality was expected to be high. It does not seem appropriate to throw away the 1000 individuals who were only observed for 1 year. If we need to do this, we had an extremely poor experimental design.

The solution to our problem is to note that we can efficiently estimate the probability of 1-year survival using both groups of people. Further using the second group we can estimate the probability of surviving the second year *conditionally upon having survived the first year*. The two estimates as percentages are

$$\text{Percent of 1-year survival} = 78.0\%,$$

$$\begin{array}{l} \text{Percent surviving year 2, given that an} \\ \text{individual survives year 1} \end{array} = 100 \times \frac{800 - 16}{800} = 98.0\%.$$

We can then combine these to get an estimate of the probability of surviving in the first year and the second year by using the concept of conditional probability. We see that the probability of 2-year survival is the probability of 1-year survival times the probability of 2-year survival given 1-year survival. Thus, the probability of 2-year survival is as follows:

$$P[A \text{ and } B] = P[A]P[B \mid A].$$

Let A = survival of 1 year, B = survival of 2 years. Then

$$P[\text{2-year survival}] = P[\text{1-year survival}]$$

$$\times P[\text{2-year survival} \mid \text{1-year survival}]$$

$$= 0.78 \times 0.98 = 0.7644.$$

For these probability calculations, note that it is more convenient to have probabilities than percents because the probabilities multiply. If we had percents it would have an extra factor of 100. For this reason the calculations on the survival curves are usually done as probabilites and then switched to percentages for graphical presentation. We will adhere to this.

Figure 16.3 presents the three estimates; for these data they are all close. The third estimate gives a consistent estimate of the curve (that is, the curve will never increase) and the estimate is efficient (because it uses all the data); it is the correct method for estimating survival.

It is possible to generalize this idea to more than two intervals.

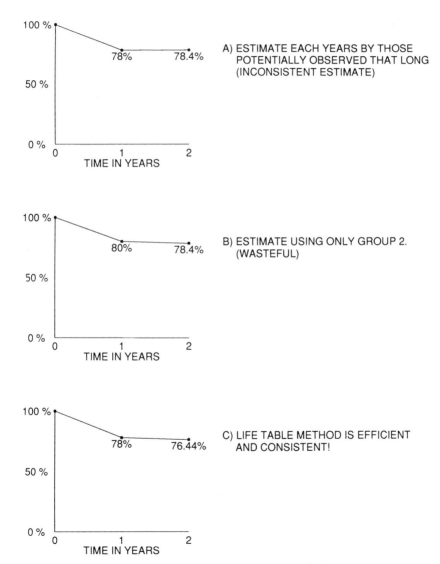

Figure 16.3. Three methods of estimating survival.

When the data are grouped into time intervals, we can estimate the survival in each interval. Let x denote the lower endpoint of each interval. (x rather than t is used here to conform to standard notation in the actuarial field. When it is necessary to index the intervals, we will use $i(x)$ to denote the inverse relationship.) Let Π denote the probability of surviving to $x(i)$ where $x(i)$ is the lower endpoint of the ith interval, that is,

$$\prod_i = S(x(i)),$$

where S is the survival curve (expressed here as the proportion surviving). Further, let π_i be the probability of living through the interval, with lower endpoint $x(i)$, conditionally upon the event of being alive at the beginning of the interval. Using the definition of a conditional probability

$$\pi_i = \frac{\prod_{i+1}}{\prod_i} = \frac{P[\text{Survive to the end of the } i\text{th interval}]}{P[\text{Survive to the end of the } i - 1\text{st interval}]}. \tag{16.4}$$

From this,

$$\prod_{i+1} = \pi_i \prod_i$$

and

$$\prod_{i+1} = \pi_1 \pi_2 \cdots \pi_i, \quad \text{where } \prod_1 = 1. \tag{16.5}$$

In graphically presenting group data, one plots points corresponding to the time of the lower endpoint of the interval and the corresponding Π_i value.

The plotted points are then joined by straight-line segments, as in Figure 16.4.

There is one further complication before we present the life table estimates. If we are periodically following people for example every 6 months or every year, it will occasionally happen that individuals cannot be located. Such individuals are called *lost to follow-up* in the study. Further, individuals may be withdrawn from the study for a variety of reasons. In clinical studies in the United States, all individuals have the right to withdraw from participation at any time. Or we might be trying to examine

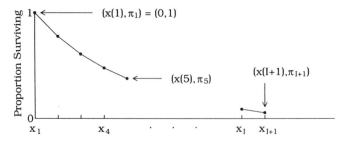

Figure 16.4. Form of the presentation of the survival curve for grouped survival data.

a medical survival in patients who could potentially be treated with surgery. Some of them may subsequently receive surgery; we could withdraw such individuals from the analysis at the time they received surgery. The rationale for this would be that after they received surgery their survival experience is potentially altered. Whatever the reason for an individual being lost to follow-up or withdrawn, this fact must be considered in the life table analysis.

To estimate the survival curve from data, the method is to estimate the π_i, and Π_i by the product of the estimates of the π_i according to Equation (16.5).

The data are usually presented in the form of Table 16.1. How might one estimate the probability of dying in the interval whose lower endpoint is x conditionally upon being alive at the beginning of the interval? At first glance one might reason that there were ℓ_x individuals of whom a (binomial) d_x died so that the estimate should be d_x/ℓ_x. The problem is that those who were lost to follow-up or withdrew during the interval might have died during the interval *after* withdrawing and this would not be counted. If such persons were equally likely to withdraw at any time during the interval, on the average they would be observed for only $1/2$ of the time. Thus, they really represent only $1/2$ a person at risk. Thus the effective number of individuals at risk, ℓ'_x, is

$$\ell'_x = \underbrace{\ell_x - (u_x + w_x)}_{\substack{\text{number} \\ \text{observed over} \\ \text{whole interval}}} + (1/2) \underbrace{(u_x + w_x)}_{\substack{\text{number} \\ \text{observed} \\ \text{over } 1/2 \\ \text{interval}}} \qquad (16.6)$$

$$= \ell_x - (1/2)(u_x + w_x).$$

The estimate of the proportion dying, q_x, is thus

$$q_x = \frac{d_x}{\ell'_x}.$$

The estimate of π_i, the probability of surviving the interval $x(i)$ to $x(i+1)$ is

$$p_{x(i)} = 1 - q_{x(i)}.$$

Table 16.1. Presentation of Life Table Data.

Interval	Number of Individuals Observed Alive at Beginning of Interval	Died During Interval	Lost to Follow-up During the Interval	Withdrawn Alive During Interval
x to $x + \Delta x$	ℓ_x	d_x	u_x	w_x
$x(1) - x(2)$	$\ell_{x(1)}$	$d_{x(1)}$	$u_{x(1)}$	$w_{x(1)}$
$x(2) - x(3)$	$\ell_{x(2)}$	$d_{x(2)}$	$u_{x(2)}$	$w_{x(2)}$
$\vdots$	$\vdots$	$\vdots$	$\vdots$	$\vdots$
$x(I) - x(I+1)$	$\ell_{x(I)}$	$d_{x(I)}$	$u_{x(I)}$	$w_{x(I)}$

Finally, the estimate of $\Pi_i = \pi_1 \pi_2 \cdots \pi_{i-1}$, $\Pi_1 = 1$ is

$$P_{x(i)} = p_{x(1)}p_{x(2)} \cdots p_{x(i-1)}, P_{x(0)} = 1. \tag{16.7}$$

Note that those who are lost to follow-up and those who are withdrawn alive are treated together; that is, in the estimates only the sum of the two is used. In many presentations such individuals are lumped together as *withdrawn* or *censored*.

Before presenting the estimates, it is also clear that an estimate of the survival curve will be more useful if some idea of its variability is given.

An estimate of the standard error of the P_x is given by Greenwood's formula (Greenwood [1926])

$$\text{SE}(P_{x(i)}) \doteq P_{x(i)} \sqrt{\sum_{j=1}^{i-1} \frac{q_{x(j)}}{\ell'_{x(j)} - d_{x(j)}}}$$

$$= P_{x(i)} \sqrt{\sum_{j=1}^{i-1} \frac{q_{x(j)}}{\ell'_{x(j)} P_{x(j)}}}. \tag{16.8}$$

This formula is a large sample approximation; *it is not reliable for small numbers of individuals.*

Example 16.1.   The method is illustrated by data of Parker et al. [1946], as discussed in Gehan [1969]. That data are from 2418 males with a diagnosis of angina pectoris (chest pain thought to be of cardiac origin) at the Mayo Clinic, between January 1, 1927 and December 31, 1936. The life table of survival time from diagnosis (in yearly intervals) is shown in Table 16.2.

Table 16.2. Life Table Analysis of 2418 Males with Angina Pectoris. Data from Gehan [1969].

x to $x + \Delta x$ (years)	ℓ_x	d_x	u_x	w_x	ℓ'_x	q_x	p_x	P_x	$\text{SE}(P_x)$
0–1	2418	456	0	0	2418	0.1886	0.8114	1.0000	—
1–2	1962	226	39	0	1942.5	0.1163	0.8837	0.8114	0.0080
2–3	1697	152	22	0	1686.0	0.0902	0.9098	0.7170	0.0092
3–4	1523	171	23	0	1511.5	0.1131	0.8869	0.6524	0.0097
4–5	1329	135	24	0	1317.0	0.1025	0.8975	0.5786	0.0101
5–6	1170	125	107	0	1116.5	0.1120	0.8880	0.5139	0.0103
6–7	938	83	133	0	871.5	0.0952	0.9048	0.4611	0.0104
7–8	722	74	102	0	671.0	0.1103	0.8897	0.4172	0.0105
8–9	546	51	68	0	512.0	0.0996	0.9004	0.3712	0.0106
9–10	427	42	64	0	395.0	0.1063	0.8937	0.3342	0.0107
10–11	321	43	45	0	298.5	0.1441	0.8559	0.2987	0.0109
11–12	233	34	53	0	206.5	0.1646	0.8354	0.2557	0.0111
12–13	146	18	33	0	129.5	0.1390	0.8610	0.2136	0.0114
13–14	95	9	27	0	81.5	0.1104	0.8896	0.1839	0.0118
14–15	59	6	23	0	47.5	0.1263	0.8737	0.1636	0.0123

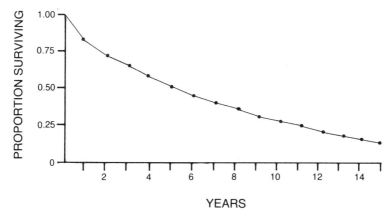

Figure 16.5. Survivorship function. Data from Table 16.2.

The survival data are given graphically in Figure 16.5. Note that in this case the proportion rather than the percent is presented.   □

As a second example, we consider individuals with the same diagnosis, angina pectoris, or chest pain thought to be of cardiac origin. These data are more recent.

Example 16.2.   Passamani *et al.* [1982] study individuals with chest pain that were studied for possible coronary artery disease. Chest pain upon exertion is often associated with coronary artery disease. The chest pain was evaluated by a physician as definite angina, probably angina, probably not angina, and definitely not angina. The definitions of these four classes were:

Definite Angina. A substantial discomfort which is preciptated by exertion, relieved by rest and/or nitroglycerin in less than ten minutes, and has a typical radiation to either shoulder, jaw, or the inner aspect of the arm. At times, definite angina may be isolated to the shoulder jaw, arm or upper abdomen.

Probable Angina. Has most of the features of definite angina but may not be entirely typical in some aspects.

Probably Not Angina. An atypical overall pattern of chest pain symptoms which does not fit the description of definite angina.

Definitely Not Angina. A pattern of chest pain symptoms which are unrelated to activity, unrelieved by nitroglycerin and/or rest, and appear clearly noncardiac in origin.

The data are plotted in Figure 16.6. Note how much improved the survival of the angina patients (definite and probable) is compared with the Mayo data of Figure 16.5. Those data had a 52% 5-year survival. These data have 91% and 85% 5-year survival! This indicates the great difficulty of using historical control data.

The lower left corner of the graph displays a statistic and *p*-value for testing for differences between the four groups. This is discussed in Section 16.7 below.

Table 16.3. Life Table for Definite Angina Patients. Time in Days.

$t(i)$	Enter	At Risk	Dead	Withdrawn Alive
0.0–90.9	2894	2894.0	44	0
91.0–181.9	2850	2850.0	28	0
182.0–272.9	2822	2822.0	22	0
273.0–363.9	2800	2799.0	25	2
364.0–454.9	2773	2773.0	23	0
455.0–545.9	2750	2750.0	23	0
546.0–636.9	2727	2727.0	23	0
637.0–727.9	2704	2563.5	12	281
728.0–818.9	2411	2394.0	17	34
819.0–909.9	2360	2359.0	19	2
910.0–1000.9	2339	2336.5	19	5
1001.0–1091.9	2315	2035.5	11	559
1092.0–1182.9	1745	1722.5	15	45
1183.0–1273.9	1685	1685.0	19	0
1274.0–1364.9	1675	1670.5	6	9
1365.0–1455.9	1660	1274.5	9	771

$t(i)$	Proportion Dead	Cumulative Survival to the end of Interval	SE	Effective Sample Size[a]
0.0–90.9	0.0152	0.9848	0.002	2893.99
91.0–181.9	0.0098	0.9751	0.003	2893.99
182.0–272.9	0.0078	0.9675	0.003	2894.00
273.0–363.9	0.0089	0.9589	0.004	2893.77
364.0–454.9	0.0083	0.9509	0.004	2893.46
455.0–545.9	0.0084	0.9430	0.004	2893.23
546.0–636.9	0.0084	0.9350	0.005	2893.06
637.0–727.9	0.0047	0.9306	0.005	2882.32
728.0–818.9	0.0071	0.9240	0.005	2850.22
819.0–909.9	0.0081	0.9166	0.005	2818.52
910.0–1000.9	0.0081	0.9091	0.005	2792.12
1001.0–1091.9	0.0054	0.9042	0.006	2753.73
1092.0–1182.9	0.0087	0.8963	0.006	2654.36
1183.0–1273.9	0.0059	0.8910	0.006	2596.11
1274.0–1364.9	0.0036	0.8878	0.006	2564.52
1365.0–1455.9	0.0071	0.8816	0.007	2449.65

[a]See Note 16.6 for a discussion of "effective sample size".

Table 16.3 gives the calculation using 91 day intervals and 4 intervals to approximate a year for one of the four groups, the definite angina patients.

As a sample calculation consider the interval from 637 to 728 days. We see that

$$\ell_x = 2704, \quad u_x + d_x = 281,$$

$$\ell'_x = 2704 - \frac{281}{2} = 2563.5,$$

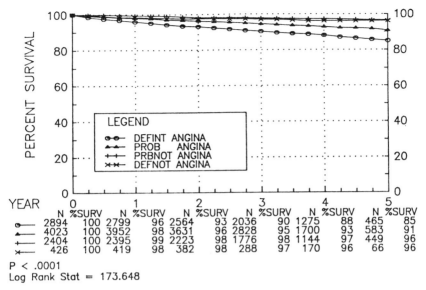

Figure 16.6. Survival by classification of chest pain. Data from Passamani *et al.* [1982].

$$q_x = \frac{12}{2563.5} = 0.0047,$$

$$p_x = 1 - 0.0047 = 0.9953,$$

$$P_x = 0.9350 \times 0.9953 = 0.9306.$$

Note in this (computer) output the cumulative survival is given to the end of the interval, rather than the beginning of the interval as in Table 16.2. Both presentations are in use.

Note also that the definite angina cases have the worst 5-year survival (85%) followed by the probable angina cases (91%). The other two categories have 96% 5-year survival (Figure 16.6).   □

As we have seen, in the life table method we have some data for which the event in question is not observed, often because at the time of the end of data collection and analysis, individuals are still alive. One term used for such data is *censoring*. The term censoring brings to mind a powerful, possibly sinister figure throwing away data to mislead one in the data analysis. In this context, it refers to the fact that although one is interested in survival times the actual survival times are not observed for all of the subjects. We have seen several sources of censored data. Subjects may be alive at the time of analysis; individuals may be lost to follow-up; individuals may refuse to participate further in research; or the individual may undergo a different therapy which removes them from estimates of the survival in a particular therapeutic group.

The *life table* or *actuarial* method that we have used above has the strength of allowing censored data and also uses the data with maximum efficiency. There is an important underlying assumption if we are to get unbiased estimates of the survival in a population from which such individuals may be considered to come. *It is necessary that the withdrawal or censoring not be associated with the endpoint.* Obviously if everyone is withdrawn because their situation deteriorates one would expect a bias in the estimation of death. Let us emphasize this again. The life table estimate gives unbiased estimates *only if the censoring mechanism is independent of the occurrence of the endpoint.*

16.4 THE HAZARD FUNCTION OR FORCE OF MORTALITY

In the analysis of survival data one is often interested in examining which periods have the highest or lowest risk of death. By risk of death, one has in mind the risk or probability among those alive at that time. For example, in very old age there is a high risk of dying each year *among* those reaching that age. The probability of any individual dying, say, in the 100th year is small because so few individuals live to be 100 years old.

   This concept is made rigorous by the idea of the hazard function or *hazard rate*. (A very precise definition of the hazard function requires ideas beyond the scope of this text and is discussed in the notes at the end of this chapter.) The hazard function is also called the *force of mortality*, the *age specific death rate, conditional failure rate* and *instantaneous death rate.*

   Definition 16.2. In a life table situation, *the (interval or actuarial) hazard rate* is the expected number dying in the interval divided by the product of the average number exposed in the interval and the interval width.

   In other words, the hazard rate, λ, is the probability of dying per unit time given survival to the time point in question. The estimate h of the hazard function is given by

$$h_x = \frac{d_x}{\ell'_x - d_x/2} \frac{1}{\Delta x}, \qquad (16.9)$$

where $\Delta x(1) = x(i+1) - x(i)$, the interval width. This is an estimate of the form

$$\frac{\text{No. dying}}{\text{Total exposure time}}.$$

ℓ'_x is an estimate of the number at risk of death. Note that this estimate is analogous to the definition in Section 15.4. Those who die will on the average have been exposed for approximately one-half of the time interval, so the number of intervals of observed time is approximately, $(\ell'_x - d_x/2) \Delta x$. Thus, the hazard rate is a death rate. If the hazard rate has a constant value λ over time the survival is exponential, that is, $S(t) = 100e^{-\lambda t}$, a point returned to later. The estimated hazard rate for Parker's data of Example 16.1 is given in Figure 16.7.

   A large sample approximation from Gehan [1969] for the se of h is

$$\text{SE}(h_x) = \left(\frac{h_x^3}{\ell_x q_x} \left(1 - \left(\frac{h_x \Delta x}{2} \right)^2 \right) \right)^{1/2}. \qquad (16.10)$$

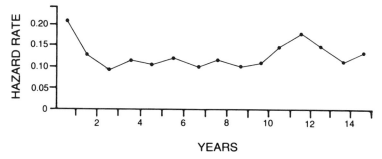

Figure 16.7. The hazard function for Example 16.1.

For the data of Example 16.1, we compute the hazard function for the second interval. We find

$$h_1 = \frac{226}{1942.5 - 226/2} \frac{1}{1} = 0.124.$$

16.5 PRESENTATION OF VITAL STATISTICS DATA

In vital statistics data for large populations the estimates are for most practical purposes, without error due to the sample size. (Other factors about the keeping of vital statistics may introduce systematic sorts of errors).

In presenting vital statistics data, the data are often presented in terms of a "fictitious" standard population of convenient size c, often 100,000. The life table presents "ℓ_x" $= cP(x)$. c is called the radix. Similarly, "d_x" $= \ell_x q_x$. Also of interest is an estimate of the expected years of life remaining for those who enter the interval with lower endpoint $x(i)$. For the interval with lower endpoint $x(j)$, let $x^*(j)$ be the midpoint of the interval. The *expected years of life remaining if one is alive at time $x(i)$* is denoted by $e^o_{x(i)}$ and estimated by

$$e^o_{x(i)} = \frac{\sum\limits_{j=i}^{\substack{last\\interval}} x^*((j+1) - x(i))d_{x(j)}}{\sum\limits_{j=i}^{\substack{last\\interval}} d_{x(j)}}. \tag{16.11}$$

Table 16.4 below presents vital statistics [1976] as prepared by the National Center for Health Statistics. (The explanation of the columns referring to a stationary population are given in the notes at the end of the chapter.)

The material of Table 16.4 for the total US population was used to plot the cumulative distribution function and survival curves of Figures 16.1 and 16.2.

Table 16.4. Abridged Life Table by Race and Sex: United States, 1974.

AGE INTERVAL	PROPORTION DYING	OF 100,000 BORN ALIVE		STATIONARY POPULATION		AVERAGE REMAINING LIFETIME
PERIOD OF LIFE BETWEEN TWO EXACT AGES STATED IN YEARS	PROPORTION OF PERSONS ALIVE AT BEGINNING OF AGE INTERVAL, DYING DURING INTERVAL	NUMBER LIVING AT BEGINNING OF AGE INTERVAL	NUMBER DYING DURING AGE INTERVAL	IN THE AGE INTERVAL	IN THIS AND ALL SUBSEQUENT AGE INTERVALS	AVERAGE NUMBER OF YEARS OF LIFE REMAINING AT BEGINNING OF AGE INTERVAL
x to $x+n$	$_nq_x$	l_x	$_nd_x$	$_nL_x$	T_x	e_x
TOTAL						
0-1	0.0167	100,000	1,675	98,503	7,191,375	71.9
1-5	.0029	98,325	287	392,633	7,092,872	72.1
5-10	.0019	98,038	185	489,693	6,700,239	68.3
10-15	.0019	97,853	190	488,864	6,210,546	63.5
15-20	.0053	97,663	519	487,138	5,721,682	58.6
20-25	.0070	97,144	677	484,049	5,234,544	53.9
25-30	.0068	96,467	660	480,685	4,750,495	49.2
30-35	.0079	95,807	756	477,227	4,269,810	44.6
35-40	.0108	95,051	1,030	472,845	3,792,583	39.9
40-45	.0168	94,021	1,579	466,417	3,319,738	35.3
45-50	.0267	92,442	2,470	456,421	2,853,321	30.9
50-55	.0397	89,972	3,574	441,480	2,396,900	26.6
55-60	.0608	86,398	5,249	419,575	1,955,420	22.6
60-65	.0904	81,149	7,339	388,244	1,535,845	18.9
65-70	.1270	73,810	9,373	346,531	1,147,601	15.5
70-75	.1908	64,437	12,293	292,426	801,070	12.4
75-80	.2790	52,144	14,549	224,856	508,644	9.8
80-85	.3629	37,595	14,394	151,407	283,788	7.5
85 AND OVER	1.0000	23,201	23,201	132,381	132,381	5.7
MALE						
0-1	0.0187	100,000	1,874	98,323	6,814,371	68.1
1-5	.0032	98,126	318	391,773	6,716,048	68.4
5-10	.0022	97,808	220	488,454	6,324,275	64.7
10-15	.0024	97,588	238	487,459	5,835,821	59.8
15-20	.0078	97,350	755	485,059	5,348,362	54.9
20-25	.0106	96,595	1,020	480,456	4,863,303	50.3
25-30	.0098	95,575	938	475,492	4,382,847	45.9
30-35	.0107	94,637	1,011	470,744	3,907,355	41.3
35-40	.0141	93,626	1,323	465,025	3,436,611	36.7
40-45	.0213	92,303	1,970	456,917	2,971,586	32.2
45-50	.0349	90,333	3,151	444,297	2,514,669	27.8
50-55	.0524	87,182	4,567	425,232	2,070,372	23.7
55-60	.0815	82,615	6,729	397,159	1,645,140	19.9
60-65	.1223	75,886	9,282	357,178	1,247,981	16.4
65-70	.1741	66,604	11,599	304,825	890,803	13.4
70-75	.2510	55,005	13,807	240,962	585,978	10.7
75-80	.3505	41,198	14,441	169,512	345,016	8.4
80-85	.4539	26,757	12,146	102,127	175,504	6.6
85 AND OVER	1.0000	14,611	14,611	73,377	73,377	5.0
FEMALE						
0-1	0.0146	100,000	1,465	98,694	7,581,958	75.8
1-5	.0026	98,535	255	393,540	7,483,264	75.9
5-10	.0015	98,280	149	490,997	7,089,724	72.1
10-15	.0014	98,131	138	490,345	6,598,727	67.2
15-20	.0028	97,993	275	489,325	6,108,382	62.3
20-25	.0034	97,718	333	487,772	5,619,057	57.5
25-30	.0039	97,385	383	485,999	5,131,285	52.7
30-35	.0052	97,002	503	483,831	4,645,286	47.9
35-40	.0077	96,499	744	480,762	4,161,455	43.1
40-45	.0124	95,755	1,190	475,992	3,680,693	36.4
45-50	.0190	94,565	1,798	468,594	3,204,701	33.9
50-55	.0278	92,767	2,581	457,748	2,736,107	29.5
55-60	.0415	90,186	3,742	442,066	2,278,359	25.3
60-65	.0617	86,444	5,330	419,622	1,836,293	21.2
65-70	.0877	81,114	7,111	388,798	1,416,671	17.5
70-75	.1438	74,003	10,639	344,843	1,027,873	13.9
75-80	.2285	63,364	14,477	281,990	683,030	10.8
80-85	.3381	48,887	16,529	203,325	401,040	8.2
85 AND OVER	1.0000	32,358	32,358	197,715	197,715	6.1

16.6   THE PRODUCT LIMIT OR KAPLAN–MEIER ESTIMATE OF THE SURVIVAL CURVE

If survival data are recorded in great detail, accuracy is preserved by placing the data into smaller rather than larger intervals. Obviously, if data are grouped, for example into 5-year intervals while the time of death is recorded to the nearest day, considerable detail is lost. The product limit or Kaplan–Meier estimate is based upon the idea of taking more and more intervals. In the limit, the intervals become arbitrarily small.

Suppose in the following that the time at which data are censored (lost to follow-up or withdrawn from the study) and the time of death (when observed) is measured to a high degree of accuracy. The *product limit or Kaplan–Meier* (see Kaplan and Meier [1958]) estimate (KM estimate) results from the actuarial or life table method of the last section as the number of intervals increases in such a way that the maximum interval width approaches zero. In this case it can be seen that the estimated survival curve is constant except for jumps at the observed time of death. The values of the survival probability before a time of death(s) is multiplied by the estimated probability of surviving past the time of death to find the new value of the survival curve.

To be more precise, suppose that n individuals are observed. Further, suppose that the time of death is observed in ℓ of the individuals at k distinct times $t_1 < t_2 < \cdots < t_k$. Let m_i be the number of deaths at time t_i. The other $n - \ell$ individuals are censored observations. If a censoring time and a death occur at the same time, it is assumed that the true time of death for the censored individual is greater than the observed censoring time. Let n_i be the number of individuals at risk of dying at time t_i. That is, $n_i = n$ minus the number of deaths prior to t_i and minus the number of individuals whose observations were censored prior to time t_i. The product limit estimate of the survival curve expressed as a proportion is

$$S(t) = 1, \quad \text{for } t < t_1,$$

$$S(t) = \prod_{j=1}^{i} \frac{n_i - m_i}{n_i}, \quad t_i \le t < t_{i+1} \; (i < k),$$

$$S(t) = 0, \quad \text{for } t_k \le t \quad \text{if } m_k = n_k \text{ (i.e., no one survives past time } t_k),$$

$$S(t) = \prod_{j=1}^{k} \frac{n_i - m_i}{n_i}, \quad \text{for } t_k \le t \le \text{largest observed censored observation.}$$

$$(16.12)$$

If $m_k < n_k$, then $S(t)$ is undefined for $t >$ largest observed censored observation.

We illustrate the method with an example.

Example 16.3.   We again use the Stanford heart transplant data discussed in Section 15.4. Suppose that we wished to estimate the medical survival of these patients. A complication is that when a donor heart becomes available, the patient has a heart transplant; we can no longer observe what the medical survival would have been. Presumably, the availability of a heart transplant, which depends on some unfortunate circumstance such as an accident to the individual who supplies the donor heart, is independent of the need for the heart by the recipient. Thus, we may use surgery for heart transplantation as a source of censoring medical survival. One *incorrect* way to analyze such data would be the following. Since we are interested in medical survival, we should not worry about individuals who have had surgery. We should go through the records and look at the survival curves only for the individuals who did not have surgery. Since such individuals by definition died awaiting the donor heart, their early survival experience would be quite poor.

Table 16.5 presents the medical survival data using surgery as the source of censoring for the Stanford heart transplant patients. The computations as described above are given.

The product limit estimate of the correct survival curve is plotted in Figure 16.8. It is given by the solid line. The line with the X's is the incorrect curve if one ignores the effect of surgery as censoring and totally eliminates such individuals from the

Table 16.5.

t(days)	Death ($*$)	n_i	$(n_i - m_i)/n_i$	$S(t)$, $t_i \leq t < t_{i+1}$
1	$*$	34	33/34	0.971
1		33		
2		32		
5	$*$	31	30/31	0.939
7	$*$	30	29/30	0.908
7		29		
11		28		
11		27		
12	$*$	26	25/26	0.873
15	$*$	25	24/25	0.838
15		24		
16		23		
17	$*$	22	21/22	0.800
17		21		
17		20		
19		19		
22		18		
24		17		
24		16		
26		15		
34	$*$	14	13/14	0.743
34		13		
35	$*$	12	11/12	0.681
36	$*$	11	10/11	0.619
36		10		
40	$*$	9	8/9	0.550
49	$*$	8	7/8	0.481
49		7		
50		6		
69		5		
81		4		
84	$*$	3	2/3	0.321
111	$*$	2	1/2	0.160
480		1		

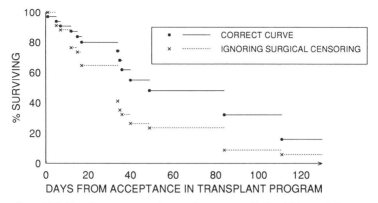

Figure 16.8. Days from acceptance in transplant program. Kaplan–Meier survival curve.

analysis. Finally, note that there was one patient who spontaneously improved under medical treatment and was reported alive at 16 months. The data of that subject is reported in the medical survival data as a 480 day survivor. As before, an asymptotic formula for the standard error of the estimate may be given. Greenwood's formula for the approximate standard error of the estimate also holds in this case. The form it takes is

$$\text{SE}(S(t)) \doteq S(t) \sqrt{\sum_{j=1}^{i} \frac{m_j}{n_j(n_j - m_j)}}, \quad \text{for } t_i \le t < t_{i+1}. \tag{16.13}$$

$\square$

16.7 THE COMPARISON OF DIFFERENT SURVIVAL CURVES: THE LOG-RANK TEST

In this section we consider a test statistic for comparing two or more survival curves for different groups of individuals. This statistic is based upon the following idea. Take a particular interval in which deaths occur, or in the case of the product limit curve, a time when one or more deaths occur. Suppose that the first group considered has one third of the individuals being observed. How many deaths would we expect in the first group if in fact the survival experience is the same for all of the different groups? We expect the number of deaths to be proportional to the fraction of the people at risk of dying in the group. That is, for the first group the expected number of deaths would be the observed number of deaths at that time divided by three. The log-rank test uses this simple fact. At each interval or time of death we take the observed number of deaths and calculate the expected number of deaths that would occur in each of the groups if all had the same risk of dying. For each group, the expected number of deaths is summed over all intervals and then compared to the observed number of deaths. Using this comparison, we get a statistic, the *log-rank statistic*, which has approximately a chi-square distribution with $k - 1$ degrees of freedom when k groups are observed. We formalize this.

Suppose that one is interested in comparing the survival experience of k populations. Suppose there are M different times at which deaths appear. For the life table method, this will usually be each interval. In the product limit approach, each observed death will be associated with a unique time. At the mth time let d_{im} be the number of deaths observed in the ith population and ℓ_{im} the number at risk of dying. (For the life table approach with withdrawls ℓ_{ij} is the appropriate ℓ'_x.) The data may be presented in M $2 \times k$ contingency tables with totals:

1	2	$\cdots$	k	
d_{1m}	d_{2m}	$\cdots$	d_{km}	D_m Dying
$\ell_{1m} - d_{1m}$	$\ell_{2m} - d_{2m}$	$\cdots$	$\ell_{km} - d_{km}$	A_m Alive
ℓ_{1m}	ℓ_{2m}	$\cdots$	ℓ_{km}	T_m Total

$$m = 1, 2, \ldots, M.$$

If all of the k populations are at equal risk of death, the probability of death will be the same in each population and conditionally upon the row and column totals,

$$E(d_{im}) = \frac{\ell_{im} D_m}{T_m}, \tag{16.14}$$

as in the chi-square test for contingency tables.

In the ith population, the total number of observed deaths is

$$O_i = \sum_{m=1}^{M} d_{im}. \tag{16.15}$$

Examining all of the times of death, the expected number of deaths in the ith population is

$$E_i = \sum_{m=1}^{M} E(d_{im}) = \sum_{m=1}^{M} \frac{\ell_{im} D_m}{T_m}. \tag{16.16}$$

The test statistic is then computed from the observed minus expected values. A note at the end of the chapter describes in detail how to compute the statistic. A simple approximate statistic (which may be used in place of the more powerful statistic that is more difficult to compute) is the following:

$$X^2 = \sum_{i=1}^{k} (O_i - E_i)^2 / E_i. \tag{16.17}$$

The statistic is written in the familiar form of the chi-square test for comparing observed and expected values. (If any $E_i = 0$, define $(O_i - E_i)^2 / E_i = 0$). Under the null hypothesis of equal survival curves in the k groups this statistic will have approximately a chi-square distribution with $k - 1$ degrees of freedom if,

1. All $E_i > 0$,
2. $A_m/(T_m - 1)$ is close to one for all m,
3. At each time of death the proportion of subjects in each group is approximately the same. That is, for each i, ℓ_m/T_m does not change much, as m varies.

Crowley and Breslow [1975] discuss this approximation. (Also see the notes at the end of this chapter.) If (1), (2), or (3) are not satisfied, the use of X^2 as a chi-square variable is conservative and the true probability of rejecting the null hypothesis, when it is true, is less than the significance level being used.

The method is illustrated by using the data of the Stanford transplant patients (Table 16.5) and comparing it with the data of Houston heart transplant patients, as reported in Messmer *et al.* [1969]. The time of survival for 15 Houston patients is read from Figure 16.9 and therefore has some inaccuracy.

Ordering both the Stanford and Houston transplant patients by their survival time after transplantation and status (dead or alive) gives Table 16.6. The dashes for the d_{im} indicate where withdrawals occur and those lines could have been omitted in the calculation. One stops when there are no future deaths at a time when members of

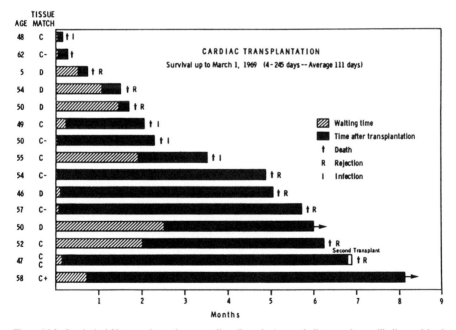

Figure 16.9. Survival of fifteen patients given a cardiac allograft. Arrows indicate patients still alive on March 1, 1969. Data from Messmer *et al.* [1969].

Table 16.6. Stanford and Houston Survival Data.

Day	Stanford		Houston			
	ℓ_{1m}	d_{1m}	ℓ_{2m}	d_{2m}	$E(d_{1m})$	$E(d_{2m})$
1	20	1	15	0	0.571	0.429
3	19	1	15	0	0.559	0.441
4	18	0	15	1	0.545	0.455
6	18	0	14	2	1.125	0.875
7	18	0	12	1	0.600	0.400
10	18	1	11	0	0.621	0.379
12	17	0	11	1	0.607	0.393
15	17	1	10	0	0.630	0.370
24	16	1	10	0	0.615	0.385
39	15	1	10	0	0.600	0.400
46	14	1	10	0	0.583	0.417
48	13	0	10	1	0.565	0.435
54	13	0	9	1	0.591	0.409
60	13	1	8	0	0.619	0.381
61	12	1	8	1	1.200	0.800
102	11	0	7	0	—	—
104	10	0	6	0	—	—
110	10	0	6	1	0.625	0.375
118	10	0	5	0	—	—
127	9	1	5	0	0.643	0.357
136	8	1	5	0	0.615	0.385
146	7	0	5	1	0.583	0.417
148	7	0	4	1	0.636	0.364
169	7	0	3	1	0.700	0.300
200	7	0	2	1	0.778	0.222

both populations are present. Summing the appropriate columns, one finds that

$$O_1 = \sum_m d_{1m} = 11,$$

$$E_1 = \sum_m D(d_{1m}) = 14.611,$$

$$O_2 = \sum_m d_{2m} = 13,$$

$$E_2 = \sum_m E(d_{2m}) = 9.389.$$

The log-rank statistic is 2.32. The simple, less powerful, approximation is $X^2 = (11 - 14.611)^2/14.611 + (13 - 9.389)^2/9.389 = 2.28$. Looking at the critical values of the chi-square distribution with one degree of freedom, there is not a statistically significant difference in the survival experience of the two populations.

Another approach is to look at the difference between survival curves at a fixed time point. Using either the life table or Kaplan–Meier product limit estimate at a fixed time T_o, one can estimate the probability of survival to T_o, say, $S(T_o)$ and the standard error of $S(T_o)$, $\text{SE}(S(T_o))$, as described in the sections above. Suppose that a subscript is used on S to denote estimates for different populations. To compare the survival experience of two populations with regard to surviving to T_o, the following statistic is $N(0, 1)$, as the sample sizes become large (when the null hypothesis of $S_1(T_o) = S_2(T_o)$ is valid.)

$$Z = \frac{S_1(T_o) - S_2(T_o)}{\sqrt{\text{SE}(S_1(T_o))^2 + \text{SE}(S_2(T_o))^2}}. \qquad (16.18)$$

One- or two-sided test may be performed depending upon the alternative hypothesis of interest.

For k groups, to compare the probability of survival to time T_o, the estimated values may be compared by constructing multiple comparison confidence intervals.

16.8 ADJUSTMENT FOR CONFOUNDING FACTORS BY STRATIFICATION

In Example 16.2, in the Coronary Artery Surgery Study (Passamani et al. [1982]), the degree of impairment due to chest pain pattern was related to survival. Individuals with pain definitely not angina had a better survival pattern than individuals with definite angina. The chest pain status is predictive of survival. These patients were studied by coronary angiography; the amount of disease in their coronary arteries as well as their left ventricular performance (the performance of the pumping part of the heart) were also evaluated. One might argue that the amount of disease is a more fundamental predictor than type of chest pain. If the pain results from coronary artery disease which affects the arteries and ventricle, then the latter affects survival more fundamentally. We might ask the question; is there additional prognostic information in the type of chest pain if one takes into account, or adjusts for, the angiographic findings?

We have used various methods of adjusting for variables. As discussed in Chapter 2, twin studies adjust for genetic variation by matching individuals with the same genetic pattern. Analogously, matched pairs studies match individuals to be (effectively) twins in the pertinent variables; this adjusts for covariates. One step up from this is the *stratified* analysis. In this case, the strata are to be quite homogeneous. Individuals in the same strata are (to a good approximation) the same with respect to the variable or variables used to define the strata. One example of stratified analysis occurred with the Mantel–Hanszel procedure for summing 2×2 tables. The point of the stratification was to adjust for the variable or variables defining the strata. In this section we consider the same approach to the analysis of the life table or actuarial method of comparing survival curves from different groups.

16.8.1 Stratification of Life Table Analyses: The Log-Rank Test

To extend the life table approach to stratification is straightforward. The first step is to perform the life table survival analysis *within each stratum*. If we do this for the four chest pain classes as discussed in Example 16.2 to adjust for angiographic data, we would use strata that depend upon the angiographic findings. This is done below. Within each of the strata, we will be comparing individuals with the same angiographic findings but different chest pain status. The log-rank statistic may be computed *separately* for each of the strata, giving us an observed and expected number of deaths for each group being studied. Somehow we want to combine the information across all of the strata. This was done for example in the Mantel–Hanszel approach to 2×2 tables. We do this by summing the values for each group of the observed and expected numbers of deaths for the different strata. These observed and expected

Table 16.7. Stratified Analysis of Survival by Chest Pain Classification.

	Stratification Variables			Deaths								
Stratum Number	# Vessels	# Prox. Vessels	Left Ventric- ular Score	Definite Angina		Probable Angina		Probably Not Angina		Definitely Not Angina		log rank statistic p-value
				Obs.	Exp.	Obs.	Exp.	Obs.	Exp.	Obs.	Exp.	
1	0	0	5-11	9	10.07	42	38.33	39	43.35	9	7.25	0.74
2	0	0	12-16	0	0.79	2	1.25	1	0.87	0	0.09	0.73
3	0	0	17-30	0	0.00	0	0.00	0	0.00	0	0.00	1.00
4	1	0	5-11	19	18.88	26	23.84	5	6.71	0	0.56	0.85
5	1	0	12-16	3	3.46	5	3.25	0	1.06	0	0.23	0.52
6	1	0	17-30	1	0.31	0	0.62	0	0.08	.	–	0.43
7	1	1	5-11	14	13.36	13	13.19	2	2.00	0	0.45	0.96
8	1	1	12-16	1	2.53	3	2.05	0	0.27	1	0.15	0.15
9	1	1	17-30	4	3.49	2	2.22	0	0.30	.	–	0.93
10	2	0	5-11	17	18.54	16	14.62	2	2.29	1	0.55	0.93
11	2	0	12-16	7	6.81	2	3.90	3	1.11	0	0.18	0.20
12	2	0	17-30	5	3.49	3	3.99	1	0.98	0	0.53	0.72
13	2	1	5-11	18	15.50	10	14.91	1	1.07	2	0.24	0.11
14	2	1	12-16	9	9.08	6	4.99	0	0.80	0	0.14	0.77
15	2	1	17-30	3	3.40	3	2.38	0	0.22	.	–	0.93
16	2	2	5-11	18	17.36	13	13.56	1	0.92	0	0.16	0.99
17	2	2	12-16	19	6.70	4	5.98	0	0.32	.	–	0.59
18	2	2	17-30	3	4.67	4	2.33	.	–	.	–	0.62
19	3	0	5-11	11	11.75	9	7.44	0	0.72	0	0.10	0.76
20	3	0	12-16	8	7.49	7	6.69	.	–	0	0.83	0.83
21	3	0	17-30	4	4.31	1	0.69	.	–	.	–	0.98
22	3	1	5-11	28	23.67	15	17.78	0	1.54	.	–	0.37
23	3	1	12-16	17	16.66	6	6.34	.	–	.	–	1.00
24	3	1	17-30	9	7.32	5	6.15	0	0.53	.	–	0.72
25	3	2	5-11	36	32.08	11	17.55	2	0.34	1	0.03	0.01
26	3	2	12-16	20	16.48	6	8.45	0	1.07	.	–	0.42
27	3	2	17-30	8	9.34	7	5.17	.	–	0	0.49	0.72
28	3	3	5-11	17	22.42	19	14.36	1	0.22	.	–	0.1
29	3	3	12-16	16	14.62	6	8.24	1	0.14	.	–	0.0
30	3	3	17-30	11	12.93	4	2.07	.	–	.	–	0.5
TOTAL				325	317.49	250	251.63	59	66.91	14	11.97	0.69

* A DASH INDICATES NO INDIVIDUALS IN THE GROUP IN THE GIVEN STRATUM. Obs.=Observed; Exp.=Expected. Log rank statistic = 1.47 with degrees of freedom.

numbers are then combined into a final log-rank statistic. Note 16.3 gives the details of the computation of the statistic. The final statistic, because it is based upon many more individuals, will be much more powerful then the log-rank statistic for any one stratum, *provided* that there is a consistent trend in the same direction within strata. We illustrate this by example.

Example 16.4. We continue with Example 16.2 of the chest pain groups. We would like to adjust for angiographic variables. A study of the angiographic variables showed that most of the prognostic information is contained within these variables:

1. The number of vessels diseased of the three major coronary vessels;
2. The number of proximal vessels diseased (that is, the number of diseased vessels where the disease is near the point where the blood pumps into the heart);
3. The left ventricular function, which will be measured by a variable called LVS-CORE.

Various combinations of these three variables were used to define 30 different strata. Table 16.7 gives the values of the variables and the strata.

Separate survival curves result in the differing strata. Figures 16.10 and 16.11 present the survival curves for two of the different strata used.

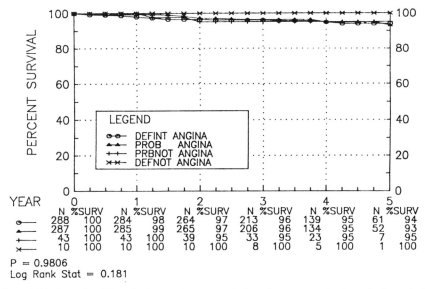

Figure 16.10. Example 16.4: survival curves for stratum 7. Cases have one proximal vessels diseased with good ventricular function (LVSCORE of 5–11).

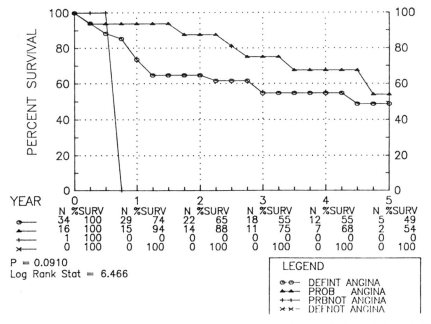

Figure 16.11. Example 16.4: survival curves for stratum 29. Cases have three proximal vessels diseased with impaired ventricular function (LVSCORE of 12–17).

Note that the overall p-value is 0.69, a result that is not statistically significant. Thus although the survival patterns differ among chest pain categories, the differences may be explained by different amounts of underlying coronary artery disease. In other words, adjustment for the arteriographic findings removed the group differences.

Note that of 30 strata one p-value is less than 0.05. Because of the multiple comparison problem this is not a worry. Further, in this strata 25, the definite angina cases have one observed and 0.03 expected deaths. As the log-rank statistic has an *asymptotic* chi-square distribution the small expected number of deaths make the asymptotic distribution inappropriate in this stratum. □

The survival curves after adjustment are sometimes presented using the method of direct adjustment.

Definition 16.3. Consider a stratified life table analysis with m strata and k groups. Let $S_{ij}(t)$ be the survival curve for the ith group in the j strata. Let n_j be the number of individuals in the jth strata (among all groups). Let $r_j = n_j / \sum_{k=1}^{m} n_k$ be the proportion of individuals in the jth strata. The *directly adjusted survival curve* for the ith group is

$$S_i(t) = \sum_{j=1}^{m} r_j S_{ij}(t). \tag{16.19}$$

Visual comparison of adjusted and unadjusted survival curves gives an idea of the effect of the adjustment.

Example 16.5. The study of Examples 16.2 and 16.4 also considered use of a type of drug called a beta blocking drug. Figures 16.12 and 16.13 present the adjusted and unadjusted survival curves. The 30 strata of Example 16.4 were used for adjustment. We see that the individuals on beta blocking drugs had a slightly worse unadjusted

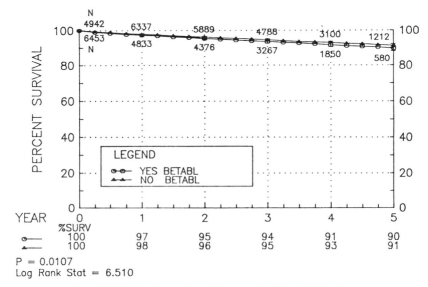

Figure 16.12. Survival curves for patients on and off beta blocking drugs.

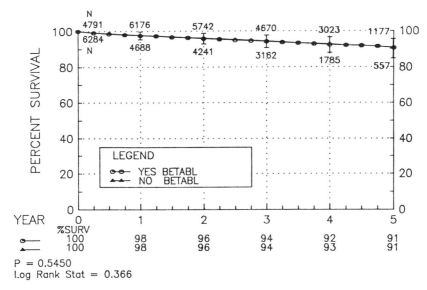

Figure 16.13. Adjusted survival curves for patients on and off beta blocking drugs.

survival ($p = 0.01$). After adjustment for the angiographic data there is no observable difference. (Of course adjustment for other variables could conceivably introduce a difference either way.) Some cases are missing from the adjusted survival curves because the data needed to make the adjustment were not available. □

16.9 THE COX PROPORTIONAL HAZARD REGRESSION MODEL

In earlier work on the life table method, we observed various ways of dealing with factors that were related to survival. One method is to plot data for different groups, where the groups were defined by different values on the factor(s) being analyzed. When we wanted to adjust for covariates, we examined stratified life table analyses. These approaches are limited, however, by the numbers involved. If we want to divide the data into strata on ten variables simultaneously, there will be so many strata that most strata will contain no individuals or at most one individual. This makes comparisons impossible. One way of getting around the number problem is to have an appropriate mathematical model with covariates. In this section we consider the *Cox proportional hazards regression model*. This model is a mathematical model of survival which allows covariate values to be taken into account. The usage of the model in survival analysis is quite similar to the multiple regression analysis of Chapter 11. We first turn to examination of the model itself.

16.9.1 The Cox Proportional Hazard Model

Suppose we want to examine the survival pattern of the two individuals, one of whom initially is at higher risk than a second individual. A natural way to quantify the idea of risk is the hazard function discussed previously. We may think of the hazard function as the instantaneous probability of dying, given that an individual has survived to a particular time. The person with the higher risk will have a higher value for the hazard function than an individual who has lower risk at the particular time. The Cox proportional hazard model works with covariates; the model expresses the hazard as a function of the covariate values. The major assumption of the model is that if the first individual has a risk of death at the initial time point that is, say, twice as high as that of a second individual, then at later times the risk of death is also twice as large. We now express this mathematically.

Suppose that at the average value of all of our covariates in the population, the hazard at time t is denoted by $h_0(t)$. Any other individual whose values on the variables being considered are not equal to the mean values, will have a hazard function proportional to $h_0(t)$. This proportionality constant varies from individual to individual depending upon the values of the variables. We develop this algebraically. There are variables $X_1, \ldots, X_p$ to be considered. Let $\mathbf{X}$ denote the values of all the X_i, that is, $\mathbf{X} = (X_1, \ldots, X_p)$.

1. If an individual has $\mathbf{X} = \overline{\mathbf{X}} = (\overline{X}_1, \ldots, \overline{X}_p)$ then the hazard function is $h_0(t)$.
2. If an individual has different values for $\mathbf{X}$ the hazard function is $h_0(t)C$, where C is a constant that depends on the values of $\mathbf{X}$. If we think of the hazard as depending upon $\mathbf{X}$, as well as t, the hazard is

$$h_0(t)C(\mathbf{X}).$$

3. For any two individuals with values of $\mathbf{X} = \mathbf{X}(1)$ and $\mathbf{X} = \mathbf{X}(2)$, respectively, the ratio of their two hazard functions is

$$\frac{h_0(t)C(\mathbf{X}(1))}{h_0(t)C(\mathbf{X}(2))} = \frac{C(\mathbf{X}(1))}{C(\mathbf{X}(2))}. \tag{16.20}$$

The hazard functions are *proportional*, the ratio does not depend upon t.

Let us reiterate this last point. Given two individuals, if one has one-half as much risk initially as a second person, then at all time points risk is one-half that of the second person. Thus the two hazard functions are proportional, and such models are called *proportional hazard models*.

Note that proportionality of the hazard function is an assumption which does not necessarily hold. For example, if two individuals were such that one is to be treated medically and the second surgically by open heart surgery, the individual being treated surgically may be at higher risk initially because of the possibility of operative mortality; later, however, the risk may be the same or even less than the equivalent individual being treated medically. In this case, if one of the covariate values indicates whether a person is treated medically or surgically, the proportional hazards model will not hold. In a given situation you need to examine the plausibility of the assumption. The model has been shown empirically to hold reasonably well for many populations over moderately long periods, say 5–10 years. Still, it is an assumption.

As currently used, one particular parametric form has been chosen for the proportionality constant $C(\mathbf{X})$. This constant, since it multiplies a hazard function, must always be positive because the resulting hazard function is an instantaneous probability of an endpoint and consequently must be nonnegative. A convenient functional form which reasonably fits many data sets is the following:

$$C(\mathbf{X}) = e^{\alpha + \beta_1 X_1 + \cdots + \beta_p X_p}, \text{ where } \alpha = -\beta_1 \overline{X}_1 - \cdots - \beta_p \overline{X}_p. \tag{16.21}$$

In this parameterization, the unknown population parameters β_i are to be estimated from a data set at hand.

With hazard $h_0(t)$, let $S_{0,pop}(t)$ be the corresponding survival curve. For an individual with covariate values $\mathbf{X} = (X_1, \ldots, X_p)$ let the survival be $S(t \mid \mathbf{X})$. Using the previous equations the survival curve is:

$$S(t \mid \mathbf{X}) = (S_{0,pop}(t))^{\exp(\alpha + \beta_1 X_1 + \cdots + \beta_p X_p)} . \tag{16.22}$$

This equation looks quite complex; the survival curve for a fixed individual is a survival curve for the entire group evaluated at the mean value, raised to an appropriate power. To estimate this quantity, the following steps are performed:

1. Estimate $S_{0,pop}$ and $\alpha, \beta_1, \ldots, \beta_p$ by $S_0(t), a, b_1, \ldots, b_p$. This is done by a computer program. The estimation is too complex to do by hand.
2. Compute $Y = a + b_1 X_1 + \cdots + b_p X_p$ (where $\mathbf{X} = (X_1, \ldots, X_p)$).
3. Compute $k = e^Y$.
4. Finally, compute $S_0(t)^k$.

The estimated survival curve is the population curve (the curve for the mean covariate values) raised to a power. If the power k is equal to 1, corresponding to e^0, the underlying curve for S_0 results. If k is greater than 1, the curve lies below S_0, and if k is less than 1 the curve lies above S_0. This is presented graphically in Figure 16.14.

Note several factors about these curves.

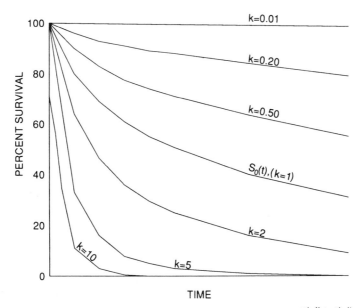

Figure 16.14. Proportional hazard survival curves as a function of $k = e^{a + b_1 X_1 + \cdots + b_p X_p}$.

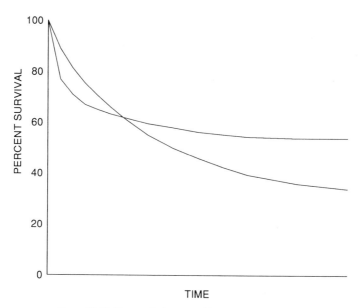

Figure 16.15. Two survival curves without proportional hazards.

1. The curves do not cross each other. This means that a procedure having a high initial mortality such as a high dose of radiation in cancer therapy, but better long term survival, as in Figure 16.15, could not be modeled by the proportional hazard model with one of the variables, say X_1, equal to 1 if the therapy were radiation and 0 if an alternative therapy were used.

2. The proportionality constant in the proportional hazard model,

$$e^{\alpha + \beta_1 X_1 + \cdots + \beta_p X_p},$$

is parametric. We have not specified the form of the underlying survival S_0. This curve is not estimated by a parametric model but by other means.

16.9.2 An Example of the Cox Proportional Hazard Regression Model

The Cox proportional hazard model is also called the Cox proportional regression model or the Cox regression model. The reason for calling this model a regression model is that the dependent variable of interest, survival, is modeled upon or "regressed upon" the values of the covariates or independent variables. The analogies between multiple regression and the Cox regression are quite good, although there is not a one to one correspondence between the techniques.

Computer software for Cox regression typically produces at least the following quantities:

Output	Description	Use of Output
a	Estimate of the constant in the hazard	The constant appears in the exponent of the covariate term for the hazard function
b_i	Estimate of the regression coefficient β_i	1. The regression coefficients allow estimation of $e^{\alpha+\beta_1 X_1+\cdots+\beta_p X_p}$ by $e^{a+b_1 X_1+\cdots+b_p X_p}$. By using this and the estimate of $S_0(t)$ we can estimate survival for any individual in terms of the values of $X_1,\ldots,X_p$ for each time t 2. The b_i give an estimate of the increase in risk (the hazard function) for different values of $X_1,\ldots,X_p$
$\text{SE}(b_i)$	The estimated standard error of b_i	1. The distribution of b_i is approximately $N(\beta_i, \text{SE}(b_i)^2)$ for large sample sizes. We can obtain $100(1-\alpha)\%$ confidence intervals for β_i as: $(b_i - z_{1-\alpha/2}\text{SE}(b_i), b_i + z_{1-\alpha/2}\text{SE}(b_i))$ 2. We test for statistical significance of β_i (in a model with the other X_j's) by rejecting $\beta_i = 0$ if $b_i^2/(\text{SE}(b_i))^2 \geq \chi^2_{1,1-\alpha}$. $\chi^2_{1,1-\alpha}$ is the $1-\alpha$ percentile of the χ^2 distribution with one degree of freedom. Sometimes the chi-square value and/or p-value are presented directly. Note that if only X_1 is in the model we may have X_1 statistically significantly related to survival, but with $X_2,\ldots,X_p$ also in the model b_1 may not differ statistically from zero. The testing of b_1 with $X_2,\ldots,X_p$ in the model is similar to testing for the partial correlation of a variable in a multiple regression situation
Model chi-square	A chi-square value for the entire model with p degrees of freedom	1. This chi-square statistic tests for *any* relationship among the $X_1,\ldots,X_p$ and the survival experience. The null hypothesis tested is $\beta_1 = \cdots = \beta_p = 0$. This is analogous to testing for zero multiple correlation between survival and $(X_1,\ldots,X_p)$ in a multiple regression setting 2. For nested models the chi-square values may be subtracted (as are the degrees of freedom) to give a chi-square test

Output	Description	Use of Output
$S_0(t)$	The estimate of the survival function for an individual with co-variate values equal to the mean of each variable	1. With $S_0(t)$, a, and the b_i we may plot the estimated survival experience of the population for any fixed value of the covariates 2. For a fixed time, say t_0, by varying the values of the covariates **X**, we may present the effect of combinations of the covariate values (see Example 16.6)

The following example illustrates the use of the Cox proportional hazards model.

Example 16.6. The left main coronary artery is a short segment of the arteries delivering blood to the heart. Two of the three major arterial systems branch off of the left main coronary artery. If this artery should close death is almost certain. Two randomized clinical trials (Veteran's Administration Study Group, Takaro *et al.* [1976] and the European Coronary Surgery Study Group [1980]) reported superior survival in individuals undergoing coronary artery bypass surgery. Chaitman *et al.* [1981] examined the observational data of the Coronary Artery Surgery Study, CASS, registry. Individuals were analyzed as being in the medical group until censored at the time of surgery. They were then entered into the surgical survival experience at the day of surgery.

A Cox model using a therapy indicator variable was used to examine the effect of therapy. Eight variables were used in this model:

CHFSCR a score for congestive heart failure (CHF). The score ranged from 0 to 4; 0 indicated no CHF symptoms. A score of 4 was indicative of severe, treated CHF.

LMCA the percent of diameter narrowing of the left main coronary artery due to atherosclerotic heart disease. By selection all cases had at least 50% narrowing of the left main coronary artery, LMCA.

LVSCR a measure of ventricular function, the pumping action of the heart. The score ranged from 5 (normal) to a potential maximum of 30 (not attained). The higher the score the worse the ventricular function.

DOM the dominance of the heart shows whether the right coronary artery carries the usual amount of blood; there is great biological variability. Individuals are classed as right or balanced dominance (DOM = 0). A left dominant individual has a higher proportion of blood flow through the LMCA, making left main disease even more important.

AGE the patient's age in years.

HYPTEN is there a history of hypertension? HYPTEN = 1 for yes and HYPTEN = 0 for no.

THRPY, THRPY = 1 for medical therapy; and THRPY = 2 for surgical therapy.

RCA this variable is one if the right coronary artery has $\geq 70\%$ stenosis and is equal to zero otherwise.

The Cox model produces the following results.

Variable	Beta	Standard Error	Chi-Square	Probability
CHFSCR	0.2985	0.0667	20.01	0.0000
LMCA	0.0178	0.0049	13.53	0.0002
LVSCR	0.1126	0.0182	38.41	0.0000
DOM	1.2331	0.3564	11.97	0.0006
AGE	0.0423	0.0098	18.75	0.0000
HYPTEN	−0.5428	0.1547	12.31	0.0005
THRPY	−1.0777	0.1668	41.77	0.0000
RCA	0.5285	0.2923	3.27	0.0706
Constant	−2.8968			

The chi-square value for CHFSCR is found by the square of beta divided by the standard error. For example $(0.2985/0.0667)^2 = 20.03$, which is the chi-square value to within the numerical accuracy.

The underlying survival curve (at the mean covariate values) has probabilities 0.944 and 0.910 of 1- and 2-year survival, respectively. The first case in the file has values: CHFSCR = 3, LMCA = 90, LVSCR = 18, DOM = 0, AGE = 49, HYPTEN = 1, THRPY = 1, and RCA = 1. What is the estimated probability of 1- and 2-year survival for this person?

$$a + b_1 X_1 + \cdots + b_n X_n = -2.8968 + (0.2985 \times 3) + (0.0178 \times 90)$$

$$+ (0.1126 \times 18) + (1.2331 \times 0) + (0.0423 \times 49)$$

$$+ (-0.5428 \times 1) + (-1.0777 \times 1) + (0.5285 \times 1)$$

$$= 2.6622$$

$$\text{Estimated probability of 1-year survival} = 0.944^{e^{2.6622}}$$

$$= 0.944^{14.328}$$

$$= 0.438$$

$$\text{Estimated probability of 2-year survival} = 0.910^{14.328}$$

$$= 0.259$$

The estimated probability of survival under medical therapy is 44% for 1 year and 26% for 2 years. This bad prognosis is largely due to the heart failure (CHFSCR) and very poor ventricular function (LVSCR). □

16.9.3 The Stepwise Cox Regression Model

Stepwise models were used in Chapters 11 and 13; for the same reasons and in the same manner we may perform a stepwise Cox regression model. In this section we consider adding variables in a stepwise fashion to the Cox regression model. The reasons are the same as for stepwise multiple regression models; we may do this to find a parsimonious set of variables to describe the relationship between a potential set of explanatory variables and the survival. We might also do this to look at the fit of a model by adding into the Cox model powers of variables and products among variables. These ideas are best illustrated by example.

Example 16.7.   The extent of coronary artery disease in coronary patients is prognostic for survival. Several different indices have been proposed to summarize the extent of the disease. These were examined in Ringqvist *et al.* [1983]. The authors present data on 15,616 patients of whom 1062 died (i.e., 1062 "events"). The details of the eight indices are not presented here. Mnemonics for the eight indices are:

DISVES the number of vessels diseased.

PRXVES the number of proximal vessels diseased.

PRXSCR a proximal vessel score.

FRIES Friesinger score.

GENSIN Gensini score.

NHCH National Heart and Chest Hospital, London, index.

MNHCH Modified NHCH index.

SEGMEN a weighted sum score of the percent of stenosis in different segments of the coronary artery tree.

In addition to the disease in the arteries the left ventricular (pumping) function of the heart is very important in prognosis. A variable, LVSCOR, measures this.

These nine variables were examined in their relationship to medical survival in the CASS (Coronary Artery Surgery Study) study. Cases were selected with at least one diseased artery. A stepwise Cox regression analysis was performed to examine whether there was independent predictive information in the indices. The computer output for the first step is given. The S-statistic is the score statistic for a variable (Kalbfleisch and Prentice [1980]). See next page for an explanation.

STEP 0. S-STATISTIC FOR VARIABLES NOT IN THE MODEL

Number	Variables	S-Statistic	Probability
1	DISVES	580.1	0.0000
2	PRXVES	602.9	0.0000
3	PRXSCR	610.6	0.0000
4	FRIES	481.8	0.0000
5	GENSEN	836.1	0.0000
6	NHCH	611.3	0.0000
7	MNHCH	559.5	0.0000
8	SEGMEN	710.5	0.0000
9	LVSCOR	1051	0.0000

STEP 1. VARIABLE LVSCOR ADDED WITH CHI-SQUARE = 1051, DF= 1
(COEFFICIENTS, CHI-SQUARE STATISTICS FOR VARIABLES IN THE MODEL)

Number	Variable	Beta	Standard Error	Chi-Square	Probability
9	LVSCOR	0.1743	0.0058	915.6	0.0000

Model chi-square statistic = 802.3, DF = 1, Probability = 0.0000

At step 0, before the first variable is entered, the values of a statistic, here called the S-statistic, is given. Under the null hypothesis of no relationship to survival, the statistic, has a chi-square distribution with one degree of freedom. The most significant variable (or the variable with the largest value of the statistic) is added to the model. In this case the variable LVSCOR is added. Below this, the estimated beta with its standard error and chi-square value, $(b_i/\text{SE}(b_i))^2$, and p-value are given. Note that the S-statistic value and chi-square value differ slightly. This is because a different method of obtaining the statistic is used when the variable is in the model.

At the next step, the output is

STEP 1. S-STATISTIC FOR VARIABLES NOT IN THE MODEL

Number	Variables	S-Statistic	Probability
1	DISVES	244.6	0.0000
2	PRXVES	233.5	0.0000
3	PRXSCR	220.1	0.0000
4	FRIES	218.2	0.0000
5	GENSEN	228.8	0.0000
6	NHCH	280.2	0.0000
7	MNHCH	248.6	0.0000
8	SEGMEN	287.6	0.0000

STEP 2. VARIABLE SEGMEN ADDED WITH CHI-SQUARE = 287.6, DF = 1
(COEFFICIENTS, CHI-SQUARE STATISTICS FOR VARIABLES IN THE MODEL)

Number	Variable	Beta	Standard Error	Chi-Square	Probability
9	LVSCOR	0.1412	0.0063	499.3	0.0000
8	SEGMEN	0.0662	0.0039	283.2	0.0000

Model chi-square statistic = 1076, DF = 2, Probability = 0.0000

Note the large drop in the S-statistic values. This shows a great overlap in prognostic information between the coronary artery indices and the ventricular function. This is expected since the ventricular function often decreases because arteries occlude (close up). Thus, the amount of disease is usually high when ventricular function is poor. Note also in steps 0 and 1 that the S-statistics are close for the eight indices; they contain roughly the same prognostic power. Also the LVSCOR chi-square drops when SEGMEN, the most significant index, is added to the model. This occurs because of their overlapping prognostic information. The model chi-square is a test of the hypothesis that all variables *in* the model have $\beta_i = 0$.

The output for the next step:

STEP 2. S-STATISTIC FOR VARIABLES NOT IN THE MODEL

Number	Variables	S-Statistic	Probability
1	DISVES	48.08	0.0000
2	PRXVES	15.23	0.0001
3	PRXSCR	13.93	0.0002
4	FRIES	41.53	0.0000
5	GENSEN	15.37	0.0001
6	NHCH	58.80	0.0000
7	MNHCH	56.17	0.0000

STEP 3. VARIABLE NHCH ADDED WITH CHI-SQUARE = 58.80, DF = 1
(COEFFICIENTS, CHI-SQUARE STATISTICS FOR VARIABLES IN THE MODEL)

Number	Variable	Beta	Standard Error	Chi-Square	Probability
9	LVSCOR	0.1368	0.0063	464.9	0.0000
8	SEGMEN	0.0339	0.0056	37.14	0.0000
6	NHCH	−0.0145	0.0019	58.58	0.0000

Model Chi-Square Statistic = 1132, DF = 3, Probability = 0.0000

For the remaining seven indices note the great drop in prognostic power (decrease in the S-statistic) in the presence of the index SEGMEN in the model. The index variables are closely related so this is expected.

This continues until the last step. The final output is given below.

STEP 4. S-STATISTIC FOR VARIABLES NOT IN THE MODEL

Number	Variables	S-Statistic	Probability
1	DISVES	7.440	0.0064
2	PRXVES	0.02875	0.8654
4	FRIES	4.749	0.0293
5	GENSEN	1.937	0.1640
7	MNHCH	5.714	0.0168

STEP 5. VARIABLE DISVES ADDED WITH CHI-SQUARE = 7.440, DF = 1)
(COEFFICIENTS, CHI-SQUARE STATISTICS FOR VARIABLES IN THE MODEL)

Number	Variable	Beta	Standard Error	Chi-Square	Probability
9	LVSCOR	0.1330	0.0064	430.1	0.0000
8	SEGMEN	0.0235	0.0062	14.35	0.0002
6	NHCH	−0.0093	0.0025	14.30	0.0002
3	PRXSCR	0.0626	0.0221	8.039	0.0046
1	DISVES	0.1613	0.0591	7.439	0.0064

Model Chi-Square Statistic = 1148, DF = 5, Probability = 0.0000

STEP 5. S-STATISTIC FOR VARIABLES NOT IN THE MODEL

Number	Variables	S-Statistic	Probability
2	PRXVES	0.05132	0.8208
4	FRIES	1.167	0.2800
5	GINSIN	1.394	0.2378
7	MNHCH	3.271	0.0705

No Variables Significant for Entry
at the 0.05 Level
Stepwise Analysis Abandoned

SUMMARY OF STEPWISE ANALYSIS

Variable	Beta
LVSCOR	0.1330
SEGMEN	0.0235
NHCH	−0.0093
PRXSCR	0.0626
DISVES	0.1613

Model Constant $= -1.4752$
(minus the sum of the products of selected variable's means and betas)

Note that four of the indices SEGMEN to DISVES entered. This shows that although the indices have much of their information in common, there are some prognostic differences. In large part, however, this results from the large number of cases (15616) and endpoints (1062). The large number of cases and great predictive power account for the extremely large values of the chi-square statistics.   □

16.9.4 Interpretation of the β_i Coefficients

In the multiple regression setting, the regression coefficients may be interpreted as the average change in the response variables if the predictive variable changes by one unit with everything else fixed. In this section we look at the interpretation of the β_i for the Cox proportional hazard model. Recall that the hazard function is proportional to the probability of failure in a short time interval. Suppose that we have two individuals whose covariate values are the same on all the p regression variables for the Cox model with the exception of the ith variable. If we take the ratio of the hazard functions for the two individuals at some time t, we have the ratio of the probability of an event in a short interval after time t. The ratio of these two probabilities is the relative risk of an event during this time period. This is also called the *instantaneous relative risk*. For the Cox proportional hazards model, we find

$$\text{Instantaneous relative risk} = RR = \frac{h_0(t)e^{\alpha+\beta_1 X_1+\cdots+\beta_i X_i(1)+\cdots+\beta_p X_p}}{h_0(t)e^{\alpha+\beta_1 X_1+\cdots+\beta_i X_i(2)+\cdots+\beta_p X_p}}$$

$$= e^{\beta_i(X_i(1)-X_i(2))}. \tag{16.23}$$

An equivalent formulation is to take the logarithm of the instantaneous relative risk

(RR). The logarithm is given by the following equation

$$\ln(RR) = \beta_i(X_i(1) - X_i(2)). \tag{16.24}$$

In words, the regression coefficients β of the Cox proportional hazard model are equal to the logarithm of the relative risk if the variable X is increased by one unit.

16.9.5 Use of the Cox Model as a Method of Adjustment

In Section 16.8 we considered stratified life table analyses to adjust for confounding factors or covariates. The Cox model may be used for the same purpose. As in the multiple linear regression model there are two ways in which we may adjust. One is to consider a variable whose effect we want to study in relationship to survival. Suppose that we want adjust for variables $X_1, \ldots, X_k$. We run the Cox proportional hazards regression model with the variable of interest and the adjustment covariates in the model. The statistical significance of the variable of interest may be tested by taking its estimated regression coefficient, dividing by its standard error and using a normal probability critical value. An equivalent approach, similar to nested hypotheses in the multiple linear regression model, is to run the Cox proportional hazards model with only the adjusting covariates. This will result in a chi-square statistic for the entire model. A second Cox proportional hazards model may be run with the variable of interest in the model in addition to the adjustment covariates. This will result in a second chi-square statistic for the model. The chi-square statistic for the second model minus the chi-square statistic for the first model will have approximately a chi-square distribution with one degree of freedom if the variable of interest has no effect upon the survival after adjustment for the covariates $X_1, \ldots, X_p$.

16.10 PARAMETRIC MODELS

16.10.1 The Exponential Model; Rates

Suppose that at each instant of time the instantaneous probability of death is the same. That is, suppose the hazard rate or force of mortality is constant. Although in human populations this is not a useful assumption over a wide time interval, it may be a valid assumption over a 5- or 10-year interval, say.

If the constant hazard rate is λ, the survival curve is $S(t) = e^{-\lambda t}$. From this expression the term *exponential survival* arises. The expected length of survival is $1/\lambda$.

If the exponential situation holds, the parameter λ is estimated by the number of events divided by total exposure time. The methods and interpretation of rates are then appropriate.

Note that if $S(t)$ is exponential, then $\log S(t) = -\lambda t$ is a straight line with slope $-\lambda$. Plotting an estimate of $S(t)$ on a logarithmic scale is one way of visually examining the appropriateness of assuming an exponential model. Figure 16.16 shows some of the patterns one might observe.

Return to the study of Bruce *et al.* [1976] discussed in Chapter 15 concerning a cardiopulmonary rehabilitation program. One of the questions of interest was the pattern with which individuals adhered to (did not drop out of) the training program. For this purpose a survival analysis was run with the endpoint the time after enrollment that a subject dropped out of the program. Thus, the endpoint of interest is the event of dropping out of the program. The data were censored in two ways:

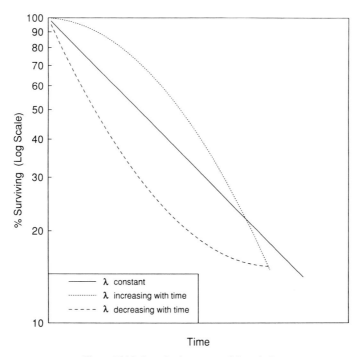

Figure 16.16. Log plot for exponential survival.

1. Many individuals were still enrolled at the time of the analysis

2. Death was a source of censoring of the time at which individuals might have dropped out.

From the plot one sees that the data do *not* look exponential (Figure 16.17). Rather it appears that the hazard of dropping out decreases as the time enrolled in the program increases.

16.10.2 Two Other Parametric Models for Survival Analysis

There are a variety of parametric models for survival distributions. In this subsection a few are mentioned. For details of the distributions and parameter estimates, the reader is referred to texts by Mann *et al.* [1974] and Gross and Clark [1975]. These texts also present a variety of models not touched upon here.

The two-parameter *Weibull Distribution* has a survival curve of the form

$$S(t) = e^{-\alpha t^\beta}, \quad \text{for } t > 0 \ (\alpha > 0, \ \beta > 0). \tag{16.25}$$

If $\beta = 1$, the Weibull distribution is the exponential model with constant hazard rate. The hazard rate decreases with time if $\beta < 1$ and increases with time if $\beta > 1$. Often, if the time of survival is measured from diagnosis of a disease, a Weibull with $\beta > 1$ will reasonably model the situation. Estimates are made by computer.

Another distribution, the *lognormal* distribution, asssumes that the logarithm of the survival time is normally distributed. If there is no censoring of data, one may

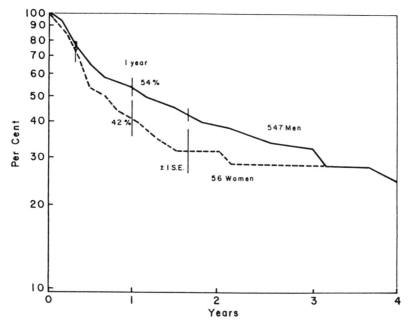

Figure 16.17. Adherence to training programs 547 men (—) and 56 women (- - - -). SE = standard error computed by Greenwood's formula.

work with the logarithm of the survival times and use methods appropriate for the normal distribution.

16.11   EXTENSIONS

16.11.1   The Cox Model with Time Dependent Covariates

Often the variables used for prediction in the Cox model may change with time. For example, in a cancer study, white cell count may be used as a predictive variable; in the cardiovascular study blood pressure may be used as a predictive variable. As additional observations are made over time the values of these quantities may change. One might then want to change the hazard putting in the new values for the covariates. This may be done using the Cox model with *time dependent covariates*. The covariates at each particular time are then substituted into the hazard function. More detail about this model is given in Kalbfleisch and Prentice [1980].

16.11.2   Stratification in the Cox Model

Other more sophisticated methods combine both stratification and the Cox model. Important covariates whose form may not be known and which must be taken into account may be used to define strata. The Cox model is then used with a different constant coefficient added to the linear term in the exponent of the covariate term for the hazard function. The constant term differs from stratum to stratum but the effect of other variables is assumed to be the same across the strata. The underlying survival curve $S_o(t)$ is allowed to differ between strata.

NOTES

Note 16.1: More on the Hazard Rate

Many of the concepts presented in this chapter are analogues of continuous quantities that are best defined in terms of calculus.

If the survival function is $S(t)$ its probability density function is,

$$f(t) = -\frac{dS(t)}{dt}.$$

The hazard rate is then

$$h(t) = \frac{f(t)}{S(t)}.$$

From this it follows that the survival is found from the hazard rate by the equation

$$S(t) = e^{-\int_0^t h(x)\,dx}.$$

Note 16.2: Stationary Populations

Table 16.4 from the US Vital Statistics Life Tables, involves columns labeled "Stationary Population". The columns show what the population would look like at long term zero population growth if the age specific death rates were as given in the table. The explanation with the life tables is as follows.

Columns 5 and 6: Stationary population ($_nL_x$ and T_x). Suppose that a group of 100,000 individuals like that assumed in columns 3 and 4 is born every year and that the proportions dying in each such group in each age interval throughout the lives of the members are exactly as shown in column 2. If there were no migration and if the births were evenly distributed over the calendar year, the survivors of these births would make up what is called a stationary population; stationary because in such a population the number of persons living in any given age group would never change. When an individual left the group, either by death or by growing older and entering the next higher age group, his or her place would immediately be taken by someone entering from the next lower age group. Thus a census taken at any time in such a stationary community would always show the same total population and the same numerical distribution of that population among the various age groups. In such a stationary population supported by 100,000 annual births, column 3 shows the number of persons who, each year, reach the birthday which marks the beginning of the age interval indicated in column 1, and column 4 shows the number of persons who die each year in the indicated age interval.

Column 5 shows the number of persons in the stationary population indicated age interval. For example, the figure given for males in the age interval 20–25 is 480, 456. This means that in a stationary population of males supported by 100,000 annual births and with proportions dying in each age group always in accordance with column 2, a census taken on any date would show 480, 456 persons between exact ages 20 and 25.

Column 6 shows the total number of persons in the stationary population (column 5) in the indicated age interval and all subsequent age intervals. For example, in the stationary population of males referred to in the last illustration, column 6 shows that there would be at any given moment a total of 4,863,303 persons who have passed their twentieth birthday. The population at all ages 0 and above (in other words, the total population of the stationary community) would be 6,814,371.

Note 16.3: The Log-Rank Statistic and the Log-Rank Statistic for Stratified Data

We present the statistic using some matrix ideas. The notation is that of Section 16.7 on the log-rank test. For the ith group at the mth time of a death (or deaths) there were d_{im} deaths and l_{im} individuals at risk. Suppose we have k groups and M times of death.

For $i, j = 1, \ldots, k$, let

$$
V_{ij} =
\begin{cases}
\sum_{m=1}^{M} \dfrac{l_{im}(T_m - l_{im})D_m(T_m - D_m)}{T_m^2(T_m - 1)}, & i = j. \\[4mm]
\sum_{m=1}^{M} \dfrac{-l_{im}l_{jm}D_m(T_m - D_m)}{T_m^2(T_m - 1)}, & i \neq j.
\end{cases}
$$

Define the $(k-1) \times (k-1)$ matrix V by

$$
V =
\begin{pmatrix}
V_{11} & V_{12} & \cdots & V_{1,k-1} \\
V_{21} & & & \vdots \\
\vdots & & & \vdots \\
V_{k-1,1} & \cdots & \cdots & V_{k-1,k-1}
\end{pmatrix}.
$$

Define vectors of observed and expected number of deaths in groups $1, 2, \ldots, k-1$ by

$$
\mathbf{O} =
\begin{pmatrix}
O_1 \\
\vdots \\
O_{k-1}
\end{pmatrix}, \quad
\mathbf{E} =
\begin{pmatrix}
E_1 \\
\vdots \\
E_{k-1}
\end{pmatrix}.
$$

The log-rank statistic is

$$
(\mathbf{O} - \mathbf{E})' V^{-1} (\mathbf{O} - \mathbf{E})
$$

where $'$ denotes a transpose and -1 a matrix inverse.

If there are $s = 1, \ldots, S$ strata, for each stratum we have $\mathbf{O}$, $\mathbf{E}$, and V. Let these values be indexed by s to denote the strata. The log-rank statistic is

$$
\left(\sum_{s=1}^{S} (\mathbf{O}_s - \mathbf{E}_s) \right)' \left(\sum_{s=1}^{S} V_s \right)^{-1} \left(\sum_{s=1}^{S} (\mathbf{O}_s - \mathbf{E}_s) \right).
$$

Note 16.4: Estimating the Probability Density Function in the Life Table Methods

The density function in the interval from $x(i)$ to $x(i+1)$ for the life table is estimated by

$$
f_i = \frac{P_i - P_{i+1}}{x(i+1) - x(i)}.
$$

The standard error of f_i is estimated by

$$\frac{p_i q_i}{\sqrt{x(i+1) - x(i)}} \left(\sum_{j=1}^{i-1} \frac{q_j}{l'_j p_j} + \frac{p_i}{l'_i q_i} \right)^{1/2}.$$

Note 16.5: Another Approach to Constructing Life Table Estimates of Survival

In the actuarial or life table situation, suppose that in each interval one can find the total exposure time. In the interval from $x(i)$ to $x(i+1)$, denote the total exposure time by E_i. (In Section 16.4 this was estimated by $l'_{x(i)} - d_{x(i)}/2$.) The hazard in the interval is estimated by

$$h_i = \frac{d_{x(i)}}{E_i \times (x(i+1) - x(i))}.$$

The probability of surviving the interval assuming a constant hazard rate in the interval and using the results of Note 16.1 above, is estimated by

$$p_i = e^{-h_i \Delta x_i}, \quad \text{where } (\Delta x_i = x(i+1) - x(i)).$$

The survival curve is then estimated by

$$P_1 = 1, \quad P_i = \prod_{j=1}^{i-1} p_i = e^{-\sum_{j=1}^{i-1} h_i \Delta x_i}.$$

An estimate of the standard error for h_i is

$$\text{SE}(h_i) \doteq \sqrt{\frac{h_i}{E_i \Delta x_i}}.$$

To find a confidence interval for $S(x(i))$ one first finds a confidence interval for $\ln S(x(i))$ estimated by $\ln P_i$. Now,

$$\text{SE}(\ln P_i) = \text{SE}\left(-\sum_{j=1}^{i-1} h_i \Delta x_i \right)$$

$$= \left(\sum_{j=1}^{i-1} \text{SE}(h_i \Delta x_i)^2 \right)^{1/2}$$

$$= \left(\sum_{j=1}^{i-1} \text{SE}(h_i)^2 \Delta x_i^2 \right)^{1/2}$$

$$\doteq \left(\sum_{j=1}^{i-1} \frac{h_i}{E_i} \Delta x_i \right)^{1/2}.$$

A $100(1 - \alpha)\%$ confidence interval, in terms of $N(0,1)$ critical values, for $S(x(i))$ is given by

$$\ln P_i \pm Z_{1-\alpha/2}\text{SE}(\ln P_i).$$

The confidence interval for $S(x_i)$ is then given by

$$P_i e^{\pm Z_{1-\alpha/2}\text{SE}(\ln P_i)}.$$

Note 16.6: Effective Sample Size in the Life Table Method

In using the life table method, Greenwood's formula allowed an estimate of the standard error of the survival probability

$$P_{x(i)} = P.$$

Let s be the estimated standard error. If the estimate of the survival probability P had arisen from a binomial situation observing n individuals until $x(i)$ the standard error would have been estimated by

$$\sqrt{\frac{P(1-P)}{n}}.$$

By equating the estimated standard error s to this quantity and "solving" for n, we have the binomial sample size that would have been needed to obtain the same precision as that given by the life table method. Recall that in this context binomial means that all n individuals could have been observed for the length of time $x(i)$. Thus, the effective sample size n is given by

$$n = \frac{P(1-P)}{s^2}.$$

Table 16.2, of Section 16.3, gives a life table for angina patients. Beginning year 10, $P = 0.2987$ and $\text{SE}(P) = s = 0.109$. The effective sample size is

$$n = \frac{0.2987(1 - 0.2987)}{0.0190^2} = 1763.13.$$

That is, the precision is the same as that estimated from 1763 individuals over this 10-year time period without any withdrawls or dropouts. Comparison of the numbers that might have been observed to $x(i)$ with the effective sample size gives an indication of the gain in statistical power from using the life table method. Table 16.3 presents effective sample sizes for all the entries.

Note 16.7: Other References Dealing with Survival Analysis and the Heart Transplant Data

The first heart transplant data has been extensively used as an illustration in the development of survival techniques. Further references are Mantel and Byar [1974], Turnbull et al. [1974], and Crowley and Hu [1977].

Note 16.8: Competing Risks

In certain situations one is interested in only certain causes of death that may be linked to the disease in question. For example, in a study of heart disease a death in a plane crash might be considered an unreasonable endpoint to attribute to the disease. Such endpoints are called *competing risks*. Kalbfleisch and Prentice [1980], Gross and Clark [1975] and Prentice *et al.* [1978] discuss such issues.

PROBLEMS

The first four problems deal with the life table or actuarial method of estimating the survival curve. In each case fill in the question marks from the other numbers given in the table.

1. Example 16.2 deals with chest pain in groups in the Coronary Artery Surgery Study, all times are in days. The life table for the individuals with chest pain thought probably not to be angina is:

GROUP 3: PRBNOT ANGINA

$t(i)$	Enter	At Risk	Dead	Withdraw Alive	Proportion Dead	Cumulative Survival	SE
0.0–90.9	2404	2404.0	2	0	0.0008	0.9992	?
91.0–181.9	2402	?	2	0	0.0008	0.9983	?
182.0–272.9	2400	2400.0	?	0	0.0021	0.9963	0.001
273.0–363.9	2395	2395.0	6	0	?	0.9938	0.002
364.0–454.9	?	2388.0	4	2	0.0017	0.9921	0.002
455.0–545.9	2383	2383.0	3	0	0.0013	?	0.002
546.0–636.9	2380	2380.0	7	0	0.0029	0.9879	0.002
637.0–727.9	2373	?	12	300	?	?	0.003
728.0–818.9	2061	2051.5	?	19	0.0015	0.9812	0.003
819.0–909.9	?	2039.0	1	0	0.0005	0.9807	0.003
910.0–1000.9	2038	2037.0	2	?	0.0010	0.9797	0.003
1001.0–1091.9	2034	?	3	517	0.0017	0.9781	0.003
1092.0–1182.9	1514	1494.0	3	40	0.0020	0.9761	0.003
1183.0–1273.9	1471	1471.0	4	0	?	0.9734	0.004
1274.0–1364.9	1467	1466.5	1	1	0.0007	0.9728	0.004
1365.0–1455.9	?	1144.0	1	642	0.0009	0.9719	0.004
1456.0–1546.9	822	777.5	1	?	0.0013	0.9707	0.004
1547.0–1637.9	732	732.0	1	0	0.0014	?	0.004
1638.0–1728.9	731	730.0	2	2	0.0027	0.9667	0.004
1729.0–1819.9	727	449.0	1	?	0.0022	0.9645	0.005

2. From Example 16.2 for the individuals with chest pain thought definitely to be angina the life table is:

GROUP 4: DEFNOT ANGINA

$t(i)$	Enter	At Risk	Dead	Withdraw Alive	Proportion Dead	Cumulative Survival	SE
0.0–90.9	426	426.0	2	?	0.0047	0.9953	0.003
91.0–181.9	?	424.0	2	0	0.0047	0.9906	?
182.0–272.9	422	?	3	0	?	?	0.006
273.0–363.9	419	419.0	0	0	0.0000	0.9836	0.006
364.0–454.9	419	419.0	1	0	0.0024	0.9812	0.007
455.0–545.9	418	417.5	?	1	0.0024	0.9789	0.007
546.0–636.9	416	416.0	1	0	0.0024	0.9765	0.007
637.0–727.9	415	382.0	0	?	0.0000	0.9765	0.007
728.0–818.9	349	343.0	0	11	0.0000	0.9765	0.007
819.0–909.9	338	338.0	1	0	0.0030	0.9736	0.008
910.0–1000.9	337	336.5	0	1	0.0000	0.9736	0.008
1001.0–1091.9	336	?	1	97	?	?	0.009
1092.0–1182.9	238	232.5	0	11	0.0000	0.9702	0.009
1183.0–1273.9	227	?	1	1	0.0044	0.9660	0.010
1274.0–1364.9	?	224.5	1	1	0.0045	0.9617	0.010
1365.0–1455.9	?	170.0	0	106	0.0000	0.9617	0.010
1456.0–1446.9	117	114.0	?	6	0.0000	0.9617	0.010
1547.0–1637.9	?	?	0	1	0.0000	0.9617	0.010
1638.0–1728.9	110	109.5	0	1	0.0000	0.9617	0.010
1729.0–1819.9	109	65.5	0	87	0.0000	0.9617	0.010

3. The beta-blocking drug data of Example 16.4 of Section 16.6.1 are used here and in the next problem. The life table for those using such drugs at enrollment is given here.

GROUP 1: YES BETABL

$t(i)$	Enter	At Risk	Dead	Withdraw Alive	Proportion Dead	Cumulative Survival	SE
0.0–90.9	4942	4942.0	?	0	0.0097	0.9903	0.001
91.0–181.9	4894	4894.0	33	0	0.0067	0.9836	0.002
182.0–272.9	4861	4861.0	?	?	0.0058	0.9779	?
273.0–363.9	4833	4832.5	28	1	0.0058	0.9723	0.002
364.0–454.9	4804	4804.0	17	0	0.0035	?	0.002
455.0–545.9	4787	4786.5	29	1	?	?	0.003
546.0–636.9	4757	4757.0	22	0	0.0046	0.9585	0.003
637.0–727.9	4735	4376.0	25	718	0.0057	0.9530	0.003
728.0–818.9	?	?	?	62	0.0043	0.9489	0.003
819.0–909.9	3913	3912.0	23	2	?	0.9434	0.003
910.0–1000.9	3888	3884.5	19	7	0.0049	0.9388	0.004
1001.0–1091.9	?	?	?	1191	0.0040	0.9350	0.004
1092.0–1182.9	2658	2624.5	14	67	0.0053	0.9300	0.004
1183.0–1273.9	2577	2576.5	11	1	0.0043	0.9261	0.004
1274.0–1364.9	2565	2561.0	15	8	?	0.9206	0.004
1365.0–1455.9	2542	1849.5	12	1385	0.0065	0.9147	0.005
1456.0–1446.9	1145	1075.0	5	?	0.0047	?	0.005
1547.0–1637.9	1000	999.0	4	2	0.0040	0.9068	0.005
1638.0–1728.9	994	989.0	4	10	0.0040	0.9031	0.006
1729.0–1819.9	980	580.0	5	800	0.0086	0.8953	0.006

4. Those not using beta blocking drugs have the following survival experience.

GROUP 1: NO BETABL

t (i)	Enter	At Risk	Dead	Withdraw Alive	Proportion Dead	Cumulative Survival	SE
0.0–90.9	6453	?	45	0	?	?	?
91.0–181.9	6408	?	28	0	?	?	?
182.0–272.9	6380	?	42	0	?	?	?
273.0–363.9	6338	?	25	2	?	?	?
364.0–454.9	6311	6310.0	24	2	0.0038	0.9746	0.002
455.0–545.9	6285	6285.0	32	0	0.0051	0.9696	0.002
546.0–636.9	6253	6253.0	?	0	0.0048	0.9650	0.002
637.0–727.9	6223	5889.0	23	668	0.0039	0.9612	0.002
728.0–818.9	?	?	23	40	0.0042	0.9572	0.003
819.0–909.9	?	5467.0	17	4	?	0.9542	0.003
910.0–1000.9	5448	5444.5	23	7	0.0042	0.9502	0.003
1001.0–1091.9	5418	4787.4	25	1261	0.0052	0.9452	0.003
1092.0–1182.9	4132	4082.0	?	100	0.0054	0.9401	0.003
1183.0–1273.9	4010	4010.0	23	0	0.0057	0.9347	0.003
1274.0–1364.9	3987	3981.0	8	?	0.0020	0.9329	0.003
1365.0–1455.9	3967	3100.0	13	1734	0.0042	0.9289	0.003
1456.0–1446.9	2220	2104.0	13	?	0.0062	0.9232	0.004
1547.0–1637.9	1975	1974.0	?	2	0.0020	0.9213	0.004
1638.0–1728.9	1969	1961.5	11	15	0.0056	0.9162	0.004
1729.0–1819.9	1943	1212.0	7	7	0.0058	0.9109	0.005

5. Take the Stanford heart transplant data of Example 16.3, Section 16.6. Place the data in a life table analysis using 50 day intervals. Plot the data over the interval from zero to 300 days. (Do not compute the Greenwood standard errors.)

6. Use the survival information from the Glasser data of Section 16.11.3. For group 1 construct a life table using 60 day intervals from 0 through 360 days. (Do not compute the Greenwood standard errors.) Plot your survival curve.

7. For Problem 16.1 compute the hazard function (in probability of dying/day) for intervals:
   a. 546–637
   b. 1092–1183
   c. 1456–1547

8. For the data of Problem 16.2 compute the hazard rate for the individuals:
   a. 0–91
   b. 91–182
   c. 819–910

9. Compute the product limit survival curve using the values for Glasser's group 1 of Section 16.11.3.

10. Compute the product limit survival curve using the values for Glasser's group 2 of Section 16.11.3. Compute the estimated standard error at the times of the first three deaths.

11. Data used by Pike [1966] are quoted in Kalbfleisch and Prentice [1980]. Two groups of rats with different pretreatment regimes were exposed to the carcinogen DBMA. The time to mortality from vaginal cancer in the two groups was:

Group 1: 143, 164, 188, 188, 190, 192, 206, 209, 213, 216, 216*, 220, 227, 230, 234, 244*, 246, 265, 304.

Group 2: 142, 156, 163, 198, 204*, 205, 232, 232, 233, 233, 233, 233, 239, 240, 261, 280, 280, 296, 296, 323, 344*.

*censored observation

a. Compute and graph the two product limit curves of the groups.

b. Compute the expected number of deaths in each group and the value of the approximation $(\sum(O - E)^2/E)$ to the log rank test. Are the survival times different in the two groups at the 5% significance level?

12. For the data of Problems 16.3 and 16.4 assume that only four 91-day intervals had been observed for each of the two groups. Using the procedure for the log rank test compute the expected number of deaths in each group (under the null hypothesis of no difference in true survival patterns in the groups). Compute the approximate log rank test. Do you reject the null hypothesis of no difference in the survival patterns at the 10%, 5% or 1% significance level?

13. Place the Glasser data for groups 1 and 2, into 50-day life table intervals (ignoring the age data). Find the observed and expected number of deaths in each group, as well as the simple log rank statistic. What can you say about the p-value for testing equality of the curves?

14. The beta blocker data of Example 16.4, Section 16.6.1 and Problems 16.3 and 16.4 when stratified into the 30 strata as discussed in the text give the following results.

	Drug Use		No Drug Usage		
Strata	**Obs.**	**Exp.**	**Obs.**	**Exp.**	**p-Value**
1	45	43.30	71	72.70	0.74
2	2	2.23	4	3.77	0.84
3	0	0.20	1	0.80	0.54
4	27	28.54	37	35.46	0.69
5	6	4.84	5	6.16	0.48

(Continued)

Strata	Drug Use		No Drug Usage		p-Value
	Obs.	Exp.	Obs.	Exp.	
6	2	0.76	1	2.24	0.08
7	20	16.87	20	23.13	0.31
8	4	5.25	10	8.75	0.49
9	3	3.17	5	4.83	0.90
10	18	16.55	21	22.45	0.63
11	5	6.68	9	7.32	0.35
12	8	4.58	1	4.42	0.02
13	21	16.04	13	17.96	0.08
14	6	8.95	16	13.05	0.19
15	2	2.63	5	4.37	0.61
16	16	16.82	20	19.81	0.78
17	5	9.86	15	10.14	0.02
18	4	4.40	5	4.60	0.78
19	7	11.48	16	11.52	0.06
20	10	8.98	8	9.02	0.62
21	4	2.89	2	3.11	0.34
22	21	19.67	24	25.33	0.68
23	13	14.59	20	18.41	0.56
24	5	6.86	11	9.14	0.32
25	35	29.64	21	26.36	0.14
26	18	14.82	13	16.18	0.24
27	7	8.89	8	6.11	0.29
28	22	17.08	18	22.92	0.10
29	11	11.24	15	14.76	0.92
30	8	9.11	8	6.89	0.52

a. What are the observed and expected numbers in the two groups? (Why do you have to add only three columns?)

b. Two strata (12 and 17) are significant with $p = 0.02$. If the true survival patterns (in the conceptual underlying populations) are the same, does this surprise you?

c. What is $\sum (O - E)^2/E$? How does this compare to the more complicated log-rank statistic shown in Figure 16.12.

15. The paper by Chaitman et al. [1981] studied patients with left main coronary artery disease, as discussed in Example 16.6. Separate Cox survival runs were performed for the medical and surgical groups. The data are presented in Table 16.11. The survival, at the mean covariate values, for 1, 2, and 3 years are given by $S_o(1)$, $S_o(2)$, and $S_o(3)$ respectively. The zero-one variables are 0 for no and 1 for yes. Consider five patients with the values on the variables as given in the next table.

Table 16.11. Significant Independent Predictors of Mortality in Patients with Greater than 50 Percent Stenosis of the Left Main Coronary Artery.

Variable	Medical Group		Surgical Group	
	X^{2a}	β_i	X^{2a}	β_i
LV score (5–30)	19.12	0.1231	18.54	0.1176
CHF score (0–4)	9.39	0.2815	8.16	0.2964
Age	14.42	0.0526	6.98	0.0402
% LMCA stenosis (50–100)	19.81	0.0293	—	
Hypertension (0–1)	9.41	0.7067	5.74	0.5455
Left dominance (0–1)	—		10.23	1.0101
Smoking (1 = never, 2 = ever, 3 = present)	7.26	0.4389	—	
MI status (0 = none, 1 = single, 2 = multiple)	4.41	−0.2842	—	
Diabetes (0–1)	—		4.67	0.5934
Total chi-square	90.97		67.11	
Degrees of freedom	7		6	
p	< 0.0001		< 0.0001	
Constant c		−7.2956		−3.7807
Estimated survival				
$S_o(1)$		0.90		0.97
$S_o(2)$		0.83		0.95
$S_o(3)$		0.76		0.93

[a] Adjusted chi-square (X^2) statistics were computed with all variables considered together. Chi-square > 6.63 corresponds to $p < 0.01$ and chi-square > 10.83 to $p < 0.001$. β, beta coefficient; CHF, congestive heart failure; LMCA, left main coronary artery; LV, left ventricular; MI, myocardial infarction. Dashes indicate a variable not in the particular model.

	Patient Number				
Variable	**1**	**2**	**3**	**4**	**5**
LV score	13	5	7	8	12
CHF score	2	0	1	0	3
Age	71	62	42	55	46
% LMCA stenosis	75	90	50	70	95
Hypertension	No	Yes	Yes	No	No
Left dominance	No	No	No	Yes	No
Smoking	Ever	Present	Ever	Ever	Present
MI status	Multiple	None	Single	None	Single
Diabetes	No	No	No	Yes	No

a. What is the estimate of the 2-year medical survival in patients 1, 2, and 3?

b. What is the estimate of the 3-year surgical survival in patients 4 and 5?

c. What is the estimated 1-year medical and 1-year surgical survival for patient 1? For patient 3?

d. What is the logarithm of the instantaneous relative risk for two individuals treated medically who differ by 20 years, but otherwise have the same values for the variables? What is the instantaneous relative risk?

e. What is the instantaneous relative risk due to diabetes (yes versus no) for surgical cases?

*__f.__ What is the standard error for the LV score coefficient for the surgical group? For the age coefficient for the medical group? Form an approximate 95% confidence interval for the age coefficient in the medical group.

16. Alderman *et al.* [1983] studied the medical and surgical survival of individuals with poor left ventricular function; that is, they studied individuals whose hearts pumped poorly. Their model (in one analysis) included the following variables:

> Impairment: impairment due to congestive heart failure (CHF); $0 =$ never had CHF; $1 =$ had CHF, but have no impairment; $2 =$ mild CHF impairment; $3 =$ moderate CHF impairment; and $4 =$ severe CHF impairment.
>
> Age: In years.
>
> LMCA: Percent of diameter narrowing of the left main coronary artery.
>
> EF: Ejection fraction, the percent of the blood in the pumping chamber (left ventricle) of the heart pumped out during heart beat.
>
> Digitalis: Does the patient use digitalis? $1 =$ yes, $2 =$ no.
>
> Therapy: $1 =$ medical; $2 =$ surgical.
>
> Vessel: Number (0–3) of vessels diseased with 70% or more stenosis.

The betas and their standard errors are:

Variable	Beta	Standard Error	Chi-Square
Impairment	0.2677	0.0505	?
Age	0.0430	0.0084	26.02
LMCA	0.0090	0.0024	?
EF	−0.0362	0.0098	?
Digitalis	−0.3802	0.1625	?
Therapy	−0.3418	0.1458	5.49
Vessel	0.2081	0.1012	4.23
Constant	−1.2873		

a. Fill in the chi-square value column where missing.

b. For which variables is $p < 0.10$? 0.05? 0.01? 0.001?

c. What is the instantaneous relative risk of 70% LMCA compared to 0% LMCA.

d. Consider three patients with the following covariate values:

	Patient Number		
Variable	1	2	3
Impairment	Severe	Mild	Moderate
Age	64	51	59
LMCA	50%	0%	0%
EF	15	32	23
Digitalis	Yes	Yes	Yes
Therapy	Medical	Surgical	Medical
Vessel	3	2	3

At the mean values of the data the 1- and 2-year survival were 88.0% and 80.16%, respectively. Find the probability of 1- and 2-year survival for these three patients.

e. With this model:

    i. Can surgery be better for one individuals and medical treatment for another? Why? What does this say about unthinking application of the model?

    ii. Under surgical therapy (considered for an individual) can the curve crossover the estimated medical survival for some individuals? For heavy surgical mortality would a proportional hazard model always seem appropriate?

17. The eight indices of Example 16.6 were run in a stepwise Cox analysis without the ventricular function variable, LVSCOR, included. (There were more cases in this analysis—as some individuals did not have a known value of the ventricular function—than in the previous analysis. This accounts for the differences at the initial step.) Portions of the output are:

STEP 0. S-STATISTIC FOR VARIABLES NOT IN THE MODEL

Number	Variables	S-Statistic	Probability
1	DISVES	597.1	0.0000
2	PRXVES	624.2	0.0000
3	PRXSCR	637.8	0.0000
4	FRIES	502.6	0.0000
5	GENSEN	872.0	0.0000
6	NHCH	627.8	0.0000
7	MNHCH	574.2	0.0000
8	SEGMEN	717.7	0.0000

STEP 1. VARIABLE GENSIN ADDED WITH CHI-SQUARE = 872.0, DF = 1
(COEFFICIENTS, CHI-SQUARE STATISTICS FOR VARIABLES IN THE MODEL)

Number	Variable	Beta	Standard Error	Chi-Square	Probability
5	GENSIN	0.0068	0.0002	828.5	0.0000

Model chi-square statistic = 654.9, DF = 1, probability = 0.0000

STEP 1. *S*-STATISTIC FOR VARIABLES NOT IN THE MODEL

Number	Variables	*S*-Statistic	Probability
1	DISVES	116.5	0.0000
2	PRXVES	69.41	0.0000
3	PRXSCR	65.85	0.0000
4	FRIES	98.11	0.0000
6	NHCH	106.3	0.0000
7	MNHCH	121.0	0.0000
8	SEGMEN	81.52	0.0000

STEP 2. VARIABLE MNHCH ADDED WITH CHI-SQUARE = 121.0, DF = 1.
(COEFFICIENTS, CHI-SQUARE STATISTICS FOR VARIABLES IN THE MODEL)

Number	Variable	Beta	Standard Error	Chi-Square	Probability
5	GENSIN	0.0042	0.0003	145.5	0.0000
7	MNHCH	−0.0496	0.0045	119.3	0.0000

Model chi-square statistic = 783.1, DF = 2, probability = 0.0000

STEP 2. *S*-STATISTIC FOR VARIABLES NOT IN THE MODEL

Number	Variables	*S*-Statistic	Probability
1	DISVES	19.60	0.0000
2	PRXVES	28.02	0.0000
3	PRXSCR	27.35	0.0000
4	FRIES	13.04	0.0003
6	NHCH	4.851	0.0276
8	SEGMEN	22.75	0.0000

Step 5. COEFFICIENTS, CHI-SQUARE STATISTICS FOR VARIABLES IN THE MODEL

Number	Variable	Beta	Standard Error	Chi-Square	Probability
5	GENSIN	0.0029	0.0003	48.36	0.0000
7	MNHCH	−0.0279	0.0061	20.69	0.0000
2	PRXVES	0.1492	0.0452	9.607	0.0020
1	DISVES	0.1846	0.0560	10.85	0.0010
8	SEGMEN	0.0140	0.0063	4.915	0.0266

Model chi-square statistic = 829.3, DF = 5, probability = 0.0000

STEP 5. *S*-STATISTIC FOR VARIABLES NOT IN THE MODEL

Number	Variables	*S*-Statistic	Probability
3	PRXSCR	1.909	0.1671
4	FRIES	1.580	0.2088
6	NHCH	0.6313	0.4269

No Variables Significant for Entry
at the 0.05 Level
Stepwise Analysis Abandoned

a. Why do the S-statistic values drop so dramatically after the first variable, GENSIN, is entered into the model?

b. When MNHCH is added at step 2, the coefficient of GENSIN drops by about a third (0.0068 to 0.0042) while the standard error increases. Does this surprise you? Using an analogy to the multiple linear regression situation, why does this occur?

c. Does the trend (in (b)) continue to the last step?

d. At the last step, the chi-square values for the individual values are quite small compared to the model chi-square. How can this occur? (Draw an analogy to the multiple linear regression situation.)

18. In the arteriographic index study of Ringqvist *et al.* [1982] mentioned above (Example 16.6), one index was constructed empirically by a stepwise Cox survival analysis. In individuals with at least one vessel disease and a normally large right coronary artery (right or balanced dominance of the heart) the percent of narrowing in 27 possible segments (as shown in Figure 16.18) were coded. The double numbered segments (e.g. 4 and 27) depend on whether the blood supply comes from the left or right side of the heart (reversed as you face the heart). Selected output from the stepwise Cox analysis on the 12,518 cases follows.

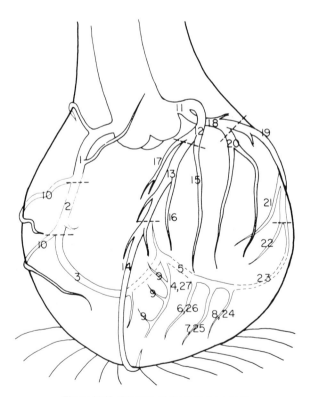

Figure 16.18. Diagram of the coronary arteries.

STEP 0. S-STATISTIC FOR VARIABLES NOT IN THE MODEL

Number	Variables	S-Statistic	Probability
1	PRXRCA	118.5	0.0000
2	MIDRCA	18.31	0.0000
3	DSTRCA	8.469	0.0036
4	RPDA	10.72	0.0011
5	RPLS	8.662	0.0033
6	RPL1	0.02630	0.8712
7	RPL2	0.4057	0.5242
8	RPL3	3.032	0.0817
9	ACMARG	14.47	0.0002
10	LMCA	161.0	0.0000
11	PRXLAD	163.4	0.0000
12	MIDLAD	33.80	0.0000
13	DSTLAD	27.54	0.0000
14	DIAG1	11.53	0.0007
15	DIAG2	6.022	0.0141
16	SEPTL1	18.89	0.0000
17	PRXCX	101.1	0.0000
18	DISTCX	72.34	0.0000
19	OBMRG1	95.49	0.0000
20	OBMRG2	19.07	0.0000
21	OBMRG3	15.83	0.0001
22	LAV	10.55	0.0012
23	LPL1	1.578	0.2090
24	LPL2	5.519	0.0188
25	LPL3	10.36	0.0013
26	LPDA	0.006011	0.9382

STEP 1. VARIABLE PRXLAD ADDED WITH CHI-SQUARE = 163.4, DF= 1.
(COEFFICIENTS, CHI-SQUARE STATISTICS FOR VARIABLES IN THE MODEL)

Number	Variable	Beta	Standard Error	Chi-Square	Probability
11	PRXLAD	0.0109	0.0009	155.7	0.0000

Model chi-square statistic = 155.9, DF = 1, probability = 0.0000

STEP 1. S-STATISTIC FOR VARIABLES NOT IN THE MODEL

Number	Variables	S-Statistic	Probability
1	PRXRCA	112.2	0.0000
2	MIDRCA	18.70	0.0000
3	DSTRCA	6.734	0.0095
4	RPDA	5.043	0.0247
5	RPLS	6.121	0.0134
6	RPL1	0.06038	0.8059
7	RPL2	0.3651	0.5457

(Continued)

Number	Variables	S-Statistic	Probability
8	RPL3	2.078	0.1494
9	ACMARG	9.801	0.0018
10	LMCA	110.4	0.0000
12	MIDLAD	59.54	0.0000
13	DSTLAD	22.23	0.0000
14	DIAG1	4.614	0.0317
15	DIAG2	6.759	0.0093
16	SEPTL1	6.649	0.0099
17	PRXCX	78.78	0.0000
18	DISTCX	65.33	0.0000
19	OBMRG1	68.71	0.0000
20	OBMRG2	16.48	0.0001
21	OBMRG3	13.87	0.0002
22	LAV	7.469	0.0063
23	LPL1	1.006	0.3159
24	LPL2	3.453	0.0632
25	LPL3	6.847	0.0089
26	LPDA	0.1186	0.7305

STEP 2. VARIABLE PRXRCA ADDED WITH CHI-SQUARE = 112.2, DF= 1
(COEFFICIENTS, CHI-SQUARE STATISTICS FOR VARIABLES IN THE MODEL)

Number	Variable	Beta	Standard Error	Chi-Square	Probability
11	PRXLAD	0.0108	0.0009	149.7	0.0000
1	PRXRCA	0.0091	0.0009	108.5	0.0000

Model chi-square statistic = 265.8, DF = 2, probability = 0.0000

STEP 9. S-STATISTIC FOR VARIABLES NOT IN THE MODEL

Number	Variables	S-Statistic	Probability
3	DSTRCA	0.3662	0.5451
4	RPDA	0.2256	0.6348
5	RPLS	0.3989	0.5276
6	RPL1	1.944	0.1633
7	RPL2	0.01976	0.8882
8	RPL3	0.3261	0.5679
9	ACMARG	0.3124	0.5762
13	DSTLAD	0.1182	0.7310
14	DIAG1	1.997	0.1576
15	DIAG2	0.2983	0.5850
16	SEPTL1	0.5156	0.4727
20	OBMRG2	0.07166	0.7889
21	OBMRG3	4.375	0.0365
23	LPL1	0.4782	0.4893
24	LPL2	1.965	0.1610
25	LPL3	2.992	0.0837
26	LPDA	0.005970	0.9384

STEP 10. VARIABLE OBMRG3 ADDED WITH CHI-SQUARE = 4.375, DF= 1
(COEFFICIENTS, CHI-SQUARE STATISTICS FOR VARIABLES IN THE MODEL)

Number	Variable	Beta	Standard Error	Chi-Square	Probability
11	PRXLAD	0.0093	0.0009	97.97	0.0000
1	PRXRCA	0.0066	0.0009	50.87	0.0000
10	LMCA	0.0103	0.0016	42.33	0.0000
19	OBMRG1	0.0036	0.0009	15.66	0.0001
12	MIDLAD	0.0038	0.0009	17.38	0.0000
17	PRXCX	0.0045	0.0010	21.32	0.0000
18	DISTCX	0.0038	0.0009	18.11	0.0000
2	MIDRCA	0.0020	0.0009	5.181	0.0228
22	LAV	0.0047	0.0022	4.635	0.0313
21	OBMRG3	0.0029	0.0014	4.354	0.0369
	Constant	−1.4826			

Model chi-square statistic $= 455.7$, DF $= 10$, probability $= 0.0000$
Maximum of 10 Variables in Model
Stepwise Analysis Terminated

The stepwise Cox program was instructed to stop adding predictive variables when no remaining statistically significant ($p \leq 0.05$) variables were left, or when 10 variables had been entered, whichever occurred first.

a. After observing the addition of the first segment (the proximal LAD), do you feel there was a large overlap in predictive information between this segment and any of the other measurements? Why or why not?

b. Consider the final model where the tenth variable is added. Which segment adds the most to the risk with a narrowing of a fixed percent? *How can it occur that this "most important" segment does not have the largest chi-square value?

c. What is the instantaneous relative risk for an individual with 70% LMCA disease compared to an individual with 0% narrowing?

*d. If someone told you they had thought the distal LAD segment was important, but it must not be because it is not in the model, what would you reply?

19. The Clark *et al.* [1971] heart transplant data were collected as follows. Individuals with failing hearts waited for a donor heart to become available; this usually occurred within 90 days. Some individuals, however, died before a donor heart became available. Figure 16.19 plots the survival curves of (1) those not transplanted (indicated by circles) and (2) the transplant patients from time of surgery (indicated by the triangles).

a. Is the survival of the nontransplanted individuals a reasonable estimate of the nonoperative survival of candidates for heart transplant? Why or why not?

b. Would you be willing to conclude from the figure (assuming a statistically significant result) that heart transplant surgery prolongs life? Why or why not?

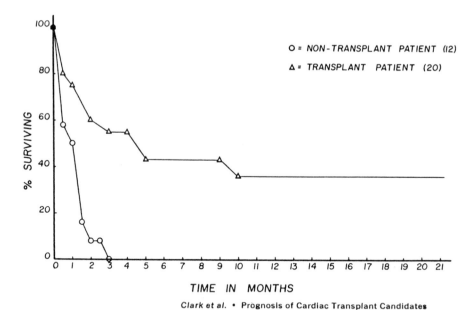

Clark et al. • Prognosis of Cardiac Transplant Candidates

Figure 16.19. Survival calculated by the life table method. Survival for transplanted patients is calculated from the time of operation; survival of nontransplanted patients is calculated from the time of selection for transplantation.

REFERENCES

Alderman, E. L., Fisher, L. D., Litwin, P., Kaiser, G. C., Myers, W. O., Maynard, C., Levine, F., and Schloss, M. [1983]. Results of coronary artery in patients with poor left ventricular function (CASS). *Circulation.* **68:** 785–789. Used with permission from the American Heart Society.

Chaitman, B. R., Fisher, L. D., Bourassa, M. G., Davis, K., Rogers, W. J., Maynard, C., Tyras, D. H., Berger, R. L. Judkins, M. P., Ringqvist, I., Mock, M. B., and Killip, T. [1981]. Effect of coronary bypass surgery on survival patterns in subsets of patients with left main coronary disease. *American Journal of Cardiology,* **48:** 765–777.

Clark, D. A., Stinson, E. B., Grieppe, R. B., Schroeder, J. S., Shumway, N. E., and Harrison, D. B. [1971]. Cardiac transplantation in man. VI. Prognosis of patients selected for cardiac transplantation. *Annals of Internal Medicine,* **75:** 15–21.

Cox, D. R. [1972]. Regression models and life tables. *Journal of the Royal Statistical Society, Series B,* **34:** 187–220.

Crowley, J. and Breslow, N. [1975]. Remarks on the conservatism of $(O - E)^2/E$ in survival data. *Biometrics.* **31:** 957–961.

Crowley, J. and Hu, M. [1977]. Covariance analysis of heart transplant survival data. *Journal of the American Statistical Association,* **72:** 27–36.

Cutler, S. J. and Ederer, F. [1958]. Maximum utilization of the life table method in analyzing survival. *Journal of Chronic Diseases,* **8:** 699–712.

European Coronary Surgery Study Group [1980]. Prospective randomized study of coronary artery bypass surgery in stable angina pectoris. Second interim report. *The Lancet,* 6 September 1980, **2:** 491–495.

Gehan, E. A. [1969]. Estimating survival functions from the life table. *Journal of Chronic Diseases*, **21:** 629–644. Copyright © 1969 by Pergamon Press, Inc. Used with permission.

Greenwood, M. [1926]. Reports on Public Health and Medical Subjects. No. 33, Appendix I. The Errors of Sampling of the Survivorship Tables. H. M. Stationary Office, London.

Kalbfleisch, J. D. and Prentice, R. L. [1980]. *The Statistical Analysis of Failure Time Data*, Wiley, New York.

Kaplan, E. L. and Meier, P. [1958]. Nonparametric estimation for incomplete observations. *Journal of the American Statistical Association*, **53:** 457–481.

Mantel, N. [1966]. Evaluation of survival data and two new rank order statistics arising in its consideration. *Cancer Chemotherapy Reports*, **50:** 163–170.

Mantel, N. and Byar, D. [1974]. Evaluation of response time 32 data involving transient states: an illustration using heart transplant data. *Journal of the American Statistical Association*, **69:** 81–86.

Mantel, N. and Haenzel, W. [1959]. Statistical aspects of the analysis of data from retrospective studies of disease. *Journal of the National Cancer Institute*, **22:** 719–748.

Messmer, B. J., Nora, J. J., Leachman, R. E., and Cooley, D. A. [1969]. Survival times after cardiac allographs. *Lancet*, May 10, 1969, **1:** 954–956.

Miller, R. G. [1981]. *Survival Analysis*. Wiley, New York.

Parker, R. L., Dry, T. J., Willius, F. A., and Gage, R. P. [1946]. Life expectancy in angina pectoris. *Journal of the American Medical Association*, **131:** 95–100.

Passamani, E. R., Fisher, L. D., Davis, K. B., Russel, R. O., Oberman, A., Rogers, W. J., Kennedy, J. W., Alderman, E., and Cohen, L. [1982]. The relationship of symptoms to severity, location and extent of coronary artery disease and mortality. Unpublished study.

Peto, R., and Peto, J. [1972]. Asymptotically efficient rank invariant test procedures. *Journal of the Royal Statistical Society, Series A*, **135:** 185–207.

Pike, M. C. [1966]. A method of analysis of a certain class of experiments in carcinogenesis. *Biometrics*, **26:** 579–581.

Prentice R. L., Kalbfleish, J. D., Peterson, A. V., Flournoy, N., Farewell, V. T., and Breslow, N. L. [1978]. The analysis of failure times in the presence of competing risks. *Biometrics*, **34:** 541–554.

Ringqvist, I., Fisher, L. D., Mock, M. B., Davis, K. B., Wedel, H., Chaitman, B. R., Passamani, E., Russell, R. O., Alderman, E. L., Kouchakas, N., Kaiser, G. C., Ryan, T. J., Killip, T., and Fray, D. [1983]. Prognostic value of angiographic indices of coronary artery disease from the coronary artery surgery study (CASS). *Journal of Clinical Investigation*, **71:** 1854–1866.

Takaro, T., Hultgren, H. N., Lipton, M. J., and Detre, K. M. and Participants in the Study Group [1976]. The veteran's administration cooperative randomized study of surgery for coronary arterial occlusive disease. II. Subgroup with significant left main lesions. *Circulation Supplement 3*, **54:** III-107–III-117.

Turnbull, B., Brown, B., and Hu, M. [1974]. Survivorship analysis of heart transplant data. *Journal of the American Statistical Association*, **69:** 74–80.

Vital Statistics of the United States, 1974. Volume II, Section 5. Life Tables [1976]. DHEW, National Center for Health Statistics. US Government Printing Office, Washington, DC, 20402.

CHAPTER 17

Sample Sizes for Observational Studies

17.1 INTRODUCTION

In this chapter we deal with the problem of calculating sample sizes in various observational settings. There is a very diverse literature on sample size calculations, dealing with many interesting areas. We can only give you a feeling for some approaches. In Note 17.2, we give a brief literature review to point you in the right direction for further study.

We start the chapter by considering the topic of screening in the context of adverse effects attributable to drug usage, trying to accommodate both the "rare disease" assumption and the multiple comparison problem. Section 17.3 discusses sample size considerations when costs of observations are not equal, or the variability is unequal; some very simple but elegant relationships are derived. Section 17.4 considers sample size consideration in the context of discriminant analysis. Three questions are considered: (1) how to select variables to be used in discriminating between two populations in the face of multiple comparisons; (2) given that m variables have been selected what sample size is needed to discriminate between two populations with satisfactory power; and (3) how large a sample size is needed to estimate the probability of correct classification with adequate precision and power. Notes, problems and references complete the chapter.

17.2 SCREENING STUDIES

A screening study is a scientific fishing expedition; for example, attempting to relate exposure to one of several drugs to presence or absence of one or more side effects (disease). For example, in Chapter 1, we discussed the relationship between the usage of post-menopausal estrogen and endometrial carcinoma. In such screening studies the number of drug categories is usually very large—500 is not uncommon—and the number of diseases is very large—50 or more is not unusual. Thus the number of combinations of disease and drug exposure can be very large—25,000 in the above example. In this section we want to consider the determination of sample size in screening studies in terms of the following considerations: many variables are tested and side effects are rare. A cohort of exposed and unexposed subjects is either followed or observed. We have looked at many diseases or exposures, and want to "protect" ourselves against a large Type I error, and want to know how many observations are to be taken. We proceed in two steps: first we derive the formula for the sample size without consideration of the multiple testing aspect, then we will incorporate the multiple testing aspect. Let

$X_1 =$ Number of occurrences of a disease of interest (per 100,000 person-years, say) in the unexposed population,

$X_2 =$ Number of occurrences (per 100,000 person-years) in the exposed population.

If X_1 and X_2 are rare events, then $X_1 \sim$ Poisson(θ_1) and $X_2 \sim$ Poisson(θ_2). Let $\theta_2 = R\theta_1$, that is, the risk in the exposed population is R times that in the unexposed population ($0 < R < \infty$). We can approximate the distributions by using the variance stabilizing transformation (discussed in Chapter 10).

$$Y_1 = \sqrt{X_1} \sim N(\sqrt{\theta_1}, \sigma^2 = 0.25),$$

$$Y_2 = \sqrt{X_2} \sim N(\sqrt{\theta_2}, \sigma^2 = 0.25).$$

Assuming independence,

$$Y_2 - Y_1 \sim N\left(\sqrt{\theta_1}(\sqrt{R} - 1), \sigma^2 = 0.5\right). \tag{17.1}$$

For specified Type I and Type II errors α, β, the *number of events* n_1 and n_2 in the unexposed and exposed groups required to detect a relative risk of R with power $1 - \beta$ are given by the equation

$$n_1 = \frac{\left(Z_{1-\alpha/2} + Z_{1-\beta}\right)^2}{2\left(\sqrt{R} - 1\right)^2}, \quad n_2 = Rn_1. \tag{17.2}$$

Equation (17.2) assumes a two-sided, two-sample test with an equal number of subjects observed in each group. It is an approximation, based upon the normality of the square root of a Poisson random variable. If the prevalence, π_1, in the unexposed population is known, the number of subjects per group, N, can be calculated by using the relationship

$$N\pi_1 = n_1 \quad \text{or} \quad N = n_1/\pi_1. \tag{17.3}$$

Example 17.1. In Section 15.4, mortality in active participants in an exercise program and dropouts were compared. Among the active participants, there were 16 deaths in 593 person-years of active participation; in dropouts there were 34 deaths in 723 person-years. Using an α of 0.05, the results were not significantly different. The relative risk, R, for dropouts is estimated by

$$R = \frac{34/723}{16/593} = 1.74.$$

Assuming equal exposure time in the active participants and dropouts, how large should the sample sizes n_1 and n_2 be to declare the relative risk, $R = 1.74$, significant at the 0.05 level with probability 0.95? In this case we use a two-tailed test and $Z_{1-\alpha/2} = 1.960$, and $Z_{1-\beta} = 1.645$, so that

$$n_1 = \frac{(1.960 + 1.645)^2}{2(\sqrt{1.74} - 1)^2} = 63.4 \doteq 64 \quad \text{and} \quad n_2 = (1.74)n_1 = 111,$$

for a total number of observed events $= n_1 + n_1 = 64 + 111 = 175$ deaths. We would need approximately $(111/34) \times 723 = 2360$ person-years exposure in the dropouts and the same number of years of exposure among the controls. The exposure years in the observed data are not split equally between the two groups. Some further discussion of this aspect can be found in Note 17.1.   □

If there is only one observational group, the group's experience perhaps being compared with that of a known population, then the sample size required is $n_1/2$, again illustrating the fact that comparing two groups requires four times more exposure-times than comparing one group with a known population.

We now turn to the second aspect of our question. Suppose the above comparison is one of a multitude of comparisons? To maintain a per-experiment significance level of α we use the Bonferroni inequality to calculate the per-comparison error rate. Table 17.1 relates the per-comparison critical values to the number of tests performed and the per-experiment error rate. It is remarkable that the critical values do not increase too rapidly with the number of tests.

Example 17.2. Suppose the FDA is screening a large number of drugs; relating 10 kinds of congenital malformations to 100 drugs that could be taken during pregnancy. A particular drug and a particular malformation is now being examined. Equal numbers of exposed and unexposed women are to be selected and a relative risk of $R = 2$ is to be detected with power 0.80 and per-experiment one-sided error rate of $\alpha = 0.05$. In this situation $\alpha^* = \alpha/1000$ and $Z_{1-\alpha^*} = Z_{1-\alpha/1000} = Z_{0.99995} = 3.891$. The required number of events in the unexposed group is

Table 17.1. Relationship Between Overall Significance Level α, Significance Level per Test, Number of Tests and Associated Z Values, Using the Bonferroni Inequality.

Number of	Overall	Required Level	Z Values	
Tests (K)	α Level	per Test (α)	One-tail	Two-tail
1	0.05	0.05	1.645	1.960
2	0.05	0.025	1.960	2.241
3	0.05	0.01667	2.128	2.394
4	0.05	0.0125	2.241	2.498
5	0.05	0.01	2.326	2.576
10	0.05	0.005	2.576	2.807
100	0.05	0.0005	3.291	3.481
1000	0.05	0.00005	3.891	4.056
10000	0.05	0.000005	4.417	4.565

$$n_1 = \frac{(3.891 + 0.842)^2}{2(\sqrt{2} - 1)^2} = \frac{22.4013}{0.343146} = 65.3 \doteq 66,$$

$$n_2 = 2n_1 = 132.$$

In total, $66 + 132 = 198$ malformations must be observed. For a particular malformation if the congenital malformation rate is of the order of $3/1000$ live births, approximately 22,000 unexposed women and 22,000 women exposed to the drug must be examined. This large sample size is not only a result of the multiple testing but, also the rarity of the disease. (The comparable number testing only once, $\alpha^* = \alpha = 0.05$ is $n_1 = \frac{1}{2}(1.645 + 0.842)^2/(\sqrt{2} - 1)^2 = 18$, or 3000 women per group.) $\qquad\square$

17.3 SAMPLE SIZE AS A FUNCTION OF COST AND AVAILABILITY

17.3.1 Equal Variance Case

Consider the comparison of means from two independent groups with the same variance σ, the standard error of the difference is

$$\sigma\sqrt{\frac{1}{n_1} + \frac{1}{n_2}}, \tag{17.4}$$

where n_1 and n_2 are the sample sizes in the two groups. As is well known, for fixed N the standard error of the difference is minimized (maximum precision) when

$$n_1 = n_2 = N,$$

that is, the sample sizes are equal. Suppose now that there is a differential cost in obtaining the observations in the two groups, then it may pay to choose n_1 and n_2 unequal, subject to the constraint that the standard error of the difference remain the same. For example,

$$\frac{1}{10} + \frac{1}{10} = \frac{1}{6} + \frac{1}{30}.$$

Two groups of equal sample size, $n_1 = n_2 = 10$ give the same precision as two groups with $n_1 = 6$ and $n_2 = 30$. Of course, the total number of observations N is larger, 20 versus 36.

In many instances sample size calculations are based on additional considerations such as:

1. Relative cost of the observations in the two groups;
2. Unequal hazard or potential hazard of treatment in the two groups;
3. A limited number of observations is available for one group.

In the last category are case-control studies where the number of cases is limited. For example, in studying the Sudden Infant Death Syndrome (SIDS), by means of a case-control study, the number of cases in a defined population is fairly well fixed, while an arbitrary number of (matching) controls can be obtained.

We now formalize the argument. Suppose there are two groups, G_1 and G_2, with costs per observations c_1 and c_2, respectively. The total cost, C, of the experiment is

$$C = c_1 n_1 + c_2 n_2, \tag{17.5}$$

where n_1 and n_2 are the number of observations in G_1 and G_2, respectively. The values of n_1 and n_2 are to be chosen to minimize (maximum precision),

$$\frac{1}{n_1} + \frac{1}{n_2},$$

subject to the constraint that the total cost is to be C. It can be shown that under these conditions the required sample sizes are

and

$$n_1 = \frac{C}{c_1 + \sqrt{c_1 c_2}} \tag{17.6}$$

$$n_2 = \frac{C}{c_2 + \sqrt{c_1 c_2}}. \tag{17.7}$$

The ratio of the two sample sizes is

$$\frac{n_2}{n_1} = \sqrt{\frac{c_1}{c_2}} = h, \quad \text{say.} \tag{17.8}$$

That is, if costs per observation in Groups G_1 and G_2, are c_1 and c_2, respectively, then choose n_1 and n_2 on the basis of the ratio of the square root of the costs. This rule has been termed the square root rule by Gail et al. [1976]; the derivation can also be found in Nam [1973] and Cochran [1962].

If the costs are equal, then $n_1 = n_2$, as before. The application of this rule can decrease the cost of an experiment although it will increase the total number of observations. Note that the population means and standard deviation need not be known to determine the ratio of the sample sizes only the costs. If the desired precision is specified—perhaps on the basis of sample size calculations assuming equal costs— the values of n_1 and n_2 can be determined. Compared with an experiment with equal sample sizes, the ratio, ρ, of the costs of the two experiments can be shown to be

$$\rho = \frac{1}{2} + \frac{h}{(1 + h^2)}. \tag{17.9}$$

If $h = 1$, then $\rho = 1$, as expected; if h is very close to zero or very large, $\rho = \frac{1}{2}$; thus no matter what the relative costs of the observations the savings can be no larger than 50%.

Example 17.3 (after Gail et al. [1976]). A new therapy, G_1, for hypertension is introduced and costs $400 per subject. The standard therapy, G_2, costs $16 per subject.

On the basis of power calculations, the precision of the experiment is to be equivalent to an experiment using 22 subjects per treatment, so that

$$\frac{1}{22} + \frac{1}{22} = 0.09091.$$

The square root rule specifies the ratio of the number of subjects in G_1 and G_2, by

$$n_2 = \sqrt{\frac{400}{16}}n_1$$

$$= 5n_1.$$

To obtain the same precision, we need to solve

$$\frac{1}{n_1} + \frac{1}{5n_1} = 0.09091$$

or

$$n_1 = 13.2 \quad \text{and} \quad n_2 = 66.0.$$

(that is, $1/13.2 + 1/66.0 = 0.09091$, the same precision).

Rounding up, we require 14 observations in G_1 and 66 observations in G_2. The costs can also be compared as in Table 17.2.

A savings of $3896 has been obtained yet the precision is the same. The total number of observations is now 80, compared to 44 in the equal sample size experiment. The ratio of the savings is

$$\rho = \frac{6656}{9152} = 0.73.$$

The value for ρ calculated from Equation (17.9) is

$$\rho = \frac{1}{2} + \frac{5}{26} = 0.69.$$

The reason for the discrepancy is the rounding of sample sizes to integers. ☐

17.3.2   Unequal Variance Case

Suppose we want to compare the means from groups with unequal variance. Again, suppose there are n_1 and n_2 observations in the two groups. Then the standard error of the difference between the two means is

Table 17.2. Costs Comparisons for Example 17.3.

	Equal Sample Size		Sample Size Determined by Cost	
	n	Cost	n	Cost
G_1	22	8800	14	5600
G_2	22	352	66	1056
Total	44	9152	80	6656

$$\sqrt{\frac{\sigma_1^2}{n_1} + \frac{\sigma_2^2}{n_2}}.$$

Let the ratio of the variances be $\eta^2 = \sigma_2^2/\sigma_1^2$. Gail et al. [1976] show that the sample size now should be allocated in the ratio

$$\frac{n_2}{n_1} = \sqrt{\frac{\sigma_2^2}{\sigma_1^2} \frac{c_1}{c_2}} = \eta h.$$

The calculations can then be carried out as before. In this case, the cost relative to the experiment with equal sample size is

$$\rho^* = \frac{(h + \eta)^2}{(1 + h^2)(1 + \eta^2)}. \tag{17.10}$$

These calculations also apply when the costs are equal but the variances unequal, as is the case in binomial sampling.

17.3.3 Rule of Diminishing Precision Gain

One of the reasons advanced at the beginning of Section 17.3 for distinguishing between the sample sizes of two groups is that a limited number of observations may be available for one group and a virtually unlimited number in the second group. Case-control studies were cited where the number of cases per population is relatively fixed. Analogous to Gail et al. [1976] we define a rule of diminishing precision gain. Suppose there are n cases and an unlimited number of controls are available. Assume that costs and variances are equal. The precision of the difference is then proportional to

$$\sigma\sqrt{\frac{1}{n} + \frac{1}{hn}},$$

where hn is the number of controls selected for the n cases.

We calculate the ratio, P_h

$$P_h = \frac{\sqrt{\dfrac{1}{n} + \dfrac{1}{hn}}}{\sqrt{\dfrac{1}{n} + \dfrac{1}{n}}}$$

$$= \sqrt{\frac{1}{2}\left(1 + \frac{1}{h}\right)}.$$

This ratio P_h is a measure of the precision of a case-control study with n and hn cases and controls, respectively, relative to the precision of a study with an equal number, n, of cases and controls. Table 17.3 presents the values of P_h and $100(P_h - P_\infty)/P_\infty$ as function of h. This table indicates that in the above context the gain in precision with, say, more than four controls per case is minimal. At $h = 4$, one

Table 17.3. Comparison of Precision of Case Control Study
with n and hn Cases and Controls, Respectively.

h	P_h	$100[(P_h - P_\infty)/P_\infty]\%$
1	1.00	41%
2	0.87	22%
3	0.82	15%
4	0.79	12%
5	0.77	10%
10	0.74	5%
∞	0.71	0%

obtains all but 12% of the precision associated with a study using an infinite number of controls. Hence, in the above situation there is little merit in obtaining more than four or five times as many controls as cases. Lubin [1980] approaches this from the point of view of the logarithm of the odds ratio and comes to a similar conclusion.

17.4 SAMPLE SIZE CALCULATIONS IN SELECTING CONTINUOUS VARIABLES TO DISCRIMINATE BETWEEN POPULATIONS

In certain situations, there is interest in examining a large number of continuous variables to explain the difference between two populations. For example, an investigator might be "fishing" for clues explaining the presence (one population) or absence (the other population) of a disease of unknown etiology. Or in a disease where a variety of factors are known to affect prognosis, the investigator may desire to find a good set of variables for predicting which individuals will survive for a fixed number of years. In this section, the determination of sample size for such studies is discussed.

There are a variety of approaches to the data analysis in this situation. With a large, say 50 or more, number of variables, we would hesitate to run stepwise discriminant analysis to select a few important variables, since (1) in typical data sets there are often many dependencies that make the method numerically unstable, that is, the results coming forth from some computers cannot be relied upon; (2) the more complex the mathematical model used, the less faith we have that it is useful in other situations, that is, the more parameters that are used and estimated, the less confidence that the result is transportable to another population in time or space; here we might be envisioning a discriminant function with a large number of variables; and (3) the multiple comparison problems inherent in considering the large number of variables at each step in the stepwise procedure make the result of doubtful value.

One approach to the analysis is to first perform a *univariate screen*. This means that variables (used singly, that is, univariately) with the most power to discriminate between the two populations are selected. Secondly, use these univariate discriminating variables in the discriminant analysis. The sample size calculations below are based upon this method of analysis. There is some danger in this approach, as variables which univariately are not important in discrimination could be important when used in conjunction with other variables. In many practical situations this is not usually the case. Before discussing the sample-size considerations, we will consider a second approach to the analysis of such data as envisioned here.

Often the discriminating variables fall naturally in smaller subsets. For example, the subsets for patients may involve data from (i) the history, (ii) physical exam, and (iii) some routine tests. In many situations the predictive information of the variables within each subset is roughly the same. This being the case, a two-step method of selecting the predictive variables is to: (1) use stepwise selection within subsets to select a few variables from each subset, and (2) combine the selected variables into a group to be used for another stepwise selection procedure to find the final subset of predictive variables.

After selecting a smaller subset of variables to use in the prediction process, one of two steps is usually taken. (1) The predictive equation is validated (tested) on a new sample to show that it has predictive power. That is, an F-test for the discriminant function is performed. Or, (2) a larger independent sample is used to provide an indication of the accuracy of the prediction. The second approach requires a larger sample size than merely establishing that there is some predictive ability as in the first approach. The next three subsections make this general discussion precise.

17.4.1 Univariate Screening of Continuous Variables

To obtain an approximate idea of the sample size needed to screen among k variables, the following is assumed: the variables are normally distributed with the same variance in each population and possibly different means. The power to classify into the two populations depends upon δ, the number of standard deviations distance between the two populations means. Some idea of the relationship of classificatory power to δ is given in Figure 17.1.

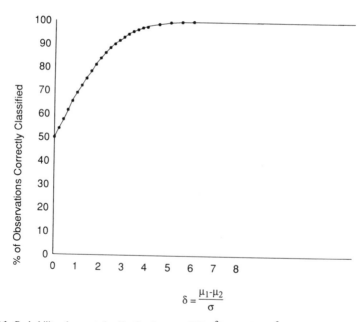

Figure 17.1. Probability of correct classification between $N(0, \sigma^2)$ and $N(\delta\sigma, \sigma^2)$ populations, assuming equal priors and $\delta\sigma/2$ as the cutoff values for classifying into the two populations.

Suppose that we are going to screen k variables and want to be sure, with probability at least $1 - \alpha$, to include all variables with $\delta \geq D$. In this case, we must be willing to accept some variables with values close to but less than D. Suppose at the same time we want probability at least $1 - \alpha$ of not including any variables with $\delta \leq fD$ where $0 < f < 1$. One approach is to look at confidence intervals for the difference in the population means. If the absolute value of the difference is greater than $fD + (1 - f)D/2$, the variable is included. If the absolute value of the difference is less than this value, the variable is not included. Figure 17.2 presents the situation.

To recap, with probability at least $1 - \alpha$, we include for use in prediction all variables with $\delta \geq D$ and do not include those with $\delta \leq fD$. In between, we are willing for either action to take place. The dividing line is placed in the middle.

Let us suppose that the number of observations, n, is large enough so that a normal approximation for confidence intervals will hold. Further, suppose that a fraction p of the data is from the first population and $1 - p$ from the second population. If we choose $1 - \alpha^*$ confidence intervals so that the probability is about $1 - \alpha$ that all intervals have half-width $\sigma(1 - f)D/2$, the result will hold.

If n is large, the pooled variance is approximately σ and the half-interval has width (in standard deviation units) of about

$$\sqrt{\frac{1}{Np} + \frac{1}{N(1-p)}} \, Z_{1-\alpha^*},$$

where $Z_{1-\alpha^*}$ is the $N(0,1)$ critical value. To make this approximately $(1 - f)D/2$, we need

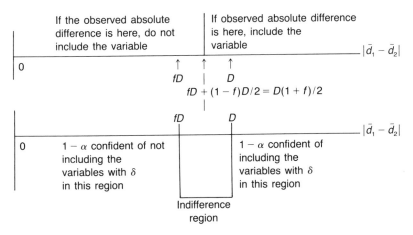

Figure 17.2. Inclusion and exclusion scheme for differences in sample means $|d_1 - d_2|$ from populations G_1 and G_2.

Table 17.4. Sample Sizes Needed for Univariate Screening when $f = 2/3$. For Each Entry the Top, Middle and Bottom Numbers are for $\alpha = 0.10, 0.05$ and 0.01, Respectively.

$D =$	$p = 0.5$			$p = 0.6$			$p = 0.7$			$p = 0.8$			$p = 0.9$		
	0.5	1	2	0.5	1	2	.5	1	2	0.5	1	2	0.5	1	2
$k = 20$	2121	527	132	2210	553	136	2525	629	157	3315	829	204	5891	1471	366
	2478	616	153	2580	642	157	2950	735	183	3872	965	238	6881	1717	429
	3289	825	204	3434	859	213	3923	978	242	5151	1288	319	9159	2287	570
$k = 100$	2920	721	179	3043	761	187	3477	867	217	4565	1139	285	8118	2028	506
	3285	820	204	3421	854	213	3910	978	242	5134	1284	319	9129	2282	570
	4118	1029	255	4288	1071	268	4905	1224	306	6435	1607	400	11445	2860	714
$k = 300$	3477	867	217	3625	905	225	4140	1033	255	5436	1356	336	9665	2414	604
	3846	961	238	4008	999	247	4577	1143	285	6010	1500	374	10685	2669	667
	4684	1169	289	4879	1220	302	5576	1394	349	7323	1828	455	13018	3251	812

$$N = \frac{4z_{1-\alpha^*}^2}{p(1-p)D^2(1-f)^2}. \qquad (17.11)$$

In Chapter 12, it was shown that $\alpha^* = \alpha/2k$ was an appropriate choice by Bonferroni's inequality. In most practical situations, the observations tend to vary together, and the probability of all the confidence statements holding is greater than $1 - \alpha$. A slight compromise is to use $\alpha^* = [1 - (1 - \alpha)^{1/k}]/2$ as if the tests are independent. This α^* was used in computing Table 17.4.

From the table, it is very clear there is a large price to be paid if the smaller population is a very small fraction of the sample. There is often no way around this if the data needs to be collected prospectively before individuals have the population membership determined (by having a heart attack or myocardial infarction, for example).

17.4.2 Sample Size to Determine that a Set of Variables Has Discriminating Power

In this section, we find the answer to the following question. Assume that a discriminant analysis is being performed at significance level α with m variables. Assume that one population has a fraction p of the observations the other population, a fraction $1 - p$ of the observations. What sample size, n, is needed so that with probability $1 - \beta$, we reject the null hypothesis of no predictive power (i.e., Mahalanobis distance equal to zero), when in fact the Mahalanobis distance is $\Delta > 0$ (where Δ is fixed and known)? (See Chapter 13 for a definition of the Mahalanobis distance.)

The procedure is to use tables for the power functions of the analysis of variance tests as given in the CRC tables, [1968] pp. 311–319. To enter the charts, first find the chart for $v_1 = m$, the number of predictive variables.

The charts are for $\alpha = 0.05$ or 0.01. It is necessary to iterate to find the correct sample size n. The method is as follows:

1. Select an estimate of n.
2. Compute

$$\phi_n = \Delta\sqrt{\frac{p(1-p)}{m+1}} \times \sqrt{n}. \qquad (17.12)$$

This quantity indexes the power curves and is a measure of the difference between the two populations, adjusting for p and m.
3. Compute $v_2 = n - 2$.
4. On the horizontal axis find ϕ and go vertically to the v_2 curve. Follow the intersection horizontally to find $1 - \beta$.
5. **a.** If $1 - \widetilde{\beta}$ is greater than $1 - \beta$, decrease the estimate of n and go back to step 2.
   b. If $1 - \widetilde{\beta}$ is less than $1 - \beta$, increase the estimate of n and go back to step 2.
   c. If $1 - \widetilde{\beta}$ is approximately equal to $1 - \beta$, stop and use the given value of n as your estimate.

Example 17.4. Working at a significance level 0.05 with 5 predictive variables, find the total sample size needed to be 90% certain of establishing predictive power when $\Delta = 1$ and $p = 0.34$. Figure 17.3 is used in the calculation.

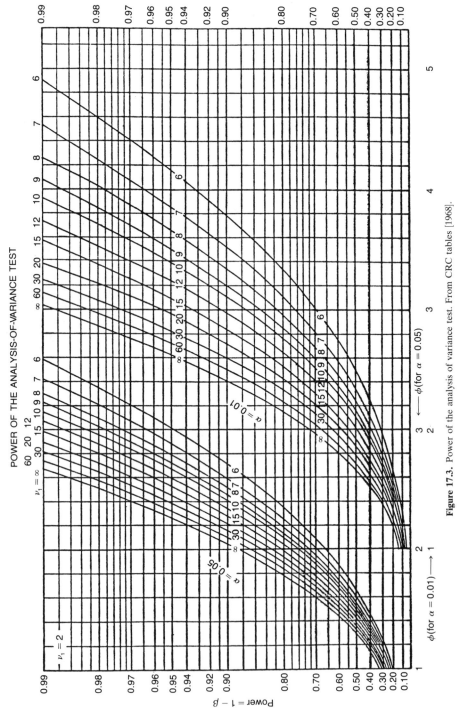

Figure 17.3. Power of the analysis of variance test. From CRC tables [1968].

We use

$$\phi_n = 1 \times \sqrt{\frac{0.3 \times 0.7}{5+1}} \sqrt{n} = 0.187\sqrt{n}$$

The method proceeds as follows:

1. Try $n = 30$, $\phi = 1.024$, $\nu_2 = 28$, $1 - \beta \doteq 0.284$.
2. Try $n = 100$, $\phi = 1.870$, $\nu_2 = 98$, $1 - \beta \doteq 0.958$.
3. Try $n = 80$, $\phi = 1.672$, $\nu_2 = 78$, $1 - \beta \doteq 0.893$.
4. Try $n = 85$, $\phi = 1.724$, $\nu_2 = 83$, $1 - \beta \doteq 0.92$.

Use $n = 83$. Note that the method is somewhat approximate due to the amount of interpolation (rough visual interpretation) needed. □

17.4.3 Quantifying the Precision of a Discrimination Method

After developing a method of classification, it is useful to validate the method on a new independent sample from the data used to find the classification algorithm. The approach of Section 17.4.2 is designed to show there is some classification power. Of more interest is to be able to make a statement on the amount of correct and incorrect classification. Suppose that one is hoping to develop a classification method that classifies correctly $100\pi\%$ of the time.

To estimate with $100(1 - \alpha)\%$ confidence the correct classification percentage to within $100\varepsilon\%$, what number of additional observations are required? The confidence interval (we'll assume n large enough for the normal approximation) will be, letting c equal the number of n trials correctly classified,

$$\frac{c}{n} \pm \sqrt{\frac{1}{n}\frac{c}{n}\left(1 - \frac{c}{n}\right)} z_{1-\alpha/2},$$

where $z_{1-\alpha/2}$ is the $N(0,1)$ critical value. We expect $c/n \doteq \pi$, so it is reasonable to choose n to satisfy $z_{1-\alpha/2} = \varepsilon\sqrt{\pi(1 - \pi)/n}$. This implies that

$$n = z_{1-\alpha/2}^2 \pi(1 - \pi)/\varepsilon^2, \tag{17.13}$$

where $\varepsilon = $ (predicted-actual) probability of misclassification.

Example 17.5. If one plans for $\pi = 90\%$ correct classification and wishes to be 99% confident of estimating the correct classification to within 2%, how many new experimental units must be allowed? From Equation (17.13) and $z_{0.995} = 2.576$, the answer is

$$n = (2.576)^2 \times 0.9(1 - 0.9)/(0.02)^2 \doteq 1493.$$ □

17.4.4   Total Sample Size for an Observational Study to Select Classfication Variables

In planning an observational study to discriminate between two populations, if the predictive variables are few in number and known, the sample size will be selected in the manner of Sections 17.4.2 or 17.4.3. The size depends upon whether the desire is to show some predictive power or to have desired accuracy of estimation of the probability of correct classification. In addition, a different sample is needed to estimate the discriminant function. Usually this is of approximately the same size.

If the predictive variables are to be culled from a large number of choices, an *additional* number of observations must be added for the selection of the predictive variables (e.g., in the manner of Section 17.4.1). Note that the method cannot be validated by application to the observations used to select the variables and to construct the discriminant function: this would lead to an exaggerated idea of the accuracy of the method. As the coefficients and variables were chosen specifically for these data, the method will work better (often considerably better) on these data than on an independent sample chosen as in Section 17.4.2 or 17.4.3.

NOTES

Note 17.1: Sample Sizes for Cohort Studies

There are four major journals which are sources for papers dealing with sample sizes in cohort and case-control studies: *Biometrics, Controlled Clinical Trials, Journal of Clinical Epidemiology,* and the *American Journal of Epidemiology.* In addition there are books by Fleiss [1981], Schlesselman [1982], and Schuster [1993].

A cohort study can be thought of as a cross-sectional study; there is no selection on case status or exposure status. The table generated is then the usual 2×2 table. Let the sample proportions be as follows:

	Exposure	**No Exposure**	
Case	p_{11}	p_{12}	$p_{1\bullet}$
Control	p_{21}	p_{22}	$p_{2\bullet}$
	$p_{\bullet 1}$	$p_{\bullet 2}$	1

If $p_{11}, p_{1\bullet}, p_{2\bullet}, p_{\bullet 1}, p_{\bullet 2}$ estimate $\pi_{11}, \pi_{1\bullet}, \pi_{2\bullet}, \pi_{\bullet 1}, \pi_{\bullet 2}$ respectively, then, approximately, the required total sample size for significance level α, and power $1 - \beta$ is

$$n = \frac{\left(Z_{1-\alpha/2} + Z_{1-\beta}\right)^2 \pi_{11}\pi_{1\bullet}\pi_{2\bullet}\pi_{\bullet 1}\pi_{\bullet 2}}{\left(\pi_{11} - \pi_{1\bullet}\pi_{\bullet 1}\right)^2}. \tag{17.14}$$

Given values of $\pi_{1\bullet}, \pi_{\bullet 1}$, and $R = (\pi_{11}/\pi_{\bullet 1})/(\pi_{12}/\pi_{\bullet 2}) = $ the relative risk, the value of π_{11} is determined by

$$\pi_{11} = \frac{R\pi_{\bullet 1}\pi_{1\bullet}}{R\pi_{\bullet 1} + \pi_{\bullet 2}}. \tag{17.15}$$

The formula for the required sample size then becomes

$$n = \left(Z_{1-\alpha/2} + Z_{1-\beta}\right)^2 \left(\frac{\pi_{\bullet 1}}{1 - \pi_{\bullet 1}}\right) \left(\frac{1 - \pi_{\bullet 1}}{\pi_{1\bullet}}\right) \left[1 + \frac{1}{\pi_{\bullet 1}(R - 1)}\right]^2. \qquad (17.16)$$

If the events are rare, the Poisson approximation derived in the text can be used. For a discussion of sample sizes in $r \times c$ contingency tables see Lachin [1977] and Cohen [1988].

Note 17.2: Sample Size Formulae for Case-Control Studies

There are a variety of sample size formulae for case-control studies. Let the data be arranged in a table as follows:

	Exposed	Not Exposed	
Case	X_{11}	X_{12}	n
Control	X_{21}	X_{22}	n

and

$$P[\text{exposure} \mid \text{case}] = \pi_1, \quad P[\text{exposure} \mid \text{control}] = \pi_2,$$

estimated by $P_1 = X_{11}/n$ and $P_2 = X_{21}/n$ (we assume $n_1 = n_2 = n$).

For a two sample, two-tailed test with,

$$P[\text{Type I error}] = \alpha \quad \text{and} \quad P[\text{Type II error}] = \beta,$$

the approximate sample size per group is

$$n = \frac{\left(Z_{1-\alpha/2}\sqrt{2\overline{\pi}(1 - \overline{\pi})} + Z_{1-\beta}\sqrt{\pi_1(1 - \pi_1) + \pi_2(1 - \pi_2)}\right)^2}{(\pi_1 - \pi_2)^2}, \qquad (17.17)$$

where $\overline{\pi} = \frac{1}{2}(\pi_1 + \pi_2)$. The total number of subjects is $2n$, of which n are cases and n are controls.

Another formula is

$$n = \frac{(\pi_1(1 - \pi) + \pi_2(1 - \pi_2))\left(Z_{1-\alpha/2} + Z_{1-\beta}\right)^2}{(\pi_1 - \pi_2)^2}. \qquad (17.18)$$

All of these formulae tend to give the same answers, and underestimate the sample sizes required. The choice of these formulae is a matter of esthetics primarily.

The formulae for sample sizes for case-control studies are approximations and several corrections are available to get closer to the exact value. Exact values for equal sample sizes have been tabulated in Haseman [1978]. Adjustment for the approximate sample size have been presented by Casagrande et al. [1978] who give a slightly more complicated and accurate formulation. See also Lachin [1981].

Two other considerations will be mentioned. First, unequal sample size. Particularly in case-control studies, it may be difficult to recruit more cases. Suppose we can select n observations from the first population and rn from the second ($0 < r < \infty$).

Following Schlesselman [1982], a very good approximation for the exact sample size for the number of cases is

$$n_1 = n \left(\frac{r+1}{2r} \right),$$ (17.19)

and for the number of controls

$$n_2 = n \left(\frac{r+1}{2} \right),$$ (17.20)

where n is determined by Equation (17.17) or (17.18). The total sample size is then $n((r+1)^2/2r)$. Note that the number of cases can never be reduced to more than $n/2$ no matter what the number of controls. This is closely related to the discussion in Section 17.3. Following Fleiss et al. [1980] a slightly improved estimate can be obtained by using

$$n_1^* = n_1 + \frac{r+1}{r\Delta} = \text{number of cases}$$

and

$$n_2^* = rn_1^* = \text{number of controls.}$$

A second consideration is cost. In Section 17.3, we considered sample sizes as a function of cost and related the sample sizes to precision. Now consider a slight reformulation of the problem in the case-control context again. Suppose enrollment of a case costs c_1 and enrollment of a control c_2. Pike and Casagrande [1979] show that a reasonable sample size approximation is

$$n_1 = n(1 + \sqrt{c_1/c_0}),$$
$$n_2 = n(1 + \sqrt{c_0/c_1}),$$

where n is defined by Equations (17.17) or (17.18).

Finally, frequently case-control study questions are put in terms of odds ratios (or relative risks). Let the odds ratio be $R = \pi_1(1 - \pi_2)/\pi_2(1 - \pi_1)$ where π_1 and π_2 are as defined at the beginning of this section. If the control group has known exposure rate π_2 that is, $P[\text{exposure} \mid \text{control}] = \pi_2$, then

$$\pi_1 = \frac{R\pi_2}{1 + \pi_2(R - 1)}.$$

To calculate sample sizes use Equation (17.17) for specified values of π_2 and R.

Mantel [1983] gives some clever suggestions for making binomial sample size tables more useful by making use of the fact that sample size is "inversely proportional to the square of the difference being sought, everything else being more or less fixed."

Gross and Clark [1975] is a good reference for sample size questions involving survival data.

Note 17.3: Power as a Function of Sample Size

Frequently the question is not "how big should my sample size be" rather, "I have 60 observations available, what kind of power do I have to detect a specified difference, relative risk or odds ratio?" The charts by Feigl as illustrated in Chapter 6 provided one answer. Basically it involves inversion of formulae such as given by Equations (17.17), (17.18), solving them for $Z_{1-\beta}$ and calculating the associated area under the normal curve. Besides Feigl, several authors have studied this problem or variations of it. Walter [1977] derived formulae for the smallest and largest relative risk, R, that can be detected as a function of sample size, Type I and Type II errors. Brittain and Schlesselman [1982], among other things, present estimates of power as a function of possibly unequal sample size and cost.

Note 17.4: Sample Size as a Function of Coefficient of Variation

Sometimes sample size questions are asked in the context of percent variability and percent changes in means. With an appropriate, natural interpretation valid answers can be provided. Specifically, assume that by "percent variability" is meant the coefficient of variation, call it V, and that the second mean differs from the first mean by a factor f.

Let two normal populations have means μ_1, μ_2 and standard deviations σ_1, σ_2. The usual sample size formula for two independent samples needed to detect a difference $\mu_1 - \mu_2$ in means with Type I error α and power $1 - \beta$ is given by

$$n = \frac{(z_{1-\alpha/2} + z_{1-\beta})^2 (\sigma_1^2 + \sigma_2^2)}{(\mu_1 - \mu_2)^2},$$

where $z_{1-\gamma}$ is the $100(1-\gamma)$th percentile of the standard normal distribution. This is the formula for a two-sided alternative; n is the number of observations per group. Now assume $\mu_1 = f\mu_2$ and $\sigma_1/\mu_1 = \sigma_2/\mu_2 = V$. Then the formula transforms to

$$n = (z_{1-\alpha/2} + z_{1-\beta})^2 V^2 [1 + 2f/(f-1)^2]. \tag{17.21}$$

The quantity V is the usual coefficient of variation and f is the ratio of means. It does not matter whether the ratio of means is defined in terms of $1/f$ rather than f.

Sometimes the problem is formulated with the variability V as specified but a percentage change between means is given. If this is interpreted as the second mean, μ_2, being a percent change from from the first mean, then this percentage change is simply $100(f-1)\%$ and the formula again applies. However, sometimes the relative status of the means cannot be specified, so an interpretation of "percent change" is needed. If we only know that $\sigma_1 = V\mu_1$ and $\sigma_2 = V\mu_2$, then the formula for sample size becomes

$$n = \frac{V^2(z_{1-\alpha/2} + z_{1-\beta})^2}{\left((\mu_1 - \mu_2)/\sqrt{\mu_1\mu_2}\right)^2}.$$

The quantity $\left((\mu_1 - \mu_2)/\sqrt{\mu_1\mu_2}\right)$ is the proportional change from μ_1 to μ_2 as a function of their geometric mean. If the questioner, therefore, can only specify a "percent change" then this interpretation is quite reasonable. Solving Equation (17.21) for $z_{1-\beta}$ allows us to calculate values for power curves:

$$z_{1-\beta} = -z_{1-\alpha/2} + \frac{\sqrt{n}|f-1|}{V\sqrt{f^2+1}}. \tag{17.22}$$

A useful set of curves as function of n and common coefficient of variation $V = 1$ can be constructed by noting that for two coefficients of variation V_1, V_2 the sample sizes $n(V_1)$ and $n(V_2)$, as functions of V_1 and V_2, are related by

$$\frac{n(V_1)}{n(V_2)} = \frac{\sigma_1^2}{\sigma_2^2}$$

for the same power and type I error. See van Belle and Martin [1993].

PROBLEMS

1. **a.** Verify that the odds ratio and relative risk are virtually equivalent for

$$P[\text{exposure}] = 0.10, \quad P[\text{disease}] = 0.01$$

in the following two situations

$$\pi_{11} = P[\text{exposed and disease}] = 0.005$$

and $\pi_{11} = 0.0025$.

   b. Using Equation (17.2) calculate the number of disease occurrences in the exposed and unexposed groups that would have to be observed to detect the relative risks calculated above with $\alpha = 0.05$ (one-tail) and $\beta = 0.10$.

   c. How many exposed persons would have to be observed? (and hence, unexposed persons as well).

   d. Calculate the sample size needed if this test is one of K tests for $K = 10, 100, 1000$.

   e. In Part (d) plot the logarithm of the sample size as a function of $\log K$. What kind of relationship is suggested? Can you state a general rule?

2. (After N. E. Breslow) Workers at all nuclear reactor facilities will be observed for a period of 10 years to determine whether they are at excess risk for leukemia. The rate in the general population is 7.5 cases per 100,000 person-years of observation. We want to be 80% sure that a doubled risk will be detected at the 0.05 level of significance.

   a. Calculate the number of leukemia cases that must be detected among the nuclear plant workers.

   b. How many workers must be observed? That is, assuming the null hypothesis holds how many workers must be observed to accrue 9.1 leukemia cases?

c. Consider this as a binomial sampling problem. Let $\pi_1 = 9.1/$answer in (b), let $\pi_2 = 2\pi_1$. Now use Equation (17.17) to calculate $n/2$ as the required sample size. How close is your answer to Part (b)?

3. (After N. E. Breslow) The rate of lung cancer for men of working age in a certain population is known to be on the order of 60 cases per 100,000 person-years of observation. A cohort study using equal numbers of exposed and unexposed persons is desired so that an increased risk of $R = 1.5$ can be detected with power $1 - \beta = 0.95$ and $\alpha = 0.01$.

a. How many cases will have to be observed in the unexposed population? The exposed population?

b. How many person-years of observation at the normal rates will be required for either of the two groups?

c. How many workers will be needed assuming a 20 year follow-up?

4. (After N. E. Breslow) A case-control study is to be designed to detect an odds ratio of three for bladder cancer associated with a certain medication which is used by about one person out of 50 in the general population.

a. For $\alpha = 0.05$ and $\beta = 0.05$ calculate the number of cases and number of controls needed to detect the increased odds ratio.

b. Use the Poisson approximation procedure to calculate the required sample sizes.

c. Four controls can be provided for each case. Use Equations (17.19) and (17.20) to calculate the sample sizes. Compare this result with the total sample size in Part (a).

5. The Sudden Infant Death Syndrome (SIDS) occurs at a rate of approximately 3 cases per 1000 live births. It is thought that smoking is a risk factor for SIDS and a case-control study is initiated to check this assumption. Since the major effort was in the selection and recruitment of cases and controls a questionnaire was developed which contained 99 additional questions.

a. Calculate the sample size needed for a case-control study using $\alpha = 0.05$, in which we want to be 95% certain of picking up an increased relative risk of 2 associated with smoking. Assume equal number of cases and controls are selected.

b. Considering smoking just one of the 100 risk factors considered, what sample sizes will be needed to maintain an $\alpha = 0.05$ per-experiment error rate?

c. Given the increased value of Z in Part (b) suppose that the sample size is not changed what is the effect on the power? What is the power now?

d. Suppose in Part (c) the power also remains fixed at 0.95, what is the minimum relative risk that can be detected?

e. Since smoking was the risk factor that precipitated the study, can an argument be made for not testing it at a reduced α level? Formulate your answer carefully.

*6. Derive the square root rule starting with Equations (17.4) and (17.5).

*7. Derive formula (17.16) from Equation (17.14).

8. It has been shown that coronary bypass surgery does not prolong life in selected patients with relatively mild angina (but may relieve the pain). A surgeon has invented a new bypass procedure that, she claims, will substantially prolong life. A trial is planned with patients randomized to surgical treatment or standard medical therapy. Currently, the 5-year survival probability of patients with relatively mild symptoms is 80%. The surgeon claims that the new technique will increase survival to 90%.

   a. Calculate the sample size needed to be 95% certain that this difference will be detected using an $\alpha = 0.05$ significance level.

   b. Suppose the cost of a coronary bypass operation is approximately $50,000; the cost of general medical care about $10,000. What is the most economical experiment under the conditions specified in Part (a)? What are the total costs of the two studies?

   c. The picture is more complicated than described in Part (b). Suppose about 25% of the patients receiving the medical treatment will go on to have a coronary bypass operation in the next 5 years. Recalculate the sample sizes under the conditions specified in Part (a).

*9. Derive the sample sizes in Table 17.4 for $D = 0.5$, $p = 0.8$, $\alpha = 0.5$, and $k = 20, 100, 300$.

*10. Consider the situation in Example 17.4.

   a. Calculate the sample size as a function of m, the number of variables by considering $m = 10$ and $m = 20$.

   b. What is the relationship of sample size and variables?

11. Two groups of rats, one young and the other old, are to be compared with respect to levels of nerve growth factor (NGF) in the cerebrospinal fluid. It is estimated that the variability in NGF from animal to animal is of the order of 60%. We want to look at a two-fold ratio in means between the two groups.

   a. Using the formula in Note 17.4 calculate the sample size per group using a two-sided alternative, $\alpha = 0.05$, and power of 0.80.

   b. Suppose the ratio of the means is really 1.6. What is the power of detecting this difference with the sample sizes calculated in Part (a)?

REFERENCES

Beyer, W. H. (ed) [1968]. *CRC Handbook of Tables for Probability and Statistics*, 2nd edition. CRC Press, Cleveland, Ohio.

Brittain, E. and Schlesselman, J. J. [1982]. Optimal allocation for the comparison of proportions. *Biometrics*, **38:** 1003–1009.

Casagrande, J. T., Pike, M. C., and Smith, P. C. [1978]. An improved approximate formula for calculating sample sizes for comparing two binomial distributions. *Biometrics*, **34:** 483–486.

Cochran, W. G. [1962]. *Sampling Techniques*, 2nd edition. Wiley, New York.

Cohen, J. [1988]. *Statistical Power Analysis for the Behavioral Sciences*, 2nd edition. Lawrence Erlbaum, Hillsdale, New Jersey.

Fleiss, J. L. [1981]. *Statistical Methods for Rates and Proportions*, 2nd edition. Wiley, New York.

Fleiss, J. L., Tytun, A., and Ury, H. K. [1980]. A simple approximation for calculating sample sizes for comparing independent proportions. *Biometrics*, **36:** 343–346.

Gail, M., Williams, R., Byar, D. P., and Brown, C. [1976]. How many controls. *Journal of Chronic Diseases*, **29:** 723–731.

Gross, A. J. and Clark, V. A. [1975]. *Survival Distributions: Reliability Applications in the Biomedical Sciences*. Wiley, New York.

Haseman, J. K. [1978]. Exact sample sizes for the use with the Fisher–Irwin test for 2×2 tables. *Biometrics*, **34:** 106–109.

Lachin, J. M. [1977]. Sample size determinations for $r \times c$ comparative trials. *Biometrics*, **33:** 315–324.

Lachin, J. M. [1981]. Introduction to sample size determination and power analysis for clinical trials. *Controlled Clinical Trials*, **2:** 93–113.

Lubin, J. H. [1980]. Some efficiency comments on group size in study design. *American Journal of Epidemiology*, **111:** 453–457.

Mantel, H. [1983]. Extended use of binomial sample-size tables. *Biometrics*, **39:** 777–779.

Nam, J. M. [1973]. Optimum sample sizes for the comparison of a control and treatment. *Biometrics*, **29:** 101–108.

Pike, M. C. and Casagrande, J. T. [1979]. Cost considerations and sample size requirements in cohort and case-control studies. *American Journal of Epidemiology*, **110:** 100–102.

Schlesselman, J. J. [1982]. Case-Control Studies: Design, Conduct, Analysis. Oxford University Press, New York.

Schuster, J. J. [1993]. *Practical Handbook of Sample Size Guidelines for Clinical Trials*. CRC Press, Boca Raton, Florida.

Ury, H. K. and Fleiss, J. R. [1980]. On approximate sample sizes for comparing two independent proportions with the use of Yates' correction. *Biometrics*, **36:** 347–351.

van Belle, G. and Martin, D. C. [1993]. Sample size as a function of coefficient of variation and ratio of means. *American Statistician* (to appear).

Walter, S. D. [1977]. Determination of significant relative risks and optimal sampling procedures in prospective and retrospective comparative studies of various sizes. *American Journal of Epidemiology*, **105:** 387–397.

CHAPTER 18

A Personal Postscript

18.1 INTRODUCTION

One of the reviewers of this text felt that it would be desirable to have a final chapter that ended the book with more interesting material than yet another statistical method. This stimulated us (the authors) to think about all the exciting, satisfying, and interesting things that had occurred in our own careers as biostatisticians. We decided to try to convey some of these feelings through our own experiences. This chapter is unabashedly written from the first person point of view. The examples do not represent a random sample of our experiences but rather the most important and/or interesting experiences of our careers. There is some duplication of background material that appears in other chapters so that this chapter may be self-contained (except for the statistical methods used). We have not made an effort to choose experiences that illustrate the use of many different statistical methods (although this would have been possible). Rather we want to entertain, and in doing so show the important collaborative role of biostatistics in biomedical research.

18.2 IS THERE TOO MUCH CORONARY ARTERY SURGERY?

The National Institutes of Health of the US Federal Government funds much of the health research in the United States. During the late 1960s and early 1970s, an exciting new technique for dealing with anginal chest pain caused by coronary artery disease was developed. Recall that coronary artery disease is caused by fibrous fatty deposits building up within the arteries that supply blood to the heart muscle (that is, the coronary arteries). As the arteries narrow the blood supply to the heart is inadequate with the increased demands of exercise and/or stress; the resulting pain is called angina. Further, the narrowed arteries tend to close with blood clots which result in the death (infarction) of the heart muscle (myocardium) whose oxygen and nutrients are supplied by the blood coming through the artery; these heart attacks are also called myocardial infarctions or Mls. Coronary artery bypass graft (CABG— pronounced "cabbage") surgery replumbs the system. Either saphenous veins from the leg or the internal mammary arteries already in the chest are used to supply blood beyond the narrowing; that is, bypassing the narrowing. Figure 18.1 schematically shows the results of bypass surgery.

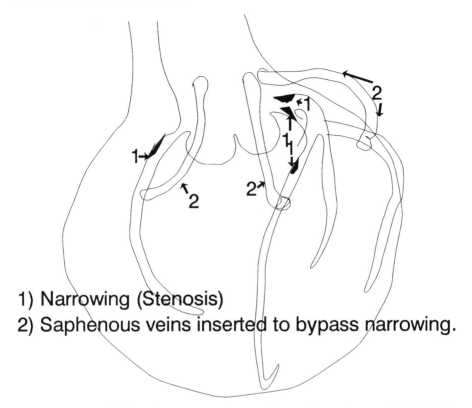

1) Narrowing (Stenosis)
2) Saphenous veins inserted to bypass narrowing.

Figure 18.1. Schematic display of coronary artery bypass graft surgery. Here saphenous veins from the leg are sewn into the aorta where the blood is pumped out of the heart and then sewn into coronary arteries beyond narrowings in order to deliver a normal blood supply. See also picture on page 838.

Because the restored blood flow should allow normal function it was conjectured that the surgery would both remove the anginal pain and also prolong life by reducing both the stress on the heart and the number of myocardial infarctions. It became clear early on that the surgery did help to relieve angina pain (although even this has been debated; see Preston [1977]). However the issue of prolonging life was more debatable. The amount of surgery had important implications for the health care budget since in the early 1970s the cost per operation ranged between $12,000 and $50,000 depending upon the location of the clinic, complexity of the surgery, and a variety of other factors. The number of surgeries by year up to 1972 is shown in Figure 18.2.

Because of the potential savings in lives and large requirement for health resources, the National Heart, Lung and Blood Institute (NHLBI; at that time the National Heart Institute) decided it was appropriate to obtain firm information about which patients have improved survival with CABG surgery. Such therapeutic comparisons are best addressed through a randomized clinical trial and that was the approach taken here with randomization to early surgery or early medical treatment. However, because not all patients could ethically be randomized it was also decided to have

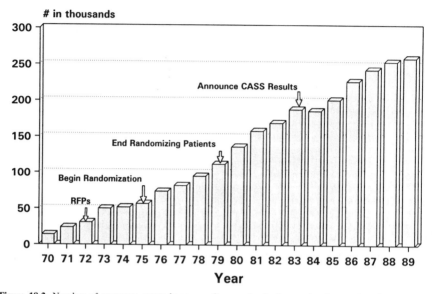

Figure 18.2. Number of coronary artery bypass graft surguries in thousands of operations by year from 1970–1989. Data courtesy of the cardiac diseases branch of the National Heart, Lung and Blood Institute from the National Hospital Discharge Survey, National Center for Health Services. Marked are some of the key timepoints in the Coronary Artery Surgery Study, CASS.

a registry of patients studied with coronary angiography so that observational data analyses could be performed on other subsets of patients to compare medical and surgical therapy. When the NHLBI has internally sponsored initiatives they are developed through a Request for Proposals (RFP) that recruits investigators to perform the collaborative research. This trial and registry called the Coronary Artery Surgery Study (CASS) had two RFPs; one was for clinical sites and the other for a Coordinating Center. The RFP for the clinical sites was issued in November, 1972 and described the proposed study, both randomized and registry components, and asked for clinics to help complete the design and to enroll patients in the randomized and registry components of the study. The Coordinating Center RFP requested applications for a center to help with the statistical design and analysis of the study, to receive and process the study forms with a resultant database, to produce reports for monitoring the progress of the study and to otherwise participate in the quality assurance of the study, and finally to collaborate in the analysis and publication of the randomized study and registry results. The organization of such a large multicenter study had a number of components: the NHLBI had a program office with medical, biostatistical and financial expertise to oversee the operation of the study; there were 15 cooperating clinical sites in the United States and Canada; the Coordinating Center was at the University of Washington under the joint direction of Lloyd Fisher and Richard Kronmal; a laboratory to read electrocardiograms (ECG lab) was at the University of Alabama. The investigators were represented by a Steering Committee with representatives from each clinic and headed by Thomas Killip, M.D. Further the NHLBl appointed a Policy Advisory and Data Monitoring Board to oversee the study, give advice to the NHLBl, and, more important, sequentially monitor the results of the study to recommend early termination if it became clear that one therapeutic arm was superior to the other.

There were 780 cases enrolled in the randomized study; the patients either had mild angina, no angina with a prior MI, and significant disease (defined as a 70% or more narrowing of the internal diameter of a coronary artery that was suitable for bypass surgery). There were a variety of other criteria for eligibility for randomization; these are detailed in CASS Principal lnvestigators and Their Associates [1981]. The registry, including the patients randomized, enrolled 24,959 patients. Extensive data were collected on all patients. The first patients were enrolled in July 1974, with randomization commencing in August 1975. Follow-up of patients within the randomized study ended in 1992. Needless to say, such a large effort cost a considerable amount of money, over $30,000,000. Below it will be shown that the investment was very cost effective.

The results of the survival analysis and indicators about the quality of life were made public in 1983 (CASS Investigators [1983a,b, 1984b]). The ejection fraction (EF) of a person is the fraction of blood in the pumping chamber of the heart, the left ventricle, that is pumped out during a heartbeat. A normal ejection fraction is greater than or equal to 0.5. The ejection fraction decreases when the heart muscle is damaged (say by an MI), or has a severely limited blood supply and cannot pump as well as the normal heart. When the heart severely fails, there are a variety of clinical signs and symptoms. The CASS randomized study did not enroll patients in heart failure but did enroll cases with mild to moderate impairment in ventricular function; EFs between 0.35 and 0.49. The survival estimates for the individuals randomized to initial medical and surgical treatment are given in Figure 18.3.

It can be seen that for patients with an EF of 0.5 or more, the survival curves were virtually identical; for individuals with lower EFs, there was a trend towards favorable mortality in the surgery group ($p = 0.085$ by the log-rank test).

A number of points were important in interpreting these data:

1. The CASS investigators agreed before the study started that the surgery was efficacious in relieving angina. Thus if a patient started to have severe angina that could not be controlled by medication, the patient was allowed to "cross over" to surgery. By year 5, 24% of the patients assigned to initial medical therapy had crossed over to the CABG surgery group. If surgery in fact is having a beneficial effect and there is much crossover, the statistical power of the comparison is reduced. Is this a bad thing? The issue is a complex one (see Peto *et al.* [1977], Weinstein and Levin [1989], Fisher *et al.* [1989, 1990]). A brief outline of some of the salient points include the following: we know that one of the benefits of randomization is that we are assured of comparable groups (on average) even with respect to unrecorded and unknown variables. If we manipulated people, or parts of their experience, between groups by using events that occurred after the time of randomization, bias can enter the analysis. Thus people should be included only in the group to which they are randomized; this is called an intent-to-treat analysis since they are counted with the group whose treatment was intended. [Does such an approach avoid bias? Does it always make biological sense?] The CASS investigators favored an intent-to-treat

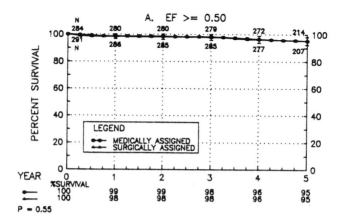

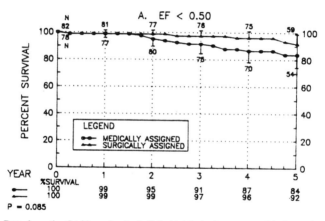

Figure 18.3. Data from the CASS randomized clinical trial; the bottom panel is for patients with ejection fractions less than 0.50; the top panel is for the patients with ejection fractions 0.50 or above. The *p*-values are the log-rank statistic for the comparison.

analysis not only because it avoided possible bias but also because of the ethical imperative to perform CABG surgery for pain relief when the pain became intractable under medical treatment. Thus including all the experience of those assigned to initial medical treatment, including CABG surgery and the subsequent events, mirrored what would happen to such a group in real life. This is the question that the trial should answer: is early surgery helpful when patients will receive it anyway when the pain becomes too severe? However, the power of such a comparison will be diminished by the crossovers. The interpretation of such intent-to-treat analyses must acknowledge that without the crossover, the results could have been substantially different.

   2. Because bypass surgery is such a big industry (e.g., 200,000 surgeries per year at $30,000 per operation adds up to 6 billion dollars/year) with many individuals' careers and professional prestige committed to the field, one could expect a counter reaction if surgery did not look beneficial. This is not to say that criticism must be considered biased on the face of it; it is only to say that one would expect considerable criticism of the study—both appropriate and inappropriate—when the results were negative. Such reactions did occur and a number of editorials, reviews, and sessions at professional meetings were given to consideration of the results. One of the authors (LF) appeared on the CBS national news as well as going to New York City to be interviewed by Mike Wallace and appearing on the TV program "60-minutes"; the program suggested there was too much CABG surgery based largely on the CASS results, a little excitement in the hum-drum biostatistical life.

3. It is important to keep the findings in context. They did not apply to all, or even most, patients. The CASS was one of three major randomized trials of CABG surgery. One study showed definitively that the surgery prolonged life in individuals with a particular coronary anatomy, left main disease (Takaro *et al.* [1976]). This study excluded patients with severe angina and thus had nothing to say about the differential survival in such patients. In fact there is observational data to suggest that early elective CABG surgery prolonged life in such patients (Kaiser *et al.* [1985], Myers *et al.* [1989]).

4. Even though the findings may apply to a *relatively* small number of patients the results could have a very substantial impact on the national health scene. Subsequent CASS papers showed that the trend in the low ejection fraction patients was real (Passamani *et al.* [1985], Alderman *et al.* [1990]). Thus suppose that we restrict ourselves to those patients with EFs of at least 0.5. This accounted for 575 of the 780 randomized patients. Suppose the randomized study had not been in effect, how many of these patients might have received early surgery? In the CASS study, there were 1315 patients who met the eligibility criteria and might have been randomized but in fact were not randomized (CASS Principle Investigators [1984a], Chaitman *et al.* [1990]); these patients were called the *randomizable* patients. The group of randomizable patients had 43% (570/1315) receiving early elective surgery. Of those who did not receive early surgery and had good ejection fractions, by 10 years 38% had received surgery. That is, 60% or so did not receive surgery. These facts and the amount of surgery in the CASS registry lead to the following speculations. Assuming that the CASS clinics were representative of the surgical practice in the country (when, in fact, they may have been more conservative than many centers because they were willing to participate in research to assess the appropriate role of bypass surgery), about 4.4% of the surgery in the United States might be prevented by applying the results of the study. In a year with 188,000 CABGs costing $30,000 each, this would lead to a savings of over 245 million dollars. Over a 4-year period over 1 billion dollars could be saved in surgical costs. However, because the patients treated medically have more anginal pain, they have higher drug costs; they might have higher hospitalization costs (but they do not; see CASS Principle Investigators [1983b] and Rogers *et al.* [1990]). Without going into detail, it is my (LF) opinion that the study saved several billion dollars in health care costs without added risk to patient lives.

5. The issues are more complex than presented here; we have not discussed the findings and integration of results with the other major randomized studies of CABG surgery. Further it is important to note that at the time of this writing (1991) a number of other proven and/or promising techniques for dealing with coronary artery disease (CAD) have been developed. These include drug and/or dietary therapy; blowing up balloons in the artery to "squish" the narrowing into the walls of the artery (percutaneous transluminal coronary angioplasty (PTCA)); introducing lasers into the coronary arteries to disintegrate the plaques that narrow the arteries; and using a roto-rooter in the arteries to replumb by grinding up the plaques. With constant medical progress, the life span of applicability of a given study is finite.

6. The surgery may improve with time as techniques and skills improve. Further it became apparent that the results of the surgery deteriorated at 10–12 years or so. The disease process at work in the coronary arteries also was at work in the grafts that bypassed the narrowed areas; thus the grafts themselves narrow and close, often requiring repeat CABG surgery. Internal mammary grafts have a longer life time and are now used more often suggesting that current long term results will be better.

In summary, the CASS study showed that, in patients with selected characteristics, CABG surgery is not needed immediately to prolong life and can often be avoided. The study was a bargain both in human and economic terms illustrating the need and benefits of careful evaluation of important health care procedures.

18.3 SCIENCE, REGULATION, AND THE STOCK MARKET

In the United States, foods, drugs, biologics, devices, and cosmetics are regulated by the Food and Drug Administration (the FDA). In order to get a new drug or biologic approved for marketing within the United States, the sponsor (usually a pharmaceutical company or biotechnology company) must perform adequate and well-controlled clinical trials that show the efficacy and safety of the product. The FDA is staffed with individuals with expertise in a number of areas including pharmacology, medicine, and biostatistics. The FDA staff reviews submitted materials and rules upon the approval or nonapproval of a product. The FDA also regulates marketing of the compounds. Marketing before approval is not allowed. The FDA uses the services of a number of advisory committees composed of experts in the areas considered. The deliberations of the advisory committees are in public, often with large audiences in attendance; the proceedings are transcribed and available for any interested parties. Advisory committee members are cautioned not to discuss upcoming issues among themselves outside of the meeting. While a new drug application (NDA) consists of many volumes [sometimes over 100] and thousands of pages of material for review by the FDA, the advisory committee members get summary material prepared by the sponsor, but agreed upon by the FDA. This material which the committee members have to review before the public meeting is typically of the order of 50–300 pages. At the meetings, the sponsor makes a presentation, usually with both company and clinical experts, and answers questions from the committee. The FDA has a presence and gets to ask questions, particularly of the advisory committee, but usually does not play a dominant role. Typically the committee at the end of its deliberations votes on whether the drug or biologic should (i) be approved; (ii) should be disapproved; or (iii) is at least temporarily disapproved because further information is needed before a final approval or disapproval is appropriate.

Both of the authors have been members of FDA advisory committees, GvB with the peripheral and central nervous system drugs advisory committee and LF with the cardiovascular and renal drugs advisory committee (both in 1981–1985 and again at present in 1991–1995). Here we discuss the consideration of one biologic: tissue plasminogen activator (tPA). A biologic is a compound that occurs naturally in the human body, while a drug is a compound that does not naturally occur but which is artificially introduced solely for therapeutic purposes. For example, insulin is a biologic while aspirin is a drug. Here we will use the term "drug" for tPA because that is the more common usage although within the FDA, drugs and biologics go to different divisions. Now we turn to the background and rationale for the use of tPA.

As discussed above, when coronary artery disease occurs it narrows the arteries changing the fluid flow properties of the blood. One of the survival mechanisms of blood is that when it is not moving, it clots and when tissues are cut they release compounds that make the blood clot. Lack of such clotting, hemophilia, is life threatening and a cause of great concern. However, when the fluid flow changes as well as the properties of the coronary artery walls, this leads to clotting within the coronary arteries. These clots then block the blood supply to the heart muscle and leads to heart attacks, or myocardial infarctions (MIs), as discussed above. The clot is composed largely of fibrin. Plasminogen when converted to plasmin converts insoluable fibrin into souluable fragments. One conceptual way to treat a heart attack would be to dissolve the blood clot thus re-establishing blood flow to the heart muscle and preventing the death of the muscle, saving the heart and often saving the life. Should a drug be approved for dissolving blood clots alone? Although biologically plausible, does this assure that the drug will work?

The purpose of giving a drug is to benefit the patient. For example, since moderately high blood pressure is asymptomatic, the lowering of blood pressure, per se, is not of benefit to the patient. However, early trials of antihypertensive drugs showed that the number of cardiovascular endpoints was reduced. Thus it was considered desirable to lower blood pressure. Blood pressure is a *surrogate* endpoint for the actual endpoints that benefit the patient. It is conceptually possible that a new antihypertensive formulation has effects not part of the original drugs tested and that in fact the new drug has an overall adverse effect. The judgment of the FDA is that reduction of blood pressure is an adequate surrogate and that drugs can be licensed based upon this information alone. Returning to the thrombolytic (that is to lyse, or breakup, the blood clot, the thrombosis) therapy tPA, it is clear that lysing the coronary arterial blood clot is a surrogate endpoint. Should this surrogate endpoint be appropriate for approving the drug? After all there is such a clear cut biological rationale: coronary artery clots cause heart attacks; heart attacks damage the heart often either impairing the heart function and thus lowering exercise capacity, or directly killing the person. But, experience has shown that very convincing biological scenarios do not always deliver the benefits expected; in the next experience we will see an important example where an obvious surrogate endpoint did not work out.

Let us return now to the tPA cardio-renal advisory committee meeting and decision. In addition to tPA, another older thrombolytic drug, streptokinase, was also being presented for approval for the same indication. Prior to the meeting there was considerable publicity over the upcoming meeting and possible approval of the drug tPA. The advisory committee meeting was to take place on Friday, May 29, 1987. On Thursday, May 28, 1987 the Wall Street Journal published an editorial entitled "The TPA Decision". The editorial read as follows:

"Profile of a heart-attack victim: 49 years old, three children, middle-manager, in seemingly good health. Cutting the grass on a Saturday afternoon, he is suddenly driven to the ground with severe chest pain. An ambulance takes him to the nearest emergency room where he receives drugs to reduce shock and pain.

At this point, he is one of approximately 4000 people who suffer a heart attack each day. If he has indeed had a heart attack, he will experience one of two possible outcomes. Either he will be dead, joining the 500,000 Americans killed each year by heart attack. Or, if he's lucky, he will join the one million others who go on to receive some form of therapy for his heart disease.

Chances of survival will depend in great part on the condition of the victim's heart, that is, how much permanent muscular damage the heart sustained during the time a clot prevented the normal flow of blood into the organ. Heart researchers have long understood that if these clots can be broken up early after a seizure's onset, the victim's chances of staying alive increase significantly. Dissolving the clot early enhances the potential benefits of such post-attack therapies as coronary bypass surgery or balloon angioplasty.

Tomorrow morning, a panel of the Food and Drug Administration will review the data on a blood-clot dissolver called TPA, for tissue-type plasminogen activator. In our mind, TPA—not any of the pharmaceutical treatments for AIDS—is the most noteworthy, unavailable drug therapy in the United States. Put another way, the FDA's new rules permitting the distribution of experimental drugs for life-threatening diseases came under pressure to do something about the AIDS epidemic. But isn't it as important for the government to move with equal speed on the epidemic of heart attacks already upon us?

This isn't to say that TPA is more important than AIDS treatments. Both have a common goal: keeping people alive. The difference is that while the first AIDS drug received final approval in about six months, TPA remains unapproved and unavailable to heart-attack victims despite the fact that the medical community has known for more than two years that it can save lives.

How many lives? Obviously no precise projection is possible, but the death toll is staggering, with about 41,000 individuals killed monthly by heart attacks.

In its April 4, 1985, issue, the New England Journal of Medicine carried the first report on the results of the National Institutes of Health's TIMI study comparing TPA's clot dissolving abilities with a drug already approved by the FDA. NIH prematurely ended that trial because TPA's results were so significantly better than the other drug.

In an accompanying editorial, the Journal's editor, Dr. Arnold Relman, said a safe and effective thrombolytic "might be of immense clinical value." In October 1985, a medical-policy committee of California's Blue Shield recommended that TPA be recognized "as acceptable medical practice." The following month at the American Heart Association's meeting, Dr. Eugene Braunwald, chairman of the department of medicine at Harvard Medical School, said, "If R-TPA were available on a wide basis, I would select that drug today." In its original TIMI report, the NIH said TPA would next be tested against a placebo; later, citing ethical reasons, the researchers dropped the placebo and now all heart patients in the TIMI trial are receiving TPA.

It is for these reasons that we call TPA the most noteworthy unavailable drug in the U.S. The FDA may believe it is already moving faster than usual with the manufacturer's new-drug application. Nonetheless, bureaucratic progess [sic] must be measured against the real-world costs of keeping this substance out of the nation's emergency rooms. The personal, social and economic consequences of heart disease in this country are immense. The American Heart Association estimates the total costs of providing medical services for all cardiovascular disease at $71 billion annually.

By now more than 4,000 patients have been treated with TPA in clinical trials. With well over a thousand Americans going to their deaths each day from heart attack, it is hard to see what additional data can justify the government's further delay in making a decision about this drug. If tomorrow's meeting of the FDA's cardio-renal advisory committee only results in more temporizing, some in Congress or at the White House should get on the phone and demand that the American public be given a reason for this delay".

The premeeting publicity also occurred in a more widely circulated newspaper. *USA Today's* lead article in the Life section of the Wednesday, May 27, 1987 edition (Findlay [1987a]) contained the following excerpts:

"Medicine is on the verge of a new era in the treatment of heart attacks, with the potential of saving tens of thousands of lives in day USA each year. . . . Doctors say the development could do for heart attacks what penicillin did for infections. It's a major breakthrough," says Dr. Eugene Braunwald. . . On Friday, after five years of intensive research, TPA will get its first official hearing before an expert advisory panel to the Food and Drug Administration. The panel will hear the evidence and recommend whether TPA should be approved for general use or undergo further study. Talk show host and USA TODAY columnist Larry King—a recent heart attack victim—is one of TPA's fans. "No complaints," King says. "They gave it to me right after they told me I was having a heart attack. It took away the pain immediately. The doctors told me TPA had a great deal to do with my recovery." . . . Also Friday, the FDA panel will hear data on streptokinase. It, too, is aimed at heart attack treatment. Both drugs— and three others under development—perform the same function. But there is intense debate on their relative safety and effectiveness. Five major studies, involving more than 12,000 people, have shown that streptokinase reduces death in the first few days after a heart attack by 35 percent to 50 percent. But there are problems. Streptokinase dissolves clots, but it also attacks other proteins in the blood, leading to other problems, particularly excessive bleeding. Also, because of the way it is made, streptokinase causes allergic reactions. . . . Enter TPA. At Friday's hearing, researchers will present evidence that TPA is more "clot specific". . . In the USA, Genentech Inc.—a San Francisco bioengineering giant—has won a hard fought race to be first to seek FDA approval. . . . Not everyone agrees that TPA will be the clotbuster of choice, however. Dr. Sol Sherry . . . says that TPA, streptokinase and other newer clotbusters each have their strengths and weaknesses".

The publicity before the meeting of the advisory committee was quite unusual since companies are prohibited from preapproval advertising; thus the impetus presumably came from other sources.

The cardio-renal advisory committee members met and considered the two thrombolytic drugs, streptokinase and tPA. They voted to recommend approval of streptokinase but felt that further data were needed before tPA could be approved. The reactions to the decision were extreme, but probably predictable given the positions expressed prior to the meeting.

The Wall Street Journal responded with an editorial on Tuesday, June 2, 1987 entitled "Human Sacrifice". It follows in its entirety.

"Last Friday an advisory panel of the Food and Drug Administration decided to sacrifice thousands of American lives on an altar of pedantry.

Under the klieg lights of a packed hearing room at the FDA, an advisory panel picked by the agency's Center for Drugs and Biologics declined to recommend approval of TPA, a drug that dissolves blood clots after heart attacks. In a 1985 multicenter study conducted by the U.S. National Heart, Lung and Blood Institute, TPA was so conclusively effective at this that the trial was stopped. The decision to withhold it from patients should be properly viewed as throwing U.S. medical research into a major crisis.

Heart disease dwarfs all other causes of death in the industrialized world, with some 500,000 Americans killed annually; by comparison, some 20,000 have died of AIDS. More than a thousand lives are being destroyed by heart attacks every day. In turning down treatment with TPA, the committee didn't dispute that TPA breaks up the blood clots impeding blood flow to the heart. But the committee asked that Genentech, which makes the genetically engineered drug, collect some more mortality data. Its submission didn't include enough statistics to prove to the panel that dissolving blood clots actually helps people with heart attacks.

Yet on Friday, the panel also approved a new procedure for streptokinase, the less effective clot dissolver—or thrombolytic agent—currently in use. Streptokinase previously had been approved for use in an expensive, specialized procedure call intracoronary infusion. An Italian study, involving 11,712 randomized heart patients at 176 coronary-care units in 1984–1985, concluded that administering streptokinase intravenously reduced deaths by 18%. So the advisory panel decided to approve intravenous streptokinase, but not approve the superior thrombolytic TPA. This is absurd.

Indeed, the panel's suggestion that it is necessary to establish the efficacy of thrombolysis stunned specialists in heart disease. Asked about the committee's justification for its decision, Dr. Eugene Braunwald, chairman of Harvard Medical School's department of medicine, told us: 'The real question is, do you accept the proposition that the proximate cause of a heart attack is a blood clot in the coronary artery? The evidence is overwhelming, *overwhelming*. It is sound, basic medical knowledge. It is in every textbook of medicine. It has been firmly established in the past decade beyond any reasonable question. If you accept the fact that a drug [TPA] is twice as effective as streptokinase in opening closed vessels, and has a good safety profile, then I find it baffling how that drug was not recommended for approval.'

Patients will die who would otherwise live longer. Medical research has allowed statistics to become the supreme judge of its inventions. The FDA, in particular its bureau of drugs under Robert Temple, has driven that system to its absurd extreme. The system now serves itself first and people later. Data supersede the dying.

The advisory panel's suggestion that TPA's sponsor conduct further mortality studies poses grave ethical questions. On the basis of what medicine already knows about TPA, what U.S. doctor will give a randomized placebo or even streptokinase? We'll put it bluntly: Are American doctors going to let people die to satisfy the bureau of drugs' chi-square studies?

Friday's TPA decision should finally alert policy makers in Washington and the medical-research community that the theories and practices now controlling drug approval in this country are significantly flawed and need to be rethought. Something has gone grievously wrong in the FDA bureaucracy. As an interim measure FDA Commissioner Frank Young, with Genentech's assent, could approve TPA under the agency's new experimental drug rules. Better still, Dr. Young should take the matter in hand, repudiate the panel's finding and force an immediate reconsideration. Moreover, it is about time Dr. Young received the clear, public support of Health and Human Services Secretary Dr. Otis Bowen in his efforts to fix the FDA.

If on the other hand Drs. Young and Bowen insist that the actions of bureaucrats are beyond challenge, then perhaps each of them should volunteer to personally administer the first randomized mortality trials of heart-attack victims receiving the TPA clot buster or nothing. Alternatively, coronary-care units receiving heart-attack victims might use a telephone hotline to ask Dr. Temple to randomize the trial himself by flipping a coin for each patient. The gods of pedantry are demanding more sacrifice."

Soon after joining the Cardiovascular and Renal Drugs Advisory Committee (LF) noticed that a number of individuals left the room at what seemed inappropriate times near the end of some advisory deliberations. I was informed that often stock analysts with expertise in the pharmaceutical industry attended meetings about key drugs; when the analysts thought they knew how the vote was going to turn out, they went out to the phones to send instructions. In the last few years I was told that they no longer left the room since electronic devices now could transmit the recommendation from the person's spot in the audience. I do not know for a fact that the information I was given was correct; however the deliberations of the advisory committee meetings do have an effect on the stock market sometimes. That was the case during the tPA deliberations (and made it particularly appropriate that the Wall Street Journal take an interest in the result). Again we convey the effect of the deliberations through quotations taken from the press. On June 1 1978, the *Wall Street Journal* had an article under the heading "FDA Panel Rejection of Anti-Clot Drug Set Genentech Back Months, Perils Stock". The article said in part:

"A Food and Drug Administration advisory panel rejected licensing the medication TPA, spoiling the summmer debut of what was touted as biotechnolgy's first billion-dollar drug. ... Genentech's stock—which reached a high in March of $64.50 following a 2-for-1 split—closed Friday at $48.25, off $2.75, in national over-the-counter trading, even before the close of the FDA panel hearing attended by more than 400 watchful analysts, scientists and competitors. Some analysts expect the shares to drop today. ... Wall Street bulls will also be rethinking their forecasts. For example, Kidder Peabody & Co.'s Peter Drake, confident of TPA's approval, last week predicted sales of $51 million in the second half of 1987, rising steeply to $205 million in 1988, $490 million in 1989 and $850 million in 1990."

The *USA Today* on Tuesday, June 2 1987, on the first page of the Money section had an articfe headed "Biotechs hit a roadblock, investors sell". The article began:

"Biotechnology stocks, buoyed more by promise than products, took one of their worst beatings Monday. Leading the bad-news pack: Biotech giant Genentech Inc., dealt a blow when its first blockbuster drug failed to get federal approval Friday. Its stock plummeted $11\frac{1}{2}$ points to $36\frac{3}{4}$, on 14.2 million shares traded—a one-day record for Genentech. "This is very serious, dramatically serious," said analyst Peter Drake, of Kidder, Peabody & Co., who Monday changed his recommendations for the group from buy to "unattractive." His reasoning: The stocks are driven by "a blend of psychology and product possibilities. And right now, the psychology is terrible." "

Biotechnology stocks as a group dropped with the Genentech panel vote. This seemed strange to me because the panel had not indicated that the drug, TPA, was bad but only that in a number of areas the data needed to be gathered and analyzed more appropriately (as described below). The panel was certainly not down on thrombolysis (as the streptokinase approval showed); it felt that the risk to benefit ratio of tPA needed to be clarified before approval could be made.

The advisory committee members replied to the Wall Street Journal editorials both individually and in groups explaining the reasons for the decision (Borer [1987], Kowey *et al.* [1988], and Fisher *et al.* [1987]). This last response to the *Wall Street Journal* was submitted with the title "The Prolongation of Human Life", however after the review of the article by the editor, the title was changed by the *Wall Street Journal* to "The FDA Cardio-Renal Committee Replies". The reply:

"The evaluation and licensing of new drugs is a topic of legitimate concern to not only the medical profession but our entire populace. Thus it is appropriate when the media, such as the Wall Street Journal, take an interest in these matters. The Food and Drug Administration recognizes the public interest by holding open meetings of advisory committees that review material presented by pharmaceutical companies, listen to expert opinions, listen to public comment from the floor and then give advice to the FDA. The Cardiovascular and Renal Drugs Advisory Committee met on May 29 to consider two drugs to dissolve blood clots causing heart attacks. The Journal published editorials prior to the meeting ("The TPA Decision," May 28) and after the meeting ("Human Sacrifice," June 2 and "The Flat Earth Committee," July 13). The second editorial began with the sentence: "Last Friday an advisory committee of the Food and Drug Administration decided to sacrifice thousands of American lives on an altar of pedantry." How can such decisions occur in our time? This reply by members of the advisory panel presents another side to the story. In part the reply is technical, although we have tried to simplify it. We first discuss drug evaluation in general and then turn to the specific issues involved in the evaluation of the thrombolytic drugs streptokinase and TPA.

The history of medicine has numerous instances of well-meaning physicians giving drugs and treatments that were harmful rather than beneficial. For example, the drug thalidomide was widely marketed in many countries—and in West Germany without a prescription—in the late 1950s and early 1960s. The drug was considered a safe and effective sleeping pill and tranquilizer. Marketing was delayed in the U.S. despite considerable pressure from the manufacturer upon the FDA. The drug was subsequently shown to cause birth defects and thousands of babies world-wide were born with grotesque malformations, including seal-like appendages and lack of limbs. The FDA physician who did not approve the drug in the U.S. received an award from President Kennedy. One can hardly argue with the benefit of careful evaluation in this case. We present this, not as a parallel to TPA, but to point out that there are two sides to the approval coin—early approval of a good drug, with minimal supporting data, looks wise in retrospect; early approval, with minimal supporting data, of a poor drug appears extremely unwise in retrospect. Without adequate and well-controlled data one cannot distinguish between the two cases. Even with the best available data, drugs are sometimes found to have adverse effects that were not anticipated. Acceptance of un-usually modest amounts of data, based on assumptions and expectations rather than actual observation is very risky. As will be explained below, the committee concluded there were major gaps in the data available to evaluate TPA.

The second editorial states that "Medical research has allowed statistics to become the supreme judge of its inventions." If this means that data are required, we agree; people evaluate new therapies with the hope that they are effective—again, before licensing, proof of effectiveness and efficacy is needed. If the editorial meant that the TPA decision turned on some arcane mathematical issue, it is incorrect. Review of the transcript shows that statistical issues played no substantial role.

We now turn to the drug of discussion, TPA. Heart attacks are usually caused by a "blood clot in an artery supplying the heart muscle with blood." The editorial quotes Dr. Eugene Braunwald, "The real question is, do you accept the proposition that the proximate cause of a heart attack is a blood clot in the coronary artery?" We accept the statement, but there is still a significant question: "What can one then do to benefit the victim?" It is not obvious that modifying the cause after the event occurs is in the patient's best interest, especially when the intervention has toxicity of its own. Blood clots cause pulmonary embolism; it is the unusual patient who requires dissolution of the clot by streptokinase. Several trials show the benefit does not outweigh the risk.

On May 29 the Cardiovascular and Renal Drugs Advisory Committee reviewed two drugs that "dissolve" blood clots. The drug streptokinase had been tested in a randomized clinical trial in Italy involving 11,806 patients. The death rate in those treated with streptokinase was 18% lower than in patients not given streptokinase; patients treated within six hours did even better. Review of 10 smaller studies, and early results of a large international study, also showed improved survival. It is important to know that the 18% reduction in death rate is a reduction of a few percent of the patients studied. The second drug considered—recombinant tissue plasminogen activator (TPA)—which also was clearly shown to dissolve blood clots was not approved. Why? At least five issues contributed, to a greater or lesser amount, to the vote not to recommend approval for TPA at this time. These issues were: the safety of the drug, the completeness and adequacy of the data presented, the dose to be used, and the mechanism of action by which streptokinase (and hopefully TPA) saves lives.

Safety was the first and most important issue concerning TPA. Two formulations of TPA were studied at various doses; the highest dose was 150 milligrams. At this dose there was an unacceptable incidence of cerebral hemorrhage (that is, bleeding in the brain), in many case leading to both severe stroke and death. The incidence may be as high as 4% or as low as 1.5% to 2% (incomplete data at the meeting made it difficult to be sure of the exact figure), but in either case it is disturbingly high; this death rate due to side effects is of the same magnitude as the lives saved by streptokinase. This finding led the National Heart, Lung and Blood Institute to stop the 150-milligram treatment in a clinical trial. It is important to realize that this finding was unexpected, as TPA was thought to be relatively unlikely to cause such bleeding. Because of bleeding, the dose of TPA recommended by Genentech was reduced to 100 milligrams. The safety profile at doses of 100 milligrams looks better, but there were questions of exactly how many patients had been treated and evaluated fully. Relatively few patients getting this dose had been reported in full. Without complete reports from the studies there could be smaller strokes not reported and uncertainty as to how patients were examined. The committee felt a substantially larger database was needed to show safety.

The TPA used to evaluate the drug was manufactured by two processes. Early studies used the double-stranded (roller bottle) form of the drug; the sponsor then changed to a predominantly single-stranded form (suspension culture method) for marketing and production reasons. The second drug differed from the first in how long the drug remained in the blood, in peak effect, in the effect on fibrinogen and in the dose needed to cause lysis of clots. Much of the data was from the early form; these data were not considered very helpful with respect to the safety of the recommended dose of the suspension method drug. This could perhaps be debated, but the intracranial bleeding makes the issue an important one. The excessive bleeding may well prove to be a simple matter of excessive dose, but this is not yet known unequivocally.

Data were incomplete in that many of the patients' data had not been submitted yet and much of the data came from treatment with TPA made by the early method of manufacture. There was uncertainty about the data used to choose the 100-milligram dose, i.e., perhaps a lower dose is adequate. When there is a serious dose-related side effect it is crucial that the dose needed for effectiveness has been well-defined and has acceptable toxicity.

Let us turn to the mechanism of action, the means by which the beneficial effect occurs. There may be a number of mechanisms. The most compelling is clot lysis (dissolution). However, experts presented data that streptokinase changes the vicosity of the blood that could improve the blood flow; the importance is uncertain. Streptokinase also lowers blood pressure, which may decrease tissue damage during a heart attack. While there is convincing evidence that TPA (at least by the first method of manufacture) dissolves clots faster than streptokinase (at least after a few hours from the onset of the heart attack), we do not have adequate knowledge to know what portion of the benefit of streptokinase comes from dissolving the clot. TPA, thus, may differ in its effect on the heart or on survival. The drugs could differ in other respects, such as how often after opening a vessel they allow reclosure, and, of course, the frequency of important adverse effects.

These issues delay possible approval. Fortunately, more data are being collected. It is our sincere hope that the drug lives up to its promise, but should the drug prove as valuable as hoped, that would not imply the decision was wrong. The decision must be evaluated as part of the overall process of drug approval.

The second editorial suggests that if the drug is not approved, Dr. Temple (director of the Bureau of Drugs, FDA), Dr. Young (FDA commissioner) and Dr. Bowen (secretary of health and human services) should administer "randomized mortality trials of heart-attack victims receiving the TPA clot buster or nothing." This indignant rhetoric seems inappropriate on several counts. First, the advisory committee has no FDA members; our votes are independent and in the past, on occasion, we have voted against the FDA's position. It is particularly inappropriate to criticize Drs. Temple and Young for the action of an independent group. The decision (by a vote of eight against approval, one for and two abstaining) was made by an independent panel of experts in cardiovascular medicine and research from excellent institutions. These unbiased experts reviewed the data presented and arrived at this decision; the FDA deserves no credit or blame. Second, we recommend approval of streptokinase; we are convinced that the drug saves lives of heart-attack victims (at least in the short term). To us it would be questionable to participate in a trial without some treatment in patients of the type shown to benefit from streptokinase. A better approach is to use streptokinase as an active control drug in a randomized trial. If it is as efficacious or better than streptokinase, we will rejoice. We have spent our adult lives in the care of patients and/or research to develop better methods for treatment. Both for our patients and our friends, our families and ourselves, we want proven beneficial drugs available.

In summary, with all good therapeutic modalities the benefits must surely outweigh the risks of treatment. In interpreting the data presented by Genentech in May 1987 the majority of the Cardiovascular and Renal Drugs Advisory Committee members could not confidently identify significant benefits without concomitant significant risk. The review was clouded by issues of safety, manufacturing process, dose size and the mechanism of action. We are hopeful these issues will be addressed quickly, allowing more accurate assessment of TPA's risk-benefit ratio with conclusive evidence that treatment can be recommended that allows us to uphold the physician's credo, *primum non nocere* (first do no harm)."

The July 28 1987, *USA Today Life* section carried an article on the first page entitled "FDA speeds approval of heart drug." The article mentioned that the FDA commissioner Frank Young was involved in the data gathering. Within a few months of the advisory committee meeting, tPA was approved for use in treating myocardial infarctions. The drug was 5–10 times more expensive than streptokinase; however it opened arteries faster and that was thought to be a potential advantage. A large randomized comparison of streptokinase and tPA was performed (ISIS 3); the preliminary results were presented at the November 1990 American Heart Association meeting. The conclusion was that the efficacy of the two drugs was essentially equivalent. Thus by approving streptokinase, even in retrospect, no period of the lack of availability of a clearly superior drug occurred because of the time delay needed to clear up the questions about tPA. This experience shows that biostatistical collaboration has consequences above and beyond the scientific and humanitarian aspects; large political and financial issues also are often involved.

18.4 IF WE STOP THE THING THAT APPEARS TO CAUSE THE DEATHS, WE MUST BE PROLONGING LIFE (OR ARE WE?)

We continue looking at research involving the heart. One of the wonders of the body is our heart; it beats steadily minute after minute, year after year. If the average number of beats is 60/min there are 86,400 beats/day or 31,536,000 beats/year. In a 65-year-old the heart may have delivered over 2 billion heart beats. The contraction of the heart muscle to force blood out into the body is triggered by electrical impulses that depolarize and thus contract the heart in a fixed pattern. As the heart muscle becomes damaged there can be problems with the electrical trigger that leads to the contraction of the heart. The electrical changes in the heart are monitored when a physician takes an electrocardiogram (ECG) of the heart. If the depolarization inappropriately starts someplace other than the usual trigger point (the sinus atrial node), then the heart can contract early; one resulting irregular heart beat, or arrhythmic beat is called a ventricular premature depolarization (VPD). While most people have occassional VPDs, after an MI patients may have many more VPDs and complex patterns of arrhythmia. The VPDs place patients at an increased risk of sudden cardiac death. To monitor the electrical activity of the heart over longer time periods, ambulatory electrocardiographic monitors (AECGMs) may be used. These units, also called Holter monitors, measure and record the electrical activity of the heart over approximately 24-h periods. In this way, the arrhythmic patterns of individuals may be monitored over time. Patients have suffered sudden cardiac death, or sudden death, while wearing these monitors and the electrical sequence of events is usually the following: patients experience numerous VPDs and then a run of VPDs that occur rapidly in succession (say at a rate greater than or equal to 120 beats/min); the runs are called ventricular tachycardia, or VT for short. Now many coronary patients have runs of VT; however before death, the VT leads to rapid, irregular, continuous electrical activity of the heart called ventricular fibrillation, or VF. Observed in a cardiac operation, VF is a fluttering, or quivering, of the heart. This irregular activity interrupts the blood flow and the patient blacks out and if not resuscitated, invariably dies. In hospital monitoring settings and cities with emergency rescue systems, the institution of cardiopulmonary resuscitation (CPR) has led to the misnomer of sudden death survivors. In a hospital setting and when emergency vehicles arrive electrical defibrillation using paddles that transmit an electrical shock is used. Individuals with high VPD counts on AECGMs are known to be at increased risk of sudden death with the risk increasing with the amount of arrhythmia.

This being the case, it was natural to try to find drugs that reduced, or even abolished, the arrhythmia in many or most patients. A number of such compounds have been developed. In patients with severe life threatening arrhythmia, if an antiarrhytmic drug can be found that controls the arrhythmia, the survival is greatly superior to the survival if the arrhythmia cannot be controlled (Graboys *et al.* [1982]). Graboys and colleagues examined the survival of patients with severe arrhythmia defined as VF (outside of the period of an MI) or VT that compromised the hemodynamics to the degree the patients were symptomatic. Figure 18.4 gives the survival from cardiac deaths in 98 patients with the arrhythmia controlled and 25 patients in whom the arrhythmia was not controlled.

Thus there was a very compelling biological scenario. Arrhythmia lead to runs of VT which lead to VF and sudden death. Drugs were developed, and could be evaluated using AECGMs, that reduced the amount of arrhythmia and even abolished arrhythmia on AECGMs in many patients. Thus these people with the reduced or abolished arrhythmia should live longer. One would then rely on the *surrogate endpoint* of the arrhytmia evaluation from an AECGM. Antiarrhythmic drugs were approved by the FDA based upon this surrogate endpoint. It is important to point out that antiarrhythmic durgs may have other benefits than preventing sudden death.

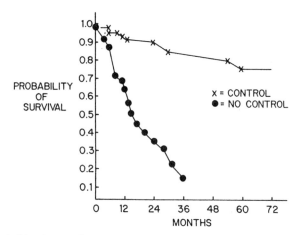

Figure 18.4. Survival free from cardiac mortality in patients with severe arrhythmia. The curves are for those whose arrhythmia was controlled by antiarrhythmic drugs and for those in whom the arrhythmia was not controlled by antiarrhythmic drugs.

For example, some patients have such severe runs of VT that they faint. Prevention of fainting spells is of direct benefit to the patient. However, asymptomatic or mildly symptomatic patients with arrhythmia were being prescribed antiarrhythmics with the faith(?), hope(?) that the drugs would prolong their life.

Why then would anyone want to perform a randomized survival trial in patients with arrhythmia? How could one ethically perfom such a trial? There were a number of reasons: (1) the patients for whom arrhythmia could be controlled by drugs have selected themselves out as biologically different; thus the survival *even without anti-arrhythmic therapy* might naturally be much better than patients for whom no drug worked. That is to say that the modification of the surrogate endpoint of arrhythmia had never been shown to improve the results of the real endpoint of interest (sudden death). (2) Some trials had disturbing results with adverse trends in mortality on antiarrhythmic drugs (IMPACT Research Group [1984] and Furberg [1983]). (3) All antiarrhythmic drugs actually produce more arrhythmia in some patients, a *proarrhythmic* effect.

The National Heart, Lung and Blood Institute decided to study the survival benefit of antiarrhythmic drugs in survivors of an MI. They issued an RFP for clinical sites and a coordinating center. One of the authors, LF, was the director of the co-ordinating center. The study began with a pilot phase to see if antiarrhythmic drugs could be found that reduced arrhythmia by a satisfactory amount. If this could be done then the randomized survival trial would begin. (Due to professional changes Dr. Al Hallstrom took over the direction of the coordinating center in 1984 and has coordinated the study since that time.) The first study was called the Cardiac Arrhythmia Pilot Study (CAPS). The CAPS (Cardiac Arrhytmia Pilot Study (CAPS) Investigators [1988]) showed that three of the drugs studied—encainide, flecainide, and moricizine—adequately suppressed arrhythmias to allow proceeding with the primary survival trial, the Cardiac Arrhythmia Suppression Trial (CAST). Patients within 6 weeks to 2 years of an MI needed 6 VPDs/h to be eligible for the study. There was an open label, dose titration period where drugs were required to reduce

VPDs by at least 80% and runs of VT by at least 90%. (For more detail see The Cardiac Arrhythmia Suppression Trial (CAST) Investigators [1989] and Echt *et al.* [1991].) Patients for whom an effective drug was found were then randomized to placebo or to the effective drug. Such was the confidence of the investigators that the drugs at least were doing no harm, that the test statistic was one-sided to stop for a drug benefit at the 0.025 significance level. The trial was not envisioned as stopping early for excess mortality in the antiarrhythmic drug groups.

The first results to appear were a tremendous shock to the cardiology community. The encainide and flecainide arms were dropped from the study because of excess mortality!! Strictly speaking, the investigators could not conclude this with their one-sided design. However, the evidence was so strong that the investigators, and almost everyone else, were convinced of the harmful effects of these two antiarrhythmic drugs in this patient population (Figure 18.5).

The results of the study have been addressed by Pratt *et al.* [1990] and Pratt [1990]; the timing of the announcement of the results is described in Bigger [1990]; this paper gives a feeling for the ethical pressure of quickly promulgating the results. Ruskin *et al.* [1989] conveys some of the impact of the trial results: "The preliminary results...have astounded most observers and challenge much of the conventional wisdom about antiarrhythmic drugs and some of the arrhythmias they are used to treat.... Although its basis is not entirely clear, this unexpected outcome is best explained as the result of the induction of lethal ventricular arrhythmias (i.e., a proarrhythmic effect) by encainide and flecainide."

This trial has saved, and will continue, to save lives by virtue of changed physician behavior. In addition, it clearly illustrates that consistent, plausible theories and changes in surrogate endpoints cannot be used to replace trials involving the endpoints of importance to the patient; at least not initially. Finally it is important to note that one should not over-extrapolate the results of a trial; the study does not directly apply to patients with other characteristics than those in the trial; it does not imply that other antiarrhythmic drugs have the same effect in this population. However it does make one more suspicious about the role of antiarrhythmic therapy

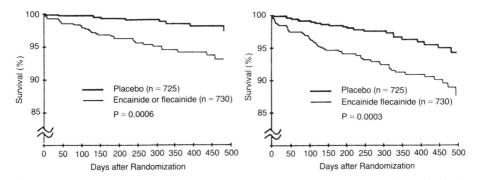

Figure 18.5. The first panel shows the survival, free of an arrhythmic death, among 1455 patients randomized to either placebo or one of encainide or flecainide. The second similar figure is based on all cause mortality. From The Cardiac Arrhythmia Suppression Trial, CAST, Investigators [1989].

with a resultant need for even more well controlled randomized data for other patient populations and/or drugs.

18.5 OH MY ACHING BACK!

One of the most common maladies in the industrialized world is the occurrence of low back problems. By the age of 50 nearly 85% of individuals can recall back symptoms; and, as someone said, the other 15% probably forgot. Among individuals in the United States, back and spine impairment are the chronic conditions that most frequently cause activity limitation. The occurrence of industrial back disability is one of the most expensive health problems afflicting industry and its employees. The cost associated with back injury in 1976 was $14,000,000,000; the costs are greatly skewed with a relatively low percent of the cost accrued by a few chronic back injury cases (Spengler *et al.* [1986]). The costs and human price associated with industrial back injury prompted the Boeing Company to contact the orthopedics department at the University of Washington to institute a collaborative study of back injury at a Boeing factory in western Washington state. Collaboration was obtained from the Boeing company management, the workers and their unions, and a research group at the University of Washington (including one of the authors, LF). The study was supported financially by the National Institutes of Health, the National Institute for Occupational Safety and Health, the Volvo Foundation, and the Boeing company. The study was designed in two phases. The first phase was a retrospective analysis of past back injury reports and insurance costs from already existing Boeing records; the second phase was a prospective study looking at a variety of possible predictors (to be described below) of industrial back injury.

The retrospective Boeing data were analyzed and presented in a series of three papers (Spengler *et al.* [1986], Bigos *et al.* [1986a,b]). The analysis covered 31,200 employees who reported 900 back injuries among 4645 claims filed by 3958 different employees. The data emphasized the cost to Boeing of this malady and, as in previous studies, showed that a small percent of the back injury reports lead to most of the cost; for example, 10% of the cases accounted for 79% of the cost. The incurred costs of back injury claims was 41% of the Boeing total although only 19% of the claims were for the back. The most expensive 10% of the back injury claims accounted for 32% of all the Boeing injury claims. Individuals were more likely to have reported an acute back injury if they had a poor employee appraisal rating from their supervisor within 6 months before the injury.

The prospective study was unique and had some very interesting findings (the investigators were awarded the highest award, the Kappa Delta award, of the American Academy of Orthopedic Surgeons for excellence in orthopedic research). Based upon previously published results and investigator conjectures, data were collected in a number of areas with potential ability to predict reports of industrial back injury. Among the information prospectively obtained from the 3020 aircraft employees who volunteered to participate in the study were the following:

Demographics: race, age, gender, total education, marital status, number in family, method, and time spent in communting to work.

Medical History: questions about treatment for back pain by physicians, by chiropractors; hospitalization for back pain; surgery for back injury; smoking status.

Physical Examination: flexibility, spinal canal size by ultrasonography, and anthropometric measures such as height and weight.

Physical Capacities: arm strength; leg strength; and aerobic capacity measured by a submaximal treadmill test.

Psychological Testing: the MMPI (Minnesota Multiphasic Inventory and its subscales); a schedule of recent life change events; a family questionnaire about interactions at home; a health locus of control questionnaire.

Job Satisfaction: subjects were asked a number of questions about their job: did they enjoy their job almost always, some of the time, hardly ever; do they get along well with their supervisor; do they get along well with their fellow employees, etc.

The details of the design and many of the study results are found in Battie *et al.* [1989, 1990a,b] and Bigos *et al.* [1991, 1992a–b]. The extensive psychological questionnaires were given to the employees to be taken home and filled out; 54% of the 3020 employees returned completed, take home, questionnaries and some data analyses were necessarily restricted to those who completed the questionnaire(s). Figure 18.6 with four panels summarizes graphically some of the important predictive results.

The results of several stepwise, step-up multivariate Cox models are presented in Table 18.1.

There are some substantial risk gradients among the employees. However, the predicitive power is not such that one can conclusively indentify individuals likely to report an acute industrial back injury report. Of more importance, given the traditional approaches to this field which have been largely biomechanical, work per-

Table 18.1. Predicting Acute Back Injury Reports.[a]

Variable	Univariate Analysis *p*-value	Multivariate Analysis *p*-value	Relative Risk	(95% Confidence Interval)
I. Entire Population (n=1326, injury=117)				
Enjoy job[b]	0.0001	0.0001	1.70	(1.31, 2.21)
MMPI 3[c]	0.0003	0.0032	1.37	(1.11, 1.68)
Prior back pain[d]	0.0010	0.0050	1.70	(1.17, 2.46)
II. Those with a history of prior back injury (n=518, inj=63)				
Enjoy job[b]	0.0003	0.0006	1.85	(1.30, 2.62)
MMPI 3[c]	0.0195	0.0286	1.34	(1.17, 1.54)
III. Those without a history of prior back pain (n=808, inj=54)				
Enjoy jobs[b]	0.0220	0.0353	1.53	(1.09, 2.29)
MMPI 3[c]	0.0334	0.0475	1.41	(1.19, 1.68)

[a]Using the Cox proportional hazards regression model.
[b]Only subjects with complete information on the enjoy job question, MMPl, and history of back pain were included in these analyses.
[c] For an increase of one unit.
[d] For an increase of 10 units.

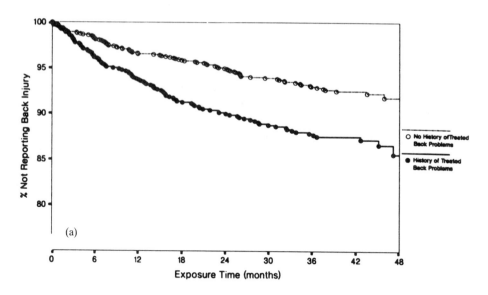

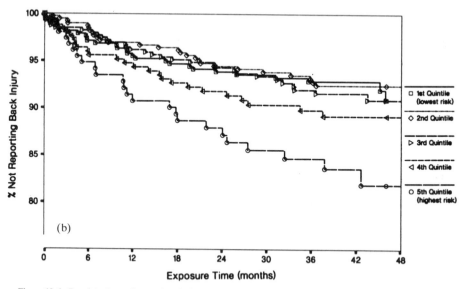

Figure 18.6. Panel A shows the product limit curves for the time to a subsequent back injury report for those reporting previous back problems and those who did not report such problems. Panel B divides the MMPI Scale 3 (hysteria) values by cutpoints taken from the quintiles of those actually reporting events. Panel C divides the subjects by their response to the question do you enjoy your job (a) almost always, (b) some of the time, or (c) hardly ever. Panel D gives the results of the multivariate Cox model of Table 18.1; the predictive equation uses the variables from the first three panels. From Bigos *et al.* [1991].

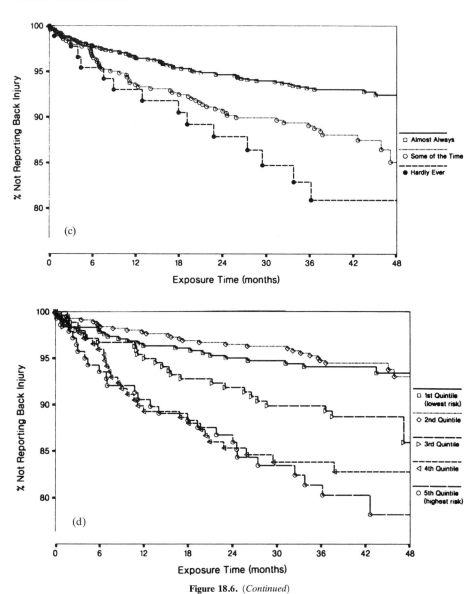

Figure 18.6. (*Continued*)

ception and psychological variables are important predictors and the problem cannot effectively be addressed with only one factor in mind. This is emphasized in Figure 18.7 which represents the amount of information (in a formal sense) in each of the categories of variables as given above.

The figure gives a Venn diagram of the estimated amount of predictive information for variables in each of the data collection areas (Fisher and Zeh [1991]). The job perception and psychological areas are about as important as the medical history and

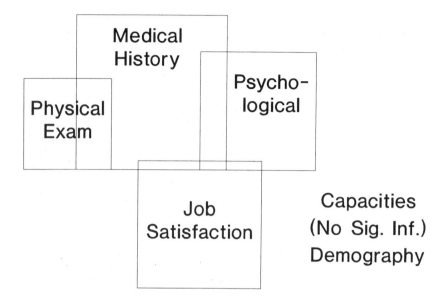

Predictive Information

Figure 18.7. Predictive information by type of variable collected. Note that the job satisfaction and psychological areas contribute the same order of magnitude as the more classical medical history and physical examination variables. The relative lack of overlap in predictive information means that at least these areas must be considered if the problem is to be fully characterized. Capacities and demography variables added no information and so have no boxes.

physical examination areas. To truly understand industrial back injury, a multifactorial approach must be used.

Among the more interesting aspects of the study is speculation on the meaning and implications of the findings. Since, as mentioned above, most individuals experience back problems at some time in their lives, could legitimate back discomfort be used as an escape if one does not enjoy their job? Can the problem be reduced by taking measures to make workers more satisfied with their employment—or are there a number of individuals who tend to be unhappy, no matter what? Is the problem a mixture of these? The results invite systematic, randomized intervention studies. Because of the magnitude of the problem, such approaches may be effective both in human and financial terms; however, this remains for the future.

18.6 IF YOU DON'T SEE THEM DOES IT MEAN THEY AREN'T THERE? DISTRIBUTION OF *GIARDIA* CYSTS IN DRINKING WATER

18.6.1 Background

The problem involved estimating the *Giardia lamblia* concentration in pristine rivers forming part of the Seattle area water supply. *Giardia lamblia* cysts are responsible for waterborne gastroenteritis by means of fecal contamination of water supply by both animals and humans. Symptoms of giardiasis include diarrhea, cramps, and weight loss. The watersheds have restricted access to keep the water supply as pure as possible. The Seattle Water Department sponsored a study to determine occurrence of *Giardia* in three watersheds: Cedar river, Green river, and Tolt river, Ongerth [1990].

The concentration of *Giardia* is usually quite low, 1 cyst per liter, so that large volumes of water must be analyzed in order to determine the concentration of cysts. You will also recognize the issue of power: if small volumes of water are analyzed the absence of cysts may simply be due to a lack of power.

The Environmental Protection Agency (EPA) has a model that assumes that water supplies are either pure or contaminated. A more reasonable model assumes that the concentration of cysts is continuous and that the problem is one of detection. In discussions with a colleague, a series of questions were formulated.

Questions

1. Do negative water samples imply no *Giardia* are present? How confident can I be? How large a sample do I have to take to ensure that the concentration of *Giardia* is less than a specified amount?
2. What is a good statistical model for this problem? What is meant by good?
3. What is the effect of less than 100% recovery in the assay (approximately 20% of cysts in samples are recovered). How does this fact affect the model?
4. It is difficult to carry large containers to and from the field. What is the effect of taking, say, 10 1-liter samples versus 1 10-liter sample?

We will discuss the answers in terms of a number of possible scenarios. This example will illustrate (we hope) that a great number of questions can fruitfully be formulated in terms of a fairly simple statistical model.

18.6.2 Poisson Model

It seemed plausible to assume that the distribution of *Giardia* followed a Poisson distribution. Remember that the Poisson distribution can be thought of as a limiting case of the binomial distribution, or derived in its own right. Here we will use the latter approach. We assume that:

a. The probability of exactly one cyst in a small volume δ is proportional to $\delta\mu$ where μ is the mean number of cysts per unit volume.

b. The probability of zero cysts in this small volume δ is $1 - \delta\mu$.

c. The probability of two or more cysts in this small volume is an order of magnitude smaller.

If X is the number of cysts per unit volume, then the probability of observing $X = x$ cysts in a volume V is

$$P[X = x] = e^{-\mu V}(\mu V)^x/x!.$$

This is the Poisson distribution of Chapter 6 with the mean expressed as a function of concentration and volume. Using this notation we get $E[X] = \text{Variance}[X] = \mu V$ and

$$P[X = 0] = e^{-\mu V}.$$

18.6.3 Rule of Threes

We meet the rule of threes, discussed in Chapter 6, again! The question is, how big a volume, V^*, must be taken to be 95% certain that it contains at least one cyst? That is,

$$P[X = 0] = 1 - 0.95 = 0.05.$$
$$= e^{-\mu V^*}.$$

Solving this equation,

$$-\mu V^* = \ln[0.05] = -2.966 \approx -3,$$

or

$$\mu V^* = 3.$$

Application 1. If there is one cyst per liter take 3 liters to be 95% certain that your sample will have at least one cyst.

Application 2. Go backwards. You observe zero cysts in a random sample of 10 liters. What is an upper bound on the number of cysts in the population? Take a 95% upper bound. In terms of our model, we require

$$\mu(10) \leq 3, \quad \mu \leq 0.3.$$

That is, the number of cysts is less than 0.3 per 10 liters. But it could be as high as 0.3 per 10 liters. So we say that 0.3 is a 95% upper confidence bound on the number of cysts per ten liters.

18.6.4 Effect of Recovery Efficiency

The assay for *Giardia* is very time consuming involving repeated filtrations and centrifuging. In the process a great number of cysts are destroyed. From studies where a known number of cysts are put in the water, the recovery is estimated to be about 20%. To approach this issue more generally, we assume that a proportion π of cysts is recovered. How does this affect the Poisson model? One way is to view this as a "compound model". Assume that X is Poisson (μV). A value of $X = x$ is generated, corresponding to the initial number of cysts in the sample. Each of the x cysts has a probability π of being recovered. A reasonable model is for Y, the observed number of cysts given that $X = x$ are in the sample originally, to be binomial (x, π). But we do not know the value of X. It can be shown that the distribution of Y, summed over all possible values of X is Poisson$(\pi\mu V)$.

Application 1. What proportion of samples will be negative as a function of the recovery efficiency, π ? The following table list the proportion of samples that will be negative as the recovery efficiency varies from 1 to 0.1 and the concentration varies from 10 to 0.1 per unit volume.

PROPORTION OF NEGATIVE SAMPLES

Number of Cysts per Unit Volume	Recovery Efficiency			
	1	0.5	0.20	0.1
10	0.000	0.007	0.135	0.368
5	0.007	0.082	0.368	0.607
2	0.135	0.368	0.670	0.819
1	0.368	0.607	0.819	0.905
0.5	0.607	0.779	0.905	0.951
0.1	0.905	0.951	0.980	0.990

If, for example, the concentration is 1 cyst per liter and the recovery efficiency is 0.2 (20%) then 0.819 or approximately 82% of the samples will be negative. This is clearly unacceptable and larger samples must be taken.

Application 2.   We can also apply the rule of threes to this situation. How large a volume must be taken to be 95% sure that at least one cyst is present when the concentration is μ and the recovery efficiency is π. Approximately then,

$$\pi \mu V = 3, \quad \mu V = 3/\pi,$$

or

$$V = 3/\pi\mu.$$

SCENARIO 1.   If there is one cyst per liter and you have a recovery efficiency of 20% you must take at $3/0.2 = 15$ liters of water to be 95% certain that you detect at least one cyst.

SCENARIO 2.   You observe 0 cysts in 10 liters and your recovery efficiency is 20% then the concentration of cysts per 10 liters could be as high as

$$3/0.20 = 15 \text{ cysts per 10 liters,}$$

or

$$1.5 \text{ cysts/liter.}$$

Application 3.   All the examples above dealt with the probabilities of observing no cysts or at least one cyst. The model can be extended to cover any number of observed cysts by using Appendix Table A.11. Let us start with one. You observe 1 cyst. What is a 95% confidence interval on the concentration. In this case the upper bound on the negative logarithm of the probability is given by

$$\pi \mu V = 4.74.$$

SCENARIO 1.   You observe one cyst in 10 liters. The recovery efficiency is 0.20. A 95% upper bound on the concentration in the water is

$$\mu V = 4.74/0.20$$
$$= 23.7 \text{ cysts per 10 liters,}$$

or

$$= 2.37 \text{ cysts per liter.}$$

Scenario 2. You take 20 10-liter samples and observe one sample with a single cyst. Your recovery efficiency is 10%. What is an upper bound on the concentration? In this case $V = 10 \times 20 = 200$, $\pi = 0.10$ and

$$\mu V = 4.74/(200 \times 0.10)$$
$$= 0.237 \text{ cysts per liter}$$
$$= 2.37 \text{ cysts per 10 liters.}$$

Application 4. From the table of exact confidence intervals for a Poisson random variable (Appendix Table A.11) we can construct the following table:

Observed Number of Cysts	Upper 95% Confidence Limit
0	3.00
1	4.74
2	6.30
3	7.75
4	9.15
5	10.51
6	11.84
7	13.15
8	14.43
9	15.70
10	16.96

Scenario. You take 20 10-liter samples; 18 are negative, in sample 19 you observe one cyst, in sample 20 four cysts. Your recovery efficiency is 25%. What is an upper bound on the cyst concentration? Total number of cysts is 5 in 200 liters. The upper limit on the Poisson number is 10.51, from the table. The upperbound is

$$\mu = 10.51/(200 \times 0.25)$$
$$= 0.2102 \text{ cysts/liter}$$
$$= 2.102 \text{ cysts per 10 liters.}$$

This example illustrates the usefulness of a simple model.

As a postscript, you may ask why we did not address Question 4: is it better to take 10 1-liter samples rather than 1 10-liter sample? The answer is implicit in the scenario: if the Poisson model holds, it does not matter since, as we discussed in Chapter 6, a sum of independent Poisson random variables is Poisson distributed with the mean equal to the sum of the individual means. Of course, there may be other, nonstatistical, reasons for preferring one sampling scheme over another.

Our colleague found these considerations quite helpful and it guided him in thinking about a feasible theoretical model, gave him some ideas about sampling and allowed him to analyze the data in a manner consistent with the model.

18.7 SYNTHESIZING INFORMATION FROM MANY STUDIES

A standard scientific activity is the review of literature. Such reviews are frequently assigned to respected experts in the field who presumably have a good sense about key issues facing the field, current research efforts, quality of work of various investigators, a feeling for what is important and what is not. Often the reviewer will identify unresolved issues and will suggest new lines of research. Perhaps one of the most famous is the review by David Hilbert at the turn of the century in the field of mathematics. His review included a statement of what would (should?) be the twenty-three key problems addressed by mathematicians in the twentieth century.

Meta-analysis may be defined as a systematic statistical method for analyzing and synthesizing results from independent studies, taking into account all pertinent information. Meta-analysis is a scientific activity that borrows from both the expert review and the methodology of multi-center trials. It is, perhaps, closer in spirit to the multi-center study (Chalmers *et al.* [1987]). Meta-analyses try to mimic multi-center studies and their validity depends on how well this can be done.

The literature of meta-analysis can be divided into three categories. First, there are the papers that deal with methodological, statistical methods. This area tends to be fairly theoretical, particularly those papers that deal with methods for detecting publication bias. Good examples are the text by Hedges and Olkin [1985] and a paper by Begg and Berlin [1988]; see also the winter 1992 issue of the Journal of Educational Statistics devoted to this issue. A second category consists of those papers actually carrying out meta-analyses. Examples are papers by Inglis and Lawson [1982] and Hyde and Linn [1988]. In the third category are review papers of meta-analysis (Jenicek [1989]).

A multi-center study is usually carried out for one or more of the following reasons. First, to increase the power of a study. Particularly in clinical trials of antiepileptic drugs one center or investigator may not have access to enough patients to detect with reasonable certainty a clinically meaningful treatment effect. Second, it may be desirable to do a study that reflects a variety of clinical practice. Third, there may be small variations in treatment and/or treatment conditions from center to center which should not impact the agent's effectiveness.

The number of centers is usually fixed by the study team or the pharmaceutical firm. A common protocol is developed and adhered to. A great deal of time and money is spent on ensuring that the study procedures are as standardized as possible. This may include sending blood samples or tissue samples to a common laboratory for analysis, and training sessions for key personnel. Patients are admitted to the study only if they satisfy clearly defined inclusion criteria. Endpoints or outcomes are measured, recorded and reported in the same way. Frequently, a data coordinating center is used to manage and monitor the study. Another characteristic is a built-in review of all activities to ensure that procedures do not deviate from the written guidelines. Particular attention is paid to dropouts and completion of the study by a patient. Everything is documented and available for inspection by the researchers. Ethical requirements usually specify one or more interim analyses in clinical trials with potential beneficial and/or harmful effects.

All the care taken in the multi-center study pays off in the analysis. With the common endpoints and standardized data collection, it is possible to pool the data at the measurement level rather than at the study level. Usually, a preliminary analysis of homogeneity of results is carried out to make sure that the results are comparable. Results are weighted by the size of the study. An example of a small multi-center study is contained in Leppik *et al.* [1986].

The overview of the multi-center study suggests that there are at least three pre-conditions for a valid meta-analysis. All the studies conducted in the area of interest must be available; that is, there must not be "publication bias": the likelihood of pub-lication must be independent of the treatment effect. Publication bias occurs when journals are more likely to accept, or authors are more likely to submit, results that are statistically significant. Recent methodological work has given some procedures for testing for publication bias. The basic idea is to consider the magnitude of a treat-ment effect as a function of the size of the study. If there is no publication bias, there should be no association (Begg and Berlin [1988]). Second, before studies are com-bined, some assessment of their quality must be made. A formal analysis of results does not guarantee that the quality of the studies is taken into account. This is where the expert review by someone knowledgeable in the field is apt to be more satisfac-tory. A third condition is that of homogeneity of results. Fortunately there are several statistical methods that allow the researcher to assess the magnitude of the problem (DerSimonian and Laird [1986]). In fact, such tests are very informative, creating the possibility of identifying groups of studies with homogeneous outcomes.

A meta-analysis is like the analysis of data from a multi-center study particularly in the requirement of common endpoints in the studies. If this is not the case it is difficult to combine the studies. A very old technique due to R. A. Fisher combines *p*-values, or significance levels, of studies (Hedges and Olkin [1985]). Sometimes, studies are excluded from a meta-analysis because there is not a common endpoint.

In comparative experiments where a new treatment may be compared with a placebo or standard treatment, an "effect size" can be calculated. This is defined as the treatment effect scaled by the amount of variability, usually the standard deviation. The effect size is the key ingredient in combining studies. We now summarize a meta-analysis of series of studies assessing the clinical effectiveness of a new antiepileptic drug. The primary purpose of this analysis is to show that studies should not be combined indiscriminately but that there sometimes are intrinsic differences between apparently similar studies.

Twenty-four reports of clinical trials of a new antiepileptic drug, progabide, span-ning the years 1980–1987 were collected van Belle [1988]. The studies dealt with a mixture of seizure types: partial seizures, primary generalized, secondary generalized, and other. The studies ranged in size from 13 patients to 187 patients. All studies were some type of cross-over, from baseline, placebo or active control. Eight studies were open, two were single blind and fourteen were double blind. The endpoint selected was "improvement" defined as a greater than 50% reduction in seizure frequency as compared to baseline in the open studies and as compared to placebo or active con-trol in the double blind, controlled studies. The predictor variables of interest were type of study (open, double blind), seizure type, and size of study.

A preliminary review determined the number of independent studies represented by the reports, that is, there were instances where the same data set apparently was used in more than one publication. It was decided that there were seventeen independent reports. Of these, fifteen presented enough detail to be available for

Table 18.2. Basic Data from Fifteen Independent Studies of the Antiepileptic Efficacy of Progabide.

Report Number	Type[a]	Number of patients		Improved[b]		Effect Size[c]
		Start	Complete	No.	(%)	
1	O	30	30	17	57	1.6
3	C	20	20	5	25	-1.8
5	C	24	20	9	45	0.2
9	C	20	17	3	18	-2.7
10	C	20	15	7	47	0.3
12	C	20	18	8	44	0.2
14	O	16	16	8	50	0.6
15	C	18	17	9	53	0.9
16	O	90	69	41	59	2.9
17	O	34	23	13	56	1.4
18	O	52	42	32	76	5.1
21	C	22	19	1	5	-7.3
22	C	64	51	12	24	-3.2
23	C	75	59	17	29	-2.3
24	O	187	151	90	60	4.3
Total		698	567	272	48	

[a]O, open trial; C, Placebo controlled, double blind (except report no. 9 which is single blind).
[b]Improved means greater than 50% reduction in seizures compared with baseline or placebo periods.
[c]Effect size, see text for definition.

meta-analysis. No formal analysis of publication bias was carried out; an informal review with knowledgeable epileptologists suggested that all studies of progabide were probably reported. Table 18.2 lists the studies that were used, the type of study, the sample size and the percentage improved.

The fifteen studies involved 698 patients, of whom 567 (81%) completed their study. Of these 567 patients, 272 were reported as improved (48%), that is, experienced a greater than 50% reduction in seizure frequency as compared with baseline. The effect size of a study relative to the average of the studies was defined as follows. Let p_i and n_i be the proportion out of n_i improved in report i. Then the weight of the study is defined as $n_i/p_i(1 - p_i)$; that is, the reciprocal of the variance of report i's estimate. The weighted average of the proportion improved is $P = \sum w_i p_i / \sum w_i$. The relative effect size for report i is then defined as $(p_i - P)\sqrt{w_i}$. A simple test of homogeneity involved a chi-square analysis of proportions of patients improved over the fifteen studies (see Chapter 6). This test was highly significant, indicating that it would be inappropriate to pool the results of the fifteen independent studies. This is perhaps the most useful feature of the meta-analysis since it forces us to begin a search for possible reasons for the heterogeneity of the results.

The first explanation is the type of study: open vs. controlled. In Figure 18.8 the percent of patients with greater than 50% reduction in seizure frequency is sorted by type of study (a plot of the relative effect size shows the same pattern, we chose the simpler scale). It is clear that the open studies, usually report much higher rates of improvement. If a homogeneity analysis is carried out within the open studies, it can be shown that they are homogeneous, that is, the variability in the results are

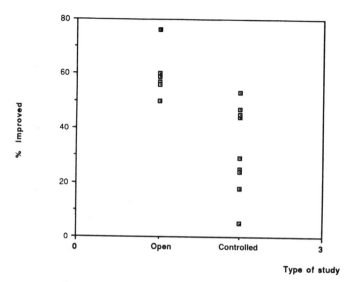

Figure 18.8. Percent of subjects improved by type of study.

consistent with random variation. This is not the case with the controlled studies; they still exhibit significant heterogeneity. If the controlled study with the lowest percent improvement is removed, the results from the remaining controlled studies are also not inconsistent with the homogeneity hypothesis.

These partitioning analyses are *ex post facto* and subject to criticism. But they can at least be considered illustrations of the usefulness of meta-analysis. A comparable analysis was carried out using seizure type, based on published data (Morselli *et al.* [1986]). Relatively little variability was associated with seizure type.

A problem that is less disposable is the following: several, but not all, of the controlled studies reported a "statistically significant treatment effect." What are we to make of those claims? A methodological response is that by following the hypothesis testing framework we are allowing for a Type I error; declaring a result significant when, in fact, it is not. We would expect one study in twenty to err in this way. If there is publication bias we would expect to see more of such studies. A second consideration is the intrinsic quality of the study. It is clear that open studies exaggerate the treatment effectiveness by incorporating an effect due to suggestion. In Table 18.3 are listed some factors that may compromise the outcome of a controlled trial (all these were gleaned from the studies under discussion).

We discuss briefly three of these factors. First, effective blinding. This is more difficult to achieve than many researchers are willing to admit. The active drug may color the urine thus giving away the treatment or may produce gastrointestinal side effects. Second, drug interactions are well known but not all studies maintained co-medications at the same plasma level. Finally, some studies had basic design flaws such as substitution of patients for patients that dropped out. The other reasons listed should also be explored. It is especially difficult to carry out controlled clinical trials

Table 18.3. Factors that May Affect the Quality and Outcome of Clinical Trials of Antiepileptic Agents.

1.	Open *vs.* controlled trial
2.	Type of patient
3.	Type/severity of seizures
4.	Ineffective blinding
5.	Co-medication
6.	Dropouts/substitution
7.	Study size (power)
8.	Plasma level monitoring
9.	Active *vs.* placebo control
10.	Period/cross-over effects
11.	Investigator quality

of antiepileptic agents. The nature of the disease and the treatment models present unique difficulties.

Meta-analysis is a useful, formal adjunct in literature review. The prerequisites for a valid meta-analysis remind us about the issues of publication bias, quality of the studies and heterogeneity of the studies. It also stresses the importance of study comparability, especially comparable outcomes. For example, there would appear to be no reason for one study to use a 75% reduction in seizure frequency as the key endpoint when all other reports use the 50% endpoint.

Meta-analysis also suggests that a series of carefully controlled clinical trials is the method of choice in determining the effectiveness of an antiepileptic agent. Open trials cannot be used in establishing antiepileptic efficacy.

One possible area where meta-analysis could be applied is in the evaluation of surgical outcomes of seizure control. The operation is such that one center can only do a limited number of procedures. A formal analysis across many centers would be very helpful in identifying unusual patterns and/or optimal strategies not apparent at the center level.

18.8 ARE TECHNICIANS AS GOOD AS PHYSICIANS?

The neuropathological diagnosis of Alzheimer's disease (AD) is time-consuming and difficult—even for experienced neuropathologists. Work in the late 1960s and early 1970s found that the presence of senile neuritic plaques in the neocortex and hippocampus justified a neuropathological diagnosis of Alzheimer's disease (Tomlinson *et al.* [1968, 1970]). Plaques are proteins associated with degenerating nerve cells in the brain, they tend to be located near the points of contact between cells. Typically, they are found in the brains of older persons.

These studies also found that large numbers of neurofibrillary tangles were often present in the neocortex and the hippocampus of brains from Alzheimer's disease victims. A tangle is another protein in the shape of a paired helical fragment found in the nerve cell. Neurofibrillary tangles are also found in other diseases. Later studies showed that plaques and tangles could be found in the brains of elderly individuals with preserved mental status. Thus, the quantity and distribution of plaques and tangles, rather than their mere presence are important in distinguishing Alzheimer's brains from the brains of normal aging individuals.

A joint conference of 1985 (Khachaturian [1985]) stressed the need for standardized clinical and neuropathological diagnoses for Alzheimers's disease.

We wanted to find out whether subjects with minimal training can count plaques and tangles in histological specimens of patients with Alzheimer's disease and controls van Belle et al. [1993]. Two experienced neuropathologists trained three student helpers to recognize plaques and tangles in slides obtained from autopsy material. After training, the students and pathologists examined coded slides from patients with Alzheimer's disease and controls. Some of the slides were repeated to provide an estimate of reproducibility. Each reader read four fields which were then averaged.

Ten sequential cases with a primary clinical and neuropathological diagnosis of Alzheimer's disease were chosen from the Alzheimer's Disease Research Center's (ADRC) brain autopsy registry. Age at death ranged from 67 years to 88 years, with a mean of 75.7 years and a standard deviation of 5.9 years.

Ten controls were examined for this study. Nine controls were selected from the ADRC registry of patients with brain autopsy, representing all subjects in the registry with no neuropathological evidence of AD. Four of these did have a clinical diagnosis of Alzheimer's disease, however. One additional control was drawn from files at the University of Washington's Department of Neuropathology. This control, aged 65 years at death, had no clinical history of Alzheimer's Disease.

For each case and control, sections from the hippocampus, and the temporal, parietal, and frontal lobes were viewed by the two neuropathologists and the three technicians. The three technicians were a first year medical school student, a graduate student in biostatistics with previous histological experience, and a pre-medical student. The technicians were briefly trained (for several hours) by a neuropathologist. The training consisted of looking at brain tissue (both Alzheimer's cases and normal brains) with a double headed microscope and at photographs of tissue. The neuropathologist trained the technicians to identify plaques and tangles in the tissue samples viewed. The training ended when the neuropathologist was satisfied that the technicians would be able to identify plaques and tangles in brain tissue samples on their own for the purposes of this study. The slides were masked to hide patient identity and were arbitrarily divided into batches of five subjects, with cases and controls mixed. Each viewer was asked to scan the entire slide to find the areas of the slide with the highest density of plaques and tangles (implied by Khachaturian [1985]). The viewer then chose the four fields on the slide which appeared to contain the highest density of plaques and tangles when viewed at 25X. Neurofibrillary tangles and senile plaques were counted in these four fields at 200X. If the field contained more than 30 plaques or tangles, the viewer scored the number of lesions in that field as 30.

The most important area in the brain for the diagnosis of Alzheimer's is the hippocampus and the results will be presented for that region. Results for other regions were similar. In addition, we will deal only with cases and plaques. Table 18.4 contains results for the estimated number of plaques per field for cases, each reading is the average of readings from four fields. The estimated number of plaques varied considerably ranging from zero to more than twenty. Inspection of Table 18.4 suggests that technician no. 3 tends to read higher than the other technicians and the neuropathologists, that is, tends to see more plaques. An analysis of variance confirms this impression:

Table 18.4. Average Number of Plaques per Field in the Hippocampus as Estimated by Three Technicians and Two Neuropathologists. Averages are Over Four Fields.

	Technician			Neuropathologist	
Case	1	2	3	A	B
1	0.75	0.00	0.00	0.00	0.00
2	7.25	6.50	7.50	4.75	3.75
3	5.50	7.25	5.50	5.75	8.75
4	5.25	8.00	14.30	5.75	6.50
5	10.00	8.25	9.00	3.50	7.75
6	7.25	7.00	21.30	13.00	8.50
7	5.75	15.30	18.80	10.30	8.00
8	1.25	4.75	3.25	3.25	4.00
9	1.75	5.00	7.25	2.50	3.SQ
10	10.50	16.00	18.30	13.80	19.00
Mean	5.25	7.80	10.50	6.26	6.98
SD	3.44	4.76	7.21	4.60	5.10

Correlations

	Technician		Neuropathologist	
	2	3	A	B
1	0.69	0.63	0.65	0.76
2		0.77	0.79	0.84
3			0.91	0.67
A				0.82

Source of Variation	d.f.	Mean Square	F
Patients	9	102.256	
Observers	4		
Technicians *vs.* neuropathologists	1	21.31	2.70
Within technicians	2	42.53	5.39
Neuropathologist A *vs.* Neuropathologist B	1	2.556	0.32
Patients × observers	36	7.888	

You will recognize from Chapter 10 the idea of partitioning the variance attributable to observers into three components; there are many ways of partitioning this variance. The above table contains one useful way of doing this. The analysis suggests that the average levels of response do not vary within neuropathologists. There is a highly significant difference among technicians. We would conclude that technician no. 3 is high, rather than technician no. 1 being low, because of the values obtained by the two neuropathologists. Note also that the residual variability is estimated to be $\sqrt{7.888} = 2.81$ plaques per patient. This represents considerable variability since the values represent averages of four readings. Using a single reading as a basis produces an estimated standard deviation of $(\sqrt{4})(2.81) = 5.6$ plaques per reading.

But how shall agreement be measured or evaluated? Equality of the mean levels suggests only that the raters tended to count the same number of plaques on the average. We need a more precise formulation of the issue. A correlation between the technicians and the neuropathologists will provide some information but is not sufficient because the correlation is invariant under changes in location and scale. In Chapter 4 we distinguished between precision and accuracy. Precision is the degree to which the observations cluster around a line, accuracy is the degree to which the observations are close to some standard. In this case the standard is the score of the neuropathologist and accuracy can be measured by the extent to which a technician's readings are from a 45° line. A paper by Lin [1989] nicely provides a framework for analyzing these data. In our case, the data are analyzed according to five criteria: location shift, scale shift, precision, accuracy, and concordance. Location shift refers to the degree to which the means of the data differ between technician and neuropathologist. A scale shift measures the differences in variability. Precision is quantified by a measure of correlation (Pearson's in our case), accuracy is estimated by the distance that the observations are from the 45° line. Concordance is defined as the product of the precision and the accuracy. In symbols, denote two raters by subscripts 1 and 2. Then we define,

$$\text{Location shift} = u = [(\mu_1 - \mu_2)]/\sqrt{(\sigma_1\sigma_2)},$$

$$\text{Scale shift} = v = \sigma_1/\sigma_2,$$

$$\text{Precision} = r,$$

$$\text{Accuracy} = A = [(v + 1/v + u^2)/2]^{-1},$$

$$\text{Concordance} = rA$$

We briefly discuss these. The location shift is a standardized estimate of the difference between the two raters. The quantity $\sqrt{(\sigma_1\sigma_2)}$ is the geometric mean of the two standard deviations. If there is no location difference between the two raters, this quantity is centered around zero. The scale shift is a ratio, if there is no scale shift, this quantity is centered around 1. The precision is the usual correlation coefficient, if the paired data fall on a straight line the correlation is 1. The accuracy is made up of a mixture of the means and the standard deviations. Note that if there is no location or scale shift the Accuracy is 1, the upper limit for this statistic. The concordance is the product of the accuracy and the precision; it is also bounded by 1. The data in Table 18.4 are analyzed according to the above criteria and displayed in Table 18.5. This table suggests that all the associations between technicians and neuropathologists are comparable. In addition, the comparisons between neuropathologists provide an internal measure of consistency. The "location shift" column indicates that, indeed, technician no. 3 tended to see more plaques than the neuropathologists. Technician no. 3 was also more variable, as indicated in the "scale shift" column. Technician no. 1 tended to be less variable than the neuropathologists. The precision of the technicians was comparable to that of the two neuropathologists compared with each other. The neuropathologists also displayed very high accuracy, almost matched by technician nos. 1 and 2. The concordance, the product of the precision and the accuracy, averaged over the two neuropathologists is comparable to their concordance. As usual,

Table 18.5. Characteristics of Ratings of Three Technicians and Two neuropathologists. Estimated Numbers of Plaques in the Hippocampus of Ten Cases. Based on Data from Table 18.4.

Technician	Pathologist	Location Shift	Scale Shift	Precision	Accuracy	Concordance
1	A	−0.18	0.75	0.95	0.94	0.89
	B	−0.35	0.68	0.76	0.88	0.67
2	A	0.33	1.03	0.79	0.95	0.75
	B	0.17	0.94	0.84	0.98	0.83
3	A	0.74	1.57	0.91	0.73	0.66
	B	0.58	1.42	0.67	0.81	0.55
A	B	−0.14	0.98	0.82	0.99	0.81

it is very important to graph the data to confirm these analytical results by a graphical display. Figure 18.9 displays the seven possible graphs. This graph was made by SYSTAT, a very comprehensive software package, Wilkinson [1989].

In summary, we conclude that it is possible to train relatively naive observers to count plaques in a manner comparable to that of experienced neuropathologists, as defined by the measures above. By this methodology, we have also been able to isolate the strengths and weaknesses of each of the technicians.

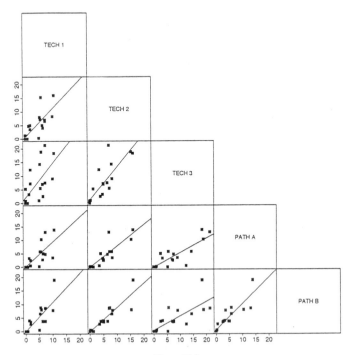

Figure 18.9.

18.9   RISKY BUSINESS

Every day of our life we meet many risks. The risk of being struck by lightning, getting into a car accident on the way to work, eating contaminated food and getting hepatitis. Many risks have moral and societal values associated with them, For example, what is the risk of being infected by AIDS through a HIV positive health practitioner? How does this risk compare with getting infectious hepatitis from an infected worker? What is the risk to the health practitioner in being identified as HIV positive? As we evaluate risks we may ignore them, in spite of being real and substantial. For example, smoking in the face of the evidence of the Surgeon General's reports. Or we may react to risks even though they are small. For example, worry about being hit by a falling airplane.

What is a risk? A risk in ordinary language (not statistics) is usually an event or the probability of the event. Thus, the risk of being hit by lightning is defined to be the probability of this event. The word risk has an unfavorable connotation. We usually do not speak of the risk of winning the lottery. For purposes of this chapter, we will relate the risk of an event to the probability of the occurrence of the event. In Chapter 3, we stated that all probabilities are conditional probabilities. When we talk about the risk of breast cancer we usually refer to its occurrence among women. Probabilities are modified as we define different groups at risk. R. A. Fisher talked about "relevant subsets", that is, what group or set of events is intended when a probability is specified.

In the course of thinking about environmental and occupational risks, one of us (GvB) wanted to develop a scale of risks similar to a Richter scale for earthquakes. The advantages of such a scale is to present risks numerically in such a way that the public would have an intuitive understanding of the risks. This, in spite of not understanding the full basis of the scale (it turns out to be fairly difficult to find a complete description of the Richter scale).

What should be the characteristics of such a scale? It became clear very quickly that the scale would have to be logarithmic. Second, it seemed that increasing risks should be associated with increasing values of the scale. It would also be nice to have the scale have roughly the same numerical range as the Richter scale. Most of its values are in the range from 3 to 7. The Risk Scale for events is defined as follows: let $P(E)$ be the probability of an event, then the Risk Units, $RU(E)$, for this event are defined to be

$$RU(E) = 10 + \log_{10}[P(E)].$$

This scale has several nice properties. First, the scale is logarithmic. Second, if the event is certain, $P(E) = 1$ and $RU(E) = 10$. Given two independent events, E_1 and E_2, the difference in their risks is

$$RU(E_1) - RU(E_2) = \log_{10}[P(E_1)/P(E_2)],$$

that is, the difference in the Risk Units is related to the relative risk of the events in a logarithmic fashion, that is, logarithm of the odds. Third, the progression is terms of powers of 10 is very simple:

Probability of Event	Risk Units
1	10
1/10	9
1/100	8
1/1000	7
1/10,000	6
1/100,000	5
1/1,000,000	4
1/10,000,000	3
1/100,000,000	2
1/1,000,000,000	1
1/10,000,000,000	0
1/100,000,000,000	−1

and so on. So a shift of two Risk Units represents a 100-fold change in probabilities. Events with Risk Units of the order of 1–4 are associated with relatively rare events. Note that the scale can go below zero.

As with the Richter scale, familiarity with common events will help you get a feeling for the scale. Let us start by considering some random events, next we deal with some common risks and locate them on the scale. Finally, we give you some risks and ask you to place them on the scale (the answers are given at the end of the chapter). The simplest case is the coin toss. The probability of, say, a head is 0.5. Hence the Risk Units associated with observing a head with a single toss of a coin is $RU(\text{Heads}) = 10 - \log_{10}(0.5) = 9.7$ (expressing Risk Units to one decimal place is usually enough). For a second example, the Risk Units of drawing at random a specified integer from the digits $(0, 1, 2, 3 \ldots, 9)$ is $1/10$ and the RU value is 9. Rolling a pair of sevens with two dice has a probability of $1/36$ and RU value is 8.4. Now consider some very small probabilities. Suppose you dial at random, what is the chance of dialing your phone number? Assume we are talking about the seven digit code and we allow all zeros as a possible number. The RU value is 3. If you throw in the area code as well you must deduct three more units to get the value $RU = 0$. There are clearly more efficient ways for making phone calls.

The idea of a logarithmic-like scale for probabilities appears in the literature quite frequently. In a delightful, little noticed book, *Risk Watch*, Urquhart and Heilmann [1984] defined the Safety Unit of an event, E as

$$\text{Safety Unit of } E = -\log_{10}[P(E)].$$

The drawback of this definition is that it calibrates events in terms of safety rather than risk. People are more inclined to think in terms of risk, they are "risk avoiders" rather than safety seekers. But it is clear than Risk Units and Safety Units very simply related:

$$RU(E) = 10 - SU(E).$$

Figure 18.10 lists the Risk Units for a series of events. Most of these probabilities were gleaned from various places in the risk literature. Beside the events mentioned already, the Risk Unit for a car accident with injury in a 1-year time interval has a

Risk Unit	Event
10	Certain event
9	Pick number 3 at random from 0 to 3
8	Car accident with injury (annual)
7	Killed in hang gliding (annual)
6	EPA action (life time risk)
5	Cancer from 4 tbsp peanut butter/day (annual)
4	Cancer from one transcontinental trip
3	Killed by falling aircraft
2	Dollar bill has specified set of eight numbers
1	Pick spot on earth at random and land within $\frac{1}{4}$ mile of your house
0	Your phone number picked at random (+ area code)
−0.5	Killed by falling meteorite (annual)

Figure 18.10. The Risk Unit scale and some associated risks.

value of 8. This corresponds to a probability of 0.01 or 1/100. The Environmental Protection Agency takes action on life time risks of Risk Unit 6. That is, if the life time probability of death is 1/10,000, the agency will take some action. This may seem rather anti-conservative but there are many risks and some selection has to be made. All these probabilities are estimates with varying degrees of precision. Crouch and Wilson [1982] include references to the data set upon which the estimate is based and also indicate whether the risk is changing. Table 18.6 describes some events for which you are ask to estimate the risk units. The answers are given at the end of the chapter.

Table 18.6. Events to be Ranked and Placed on Risk Units Scale. All Risks are Annual Unless Otherwise Indicated. Events NOT Ordered by Risk.

a.	Accidental drowning
b.	Amateur pilot death
c.	Appear on the Johnny Carson show (1991)
d.	Death due to smoking
e.	Die in mountain climbing accident
f.	Fatality due to insect bite or sting
g.	Hit by lightning (in life time)
h.	Killed in college football
i.	Life time risk of cancer due to chlorination
j.	Cancer from one diet cola/day with saccharin
k.	Ace of spades in one draw from 52 card deck
l.	Win the Reader's Digest Sweepstakes
m.	Win the Washington State lottery grand prize (with one ticket)

How do we evaluate risks? Why do we take action on some risks but not on others? The study of risks has been become a separate science with its own journals and society. The Borgen [1990] and Slovic [1986] articles are in a journal Risk Analysis that is worth examining. The following dimensions about evaluating risks have been mentioned in the literature:

Voluntary	Involuntary
Immediate effect	Delayed effect
Exposure essential	Exposure a luxury
Common hazard	"Dread" hazard
Affects average person	Affects special group
Reversible	Irreversible

We discuss these briefly. Recreational scuba diving has an annual probability of death of 4/10,000 or Risk Unit of 6.6 (Crouch and Wilson, Table 7.4). Compare this with some of the risks in Figure 18.10. Another dimension is the timing of the effect. If the effect is delayed we are usually willing to take a bigger risk, the most obvious example is smoking (which also is a voluntary behavior). If the exposure is essential, as part of one's occupation, then, again, larger risks are acceptable. A "dread" hazard is often perceived as of greater risk than a common hazard. The most conspicuous example is an airplane crash versus an automobile accident. But, perversely, we are less likely to be concerned about hazards that affect special groups we are not immediately linked to. For example, migrant workers have high exposures to pesticides and resultant increased immediate risks of neurological damage and long term risks of cancer. As a society, we are not vigorous in reducing those risks. Finally, if the effects of a risk are reversible we are willing to take larger risks.

Table 18.7 lists some risks with the same estimated value: each one increases the annual risk of death by one in a million, that is, All events have a Risk Unit value of 4. These examples, taken from Wynne [1991], illustrate that we do not judge risks to be the same even though the probabilities are equal. Some of the risks are avoidable others may not be. For example, flight attendants on international flights cannot escape passive smoking. It may be possible to avoid drinking Miami drinking water by drinking bottled water, or moving to Alaska. Most of the people who live in New York or Boston are not aware of the risk of living in those cities. But even if they did it is unlikely that they would move. A risk of one in a million is too small to act on.

How can risks be ranked? There are many ways. The primary one is by the prob-

Table 18.7. Activities Estimated to Increase the Annual Probability of Death by One in a Million (All Events Have a Risk Unit value of 4).

Activity	Cause of death
Smoking 1.4 cigarettes	Cancer, heart disease
Drinking 0.5 liters of wine	Cirrhosis of the liver
Living 2 days in New York or Boston	Air pollution
Travelling 10 miles by bicycle	Accident
Living 2 months with a cigarette smoker	Cancer, heart disease
Drinking Miami drinking water for 1 year	Cancer from chloroform
Living 150 years within 5 miles of a nuclear power plant	Cancer from radiation
Eating 100 charcoal-broiled steaks	Cancer from benzopyrene

[a]Condensed from Wynne (1991).

Answers (in Risk Units) to Risks in Table 18.6.

	Risk Units	Source/comments
a.	5.6	Crouch and Wilson [1982], Table 7.2
b.	7.0	Crouch and Wilson [1982], Table 7.4
c.	4.3	Siskin *et al.* [1990]
d.	7.5	Slovic [1986], Table 1
e.	6.8	Crouch and Wilson [1982], Table 7.4
f.	3.4	Crouch and Wilson [1982], Table 7.2
g.	4.2	Siskin *et al.* [1990]
h.	5.5	Crouch and Wilson [1982], Table 7.4
i.	4.0	Crouch and Wilson [1982], Table 7.5 and text, pp. 186–187
j.	5.0	Slovic [1986], Table 1
k.	8.3	$10+\log(1/52)$
l.	1.6	From back of announcement; $10 + \log(1/250,000,000)$
m.	3.0	From back of lottery ticket; $10 + \log(1/10,000,000)$

ability of occurrence as we have discussed so far. Another is by the expected loss (or gain). For example, the probability of a fire destroying your home is fairly small but the loss is so great that it pays to make the unfair bet with the insurance company. An unfair bet is one where the expected gain is negative. Another example is the lottery. A typical state lottery takes more than 50 cents from every dollar that is bet (compared to about 4 cents for roulette play in a casino). But the reward is so large (and the investment apparently small) that many people gladly play this unfair game.

How can risks be changed? It is clearly possible to stop smoking, to give up scuba diving, quit the police force, never drive a car. Many risks are associated with specific behaviors and changing those behaviors will change the risks. In the language of probability we have moved to another subset. Some changes will not completely remove the risks because of lingering effects of the behavior. But a great deal of risk reduction can be effected by changes in behavior. It behooves each one of us to assess the risks we take and to decide whether they are worth it.

REFERENCES

Alderman, E. L., Bourassa, M. G., Cohen, L. S., Davis, K. B., Kaiser, G. C., Killip, T., Mock, M. B., Pettinger, M., and Robertson, T. L. [1990]. Ten-year follow-up of survival and myocardial infarction in the randomized Coronary Artery Surgery Study. *Circulation*, **82:** 1629–1646.

Battie, M. C., Bigos, S. J., Fisher, L. D., Hansson, T. H., Nachemson, A. L., Spengler, D. M., Wortley, M. D., and Zeh, J. [1989]. A prospective study of the role of cardiovascular risk factors and fitness in industrial back pain complaints. *Spine*, **14:** 141–147.

Battie, M. C., Bigos, S. J., Fisher, L. D., Spengler, D. M., Hansson, T. H., Nachemson, A. L., and Wortley, M. D. [1990a]. Anthropometric and clinical measures as predictors of back pain complaints in industry: a prospective study. *Journal of Spinal Disorders*, **3:** 195–204.

Battie, M. C., Bigos, S. J., Fisher, L. D., Spengler, D. M., Hansson, T. H., Nachemson, A. L., and Wortley, M. D. [1990b]. The role of spinal flexibility in back pain complaints within industry: a prospective study. *Spine*, **15:** 768–773.

Begg, C. B. and Berlin, A. [1988]. Publication bias: a problem in interpreting medical data. *Journal of the Royal Statistical Society*, **151:** 419–463.

Bigger, Jr., J. T. [1990]. Editorial: the events surrounding the removal of encainide and flecainide from the Cardiac Arrhythmia Suppression Trial (CAST) and why CAST is continuing with moricizine. *Journal of the American College of Cardiology*, **15**: 243–245.

Bigos, S. J., Spengler, D. M., Martin, N. A., Zeh, J., Fisher, L., Nachemson, A., and Wang, M. H. [1986a]. Back injuries in industry: a retrospective study II. Injury factors. *Spine*, **11**: 246–251.

Bigos, S. J., Spengler, D. M., Martin, N. A., Zeh, J., Fisher, L., Nachemson, A., and Wang, M. H. [1986b]. Back injuries in industry: a retrospective study III. Employee-related factors. *Spine*, **11**: 252–256.

Bigos, S. J., Battie, M. C., Spengler, D. M., Fisher, L. D., Fordyce, W. E., Hansson, T. H., Nachemson, A. L., and Wortley, M. D. [1991]. A prospective study of work perceptions and psychosocial factors affecting the report of back injury. *Spine*, **16**: 1–6.

Bigos, S. J., Battie, M. C., Fisher, L. D., Fordyce, W. E., Hansson, T. H., Nachemson, A. L., and Spengler, D. M. [1992a]. A longitudinal, prospective study of industrial back injury reporting in industry. *Clinical Orthopaedics*, **279**: 21–34.

Bigos, S. J., Battie, M. C., Fisher, L. D., Hansson, T. H., Spengler, D. M., and Nachemson, A. L. [1992b]. A prospective evaluation of commonly used pre-employment screening tools for acute industrial back pain. *Spine*, **17**: 922–926.

Borer, J. S. [1987]. t-PA and the principles of drug approval. (Editorial) *New England Journal of Medicine*, **317**: 1659–1661.

Borgen, K. T. [1990]. Of apples, alcohol, and unacceptable risks. *Risk Analysis*, **10**: 199–200.

Cardiac Arrhythmia Pilot Study (CAPS) Investigators [1988]. Effect of encainide, flecainide, imipramine and moricizine on ventricular arrhythmias during the year after acute myocardial infarction: the CAPS. *American Journal of Cardiology*, **61**: 501–509.

Cardiac Arrhythmia Trial (CAST) Investigators [1989]. Preliminary Report: effect of encainide and flecainide on mortality in a randomized trial of arrhythmia suppression after myocardial infarction. *New England Journal of Medicine*, **321**: 406–412.

CASS Principal Investigators and Their Associates: Coronary Artery Surgery Study (CASS). [1983a]. A randomized trial of coronary artery bypass surgery: Survival data. *Circulation*, **68**: 939–950.

CASS Principal Investigators and Their Associates: Coronary Artery Surgery Study (CASS). [1983b]. A randomized trial of coronary artery bypass surgery: quality of life in patients randomly assigned to treatment groups. *Circulation*, **68**: 951–960.

CASS Principal Investigators and Their Associates: Coronary Artery Surgery Study (CASS). [1984a]. A randomized trial of coronary artery bypass surgery. comparability of entry characteristics and survival in randomized patients and nonrandomized patient meeting randomization criteria. *Journal of the American College of Cardiology*, **3**: 114–128.

CASS Principal Investigators and Their Associates. [1984b]. Myocardial infarction and mortality in the Coronary Artery Surgery Study (CASS) randomized trial. *New England Journal of Medicine*, **310**: 750–758.

Chaitman, B. R., Ryan, T. J., Kronmal, R. A., Foster, E. D., Frommer, P. L., Killip, T., and the CASS Investigators. [1990]. Coronary Artery Surgery Study (CASS): comparability of 10 year survival in randomized and randomizable patients. *Journal of the American College of Cardiology*, **16**: 1071–1078.

Chalmers, T. C., Levin, H., Sacks, H. S., Reitman, D., Berrier, J., and Nagalingam, R. [1987]. Meta-analysis of clinical trials as a scientific discipline. I. Control of bias and comparison with large cooperative trials. *Statistics in Medicine*, **6**: 315–325.

Chase, M. [1987]. FDA Panel rejection of anti-clot drug sets Genentech back months, perils stock. *Wall Street Journal*, **June 1**: 26.

Crouch, E. A. C., and Wilson, R. [1982]. *Risk Benefit Analysis*. Ballinger, Cambridge, MA.

DerSimonian, R. and Laird, N. [1986]. Meta-analysis in clinical trials. *Controlled Clinical Trials*, **7**: 177–178.

Echt, D. S., Liebson, P. R., Mitchell, B., Peters, R. W., Obias-Manno, D., Barker, A. H., Arensberg, D., Baker, A., Friedman, L., Greene, H. L., Huther, M. L., Richardson, D. W., and the CAST Investigators. [1991]. Mortality and morbidity in patients receiving encainide, flecainide, or placebo: the Cardiac Arrhythmia Suppression Trial. *New England Journal of Medicine*, **324:** 781–788.

Findlay S. [1987a]. Cover story: Heart attack victims get new hope. Subheading: On Friday, the FDA will scrutinize two drugs that could save many lives. *USA Today,* May 27: 1D, 2D.

Findlay, S. [1987b]. FDA speeds approval of heart drug. *USA Today,* July 28: 1D.

Fisher, L. D., Giardina, E.-G., Kowy, P. R., Leier, C. V., Lowenthal, D. T., Messerli, F. H., Pratt, C. M., and Ruskin, J. [1987]. The FDA Cardio-Renal Committee replies, Letter to the Editor. *Wall Street Journal*, Wednesday, August 12: 19.

Fisher L. D., Kaiser, G. C., Davis, K. B., and Mock, M. [1989]. Crossovers in coronary bypass grafting trials: Desirable, undesirable, or both? *Annals of Thoracic Surgery*, **48:** 465–466.

Fisher, L. D., Dixon, D. O., Herson, J., and Frankowski, R. F. [1990]. Analysis of randomized clinical trials: intention to treat. In *Statistical Issues in Drug Research and Development*. K. E. Peace, (ed.). Marcel Dekker, New York, pp. 331–344.

Fisher, L. D., and Zeh, J. [1991]. An information theory approach to presenting predictive value in the Cox proportional hazards regression model. (Unpublished).

Furburg, C. D. [1983]. Effect of antiarrhythmic drugs on mortality after myocardial infarction. *American Journal of Cardiology*, **52:** 32C–36C.

Glass, G. V. [1976]. Primary, secondary, and meta-analysis of research. *Educational Researcher*, **5:** 3–8.

Graboys, T. B., Lown, B., Podrid, P. J. and DeSilva, R. [1982]. Long-term survival of patients with malignant ventricular arrhythmia treated with antiarrhythmic drugs. *American Journal of Cardiology*, **50:** 437–443.

Hedges, L. V. and Olkin, I. [1985]. *Statistical Methods for Meta-analysis*. Academic Press, Orlando, FL.

Hyde, J. S. and Linn, M. C. [1988]. Gender differences in verbal ability: a meta-analysis. *Psychological Bulletin*, **1:** 53–69.

IMPACT Research Group [1984]. International mexiletine and placebo antiarrhythmic coronary trial: I. Report on arrhythmias and other findings. *Journal of the American College of Cardiology*, **4:** 1148–1163.

Inglis J. and Lawson J. S. [1982]. Meta-analysis of sex differences in the effects of unilateral brain damage on intelligence test results. *Canadian Journal of Psychology*, **36:** 670–683.

Jenicek, M. [1989]. Meta-analysis in medicine. *Journal of Clinical Epidemiology*, **42:** 35–44.

Kaiser, G. C., Davis, K. B., Fisher, L. D., Myers, W. O., Foster, E. D., Passamani, E. R., and Gillespie, M. J. [1985]. Survival following coronary artery bypass grafting in patients with severe angina pectoris (CASS), (with discussion). *Journal of Thoracic and Cardiovascular Surgery*, **89:** 513–524.

Khachaturian, Z. S. [1985]. Diagnosis of Alzheimer's Disease. *Archives of Neurology*, **42:** 1097–1105.

Kowey, P. R., Fisher, L. D., Giardina, E-G., Leier, C. V., Lowenthal, D. T., Messerli, F. H., and Pratt, C. M. [1988] The TPA controversy and the drug approval process; the view of the Cardiovascular and Renal Drugs Advisory Committee. *Journal of the American Medical Association*, **260:** 2250–2252.

Leppik, I. E., Brundage, R. C., Krall, R., Cloyd, J. C., Bowman-Cloyd, T., and Jacobs, M. P. [1986]. Double-blind withdrawal of phenytoin and carbamazapine in patients treated with progabide for partial seizures. *Epilepsia*, **27:** 563–568.

Lin, L. I. [1989]. A concordance correlation coefficient to evaluate reproducibility. *Biometrics*, **45:** 255–268.

Moolgavkar, S. H. (ed.) [1990]. *Scientific Issues in Quantitative Cancer Risk Assessment.* Birkhauser, Boston, Secaucus, NJ.

Morselli, P. L., Bartholini, G., and Lloyd, K. G. [1986]. Progabide. In *New Anticonvulsant Drugs.* R. J. Porter and B. S. Meldrum (eds.). John Libbey, London, pp. 237–253.

Myers, W. O., Schaff H. V., Gersh, B. J., Fisher, L. D., Kosinski, A. S., Mock, M. B., Holmes, D. R., Ryan, T. J., Kaiser, G. C., and CASS Investigators. [1989]. Improved survival of surgically treated patients with triple vessel coronary disease and severe angina pectoris: a report from the Coronary Artery Surgery Study (CASS) registry. *Journal of Thoracic and Cardiovascular Surgery,* **97:** 487–495.

Ongerth, J. E. [1990]. Evaluation of treatment for removing Giardia cysts. *Journal American Water Works Association,* **82:** 51–61.

Passamani, E., Davis, K. B., Gillespie, M. J., Killip, T., and the CASS Principal Investigators and Their Associates. [1985]. A randomized trial of coronary artery bypass surgery: survival of patients with a low ejection fraction. *New England Journal of Medicine,* **312:** 1665–1671.

Peto, R., Pike, M. C., Armitage, P., Breslow, N. L., Cox, D. R., Howard, S. V., Mantel, N., McPherson, K., Peto, J., and Smith, P. G. [1977]. Design and analysis of randomized clinical trials requiring prolonged observation of each patient. II. Analysis and examples. *British Journal of Cancer,* **35:** 1–39.

Pettinger, M. B. [1989]. Examination of meta-analysis: an evaluation with Monte Carlo studies and a partially annotated biliography. Unpublished M. S. thesis. University of Washington.

Pratt, C. M. (ed.) [1990]. A symposium: The Cardiac Arrhythmia Suppression Trial — does it alter our concepts of and approaches to ventricular arrhythmias? *American Journal of Cardiology,* **65:** 1B–42B.

Pratt, C. M., Brater, D. C., Harrell, Jr., F. E., Kowey, P. R., Leier, C. V., Lowenthal, D. T., Messerli, F., Packer, M., Pritchett, E. L. C., and Ruskin, J. N. [1990]. Clinical and regulatory implications of the Cardiac Arrhythmia Suppression Trial. *American Journal of Cardiology,* **65:** 103–105.

Preston, T. A. [1977]. *Coronary Artery Surgery: A Critical Review.* Raven Press, New York.

Principal Investigators of CASS and their Associates. [1981]. National Heart, Lung, Blood Institute Coronary Artery Surgery Study. T. Killip, L. D. Fisher, and M. B. Mock (eds.). American Heart Association Monograph Number 79. *Circulation,* **63**(part II): I-1–I-81.

Rogers, W. J., Coggin, C. J., Gersh, B. J., Fisher, L. D., Myers, W. O., Oberman, A., and Sheffield, L. T. [1990]. Ten-year follow-up of quality of life in patients randomized to receive medical therapy or coronary artery bypass graft surgery. *Circulation,* **82:** 1647–1658.

Rebello, K. [1987]. Biotechs hit a roadblock, investors sell. *USA Today,* June 2: 1B.

Ruskin, J. N. [1989]. The Cardiac Arrhythmia Suppression Trial (CAST) (Editorial). *New England Journal of Medicine,* **321:** 386–388.

Siskin, B., Staller, J., and Rornik, D. [1990]. What are the chances? Risk, odds and likelihood in everyday life. Crown publishers, Inc. New York.

Slovic, P. [1986]. Informing and educating the public about risk. *Risk Analysis,* **6:** 403–415.

Spengler, D. M., Bigos, S. J., Martin, N. A., Zeh, J., Fisher, L. D., and Nachemson, A. [1986]. Back injuries in industry: a retrospective study I. Overview and cost analysis. *Spine,* **11:** 241–245.

Takaro, T., Hultgren, H., Lipton, M., Detre, K. and participants in the Veterans Administration Cooperative Study Group. [1976]. VA cooperative randomized study for coronary arterial occlusive disease. II. Left main disease. *Circulation,* **54**(suppl 3): III-107.

Tomlinson, B. E. [1982]. Plaques, tangles and Alzheimer's disease, *Psychological Medicine,* **12:** 449–459.

Tomlinson, B. E., Blessed, G., and Roth, M. [1968]. Observations on the brains of non-demented old people. *Journal of Neurological Science,* **7:** 331–356.

Tomlinson, B. E., Blessed, G., and Roth, M. [1970]. Observations on the brains of demented old people. *Journal of Neurological Science*, **11:** 205–242.

Urquhart, J., and Heilmann, K. [1984]. *Risk Watch: The Odds of Life.* Facts on File Publications, New York.

van Belle, G. [1992]. A meta-analysis of the effectiveness of progabide as an antiepileptic drug. Manuscript.

van Belle, G., Gibson, K., Nochlin, D., Sumi, M., and Larson, E. B. [1993]. Counting plaques and tangles in Alzheimer's disease: concordance of technicians and pathologists. Submitted.

Wall Street Journal [1987a]. The TPA Decision. (Editorial). *Wall Street Journal,* Thursday, May 28: 26.

Wall Street Journal [1987b]. Human Sacrifice. (Editorial). *Wall Street Journal,* Tuesday, June 2: 30.

Weinstein, G. S., and Levin, B. [1989]. Effect of crossover on the statistical power of randomized studies. *Annals of Thoracic Surgery*, **48:** 490–495.

Wilkinson, L. [1989]. *SYGRAPH: The System for Graphics.* Evanston, IL, SYSTAT, Inc.

Wynne, B. [1991]. Public perception and communication of risk: what do we know? *NIH Journal of Health*, **3:** 65–71.

Appendix

Table A.1. Standard normal distribution.

Let Z be a normal random variable with mean zero and variance one. For selected values of Z three values are tabled: (1) the two-sided p-value, or $P[|Z| \geq z]$; (2) the one-sided p-value, or $P[Z \geq z]$; and (3) the cumulative distribution function at Z, or $P[Z \leq z]$.

z	Two-sided	One-sided	Cum. dist.	z	Two-sided	One-sided	Cum. dist.	z	Two-sided	One-sided	Cum. dist.
0.00	1.0000	.5000	.5000	1.30	.1936	.0968	.9032	1.80	.0719	.0359	.9641
0.05	.9601	.4801	.5199	1.31	.1902	.0951	.9049	1.81	.0703	.0351	.9649
0.10	.9203	.4602	.5398	1.32	.1868	.0934	.9066	1.82	.0688	.0344	.9656
0.15	.8808	.4404	.5596	1.33	.1835	.0918	.9082	1.83	.0673	.0336	.9664
0.20	.8415	.4207	.5793	1.34	.1802	.0901	.9099	1.84	.0658	.0329	.9671
0.25	.8026	.4013	.5987	1.35	.1770	.0885	.9115	1.85	.0643	.0322	.9678
0.30	.7642	.3821	.6179	1.36	.1738	.0869	.9131	1.86	.0629	.0314	.9686
0.35	.7263	.3632	.6368	1.37	.1707	.0853	.9147	1.87	.0615	.0307	.9693
0.40	.6892	.3446	.6554	1.38	.1676	.0838	.9162	1.88	.0601	.0301	.9699
0.45	.6527	.3264	.6736	1.39	.1645	.0823	.9177	1.89	.0588	.0294	.9706
0.50	.6171	.3085	.6915	1.40	.1615	.0808	.9192	1.90	.0574	.0287	.9713
0.55	.5823	.2912	.7088	1.41	.1585	.0793	.9207	1.91	.0561	.0281	.9719
0.60	.5485	.2743	.7257	1.42	.1556	.0778	.9222	1.92	.0549	.0274	.9726
0.65	.5157	.2578	.7422	1.43	.1527	.0764	.9236	1.93	.0536	.0268	.9732
0.70	.4839	.2420	.7580	1.44	.1499	.0749	.9251	1.94	.0524	.0262	.9738
0.75	.4533	.2266	.7734	1.45	.1471	.0735	.9265	1.95	.0512	.0256	.9744
0.80	.4237	.2119	.7881	1.46	.1443	.0721	.9279	1.96	.0500	.0250	.9750
0.85	.3953	.1977	.8023	1.47	.1416	.0708	.9292	1.97	.0488	.0244	.9756
0.90	.3681	.1841	.8159	1.48	.1389	.0694	.9306	1.98	.0477	.0239	.9761
0.95	.3421	.1711	.8289	1.49	.1362	.0681	.9319	1.99	.0466	.0233	.9767
1.00	.3173	.1587	.8413	1.50	.1336	.0668	.9332	2.00	.0455	.0228	.9772
1.01	.3125	.1562	.8438	1.51	.1310	.0655	.9345	2.01	.0444	.0222	.9778
1.02	.3077	.1539	.8461	1.52	.1285	.0643	.9357	2.02	.0434	.0217	.9783
1.03	.3030	.1515	.8485	1.53	.1260	.0630	.9370	2.03	.0424	.0212	.9788
1.04	.2983	.1492	.8508	1.54	.1236	.0618	.9382	2.04	.0414	.0207	.9793
1.05	.2937	.1469	.8531	1.55	.1211	.0606	.9394	2.05	.0404	.0202	.9798
1.06	.2891	.1446	.8554	1.56	.1188	.0594	.9406	2.06	.0394	.0197	.9803
1.07	.2846	.1423	.8577	1.57	.1164	.0582	.9418	2.07	.0385	.0192	.9808
1.08	.2801	.1401	.8599	1.58	.1141	.0571	.9429	2.08	.0375	.0188	.9812
1.09	.2757	.1379	.8621	1.59	.1118	.0559	.9441	2.09	.0366	.0183	.9817
1.10	.2713	.1357	.8643	1.60	.1096	.0548	.9452	2.10	.0357	.0179	.9821
1.11	.2670	.1335	.8665	1.61	.1074	.0537	.9463	2.11	.0349	.0174	.9826
1.12	.2627	.1314	.8686	1.62	.1052	.0526	.9474	2.12	.0340	.0170	.9830
1.13	.2585	.1292	.8708	1.63	.1031	.0516	.9484	2.13	.0332	.0166	.9834
1.14	.2543	.1271	.8729	1.64	.1010	.0505	.9495	2.14	.0324	.0162	.9838
1.15	.2501	.1251	.8749	1.65	.0989	.0495	.9505	2.15	.0316	.0158	.9842
1.16	.2460	.1230	.8770	1.66	.0969	.0485	.9515	2.16	.0308	.0154	.9846
1.17	.2420	.1210	.8790	1.67	.0949	.0475	.9525	2.17	.0300	.0150	.9850
1.18	.2380	.1190	.8810	1.68	.0930	.0465	.9535	2.18	.0293	.0146	.9854
1.19	.2340	.1170	.8830	1.69	.0910	.0455	.9545	2.19	.0285	.0143	.9857
1.20	.2301	.1151	.8849	1.70	.0891	.0446	.9554	2.20	.0278	.0139	.9861
1.21	.2263	.1131	.8869	1.71	.0873	.0436	.9564	2.21	.0271	.0136	.9864
1.22	.2225	.1112	.8888	1.72	.0854	.0427	.9573	2.22	.0264	.0132	.9868
1.23	.2187	.1093	.8907	1.73	.0836	.0418	.9582	2.23	.0257	.0129	.9871
1.24	.2150	.1075	.8925	1.74	.0819	.0409	.9591	2.24	.0251	.0125	.9875
1.25	.2113	.1056	.8944	1.75	.0801	.0401	.9599	2.25	.0244	.0122	.9878
1.26	.2077	.1038	.8962	1.76	.0784	.0392	.9608	2.26	.0238	.0119	.9881
1.27	.2041	.1020	.8980	1.77	.0767	.0384	.9616	2.27	.0232	.0116	.9884
1.28	.2005	.1003	.8997	1.78	.0751	.0375	.9625	2.28	.0226	.0113	.9887
1.29	.1971	.0985	.9015	1.79	.0735	.0367	.9633	2.29	.0220	.0110	.9890

Table A.1 *continued*. Standard normal distribution.											
z	*Two-sided*	*One-sided*	*Cum. dist.*	*z*	*Two-sided*	*One-sided*	*Cum. dist.*	*z*	*Two-sided*	*One-sided*	*Cum. dist.*
2.30	.0214	.0107	.9893	2.80	.0051	.0026	.9974	3.30	.0010	.0005	.9995
2.31	.0209	.0104	.9896	2.81	.0050	.0025	.9975	3.31	.0009	.0005	.9995
2.32	.0203	.0102	.9898	2.82	.0048	.0024	.9976	3.32	.0009	.0005	.9995
2.33	.0198	.0099	.9901	2.83	.0047	.0023	.9977	3.33	.0009	.0004	.9996
2.34	.0193	.0096	.9904	2.84	.0045	.0023	.9977	3.34	.0008	.0004	.9996
2.35	.0188	.0094	.9906	2.85	.0044	.0022	.9978	3.35	.0008	.0004	.9996
2.36	.0183	.0091	.9909	2.86	.0042	.0021	.9979	3.36	.0008	.0004	.9996
2.37	.0178	.0089	.9911	2.87	.0041	.0021	.9979	3.37	.0008	.0004	.9996
2.38	.0173	.0087	.9913	2.88	.0040	.0020	.9980	3.38	.0007	.0004	.9996
2.39	.0168	.0084	.9916	2.89	.0039	.0019	.9981	3.39	.0007	.0003	.9997
2.40	.0164	.0082	.9918	2.90	.0037	.0019	.9981	3.40	.0007	.0003	.9997
2.41	.0160	.0080	.9920	2.91	.0036	.0018	.9982	3.41	.0006	.0003	.9997
2.42	.0155	.0078	.9922	2.92	.0035	.0018	.9982	3.42	.0006	.0003	.9997
2.43	.0151	.0075	.9925	2.93	.0034	.0017	.9983	3.43	.0006	.0003	.9997
2.44	.0147	.0073	.9927	2.94	.0033	.0016	.9984	3.44	.0006	.0003	.9997
2.45	.0143	.0071	.9929	2.95	.0032	.0016	.9984	3.45	.0006	.0003	.9997
2.46	.0139	.0069	.9931	2.96	.0031	.0015	.9985	3.46	.0005	.0003	.9997
2.47	.0135	.0068	.9932	2.97	.0030	.0015	.9985	3.47	.0005	.0003	.9997
2.48	.0131	.0066	.9934	2.98	.0029	.0014	.9986	3.48	.0005	.0003	.9997
2.49	.0128	.0064	.9936	2.99	.0028	.0014	.9986	3.49	.0005	.0002	.9998
2.50	.0124	.0062	.9938	3.00	.0027	.0013	.9987	3.50	.0005	.0002	.9998
2.51	.0121	.0060	.9940	3.01	.0026	.0013	.9987	3.51	.0004	.0002	.9998
2.52	.0117	.0059	.9941	3.02	.0025	.0013	.9987	3.52	.0004	.0002	.9998
2.53	.0114	.0057	.9943	3.03	.0024	.0012	.9988	3.53	.0004	.0002	.9998
2.54	.0111	.0055	.9945	3.04	.0024	.0012	.9988	3.54	.0004	.0002	.9998
2.55	.0108	.0054	.9946	3.05	.0023	.0011	.9989	3.55	.0004	.0002	.9998
2.56	.0105	.0052	.9948	3.06	.0022	.0011	.9989	3.56	.0004	.0002	.9998
2.57	.0102	.0051	.9949	3.07	.0021	.0011	.9989	3.57	.0004	.0002	.9998
2.58	.0099	.0049	.9951	3.08	.0021	.0010	.9990	3.58	.0003	.0002	.9998
2.59	.0096	.0048	.9952	3.09	.0020	.0010	.9990	3.59	.0003	.0002	.9998
2.60	.0093	.0047	.9953	3.10	.0019	.0010	.9990	3.60	.0003	.0002	.9998
2.61	.0091	.0045	.9955	3.11	.0019	.0009	.9991	3.61	.0003	.0002	.9998
2.62	.0088	.0044	.9956	3.12	.0018	.0009	.9991	3.62	.0003	.0001	.9999
2.63	.0085	.0043	.9957	3.13	.0017	.0009	.9991	3.63	.0003	.0001	.9999
2.64	.0083	.0041	.9959	3.14	.0017	.0008	.9992	3.64	.0003	.0001	.9999
2.65	.0080	.0040	.9960	3.15	.0016	.0008	.9992	3.65	.0003	.0001	.9999
2.66	.0078	.0039	.9961	3.16	.0016	.0008	.9992	3.66	.0003	.0001	.9999
2.67	.0076	.0038	.9962	3.17	.0015	.0008	.9992	3.67	.0002	.0001	.9999
2.68	.0074	.0037	.9963	3.18	.0015	.0007	.9993	3.68	.0002	.0001	.9999
2.69	.0071	.0036	.9964	3.19	.0014	.0007	.9993	3.69	.0002	.0001	.9999
2.70	.0069	.0035	.9965	3.20	.0014	.0007	.9993	3.70	.0002	.0001	.9999
2.71	.0067	.0034	.9966	3.21	.0013	.0007	.9993	3.71	.0002	.0001	.9999
2.72	.0065	.0033	.9967	3.22	.0013	.0006	.9994	3.72	.0002	.0001	.9999
2.73	.0063	.0032	.9968	3.23	.0012	.0006	.9994	3.73	.0002	.0001	.9999
2.74	.0061	.0031	.9969	3.24	.0012	.0006	.9994	3.74	.0002	.0001	.9999
2.75	.0060	.0030	.9970	3.25	.0012	.0006	.9994	3.75	.0002	.0001	.9999
2.76	.0058	.0029	.9971	3.26	.0011	.0006	.9994	3.76	.0002	.0001	.9999
2.77	.0056	.0028	.9972	3.27	.0011	.0005	.9995	3.77	.0002	.0001	.9999
2.78	.0054	.0027	.9973	3.28	.0010	.0005	.9995	3.78	.0002	.0001	.9999
2.79	.0053	.0026	.9974	3.29	.0010	.0005	.9995	3.79	.0002	.0001	.9999

Table A.2. Critical values (percentiles) for the standard normal distribution.							
The fourth column is the $N(0,1)$ percentile for the percent given in column one. It is also the upper one-sided $N(0,1)$ critical value and two-sided $N(0,1)$ critical value for the significance levels given in columns two and three, respectively.							
Percent	*1-sided*	*2-sided*	*z*	*Percent*	*1-sided*	*2-sided*	*z*
50	.50	1.00	0.00	99.59	.0041	.0082	2.64
55	.45	.90	0.13	99.60	.0040	.0080	2.65
60	.40	.80	0.25	99.61	.0039	.0078	2.66
65	.35	.70	0.39	99.62	.0038	.0076	2.67
70	.30	.60	0.52	99.63	.0037	.0074	2.68
75	.25	.50	0.67	99.64	.0036	.0072	2.69
80	.20	.40	0.84	99.65	.0035	.0070	2.70
85	.15	.30	1.04	99.66	.0034	.0068	2.71
90	.10	.20	1.28	99.67	.0033	.0066	2.72
91	.09	.18	1.34	99.68	.0032	.0064	2.73
92	.08	.16	1.41	99.69	.0031	.0062	2.74
93	.07	.14	1.48	99.70	.0030	.0060	2.75
94	.06	.12	1.55	99.71	.0029	.0058	2.76
95	.05	.10	1.64	99.72	.0028	.0056	2.77
95.5	.045	.090	1.70	99.73	.0027	.0054	2.78
96.0	.040	.080	1.75	99.74	.0026	.0052	2.79
96.5	.035	.070	1.81	99.75	.0025	.0050	2.81
97.0	.030	.060	1.88	99.76	.0024	.0048	2.82
97.5	.025	.050	1.96	99.77	.0023	.0046	2.83
98.0	.020	.040	2.05	99.78	.0022	.0044	2.85
98.5	.015	.030	2.17	99.79	.0021	.0042	2.86
99.0	.010	.020	2.33	99.80	.0020	.0040	2.88
99.05	.0095	.0190	2.35	99.81	.0019	.0038	2.89
99.10	.0090	.0180	2.37	99.82	.0018	.0036	2.91
99.15	.0085	.0170	2.39	99.83	.0017	.0034	2.93
99.20	.0080	.0160	2.41	99.84	.0016	.0032	2.95
99.25	.0075	.0150	2.43	99.85	.0015	.0030	2.97
99.30	.0070	.0140	2.46	99.86	.0014	.0028	2.99
99.35	.0065	.0130	2.48	99.87	.0013	.0026	3.01
99.40	.0060	.0120	2.51	99.88	.0012	.0024	3.04
99.45	.0055	.0110	2.54	99.89	.0011	.0022	3.06
99.50	.0050	.0100	2.58	99.90	.0010	.0020	3.09
99.51	.0049	.0098	2.58	99.91	.0009	.0018	3.12
99.52	.0048	.0096	2.59	99.92	.0008	.0016	3.16
99.53	.0047	.0094	2.60	99.93	.0007	.0014	3.19
99.54	.0046	.0092	2.60	99.94	.0006	.0012	3.24
99.55	.0045	.0090	2.61	99.95	.0005	.0010	3.29
99.56	.0044	.0088	2.62	99.96	.0004	.0008	3.35
99.57	.0043	.0086	2.63	99.97	.0003	.0006	3.43
99.58	.0042	.0084	2.64	99.98	.0002	.0004	3.54
				99.99	.0001	.0002	3.72

Table A.3. Critical values (percentiles) for the chi-square distribution.

For each degree of freedom (D) in the first column, the table entries are the critical values for the upper one-sided significance levels in the column headings or, equivalently, the percentiles for the corresponding percentages.

	Percentage								
	2.5	5	50	75	90	95	97.5	99	99.9
D	Upper one-sided α								
	.975	.95	.50	.25	.10	.05	.025	.01	.001
1	.001	.004	.455	1.32	2.71	3.84	5.02	6.63	10.83
2	.051	.103	1.39	2.77	4.61	5.99	7.38	9.21	13.82
3	.216	.352	2.37	4.11	6.25	7.82	9.35	11.34	16.27
4	.484	.711	3.36	5.39	7.78	9.49	11.14	13.28	18.47
5	.831	1.15	4.35	6.63	9.24	11.07	12.83	15.09	20.52
6	1.24	1.64	5.35	7.84	10.64	12.59	14.45	16.81	22.46
7	1.69	2.17	6.35	9.04	12.02	14.07	16.01	18.47	24.32
8	2.18	2.73	7.34	10.22	13.36	15.51	17.53	20.09	26.12
9	2.70	3.33	8.34	11.39	14.68	16.92	19.02	21.67	27.88
10	3.25	3.94	9.34	12.55	15.99	18.31	20.48	23.21	29.59
11	3.82	4.57	10.34	13.70	17.27	19.68	21.92	24.72	31.26
12	4.40	5.23	11.34	14.85	18.55	21.03	23.34	26.22	32.91
13	5.01	5.89	12.34	15.98	19.81	22.36	24.74	27.69	34.53
14	5.63	6.57	13.34	17.12	21.06	23.68	26.12	29.14	36.12
15	6.26	7.26	14.34	18.25	22.31	25.00	27.49	30.58	37.70
16	6.91	7.96	15.34	19.37	23.54	26.30	28.85	32.00	39.25
17	7.56	8.67	16.34	20.49	24.77	27.59	30.19	33.41	40.79
18	8.23	9.39	17.34	21.60	25.99	28.87	31.53	34.81	42.31
19	8.91	10.12	18.34	22.72	27.20	30.14	32.85	36.19	43.82
20	9.59	10.85	19.34	23.83	28.41	31.41	34.17	37.57	45.31
21	10.28	11.59	20.34	24.93	29.62	32.67	35.48	38.93	46.80
22	10.98	12.34	21.34	26.04	30.81	33.92	36.78	40.29	48.27
23	11.69	13.09	22.34	27.14	32.01	35.17	38.08	41.64	49.73
24	12.40	13.85	23.34	28.24	33.20	36.42	39.36	42.98	51.18
25	13.12	14.61	24.34	29.34	34.38	37.65	40.65	44.31	52.62
26	13.84	15.38	25.34	30.43	35.56	38.89	41.92	45.64	54.05
27	14.57	16.15	26.34	31.53	36.74	40.11	43.19	46.96	55.48
28	15.31	16.93	27.34	32.62	37.92	41.34	44.46	48.28	56.89
29	16.05	17.71	28.34	33.71	39.09	42.56	45.72	49.59	58.30
30	16.79	18.49	29.34	34.80	40.26	43.77	46.98	50.89	59.70
35	20.57	22.47	34.34	40.22	46.06	49.80	53.20	57.34	66.62
40	24.43	26.51	39.34	45.62	51.81	55.76	59.34	63.69	73.40
45	28.37	30.61	44.34	50.98	57.51	61.66	65.41	69.96	80.08
50	32.36	34.76	49.33	56.33	63.17	67.50	71.42	76.15	86.66
55	36.40	38.96	54.33	61.66	68.80	73.31	77.38	82.29	93.17
60	40.48	43.19	59.33	66.98	74.40	79.08	83.30	88.38	99.61
65	44.60	47.45	64.33	72.28	79.97	84.82	89.18	94.42	105.99
70	48.76	51.74	69.33	77.58	85.53	90.53	95.02	100.43	112.32
75	52.94	56.05	74.33	82.86	91.06	96.22	100.84	106.39	118.60
80	57.15	60.39	79.33	88.13	96.58	101.88	106.63	112.33	124.84
85	61.39	64.75	84.33	93.39	102.08	107.52	112.39	118.24	131.04
90	65.65	69.13	89.33	98.65	107.57	113.15	118.14	124.12	137.21
95	69.92	73.52	94.33	103.90	113.04	118.75	123.86	129.97	143.34
100	74.22	77.93	99.33	109.14	118.50	124.34	129.56	135.81	149.45

For more than 100 degrees of freedom chi-square critical values may be found in terms of the degrees of freedom D and the corresponding two-sided critical value for a standard normal deviate Z by the equation $X^2 = 0.5 \cdot (Z + \sqrt{2 \cdot D - 1})^2$.

Table A.4. *p*-values for the chi-square distribution divided by degrees of freedom.

This table gives the *p*-value for a chi-square variable divided by it's degrees of freedom. Values with $p \leq 0.05$ are tabled. The column labelled "ratio" is the value of the chi-square statistic divided by it's degrees of freedom. The other columns, to the right, are headed by the degrees of freedom. For 1 degree of freedom the two-sided standard normal *p*-value (Table 2) for the square root of the chi-square statistic should be used. For 2 degrees of freedom, if Y is the value of the chi-square statistic, then $\exp(-Y/2)$ is the *p*-value. For more than 100 degrees of freedom use the two-sided standard normal *p*-value (Table 2) for $Z = \sqrt{Y} - \sqrt{2 \cdot DF - 1}$, where DF is the degrees of freedom.

	Degrees of freedom					Degrees of freedom		
Ratio	3	4	5	Ratio	6	7	8	9
2.2	.0858	.0663	.0514	1.85	.0853	.0733	.0632	.0545
2.3	.0752	.0563	.0423	1.90	.0768	.0651	.0554	.0472
2.4	.0658	.0477	.0348	1.95	.0690	.0578	.0485	.0408
2.5	.0576	.0404	.0285	2.00	.0620	.0512	.0424	.0352
2.6	.0503	.0342	.0234	2.05	.0556	.0453	.0370	.0303
2.7	.0440	.0289	.0191	2.10	.0498	.0400	.0323	.0261
2.8	.0384	.0244	.0156	2.15	.0447	.0354	.0281	.0224
2.9	.0336	.0206	.0127	2.20	.0400	.0312	.0244	.0192
3.0	.0293	.0174	.0104	2.25	.0357	.0275	.0212	.0164
3.1	.0256	.0146	.0084	2.30	.0320	.0242	.0184	.0141
3.2	.0223	.0123	.0068	2.35	.0285	.0213	.0160	.0120
3.3	.0194	.0103	.0056	2.40	.0255	.0187	.0138	.0102
3.4	.0169	.0087	.0045	2.45	.0227	.0165	.0120	.0087
3.5	.0148	.0073	.0036	2.50	.0203	.0144	.0103	.0074
3.6	.0129	.0061	.0029	2.55	.0180	.0127	.0089	.0063
3.7	.0112	.0051	.0024	2.60	.0161	.0111	.0077	.0054
3.8	.0097	.0043	.0019	2.65	.0143	.0097	.0066	.0045
3.9	.0085	.0036	.0016	2.70	.0127	.0085	.0057	.0039
4.0	.0074	.0030	.0012	2.75	.0113	.0074	.0049	.0033
4.1	.0064	.0025	.0010	2.80	.0100	.0065	.0042	.0028
4.2	.0056	.0021	.0008	2.85	.0089	.0057	.0036	.0023
4.3	.0049	.0018	.0007	2.90	.0079	.0050	.0031	.0020
4.4	.0042	.0015	.0005	2.95	.0070	.0043	.0027	.0017
4.5	.0037	.0012	.0004	3.00	.0062	.0038	.0023	.0014
4.6	.0032	.0010	.0003	3.05	.0055	.0033	.0020	.0012
4.7	.0028	.0009	.0003	3.10	.0049	.0029	.0017	.0010
4.8	.0024	.0007	.0002	3.15	.0043	.0025	.0014	.0008
4.9	.0021	.0006	.0002	3.20	.0038	.0022	.0012	.0007
5.0	.0018	.0005	.0001	3.25	.0034	.0019	.0011	.0006
5.1	.0016	.0004	.0001	3.30	.0030	.0016	.0009	.0005
5.2	.0014	.0003	.0001	3.35	.0027	.0014	.0008	.0004
5.3	.0012	.0003	.0001	3.40	.0024	.0012	.0007	.0003
5.4	.0010	.0002	.0001	3.45	.0021	.0011	.0006	.0003
5.5	.0009	.0002	.0000	3.50	.0018	.0009	.0005	.0002
5.6	.0008	.0002	.0000	3.55	.0016	.0008	.0004	.0002
5.7	.0007	.0001	.0000	3.60	.0014	.0007	.0003	.0002
5.8	.0006	.0001	.0000	3.65	.0013	.0006	.0003	.0001
5.9	.0005	.0001	.0000	3.70	.0011	.0005	.0002	.0001
6.0	.0004	.0001	.0000	3.75	.0010	.0005	.0002	.0001
6.1	.0004	.0001	.0000	3.80	.0009	.0004	.0002	.0001
6.2	.0003	.0001	.0000	3.85	.0008	.0003	.0002	.0001
6.3	.0003	.0000	.0000	3.90	.0007	.0003	.0001	.0001
6.4	.0002	.0000	.0000	3.95	.0006	.0003	.0001	.0000
6.5	.0002	.0000	.0000	4.00	.0005	.0002	.0001	.0000
6.6	.0002	.0000	.0000	4.05	.0005	.0002	.0001	.0000
6.7	.0002	.0000	.0000	4.10	.0004	.0002	.0001	.0000
6.8	.0001	.0000	.0000	4.15	.0004	.0001	.0001	.0000
				4.20	.0003	.0001	.0000	.0000
				4.25	.0003	.0001	.0000	.0000
				4.30	.0002	.0001	.0000	.0000
				4.35	.0002	.0001	.0000	.0000
				4.40	.0002	.0001	.0000	.0000
				4.45	.0002	.0001	.0000	.0000
				4.50	.0001	.0001	.0000	.0000

				Degrees of freedom					
Ratio	10	11	12	13	14	15	16	17	18	19
1.46	.1473	.1389	.1311	.1237	.1169	.1105	.1044	.0988	.0935	.0886
1.48	.1395	.1311	.1232	.1159	.1090	.1027	.0967	.0912	.0860	.0811
1.50	.1321	.1236	.1157	.1084	.1016	.0953	.0895	.0841	.0790	.0743
1.52	.1249	.1164	.1086	.1014	.0947	.0885	.0827	.0774	.0725	.0679
1.54	.1181	.1097	.1019	.0947	.0881	.0820	.0764	.0712	.0664	.0620
1.56	.1117	.1032	.0955	.0884	.0820	.0760	.0705	.0655	.0608	.0566
1.58	.1055	.0971	.0895	.0825	.0762	.0704	.0650	.0602	.0557	.0515
1.60	.0996	.0913	.0838	.0770	.0708	.0651	.0599	.0552	.0509	.0469
1.62	.0940	.0859	.0785	.0718	.0657	.0602	.0552	.0506	.0465	.0427
1.64	.0887	.0807	.0734	.0668	.0609	.0556	.0507	.0464	.0424	.0388
1.66	.0837	.0757	.0686	.0622	.0565	.0513	.0466	.0424	.0386	.0352
1.68	.0789	.0711	.0641	.0579	.0523	.0473	.0428	.0388	.0352	.0319
1.70	.0744	.0667	.0599	.0538	.0484	.0436	.0393	.0355	.0320	.0289
1.72	.0701	.0625	.0559	.0500	.0448	.0402	.0361	.0324	.0291	.0262
1.74	.0660	.0586	.0522	.0465	.0414	.0370	.0330	.0295	.0264	.0237
1.76	.0621	.0549	.0487	.0431	.0383	.0340	.0303	.0269	.0240	.0214
1.78	.0584	.0514	.0454	.0400	.0354	.0313	.0277	.0245	.0217	.0193
1.80	.0550	.0482	.0423	.0371	.0326	.0287	.0253	.0223	.0197	.0174
1.82	.0517	.0451	.0394	.0344	.0301	.0264	.0231	.0203	.0178	.0157
1.84	.0486	.0422	.0366	.0319	.0278	.0242	.0211	.0185	.0161	.0141
1.86	.0456	.0394	.0341	.0295	.0256	.0222	.0193	.0168	.0146	.0127
1.88	.0429	.0368	.0317	.0273	.0236	.0204	.0176	.0152	.0132	.0114
1.90	.0403	.0344	.0295	.0253	.0217	.0186	.0160	.0138	.0119	.0103
1.92	.0378	.0321	.0274	.0234	.0200	.0171	.0146	.0125	.0107	.0092
1.94	.0355	.0300	.0254	.0216	.0184	.0156	.0133	.0113	.0097	.0083
1.96	.0333	.0280	.0236	.0199	.0169	.0143	.0121	.0103	.0087	.0074
1.98	.0312	.0261	.0219	.0184	.0155	.0131	.0110	.0093	.0078	.0066
2.00	.0293	.0244	.0203	.0170	.0142	.0119	.0100	.0084	.0071	.0059
2.02	.0274	.0227	.0189	.0157	.0131	.0109	.0091	.0076	.0063	.0053
2.04	.0257	.0212	.0175	.0145	.0120	.0099	.0082	.0069	.0057	.0047
2.06	.0241	.0197	.0162	.0133	.0110	.0091	.0075	.0062	.0051	.0042
2.08	.0225	.0184	.0150	.0123	.0101	.0083	.0068	.0056	.0046	.0038
2.10	.0211	.0171	.0139	.0113	.0092	.0075	.0061	.0050	.0041	.0034
2.12	.0197	.0159	.0129	.0104	.0084	.0069	.0056	.0045	.0037	.0030
2.14	.0185	.0148	.0119	.0096	.0077	.0062	.0050	.0041	.0033	.0027
2.16	.0173	.0138	.0110	.0088	.0071	.0057	.0046	.0037	.0030	.0024
2.18	.0162	.0128	.0102	.0081	.0065	.0052	.0041	.0033	.0026	.0021
2.20	.0151	.0119	.0094	.0075	.0059	.0047	.0037	.0030	.0024	.0019
2.22	.0141	.0111	.0087	.0069	.0054	.0043	.0034	.0027	.0021	.0017
2.24	.0132	.0103	.0080	.0063	.0049	.0039	.0030	.0024	.0019	.0015
2.26	.0123	.0096	.0074	.0058	.0045	.0035	.0027	.0022	.0017	.0013
2.28	.0115	.0089	.0069	.0053	.0041	.0032	.0025	.0019	.0015	.0012
2.30	.0107	.0082	.0063	.0049	.0038	.0029	.0022	.0017	.0013	.0010
2.32	.0100	.0076	.0058	.0045	.0034	.0026	.0020	.0016	.0012	.0009
2.34	.0094	.0071	.0054	.0041	.0031	.0024	.0018	.0014	.0011	.0008
2.36	.0087	.0066	.0050	.0038	.0028	.0022	.0016	.0012	.0009	.0007
2.38	.0081	.0061	.0046	.0034	.0026	.0020	.0015	.0011	.0008	.0006
2.40	.0076	.0057	.0042	.0032	.0024	.0018	.0013	.0010	.0007	.0006
2.42	.0071	.0052	.0039	.0029	.0021	.0016	.0012	.0009	.0007	.0005
2.44	.0066	.0049	.0036	.0026	.0020	.0014	.0011	.0008	.0006	.0004
2.46	.0062	.0045	.0033	.0024	.0018	.0013	.0010	.0007	.0005	.0004
2.48	.0057	.0042	.0030	.0022	.0016	.0012	.0009	.0006	.0005	.0003
2.50	.0053	.0039	.0028	.0020	.0015	.0011	.0008	.0006	.0004	.0003
2.52	.0050	.0036	.0026	.0019	.0013	.0010	.0007	.0005	.0004	.0003
2.54	.0046	.0033	.0024	.0017	.0012	.0009	.0006	.0005	.0003	.0002
2.56	.0043	.0031	.0022	.0015	.0011	.0008	.0006	.0004	.0003	.0002
2.58	.0040	.0028	.0020	.0014	.0010	.0007	.0005	.0004	.0003	.0002
2.60	.0037	.0026	.0018	.0013	.0009	.0006	.0005	.0003	.0002	.0002
2.62	.0035	.0024	.0017	.0012	.0008	.0006	.0004	.0003	.0002	.0001
2.64	.0032	.0022	.0016	.0011	.0007	.0005	.0004	.0003	.0002	.0001

Table A.4 *continued*. *p*-values for the chi-square distribution divided by degrees of freedom.

Table A.4 *continued.* *p*-values for the chi-square distribution divided by degrees of freedom.

| | Degrees of freedom | | | | | | | | | |
Ratio	20	21	22	23	24	25	26	27	28	29
1.86	.0111	.0097	.0084	.0074	.0064	.0056	.0049	.0043	.0038	.0033
1.88	.0099	.0086	.0075	.0065	.0056	.0049	.0043	.0037	.0032	.0028
1.90	.0089	.0076	.0066	.0057	.0049	.0043	.0037	.0032	.0028	.0024
1.92	.0079	.0068	.0058	.0050	.0043	.0037	.0032	.0028	.0024	.0021
1.94	.0071	.0060	.0052	.0044	.0038	.0032	.0028	.0024	.0021	.0018
1.96	.0063	.0054	.0046	.0039	.0033	.0028	.0024	.0021	.0018	.0015
1.98	.0056	.0047	.0040	.0034	.0029	.0025	.0021	.0018	.0015	.0013
2.00	.0050	.0042	.0035	.0030	.0025	.0021	.0018	.0015	.0013	.0011
2.02	.0044	.0037	.0031	.0026	.0022	.0018	.0016	.0013	.0011	.0009
2.04	.0040	.0033	.0027	.0023	.0019	.0016	.0013	.0011	.0009	.0008
2.06	.0035	.0029	.0024	.0020	.0017	.0014	.0012	.0010	.0008	.0007
2.08	.0031	.0026	.0021	.0018	.0014	.0012	.0010	.0008	.0007	.0006
2.10	.0028	.0023	.0019	.0015	.0013	.0010	.0009	.0007	.0006	.0005
2.12	.0025	.0020	.0016	.0013	.0011	.0009	.0007	.0006	.0005	.0004
2.14	.0022	.0018	.0014	.0012	.0009	.0008	.0006	.0005	.0004	.0003
2.16	.0019	.0016	.0013	.0010	.0008	.0007	.0005	.0004	.0004	.0003
2.18	.0017	.0014	.0011	.0009	.0007	.0006	.0005	.0004	.0003	.0002
2.20	.0015	.0012	.0010	.0008	.0006	.0005	.0004	.0003	.0003	.0002
2.22	.0013	.0011	.0008	.0007	.0005	.0004	.0003	.0003	.0002	.0002
2.24	.0012	.0009	.0007	.0006	.0005	.0004	.0003	.0002	.0002	.0001
2.26	.0010	.0008	.0006	.0005	.0004	.0003	.0002	.0002	.0002	.0001
2.23	.0009	.0007	.0006	.0004	.0003	.0003	.0002	.0002	.0001	.0001
2.30	.0008	.0006	.0005	.0004	.0003	.0002	.0002	.0001	.0001	.0001
2.32	.0007	.0005	.0004	.0003	.0003	.0002	.0002	.0001	.0001	.0001
2.34	.0006	.0005	.0004	.0003	.0002	.0002	.0001	.0001	.0001	.0001
2.36	.0006	.0004	.0003	.0002	.0002	.0001	.0001	.0001	.0001	.0000
2.38	.0005	.0004	.0003	.0002	.0002	.0001	.0001	.0001	.0001	.0000
2.40	.0004	.0003	.0002	.0002	.0001	.0001	.0001	.0001	.0000	.0000
2.42	.0004	.0003	.0002	.0002	.0001	.0001	.0001	.0001	.0000	.0000
2.44	.0003	.0002	.0002	.0001	.0001	.0001	.0001	.0000	.0000	.0000
2.46	.0003	.0002	.0002	.0001	.0001	.0001	.0000	.0000	.0000	.0000
2.48	.0003	.0002	.0001	.0001	.0001	.0001	.0000	.0000	.0000	.0000
2.50	.0002	.0002	.0001	.0001	.0001	.0000	.0000	.0000	.0000	.0000
2.52	.0002	.0001	.0001	.0001	.0001	.0000	.0000	.0000	.0000	.0000

| | Degrees of Freedom | | | | | | | |
Ratio	30	35	40	45	50	55	60	65
1.30	.1257	.1102	.0968	.0853	.0754	.0667	.0591	.0525
1.31	.1192	.1037	.0906	.0793	.0696	.0612	.0539	.0476
1.32	.1128	.0976	.0846	.0736	.0642	.0561	.0491	.0430
1.33	.1068	.0917	.0790	.0683	.0591	.0513	.0446	.0389
1.34	.1010	.0862	.0737	.0633	.0544	.0469	.0405	.0351
1.35	.0955	.0809	.0687	.0586	.0500	.0429	.0368	.0316
1.36	.0903	.0759	.0640	.0542	.0460	.0391	.0333	.0284
1.37	.0853	.0711	.0596	.0501	.0422	.0356	.0301	.0255
1.38	.0805	.0666	.0554	.0462	.0387	.0324	.0272	.0229
1.39	.0759	.0624	.0515	.0426	.0354	.0295	.0246	.0205
1.40	.0716	.0584	.0478	.0393	.0324	.0267	.0221	.0184
1.41	.0674	.0546	.0444	.0362	.0296	.0243	.0199	.0164
1.42	.0635	.0510	.0411	.0333	.0270	.0220	.0179	.0146
1.43	.0598	.0476	.0381	.0306	.0246	.0199	.0161	.0130
1.44	.0563	.0444	.0353	.0281	.0224	.0180	.0144	.0116
1.45	.0529	.0415	.0326	.0258	.0204	.0162	.0129	.0103
1.46	.0497	.0386	.0302	.0236	.0186	.0146	.0116	.0092
1.47	.0467	.0360	.0279	.0217	.0169	.0132	.0103	.0081
1.48	.0439	.0335	.0257	.0198	.0153	.0119	.0092	.0072
1.49	.0412	.0312	.0237	.0181	.0139	.0107	.0082	.0064
1.50	.0386	.0290	.0219	.0166	.0126	.0096	.0073	.0056
1.51	.0362	.0269	.0202	.0151	.0114	.0086	.0065	.0050
1.52	.0339	.0250	.0185	.0138	.0103	.0077	.0058	.0044
1.53	.0318	.0232	.0171	.0126	.0093	.0069	.0052	.0038
1.54	.0297	.0215	.0157	.0115	.0084	.0062	.0046	.0034

Table A.4 *continued*. p-values for the chi-square distribution divided by degrees of freedom.

					Degrees of freedom					
Ratio	10	11	12	13	14	15	16	17	18	19
2.66	.0030	.0021	.0014	.0010	.0007	.0005	.0003	.0002	.0C02	.0001
2.68	.0028	.0019	.0013	.0009	.0006	.0004	.0003	.0002	.0001	.0001
2.70	.0026	.0018	.0012	.0008	.0006	.0004	.0003	.0002	.0001	.0001
2.72	.0024	.0016	.0011	.0007	.0005	.0003	.0002	.0002	.0001	.0001
2.74	.0023	.0015	.0010	.0007	.0005	.0003	.0002	.0001	.0001	.0001
2.76	.0021	.0014	.0009	.0006	.0004	.0003	.0002	.0001	.0001	.0001
2.78	.0019	.0013	.0009	.0006	.0004	.0002	.0002	.0001	.0001	.0000
2.80	.0018	.0012	.0008	.0005	.0003	.0002	.0001	.0001	.0001	.0000
2.82	.0017	.0011	.0007	.0005	.0003	.0002	.0001	.0001	.0001	.0000
2.84	.0016	.0010	.0007	.0004	.0003	.0002	.0001	.0001	.0001	.0000
2.86	.0014	.0009	.0006	.0004	.0003	.0002	.0001	.0001	.0000	.0000
2.88	.0013	.0009	.0006	.0004	.0002	.0001	.0001	.0001	.0000	.0000
2.90	.0012	.0008	.0005	.0003	.0002	.0001	.0001	.0001	.0000	.0000
2.92	.0012	.0007	.0005	.0003	.0002	.0001	.0001	.0000	.0000	.0000
2.94	.0011	.0007	.0004	.0003	.0002	.0001	.0001	.0000	.0000	.0000
2.96	.0010	.0006	.0004	.0002	.0002	.0001	.0001	.0000	.0000	.0000
2.98	.0009	.0006	.0004	.0002	.0001	.0001	.0001	.0000	.0000	.0000
3.00	.0009	.0005	.0003	.0002	.0001	.0001	.0000	.0000	.0000	.0000
3.02	.0008	.0005	.0003	.0002	.0001	.0001	.0000	.0000	.0000	.0000
3.04	.0007	.0004	.0003	.0002	.0001	.0001	.0000	.0000	.0000	.0000
3.06	.0007	.0004	.0002	.0001	.0001	.0001	.0000	.0000	.0000	.0000
3.08	.0006	.0004	.0002	.0001	.0001	.0000	.0000	.0000	.0000	.0000
3.10	.0006	.0003	.0002	.0001	.0001	.0000	.0000	.0000	.0000	.0000
3.12	.0005	.0003	.0002	.0001	.0001	.0000	.0000	.0000	.0000	.0000
3.14	.0005	.0003	.0002	.0001	.0001	.0000	.0000	.0000	.0000	.0000
3.16	.0005	.0003	.0002	.0001	.0001	.0000	.0000	.0000	.0000	.0000
3.18	.0004	.0002	.0001	.0001	.0000	.0000	.0000	.0000	.0000	.0000
3.20	.0004	.0002	.0001	.0001	.0000	.0000	.0000	.0000	.0000	.0000
3.22	.0004	.0002	.0001	.0001	.0000	.0000	.0000	.0000	.0000	.0000
3.24	.0003	.0002	.0001	.0001	.0000	.0000	.0000	.0000	.0000	.0000
3.26	.0003	.0002	.0001	.0001	.0000	.0000	.0000	.0000	.0000	.0000
3.28	.0003	.0002	.0001	.0001	.0000	.0000	.0000	.0000	.0000	.0000
3.30	.0003	.0002	.0001	.0000	.0000	.0000	.0000	.0000	.0000	.0000
3.32	.0003	.0001	.0001	.0000	.0000	.0000	.0000	.0000	.0000	.0000
3.34	.0002	.0001	.0001	.0000	.0000	.0000	.0000	.0000	.0000	.0000
3.36	.0002	.0001	.0001	.0000	.0000	.0000	.0000	.0000	.0000	.0000
3.38	.0002	.0001	.0001	.0000	.0000	.0000	.0000	.0000	.0000	.0000
3.40	.0002	.0001	.0001	.0000	.0000	.0000	.0000	.0000	.0000	.0000
3.42	.0002	.0001	.0000	.0000	.0000	.0000	.0000	.0000	.0000	.0000
3.44	.0002	.0001	.0000	.0000	.0000	.0000	.0000	.0000	.0000	.0000

					Degrees of freedom					
Ratio	20	21	22	23	24	25	26	27	28	29
1.46	.0839	.0795	.0754	.0715	.0678	.0643	.0611	.0580	.0551	.0523
1.48	.0766	.0723	.0683	.0646	.0611	.0577	.0546	.0517	.0489	.0463
1.50	.0699	.0657	.0619	.0583	.0549	.0517	.0488	.0460	.0434	.0409
1.52	.0636	.0597	.0559	.0525	.0493	.0463	.0435	.0408	.0384	.0361
1.54	.0579	.0541	.0505	.0472	.0442	.0413	.0387	.0362	.0339	.0317
1.56	.0526	.0489	.0456	.0424	.0395	.0368	.0343	.0320	.0299	.0279
1.58	.0478	.0443	.0410	.0381	.0353	.0328	.0305	.0283	.0263	.0245
1.60	.0433	.0400	.0369	.0341	.0315	.0292	.0270	.0250	.0231	.0214
1.62	.0392	.0361	.0332	.0305	.0281	.0259	.0239	.0220	.0203	.0187
1.64	.0355	.0325	.0298	.0273	.0250	.0230	.0211	.0193	.0178	.0163
1.66	.0321	.0293	.0267	.0244	.0222	.0203	.0186	.0170	.0155	.0142
1.68	.0290	.0263	.0239	.0217	.0198	.0180	.0164	.0149	.0136	.0124
1.70	.0261	.0236	.0214	.0193	.0175	.0159	.0144	.0130	.0118	.0107
1.72	.0235	.0212	.0191	.0172	.0155	.0140	.0126	.0114	.0103	.0093
1.74	.0212	.0190	.0170	.0153	.0137	.0123	.0111	.0100	.0090	.0081
1.76	.0191	.0170	.0152	.0136	.0121	.0108	.0097	.0087	.0078	.0070
1.78	.0171	.0152	.0135	.0120	.0107	.0095	.0085	.0076	.0067	.0060
1.80	.0154	.0136	.0120	.0107	.0094	.0084	.0074	.0066	.0058	.0052
1.82	.0138	.0121	.0107	.0094	.0083	.0073	.0065	.0057	.0050	.0045
1.84	.0124	.0108	.0095	.0083	.0073	.0064	.0056	.0050	.0044	.0038

Table A.4 *continued.* *p*-values for the chi-square distribution divided by degrees of freedom.

Ratio	30	35	40	45	50	55	60	65
1.55	.0278	.0200	.0144	.0104	.0076	.0055	.0041	.0030
1.56	.0260	.0185	.0132	.0095	.0068	.0049	.0036	.0026
1.57	.0243	.0171	.0121	.0086	.0062	.0044	.0032	.0023
1.58	.0227	.0159	.0111	.0078	.0055	.0039	.0028	.0020
1.59	.0212	.0147	.0102	.0071	.0050	.0035	.0025	.0017
1.60	.0198	.0136	.0093	.0065	.0045	.0031	.0022	.0015
1.61	.0185	.0125	.0086	.0059	.0040	.0028	.0019	.0013
1.62	.0173	.0116	.0078	.0053	.0036	.0025	.0017	.0012
1.63	.0161	.0107	.0071	.0048	.0032	.0022	.0015	.0010
1.64	.0150	.0099	.0065	.0043	.0029	.0019	.0013	.0009
1.65	.0140	.0091	.0060	.0039	.0026	.0017	.0011	.0008
1.66	.0130	.0084	.0054	.0035	.0023	.0015	.0010	.0007
1.67	.0121	.0077	.0050	.0032	.0021	.0013	.0009	.0006
1.68	.0113	.0071	.0045	.0029	.0019	.0012	.0008	.0005
1.69	.0105	.0065	.0041	.0026	.0017	.0011	.0007	.0004
1.70	.0097	.0060	.0037	.0023	.0015	.0009	.0006	.0004
1.71	.0091	.0055	.0034	.0021	.0013	.0008	.0005	.0003
1.72	.0084	.0051	.0031	.0019	.0012	.0007	.0004	.0003
1.73	.0078	.0047	.0028	.0017	.0010	.0006	.0004	.0002
1.74	.0072	.0043	.0026	.0015	.0009	.0006	.0003	.0002
1.75	.0067	.0039	.0023	.0014	.0008	.0005	.0003	.0002
1.76	.0062	.0036	.0021	.0012	.0007	.0004	.0003	.0002
1.77	.0058	.0033	.0019	.0011	.0006	.0004	.0002	.0001
1.78	.0054	.0030	.0017	.0010	.0006	.0003	.0002	.0001
1.79	.0050	.0028	.0016	.0009	.0005	.0003	.0002	.0001
1.80	.0046	.0025	.0014	.0008	.0004	.0003	.0001	.0001
1.81	.0043	.0023	.0013	.0007	.0004	.0002	.0001	.0001
1.82	.0039	.0021	.0012	.0006	.0004	.0002	.0001	.0001
1.83	.0036	.0020	.0011	.0006	.0003	.0002	.0001	.0001
1.84	.0034	.0018	.0010	.0005	.0003	.0001	.0001	.0000
1.85	.0031	.0016	.0009	.0005	.0002	.0001	.0001	.0000
1.86	.0029	.0015	.0008	.0004	.0002	.0001	.0001	.0000
1.87	.0027	.0014	.0007	.0004	.0002	.0001	.0001	.0000
1.88	.0025	.0012	.0006	.0003	.0002	.0001	.0000	.0000
1.89	.0023	.0011	.0006	.0003	.0001	.0001	.0000	.0000
1.90	.0021	.0010	.0005	.0003	.0001	.0001	.0000	.0000
1.91	.0019	.0009	.0005	.0002	.0001	.0001	.0000	.0000
1.92	.0018	.0009	.0004	.0002	.0001	.0000	.0000	.0000
1.93	.0016	.0008	.0004	.0002	.0001	.0000	.0000	.0000
1.94	.0015	.0007	.0003	.0002	.0001	.0000	.0000	.0000
1.95	.0014	.0006	.0003	.0001	.0001	.0000	.0000	.0000
1.96	.0013	.0006	.0003	.0001	.0001	.0000	.0000	.0000
1.97	.0012	.0005	.0002	.0001	.0001	.0000	.0000	.0000
1.98	.0011	.0005	.0002	.0001	.0000	.0000	.0000	.0000
1.99	.0010	.0004	.0002	.0001	.0000	.0000	.0000	.0000
2.00	.0009	.0004	.0002	.0001	.0000	.0000	.0000	.0000
2.01	.0008	.0004	.0002	.0001	.0000	.0000	.0000	.0000
2.02	.0008	.0003	.0001	.0001	.0000	.0000	.0000	.0000
2.03	.0007	.0003	.0001	.0001	.0000	.0000	.0000	.0000
2.04	.0007	.0003	.0001	.0000	.0000	.0000	.0000	.0000
2.05	.0006	.0002	.0001	.0000	.0000	.0000	.0000	.0000
2.06	.0006	.0002	.0001	.0000	.0000	.0000	.0000	.0000
2.07	.0005	.0002	.0001	.0000	.0000	.0000	.0000	.0000
2.08	.0005	.0002	.0001	.0000	.0000	.0000	.0000	.0000
2.09	.0004	.0002	.0001	.0000	.0000	.0000	.0000	.0000
2.10	.0004	.0002	.0001	.0000	.0000	.0000	.0000	.0000
2.11	.0004	.0001	.0001	.0000	.0000	.0000	.0000	.0000
2.12	.0003	.0001	.0000	.0000	.0000	.0000	.0000	.0000
2.13	.0003	.0001	.0000	.0000	.0000	.0000	.0000	.0000
2.14	.0003	.0001	.0000	.0000	.0000	.0000	.0000	.0000

Table A.4 *continued.* p-values for the chi-square distribution divided by degrees of freedom.

Ratio	\multicolumn{8}{c}{Degrees of freedom}

Ratio	30	35	40	45	50	55	60	65
2.15	.0003	.0001	.0000	.0000	.0000	.0000	.0000	.0000
2.16	.0002	.0001	.0000	.0000	.0000	.0000	.0000	.0000
2.17	.0002	.0001	.0000	.0000	.0000	.0000	.0000	.0000
2.18	.0002	.0001	.0000	.0000	.0000	.0000	.0000	.0000
2.19	.0002	.0001	.0000	.0000	.0000	.0000	.0000	.0000
2.20	.0002	.0001	.0000	.0000	.0000	.0000	.0000	.0000
2.21	.0001	.0000	.0000	.0000	.0000	.0000	.0000	.0000

Ratio	\multicolumn{7}{c}{Degrees of freedom}

Ratio	70	75	80	85	90	95	100
1.24	.0845	.0778	.0718	.0662	.0611	.0565	.0522
1.25	.0769	.0704	.0646	.0593	.0544	.0500	.0460
1.26	.0698	.0636	.0580	.0529	.0483	.0442	.0404
1.27	.0633	.0573	.0520	.0472	.0428	.0389	.0354
1.28	.0573	.0516	.0465	.0420	.0379	.0343	.0310
1.29	.0517	.0463	.0415	.0372	.0334	.0300	.0270
1.30	.0467	.0415	.0370	.0330	.0294	.0263	.0235
1.31	.0420	.0372	.0329	.0292	.0259	.0230	.0204
1.32	.0378	.0332	.0292	.0257	.0227	.0200	.0177
1.33	.0339	.0296	.0259	.0227	.0198	.0174	.0153
1.34	.0304	.0264	.0229	.0199	.0173	.0151	.0132
1.35	.0272	.0234	.0202	.0175	.0151	.0131	.0113
1.36	.0243	.0208	.0178	.0153	.0131	.0113	.0097
1.37	.0217	.0184	.0157	.0134	.0114	.0097	.0083
1.38	.0193	.0163	.0138	.0117	.0099	.0084	.0071
1.39	.0172	.0144	.0121	.0101	.0085	.0072	.0060
1.40	.0153	.0127	.0106	.0088	.0074	.0061	.0051
1.41	.0135	.0112	.0092	.0076	.0063	.0053	.0044
1.42	.0120	.0098	.0081	.0066	.0054	.0045	.0037
1.43	.0106	.0086	.0070	.0057	.0047	.0038	.0031
1.44	.0094	.0075	.0061	.0049	.0040	.0032	.0026
1.45	.0082	.0066	.0053	.0043	.0034	.0027	.0022
1.46	.0073	.0058	.0046	.0037	.0029	.0023	.0019
1.47	.0064	.0050	.0040	.0031	.0025	.0020	.0016
1.48	.0056	.0044	.0034	.0027	.0021	.0017	.0013
1.49	.0049	.0038	.0030	.0023	.0018	.0014	.0011
1.50	.0043	.0033	.0025	.0020	.0015	.0012	.0009
1.51	.0038	.0029	.0022	.0017	.0013	.0010	.0008
1.52	.0033	.0025	.0019	.0014	.0011	.0008	.0006
1.53	.0029	.0022	.0016	.0012	.0009	.0007	.0005
1.54	.0025	.0019	.0014	.0010	.0008	.0006	.0004
1.55	.0022	.0016	.0012	.0009	.0006	.0005	.0004
1.56	.0019	.0014	.0010	.0007	.0005	.0004	.0003
1.57	.0016	.0012	.0009	.0006	.0005	.0003	.0002
1.58	.0014	.0010	.0007	.0005	.0004	.0003	.0002
1.59	.0012	.0009	.0006	.0004	.0003	.0002	.0002
1.60	.0011	.0008	.0005	.0004	.0003	.0002	.0001
1.61	.0009	.0006	.0004	.0003	.0002	.0002	.0001
1.62	.0008	.0005	.0004	.0003	.0002	.0001	.0001
1.63	.0007	.0005	.0003	.0002	.0002	.0001	.0001
1.64	.0006	.0004	.0003	.0002	.0001	.0001	.0001
1.65	.0005	.0003	.0002	.0002	.0001	.0001	.0000
1.66	.0004	.0003	.0002	.0001	.0001	.0001	.0000
1.67	.0004	.0002	.0002	.0001	.0001	.0000	.0000
1.68	.0003	.0002	.0001	.0001	.0001	.0000	.0000
1.69	.0003	.0002	.0001	.0001	.0000	.0000	.0000
1.70	.0002	.0001	.0001	.0001	.0000	.0000	.0000
1.71	.0002	.0001	.0001	.0001	.0000	.0000	.0000
1.72	.0002	.0001	.0001	.0000	.0000	.0000	.0000
1.73	.0001	.0001	.0001	.0000	.0000	.0000	.0000

Table A.5. Critical values (percentiles) for the t distribution.

The table entries are the critical values (percentiles) for the t distribution. The column headed df (degrees of freedom) gives the degrees of freedom for the values in that row. The columns are labeled by "Percent","One-sided" and "Two-sided". Percent is 100 × cumulative distribution function - the table entry is the corresponding percentile. One-sided is the significance level for the one-sided upper critical value - the table entry is the critical value. Two-sided gives the two-sided significance level - the table entry is the corresponding two-sided critical value.

							Percent					
	75	90	95	97.5	99	99.5	99.75	99.9	99.95	99.975	99.99	99.995
							One-sided α					
	.25	.10	.05	.025	.01	.005	.0025	.001	.0005	.00025	.0001	.00005
							Two-sided α					
	.50	.20	.10	.05	.02	.01	.005	.002	.001	.0005	.0002	.0001
df												
1	1.00	3.08	6.31	12.71	31.82	63.66	127.32	318.31	636.62	1273.24	3183.10	6366.20
2	.82	1.89	2.92	4.30	6.96	9.??	14.09	22.33	31.60	44.70	70.70	99.99
3	.76	1.64	2.35	3.18	4.54	5.84	7.45	10.21	12.92	16.33	22.20	28.00
4	.74	1.53	2.13	2.78	3.75	4.60	5.60	7.17	8.61	10.31	13.03	15.54
5	.73	1.48	2.02	2.57	3.37	4.03	4.77	5.89	6.87	7.98	9.68	11.18
6	.72	1.44	1.94	2.45	3.14	3.71	4.32	5.21	5.96	6.79	8.02	9.08
7	.71	1.42	1.90	2.37	3.00	3.50	4.03	4.79	5.41	6.08	7.06	7.88
8	.71	1.40	1.86	2.31	2.90	3.36	3.83	4.50	5.04	5.62	6.44	7.12
9	.70	1.38	1.83	2.26	2.82	3.25	3.69	4.30	4.78	5.29	6.01	6.59
10	.70	1.37	1.81	2.23	2.76	3.17	3.58	4.14	4.59	5.05	5.69	6.21
11	.70	1.36	1.80	2.20	2.72	3.11	3.50	4.03	4.44	4.86	5.45	5.92
12	.70	1.36	1.78	2.18	2.68	3.06	3.43	3.93	4.32	4.72	5.26	5.69
13	.69	1.35	1.77	2.16	2.65	3.01	3.37	3.85	4.22	4.60	5.11	5.51
14	.69	1.35	1.76	2.15	2.63	2.98	3.33	3.79	4.14	4.50	4.99	5.36
15	.69	1.34	1.75	2.13	2.60	2.95	3.29	3.73	4.07	4.42	4.88	5.24
16	.69	1.34	1.75	2.12	2.58	2.92	3.25	3.69	4.02	4.35	4.79	5.13
17	.69	1.33	1.74	2.11	2.57	2.90	3.22	3.65	3.97	4.29	4.71	5.04
18	.69	1.33	1.73	2.10	2.55	2.88	3.20	3.61	3.92	4.23	4.65	4.97
19	.69	1.33	1.73	2.09	2.54	2.86	3.17	3.58	3.88	4.19	4.59	4.90
20	.69	1.33	1.73	2.09	2.53	2.85	3.15	3.55	3.85	4.15	4.54	4.84
21	.69	1.32	1.72	2.08	2.52	2.83	3.14	3.53	3.82	4.11	4.49	4.78
22	.69	1.32	1.72	2.07	2.51	2.82	3.12	3.51	3.79	4.08	4.45	4.74
23	.68	1.32	1.71	2.07	2.50	2.81	3.10	3.49	3.77	4.05	4.42	4.69
24	.68	1.32	1.71	2.06	2.49	2.80	3.09	3.47	3.75	4.02	4.38	4.65
25	.68	1.32	1.71	2.06	2.49	2.79	3.08	3.45	3.73	4.00	4.35	4.62
26	.68	1.32	1.71	2.06	2.48	2.78	3.07	3.44	3.71	3.97	4.32	4.59
27	.68	1.31	1.70	2.05	2.47	2.77	3.06	3.42	3.69	3.95	4.30	4.56
28	.68	1.31	1.70	2.05	2.47	2.76	3.05	3.41	3.67	3.94	4.28	4.53
29	.68	1.31	1.70	2.05	2.46	2.76	3.04	3.40	3.66	3.92	4.25	4.51
30	.68	1.31	1.70	2.04	2.46	2.75	3.03	3.39	3.65	3.90	4.23	4.48
35	.68	1.31	1.69	2.03	2.44	2.72	3.00	3.34	3.59	3.84	4.15	4.39
40	.68	1.30	1.68	2.02	2.42	2.70	2.97	3.31	3.55	3.79	4.09	4.32
45	.68	1.30	1.68	2.01	2.41	2.69	2.95	3.28	3.52	3.75	4.05	4.27
50	.68	1.30	1.68	2.01	2.40	2.68	2.94	3.26	3.50	3.72	4.01	4.23
55	.68	1.30	1.67	2.00	2.40	2.67	2.93	3.25	3.48	3.70	3.99	4.20
60	.68	1.30	1.67	2.00	2.39	2.66	2.91	3.23	3.46	3.68	3.96	4.17
65	.68	1.29	1.67	2.00	2.39	2.65	2.91	3.22	3.45	3.66	3.94	4.15
70	.68	1.29	1.67	1.99	2.38	2.65	2.90	3.21	3.44	3.65	3.93	4.13
75	.68	1.29	1.67	1.99	2.38	2.64	2.89	3.20	3.43	3.64	3.91	4.11
80	.68	1.29	1.66	1.99	2.37	2.64	2.89	3.20	3.42	3.63	3.90	4.10
85	.68	1.29	1.66	1.99	2.37	2.64	2.88	3.19	3.41	3.62	3.89	4.08
90	.68	1.29	1.66	1.99	2.37	2.63	2.88	3.18	3.40	3.61	3.88	4.07
95	.68	1.29	1.66	1.99	2.37	2.63	2.87	3.18	3.40	3.60	3.87	4.06
100	.68	1.29	1.66	1.98	2.36	2.63	2.87	3.17	3.39	3.60	3.86	4.05
200	.68	1.29	1.65	1.97	2.35	2.60	2.84	3.13	3.34	3.54	3.79	3.97
500	.68	1.28	1.65	1.97	2.33	2.59	2.82	3.11	3.31	3.50	3.75	3.92
∞	.67	1.28	1.65	1.96	2.33	2.58	2.81	3.10	3.30	3.49	3.73	3.91

Table A.6. Two-sided p-values for the t distribution.

For each observed value of the t statistic in column one, table entries correspond to the two-sided p-value for the degrees of freedom in the column heading.

t	1	2	3	4	5	6	7	8	9	10	11	12	13	14	15
.5	.705	.667	.651	.643	.638	.635	.632	.631	.629	.628	.627	.626	.625	.625	.624
1.0	.500	.423	.391	.374	.363	.356	.351	.347	.343	.341	.339	.337	.336	.334	.333
1.5	.374	.272	.231	.208	.194	.184	.177	.172	.168	.165	.162	.159	.158	.156	.154
1.6	.356	.251	.208	.185	.170	.161	.154	.148	.144	.141	.138	.136	.134	.132	.130
1.7	.339	.231	.188	.164	.150	.140	.133	.128	.123	.120	.117	.115	.113	.111	.110
1.8	.323	.214	.170	.146	.132	.122	.115	.110	.105	.102	.099	.097	.095	.093	.092
1.9	.308	.198	.154	.130	.116	.106	.099	.094	.090	.087	.084	.082	.080	.078	.077
2.0	.295	.184	.139	.116	.102	.092	.086	.081	.077	.073	.071	.069	.067	.065	.064
2.1	.283	.171	.127	.104	.090	.080	.074	.069	.065	.062	.060	.058	.056	.054	.053
2.2	.272	.159	.115	.093	.079	.070	.064	.059	.055	.052	.050	.048	.046	.045	.044
2.3	.261	.148	.105	.083	.070	.061	.055	.050	.047	.044	.042	.040	.039	.037	.036
2.4	.251	.138	.096	.074	.062	.053	.047	.043	.040	.037	.035	.034	.032	.031	.030
2.5	.242	.130	.088	.067	.054	.047	.041	.037	.034	.031	.030	.028	.027	.025	.025
2.6	.234	.122	.080	.060	.048	.041	.035	.032	.029	.026	.025	.023	.022	.021	.020
2.7	.226	.114	.074	.054	.043	.036	.031	.027	.024	.022	.021	.019	.018	.017	.016
2.8	.218	.107	.068	.049	.038	.031	.027	.023	.021	.019	.017	.016	.015	.014	.013
2.9	.211	.101	.063	.044	.034	.027	.023	.020	.018	.016	.014	.013	.012	.012	.011
3.0	.205	.095	.058	.040	.030	.024	.020	.017	.015	.013	.012	.011	.010	.010	.009
3.1	.199	.090	.053	.036	.027	.021	.017	.015	.013	.011	.010	.009	.008	.008	.007
3.2	.193	.085	.049	.033	.024	.019	.015	.013	.011	.009	.008	.008	.007	.006	.006
3.3	.187	.081	.046	.030	.021	.016	.013	.011	.009	.008	.007	.006	.006	.005	.005
3.4	.182	.077	.042	.027	.019	.014	.011	.009	.008	.007	.006	.005	.005	.004	.004
3.5	.177	.073	.039	.025	.017	.013	.010	.008	.007	.006	.005	.004	.004	.004	.003
3.6	.172	.069	.037	.023	.016	.011	.009	.007	.006	.005	.004	.004	.003	.003	.003
3.7	.168	.066	.034	.021	.014	.010	.008	.006	.005	.004	.004	.003	.003	.002	.002
3.8	.164	.063	.032	.019	.013	.009	.007	.005	.004	.003	.003	.003	.002	.002	.002
3.9	.160	.060	.030	.018	.011	.008	.006	.005	.004	.003	.002	.002	.002	.002	.001
4.0	.156	.057	.028	.016	.010	.007	.005	.004	.003	.003	.002	.002	.002	.001	.001
4.1	.152	.055	.026	.015	.009	.006	.005	.003	.003	.002	.002	.001	.001	.001	.001
4.2	.149	.052	.025	.014	.008	.006	.004	.003	.002	.002	.001	.001	.001	.001	.001
4.3	.145	.050	.023	.013	.008	.005	.004	.003	.002	.002	.001	.001	.001	.001	.001
4.4	.142	.048	.022	.012	.007	.005	.003	.002	.002	.001	.001	.001	.001	.001	.001
4.5	.139	.046	.020	.011	.006	.004	.003	.002	.001	.001	.001	.001	.001	.000	.000
4.6	.136	.044	.019	.010	.006	.004	.002	.002	.001	.001	.001	.001	.000	.000	.000
4.7	.133	.042	.018	.009	.005	.003	.002	.002	.001	.001	.001	.001	.000	.000	.000
4.8	.131	.041	.017	.009	.005	.003	.002	.001	.001	.001	.001	.000	.000	.000	.000
4.9	.128	.039	.016	.008	.004	.003	.002	.001	.001	.001	.000	.000	.000	.000	.000
5.0	.126	.038	.015	.007	.004	.002	.002	.001	.001	.001	.000	.000	.000	.000	.000
5.2	.121	.035	.014	.007	.003	.002	.001	.001	.001	.000	.000	.000	.000	.000	.000
5.4	.117	.033	.012	.006	.003	.002	.001	.001	.000	.000	.000	.000	.000	.000	.000
5.6	.112	.030	.011	.005	.003	.001	.001	.001	.000	.000	.000	.000	.000	.000	.000
5.8	.109	.028	.010	.004	.002	.001	.001	.000	.000	.000	.000	.000	.000	.000	.000
6.0	.105	.027	.009	.004	.002	.001	.001	.000	.000	.000	.000	.000	.000	.000	.000
6.2	.102	.025	.008	.003	.002	.001	.000	.000	.000	.000	.000	.000	.000	.000	.000
6.4	.099	.024	.008	.003	.001	.001	.000	.000	.000	.000	.000	.000	.000	.000	.000
6.6	.096	.022	.007	.003	.001	.001	.000	.000	.000	.000	.000	.000	.000	.000	.000
6.8	.093	.021	.007	.002	.001	.000	.000	.000	.000	.000	.000	.000	.000	.000	.000
7.0	.090	.020	.006	.002	.001	.000	.000	.000	.000	.000	.000	.000	.000	.000	.000
7.2	.088	.019	.006	.002	.001	.000	.000	.000	.000	.000	.000	.000	.000	.000	.000
7.4	.086	.018	.005	.002	.001	.000	.000	.000	.000	.000	.000	.000	.000	.000	.000
7.6	.083	.017	.005	.002	.001	.000	.000	.000	.000	.000	.000	.000	.000	.000	.000
7.8	.081	.016	.004	.001	.001	.000	.000	.000	.000	.000	.000	.000	.000	.000	.000
8.0	.079	.015	.004	.001	.000	.000	.000	.000	.000	.000	.000	.000	.000	.000	.000
8.5	.075	.014	.003	.001	.000	.000	.000	.000	.000	.000	.000	.000	.000	.000	.000
9.0	.070	.012	.003	.001	.000	.000	.000	.000	.000	.000	.000	.000	.000	.000	.000
9.5	.067	.011	.002	.001	.000	.000	.000	.000	.000	.000	.000	.000	.000	.000	.000
10.0	.063	.010	.002	.001	.000	.000	.000	.000	.000	.000	.000	.000	.000	.000	.000
10.5	.060	.009	.002	.000	.000	.000	.000	.000	.000	.000	.000	.000	.000	.000	.000
11.0	.058	.008	.002	.000	.000	.000	.000	.000	.000	.000	.000	.000	.000	.000	.000
11.5	.055	.007	.001	.000	.000	.000	.000	.000	.000	.000	.000	.000	.000	.000	.000
12.0	.053	.007	.001	.000	.000	.000	.000	.000	.000	.000	.000	.000	.000	.000	.000
12.5	.051	.006	.001	.000	.000	.000	.000	.000	.000	.000	.000	.000	.000	.000	.000
13.0	.049	.006	.001	.000	.000	.000	.000	.000	.000	.000	.000	.000	.000	.000	.000

						Degrees of freedom									
t	16	17	18	19	20	21	22	23	24	25	26	27	28	29	30
.50	.624	.623	.623	.623	.623	.622	.622	.622	.622	.621	.621	.621	.621	.621	.621
1.00	.332	.331	.331	.330	.329	.329	.328	.328	.327	.327	.327	.326	.326	.326	.325
1.50	.153	.152	.151	.150	.149	.148	.148	.147	.147	.146	.146	.145	.145	.144	.144
1.60	.129	.128	.127	.126	.125	.125	.124	.123	.123	.122	.122	.121	.121	.120	.120
1.65	.118	.117	.116	.115	.115	.114	.113	.113	.112	.111	.111	.111	.110	.110	.109
1.70	.108	.107	.106	.105	.105	.104	.103	.103	.102	.102	.101	.101	.100	.100	.099
1.75	.099	.098	.097	.096	.095	.095	.094	.093	.093	.092	.092	.091	.091	.091	.090
1.80	.091	.090	.089	.088	.087	.086	.086	.085	.084	.084	.083	.083	.083	.082	.082
1.85	.083	.082	.081	.080	.079	.078	.078	.077	.077	.076	.076	.075	.075	.075	.074
1.90	.076	.075	.074	.073	.072	.071	.071	.070	.070	.069	.069	.068	.068	.067	.067
1.95	.069	.068	.067	.066	.065	.065	.064	.063	.063	.062	.062	.062	.061	.061	.061
2.00	.063	.062	.061	.060	.059	.059	.058	.057	.057	.056	.056	.056	.055	.055	.055
2.05	.057	.056	.055	.054	.054	.053	.052	.052	.051	.051	.051	.050	.050	.050	.049
2.10	.052	.051	.050	.049	.049	.048	.047	.047	.046	.046	.046	.045	.045	.045	.044
2.15	.047	.046	.045	.045	.044	.043	.043	.042	.042	.041	.041	.041	.040	.040	.040
2.20	.043	.042	.041	.040	.040	.039	.039	.038	.038	.037	.037	.037	.036	.036	.036
2.25	.039	.038	.037	.036	.036	.035	.035	.034	.034	.033	.033	.033	.032	.032	.032
2.30	.035	.034	.034	.033	.032	.032	.031	.031	.030	.030	.030	.029	.029	.029	.029
2.35	.032	.031	.030	.030	.029	.029	.028	.028	.027	.027	.027	.026	.026	.026	.026
2.40	.029	.028	.027	.027	.026	.026	.025	.025	.025	.024	.024	.024	.023	.023	.023
2.45	.026	.025	.025	.024	.024	.023	.023	.022	.022	.022	.021	.021	.021	.021	.020
2.50	.024	.023	.022	.022	.021	.021	.020	.020	.020	.019	.019	.019	.019	.018	.018
2.55	.021	.021	.020	.020	.019	.019	.018	.018	.018	.017	.017	.017	.017	.016	.016
2.60	.019	.019	.018	.018	.017	.017	.016	.016	.016	.015	.015	.015	.015	.015	.014
2.65	.017	.017	.016	.016	.015	.015	.015	.014	.014	.014	.014	.013	.013	.013	.013
2.70	.016	.015	.015	.014	.014	.013	.013	.013	.013	.012	.012	.012	.012	.011	.011
2.75	.014	.014	.013	.013	.012	.012	.012	.011	.011	.011	.011	.011	.010	.010	.010
2.80	.013	.012	.012	.011	.011	.011	.010	.010	.010	.010	.010	.009	.009	.009	.009
2.85	.012	.011	.011	.010	.010	.010	.009	.009	.009	.009	.008	.008	.008	.008	.008
2.90	.010	.010	.010	.009	.009	.009	.008	.008	.008	.008	.007	.007	.007	.007	.007
2.95	.009	.009	.009	.008	.008	.008	.007	.007	.007	.007	.007	.006	.006	.006	.006
3.00	.008	.008	.008	.007	.007	.007	.007	.006	.006	.006	.006	.006	.006	.005	.005
3.05	.008	.007	.007	.007	.006	.006	.006	.006	.006	.005	.005	.005	.005	.005	.005
3.10	.007	.007	.006	.006	.006	.005	.005	.005	.005	.005	.005	.004	.004	.004	.004
3.15	.006	.006	.006	.005	.005	.005	.005	.004	.004	.004	.004	.004	.004	.004	.004
3.20	.006	.005	.005	.005	.004	.004	.004	.004	.004	.004	.004	.003	.003	.003	.003
3.25	.005	.005	.004	.004	.004	.004	.004	.004	.003	.003	.003	.003	.003	.003	.003
3.30	.005	.004	.004	.004	.004	.003	.003	.003	.003	.003	.003	.003	.003	.003	.002
3.35	.004	.004	.004	.003	.003	.003	.003	.003	.003	.003	.002	.002	.002	.002	.002
3.40	.004	.003	.003	.003	.003	.003	.003	.002	.002	.002	.002	.002	.002	.002	.002
3.45	.003	.003	.003	.003	.003	.002	.002	.002	.002	.002	.002	.002	.002	.002	.002
3.50	.003	.003	.003	.002	.002	.002	.002	.002	.002	.002	.002	.002	.002	.002	.001
3.60	.002	.002	.002	.002	.002	.002	.002	.002	.001	.001	.001	.001	.001	.001	.001
3.70	.002	.002	.002	.002	.001	.001	.001	.001	.001	.001	.001	.001	.001	.001	.001
3.80	.002	.001	.001	.001	.001	.001	.001	.001	.001	.001	.001	.001	.001	.001	.001
3.90	.001	.001	.001	.001	.001	.001	.001	.001	.001	.001	.001	.001	.001	.001	.001
4.00	.001	.001	.001	.001	.001	.001	.001	.001	.001	.000	.000	.000	.000	.000	.000
4.10	.001	.001	.001	.001	.001	.001	.000	.000	.000	.000	.000	.000	.000	.000	.000
4.20	.001	.001	.001	.000	.000	.000	.000	.000	.000	.000	.000	.000	.000	.000	.000
4.30	.001	.000	.000	.000	.000	.000	.000	.000	.000	.000	.000	.000	.000	.000	.000
4.40	.000	.000	.000	.000	.000	.000	.000	.000	.000	.000	.000	.000	.000	.000	.000

Table A.6 continued. Two-sided p-values for the t distribution.

t	35	40	45	50	55	60	65	70	80	90	100	200	∞
						Degrees of freedom							
.50	.620	.620	.620	.619	.619	.619	.619	.619	.618	.618	.618	.618	.617
1.00	.324	.323	.323	.322	.322	.321	.321	.321	.320	.320	.320	.319	.318
1.60	.119	.117	.117	.116	.115	.115	.114	.114	.114	.113	.113	.111	.110
1.65	.108	.107	.106	.105	.105	.104	.104	.103	.103	.102	.102	.101	.099
1.70	.098	.097	.096	.095	.095	.094	.094	.094	.093	.093	.092	.091	.089
1.75	.089	.088	.087	.086	.086	.085	.085	.085	.084	.084	.083	.082	.080
1.80	.080	.079	.079	.078	.077	.077	.077	.076	.076	.075	.075	.073	.072
1.85	.073	.072	.071	.070	.070	.069	.069	.069	.068	.068	.067	.066	.065
1.90	.066	.065	.064	.063	.063	.062	.062	.062	.061	.061	.060	.059	.058
1.95	.059	.058	.057	.057	.056	.056	.055	.055	.055	.054	.054	.053	.051
1.98	.056	.055	.054	.053	.053	.052	.052	.052	.051	.051	.050	.049	.048
2.00	.053	.052	.052	.051	.050	.050	.050	.049	.049	.049	.048	.047	.046
2.02	.051	.050	.049	.049	.048	.048	.048	.047	.047	.046	.046	.045	.044
2.04	.049	.048	.047	.047	.046	.046	.045	.045	.045	.044	.044	.043	.042
2.06	.047	.046	.045	.045	.044	.044	.043	.043	.043	.042	.042	.041	.040
2.08	.045	.044	.043	.043	.042	.042	.041	.041	.041	.040	.040	.039	.038
2.10	.043	.042	.041	.041	.040	.040	.040	.039	.039	.039	.038	.037	.036
2.12	.041	.040	.040	.039	.039	.038	.038	.038	.037	.037	.036	.035	.034
2.14	.039	.039	.038	.037	.037	.036	.036	.036	.035	.035	.035	.034	.033
2.16	.038	.037	.036	.036	.035	.035	.034	.034	.034	.033	.033	.032	.031
2.18	.036	.035	.035	.034	.034	.033	.033	.033	.032	.032	.032	.030	.029
2.20	.035	.034	.033	.032	.032	.032	.031	.031	.031	.030	.030	.029	.028
2.22	.033	.032	.031	.031	.031	.030	.030	.030	.029	.029	.029	.028	.027
2.24	.032	.031	.030	.030	.029	.029	.029	.028	.028	.028	.027	.026	.025
2.26	.030	.029	.029	.028	.028	.027	.027	.027	.027	.026	.026	.025	.024
2.28	.029	.028	.027	.027	.027	.026	.026	.026	.025	.025	.025	.024	.023
2.30	.028	.027	.026	.026	.025	.025	.025	.024	.024	.024	.024	.022	.022
2.32	.026	.026	.025	.024	.024	.024	.023	.023	.023	.023	.022	.021	.021
2.34	.025	.024	.024	.023	.023	.023	.022	.022	.022	.021	.021	.020	.019
2.36	.024	.023	.023	.022	.022	.022	.021	.021	.021	.020	.020	.019	.018
2.38	.023	.022	.022	.021	.021	.021	.020	.020	.020	.019	.019	.018	.017
2.40	.022	.021	.021	.020	.020	.020	.019	.019	.019	.018	.018	.017	.017
2.42	.021	.020	.020	.019	.019	.019	.018	.018	.018	.018	.017	.016	.016
2.44	.020	.019	.019	.018	.018	.018	.017	.017	.017	.017	.016	.016	.015
2.46	.019	.018	.018	.017	.017	.017	.017	.016	.016	.016	.016	.015	.014
2.48	.018	.017	.017	.017	.016	.016	.016	.016	.015	.015	.015	.014	.013
2.50	.017	.017	.016	.016	.015	.015	.015	.015	.014	.014	.014	.013	.013
2.45	.019	.019	.018	.018	.017	.017	.017	.017	.016	.016	.016	.015	.014
2.50	.017	.017	.016	.016	.015	.015	.015	.015	.014	.014	.014	.013	.013
2.55	.015	.015	.014	.014	.014	.013	.013	.013	.013	.013	.012	.012	.011
2.60	.014	.013	.013	.012	.012	.012	.012	.011	.011	.011	.011	.010	.009
2.65	.012	.011	.011	.011	.010	.010	.010	.010	.010	.010	.009	.009	.008
2.70	.011	.010	.010	.009	.009	.009	.009	.009	.008	.008	.008	.008	.007
2.75	.009	.009	.009	.008	.008	.008	.008	.008	.007	.007	.007	.007	.006
2.80	.008	.008	.008	.007	.007	.007	.007	.007	.006	.006	.006	.006	.005
2.85	.007	.007	.007	.006	.006	.006	.006	.006	.006	.005	.005	.005	.004
2.90	.006	.006	.006	.006	.005	.005	.005	.005	.005	.005	.005	.004	.004
2.95	.006	.005	.005	.005	.005	.005	.004	.004	.004	.004	.004	.004	.003
3.00	.005	.005	.004	.004	.004	.004	.004	.004	.004	.003	.003	.003	.003
3.05	.004	.004	.004	.004	.004	.003	.003	.003	.003	.003	.003	.003	.002
3.10	.004	.004	.003	.003	.003	.003	.003	.003	.003	.003	.003	.002	.002
3.15	.003	.003	.003	.003	.003	.003	.002	.002	.002	.002	.002	.002	.002
3.20	.003	.003	.003	.002	.002	.002	.002	.002	.002	.002	.002	.002	.001
3.25	.003	.002	.002	.002	.002	.002	.002	.002	.002	.002	.002	.001	.001
3.30	.002	.002	.002	.002	.002	.002	.002	.002	.001	.001	.001	.001	.001
3.35	.002	.002	.002	.002	.001	.001	.001	.001	.001	.001	.001	.001	.001
3.40	.002	.002	.001	.001	.001	.001	.001	.001	.001	.001	.001	.001	.001
3.45	.001	.001	.001	.001	.001	.001	.001	.001	.001	.001	.001	.001	.001
3.50	.001	.001	.001	.001	.001	.001	.001	.001	.001	.001	.001	.001	.000
3.60	.001	.001	.001	.001	.001	.001	.001	.001	.001	.001	.000	.000	.000
3.70	.001	.001	.001	.001	.000	.000	.000	.000	.000	.000	.000	.000	.000

Table A.7a. Critical values (percentiles) for the F distribution.

Upper one-sided 0.10 significance levels; two-sided 0.20 significance levels; 90 percent percentiles.

Tabulated are critical values for the F distribution. The column headings give the numerator degrees of freedom and the row headings the denominator degrees of freedom. Lower one-sided critical values may be found from these tables by reversing the degrees of freedom and using the reciprocal of the tabled value at the same significance level (100 minus the percent for the percentile).

	Numerator degrees of freedom																		
	1	2	3	4	5	6	7	8	9	10	11	12	13	14	15	16	17	18	19
1	39.86	49.50	53.59	55.83	57.24	58.20	58.91	59.44	59.86	60.19	60.71	61.22	61.74	62.00	62.26	62.53	62.79	63.06	63.33
2	8.53	9.00	9.16	9.24	9.29	9.33	9.35	9.37	9.38	9.39	9.41	9.42	9.44	9.45	9.46	9.47	9.47	9.48	9.49
3	5.54	5.46	5.39	5.34	5.31	5.28	5.27	5.25	5.24	5.23	5.22	5.20	5.18	5.18	5.17	5.16	5.15	5.14	5.13
4	4.54	4.32	4.19	4.11	4.05	4.01	3.98	3.95	3.94	3.92	3.90	3.87	3.84	3.83	3.82	3.80	3.79	3.78	3.76
5	4.06	3.78	3.62	3.52	3.45	3.40	3.37	3.34	3.32	3.30	3.27	3.24	3.21	3.19	3.17	3.16	3.14	3.12	3.10
6	3.78	3.46	3.29	3.18	3.11	3.05	3.01	2.98	2.96	2.94	2.90	2.87	2.84	2.82	2.80	2.78	2.76	2.74	2.72
7	3.59	3.26	3.07	2.96	2.88	2.83	2.78	2.75	2.72	2.70	2.67	2.63	2.59	2.58	2.56	2.54	2.51	2.49	2.47
8	3.46	3.11	2.92	2.81	2.73	2.67	2.62	2.59	2.56	2.54	2.50	2.46	2.42	2.40	2.38	2.36	2.34	2.32	2.29
9	3.36	3.01	2.81	2.69	2.61	2.55	2.51	2.47	2.44	2.42	2.38	2.34	2.30	2.28	2.25	2.23	2.21	2.18	2.16
10	3.29	2.92	2.73	2.61	2.52	2.46	2.41	2.38	2.35	2.32	2.28	2.24	2.20	2.18	2.16	2.13	2.11	2.08	2.06
11	3.23	2.86	2.66	2.54	2.45	2.39	2.34	2.30	2.27	2.25	2.21	2.17	2.12	2.10	2.08	2.05	2.03	2.00	1.97
12	3.18	2.81	2.61	2.48	2.39	2.33	2.28	2.24	2.21	2.19	2.15	2.10	2.06	2.04	2.01	1.99	1.96	1.93	1.90
13	3.14	2.76	2.56	2.43	2.35	2.28	2.23	2.20	2.16	2.14	2.10	2.05	2.01	1.98	1.96	1.93	1.90	1.88	1.85
14	3.10	2.73	2.52	2.39	2.31	2.24	2.19	2.15	2.12	2.10	2.05	2.01	1.96	1.94	1.91	1.89	1.86	1.83	1.80
15	3.07	2.70	2.49	2.36	2.27	2.21	2.16	2.12	2.09	2.06	2.02	1.97	1.92	1.90	1.87	1.85	1.82	1.79	1.76
16	3.05	2.67	2.46	2.33	2.24	2.18	2.13	2.09	2.06	2.03	1.99	1.94	1.89	1.87	1.84	1.81	1.78	1.75	1.72
17	3.03	2.64	2.44	2.31	2.22	2.15	2.10	2.06	2.03	2.00	1.96	1.91	1.86	1.84	1.81	1.78	1.75	1.72	1.69
18	3.01	2.62	2.42	2.29	2.20	2.13	2.08	2.04	2.00	1.98	1.93	1.89	1.84	1.81	1.78	1.75	1.72	1.69	1.66
19	2.99	2.61	2.40	2.27	2.18	2.11	2.06	2.02	1.98	1.96	1.91	1.86	1.81	1.79	1.76	1.73	1.70	1.67	1.63
20	2.97	2.59	2.38	2.25	2.16	2.09	2.04	2.00	1.96	1.94	1.89	1.84	1.79	1.77	1.74	1.71	1.68	1.64	1.61
21	2.96	2.57	2.36	2.23	2.14	2.08	2.02	1.98	1.95	1.92	1.87	1.83	1.78	1.75	1.72	1.69	1.66	1.62	1.59
22	2.95	2.56	2.35	2.22	2.13	2.06	2.01	1.97	1.93	1.90	1.86	1.81	1.76	1.73	1.70	1.67	1.64	1.60	1.57
23	2.94	2.55	2.34	2.21	2.11	2.05	1.99	1.95	1.92	1.89	1.84	1.80	1.74	1.72	1.69	1.66	1.62	1.59	1.55
24	2.93	2.54	2.33	2.19	2.10	2.04	1.98	1.94	1.91	1.88	1.83	1.78	1.73	1.70	1.67	1.64	1.61	1.57	1.53
25	2.92	2.53	2.32	2.18	2.09	2.02	1.97	1.93	1.89	1.87	1.82	1.77	1.72	1.69	1.66	1.63	1.59	1.56	1.52
26	2.91	2.52	2.31	2.17	2.08	2.01	1.96	1.92	1.88	1.86	1.81	1.76	1.71	1.68	1.65	1.61	1.58	1.54	1.50
27	2.90	2.51	2.30	2.17	2.07	2.00	1.95	1.91	1.87	1.85	1.80	1.75	1.70	1.67	1.64	1.60	1.57	1.53	1.49
28	2.89	2.50	2.29	2.16	2.06	2.00	1.94	1.90	1.87	1.84	1.79	1.74	1.69	1.66	1.63	1.59	1.56	1.52	1.48
29	2.89	2.50	2.28	2.15	2.06	1.99	1.93	1.89	1.86	1.83	1.78	1.73	1.68	1.65	1.62	1.58	1.55	1.51	1.47
30	2.88	2.49	2.28	2.14	2.05	1.98	1.93	1.88	1.85	1.82	1.77	1.72	1.67	1.64	1.61	1.57	1.54	1.50	1.46
40	2.84	2.44	2.23	2.09	2.00	1.93	1.87	1.83	1.79	1.76	1.71	1.66	1.61	1.57	1.54	1.51	1.47	1.42	1.38
60	2.79	2.39	2.18	2.04	1.95	1.87	1.82	1.77	1.74	1.71	1.66	1.60	1.54	1.51	1.48	1.44	1.40	4.35	1.29
120	2.75	2.35	2.13	1.99	1.90	1.82	1.77	1.72	1.68	1.65	1.60	1.55	1.48	1.45	1.41	1.37	1.32	1.26	1.19

926

Table A.7b. Critical values (percentiles) for the F distribution.
Upper one-sided 0.05 significance levels; two-sided 0.10 significance levels; 95 percent percentiles.

	Numerator degrees of freedom																		
	1	2	3	4	5	6	7	8	9	10	11	12	13	14	15	16	17	18	19
1	161.4	199.5	215.7	224.6	230.2	234.0	236.8	238.9	240.5	241.9	243.9	245.9	248.0	249.1	250.1	251.1	252.2	253.3	254.3
2	18.51	19.00	19.16	19.25	19.30	19.33	19.35	19.37	19.38	19.40	19.41	19.43	19.45	19.45	19.46	19.47	19.48	19.49	19.50
3	10.13	9.55	9.28	9.12	9.01	8.94	8.89	8.85	8.81	8.79	8.74	8.70	8.66	8.64	8.62	8.59	8.57	8.55	8.53
4	7.71	6.94	6.59	6.39	6.26	6.16	6.09	6.04	6.00	5.96	5.91	5.86	5.80	5.77	5.75	5.72	5.69	5.66	5.63
5	6.61	5.79	5.41	5.19	5.05	4.95	4.88	4.82	4.77	4.74	4.68	4.62	4.56	4.53	4.50	4.46	4.43	4.40	4.36
6	5.99	5.14	4.76	4.53	4.39	4.28	4.21	4.15	4.10	4.06	4.00	3.94	3.87	3.84	3.81	3.77	3.74	3.70	3.67
7	5.59	4.74	4.35	4.12	3.97	3.87	3.79	3.73	3.68	3.64	3.57	3.51	3.44	3.41	3.38	3.34	3.30	3.27	3.23
8	5.32	4.46	4.07	3.84	3.69	3.58	3.50	3.44	3.39	3.35	3.28	3.22	3.15	3.12	3.08	3.04	3.01	2.97	2.93
9	5.12	4.26	3.86	3.63	3.48	3.37	3.29	3.23	3.18	3.14	3.07	3.01	2.94	2.90	2.86	2.83	2.79	2.75	2.71
10	4.96	4.10	3.71	3.48	3.33	3.22	3.14	3.07	3.02	2.98	2.91	2.85	2.77	2.74	2.70	2.66	2.62	2.58	2.54
11	4.84	3.98	3.59	3.36	3.20	3.09	3.01	2.95	2.90	2.85	2.79	2.72	2.65	2.61	2.57	2.53	2.49	2.45	2.40
12	4.75	3.89	3.49	3.26	3.11	3.00	2.91	2.85	2.80	2.75	2.69	2.62	2.54	2.51	2.47	2.43	2.38	2.34	2.30
13	4.67	3.81	3.41	3.18	3.03	2.92	2.83	2.77	2.71	2.67	2.60	2.53	2.46	2.42	2.38	2.34	2.30	2.25	2.21
14	4.60	3.74	3.34	3.11	2.96	2.85	2.76	2.70	2.65	2.60	2.53	2.46	2.39	2.35	2.31	2.27	2.22	2.18	2.13
15	4.54	3.68	3.29	3.06	2.90	2.79	2.71	2.64	2.59	2.54	2.48	2.40	2.33	2.29	2.25	2.20	2.16	2.11	2.07
16	4.49	3.63	3.24	3.01	2.85	2.74	2.66	2.59	2.54	2.49	2.42	2.35	2.28	2.24	2.19	2.15	2.11	2.06	2.01
17	4.45	3.59	3.20	2.96	2.81	2.70	2.61	2.55	2.49	2.45	2.38	2.31	2.23	2.19	2.15	2.10	2.06	2.01	1.96
18	4.41	3.55	3.16	2.93	2.77	2.66	2.58	2.51	2.46	2.41	2.34	2.27	2.19	2.15	2.11	2.06	2.02	1.97	1.92
19	4.38	3.52	3.13	2.90	2.74	2.63	2.54	2.48	2.42	2.38	2.31	2.23	2.16	2.11	2.07	2.03	1.98	1.93	1.88
20	4.35	3.49	3.10	2.87	2.71	2.60	2.51	2.45	2.39	2.35	2.28	2.20	2.12	2.08	2.04	1.99	1.95	1.90	1.84
21	4.32	3.47	3.07	2.84	2.68	2.57	2.49	2.42	2.37	2.32	2.25	2.18	2.10	2.05	2.01	1.96	1.92	1.87	1.81
22	4.30	3.44	3.05	2.82	2.66	2.55	2.46	2.40	2.34	2.30	2.23	2.15	2.07	2.03	1.98	1.94	1.89	1.84	1.78
23	4.28	3.42	3.03	2.80	2.64	2.53	2.44	2.37	2.32	2.27	2.20	2.13	2.05	2.01	1.96	1.91	1.86	1.81	1.76
24	4.26	3.40	3.01	2.78	2.62	2.51	2.42	2.36	2.30	2.25	2.18	2.11	2.03	1.98	1.94	1.89	1.84	1.79	1.73
25	4.24	3.39	2.99	2.76	2.60	2.49	2.40	2.34	2.28	2.24	2.16	2.09	2.01	1.96	1.92	1.87	1.82	1.77	1.71
26	4.23	3.37	2.98	2.74	2.59	2.47	2.39	2.32	2.27	2.22	2.15	2.07	1.99	1.95	1.90	1.85	1.80	1.75	1.69
27	4.21	3.35	2.96	2.73	2.57	2.46	2.37	2.31	2.25	2.20	2.13	2.06	1.97	1.93	1.88	1.84	1.79	1.73	1.67
28	4.20	3.34	2.95	2.71	2.56	2.45	2.36	2.29	2.24	2.19	2.12	2.04	1.96	1.91	1.87	1.82	1.77	1.71	1.65
29	4.18	3.33	2.93	2.70	2.55	2.43	2.35	2.28	2.22	2.18	2.10	2.03	1.94	1.90	1.85	1.81	1.75	1.70	1.64
30	4.17	3.32	2.92	2.69	2.53	2.42	2.33	2.27	2.21	2.16	2.09	2.01	1.93	1.89	1.84	1.79	1.74	1.68	1.62
40	4.08	3.23	2.84	2.61	2.45	2.34	2.25	2.18	2.12	2.08	2.00	1.92	1.84	1.79	1.74	1.69	1.64	1.58	1.51
60	4.00	3.15	2.76	2.53	2.37	2.25	2.17	2.10	2.04	1.99	1.92	1.84	1.75	1.70	1.65	1.59	1.53	1.47	1.39
120	3.92	3.07	2.68	2.45	2.29	2.17	2.09	2.02	1.96	1.91	1.83	1.75	1.66	1.61	1.55	1.50	1.43	1.35	1.25
∞	3.84	3.00	2.60	2.37	2.21	2.10	2.01	1.94	1.88	1.83	1.75	1.67	1.57	1.52	1.46	1.39	1.32	1.22	1.00

Table A.7c. Critical values (percentiles) for the F distribution.
Upper one-sided 0.025 significance levels; two-sided 0.05 significance levels; 97.5 percent percentiles.

	Numerator degrees of freedom																		
	1	2	3	4	5	6	7	8	9	10	11	12	13	14	15	16	17	18	19
1	647.8	799.5	864.2	899.6	921.8	937.1	948.2	956.7	963.3	968.6	976.7	984.9	993.1	997.2	1001	1006	1010	1014	1018
2	38.51	39.00	39.17	39.25	39.30	39.33	39.36	39.37	39.39	39.40	39.41	39.43	39.45	39.46	39.46	39.47	39.48	39.49	39.50
3	17.44	16.04	15.44	15.10	14.88	14.73	14.62	14.54	14.47	14.42	14.34	14.25	14.17	14.12	14.08	14.04	13.99	13.95	13.90
4	12.22	10.65	9.98	9.60	9.36	9.20	9.07	8.98	8.90	8.84	8.75	8.66	8.56	8.51	8.46	8.41	8.36	8.31	8.26
5	10.01	8.43	7.76	7.39	7.15	6.98	6.85	6.76	6.68	6.62	6.52	6.43	6.33	6.28	6.23	6.18	6.12	6.07	6.02
6	8.81	7.26	6.60	6.23	5.99	5.82	5.70	5.60	5.52	5.46	5.37	5.27	5.17	5.12	5.07	5.01	4.96	4.90	4.85
7	8.07	6.54	5.89	5.52	5.29	5.12	4.99	4.90	4.82	4.76	4.67	4.57	4.47	4.42	4.36	4.31	4.25	4.20	4.14
8	7.57	6.06	5.42	5.05	4.82	4.65	4.53	4.43	4.36	4.30	4.20	4.10	4.00	3.95	3.89	3.84	3.78	3.73	3.67
9	7.21	5.71	5.08	4.72	4.48	4.32	4.20	4.10	4.03	3.96	3.87	3.77	3.67	3.61	3.56	3.51	3.45	3.39	3.33
10	6.94	5.46	4.83	4.47	4.24	4.07	3.95	3.85	3.78	3.72	3.62	3.52	3.42	3.37	3.31	3.26	3.20	3.14	3.08
11	6.72	5.26	4.63	4.28	4.04	3.88	3.76	3.66	3.59	3.53	3.43	3.33	3.23	3.17	3.12	3.06	3.00	2.94	2.88
12	6.55	5.10	4.47	4.12	3.89	3.73	3.61	3.51	3.44	3.37	3.28	3.18	3.07	3.02	2.96	2.91	2.85	2.79	2.72
13	6.41	4.97	4.35	4.00	3.77	3.60	3.48	3.39	3.31	3.25	3.15	3.05	2.95	2.89	2.84	2.78	2.72	2.66	2.60
14	6.30	4.86	4.24	3.89	3.66	3.50	3.38	3.29	3.21	3.15	3.05	2.95	2.84	2.79	2.73	2.67	2.61	2.55	2.49
15	6.20	4.77	4.15	3.80	3.58	3.41	3.29	3.20	3.12	3.06	2.96	2.86	2.76	2.70	2.64	2.59	2.52	2.46	2.40
16	6.12	4.69	4.08	3.73	3.50	3.34	3.22	3.12	3.05	2.99	2.89	2.79	2.68	2.63	2.57	2.51	2.45	2.38	2.32
17	6.04	4.62	4.01	3.66	3.44	3.28	3.16	3.06	2.98	2.92	2.82	2.72	2.62	2.56	2.50	2.44	2.38	2.32	2.25
18	5.98	4.56	3.95	3.61	3.38	3.22	3.10	3.01	2.93	2.87	2.77	2.67	2.56	2.50	2.44	2.38	2.32	2.26	2.19
19	5.92	4.51	3.90	3.56	3.33	3.17	3.05	2.96	2.88	2.82	2.72	2.62	2.51	2.45	2.39	2.33	2.27	2.20	2.13
20	5.87	4.46	3.86	3.51	3.29	3.13	3.01	2.91	2.84	2.77	2.68	2.57	2.46	2.41	2.35	2.29	2.22	2.16	2.09
21	5.83	4.42	3.82	3.48	3.25	3.09	2.97	2.87	2.80	2.73	2.64	2.53	2.42	2.37	2.31	2.25	2.18	2.11	2.04
22	5.79	4.38	3.78	3.44	3.22	3.05	2.93	2.84	2.76	2.70	2.60	2.50	2.39	2.33	2.27	2.21	2.14	2.08	2.00
23	5.75	4.35	3.75	3.41	3.18	3.02	2.90	2.81	2.73	2.67	2.57	2.47	2.36	2.30	2.24	2.18	2.11	2.04	1.97
24	5.72	4.32	3.72	3.38	3.15	2.99	2.87	2.78	2.70	2.64	2.54	2.44	2.33	2.27	2.21	2.15	2.08	2.01	1.94
25	5.69	4.29	3.69	3.35	3.13	2.97	2.85	2.75	2.68	2.61	2.51	2.41	2.30	2.24	2.18	2.12	2.05	1.98	1.91
26	5.66	4.27	3.67	3.33	3.10	2.94	2.82	2.73	2.65	2.59	2.49	2.39	2.28	2.22	2.16	2.09	2.03	1.95	1.88
27	5.63	4.24	3.65	3.31	3.08	2.92	2.80	2.71	2.63	2.57	2.47	2.36	2.25	2.19	2.13	2.07	2.00	1.93	1.85
28	5.61	4.22	3.63	3.29	3.06	2.90	2.78	2.69	2.61	2.55	2.45	2.34	2.23	2.17	2.11	2.05	1.98	1.91	1.83
29	5.59	4.20	3.61	3.27	3.04	2.88	2.76	2.67	2.59	2.53	2.43	2.32	2.21	2.15	2.09	2.03	1.96	1.89	1.81
30	5.57	4.18	3.59	3.25	3.03	2.87	2.75	2.65	2.57	2.51	2.41	2.31	2.20	2.14	2.07	2.01	1.94	1.87	1.79
40	5.42	4.05	3.46	3.13	2.90	2.74	2.62	2.53	2.45	2.39	2.29	2.18	2.07	2.01	1.94	1.88	1.80	1.72	1.64
60	5.29	3.93	3.34	3.01	2.79	2.63	2.51	2.41	2.33	2.27	2.17	2.06	1.94	1.88	1.82	1.74	1.67	1.58	1.48
120	5.15	3.80	3.23	2.89	2.67	2.52	2.39	2.30	2.22	2.16	2.05	1.94	1.82	1.76	1.69	1.61	1.53	1.43	1.31
∞	5.02	3.69	3.12	2.79	2.57	2.41	2.29	2.19	2.11	2.05	1.94	1.83	1.71	1.64	1.57	1.48	1.39	1.27	1.00

Table A.7d. Critical values (percentiles) for the F distribution.
Upper one-sided 0.01 significance levels; two-sided 0.02 significance levels; 99 percent percentiles.

Numerator degrees of freedom

	1	2	3	4	5	6	7	8	9	10	11	12	13	14	15	16	17	18	19
1	4052	5000	5403	5625	5764	5859	5928	5982	6022	6056	6106	6157	6209	6235	6261	6287	6313	6339	6366
2	98.50	99.00	99.17	99.25	99.30	99.33	99.36	99.37	99.39	99.40	99.42	99.43	99.45	99.46	99.47	99.47	99.48	99.49	99.50
3	34.12	30.82	29.46	28.71	28.24	27.91	27.67	27.49	27.35	27.23	27.05	26.87	26.69	26.60	26.50	26.41	26.32	26.22	26.13
4	21.20	18.00	16.69	15.98	15.52	15.21	14.98	14.80	14.66	14.55	14.37	14.20	14.02	13.93	13.84	13.75	13.65	13.56	13.46
5	16.26	13.27	12.06	11.39	10.97	10.67	10.46	10.29	10.16	10.05	9.89	9.72	9.55	9.47	9.38	9.29	9.20	9.11	9.02
6	13.75	10.92	9.78	9.15	8.75	8.47	8.26	8.10	7.98	7.87	7.72	7.56	7.40	7.31	7.23	7.14	7.06	6.97	6.88
7	12.25	9.55	8.45	7.85	7.46	7.19	6.99	6.84	6.72	6.62	6.47	6.31	6.16	6.07	5.99	5.91	5.82	5.74	5.65
8	11.26	8.65	7.59	7.01	6.63	6.37	6.18	6.03	5.91	5.81	5.67	5.52	5.36	5.28	5.20	5.12	5.03	4.95	4.86
9	10.56	8.02	6.99	6.42	6.06	5.80	5.61	5.47	5.35	5.26	5.11	4.96	4.81	4.73	4.65	4.57	4.48	4.40	4.31
10	10.04	7.56	6.55	5.99	5.64	5.39	5.20	5.06	4.94	4.85	4.71	4.56	4.41	4.33	4.25	4.17	4.08	4.00	3.91
11	9.65	7.21	6.22	5.67	5.32	5.07	4.89	4.74	4.63	4.54	4.40	4.25	4.10	4.02	3.94	3.86	3.78	3.69	3.60
12	9.33	6.93	5.95	5.41	5.06	4.82	4.64	4.50	4.39	4.30	4.16	4.01	3.86	3.78	3.70	3.62	3.54	3.45	3.36
13	9.07	6.70	5.74	5.21	4.86	4.62	4.44	4.30	4.19	4.10	3.96	3.82	3.66	3.59	3.51	3.43	3.34	3.25	3.17
14	8.86	6.51	5.56	5.04	4.69	4.46	4.28	4.14	4.03	3.94	3.80	3.66	3.51	3.43	3.35	3.27	3.18	3.09	3.00
15	8.68	6.36	5.42	4.89	4.56	4.32	4.14	4.00	3.89	3.80	3.67	3.52	3.37	3.29	3.21	3.13	3.05	2.96	2.87
16	8.53	6.23	5.29	4.77	4.44	4.20	4.03	3.89	3.78	3.69	3.55	3.41	3.26	3.18	3.10	3.02	2.93	2.84	2.75
17	8.40	6.11	5.18	4.67	4.34	4.10	3.93	3.79	3.68	3.59	3.46	3.31	3.16	3.08	3.00	2.92	2.83	2.75	2.65
18	8.29	6.01	5.09	4.58	4.25	4.01	3.84	3.71	3.60	3.51	3.37	3.23	3.08	3.00	2.92	2.84	2.75	2.66	2.57
19	8.18	5.93	5.01	4.50	4.17	3.94	3.77	3.63	3.52	3.43	3.30	3.15	3.00	2.92	2.84	2.76	2.67	2.58	2.49
20	8.10	5.85	4.94	4.43	4.10	3.87	3.70	3.56	3.46	3.37	3.23	3.09	2.94	2.86	2.78	2.69	2.61	2.52	2.42
21	8.02	5.78	4.87	4.37	4.04	3.81	3.64	3.51	3.40	3.31	3.17	3.03	2.88	2.80	2.72	2.64	2.55	2.46	2.36
22	7.95	5.72	4.82	4.31	3.99	3.76	3.59	3.45	3.35	3.26	3.12	2.98	2.83	2.75	2.67	2.58	2.50	2.40	2.31
23	7.88	5.66	4.76	4.26	3.94	3.71	3.54	3.41	3.30	3.21	3.07	2.93	2.78	2.70	2.62	2.54	2.45	2.35	2.26
24	7.82	5.61	4.72	4.22	3.90	3.67	3.50	3.36	3.26	3.17	3.03	2.89	2.74	2.66	2.58	2.49	2.40	2.31	2.21
25	7.77	5.57	4.68	4.18	3.85	3.63	3.46	3.32	3.22	3.13	2.99	2.85	2.70	2.62	2.54	2.45	2.36	2.27	2.17
26	7.72	5.53	4.64	4.14	3.82	3.59	3.42	3.29	3.18	3.09	2.96	2.81	2.66	2.58	2.50	2.42	2.33	2.23	2.13
27	7.68	5.49	4.60	4.11	3.78	3.56	3.39	3.26	3.15	3.06	2.93	2.78	2.63	2.55	2.47	2.38	2.29	2.20	2.10
28	7.64	5.45	4.57	4.07	3.75	3.53	3.36	3.23	3.12	3.03	2.90	2.75	2.60	2.52	2.44	2.35	2.26	2.17	2.06
29	7.60	5.42	4.54	4.04	3.73	3.50	3.33	3.20	3.09	3.00	2.87	2.73	2.57	2.49	2.41	2.33	2.23	2.14	2.03
30	7.56	5.39	4.51	4.02	3.70	3.47	3.30	3.17	3.07	2.98	2.84	2.70	2.55	2.47	2.39	2.30	2.21	2.11	2.01
40	7.31	5.18	4.31	3.83	3.51	3.29	3.12	2.99	2.89	2.80	2.66	2.52	2.37	2.29	2.20	2.11	2.02	1.92	1.80
60	7.08	4.98	4.13	3.65	3.34	3.12	2.95	2.82	2.72	2.63	2.50	2.35	2.20	2.12	2.03	1.94	1.84	1.73	1.60
120	6.85	4.79	3.95	3.48	3.17	2.96	2.79	2.66	2.56	2.47	2.34	2.19	2.03	1.95	1.86	1.76	1.66	1.53	1.38
∞	6.63	4.61	3.78	3.32	3.02	2.80	2.64	2.51	2.41	2.32	2.18	2.04	1.88	1.79	1.70	1.59	1.47	1.32	1.00

Table A.7e. Critical values (percentiles) for the F distribution.

Upper one-sided 0.005 significance levels; two-sided 0.01 significance levels; 99.5 percent percentiles.

| | \multicolumn{19}{c}{Numerator degrees of freedom} |
	1	2	3	4	5	6	7	8	9	10	11	12	13	14	15	16	17	18	19
1	16211	20000	21615	22500	23056	23437	23715	23925	24091	24224	24426	24630	24836	24940	25044	25148	25253	25359	25465
2	198.5	199.0	199.2	199.2	199.3	199.3	199.4	199.4	199.4	199.4	199.4	199.4	199.4	199.5	199.5	199.5	199.5	199.5	199.5
3	55.55	49.80	47.47	46.19	45.39	44.84	44.43	44.13	43.88	43.69	43.39	43.08	42.78	42.62	42.47	42.31	42.15	41.99	41.83
4	31.33	26.28	24.26	23.15	22.46	21.97	21.62	21.35	21.14	20.97	20.70	20.44	20.17	20.03	19.89	19.75	19.61	19.47	19.32
5	22.78	18.31	16.53	15.56	14.94	14.51	14.20	13.96	13.77	13.62	13.38	13.15	12.90	12.78	12.66	12.53	12.40	12.27	12.14
6	18.63	14.54	12.92	12.03	11.46	11.07	10.79	10.57	10.39	10.25	10.03	9.81	9.59	9.47	9.36	9.24	9.12	9.00	8.88
7	16.24	12.40	10.88	10.05	9.52	9.16	8.89	8.68	8.51	8.38	8.18	7.97	7.75	7.65	7.53	7.42	7.31	7.19	7.08
8	14.69	11.04	9.60	8.81	8.30	7.95	7.69	7.50	7.34	7.21	7.01	6.81	6.61	6.50	6.40	6.29	6.18	6.06	5.95
9	13.61	10.11	8.72	7.96	7.47	7.13	6.88	6.69	6.54	6.42	6.23	6.03	5.83	5.73	5.62	5.52	5.41	5.30	5.19
10	12.83	9.43	8.08	7.34	6.87	6.54	6.30	6.12	5.97	5.85	5.66	5.47	5.27	5.17	5.07	4.97	4.86	4.75	4.64
11	12.23	8.91	7.60	6.88	6.42	6.10	5.86	5.68	5.54	5.42	5.24	5.05	4.86	4.76	4.65	4.55	4.44	4.34	4.23
12	11.75	8.51	7.23	6.52	6.07	5.76	5.52	5.35	5.20	5.09	4.91	4.72	4.53	4.43	4.33	4.23	4.12	4.01	3.90
13	11.37	8.19	6.93	6.23	5.79	5.48	5.25	5.08	4.94	4.82	4.64	4.46	4.27	4.17	4.07	3.97	3.87	3.76	3.65
14	11.06	7.92	6.68	6.00	5.56	5.26	5.03	4.86	4.72	4.60	4.43	4.25	4.06	3.96	3.86	3.76	3.66	3.55	3.44
15	10.80	7.70	6.48	5.80	5.37	5.07	4.85	4.67	4.54	4.42	4.25	4.07	3.88	3.79	3.69	3.58	3.48	3.37	3.26
16	10.58	7.51	6.30	5.64	5.21	4.91	4.69	4.52	4.38	4.27	4.10	3.92	3.73	3.64	3.54	3.44	3.33	3.22	3.11
17	10.38	7.35	6.16	5.50	5.07	4.78	4.56	4.39	4.25	4.14	3.97	3.79	3.61	3.51	3.41	3.31	3.21	3.10	2.98
18	10.22	7.21	6.03	5.37	4.96	4.66	4.44	4.28	4.14	4.03	3.86	3.68	3.50	3.40	3.30	3.20	3.10	2.99	2.87
19	10.07	7.09	5.92	5.27	4.85	4.56	4.34	4.18	4.04	3.93	3.76	3.59	3.40	3.31	3.21	3.11	3.00	2.89	2.78
20	9.94	6.99	5.82	5.17	4.76	4.47	4.26	4.09	3.96	3.85	3.68	3.50	3.32	3.22	3.12	3.02	2.92	2.81	2.69
21	9.83	6.89	5.73	5.09	4.68	4.39	4.18	4.01	3.88	3.77	3.60	3.43	3.24	3.15	3.05	2.95	2.84	2.73	2.61
22	9.73	6.81	5.65	5.02	4.61	4.32	4.11	3.94	3.81	3.70	3.54	3.36	3.18	3.08	2.98	2.88	2.77	2.66	2.55
23	9.63	6.73	5.58	4.95	4.54	4.26	4.05	3.88	3.75	3.64	3.47	3.30	3.12	3.02	2.92	2.82	2.71	2.60	2.48
24	9.55	6.66	5.52	4.89	4.49	4.20	3.99	3.83	3.69	3.59	3.42	3.25	3.06	2.97	2.87	2.77	2.66	2.55	2.43
25	9.48	6.60	5.46	4.84	4.43	4.15	3.94	3.78	3.64	3.54	3.37	3.20	3.01	2.92	2.82	2.72	2.61	2.50	2.38
26	9.41	6.54	5.41	4.79	4.38	4.10	3.89	3.73	3.60	3.49	3.33	3.15	2.97	2.87	2.77	2.67	2.56	2.45	2.33
27	9.34	6.49	5.36	4.74	4.34	4.06	3.85	3.69	3.56	3.45	3.28	3.11	2.93	2.83	2.73	2.63	2.52	2.41	2.25
28	9.28	6.44	5.32	4.70	4.30	4.02	3.81	3.65	3.52	3.41	3.25	3.07	2.89	2.79	2.69	2.59	2.48	2.37	2.29
29	9.23	6.40	5.28	4.66	4.26	3.98	3.77	3.61	3.48	3.38	3.21	3.04	2.86	2.76	2.66	2.56	2.45	2.33	2.24
30	9.18	6.35	5.24	4.62	4.23	3.95	3.74	3.58	3.45	3.34	3.18	3.01	2.82	2.73	2.63	2.52	2.42	2.30	2.18
40	8.83	6.07	4.98	4.37	3.99	3.71	3.51	3.35	3.22	3.12	2.95	2.78	2.60	2.50	2.40	2.30	2.18	2.06	1.93
60	8.49	5.79	4.73	4.14	3.76	3.49	3.29	3.13	3.01	2.90	2.74	2.57	2.39	2.29	2.19	2.08	1.96	1.83	1.69
120	8.18	5.54	4.50	3.92	3.55	3.28	3.09	2.93	2.81	2.71	2.54	2.37	2.19	2.09	1.98	1.87	1.75	1.61	1.43
∞	7.88	5.30	4.28	3.72	3.35	3.09	2.90	2.74	2.62	2.52	2.36	2.19	2.00	1.90	1.79	1.67	1.53	1.36	1.00

Table A.7f. Critical values (percentiles) for the F distribution.

Upper one-sided 0.001 significance levels; two-sided 0.002 significance levels; 99.9 percent percentiles.

Numerator degrees of freedom

	1	2	3	4	5	6	7	8	9	10	11	12	13	14	15	16	17	18	19
1	4053*	5000*	5404*	5625*	5764*	5859*	5929*	5981*	6023*	6056*	6107*	6158*	6209*	6235*	6261*	6287*	6313*	6340*	6366*
2	998.5	999.0	999.2	999.2	999.3	999.3	999.4	999.4	999.4	999.4	999.4	999.4	999.4	999.4	999.5	999.5	999.5	999.5	999.5
3	167.0	148.5	141.1	137.1	134.6	132.8	131.6	130.6	129.9	129.2	128.3	127.4	126.4	125.9	125.0	125.0	124.5	124.0	123.5
4	74.14	61.25	56.18	53.44	51.71	50.53	49.66	49.00	48.47	48.05	47.41	46.76	46.10	45.77	45.43	45.09	44.75	44.40	44.05
5	47.18	37.12	33.20	31.09	29.75	28.84	28.16	27.64	27.24	26.92	26.42	25.91	25.39	25.14	24.87	24.60	24.33	24.06	23.79
6	35.51	27.00	23.70	21.92	20.81	20.03	19.46	19.03	18.69	18.41	17.99	17.56	17.12	16.89	16.67	16.44	16.21	15.99	15.75
7	29.25	21.69	18.77	17.19	16.21	15.52	15.02	14.63	14.33	14.08	13.71	13.32	12.93	12.73	12.53	12.33	12.12	11.91	11.70
8	25.42	18.49	15.83	14.39	13.49	12.86	12.40	12.04	11.77	11.54	11.19	10.84	10.48	10.30	10.11	9.92	9.73	9.53	9.33
9	22.86	16.39	13.90	12.56	11.71	11.13	10.70	10.37	10.11	9.89	9.57	9.24	8.90	8.72	8.55	8.37	8.19	8.00	7.81
10	21.04	14.91	12.55	11.28	10.48	9.92	9.52	9.20	8.96	8.75	8.45	8.13	7.80	7.64	7.47	7.30	7.12	6.94	6.76
11	19.69	13.81	11.56	10.35	9.58	9.05	8.66	8.35	8.12	7.92	7.63	7.32	7.01	6.85	6.68	6.62	6.35	6.17	6.00
12	18.64	12.97	10.80	9.63	8.89	8.38	8.00	7.71	7.48	7.29	7.00	6.71	6.40	6.25	6.09	5.93	5.76	5.59	5.42
13	17.81	12.31	10.21	9.07	8.35	7.86	7.49	7.21	6.98	6.80	6.52	6.23	5.93	5.78	5.63	5.47	5.30	5.14	4.97
14	17.14	11.78	9.73	8.62	7.92	7.43	7.08	6.80	6.58	6.40	6.13	5.85	5.56	5.41	5.25	5.10	4.94	4.77	4.60
15	16.59	11.34	9.34	8.25	7.57	7.09	6.74	6.47	6.26	6.08	5.81	5.54	5.25	5.10	4.95	4.80	4.64	4.47	4.31
16	16.12	10.97	9.00	7.94	7.27	6.81	6.46	6.19	5.98	5.81	5.55	5.27	4.99	4.85	4.70	4.54	4.39	4.23	4.06
17	15.72	10.66	8.73	7.68	7.02	6.56	6.22	5.96	5.75	5.58	5.32	5.05	4.78	4.63	4.48	4.33	4.18	4.02	3.85
18	15.38	10.39	8.49	7.46	6.81	6.35	6.02	5.76	5.56	5.39	5.13	4.87	4.59	4.45	4.30	4.15	4.00	3.84	3.67
19	15.08	10.16	8.28	7.26	6.62	6.18	5.85	5.59	5.39	5.22	4.97	4.70	4.43	4.29	4.14	3.99	3.84	3.68	3.51
20	14.82	9.95	8.10	7.10	6.46	6.02	5.69	5.44	5.24	5.08	4.82	4.56	4.29	4.15	4.00	3.86	3.70	3.54	3.38
21	14.59	9.77	7.94	6.95	6.32	5.88	5.56	5.31	5.11	4.95	4.70	4.44	4.17	4.03	3.88	3.74	3.58	3.42	3.26
22	14.38	9.61	7.80	6.81	6.19	5.76	5.44	5.19	4.99	4.83	4.58	4.33	4.06	3.92	3.78	3.63	3.48	3.32	3.15
23	14.19	9.47	7.67	6.69	6.08	5.65	5.33	5.09	4.89	4.73	4.48	4.23	3.96	3.82	3.68	3.53	3.38	3.22	3.05
24	14.03	9.34	7.55	6.59	5.98	5.55	5.23	4.99	4.80	4.64	4.39	4.14	3.87	3.74	3.59	3.45	3.29	3.14	2.97
25	13.88	9.22	7.45	6.49	5.88	5.46	5.15	4.91	4.71	4.56	4.31	4.06	3.79	3.66	3.52	3.37	3.22	3.06	2.89
26	13.74	9.12	7.36	6.41	5.80	5.38	5.07	4.83	4.64	4.48	4.24	3.99	3.72	3.59	3.44	3.30	3.15	2.99	2.82
27	13.61	9.02	7.27	6.33	5.73	5.31	5.00	4.76	4.57	4.41	4.17	3.92	3.66	3.52	3.38	3.23	3.08	2.92	2.75
28	13.50	8.93	7.19	6.25	5.66	5.24	4.93	4.69	4.50	4.35	4.11	3.86	3.60	3.46	3.32	3.18	3.02	2.86	2.69
29	13.39	8.85	7.12	6.19	5.59	5.18	4.87	4.64	4.45	4.29	4.05	3.80	3.54	3.41	3.27	3.12	2.97	2.81	2.64
30	13.29	8.77	7.05	6.12	5.53	5.12	4.82	4.58	4.39	4.24	4.00	3.75	3.49	3.36	3.22	3.07	2.92	2.76	2.59
40	12.61	8.25	6.60	5.70	5.13	4.73	4.44	4.21	4.02	3.87	3.64	3.40	3.15	3.01	2.87	2.73	2.57	2.41	2.23
60	11.97	7.76	6.17	5.31	4.76	4.37	4.09	3.87	3.69	3.54	3.31	3.08	2.83	2.69	2.55	2.41	2.25	2.08	1.89
120	11.38	7.32	5.79	4.95	4.42	4.04	3.77	3.55	3.38	3.24	3.02	2.78	2.53	2.40	2.26	2.11	1.95	1.76	1.54
∞	10.83	6.91	5.42	4.62	4.10	3.74	3.47	3.27	3.10	2.96	2.74	2.51	2.27	2.13	1.99	1.84	1.66	1.45	1.00

* multiply these entries by 100.

Table A.8. Binomial individual term and tail probabilities.

Under each value of π the first column gives $b(k;n,\pi)$, the probability of k successes in n independent trials with probability of success π in each trial. The second column gives $P[X \leq k]$ or $P[X \geq k]$ if one of these is less than 0.5. A missing entry indicates that neither is less than 0.5.

n	k	$\pi = 0.1$		$\pi = 0.2$		$\pi = 0.3$		$\pi = 0.4$		$\pi = 0.5$	
2	0	.8100	---	.6400	---	.4900	.4900	.3600	.3600	.2500	.2500
	1	.1800	.1900	.3200	.3600	.4200	---	.4800	---	.5000	---
	2	.0100	.0100	.0400	.0400	.0900	.0900	.1600	.1600	.2500	.2500
3	0	.7290	---	.5120	---	.3430	.3430	.2160	.2160	.1250	.1250
	1	.2430	.2710	.3840	.4880	.4410	---	.4320	---	.3750	---
	2	.0270	.0280	.0960	.1040	.1890	.2160	.2880	.3520	.3750	---
	3	.0010	.0010	.0080	.0080	.0270	.0270	.0640	.0640	.1250	.1250
4	0	.6561	---	.4096	.4096	.2401	.2401	.1296	.1296	.0625	.0625
	1	.2916	.3439	.4096	---	.4116	---	.3456	.4752	.2500	.3125
	2	.0486	.0523	.1536	.1808	.2646	.3483	.3456	---	.3750	---
	3	.0036	.0037	.0256	.0272	.0756	.0837	.1536	.1792	.2500	.3125
	4	.0001	.0001	.0016	.0016	.0081	.0081	.0256	.0256	.0625	.0625
5	0	.5905	---	.3277	.3277	.1681	.1681	.0778	.0778	.0313	.0313
	1	.3281	.4095	.4096	---	.3602	---	.2592	.3370	.1563	.1875
	2	.0729	.0815	.2048	.2627	.3087	.4718	.3456	---	.3125	---
	3	.0081	.0086	.0512	.0579	.1323	.1631	.2304	.3174	.3125	---
	4	.0005	.0005	.0064	.0067	.0284	.0308	.0768	.0870	.1563	.1875
	5	.0000	.0000	.0003	.0003	.0024	.0024	.0102	.0102	.0313	.0313
6	0	.5314	---	.2621	.2621	.1176	.1176	.0467	.0467	.0156	.0156
	1	.3543	.4686	.3932	---	.3025	.4202	.1866	.2333	.0938	.1094
	2	.0984	.1143	.2458	.3446	.3241	---	.3110	---	.2344	.3438
	3	.0146	.0159	.0819	.0989	.1852	.2557	.2765	.4557	.3125	---
	4	.0012	.0013	.0154	.0170	.0595	.0705	.1382	.1792	.2344	.3438
	5	.0001	.0001	.0015	.0016	.0102	.0109	.0369	.0410	.0938	.1094
	6	.0000	.0000	.0001	.0001	.0007	.0007	.0041	.0041	.0156	.0156
7	0	.4783	.4783	.2097	.2097	.0824	.0824	.0280	.0280	.0078	.0078
	1	.3720	---	.3670	---	.2471	.3294	.1306	.1586	.0547	.0625
	2	.1240	.1497	.2753	.4233	.3177	---	.2613	.4199	.1641	.2266
	3	.0230	.0257	.1147	.1480	.2269	.3529	.2903	---	.2734	---
	4	.0026	.0027	.0287	.0333	.0972	.1260	.1935	.2898	.2734	---
	5	.0002	.0002	.0043	.0047	.0250	.0288	.0774	.0963	.1641	.2266
	6	.0000	.0000	.0004	.0004	.0036	.0038	.0172	.0188	.0547	.0625
	7	.0000	.0000	.0000	.0000	.0002	.0002	.0016	.0016	.0078	.0078
8	0	.4305	.4305	.1678	.1678	.0576	.0576	.0168	.0168	.0039	.0039
	1	.3826	---	.3355	---	.1977	.2553	.0896	.1064	.0313	.0352
	2	.1488	.1869	.2936	.4967	.2965	---	.2090	.3154	.1094	.1445
	3	.0331	.0381	.1468	.2031	.2541	.4482	.2787	---	.2188	.3633
	4	.0046	.0050	.0459	.0563	.1361	.1941	.2322	.4059	.2734	---
	5	.0004	.0004	.0092	.0104	.0467	.0580	.1239	.1737	.2188	.3633
	6	.0000	.0000	.0011	.0012	.0100	.0113	.0413	.0498	.1094	.1445
	7	.0000	.0000	.0001	.0001	.0012	.0013	.0079	.0085	.0313	.0352
	8	.0000	.0000	.0000	.0000	.0001	.0001	.0007	.0007	.0039	.0039
9	0	.3874	.3874	.1342	.1342	.0404	.0404	.0101	.0101	.0020	.0020
	1	.3874	---	.3020	.4362	.1556	.1960	.0605	.0705	.0176	.0195
	2	.1722	.2252	.3020	---	.2668	.4628	.1612	.2318	.0703	.0898
	3	.0446	.0530	.1762	.2618	.2668	---	.2508	.4826	.1641	.2539
	4	.0074	.0083	.0661	.0856	.1715	.2703	.2508	---	.2461	---
	5	.0008	.0009	.0165	.0196	.0735	.0988	.1672	.2666	.2461	---
	6	.0001	.0001	.0028	.0031	.0210	.0253	.0743	.0994	.1641	.2539
	7	.0000	.0000	.0003	.0003	.0039	.0043	.0212	.0250	.0703	.0898
	8	.0000	.0000	.0000	.0000	.0004	.0004	.0035	.0038	.0176	.0195
	9	.0000	.0000	.0000	.0000	.0000	.0000	.0003	.0003	.0020	.0020
10	0	.3487	.3487	.1074	.1074	.0282	.0282	.0060	.0060	.0010	.0010
	1	.3874	---	.2684	.3758	.1211	.1493	.0403	.0464	.0098	.0107
	2	.1937	.2639	.3020	---	.2335	.3828	.1209	.1673	.0439	.0547
	3	.0574	.0702	.2013	.3222	.2668	---	.2150	.3823	.1172	.1719
	4	.0112	.0128	.0881	.1209	.2001	.3504	.2508	---	.2051	.3770

Table A.8 *continued*. Binomial individual term and tail probabilities.											
		$\pi = 0.1$		$\pi = 0.2$		$\pi = 0.3$		$\pi = 0.4$		$\pi = 0.5$	
n	k										
10	5	.0015	.0016	.0264	.0328	.1029	.1503	.2007	.3669	.2461	---
	6	.0001	.0001	.0055	.0064	.0368	.0473	.1115	.1662	.2051	.3770
	7	.0000	.0000	.0008	.0009	.0090	.0106	.0425	.0548	.1172	.1719
	8	.0000	.0000	.0001	.0001	.0014	.0016	.0106	.0123	.0439	.0547
	9	.0000	.0000	.0000	.0000	.0001	.0001	.0016	.0017	.0098	.0107
	10	.0000	.0000	.0000	.0000	.0000	.0000	.0001	.0001	.0010	.0010
11	0	.3138	.3138	.0859	.0859	.0198	.0198	.0036	.0036	.0005	.0005
	1	.3835	---	.2362	.3221	.0932	.1130	.0266	.0302	.0054	.0059
	2	.2131	.3026	.2953	---	.1998	.3127	.0887	.1189	.0269	.0327
	3	.0710	.0896	.2215	.3826	.2568	---	.1774	.2963	.0806	.1133
	4	.0158	.0185	.1107	.1611	.2201	.4304	.2365	---	.1611	.2744
	5	.0025	.0028	.0388	.0504	.1321	.2103	.2207	.4672	.2256	---
	6	.0003	.0003	.0097	.0117	.0566	.0782	.1471	.2465	.2256	---
	7	.0000	.0000	.0017	.0020	.0173	.0216	.0701	.0994	.1611	.2744
	8	.0000	.0000	.0002	.0002	.0037	.0043	.0234	.0293	.0806	.1133
	9	.0000	.0000	.0000	.0000	.0005	.0006	.0052	.0059	.0269	.0327
	10	.0000	.0000	.0000	.0000	.0000	.0000	.0007	.0007	.0054	.0059
	11	.0000	.0000	.0000	.0000	.0000	.0000	.0000	.0000	.0005	.0005
12	0	.2824	.2824	.0687	.0687	.0138	.0138	.0022	.0022	.0002	.0002
	1	.3766	---	.2062	.2749	.0712	.0850	.0174	.0196	.0029	.0032
	2	.2301	.3410	.2835	---	.1678	.2528	.0639	.0834	.0161	.0193
	3	.0852	.1109	.2362	.4417	.2397	.4925	.1419	.2253	.0537	.0730
	4	.0213	.0256	.1329	.2054	.2311	---	.2128	.4382	.1208	.1938
	5	.0038	.0043	.0532	.0726	.1585	.2763	.2270	---	.1934	.3872
	6	.0005	.0005	.0155	.0194	.0792	.1178	.1766	.3348	.2256	---
	7	.0000	.0001	.0033	.0039	.0291	.0386	.1009	.1582	.1934	.3872
	8	.0000	.0000	.0005	.0006	.0078	.0095	.0420	.0573	.1208	.1938
	9	.0000	.0000	.0001	.0001	.0015	.0017	.0125	.0153	.0537	.0730
	10	.0000	.0000	.0000	.0000	.0002	.0002	.0025	.0028	.0161	.0193
	11	.0000	.0000	.0000	.0000	.0000	.0000	.0003	.0003	.0029	.0032
	12	.0000	.0000	.0000	.0000	.0000	.0000	.0000	.0000	.0002	.0002
13	0	.2542	.2542	.0550	.0550	.0097	.0097	.0013	.0013	.0001	.0001
	1	.3672	---	.1787	.2336	.0540	.0637	.0113	.0126	.0016	.0017
	2	.2448	.3787	.2680	---	.1388	.2025	.0453	.0579	.0095	.0112
	3	.0997	.1339	.2457	.4983	.2181	.4206	.1107	.1686	.0349	.0461
	4	.0277	.0342	.1535	.2527	.2337	---	.1845	.3530	.0873	.1334
	5	.0055	.0065	.0691	.0991	.1803	.3457	.2214	---	.1571	.2905
	6	.0008	.0009	.0230	.0300	.1030	.1654	.1968	.4256	.2095	---
	7	.0001	.0001	.0058	.0070	.0442	.0624	.1312	.2288	.2095	---
	8	.0000	.0000	.0011	.0012	.0142	.0182	.0656	.0977	.1571	.2905
	9	.0000	.0000	.0001	.0002	.0034	.0040	.0243	.0321	.0873	.1334
	10	.0000	.0000	.0000	.0000	.0006	.0007	.0065	.0078	.0349	.0461
	11	.0000	.0000	.0000	.0000	.0001	.0001	.0012	.0013	.0095	.0112
	12	.0000	.0000	.0000	.0000	.0000	.0000	.0001	.0001	.0016	.0017
	13	.0000	.0000	.0000	.0000	.0000	.0000	.0000	.0000	.0001	.0001
14	0	.2288	.2288	.0440	.0440	.0068	.0068	.0008	.0008	.0001	.0001
	1	.3559	---	.1539	.1979	.0407	.0475	.0073	.0081	.0009	.0009
	2	.2570	.4154	.2501	.4481	.1134	.1608	.0317	.0398	.0056	.0065
	3	.1142	.1584	.2501	---	.1943	.3552	.0845	.1243	.0222	.0287
	4	.0349	.0441	.1720	.3018	.2290	---	.1549	.2793	.0611	.0898
	5	.0078	.0092	.0860	.1298	.1963	.4158	.2066	.4859	.1222	.2120
	6	.0013	.0015	.0322	.0439	.1262	.2195	.2066	---	.1833	.3953
	7	.0002	.0002	.0092	.0116	.0618	.0933	.1574	.3075	.2095	---
	8	.0000	.0000	.0020	.0024	.0232	.0315	.0918	.1501	.1833	.3953
	9	.0000	.0000	.0003	.0004	.0066	.0083	.0408	.0583	.1222	.2120
	10	.0000	.0000	.0000	.0000	.0014	.0017	.0136	.0175	.0611	.0898
	11	.0000	.0000	.0000	.0000	.0002	.0002	.0033	.0039	.0222	.0287
	12	.0000	.0000	.0000	.0000	.0000	.0000	.0005	.0006	.0056	.0065
	13	.0000	.0000	.0000	.0000	.0000	.0000	.0001	.0001	.0009	.0009
	14	.0000	.0000	.0000	.0000	.0000	.0000	.0000	.0000	.0001	.0001

Table A.8 continued. Binomial individual term and tail probabilities.										
		$\pi = 0.1$		$\pi = 0.2$		$\pi = 0.3$		$\pi = 0.4$		$\pi = 0.5$
n	k									
15	0	.2059 .2059		.0352 .0352		.0047 .0047		.0005 .0005		.0000 .0000
	1	.3432 ---		.1319 .1671		.0305 .0353		.0047 .0052		.0005 .0005
	2	.2669 .4510		.2309 .3980		.0916 .1268		.0219 .0271		.0032 .0037
	3	.1285 .1841		.2501 ---		.1700 .2969		.0634 .0905		.0139 .0176
	4	.0428 .0556		.1876 .3518		.2186 ---		.1268 .2173		.0417 .0592
	5	.0105 .0127		.1032 .1642		.2061 .4845		.1859 .4032		.0916 .1509
	6	.0019 .0022		.0430 .0611		.1472 .2784		.2066 ---		.1527 .3036
	7	.0003 .0003		.0138 .0181		.0811 .1311		.1771 .3902		.1964 ---
	8	.0000 .0000		.0035 .0042		.0348 .0500		.1181 .2131		.1964 ---
	9	.0000 .0000		.0007 .0008		.0116 .0152		.0612 .0950		.1527 .3036
	10	.0000 .0000		.0001 .0001		.0030 .0037		.0245 .0338		.0916 .1509
	11	.0000 .0000		.0000 .0000		.0006 .0007		.0074 .0093		.0417 .0592
	12	.0000 .0000		.0000 .0000		.0001 .0001		.0016 .0019		.0139 .0176
	13	.0000 .0000		.0000 .0000		.0000 .0000		.0003 .0003		.0032 .0037
	14	.0000 .0000		.0000 .0000		.0000 .0000		.0000 .0000		.0005 .0005
	15	.0000 .0000		.0000 .0000		.0000 .0000		.0000 .0000		.0000 .0000
16	0	.1853 .1853		.0281 .0281		.0033 .0033		.0003 .0003		.0000 .0000
	1	.3294 ---		.1126 .1407		.0228 .0261		.0030 .0033		.0002 .0003
	2	.2745 .4853		.2111 .3518		.0732 .0994		.0150 .0183		.0018 .0021
	3	.1423 .2108		.2463 ---		.1465 .2459		.0468 .0651		.0085 .0106
	4	.0514 .0684		.2001 .4019		.2040 .4499		.1014 .1666		.0278 .0384
	5	.0137 .0170		.1201 .2018		.2099 ---		.1623 .3288		.0667 .1051
	6	.0028 .0033		.0550 .0817		.1649 .3402		.1983 ---		.1222 .2272
	7	.0004 .0005		.0197 .0267		.1010 .1753		.1889 .4728		.1746 .4018
	8	.0001 .0001		.0055 .0070		.0487 .0744		.1417 .2839		.1964 ---
	9	.0000 .0000		.0012 .0015		.0185 .0257		.0840 .1423		.1746 .4018
	10	.0000 .0000		.0002 .0002		.0056 .0071		.0392 .0583		.1222 .2272
	11	.0000 .0000		.0000 .0000		.0013 .0016		.0142 .0191		.0667 .1051
	12	.0000 .0000		.0000 .0000		.0002 .0003		.0040 .0049		.0278 .0384
	13	.0000 .0000		.0000 .0000		.0000 .0000		.0008 .0009		.0085 .0106
	14	.0000 .0000		.0000 .0000		.0000 .0000		.0001 .0001		.0018 .0021
	15	.0000 .0000		.0000 .0000		.0000 .0000		.0000 .0000		.0002 .0003
	16	.0000 .0000		.0000 .0000		.0000 .0000		.0000 .0000		.0000 .0000
17	0	.1668 .1668		.0225 .0225		.0023 .0023		.0002 .0002		.0000 .0000
	1	.3150 .4818		.0957 .1182		.0169 .0193		.0019 .0021		.0001 .0001
	2	.2800 ---		.1914 .3096		.0581 .0774		.0102 .0123		.0010 .0012
	3	.1556 .2382		.2393 ---		.1245 .2019		.0341 .0464		.0052 .0064
	4	.0605 .0826		.2093 .4511		.1868 .3887		.0796 .1260		.0182 .0245
	5	.0175 .0221		.1361 .2418		.2081 ---		.1379 .2639		.0472 .0717
	6	.0039 .0047		.0680 .1057		.1784 .4032		.1839 .4478		.0944 .1662
	7	.0007 .0008		.0267 .0377		.1201 .2248		.1927 ---		.1484 .3145
	8	.0001 .0001		.0084 .0109		.0644 .1046		.1606 .3595		.1855 ---
	9	.0000 .0000		.0021 .0026		.0276 .0403		.1070 .1989		.1855 ---
	10	.0000 .0000		.0004 .0005		.0095 .0127		.0571 .0919		.1484 .3145
	11	.0000 .0000		.0001 .0001		.0026 .0032		.0242 .0348		.0944 .1662
	12	.0000 .0000		.0000 .0000		.0006 .0007		.0081 .0106		.0472 .0717
	13	.0000 .0000		.0000 .0000		.0001 .0001		.0021 .0025		.0182 .0245
	14	.0000 .0000		.0000 .0000		.0000 .0000		.0004 .0005		.0052 .0064
	15	.0000 .0000		.0000 .0000		.0000 .0000		.0001 .0001		.0010 .0012
	16	.0000 .0000		.0000 .0000		.0000 .0000		.0000 .0000		.0001 .0001
	17	.0000 .0000		.0000 .0000		.0000 .0000		.0000 .0000		.0000 .0000
	0	.1501 .1501		.0180 .0180		.0016 .0016		.0001 .0001		.0000 .0000
	1	.3002 .4503		.0811 .0991		.0126 .0142		.0012 .0013		.0001 .0001
	2	.2835 ---		.1723 .2713		.0458 .0600		.0069 .0082		.0006 .0007
	3	.1680 .2662		.2297 ---		.1046 .1646		.0246 .0328		.0031 .0038
	4	.0700 .0982		.2153 .4990		.1681 .3327		.0614 .0942		.0117 .0154
	5	.0218 .0282		.1507 .2836		.2017 ---		.1146 .2088		.0327 .0481
	6	.0052 .0064		.0816 .1329		.1873 .4656		.1655 .3743		.0708 .1189
	7	.0010 .0012		.0350 .0513		.1376 .2783		.1892 ---		.1214 .2403
	8	.0002 .0002		.0120 .0163		.0811 .1407		.1734 .4366		.1669 .4073
	9	.0000 .0000		.0033 .0043		.0386 .0596		.1284 .2632		.1855 ---

		$\pi = 0.1$		$\pi = 0.2$		$\pi = 0.3$		$\pi = 0.4$		$\pi = 0.5$	
n	k										
18	10	.0000	.0000	.0008	.0009	.0149	.0210	.0771	.1347	.1669	.4073
	11	.0000	.0000	.0001	.0002	.0046	.0061	.0374	.0576	.1214	.2403
	12	.0000	.0000	.0000	.0000	.0012	.0014	.0145	.0203	.0708	.1189
	13	.0000	.0000	.0000	.0000	.0002	.0003	.0045	.0058	.0327	.0481
	14	.0000	.0000	.0000	.0000	.0000	.0000	.0011	.0013	.0117	.0154
	15	.0000	.0000	.0000	.0000	.0000	.0000	.0002	.0002	.0031	.0038
	16	.0000	.0000	.0000	.0000	.0000	.0000	.0000	.0000	.0006	.0007
	17	.0000	.0000	.0000	.0000	.0000	.0000	.0000	.0000	.0001	.0001
	18	.0000	.0000	.0000	.0000	.0000	.0000	.0000	.0000	.0000	.0000
19	0	.1351	.1351	.0144	.0144	.0011	.0011	.0001	.0001	.0000	.0000
	1	.2852	.4203	.0685	.0829	.0093	.0104	.0008	.0008	.0000	.0000
	2	.2852	---	.1540	.2369	.0358	.0462	.0046	.0055	.0003	.0004
	3	.1796	.2946	.2182	.4551	.0869	.1332	.0175	.0230	.0018	.0022
	4	.0798	.1150	.2182	---	.1491	.2822	.0467	.0696	.0074	.0096
	5	.0266	.0352	.1636	.3267	.1916	.4739	.0933	.1629	.0222	.0318
	6	.0069	.0086	.0955	.1631	.1916	---	.1451	.3081	.0518	.0835
	7	.0014	.0017	.0443	.0676	.1525	.3345	.1797	.4878	.0961	.1796
	8	.0002	.0003	.0166	.0233	.0981	.1820	.1797	---	.1442	.3238
	9	.0000	.0000	.0051	.0067	.0514	.0839	.1464	.3325	.1762	---
	10	.0000	.0000	.0013	.0016	.0220	.0326	.0976	.1861	.1762	---
	11	.0000	.0000	.0003	.0003	.0077	.0105	.0532	.0885	.1442	.3238
	12	.0000	.0000	.0000	.0000	.0022	.0028	.0237	.0352	.0961	.1796
	13	.0000	.0000	.0000	.0000	.0005	.0006	.0085	.0116	.0518	.0835
	14	.0000	.0000	.0000	.0000	.0001	.0001	.0024	.0031	.0222	.0318
	15	.0000	.0000	.0000	.0000	.0000	.0000	.0005	.0006	.0074	.0096
	16	.0000	.0000	.0000	.0000	.0000	.0000	.0001	.0001	.0018	.0022
	17	.0000	.0000	.0000	.0000	.0000	.0000	.0000	.0000	.0003	.0004
	18	.0000	.0000	.0000	.0000	.0000	.0000	.0000	.0000	.0000	.0000
	19	.0000	.0000	.0000	.0000	.0000	.0000	.0000	.0000	.0000	.0000
20	0	.1216	.1216	.0115	.0115	.0008	.0008	.0000	.0000	.0000	.0000
	1	.2702	.3917	.0576	.0692	.0068	.0076	.0005	.0005	.0000	.0000
	2	.2852	---	.1369	.2061	.0278	.0355	.0031	.0036	.0002	.0002
	3	.1901	.3231	.2054	.4114	.0716	.1071	.0123	.0160	.0011	.0013
	4	.0898	.1330	.2182	---	.1304	.2375	.0350	.0510	.0046	.0059
	5	.0319	.0432	.1746	.3704	.1789	.4164	.0746	.1256	.0148	.0207
	6	.0089	.0113	.1091	.1958	.1916	---	.1244	.2500	.0370	.0577
	7	.0020	.0024	.0545	.0867	.1643	.3920	.1659	.4159	.0739	.1316
	8	.0004	.0004	.0222	.0321	.1144	.2277	.1797	---	.1201	.2517
	9	.0001	.0001	.0074	.0100	.0654	.1133	.1597	.4044	.1602	.4119
	10	.0000	.0000	.0020	.0026	.0308	.0480	.1171	.2447	.1762	---
	11	.0000	.0000	.0005	.0006	.0120	.0171	.0710	.1275	.1602	.4119
	12	.0000	.0000	.0001	.0001	.0039	.0051	.0355	.0565	.1201	.2517
	13	.0000	.0000	.0000	.0000	.0010	.0013	.0146	.0210	.0739	.1316
	14	.0000	.0000	.0000	.0000	.0002	.0003	.0049	.0065	.0370	.0577
	15	.0000	.0000	.0000	.0000	.0000	.0000	.0013	.0016	.0148	.0207
	16	.0000	.0000	.0000	.0000	.0000	.0000	.0003	.0003	.0046	.0059
	17	.0000	.0000	.0000	.0000	.0000	.0000	.0000	.0000	.0011	.0013
	18	.0000	.0000	.0000	.0000	.0000	.0000	.0000	.0000	.0002	.0002
	19	.0000	.0000	.0000	.0000	.0000	.0000	.0000	.0000	.0000	.0000
	20	.0000	.0000	.0000	.0000	.0000	.0000	.0000	.0000	.0000	.0000
22	0	.0985	.0985	.0074	.0074	.0004	.0004	.0000	.0000	.0000	.0000
	1	.2407	.3392	.0406	.0480	.0037	.0041	.0002	.0002	.0000	.0000
	2	.2808	---	.1065	.1545	.0166	.0207	.0014	.0016	.0001	.0001
	3	.2080	.3800	.1775	.3320	.0474	.0681	.0060	.0076	.0004	.0004
	4	.1098	.1719	.2108	---	.0965	.1645	.0190	.0266	.0017	.0022
	5	.0439	.0621	.1898	.4571	.1489	.3134	.0456	.0722	.0063	.0085
	6	.0138	.0182	.1344	.2674	.1808	.4942	.0862	.1584	.0178	.0262
	7	.0035	.0044	.0768	.1330	.1771	---	.1314	.2898	.0407	.0669
	8	.0007	.0009	.0360	.0561	.1423	.3287	.1642	.4540	.0762	.1431
	9	.0001	.0001	.0140	.0201	.0949	.1865	.1703	---	.1186	.2617

Table A.8 *continued*. Binomial individual term and tail probabilities.

		$\pi = 0.1$		$\pi = 0.2$		$\pi = 0.3$		$\pi = 0.4$		$\pi = 0.5$	
n	k										
22	10	.0000	.0000	.0046	.0061	.0529	.0916	.1476	.3756	.1542	.4159
	11	.0000	.0000	.0012	.0016	.0247	.0387	.1073	.2280	.1682	---
	12	.0000	.0000	.0003	.0003	.0097	.0140	.0656	.1207	.1542	.4159
	13	.0000	.0000	.0001	.0001	.0032	.0043	.0336	.0551	.1186	.2617
	14	.0000	.0000	.0000	.0000	.0009	.0011	.0144	.0215	.0762	.1431
	15	.0000	.0000	.0000	.0000	.0002	.0002	.0051	.0070	.0407	.0669
	16	.0000	.0000	.0000	.0000	.0000	.0000	.0015	.0019	.0178	.0262
	17	.0000	.0000	.0000	.0000	.0000	.0000	.0004	.0004	.0063	.0085
	18	.0000	.0000	.0000	.0000	.0000	.0000	.0001	.0001	.0017	.0022
	19	.0000	.0000	.0000	.0000	.0000	.0000	.0000	.0000	.0004	.0004
	20	.0000	.0000	.0000	.0000	.0000	.0000	.0000	.0000	.0001	.0001
	21	.0000	.0000	.0000	.0000	.0000	.0000	.0000	.0000	.0000	.0000
	22	.0000	.0000	.0000	.0000	.0000	.0000	.0000	.0000	.0000	.0000
24	0	.0798	.0798	.0047	.0047	.0002	.0002	.0000	.0000	.0000	.0000
	1	.2127	.2925	.0283	.0331	.0020	.0022	.0001	.0001	.0000	.0000
	2	.2718	---	.0815	.1145	.0097	.0119	.0006	.0007	.0000	.0000
	3	.2215	.4357	.1493	.2639	.0305	.0424	.0028	.0035	.0001	.0001
	4	.1292	.2143	.1960	.4599	.0687	.1111	.0099	.0134	.0006	.0008
	5	.0574	.0851	.1960	---	.1177	.2288	.0265	.0400	.0025	.0033
	6	.0202	.0277	.1552	.3441	.1598	.3886	.0560	.0960	.0080	.0113
	7	.0058	.0075	.0998	.1889	.1761	---	.0960	.1919	.0206	.0320
	8	.0014	.0017	.0530	.0892	.1604	.4353	.1360	.3279	.0438	.0758
	9	.0003	.0003	.0236	.0362	.1222	.2750	.1612	.4891	.0779	.1537
	10	.0000	.0001	.0088	.0126	.0785	.1528	.1612	---	.1169	.2706
	11	.0000	.0000	.0028	.0038	.0428	.0742	.1367	.3498	.1488	.4194
	12	.0000	.0000	.0008	.0010	.0199	.0314	.0988	.2130	.1612	---
	13	.0000	.0000	.0002	.0002	.0079	.0115	.0608	.1143	.1488	.4194
	14	.0000	.0000	.0000	.0000	.0026	.0036	.0318	.0535	.1169	.2706
	15	.0000	.0000	.0000	.0000	.0008	.0010	.0141	.0217	.0779	.1537
	16	.0000	.0000	.0000	.0000	.0002	.0002	.0053	.0075	.0438	.0758
	17	.0000	.0000	.0000	.0000	.0000	.0000	.0017	.0022	.0206	.0320
	18	.0000	.0000	.0000	.0000	.0000	.0000	.0004	.0005	.0080	.0113
	19	.0000	.0000	.0000	.0000	.0000	.0000	.0001	.0001	.0025	.0033
	20	.0000	.0000	.0000	.0000	.0000	.0000	.0000	.0000	.0006	.0008
	21	.0000	.0000	.0000	.0000	.0000	.0000	.0000	.0000	.0001	.0001
	22	.0000	.0000	.0000	.0000	.0000	.0000	.0000	.0000	.0000	.0000
	23	.0000	.0000	.0000	.0000	.0000	.0000	.0000	.0000	.0000	.0000
	24	.0000	.0000	.0000	.0000	.0000	.0000	.0000	.0000	.0000	.0000
26	0	.0646	.0646	.0030	.0030	.0001	.0001	.0000	.0000	.0000	.0000
	1	.1867	.2513	.0196	.0227	.0010	.0011	.0000	.0000	.0000	.0000
	2	.2592	---	.0614	.0841	.0056	.0067	.0002	.0003	.0000	.0000
	3	.2304	.4895	.1228	.2068	.0192	.0260	.0013	.0016	.0000	.0000
	4	.1472	.2591	.1765	.3833	.0473	.0733	.0050	.0066	.0002	.0003
	5	.0720	.1118	.1941	---	.0893	.1626	.0148	.0214	.0010	.0012
	6	.0280	.0399	.1699	.4225	.1339	.2965	.0345	.0559	.0034	.0047
	7	.0089	.0119	.1213	.2526	.1640	.4605	.0657	.1216	.0098	.0145
	8	.0023	.0030	.0720	.1313	.1669	---	.1040	.2255	.0233	.0378
	9	.0005	.0006	.0360	.0592	.1431	.3726	.1386	.3642	.0466	.0843
	10	.0001	.0001	.0153	.0232	.1042	.2295	.1571	---	.0792	.1635
	11	.0000	.0000	.0056	.0079	.0650	.1253	.1524	.4787	.1151	.2786
	12	.0000	.0000	.0017	.0023	.0348	.0603	.1270	.3263	.1439	.4225
	13	.0000	.0000	.0005	.0006	.0161	.0255	.0912	.1993	.1550	---
	14	.0000	.0000	.0001	.0001	.0064	.0094	.0564	.1082	.1439	.4225
	15	.0000	.0000	.0000	.0000	.0022	.0030	.0301	.0518	.1151	.2786
	16	.0000	.0000	.0000	.0000	.0006	.0009	.0138	.0217	.0792	.1635
	17	.0000	.0000	.0000	.0000	.0002	.0002	.0054	.0079	.0466	.0843
	18	.0000	.0000	.0000	.0000	.0000	.0000	.0018	.0025	.0233	.0378
	19	.0000	.0000	.0000	.0000	.0000	.0000	.0005	.0007	.0098	.0145
	20	.0000	.0000	.0000	.0000	.0000	.0000	.0001	.0001	.0034	.0047
	21	.0000	.0000	.0000	.0000	.0000	.0000	.0000	.0000	.0010	.0012
	22	.0000	.0000	.0000	.0000	.0000	.0000	.0000	.0000	.0002	.0003
	23	.0000	.0000	.0000	.0000	.0000	.0000	.0000	.0000	.0000	.0000
	24	.0000	.0000	.0000	.0000	.0000	.0000	.0000	.0000	.0000	.0000

Table A.8 *continued*. Binomial individual term and tail probabilities.

Table A.8 *continued*. Binomial individual term and tail probabilities.										
		$\pi = 0.1$		$\pi = 0.2$		$\pi = 0.3$		$\pi = 0.4$		$\pi = 0.5$
n	k									
26	25	.0000 .0000		.0000 .0000		.0000 .0000		.0000 .0000		.0000 .0000
	26	.0000 .0000		.0000 .0000		.0000 .0000		.0000 .0000		.0000 .0000
28	0	.0523 .0523		.0019 .0019		.0000 .0000		.0000 .0000		.0000 .0000
	1	.1628 .2152		.0135 .0155		.0006 .0006		.0000 .0000		.0000 .0000
	2	.2442 .4594		.0457 .0612		.0032 .0038		.0001 .0001		.0000 .0000
	3	.2352 - - -		.0990 .1602		.0119 .0157		.0006 .0007		.0000 .0000
	4	.1633 .3054		.1547 .3149		.0318 .0474		.0025 .0032		.0001 .0001
	5	.0871 .1421		.1856 - - -		.0654 .1128		.0079 .0111		.0004 .0005
	6	.0371 .0550		.1779 .4995		.1074 .2202		.0203 .0315		.0014 .0019
	7	.0130 .0179		.1398 .3216		.1446 .3648		.0426 .0740		.0044 .0063
	8	.0038 .0050		.0917 .1818		.1627 - - -		.0745 .1485		.0116 .0178
	9	.0009 .0012		.0510 .0900		.1550 .4725		.1103 .2588		.0257 .0436
	10	.0002 .0002		.0242 .0391		.1262 .3175		.1398 .3986		.0489 .0925
	11	.0000 .0000		.0099 .0149		.0885 .1913		.1525 - - -		.0800 .1725
	12	.0000 .0000		.0035 .0050		.0537 .1028		.1440 .4490		.1133 .2858
	13	.0000 .0000		.0011 .0015		.0283 .0491		.1181 .3050		.1395 .4253
	14	.0000 .0000		.0003 .0004		.0130 .0208		.0844 .1868		.1494 - - -
	15	.0000 .0000		.0001 .0001		.0052 .0077		.0525 .1025		.1395 .4253
	16	.0000 .0000		.0000 .0000		.0018 .0025		.0284 .0499		.1133 .2858
	17	.0000 .0000		.0000 .0000		.0005 .0007		.0134 .0215		.0800 .1725
	18	.0000 .0000		.0000 .0000		.0001 .0002		.0055 .0081		.0489 .0925
	19	.0000 .0000		.0000 .0000		.0000 .0000		.0019 .0027		.0257 .0436
	20	.0000 .0000		.0000 .0000		.0000 .0000		.0006 .0008		.0116 .0178
	21	.0000 .0000		.0000 .0000		.0000 .0000		.0001 .0002		.0044 .0063
	22	.0000 .0000		.0000 .0000		.0000 .0000		.0000 .0000		.0014 .0019
	23	.0000 .0000		.0000 .0000		.0000 .0000		.0000 .0000		.0004 .0005
	24	.0000 .0000		.0000 .0000		.0000 .0000		.0000 .0000		.0001 .0001
	25	.0000 .0000		.0000 .0000		.0000 .0000		.0000 .0000		.0000 .0000
	26	.0000 .0000		.0000 .0000		.0000 .0000		.0000 .0000		.0000 .0000
	27	.0000 .0000		.0000 .0000		.0000 .0000		.0000 .0000		.0000 .0000
	28	.0000 .0000		.0000 .0000		.0000 .0000		.0000 .0000		.0000 .0000
30	0	.0424 .0424		.0012 .0012		.0000 .0000		.0000 .0000		.0000 .0000
	1	.1413 .1837		.0093 .0105		.0003 .0003		.0000 .0000		.0000 .0000
	2	.2277 .4114		.0337 .0442		.0018 .0021		.0000 .0000		.0000 .0000
	3	.2361 - - -		.0785 .1227		.0072 .0093		.0003 .0003		.0000 .0000
	4	.1771 .3526		.1325 .2552		.0208 .0302		.0012 .0015		.0000 .0000
	5	.1023 .1755		.1723 .4275		.0464 .0766		.0041 .0057		.0001 .0002
	6	.0474 .0732		.1795 - - -		.0829 .1595		.0115 .0172		.0006 .0007
	7	.0180 .0258		.1538 .3930		.1219 .2814		.0263 .0435		.0019 .0026
	8	.0058 .0078		.1106 .2392		.1501 .4315		.0505 .0940		.0055 .0081
	9	.0016 .0020		.0676 .1287		.1573 - - -		.0823 .1763		.0133 .0214
	10	.0004 .0005		.0355 .0611		.1416 .4112		.1152 .2915		.0280 .0494
	11	.0001 .0001		.0161 .0256		.1103 .2696		.1396 .4311		.0509 .1002
	12	.0000 .0000		.0064 .0095		.0749 .1593		.1474 - - -		.0806 .1808
	13	.0000 .0000		.0022 .0031		.0444 .0845		.1360 .4215		.1115 .2923
	14	.0000 .0000		.0007 .0009		.0231 .0401		.1101 .2855		.1354 .4278
	15	.0000 .0000		.0002 .0002		.0106 .0169		.0783 .1754		.1445 - - -
	16	.0000 .0000		.0000 .0001		.0042 .0064		.0489 .0971		.1354 .4278
	17	.0000 .0000		.0000 .0000		.0015 .0021		.0269 .0481		.1115 .2923
	18	.0000 .0000		.0000 .0000		.0005 .0006		.0129 .0212		.0806 .1808
	19	.0000 .0000		.0000 .0000		.0001 .0002		.0054 .0083		.0509 .1002
	20	.0000 .0000		.0000 .0000		.0000 .0000		.0020 .0029		.0280 .0494
	21	.0000 .0000		.0000 .0000		.0000 .0000		.0006 .0009		.0133 .0214
	22	.0000 .0000		.0000 .0000		.0000 .0000		.0002 .0002		.0055 .0081
	23	.0000 .0000		.0000 .0000		.0000 .0000		.0000 .0000		.0019 .0026
	24	.0000 .0000		.0000 .0000		.0000 .0000		.0000 .0000		.0006 .0007
	25	.0000 .0000		.0000 .0000		.0000 .0000		.0000 .0000		.0001 .0002
	26	.0000 .0000		.0000 .0000		.0000 .0000		.0000 .0000		.0000 .0000
	27	.0000 .0000		.0000 .0000		.0000 .0000		.0000 .0000		.0000 .0000
	28	.0000 .0000		.0000 .0000		.0000 .0000		.0000 .0000		.0000 .0000
	29	.0000 .0000		.0000 .0000		.0000 .0000		.0000 .0000		.0000 .0000
	30	.0000 .0000		.0000 .0000		.0000 .0000		.0000 .0000		.0000 .0000

Table A.9. Fisher's exact test for 2 × 2 tables.

Consider a 2 by 2 table: $\begin{array}{c|c} a & A\text{-}a \\ \hline b & B\text{-}b \end{array} \begin{array}{c} A \\ B \end{array}$

with rows and/or columns exchanged so that (1) $A \geqq B$ and (2) $(a/A) \geqq (b/B)$. The table entries are ordered lexicographically by A (ascending), B (descending) and a (descending). For each triple (A,B,a) the table presents critical values for one-sided tests of the hypothesis that the true proportion corresponding to a/A is greater than the true proportion corresponding to b/B. Significance levels of 0.05, 0.025 and 0.01 are considered. For $A \leqq 15$ all values where critical values exist are tabulated. For each significance level two columns give: (1) the nominal critical value for b (that is, reject the null hypothesis if the observed b is less than or equal to the table entry) and (2) the p-value corresponding to the critical value (this is less than the nominal significance level in most cases due to the discreteness of the distribution).

A	B	a	b	p	b	p	b	p	A	B	a	b	p	b	p	b	p
3	3	3	0	.050	-	---	-	---	8	5	8	2	.035	1	.007	1	.007
4	4	4	0	.014	0	.014	-	---	8	5	7	1	.032	0	.005	0	.005
4	3	4	0	.029	-	---	-	---	8	5	6	0	.016	0	.016	-	---
5	5	5	1	.024	1	.024	0	.004	8	5	5	0	.044	-	---	-	---
5	5	4	0	.024	0	.024	-	---	8	4	8	1	.018	1	.018	0	.002
5	4	5	1	.048	0	.008	0	.008	8	4	7	0	.010	0	.010	-	---
5	4	4	0	.040	-	---	-	---	8	4	6	0	.030	-	---	-	---
5	3	5	0	.018	0	.018	-	---	8	3	8	0	.006	0	.006	0	.006
5	2	5	0	.048	-	---	-	---	8	3	7	0	.024	0	.024	-	---
6	6	6	2	.030	1	.008	1	.008	8	2	8	0	.022	0	.022	-	---
6	6	5	1	.040	0	.008	0	.008	9	9	9	5	.041	4	.015	3	.005
6	6	4	0	.030	-	---	-	---	9	9	8	3	.025	3	.025	2	.008
6	5	6	1	.015	1	.015	0	.002	9	9	7	2	.028	1	.008	1	.008
6	5	5	0	.013	0	.013	-	---	9	9	6	1	.025	1	.025	0	.005
6	5	4	0	.045	-	---	-	---	9	9	5	0	.015	0	.015	-	---
6	4	6	1	.033	0	.005	0	.005	9	9	4	0	.041	-	---	-	---
6	4	5	0	..024	0	.024	-	---	9	8	9	4	.029	3	.009	3	.009
6	3	6	0	.012	0	.012	-	---	9	8	8	3	.043	2	.013	1	.003
6	3	5	0	.048	-	---	-	---	9	8	7	2	.044	1	.012	0	.002
6	2	6	0	.036	-	---	-	---	9	8	6	1	.036	0	.007	0	.007
7	7	7	3	.035	2	.010	1	.002	9	8	5	0	.020	0	.020	-	---
7	7	6	1	.015	1	.015	0	.002	9	7	9	3	.019	3	.019	2	.005
7	7	5	0	.010	0	.010	-	---	9	7	8	2	.024	2	.024	1	.006
7	7	4	0	.035	-	---	-	---	9	7	7	1	.020	1	.020	0	.003
7	6	7	2	.021	2	.021	1	.005	9	7	6	0	.010	0	.010	-	---
7	6	6	1	.025	0	.004	0	.004	9	7	5	0	.029	-	---	-	---
7	6	5	0	.016	0	.016	-	---	9	6	9	3	.044	2	.011	1	.002
7	6	4	0	.049	-	---	-	---	9	6	8	2	.047	1	.011	0	.001
7	5	7	2	.045	1	.010	0	.001	9	6	7	1	.035	0	.006	0	.006
7	5	6	1	.045	0	.008	0	.008	9	6	6	0	.017	0	.017	-	---
7	5	5	0	.027	-	---	-	---	9	6	5	0	.042	-	---	-	---
7	4	7	1	.024	1	.024	0	.003	9	5	9	2	.027	1	.005	1	.005
7	4	6	0	.015	0	.015	-	---	9	5	8	1	.023	1	.023	0	.003
7	4	5	0	.045	-	---	-	---	9	5	7	0	.010	0	.010	-	---
7	3	7	0	.008	0	.008	0	.008	9	5	6	0	.028	-	---	-	---
7	3	6	0	.033	-	---	-	---	9	4	9	1	.014	1	.014	0	.001
7	2	7	0	.028	-	---	-	---	9	4	8	0	.007	0	.007	0	.007
8	8	8	4	.038	3	.013	2	.003	9	4	7	0	.021	0	.021	-	---
8	8	7	2	.020	2	.020	1	.005	9	4	6	0	.049	-	---	-	---
8	8	6	1	.020	1	.020	0	.003	9	3	9	1	.045	0	.005	0	.005
8	8	5	0	.013	0	.013	-	---	9	3	8	0	.018	0	.018	-	---
8	8	4	0	.038	-	---	-	---	9	3	7	0	.045	-	---	-	---
8	7	8	3	.026	2	.007	2	.007	9	2	9	0	.018	0	.018	-	---
8	7	7	2	.035	1	.009	1	.009	10	10	10	6	.043	5	.016	4	.005
8	7	6	1	.032	0	.006	0	.006	10	10	9	4	.029	3	.010	3	.010
8	7	5	0	.019	0	.019	-	---	10	10	8	3	.035	2	.012	1	.003
8	6	8	2	.015	2	.015	1	.003	10	10	7	2	.035	1	.010	1	.010
8	6	7	1	.016	1	.016	0	.002	10	10	6	1	.029	0	.005	0	.005
8	6	6	0	.009	0	.009	0	.009	10	10	5	0	.016	0	.016	-	---
8	6	5	0	.028	-	---	-	---	10	10	4	0	.043	-	---	-	---

Table A.9 *continued*. Fisher's exact test for 2 × 2 tables.

A	B	a	b	p	b	p	b	p	A	B	a	b	p	b	p	b	p
10	9	10	5	.033	4	.011	3	.003	11	8	10	3	.024	3	.024	2	.006
10	9	9	4	.050	3	.017	2	.005	11	8	9	2	.022	2	.022	1	.005
10	9	8	2	.019	2	.019	1	.004	11	8	8	1	.015	1	.015	0	.002
10	9	7	1	.015	1	.015	0	.002	11	8	7	1	.037	0	.007	0	.007
10	9	6	1	.040	0	.008	0	.008	11	8	6	0	.017	0	.017	–	—
10	9	5	0	.022	0	.022	–	—	11	8	5	0	.040	–	—	–	—
10	8	10	4	.023	4	.023	3	.007	11	7	11	4	.043	3	.011	2	.002
10	8	9	3	.032	2	.009	2	.009	11	7	10	3	.047	2	.013	1	.002
10	8	8	2	.031	1	.008	1	.008	11	7	9	2	.039	1	.009	1	.009
10	8	7	1	.023	1	.023	0	.004	11	7	8	1	.025	1	.025	0	.004
10	8	6	0	.011	0	.011	–	—	11	7	7	0	.010	0	.010	–	—
10	8	5	0	.029	–	—	–	—	11	7	6	0	.025	0	.025	–	—
10	7	10	3	.015	3	.015	2	.003	11	6	11	3	.029	2	.006	2	.006
10	7	9	2	.018	2	.018	1	.004	11	6	10	2	.028	1	.005	1	.005
10	7	8	1	.013	1	.013	0	.002	11	6	9	1	.018	1	.018	0	.002
10	7	7	1	.036	0	.006	0	.006	11	6	8	1	.043	0	.007	0	.007
10	7	6	0	.017	0	.017	–	—	11	6	7	0	.017	0	.017	–	—
10	7	5	0	.041	–	—	–	—	11	6	6	0	.037	–	—	–	—
10	6	10	3	.036	2	.008	2	.008	11	5	11	2	.018	2	.018	1	.003
10	6	9	2	.036	1	.008	1	.008	11	5	10	1	.013	1	.013	0	.001
10	6	8	1	.024	1	.024	0	.003	11	5	9	1	.036	0	.005	0	.005
10	6	7	0	.010	0	.010	–	—	11	5	8	0	.013	0	.013	–	—
10	6	6	0	.026	–	—	–	—	11	5	7	0	.029	–	—	–	—
10	5	10	2	.022	2	.022	1	.004	11	4	11	1	.009	1	.009	1	.009
10	5	9	1	.017	1	.017	0	.002	11	4	10	1	.033	0	.004	0	.004
10	5	8	1	.047	0	.007	0	.007	11	4	9	0	.011	0	.011	–	—
10	5	7	0	.019	0	.019	–	—	11	4	8	0	.026	–	—	–	—
10	5	6	0	.042	–	—	–	—	11	3	11	1	.033	0	.003	0	.003
10	4	10	1	.011	1	.011	0	.001	11	3	10	0	.011	0	.011	–	—
10	4	9	1	.041	0	.005	0	.005	11	3	9	0	.027	–	—	–	—
10	4	8	0	.015	0	.015	–	—	11	2	11	0	.013	0	.013	–	—
10	4	7	0	.035	–	—	–	—	11	2	10	0	.038	–	—	–	—
10	3	10	1	.038	0	.003	0	.003	12	12	12	8	.047	7	.019	6	.007
10	3	9	0	.014	0	.014	–	—	12	12	11	6	.034	5	.014	4	.005
10	3	8	0	.035	–	—	–	—	12	12	10	5	.045	4	.018	3	.006
10	2	10	0	.015	0	.015	–	—	12	12	9	4	.050	3	.020	2	.006
10	2	9	0	.045	–	—	–	—	12	12	8	3	.050	2	.018	1	.005
11	11	11	7	.045	6	.018	5	.006	12	12	7	2	.045	1	.014	0	.002
11	11	10	5	.032	4	.012	3	.004	12	12	6	1	.034	0	.007	0	.007
11	11	9	4	.040	3	.015	2	.004	12	12	5	0	.019	0	.019	–	—
11	11	8	3	.043	2	.015	1	.004	12	12	4	0	.047	–	—		
11	11	7	2	.040	1	.012	0	.002	12	11	12	7	.037	6	.014	5	.005
11	11	6	1	.032	0	.006	0	.006	12	11	11	5	.024	5	.024	4	.008
11	11	5	0	.018	0	.018	–	—	12	11	10	4	.029	3	.010	2	.003
11	11	4	0	.045	–	—			12	11	9	3	.030	2	.009	2	.009
11	10	11	6	.035	5	.012	4	.004	12	11	8	2	.026	1	.007	1	.007
11	10	10	4	.021	4	.021	3	.007	12	11	7	1	.019	1	.019	0	.003
11	10	9	3	.024	3	.024	2	.007	12	11	6	1	.045	0	.009	0	.009
11	10	8	2	.023	2	.023	1	.006	12	11	5	0	.024	0	.024	–	—
11	10	7	1	.017	1	.017	0	.003	12	10	12	6	.029	5	.010	5	.010
11	10	6	1	.043	0	.009	0	.009	12	10	11	5	.043	4	.015	3	.005
11	10	5	0	.023	0	.023	–	—	12	10	10	4	.048	3	.017	2	.005
11	9	11	5	.026	4	.008	4	.008	12	10	9	3	.046	2	.015	1	.004
11	9	10	4	.038	3	.012	2	.003	12	10	8	2	.038	1	.010	0	.002
11	9	9	3	.040	2	.012	1	.003	12	10	7	1	.026	0	.005	0	.005
11	9	8	2	.035	1	.009	1	.009	12	10	6	0	.012	0	.012	–	—
11	9	7	1	.025	1	.025	0	.004	12	10	5	0	.030	–	—	–	—
11	9	6	0	.012	0	.012	–	—	12	9	12	5	.021	5	.021	4	.006
11	9	5	0	.030	–	—	–	—	12	9	11	4	.029	3	.009	3	.009
11	8	11	4	.018	4	.018	3	.005	12	9	10	3	.029	2	.008	2	.008

Table A.9 *continued*. Fisher's exact test for 2 × 2 tables.

A	B	a	b	p	b	p	b	p	A	B	a	b	p	b	p	b	p
12	9	9	2	.024	2	.024	1	.006	13	12	6	1	.046	0	.010	0	.010
12	9	8	1	.016	1	.016	0	.002	13	12	5	0	.024	0	.024	—	—
12	9	7	1	.037	0	.007	0	.007	13	11	13	7	.031	6	.011	5	.003
12	9	6	0	.017	0	.017	—	—	13	11	12	6	.048	5	.018	4	.006
12	9	5	0	.039	—	—	—	—	13	11	11	4	.021	4	.021	3	.007
12	8	12	5	.049	4	.014	3	.004	13	11	10	3	.021	3	.021	2	.006
12	8	11	3	.018	3	.018	2	.004	13	11	9	3	.050	2	.017	1	.004
12	8	10	2	.015	2	.015	1	.003	13	11	8	2	.040	1	.011	0	.002
12	8	9	2	.040	1	.010	1	.010	13	11	7	1	.027	0	.005	0	.005
12	8	8	1	.025	1	.025	0	.004	13	11	6	0	.013	0	.013	—	—
12	8	7	0	.010	0	.010	—	—	13	11	5	0	.030	—	—	—	—
12	8	6	0	.024	0	.024	—	—	13	10	13	6	.024	6	.024	5	.007
12	7	12	4	.036	3	.009	3	.009	13	10	12	5	.035	4	.012	3	.003
12	7	11	3	.038	2	.010	2	.010	13	10	11	4	.037	3	.012	2	.003
12	7	10	2	.029	1	.006	1	.006	13	10	10	3	.033	2	.010	1	.002
12	7	9	1	.017	1	.017	0	.002	13	10	9	2	.026	1	.006	1	.006
12	7	8	1	.040	0	.007	0	.007	13	10	8	1	.017	1	.017	0	.003
12	7	7	0	.016	0	.016	—	—	13	10	7	1	.038	0	.007	0	.007
12	7	6	0	.034	—	—	—	—	13	10	6	0	.017	0	.017	—	—
12	6	12	3	.025	3	.025	2	.005	13	10	5	0	.038	—	—	—	—
12	6	11	2	.022	2	.022	1	.004	13	9	13	5	.017	5	.017	4	.005
12	6	10	1	.013	1	.013	0	.002	13	9	12	4	.023	4	.023	3	.007
12	6	9	1	.032	0	.005	0	.005	13	9	11	3	.022	3	.022	2	.006
12	6	8	0	.011	0	.011	—	—	13	9	10	2	.017	2	.017	1	.004
12	6	7	0	.025	0	.025	—	—	13	9	9	2	.040	1	.010	0	.001
12	6	6	0	.050	—	—	—	—	13	9	8	1	.025	1	.025	0	.004
12	5	12	2	.015	2	.015	1	.002	13	9	7	0	.010	0	.010	—	—
12	5	11	1	.010	1	.010	1	.010	13	9	6	0	.023	0	.023	—	—
12	5	10	1	.028	0	.003	0	.003	13	9	5	0	.049	—	—	—	—
12	5	9	0	.009	0	.009	0	.009	13	8	13	5	.042	4	.012	3	.003
12	5	8	0	.020	0	.020	—	—	13	8	12	4	.047	3	.014	2	.003
12	5	7	0	.041	—	—	—	—	13	8	11	3	.041	2	.011	1	.002
12	4	12	2	.050	1	.007	1	.007	13	8	10	2	.029	1	.007	1	.007
12	4	11	1	.027	0	.003	0	.003	13	8	9	1	.017	1	.017	0	.002
12	4	10	0	.008	0	.008	0	.008	13	8	8	1	.037	0	.006	0	.006
12	4	9	0	.019	0	.019	—	—	13	8	7	0	.015	0	.015	—	—
12	4	8	0	.038	—	—	—	—	13	8	6	0	.032	—	—	—	—
12	3	12	1	.029	0	.002	0	.002	13	7	13	4	.031	3	.007	3	.007
12	3	11	0	.009	0	.009	0	.009	13	7	12	3	.031	2	.007	2	.007
12	3	10	0	.022	0	.022	—	—	13	7	11	2	.022	2	.022	1	.004
12	3	9	0	.044	—	—	—	—	13	7	10	1	.012	1	.012	0	.002
12	2	12	0	.011	0	.011	—	—	13	7	9	1	.029	0	.004	0	.004
12	2	11	0	.033	—	—	—	—	13	7	8	0	.010	0	.010	—	—
13	13	13	9	.048	8	.020	7	.007	13	7	7	0	.022	0	.022	—	—
13	13	12	7	.037	6	.015	5	.006	13	7	6	0	.044	—	—	—	—
13	13	11	6	.048	5	.021	4	.008	13	6	13	3	.021	3	.021	2	.004
13	13	10	4	.024	4	.024	3	.008	13	6	12	2	.017	2	.017	1	.003
13	13	9	3	.024	3	.024	2	.008	13	6	11	2	.046	1	.010	1	.010
13	13	8	2	.021	2	.021	1	.006	13	6	10	1	.024	1	.024	0	.003
13	13	7	2	.048	1	.015	0	.003	13	6	9	1	.050	0	.008	0	.008
13	13	6	1	.037	0	.007	0	.007	13	6	8	0	.017	0	.017	—	—
13	13	5	0	.020	0	.020	—	—	13	6	7	0	.034	—	—	—	—
13	13	4	0	.048	—	—	—	—	13	5	13	2	.012	2	.012	1	.002
13	12	13	8	.039	7	.015	6	.005	13	5	12	2	.044	1	.008	1	.008
13	12	12	6	.027	5	.010	5	.010	13	5	11	1	.022	1	.022	0	.002
13	12	11	5	.033	4	.013	3	.004	13	5	10	1	.047	0	.007	0	.007
13	12	10	4	.036	3	.013	2	.004	13	5	9	0	.015	0	.015	—	—
13	12	9	3	.034	2	.011	1	.003	13	5	8	0	.029	—	—	—	—
13	12	8	2	.029	1	.008	1	.008	13	4	13	2	.044	1	.006	1	.006
13	12	7	1	.020	1	.020	0	.004	13	4	12	1	.022	1	.022	0	.002

Table A.9 *continued*. Fisher's exact test for 2 × 2 tables.

A	B	a	b	p	b	p	b	p	A	B	a	b	p	b	p	b	p
13	4	11	0	.006	0	.006	0	.006	14	9	14	6	.047	5	.014	4	.004
13	4	10	0	.015	0	.015	-	---	14	9	13	4	.018	4	.018	3	.005
13	4	9	0	.029	-	---	-	---	14	9	12	3	.017	3	.017	2	.004
13	3	13	1	.025	1	.025	0	.002	14	9	11	3	.042	2	.012	1	.002
13	3	12	0	.007	0	.007	0	.007	14	9	10	2	.029	1	.007	1	.007
13	3	11	0	.018	0	.018	-	---	14	9	9	1	.017	1	.017	0	.002
13	3	10	0	.036	-	---	-	---	14	9	8	1	.036	0	.006	0	.006
13	2	13	0	.010	0	.010	0	.010	14	9	7	0	.014	0	.014	-	---
13	2	12	0	.029	-	---	-	---	14	9	6	0	.030	-	---	-	---
14	14	14	10	.049	9	.020	8	.008	14	8	14	5	.036	4	.010	4	.010
14	14	13	8	.038	7	.016	6	.006	14	8	13	4	.039	3	.011	2	.002
14	14	12	6	.023	6	.023	5	.009	14	8	12	3	.032	2	.008	2	.008
14	14	11	5	.027	4	.011	3	.004	14	8	11	2	.022	2	.022	1	.005
14	14	10	4	.028	3	.011	2	.003	14	8	10	2	.048	1	.012	0	.002
14	14	9	3	.027	2	.009	2	.009	14	8	9	1	.026	0	.004	0	.004
14	14	8	2	.023	2	.023	1	.006	14	8	8	0	.009	0	.009	0	.009
14	14	7	1	.016	1	.016	0	.003	14	8	7	0	.020	0	.020	-	---
14	14	6	1	.038	0	.008	0	.008	14	8	6	0	.040	-	---	-	---
14	14	5	0	.020	0	.020	-	---	14	7	14	4	.026	3	.006	3	.006
14	14	4	0	.049	-	---	-	---	14	7	13	3	.025	2	.006	2	.006
14	13	14	9	.041	8	.016	7	.006	14	7	12	2	.017	2	.017	1	.003
14	13	13	7	.029	6	.011	5	.004	14	7	11	2	.041	1	.009	1	.009
14	13	12	6	.037	5	.015	4	.005	14	7	10	1	.021	1	.021	0	.003
14	13	11	5	.041	4	.017	3	.006	14	7	9	1	.043	0	.007	0	.007
14	13	10	4	.041	3	.016	2	.005	14	7	8	0	.015	0	.015	-	---
14	13	9	3	.038	2	.013	1	.003	14	7	7	0	.030	-	---	-	---
14	13	8	2	.031	1	.009	1	.009	14	6	14	3	.018	3	.018	2	.003
14	13	7	1	.021	1	.021	0	.004	14	6	13	2	.014	2	.014	1	.002
14	13	6	1	.048	0	.010	-	---	14	6	12	2	.037	1	.007	1	.007
14	13	5	0	.025	0	.025	-	---	14	6	11	1	.018	1	.018	0	.002
14	12	14	8	.033	7	.012	6	.004	14	6	10	1	.038	0	.005	0	.005
14	12	13	6	.021	6	.021	5	.007	14	6	9	0	.012	0	.012	-	---
14	12	12	5	.025	4	.009	4	.009	14	6	8	0	.024	0	.024	-	---
14	12	11	4	.026	3	.009	3	.009	14	6	7	0	.044	-	---	-	---
14	12	10	3	.024	3	.024	2	.007	14	5	14	2	.010	2	.010	1	.001
14	12	9	2	.019	2	.019	1	.005	14	5	13	2	.037	1	.006	1	.006
14	12	8	2	.042	1	.012	0	.002	14	5	12	1	.017	1	.017	0	.002
14	12	7	1	.028	0	.005	0	.005	14	5	11	1	.038	0	.005	0	.005
14	12	6	0	.013	0	.013	-	---	14	5	10	0	.011	0	.011	-	---
14	12	5	0	.030	-	---	-	---	14	5	9	0	.022	0	.022	-	---
14	11	14	7	.026	6	.009	6	.009	14	5	8	0	.040	-	---	-	---
14	11	13	6	.039	5	.014	4	.004	14	4	14	2	.039	1	.005	1	.005
14	11	12	5	.043	4	.016	3	.005	14	4	13	1	.019	1	.019	0	.002
14	11	11	4	.042	3	.015	2	.004	14	4	12	1	.044	0	.005	0	.005
14	11	10	3	.036	2	.011	1	.003	14	4	11	0	.011	0	.011	-	---
14	11	9	2	.027	1	.007	1	.007	14	4	10	0	.023	0	.023	-	---
14	11	8	1	.017	1	.017	0	.003	14	4	9	0	.041	-	---	-	---
14	11	7	1	.038	0	.007	0	.007	14	3	14	1	.022	1	.022	0	.001
14	11	6	0	.017	0	.017	-	---	14	3	13	0	.006	0	.006	0	.006
14	11	5	0	.038	-	---	-	---	14	3	12	0	.015	0	.015	-	---
14	10	14	6	.020	6	.020	5	.006	14	3	11	0	.029	-	---	-	---
14	10	13	5	.028	4	.009	4	.009	14	2	14	0	.008	0	008	0	.008
14	10	12	4	.028	3	.009	3	.009	14	2	13	0	.025	0	.025	-	---
14	10	11	3	.024	3	.024	2	.007	14	2	12	0	.050	-	---	-	---
14	10	10	2	.018	2	.018	1	.004	15	15	15	11	.050	10	.021	9	.008
14	10	9	2	.040	1	.011	0	.002	15	15	14	9	.040	8	.018	7	.007
14	10	8	1	.024	1	.024	0	.004	15	15	13	7	.025	6	.010	5	.004
14	10	7	0	.010	0	.010	0	.010	15	15	12	6	.030	5	.013	4	.005
14	10	6	0	.022	0	.022	-	---	15	15	11	5	.033	4	.013	3	.005
14	10	5	0	.047	-	---	-	---	15	15	10	4	.033	3	.013	2	.004

Table A.9 *continued*. Fisher's exact test for 2 × 2 tables.													
A	*B*	*a*	*b*	*p*	*b*	*p*	*b*	*p*	*A*	*B*	*a*	*b*	*p*	*b*	*p*	*b*	*p*

A	B	a	b	p	b	p	b	p	A	B	a	b	p	b	p	b	p
15	15	9	3	.030	2	.010	1	.003	15	9	15	6	.042	5	.012	4	.003
15	15	8	2	.025	1	.007	1	.007	15	9	14	5	.047	4	.015	3	.004
15	15	7	1	.018	1	.018	0	.003	15	9	13	4	.042	3	.013	2	.003
15	15	6	1	.040	0	.008	0	.008	15	9	12	3	.032	2	.009	2	.009
15	15	5	0	.021	0	.021	-	---	15	9	11	2	.021	2	.021	1	.005
15	15	4	0	.050	-	---	-	---	15	9	10	2	.045	1	.011	0	.002
15	14	15	10	.042	9	.017	8	.006	15	9	9	1	.024	1	.024	0	.004
15	14	14	8	.031	7	.013	6	.005	15	9	8	1	.048	0	.009	0	.009
15	14	13	7	.041	6	.017	5	.007	15	9	7	0	.019	0	.019	-	---
15	14	12	6	.046	5	.020	4	.007	15	9	6	0	.037	-	---	-	---
15	14	11	5	.048	4	.020	3	.007	15	8	15	5	.032	4	.008	4	.008
15	14	10	4	.046	3	.018	2	.006	15	8	14	4	.033	3	.009	3	.009
15	14	9	3	.041	2	.014	1	.004	15	8	13	3	.026	2	.006	2	.006
15	14	8	2	.033	1	.009	1	.009	15	8	12	2	.017	2	.017	1	.003
15	14	7	1	.022	1	.022	0	.004	15	8	11	2	.037	1	.008	1	.008
15	14	6	1	.049	0	.011	-	---	15	8	10	1	.019	1	.019	0	.003
15	14	5	0	.025	-	---	-	---	15	8	9	1	.038	0	.006	0	.006
15	13	15	9	.035	8	.013	7	.005	15	8	8	0	.013	0	.013	-	---
15	13	14	7	.023	7	.023	6	.009	15	8	7	0	.026	-	---	-	---
15	13	13	6	.029	5	.011	4	.004	15	8	6	0	.050	-	---	-	---
15	13	12	5	.031	4	.012	3	.004	15	7	15	4	.023	4	.023	3	.005
15	13	11	4	.030	3	.011	2	.003	15	7	14	3	.021	3	.021	2	.004
15	13	10	3	.026	2	.008	2	.008	15	7	13	2	.014	2	.014	1	.002
15	13	9	2	.020	2	.020	1	.005	15	7	12	2	.032	1	.007	1	.007
15	13	8	2	.043	1	.013	0	.002	15	7	11	1	.015	1	.015	0	.002
15	13	7	1	.029	0	.005	0	.005	15	7	10	1	.032	0	.005	0	.005
15	13	6	0	.013	0	.013	-	---	15	7	9	0	.010	0	.010	-	---
15	13	5	0	.031	-	---	-	---	15	7	8	0	.020	0	.020	-	---
15	12	15	8	.028	7	.010	7	.010	15	7	7	0	.038	-	---	-	---
15	12	14	7	.043	6	.016	5	.006	15	6	15	3	.015	3	.015	2	.003
15	12	13	6	.049	5	.019	4	.007	15	6	14	2	.011	2	.011	1	.002
15	12	12	5	.049	4	.019	3	.006	15	6	13	2	.031	1	.006	1	.006
15	12	11	4	.045	3	.017	2	.005	15	6	12	1	.014	1	.014	0	.002
15	12	10	3	.038	2	.012	1	.003	15	6	11	1	.029	0	.004	0	.004
15	12	9	2	.028	1	.007	1	.007	15	6	10	0	.009	0	.009	0	.009
15	12	8	1	.018	1	.018	0	.003	15	6	9	0	.017	0	.017	-	---
15	12	7	1	.038	0	.007	0	.007	15	6	8	0	.032	-	---	-	---
15	12	6	0	.017	0	.017	-	---	15	5	15	2	.009	2	.009	2	.009
15	12	5	0	.037	-	---	-	---	15	5	14	2	.032	1	.005	1	.005
15	11	15	7	.022	7	.022	6	.007	15	5	13	1	.014	1	.014	0	.001
15	11	14	6	.032	5	.011	4	.003	15	5	12	1	.031	0	.004	0	.004
15	11	13	5	.034	4	.012	3	.003	15	5	11	0	.008	0	.008	0	.008
15	11	12	4	.032	3	.010	2	.003	15	5	10	0	.016	0	.016	-	---
15	11	11	3	.026	2	.008	2	.008	15	5	9	0	.030	-	---	-	---
15	11	10	2	.019	2	.019	1	.004	15	4	15	2	.035	1	.004	1	.004
15	11	9	2	.040	1	.011	0	.002	15	4	14	1	.016	1	.016	0	.001
15	11	8	1	.024	1	.024	0	.004	15	4	13	1	.037	0	.004	0	.004
15	11	7	1	.049	0	.010	0	.010	15	4	12	0	.009	0	.009	0	.009
15	11	6	0	.022	0	.022	-	---	15	4	11	0	.018	0	.018	-	---
15	11	5	0	.046	-	---	-	---	15	4	10	0	.033	-	---	-	---
15	10	15	6	.017	6	.017	5	.005	15	3	15	1	.020	1	.020	0	.001
15	10	14	5	.023	5	.023	4	.007	15	3	14	0	.005	0	.005	0	.005
15	10	13	4	.022	4	.022	3	.007	15	3	13	0	.012	0	.012	-	---
15	10	12	3	.018	3	.018	2	.005	15	3	12	0	.025	0	.025	-	---
15	10	11	3	.042	2	.013	1	.003	15	3	11	0	.043	-	---	-	---
15	10	10	2	.029	1	.007	1	.007	15	2	15	0	.007	0	.007	0	.007
15	10	9	1	.016	1	.016	0	.002	15	2	14	0	.022	0	.022	-	---
15	10	8	1	.034	0	.006	0	.006	15	2	13	0	.044	-	---	-	---
15	10	7	0	.013	0	.013	-	---									
15	10	6	0	.028	-	---	-	---	23	10	21	5	.016	5	.016	4	.004
									32	13	32	10	.020	10	.020	9	.005

Table A.10. Sample sizes for comparing two proportions with a one-sided Fisher's exact test in 2 × 2 tables.

Let P_A and P_B be the true proportions in two populations. The sample size, N, for two equally sized groups is tabulated for one-sided significance level α and probability β of not rejecting the null hypothesis. Each rectangular portion of the table contains sample sizes for *two* pairs of α and β values, one above the diagonal and one below it. The arcsine approximation was used to estimate N.

P_A \ P_B	.001	.01	.05	.10	.15	.20	.25	.30	.40	.50	.60	.70	.80	.90
						$\alpha = .01$ and $\beta = .01$								
.001	---	2305	288	129	81	58	45	37	26	20	15	12	10	8
.01	1679	---	689	221	123	82	61	48	32	24	18	14	11	9
.05	210	502	---	1169	366	191	122	87	52	35	25	19	14	11
.10	94	161	852	---	1877	538	266	163	83	51	34	25	18	13
.15	59	90	266	1368	---	2489	683	327	132	73	46	31	22	15
.20	43	60	140	392	1814	---	3012	805	222	105	61	39	27	18
.25	33	44	89	194	498	2194	---	3447	417	158	83	50	32	21
.30	27	35	63	119	239	587	2511	---	981	256	116	64	39	25
.40	19	24	38	60	96	162	304	715	---	1068	267	116	61	34
.50	14	17	26	37	53	77	116	187	778	---	1068	256	105	51
.60	11	13	19	25	34	45	61	84	195	778	---	981	222	83
.70	9	10	14	18	23	29	37	47	84	187	715	---	805	163
.80	7	8	11	13	16	20	24	29	45	77	162	587	---	538
.90	6	6	8	10	11	13	15	18	25	37	60	119	392	---

$\alpha = .01$ and $\beta = .05$ (or $\alpha = .05$ and $\beta = .01$)

P_A \ P_B	.001	.01	.05	.10	.15	.20	.25	.30	.40	.50	.60	.70	.80	.90
						$\alpha = .025$ and $\beta = .05$ (or $\alpha = .05$ and $\beta = .025$)								
.001	---	1384	173	78	49	35	27	22	16	12	9	8	6	5
.01	1119	---	414	133	74	50	37	29	20	14	11	9	7	5
.05	140	335	---	702	220	115	74	52	31	21	15	12	9	7
.10	63	108	568	---	1127	323	160	98	50	31	21	15	11	8
.15	40	60	178	911	---	1494	410	197	79	44	28	19	13	9
.20	29	40	93	261	1208	---	1808	483	133	63	37	24	16	11
.25	22	30	60	129	332	1462	---	2069	251	95	50	30	20	13
30	18	23	42	79	159	391	1673	---	589	154	70	39	24	15
.40	13	16	25	40	64	108	203	476	---	641	161	70	37	21
.50	10	12	17	25	35	51	77	125	519	---	641	154	63	31
.60	8	9	13	17	23	30	40	56	130	519	---	589	133	50
.70	6	7	9	12	15	19	25	32	56	125	476	---	483	98
.80	5	6	7	9	11	13	16	19	30	51	108	391	---	323
.90	4	4	6	7	8	9	10	12	17	25	40	79	261	---

$\alpha = .025$ and $\beta = .10$ (or $\alpha = .10$ and $\beta = .025$)

P_A \ P_B	.001	.01	.05	.10	.15	.20	.25	.30	.40	.50	.60	.70	.80	.90
						$\alpha = .05$ and $\beta = .05$								
.001	---	1152	144	65	41	29	23	19	13	10	8	6	5	4
.01	912	---	345	111	62	41	31	24	16	12	9	7	6	5
.05	114	273	---	585	183	96	61	44	26	18	13	10	7	6
.10	51	88	463	---	939	269	133	82	42	26	17	13	9	7
.15	32	49	145	743	---	1245	342	164	66	36	23	16	11	8
.20	23	33	76	213	985	---	1506	403	111	53	31	20	14	9
.25	18	24	49	106	271	1192	---	1723	209	79	42	25	16	11
.30	15	19	35	65	130	319	1364	---	491	128	58	32	20	13
.40	11	13	21	33	52	88	165	388	---	534	134	58	31	17
.50	8	10	14	20	29	42	63	102	423	---	534	128	53	26
.60	6	7	10	14	18	24	33	46	106	423	---	491	111	42
.70	5	6	8	10	13	16	20	26	46	102	388	---	403	82
.80	4	5	6	.7	9	11	13	16	24	42	88	319	---	269
.90	3	4	5	5	6	7	9	10	14	20	33	65	213	---

$\alpha = .05$ and $\beta = .10$ (or $\alpha = .10$ and $\beta = .05$)

P_A	P_B													
	.001	.01	.05	.10	.15	.20	.25	.30	.40	.50	.60	.70	.80	.90
.001	---	700	88	40	25	18	14	11	8	6	5	4	3	3
.01	480	---	210	67	38	25	19	15	10	7	6	5	4	3
.05	60	144	---	355	111	58	37	27	16	11	8	6	5	4
.10	27	46	244	---	570	164	81	50	25	16	11	8	6	4
.15	17	26	77	391	---	756	208	100	40	22	14	10	7	5
.20	13	18	40	112	519	---	914	245	68	32	19	12	8	6
.25	10	13	26	56	143	628	---	1046	127	48	25	16	10	7
.30	8	10	18	34	69	168	718	---	298	78	35	20	12	8
.40	6	7	11	18	28	47	87	205	---	325	82	35	19	11
.50	4	5	8	11	15	22	33	54	223	---	325	78	32	16
.60	4	4	6	8	10	13	18	25	56	223	---	298	68	25
.70	3	3	4	6	7	9	11	14	25	54	205	---	245	50
.80	2	3	3	4	5	6	7	9	13	22	47	168	---	164
.90	2	2	3	3	4	4	5	6	8	11	18	34	112	---

Table A.10 *continued*. **Sample sizes for Fisher's exact test.**

$\alpha = .10$ and $\beta = .10$

$\alpha = .10$ and $\beta = .20$ (or $\alpha = .20$ and $\beta = .10$)

Table A.11. Confidence intervals for the expected value (parameter) of a Poisson random variable.

Suppose that one observation, N, is obtained from a Poisson distribution with expected value λ. Ninety percent, 95 percent and 99 percent confidence intervals for the parameter λ are given. For $N > 100$ a normal approximation with variance N may be used.

| | Percent confidence | | | | | | | Percent confidence | | | | | |
| | 90 | | 95 | | 99 | | | 90 | | 95 | | 99 | |
N	Lower limit	Upper limit	Lower limit	Upper limit	Lower limit	Upper limit	N	Lower limit	Upper limit	Lower limit	Upper limit	Lower limit	Upper limit
0	0.00	3.00	0.00	3.69	0.00	5.30	50	38.96	63.29	37.11	65.92	33.66	71.27
1	0.05	4.74	0.03	5.57	0.01	7.43	51	39.85	64.40	37.97	67.06	34.48	72.45
2	0.36	6.30	0.24	7.22	0.10	9.27	52	40.73	65.52	38.84	68.19	35.30	73.62
3	0.82	7.75	0.62	8.77	0.34	10.98	53	41.62	66.63	39.70	69.33	36.13	74.80
4	1.37	9.15	1.09	10.24	0.67	12.59	54	42.51	67.74	40.57	70.46	36.95	75.97
5	1.97	10.51	1.62	11.67	1.08	14.15	55	43.40	68.85	41.43	71.59	37.78	77.15
6	2.61	11.84	2.20	13.06	1.54	15.66	56	44.29	69.96	42.30	72.72	38.60	78.32
7	3.29	13.15	2.81	14.42	2.04	17.13	57	45.18	71.07	43.17	73.85	39.43	79.49
8	3.98	14.43	3.45	15.76	2.57	18.58	58	46.07	72.18	44.04	74.98	40.26	80.66
9	4.70	15.71	4.12	17.08	3.13	20.00	59	46.96	73.28	44.91	76.11	41.09	81.82
10	5.43	16.96	4.80	18.39	3.72	21.40	60	47.85	74.39	45.79	77.23	41.93	82.99
11	6.17	18.21	5.49	19.68	4.32	22.78	61	48.75	75.49	46.66	78.36	42.76	84.15
12	6.92	19.44	6.20	20.96	4.94	24.14	62	49.64	76.60	47.54	79.48	43.60	85.32
13	7.69	20.67	6.92	22.23	5.58	25.50	63	50.54	77.70	48.41	80.60	44.43	86.48
14	8.46	21.89	7.65	23.49	6.23	26.84	64	51.43	78.80	49.29	81.73	45.27	87.64
15	9.25	23.10	8.40	24.74	6.89	28.16	65	52.33	79.91	50.17	82.85	46.11	88.80
16	10.04	24.30	9.15	25.98	7.57	29.48	66	53.23	81.01	51.04	83.97	46.95	89.96
17	10.83	25.50	9.90	27.22	8.25	30.79	67	54.13	82.11	51.92	85.09	47.79	91.11
18	11.63	26.69	10.67	28.45	8.94	32.09	68	55.03	83.21	52.80	86.21	48.64	92.27
19	12.44	27.88	11.44	29.67	9.64	33.38	69	55.93	84.31	53.69	87.32	49.48	93.42
20	13.25	29.06	12.22	30.89	10.35	34.67	70	56.83	85.40	54.57	88.44	50.33	94.58
21	14.07	30.24	13.00	32.10	11.07	35.95	71	57.73	86.50	55.45	89.56	51.17	95.73
22	14.89	31.41	13.79	33.31	11.79	37.22	72	58.63	87.60	56.34	90.67	52.02	96.88
23	15.72	32.59	14.58	34.51	12.52	38.48	73	59.54	88.69	57.22	91.79	52.87	98.03
24	16.55	33.75	15.38	35.71	13.26	39.74	74	60.44	89.79	58.11	92.90	53.72	99.18
25	17.38	34.92	16.18	36.90	14.00	41.00	75	61.35	90.89	58.99	94.01	54.57	100.33
26	18.22	36.08	16.98	38.10	14.74	42.25	76	62.25	91.98	59.88	95.13	55.42	101.48
27	19.06	37.23	17.79	39.28	15.49	43.50	77	63.16	93.07	60.77	96.24	56.28	102.62
28	19.90	38.39	18.61	40.47	16.25	44.74	78	64.06	94.17	61.66	97.35	57.13	103.77
29	20.75	39.54	19.42	41.65	17.00	45.98	79	64.97	95.26	62.55	98.46	57.98	104.91
30	21.59	40.69	20.24	42.83	17.77	47.21	80	65.88	96.35	63.44	99.57	58.84	106.06
31	22.44	41.84	21.06	44.00	18.53	48.44	81	66.79	97.44	64.33	100.68	59.70	107.20
32	23.30	42.98	21.89	45.17	19.30	49.67	82	67.70	98.53	65.22	101.78	60.55	108.34
33	24.15	44.13	22.72	46.34	20.08	50.89	83	68.60	99.62	66.11	102.89	61.41	109.48
34	25.01	45.27	23.55	47.51	20.86	52.11	84	69.51	100.71	67.00	104.00	62.27	110.62
35	25.87	46.40	24.38	48.68	21.64	53.32	85	70.42	101.80	67.89	105.10	63.13	111.76
36	26.73	47.54	25.21	49.84	22.42	54.54	86	71.34	102.89	68.79	106.21	63.99	112.90
37	27.59	48.68	26.05	51.00	23.21	55.75	87	72.25	103.98	69.68	107.31	64.85	114.04
38	28.46	49.81	26.89	52.16	24.00	56.96	88	73.16	105.06	70.58	108.42	65.72	115.17
39	29.33	50.94	27.73	53.31	24.79	58.16	89	74.07	106.15	71.47	109.52	66.58	116.31
40	30.20	52.07	28.58	54.47	25.59	59.36	90	74.98	107.24	72.37	110.63	67.44	117.45
41	31.07	53.20	29.42	55.62	26.38	60.56	91	75.90	108.32	73.27	111.73	68.31	118.58
42	31.94	54.32	30.27	56.77	27.18	61.76	92	76.81	109.41	74.16	112.83	69.17	119.71
43	32.81	55.45	31.12	57.92	27.99	62.96	93	77.73	110.50	75.06	113.93	70.04	120.85
44	33.69	56.57	31.97	59.07	28.79	64.15	94	78.64	111.58	75.96	115.03	70.91	121.98
45	34.56	57.69	32.82	60.21	29.60	65.34	95	79.56	112.66	76.86	116.13	71.77	123.11
46	35.44	58.82	33.68	61.36	30.41	66.53	96	80.47	113.75	77.76	117.23	72.64	124.24
47	36.32	59.94	34.53	62.50	31.22	67.72	97	81.39	114.83	78.66	118.33	73.51	125.37
48	37.20	61.05	35.39	63.64	32.03	68.90	98	82.30	115.91	79.56	119.43	74.38	126.50
49	38.08	62.17	36.25	64.78	32.85	70.08	99	83.22	117.00	80.46	120.53	75.25	127.63
							100	84.14	118.08	81.36	121.63	76.12	128.76

Table A.12. Critical values for the signed ranks test.

For the given n, critical values for the signed ranks test are tabled corresponding to the upper one-sided and two-sided significance levels in the column headings.

						One-sided α							
.05	.025	.01	.005		.05	.025	.01	.005		.05	.025	.01	.005
						Two-sided α							
.10	.05	.02	.01		.10	.05	.02	.01		.10	.05	.02	.01
n				n					n				
5	1	—	—	—	20	60	52	43	37	35	214	195	174	160
6	2	1	—	—	21	68	59	49	43	36	228	208	186	171
7	4	2	0	—	22	75	66	56	49	37	242	222	198	183
8	6	4	2	0	23	83	73	62	55	38	256	235	211	195
9	8	6	3	2	24	92	81	69	61	39	271	250	224	208
10	11	8	5	3	25	101	90	77	68	40	287	264	238	221
11	14	11	7	5	26	110	98	85	76	41	303	279	252	234
12	17	14	10	7	27	120	107	93	84	42	319	295	267	248
13	21	17	13	10	28	130	117	102	92	43	336	311	281	262
14	26	21	16	13	29	141	127	111	100	44	353	327	297	277
15	30	25	20	16	30	152	137	120	109	45	371	344	313	292
16	36	30	24	19	31	163	148	130	118	46	389	361	329	307
17	41	35	28	23	32	175	159	141	128	47	408	379	345	323
18	47	40	33	28	33	188	171	151	138	48	427	397	362	339
19	54	46	38	32	34	201	183	162	149	49	446	415	380	356
										50	466	434	398	373

Table A.13. Critical values for the Mann-Whitney (Wilcoxon) statistic.

This table presents upper one-sided and two-sided critical values for the Mann-Whitney U statistic. Lower one-sided critical values are computed from the upper one-sided critical value (at the same significance level) as $(M \cdot N) - U$. The Wilcoxon two-sample statistic, W, is related to U by the equation $W = (M \cdot N) + (M \cdot (M+1)/2) - U$, where W is the sum of the ranks of the sample of size M in the combined sample.

One-sided α: .10 .05 .025 .01 .005 .001
Two-sided α: .20 .10 .05 .02 .01 .002

n	m	.10 / .20	.05 / .10	.025 / .05	.01 / .02	.005 / .01	.001 / .002	n	m	.10 / .20	.05 / .10	.025 / .05	.01 / .02	.005 / .01	.001 / .002
3	2	6	--	--	--	--	--	11	1	11	--	--	--	--	--
3	3	8	9	--	--	--	--	11	2	19	21	22	--	--	--
								11	3	26	28	30	32	33	--
4	2	8	--	--	--	--	--	11	4	33	36	38	40	42	44
4	3	11	12	--	--	--	--	11	5	40	43	46	48	50	53
4	4	13	15	16	--	--	--	11	6	47	50	53	57	59	62
								11	7	54	58	61	65	67	71
5	2	9	10	--	--	--	--	11	8	61	65	69	73	75	80
5	3	13	14	15	--	--	--	11	9	68	72	76	81	83	89
5	4	16	18	19	20	--	--	11	10	74	79	84	88	92	98
5	5	20	21	23	24	25	--	11	11	81	87	91	96	100	106
								12	1	12	--	--	--	--	--
6	2	11	12	--	--	--	--	12	2	20	22	23	--	--	--
6	3	15	16	17	--	--	--	12	3	28	31	32	34	35	--
6	4	19	21	22	23	24	--	12	4	36	39	41	43	45	48
6	5	23	25	27	28	29	--	12	5	43	47	49	52	54	58
6	6	27	29	31	33	34	--	12	6	51	55	58	61	63	68
7	2	13	14	--	--	--	--	12	7	58	63	66	70	72	77
7	3	17	19	20	21	--	--	12	8	66	70	74	79	81	87
7	4	22	24	25	27	28	--	12	9	73	78	82	87	90	96
7	5	27	29	30	32	34	--	12	10	81	86	91	96	99	106
7	6	31	34	36	38	39	42	12	11	88	94	99	104	108	115
7	7	36	38	41	43	45	48	12	12	95	102	107	113	117	124
8	2	14	15	16	--	--	--	13	1	13	--	--	--	--	--
8	3	19	21	22	24	--	--	13	2	22	24	25	26	--	--
8	4	25	27	28	30	31	--	13	3	30	33	35	37	38	--
8	5	30	32	34	36	38	40	13	4	39	42	44	47	49	51
8	6	35	38	40	42	44	47	13	5	47	50	53	56	58	62
8	7	40	43	46	49	50	54	13	6	55	59	62	66	68	73
8	8	45	49	51	55	57	60	13	7	63	67	71	75	78	83
								13	8	71	76	80	84	87	93
9	1	9	--	--	--	--	--	13	9	79	84	89	94	97	103
9	2	16	17	18	--	--	--	13	10	87	93	97	103	106	113
9	3	22	23	25	26	27	--	13	11	95	101	106	112	116	123
9	4	27	30	32	33	35	--	13	12	103	109	115	121	125	133
9	5	33	36	38	40	42	44	13	13	111	118	124	130	135	143
9	6	39	42	44	47	49	52								
9	7	45	48	51	54	56	60	14	1	14	--	--	--	--	--
9	8	50	54	57	61	63	67	14	2	23	25	27	28	--	--
9	9	56	60	64	67	70	74	14	3	32	35	37	40	41	--
								14	4	41	45	47	50	52	55
10	1	10	--	--	--	--	--	14	5	50	54	57	60	63	67
10	2	17	19	20	--	--	--	14	6	59	63	67	71	73	78
10	3	24	26	27	29	30	--	14	7	67	72	76	81	83	89
10	4	30	33	35	37	38	40	14	8	76	81	86	90	94	100
10	5	37	39	42	44	46	49	14	9	85	90	95	100	104	111
10	6	43	46	49	52	54	57	14	10	93	99	104	110	114	121
10	7	49	53	56	59	61	65	14	11	102	108	114	120	124	132
10	8	56	60	63	67	69	74	14	12	110	117	123	130	134	143
10	9	62	66	70	74	77	82	14	13	119	126	132	139	144	153
10	10	68	73	77	81	84	90	14	14	127	135	141	149	154	164

Table A.13 continued. Mann-Whitney critical values.

Left block

		One-sided α .10	.05	.025	.01	.005	.001
		Two-sided α .20	.10	.05	.02	.01	.002
n	m						
15	1	15	--	--	--	--	--
15	2	25	27	29	30	--	--
15	3	35	38	40	42	43	--
15	4	44	48	50	53	55	59
15	5	53	57	61	64	67	71
15	6	63	67	71	75	78	83
15	7	72	77	81	86	89	95
15	8	81	87	91	96	100	106
15	9	90	96	101	107	111	118
15	10	99	106	111	117	121	129
15	11	108	115	121	128	132	141
15	12	117	125	131	138	143	152
15	13	127	134	141	148	153	163
15	14	136	144	151	159	164	174
15	15	145	153	161	169	174	185
16	1	16	--	--	--	--	--
16	2	27	29	31	32	--	--
16	3	37	40	42	45	46	--
16	4	47	50	53	57	59	62
16	5	57	61	65	68	71	75
16	6	67	71	75	80	83	88
16	7	76	82	86	91	94	101
16	8	86	92	97	102	106	113
16	9	96	102	107	113	117	125
16	10	106	112	118	124	129	137
16	11	115	122	129	135	140	149
16	12	125	132	139	146	151	161
16	13	134	143	149	157	163	173
16	14	144	153	160	168	174	185
16	15	154	163	170	179	185	197
16	16	163	173	181	190	196	208
17	1	17	--	--	--	--	--
17	2	28	31	32	34	--	--
17	3	39	42	45	47	49	51
17	4	50	53	57	60	62	66
17	5	60	65	68	72	75	80
17	6	71	76	80	84	87	93
17	7	81	86	91	96	100	106
17	8	91	97	102	108	112	119
17	9	101	108	114	120	124	132
17	10	112	119	125	132	136	145
17	11	122	130	136	143	148	158
17	12	132	140	147	155	160	170
17	13	142	151	158	166	172	183
17	14	153	161	169	178	184	195
17	15	163	172	180	189	195	208
17	16	173	183	191	201	207	220
17	17	183	193	202	212	219	232
18	1	18	--	--	--	--	--
18	2	30	32	34	36	--	--
18	3	41	45	47	50	52	54
18	4	52	56	60	63	66	69

Right block

		One-sided α .10	.05	.025	.01	.005	.001
		Two-sided α .20	.10	.05	.02	.01	.002
n	m						
18	5	63	68	72	76	79	84
18	6	74	80	84	89	92	98
18	7	85	91	96	102	105	112
18	8	96	103	108	114	118	126
18	9	107	114	120	126	131	139
18	10	118	125	132	139	143	153
18	11	129	137	143	151	156	166
18	12	139	148	155	163	169	179
18	13	150	159	167	175	181	192
18	14	161	170	178	187	194	206
18	15	172	182	190	200	206	219
18	16	182	193	202	212	218	232
18	17	193	204	213	224	231	245
18	18	204	215	225	236	243	258
19	1	18	19	--	--	--	--
19	2	31	34	36	37	38	--
19	3	43	47	50	53	54	57
19	4	55	59	63	67	69	73
19	5	67	72	76	80	83	88
19	6	78	84	89	94	97	103
19	7	90	96	101	107	111	118
19	8	101	108	114	120	124	132
19	9	113	120	126	133	138	146
19	10	124	132	138	146	151	161
19	11	136	144	151	159	164	175
19	12	147	156	163	172	177	188
19	13	158	167	175	184	190	202
19	14	169	179	188	197	203	216
19	15	181	191	200	210	216	230
19	16	192	203	212	222	230	244
19	17	203	214	224	235	242	257
19	18	214	226	236	248	255	271
19	19	226	238	248	260	268	284
20	1	19	20	--	--	--	--
20	2	33	36	38	39	40	--
20	3	45	49	52	55	57	60
20	4	58	62	66	70	72	77
20	5	70	75	80	84	87	93
20	6	82	88	93	98	102	108
20	7	94	101	106	112	116	124
20	8	106	113	119	126	130	139
20	9	118	126	132	140	144	154
20	10	130	138	145	153	158	168
20	11	142	151	158	167	172	183
20	12	154	163	171	180	186	198
20	13	166	176	184	193	200	212
20	14	178	188	197	207	213	226
20	15	190	200	210	220	227	241
20	16	201	213	222	233	241	255
20	17	213	225	235	247	254	270
20	18	225	237	248	260	268	284
20	19	237	250	261	273	281	298
20	20	249	262	273	286	295	312

Table A.14. Fisher's Z-transform. $Z = \tanh^{-1} r$.

The row value plus the column value gives the value of r. The table entry is Z. For example: if $r = .34$, then $Z = .3541$; if $Z = 1.7736$, then $r = .944$. For bivariate normal samples and large N, Z is approximately normally distributed with mean equal to the inverse hyperbolic tangent of the population correlation coefficient and variance equal to $1/(N-3)$.

	.00	.01	.02	.03	.04	.05	.06	.07	.08	.09
.0	.0000	.0100	.0200	.0300	.0400	.0500	.0601	.0701	.0802	.0902
.1	.1003	.1104	.1206	.1307	.1409	.1511	.1614	.1717	.1820	.1923
.2	.2027	.2132	.2237	.2342	.2448	.2554	.2661	.2769	.2877	.2986
.3	.3095	.3205	.3316	.3428	.3541	.3654	.3769	.3884	.4001	.4118
.4	.4236	.4356	.4477	.4599	.4722	.4847	.4973	.5101	.5230	.5361
.5	.5493	.5627	.5763	.5901	.6042	.6184	.6328	.6475	.6625	.6777
.6	.6931	.7089	.7250	.7414	.7582	.7753	.7928	.8107	.8291	.8480
.7	.8673	.8872	.9076	.9287	.9505	.9730	.9962	1.0203	1.0454	1.0714
.8	1.0986	1.1270	1.1568	1.1881	1.2212	1.2562	1.2933	1.3331	1.3758	1.4219

	.000	.001	.002	.003	.004	.005	.006	.007	.008	.009
.90	1.4722	1.4775	1.4828	1.4882	1.4937	1.4992	1.5047	1.5103	1.5160	1.5217
.91	1.5275	1.5334	1.5393	1.5453	1.5513	1.5574	1.5636	1.5698	1.5762	1.5826
.92	1.5890	1.5956	1.6022	1.6089	1.6157	1.6226	1.6296	1.6366	1.6438	1.6510
.93	1.6584	1.6658	1.6734	1.6811	1.6888	1.6967	1.7047	1.7129	1.7211	1.7295
.94	1.7380	1.7467	1.7555	1.7645	1.7736	1.7828	1.7923	1.8019	1.8117	1.8216
.95	1.8318	1.8421	1.8527	1.8635	1.8745	1.8857	1.8972	1.9090	1.9210	1.9333
.96	1.9459	1.9588	1.9721	1.9857	1.9996	2.0139	2.0287	2.0439	2.0595	2.0756
.97	2.0923	2.1095	2.1273	2.1457	2.1649	2.1847	2.2054	2.2269	2.2494	2.2729
.98	2.2976	2.3235	2.3507	2.3796	2.4101	2.4427	2.4774	2.5147	2.5550	2.5987
.99	2.6467	2.6996	2.7587	2.8257	2.9031	2.9945	3.1063	3.2504	3.4534	3.8002

Table A.15. Critical values of the bivariate normal sample correlation coefficient ρ.

When $\rho=0$ the distribution is symmetric about zero; thus one-sided lower critical values are -1 times the tabled one-sided upper critical values. Column headings are also labeled for the corresponding two-sided significance level and the percentage of the distribution less than the tabled value. N is the number of observations; the degrees of freedom is two less than this.

	Percent								Percent						
	90	95	97.5	99	99.5	99.9	99.95		90	95	97.5	99	99.5	99.9	99.95
			One-sided α								One-sided α				
	.10	.05	.025	.01	.005	.001	.0005		.10	.05	.025	.01	.005	.001	.0005
			Two-sided α								Two-sided α				
	.20	.10	.05	.02	.01	.002	.001		.20	.10	.05	.02	.01	.002	.001
N								N							
3	.951	.988	.997	1.000	1.000	1.000	1.000	20	.299	.378	.444	.516	.562	.648	.679
4	.800	.900	.950	.980	.990	.998	.999	25	.265	.337	.396	.462	.505	.588	.618
5	.687	.805	.878	.934	.959	.986	.991	30	.241	.306	.361	.423	.463	.542	.570
6	.608	.729	.811	.882	.917	.963	.974								
7	.551	.669	.755	.833	.875	.935	.951	35	.222	.283	.334	.392	.430	.505	.532
								40	.207	.264	.312	.367	.403	.474	.501
8	.507	.622	.707	.789	.834	.905	.925	45	.195	.248	.294	.346	.380	.449	.474
9	.472	.582	.666	.750	.798	.875	.898	50	.184	.235	.279	.328	.361	.427	.451
10	.443	.549	.632	.716	.765	.847	.872	55	.176	.224	.266	.313	.345	.408	.432
11	.419	.522	.602	.685	.735	.820	.847								
12	.398	.497	.576	.658	.708	.795	.823	60	.168	.214	.254	.300	.330	.391	.414
								65	.161	.206	.244	.288	.317	.376	.399
13	.380	.476	.553	.634	.684	.772	.801	70	.155	.198	.235	.278	.306	.363	.385
14	.365	.458	.533	.612	.661	.750	.780	75	.150	.191	.227	.268	.296	.351	.372
15	.351	.441	.514	.592	.641	.730	.760	80	.145	.185	.220	.260	.286	.341	.361
16	.338	.426	.497	.574	.623	.711	.742								
17	.327	.412	.482	.558	.606	.694	.725	85	.140	.180	.213	.252	.278	.331	.351
								90	.136	.175	.207	.245	.270	.322	.341
18	.317	.400	.468	.543	.590	.678	.708	95	.133	.170	.202	.238	.263	.313	.332
19	.308	.389	.456	.529	.575	.662	.693	100	.129	.165	.197	.232	.257	.305	.324

Table A.16. Critical values for Spearman's rank correlation coefficient.

For a sample of size n, two-sided critical values are given for significance levels .10, .05, and .01. Reject the null hypothesis of independence if the absolute value of the sample Spearman correlation coefficient exceeds the tabled value.

			Two-sided α				
	.10	.05	.01		.10	.05	.01
n				n			
5	.900	---	---	19	.388	.462	.608
6	.829	.886	---				
7	.714	.786	.929	20	.377	.450	.591
8	.643	.738	.881	21	.368	.438	.576
9	.600	.700	.833	22	.359	.428	.562
				23	.351	.418	.549
10	.564	.648	.794	24	.343	.409	.537
11	.536	.618	.818				
12	.497	.591	.780	25	.336	.400	.526
13	.475	.566	.745	26	.329	.392	.515
14	.457	.545	.716	27	.323	.385	.505
				28	.317	.377	.496
15	.441	.525	.689	29	.311	.370	.487
16	.425	.507	.666				
17	.412	.490	.645	30	.305	.364	.478
18	.399	.476	.625				

Table A.17. Critical values for Kendall's rank correlation coefficient.

For a bivariate sample of size N let K be the number of pairs of observations where the first member of the pair has both X and Y values larger than the second member of the pair. The table entry gives the smallest integer k for which $P[K \leq k]$ is less than or equal to one minus the cumulative distribution as specified. Equivalently the table entry is the corresponding critical value. Kendall's rank correlation coefficient is $(4 \cdot K / (N \cdot (N-1))) - 1$.

	Cumulative distribution									
	.7	.75	.80	.85	.90	.95	.99	.975	.995	.999
	Significance level									
N	.3	.25	.20	.15	.10	.05	.01	.025	.005	.001
3	3	3	3	–	–	–	–	–	–	–
4	5	5	5	6	6	6	–	–	–	–
5	7	7	8	8	9	9	10	10	–	–
6	10	10	11	11	12	13	14	14	15	–
7	13	14	14	15	16	17	18	19	20	21
8	17	18	18	19	20	22	23	24	25	26
9	22	22	23	24	25	27	28	30	31	33
10	27	27	28	29	31	33	34	36	37	40
11	32	33	34	35	37	39	41	43	44	47
12	38	39	40	42	43	46	48	51	52	55
13	44	46	47	49	51	53	56	59	61	64
14	51	53	54	56	58	62	64	67	69	73
15	59	60	62	64	67	70	73	77	79	83
16	67	69	70	73	75	79	83	86	89	94
17	75	77	79	82	85	89	93	97	100	105
18	85	87	89	91	95	99	103	108	111	117
19	94	96	99	101	105	110	114	119	123	129
20	104	107	109	112	116	121	126	131	135	142
21	115	117	120	123	127	133	138	144	148	156
22	126	129	132	135	139	146	151	157	161	170
23	138	140	144	147	152	159	164	171	176	184
24	150	153	156	160	165	172	178	185	190	200
25	162	166	169	173	179	186	193	200	205	216

Table A.18. Critical values for the Studentized range.

Tabulated are the upper 0.05 and 0.01 critical values for the Studentized range. The first column gives the degrees of freedom for the estimate of the variance used in the denominator. The corresponding table entries give the critical values for 2 to 10 groups, or means, in the numerator.

	Upper 0.05 critical values								
df	k = sample size for the range = number of groups								
	2	3	4	5	6	7	8	9	10
1	17.97	26.98	32.82	37.08	40.41	43.12	45.40	47.36	49.07
2	6.08	8.33	9.80	10.88	11.74	12.44	43.03	13.54	13.99
3	4.50	5.91	6.82	7.50	8.04	8.48	8.85	9.18	9.46
4	3.93	5.04	5.76	6.29	6.71	7.05	7.35	7.60	7.83
5	3.64	4.60	5.22	5.67	6.03	6.33	6.58	6.80	6.99
6	3.46	4.34	4.90	5.30	5.63	5.90	6.12	6.32	6.49
7	3.34	4.16	4.68	5.06	5.36	5.61	5.82	6.00	6.16
8	3.26	4.04	4.53	4.89	5.17	5.40	5.60	5.77	5.92
9	3.20	3.95	4.41	4.76	5.02	5.24	5.43	5.59	5.74
10	3.15	3.88	4.33	4.65	4.91	5.12	5.30	5.46	5.60
11	3.11	3.82	4.26	4.57	4.82	5.03	5.20	5.35	5.49
12	3.08	3.77	4.20	4.51	4.75	4.95	5.12	5.27	5.39
13	3.06	3.73	4.15	4.45	4.69	4.88	5.05	5.19	5.32
14	3.03	3.70	4.11	4.41	4.64	4.83	4.99	5.13	5.25
15	3.01	3.67	4.08	4.37	4.59	4.78	4.94	5.08	5.20
16	3.00	3.65	4.05	4.33	4.56	4.74	4.90	5.03	5.15
17	2.98	3.63	4.02	4.30	4.52	4.70	4.86	4.99	5.11
18	2.97	3.61	4.00	4.28	4.49	4.67	4.82	4.96	5.07
19	2.96	3.59	3.98	4.25	4.47	4.65	4.79	4.92	5.04
20	2.95	3.58	3.96	4.23	4.45	4.62	4.77	4.90	5.01
24	2.92	3.53	3.90	4.17	4.37	4.54	4.68	4.81	4.92
30	2.89	3.49	3.85	4.10	4.30	4.46	4.60	4.72	4.82
40	2.86	3.44	3.79	4.04	4.23	4.39	4.52	4.63	4.73
6C	2.83	3.40	3.74	3.98	4.16	4.31	4.44	4.47	4.65
120	2.80	3.36	3.69	3.92	4.10	4.24	4.36	4.39	4.56
∞	2.77	3.31	3.63	3.86	4.03	4.17	4.29	4.39	4.47
	Upper 0.01 critical values								
1	90.03	135.0	164.3	185.6	202.2	215.8	227.2	237.0	245.6
2	14.04	19.02	22.29	24.72	26.63	28.20	29.53	30.68	31.69
3	8.26	10.62	12.17	13.33	14.24	15.00	15.64	16.20	16.69
4	6.51	8.12	9.17	9.96	10.58	11.10	11.55	11.93	12.27
5	5.70	6.98	7.80	8.42	8.91	9.32	9.67	9.97	10.24
6	5.24	6.33	7.03	7.56	7.97	8.32	8.61	8.87	9.10
7	4.95	5.92	6.54	7.01	7.37	7.68	7.94	8.17	8.37
8	4.75	5.64	6.20	6.62	6.96	7.24	7.47	7.68	7.86
9	4.60	5.43	5.96	6.35	6.66	6.91	7.13	7.33	7.49
10	4.48	5.27	5.77	6.14	6.43	6.67	6.87	7.05	7.21
11	4.39	5.15	5.62	5.97	6.25	6.48	6.67	6.84	6.99
12	4.32	5.05	5.50	5.84	6.10	6.32	6.51	6.67	6.81
13	4.26	4.96	5.40	5.73	5.98	6.19	6.37	6.53	6.67
14	4.21	4.89	5.32	5.63	5.88	6.08	6.26	6.41	6.54
15	4.17	4.84	5.25	5.56	5.80	5.99	6.16	6.31	6.44
16	4.13	4.79	5.19	5.49	5.72	5.92	6.08	6.22	6.35
17	4.10	4.74	5.14	5.43	5.66	5.85	6.01	6.15	6.27
18	4.07	4.70	5.09	5.38	5.60	5.79	5.94	6.08	6.20
19	4.05	4.67	5.05	5.33	5.55	5.73	5.89	6.02	6.14
20	4.02	4.64	5.02	5.29	5.51	5.69	5.84	5.97	6.09
24	3.96	4.55	4.91	5.17	5.37	5.54	5.69	5.81	5.92
30	3.89	4.45	4.80	5.05	5.24	5.40	5.54	5.65	5.76
40	3.82	4.37	4.70	4.93	5.11	5.26	5.39	5.50	5.60
60	3.76	4.28	4.59	4.82	4.99	5.13	5.25	5.36	5.45
120	3.70	4.20	4.50	4.71	4.87	5.01	5.12	5.21	5.30
∞	3.64	4.12	4.40	4.60	4.76	4.88	4.99	5.08	5.16

Table A.19. Critical values for the maximum F ratio.

Consider K independent mean squares, each with df degrees of freedom. Under the null hypothesis each has a chi-square distribution divided by the degrees of freedom times the same fixed constant. The test statistic is the largest mean square divided by the smallest mean square. The column headings are values of K; the table entries are the upper critical values.

df	Upper 0.05 significance level critical values										
	2	3	4	5	6	7	8	9	10	11	12
2	39.0	87.5	142	202	266	333	403	475	550	626	704
3	15.4	27.8	39.2	50.7	62.0	72.9	83.5	93.9	104	114	124
4	9.60	15.5	20.6	25.2	29.5	33.6	37.5	41.1	44.6	48.0	51.4
5	7.15	10.8	13.7	16.3	18.7	20.8	22.9	24.7	26.5	28.2	29.9
6	5.82	8.38	10.4	12.1	13.7	15.0	16.3	17.5	18.6	19.7	20.7
7	4.99	6.94	8.44	9.70	10.8	11.8	12.7	13.5	14.3	15.1	15.8
8	4.43	6.00	7.18	8.12	9.03	9.78	10.5	11.1	11.7	12.2	12.7
9	4.03	5.34	6.31	7.11	7.80	8.41	8.95	9.45	9.91	10.3	10.7
10	3.72	4.85	5.67	6.34	6.92	7.42	7.87	8.28	8.66	9.01	9.34
12	3.28	4.16	4.79	5.30	5.72	6.09	6.42	6.72	7.00	7.25	7.48
15	2.86	3.54	4.01	4.37	4.68	4.95	5.19	5.40	5.59	5.77	5.93
20	2.46	2.95	3.29	3.54	3.76	3.94	4.10	4.24	4.37	4.49	4.59
30	2.07	2.40	2.61	2.78	2.91	3.02	3.12	3.21	3.29	3.36	3.39
∞	1.67	1.85	1.96	2.04	2.11	2.17	2.22	2.26	2.30	2.33	2.36

df	Upper 0.01 significance level critical values[*]										
	2	3	4	5	6	7	8	9	10	11	12
2	199	448	729	1036	1362	1705	2063	2432	2813	3204	3605
3	47.5	85	120	151	184	21(6)	24(9)	28(1)	31(0)	33(7)	36(1)
4	23.2	37	49	59	69	79	89	97	106	113	120
5	14.9	22	28	33	38	42	46	50	54	57	60
6	11.1	15.5	19.1	22	25	27	30	32	34	36	37
7	8.89	12.1	14.5	16.5	18.4	20	22	23	24	26	27
8	7.50	9.9	11.7	13.2	14.5	15.8	16.9	17.9	18.9	19.8	21
9	6.54	8.5	9.9	11.1	12.1	13.1	13.9	14.7	15.3	16.0	16.6
10	5.85	7.4	8.6	9.6	10.4	11.1	11.8	12.4	12.9	13.4	13.9
12	4.91	6.1	6.9	7.6	8.2	8.7	9.1	9.5	9.9	10.2	10.6
15	4.07	4.9	5.5	6.0	6.4	6.7	7.1	7.3	7.5	7.8	8.0
20	3.32	3.8	4.3	4.6	4.9	5.1	5.3	5.5	5.6	5.8	5.9
30	2.63	3.0	3.3	3.4	3.6	3.7	3.8	3.9	4.0	4.1	4.2
∞	1.96	2.2	2.3	2.4	2.4	2.5	2.5	2.6	2.6	2.7	2.7

[*]The figures in parentheses are uncertain. In general the third digit may be in error by a few units.

Table A.20. Critical values for Cochran's test.

Tabled are critical values for the ratio of the largest of K independent estimates of a variance, each based on N observations, to the sum of the estimates of the variances. Normality of the observations is assumed. The null hypothesis is that all K variances are the same; the test for equality of variances is Cochran's test. The column headings are the values of N.

					Upper 0.01 significance level critical values									
df	2	3	4	5	6	7	8	9	10	11	17	37	145	∞
2	.9999	.9950	.9794	.9586	.9373	.9172	.8988	.8823	.8674	.8539	.7949	.7067	.6062	.5000
3	.9933	.9423	.8831	.8335	.7933	.7606	.7335	.7107	.6912	.6743	.6059	.5153	.4230	.3333
4	.9676	.8643	.7814	.7212	.6761	.6410	.6129	.5897	.5702	.5536	.4884	.4057	.3251	.2500
5	.9279	.7885	.6957	.6329	.5875	.5531	.5259	.5037	.4854	.4697	.4094	.3351	.2644	.2000
6	.8828	.7218	.6258	.5635	.5195	.4866	.4608	.4401	.4229	.4084	.3529	.2858	.2229	.1667
7	.8376	.6644	.5685	.5080	.4659	.4347	.4105	.3911	.3751	.3616	.3105	.2494	.1929	.1429
8	.7945	.6152	.5209	.4627	.4226	.3932	.3704	.3522	.3373	.3248	.2779	.2214	.1700	.1250
9	.7544	.5727	.4810	.4251	.3870	.3592	.3378	.3207	.3067	.2950	.2514	.1992	.1521	.1111
10	.7175	.5358	.4469	.3934	.3572	.3308	.3106	.2945	.2813	.2704	.2297	.1811	.1376	.1000
12	.6528	.4751	.3919	.3428	.3099	.2861	.2680	.2535	.2419	.2320	.1961	.1535	.1157	.0833
15	.5747	.4069	.3317	.2882	.2593	.2386	.2228	.2104	.2002	.1918	.1612	.1251	.0934	.0667
20	.4799	.3297	.2654	.2288	.2048	.1877	.1748	.1646	.1567	.1501	.1248	.0960	.0709	.0500
24	.4247	.2871	.2295	.1970	.1759	.1608	.1495	.1406	.1338	.1283	.1060	.0810	.0595	.0417
30	.3632	.2412	.1913	.1635	.1454	.1327	.1232	.1157	.1100	.1054	.0867	.0658	.0480	.0333
40	.2940	.1915	.1508	.1281	.1135	.1033	.0957	.0898	.0853	.0816	.0668	.0503	.0363	.0250
60	.2151	.1371	.1069	.0902	.0796	.0722	.0668	.0625	.0594	.0567	.0461	.0344	.0245	.0167
120	.1225	.0759	.0585	.0489	.0429	.0387	.0357	.0334	.0316	.0302	.0242	.0178	.0125	.0083
∞	.0000	.0000	.0000	.0000	.0000	.0000	.0000	.0000	.0000	.0000	.0000	.0000	.0000	.0000

					Upper 0.05 significance level critical values									
df	2	3	4	5	6	7	8	9	10	11	17	37	145	∞
2	.9985	.9750	.9392	.9057	.8772	.8534	.8332	.8159	.8010	.7880	.7341	.6602	.5813	.5000
3	.9669	.8709	.7977	.7457	.7071	.6771	.6530	.6333	.6167	.6025	.5466	.4748	.4031	.3333
4	.9065	.7679	.6841	.6287	.5895	.5598	.5365	.5175	.5017	.4884	.4366	.3720	.3093	.2500
5	.8412	.6838	.5981	.5441	.5065	.4783	.4564	.4387	.4241	.4118	.3645	.3066	.2513	.2000
6	.7808	.6161	.5321	.4803	.4447	.4184	.3980	.3817	.3682	.3568	.3135	.2612	.2119	.1667
7	.7271	.5612	.4800	.4307	.3974	.3726	.3535	.3384	.3259	.3154	.2756	.2278	.1833	.1429
8	.6798	.5157	.4377	.3910	.3595	.3362	.3185	.3043	.2926	.2862	.2462	.2022	.1616	.1250
9	.6385	.4775	.4027	.3584	.3286	.3067	.2901	.2768	.2659	.2568	.2226	.1820	.1446	.1111
10	.6020	.4450	.3733	.3311	.3029	.2823	.2666	.2541	.2439	.2353	.2032	.1655	.1308	.1000
12	.5410	.3924	.3264	.2880	.2624	.2439	.2299	.2187	.2098	.2020	.1737	.1403	.1100	.0833
15	.4709	.3346	.2758	.2419	.2195	.2034	.1911	.1815	.1736	.1671	.1429	.1144	.0889	.0667
20	.3894	.2705	.2205	.1921	.1735	.1602	.1501	.1422	.1357	.1303	.1108	.0879	.0675	.0500
24	.3434	.2354	.1907	.1656	.1493	.1374	.1286	.1216	.1160	.1113	.0942	.0743	.0567	.0417
30	.2929	.1980	.1593	.1377	.1237	.1137	.1061	.1002	.0958	.0921	.0771	.0604	.0457	.0333
40	.2370	.1576	.1259	.1082	.0968	.0887	.0827	.0780	.0745	.0713	.0595	.0462	.0347	.0250
60	.1737	.1131	.0895	.0765	.0682	.0623	.0583	.0552	.0520	.0497	.0411	.0316	.0234	.0167
120	.0998	.0632	.0495	.0419	.0371	.0337	.0312	.0292	.0279	.0266	.0218	.0165	.0120	.0083
∞	.0000	.0000	.0000	.0000	.0000	.0000	.0000	.0000	.0000	.0000	.0000	.0000	.0000	.0000

Table A.21. Expected values of normal order statistics.

A sample of N N(0,1) observations is ranked from largest (rank 1) to smallest (rank N). The expected values of the order statistics (the ranked values) are given. Only the expected values for the upper half of the order statistics are given since the expected values are symmetric about zero. The column headings give the size of the sample and the row headings the rank of the order statistic.

Rank	Sample size 2	3	4	5	6	7	8	9	10	11	12	13	14
1	.56419	.34628	1.02938	1.16296	1.26721	1.35218	1.42360	1.48501	1.53875	1.58644	1.62923	1.66799	1.70338
2		.00000	.29701	.49502	.64176	.75737	.85222	.93230	1.00136	1.06192	1.11573	1.16408	1.20790
3			.00000	.20155	.35271	.47282	.57197	.65606	.72884	.79284	.84983	.90113
4					.00000	.15251	.27453	.37576	.46198	.53684	.60285	.66176
5							.00000	.12267	.22489	.31225	.38833	.45557
6									.00000	.10259	.19052	.26730
7											.00000	.08816

Rank	Sample size 15	16	17	18	19	20	21	22	23	24	25	26	27
1	1.73591	1.76599	1.79394	1.82003	1.84448	1.86748	1.88917	1.90969	1.92916	1.94767	1.96531	1.98216	1.99827
2	1.24794	1.28474	1.31878	1.35041	1.37994	1.40760	1.43362	1.45816	1.48137	1.50338	1.52430	1.54423	1.56326
3	.94769	.99027	1.02946	1.06573	1.09945	1.13095	1.16047	1.18824	1.21445	1.23924	1.26275	1.28511	1.30641
4	.71488	.76317	.80738	.84812	.88586	.92098	.95380	.98459	1.01356	1.04091	1.06679	1.09135	1.11471
5	.51570	.57001	.61946	.66479	.70661	.74538	.78150	.81527	.84697	.87682	.90501	.93171	.95705
6	.33530	.39622	.45133	.50158	.54771	.59030	.62982	.66667	.70115	.73354	.76405	.79289	.82021
7	.16530	.23375	.29519	.35084	.40164	.44833	.49148	.53157	.56896	.60299	.63690	.66794	.69727
8	.00000	.07729	.14599	.20774	.26374	.31493	.36203	.40559	.44609	.48391	.51935	.55267	.58411
9			.00000	.06880	.13072	.18698	.23841	.28579	.32965	.37047	.40860	.44436	.47801
10					.00000	.06200	.11836	.16997	.21755	.26163	.30268	.34105	.37706
11							.00000	.05642	.10813	.15583	.20006	.24128	.27983
12									.00000	.05176	.09953	.14387	.18520
13											.00000	.04781	.09220
14													.00000

Rank	Sample size 28	29	30	31	32	33	34	35	36	37	38	39
1	2.01371	2.02852	2.04276	2.05646	2.06967	2.08241	2.09471	2.10661	2.11812	2.12928	2.14009	2.15059
2	1.58145	1.59888	1.61560	1.63166	1.64712	1.66200	1.67636	1.69023	1.70362	1.71659	1.72914	1.74131
3	1.32674	1.34619	1.36481	1.38268	1.39985	1.41637	1.43228	1.44762	1.46244	1.47676	1.49061	1.50402
4	1.13697	1.15822	1.17855	1.19803	1.21672	1.23468	1.25196	1.26860	1.28466	1.30016	1.31514	1.32964
5	.98115	1.00414	1.02609	1.04709	1.06721	1.08652	1.10509	1.12295	1.14016	1.15677	1.17280	1.18830
6	.84815	.87084	.89439	.91688	.93841	.95905	.97886	.99790	1.01624	1.03390	1.05095	1.06741
7	.72508	.75150	.77666	.80066	.82359	.84555	.86660	.88681	.90625	.92496	.94300	.96041
8	.61385	.64205	.66885	.69438	.71875	.74204	.76435	.78574	.80629	.82605	.84508	.86343
9	.50977	.53982	.56834	.59545	.62129	.64596	.66954	.69214	.71382	.73465	.75468	.77398
10	.41096	.44298	.47329	.50206	.52943	.55552	.58043	.60427	.62710	.64902	.67009	.69035
11	.31603	.35013	.38235	.41287	.44185	.46942	.49572	.52084	.54488	.56793	.59005	.61131
12	.22389	.26023	.29449	.32686	.35755	.38669	.41444	.44091	.46620	.49042	.51363	.53592
13	.13361	.17240	.20885	.24322	.27573	.30654	.33582	.36371	.39032	.41576	.44012	.46348
14	.04442	.08588	.12473	.16126	.19572	.22832	.25924	.28863	.31663	.34338	.36892	.39340
15		.00000	.04148	.08037	.11695	.15147	.18415	.21515	.24463	.27272	.29954	.32520
16				.00000	.03890	.07552	.11009	.14282	.17388	.20342	.23159	.25849
17						.00000	.03663	.07123	.10399	.13509	.16469	.19292
18								.00000	.03461	.06739	.09853	.12817
19										.00000	.03280	.06395
20												.00000

Example Index

Symbol Index

Author Index

Subject Index

— T —

*Now available in a lower priced paperback edition in the Wiley Classics Library.

*Now available in a lower priced paperback edition in the Wiley Classics Library.